Springer Collected Works in Mathematics

For further volumes:
http://www.springer.com/series/11104

Spring 1980
at the Institute for Advanced Study, Princeton

Armand Borel

Oeuvres - Collected Papers III

1969 – 1982

Reprint of the 1983 Edition

 Springer

Armand Borel (1923 – 2003)
The Institute for Advanced Study
Princeton, NJ
USA

ISSN 2194-9875
ISBN 978-3-662-44313-2   (Softcover)
       978-3-540-12126-8   (Hardcover)
DOI 10.1007/978-3-662-44314-9
Springer Heidelberg New York Dordrecht London

Library of Congress Control Number: 2012954381

Printed on acid-free paper

Springer is part of Springer Science+Business Media (www.springer.com)

Préface

Ces trois volumes contiennent tous les articles que j'ai publiés, seul ou en collaboration, jusqu'en 1982, à l'exception de quelques exposés de Séminaire. N'y figurent pas non plus, bien entendu, les livres ou Notes de cours (à part une introduction). J'ai par contre inclus deux manuscrits restés inédits, ne serait-ce que pour fournir une référence originale à des travaux qui avaient été indirectement publiés, au moins en partie, à l'époque.

Les articles sont pour la plupart reproduits par photocopie. Quelques-uns d'entre eux cependant ont fait l'objet d'une composition typographique. Des corrections mineures ont été faites directement sur le texte. D'autres, plus longues, figurent dans les «Commentaires et corrections» à la fin de chaque volume. Ces derniers fournissent aussi des références à des résultats ultérieurs qui complètent ou généralisent des propositions du texte, ou encore répondent à des questions qui y sont posées. Quelques problèmes encore ouverts à ma connaissance sont aussi signalés. Ni ces renseignements ni les corrections ne prétendent à être exhaustifs.

On remarquera la forte proportion d'articles écrits en collaboration. Ils furent une source de discussions et échanges dont j'ai tiré grand profit, aussi suis-je heureux de saisir cette occasion pour exprimer ma gratitude aux coauteurs de ces travaux, non seulement pour m'avoir autorisé à les reproduire ici, mais avant tout pour la collaboration elle-même.

Enfin, je remercie vivement Springer-Verlag pour la proposition, si flatteuse, de faire figurer mes travaux dans sa série de «Collected Papers» et pour avoir mené à bien cette publication avec son habileté coutumière, en accédant volontiers à toutes mes demandes.

Princeton, Décembre 1982   A. Borel

Curriculum vitae

Né à la Chaux-de-Fonds, Neuchâtel, Suisse le 21 mai 1923

Études à la section de Mathématiques et Physique de l'École Polytechnique Fédérale, Zürich, 1942–1947. Diplôme en mathématiques au printemps 1947
Assistant à l'E.P.F., 1947–1949
Boursier du C.N.R.S., Paris, 1949–50
Remplaçant du professeur d'algèbre à l'Université de Genève, 1950–1952
Doctorat ès Sciences, Université de Paris, 1952
Membre de l'Institute for Advanced Study, Princeton, 1952–54
Visiting lecturer, University of Chicago, 1954–55
Professeur à l'E.P.F., Zürich, 1955–57
Professeur à l'Institute for Advanced Study, depuis 1957

Invitations de un mois ou plus:
MIT, Cambridge, Mass., printemps 1958, automne 1969
Tata Institute of Fundamental Research, Bombay, janvier-mars 1961, janvier 1968
Université de Paris, janvier-juin 1964
Université de Genève, printemps 1966
Cours à Columbia University, New York, printemps 1968
University of Utrecht, The Netherlands, printemps 1971
University of Buenos Aires, Argentina, août 1973
Institut des Hautes Etudes Scientifiques, Bures-sur-Yvette, automne 1973, juin 1978
Cours à Princeton University, octobre 1974–janvier 1975
University of California at Berkeley, printemps 1975
University of Chicago, printemps 1976
University of Washington, Seattle (Walker Ames Professorship), été 1976
Collège de France, Paris, mai 1977
Yale University, automne 1978
University of Amsterdam, mai 1978
University of Mexico, été 1953, été 1979
Mathematics Institute, Academia Sinica, Beijing, Chine, mai-juin 1981

Membre du Comité de rédaction: Annals of Mathematics 1962–1979; Inventiones Math., depuis 1979

Table des matières

Volume III

81.

On the automorphisms of certain subgroups
of semi-simple Lie groups

Proc. Colloquium on Algebraic Geometry 1968, Tata Institute, Bombay (1969) 43–73

LET L be a group. We denote by $E(L)$ the quotient group Aut L/Int L of the group Aut L of automorphisms of L by the group Int L of inner automorphisms Int $a: x \to a.x.a^{-1}(a, x \in L)$ of L. Our first aim is to show that $E(L)$ is finite if L is arithmetic, S-arithmetic (see 3.2) or uniform in a semi-simple Lie group (with some exceptions, see Theorem 1.5, Theorem 3.6 and Theorem 5.2 for the precise statements). A slight variant of the proof also shows that in these cases L is not isomorphic to a proper subgroup of finite index. As a consequence, a Riemannian symmetric space with negative curvature, and no flat component, has infinitely many non-homeomorphic compact Clifford-Klein forms, Theorem 6.2.

Further information on $E(L)$ is obtained when L is an S-arithmetic group of a semi-simple k-group G (with some conditions on G and S). If L contains the center of G_k, and G is simply connected, then $E(L)$ is essentially generated by four kinds of automorphisms: exterior automorphisms of G, automorphisms deduced from certain automorphisms of k, automorphisms of the form $x \to f(x). x$ where f is a suitable homomorphism of L into the center of G_k, and automorphisms induced by the normalizer $N(L)$ of L in G (see Lemma 1.8, Remark in 1.9). In the case where G is split and L is the group of $\mathfrak{o}(S)$-points of G for its canonical integral structure, these results are made more precise (Theorem 2.2, Theorem 4.3), and $N(L)/L$ is put into relation with the S-ideal class group of k and S-units (see Lemma 2.3, Lemma 4.5; these results overlap with those of Allan [1, 2]). As an illustration, we discuss Aut L for some classical groups (Examples 2.6, 4.6). The results are related to those of O'Meara [24] if $G = \mathbf{SL}_n$, and of Hua-Reiner [12, 13] and Reiner [27] if $G = \mathbf{SL}_n, \mathbf{Sp}_{2n}$, and $k = \mathbf{Q}, \mathbf{Q}(i)$.

The finiteness of $E(L)$ follows here rather directly from rigidity theorems [25], [26], [32], [33]. The connection between the two is established by Lemma 1.1.

NOTATION.   In this paper, all algebraic groups are linear, and we follow in general the notations and conventions of [9]. In particular, we make no notational distinction between an algebraic group G over a field k and its set of points in an algebraic closure $\bar{k}$ of the field of definition (usually $\mathbf{C}$ here). The Lie algebra of a real Lie group or of an algebraic group is denoted by the corresponding German letter.

If L is a group and V a L-module, then $H^1(L, V)$ is the 1st cohomology group of L with coefficients in V. In particular, if L is a subgroup of the Lie group G, then $H^1(L, \mathfrak{g})$ is the 1st cohomology group of L with coefficients in the Lie algebra $\mathfrak{g}$ of G, on which L operates by the adjoint representation.

A closed subgroup L of a topological group G is *uniform* if G/L is compact.

1.   Uniform or arithmetic subgroups.

LEMMA 1.1.   *Let G be an algebraic group over* $\mathbf{R}$, *L a finitely generated subgroup of $G_{\mathbf{R}}$ and N the normalizer of L in $G_{\mathbf{R}}$.   Assume that $H^1(L, \mathfrak{g}) = 0$.   Then the group of automorphisms of L induced by elements of N has finite index in* Aut L.

Let L_0 be a group isomorphic to L and ι an isomorphism of L_0 onto L. For $M = G_{\mathbf{R}}$, G, let $R(L_0, M)$ be the set of homomorphisms of L_0 into M. Let (x_i) $(1 \leqslant i \leqslant q)$ be a generating set of elements of L. Then $R(L_0, M)$ may be identified with a subset of M^q, namely, the set of m-uples (y_i) which satisfy a set of defining relations for L_0 in the x_i, x_i^{-1}. In particular $R(L_0, G)$ is an affine algebraic set over $\mathbf{R}$, whose set of real points is $R(L_0, G_{\mathbf{R}})$. The group M operates on $R(L_0, M)$, by composition with inner automorphisms, and G is an algebraic transformation group of $R(L_0, G)$, with action defined over $\mathbf{R}$.

To $\alpha \in \mathrm{Aut}\, L$, let us associate the element $j(\alpha) = \alpha \circ \iota$ of $R(L_0, G)$. The map j is then a bijection of $\mathrm{Aut}\, L$ onto the set $I(L_0, L) \subset R(L_0, G_{\mathbf{R}})$ of isomorphisms of L_0 onto L. If $a, b \in I(L_0, L)$, then $b \in G_{\mathbf{R}}(a)$ if and only if there exists $n \in N$ such that $b = (\mathrm{Int}\, n) \circ a$. Our assertion is therefore equivalent to: "$I(L_0, L)$ is contained in finitely many orbits of $G_{\mathbf{R}}$," which we now prove.

Let $b \in I(L_0, L)$. Since

$$H^1(b\,(L^0), \mathfrak{g}) = H^1(L, \mathfrak{g}) = H^1(L, \mathfrak{g}_{\mathbf{R}}) \otimes \mathbf{C},$$

we also have $H^1(b(L_0), \mathfrak{g}) = 0$. By the lemma of [33], it follows that the orbit $G(b)$ contains a Zariski-open subset of $R(L_0, G)$. Since the latter is the union of finitely many irreducible components, this shows that $I(L_0, L)$ is contained in finitely many orbits of G. But an orbit of G containing a real point can be identified to a homogeneous space G/H where H is an algebraic subgroup of G, defined over $\mathbf{R}$. Therefore its set of real points is the union of finitely many orbits of $G_{\mathbf{R}}([8], \S6.4)$, whence our contention.

REMARKS. 1.2. (i) The lemma and its proof remain valid if $\mathbf{R}$ and $\mathbf{C}$ are replaced by a locally compact field of characteristic zero K and an algebraically closed extension of K.

(ii) The group $\mathbf{SL}(2, \mathbf{Z})$ has a subgroup of finite index L isomorphic to the free group on m generators, where $m \geqslant 2$ (and in fact may be taken arbitrarily large). The group $E(L)$ has the group $\mathbf{GL}(m, \mathbf{Z}) = \mathrm{Aut}\,(L/(L, L))$ as a quotient, hence is infinite. On the other hand, L has finite index in its normalizer in $\mathbf{SL}(2, \mathbf{C})$, as is easily checked (and follows from Proposition 3.3 (d)). Thus, 1.1 implies that $H^1(L, \mathfrak{g}) \neq 0$, as is well known. Similarly, taking (i) into account, we see that the free uniform subgroups of $\mathbf{PSL}(2, \mathbf{Q}_p)$ constructed by Ihara [15] have non-zero first cohomology group with coefficients in $\mathfrak{g}$.

LEMMA 1.3. *Let G be an algebraic group over $\mathbf{R}$, L a finitely generated discrete subgroup of $G_{\mathbf{R}}$ such that $G_{\mathbf{R}}/L$ has finite invariant measure. Assume that $H^1(L', \mathfrak{g}) = 0$ for all subgroups of finite index L' of L. Then L is not isomorphic to a proper subgroup of finite index.*

Define L_0, M, ι, $R(L_0, M)$ and the action of M on $R(L_0, M)$ as
in the proof of Lemma 1.1. Let C be the set of monomorphisms of L_0
onto subgroups of finite index of L. Then, $j: \alpha \longmapsto \alpha \circ \iota$ is a bijection
of C onto a subset J of $R(L_0, G_\mathbf{R})$, and the argument of Lemma 1.1
shows that J is contained in the union of finitely many orbits of $G_\mathbf{R}$.
Fix a Haar measure on $G_\mathbf{R}$, and hence on all quotients of $G_\mathbf{R}$ by
discrete subgroups. The total measure $m(G_\mathbf{R}/L')$ is finite for every
subgroup of finite index of L, since $m(G_\mathbf{R}/L)$ is finite. If $b, c \in C$, and
$b = \mathrm{Int}\, g \circ c$, $(g \in G_\mathbf{R})$, then $m(G_\mathbf{R}/b(L)) = m(G_\mathbf{R}/c(L))$. Consequently,
$m(G_\mathbf{R}/L')$ takes only finitely many values, as L' runs through the
subgroups of finite index of L, isomorphic to L. But, if there is one
such group $L' \neq L$, then there is one of arbitrary high index in L, a
contradiction.

LEMMA 1.4. *Let L be a finitely generated group, M a normal
subgroup of finite index, whose center is finitely generated, N a characteristic finite subgroup of L.*

(a) *If $E(M)$ is finite, then $E(L)$ is finite.*

(b) *If $E(L/N)$ is finite, then $E(L)$ is finite.*

(a)   It is well known and elementary that a finitely generated
group contains only finitely many subgroups of a given finite index
(see e. g. [11]). Therefore, the group Aut (L, M) of automorphisms of
L leaving M stable has finite index in Aut(L). Since L/M is finite,
the subgroup Q of elements of Aut(L, M) inducing the identity on
L/M has also finite index. Let $r: Q \to \mathrm{Aut}\, M$ be the restriction map.
Our assumption implies that $r^{-1}(\mathrm{Int}\, M)$ has finite index in Q, hence
that $\mathrm{Int}\, L.\ker r$ is a subgroup of finite index of $Q.\mathrm{Int}\, L$. It suffices
therefore to show that $\ker r \cap \mathrm{Int}\, L$ has finite index in $\ker r$. Let
$b \in \ker r$. Write

$$b(x) = u_x.x \quad (x \in L).$$

Then $u_x \in M$ and routine checking shows: the map $u: x \longmapsto u_x$ is a
1-cocycle on L, with coefficients in the center C of M, which is
constant on the cosets mod M, and may consequently be viewed as a
1-cocycle of L/M with coefficients in C; furthermore, two cocycles
thus associated to elements $b, c \in Q$ are cohomologous if and only if

there exists $n \in C$ such that $b = \mathrm{Int}\, n \circ c$, and any such cocycle is associated to an automorphism. Therefore

$$\ker r/\mathrm{Int}_L C \cong H^1(L/M, C).$$

By assumption, L/M is finite, and C is finitely generated. Hence the right hand side is finite, which implies our assertion.

(b)   The group N being characteristic, we have a natural homomorphism $\pi : \mathrm{Aut}\, L \to \mathrm{Aut}\, L/N$. The finiteness of $E(L/N)$ implies that $\mathrm{Int}\, L.\ker \pi$ has finite index in $\mathrm{Aut}\, L$. Moreover $\ker \pi$ consists of automorphisms of the form $x \longmapsto x.v_x$, $(x \in L,\, v_x \in N)$, and is finite, since N is finite and L is finitely generated.

THEOREM 1.5.   *Let G be a semi-simple Lie group, with finitely many connected components, whose identity component G^0 has a finite center, and L a discrete subgroup of G. Then $E(L)$ is finite if one of the two following conditions is fulfilled:*

(a) G/L is compact, G^0 has no non-compact three-dimensional factor;

(b) $\mathrm{Aut}(\mathfrak{g} \otimes \mathbf{C})$ may be identified with an algebraic group G' over $\mathbf{Q}$, such that the image L' of $L \cap G$ in G' by the natural projection is an arithmetic subgroup of G', and $G'_{\mathbf{R}}$ has no factor locally isomorphic to $\mathbf{SL}(2, \mathbf{R})$ on which the projection of L' is discrete.

(a)   Let A be the greatest compact normal subgroup of G^0 and $\pi: G^0 \to G^0/A$ the canonical projection. Since the center of G^0 is finite, it is contained in A, and G^0/A is the direct product of non-compact simple groups with center reduced to $\{e\}$. The group $\pi(L \cap G^0)$ is discrete and uniform in G^0/A. By density [4], its center is contained in the center of G^0/A, hence is reduced to $\{e\}$. Consequently, the center of $L \cap G^0$ is contained in $A \cap L$, hence is finite. We may then apply Lemma 1.4(a), which reduces us to the case where G is connected. Moreover, by [4], any finite normal subgroup of $\pi(L)$ is central in G/A, hence reduced to $\{e\}$. Therefore $A \cap L$ is the greatest finite normal subgroup of L, and is characteristic. By Lemma 1.4(b), it suffices to show that $E(\pi(L))$ is finite. We are thus reduced to the case where G has no center, and is a direct product of non-compact simple groups of dimension > 3. In particular, G is of finite index in the group of real points of an algebraic group defined over $\mathbf{R}$, namely Aut $(\mathfrak{g} \otimes \mathbf{C})$. By a theorem of Weil [33], $H^1(L, \mathfrak{g}) = 0$, hence by Lemma 1.1, it is enough to show that L has finite index in its normalizer $N(L)$. By [4], $N(L)$ is discrete. Since G/L is fibered by $N(L)/L$, and is compact, $N(L)/L$ is finite.

(b) Let A be the greatest normal $\mathbf{Q}$-subgroup of G'^0, whose group of real points is compact, and let π be the composition of the natural homomorphisms

$$G^0 \to \operatorname{Ad} \mathfrak{g} \to (G'^0)_{\mathbf{R}}/A_{\mathbf{R}}.$$

G'/A is a $\mathbf{Q}$-group without center, which is a product of $\mathbf{Q}$-simple groups, each of which has dimension > 3 and a non-compact group of real points. $\pi(L)$ is arithmetic in G'/A ([6], Theorem 6) and $L \cap \ker \pi$ is finite. By Zariski-density ([6], Theorem 1), any finite normal subgroup of $\pi(L)$ is central in G'/A, hence reduced to $\{e\}$. Thus $L \cap \ker \pi$ is the greatest finite normal subgroup of L, and is characteristic. Also the center of $\pi(L \cap G^0)$ is central in Ad G', hence reduced to $\{e\}$, and the center of $L \cap G^0$ is compact, and therefore finite. By Lemma 1.4, we are thus reduced to the case where G, G' are connected, $A = \{e\}$, and L is arithmetic in G'. Let $G_1, \ldots, G_q$ be the simple $\mathbf{Q}$-factors of G'. The group L is commensurable with the product of the intersections $L_i = L \cap G_i$, which are arithmetic ([7], 6.3). If $G_i \cap L$ is uniform in $G_{i\mathbf{R}}$, then $H^1(L_i, \mathfrak{g}_{\mathbf{R}}) = 0$ by [33]. If not, then $rk_{\mathbf{Q}} G_i \geqslant 1$, and $H^1(L_i, \mathfrak{g}_i) = 0$ by theorems of Raghunathan [25], [26]. Consequently, $H^1(L \cap G', \mathfrak{g}'_{\mathbf{R}}) = 0$. Moreover, $L \cap G'$ is of finite index in its normalizer N. In fact, N is closed, belongs to $G'_{\mathbf{Q}}$ ([6], Theorem 2), hence to $G'_{\mathbf{R}}$, and $N/(L \cap G')$ is compact since $G'_{\mathbf{R}}/(L \cap G')$ has finite invariant measure [7]. The conclusion now follows from Lemma 1.1 and Lemma 1.4.

LEMMA 1.6. *Let L be a finitely generated group, N a normal subgroup of finite index. Assume L is isomorphic to a proper subgroup of finite index. Then N has a subgroup of finite index which is isomorphic to a proper subgroup of finite index.*

The assumption implies the existence of a strictly decreasing sequence (L_i) $(i = 1, 2, \dots)$ of subgroups of finite index of L and of isomorphisms $f_i : L \xrightarrow{\sim} L_i$ $(i = 1, 2, \dots)$. Let a be the index of N in L. Then $L_i \cap N$ has index $\leqslant a$ in L_i, hence $M_i = f_i^{-1}(L_i \cap N)$ has index $\leqslant a$. Passing to a subsequence if necessary, we may assume that M_i is independent of i. Then, $L_i \cap N$ is isomorphic to a proper subgroup of finite index.

PROPOSITION 1.7. *Let G and L be as in Theorem 1.5. Assume one of the conditions (a), (b) of Theorem 1.5 to be fulfilled. Then L is not isomorphic to a proper subgroup of finite index.*

By use of Lemma 1.6, the proof is first reduced to the case where G is connected. Let π be as in the proof of (a) or (b) in Theorem 1.5. Then $L \cap \ker \pi$ is the greatest finite normal subgroup of L. Similarly $L' \cap \ker \pi$ is the greatest finite normal subgroup of L', if L' has finite index in L. Therefore, if L' is isomorphic to L, the groups $\ker \pi \cap L$ and $\ker \pi \cap L'$ are equal, and are mapped onto each other by any isomorphism of L onto L'; hence $\pi(L) \simeq \pi(L')$, and $\pi(L) \neq \pi(L')$ if $L \neq L'$. We are thus reduced to the case where the group A of (a) or (b) in Lemma 1.4 is $= \{e\}$. Moreover, in case (b), it suffices to consider $L \cap G'$ in view of Lemma 1.6. Our assertion then follows from the rigidity theorems of Weil and Raghunathan and from Lemma 1.3.

We shall need the following consequence of a theorem of Raghunathan:

LEMMA 1.8. *Let G, G' be connected semi-simple $\mathbf{Q}$-groups, which are almost simple over $\mathbf{Q}$. Let L be a subgroup of $G_{\mathbf{Q}}$ containing an arithmetic subgroup L_0 of G, and s an isomorphism of L onto a subgroup of $G'_{\mathbf{Q}}$ which maps L_0 onto a Zariski-dense subgroup of G'. Assume that $rk_{\mathbf{Q}}(G) \geqslant 2$.*

(i) *G and G' are isogeneous over $\mathbf{Q}$.*

(ii) *If G is simply connected, or G' is centerless, there exists a $\mathbf{Q}$-isogeny s' of G onto G', and a homomorphism g of L into the center of $G'_{\mathbf{Q}}$ such that $s(x) = s'(x) \cdot g(x)$ $(x \in L)$.*

Let $\widetilde{G}$ be the universal covering group of G, $\pi : \widetilde{G} \to G$ the canonical projection and $\widetilde{L} = \pi^{-1}(L) \cap G_{\mathbf{Q}}$. The group $\widetilde{L}_0 = \pi^{-1}(L_0) \cap \widetilde{L}$ is arithmetic, as follows e.g. from ([7], §6.11); in particular, $\pi(\widetilde{L}_0)$ has finite index in L_0, and $L_0' = s \circ \pi(\widetilde{L}_0)$ is Zariski-dense in G'.

We identify G' with a $\mathbf{Q}$-subgroup of $\mathbf{GL}(n, \mathbf{C})$, for some n. The map $r = s \circ \pi$ may be viewed as a linear representation of $\widetilde{L}$ into $\mathbf{GL}(n, \mathbf{Q})$. By Theorem 1 of [25], there exists a normal subgroup $\widetilde{N}$ of $\widetilde{L}_0$, Zariski-dense in $\widetilde{G}$, and a morphism $t : \widetilde{G} \to \mathbf{GL}(n, \mathbf{C})$ which coincides with r on $\widetilde{N}$. Let C be the Zariski-closure of $t(\widetilde{N})$. It is an algebraic subgroup contained in $t(\widetilde{G}) \cap G'$. Since $t(\widetilde{N})$ is normal in L_0', and L_0' is Zariski-dense, the group C is normal in G'. However, ([9], §6.21(ii)), the group G' is isogeneous to a group $R_{k/\mathbf{Q}}H$, where k is a number field, H an absolutely simple k-group, and $R_{k/\mathbf{Q}}$ denotes restriction of the scalars ([31], Chap. I). Consequently, an infinite subgroup of $G_{\mathbf{Q}}'$ is not contained in a proper direct factor of G', whence $C = G' = t(\widetilde{G})$. If f is a regular function defined over $\mathbf{Q}$ on G', then $f \circ t$ is a regular function on $\widetilde{G}$, which takes rational values on the dense set $\widetilde{N}$. It follows immediately that $f \circ t$ is defined over $\mathbf{Q}$, hence t is defined over $\mathbf{Q}$. Its kernel is a proper normal $\mathbf{Q}$-subgroup of G', hence is finite, and t is a $\mathbf{Q}$-isogeny. This implies (i).

If G' is centerless, then $Z(\widetilde{G})$ belongs to the kernel of t. Thus, if G is simply connected, or G' centerless, t defines a $\mathbf{Q}$-isogeny s' of G onto G', which coincides with s on the Zariski-dense subgroup $N = \pi(\widetilde{N})$. The group $s(N)$ is then Zariski-dense in G'. Let $x \in L$, $y \in N$. Then $x . y . x^{-1} \in N$, hence $s(x) . s'(x)^{-1}$ centralizes $s(N)$, and therefore also G'. Consequently, $s(x).S'(x)^{-1}$ belongs to $G_{\mathbf{Q}}' \cap Z(G')$, and $f : x \longmapsto s(x).s'(x)^{-1}$, and s' fulfil our conditions.

THEOREM 1.9. *Let G and L be as in Lemma 1.8. Assume G to be centerless and L to be equal to its normalizer in $G_{\mathbf{Q}}$. Then $E(L)$ may be identified with a subgroup of $(\mathrm{Aut}\ G)_{\mathbf{Q}}/(\mathrm{Int}\ G)_{\mathbf{Q}}$.*

If G is centerless, and M is a Zariski-dense subgroup of $G_{\mathbf{Q}}$, then the normalizer of M in G belongs to $G_{\mathbf{Q}}$. This follows from

Theorem 2 in [6] if M is arithmetic, but the proof yields this more general statement, as well as Proposition 3.3(b) below. In view of this, Theorem 1.9 is a consequence of Lemma 1.8.

REMARK. It is no great loss in generality to assume that L contains the center of G_Q, and this assumption will in fact be fulfilled in the cases to be considered below. In this case, Aut L is generated by three kinds of automorphisms: (a) exterior automorphisms of G leaving L stable, (b) automorphisms $x \longmapsto f(x).x$, where f is a homomorphism of L into its center, (c) automorphisms of the form $x \longmapsto y.x.y^{-1}$, where y belongs to the normalizer of L in G.

Using some information on these three items, we shall in the following paragraph give a more precise description of Aut L, when G is a split group.

2. Arithmetic subgroups of split groups over Q. *In this paragraph G is a connected semi-simple and almost simple Q-group, which is split, of Q-rank $\geqslant 2$; L is the group of integral points of G for the canonical Z-structure associated to a splitting of G [10], [18], and $N(L)$ the normalizer of L in G_C.*

2.1. The group L is equal to its normalizer in G_Q, and also to its normalizer in G if G has no center. To see this, we first notice that L has finite index in its normalizer in G. In fact, since the image of L in Int G is arithmetic ([7], §6.11), it suffices to show that if G is centerless, any arithmetic subgroup of G is of finite index in its normalizer, which follows from the end argument of Theorem 1.5(b). Our assertion is then a consequence of ([6], Theorem 7). Another proof will be given below (Theorem 4.3).

THEOREM 2.2. *If G is centerless, then* Aut L *is a split extension of $E(G) = $ Aut $G/$Int G by L. If G is simply connected, then* Aut L *is a split extension of $E(G)$ by the subgroup A of automorphisms of the form $x \longmapsto f(x).y.x.y^{-1}$ where $y \in N(L)$ and f is a homomorphism of L into its center.*

If G is centerless or simply connected, the standard construction of automorphisms of G leaving stable the splitting of G yields a subgroup $E'(G)$ of (Aut $G)_{\mathbf{Q}}$, isomorphic to $E(G)$ under the canonical projection, and leaving the $\mathbf{Z}$-structure of G invariant. Thus $E(G)$ may be identified with a subgroup of Aut L. If G is centerless, the theorem follows then from Theorem 1.9, and 2.1. Let now G be simply connected. Let $s \in$ Aut L. By Lemma 1.8, we can find $s' \in$ (Aut $G)_{\mathbf{Q}}$ and $f \in$ Hom $(L, Z(G)_{\mathbf{Q}})$ such that $s(x) = f(x).s'(x)$ $(x \in L)$. However, L contains $Z(G)_{\mathbf{Q}}$ by 2.1. Therefore, f maps L into $Z(G) \cap L$. But $Z(G) \cap L$ is equal to the center of L, since L is Zariski-dense in G, whence our assertion.

If G has a non-trivial center the image of $N(L)$ in Aut L is in general different from Int L. The quotient $N(L)/L$ has been studied in various cases, including those of Examples 2.6(1), (2), notably by Maass, Ramanathan, Allan (see [1], where references to earlier work are also given). We shall discuss it here and in §4 from a somewhat different point of view. In the following statement, the group $G' =$ Int G is endowed with the $\mathbf{Z}$-structure associated to the splitting defined by the given splitting of G.

LEMMA 2.3. *Let $\pi: G \to G' =$ Int G be the canonical projection, T the maximal torus given by the splitting of G, and $T' = \pi(T)$. Then*

$$\pi(N(L)) = G'_{\mathbf{Z}}, \tag{1}$$

$$\pi(N(L))/\text{Int } L \simeq T'_{\mathbf{Z}}/\pi(T_{\mathbf{Z}}) \simeq Z(L). \tag{2}$$

The group L is the normalizer in $G_{\mathbf{Q}}$ of a Chevalley lattice $\mathfrak{g}_{\mathbf{Z}}$ in $\mathfrak{g}$ as follows from 2.17 in [16]. Moreover, a Chevalley lattice is spanned by the logarithms of the unipotent elements in L. Consequently $N(L)$ is the normalizer in G of $\mathfrak{g}_{\mathbf{Z}}$. From this (1) follows.

Let B be the maximal solvable subgroup of G corresponding to the positive roots in the given splitting of G and U its unipotent radical. Then $B = T.U$ (semi-direct). Let $x \in N(L)$. Since Int x preserves the $\mathbf{Q}$-structure of G, the group $x.B.x^{-1}$ is a maximal connected solvable subgroup defined over $\mathbf{Q}$, hence ([9], §4.13) there exists $z \in G_{\mathbf{Q}}$ such that $z.B.z^{-1} = x.B.x^{-1}$. But we have $G_{\mathbf{Q}} = L.B_{\mathbf{Q}}$ ([6], Lemma 1). Since B is equal to its normalizer, it follows that $N(L) = L.(N(L) \cap B)$. Let now $x \in N(L) \cap B$. Write $x = t.v$ $(t \in T, v \in U)$. We have $\pi(x) \in L$

(see 2.1). But, with respect to a suitable basis of a Chevalley lattice in $\mathfrak{g}_\mathbf{Q}$, $\pi(t)$ is diagonal, and $\pi(u)$ upper triangular, unipotent, therefore $\pi(t)$, $\pi(u) \in L$. However [10], π defines a $\mathbf{Z}$-isomorphism of U onto $\pi(U)$, hence $u \in L$, which shows that

$$N(L) = L.(N(L) \cap T). \tag{3}$$

The kernel of π is contained in T, therefore (1) implies that $N(L) \cap T$ is the full inverse image of $T'_\mathbf{Z}$, which yields the first equality of (2). The groups $T_\mathbf{Z}$ and $T'_\mathbf{Z}$ consist of the elements of order 2 of T and T' respectively and are both isomorphic to $(\mathbf{Z}/2\mathbf{Z})^l$, where l is the rank of G. Consequently $T'_\mathbf{Z}/\pi(T_\mathbf{Z})$ is isomorphic to the kernel of $\pi : T_\mathbf{Z} \to T'_\mathbf{Z}$, i.e. to $Z(L)$, which ends the proof of (2).

The determination of Aut L/Int L is thus to a large extent reduced to that of the center $Z(L)$ of L, and of the quotient of L by its commutator subgroup (L,L). We now make some remarks on these two groups.

2.4. The center $Z(L)$ of L is of order two if G is simply connected of type $\mathbf{A}_n$ (n odd), $\mathbf{B}_n$, $\mathbf{C}_n$ ($n \geqslant 1$), $\mathbf{D}_n$ ($n \geqslant 3$, n odd), $\mathbf{E}_7$, of type $(2, 2)$ if $G = \mathbf{Spin}\ 4m$ (m positive integer), of order one in the other cases.

In fact the $\mathbf{Z}$-structure on G may be defined by means of an admissible lattice in the representation space of a faithful representation defined over $\mathbf{Q}$. If we assume $G \subset \mathbf{GL}(n, \mathbf{C})$ and $\mathbf{Z}^n$ to be an admissible lattice, then $Z(L)$ is represented by diagonal matrices with integral coefficients, which shows first that $Z(L)$ is an elementary abelian 2-group. All almost simple simply connected groups have faithful irreducible representations, except for the type $\mathbf{D}_{2m}$. Thus, except in that case, $Z(L)$ is of order 2 (resp. 1) if $Z(G)$ has even (resp. odd) order, whence our contention. The case of $\mathbf{D}_{2m}$ is settled by considering the sum of the two half-spinor representations.

2.5. It is well known that $\mathbf{SL}(n, \mathbf{Z})$ is equal to its commutator subgroup if $n \geqslant 3$ (see [3] e.g.). Also the commutator subgroup of $\mathbf{Sp}(2n, \mathbf{Z})$ is equal to $\mathbf{Sp}(2n, \mathbf{Z})$ if $n \geqslant 3$, has index two if $n = 2$

[3], [28]. More generally, if the congruence subgroup theorem holds, which is the case if $rkG \geqslant 2$ and G is simply connected, according to [22], then $L/(L, L)$ is the product of the corresponding local groups $G_{0_p}/(G_{0_p}, G_{0_p})$. Serre has pointed out to me that, using this, one can show that $L = (L, L)$ if G has rank $\geqslant 3$, and is simply connected. Another more direct proof was mentioned to me by R. Steinberg, who also showed that $L/(L,L)$ is of order two if $G = \mathbf{G}_2$. He uses known commutation rules among unipotent elements of L, and the fact that they generate L.

EXAMPLES. 2.6. (1) $G = \mathbf{SL}(n, \mathbf{C})$, $L = \mathbf{SL}(n, \mathbf{Z})$, $(n \geqslant 3)$. In this case, $E(G)$ is of order two, generated by the automorphism $\sigma : x \longmapsto {}^t x^{-1}$. By Lemma 2.3 and 2.4, Int L has index one (resp. two) in the image of $N(L)$ if n is odd (resp. even). Furthermore, it is easily seen, and will follow from Lemma 4.5, that, in the even dimensional case, the non-interior automorphisms defined by $N(L)$ are of the form $x \longmapsto y.x.y^{-1}(y \in \mathbf{GL}(n, \mathbf{Z}), \det y = -1)$. Thus, taking 2.5 into account, we see that Aut L is generated by Int L, σ, and, for n even, by one further automorphism induced by an element of $\mathbf{GL}(n, \mathbf{Z})$ of determinant -1. This is closely related to results of Hua-Reiner [12], [13].

(2) $G = \mathbf{Sp}(2n, \mathbf{C})$, $L = \mathbf{Sp}(2n, \mathbf{Z})$, $(n \geqslant 2)$. Here, $E(G)$ is reduced to the identity. Thus, by the above, Int L has index two in Aut L if $n \geqslant 3$, index four if $n = 2$. The non-trivial element of $N(L)/L$ is represented by an automorphism of the form $x \longmapsto y.x.y^{-1}$ where y is an element of $\mathbf{GL}(2n, \mathbf{Z})$ which transforms the bilinear form underlying the definition of $\mathbf{Sp}(2n, \mathbf{C})$ into its opposite (see Examples 4.6). For $n = 2$, one has to add the automorphism $x \to \chi(x).x$, where χ is the non-trivial character of L. This result is due to Reiner [27].

(3) G is simply connected, of type $\mathbf{D}_{2m}$. Then we have a composition series

$$\text{Aut } L \supset A \supset \text{Int } L,$$

where Aut L/A is of order two, and $A/\text{Int } L$ has order two if m is odd, is of type $(2,2)$ if m is even. This follows from (2.2), (2.3), (2.4), (2.5). The other simple groups of rank $\geqslant 3$ are discussed similarly.

3. S-arithmetic groups over number fields. 3.1. Throughout the rest of this paper, k is an algebraic number field of finite degree over $\mathbf{Q}$, $\mathfrak{o}$ its ring of integers, V the set of primes of k, V_∞ the set of infinite primes of k, S a finite subset of V containing V_∞, and $\mathfrak{o}(S)$ the subring of $x \in k$ which are integral outside S. We let $I(k,S)$ be the S-ideal class group of k, i.e. the quotient of the group of fractional $\mathfrak{o}(S)$-ideals by the group of principal $\mathfrak{o}(S)$-ideals. We follow the notation of [5]. In particular k_v is the completion of k at $v \in S$, $\mathfrak{o}_v$ the ring of integers of k_v. If G is a k-group, then G^0 is its identity component, and

$$G_v = G_{k_v}(v \in S),\ G_S = \prod_{v \in S} G_v,\ G_\infty = \prod_{v \in V_\infty} G_v.$$

Moreover $G' = R_{k/\mathbf{Q}}G$ is the group obtained from G by restriction of the groundfield from k to $\mathbf{Q}$([31], Chap. I), and we let μ denote the canonical isomorphism of G_k onto $G'_{\mathbf{Q}}$.

If A is an abelian group, and q a positive integer, we let $_qA$ and $A^{(q)}$ denote the kernel and the image of the homomorphism $x \longmapsto x^q$.

3.2. Let G be a k-group. A subgroup L of G_k is S-arithmetic if there is a faithful k-morphism $r: G \to \mathbf{GL}_n$ such that $r(L)$ is commensurable with $r(G)_{\mathfrak{o}(S)}$.

If S' is a finite set of primes of $\mathbf{Q}$, including ∞, and S is the set of primes dividing some element of S', then $\nu: G_k \xrightarrow{\ \sim\ } G'_{\mathbf{Q}}$ induces a bijection between S-arithmetic subgroups of G and S'-arithmetic subgroups of G'. This follows directly from the remarks made in ([5], § 1).

In the following proposition, we collect some obvious generalizations of known facts.

PROPOSITION 3.3. *Let G be a semi-simple k-group, L a S-arithmetic subgroup of G, N the greatest normal k-subgroup of G^0 such that N_∞ is compact, and $\pi: G \to G/N$ the natural projection.*

(a) *If N is finite and G is connected, L is Zariski-dense in G.*

(b) *If G is connected, the commensurability group $C(L)$ of L in G is equal to $\pi^{-1}((G/N)_k)$.*

(c) *If $\sigma:G\to H$ is a surjective k-morphism, $\sigma(L)$ is S-arithmetic in H.*

(d) *If N is finite, L has finite index in its normalizer in G.*

(a) follows from ([6], Theorem 3), and from the fact that L contains an arithmetic subgroup of G.

(b) We recall that $C(L)$ is the group of elements $x \in G$ such that $x.L.x^{-1}$ is commensurable with L. The proof of (b) is the same as that of Theorem 2 in [6]. In fact, this argument shows that if G is centerless, then $C(M) \subset G_k$ whenever M is a subgroup of G_k Zariski-dense in G.

(c) If σ is an isomorphism, the argument is the same as that of ([7], §6.3). If σ is an isogeny, this has been proved in ([5], §8.12). From there, the extension to the general case proceeds exactly in the same way as in the case $S = V_{\infty}$([6], Theorem 6).

(d) We may assume G to be connected and $N = \{e\}$. Then by (c), $N(L) \subset G_k$. We view G_k and L as diagonally embedded in G_S. Then L is discrete in G_S.

The group L has a finite system of generators [17], say $(x_i)_{1 \leqslant i \leqslant q}$. Since L is discrete, there exists a neighbourhood U of e in G_S such that if $x \in N(L) \cap U$, then x centralizes the x_i's, hence L. The latter being Zariski-dense in G, this implies that the component x_v of x in G_v $(v \in S)$ is central in G_v, whence $x_v = e$, which shows that $N(L)$ is discrete in G_S. But G_S/L has finite invariant volume ([5], §5.6) and is fibered by $N(L)/L$, hence $N(L)/L$ is finite.

PROPOSITION 3.4. *Let G be a semi-simple k-group, L a subgroup of G_k which is Zariski-dense, and is equal to its normalizer in G_k, and $N(L)$ the normalizer of L in G. Then $N(L)/L$ is a commutative group whose exponent divides the order m of the center $Z(G)$ of G.*

We show first that if $x \in N(L)$, then $x^m \in L$. In view of the assumption, it suffices to prove that $x^m \in G_k$. Let $\pi : G \to G/Z(G)$ be the canonical projection. The fiber $F_x = \pi^{-1}(\pi(x))$ of x consists of the elements $x.z_i(1 \leqslant i \leqslant m)$, where z_i runs through $Z(G)$, and belongs to $N(L)$. By the remark made in Proposition 3.3 (b), $\pi(x)$ is rational over k, hence F_x is defined over k, and its points are permuted by the Galois group of $\bar{k}$ over k. Since the z_i's are central, the product of the xz_i is equal to $x^m . z_1...z_m$ and is rational over k. Similarly the product of the z_i's is rational over k, whence our assertion.

It is possible to embed $Z(G)$ as a k-subgroup in a k-torus T' whose first Galois cohomology group is zero (see Ono, Annals of Math. (2), 82 (1965), p. 86). Let $H = (G \times T')/Z(G)$ where $Z(G)$ is embedded diagonally in $G \times T'$. Then $G/Z(G)$ may be identified with H/T'. Let $x \in N(L)$. We have already seen that $\pi(x)$ is rational over k. But, since T' has trivial first Galois-cohomology group, the map $H_k \to (H/T')_k$ is surjective. There exists therefore $d \in T'$ such that $d.x \in H_k$. Thus, if $x,y \in N(L)$, we can find two elements x', $y' \in H_k$, which normalize L, whose commutator (x', y') is equal to (x, y). But, obviously, $G = (H, H)$, therefore $(x, y) \in N(L)_k$ hence $(x, y) \in L$, and $(N(L), N(L)) \subset L$. This argument also proves that

$$N(L) = G \cap (N_H(L)_k . T'), \tag{1}$$

where $N_H(L)$ is the normalizer of L in H, and $N_H(L)_k = N_H(L) \cap H_k$.

For the sake of reference, we state as a lemma a remark made by Ihara in ([14], p. 269).

LEMMA 3.5. *Let A be a group, B a subgroup, and V a A-module. Assume that for any $a \in A$, there is no non-zero element of V fixed under $a.B.a^{-1} \cap B$. Then the restriction map $r: H^1(A, V) \to H^1(B, V)$ is injective.*

It suffices to show that if z is a 1-cocycle of A which is zero on B, then z is zero. Let $a \in A$, $b \in B$ be such that $a.b.a^{-1} = b' \in B$. We have then

$$z(a.b) = z(a) = z(b'.a) = b'.z(a),$$

which shows that $z(a)$ is fixed under $a.B.a^{-1} \cap B$, hence is zero.

THEOREM 3.6. *Let G be a semi-simple k-group and L a S-arithmetic subgroup. Then $E(L)$ is finite if one of the following conditions is fulfilled :*

(a) G has no normal k-subgroup N such that N_∞ has a non-compact factor of type $\mathbf{SL}(2, \mathbf{R})$, or also of type $\mathbf{SL}(2,\mathbf{C})$ if G_S/L is not compact;

(b) *G is of type $\mathbf{SL}_2$ over k, and S has at least two elements.*

By Lemma 1.4(b), we may assume G to be connected. Let N be the greatest normal k-subgroup of G such that N_∞ is compact and $\pi: G \to G/N$ the natural projection.

(a) Arguing as in Lemma 1.4, we see that it suffices to show that $E(\pi(L))$ is finite, which reduces us to the case where G is a direct product of simple k-groups G_i. The group $L_i = G_i \cap L$ is S-arithmetic in G_i and the product of the L_i is normal of finite index in L. By Lemma 1.4, we may therefore assume L to be the product of its intersection with the G_i's. The group L has finite index in its normalizer (Proposition 3.3) and is finitely generated [17], so that, in order to deduce our assertion from Lemma 1.1, applied to L and G_∞, it suffices to show that $H^1(L, \mathfrak{g}_\infty) = 0$. Since this group is isomorphic to the product of the groups $H^1(L_i, \mathfrak{g}_{i\infty})$, we may assume G to be simple over k. Let $L_0 = L \cap G_0$.

The group L_0 is arithmetic, and therefore so is $x.L_0.x^{-1} \cap L_0 = L_{0,x}$ ($x \in L$). Consequently, $L_{0,x}$ is Zariski-dense in G ([6], Theorem 1), and has no non-zero fixed vector in $\mathfrak{g}_\infty$. By Lemma 3.5, the restriction map : $H^1(L, \mathfrak{g}_\infty) \to H^1(L_0, \mathfrak{g}_\infty)$ is injective. But $H^1(L_0, \mathfrak{g}_\infty) = 0$: if $\mathrm{rk}_k G > 1$, this follows from [25], [26]. Let now $\mathrm{rk}_k G = 0$. Then G_∞/L_0 is compact ([4], §11.6). In view of 3.2, we may further assume G to be almost absolutely simple over k. Let J be the set of $v \in V_\infty$ such that G_v is not compact and H the subgroup of G generated by the G_v's ($v \in J$). Then, by Weil's theorem ([32], [33]),

$$H^1(\Gamma_0, \mathfrak{h}) = 0. \tag{1}$$

But we have

$$H^1(\Gamma_0, \mathfrak{g}_v) = H^1(\Gamma_0, {}^v\mathfrak{g}_k) \otimes_{v(k)} k_v, \quad (v \in V_\infty), \tag{2}$$

$$H^1(\Gamma_0, \mathfrak{g}) = \prod_{v \in V_\infty} H^1(\Gamma_0, \mathfrak{g}_v), \tag{3}$$

$$H^1(\Gamma_0, \mathfrak{h}) = \prod_{v \in J} H^1(\Gamma_0, \mathfrak{g}_v), \tag{4}$$

whence $H^1(\Gamma_0, \mathfrak{g}_\infty) = 0$.

(b) N is finite, and therefore, L has finite index in its normalizer (Proposition 3.3). Again, there remains to show that $H^1(L, \mathfrak{g}_\infty) = 0$. Let $\mathbf{SL}_2 \to G$ be the covering map, and L' the inverse image of L in $\mathbf{SL}(2, k)$. The homomorphism $H^1(L, \mathfrak{g}_\infty) \to H^1(L', \mathfrak{g}_\infty)$ is injective, hence we may assume $G = \mathbf{SL}_2$. But then the vanishing of H^1 follows from [29].

REMARK. It is also true that in both cases of Theorem 3.6, the group L is not isomorphic to a proper subgroup of finite index. This is seen by modifying the proof of Theorem 3.6, in the same way as Proposition 1.7 was obtained from Theorem 1.5.

If G is almost simple over k, of k-rank > 2, then we may apply Theorem 1.9 to G'. Thus, in that case, we see that, if L contains $Z(G)_k$, the determination of Aut L is essentially reduced to that of the normalizer of L in G, of the homomorphisms of L into its center, and of the exterior automorphisms of G' leaving stable. We shall use this in §4 to get more explicit information when G is a split group. Here, we mention another consequence of Theorem 1.9.

PROPOSITION 3.7. *Let G be an almost absolutely simple k-group, of k-rank $\geqslant 2$, k' a number field, and G' an almost absolutely simple k'-group. Let L be an arithmetic subgroup of G_k, and s an isomorphism of L onto an arithmetic subgroup of G'. Then there is an isomorphism ϕ of k' onto k and the k-group $^\phi G'$ obtained from G' by change of the groundfield ϕ is k-isogeneous to G.*

Let $H = R_{k/\mathbf{Q}} G$, $H' = R_{k'/\mathbf{Q}} G'$, and M, M' the images of L and $L' = s(L)$ under the canonical isomorphisms $G_k \xrightarrow{\sim} H_{\mathbf{Q}}$ and $G'_{k'} \xrightarrow{\sim} H'_{\mathbf{Q}}$.

Then s may be viewed as an isomorphism of M onto M'. The group M' is infinite, hence $H'_{\mathbf{R}}$ is not compact, and M' is Zariski-dense in H' ([6], Theorem 1). By Lemma 1.8, H and H' are $\mathbf{Q}$-isogeneous. There exists therefore an isomorphism α of $\mathfrak{h}_{\mathbf{Q}}$ onto $\mathfrak{h}'_{\mathbf{Q}}$. But the commuting algebra of $\mathrm{ad}\,\mathfrak{h}_{\mathbf{Q}}$ (resp. $\mathrm{ad}\,\mathfrak{h}'_{\mathbf{Q}}$) in the ring of linear transformations of $\mathfrak{h}_{\mathbf{Q}}$ (resp. $\mathfrak{h}'_{\mathbf{Q}}$) into itself is isomorphic to k (resp. k'). Hence α induces an isomorphism $\phi: k' \xrightarrow{\sim} k$. Let $\mathfrak{g}''_k = {}^\phi \mathfrak{g}_{k'}$ be the Lie algebra over k obtained from $\mathfrak{g}'$ by the change of groundfield ϕ. Then, it is clear from the definition of ϕ that $\alpha = R_{k/\mathbf{Q}}\,\beta$, where β is a k-isomorphism of $\mathfrak{g}$ onto $\mathfrak{g}''$. This isomorphism is then the differential of a k-isogeny of the universal covering of G onto the k-group $^{\phi}G'$.

3.8. We need some relations between $(\mathrm{Aut}\ G)_k$ and $(\mathrm{Aut}\ G')_{\mathbf{Q}}$. For simplicity, we establish them in the context of Lie algebras, and assume G to be almost simple over $\overline{k}$. The Lie algebra $\mathfrak{g}'_{\mathbf{Q}}$ is just $\mathfrak{g}_k$, viewed as a Lie algebra over $\mathbf{Q}$. Since $\mathfrak{g}_k$ is absolutely simple, the commuting algebra of $\mathrm{ad}\mathfrak{g}'_{\mathbf{Q}}$ in $\mathfrak{gl}(\mathfrak{g}'_{\mathbf{Q}})$ may be identified to k. Let $a \in \mathrm{Aut}\ \mathfrak{g}'_{\mathbf{Q}}$. Then a defines an automorphism of $\mathfrak{gl}(\mathfrak{g}'_{\mathbf{Q}})$ leaving $\mathrm{ad}\mathfrak{g}'_{\mathbf{Q}}$ stable, and therefore an automorphism $\beta(a)$ of k. If β is the identity, this means that a is a k-linear map of $\mathfrak{g}'_{\mathbf{Q}}$, hence comes from an automorphism of $\mathfrak{g}_k$. We have therefore an exact sequence

$$1 \to \mathrm{Aut}\ \mathfrak{g}_k \to \mathrm{Aut}\ \mathfrak{g}'_{\mathbf{Q}} \to \mathrm{Aut}\ k. \tag{1}$$

Let k_0 be the fixed field of $\mathrm{Aut}\ k$ in k. Assume that $\mathfrak{g}_k = \mathfrak{g}_0 \otimes_{k_0} k$, where $\mathfrak{g}_0$ is a Lie algebra over k_0. Then, for $s \in \mathrm{Aut}\ k$, $^s\mathfrak{g}_k = \mathfrak{g}_k$, and s, acting by conjugation with respect to $\mathfrak{g}_0$, defines a s-linear automorphism of $\mathfrak{g}_k$, and therefore an automorphism a of $\mathrm{Aut}\ \mathfrak{g}_{\mathbf{Q}}$ such that $\beta(a) = s$. Thus, in this case, the sequence

$$1 \to \mathrm{Aut}\ \mathfrak{g}_k \to \mathrm{Aut}\ \mathfrak{g}'_{\mathbf{Q}} \to \mathrm{Aut}\ k \to 1 \tag{2}$$

is exact and split. Translated into group terms, this yields the following lemma:

LEMMA 3.9. *Let G be absolutely almost simple over k. Then we have an exact sequence*

$$1 \to (\mathrm{Aut}\ G)_k \to (\mathrm{Aut}\ G')_{\mathbf{Q}} \to \mathrm{Aut}\ k. \tag{1}$$

Let k_0 be the fixed field of $\mathrm{Aut}\ k$ and assume that G is obtained by extension of the field of definition from a k_0-group G_0. Then the sequence

$$1 \to (\mathrm{Aut}\ G)_k \to (\mathrm{Aut}\ G')_{\mathbf{Q}} \to \mathrm{Aut}\ k \to 1 \tag{2}$$

is exact and split. On G_k, identified with $G_{0,k}$, the group $\mathrm{Aut}\ k$ acts by conjugation.

Strictly speaking, the sequences 3.8 (1), (2) give Lemma 3.9 (1), (2) if G is centerless or simply connected (the only cases of interest below). But in the general case, we may argue in the same way as above, replacing Aut $\mathfrak{g}_k$ and Aut $\mathfrak{g}'_\mathbf{Q}$ by the images of (Aut $G)_k$ and (Aut $G')_\mathbf{Q}$ in those groups. The proof can also be carried out directly in G and G', using the structure of $R_{k/\mathbf{Q}}\, G$, and is then valid if $\mathbf{Q}$, k and Aut k are replaced by a field K, a finite separable extension K' of K, and Aut(K'/K).

REMARK. The above lemma was obtained with the help of Serre, who has also given examples where G has no k_0-form and (2) is not exact.

4. Split groups over number fields. In this paragraph, G is a connected almost simple k-split group. G is viewed as obtained by extension of the groundfield from a $\mathbf{Q}$-split group G_0, endowed with the $\mathbf{Z}$-structure associated to a splitting over $\mathbf{Q}$. G is then endowed with an $\mathfrak{o}$-structure associated to its given splitting, and G_B is well defined for any $\mathfrak{o}$-algebra B. We shall be interested mainly in the canonical S-arithmetic subgroup $G_{\mathfrak{o}(S)}$.

LEMMA 4.1. *Let G be split over k, almost simple over k, and $L = G_{\mathfrak{o}(S)}$.*

(i) *L is equal to its normalizer in G_k. The image in $G/Z(G)$ of the normalizer $N(L)$ of L in G is equal to $(G/Z(G))_{\mathfrak{o}(S)}$. In particular, $L = N(L)$ if G is centerless.*

(ii) *The group $N(L)/L$ is a finite commutative group whose exponent divides the order m of $Z(G)$.*

(i) Let Γ be a Chevalley lattice in $\mathfrak{g}_{0,\mathbf{Q}}$. Then ([16], 2.17) shows that $G_{\mathfrak{o}(S)}$ is the stabilizer of $\mathfrak{o}(S).\Gamma$ in G_k, operating on $\mathfrak{g}$ by the adjoint representation. The lattice Γ is spanned by the logarithms of the unipotent elements in $G_{0,\mathbf{Z}}$, hence $\mathfrak{o}(S).\Gamma$ is spanned by the logarithms of unipotent elements in $G_{\mathfrak{o}(S)}$. It is then clear that if $x \in G$ normalizes $G_{\mathfrak{o}(S)}$, then Ad x normalizes $\mathfrak{o}(S).\Gamma$. If moreover $x \in G_k$, then $x \in G_{\mathfrak{o}(S)}$, which proves the first assertion. Together with Proposition 3.3, this proves (i).

(ii)   The group $N(L)/L$ is finite by Proposition 3.3. The other assertions of (ii) follow from (i) and Proposition 3.4.

LEMMA. 4.2.   *Let $G = \mathbf{SL}_2, \mathbf{PSL}_2$ and L a S-arithmetic subgroup of G. Assume that S has at least two elements. Let s be an automorphism of L. There exists an automorphism s' of G', defined over $\mathbf{Q}$, and a homomorphism f of L into $Z(G')_\mathbf{Q}$ such that $s(x) = f(x) \cdot s'(x)$ $(x \in L)$.*

Let $\widetilde{G} = \mathbf{SL}_2$, $\pi : \widetilde{G} \to G$ the natural homomorphism and $\widetilde{L} = \pi^{-1}(L) \cap G_k$. Then $\widetilde{L}$ is S-arithmetic in $\widetilde{G}$. The map $s \circ \pi$ defines a homomorphism of $\widetilde{L}$ into $G'_\mathbf{Q}$. It follows from [29] that there exists a $\mathbf{Q}$-morphism $t : R_{k/\mathbf{Q}} \widetilde{G} \to G'$, which coincides with $s \circ \pi$ on a normal subgroup of finite index of $\widetilde{L}$. The end of the argument is then the same as in Lemma 1.8.

THEOREM 4.3.   *Let $\mathrm{Aut}\,(k,S)$ be the subgroup of $\mathrm{Aut}\,k$ leaving S stable. Assume either $\mathrm{rk}_k G \geqslant 2$ or $\mathrm{rk}_k G = 1$ and $\mathrm{Card}\,S \geqslant 2$. Let $L = G_{\mathfrak{o}(S)}$.*

(i)   *If G is centerless, $\mathrm{Aut}\,L$ is generated by $E(G)$, the group $\mathrm{Aut}(k,S)$ acting by conjugation, and $\mathrm{Int}\,L$.*

(ii)   *If G is simply connected, $\mathrm{Aut}\,L$ is generated by $E(G)$, $\mathrm{Aut}(k,S)$, and automorphisms of the form $x \longmapsto f(x) \cdot y \cdot x \cdot y^{-1}$ where f is a homomorphism of L into its center, and y belongs to the normalizer of L in G.*

By Lemma 4.1, L contains $Z(G)_k$. Let $s \in \mathrm{Aut}\,L$. By Lemma 1.8 and Lemma 4.2 we may write $s(x) = f(x) \cdot s'(x)$ where s' is a $\mathbf{Q}$-automorphism of G' and f a homomorphism of L into $Z(G')_\mathbf{Q} \cong Z(G)_k$, hence of L into its center.

The group G comes by extension of the groundfield from a split $\mathbf{Q}$-group G_0. Therefore Lemma 3.9 obtains. After having modified s by a field automorphism J, we may consequently assume s' to belong to $(\mathrm{Aut}\,G)_k$. In both cases (i), (ii) $(\mathrm{Aut}\,G)_k$ is a split extension of $E(G)$ by $(\mathrm{Int}\,G)_k$; moreover, the representative $E'(G)$ of $E(G)$ alluded to in Theorem 2.2 leaves L stable. Thus, after having multiplied s' by an element of $E(G)$, we may assume $s' \in$ (Int $G)_k$, hence $s' = $ Int y, $(y \in N(L))$.

4.4. Let G have a non-trivial center. We assume that the underlying $\mathbf{Q}$-split group G_0 may be (and is) identified with a $\mathbf{Q}$-subgroup of $\mathbf{GL}_n$ by means of an irreducible representation all of whose weights are extremal, i.e. form one orbit under the Weyl group, in such a way that $\mathbf{Z}^n$ is an admissible lattice, in the sense of [10]. (This assumption is fulfilled in all cases, except for the one of the spinor group in a number of variables multiple of four.)

Let D be the group of scalar multiples of the identity in $\mathbf{GL}_n$, and $H = D.G$. The group H is the identity component of the normalizer of G in $\mathbf{GL}_n$. The group D is a one-dimensional split torus. In particular, its first Galois cohomology group is zero. We have $G \cap D = Z(G)$, and $G \subset \mathbf{SL}_n$, therefore the order m of $Z(G)$ divides n, and Proposition 3.4 (i) yields

$$N(L) = G \cap N_H(L)_k.D. \tag{1}$$

LEMMA 4.5. *We keep the assumptions of 4.4. Let m be the order of $Z(G)$. Let A and B be the images of $N(L)$ and $H_{\mathfrak{o}(S)}$ in* Aut L.

(i) *The enveloping algebra M of L over $\mathfrak{o}(S)$ is $\mathbf{M}(n, \mathfrak{o}(S))$.*

(ii) *A/B is isomorphic to a subgroup of $_m I(k,S)$ and B to $\mathfrak{o}(S)^*/\mathfrak{o}(S)^{*(m)}$.*

(i) In view of the definition of admissible lattices [10], the maximal k-split torus T of the given splitting G may be assumed to be diagonal and the $\mathfrak{o}(S)$-lattice $\Gamma_0 = \mathfrak{o}(S)^n$ is the direct sum of its intersections with the eigenspaces of T. Our assumption on the weights implies further that these eigenspaces are one-dimensional, permuted transitively by the normalizer $N(T)$ of T.

Given a prime ideal $v \in V = V_\infty$, we denote by F_v the residue field $\mathfrak{o}/v$ and by $\overline{F}_v$ an algebraic closure of F_v. By [10], reduction mod v of G, (endowed with its canonical $\mathfrak{o}$-structure), yields a F_v-subgroup $G_{(v)}$ of $\mathbf{GL}(n, \overline{F}_v)$ which is connected, almost simple, has the same Dynkin diagram as G, and is simply connected if

G is. The reduction mod v also defines an isomorphism of the character group $X^*(T)$ of T onto the character group $X^*(T_{(v)})$ of the reduction mod v of T, which induces a bijection of the weights of the identity representation of G onto those of the identity representation of $G_{(v)}$. Thus the eigenspaces of $T_{(v)}$ are one-dimensional, and permuted transitively by the normalizer of $T_{(v)}$. Consequently, the identity representation of $G_{(v)}$ is irreducible.

The given splitting of G defines one of the universal covering $\widetilde{G}$ of G, hence an $\mathfrak{o}$-structure on $\widetilde{G}$. The reduction mod v $\widetilde{G}_{(v)}$ of G is the universal covering group of $G_{(v)}$ and the identity representation may be viewed as a irreducible representation of $\widetilde{G}_{(v)}$, say $f_{(v)}$. But $f_{(v)}$ has only extremal weights, therefore is a fundamental representation. It follows then from results of Steinberg ([30]; 1.3, 7.4) that the representation $f_{(v)}$ of the finite group $\widetilde{G}_{(v),\,F_v}$ is absolutely irreducible. Now, since reduction mod v is good, $\widetilde{G}_{(v),\,F_v}$ is the reduction of $\widetilde{G}_{\mathfrak{o}_v}$. Moreover, $\widetilde{G}$ being split and simply connected, strong approximation is valid in $\widetilde{G}$, hence $\widetilde{G}_{\mathfrak{o}}$ is dense in $\widetilde{G}_{\mathfrak{o}_v}$, which implies that reduction mod v maps $\widetilde{G}_{\mathfrak{o}}$ onto $\widetilde{G}_{(v),F_v}$. But the canonical projection of $\widetilde{G}$ onto G maps $\widetilde{G}_{\mathfrak{o}}$ into $\widetilde{G}_{\mathfrak{o}}$. Consequently, the image of $G_{\mathfrak{o}}$ in $G_{(v)}$ by reduction is a subgroup which contains $f_{(v)}(\widetilde{G}_{(v),\,F_v})$, hence is irreducible. Therefore

$$M \otimes F_v = \mathbf{M}(n,\, F_v), \quad (v \in V - S).$$

This shows that the index of M in $\mathbf{M}(n,\mathfrak{o}(S))$ is prime to all elements in $V - S$, whence (i).

(ii) By 4.4 (1), the image of $N(L)$ in Aut L is the same as that of $N' = N_H(L)_k$. Let $x \in N'$ and $\Gamma = x.\Gamma_0$ be the transform under x of the standard lattice $\Gamma_0 = \mathfrak{o}(S)^n$. This is a $\mathfrak{o}(S)$-lattice stable under L hence, by (i), also stable under $\mathbf{GL}(n,\mathfrak{o}(S))$. For $v \in V - S$, the local lattice $\mathfrak{o}_v.\Gamma$ in k_v^n is then stable under $\mathbf{GL}(n,\mathfrak{o}_v)$. There exists therefore a power $v^{a(v)}(a(v) \in Z)$ of v such that $\mathfrak{o}_v.\Gamma = v^{a(v)}.\mathfrak{o}_v^n$. We have then also $\mathfrak{o}_v.(\det x) = v^{n.a(v)}$. In view of the relation between a lattice and its localizations, we have then $\Gamma = \mathfrak{a} . \Gamma_0$ with $\mathfrak{a} = \Pi \mathfrak{v}^{a(v)}$, and moreover $\mathfrak{a}^n . \mathfrak{o}(S) = \mathfrak{o}(S).(\det x)$. By assigning to x the image of $\mathfrak{a} . \mathfrak{o}(S)$ in $I(k,S)$, we define therefore a map α of N' into ${}_n I(k,S)$, which is obviously a homomorphism. If $d \in D$, then $\alpha(d.x) = \alpha(x)$, whence a homomorphism of A into ${}_n I(k, S)$, to be denoted also by α. Clearly, $H_{\mathfrak{o}(S)} \subset \ker \alpha$. Conversely, assume that $x \in \ker \alpha$. Then $x.\Gamma_0$ is homothetic to Γ_0, and there exists $d \in k^*$ such that $d.x$ leaves Γ_0 stable. But then $d.x \in H_{\mathfrak{o}(S)}$, so that the image of x in $\mathrm{Aut}\ L$ belongs to B. Thus, A/B is isomorphic to a subgroup of ${}_n I(k, S)$. But $N(L)/L$ is of exponent m by Lemma 4.1, and m divides n, therefore α maps A/B into a subgroup of ${}_m I(k,S)$.

Let $\sigma : H \to H/G$ be the canonical projection. Its restriction to D is the projection $D \to D/Z(G)$, and $H/G = D/Z(G)$. If an element $x \in H_{\mathfrak{o}(S)}$ defines an inner automorphism of L, then $x \in D_{\mathfrak{o}(S)} . L$, and $\sigma(x) \in \sigma(D_{\mathfrak{o}(S)})$. Since elements of $D_{\mathfrak{o}(S)}$ define trivial automorphisms of L, we see that

$$B/\mathrm{Int}\ L \cong \sigma(H_{\mathfrak{o}(S)})/\sigma(D_{\mathfrak{o}(S)}). \tag{1}$$

Identify $D/Z(G)$ to $\mathbf{GL}_1$. Then $\sigma(H_{\mathfrak{o}(S)})$ is an S-arithmetic subgroup of $\mathbf{GL}_1$ hence a subgroup of finite index of $\mathfrak{o}(S)^*$. The group $Z(G)$ is cyclic of order m, therefore the projection $D = \mathbf{GL}_1 \to D'$ is either $x \longmapsto x^m$ or $x \longmapsto x^{-m}$, hence $\sigma(D_{\mathfrak{o}(S)}) \cong \mathfrak{o}(S)^{*(m)}$, so that $B/\mathrm{Int}\ L$ may be identified to a subgroup of $\mathfrak{o}(S)^*/\mathfrak{o}(S)^{*(m)}$. Thus (1) yields an injective homomorphism $\tau : B/\mathrm{Int}\ L \to \mathfrak{o}(S^*)/\mathfrak{o}(S)^{*(m)}$. There remains to show that τ is surjective.

Let $\pi : H \to H/D = G/Z(G) = \mathrm{Int}\ G$ be the canonical projection, T the maximal torus given by the splitting of G and $T' = \pi(T)$. We have already remarked that $x \in H_{\mathfrak{o}(S)}$ defines an inner automorphism of L if and only if $x \in D_{\mathfrak{o}(S)} . L$, so $B/\mathrm{Int}\ L \cong \pi(H_{\mathfrak{o}(S)})/\pi(L)$. By Lemma 4.1, $\pi(N(L)) \cong (G/Z(G))_{\mathfrak{o}(S)}$. On the other hand, since TD is split, D is a direct factor over k; this implies immediately that $\pi : (TD)_{\mathfrak{o}(S)} \to T'_{\mathfrak{o}(S)}$ is surjective, hence $\pi(H_{\mathfrak{o}(S)}) \cap T' = T'_{\mathfrak{o}(S)}$. We have $\pi(D_{\mathfrak{o}(S)} . L) = \pi(L)$, and consequently, since $\ker \pi \cap G \subset T$,

$$\pi(D_{\mathfrak{o}(S)} . L) \cap T' = \pi(L \cap T)\pi = (T_{\mathfrak{o}(S)});$$

hence $B/\mathrm{Int}\ L$ contains a subgroup isomorphic to $T'_{\mathfrak{o}(S)}/\pi(T_{\mathfrak{o}(S)})$. However, the kernel of $\pi\colon T \to T'$ is a cyclic group of order m. It is then elementary that we can write $T = T_1 \times T_2$, over k, with T_1 containing $Z(G)$ of dimension one. This implies

$$T'_{\mathfrak{o}(S)}/\pi(T_{\mathfrak{o}(S)}) \cong \pi(T_1)_{\mathfrak{o}(S)}/\pi(T_{1,\mathfrak{o}(S)}) = \mathfrak{o}(S)^*/\mathfrak{o}(S)^{*(m)};$$

this shows that the order of $B/\mathrm{Int}\ L$ exceeds that of $\mathfrak{o}(S)^*/\mathfrak{o}(S)^{*(m)}$. Therefore T is surjective.

EXAMPLES. 4.6. (1) $G = \mathbf{SL}_n$. $H = \mathbf{GL}_n$, $(n \geqslant 3)$. The group $L = \mathbf{SL}\ (n, \mathfrak{o}(S))$ is equal to its derived group [3], Corollary 4.3. By Lemma 4.1 Aut L is generated by Aut $(k,\ S)$, acting by conjugation on the coefficients, by the automorphism $x \longmapsto {}^t x^{-1}$, and by the image A in Aut L of $N(L)$.

If $\mathfrak{a}$ is an $\mathfrak{o}(S)$-ideal, then $\mathfrak{a}.\Gamma_0$ is isomorphic to $\mathfrak{a}^n \oplus \mathfrak{o}(S)^{n-1}$ by standard facts on lattices. Therefore, if $\mathfrak{a}^n$ is principal, then $\mathfrak{a}.\Gamma_0$ is isomorphic to Γ_0 and there exists $g \in \mathbf{GL}(n,k)$ such that $g.\Gamma_0 = \mathfrak{a}.\Gamma_0$. But the stabilizer of Γ_0 in G is the same as that of $\mathfrak{a}.\Gamma_0$, hence $g \in N_H(L)_k$, which shows that, in this case, the monomorphism $A/B \to {}_nI(k,S)$ is an isomorphism. We have therefore a composition series

$$\mathrm{Aut}\ L \supset A' \supset A \supset B \supset \mathrm{Int}\ L,$$

whose successive quotients are isomorphic to $\mathrm{Aut}(k,S)$, $\mathbf{Z}/2\mathbf{Z}$, ${}_nI(k,S)$ and $\mathfrak{o}(S)^*/\mathfrak{o}(S)^{*(n)}$.

This result is contained in [24], where Aut $\mathbf{SL}(n,Q)$ is determined for any commutative integral domain Q, except for the fact that the structure of the subgroup corresponding to $A/\mathrm{Int}\ L$ is not discussed there. For $\mathfrak{o}(S) = \mathfrak{o}$, it is related to those of [19] if k has class number one, and of [20] if $k = \mathbf{Q}(i)$.

(2) $G = \mathbf{SL}_2$, card $S \geqslant 2$. The above discussion of $A/\mathrm{Int}\ L$ is still valid, (without restriction on S, in fact). Furthermore, the contragredient mapping $x \longmapsto {}^t x^{-1}$ is an inner automorphism for $n = 2$. However, in general, L is not equal to its commutator subgroup, and $L/(L,L)$ has a non-trivial 2-primary component. Therefore there may be non-trivial automorphisms of the form $x \to f(x).x$

where f is a character of order two of L. Clearly, such a homomorphism of L into itself is bijective if and only if $\chi(-1)=1$. It follows from Lemma 4.2 that Aut L is generated by automorphisms of the previous type, field automorphisms, and elements of A.

We note that this conclusion does not hold true without some restriction on k, S. For instance, there is one further automorphism if $k=\mathbf{Q}(i)$, $\mathfrak{o}(S)=\mathbf{Z}(i)$, (see [20], and also [21] for a further discussion of the case $n=2$).

(3) $G=\mathbf{Sp}_{2n}$, $L=\mathbf{Sp}(2n, \mathfrak{o}(S))$. The commutator subgroup of L is equal to L if $n \geqslant 3$, and has index a power of two if $n=2$ ([3], Remark to 12.5). The group G has no outer automorphisms, therefore, if $n \geqslant 3$, Theorem 4.3, and Lemma 4.5 show that we have a composition series

$$\text{Aut } L \supset A \supset B \supset \text{Int } L,$$

with

$$\text{Aut } L/A \simeq \text{Aut } (k,S), \qquad B/\text{Int } L \simeq \mathfrak{o}(S)^*/\mathfrak{o}(S)^{*(2)},$$

and A/B isomorphic to a subgroup of $_2I(k,S)$. We claim that in fact

$$A/B \simeq {}_2I(k,S).$$

We write the elements of $\mathbf{GL}_{2n}$ as 2×2 matrices whose entries are $n \times n$ matrices. $\mathbf{Sp}_{2n}$ is the group of elements in $\mathbf{GL}_{2n}$ leaving $J=\begin{pmatrix} 0 & 1 \\ -1 & 0 \end{pmatrix}$ invariant, and its normalizer H in $\mathbf{GL}_{2n}$ is the group of similitudes of J. Let $\mathfrak{a}$ be an $\mathfrak{o}(S)$-ideal such that $\mathfrak{a}^2$ is principal. As remarked above, there exists $x \in \mathbf{GL}(2,k)$ such that $x.\mathfrak{o}(S)^2 = \mathfrak{a}.\mathfrak{o}(S)^2$. Let y be the element of $\mathbf{GL}(2n,k)$ which acts via x on the space spanned by the i-th and $(n+i)$-th canonical basis vectors $(i=1,\dots,n)$. Then $y \in H_k$, and $y.\mathfrak{o}(S)^{2n}=\mathfrak{a}.\mathfrak{o}(S)^{2n}$. Thus, y is an element of $N_H(L)_k$ which is mapped onto the image of $\mathfrak{a}$ in $_2I(k,S)$ by the homomorphism $\alpha\colon A/B \to {}_2I(k,S)$ of Lemma 4.5. Hence, α is also surjective.

If $n=2$, Aut L is obtained by combining automorphisms of the above types with those of the form $x \longmapsto f(x).x$, where f is a homomorphism of L into ± 1 whose kernel contains -1.

REMARK 4.7. It was noticed in Lemma 4.1 that L is equal to its normalizer in G_k. Since L has finite index in its normalizer in G, (Proposition 3.3 (d)), this means that L is not a proper normal subgroup of an arithmetic subgroup of G. More generally, we claim that L is maximal among arithmetic subgroups, i.e. that no subgroup M of G_k contains L as a proper subgroup of finite index. This was proved by Matsumoto [23] when $S = V_\infty$, and his proof extends immediately to the present case. In fact, the argument in the proof of Theorem 1 of [23] shows that if L has finite index in $M \subset G_k$, then the closure of M in G_v ($v \in V$, $v \notin S$) is contained in G_{0_v}, whence $M \subset L$.

5. Uniform subgroups in G_S. In §1, we proved the finiteness of $E(L)$ for subgroups which are either uniform or arithmetic. In §3 the arithmetic case was extended to S-arithmetic groups. Now a S-arithmetic group may be viewed as a discrete subgroup of G_S, which is *irreducible* in the sense that its intersection with any proper partial product of the G_v's ($v \in S$) reduces to the identity. We wish to point out here that there is also a generalization to G_S of the uniform subgroup case. We assume that $S \neq V_\infty$. Such groups have been considered by Ihara [14] for $G = \mathbf{SL}_2$, and Lemma 5.1 is an easy extension of results of his.

LEMMA 5.1. *Let G be a connected semi-simple, almost simple k-group, L a uniform irreducible subgroup of G_S, and L' its projection on G_∞.*

(i) *L is finitely generated.*

(ii) *If* rank $G \geqslant 2$ *and G_∞ has no compact or three-dimensional factor or $k = \mathbf{Q}$, $G = \mathbf{SL}_2$, then $H^1(L', \mathfrak{g}_\infty) = 0$.*

(i) Let $S' = S - V_\infty$, $G_{S'} = \prod_{v \in S'} G_v$, and $K = G_\infty \times \prod_{v \in S'} G_{0_v}$. The latter is an open subgroup of G_S. The orbits of K in G_S/L are open, hence closed, hence compact. Therefore $L_0 = L \cap K$ is uniform in K. Since K is the product of G_∞ by a compact group, the projection L'_0 of L_0 in G_∞ is a discrete uniform subgroup of G_∞. But G_∞ is a real Lie group with a finite number of connected components. Therefore the standard topological argument shows that L_0 is finitely generated. Let L'' be the projection of L on $G_{S'}$. Since L is uniform in G_S, there exists a compact subset C of $G_{S'}$ such that $G_S = L''.C$. On the other hand, it follows from ([9], 13.4) that $G_{S'}$ has a compact set of generators, say D. Then the standard Schreier-Reidemeister procedure to find generators for a subgroup shows that L is generated by $L \cap (G_\infty \times D.C.D^{-1})$ and consequently by L_0 and finitely many elements. (This argument is quite similar to the one used by Kneser [17] to prove the finite generation of G_S.)

(ii) We first notice that the restriction map $r : H^1(L', \mathfrak{g}_\infty) \to H^1(L'_0, \mathfrak{g}_\infty)$ is injective. The argument is the same as one of Ihara's ([14], p. 269) in the case $G = \mathbf{SL}_2$: if $x \in G_S$, then $x.K.x^{-1}$ is commensurable with K, hence, if $x \in L$, the group $L_{0,x} = x.L_0.x^{-1} \cap L_0$ has finite index in L_0. In particular, $L'_{0,x}$ is uniform in G_∞, hence, by density [4], has no fixed vector $\neq 0$ in $\mathfrak{g}_\infty$. This implies by Lemma 3.5 that $\ker r = 0$. If $\mathrm{rk}(G) \geqslant 2$, then $H^1(L'_0, \mathfrak{g}_\infty) = 0$ by [32] and [33], whence our assertion in this case. If $G = \mathbf{SL}_2$, $k = \mathbf{Q}$, the vanishing of $H^1(L', \mathfrak{g}_\infty)$ has been proved by Ihara, loc. cit. (it is stated there only in the case where S consists of ∞ and one prime, but the proof is *a fortiori* valid in the more general case).

THEOREM 5.2. *Let G and L be as in Lemma 5.1 (ii). Then $E(L)$ is finite, and L is not isomorphic to a proper subgroup of finite index.*

Identify L to its projection L' in G_∞. Then the theorem follows from Lemma 1.1 and Lemma 1.3 in the same way as in the case $S = V_\infty$.

APPENDIX

6. On compact Clifford-Klein forms of symmetric spaces with negative curvature.

6.1. Let M be a simply connected and connected Riemannian symmetric space of negative curvature, without flat component. A Clifford-Klein form of M is the quotient M/L of M by a properly discontinuous group of isometries acting freely, endowed with the metric induced from the given metric on M. In an earlier paper (Topology 2 (1963), 111-122), it was proved that M

always has at least one compact Clifford-Klein form. In answer to
a question of H. Hopf, we point out here that M always has
infinitely many different compact forms. More precisely:

THEOREM 6.2. *Let M be as in 6.1. Then M has infinitely many
compact Clifford-Klein forms with non-isomorphic fundamental
groups.*

M is the direct product of irreducible symmetric spaces. We may
therefore assume M to be irreducible. Then $M = G/K$, where G is a
connected simple non-compact Lie group, with center reduced to
$\{e\}$, and K is a maximal compact subgroup of G. Moreover, G is the
identity component of the group of isometries of M. Let L be a
discrete uniform subgroup of G, without elements of finite order $\neq e$.
Then L operates freely, in a properly discontinuous manner, on M,
and M/L is compact. Moreover, by a known result of Selberg
(see e.g. loc. cit., Theorem B), L has subgroups of arbitrary high
finite index. Since M is homeomorphic to euclidean space, L is
isomorphic to the fundamental group of M/L; it suffices therefore
to show that L is not isomorphic to any proper subgroup L' of
finite index. If dim $G = 3$, then M is the upper half-plane, and
this is well known. It follows for instance from the relations

$$\chi(M/L') = [L : L'] . \chi(M/L) \neq 0, \tag{1}$$

where $[L : L']$ is the index of L' in L, and $\chi(X)$ denotes the Euler-
Poincaré-characteristic of the space X. If dim $G > 3$, our assertion
is a consequence of Proposition 1.7. If $\chi(M/L) \neq 0$, which is the
case if and only if G and K have the same rank, one can of course
also use (1).

REFERENCES

1. N. ALLAN : The problem of maximality of arithmetic groups,
     Proc. Symp. pur. math. 9, Algebraic groups and discontinuous
     subgroups, *A.M.S., Providence, R. I.*, (1966), 104-109.

2. N. ALLAN : Maximality of some arithmetic groups, *Annals of
     the Brazilian Acad. of Sci.* [38 (1966), 234-244].

3. H. Bass, J. Milnor and J.-P. Serre : Solution of the congruence subgroup problem for $\mathbf{SL}_n$ $(n \geq 3)$ and $\mathbf{Sp}_{2n}$ $(n > 2)$, *Publ. Math. I.H.E.S.* [33 (1967), 421-499].

4. A. Borel : Density properties for certain subgroups of semi-simple groups without compact factors, *Annals of Math.* 72 (1960), 179-188.

5. A. Borel : Some finiteness properties of adele groups over number fields, *Publ. Math. I.H.E.S.* 16 (1963), 5-30.

6. A. Borel : Density and maximality of arithmetic groups, *J. f. reine u. ang. Mathematik* 224 (1966), 78-89.

7. A. Borel and Harish-Chandra, Arithmetic subgroups of algebraic groups, *Annals of Math.* (2) 75 (1962), 485-535.

8. A. Borel and J-P. Serre : Théorèmes de finitude en cohomologie galoisienne, *Comm. Math. Helv.* 39 (1964), 111-164.

9. A. Borel and J. Tits : Groupes réductifs, *Publ. Math. I. H. E. S.* 27 (1965), 55-150.

10. C. Chevalley : Certains schémas de groupes semi-simples, *Sém. Bourbaki* (1961), *Exp.* 219.

11. M. Hall : A topology for free groups and related groups, *Annals of Math.* (2) 52 (1950), 127-139.

12. L. K. Hua and I. Reiner : Automorphisms of the unimodular group, *Trans. A. M. S.* 71 (1955), 331-348.

13. L. K. Hua and I. Reiner : Automorphisms of the projective unimodular group, *Trans. A. M. S.* 72 (1952), 467-473.

14. Y. Ihara : Algebraic curves mod p and arithmetic groups, *Proc. Symp. pure math.* 9, Algebraic groups and discontinuous subgroups, *A.M.S., Providence, R.I.* (1966), 265-271.

15. Y. Ihara : On discrete subgroups of the two by two projective linear group over p-adic fields, *Jour. Math. Soc. Japan* 18 (1966), 219-235.

16. N. Iwahori and H. Matsumoto : On some Bruhat decompositions and the structure of the Hecke rings of p-adic Chevalley groups, *Publ. Math. I.H.E.S.* 25 (1965), 5-48.

17. M. KNESER : Erzeugende und Relationen verallgemeinerter Einheitsgruppen, *Jour. f. reine u. ang. Mat.* 214-15 (1964), 345-349.

18. B. KOSTANT : Groups over **Z**, *Proc. Symp. pure mat.* 9, Algebraic groups and discontinuous subgroups, *A. M. S., Providence, R.I.,* 1966, 90-98.

19. J. LANDIN and I. REINER : Automorphisms of the general linear group over a principal ideal domain, *Annals of Math.* (2) 65 (1957), 519-526.

20. J. LANDIN and I. REINER : Automorphisms of the Gaussian modular group, *Trans. A. M. S.* 87 (1958), 76-89.

21. J. LANDIN and I. REINER : Automorphisms of the two-dimensional general linear group over a Euclidean ring, *Proc. A.M.S.* 9 (1958), 209-216.

22. H. MATSUMOTO : Subgroups of finite index in certain arithmetic groups, *Proc. Symp. pure math.* 9, Algebraic groups and discontinuous subgroups, *A.M.S., Providence, R.I.* (1966), 99-103.

23. H. MATSUMOTO : Sur les groupes semi-simples déployés sur un anneau principal, *C. R. Acad. Sci. Paris* 262 (1966), 1040-1042.

24. O. T. O'MEARA : The automorphisms of the linear groups over any integral domain, *Jour. f. reine u. ang. Mat.* 223 (1966), 56-100.

25. M. S. RAGHUNATHAN : Cohomology of arithmetic subgroups of algebraic groups I, *Annals of Math.* (2) 86 (1967), 409-424.

26. M. S. RAGHUNATHAN : Cohomology of arithmetic subgroups of algebraic groups II, 87 (1968), 279-304.

27. I. REINER : Automorphisms of the symplectic modular group, *Trans. A.M.S.* 80 (1955), 35-50.

28. I. REINER : Real linear characters of the symplectic unimodular group, *Proc. A. M. S.* 6 (1955), 987-990.

29. J. P. SERRE : Le problème des groupes de congruence pour SL_2, *Annals of Math.* 92 (1970), 489-527.

30. R. Steinberg : Representations of algebraic groups, *Nagoya M.J.* 22 (1963), 33-56.

31. A. Weil : Adeles and algebraic groups, *Notes, The Institute for Advanced Study, Princeton, N. J.* 1961.

32. A. Weil : On discrete subgroups of Lie groups II, *Annals of Math.* (2) 75 (1962), 578-602.

33. A. Weil : Remarks on the cohomology of groups, (ibid), (2) 80 (1964), 149-177.

The Institute for Advanced Study,
Princeton, N.J.

82.

(avec J. Tits)

On 'abstract' homomorphisms of simple algebraic groups

Proc. Colloquium on Algebraic Geometry 1968, Tata Institute, Bombay (1969) 75–82

THIS Note describes some results pertaining chiefly to homomorphisms of groups of rational points of semi-simple algebraic groups, and gives an application to a conjecture of Steinberg's [9] on irreducible projective representations. Some proofs are sketched. Full details will be given elsewhere.

NOTATION. The notation and conventions of [1] are used. In particular, all algebraic groups are affine, k is a commutative field, $\bar{k}$ an algebraic closure of k, p its characteristic, and G is a k-group. In this Note, G is moreover assumed to be *connected*. k' also denotes a commutative field.

Let $\phi : k \to k'$ be a (non-zero) homomorphism. We let $^\phi G$ be the k'-groups $G \otimes_k k'$ obtained from G by the change of basis ϕ, and ϕ_0 be the canonical homomorphism $G_k \to {}^\phi G_{k'}$ associated to ϕ.

If $p \neq 0$, then Fr^i denotes the p^i-th power homomorphism $\lambda \to \lambda^{p^i}$ of a field of characteristic p ($i = 0, 1, 2, \ldots$). If $p = 0$, Fr^i is the identity.

A connected semi-simple k-group H is *adjoint* if it is isomorphic to its image under the adjoint representation, *almost simple* (resp. *simple*) over k if it has no proper normal k-subgroup of strictly positive dimension (resp. $\neq \{e\}$).

1. Homomorphisms. 1.1. Let G be semi-simple. G^+ will denote the subgroup of G_k generated by the groups U_k, where U runs through the unipotent radicals of the parabolic k-subgroups of G. The group G^+ is normal in G_k; it is $\neq \{e\}$ if and only if $\mathrm{rk}_k(G) > 0$. If, moreover, G is almost simple over k, then G^+ is Zariski-dense in G, and the quotient of G^+ by its center is simple except in finitely many cases where k has two or three elements [10]. If $f: G \to H$ is a central k-isogeny, then $f(G^+) = H^+$. The group G^+ is equal to G_k if $k = \bar{k}$, or if G is k-split and simply connected; it is

32

conjectured to be equal to G_k if G is simply connected and $\mathrm{rk}_k(G) > 0$ [10]. It is always equal to its commutator subgroup.

THEOREM 1.2. *Assume k to be infinite, and G to be almost absolutely simple, of strictly positive k-rank. Let H be a subgroup of G_k containing G^+. Let k' be a commutative field, G' a connected almost absolutely simple k'-group, and $\alpha \colon H \to G'_{k'}$ a homomorphism whose kernel does not contain G^+, and whose image contains G'^+. Assume finally that either G is simply connected or G' is adjoint. Then there exists an isomorphism $\phi \colon k \xrightarrow{\sim} k'$, a k'-isogeny $\beta \colon {}^\phi G \to G'$, and a homomorphism γ of H into the center of $G'_{k'}$ such that $\alpha(x) = \beta(\phi_0(x)) \cdot \gamma(x)$ ($x \in H$). Moreover, β is central, except possibly in the cases: $p = 3$, G, G' split of type $\mathbf{G}_2$; $p = 2$, G, G' split of type $\mathbf{F}_4$; $p = 2$, G, G' split of type $\mathbf{B}_n$, $\mathbf{C}_n$, where β may be special.*

(The special isogenies are those discussed in [3, Exp. 21-24].) In the following corollary, G and G' need not satisfy the last assumption of the theorem.

COROLLARY 1.3. *Assume G_k is isomorphic to $G'_{k'}$. Then k is isomorphic to k', and G, G' are of the same isogeny class.*

Let $\overline{G}$ and $\overline{G}'$ be the adjoint groups of G and G'. The assumption implies the existence of an isomorphism $\alpha \colon \overline{G}^+ \xrightarrow{\sim} \overline{G}'^+$. By the theorem there is an isomorphism ϕ of k onto k' and an isogeny μ of ${}^\phi \overline{G}$ onto $\overline{G}'$, whence our assertion.

REMARKS 1.4. (i) It may be that the homomorphism γ in (1.2) is always trivial. It is obviously so if G' is adjoint, or if H is equal to its commutator subgroup. Since G^+ is equal to its commutator group, this condition will be fulfilled if G is simply connected and the conjecture $G_k = G^+$ of [10] is true, thus in particular if G splits over k. Moreover, in that case the assumption $G^+ \subset \ker \alpha$ would be superfluous.

(ii) The theorem has been known in many special cases, starting with the determination of the automorphism group of the projective linear group [7]. We refer to Dieudonné's survey [4] for the automorphisms of the classical groups. For split groups over infinite fields, see also [6].

(iii) Assume $k = k'$, $G = G'$, G adjoint, and k not to have any automorphism $\neq$ id. Theorem 1.2 implies then that every automorphism of G_k is the restriction of an automorphism of G, which is then necessarily defined over k. In particular, if k is the field of real numbers $\mathbf{R}$, every automorphism of G_k is continuous in the ordinary topology, as was proved first by Freudenthal [5].

(iv) The assumption $\mathrm{rk}_k\, G > 0$ is essential for our proof, but it seems rather likely that similar results are valid for anisotropic groups. This is the case for many classical groups [4]. Also, Freudenthal's proof is valid for compact groups. In fact, the continuity of any abstract-group automorphism of a compact semi-simple Lie group had been proved earlier, independently, by E. Cartan [2] and van der Waerden [11]. We note also that van der Waerden's proof remains valid in the p-adic case.

(v) The group Aut G_k has also been studied when k is finite. See [4] for the classical groups, and [8] for the general case.

THEOREM 1.5. *Assume k to be infinite, and G to be almost simple, split over k. Let G' be a semi-simple split k'-group, G'_i $(1 \leqslant i \leqslant s)$ the almost simple normal subgroups of G', and $\alpha : G_k \to G'_{k'}$ a homomorphism whose image is Zariski-dense. If $G_k = G^+$, then G' is connected. Assume G' to be connected and either G simply connected or G' adjoint. Then there exist homomorphisms $\phi_i : k \to k'$ and k'-isogenies $\beta_i : \phi_i G \to G'_i$ $(1 \leqslant i \leqslant s)$, which are either central or special, such that*

$$\alpha(x) = \prod_i (\beta_i \circ \phi_{i,0})\,(x), \quad (x \in G_k).$$

Moreover, $Fr^a \circ \phi_i \neq Fr^b \circ \phi_j$ if $p = 0$ and $i \neq j$, or if $p \neq 0$ and $(a, i) \neq (b, j)$ $(1 \leqslant i, j \leqslant s;\, a, b = 0, 1, 2, ...)$.

The proof of Theorem 1.5 goes more or less along the same lines as that of Theorem 1.2. In fact, it seems not unlikely that Theorem 1.5 can be generalized so as to contain Theorem 1.2. We hope to come back to this question on another occasion.

EXAMPLE 1.6. The following example, which admits obvious generalizations, shows that the assumption of semi-simplicity made on G' in Theorem 1.5 cannot be dropped.

Let $G = \mathbf{SL}_2$, and N be the additive group of 2×2 matrices over $\bar{k}$, of trace zero. Let d be a non-trivial derivation of k. Extend it to a derivation of N_k by letting it operate on the coefficients, and define $h \colon G_k \to N_k$ by $h(g) = g^{-1}.dg$. Let $G' = G.N$ be the semi-direct product of G and N, where G acts on N by the adjoint representation. Then $g \to (g,\, h(g))$ is easily checked to be a homomorphism of G_k into G'_k with dense image; clearly, it defines an "abstract" Levi section of G'_k.

2. Projective representations.

2.1. Assume $p \neq 0$. For G semi-simple, let $\mathscr{R}$ or $\mathscr{R}(G)$ be the set of p^l $(l = \operatorname{rank} G)$ irreducible projective representations whose highest weight is a linear combination of the fundamental highest weights with coefficients between 0 and $p - 1$. The following theorem, in a slightly different formulation, was conjectured by R. Steinberg [9], for $k = \bar{k}$. We show below how it follows from Theorem 1.5 and [9], (Theorem 1.1).

THEOREM 2.2. *Assume k to be infinite, $p \neq 0$, and G k-split, simple, adjoint. Let $\pi \colon G^+ \to \mathbf{PGL}(n,\, \bar{k})$ be an irreducible (not necessarily rational) projective representation of G^+. Then there exist distinct homomorphisms $\phi_j \colon k \to \bar{k}$, and elements $\pi_j \in \mathscr{R}(^{\phi_j}G)$ $(1 \leqslant j \leqslant t)$, such that $\pi = \prod\limits_j \pi_j \circ \phi_{j,0}$.*

PROOF. Let G' be the Zariski-closure of $\pi(G_k)$ in $\mathbf{PGL}_n$. It is also an irreducible projective linear group, hence its center and also its centralizer in $\mathbf{PGL}_n$, or in the Lie algebra of $\mathbf{PGL}_n$, are reduced to $\{e\}$. Thus G' is semi-simple, and its identity component is adjoint. Moreover, by Theorem 1.5, G' is connected. By Theorem 1.1 of [9], there exist elements $\pi_a \in \mathscr{R}(G')$, $(1 \leqslant a \leqslant q)$ such that the identity representation of G' is equal to $\prod\limits_a \pi_a \circ (Fr^a)$. Let G'_i $(1 \leqslant i \leqslant s)$ be the simple factors of G'.

The tensor product defines a bijection of $\mathscr{R}(G'_1) \times \ldots \times \mathscr{R}(G'_s)$ onto $\mathscr{R}(G')$. We may therefore write

$$\pi_a = \prod_{1 \leqslant i \leqslant s} \pi_{a,i}, \quad (\pi_{ai} \in \mathscr{R}(G_i'); 1 \leqslant a \leqslant q).$$

Let now $\phi_i \colon k \to \bar{k}$ and $\beta_i \colon {}^{\phi_i}G \to G_i$ be as in Theorem 1.5 (with $\bar{k} = k'$). We have then

$$\pi = \prod_{a,i} \pi_{a,i} \circ (Fr^a)_0 \circ \beta_i \circ \phi_{i,0}. \tag{1}$$

But $(Fr^a)_0 \circ \beta_i = \beta_{a,i} \circ (Fr^a)_0$, where $\beta_{a,i}$ is the transform of β_i under Fr^a. Let $\phi_{a,i} = Fr^a \circ \phi_i$. Since G, G_i are adjoint, the morphisms $\beta_{a,i}$ are either isomorphisms or special isogenies. Therefore, taking ([9], §11) into account, we see that

$$\pi'_{a,i} = \pi_{a,i} \circ \beta_{a,i} \in \mathscr{R}\left({}^{\phi_{a,i}}(G)\right), \qquad (1 \leqslant i \leqslant s; 1 \leqslant a \leqslant q),$$

and (1) yields

$$\pi = \prod_{a,i} \pi'_{a,i} \circ (\phi_{a,i})_0, \tag{2}$$

which proves the theorem, in view of the fact that the $\phi_{a,i}$ are distinct by Theorem 1.5.

3. Sketch of the proof of Theorem 1.2.

In this paragraph, k is infinite and G is semi-simple, of strictly positive k-rank.

The two following propositions are the starting point of the proofs of Theorem 1.2 and Theorem 1.5.

PROPOSITION 3.1. *Let G' be a k'-group, and $\alpha \colon G^+ \to G'_{k'}$ a non-trivial homomorphism. Let P be a minimal parabolic k-subgroup of G and U its unipotent radical. Then $\alpha(U_k)$ is a unipotent subgroup contained in the identity component of G' and $\alpha(G^+) \subset G'^0$. The field k' is also of characteristic p.*

Let S be a maximal k-split torus of P. It is easily seen that $S^+ = S \cap G^+$ is dense in S. It follows then from ([1], §11.1) that any subgroup of finite index of $S^+.U_k$ contains elements $s \in S^+$ such that $(s, U_k) = U_k$. From this we deduce first that U_k is contained in any normal subgroup of finite index of $S^+.U_k$, and then, that it is also contained in the commutator subgroup of any such subgroup. It follows that $\alpha(U_k)$ is contained in the derived group of the identity component of the Zariski closure of $\alpha(S^+.U_k)$. The latter being solvable, this implies that $\alpha(U_k)$ is unipotent.

Let $p' = \operatorname{char.} k'$. If $p \neq 0$, then U_k is a p-group. Its image is a p-group and is $\neq \{e\}$ since α is non-trivial, and G^+ is generated by the conjugates of U_k; hence $p = p'$. If $p = 0$ and $p' \neq 0$, then $\ker \alpha \cap U_k$ has finite index in U, whence easily a contradiction with the main theorem of [10].

PROPOSITION 3.2. *Let G' be a connected semi-simple k'-group. Let P, S, U be as above, P^- the parabolic k-subgroup opposed to P and containing $\mathscr{Z}(S)$, and $U^- = R_u(P^-)$. Let H be a subgroup of G_k containing G^+ and $\alpha\colon H \to G'_{k'}$ be a homomorphism with dense image. Then the Zariski-closures Q, Q^- of $\alpha(P \cap H)$ and $\alpha(P^- \cap H)$ are two opposed parabolic k'-subgroups, and $Q \cap Q^-$, $R_u(Q)$, $R_u(Q^-)$ are the Zariski-closures of $\alpha(Z(S) \cap H)$, $\alpha(U_k)$ and $\alpha(U_k^-)$ respectively.*

Let M, V, V^- be the Zariski-closures of $\alpha(\mathscr{Z}(S) \cap H)$, $\alpha(U_k)$ and $\alpha(U_k^-)$ respectively. The groups V, V^- are unipotent, by Proposition 3.1. The group G is the union of finitely many left translates of $U^-.P$. Since $\alpha(H)$ is dense, this implies that $V^-.M.V$ contains a non-empty open subset of G'. Let T be a maximal torus of M and Y, Y^- be two maximal unipotent subgroups of M^0 normalized by T such that $Y^-.\ T.\ Y$ is open in M^0 (see [1], §2.3, Remarque). Then $V^-.\ Y^-$ and $Y.\ V$ are unipotent subgroups of G' normalized by T and $V^-.\ Y^-.\ T.\ Y.\ V$ contains a non-empty open set of G'. Consequently ([1], §2.3), T is a maximal torus of G', and $V^-.\ Y^-$, $Y.\ V$ are two opposed maximal unipotent subgroups. This shows that Q, Q^- are parabolic subgroups, M is reductive, connected, and $V = R_u(Q)$, (resp. $V^- = R_u(Q^-)$). The groups Q, Q^- are obviously k'-closed. Arguing as in Proposition 3.1, we may find $s \in S \cap H$ such that $(s, U_k) = U_k$, $(s, U_k^-) = U_k^-$. It follows then from ([1], §11.1) that $\mathscr{Z}(\alpha(s))^0 = M$. Hence M is defined over k'([1], §10.3). By Grothendieck's theorem ([1], §2.14), it contains a maximal torus defined over k'. Hence ([1], §3.13), Q, Q^-, V, V^- are defined over k'.

3.3. We now sketch the proof of Theorem 1.2, assuming for simplicity that G, G' are adjoint and $H = G_k$. Then α is injective.

Proposition 3.2, applied to α and α^{-1}, shows that Q, Q^- are two opposed minimal parabolic k'-subgroups of G'. Consequently, α induces an isomorphism of $\mathcal{N}(S)/\mathcal{Z}(S)$ onto $\mathcal{N}(M)/M$, i.e. of $_kW(G) = {}_kW$ onto $_{k'}W' = {}_{k'}W(G')$. For $a \in {}_k\Phi(G)$, let $U_a = U_{(a)}/U_{(2a)}$, where we put $U_{(2a)} = \{e\}$ if $2a \notin {}_k\Phi$. It may be shown that $U_{(2a)}$ is the center of $U_{(a)}$. The groups $U_{(a)}^-$ may be characterized as minimal among the intersections $U \cap w(P)$ $(w \in {}_kW)$ not reduced to $\{e\}$. It then follows that α induces a bijection $\alpha_* : {}_k\Phi(G) \to {}_{k'}\Phi(G')$ preserving the angles, and isomorphisms $U_{a,k} \xrightarrow{\sim} V_{\alpha_*(a),k'}$. The group U_a (resp. $V_{\alpha_*(a)}$) may be endowed canonically with a vector space structure such that S (resp. a maximal k'-split torus S' of M) acts on it by dilatations. The next step is to show that $\alpha : U_{a,k} \xrightarrow{\sim} V_{\alpha_*(a),k'}$ induces a bijection ϕ_a between the algebras of dilatations. Let L_a be the subgroup of G generated by $U_{(a)}$ and $U_{(-a)}$. The assumption that G is almost absolutely simple is equivalent to the existence of one element $a \in {}_k\Phi$ such that the intersection X_a of L_a with the center C of $\mathcal{Z}(S)$ is one-dimensional, hence such that $X_a^0 \subset S$. This is the main tool used in showing that $\alpha(S_k) \subset S'_{k'}$, hence that α maps dilatations by elements of $(k^*)^2$ into dilatations. If $p \neq 2$, this suffices to yield the existence of $\phi_a : k \xrightarrow{\sim} k'$. In characteristic two, some further argument, based on properties of groups of rank one, is needed. It is clear that $\phi_a = \phi_b$ if $b \in {}_kW(a)$. Using further some facts about commutators, it is then easily proved that $\phi_a = \phi_b$ $(a, b \in {}_k\Phi)$ if α_* preserves the lengths. If not, we show that we are in one of the exceptional cases listed in the theorem, and we reduce it to the preceding one by use of a special isogeny. Write then ϕ instead of ϕ_a. Replacing G by $^\phi G$, we may assume $k = k'$, $\phi = \mathrm{id}$. It is then shown that $\alpha : U_k \xrightarrow{\sim} V_k$ is the restriction of a k-isomorphism of varieties. On the other hand, since G' is adjoint, $\mathcal{Z}(S')$ is isomorphic to its image in $GL(\mathbf{b})$ under the adjoint representation, where $\mathbf{b}$ is the sum of Lie algebras of the $V_{a'}(a' \in {}_k\Phi(G'))$. This implies readily that the restriction of α to $U_k^- . P_k$ is the restriction of a k-isomorphism of varieties of $U^- . P$ onto $V^- . Q$. The conclusion then follows readily from the fact that G is a finite union of translates $x.U^- . P$ $(x \in G_k)$.

REFERENCES

1. A. BOREL and J. TITS :   Groupes réductifs, *Publ. Math. I.H.E.S.*
    27 (1965), 55-150.

2. E. CARTAN :   Sur les représentations linéaires des groupes clos,
    Comm. Math. Helv. 2 (1930), 269-283.

3. C. CHEVALLEY :   *Séminaire sur la classification des groupes de Lie
    algébriques*, 2 vol., Paris 1958 (mimeographed Notes).

4. J. DIEUDONNÉ :   *La géométrie des groupes classiques*, Erg. d.
    Math. u. Grenzg. Springer Verlag, 2nd edition, 1963.

5. H. FREUDENTHAL :   Die Topologie der Lieschen Gruppen als alge-
    braisches Phänomen I, *Annals of Math.* (2) 42 (1941), 1051-1074.
    Erratum *ibid.* 47 (1946), 829-830.

6. J. HUMPHREYS :   On the automorphisms of infinite Chevalley
    groups (preprint).

7. O. SCHREIER und B. L. v. d. WAERDEN :   Die Automorphismen
    der projektiven Gruppen, *Abh. Math. Sem. Hamburg Univ.* 6
    (1928), 303-322.

8. R. STEINBERG :   Automorphisms of finite linear groups, *Canadian
    J. M.* 12 (1960), 606-615.

9. R. STEINBERG :   Representations of algebraic groups, *Nagoya
    Math. J.* 22 (1963), 33-56.

10. J. TITS :   Algebraic and abstract simple groups, *Annals of Math.*
    (2) 80 (1964), 313-329.

11. B. L. v. d. WAERDEN :   Stetigkeitssätze für halb-einfache Liesche
    Gruppen, *Math. Zeit.* 36 (1933), 780-786.

The Institute for Advanced Study, Princeton, N. J.
Universität Bonn.

83.

Injective endomorphisms of algebraic varieties

Arch. Math. **20** (1969) 531–537

Recently J. Ax proved the somewhat unexpected fact that an injective morphism of an algebraic variety into itself is surjective ([2], § 4). His proof uses ultraproducts, to reduce to the analogous obvious property of finite fields. Later, G. SHIMURA gave a proof by means of reduction mod $\mathfrak{p}$. This paper outlines a third one, based on cohomology with compact supports, which carries over to real varieties. We discuss first the complex case, where cohomology is the Alexander-Spanier cohomology, (2.3). It is likely that the argument carries over to arbitrary algebraically closed ground-fields, using Grothendieck's cohomology with proper supports [1, 10], at least for quasi-projective varieties (see section 3). Section 4 is devoted to a real analogue (4.4) of (2.3), which generalizes results of BIALYNICKI-BIRULA and ROSENLICHT [3][1].

Besides standard facts in algebraic geometry, the main points in the proofs of 2.3, 4.4 are Lemma 1.2 and some known topological properties of algebraic sets (2.1, 4.2).

1. Cohomology with compact supports.

1.1. For the cohomology theory used in this paper (except in section 3), see e.g. [6, 9]. $H_c^i(X; A)$ denotes the i-th Alexander-Spanier cohomology group with compact supports of the locally compact space X, with coefficients in the commutative group A, and $H_c^*(X; A)$ is the direct sum of the $H_c^i(X; A)$. We recall that if Y is a closed subspace of X, there is an exact cohomology sequence of X mod Y:

$$(1) \qquad \cdots \to H_c^i(X - Y; A) \to H_c^i(X; A) \to H_c^i(Y; A) \to H_c^{i+1}(X - Y; A) \to \cdots$$

1.2. Lemma. *Let L be a field, X be a locally compact space, and f a homeomorphism of X onto an open subspace. Let $Y_q = X - f^q(X)$, $(q = 1, 2, \ldots)$. Assume that $\dim H_c^*(X; L)$ is finite. Then there exists a constant c such that $\dim H_c^*(Y_q; L) \leqq c$ $(q = 1, 2, \ldots)$.*

---

[1]) After this paper was completed, M. RAYNAUD drew my attention on a result of A. GROTHEN-DIECK (Éléments de Géométrie Algébrique, Chap. IV, § 17.9.6), which implies that a monomorphism $f: V \to V$ of an algebraic variety into itself is an isomorphism. Using this, M. RAYNAUD proved more generally that if f has finite fibres, and is an immersion on a dense open subset, then f is an isomorphism. These results contain 2.3.

40

The q-th iterate f^q is a homeomorphism of X onto $X - Y_q$, hence

$$\dim H_c^*(X - Y_q; L) = \dim H_c^*(X; L).$$

Our assertion is then an obvious consequence of 1.1.(1), for X and Y_q.

2. Complex varieties. Our algebraic varieties over an algebraically closed field k are those of Serre's [14]. This notion is equivalent to that of reduced scheme of finite type over k. We make no notational distinction between a variety over k and its set of k-points. In this section, $k = \mathbf{C}$.

The following lemma definitely belongs to the "well-known" variety, and is included only for convenience.

2.1. Lemma. *Let X be a complex algebraic variety, $n = \dim_{\mathbf{C}} X$, and L a noetherian ring. X has cohomological dimension $2n$, $H_c^i(X; L)$ is a finitely generated L-module for all i, is zero for $i > 2n$, and $H_c^{2n}(X; L)$ is a free L-module with s generators, where s is the number of irreducible components of X of complex dimension n.*

Let S be the union of the singular set and of the irreducible components of dimension $< n$ of X. Then S is algebraic, of complex dimension $\leq n - 1$, and $X - S$ is the union of s smooth complex n-dimensional varieties, each of which is Zariski-open in an irreducible component of X, whence our assertion on the cohomological dimension of X and on $H_c^i(X; L)$ for $i \geq 2n$.

Let X be compact. Then, by the triangulation theorem (see e.g. [12]), it is a finite polyhedron, hence $H_c^*(X; L)$ is finitely generated. By 1.1 (1), this is then also true if X is Zariski-open in a compact variety, in particular if X is quasi-projective. In general, X admits a finite covering by open quasi-projective subvarieties. Using induction on the number of elements of such a covering, we are reduced to the case where $X = U \cup V$, with U, V Zariski-open, V quasi-projective, $H_c^*(U; L)$ finitely generated. Then $U \cap V$ is quasi-projective, and our assertion follows from the Mayer-Vietoris sequence [6, p. 65]:

$$\cdots \to H_c^i(U; L) \oplus H_c^i(V; L) \to H_c^i(X; L) \to H_c^{i+1}(U \cap V; L) \to \cdots$$

2.2. Lemma. *Let X be a complex algebraic variety, Y_q $(q = 1, 2, \ldots)$ an increasing sequence of Zariski-closed subsets of X, and L a field. If $\dim H_c^*(Y_q; L)$ is uniformly bounded in q, then the sequence (Y_q) is stationary.*

It suffices to show that, for every positive integer s, the number of irreducible components of dimension s of Y_q is uniformly bounded. This being clear if $s \geq \dim X$, we use descending induction on s. Write $Y_q = W_q \cup Z_q$, where W_q (resp. Z_q) is the union of the irreducible components of Y_q of dimension $\leq s$ (resp. $> s$). By induction (Z_q) is stationary, hence $\dim H_c^*(Z_q; L)$ is uniformly bounded. But the number of irreducible components of dimension s of $W_q' = W_q - Z_q \cap W_q$ is equal to that of W_q; and, by the cohomology sequence of Y_q mod Z_q, and 2.1, it is bounded by $\dim H_c^{2s-1}(Z_q; L) + \dim H_c^{2s}(Y_q; L)$, hence is uniformly bounded.

2.3. Theorem. *Let X be a complex algebraic variety, and $f: X \to X$ an injective morphism. Then f is surjective.*

(a) *X is irreducible, normal.* The map f is dominant and, since we are in characteristic zero, is birational. Then so is f^q $(q = 1, 2, \ldots)$, and, by Zariski's main theorem, f^q is an isomorphism of X onto a Zariski-open subset $X - Y_q$ of X, where (Y_q) is an increasing sequence of Zariski-closed subsets of X $(q = 1, 2, \ldots)$. Assume $Y_1 \neq \emptyset$. Then Y_q is the disjoint union of Y_1 and of the locally closed subsets $f^i(Y_1)$ $(i = 1, \ldots, q-1)$, and is not stationary. Let L be a field. In view of 2.2, $\dim H_c^*(Y_q; L)$ is not bounded, whence a contradiction with 1.2.

(b) *X is irreducible.* Let $\tilde{X}$ be the normalization of X and $\pi\colon \tilde{X} \to X$ the canonical projection. It is a covering in the sense of [8, p. 165]. The automorphism f^0 of the field $\mathbf{C}(X)$ of rational functions on X induced by f induces a birational map $\tilde{f}\colon \tilde{X} \to \tilde{X}$ and we have $f \circ \pi = \pi \circ \tilde{f}$, (as rational maps, at first). But, then, $\pi \circ \tilde{f}$ is everywhere defined, hence so is $\tilde{f}$ [7, Cor. 2, p. 180]. Consequently $\tilde{f}$ is a birational morphism with finite fibres. By Zariski's main theorem, $\tilde{f}$ is an isomorphism of $\tilde{X}$ onto an open subset of $\tilde{X}$. In particular, $\tilde{f}$ is injective. By (a), $\tilde{f}$ is then surjective, hence f is surjective, too.

(c) *General case.* Let X_i $(1 \leq i \leq t)$ be the irreducible components of X, indexed in such a way that $\dim_{\mathbf{C}} X_i \leq \dim_{\mathbf{C}} X_j$ if $i \geq j$. Assume that $X_1, \ldots, X_s$ are the components of dimension n. Given $i \leq s$, we have then $f(X_i) \subset X_{j(i)}$ with $j(i) \leq s$, and $f(X_i)$ contains a non-empty Zariski-open subset of $X_{j(i)}$. Since f is injective, we must have

$$(1) \qquad\qquad j(i) \neq j(k) \qquad (i \neq k).$$

Fix i. Let $i_1 = j(i)$, $i_2 = j(i_1)$, ..., $i_p = j(i_{p-1})$, ... There exist a, b, $(a < b)$ such that $i_a = i_b$. But, then, by (1), $i_{b-a} = i$ so that f^{b-a} yields an injective morphism of X_i into X_i. By (b) it is surjective. But, then, $X_i \to X_{j(i)}$ is also surjective.

Thus $\operatorname{Im} f$ contains X_i $(i \leq s)$. Since f is injective, it follows that f maps the set of components of dimension $< n$ into itself. We then argue by induction on $\dim X$.

3. Varieties in non-zero characteristic. Let now X be a variety over an algebraically closed groundfield k of characteristic $p \neq 0$. Assume X to be quasi-projective. Then it seems that 2.3 remains true over k, and that the above proof goes over with little change if $H_c^i(X; L)$ is meant to be the i-th cohomology group of X, with proper supports, and coefficients in a finite field of characteristic $q \neq p$, in the sense of GROTHENDIECK [1, 10]. In fact, this cohomology theory also has the cohomology sequence 1.1 (1) as well as all the formal properties needed to prove the analogue of 2.1. Moreover, it is invariant under purely inseparable morphisms. According to "well-informed" sources, all this will be found in Chap. XVII of [1], not yet published. Granting this, we briefly outline how to prove 2.3 over k.

Assume first X to be irreducible, normal. Then f is open ([7, Exp. 5, App. II] or [8, pp. 195—196]), hence $f(X) = X - F$, with F Zariski-closed. However, f need not be birational, but is at any rate purely inseparable, hence f^* is an isomorphism. From this it is clear that the argument of 2.3 (a) remains valid.

Let now X be irreducible, $\tilde{X}$ its normalization, and $\pi\colon \tilde{X} \to X$ the canonical map. As in 2.3 the map $f^0\colon k(X) \to k(X)$ induced by f defines a rational map $\tilde{f}\colon \tilde{X} \to \tilde{X}$, and it follows again from [8, Cor. 2, p. 180] that $\tilde{f}$ is a morphism. It satisfies $\pi \circ \tilde{f} =$ $= f \circ \pi$. The map $\tilde{f}$ is purely inseparable, hence is injective [7, Exp. V, Cor. 2 to Thm. 2]. It is then surjective by the first part of the proof, hence f is surjective, too. Finally, 2.3(c) goes over without change.

4. Real algebraic varieties.

4.1. By a *real algebraic variety* V we mean the variety $X(\mathbf{R})$ of real points of a complex algebraic variety X defined over $\mathbf{R}$, in which V is Zariski-dense. We say that X is a complexification of V. A morphism $f\colon V = X(\mathbf{R}) \to W = Y(\mathbf{R})$ of real algebraic varieties is the restriction of an $\mathbf{R}$-morphism $g\colon U \to Y$ of a Zariski $\mathbf{R}$-open subset U of X, containing V, into Y. As long as we are only interested in V, we may of course replace X by U, and there is therefore no restriction of generality in considering morphisms which are restrictions of $\mathbf{R}$-morphisms of X into Y.

We say that V is smooth (resp. normal) if it consists of simple (resp. normal) points of X. Replacing X by a Zariski-open subset, if needed, we may then assume X itself to be smooth (resp. normal) without loss of generality.

V carries the Zariski-topology and the (finer) ordinary topology. With respect to the latter, it is a locally compact space, and its cohomological dimension is equal to $\dim_{\mathbf{C}} X$ (see e.g. [4, 3.5]). It also admits a triangulation [12].

4.2. Lemma. *Let V be a real algebraic variety of dimension n, L a noetherian ring. Then V has finite cohomological dimension, $H_c^*(V; L)$ is finitely generated. If L is a field of characteristic two, then $\dim H_c^n(V; L) \geq$ the number of irreducible components of dimension n of V.*

The proof of the finite generation of $H_c^*(V; L)$ is the same as in the complex case. The space V carries a fundamental class mod 2 [4, 3.7, 3.8], which implies that its n-th homology group with closed supports $H_n(V; L)$ is $\neq 0$, if L is a field of characteristic two. But this space is the dual of $H_c^n(V; L)$, hence the latter is $\neq 0$, too.

Let m be the number of irreducible components of dimension n of V. Let A be one such component, and B the union of the other irreducible components of V. Then $V = A \cup B$, and $C = A \cap B$ is an algebraic subset of dimension $< n$. Let

$$d'\colon H_c^{n-1}(C; L) \to H_c^n(A - C; L),$$
$$d''\colon H_c^{n-1}(C; L) \to H_c^n(B - C; L),$$

be the coboundary homomorphisms in the cohomology sequences of A mod C and B mod C respectively. Since $H_c^n(C; L) = 0$, the image of d' has codimension ≥ 1 by the above, and, arguing by induction on the number of irreducible components of dimension n, we may assume that $\operatorname{Im} d''$ has codimension $\geq m - 1$. But $V - C$ is the disjoint union of the open subsets $A - C$ and $B - C$, and it is clear from the construction of the cohomology sequence that the coboundary homomorphism d in the cohomology sequence

$$H_c^{n-1}(C; L) \xrightarrow{d} H_c^n(A - C; L) \oplus H_c^n(B - C; L) \to H_c^n(V; L) \to H_c^n(C; L)$$

is the sum of d' and d''. Hence the codimension of $\operatorname{Im} d$ is $\geq m$, whence our contention.

4.3. For the sake of reference, we recall here that if X, Y are normal complex irreducible algebraic varieties, and $f\colon X \to Y$ is a dominant morphism all of whose fibres have the same dimension, then f is open in the Zariski-topology and in the ordinary topology.

In fact, f is open in the Zariski-topology by a result of CHEVALLEY (valid in arbitrary characteristic) [8, pp. 195—196]. If the fibres are finite, the only case of interest here, this is also contained in Appendix II of [7, Exp. 5].

The spaces X, Y are complex analytic spaces, being normal as algebraic varieties, they are locally analytically irreducible (and in fact analytically normal) by well-known results of ZARISKI. The openness of f in the ordinary topology then follows from [13, Chap. VII, Prop. 4, p. 132].

4.4. Theorem. *Let V be a real algebraic variety and $f\colon V \to V$ an injective morphism. Assume either that V is smooth, or that V is normal and f is the restriction of an* **R**-*endomorphism g with finite fibres of a complexification X of V. Then f is surjective.*

The image $f(V)$ of V contains a dense Zariski-open subset of V. Furthermore if X is irreducible, g is a dominant morphism of odd degree N [3, § 2]. We now distinguish two cases:

(a) *X is irreducible.* We want to prove first that f *is open*. This is clear, by the invariance of domain, if V is smooth. So assume now that V is normal and that all fibres of g are finite. Then g is open in both the Zariski and the ordinary topology (4.3).

The morphism g induces a monomorphism g^0 of the field $E = \mathbf{C}(X)$ of rational functions of X into itself. Let F be its image. Then $\mathbf{C}(X)$ is an extension of degree N of F. Viewing F as the function field of X, we consider the normalization M of X in E ([11], Chap. V, Thm. 3, p. 131, [8], Thm. 2, p. 180). Let $\pi\colon M \to X$ be the canonical projection. M and π are defined over $\mathbf{R}$. The fibres of π are finite; π is proper in the Zariski-topology [8], hence in the ordinary topology ([5], Prop. 5 et Remarque), and is open by 4.3. By construction, there is a birational map $h\colon X \to M$ defined over $\mathbf{R}$ such that $g = \pi \circ h$. It follows easily from Zariski's main theorem that h is an isomorphism of X onto a Zariski open subset Z of M [7, Exp. 5, Cor. 1 to Thm. 2]. It suffices therefore to show that the restriction of π to $Z(\mathbf{R})$ is an open map of $Z(\mathbf{R})$ into V. It is clearly injective.

There is a non-empty Zariski $\mathbf{R}$-open subset U of X such that every fibre of g over U consists of exactly N points [7, Exp. 5, Cor. 3 to Thm. 1]. Then $U(\mathbf{R})$ is open and Zariski-dense in V. The fibres over U will be called the regular fibres. Since the fibres of π have at most N points, we have

$$(1) \qquad \qquad \pi^{-1}(U) \subset Z.$$

For $y \in M$, let r_y be the order of ramification of π at y. It may be defined as the upper bound of the number of elements in the intersection of W_y with regular fibres, as W_y runs through a fundamental set of neighborhoods of y. Since π is proper, the sum of the indices of the points in an arbitrary fibre is equal to N. Two complex conjugate points have clearly the same ramification index.

Let now $z \in Z(\mathbf{R})$. We claim that r_z is odd. Assume it is not. The fibre over $\pi(z)$ consists of pairs of complex conjugate points, and of real points. The ramification index being invariant under complex conjugation, the sum of the indices of the complex points is even, and there exists $z' \in M(\mathbf{R}) \cap \pi^{-1}(\pi(z))$ with odd index. If $U'_{z'}$ is a neighborhood of z' stable under complex conjugation, it intersects the fibre of a regular point r close to $\pi(z)$ in an odd number of points. This intersection is stable under complex conjugation, hence contains at least one real point. This point belongs to $Z(\mathbf{R})$ by (1). On the other hand $h(g^{-1}(U(\mathbf{R})))$ is dense in $Z(\mathbf{R})$; given a neighborhood U_z of z, we may therefore find a regular fibre $Q = \pi^{-1}(q)$, with $q \in V$ close to $\pi(z)$, which has a real point in U_z. But we already remarked that it has also a real point close to z', whence a contradiction with the injectivity of f.

We know that if W is a neighborhood of z in Z, then $\pi(W)$ is a neighborhood of $\pi(z)$. In order to show that $\pi: Z(\mathbf{R}) \to V$ is open at z, it suffices to show that $\pi(W(\mathbf{R}))$ contains a neighborhood of $\pi(z)$ in V. If $q \in V$ is sufficiently close to $\pi(z)$, then the sum of the ramification indices of the points in $\pi^{-1}(q) \cap W$ is equal to m_z. It is therefore odd, which implies immediately that $\pi^{-1}(q) \cap W$ contains a real point.

Thus f is open. Similarly, f^q is open $(q = 1, 2, \ldots)$. The Y_q's form an increasing sequence of closed subsets of V. Assume that $Y_1 \neq \emptyset$. Let $a = \dim Y_1$ and L be a field of characteristic two. In order to get a contradiction, it suffices, in view of 1.2, 4.2, to show that

$$(1) \qquad \dim H^a(Y_q; L) \geqq q, \qquad (q = 1, 2, \ldots).$$

The set Y_q is not necessarily algebraic. However, if F_q is its Zariski-closure in V, then $Y_q = F_q - f^q(M_q)$, with M_q algebraic, $\dim M_q < \dim F_q = a$ $(q = 1, 2, \ldots)$.

This was proved in [3] when $V = \mathbf{R}^n$. We repeat the proof. Let $M_q = (f^q)^{-1}(F_q)$. Then M_q is algebraic and, obviously, $Y_q = F_q - f^q(M_q)$. Let Z_q be the Zariski-closure of $f^q(M_q)$. Then $f^q(M_q)$ contains a dense Zariski-open subset of Z_q, and

$$\dim(Z_q - f^q(M_q)) < \dim Z_q.$$

If now Z_q had dimension a, it would contain a top dimensional component C of F_q, and then C would be in $f^q(M_q)$, so that the union of the other components of F_q would be the Zariski-closure of Y_q, a contradiction. Therefore $\dim Z_q < a$, whence also $\dim f^q(M_q) < a$. Moreover, all simple points of F_q belong to $F_q - Z_q$, which is contained in Y_q, hence $\dim F_q = a$. The cohomology sequence of $F_q \bmod Y_q$ then yields

$$(2) \qquad H_c^a(F_q; L) \cong H_c^a(Y_q; L), \qquad (q = 1, 2, \ldots).$$

In order to prove (1), it is therefore enough, by 4.2, to show that, if m_q is the number of irreducible components of dimension a of F_q, then

$$(3) \qquad m_q \geqq q, \qquad (q = 1, 2, \ldots).$$

For $q = 1$, this is clear, so we use induction on q. We have

$$(4) \qquad Y_q = Y_{q-1} \cup f^{q-1}(Y_1), \qquad Y_{q-1} \cap f^{q-1}(Y_1) = \emptyset, \qquad (q = 2, 3, \ldots),$$

with Y_{q-1} closed and $f^{q-1}(Y_1)$ open in Y_q, whence $F_q = F_{q-1} \cup N_{q-1}$ where N_{q-1} is the Zariski-closure of $f^{q-1}(Y_1)$. We claim that

$$(5) \qquad \dim N_q = a, \qquad \dim(N_q - f^q(Y_1)) < a, \qquad (q = 1, 2, \ldots).$$

In fact, $f^q(F_1)$ contains a dense Zariski-open subset of its Zariski-closure P_q, hence $\dim P_q = a$ and $\dim(P_q - f^q(F_1)) < a$. Since

$$\dim f^q(F_1 - Y_1) = \dim(F_1 - Y_1) = \dim f(M_1) < a$$

by the above, we have $\dim(P_q - f^q(Y_1)) < a$, whence (5) immediately. Since Y_{q-1} and $f^{q-1}(Y_1)$ are disjoint, it follows then that $\dim(V_{q-1} \cap N_{q-1}) < a$, whence $m_q \geq m_{q-1} + 1 \geq q$.

(b) In the general case, X consists of disjoint irreducible components X_i, $(1 \leq i \leq q)$. Assume $X_1, \ldots, X_s$ are the components of top dimension. Then $f(X_i(\mathbf{R})) \subset X_{j(i)}(\mathbf{R})$, $(1 \leq j(i) \leq s)$, and $f(X_i(\mathbf{R}))$ contains a dense Zariski-open subset of $X_{j(i)}(\mathbf{R})$, [3, § 2]. Consequently, $j(i) \neq j(k)$ if $i \neq k$. We may then argue as in the complex case.

4.5. Remarks. 1. The proof of the finite generation of the cohomology with compact supports of a real or complex algebraic variety was reduced in 2.1, 4.2 to the compact case, and there deduced from the triangulation theorem. However, the latter is not needed in full force. It would suffice e.g. to show that a real or complex algebraic variety is clc$^\infty$, e.g. that it is locally contractible, and then use [6, § 16.5].

2. The assumptions made on V and g were used to show that f is open. I do not know whether they are necessary for the conclusion to hold.

References

[1] M. ARTIN and A. GROTHENDIECK, Cohomologie étale des schémas. Séminaire de géométrie algébrique de l'I.H.E.S. 1963—64.

[2] J. AX, The elementary theory of finite fields. Ann. of Math., II. Ser. 88, 239—271 (1968).

[3] A. BIALYNICKI-BIRULA and M. ROSENLICHT, Injective morphisms of real algebraic varieties. Proc. Amer. Math. Soc. 13, 200—203 (1962).

[4] A. BOREL et A. HAEFLIGER, La classe d'homologie fondamentale d'un espace analytique. Bull. Soc. Math. France 89, 461—513 (1961).

[5] A. BOREL et J.-P. SERRE, Le théorème de Riemann-Roch. Bull. Soc. Math. France 86, 97—136 (1958).

[6] G. BREDON, Sheaf Theory. New York 1967.

[7] C. CHEVALLEY, Séminaire sur la classification des groupes de Lie algébriques, mimeographed notes. Paris 1956—58.

[8] C. CHEVALLEY, Fondements de la géométrie algébrique, mimeographed notes. Paris 1958.

[9] R. GODEMENT, Topologie algébrique et théorie des faisceaux. Paris 1958.

[10] A. GROTHENDIECK, Formule de Lefschetz et rationalité des fonctions L. Sém. Bourbaki. Exp. 279 (1964—65).

[11] S. LANG, Introduction to algebraic geometry. New York 1958.

[12] S. ŁOJASIEWICZ, Triangulation of semi-analytic sets. Ann. Scuola Norm. Sup. Pisa, Ser. III 18, 449—473 (1964).

[13] R. NARASIMHAN, Introduction to the theory of analytic spaces. Springer Lecture Notes 25 (1967).

[14] J.-P. SERRE, Faisceaux algébriques cohérents. Ann. of Math., II. Ser. 61, 197—278 (1955).

Eingegangen am 9. 9. 1968

Anschrift des Autors:

A. Borel
The Institute for Advanced Study, Princeton, N. J., USA

86.

Sous-groupes discrets de groupes semi-simples
(d'après D. A. Kajdan et G. A. Margoulis)

Séminaire Bourbaki, Exp. 358 (1968/69)
Lect. Notes Math. **179** (1971) 199−216

§ 1. Enoncé des résultats

G désignera toujours un groupe de Lie connexe, semi-simple. Pour abréger on appellera *facteur compact* de G tout sous-groupe compact connexe distingué, et on dira que G est *sans facteur compact* s'il ne possède pas de facteur compact $\neq \{e\}$.

Dans ce paragraphe, on énonce les résultats obtenus dans [6] (à cela près que [6] suppose G linéaire, sans facteur compact, mais le passage de là au cas un peu plus général considéré ici est immédiat).

Théorème 1. *Il existe un voisinage W de l'élément neutre e de G tel que si Γ est un sous-groupe discret de G dont l'intersection avec tout facteur compact de G est dans le centre de G, alors il existe $g \in G$ tel que*

$$ g \cdot \Gamma \cdot g^{-1} \cap W = \{e\}. $$

Soit m une mesure de Haar sur G. Elle est bi-invariante, et définit canoniquement une mesure invariante sur G/Γ pour tout sous-groupe discret Γ de G. On notera $m(A)$ la mesure d'une partie mesurable de G/Γ par rapport à cette mesure.

Corollaire. *Il existe une constante $d_G > 0$ telle que si Γ est comme dans le théorème, alors $m(G/\Gamma) \geq d_G$.*

En effet, si W_0 est un voisinage symétrique de e tel que $W_0 \cdot W_0 \subset W$, alors la restriction de la projection canonique $\pi : G \to G/(g \cdot \Gamma \cdot g^{-1})$ à W_0 est injective donc $m(G/\Gamma) = m(G/(g \cdot \Gamma \cdot g^{-1})) \geq m(W_0)$.

On voit en particulier que *si G n'a pas de facteur compact, alors les conclusions du Théor. 1 et de son corollaire valent pour tout sous-groupe discret de G.* Cela est évidemment faux si G est compact.

Ce corollaire était connu auparavant pour $G = \mathbf{SL}(2, \mathbf{R})$. En effet, Siegel [7] a montré que si X est le disque unité $\{|z| < 1\}$ du plan complexe, muni de l'élément de volume $dx \wedge dy \cdot (1 - |z|^2)^{-1}$, et si Γ est un groupe d'automorphismes proprement discontinu de X, alors $m(X/\Gamma) \geq \pi/21$ (la borne inférieure étant atteinte). Récemment, H. C. Wang [8] a donné une autre démonstration du Théor. 1, qui fournit aussi des renseignements quantitatifs sur W.

47

Théorème 2. *Soit Γ un sous-groupe discret de G. Supposons G/Γ non compact et $m(G/\Gamma) < \infty$. Alors il existe $x \in \Gamma$ dont la classe de conjugaison $C(x) = \{g \cdot x \cdot g^{-1}, g \in G\}$ n'est pas fermée. Si G n'a pas de facteur compact, on peut trouver $x \in \Gamma - \{e\}$ tel que e soit contenu dans l'adhérence $\overline{C(x)}$ de $C(x)$.*

Il est élémentaire que si L est un groupe localement compact et H un sous-groupe discret tel que L/H soit compact, alors les classes de conjugaison dans L des éléments de H sont fermées (voir p. ex. [2, § 11.2]). Le théorème 2 fournit donc une réciproque (lorsque $m(G/\Gamma) < \infty$).

Supposons G linéaire. Il est alors d'indice fini dans le groupe des points réels d'un groupe algébrique sur **R**. On sait que $C(g)$ est fermée (resp. admet e comme point d'accumulation) si et seulement si g est semi-simple (resp. unipotent, i. e. g-e est nilpotent). Le théorème 2 entraîne donc le:

Corollaire. *Soit Γ comme dans le Théor. 2 et supposons G linéaire. Alors Γ possède un élément $x \neq e$ non semi-simple (resp. unipotent si G est sans facteur compact).*

Ce corollaire, classique pour les groupes fuchsiens, avait été conjecturé en général par A. Selberg. Il était connu dans plusieurs cas particuliers, notamment si Γ est arithmétique (conjecture de Godement, cf. [2, § 11]), si G est produit de groupes **SL** $(2, \mathbf{R})$ et **SL** $(2, \mathbf{C})$ (Selberg), ou si G est simple, de rang réel 1 (Garland).

Fixons un ds^2 riemannien invariant à droite sur G, et soit $d(,)$ la distance associée. Elle est invariante à droite, et fait de G un espace métrique *complet* (puisque le ds^2 est complet, G étant riemannien homogène). On posera

$$|x| = d(e, x), \quad D_r = \{g \in G \mid |g| < r\}, \quad (x \in G; r > 0).$$

Comme la métrique d est complète, D_r est relativement compact. d étant symétrique, invariante à droite, on a

$$(1) \qquad\qquad |x| = |x^{-1}|, \quad d(x, y) = |x \cdot y^{-1}|, \quad (x, y \in G).$$

De $|x| = d(xy^{-1}, y^{-1})$, on tire aussi

$$(2) \qquad\qquad |x| \leq |y| + d(x, y), \quad |x| \leq |y| + |y \cdot x|, \quad (x, y \in G).$$

Théorème A. *Supposons G sans facteur compact. Il existe un voisinage V de e dans G, et des constantes $b > 0$, $c > 1$ (dépendant de G) ayant la propriété suivante: si Γ est un sous-groupe discret de G, alors on peut trouver $g \in D_b$ tel que*

$$|g \cdot x \cdot g^{-1}| \geq c|x|, \quad (x \in \Gamma \cap V).$$

Les théor. 1 et 2 se déduisent par des raisonnements essentiellement élémentaires, mais extrêmement ingénieux, du Théor. A, qui est donc en fait le résultat central de [6]. Ce dernier sera établi au § 4, les §§ 2, 3 étant consacrés à la démonstration des Théor. 1 et 2 à partir du Théorème A. Bien qu'il s'agisse d'un rapport sur un article publié, on a donné des démonstrations complètes. Suivant l'exemple de [3], on renvoie au Mémoire original le lecteur qui préfère un exposé plus bref, non alourdi de détails techniques.

Notations. Outre celles introduites plus haut, on utilisera les suivantes:

Si $g \in G$ et $A \subset G$, on écrira quelquefois $^g A$ pour $g \cdot A \cdot g^{-1}$.

Si G est sans facteur compact, et Γ est un sous-groupe discret de G, un élément $g \in D_b$ qui vérifie la condition du Théor. A pour Γ sera dit être *adapté à* Γ.

Enfin, l'algèbre de Lie d'un groupe de Lie G, H, P, ... est notée par la minuscule gothique correspondante $\mathfrak{g}$, $\mathfrak{h}$, $\mathfrak{p}$, ...

§ 2. Démonstration du Théorème 1

Dans ce paragraphe et le suivant, on admet le Théor. A. Si G est sans facteur compact, on choisit a $(0 < a \leq 1)$ tel que

$$(1) \qquad g^{-1} \cdot D_a \cdot g \subset V, \qquad (g \in D_b).$$

Pour la commodité des références, on met en lemme le point essentiel de la démonstration du Théor. 1.

2.1. Lemme. *Supposons G sans facteur compact, et soit Γ un sous-groupe discret de G. Soient n un entier ≥ 1 et $g_i \in D_b$ $(1 \leq i \leq n)$ tels que g_1 soit adapté à $\Gamma = \Gamma_1$, et g_i à $\Gamma_i = g_{i-1} \cdot \Gamma_{i-1} \cdot g_{i-1}^{-1}$ $(2 \leq i \leq n)$. Soient $x \in \Gamma$, $h_i = g_i \cdot \ldots \cdot g_1$ et $x_i = h_i \cdot x \cdot h_i^{-1}$ $(1 \leq i \leq n)$. Supposons $x_n \in D_a$. Alors $x_i \in D_a$ et $|x_i| \geq c|x_{i-1}|$, $(1 \leq i \leq n; x_0 = x)$. En particulier $|x| \leq c^{-n} \cdot a$.*

Supposons $x_i \in D_a$. Alors, vu (1),

$$x_{i-1} = g_{i-1}^{-1} \cdot x_i \cdot g_{i-1} \in V \cap \Gamma_{i-1},$$

donc, puisque g_{i-1} est adapté à Γ_{i-1}, $|x_i| \geq c|x_{i-1}|$ et $x_{i-1} \in D_a$, ce qui établit notre assertion par récurrence descendante sur i.

2.2. *Démonstration du Théor. 1.* Supposons tout d'abord G sans facteur compact, et montrons que $W = D_a$ vérifie le théorème. Soit q le minimum de $|x|$ sur $\Gamma - \{e\}$, et soit $n \geq 1$ tel que $c^n \cdot q > a$. D'après le Théor. A, on peut trouver une suite d'éléments $g_i \in D_b$ $(1 \leq i \leq n)$ tels que g_1 soit adapté à $\Gamma = \Gamma_1$ et g_i à $\Gamma_i = g_{i-1} \cdot \Gamma_{i-1} \cdot g_{i-1}^{-1}$ $(l \geq 2)$. Posons $h_n = g_n \cdot \ldots \cdot g_1$. Soit $x \in \Gamma$. Alors, si $h_n \cdot x \cdot h_n^{-1} \in W$, le lemme 2.1 montre que

$$|x| \leq c^{-n} \cdot a < q,$$

donc $x = e$, et ainsi Γ possède un conjugué, Γ_n, dont l'intersection avec W est réduite à $\{e\}$.

Dans le cas général, soient Z le centre de G, N le plus grand facteur compact de G et $\pi: G \to G/N = G'$ la projection canonique. Alors G' est sans facteur compact. Soit W un voisinage de e dans G tel que $W \cap Z = \{e\}$ et que $\pi(W)$ vérifie le Théor. 1 pour G'. Soit Γ un sous-groupe discret de G tel que $\Gamma \cap N \subset Z$. Comme N est compact $\Gamma' = \pi(\Gamma)$ est discret et il existe donc $g \in G$ tel que $^{\pi(g)}\Gamma' \cap \pi(W) = \{e\}$. On a alors $^g \Gamma \cap W \subset N$; mais, comme $\Gamma \cap N \subset Z$, et N est distingué dans G, on a $^g \Gamma \cap N = {}^g(\Gamma \cap N) \subset Z$, donc $^g \Gamma \cap W \subset Z \cap W = \{e\}$.

<h3 align="center">§ 3. Démonstration du Théorème 2</h3>

3.1. Lemme. *Soit Γ un sous-groupe discret de G tel que G/Γ soit non compact, de mesure $m(G/\Gamma)$ finie, et soit $\varepsilon > 0$. Alors il existe $x \in \Gamma$, $x \neq e$, et $g \in G$ tels que $|^{g}x| \leqq \varepsilon$.*

Soit $\pi : G \to G/\Gamma$ la projection canonique. La métrique d définit sur G/Γ une métrique $\bar{d}$ caractérisée par

$$\bar{d}(\pi(u), \pi(v)) = \min_{x \in \Gamma} d(u, v \cdot x) \qquad (u, v \in G).$$

L'espace G/Γ étant non compact, de mesure finie, on peut trouver $r > 0$ et $g_0 \in G$ tels que

$$(1) \qquad m(G/\Gamma - \pi(D_r)) < m(D_{\varepsilon/2}), \quad \bar{d}(\pi(e), \pi(g_0)) > r + \varepsilon.$$

Vu (2) du § 1, on a alors

$$(2) \qquad \pi(D_{\varepsilon/2} \cdot g_0) \cap \pi(D_r) = \emptyset,$$

d'où

$$m(\pi(D_{\varepsilon/2} \cdot g_0)) < m(G/\Gamma - \pi(D_r)) < m(D_{\varepsilon/2}) = m(D_{\varepsilon/2} \cdot g_0),$$

ce qui prouve que π n'est pas injective sur $D_{\varepsilon/2} \cdot g_0$. Soient donc $u \in D_{\varepsilon/2}$ et $x \in \Gamma - \{e\}$ tels que $u \cdot g_0 \cdot x \in D_{\varepsilon/2} \cdot g_0$. On a alors $d(u \cdot g_0, u \cdot g_0 \cdot x) \leqq \varepsilon$, donc $|u \cdot g_0 \cdot x \cdot (u \cdot g_0)^{-1}| \leqq \varepsilon$.

3.2. Dans la suite de ce paragraphe, C désigne une constante $\geqq c$ telle que

$$(3) \qquad |^{g}x| \leqq C \cdot |x|, \quad (g \in D_b; x \in D_a).$$

[L'existence de C résulte du fait que $|\cdot|$ est associée à une métrique riemannienne, cf. 4.1.]

Supposons G *sans facteur compact*. On fixe un voisinage symétrique W_0 de e tel que $W_0 \cdot W_0 \subset D_a$, et $r > 0$ tel que

$$(4) \qquad m(G/\Gamma - \pi(D_r)) < m(W_0).$$

Soit

$$S = \{x \in \Gamma \mid D_{r+a+c} \cdot x \cap D_{r+a+c} \neq \emptyset\}.$$

Nous voulons montrer que S contient un élément $s \neq e$ tel que $e \in \overline{C(s)}$, ce qui établira le Théor. 2 lorsque G est sans facteur compact. Comme S est fini (car D_{r+a+c} est relativement compact), il suffira pour cela de prouver:

(*) *Soit $\varepsilon > 0$. Alors il existe $s \in S - \{e\}$ et $g \in G$ tels que $|^{g}s| \leqq \varepsilon$.*

50

On suppose $\varepsilon \leqq a$. On fixe un entier $n \geqq 1$ et un nombre réel d tels que

$$(5) \qquad c^n \cdot \varepsilon \geqq a, \quad 0 < d < (c/C)^n \cdot C^{-1} \cdot \varepsilon \leqq \varepsilon.$$

(Rappelons que $C \geqq c > 1$.) D'après 3.1, il existe $g_0 \in G$ et $x_0 \in \Gamma - \{e\}$ tels que

$$(6) \qquad |g_0 \cdot x \cdot g_0^{-1}| < d.$$

A l'aide du Théor. A, on peut trouver $g_1 \in D_b$ adapté à $\Gamma_1 = g_0 \cdot \Gamma \cdot g_0^{-1}$, puis g_i adapté à $\Gamma_i = g_{i-1} \cdot \Gamma_{i-1} \cdot g_{i-1}^{-1}$, $(i = 2, 3, \ldots)$. Vu (5), (6) on a $\Gamma_1 \cap D_a \neq \{e\}$. Comme on l'a vu en démontrant le Théor. 1, 2.1 entraîne alors l'existence de $m \geqq 1$ tel que

$$(7) \qquad \Gamma_m \cap D_a \neq \{e\}, \quad \Gamma_{m+1} \cap D_a = \{e\}.$$

Posons encore $h_i = g_i \cdot \ldots \cdot g_0 \, (0 \leqq i \leqq m + 1)$.

Lemme. *On conserve les notations et hypothèses précédentes. Alors $\pi(D_a \cdot h_{m+1})$ $\cap \pi(D_r) \neq \emptyset$. Si $y \in \Gamma - \{e\}$ est tel que $h_m \cdot y \cdot h_m^{-1} \in D_a$, alors $|g_0 \cdot y \cdot g_0^{-1}| \leqq \varepsilon$.*

Admettons provisoirement ce lemme. Il existe alors $x \in \Gamma - \{e\}$ et $u \in D_a$ tels que $u \cdot h_{m+1} \cdot x \in D_r$. Vu (2) du § 1, on a

$$(8) \qquad \begin{aligned} |h_{m+1} \cdot x| &\leqq |u \cdot h_{m+1} \cdot x| + |u|, \\ |h_{m+1} \cdot x| &\leqq r + a. \end{aligned}$$

Soit $y \in \Gamma - \{e\}$ tel que $h_m \cdot y \cdot h_m^{-1} \in D_a$. On a, vu (3),

$$d(h_{m+1} \cdot y \cdot x, h_{m+1} \cdot x) = d(h_{m+1} \cdot y, h_{m+1})$$
$$= |h_{m+1} \cdot y \cdot h_{m+1}^{-1}| \leqq C \cdot |h_m \cdot y \cdot h_m^{-1}|.$$

Comme $h_m \cdot y \cdot h_m^{-1} \in D_a$ et $a \leqq 1$, le dernier terme est $\leqq C$. Vu (8), et (2) du § 1, cela donne

$$(9) \qquad |h_{m+1} \cdot y \cdot x| \leqq C + r + a.$$

On a par suite

$$h_{m+1} \cdot y \cdot x = h_{m+1} \cdot x \cdot x^{-1} \cdot y \cdot x \in D_{r+a+C} \cap D_{r+a} \cdot x^{-1} \cdot y \cdot x,$$

d'où $x^{-1} \cdot y \cdot x \in S - \{e\}$. Mais $C(x^{-1} \cdot y \cdot x) = C(y)$, donc (*) résulte de la deuxième assertion du lemme.

3.3. Démontrons maintenant le lemme de 3.2. Comme $h_{m+1} \cdot \Gamma \cdot h_{m+1}^{-1} \cap D_a$ $= \{e\}$ (cf. (7)), la projection π est injective sur $W_0 \cdot h_{m+1}$. On a alors, vu (4)

$$m(\pi(W_0 \cdot h_{m+1})) = m(W_0 \cdot h_{m+1}) = m(W_0) > m(G/\Gamma - \pi(D_r)),$$

d'où $\pi(W_0 \cdot h_{m+1}) \cap \pi(D_r) \neq \emptyset$, et la première assertion du lemme, puisque $W_0 \subset D_a$.

Soit $y \in \Gamma - \{e\}$ tel que $|h_m \cdot y \cdot h_m^{-1}| < a$. D'après 2.1, on a

$$|g_0 \cdot y \cdot g_0^{-1}| < c^{-m} \cdot a,$$

donc, si $m \geq n$, on a, vu (5) et $c > 1$:

$$|g_0 \cdot y \cdot g_0^{-1}| < c^{-n} \leq \varepsilon.$$

Il reste à considérer le cas où $n > m$. Soit m' le plus grand entier $\leq m$ tel que

$$\left.\begin{array}{l}
h_{m'+1} \cdot x_0 \cdot h_{m'+1}^{-1} \notin D_a, \qquad h_j \cdot x_0 \cdot h_j^{-1} \in D_a \quad (0 \leq j \leq m'). \\[2mm]
\text{On a alors, vu (3),} \\[2mm]
\qquad a \leq |h_{m'+1} \cdot x_0 \cdot h_{m'+1}^{-1}| \leq C^{m'+1} \cdot |g_0 \cdot x_0 \cdot g_0^{-1}|. \\[2mm]
\text{D'autre part, 2.1 donne} \\[2mm]
\qquad |g_0 \cdot y \cdot g_0^{-1}| \leq c^{-m'} |h_{m'} \cdot y \cdot h_{m'}^{-1}| \leq c^{-m'} \cdot a. \\[2mm]
\text{Par conséquent, vu (5), (6)} \\[2mm]
\qquad |g_0 \cdot y \cdot g_0^{-1}| \leq c^{-m'} \cdot C^{m'+1} |g_0 \cdot x_0 \cdot g_0^{-1}| \leq (c/C)^{-m'+n} \cdot \varepsilon.
\end{array}\right\} \quad m' = m'$$

Comme $m' \leq m < n$ et $c/C \leq 1$, cela entraîne $|g_0 \cdot y \cdot g_0^{-1}| \leq \varepsilon$, et termine la démonstration du lemme, et par conséquent du Théor. 2, lorsque G est sans facteur compact.

3.4. Soient N le plus grand facteur compact de G et $\pi : G \to G' = G/N$ la projection canonique. Comme N est compact, le groupe $N \cdot \Gamma$ est fermé dans G, l'application π est propre, et $\pi(\Gamma) = \Gamma'$ est discret dans G'. Le quotient G/Γ est fibré, de fibre type $N/N \cap \Gamma$ compacte, base $G/N \cdot \Gamma$, donc $G/N \cdot \Gamma$ est non-compact, de mesure invariante finie. Comme $G'/\Gamma' \cong G/N \cdot \Gamma$, il en est de même pour G'/Γ'. Le groupe G' étant sans facteur compact, il existe d'après ce qui précède $x \in \Gamma - \{e\}$ tel que $C(\pi(x))$ ne soit pas fermée dans G'. Comme $C(\pi(x)) = \pi(C(x))$ et que π est propre, il s'ensuit que $C(x)$ n'est pas fermée dans G.

§ 4. Démonstration du Théorème A

4.1. Soit U un voisinage ouvert relativement compact de e dans G admettant des coordonnées canoniques (x_i). Cela signifie qu'il existe un voisinage U_0 de 0 dans $\mathfrak{g}$ tel que l'application exponentielle $\exp: \mathfrak{g} \to G$ soit un difféomorphisme de U_0 sur U, et que (x_i) est transporté par l'exponentielle d'un système de coordonnées sur $\mathfrak{g}$.

On fixe une norme euclidienne $|\cdot|$ sur $\mathfrak{g}$, et on note de la même manière le transporté de $|\cdot|$ à U par l'exponentielle. Comme $|\cdot|$ est associée à une métrique riemannienne, on voit facilement qu'il existe des constantes $d, d' > 0$ telles que

$$(1) \qquad\qquad d \cdot |x|_0 \leqq |x| \leqq d' \cdot |x|_0, \qquad (x \in U).$$

L'existence de la constante C de 3.2 (3) s'en déduit immédiatement. En effet, soit a' tel que ${}^g D_{a'} \subset U$ pour $g \in D_b$ $(0 < a' \leqq a)$. Alors, si $a' \leqq |x| \leqq |a|$, l'existence de C est évidente par compacité et continuité, et si $|x| \leqq a'$, elle résulte de (1) et du fait que $x \mapsto {}^g x$ est linéaire en coordonnées canoniques.

4.2. On sait que si U est suffisamment petit, alors $\Gamma \cap U$ est contenu dans un sous-groupe analytique nilpotent de G quel que soit le sous-groupe discret Γ de G (cf. [9], et, pour des résultats plus précis, [8]). Rappelons-en brièvement une démonstration: d'après la formule de Campbell-Hausdorff, on a, pour $t \in \mathbf{R}$ suffisamment petit

$$\log (\exp t \cdot x, \exp t \cdot y) = t^2 \cdot [x, y] + \text{termes d'ordre} \geqq 3, \qquad (x, y \in \mathfrak{g}),$$

où $(u, v) = u \cdot v \cdot u^{-1} \cdot v^{-1}$, $(u, v \in G)$. Il s'ensuit que, pour U convenable, on a

$$|(u, v)|_0 < \min (|u|_0, |v|_0) \qquad (u, v \in U - \{e\}).$$

Si maintenant Γ est un sous-groupe discret, on peut écrire

$$\Gamma \cap U = \{e, a_1, \ldots, a_m\}, \, (|a_1|_0 \leqq |a_2|_0 \leqq \ldots \leqq |a_m|_0),$$

et l'on déduit de l'inégalité précédente que les groupes à un paramètre $\exp(\mathbf{R} \log a_i)$, $(1 \leqq i \leqq m)$ engendrent un groupe analytique nilpotent.

Cela montre que le Théorème A est conséquence du

Théorème B. *Supposons G sans facteur compact. Il existe un voisinage V de e dans U, et des constantes $b > 0$, $c < 1$ (dépendant de G) ayant la propriété suivante: si N est un sous-groupe analytique nilpotent de G, alors il existe $g \in D_b$ tel que $|{}^g x| \geqq c |x|$ quel que soit $x \in N \cap V$.*

4.3. Soient R_p l'ensemble des sous-groupes analytiques nilpotents de dimension p de G et $L R_p$ celui de leurs algèbres de Lie. Il est immédiat que $L R_p$ est fermé dans la grassmannienne des p-plans de $\mathfrak{g}$ donc $L R_p$, muni de la topologie induite, est compact.

Pour tout sous-groupe analytique H de G, posons $H_0 = \exp (\mathfrak{h} \cap U_0)$, où $U_0 = \log U$. C'est un voisinage de e dans H pour sa topologie de groupe de Lie. (Rappelons que si H est fermé, cette dernière est la topologie induite par celle de G).

Soit N un sous-groupe analytique nilpotent. Soient $U' \subset U$ un voisinage de e, et $r > 0$ assez petit pour que la boule fermée $\{|x| \leqq r\}$ soit contenue dans U'. Notons S_r la sphère $\{|x| = r\}$ de rayon r, A un voisinage ouvert de $N_0 \cap S_r$ dans S_r, et A' le cône de sommet e, et base A. Il est clair qu'il existe un voisinage C de $\mathfrak{n}$ dans $L R_p$ tel que si $\mathfrak{m} \in C$, et si M est le groupe analytique correspondant, alors $M_0 \cap A$ est un voisinage de e dans M_0. Comme l'adhérence d'un sous-groupe analytique nilpotent est un groupe nilpotent, on voit, par compacité, que, étant donné N, il suffit d'exhiber U', A', $c_N > 1$ et $g \in G$ tels que

$$(1) \qquad\qquad |{}^g x| \geqq c_N \cdot |x|, \qquad (x \in A').$$

4.4. Lemme. *Soit* $\mathfrak{n}$ *une sous-algèbre de Lie nilpotente de* $\mathfrak{g}$ *et soit* c' *une constante* > 0. *Alors il existe* $g \in G$ *tel que*

$$|\operatorname{Ad} g\,(x)|_0 \geqq c'\,|x|_0, \qquad (x \in \mathfrak{n}).$$

Montrons tout d'abord que si $c' > 2 \cdot d'/d$ (où d, d' sont comme dans 4.1 (1)), alors 4.4 entraîne la dernière assertion de 4.3. Soit $r > 0$ tel que $\operatorname{Ad} g\,(x) \in U_0$ si $|x| \leqq r$. Il existe alors, par continuité, un voisinage A de $N_0 \cap S_r$ dans S_r tel que

$$(2) \qquad\qquad {}^g x \in U, \quad |{}^g x|_0 \geqq (c'/2) \cdot |x|_0, \qquad (x \in A).$$

Comme $|\cdot|_0$ provient d'une métrique euclidienne en coordonnées canoniques, (2) reste vraie, par homogénéité, si x parcourt le cône A' de sommet e et base A. Mais, d'après 4.1 (1), on a alors

$$|{}^g x| \geqq d \cdot |{}^g x|_0 \geqq (c' \cdot d/2) \cdot |x|_0 \geqq (c' \cdot d/2 \cdot d') \cdot |x|, \qquad (x \in A'),$$

ce qui donne bien 4.3 (1) si $c' > 2 \cdot d' \cdot d^{-1}$.

4.5. Il reste à démontrer 4.4. Il est clair que l'on peut remplacer G par un groupe localement isomorphe, par exemple par son groupe adjoint $\operatorname{Ad} G$, ce qui permet de supposer que G est la composante connexe de e dans le groupe des points réels d'un groupe algébrique $G_{\mathbf{C}}$ défini sur $\mathbf{R}$, et a un centre réduit à $\{e\}$. Une partie (resp. un sous-groupe) de G sera dite (resp. dit) algébrique si c'est l'intersection de G avec une partie (resp. un sous-groupe) algébrique de $G_{\mathbf{C}}$. Pour les propriétés des groupes algébriques (resp. semi-simples réels) utilisées ci-dessous, *voir* par exemple [1] (resp. [5]).

La plus petite sous-algèbre de Lie algébrique de $\mathfrak{g}$ contenant $\mathfrak{n}$ est aussi nilpotente, donc, quitte à agrandir $\mathfrak{n}$, on peut supposer $\mathfrak{n}$ algébrique. Soit N le sous-groupe algébrique correspondant de G. On a $N = S \times N_u$, où S est un tore algébrique (ou, plus précisément, l'intersection de G avec un tore algébrique de $G_{\mathbf{C}}$) et N_u est unipotent.

Soit $G = K \cdot A \cdot U$ une décomposition d'Iwasawa de G (K sous-groupe compact maximal, U sous-groupe unipotent maximal, A sous-groupe diagonalisable sur $\mathbf{R}$ connexe maximal normalisant U). Soient $P = M \cdot A \cdot U$ le normalisateur de U ($M = P \cap K$). C'est donc un $\mathbf{R}$-sous-groupe parabolique minimal, et $M \cdot A$ est le centralisateur $Z(A)$ de A. Soit $P^- = Z(A) \cdot U^-$ le sous-groupe parabolique opposé à P contenant $Z(A)$. ($U^- \cap U = \{e\}$.) Nous voulons montrer tout d'abord

$$(*) \quad \textit{Il existe } h \in G \textit{ tel que } \operatorname{Ad} h\,(\mathfrak{p}) \cap \mathfrak{n} = \dot{0}.$$

Si $h \in G$, ${}^h P \cap N$ est un sous-groupe algébrique de N, donc sa composante neutre est produit de ses intersections avec S et N_u, la première étant un tore. Il suffit donc

54

d'exhiber $h \in G$ tel que

$$\mathrm{Ad}\, h\,(\mathfrak{p}) \cap \mathfrak{s} = \mathrm{Ad}\, h\,(\mathfrak{p}) \cap \mathfrak{n}_u = 0.$$

Comme un ensemble algébrique propre de G est sans point intérieur, l'existence de h résulte (par Baire) des trois faits suivants:

(i) l'ensemble L des $h \in G$ tels que ${}^h P \cap N_u = \{e\}$ contient un ouvert dense;

(ii) si S' est un sous-tore de S, l'ensemble $Q\,(S')$ des $h \in G$ tels que $\mathrm{Ad}\, h\,(\mathfrak{p}) \supset \mathfrak{s}'$ est contenu dans un sous-ensemble algébrique propre;

(iii) l'ensemble des sous-tores de S est dénombrable.

Démontrons ces trois assertions:

(i) il existe $a \in G$ tel que ${}^a U \supset N_u$, donc L contient les éléments $x \cdot a$, où x est tel que ${}^{xa}P$ soit opposé à ${}^a P$, donc contient un ouvert de Zariski non vide de G.

(ii) l'ensemble Q' des sous-espaces de dimension $p = \dim P$ de $\mathfrak{g}_{\mathbf{C}} = \mathfrak{g} \otimes_{\mathbf{R}} \mathbf{C}$ qui contiennent $\mathfrak{s}'$ est algébrique dans la grassmannienne $G_p\,(\mathfrak{g}_{\mathbf{C}})$ des sous-espaces vectoriels de dimension p de $\mathfrak{g}_{\mathbf{C}}$. L'application $s : h \mapsto \mathrm{Ad}\, h\,(\mathfrak{p})$ est un morphisme de variétés algébriques de $G_{\mathbf{C}}$ dans $G_p\,(\mathfrak{g}_{\mathbf{C}})$, donc $Q\,(S')$ est contenu dans $s^{-1}\,(Q') \cap G$, qui est algébrique. D'autre part, l'intersection des sous-groupes ${}^h P\ (h \in G)$ est contenue dans $M \cdot A$. Sa composante connexe de e est un sous-groupe distingué, donc semi-simple, donc contenu dans M, donc réduit à $\{e\}$ puisque G est sans facteur compact, d'où $s^{-1}\,(Q') \cap G \neq G$.

(iii) les sous-tores de S correspondent biunivoquement aux sous-groupes facteurs directs d'un groupe abélien libre de rang égal à $\dim S$ (le groupe des caractères rationnels de S).

On peut donc supposer que $\mathfrak{p} \cap \mathfrak{n} = (0)$. On munit le système $\{\alpha\}$ des racines de G par rapport à A d'un ordre associé à P. On a donc

$$\mathfrak{g} = \mathfrak{u}^{-} \oplus \mathfrak{p}, \quad \mathfrak{u}^{-} = \bigoplus_{\alpha < 0} \mathfrak{g}_\alpha, \quad \mathfrak{g}_\alpha = \{x \in \mathfrak{g} \,|\, \mathrm{Ad}\, a\,(x) = \alpha\,(a) \cdot x,\ (a \in A)\}.$$

Soit $x \in \mathfrak{n}$. Il s'écrit de manière unique

$$x = \sum_{\alpha < 0} x_\alpha + z_x \quad (x_\alpha \in \mathfrak{g}_\alpha;\, z_x \in \mathfrak{p}).$$

Comme $\mathfrak{n} \cap \mathfrak{p} = (0)$, il existe une constante $r > 0$ telle que

$$\Big| \sum_{\alpha < 0} x_\alpha \Big|_0 \geqq r \cdot |x|_0, \quad (x \in \mathfrak{n}).$$

Soit $h \in A$. On a

$$\mathrm{Ad}\, h\,(x) = \sum_{\alpha < 0} \alpha\,(h) \cdot x_\alpha + \mathrm{Ad}\, h\,(z_x), \quad \mathrm{Ad}\, h\,(z_x) \in \mathfrak{p},\ (x \in \mathfrak{n}).$$

Soit $q > 0$. On peut trouver $h \in A$ tel que $\alpha\,(h) > q$ pour tout $\alpha < 0$. En supposant, ce qui est loisible, les $\mathfrak{g}_\alpha\ (\alpha < 0)$ et $\mathfrak{p}$ mutuellement orthogonaux, on obtient

$$|\mathrm{Ad}\, h\,(x)|_0^2 \geqq q^2 \cdot \sum_{\alpha < 0} |x_\alpha|_0^2 = q^2 \cdot \Big| \sum_{\alpha < 0} x_\alpha \Big|_0^2 \geqq q^2 \cdot r^2 \cdot |x|_0^2$$

ce qui démontre le lemme, si l'on choisit $q > r^{-1} \cdot c'$.

§ 5. Groupes de rang réel un

Dans ce paragraphe, on décrit brièvement des résultats de H. Garland et M. S. Raghunathan [4], qui ont leur point de départ dans le cor. au Théor. 2, et, entre autres choses, le renforcent considérablement dans un cas particulier. Dans ce paragraphe, G est simple, linéaire, de rang réel égal à un (i.e. le A de la décomposition d'Iwasawa $g = K \cdot A \cdot U$ de 4.5 est de dimension un), et Γ est un sous-groupe discret de G tel que $m\,(G/\Gamma) < \infty$. Bien que Γ ne soit pas nécessairement définissable arithmétiquement (Makarov, Vinberg); [4] montre que Γ a plusieurs propriétés importantes en commun avec les sous-groupes de ce type de G. On suppose G/Γ non compact (sinon ce qui suit est évident, ou connu ou trivialement faux).

Soit α la racine simple de G par rapport à A, positive pour l'ordre associé à U, et, pour $t > 0$, soit $A_t = \{a \in A \mid \alpha\,(a) \leqq t\}$. Un ensemble de Siegel $\mathfrak{S}_{t,\eta}$ de G (par rapport à K, A, U) est un ensemble de la forme

$$\mathfrak{S}_{t,\eta} = K \cdot A_t \cdot \eta, \qquad (\eta \text{ voisinage relativement compact de } e \text{ dans } U).$$

Le résultat principal de [4] affirme que l'on peut trouver $\mathfrak{S} = \mathfrak{S}_{t,\eta}$ et une partie finie non vide C de G ayant les propriétés suivantes:

(i) $^{c}U/(^{c}U \cap \Gamma)$ est compact pour tout $c \in C^{-1}$;

(ii) $G = \mathfrak{S} \cdot C \cdot \Gamma$;

(iii) $\{x \in \Gamma \mid \mathfrak{S} \cdot C \cdot x \cap \mathfrak{S} \cdot C \neq \emptyset\}$ est fini;

(iv)   pour t' suffisamment grand:

$$\mathfrak{S}_{t',\eta} \cdot c \cdot x \cap \mathfrak{S}_{t,\eta} \cdot c' \neq \emptyset, \qquad (c, c' \in C; \ x \in \Gamma),$$

entraîne $c = c'$ et $c \cdot x \cdot c^{-1} \in Z\,(A) \cdot U$.

De plus, Γ est de présentation finie, G/Γ est difféomorphe à l'intérieur d'une variété à bord compacte et, si G n'est pas localement isomorphe à $\mathbf{SL}\,(2, \mathbf{R})$ ou $\mathbf{SL}\,(2, \mathbf{C})$, alors Γ est rigide (i.e. $H^1\,(\Gamma, \mathrm{Ad}) = 0$).

Bibliographie

1. Borel, A.: Linear algebraic groups, Mathematical Lecture Notes Series, Benjamin, New York (1969)
2. Borel, A., Harish-Chandra: Arithmetic subgroups of algebraic groups, Annals of Math. (2) **75** (1962), 485−535
3. Delaroche, C., Kirillov, A.: Sur les relations entre l'espace dual d'un groupe et la structure de ses sous-groupes fermés (d'après D. A. Kajdan), Sém. Bourbaki, 20e année, 1967/68, Exp. 343
4. Garland, H., Raghunathan, M. S.: Fundamental domains for lattices in rank one semi-simple Lie groups [Annals of Math. (2) **92** (1970), 279−326]
5. Helgason, S.: Differential geometry and symmetric spaces, Academic Press, New York (1962)
6. Kajdan, D. A., Margoulis, G. A.: Démonstration d'une conjecture de Selberg, Math. Sbornik N. S. **75** (1968), 163−168 (en russe)
7. Siegel, C. L.: Some remarks on discontinuous groups, Annals of Math. (2) **46** (1945), 124−132. Gesammelte Abhandlungen, Bd. III, 67−77
8. Wang, H. C.: On discrete nilpotent subgroups of Lie groups [Jour. Diff. Geometry 3 (1969), 481−492]
9. Zassenhaus, H.: Beweis eines Satzes über diskrete Gruppen, Abh. math. Sem. Hamburg **19** (1938), 289−312

87.

On periodic maps of certain $K(\pi, 1)$

(unpublished, 1969)

1   The main purpose of this Note is to show that if a compact orientable cohomology manifold M is a $K(\pi, 1)$, where the center of π is a torsion group, then the identity component of the group Aut M of homeomorphisms of M is torsion-free. In particular, it does not contain any compact Lie group $\neq \{e\}$. In fact, the proof is cohomological, and this result will be a corollary of somewhat more general statements.

In this Note, X is a locally compact space, which is arcwise connected and locally connected, and simply connected, L a finitely generated group of homeomorphisms of X which operates freely and properly [5, § 4, Nos. 3, 4], $M = X/L$ and $\pi\colon X \to M$ the canonical projection. For $x \in X$, let α_x be the canonical isomorphism of L onto the fundamental group $\pi_1(M, \pi(x))$ of M based at $\pi(x)$. It assigns to $g \in L$ the class of the image of any arc joining x to $x \cdot g$. Of course, $\alpha_{x \cdot u} = \alpha_x \circ \operatorname{Int} u\, (u \in L)$, where $\operatorname{Int} u$ denotes the inner automorphism $g \mapsto u \cdot g \cdot u^{-1}$.

We denote by $E(L)$ the group of outer automorphisms of L, i.e. the quotient of the group Aut L of automorphisms of L by the group $\operatorname{Int} L$ of inner automorphisms. Let Aut M be the group of homeomorphisms of M, endowed with the compact open topology. There is a canonical homomorphism Aut $M \to E(L)$, to be denoted μ. If $m \in M$, and $f \in$ Aut M, the class of $\mu(f)$ is given by the natural isomorphism $\pi_1(M, m) \overset{\sim}{\to} \pi_1(M, f(m))$, induced by f, via the above isomorphisms α_x. It can also be described in the following way: given $u \in X$ and $v \in \pi^{-1}(f(\pi(x)))$, there is a unique $\tilde{f} \in$ Aut X of f such that $\tilde{f}(u) = v$ and $f \circ \pi = \pi \circ \tilde{f}$. It will be called a *lifting* of f. There is then an automorphism q of L such that $\tilde{f}(x \cdot g) = \tilde{f}(x) \cdot q(g)$ $(x \in X;\, g \in L)$. Then $\mu(f)$ is the class of q in $E(L)$. Clearly, if we replace u by $u \cdot h\,(h \in L)$, then q is replaced by $q \circ \operatorname{Int} h$. In particular, *if $\mu(f) = e$, there is a lifting $\tilde{f}$ of f which commutes with L.*

Since L is finitely generated, it is clear that μ is continuous. In particular, ker μ contains the identity component $(\text{Aut } M)^0$ of Aut M. In the sequel M is always assumed to be *compact*.

Theorem 1. *We keep the previous notation. Let p be a prime. Assume that X is an orientable and acyclic cohomology manifold* mod p [1, Chap. I] *and that L keeps an orientation of X fixed. Assume that the center of L is a torsion-group. Let G be a finite p-group in* Aut M. *Then $\mu\colon G \to E(L)$ is injective.*

It suffices to show that if $f \in$ Aut M has order p, and $\mu(f) = e$, then f is the identity. We choose a lifting $\tilde{f}$ of f which commutes with L. Since $\tilde{f}^p = \operatorname{id}.$, we have

57

$\tilde{f}^p(x) = x \cdot g_x(x \in X, g_x \in L)$. By continuity $g_x = h$ is independent of x. Since $\tilde{f}$ commutes with L, the element h is central in L, hence has finite order. Thus $\tilde{f}$ has finite order s divisible by p. Let $s = r \cdot p$. To get a contradiction, it is enough to prove that $\tilde{f}^r = e$. Let Y be a fixed point set of $\tilde{f}^r$. Since $\tilde{f}^r$ has order p, Y, which is acyclic mod p by well-known results of Smith theory (see, e.g., [1], Chap. III, Cor. 4.5, Chap. V, Thm. 3.2) is not empty. Since $\tilde{f}$ commutes with L, the subspace Y is stable under L. By the standard facts on Eilenberg-MacLane cohomology of groups, we have therefore

$$H^*(L; K_p) \cong H^*(M; K_p) = H^*(Y/L; K_p) ,$$

where K_p is a field of characteristic p, and the left-hand side denotes the cohomology of the discrete group L, with coefficients in K_p, and trivial action of L on K_p. Let $n = \dim X$. Since M is a compact orientable cohomology n-manifold, we have $H^m(M; K) \neq 0$, hence $\dim Y = \dim X = n$. But then $Y = X$.

Corollary 1. *The identity component* $(\mathrm{Aut}\, M)^0$ *of* $\mathrm{Aut}\, M$ *has no p-torsion. In particular, no compact connected Lie group $G \neq \{e\}$ can act effectively on M.*

The first assertion follows since $\ker \mu \supset (\mathrm{Aut}\, M)^0$, and the second since G has non trivial elements of order p.

Corollary 2. *Assume that M is an orientable and acyclic cohomology manifold over $\mathbf{Z}$ and that L keeps an orientation of X fixed. Assume the center of L to be a torsion group and let G be a torsion subgroup of $\mathrm{Auf}\, M$. Then $G \to E(L)$ is injective.*

In this case, the assumptions of the theorem are fulfilled for each prime p, whence the corollary.

Theorem 2. *We keep the assumption of 1. Assume that X is an orientable and acyclic cohomology manifold over $\mathbf{Q}$, and that L keeps an orientation of X fixed. If the center of L is a torsion group, then M does not admit an effective action of the circle group.*

Let G be the circle group, $\tilde{G}$ its universal covering, i.e. the additive group of the reals, and $\sigma: \tilde{G} \to G$ the canonical projection. Assume that we are given a continuous action of G on M. We claim that it can be lifted to an action of $\tilde{G}$ on X. In fact, let the given action be described by a morphism.

$$\varphi: G \times M \to M .$$

Fix $x \in X$. Since $\tilde{G} \times X$ is the universal covering of $G \times M$, the map φ can be uniquely lifted to a continuous map $\tilde{\varphi}: \tilde{G} \times X \to X$ sending (e, x) to x. We thus have a commutative diagram

$$
\begin{array}{ccc}
\tilde{G} \times X & \xrightarrow{\tilde{\varphi}} & X \\
\downarrow{\scriptstyle \sigma} \quad \downarrow{\scriptstyle \pi} & & \downarrow{\scriptstyle \pi} \\
G \times M & \xrightarrow{\varphi} & M
\end{array}
$$

We have to check that $\tilde{\varphi}$ defines an action of $\tilde{G}$ on X. Since $e \cdot m = m$ for all $m \in M$, it is clear by continuity that $\tilde{\varphi}(e, x) = x$ for all $x \in X$. There remains to prove that

$$\tilde{\varphi}(g\,h, x) = \tilde{\varphi}(g, \tilde{\varphi}(h, x)) \qquad (g, h \in \tilde{G}; \, x \in X)\,.$$

This amounts to the commutativity of the diagram

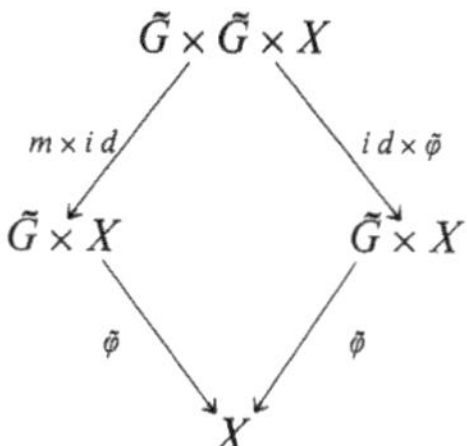

where $m\colon \tilde{G} \times \tilde{G} \to \tilde{G}$ is the product mapping. This follows immediately from the commutativity of the corresponding diagram for G and M, and the uniqueness of liftings to universal coverings.

Next we want to prove that some element $g \neq e$ in $\tilde{G}$ acts trivially on X. It is clear, by continuity, that G commutes with L. Thus, if $g \in \ker \sigma$ there exists $u \in L$ such that $g \cdot x = x \cdot u \, (x \in X)$, and u is central in L, hence some power of g is the identity. Thus $\tilde{\varphi}$ defines in fact an action of the circle group on X, which commutes with L. The argument is now as above. Let Y be the fixed point set of $\tilde{G}$. It is an acyclic cohomology manifold over $\mathbf{Q}$, stable under L, on which L acts freely. We have again

$$H^*(L; \mathbf{Q}) \cong H^*(M; \mathbf{Q}) \cong H^*(Y/L; \mathbf{Q})\,,$$

and the above dimension argument again gives $Y = X$.

Some Examples. Let G be a connected non-compact semi-simple group with center reduced to e, no compact factor $\neq \{e\}$, K a maximal compact subgroup of G and $X = G/K$. Then X is a Riemannian symmetric space of non-compact type, and is homeomorphic to euclidean space. Let L be a discrete subgroup of G which is cocompact (i.e., G/L is compact) and torsion free. Then L acts freely on X and X/L is a compact orientable manifold. Furthermore, the center of L is reduced to $\{e\}$, [2]. Thus the assumptions of Corollary 2 or Theorem 2 obtain. We refer to [3] for the construction of such groups. If, moreover, G has no factor locally isomorphic to $\mathbf{S L}(2, \mathbf{R})$, then L is rigid and $E(L)$ is finite [4, Thm. 1.5]. Thus, in that case, the torsion subgroups of Aut M are finite, with bounded orders. The author does not know whether they form finitely many conjugacy classes, nor whether one can find an L with no outer automorphism.

Remark. The fact that a $K(\pi, 1)$ with non-zero Euler characteristic admits no effective circle action was proved by Conner and Montgomery [6]. However, among the above examples, there also occur manifolds with zero Euler-characteristic.

References

1. Borel, A., et al.: Seminar on transformation groups, Annals of Math. Studies **46,** Princeton University Press, 1960
2. Borel, A.: Density properties for certain subgroups of semi-simple groups without compact factors, Annals of Math. (2) **72** (1960), 179–188
3. Borel, A.: On compact Clifford-Klein forms of symmetric spaces, Topology **2** (1963), 111–122
4. Borel, A.: On the automorphisms of certain subgroups of semi-simple Lie groups, Proc. Bombay Colloquium on algebraic geometry, 1968, Oxford U. Press 1969, 43–73
5. Bourbaki, N.: Topologie générale, Chap. III, 3ᵉ éd. Hermann Paris 1960
6. Conner, P., Montgomery, D.: Transformation groups on a $K(\pi, 1)$ I, Michigan J. M. **6** (1959), 405–412

The Institute for Advanced Study
Princeton, N. J.

88.

Pseudo-concavité et groupes arithmétiques

Essays on Topology and Related Topics, Mémoires dédiés à G. de Rham,
Springer 1970, 70–84

Introduction

Soient $H_n = \{Z \in \mathbf{M}(n, \mathbf{C}) \mid {}^t Z = Z, \operatorname{Im} Z > 0\}$ le demi-plan, et $\Gamma_n = \mathbf{Sp}(n, \mathbf{Z})/\{\pm 1\}$ le groupe modulaire de Siegel, de degré n. Le groupe discret Γ_n opère proprement sur H_n. Si $n = 1$, on obtient le demi-plan de Poincaré et le groupe modulaire classique. On sait qu'en théorie des formes ou fonctions modulaires, les cas $n = 1$ et $n \geq 2$ présentent des différences importantes. Elles apparurent pour la première fois lorsque Koecher [6] montra que les conditions de régularité à l'infini, imposées par Siegel pour $n \geq 2$ en analogie avec le cas $n = 1$, sont vérifiées d'elles-mêmes. Dans le même ordre d'idées, le corps de *toutes* les fonctions méromorphes invariantes par Γ_n est un corps de fonctions algébrique si $n \geq 2$ (mais pas si $n = 1$). Cela a été rattaché plus tard à deux phénomènes de nature plus générale. D'une part, Baily [2] a montré que la compactification de Satake V_n^* de $V_n = H_n/\Gamma_n$ s'identifie à une variété projective et projectivement normale, dont la structure analytique prolonge la structure analytique quotient naturelle de V_n, et telle que $V_n^* - V_n$ soit une sous-variété algébrique de codimension n. Les faits mentionnés plus haut trouvent alors leur explication dans des propriétés de prolongement de fonctions méromorphes, ou de sections holomorphes de fibrés analytiques, définies en dehors d'un sous-ensemble analytique de codimension ≥ 2. D'autre part, Andreotti et Grauert [1] ont introduit une notion d'espace analytique pseudo-concave, montré qu'un tel espace a plusieurs propriétés importantes en commun avec les espaces analytiques compacts, et que V_n est pseudo-concave si $n \geq 2$. Le premier résultat a été étendu au cas général d'un groupe arithmétique opérant sur un domaine borné symétrique X dans [3]. Nous nous proposons ici de faire de même pour le deuxième, en prouvant que X/Γ est pseudo-concave si, en gros, il n'est pas produit d'un quotient similaire X'/Γ' par H_1/Γ'', ou Γ'' est un sous-groupe d'indice fini de Γ_1 (cf. 4.3 pour l'énoncé précis). En fait, cela n'est pas nouveau à proprement parler: on peut le déduire de l'existence d'une compactification soit par des calculs explicites [7], soit en utilisant le fait,

61

qui m'a été signalé par H. Grauert, R. Narasimhan, que dans une variété projective irréductible, le complément d'une sous-variété de codimension ≥ 2 est pseudo-concave. Mais notre but est d'en donner une démonstration plus directe, qui n'utilise pas l'existence d'une compactification. Elle est dans son principe analogue à celle de [1], une fois cette dernière formulée en termes plus géométriques, tout en présentant quelques complications techniques. En bref, il s'agit de construire sur X/Γ une fonction réelle continue q telle que, pour un nombre réel convenable r, l'ensemble $Y = \{y \in X/\Gamma \,|\, q(y) < r\}$ soit une ouvert relativement compact non vide, et, si $q(y) = r$, il passe par y une sous-variété locale analytique M_y de dimension complexe ≥ 2 sur laquelle la restriction de q est deux fois continûment différentiable et a une forme de Levi négative non dégénérée (cf. 4.1). Ici M_y proviendra d'une fibre de la projection σ_1 de X sur une composante à la frontière rationnelle F_1 de dimension *maximum*. Pour construire q, on part d'une fonction μ_0 associée au déterminant fonctionnel dans la réalisation de X comme «domaine de Siegel de troisième espèce» adaptée à une composante à la frontière rationnelle F_s de dimension *minimum*. Si F_s n'est pas réduite à un point, on forme à partir de μ_0 une série de Poincaré τ_0 par rapport à un quotient du sous-groupe de Γ normalisant F_s. La fonction q en $x \cdot \Gamma$ est alors définie comme le maximum de $\log \tau_0(x \cdot u)$, où u parcourt un ensemble $\Gamma \cdot C$, où C est une certaine partie finie du groupe des automorphismes de X, qui est liée aux «pointes» d'un ensemble fondamental pour Γ dans X.

Le § 1 rappelle quelques propriétés de X relatives à ses composantes à la frontière rationnelles. Le § 2 introduit les fonctions μ_0, τ_0 et montre que $\log \mu_0$ a une forme de Levi négative non-dégénérée. Le § 3 étudie quelques propriétés de $\max \log \tau_0(x \cdot u) \,(u \in \Gamma \cdot C)$, et de sa forme de Levi, en utilisant les principaux résultats de la théorie de la réduction par rapport à un groupe arithmétique [4]. Enfin, le § 4 rappelle la définition d'un espace pseudo-concave, un critère de [1], et démontre le théorème de pseudo-concavité.

La méthode de Andreotti-Grauert avait déjà été étendue à quelques cas classiques par Gundlach, Ramanathan (non publié), et Spilker [8].

Notations et conventions. Nous utilisons en général celles de [3, 4]. En particulier, les groupes algébriques considérés s'identifient à des sous-groupes de $\mathbf{GL}(n, \mathbf{C})$. Si H est un tel groupe, et L un sous-anneau de $\mathbf{C}$, alors $H_L = H \cap \mathbf{GL}(n, L)$ est le groupe des éléments de H à coefficients dans L, dont le déterminant est inversible dans L. Si H est défini sur $\mathbf{Q}$, un *sous-groupe arithmétique* Γ de H est un sous-groupe de $H_\mathbf{Q}$ commensurable à $H_\mathbf{Z}$ (i. e. tel que $H_\mathbf{Z} \cap \Gamma$ soit d'indice fini dans $H_\mathbf{Z}$ et dans Γ).

Dans tout ce travail, G est un groupe algébrique connexe semi-simple défini sur $\mathbf{Q}$, et K un sous-groupe compact maximal de $G_{\mathbf{R}}$. On suppose que le quotient $X = K\backslash G_{\mathbf{R}}$ des classes à droite de $G_{\mathbf{R}}$ modulo K s'identifie à un domaine borné symétrique, dont le groupe d'automorphismes $\operatorname{Aut} X$ contient la composante connexe $G_{\mathbf{R}}^0$ de e dans $G_{\mathbf{R}}$ (pour la topologie ordinaire). Γ désigne un sous-groupe arithmétique de G, qui fait partie de $\operatorname{Aut} X$. On note s le $\mathbf{Q}$-rang de G, i. e. la dimension des plus grands tores algébriques de G qui sont définis sur $\mathbf{Q}$ et diago-nalisables sur $\mathbf{Q}$. Si $s = 0$, alors X/Γ est compact, et la pseudo-concavité est évidente. Nous supposerons donc $s \geq 1$.

Dans les §§ 1, 2, 3, on admet du plus que Γ fait partie de $G_{\mathbf{R}}^0$, et que G est simple sur $\mathbf{Q}$ (i. e. ne contient pas de sous-groupe algébrique distingué non trivial défini sur $\mathbf{Q}$).

§ 1. Réalisations et fibrations de X associés aux composantes à la frontière rationnelles

Nous rappelons ici quelques résultats démontrés ou énoncés (avec références) dans $[3, \S\S 1, 2, 3]$.

Il existe des réalisations S_b de X comme domaine de $\mathbf{C}^n (n = \dim_{\mathbf{C}} X)$ $(0 \leq b \leq s)$, des domaines bornés symétriques F_b, des applications holomorphes $\sigma_b : X \to F_b$ submersives ayant, entre autres, les propriétés suivantes:

1.1. $S_0 = F_0$ est la réalisation de Harish-Chandra de X comme domaine borné symétrique. L'action de $G_{\mathbf{R}}^0$ sur S_0 se prolonge de manière continue à l'adhérence $\overline{S_0}$ de S_0. L'espace F_b s'identifie à un domaine borné d'un sous-espace affine L_b et est l'intersection de L_b et $\overline{S_0}$. Soient

$$
\begin{aligned}
&N_b = N(F_b) = \{g \in G_{\mathbf{R}}^0 \mid F_b \cdot g = F_b\}, \\
(1) \qquad &Z_b = Z(F_b) = \{g \in N_b \mid f \cdot g = f \quad (f \in F_b)\}, \\
&\Gamma_b = \Gamma \cap N_b, \quad G(F_b) = N_b/Z_b, \quad \Gamma(F_b) = \Gamma_b/(Z_b \cap \Gamma),
\end{aligned}
$$

et $\zeta_b : N_b \to G(F_b)$ la projection canonique. Alors N_b est l'intersection de $G_{\mathbf{R}}^0$ avec un $\mathbf{Q}$-sous-groupe parabolique propre maximal P_b de G. Les éléments de N_b commutent à σ_b et définissent des automorphismes de F_b $[3, \S\S 1.5, 3.3]$. Le groupe $\Gamma(F_b)$ opère proprement sur F_b, et est de type arithmétique $[3, \S 3.7]$. On a $F_{b+1} \subset \overline{F_b}$, et, pour $b \leq c$, une factorisation

$$
(2) \qquad\qquad \sigma_c = \sigma_{b,c} \circ \sigma_b,
$$

où $\sigma_{b,c}: F_b \to F_c$ est holomorphe. En fait $F_{b+1}, \ldots, F_s$ jouent pour F_b le rôle de $F_1, \ldots, F_s$ pour X, et $\sigma_{b,c}$ est la projection canonique correspondante. En particulier, si $b \leq c$ *toute fibre de σ_b est contenue dans une fibre de σ_c.* (Voir [3, §§ 1.5, 1.7, 3.3]; l'indice b est gras dans [3] pour éviter des confusions possibles avec les notations de [3, § 1].)

1.2. Il existe un isomorphisme canonique $v_b: X \to S_b$, qui envoie les fibres de σ_b sur l'ensemble des intersections de S_b avec une famille de plans parallèles. Dans la réalisation S_b, le groupe Z_b opère par l'intermédiaire de transformations affines. On note $J_b(x, \varphi)$ le déterminant fonctionnel en x d'un automorphisme φ de X, par rapport aux coordonnées de S_b, et $j_b(y, \psi)$ la valeur en y du déterminant fonctionnel d'un automorphisme ψ de F_b, dans les coordonnées de la réalisation bornée canonique de F_b.

1.3. Si G est absolument simple, alors X et les F_b sont des domaines bornés symétriques irréductibles. Dans le cas général, il existe un corps de nombres totalement réel k de degré fini sur $\mathbf{Q}$ et un groupe G' absolument simple défini sur k, tels que $G = R_{k/\mathbf{Q}} G'$, où $R_{k/\mathbf{Q}}$ est le foncteur de restriction des scalaires [3, §§ 3.2, 3.3]. Si $\rho_1, \ldots, \rho_d$ sont les monomorphismes de k dans $\mathbf{R}$ $(d = [k: \mathbf{Q}])$, alors $G_{\mathbf{R}}$ est produit de d groupes $G_i = ({}^{\rho_i}G')_{\mathbf{R}}$, l'espace X est produit de d domaines bornés symétriques X_i, où X_i est le quotient de G_i par un sous-groupe compact maximal. De même F_b est produit de domaines $F_{b,i}$, où $F_{b,i}$ est une composante à la frontière de X_i. Si $x \in X$, $g \in G_{\mathbf{R}}^0$, on peut écrire

$$x = (x_1, \ldots, x_d), \quad g = (g_1, \ldots, g_d), \quad (x_i \in X_i, g_i \in G_i),$$

et l'on a

$$J_b(x, g) = \prod_{1 \leq i \leq d} J_{b,i}(x_i, g_i),$$

où $J_{b,i}$ désigne le déterminant fonctionnel de X_i, pour la réalisation associée à $F_{b,i}$. Etant donné un d-uple $q = (q_1, \ldots, q_d)$ de nombres réels, on convient de poser

$$|J_b(x, g)|^q = \prod_i |J_{b,i}(x, g)|^{q_i}.$$

De même j_b se décompose en produit, et on adopte une convention similaire pour $|j_b|^q$.

1.4. Soit $\mathfrak{u}_b$ l'algèbre de Lie du radical unipotent $U_b = U(F_b)$ de $N(F_b)$, et soit

$$(1) \qquad\qquad \chi_b(g) = \det(\mathrm{Ad}\, g)|_{\boldsymbol{u}_b}, \qquad (g \in N(F_b)),$$

où Ad dénote la représentation adjointe de G. Il existe un d-uple $\mathbf{q}_b$ de nombres rationnels >0, et un nombre rationnel $n_b>0$ tels que

$$(2) \qquad |J_b(x,g)| = |j_b(\sigma_b(x),g)|^{\mathbf{q}_b} \cdot |\chi_b(g)|^{-n_b}, \qquad (x\in S_b; g\in N_b),$$

[3, § 3.12]. Cela montre en particulier: si $g\in N_b$, alors la fonction $x\mapsto J_b(x,g)$ est constante sur les fibres de σ_b; si $g\in Z_b$, alors $|J_b(x,g)|$ est indépendant de x, et $g\mapsto |J_b(x,g)|$ est la restriction à Z_b d'une puissance du module d'un caractère rationnel, défini sur $\mathbf{Q}$, de P_b.

1.5. Lemme. *Soient b, c tels que $(1\leqq b\leqq c\leqq s)$ et soit $g\in N_b$. Alors la fonction $x\mapsto J_c(x,g)$, $(x\in X)$ est constante sur les fibres de σ_b.*

Soit v le composé des isomorphismes canoniques $S_b\to X\to S_c$ et soit $j(x,v)$ le déterminant fonctionnel de v en $x\in S_b$. Alors, évidemment

$$j(x,v)\cdot J_c(v(x),g)=J_b(x,g)\cdot j(x\cdot g,v), \qquad (x\in S_b).$$

On a remarqué en 1.4 que $J_b(x,g)$ est constant sur les fibres de σ_b. D'autre part, d'après [3, 1.18(ii)], appliqué à chaque facteur X_i de X, j est aussi constant sur les fibres de σ_b, d'où le lemme.

§ 2. Les fonctions μ_0 et τ_0 sur X

2.1. Soient $x_0\in X$ le point fixe par K et $\pi: G_{\mathbf{R}}\to X$ la projection canonique. Etant donné un entier positif a, on définit une fonction μ sur $G_{\mathbf{R}}^0$ par

$$(1) \qquad \mu(g)=|J_s(x_0,g)|^a= \prod_{1\leqq i\leqq d} |J_{b,i}(x_{0i},g_i)|^a,$$

(cf. 1.3). L'identité du cocycle

$$(2) \qquad J_s(x,g\cdot g')=J_s(x,g)\cdot J_s(x\cdot g,g'), \qquad (x\in X; g,g'\in G_{\mathbf{R}}^0),$$

entraine

$$(3) \qquad \mu(g\cdot g')=\mu(g)\cdot |J_s(x_0\cdot g,g')|^a, \qquad (g,g'\in G_{\mathbf{R}}^0).$$

La fonction μ est invariante à gauche par $K\cap G_{\mathbf{R}}^0$ et définit donc une fonction sur X, qui sera notée μ_0. La relation (3) s'écrit alors

$$(4) \qquad \mu_0(x\cdot g)=\mu_0(x)\cdot |J_s(x,g)|^a, \qquad (x\in X; g\in G_{\mathbf{R}}^0).$$

Nous supposons a tel que les nombres $a\cdot n_s$ et $a\cdot q_{s,i}$ (cf. 1.4) soient des entiers >0 pairs $(1\leqq i\leqq d)$. Comme χ_b est un caractère défini sur $\mathbf{Q}$, il prend les valeurs ± 1 sur tout sous-groupe arithmétique de N_s, donc

$$(5) \qquad \chi_s(\gamma)^{an_s}=1, \qquad (\gamma\in \Gamma_s).$$

Soit $\Gamma_\infty = \Gamma_s \cap Z_s$. Si $\gamma \in \Gamma_\infty$, alors, il opère trivialement sur F_s, donc, vu (5) et 1.4 (2), $|J_s(x,\gamma)| = 1$, et (3), (4) montrent que μ et μ_0 sont invariantes par γ. On peut ainsi former la série

$$(6) \qquad \tau(g) = \sum_{\gamma \in \Gamma_s/\Gamma_\infty} \mu(g \cdot \gamma),$$

qui, vue comme série de fonctions sur X, s'écrit

$$(7) \qquad \tau_0(x) = \sum_{\gamma \in \Gamma_s/\Gamma_\infty} \mu_0(x \cdot \gamma).$$

Soit

$$(8) \qquad P(y) = \sum_{\gamma \in \Gamma_s/\Gamma_\infty} |j_s(y,\gamma)|^{q_s \cdot a}, \qquad (y \in F_s).$$

C'est une série de Poincaré, ou plutôt la somme des valeurs absolues des termes d'une série de Poincaré sur F_s. Vu les conditions imposées à a, elle converge uniformément sur tout compact en vertu de résultats connus (*voir* p. ex [3, § 5.10], ou [5]). Compte tenu de (4), on a

$$(9) \qquad \tau_0(x) = \mu_0(x) \cdot \sum_{\gamma \in \Gamma_s/\Gamma_\infty} |J_s(x,\gamma)|^a,$$

ce qui, vu (5) et 1.4 (2), donne

$$(10) \qquad \tau_0(x) = \mu_0(x) \cdot P(\sigma_s(x)), \qquad (x \in X),$$

et montre en particulier que τ_0 converge. On a de même

$$(11) \qquad \tau(g) = \mu(g) \cdot P(\sigma_s(x_0 \cdot g)), \qquad (g \in G_\mathbf{R}^0).$$

2.2. Lemme. *On a*

$$(1) \qquad \tau(g \cdot \gamma) = \tau(g), \qquad (g \in G_\mathbf{R}^0, \ \gamma \in \Gamma_s).$$

$$(2) \qquad \tau(g \cdot h) = \tau(g)|\chi_s(h)|^{-a \cdot n_s}, \qquad (g \in G_\mathbf{R}^0; \ h \in Z_s).$$

La première relation est claire par construction. Soient $g \in G_\mathbf{R}^0$, $h \in Z_s$, $q \in N_s$. Alors, d'après 2.1(3)

$$\mu(g \cdot h \cdot q) = \mu(g \cdot q \cdot q^{-1} \cdot h \cdot q) = \mu(g \cdot q) \cdot |J_s(x_0 \cdot g \cdot q, \ q^{-1} \cdot h \cdot q)|^a.$$

Mais Z_s est distingué dans N_s, et le caractère χ_s est évidemment constant sur une classe de conjugaison de N_s. Il résulte alors de 1.4(2) que l'on a

$$\mu(g \cdot h \cdot q) = \mu(g \cdot q) \cdot |\chi_s(h)|^{-a \cdot n_s},$$

d'où (2) par sommation sur $q \in \Gamma_s/\Gamma_\infty$.

2.3. Soient M une variété analytique complexe, $m \in M$ et f une fonction deux fois continûment différentiable définie au voisinage de m. Rappelons que la forme de Levi $L(f)$ de f en m est la forme hermitienne sur l'espace tangent en m donnée par

$$L(f) = \sum_{i,j} \frac{\partial^2 f}{\partial z_i \partial \overline{z}_j}(m) \cdot d z_i \cdot d \overline{z}_j,$$

où (z_i) sont des coordonnées locales au voisinage de m. Il est immédiat, et bien connu, que $L(f)$ ne dépend pas des coordonnées locales.

2.4. Lemme. *La forme de Levi $L(\log \mu_0)$ de $\log \mu_0$ est négative non degénérée en tout point de X.*

Soit $\mathcal{K}_0$ le noyau de Bergmann de X, identifié au domaine borné S_0, et soit $\mathcal{K}$ le transporté de $\mathcal{K}_0$ à S_b par l'isomorphisme canonique $S_0 \to X \to S_b$. Alors

$$\omega_x = \mathcal{K}(x) \cdot d z_1 \wedge d \overline{z}_1 \wedge \dots \wedge d z_n \wedge d \overline{z}_n,$$

est un élément de volume invariant sur S_b et l'on a

(1) $\qquad\qquad \mathcal{K}(x) = \mathcal{K}(x \cdot g) \cdot |J_s(x \cdot g)|^2, \qquad (x \in S_b; \, g \in G_{\mathbf{R}}^0).$

Comme $G_{\mathbf{R}}^0$ est transitif sur X, (1) et 2.1(4) montrent que le produit $\mu_0^2 \cdot \mathcal{K}^a$ est une constante sur X. Par suite

(2) $\qquad\qquad 2 \cdot L(\log \mu_0) + a \cdot L(\log \mathcal{K}) = 0.$

Mais $L(\log \mathcal{K})$ n'est autre que la métrique de Bergmann, donc est positive non dégénérée, d'où le lemme.

§ 3. Une fonction invariante par Γ

3.1. *Notation.* Soient f, g deux fonctions à valeurs réelles > 0 sur un espace M. Comme dans [4], on écrira $f \prec g$ s'il existe une constante $c > 0$ telle que $f(m) \leqq c \cdot g(m)$ pour tout $m \in M$, $f \succ g$ si $g \prec f$, et enfin $f \asymp g$ si l'on a simultanément $f \prec g$ et $g \prec f$. Dans ce dernier cas, on dira que f et g sont *comparables* sur M.

3.2. Soit P le $\mathbf{Q}$-sous-groupe parabolique minimal de G sous-jacent à la définition des F_b. On pose $P = M \cdot S \cdot U$, où U est le radical unipotent de P, S un tore algébrique déployé sur $\mathbf{Q}$ maximal de P, M un $\mathbf{Q}$-sous-groupe connexe réductif, de $\mathbf{Q}$-rang égal à zéro, tel que $M \cdot S$ soit le centralisateur $Z(S)$ de S dans G. Soit encore $_\mathbf{Q}\varDelta$ l'ensemble des $\mathbf{Q}$-racines simples de G par rapport à S, pour un ordre associé à P (pour ces notions, voir [4, § 11]).

On notera A la composante connexe de e, en topologie ordinaire, dans le groupe $S_{\mathbf{R}}$. Les sous-groupe paraboliques $P_b (1 \leq b \leq s)$ contiennent P, et sont tous les $\mathbf{Q}$-sous-groupes paraboliques propres maximaux contenant P [3, § 3.7]. De plus [3, § 3.8]:

$$(1) \qquad\qquad Z(F_s) \supset A \cdot U_{\mathbf{R}},$$

et $F_s/\Gamma(F_s)$ est compact.

Dans la suite, nous appliquerons au groupe $G_{\mathbf{R}}^0$ quelques théorèmes de [4], qui sont formulés dans [4] pour le groupe $H_{\mathbf{R}}$ de *tous* les points réels d'un groupe réductif H défini sur $\mathbf{Q}$. Cela ne présente pas de difficulté, car il est clair que les démonstrations subsistent sans changement, si ce n'est de notation, si l'on remplace $H_{\mathbf{R}}$ par un sous-groupe ouvert H'. On peut aussi remarquer que ces théorèmes pour $H_{\mathbf{R}}$ (resp. H') sont équivalents à des énoncés similaires portant sur l'espace symétrique non compact $H_{\mathbf{R}}/L$ (resp. $H'/(L \cap H')$) quotient de $H_{\mathbf{R}}$ (resp. H') par un sous-groupe compact maximal et que l'on a $H_{\mathbf{R}}/L = H'/(L \cap H')$.

3.3. Nous voulons maintenant montrer que τ vérifie les hypothèses du Théor. 16.9 de [4]. Si l'on numérote les éléments $\alpha_1,\ldots,\alpha_s$ de ${}_{\mathbf{Q}}\varDelta$ comme dans [3], alors P_s est le groupe parabolique P_θ associé à $\theta = {}_{\mathbf{Q}}\varDelta - \{\alpha_s\}$. Vu [3, § 2.3], χ_s est un multiple strictement positif du poids dominant fondamental d_{α_s}, et est orthogonal aux éléments de θ. De plus, comme F_s est la plus petite composante à la frontière rationnelle, le sous-groupe L_θ (notation de [3, § 2.2]) fait partie du groupe dérivé de $Z(F_s)$, donc, vu (2.2(2)),

$$(1) \qquad\qquad \tau(g \cdot h) = \tau(g), \qquad (g \in G_{\mathbf{R}}^0,\ h \in L_{\theta,\mathbf{R}} \cap G_{\mathbf{R}}^0).$$

Une fonction continue ξ réelle positive sur $G_{\mathbf{R}}$ est dite de type (P, χ_s^c) si

$$\xi(g \cdot p) = \xi(g) \cdot |\chi_s(p)|^c, \qquad (g \in G_{\mathbf{R}};\ p \in P_{\mathbf{R}})$$

[4, § 14]. Montrons que τ^{-1} est comparable à une fonction $\xi > 0$ de type $(P, \chi_s^{an_s})$ sur $G_{\mathbf{R}}^0$. Le groupe M étant réductif, de $\mathbf{Q}$-rang égal à zéro, le quotient $M_{\mathbf{R}}/(M_{\mathbf{R}} \cap \Gamma)$ est compact [4, § 8]. Il existe donc un compact D de $M_{\mathbf{R}}$ tel que

$$G_{\mathbf{R}} = K \cdot P_{\mathbf{R}}^0 = K \cdot D \cdot A \cdot U_{\mathbf{R}}(\Gamma \cap M).$$

Soit $D' = K \cdot D$, et écrivons $g \in G_{\mathbf{R}}$ sous la forme

$$g = h \cdot a \cdot u \cdot \gamma, \qquad (h \in D',\ a \in A,\ u \in U_{\mathbf{R}},\ \gamma \in \Gamma \cap M).$$

Comme $\Gamma \cap P \subset \Gamma_s$, et $\chi_s(\gamma)^{an_s} = 1$ pour $\gamma \in \Gamma_s$, on a

$$\xi(g) = \xi(h) \cdot |\chi_b(a \cdot u)|^{an_s}.$$

Mais $A \cdot U_{\mathbf{R}} \subset Z_s$, donc, vu 2.2:

$$\tau(g) = \tau(h) \cdot |\chi_s(a \cdot u)|^{-a n_s},$$

et $(\tau \cdot \xi)(g) = (\tau \cdot \xi)(h)$. Comme h varie dans un compact, il s'ensuit que $\tau \cdot \xi$ est compris entre deux bornes strictement positives, d'où notre assertion. La fonction τ étant invariante à droite par $\Gamma \cap P \subset \Gamma_s$, on voit que τ^{-1} vérifie les conditions imposées à Φ dans [4, § 16.9].

3.4. Soit $\mathfrak{S}$ un ensemble de Siegel ouvert de $G_{\mathbf{R}}$ par rapport à K, P, S [4, § 12]. On a donc

$$(1) \qquad\qquad \mathfrak{S} = K \cdot A_{(t)} \cdot \omega,$$

où ω est un voisinage ouvert relativement compact de e dans $(M \cdot U)_{\mathbf{R}}$, t est réel >0 et

$$(2) \qquad\qquad A_{(t)} = \{ a \in A \,|\, a^\alpha < t, \, (\alpha \in_{\mathbf{Q}} \Delta) \}.$$

Il existe une partie finie C de $G_{\mathbf{Q}}$ et un ensemble de Siegel $\mathfrak{S}$ tels que

$$(3) \qquad\qquad G_{\mathbf{R}} = \mathfrak{S} \cdot C^{-1} \cdot \Gamma,$$

[4, Théor. 13.1]. De plus, d'après le Théor. 16.9 de [4] appliqué à τ^{-1}, ce qui est possible vu 3.3, on peut trouver C de sorte que pour tout $g \in G_{\mathbf{R}}$, la fonction $u \mapsto \tau(g \cdot u)(u \in \Gamma \cdot C)$ atteigne son maximum en un point de $\mathfrak{S} \cap g \cdot \Gamma \cdot C$. En particulier, ce maximum existe, et nous pouvons définir

$$(4) \qquad\qquad \Phi(g) = \max_{u \in \Gamma \cdot C} \log \tau(g \cdot u).$$

Cette fonction est invariante à gauche par K, d'où aussi une fonction Φ_0 sur X, donnée par

$$(5) \qquad\qquad \Phi_0(x) = \max_{u \in \Gamma \cdot C} \log \tau_0(x \cdot u).$$

Pour tout $\delta > 0$, posons

$$X_\delta = \{ x \in X \,|\, \Phi_0(x) < \delta \}.$$

On désigne par $\mathfrak{S}'$ l'image $x_0 \cdot \mathfrak{S}$ de $\mathfrak{S}$ dans X par π. C'est, par définition, un ensemble de Siegel ouvert de X. On a évidemment

$$(6) \qquad\qquad X = \mathfrak{S}' \cdot C^{-1} \cdot \Gamma$$

et, pour tout $x \in X$, la fonction $u \mapsto \log \tau_0(x \cdot u)(u \in \Gamma \cdot C)$ atteint son maximum en un point de $\mathfrak{S}' \cap x \cdot \Gamma \cdot C$.

3.5. Lemme. *La fonction Φ_0 est continue, invariante par Γ. Pour tout $c \in C$ et tout $\delta > 0$, l'ensemble $X_\delta \cap \mathfrak{S}' \cdot c^{-1}$ est relativement compact.*

Φ_0 est invariante par Γ par construction, et continue d'après [4, 16.10]. En ce qui concerne la deuxième assertion, il revient au même de montrer que

$$\mathfrak{S} \cdot c^{-1} \cap G_\delta, \quad (G_\delta = \{g \in G_{\mathbf{R}} | \Phi(g) < \delta\}),$$

est relativement compact.

Soit $g \in \mathfrak{S} \cdot c^{-1} \cap G_\delta$. On a donc $g \cdot c \in \mathfrak{S}$. Écrivons $h = g \cdot c$ sous la forme $h = k_h \cdot a_h \cdot u_h$ ($k_h \in K$, $q_h \in A_{(t)}$, $u_h \in \omega$). Comme k_h et u_h varient dans des ensembles relativement compacts, il suffit de faire voir qu'il en est de même pour a_h. Les racines simples formant un système de coordonnées sur A, on est ramené à montrer:

$$(1) \qquad\qquad a_h^\alpha \asymp 1 \quad (\alpha \in {}_{\mathbf{Q}}\varDelta; h \in \mathfrak{S}; h \cdot c^{-1} \in G_\delta).$$

Par définition de Φ, on a

$$(2) \qquad\qquad \tau(h) = \tau(g \cdot c) \leqq \exp \Phi(g) < e^\delta.$$

Mais $\tau(k_h \cdot a_h \cdot u_h) = \tau(a_h \cdot u_h) = \tau(a_h \cdot u_h \cdot a_h^{-1} \cdot a_h)$. Vu 2.2 et le fait que $A \subset Z_s$, on a donc

$$\tau(h) = \tau(a_h \cdot u_h \cdot a_h^{-1}) \cdot |\chi_s(a_h)|^{-ans}.$$

L'élément $a_h \cdot u_h \cdot a_h^{-1}$ parcourt un ensemble relativement compact si h varie dans $\mathfrak{S}$ [4, lemme 12.2], donc

$$\tau(h) \asymp |\chi_s(a_h)|^{-ans}, \quad (h \in \mathfrak{S}).$$

Si $h \cdot c^{-1} \in G_\delta$, on a $\tau(h) \leqq e^\delta$ d'après (2), donc

$$|\chi_s(a_h)| > 1, \quad (h \in \mathfrak{S}, h \cdot c^{-1} \in G_\delta).$$

On a remarqué plus haut (3.3) que χ_s est un multiple positif d'un poids dominant fondamental. En vertu de faits simples sur les systèmes de racines, cela entraine

$$\chi_s = \sum_{\alpha \in {}_{\mathbf{Q}}\varDelta} c_\alpha \cdot \alpha, \quad (c_\alpha > 0; \alpha \in {}_{\mathbf{Q}}\varDelta).$$

On a donc

$$(3) \qquad\qquad \prod_{a \in {}_{\mathbf{Q}}\varDelta} a_h^{c_a \cdot a} > 1, \quad (h \in \mathfrak{S} \cap G_\delta \cdot c).$$

Mais, vu la définition de $\mathfrak{S}$ (3.4), on a évidemment

$$(4) \qquad\qquad a_h^\alpha < 1, \quad (\alpha \in {}_{\mathbf{Q}}\varDelta; h \in \mathfrak{S} \cap G_\delta \cdot c),$$

et (1) résulte de (3), (4).

3.6. Lemme. *Il existe une partie finie E de Γ et une constante $r > 0$ ayant la propriété suivante: si $x \in \mathfrak{S}'$, $c, d \in C$ et $\gamma \in \Gamma$ sont tels que $\Phi_0(x \cdot c^{-1}) \geqq r - 1$, et $x \cdot c^{-1} \in \mathfrak{S}' \cdot d^{-1} \cdot \gamma$, alors $\gamma \in E$ et $d^{-1} \cdot \gamma \cdot c \in \bigcup_{1 \leqq b \leqq s} N_b$.*

D'après la propriété de Siegel [4, 15.4],

$$E = \{\gamma \in \Gamma \,|\, \mathfrak{S}' \cdot C^{-1} \cap \mathfrak{S} \cdot C^{-1} \cdot \gamma \neq \phi\},$$

est fini. On peut donc trouver une constante r telle que si $\mathfrak{S} \cdot c^{-1} \cap \mathfrak{S} \cdot d^{-1} \cdot \gamma$ $(c, d \in C, \gamma \in \Gamma)$ est non vide, relativement compact, alors il est contenu dans G_{r-2}. Autrement dit, si $x \in \mathfrak{S}'$, $c, d \in C$ sont tels que $\Phi_0(x \cdot c^{-1}) \geq r - 2$ $x \cdot c^{-1} \in \mathfrak{S}' \cdot d^{-1} \cdot \gamma$, alors $\mathfrak{S}' \cdot c^{-1} \cap \mathfrak{S}' \cdot d^{-1} \cdot \gamma$ est non relativement compact et $\gamma \in E$. Mais alors $d^{-1} \cdot \gamma \cdot c$ appartient à un $\mathbf{Q}$-sous-groupe parabolique propre contenant P [4, 12.6]. Or les groupes paraboliques $P_b (1 \leq b \leq s)$ sont tous les $\mathbf{Q}$-sous-groupes paraboliques propres maximaux de G contenant P (3.2) donc $d^{-1} \cdot \gamma \cdot c \in \bigcup_b N_b$.

3.7. Lemme. *Soit r comme dans 3.6. Soient $c \in C$, $x \in \mathfrak{S}'$, L_x la fibre de σ_1 contenant x, et supposons que $\Phi_0(x \cdot c^{-1}) \geq r$. Il existe un voisinage V de x dans L_x tel que la restriction de Φ_0 à $V \cdot c^{-1}$ soit la somme d'une constante et de la fonction $z \mapsto \log \mu_0(z \cdot c)\,(z \in V \cdot c^{-1})$. La restriction de Φ_0 à $V \cdot c^{-1}$ est indéfiniment différentiable, et sa forme de Levi sur $V \cdot c^{-1}$ est négative non dégénérée.*

Comme Φ_0 est continue, on peut trouver un voisinage U de x dans $\mathfrak{S}'$ tel que $\Phi_0(y \cdot c^{-1}) \geq r - 1\,(y \in U)$. On pose $V = L_x \cap U$.

Par définition $\Phi_0(z)$ est le maximum de $\log \tau_0(z \cdot u)$ pour $u \in \Gamma \cdot C$, et l'on sait que ce maximum est atteint en un point de $z \cdot \Gamma \cdot C \cap \mathfrak{S}'$. On peut donc écrire en fait

$$\Phi_0(z) = \max \log \tau_0(z \cdot u), \quad (u \in \Gamma \cdot C, z \cdot u \in \mathfrak{S}').$$

Supposons maintenant que z soit de la forme $y \cdot c^{-1}(y \in U)$. Alors, vu 3.6, il existe un ensemble fini $F \subset \Gamma \cdot C$ tel que

$$(1) \qquad\qquad\qquad F^{-1} \cdot c \subset \bigcup_{1 \leq b \leq s} N_b,$$

$$(2) \qquad\qquad \Phi_0(y \cdot c^{-1}) = \max_{u \in F} \log \tau_0(y \cdot c^{-1} \cdot u), \quad (y \in V).$$

Fixons $u \in F$ et soit b tel que $u^{-1} \cdot c \in N_b$. D'après 2.1 (10):

$$(3) \qquad \tau_0(y \cdot c^{-1} \cdot u) = \mu_0(y \cdot c^{-1} \cdot u) \cdot P(\sigma_s(y \cdot c^{-1} \cdot u)), \quad (y \in V).$$

V est dans une fibre de σ_1, donc dans une fibre de σ_b. Par conséquent $V \cdot c^{-1} \cdot u$ est aussi dans une fibre de σ_b, et *a fortiori* dans une fibre de σ_s, (1.1). Le deuxième facteur de droite dans (3) est donc constant. L'égalité 2.1 (4) donne

$$(4) \qquad\qquad \mu_0(y \cdot c^{-1} \cdot u) = \mu_0(y) \cdot |J_s(y, c^{-1} \cdot u)|^a.$$

Mais, comme $c^{-1} \cdot u \in N_b$ et $b \leqq s$, le deuxième facteur de droite est constant sur les fibres de σ_b, donc *a fortiori* sur celles de σ_1, en vertu de 1.5. Il existe par suite une constante $c(u) > 0$ telle que

$$(5) \qquad \tau_0(y \cdot c^{-1} \cdot u) = c(u) \cdot \mu_0(y), \qquad (y \in V).$$

Vu (2), on obtient donc

$$(6) \qquad \Phi_0(y \cdot c^{-1}) = \log \mu_0(y) + \max_{u \in F} \log c(u), \qquad (y \in V)$$

ce qui démontre la première assertion. La deuxième résulte alors de 2.4.

§ 4. Pseudo-concavité

4.1. Soient Z un espace analytique irréductible (réduit, sur $\mathbf{C}$), Y un ouvert relativement compact de Z et z un point de la frontière ∂Y de Y dans Z. On dit que Y est *pseudo-concave dans Z, au point z*, si, étant donné une fonction holomorphe f définie au voisinage de z, il existe un voisinage V de z tel que le maximum de $|f|$ sur $V \cap \bar{Y}$ soit atteint en un point de $V \cap Y$, [1].

Si Z contient un ouvert relativement compact non vide Y qui soit pseudo-concave dans Z en tout point de son bord, on dit que Z *est pseudo-concave*.

En vertu du principe du maximum, il est immédiat que Z est pseudo-concave s'il contient un ouvert relativement compact non vide Y tel que pour tout $z \in \partial Y$ on puisse trouver une application holomorphe $\alpha : U \to Z$, où U est un voisinage ouvert de l'origine 0 dans $\mathbf{C}$, telle que $\alpha(0) = z$ et $f(U - \{0\}) \subset Y$.

Cette condition est à son tour entrainée par la suivante [1, Satz 1]:

(*) Il existe une fonction continue réelle q sur Z et un nombre réel r ayant les propriétés suivantes: $Y = \{z \in Z \mid q(z) < r\}$ est un ouvert relativement compact non vide; si $z \in q^{-1}(r)$, il existe une application holomorphe $\alpha : V \to Z$, où V est un voisinage de l'origine 0 dans $\mathbf{C}^2$, appliquant 0 sur z, et telle que $q \circ \alpha$ soit deux fois continûment différentiable et ait une forme de Levi négative non dégénérée à l'origine.

Nous rappelons brièvement la démonstration. Soient t_1, t_2 les coordonnées canoniques de $\mathbf{C}^2$, et posons $p = q \circ \alpha$. On peut écrire, au voisinage de 0:

$$p(t) = p(0) + \Sigma a_i \cdot t_i + \Sigma \bar{a}_i \cdot \bar{t}_i + \Sigma b_{ij} t_i t_j + \Sigma \bar{b}_{ij} \bar{t}_i \bar{t}_j + \Sigma c_{ij} t_i \bar{t}_j + s(t),$$

où $t = (t_1, t_2)$, $(1 \leq i, j \leq 2)$, les a_i, b_{ij}, c_{ij} sont des dérivées partielles de p à l'origine, s et ses dérivées partielles d'ordre $\leqq 2$ sont nulles à l'origine. Soit F la sous-variété analytique donnée par

$$\Sigma a_i \cdot t_i + \Sigma b_{ij} t_i \cdot t_j = 0.$$

Comme les c_{ij} sont les coefficients de la forme de Levi de p à l'origine, on a alors, pour $t \in F - \{0\}$, voisin de l'origine, $p(t) < p(0)$, donc $\alpha(t) \in Y$. On voit immédiatement que F contient une courbe passant par 0, non singulière en 0, donc la condition mentionnée plus haut est vérifiée.

4.2. Soit, comme précédemment, Z un espace analytique irréductible. Il est immédiat à partir de la définition que Z est pseudo-concave s'il est compact, ou s'il est le produit, en tant qu'espace analytique, d'espaces pseudo-concaves. Si Z est pseudo-concave et possède m fonctions méromorphes f_i algébriquement indépendantes $(m = \dim_{\mathbf{C}} Z)$, alors le corps des fonctions méromorphes $\mathbf{C}(Z)$ sur Z est un corps de fonctions algébrique de degré de transcendance m sur $\mathbf{C}$ [1, Satz 3]. Si $Z = X/\Gamma$, il est bien connu que la deuxième condition est satisfaite donc, si l'on est dans le cas de 4.3, $\mathbf{C}(X/\Gamma)$ est un corps de fonctions algébrique de degré de transcendance $\dim_{\mathbf{C}} X/\Gamma$. De plus, comme on peut prendre pour f_i un quotient de deux séries de Poincaré $(1 \leqq i \leqq m)$ il s'ensuit que tout élément de $\mathbf{C}(X/\Gamma)$ est quotient de deux formes automorphes de même poids.

4.3. Théorème. *Supposons que G n'ait pas de sous-groupe distingué défini sur $\mathbf{Q}$ et localement isomorphe sur $\mathbf{Q}$ à $\mathbf{SL}(2,\mathbf{C})$. Alors X/Γ est pseudo-concave.*

Remarquons tout d'abord que si Γ_1, Γ_2 sont deux sous-groupes discrets commensurables du groupe $\mathrm{Aut}\, X$ des automorphismes de X, alors X/Γ_1 est pseudo-concave si et seulement si X/Γ_2 l'est [1, Satz 2]. Il est donc loisible de remplacer Γ par un sous-groupe d'indice fini. En particulier, on peut supposer $\Gamma \subset G_{\mathbf{R}}^0$.

Soit Z le centre de $G_{\mathbf{R}}^0$. C'est aussi le centre de $G_{\mathbf{C}}$ [3, §11.5]. Il est donc défini sur $\mathbf{Q}$, et G/Z s'identifie au groupe adjoint $\mathrm{Ad}\, G$ de G. Comme Z opère trivialement sur X, et que l'image de Γ dans $\mathrm{Ad}\, G$ est un groupe arithmétique [4, §8.9], on peut supposer que le centre de G est réduit à $\{e\}$. Le groupe G est alors produit direct sur $\mathbf{Q}$ de groupes $\mathbf{Q}$-simples G_i, de centre réduit à $\{e\}$. Pour tout $i, \Gamma_i = \Gamma \cap G_i$ est arithmétique dans G_i, et Γ est commensurable au produit des groupes Γ_i [4, §8.10]. D'autre part X s'identifie au produit des quotients $G_{i\mathbf{R}}/(K \cap G_i)$, qui sont des domaines bornés symétriques. En remplaçant Γ par le produit des Γ_i, et en utilisant la remarque initiale de 4.2, on voit que l'on est ramené au cas où G est simple sur $\mathbf{Q}, \Gamma \subset G_{\mathbf{R}}^0$, et où X/Γ est non-compact, donc où le $\mathbf{Q}$-rang de G est $\geqq 1$. C'est celui qui est considéré dans les paragraphes précédents, dont nous reprenons les notations.

La fonction Φ_0 du §3 est invariante à droite par Γ, donc définit une fonction q sur X/Γ, qui est réelle, continue (3.5). Vu 3.4 (6), la projection $\alpha: X \to X/\Gamma$ applique $\mathfrak{S}' \cdot C^{-1}$ sur X/Γ. Il résulte alors de 3.5 que pour tout $r \in \mathbf{R}$

$$(X/\Gamma)_r = \{y \in X/\Gamma \mid q(y) < r\},$$

est relativement compact. Prenons maintenant r comme dans 3.6.

Soit $y \in X/\Gamma$ tel que $q(y) = r$. Il existe $x \in \mathfrak{S}'$ et $c \in C$ tels que $y = \alpha(x \cdot c^{-1})$. Soit L_x la fibre de σ_1 passant par x. C'est une sous-variété et 3.7 entraine que la restriction de $\Phi_0 = q \circ \alpha$ à $(L_x \cdot c^{-1})$ est indéfiniment différentiable et possède une forme de Levi négative non dégénérée au voisinage de y. Mais, comme $\dim G > 3$, L_x est de dimension complexe ≥ 2 d'après [3, § 3.15], donc la condition (*) de 4.1 est remplie.

4.4. Supposons que $X = H_n$ et $\Gamma = \Gamma_n$ soient le demi-plan et le groupe modulaire de Siegel de degré n. Alors $G_{\mathbf{R}} = \mathbf{Sp}(n, \mathbf{R})/\{\pm 1\}$. La composante F_s est réduite à un point, et la réalisation correspondante est celle donnée à la première ligne de l'introduction. Si l'on écrit comme d'habitude $g \in \mathbf{Sp}(n, \mathbf{R})$ sous la forme

$$(1) \qquad g = \begin{pmatrix} A & B \\ C & D \end{pmatrix}, \quad (A, B, C, D \in \mathbf{M}(n, \mathbf{R})),$$

et que l'on fait opérer $G_{\mathbf{R}}$ à gauche sur H_n par

$$Z \mapsto (A \cdot Z + B) \cdot (C \cdot Z + D)^{-1},$$

alors

$$J_s(Z, g) = (\det(C \cdot Z + D))^{-(n+1)},$$

et l'on sait que

$$\det \operatorname{Im}(g \cdot Z) = (\det \operatorname{Im} Z) \cdot |\det(C \cdot Z + D)|^{-2}.$$

Prenons comme groupe compact maximal K le quotient par $\{\pm 1\}$ du groupe des éléments de $\mathbf{Sp}(n, \mathbf{R})$ satisfaisant, dans les notations de (1), aux conditions:

$$C = -B, \quad A = D, \quad A + i \cdot B \in \mathbf{U}(n).$$

Alors le point fixe de K est le produit par i de la matrice unité d'ordre n. En le notant par i, on a donc, puisque Γ_s est réduit à $\{e\}$:

$$\tau(g) = \mu(g) = |\det(C \cdot i + D)|^{-a(n+1)},$$
$$\tau_0(Z) = \mu_0(Z) = (\det \operatorname{Im} Z)^{a(n+1)/2}.$$

La composante à la frontière de dimension maximum est isomorphe au demi-plan de Siegel de degré $n-1$ et s'identifie de manière naturelle aux matrices $Z' = (z_{ij})$, $(1 \leq i, j \leq n-1)$, symétriques, à partie imaginaire > 0. Les fibres de σ_1 sont les plans de coordonnées $z_{in}(1 \leq i \leq n)$. On retrouve bien la démonstration de [1].

Bibliographie

[1] Andreotti, A., u. H. Grauert: Algebraische Körper von automorphen Funktionen. Nachr. Akad. der Wiss. zu Göttingen 1961, Nr. 3, 39—48.

[2] Baily, W.: On Satake's compactification of V_n. Amer. J. Math. **80**, 348—364 (1958).

[3] —, and A. Borel: Compactification of arithmetic quotients of bounded symmetric domains. Annals of Math. (2) **84**, 442—528 (1966).

[4] Borel, A.: Introduction aux groupes arithmétiques. Act. Sci. Ind. Paris: Hermann éd. 1969.

[5] Cartan, H.: Fonctions automorphes et séries de Poincaré. Jour. Analyse math. **6**, 169—175 (1958).

[6] Koecher, M.: Zur Theorie der Modulfunktionen n-ten Grades I. Math. Zeitschrift **59**, 399—416 (1954).

[7] Pyateckii-Shapiro, I. I.: Arithmetic groups on complex domains. Uspehi Mat. Nauk **XIX** (6), 93—121 (1964); [English translation in Russian Math. Surveys **XIX** (6), pp. 83ff].

[8] Spilker, J.: Algebraische Körper von automorphen Funktionen. Math. Annalen **149**, 341—360 (1963).

89.

Properties and linear representations of Chevalley groups

Seminar on algebraic groups and related finite groups,
Lect. Notes Math. **131** (1970) 1−55

This part of the Notes is devoted to a survey, in part with proofs, of some of the main results on Chevalley groups and their irreducible representations.

The construction of Chevalley groups relies on some properties of forms over $\mathbf{Z}$ of a complex semi-simple Lie algebra and of its universal enveloping algebra, which are stated or proved in §§ 1, 2. In conformity with one aim of this seminar, we have tried to state the main results of Chevalley groups with a minimum of prerequisites (§ 3). For their proof in the same spirit, we refer to Steinberg's Notes [12]. Here we have limited ourselves, in § 4, to give the proof of one of them, in a sense the strongest one from the point of view of algebraic group theory, about schemes over $\mathbf{Z}$ associated to admissible lattices. § 5 gives the irreducible rational representations of a Chevalley group over an algebraically closed field. The main results are due to Chevalley [4]. The presentation here follows essentially [12]. §§ 6, 7 are devoted to some results of Curtis [6, 7] and Steinberg [11] in characteristic $p > 0$. In particular, they describe the irreducible representations of a Chevalley group G which are also irreducible for the Lie algebra (6.4), prove that they remain irreducible when restricted to the (finite) group of rational points over the prime field (7.3), and show how to construct all irreducible representations of G from those (7.5). The main difference between their treatment and the one given here consists in the systematic use of the transformations $X_{a,j}$. This allows one in particular to give a somewhat more direct proof of 6.4.

For the facts on complex semi-simple Lie algebras (resp. linear algebraic groups) used without further comment, we refer to [8; 10] (resp. [1]).

I wish to thank P. Cartier, J. E. Humphreys, and R. Steinberg, who read an earlier draft, for a number of corrections and suggestions for improvement.

Notation. $\mathbf{Z}$ is the ring of integers, $\mathbf{N}$ the monoid of positive integers, $\mathbf{Q}$ (resp. $\mathbf{R}$, resp. $\mathbf{C}$) the field of rational (resp. real, resp. complex) numbers, $\mathbf{F}_q$ the field with q elements (q a power of a prime).

Throughout, k is a commutative field, k_0 its prime subfield, K an algebraically closed extension of k, and K_p an algebraically closed field of characteristic p.

76

§ 1. Z-forms of a complex semi-simple Lie algebra
and of its universal enveloping algebra

1.1. $\mathfrak{g}$ is a complex finite dimensional semi-simple Lie algebra, $\mathfrak{h}$ a Cartan sub-algebra, Φ the set of roots of $\mathfrak{g}$ with respect to $\mathfrak{h}$, Φ^+ (resp. Δ) the set of positive (resp. simple) roots in Φ for some fixed ordering, and

$$\mathfrak{g}_a = \{x \in \mathfrak{g} \mid [h, x] = a(h) \cdot x, \ (h \in \mathfrak{h})\} \qquad (a \in \Phi).$$

The space $\mathfrak{g}_a$ is one-dimensional and $\mathfrak{g} = \mathfrak{h} \oplus \left(\bigoplus_{a \in \Phi} \mathfrak{g}_a \right)$. The restriction to $\mathfrak{h}$ of the Killing form $B(x, y) = \mathrm{tr}(\mathrm{ad}\, x \circ \mathrm{ad}\, y)$ of $\mathfrak{g}$ is non-degenerate and allows one to define in the usual way a scalar product $(,)$ on $\mathfrak{h}^*$. For a, $b \in \mathfrak{h}^*$, we let $n_{ab} = 2 \cdot (a, b) \cdot (b, b)^{-1}$. If $a, b \in \Phi$, then $n_{a,b} \in \mathbf{Z}$. We put

$$\mathfrak{n} = \mathfrak{n}^+ = \sum_{a > 0} \mathfrak{g}_a, \quad \mathfrak{n}^- = \sum_{a < 0} \mathfrak{g}_a,$$

$$\mathfrak{b} = \mathfrak{b}^+ = \mathfrak{h} + \mathfrak{n}, \quad \mathfrak{b}^- = \mathfrak{h} + \mathfrak{n}^-.$$

$\mathfrak{n}^{\pm}$ (resp. $\mathfrak{b}^{\pm}$) is a maximal nilpotent (resp. solvable) subalgebra of $\mathfrak{g}$.

1.2. *Chevalley basis of* $\mathfrak{g}$. For $a, b \in \Phi \, (a \neq \pm b)$, let $p(a, b)$ be the greatest integer such that $a - p(a, b) \cdot b \in \Phi$. According to Chevalley ([3], see also [12, Thm. 1, p. 6]) it is possible to choose $x_a \in \mathfrak{g}_a - \{0\}$ such that if we put $h_a = [x_a, x_{-a}]$, then we have

$$[x_a, x_b] = 0, \qquad\qquad (a, b \in \Phi, a + b \notin \Phi, a + b \neq 0),$$

$$[x_a, x_b] = \pm (p(a, b) + 1) \cdot x_{a+b}, \quad (a, b, a + b \in \Phi),$$

$$[h_a, x_b] = n_{ba} \cdot x_b, \qquad\qquad (a, b \in \Phi).$$

The elements $h_a (a \in \Delta)$, $x_b (b \in \Phi)$ form the basis of a $\mathbf{Z}$-form of $\mathfrak{g}$, to be denoted $\mathfrak{g}_{\mathbf{Z}}$.

1.3. *A $\mathbf{Z}$-form of $U(\mathfrak{g})$.* We let $U(\mathfrak{g})$ or U denote the universal enveloping algebra of $\mathfrak{g}$. By definition, $U(\mathfrak{g})$ is the quotient of the tensor algebra $T(\mathfrak{g}) = \sum_{n \geq 0} T^n \mathfrak{g}$ by the ideal J generated by the elements $(x \otimes y - y \otimes x - [x, y])$ $(x, y \in \mathfrak{g})$. Any representation $\pi \colon \mathfrak{g} \to \mathfrak{gl}(V)$ of $\mathfrak{g}$ into the Lie algebra of endomorphisms of a complex vector space V extends canonically to a homomorphism of $U(\mathfrak{g})$ into the associative algebra of endomorphisms of V. The extension will also be denoted π.

Let $q \colon T \to U$ be the canonical projection, and $U_j = q \left(\sum_{0 \leq i \leq j} T^i(\mathfrak{g}) \right)$. The U_j's define an increasing filtration of U. Let $\mathrm{Gr}\, U = \sum_j U_j / U_{j-1}$ be the associated graded algebra. By the Poincaré-Birkhoff-Witt theorem [2; 8], the natural map $\mathfrak{g} \to U(\mathfrak{g})$ is injective, and extends to an isomorphism of graded algebras of the symmetric algebra $S(\mathfrak{g})$ of $\mathfrak{g}$ onto $\mathrm{Gr}\, U$. If $\mathfrak{g}$ is commutative then $U(\mathfrak{g})$ is isomorphic to $S(\mathfrak{g})$. A homomorphism $f \colon \mathfrak{g} \to \mathfrak{g}'$ induces a homomorphism $U(f) \colon U(\mathfrak{g}) \to U(\mathfrak{g}')$ which is injective (resp. surjective) if f is so. If $\mathfrak{m}$ is a subalgebra of $\mathfrak{g}$, we shall identify $U(\mathfrak{m})$ to a subalgebra of $U(\mathfrak{g})$ via the inclusion map. Let us write $\Phi^+ = \{a_1, \ldots, a_m\}$. For $Q = (q_1, \ldots, q_m) \in \mathbf{N}^m$ we define elements $e_{\pm Q}$ of U by

$$e_{\pm Q} = \prod_i (x_{\pm a_i})^{q_i}/q_i! \, .$$

Given $x \in \mathfrak{g}$ and $s \in \mathbf{N}$, we put, as usual, in $U(\mathfrak{g})$,

$$\binom{x}{s} = x(x-1) \ldots (x-s+1)/s! \, .$$

Let $l = \dim \mathfrak{h}$. To an l-uple $R = (r_a)_{a \in \varDelta}$ of elements of $\mathbf{N}$, we associate the element of $U(\mathfrak{h})$ given by

$$h_R = \prod_{a \in \varDelta} \binom{h_a}{r_a} \, .$$

1.4. Theorem. *The elements* $e_{-Q} \cdot h_R \cdot e_S (Q, S \in \mathbf{N}^m; R \in \mathbf{N}^l)$ *form a basis of a* $\mathbf{Z}$-*form* $U_{\mathbf{Z}}$ *of* $U(\mathfrak{g})$. *The elements* $e_{\pm Q}$ *(resp.* h_R*) form a basis of a* $\mathbf{Z}$-*form* $U(\mathfrak{n}^{\pm})_{\mathbf{Z}}$ *(resp.* $U(\mathfrak{h})_{\mathbf{Z}})$ *of* $U(\mathfrak{n}^{\pm})$, *(resp.* $U(\mathfrak{h}))$. *The ring* $U_{\mathbf{Z}}$ *is generated by the elements* e_{-Q}, $e_Q (Q \in \mathbf{N}^m)$.

This theorem is fundamental for the sequel. It underlies the construction of schemes over $\mathbf{Z}$ (see § 4) originally due to Chevalley [5]. It is not mentioned in [5], but the existence of "admissible" lattices in any finite dimensional $\mathfrak{g}$-module, which is stated there, is a closely related result. 1.4 itself is proved in ([9], p. 95−97, [12], § 2), to which we refer for the details. We mention only some commutation relations in $U(\mathfrak{g})$, which are the starting point of the proof, and will be used later. They are easily derived from (1.2) by induction on j.

$$(1) \quad (x_a^j/j!) \cdot x_b = x_b \cdot (x_a^j/j!) + \sum_{0 < i \leqq j} c_i \cdot x_{ia+b} \cdot (x_a^{j-i}/(j-i)!) \, ,$$

$$(a, b \in \Phi, a \neq \pm b, j \in \mathbf{N}; c_i \in \mathbf{Z}) \, .$$

$$(2) \quad (x_a^j/j!) \cdot x_{-a} - x_{-a} \cdot (x_a^j/j!) = h_a \cdot x_a^{j-1}/(j-1)! - x_a^{j-1}/(j-2)! \, ,$$

$$(a \in \Phi; j \in \mathbf{N}) \, .$$

$$(3) \quad (x_a^j/j!) \cdot U(\mathfrak{h})_{\mathbf{Z}} = U(\mathfrak{h})_{\mathbf{Z}} \cdot (x_a^j/j!) \, , \qquad\qquad (a \in \Phi; j \in \mathbf{N}) \, .$$

We note also that, as a $\mathbf{Z}$-module, $U_{\mathbf{Z}}$ is isomorphic to $U(\mathfrak{n}^-)_{\mathbf{Z}} \otimes U(\mathfrak{h})_{\mathbf{Z}} \otimes U(\mathfrak{n})_{\mathbf{Z}}$.

§ 2. Admissible Z-forms in g-modules

2.1. We first recall some facts on representations of $\mathfrak{g}$ (see [8; 10]). An element $m \in \mathfrak{h}^*$ is a *weight* if and only if it takes integral values on the elements $h_a (a \in \Phi)$ of 1.2. It suffices that this be the case for $a \in \varDelta$. The weights form a $\mathbf{Z}$-form of $\mathfrak{h}^*$, to be denoted $\varGamma_{\mathrm{sc}}$. The elements l_a defined by $l_a(h_b) = \delta_{ab} (a, b \in \varDelta)$ form a basis of $\varGamma_{\mathrm{sc}}$. A weight is *dominant* if $m(h_a) \geqq 0$ for $a > 0$. The dominant weights are the linear combinations, with coefficients in $\mathbf{N}$, of the l_a; the latter are the *fundamental dominant weights*. The subgroup $\varGamma_{\mathrm{ad}}$ generated by the roots has finite index in $\varGamma_{\mathrm{sc}}$.

Let (π, E) be a finite dimensional $\mathfrak{g}$-module. For $m \in \mathfrak{h}^*$, let

$$E_m = \{v \in E \mid \pi(h) \cdot v = m(h) \cdot v, \quad (h \in \mathfrak{h})\} .$$

The m's for which $E_m \neq 0$ are the weights of π. The space E is the direct sum of the E_m's, where m runs through the set $P(\pi)$ of the weights of π. We have

$$(1) \qquad \pi(\mathfrak{g}_a) \cdot E_m \subset E_{m+a}, \qquad (a \in \Phi, m \in P(\pi)) .$$

We let Γ_π denote the subgroup of $\mathfrak{h}^*$ generated by the weights of π. If π is *faithful*, then

$$(2) \qquad \Gamma_{\mathrm{sc}} \supset \Gamma_\pi \supset \Gamma_{\mathrm{ad}} .$$

Let now π be *irreducible*. The subspace of E annihilated by $\mathfrak{n}$ is one-dimensional, and is a weight space E_{l_π}. The weight l_π is dominant, and is called the *highest weight* of π; a non-zero element of E_{l_π} is a *highest weight vector*. The correspondence $\pi \mapsto l_\pi$ defines a bijection between equivalence classes of irreducible finite dimensional $\mathfrak{g}$-modules and dominant weights. We have

$$(3) \qquad E = \pi(U(\mathfrak{n}^-)) \cdot E_{l_\pi} .$$

The weights m of π are of the form

$$(4) \qquad m = l_\pi - \sum_{a \in \Delta} c_a \cdot a, \qquad (c_a \in \mathbf{N}) .$$

If l is a dominant weight, we shall also denote by Γ_l the subgroup of Γ_{sc} generated by the weights of an irreducible representation with highest weight l.

2.2. Definition. Let (π, E) be a $\mathfrak{g}$-module. An *admissible* **Z**-*form of E* is a **Z**-form which is stable under $\pi(U_{\mathbf{Z}})$.

Since $U_{\mathbf{Z}}$ is generated by the elements $x_a^j/j!$, $(a \in \Phi; j \in \mathbf{N})$, it suffices to ask that the **Z**-form be stable under the endomorphisms $\pi(x_a^j/j!)$, which is one of the two conditions imposed by Chevalley [5] (the other one follows from 2.3).

2.3. Lemma. *Let (π, E) be a finite dimensional $\mathfrak{g}$-module, and let L be a subgroup of E stable under $U_{\mathbf{Z}}$. Then $L = \bigoplus_{m \in P(\pi)} (L \cap E_m)$.*

Let $x \in L$. We may write

$$(1) \qquad x = \sum_{m \in P(\pi)} x_m, \qquad (x_m \in E_m) ,$$

and we have to show that $x_m \in L$ for every m.

Since $\mathfrak{h}$ is commutative, we may identify $U(\mathfrak{h})$ to $S(\mathfrak{h})$, and hence to the algebra of complex valued polynomials on $\mathfrak{h}^*$. With this identification we have

$$(2) \qquad \pi(F) \cdot y = F(m) \cdot y, \qquad (F \in U(\mathfrak{h}); m \in P(\pi); y \in E_m) .$$

On the other hand, it follows from the definitions, and known facts about integral valued polynomials on lattices, that the previous identification maps $U(\mathfrak{h})_{\mathbf{Z}}$ onto the ring of polynomials on $\mathfrak{h}^*$ which take integral values on Γ_{sc}. Fix $m \in P(\pi)$. We may then find $F \in U(\mathfrak{h})_{\mathbf{Z}}$ such that $F(m) = 1$, $F(m') = 0$ if $m' \in P(\pi)$, $m' \neq m$. We have then, by (1), (2), $\pi(F) \cdot x = x_m$, hence $x_m \in L$.

2.4. Proposition. *Let (π, E) be a finite dimensional g-module. Then E has an admissible $\mathbf{Z}$-form. If π is irreducible, and e_0 is a highest weight vector, then $U_{\mathbf{Z}} \cdot e_0$ is equal to $U(\mathfrak{n}^-)_{\mathbf{Z}} \cdot e_0$ and is an admissible $\mathbf{Z}$-form of E.*

Since π is fully reducible, it suffices to prove the second assertion. The space $\mathbf{C} \cdot e_0$ is annihilated by $\mathfrak{n}$, hence $U(\mathfrak{n})_{\mathbf{Z}} \cdot e_0 = \mathbf{Z} \cdot e_0$. By the remarks made in 2.3, $U(\mathfrak{h})_{\mathbf{Z}} \cdot e_0 \subset \mathbf{Z} \cdot e_0$, therefore $L = U(\mathfrak{n}^-)_{\mathbf{Z}} \cdot e_0$ is equal to $U_{\mathbf{Z}} \cdot e_0$, hence is stable under $U_{\mathbf{Z}}$. There remains to show that L is a $\mathbf{Z}$-form of E.

It follows immediately from 2.1 (1) that there exists $q \geq 1$ such that $\pi(x_a^q) = 0$ for all $a \in \Phi$. As a consequence, L is finitely generated. There remains to show that if $(v_i)_{l \leq i \leq r} \in L$ are linearly independent over $\mathbf{Z}$, then they are independent over $\mathbf{C}$. For this, it is enough to show that if we have a relation

$$(1) \qquad \sum_{1 \leq i \leq r} c_i \cdot v_i = 0 , \qquad (c_i \in \mathbf{C} - \{0\}, 1 \leq i \leq r) ,$$

then there exists a relation

$$(2) \qquad \sum m_i \cdot c_i = 0 , \qquad (m_i \in \mathbf{Z}; 1 \leq i \leq r, \text{ not all zero}).$$

In fact, if (2) holds true, we have, assuming $m_1 \neq 0$,

$$\sum_{2 \leq i \leq r} c_i (m_1 \cdot v_i - m_i \cdot v_1) = 0 ,$$

with $m_1 v_i - m_i v_1 (2 \leq i \leq r)$ independent over $\mathbf{Z}$, and our assertion follows by induction on r.

Let $(v)_0$ denote the projection of $v \in E$ into $\mathbf{C} \cdot e_0$, parallel to the E_m's, $m \neq l_\pi$. From (1), we have, for any $x \in U_{\mathbf{Z}}$

$$\sum_{1 \leq i \leq r} c_i \cdot (x \cdot v_i)_0 = 0 .$$

Moreover, by irreducibility, we may choose x such that $(x \cdot v_i)_0 \neq 0$ for some i. It suffices therefore to show that $(x \cdot v_i)_0 \in \mathbf{Z} \cdot v_0$ for every $x \in U_{\mathbf{Z}}$. Since $v_i \in U_{\mathbf{Z}} \cdot e_0$, we are reduced to showing that $(x \cdot e_0)_0 \in \mathbf{Z} \cdot e_0$ for every $x \in U_{\mathbf{Z}}$. By 1.4, we may limit ourselves to elements x of the form $e_{-\varrho} \cdot h_R \cdot e_S$. If $e_S \neq 1$, then $e_S \cdot v_0 = 0$. If $e_{-\varrho} \neq 1$, then $(x \cdot v)_0 = 0$ by 2.1 (1). Thus we may assume $x = h_R$, and then $x \cdot v_0 \in \mathbf{Z} \cdot v_0$ by 2.3 (2).

2.5. Notation. If Γ is a $\mathbf{Z}$-form of $\mathfrak{h}^*$, we let $\hat{\Gamma}$ denote the dual $\mathbf{Z}$-form of $\mathfrak{h}$, i.e.

$$\hat{\Gamma} = \{h \in \mathfrak{h} \,|\, a(h) \in \mathbf{Z} \,(a \in \Gamma)\} ,$$

and put

$$\mathfrak{g}_\Gamma = \hat{\Gamma} + \sum_{a \in \Phi} \mathbf{Z} \cdot x_a.$$

If $\Gamma = \Gamma_\pi$ (π faithful representation of $\mathfrak{g}$), then we also write $\mathfrak{g}_\pi$ for $\mathfrak{g}_\Gamma$. Note that $\hat{\Gamma}_{sc}$ is the group generated by the h_a's, hence

$$\mathfrak{g}_{\mathbf{Z}} = \hat{\Gamma}_{sc} + \sum_a \mathbf{Z} \cdot x_a = \mathfrak{g}_{sc}.$$

2.6. Proposition [12, p. 18]. *We keep the previous notation, and view $\mathfrak{g}$ as a $\mathfrak{g}$-module under the adjoint representation.*

(i) *The admissible $\mathbf{Z}$-forms of $\mathfrak{g}$ (as a Lie algebra) containing $\mathfrak{g}_{\mathbf{Z}}$ are the $\mathfrak{g}_\Gamma$, where Γ runs through the lattices between Γ_{sc} and Γ_{ad}.*

(ii) *Let (π, E) be a faithful $\mathfrak{g}$-module and L an admissible $\mathbf{Z}$-form of E. Then $\mathfrak{g}_\pi = \{X \in \mathfrak{g} \mid \pi(X) \cdot L \subset L\}$.*

(i) The elements of $\hat{\Gamma}_{ad}$ are the $h \in \mathfrak{h}$ on which the roots take integral values. Hence $[h, x_a] \subset \mathbf{Z} \cdot x_a$ for any $h \in \hat{\Gamma}_{ad}$. From this and 1.2, we see that $\mathfrak{g}_\Gamma$ is a Lie algebra over $\mathbf{Z}$, hence a $\mathbf{Z}$-form of $\mathfrak{g}$. To prove that it is admissible we have to show that it is stable under $(\operatorname{ad} x_a^j)/j!$ ($a \in \Phi, j \in \mathbf{N}$). Let $b \in \Phi$, $b \neq \pm a$. If $a + b \notin \Phi$, then $\operatorname{ad} x_a(x_b) = 0$. In general, we have, since the set of i's for which $a + i\,b \in \Phi$ is the set of all integers in some interval, $p(a + b, a) = p(b, a) + 1$. From this and 1.2 we deduce immediately

$$(1) \qquad (\operatorname{ad} x_a^j/j!)\,(x_b) = \pm \binom{p(a, b) + j}{j} x_{b+j \cdot a}, \qquad (j \in \mathbf{N}).$$

By 1.2, we have also

$$(2) \qquad \begin{aligned} (\operatorname{ad} x_a) \cdot x_{-a} &= h_a, \\ (\operatorname{ad} x_a^2) \cdot x_{-a} &= [x_a, h_a] = -2x_a, \\ (\operatorname{ad} x_a^3) \cdot x_{-a} &= 0. \end{aligned}$$

Finally, if $h \in \hat{\Gamma}_{ad}$, then $\operatorname{ad} x_a \cdot h = a(h) \cdot x_a \in \mathbf{Z} \cdot x_a$ and $(\operatorname{ad} x_a^2) \cdot h = 0$. Consequently, $(\operatorname{ad} x_a^j)/j!$ leaves $\mathfrak{g}_\Gamma$ stable.

Let now M be an admissible $\mathbf{Z}$-form of $\mathfrak{g}$ containing $\mathfrak{g}_{\mathbf{Z}}$. By 2.3

$$M = M \cap \mathfrak{h} \oplus \sum_{a \in \Phi} (M \cap \mathfrak{g}_a).$$

We want to prove first that $M \cap \mathfrak{g}_a = \mathbf{Z} \cdot x_a (a \in \Phi)$. The group $M \cap \mathfrak{g}_\alpha$ is infinite cyclic. Let $c \in \mathbf{C}$ be such that $c \cdot x_a$ generates $M \cap \mathfrak{g}_a$. We have to show that $c \in \mathbf{Z}$. We have

$$c \cdot h_a = -[x_{-a}, c \cdot x_a] = -\operatorname{ad} x_{-a}(c \cdot x_a) \in M,$$

hence

$$[c \cdot h_a, c \cdot x_a] = 2 \cdot c^2 \cdot x_a \in M$$

and therefore $2 \cdot c \in \mathbf{Z}$. We have also

$$(\operatorname{ad} x^2_{-a}/2!)(c \cdot x_a) = - c \cdot x_{-a} \in M ,$$

and therefore

$$[c \cdot x_a, c \cdot x_{-a}] = c^2 \cdot h_a \in M .$$

But now $[c^2 \cdot h_a, x_a] = 2 \cdot c^2 \cdot x_a$ and $[c^2 \cdot h_a, c \cdot x_a] = 2 \cdot c^3 \cdot x_a$ belong to M, and therefore $2c$, $2c^2 \in \mathbf{Z}$. This implies $c \in \mathbf{Z}$.

$\mathfrak{h} \cap M$ is a $\mathbf{Z}$-form of $\mathfrak{h}$. What has already been shown proves that if $h \in \mathfrak{h} \cap M$, then $[h, x_a] \in \mathbf{Z} \cdot x_a$, hence $a(h) \in \mathbf{Z}$, and $h \in \hat{\Gamma}_{\mathrm{ad}}$.

(ii) Let $M = \{X \in \mathfrak{g} \mid \pi(X) \cdot L \subset L\}$. It is clear that $M \supset \mathfrak{g}_{\mathbf{Z}}$, $[M, M] \subset M$, and that M is stable under $U_{\mathbf{Z}}$. There remains to show that M is a $\mathbf{Z}$-form of $\mathfrak{g}$, as a vector space. By 2.3

$$M = M \cap \mathfrak{h} \oplus \sum_{a \in \Phi} M \cap \mathfrak{g}_a.$$

Let $h \in \mathfrak{h}$. For any weight m of π, $L \cap E_m$ is a $\mathbf{Z}$-form of E_m. On E_m, we have $\pi(h) = m(h) \cdot I$. Therefore $\pi(h)(L \cap E_m) \subset E_m$ if and only if $m(h) \in \mathbf{Z}$, hence $M \cap \mathfrak{h} = \hat{\Gamma}_\pi$. In particular, $M \cap \mathfrak{h}$ is a $\mathbf{Z}$-form of $\mathfrak{h}$. Let $a \in \Phi$. Then $x \mapsto [x_{-a}, x]$ maps $M \cap \mathfrak{g}_a$ injectively into a group of elements in $M \cap \mathfrak{h}$ which are proportional to h_{-a}. Since $M \cap \mathfrak{h}$ is a $\mathbf{Z}$-form of $\mathfrak{h}$ containing h_{-a}, this group is infinite cyclic, hence so is $M \cap \mathfrak{g}_a$.

Remark. It is known from representation theory that given Γ between Γ_{sc} and Γ_{ad}, there always exists a faithful representation π of $\mathfrak{g}$ such that $\Gamma = \Gamma_\pi$.

2.7. Example. Let $\mathfrak{g} = \mathfrak{s}\,\mathfrak{l}_2$. Take for $\mathfrak{h}$ the algebra of diagonal matrices of trace zero. The elements

$$h = \begin{pmatrix} 1 & 0 \\ 0 & -1 \end{pmatrix}, \quad x = \begin{pmatrix} 0 & 1 \\ 0 & 0 \end{pmatrix}, \quad y = \begin{pmatrix} 0 & 0 \\ 1 & 0 \end{pmatrix}$$

satisfy the relations:

$$[h, x] = 2 \cdot x, \quad [h, y] = - 2 \cdot y, \quad [x, y] = h,$$

and form a Chevalley basis. As the positive root, we take the element $a \in \mathfrak{h}^*$ having value two on h. Then $\mathfrak{n} = \mathbf{C} \cdot x$, $\mathfrak{n}^- = \mathbf{C} \cdot y$. Furthermore

$$\Gamma_{\mathrm{ad}} = \mathbf{Z} \cdot a, \quad \Gamma_{\mathrm{sc}} = \mathbf{Z} \cdot (a/2) .$$

For each $n \geq 0$, $\mathfrak{g}$ has one equivalence class of irreducible representations of degree $n + 1$. It contains the natural representation π_n of $\mathfrak{g}$ in the n-th symmetric power E_{n+1} of $\mathbf{C}^2$. Let (e_1, e_2) be the canonical basis of $\mathbf{C}^2$. Then $e_1^{n-i} \cdot e_2^i$ (symmetric product) spans the eigenspace of $\mathfrak{h}$ of weight $(n - 2i)\,a/2$ $(0 \leq i \leq n)$. The element $e_0 = e_1^n$ is a highest weight vector. It is readily checked that the admissible $\mathbf{Z}$-form

$U_{\mathbf{Z}} \cdot e_0$ is spanned by the elements $\binom{n}{i} \cdot e_1^{n-i} \cdot e_2^{i} \, (0 \leq i \leq n)$. Moreover $\{e_1^{n-i} \cdot e_2^{i}\}_{0 \leq i \leq n}$ is also the basis of an admissible $\mathbf{Z}$-form.

The corresponding lattice $\varGamma_{\pi_n}$ is $\varGamma_{\mathrm{ad}}$ if n is odd, $\varGamma_{\mathrm{sc}}$ if n is even. The $\mathbf{Z}$-forms of $\mathfrak{g}$ given by 2.6 are $\mathfrak{g}_{\mathbf{Z}} = \mathfrak{g}_{\mathrm{sc}}$, spanned by h, x, y, and $\mathfrak{g}_{\mathrm{ad}}$, spanned by $h/2$, x, y.

§ 3. Definition and properties of Chevalley groups

3.1. Let $\pi\colon \mathfrak{g} \to \mathfrak{gl}(V)$ be a representation of $\mathfrak{g}$. Identify V to $\mathbf{C}^n$ by means of a basis of an admissible $\mathbf{Z}$-form $V_{\mathbf{Z}}$, which consists of eigenvectors of $\mathfrak{t}$. Then

$$\pi(x_a)^j/j! = \pi(x_a^j/j!) \in \mathbf{M}(n, \mathbf{Z}), \qquad (a \in \varPhi; j \in \mathbf{N}).$$

Since $\pi(x_a)$ is nilpotent, as follows from 2.1 (1), the series $\exp t\,\pi(x_a)$, where t is an indeterminate, has only finitely many non-zero terms, hence

$$\exp t \cdot \pi(x_a) \in \mathbf{SL}(n, \mathbf{Z}[t]).$$

Let k be a commutative field. Then $\exp(t \cdot \pi(x_a)) = x_a(t)$ defines an automorphism of $V_{(k)} = V_{\mathbf{Z}} \otimes k$ for every $t \in k$. By definition, the Chevalley group associated to π and k is the group

$$G_{\pi, k} = \langle x_a(t), (a \in \varPhi, t \in k) \rangle,$$

generated by the $x_a(t)$ $(a \in \varPhi, t \in k)$. It is a subgroup of $GL(V_{\mathbf{Z}} \otimes k)$.

If the context requires it, we shall also write $x_a^{(\pi)}(t)$ for $x_a(t)$. We shall now list some of the main properties of Chevalley groups, from four different points of view, in order of increasing use of algebraic geometry.

3.2. $G_{\pi, k}$ *as an abstract group*

(1) For each $a \in \varPhi$, there is a unique homomorphism $\mu_a\colon \mathbf{SL}(2, k) \to G_{\pi, k}$, which maps $\begin{pmatrix} 1 & t \\ 0 & 1 \end{pmatrix} \left(\text{resp. } \begin{pmatrix} 1 & 0 \\ t & 1 \end{pmatrix}\right)$ onto $x_a(t)$ (resp. $x_{-a}(t)$) for every $t \in k$. Let $h_a(t)$ be the image of $\begin{pmatrix} t & 0 \\ 0 & t^{-1} \end{pmatrix}$ $(t \in k^*)$ under μ_a. The elements $h_a(t)$ belong to the group of diagonal matrices of $G_{\pi, k}$, to be denoted H. The group H is generated by the elements $h_a(t)$, $(t \in k^*, a \in \varDelta)$. If $m \in P(\pi)$ is a weight of π, then, putting $V_{(k), m} = (V_{\mathbf{Z}} \cap V_m) \otimes_{\mathbf{Z}} k$, we have

$$h_a(t) \cdot x = t^{m(h_a)} \cdot x, \qquad (x \in V_{(k), m}; \; t \in k^*; \; a \in \varPhi).$$

If $\varGamma_\pi = \varGamma_{\mathrm{sc}}$, then μ_a is injective $(a \in \varPhi)$, and H is the direct product of the groups $\{h_a(t)\}$ $(a \in \varDelta)$.

To m there is associated a homomorphism of H into k^* characterized by

$$m\left(\prod_{a \in \varDelta} h_a(t_a)\right) = \prod_{a \in \varDelta} t_a^{m(h_a)}.$$

(2) Let w_a be the image of $\begin{pmatrix} 0 & 1 \\ -1 & 0 \end{pmatrix}$ under $\mu_a (a \in \Phi)$. The element w_a normalizes H. We let $N = \langle H, w_a (a \in \Delta) \rangle$. Let

$$U = U^+ = \langle x_a(t), (t \in k, a > 0) \rangle, \qquad U^- = \langle x_a(t), (t \in k, a < 0) \rangle .$$

The group $U^{\pm}$ is unipotent. Let $a_1, \ldots, a_m$ be the positive roots, written in any order. Then

$$v_{\pm} : (t_1, \ldots, t_m) \mapsto x_{\pm a_1}(t_1) \ldots x_{\pm a_m}(t_m) ,$$

is a bijection of k^m onto $U^{\pm}$.

The group $G_{\pi, k}$ is generated by the groups $\{x_a(t)\}$, $(a \in \Delta \cup (-\Delta))$. If Card $k \geq 4$, the group U (resp. U^-) is generated by the groups $\{x_a(t)\}$, $(a \in \Delta$, resp. $a \in -\Delta)$.

The group $U^{\pm}$ is normalized by H. Let $B = B^+ = H \cdot U$, $B^- = H \cdot U^-$. Then (B, N) is a $B \cdot N$-pair, or Tits system, in $G_{\pi, k}$, with root system Φ.

(3) $G_{\pi, k}$ is its own commutator group. Its center Z is finite, and if $\mathfrak{g}$ is simple, $G_{\pi, k}/Z$ is simple except in a few cases where k has two or three elements [3; 13].

(4) Let ϱ be another representation of $\mathfrak{g}$. If $\Gamma_\pi \supset \Gamma_\varrho$, then the map

$$x_a^{(\pi)}(t) \mapsto x_a^{(\varrho)}(t) , \qquad (t \in k; a \in \Phi) ,$$

extends to a surjective homomorphism $\lambda_{\varrho, \pi} : G_{\pi, k} \to G_{\varrho, k}$. In particular, if $\Gamma_\pi = \Gamma_\varrho$, then $\lambda_{\varrho, \pi}$ is an isomorphism.

(5) We refer to [12, § 6] for a presentation of $G_{\pi, k}$ by generators and relations, when $\Gamma_\pi = \Gamma_{\mathrm{sc}}$.

3.3. *Algebraic Group Properties*

(1) $G_{\pi, K}$ is the group $G_\pi(K)$ of points rational over K of an algebraic subgroup G_π of $\mathbf{G L}_n$ defined over k_0. The group G_π is connected, semi-simple. If π is faithful, the Lie algebra $L(G_\pi)$ of $G_{\pi, K}$ is isomorphic to $\mathfrak{g}_\pi \otimes_{\mathbf{Z}} K$ (notation of 2.5). More precisely, for any subfield k' of K,

$$L(G_\pi)(k') = \mathfrak{g}_\pi \otimes_{\mathbf{Z}} k' .$$

(2) For $k = K$, the map $\lambda_{\varrho, \pi}$ of 3.2 (4) is a surjective morphism, defined over k_0. Let π, ϱ be faithful. Then the kernel of $\lambda_{\varrho, \pi} : G_\pi(K) \to G_\varrho(K)$ is isomorphic to $\hat{\Gamma}_\varrho / \hat{\Gamma}_\pi$ mod p-torsion. The kernel of the differential $d\lambda_{\varrho, \pi}$ of $\lambda_{\varrho, \pi}$ is isomorphic to $(\hat{\Gamma}_\varrho / \hat{\Gamma}_\pi) \otimes K$. The homomorphism $\mu_a : \mathbf{S L}(2, K) \to G_\pi(K)$ is a k_0-morphism, with finite kernel. If $\Gamma_{\mathrm{sc}} = \Gamma_\pi$, then μ_a is an isomorphism of $\mathbf{S L}_2$ onto its image.

(3) If $k = K$, H is the group of rational points of a maximal torus of G_π which is defined over k_0 and splits over k_0. In particular, G_π splits over k_0.

For $m \in P(\pi)$, the homomorphism of H into K^* associated to m by 3.2 (1) is a rational character of H. The map $P(\pi) \mapsto X(H)$ thus defined, where $X(H)$ is the group Mor $(H, \mathbf{G L}_1)$ of rational characters of H, induces an isomorphism of Γ_π onto $X(H)$. We shall often identify $X(H)$ to Γ_π by this isomorphism.

If $\Gamma_\pi = \Gamma_{sc}$, then H is isomorphic over k_0 to the product of the groups $\{h_a(t)\}$ $(a \in \Delta)$, and $t \to h_a(t)$ is a k_0-isomorphism of $\mathbf{G}\,\mathbf{L}_1$ onto $\{h_a(t)\}$ for all $a \in \Phi$.

The group $U^\pm$ is maximal unipotent, defined over k_0, and the map $v_\pm$ of 3.2 (1) is a k_0-isomorphism of varieties.

(4) If $\Gamma_\pi = \Gamma_{sc}$, the group G_π is simply connected, in the sense that every rational projective representation of G_π can be raised to a linear representation in the underlying vector space. Equivalently, if H is a connected K-group, and $f: H \mapsto G_\pi$ is an isogeny which is an isomorphism on unipotent subgroups, then f is an isomorphism.

If $K = \mathbf{C}$, then $G_\pi(K)$ is a complex Lie group, connected in the usual topology, Γ_{sc}/Γ_π is isomorphic to the fundamental group of $G_\pi(K)$.

(5) The group $G_{\pi,k}$ is the derived group of the group $G_\pi(k)$ of points of G_π rational over k. If $\Gamma_\pi = \Gamma_{sc}$, then $G_{\pi,k} = G_\pi(k)$.

(6) Every connected semi-simple algebraic group over K is isomorphic to one of the groups $G_{\pi,K}$ [4]. The elements UwB of the Bruhat decomposition of $G_{\pi,K}$ associated to the (B, N)-pair structure of 3.2 are locally closed.

Remarks. (1) The Weyl group W acts in the familiar way on $\mathfrak{h}, \mathfrak{h}^*$ and on $P(\pi)$. We have

$$(1) \qquad \langle w(m), w(h) \rangle = \langle m, h \rangle, \qquad (m \in \mathfrak{h}^*, h \in \mathfrak{h}, w \in W),$$

where $\langle,\rangle$ is the canonical pairing. The group W acts on H, by inner automorphisms, and on $X(H)$, and there is a formula similar to (1). The canonical isomorphism $\Gamma_\pi \xrightarrow{\sim} X(H)$ mentioned in 3.3 (3) is then equivariant. In fact, if $n \in N(H)$ represents w, we have $n \cdot h \cdot x = n \cdot h \cdot n^{-1} \cdot n \cdot x$ $(x \in V)$, which implies immediately that $n \cdot V_m = V_{m'}$ where m' is characterized by $m'(h) = m(n \cdot h \cdot n^{-1})$.

(2) Let $m \in X(H)$. Its differential dm maps the Lie algebra $L(H)$ of H into K, hence can be identified with a linear form on $L(H)$. We have the canonical isomorphisms

$$\alpha\colon L(H) \xrightarrow{\sim} \hat{\Gamma}_\pi \otimes_{\mathbf{z}} K, \qquad \beta\colon L(H)^* = \Gamma_\pi \otimes_{\mathbf{z}} K,$$

which are compatible with the canonical pairings of $L(H)$ and $L(H)^*$, and of $\hat{\Gamma}_\pi$ and Γ_π. In particular:

$$(2) \qquad dm = \beta^{-1}(m \otimes 1).$$

If $p \neq 0$, then $dm = 0$ if and only if m is divisible by p in Γ_π.

3.4. *Reduction* mod p. Let V be an affine variety over k, and $k[V]$ its coordinate ring. If $V \subset K^n$, then $k[V] = k(X_1, \ldots, X_n)/J$, where J is the ideal of polynomials in $k[X_1, \ldots, X_n]$ vanishing on V. We recall that the assignment $V \mapsto k[V]$ defines an equivalence between the category of affine k-varieties and k-morphisms and that of finitely generated algebras over k which are *reduced* (no nilpotent element $\neq 0$). For any extension k' of k in K, the set $V(k')$ of k'-rational points of V is $\mathrm{Hom}_k(k[V], k')$.

Let now $k = \mathbf{Q}$, $K = \mathbf{C}$. Fix a $\mathbf{Z}$-form $\mathbf{Z}[V]$ of $\mathbf{Q}[V]$. This amounts to choosing an embedding $V \subset \mathbf{C}^m$ over $\mathbf{Q}$, such that $\mathbf{Z}[V] = \mathbf{Z}[X_1, \ldots, X_m]/J$ where J is the ideal of polynomials with integral coefficients vanishing on V. For a prime p, let $V_{(p)}$ be the set of zeros of $J \otimes \mathbf{F}_p$ in K_p^m. This is an affine variety defined over $\mathbf{F}_p$. It depends on the choice of the $\mathbf{Z}$-form, however two given $\mathbf{Z}$-forms yield isomorphic varieties for almost all primes. We shall call $V_{(p)}$ the reduction mod p of V, in spite of this ambiguity.

The ideal $J(V_{(p)})$ in $\mathbf{F}_p[X_1, \ldots, X_n]$ contains $J \otimes \mathbf{F}_p$, obviously, but it may be bigger. By the Nullstellensatz, however, the image of $J(V_{(p)})$ in $\mathbf{Z}[V] \otimes \mathbf{F}_p = \mathbf{Z}[V]/(J \otimes \mathbf{F}_p)$ is the nilradical of that ring (ideal of nilpotent elements). If the latter is zero, i.e. if $\mathbf{Z}[V] \otimes \mathbf{F}_p$ is reduced, then we say that *the reduction is good.*

The variety V is irreducible if and only if $\mathbf{Q}[V]$ is an integral domain. If $\mathbf{Z}[V] \otimes \mathbf{F}_p$ is an integral domain, then it is reduced. In that case, reduction is good, and moreover $V_{(p)}$ is irreducible. For this, it is necessary and sufficient that $p \cdot J$ be a prime ideal in $\mathbf{Z}[V]$.

There is always a canonical surjective morphism of $\mathbf{Z}[V] \otimes \mathbf{F}_p$ onto $\mathbf{F}_p[V_{(p)}]$. It is bijective if and only if the reduction is good.

Let V' be an affine variety over $\mathbf{Q}$, $\mathbf{Z}[V']$ a $\mathbf{Z}$-form of $\mathbf{Q}[V']$, and $\pi\colon V \to V'$ a $\mathbf{Q}$-morphism. Assume that the associated comorphism $\pi_0\colon \mathbf{Q}[V'] \to \mathbf{Q}[V]$, defined by $f \mapsto f \circ \pi$, maps $\mathbf{Z}[V']$ into $\mathbf{Z}[V]$. We then say that π is defined over $\mathbf{Z}$. Then π_0 induces a homomorphism of $\mathbf{Z}[V'] \otimes \mathbf{F}_p$ into $\mathbf{Z}[V] \otimes \mathbf{F}_p$. It maps the nilradical into the nilradical, whence a map $\pi_{(p),0}\colon \mathbf{F}_p[V'] \to \mathbf{F}_p[V]$, which in turn defines a $\mathbf{F}_p$-morphism $V_{(p)} \to V'_{(p)}$, to be denoted $\pi_{(p)}$, and to be called the reduction mod p of π.

Assume $V = G$ to be an algebraic subgroup of $\mathbf{GL}(n, \mathbf{C})$ defined over $\mathbf{Q}$. Then $\mathbf{Q}[G] = \mathbf{Q}[g_{11}, g_{12}, \ldots, g_{nn}, \det^{-1}]/J'$. [Recall that G is open in a affine variety of $\mathbf{M}(n, \mathbf{C})$, but can be identified to an affine variety of $\mathbf{M}(n+1, \mathbf{Z})$ by the familiar embedding $g \mapsto (g, \deg g^{-1})$ of $\mathbf{GL}(n, \mathbf{C})$ into $\mathbf{SL}(n+1, \mathbf{C})$, whence the $\det^{-1}$.] Take then $\mathbf{Z}[g_{11}, \ldots, g_{nn}, \det^{-1}]/J$ as a $\mathbf{Z}$-form $(J = J' \cap \mathbf{Z}[g_{11}, \ldots, g_{nn}, \det^{-1}])$. Then the morphism $m\colon G \times G \to G$ given by the product and the inverse map $x \mapsto x^{-1}$ are defined over $\mathbf{Z}$, from which it follows immediately that $G_{(p)}$ is an algebraic group over $\mathbf{F}_p$, with product $m_{(p)}$.

We now come back to the situation of 3.3 with $k = \mathbf{Q}$, and identify G_π to a subgroup of $\mathbf{GL}(n, \mathbf{C})$ via an admissible lattice. Then $\mathbf{Q}[G_\pi]$ has a natural $\mathbf{Z}$-form defined by $\mathbf{Z}[g_{11}, g_{12}, \ldots, g_{nn}]/J$. (Since G_π is connected, semi-simple, its elements have determinant one, so we do not have to put $\det^{-1}$.) Now the main result about reduction of Chevalley groups is that for any prime p, the reduction $G_{\pi,(p)}$ of G_π may be canonically identified with the group G_{π, K_p} of 3.1. In particular, the *reduction is good, and irreducible for every prime p.* Moreover the homomorphisms μ_a and $\lambda_{\varrho, \pi}$ are defined over $\mathbf{Z}$.

3.5. *Schemes over $\mathbf{Z}$.* From the above, it is clear that to $\mathfrak{g}$, π and an admissible lattice, there is somehow associated an object over $\mathbf{Z}$, which yields $G_{\pi, L}$ for any algebraically closed field L by a process of reduction. As such an object we could take $\mathbf{Z}[G_\pi]$. But it is better to say that $\mathbf{Z}[G_\pi]$ represents this object, and discuss this in the framework of schemes. From this point of view 3.3 may be expressed by saying that $\mathbf{Z}[G_\pi]$ represents a scheme over $\mathbf{Z}$, whose fibers are the groups $G_{\pi, \mathbf{C}}$ and G_{π, K_p}

and are irreducible, and that the morphisms $\lambda_{\varrho,\pi}, \mu_a$ are associated to $\mathbf{Z}$-morphisms of schemes over $\mathbf{Z}$. We shall discuss this in some more detail in the next section.

§ 4. Group schemes over Z

In this section, all rings have a unit, and all ring homomorphisms and modules are unitary. R is a commutative ring.

4.1. We recall that an *affine scheme* M over R is a topological ringed space, whose set of points is the set of prime ideals of a commutative R-algebra A. It is reduced, or of finite type, if A is so. Given a commutative R-algebra L, the set $M(L)$ of points of M, with values in L, is the set $\mathrm{Mor}_{R\text{-alg}}(A, L)$ of R-algebra morphisms of A into L. Thus M gives rise to a functor from commutative R-algebras to sets, represented by A. For our needs it is as well, and in some respects more convenient, to view directly the functor as a scheme, and we shall do so.

Assume now A is endowed with an R-algebra morphism $d\colon A \to A \otimes A$, to be called a diagonal map. Then $M(L)$ admits a product, which assigns to $u, v \in M(L)$ the composite of the maps

$$A \xrightarrow{d} A \otimes A \xrightarrow{u \otimes v} L \otimes L \xrightarrow{m} L \,,$$

where m is given by $m(a \otimes b) = a \cdot b$. If moreover there is a morphism $e_0\colon A \to L$ and an automorphism $i\colon A \to A$, such that d, e_0, i satisfy the conditions imposed on μ_0, p_0, i_0 in [1, p. 89−90], then $M(L)$ is a group, where the inverse of u is $u \circ i$, and the unit element is e_0 followed by the canonical inclusion of R into L. With some mild abuse of notation, these conditions can be stated as follows:

$$
\begin{array}{lll}
& \text{(Ass).} & (d \otimes 1) \circ d = (1 \otimes d) \circ d \\[4pt]
(1) & \text{(Id)} & (1 \otimes e_0) \circ d = (e_0 \otimes 1) \circ d = \mathrm{Id} \\[4pt]
& \text{(Inv.)} & (1 \otimes i) \circ d = (i \otimes 1) \circ d = e_0 \,.
\end{array}
$$

In this case, $L \mapsto M(L)$ is a functor from commutative R-algebras to groups, and M, or the functor; is said to be an *affine group scheme* over R. [In [1], loc. cit. R is a field, and A is reduced, because of a standing assumption, but these conditions are not needed to derive from (1) that $M(L)$ is a group.]

If $R = \mathbf{Z}$, and A is a $\mathbf{Z}$-form of the coordinate ring over $\mathbf{Q}$ of an affine algebraic group G defined over $\mathbf{Q}$, then A has the properties mentioned above (d being induced by the product in $G(\mathbf{C})$, and i, e_0 being defined by $i(a)(g) = a(g^{-1})$, $e_0 a = a(e)$, $(a \in A, g \in G(\mathbf{C}))$, and defines an affine group scheme over $\mathbf{Z}$ (reduced, of finite type). If L is an algebraically closed field, and p the characteristic of L, then $(A \otimes L)/\text{nilradical}$ is the coordinate ring of the reduction mod p, $G_{(p)}$, of G introduced in 3.4, and $G_{(p)}(L)$ can be canonically identified with $M(L)$.

4.2. Assume now R to be an algebraically closed field, and A to be of finite type. Then, by the Nullstellensatz, $M(R)$ can be canonically identified with the set maxspec (A) of maximal ideals of A. To an element $a \in A$ there is associated a function $\bar{a}$ on $M(R)$ defined by $\bar{a}(u) = u(a)$ $(u \in M(r))$. The kernel of the homomorphism $v\colon a \mapsto \bar{a}$ is the nilradical $\mathfrak{n}$ of A (since, under our assumptions, every prime ideal of A is an intersection of maximal ideals). To $u \in M(R)$ we associate the map

$$(1) \qquad\qquad \varrho_u = (1 \otimes u) \circ d$$

of A into A. It is a morphism, and it follows in an elementary way from 4.1 (1) that $\varrho_u \circ \varrho_v = \varrho_{vu}$, $\varrho_e = \mathrm{Id}$. In particular, ϱ_u is an automorphism of A. Moreover, in view of 4.1 and our conventions, we have

$$(2) \qquad\qquad \overline{\varrho_u(a)}\,(x) = \bar{a}(x \cdot u)\,, \qquad (x, u \in M(R); a \in A)\,.$$

Thus, ϱ_u defines the automorphism of $A/\mathfrak{n} = \bar{A}$ associated to the right translation $x \mapsto x \cdot u$ on $M(R)$. However, we shall have to consider the ϱ_u's in a case where we do not know that A is reduced, and therefore we have to start from the less intuitive definition (1).

Lemma. *Let $a \in A$, $a \notin \mathfrak{n}$. Then the ideal I_a generated by the elements $\varrho_u \cdot a$ $(u \in M(R))$ is A.*

Proof. For $h \in A$, let $M_h = \{u \in M(R) \mid u(h) \neq 0\}$. It follows from (2) that

$$(3) \qquad\qquad M_{\varrho_u \cdot h} = M_h \cdot u^{-1}, \qquad (h \in A; u \in M(R))\,.$$

By our assumption, and the fact that $\mathfrak{n} = \ker v$, the set M_a is not empty. Since $M(R)$ is a group, (3) implies that the sets $M_{\varrho_u \cdot a}$ $(u \in M(R))$ form a covering of $M(R)$. Hence I_a has no zero on $M(R)$, i.e., I_a is not contained in any maximal ideal of A, whence the lemma.

4.3. From now on, let $G = G_\pi$ and $\mathbf{Z}[G]$ be as at the end of 3.4. The $\mathbf{Z}$-structure of $\mathbf{SL}_2$ will be the obvious one given by the coordinate ring $\mathbf{Z}[a, b, c, d]/(ad-bc-1)$ generated over $\mathbf{Z}$ by the coefficients of the identity representation of $\mathbf{SL}_2$.

The main result of [5] is then: for each prime p, the ring $\mathbf{Z}[G] \otimes K_p$ is a domain of integrity, $\mathrm{Mor}(\mathbf{Z}[G], K_p)$ may be canonically identified with the Chevalley group G_{π, K_p} and is a connected semi-simple group defined over $\mathbf{F}_p$. Moreover, the morphisms $\lambda_{\varrho, \pi}$ and $\mu_r (r \in \Phi)$ are morphisms of schemes over $\mathbf{Z}$.

This is equivalent to the following set of assertions:

(i) for each prime p, the ideal $p \cdot \mathbf{Z}[G]$ is prime in $\mathbf{Z}[G]$;

(ii) let (ϱ, V) be a representation of $\mathfrak{g}$ such that $\Gamma_\varrho \subset \Gamma_\pi$. Identify V to $\mathbf{C}^q$ $(q = \dim V)$ via a basis of an admissible $\mathbf{Z}$-form of V. Then $g \mapsto (\varrho(g))_{ij} (1 \leq i, j \leq q)$ is an element of $\mathbf{Z}[G]$.

(iii) For every $r \in \Phi$, the comorphism $\mu_{r,0}$ associated to μ_r maps $\mathbf{Z}[G]$ into $\mathbf{Z}[\mathbf{SL}_2]$.

The rest of § 4 is devoted to the proof of these results. We shall assume that π is a faithful representation of $\mathfrak{g}$, which is clearly no loss in generality.

4.4. We define a $\mathbf{Z}$-structure on H by letting $\mathbf{Z}[H]$ be the group ring of the character group $X(H) \cong \Gamma_\pi$ of H. Let l be the dimension of H and $(h'_j)_{1 \leq j \leq l}$ be a basis of $X(H)$. Then $\mathbf{Z}[H] = \otimes_j \mathbf{Z}[h'_j, h'^{-1}_j]$ where $\mathbf{Z}[T, T^{-1}]$ denotes the ring of polynomials, with negative exponents allowed, in the indeterminate T, and coefficients in $\mathbf{Z}$. If $\Gamma_\pi = \Gamma_{sc}$, then we may take for $\{h'_j\}$ the set $\{l_a\}_{a \in \Delta}$ of fundamental highest weights.

Let $\zeta_{\pm i}$ $(1 \leq i \leq m)$ be the elements of $\mathbf{Q}[U^\pm]$ such that, in the notation of 3.2:

$$v^\pm(\zeta_{\pm 1}(u), \ldots, \zeta_{\pm m}(u)) = u , \qquad (u \in U^\pm) ,$$

i.e., such that $u = \prod_i x_{\pm a_i}(\zeta_{\pm i}(u))$. We define a $\mathbf{Z}$-form of $\mathbf{Q}[U^\pm]$ by putting

$$\mathbf{Z}[U^\pm] = \mathbf{Z}[\zeta_{\pm 1}, \ldots, \zeta_{\pm m}] .$$

Obviously

$$(1) \qquad \mathbf{Z}[H] \otimes k = \prod_a k[h'_a, h'^{-1}_a], \quad \mathbf{Z}[U^\pm] \otimes k = k[\zeta_{\pm 1}, \ldots, \zeta_{\pm m}] .$$

Let $\Omega = U^- \cdot H \cdot U$ and $\varphi \colon U^- \times H \times U \to \Omega$ be the morphism $(x, y, z) \mapsto x \cdot y \cdot z$ $(x \in U^-, y \in H, z \in U)$. The set Ω is open in G, and φ is an isomorphism of complex manifolds. We shall soon see that Ω is Zariski-open in G. It will then follow that φ is in fact an isomorphism over Q of algebraic varieties, Ω being given the induced structure from that of G. Meanwhile, we define $\mathbf{Z}[\Omega]$ to be the image of $\mathbf{Z}[U^-] \otimes \mathbf{Z}[H] \otimes \mathbf{Z}[U]$ under the comorphism associated to φ.

Let (h''_j) be the basis of $\hat\Gamma_\pi$ dual to (h'_j), and identify it with a basis of $\mathfrak{h}$ by means of the isomorphism $\mathfrak{h} \cong \hat\Gamma_\pi \otimes_\mathbf{Z} \mathbf{C}$. Take as a basis (y_k) of the m-th exterior power $\Lambda^m \mathfrak{g}$ of $\mathfrak{g}$ the exterior products of m elements in $S = \{x_{\pm a_i}, h''_j, (i = 1, \ldots, m, j = 1, \ldots, l)\}$. Let $\sigma = \Lambda^m \mathrm{Ad}$ be the m-th exterior power of the adjoint representation Ad of G_π. Since S spans an admissible $\mathbf{Z}$-form of $\mathfrak{g}$ (2.6), the coefficients of Ad, and hence also those of σ, belong to $\mathbf{Z}[G]$. Take $y_1 = x_{a_1} \wedge \ldots \wedge x_{a_m}$ as the first basis vector, and let $d \colon g \mapsto (\sigma(g))_{11}$ be the corresponding matrix coefficient.

4.5. Lemma. *We have $d(e) = 1$. The set Ω is the set of points $G_{\pi, \mathbf{C}}$ on which d is not zero.* $\mathbf{Z}[\Omega] \cong \mathbf{Z}[G][d^{-1}]$.

This lemma is proved in ([5], p. 9, 10 and top of p. 11). For the sake of completeness, we summarize the argument. The first assertion is clear. Let q be the sum of the positive roots. Then

$$(1) \qquad d(u' \cdot h \cdot u) = h^q, \qquad (u' \in U^-; h \in H; u \in U) .$$

Let (n_w) be a set of representatives in the normalizer $N(H)$ of H of the cosets modulo H. If $n_w \notin H$, then there exists an index $r \neq 1$ such that $\sigma(n_w) \cdot y_1$ is a multiple of y_r, and y_r is a product of elements x_a, where a runs through a set ψ of roots such that

$$\psi \cap -\psi = \varphi, \ \psi \cup -\psi = \Phi , \qquad \psi \neq \Phi^+ .$$

Then, any $u' \in U^-$ transforms y_j onto a linear combination of y_i's which are all different from y_1. Hence

$$d(u' \cdot n_w \cdot h \cdot u) = 0 , \qquad (u' \in U^-; h \in H; u \in U; n_w \in N(H) - H) .$$

Since, by the Bruhat decomposition, $G_{\pi,\mathbf{C}} = \bigcup_w U^- \cdot n_w \cdot H \cdot U$, this proves the second assertion.

For $a \in \Phi^+$, and j between 1 and l, let $m(a,j)$ be the integer such that $y_{m(a,j)}$ is obtained from y_1 by replacing x_a by h_j''. Let moreover $c_{a,j}(1 \leq j \leq l)$ be the integers such that

$$h_a = \sum_{1 \leq j \leq l} c_{a,j}\, h_j'' \,.$$

If $a, b \in \Phi$ are not proportional, then, by 1.2

$$(2) \qquad\qquad \mathrm{Ad}\, x_a(t) \cdot x_b = x_b + t\, N_{ab} \cdot x_{a+b} + (t^2/2)\, N_{a,a+b}\, x_{2a+b} + \ldots$$

where $N_{a,ia+b} \in \mathbf{Z}$. By 2.6 (2):

$$(3) \qquad\qquad \mathrm{Ad}\, \mathrm{x}_a(t) \cdot x_{-a} = x_{-a} + t \cdot h_a - 2t \cdot x_a.$$

Fix $a \in \Phi^+$, and j between 1 and l. Let $s = u' \cdot h \cdot u\ (u' \in U^-;\ h \in H;\ u \in U)$. We have

$$\sigma(h \cdot u) \cdot y_1 = d(h \cdot u) \cdot y_1 = d(s) \cdot y_1$$

hence

$$\sigma(s) \cdot y_1 = d(s) \cdot \sigma(u') \cdot y_1\,.$$

If $b = \pm a_i$, we also write ζ_b for the above $\zeta_{\pm i}$. Using (2) and (3), one shows easily that there exist polynomials Q_a, Q_a' in indeterminates $Z_b(0 < b < a;\ b \in \Phi^+)$, such that

$$(4) \qquad\qquad \sigma(s)_{m(a,j),1} = d(s)\,(- \zeta_a \cdot c_{a,j} + Q_a((\zeta_{-b})_{0<b<a}))\,,$$

$$(5) \qquad\qquad \sigma(s)_{1,m(a,j)} = d(s)\,(- a(h_j'')) \cdot \zeta_a + Q_a'((\zeta_b)_{0<b<a})\,.$$

Now we have

$$2 = a(h_a) = \sum_{1 \leq j \leq l} c_{a,j} \cdot a(h_j)\,,$$

hence either the $c_{a,j}$ or the $a(h_j'')\ (1 \leq j \leq l)$ have greatest common divisor one. In the former (resp. latter) case, the relations (4) (resp. (5)) show by induction on a that the ζ_{-a} (resp. ζ_a), $(a \in \Phi^+)$, belong to $\mathbf{Z}[G][d^{-1}]$. However, starting from the product of the $x_a\ (a < 0)$, one gets in the same way relations similar to (4) and (5), in which U^- and U are permuted. It follows then that both ζ_a and ζ_{-a} belong to $\mathbf{Z}[G][d^{-1}]$.

4.6. Lemma. *The algebra $\mathbf{Z}[G] \otimes k$ is reduced.*

For the proof we may (and shall) assume k to be algebraically closed. Let $A = \mathbf{Z}[G] \otimes k$, and $\mathfrak{n}$ be the nilradical of A. Let $\bar{A} = A/\mathfrak{n}$ and $\tau\colon A \to \bar{A}$ the canonical morphism. From 4.3, it is clear that $\mathbf{Z}[\Omega] \otimes k$ is reduced. Hence, by 4.5, $\mathbf{Z}[G][d^{-1}] \otimes k = A[d^{-1} \otimes 1]$ is reduced. Consequently, the canonical morphism $A \to A[d^{-1} \otimes 1]$ annihilates $\mathfrak{n}$. By the definition of localization, this means that, given $x \in \mathfrak{n}$, there exists $q \in \mathbf{N}$ such that $x \cdot d^q = 0$. Since $\mathfrak{n}$ is finitely generated as an ideal we may choose q such that

$$\mathfrak{n} \cdot d^q = 0 .$$

The automorphisms ϱ_u ($u \in \mathrm{maxspec}\, A$) of 4.2 leave $\mathfrak{n}$ invariant. Hence $\mathfrak{v}$ is annihilated by the ideal I_h generated in A by all the transforms of $h = d^q$ under the automorphisms ϱ_u. But $d^q(e) = e_0(d^q) = (e_0(d))^q = 1$, hence $d^q \notin \mathfrak{n}$, and, by the lemma in 4.2, $L_h = A$, whence $\mathfrak{n} = 0$.

4.7. Let p be a prime. By 4.3 (1), $\mathbf{Z}[H] \otimes \mathbf{F}_p$ and $\mathbf{Z}[U^\pm] \otimes \mathbf{F}_p$ are domains of integrity, hence the reductions mod p of H and of $U^\pm$ are irreducible groups. The first one $H_{(p)}$ is a torus; the group $U^\pm_{(p)}$ is unipotent, and $H_{(p)}$ normalizes $U^\pm_{(p)}$ (since this property is preserved by reduction mod p). The groups $B^\pm_{(p)} = H_{(p)} \cdot U^\pm_{(p)}$ are then connected solvable subgroups of $G_{(p)}$. The Lie algebras of $H_{(p)}$ and $U^\pm_{(p)}$ are

(1)
$$L(H_{(p)}) = \mathfrak{h}_\mathbf{Z} \otimes_\mathbf{Z} K_p, \qquad L(U_{(p)}) = \mathfrak{u}^\pm_\mathbf{Z} \otimes_\mathbf{Z} K_p ,$$

hence

(2)
$$L(G_{(p)}) = \mathfrak{g}_\mathbf{Z} \otimes_\mathbf{Z} K_p = L(U^-_{(p)}) \otimes L(H_{(p)}) \otimes L(U_{(p)}) .$$

It follows from 4.4 (1) that $\chi \mapsto \chi \otimes 1$ defines an isomorphism of $X(H)$ onto $X(H_{(p)})$. We denote an element of $X(H)$ and its image in the same way. In particular $\Phi \subset X(H_{(p)})$, and in fact, $a \in \Phi$ is the character of $H_{(p)}$ in $\mathfrak{g}_{a,(p)} = \mathfrak{g}_{a,\mathbf{Z}} \otimes_\mathbf{Z} K_p$. We can also view the Weyl group $W(G)$ as a group of automorphisms of $X(H_{(p)})$.

Lemma. (i) *The automorphisms of $X(H_{(p)})$ defined by $W(G)$ are induced by inner automorphisms of the identity component $(G_{(p)})^0$ of $G_{(p)}$ leaving $H_{(p)}$ stable.*

(ii) *$U_{(p)}$ is a maximal connected unipotent subgroup of $(G_{(p)})^0$.*

In $\mathbf{SL}_2$ we have

(3)
$$\begin{pmatrix} 1 & 1 \\ 0 & 1 \end{pmatrix} \cdot \begin{pmatrix} 1 & 0 \\ -1 & 1 \end{pmatrix} \cdot \begin{pmatrix} 1 & 1 \\ 0 & 1 \end{pmatrix} = \begin{pmatrix} 0 & 1 \\ -1 & 0 \end{pmatrix} .$$

The elements of $x_a(t)$ ($t \in \mathbf{Z}$) leave stable the given admissible lattice in the representation space of π, hence define elements of $G_\pi(\mathbf{Z}) = G_\pi \cap \mathbf{SL}(n, \mathbf{Z})$. By (3), it follows then that $\mu_b \begin{pmatrix} 0 & 1 \\ -1 & 0 \end{pmatrix} \in G_\pi(\mathbf{Z})$. This element induces the reflection w_b on $X(H)$. By reduction mod p, $G_\pi(\mathbf{Z})$ is mapped homomorphically into $G_\pi(\mathbf{F}_p)$, whence (i).

To prove (ii) it suffices, by standard conjugacy theorems [1, § 11], to show that $U_{(p)}$ coincides with the unipotent radical V of some maximal connected solvable subgroup of $(G_{(p)})^0$ containing $H_{(p)} \cdot U_{(p)}$.

The Lie algebra $L(G_{(p)})$ of $G_{(p)}$ is direct sum of $L(H_{(p)})$ and of one-dimensional eigenspaces $\mathfrak{g}_{a,(p)}$ of $H_{(p)}$ corresponding to the roots. Since V is normalized by $H_{(p)}$, its Lie algebra $L(V)$ is direct sum of its intersections with the eigenspaces of $H_{(p)}$. But V is unipotent, therefore $L(V)$ consists of nilpotent elements, and $L(V) \cap L(H_{(p)})$

$= \{0\}$. Thus $L(V)$ is direct sum of some of the eigenspaces $\mathfrak{g}_{a,(p)}$. If $V \neq U_{(p)}$, then $L(V)$ contains $\mathfrak{g}_{b,(p)}$ for some $b \in \Phi$, $b < 0$. But then it contains the group $U_{b,(p)}$, where U_b is the group generated by the $x_b(t)$, $(t \in \mathbf{C})$. This follows from the uniqueness assertion of [1, Theorem in (13.18), part (d)], applied to the quotient of $(G_{(p)})^0$ by its unipotent radical. The proof of (i) now shows that the subgroup generated by $U_{b,(p)}$ and $U_{-b,(p)}$ contains an element n_b which normalizes, but does not centralize, $H_{(p)}$. But this is a contradiction, because $H_{(p)} \cdot V$ would be a connected solvable group in which the normalizer of a torus would differ from its centralizer, which is impossible [1; 10.6 (5)]. Hence $V = U_{(p)}$.

4.8. By reduction mod p, σ defines a representation of $G_{(p)}$ and the function d of 4.4 defines an element of $\mathbf{F}_p[G_{(p)}]$, which we also denote respectively σ, d. We claim that *d does not vanish identically on any irreducible component of* $G_{(p)}$.

Let C be an irreducible component of $G_{(p)}$ and let $x \in C$. Since, by 4.7, $U_{(p)}$ is maximal connected unipotent in $(G_{(p)})^0$, there exists $y \in (G_{(p)})^0$ such that $y \cdot x$ normalizes U [1; 11.3]. But then $\sigma(y \cdot x)(y_1)$ is a non-zero multiple of y_1, (notation of 4.5), hence $d(y \cdot x) \neq 0$.

4.9. Lemma. *We have* $\mathbf{Z}[G] = \mathbf{Q}[G] \cap \mathbf{Z}[\Omega]$. *The restriction map identifies* $\mathbf{Z}[G]$ *to a direct summand of* $\mathbf{Z}[\Omega]$.

(In this statement, and in the sequel, we make no notational distinction between $f \in \mathbf{Q}[G]$ and its restriction to Ω.)

In view of the definitions of the various $\mathbf{Z}$-structure involved here, it is clear that the restriction mapping sends $\mathbf{Z}[G]$ into $\mathbf{Z}[\Omega]$. Let then $f \in \mathbf{Z}[\Omega]$ be such that $\mathbf{Z} \cdot f \cap \mathbf{Z}[G] \neq \{0\}$, and a be the smallest positive integer for which $a \cdot f \in \mathbf{Z}[G]$. We have to prove that $a = 1$.

By 4.5, there exists a positive integer s such that $d^s \cdot f \in \mathbf{Z}[G]$. We have then $d_s \cdot a \cdot f \in a \cdot \mathbf{Z}[G]$. Assume $a \neq 1$, and let p be a prime dividing a. The canonical image of $d^s \cdot a \cdot f$ in $\mathbf{F}_p[G_{(p)}]$ is then zero. Since, by 4.6, d^s is not identically zero on any irreducible component of $G_{(p)}$, it follows that $a \cdot f$ is identically zero on $G_{(p)}$. But, by 4.6, $\mathbf{F}_p[G_{(p)}] = \mathbf{Z}[G] \otimes \mathbf{F}_p$. Since the kernel of $\mathbf{Z}[G] \to \mathbf{Z}[G] \otimes \mathbf{F}_p$ is $p \cdot \mathbf{Z}[G]$, we see that $a \cdot f \in p \cdot \mathbf{Z}[G]$, which then gives $(a/p) \cdot f \in \mathbf{Z}[G]$, (since $\mathbf{Z}[G]$ is torsion free, as an additive group). This contradicts the minimality of a, therefore $a = 1$.

This proves the second assertion of the lemma. The first one is an obvious consequence (in fact, it is elementary that both assertions are equivalent).

4.10. The proofs of 4.2 (i), (ii), (iii) are now immediate. The ring $\mathbf{Z}[G]$ being a direct summand of $\mathbf{Z}[\Omega]$, the natural map $\mathbf{Z}[G] \otimes K \to \mathbf{Z}[\Omega] \otimes k$ is injective. But $\mathbf{Z}[\Omega] \otimes k$ is a domain of integrity (4.3), hence so is $\mathbf{Z}[G] \otimes k$, and the kernel of $\mathbf{Z}[G] \to \mathbf{Z}[G] \otimes k$ is a prime ideal. For $k = \mathbf{F}_p$, this gives (i).

The coefficient function $g \mapsto \varrho(g)_{ij}$ is obviously an element of $\mathbf{Q}[G]$. In order to prove that it belongs to $\mathbf{Z}[G]$, it suffices, by 4.9, to show that its restriction to Ω belongs to $\mathbf{Z}[\Omega]$. Since the map $(x, y, z) \mapsto x \cdot y \cdot z$ induces an isomorphism of $\mathbf{Z}[U^-] \otimes \mathbf{Z}[H] \otimes \mathbf{Z}[U]$ onto $\mathbf{Z}[\Omega]$, it is then enough to prove that the restriction of $\varrho(g)_{ij}$ to $M = U^\pm$, H belongs to $\mathbf{Z}[M]$. For $M = U^\pm$ this is clear from the definitions (see 3.1, 3.4). By 2.3, we may assume that the chosen basis of V consists of eigenvectors of H. Then $\varrho(H)$ is diagonal. Its diagonal coefficients are $h \mapsto m(h)$, where m is a weight of ϱ. Since $\Gamma_\varrho \subset \Gamma_\pi$ and $\mathbf{Z}[H] = \mathbf{Z}[\Gamma_\pi]$, this gives $m \in \mathbf{Z}[H]$, and ends the proof of (ii).

Let $a \in \Phi$. It is clear from the definition of an admissible $\mathbf{Z}$-form that if (ϱ, V) is as in (ii), an admissible $\mathbf{Z}$-form for G in V is also admissible for $\mathbf{SL}_2$, with respect to the representation $\varrho \circ \mu_a$. Thus (iii) is a special case of (ii).

§ 5. Representations of Chevalley groups over algebraically closed fields

In this section we fix a Chevalley group $G = G_{\varrho, K}$, view it as an algebraic group, and want to determine its irreducible rational representations. The main results are due to Chevalley [4].

5.1. Let (π, E) be a rational representation of G. Then $\pi(H)$ can be put in diagonal form. If we let, for $m \in X(H)$

$$E_m = \{x \in E \mid \pi(h) \cdot x = m(h) \cdot x, \ (h \in H)\},$$

then $E = \oplus E_m$. The m's for which $E_m \neq 0$ are the weights of π. We let again $P(\pi)$ denote the set of weights of π. As in the remark (1) in 3.3, we see that if $n \in N(H)$ represents $w \in W$, then $\pi(n) \cdot E_m = E_{w(m)}$. In particular $P(\pi)$ *is stable under* W.

5.2. Lemma. *Let* $\pi \colon G \to GL(E)$ *be a rational representation, m a weight of π and* $v \in E_m$. *Let a be a root of G. Then there exist $e_i \in E_{m+ia}$ $(i = 0, 1, 2, \ldots)$, $e_0 = v$, such that* $\pi(x_a(t)) \cdot v = \sum_i t^i \cdot e_i (t \in K)$.

Let $h \in H$. Then $h \cdot x_a(t) \cdot h^{-1} = x_a(h^a \cdot t)$ and, if q is a weight, $\pi(h) \cdot x = h^q \cdot x \ (x \in E_q)$.

The map $t \mapsto \pi(x_a(t)) \cdot v$ is a regular function on K, with values in E, hence is a polynomial function on K and we may find $e_i \in E$ such that

$$\pi(x_a(t)) \cdot v = \sum_i t^i \cdot e_i.$$

The above relations imply

$$\sum_i t^i \cdot h^{ia} \cdot e_i = \sum_i h^{-m} \cdot t^i \cdot h(e_i).$$

Equating the coefficients of t^i for each i yields $e_i \in E_{m+ia}$ and, for $t = 0$, $e_0 = v$.

5.3. Theorem. *Let* (π, E) *be an irreducible rational representation of G.*

(i) There exists a unique line $D \subset E$ stable under B; the line D is the space of all vectors fixed under U; the corresponding weight l_π of H in D is dominant. All other weights m in E are of the form

$$(1) \qquad\qquad l_\pi - \sum_{a \in \Delta} c_a \cdot a, \qquad (c_a \in \mathbf{N}).$$

(ii) (π, E) is equivalent to (π', E') if and only if $l_\pi = l_{\pi'}$.

(iii) Every dominant weight l such that $\Gamma_l \subset \Gamma_\varrho$ is the highest weight of some irreducible representation of G.

(i) The set $P(\pi)$ is finite. There exists therefore $l \in P(\pi)$ such that $l + i \cdot a$ is not a weight for any $a \in \Delta$ and $i \geq 1$. The group W operates on $P(\pi)$, and we have, for $a \in \Phi$

$$w_a(m) = m - 2(a, m) \cdot (a, a)^{-1} \cdot a.$$

Consequently, $(l, a) \geq 0$ for $a \in \Delta$, hence l is dominant. Let $e_0 \in E_l - \{0\}$. Iterated application of 5.2 shows that

$$(2) \qquad\qquad U^- \cdot e_0 \subset e_0 + \sum E_m,$$

where m runs through a set P' of weights distinct from l, which have the form indicated in the theorem. We have $B \cdot e_0 = H \cdot e_0 = K \cdot e_0$, hence $U^- \cdot B \cdot e_0 = K \cdot U^- \cdot e_0$. But $U^- \cdot B$ is open, dense, in G, hence $U^- \cdot e_0$ spans the same vector space as $G \cdot e_0$, which is E by irreducibility. We get then from (2)

$$(3) \qquad\qquad E = K \cdot e_0 + \sum_{m \in P'} E_m.$$

This shows that E_l is one dimensional, that $P' = P(\pi)$, whence also (1) of (i) with $l_\pi = l$. To conclude the proof of (i), there remains to show that E_l is the space of all vectors fixed under U. Let F be that space. It is stable under H, hence direct sum of its intersections with the E_m's. If $F \neq E_l$, then F contains a vector $e_1 \neq 0$ of some weight $m_1 \neq l$. But then the above argument would prove that any other weight is of the form $m = m_1 - \sum c_a \cdot a$, $(c_a \in \mathbf{N})$, which is absurd.

In the sequel, l_π is called the *highest weight of* π, and an element of $E_l - \{0\}$ is a *highest weight vector of* π.

(ii) Let (π, E) and (π', E') be irreducible, with the same highest weight l. Let $F = E \oplus E'$, and $\sigma = \pi \oplus \pi'$. Fix highest weight vectors $e_0 \in E$, $e_0' \in E'$ and let $f = e_0 + e_0'$. Let E'' be the smallest invariant subspace containing f. The previous argument shows that E'' is spanned by $U^- \cdot f$ and that $E'' \cap (E_l + E_l') = K \cdot f$. Thus $E'' \cap E \neq E$, $E'' \cap E' \neq E'$ and, by irreducibility, $E'' \cap E = E'' \cap E = \{0\}$. The projection of E'' onto E (resp. E') is then an isomorphism of G-modules. Hence (π, E), (π', E') are equivalent to the restriction of σ to E''.

(iii) Given l, we first consider an irreducible representation (σ, V) of $\mathfrak{o}$ with highest weight l and let $V_{\mathbf{Z}}$ be an admissible $\mathbf{Z}$-form of V. We have then a Chevalley group $G_{\sigma, K} \subset GL(V_{\mathbf{Z}} \otimes K)$, and a canonical homomorphism $\lambda_{\varrho, \sigma}$ (3.3). The latter defines then a rational representation of G into $GL(V_{\mathbf{Z}} \otimes K)$. It need not be irreducible, however. Let V' be the smallest invariant subspace of $V_{\mathbf{Z}} \otimes K$ containing $(V_l \cap V_{\mathbf{Z}}) \otimes K$, and let V'' be a maximal proper invariant subspace of V'. Then V'/V'' is irreducible, and it contains a line stable under B, in which the weight of H is l. Hence, by (i), l is its highest weight.

5.4. Proposition. *Let* $l = \sum_{a \in \Delta} c_a \cdot l_a (c_a \in \mathbf{N})$ *be a dominant weight contained in* Γ_ϱ. *Let* (π_0, V) *(resp.* (π, E)*) be an irreducible representation of* $\mathfrak{g}$ *over* $\mathbf{C}$ *(resp. of G over K) with highest weight l. Then* $P(\pi) \subset P(\pi_0)$. *Let* $a \in \Delta$, *and* $i \in \mathbf{N}$ *be such that* $l - i \cdot a$ *is a weight. Then* $0 \leq i \leq c_a$.

As in the proof of 5.3 (iii), we have a canonical homomorphism $\lambda: G \to GL(V_{\mathbf{Z}} \otimes K)$, where $V_{\mathbf{Z}}$ is an admissible $\mathbf{Z}$-form of V. The $\mathbf{Z}$-form $V_{\mathbf{Z}}$ is direct sum of its intersections with the weight spaces V_m ($m \in P(\pi_0)$) (2.3), hence

$$V_{\mathbf{Z}} \otimes K = \oplus \, (V_{\mathbf{Z}} \cap V_m) \otimes_{\mathbf{Z}} K \,.$$

From (3.3) it follows that weight spaces of λ are the terms on the right hand side, and that the identification $\Gamma_\varrho \xrightarrow{\sim} X(H)$ maps m onto the weight of λ in $(V_{\mathbf{Z}} \otimes V_m) \otimes K$. Therefore $P(\pi_0) = P(\lambda)$. But the irreducible representation of weight l has been constructed in 5.3 (iii) as a quotient of a subspace of $V_{\mathbf{Z}} \otimes K$, whence the first assertion. Let $w_a \in W$ be the reflection associated to a. Clearly, $w_a(l) = l - c_a \cdot a$. Let $m = l - i \cdot a$ be a weight of π. Then $i \geq 0$ by 5.3 (i). Moreover, $w_a(m) = l - c_a \cdot a + i \cdot a$. Since $w_a(m)$ is also a weight π by 5.1, Theor. 5.3 (i) yields $i \leq c_a$.

5.5. To prove 5.3 (iii), we have made full use of the characteristic zero theory, and of the existence of admissible lattices. However, it is possible to give another proof purely in the context of algebraic groups. It allows one to construct in a canonical way an irreducible representation with a given highest weight. In the rest of this section, we outline this construction.

5.6. Let H be an algebraic group defined over k. It acts on the algebra $K[H]$ of regular functions on H by right translations. We let

$$r_h \cdot f(x) = f(x \cdot h) \,, \qquad (f \in K[H]; \ x, h \in H) \,.$$

This defines a representation to be denoted $(r, K[H])$. It is infinite dimensional (unless H is finite), however, every element belongs to a finite dimensional invariant subspace, and the restriction of r to any stable finite dimensional subspace is rational. In particular, every irreducible subspace is finite dimensional.

Let (π, E) be a rational representation of H. To $x \in E$, $y \in E^*$, we associate the function

$$c_{y,x}: h \mapsto y(\pi(h) \cdot x) \,, \qquad (h \in H) \,,$$

on H. It is regular. The $c_{y,x}$'s are the "coefficients" of π. Routine checking shows that

$$c_{y,\pi(h) \cdot x} = r_h \cdot c_{y,x}, \qquad (x \in E, y \in E^*, h \in H) \,.$$

Thus, for a fixed y, the map $\lambda_y: x \mapsto c_{y,x}$ is an equivariant morphism of E into $K[H]$. If π is irreducible, and $y \neq 0$, then $\mathrm{Im}\,\lambda_y \neq 0$ and λ_y is an isomorphism of G-modules of E onto $\lambda_y(E)$. In the general case, let $\{y_i\}_{1 \leq i \leq m}$ be a basis of E^*. Then

$$x \mapsto (c_{y_1,x}, \ldots, c_{y_m,x})$$

is an equivariant homomorphism of E into $K[H]^m$, which is clearly injective. As a consequence, every (finite dimensional) H-module is isomorphic to a submodule of the direct sum of a certain number of copies of $(r, K[H])$.

5.7. Let now π be irreducible, with highest weight l. Fix a highest weight vector e_0. We have the decomposition $E = E_l \oplus E'$, where E' is the sum of the weight spaces $E_m (m \in P(\pi), m \neq l)$. Let b be the linear form on E which is one on e_0 and is zero on E', and let $c_\pi = c_{b,e_0}$. We have then

$$(1) \qquad \pi(g) \cdot e_0 = c_\pi(g) \cdot e_0 \pmod{E'} .$$

This implies

$$(2) \qquad c_\pi(x \cdot y \cdot z) = l(x) \, c_\pi(y) \cdot l(z) , \qquad (x \in B^-; y \in G; z \in B) .$$

in fact,

$$\pi(x \cdot y \cdot z) \cdot e_0 = l(z) \cdot \pi(x \cdot y) \cdot e_0 = l(z) \cdot \pi(x) (c_\pi(y) \cdot e_0 + e'(y)) ,$$

with $e'(y) \in E'$. But 5.2 and 5.3 show that $\pi(x) \cdot e_0 = l(x) e_0 \bmod E'$ and that E' is stable under B^-, whence (2). In particular, we have

$$(3) \qquad c_\pi(x \cdot y) = l(y) , \qquad (x \in U^-; y \in B) .$$

Since c_π is regular, and $U^- \cdot B$ is dense, this determines c_π completely. In particular c_π depends only on l. We shall also denote it by c_l. The value of c_π on any element can also easily be described in terms of the Bruhat decomposition of G. Let

$$J = \{a \in \Delta \mid w_a(l) = l\} .$$

(This is the set of a's for which l_a has coefficient zero in l, written as sum of the l_a's.) Let $W(J)$ be the subgroup of W generated by the $w_a(a \in J)$. Then the parabolic subgroup $P_J = B \cdot W(J) \cdot B$ is easily seen to be the full stability group of E_l. Write the Bruhat decomposition of G in the form $G = \bigcup_w U^- \cdot w \cdot B$. Since $w(e_0) = e_0$ if $w \in W(J)$, and $w(e_0) \in E'$ if $w \notin W(J)$, we get, for $x \in U^-, y \in B$,

$$(4) \qquad \begin{aligned} c_\pi(x \cdot w \cdot y) &= l(y) , \qquad (w \in W(J)) \\ c_\pi(x \cdot w \cdot y) &= 0 , \qquad (w \notin W(J)) . \end{aligned}$$

By 5.6 we know that (π, E) is isomorphic to the restriction of r to the smallest invariant subspace of $(r, K[G])$ containing c_π.

5.8. Let l be a dominant weight. We put

$$F_l = \{f \in K[G] \mid f(h \cdot x) = l(h) \cdot f(x) , \qquad (h \in B^-, x \in G)\} .$$

This is a subspace invariant under right translations. It has a wellknown interpretation in terms of line bundles (which we shall not need however): Let $X = G/B^-$ and $q : G \to X$ the natural projection. The cross product $G \times_{B^-} K$, where K is viewed as a B^--module via l, is the total space of a line bundle ξ over X. Let σ be its projection map. To $f \in F_l$ we may assign a regular cross-section s of ξ by

$$s\,(q\,(g)) = \sigma\,(x, f\,(x))\,.$$

This map is known to be an isomorphism of F_l onto the space of regular cross-sections of ξ. In particular it is finite-dimensional, a fact we shall prove in a different way below.

5.9. Lemma. *Let $M \neq \{0\}$ be an irreducible subspace of F_l. Then M is the span of* $G(c_l)$.

M is finite dimensional, as was recalled in 5.6. Let z_0 be a highest weight vector and l_0 the highest weight of (r, M). Then

$$(1) \qquad\qquad z_0\,(x \cdot b) = l_0\,(b) \cdot z_0\,(x)\,, \qquad (x \in G; b \in B)\,.$$

But

$$z_0\,(e) = z_0\,(h \cdot e \cdot h^{-1}) = l\,(h) \cdot z_0\,(e) \cdot l_0\,(h)^{-1}, \qquad (h \in H)\,,$$
$$z_0\,(x \cdot b) = z_0\,(e) \cdot l_0\,(b)\,, \qquad (x \in U^-; b \in B)\,.$$

The last relation shows that $z_0\,(e) \neq 0$, since $z_0 \neq 0$. The previous one then gives $l = l_0$, hence $z_0 = z_0\,(e) \cdot c_l$ by (1). Since $z_0\,(e) \neq 0$, this yields $c_l \in M$, whence the lemma.

5.10. In order to get an independent proof of the existence of an irreducible representation with highest weight a given dominant weight l, it suffices therefore to show that $F_l \neq \{0\}$. We sketch the proof. Since l is a linear combination with coefficients in $\mathbf{N}$ of fundamental dominant weights, it suffices to consider the case where $l = l_a (a \in \varDelta)$ is fundamental. Let $J = \varDelta - \{a\}$. If f is a rational function on G, and f^m is regular for some $m \in \mathbf{N}$, then f is also regular (since G, being smooth, is normal). In view of 5.7 (4), it suffices then to show the existence of one irreducible representation (π, E) in which the highest weight line is stable under the maximal proper parabolic subgroup P_J. By a lemma of Chevalley ([4, Exp. 10], or [1, § 5.1]), there exists a rational representation (σ, V) of G and a line $D \subset V$ whose stability group is P_J. Let W be the smallest stable subspace containing D, and W' the quotient of W by a maximal proper stable subspace. Then the induced representation in W' has the required properties.

5.11. Consider the group G_ϱ with its canonical $\mathbf{Z}$-structure, which gives rise to $G_{\varrho,K}$ by reduction. We have $\mathbf{Z}[G_\varrho] \otimes K = K[G_{\varrho,K}]$. We note that $\mathbf{Z}[G_\varrho]$ is an admissible $\mathbf{Z}$-form of $\mathbf{C}[G_\varrho]$ for the representation defined by right (resp. left) translations. In fact, the identification of G_ϱ with a subgroup of $\mathbf{G\,L}\,(n, \mathbf{C})$ by means of an admissible $\mathbf{Z}$-form yields a homomorphism of $U\,(\mathfrak{g})$ into $\mathbf{M}\,(n, \mathbf{C})$ which maps $U_\mathbf{Z}$ into $\mathbf{M}\,(n, \mathbf{Z})$, and such that the action of $U\,(\mathfrak{g})$ on $\mathbf{C}[G_\varrho]$ is the transpose of the matrix multiplication on the right (resp. left). Since $U_\mathbf{Z}$ is represented by elements of $\mathbf{M}\,(n, \mathbf{Z})$, this action preserves $\mathbf{Z}[G_\varrho]$. Let l be a dominant weight and $F_{l,0}$ be the space F_l of 5.8, defined in $\mathbf{C}[G_\varrho]$. Then $F_{l,\mathbf{Z}} = F_{l,0} \cap \mathbf{Z}[G]$ is a direct summand. This follows easily from the fact that $\mathbf{Z}[G]$ is an admissible $\mathbf{Z}$-form of $\mathbf{C}[G]$ (for the representation defined by right translations) and from 2.3. From this one sees that $F_l = F_{l,\mathbf{Z}} \otimes K$ has the same dimension as $F_{l,0}$. Moreover, by full reducibility in characteristic zero, and 5.9, the representation of G in $F_{l,0}$ is the irreducible representation of highest weight l. $\dim F_l = \dim F_{l,0}$ is finite, the weights of G in F_l are independent of the characteristic, but the representation may be reducible for certain characteristics. As an example, consider the natural representation of $\mathbf{S\,L}_2$ in the m-th symmetric power of $\mathbf{C}^2$, say endowed with the admissible lattice of 2.7. If the prime p divides m, then the p-th powers of the elements of $\mathbf{C}^2$ form an invariant irreducible subspace.

5.12. Let p be a prime, (π, E) a representation of $\mathfrak{g}$ in E, and $E_{\mathbf{Z}}$ an admissible $\mathbf{Z}$-form of E. Then, we get a representation of G_{π, K_p} in $E_{\mathbf{Z}} \otimes K_p$. However, its equivalence class may vary with the admissible $\mathbf{Z}$-form. For instance, consider the adjoint representation of $\mathfrak{s}\,\mathfrak{l}_2$ in its Lie algebra. We have two admissible lattices $\mathfrak{g}_{sc}$ and $\mathfrak{g}_{ad}$ (see 2.6). If $p = 2$, the first (resp. second) one yields a non-fully reducible representation of $\mathbf{S\,L}(2, K_2)$ with a 1-dimensional (resp. 2-dimensional) invariant subspace.

If all weights of an irreducible representation in characteristic zero are extremal (i.e. are transforms of the highest weight under the Weyl group) then reduction mod p always yields an irreducible representation. To see this, note first that the weights of the reduction mod p are the same, modulo canonical identifications, as those of the given representation, with the same multiplicities (see 5.4 and its proof). Moreover, (4.5) the Weyl group may be identified with the Weyl group of the reduction mod p of the group, and this is compatible with the action on weights. It follows that the representation space in characteristic p is also sum of one-dimensional eigenspaces which are permuted transitively by the Weyl group, whence the irreducibility. As is knwon, every almost simple group with non-trivial center, not isomorphic to a group $\mathrm{Spin}\,(4m)$ $(m = 1, 2, \ldots)$, has such a representation.

5.13. Proposition. *Let p be the characteristic of K. Let (π, E) be a rational representation of $G_{\varrho, K}$. Then for each $a \in \Phi$ and $j \in \mathbf{N}$ there exists an endomorphism $X_{a,j}$ of E with the following properties:*

(a) $X_{a,j}$ is zero for all sufficiently large j, commutes with $X_{a,i}$ for all $i, j \in \mathbf{N}$, and is the identity for $j = 0$.

(b) $\pi(x_a(t)) = \sum_i t^i X_{a,i}, \qquad (a \in \Phi; t \in K)\,.$

(c) $X_{a,j}(E_m) \subset E_{m+j \cdot a}, \qquad (m \in P(\pi), a \in \Phi, j \in \mathbf{N})\,.$

(d) $X_{a,j} = (j!)^{-1}(d\pi(x_a)^j), \qquad (0 \leq j < p$ if $p \neq 0; j \in \mathbf{N}$ if $p = 0)\,.$

(e) If $p \neq 0$, $X_{a,p \cdot s+j} = X_{a,s \cdot p} \cdot X_{a,j}, \qquad (0 < j < p)\,.$

(f) $X_{a,j}$ stabilizes every G-invariant subspace of E $(a \in \Phi, j \in \mathbf{N})$.

It follows from 5.2 that there are endomorphisms $X_{a,j}$ verifying (b), (c). They are then unique. Let F be an invariant subspace. By the remark just made, we can choose $X'_{a,j}$ verifying (b), (c) in F. By the uniqueness, the restriction of $X_{a,j}$ to F has to coincide with $X'_{a,j}$, therefore $X_{a,j}$ satisfies (f). As a consequence, if the $X_{a,j}$ verify our conditions in E, they will do so in every stable subspace. Turning this around, we see that it suffices to show the existence of a representation (π', E') in which our conditions are fulfilled, and whose restriction to some invariant subspace is isomorphic to (π, E). By 5.6, (π, E) is isomorphic to a subrepresentation of $(r, K[G])^m$. But we have $K[G] = \mathbf{Z}[G] \otimes K$. Therefore, $K[G]$ is an increasing union of stable subspaces of the form $(M \cap \mathbf{Z}[G]) \otimes K$, where M is a subspace in $\mathbf{Q}[G]$, such that $M \otimes \mathbf{C}$ is stable under the complex group $G_{\varrho,\mathbf{C}}$ associated to the representation ϱ and the admissible lattice defining $G_{\varrho,K}$. We are therefore reduced to the case of a representation obtained starting from a representation (σ, V) of $\mathfrak{g}$ such that $\Gamma_\sigma \subset \Gamma_\varrho$, taking an admissible $\mathbf{Z}$-form $V_{\mathbf{Z}}$ of V, and letting $E = V_{\mathbf{Z}} \otimes K$ and π be the natural representation of $G_{\varrho,K}$ in E. In this case, however, our conditions are all fulfilled, because we have

$$(1) \qquad X_{a,j} = (x_a^j/j!) \otimes 1\,, \qquad (a \in \Phi, j \in \mathbf{N})\,.$$

Remark. R. Steinberg has pointed out the following simpler proof of (d), (e).

Comparing the coefficients of $t^i \cdot u^j$ in $x_a(t + u) = x_a(t) \cdot x_a(u)$, we get

$$(2) \qquad \binom{i+j}{i} \cdot X_{a,i+j} = X_{a,i} \cdot X_{a,j}\,, \qquad (a \in \Phi, i, j \in \mathbf{N})\,.$$

From this, (d) follows by induction on j, and (e) by substituting $i = p\,s$.

5.14. Lemma. *The operators $X_{a,j}$ of 5.13 verify the following commutation relations, where X_a stands for $X_{a,1}$ if $a \in \Phi$, for 0 otherwise.*

$$(1) \qquad X_{a,j} \cdot X_b = X_b \cdot X_{a,j} + \sum_{0 < i \leqq j} c_i \cdot X_{ia+b} \cdot X_{a,j-i}\,,$$

$(a, b \in \Phi, a \neq \pm b, j \in \mathbf{N}; c_i \in \mathbf{F}_p, (1 \leqq i \leqq j))$.

$$(2) \qquad X_{a,j} \cdot X_{-a} = X_{-a} \cdot X_{a,j} + h_a \cdot X_{a,j-1} - (j-1)\, X_{a,j-1}\,,$$

$(a \in \Phi; j \in \mathbf{N}; X_{a,i} = 0$ if $i < 0)$.

$$(3) \qquad X_{a,j} \cdot \bar{U}(\mathfrak{h}) = \bar{U}(\mathfrak{h}) \cdot X_{a,j}\,, \qquad (a \in \Phi; j \in \mathbf{N})\,.$$

Here again, it is enough to prove this in some extension of (π, E) and we may assume $E = V_{\mathbf{Z}} \otimes K$ as at the end of 5.13. But then, the lemma follows from 5.13 (1) and 1.4 (1) (2) (3).

§ 6. Infinitesimally irreducible representations

Let H be an algebraic group over k and $\pi\colon H \to GL(E)$ a rational representation. Its differential $d\pi$ defines a representation $L(H) \to \mathfrak{g}\,\mathfrak{l}(E)$ of the Lie algebra $L(H)$ of H. If $d\pi$ is irreducible, we say that π is infinitesimally irreducible. Every subspace invariant under H is obviously invariant under $L(H)$, hence, if π is infinitesimally irreducible, it is *a fortiori* irreducible. If H is connected, the converse is true in characteristic zero, but not necessarily otherwise. In fact, we

99

shall see that a Chevalley group has only finitely many infinitesimally irreducible representations, up to equivalence, in non-zero characteristic. We assume $p = \operatorname{char} k > 0$, and G to be "simply connected"(i.e. $G = G_{\varrho,K}$, with ϱ such that Γ_ϱ is the group of all weights, see 3.3 (4)).

We shall have to deal extensively with the Lie algebra of G, whereas the complex Lie algebra which gives rise to G will hardly occur anymore. To simplify, we shall now write $\mathfrak{g}$ for $L(G)$ and use the previous notation $\mathfrak{u}^\pm$, $\mathfrak{h}$, x_a, etc. to denote the objects in $\mathfrak{g}$ corresponding to those defined in § 1.

6.1. The Lie algebra $\mathfrak{g}$ of G is *restricted:* it has a p-power operation $x \mapsto x^{[p]}$, which, with $\mathfrak{g}$ identified to a subalgebra of $\mathfrak{gl}_n$, is given by $x^{[p]} = x^p$. In particular

$$(1) \qquad x_a^{[p]} = 0, \quad h_a^{[p]} = h_a, \qquad (a \in \varDelta) .$$

The *restricted universal enveloping algebra* $\bar{U}(\mathfrak{g})$ of $\mathfrak{g}$ is, by definition, the quotient of $U(\mathfrak{g})$ by the ideal generated by the elements $(x^{[p]} - x^p)$, $(x \in \mathfrak{g})$. The "restricted" analogue of the Poincaré-Birkhoff-Witt theorem shows that the monomials

$$(2) \qquad \left(\prod_i x_{-a_i}^{q_i} \right) \cdot \left(\prod_{b \in \varDelta} h_b^{s_b} \right) \cdot \left(\prod_i x_{a_i}^{t_i} \right), \qquad (0 \le q_i, s_b, t_i < p)$$

form a vector space basis of $\bar{U}(\mathfrak{g})$. (For all this, see [8].) In particular

$$(3) \qquad \bar{U}(\mathfrak{g}) = \bar{U}(\mathfrak{u}^-) \otimes \bar{U}(\mathfrak{h}) \otimes \bar{U}(\mathfrak{u}) ,$$

and, by (1):

$$(4) \qquad \bar{U}(\mathfrak{h}) = \bigotimes_{a \in \varDelta} K[h_a]/(h_a^p - h_a) .$$

6.2. Lemma. *Let (π, E) be an infinitesimally irreducible rational representation of G, and l its highest weight. Then*

(i) $E = \bar{U}(\mathfrak{u}^-) \cdot E_l$.
(ii) E_l *is the only subspace of E annihilated by* $\mathfrak{u}$.

The space E_l is annihilated by $\mathfrak{u}$, stable under $\mathfrak{h}$, hence is also stable under $\bar{U}(\mathfrak{u})$ and $\bar{U}(\mathfrak{h})$. Consequently, $\bar{U}(\mathfrak{u}^-) \cdot E = \bar{U}(\mathfrak{g}) \cdot E_l$ is stable under $\mathfrak{g}$, and not zero, whence (i).

Let F be the zero-space of $\mathfrak{u}$. It is stable under H, hence

$$F = \sum_{m \in P(\pi)} F_m, \qquad (F = E_m \cap F).$$

Let $F' = \sum_{m \ne l} F_m$. It is annihilated by $\mathfrak{u}$, stable under $\mathfrak{h}$, hence again

$$\bar{U}(\mathfrak{g}) \cdot F' = \bar{U}(\mathfrak{u}^-) \cdot F'.$$

But $\bar{U}(\mathfrak{u}^-)$ leaves the sum of the $E_m (m \ne l)$ stable, hence $U(\mathfrak{g}) \cdot F'$ is a proper invariant subspace, whence $F' = (0)$.

6.3. Definition. We recall that the dominant weights are linear combinations with coefficients in $\mathbf{N}$ of the fundamental dominant weights $l_a (a \in \varDelta)$ (2.1). We let $M(G)$ denote the set of classes of irreducible representations of G whose highest weight l is of the form $l = \sum_a c_a \cdot l_a$ with $0 \leq c_a < p$. There are p^r $(r = \mathrm{Card}\ \varDelta)$ such classes.

6.4. Theorem (Curtis [6]). *The elements of $M(G)$ are infinitesimally irreducible.*

Let (π, E) be a representation whose class is in $M(G)$ and $l = \sum c_a \cdot l_a$ its highest weight. Let e_0 be a highest weight vector. As usual we have

$$(1) \qquad X_{a,j} \cdot e_0 = 0 , \qquad (a \in \varPhi^+, j > 0).$$

By 5.4, $l - ia$ is not a weight if $i > c_a$. Since $c_a < p$, this shows

$$(2) \qquad X_{-a,j} \cdot e_0 = 0 , \qquad (a \in \varDelta; j \geq p) .$$

We now prove

$$(3) \qquad E = \bar{U}(\mathfrak{g}) \cdot e_0.$$

For this, it suffices to show $F = \bar{U}(\mathfrak{g}) \cdot e_0$ is stable under G. Since G is generated by the groups $x_{\pm a}(t)$ $(a \in \varDelta)$, (3.2 (3)) this, by 5.13, amounts to proving:

$$(4) \qquad X_{\pm a,j} \cdot v \subset F , \qquad (v \in F; a \in \varDelta, j \in \mathbf{N}) .$$

By (1), (2), this is true for $v = e_0$. On the other hand, $F = \bar{U}(\mathfrak{u}^-) \cdot e_0$. Using induction on the degree of basis elements of $\bar{U}(\mathfrak{u}^-)$, we are then reduced to showing that if $v \in F$ verifies (4), and if b is a positive root, then $x_{-b} \cdot v$ also verifies (4). Now this follows from the commutation rules (5.14).

There remains to show that if F is a non-zero subspace of E, invariant under $\bar{U}(\mathfrak{g})$, then $F = E$. By 5.13 (c), F contains a line annihilated by $\mathfrak{u}$. In view of (3) it suffices therefore to prove that E_l is the subspace of all vectors annihilated by $\mathfrak{u}$. Let P be the latter space. The Lie algebra $\mathfrak{u}$ is invariant under H, acting via the adjoint representation, therefore P is stable under H, and we may write

$$(5) \qquad P = E_l + \sum_{m \neq l} (P \cap E_m) .$$

It is clear, by induction on $j \in \mathbf{N}$, that 5.14 (1) can also be written

$$(6) \qquad X_b \cdot X_{a,j} = X_{a,j} \cdot X_b = \sum_{0 < i \leq j} c'_i, X_{a,j-i} \cdot X_{ia+b} ,$$

$(a, b \in \varPhi, a \neq \pm b; c'_i \in \mathbf{F}_p)$. From this it follows that

$$(7) \qquad X_{a,j}(P) \subset P , \qquad (a \in \varPhi^+; j \in \mathbf{N}) .$$

By the degree $d^0 a$ of a positive root is meant the sum of the coefficients of a, expressed as a linear combination of simple roots. The degree $d^0 m$ of a weight

$m = l - \sum_{a \in \Delta} m_a \cdot a$ is the sum of the m_a's. Assume now $E_l \neq P$. Using (5), choose $m \neq l$ of smallest possible degree such that $P \cap E_m \neq 0$, and let $x \in P \cap E_m$. We claim that

$$(8) \qquad\qquad X_{a,j} \cdot x = 0 \,, \qquad (a \in \Delta; j \geq 1) \,.$$

We have $X_{a,j}(E_m) \subset E_{m+j \cdot a}$, and $d^0(m + ja) = d^0 m - j d^0 a \; (a \in \Phi^+)$. Assume (8) to be false. Then, by our minimality assumption on $d^0 m$, we must have $m + ja = l$. But, by 5.4, and the condition on π, $j < p$, hence (5.13) $X_{a,j} = (j!)^{-1} X_a$, and consequently $X_{a,j} x = 0$ since $x \in P$. This contradiction proves (8). By 5.13, we see that x is fixed under all transformations $\pi(x_a(t)) \; (a \in \Delta, t \in K)$. However, these $x_a(t)$ generate U, as was stated in 3.2 (2). Thus x is fixed under U, hence $x \in E_l$ by 5.3, and therefore $P = E_l$.

Remark. The above proof of the second part is different from the original one. The change was suggested in part by some remarks of Steinberg.

6.5. We shall see in §7 that $M(G)$ is the set of all classes of infinitesimally irreducible representations of G. To conclude this section, we prove that the differentials of the elements of $M(G)$ yield all equivalence classes of irreducible representations of $\mathfrak{g}$ (as a *restricted* Lie algebra) (result due to Curtis [6]).

Let $\pi: \mathfrak{g} \to \mathfrak{gl}(E)$ be an irreducible restricted representation. By the Engel-Jacobson theorem ([8], Thm. 1′, p. 34), the space E_0 annihilated by $\mathfrak{u}$ is $\neq \{0\}$. It is clearly stable under $\mathfrak{h}$. We can then put $\mathfrak{h}$ there in triangular form and find an element $e_0 \neq 0$ stable under $\mathfrak{h}$. There exists then $l \in \mathfrak{g}^*$ such that

$$h \cdot e_0 = l(h) \cdot e_0 \,, \qquad (h \in \mathfrak{h}).$$

From 6.1 (3) and

$$\bar{U}(\mathfrak{u}) \cdot e_0 = 0, \qquad \bar{U}(\mathfrak{h}) \cdot e_0 \subset K \cdot e_0 \,,$$

we get again

$$E = \bar{U}(\mathfrak{g}) \cdot e_0 = \bar{U}(\mathfrak{u}^-) \cdot e_0 \,.$$

By the standard Lie algebra argument, it follows that E is spanned by eigenvectors of $\mathfrak{h}$, and that any other weight of $\mathfrak{h}$ is of the form

$$l - \sum_{a \in \Delta} c_a \cdot da \,, \qquad (c_a \in \mathbf{F}_p) \,,$$

where da is the differential of a. We note however that it cannot be asserted that l has multiplicity one, because it may happen that da is zero on $\mathfrak{h}$ for some, and sometimes even for all, $a \in \Delta$. Nevertheless, l is well determined by π, and will also be called its highest weight. Then one shows, exactly as in 5.3, that two irreducible representations are equivalent if and only if they have the same highest weight. Now the highest weight is determined by its values d_a on h_a, where a runs through Δ. Furthermore, since $h_a^p = h_a$, we have $d_a \in \mathbf{F}_p$. Thus we have an injective map from classes of irreducible representations to $(\mathbf{F}_p)^r$, and there are at most p^r such classes. On the other hand let (σ, V) *belong to an element of* $M(G)$, and let

$l = \sum d_a \cdot l_a$ be its highest weight $(0 \leq d_a < p)$. Then $dl(h_a) = d_a \cdot l_a$. Since σ is infinitesimally irreducible (6.4), we see that the above map is surjective, and also that the differentials of two representations of G belonging to different elements of $M(G)$ are inequivalent representations of $\mathfrak{g}$.

§ 7. The tensor product theorem

7.1. Let V_k be a vector space over k, k' a field and $\mu: k \to k'$ a homomorphism. Let $V_{k'} = V_k \otimes_k k'$ be the vector space over k' obtained from V_k by extension of the groundfield via μ. Let L be a group and $\pi: L \to GL(V_k)$ a linear representation. Then $x \mapsto \pi(x) \otimes 1$ is a representation of L in $GL(V_{k'})$, to be denoted π^μ. If we identify V_k with k^n, then $V_{k'}$ is canonically identified with k'^n, the matrix $\pi^\mu(x)$ is obtained by applying μ to the coefficients of $\pi(x)$.

In the sequel, we keep the assumptions of 6.1. Since we want to emphasize rational points we write G_K for G, and let G_K denote the Chevalley group $G_{\mathbb{Q},k}$.

7.2. Theorem (Steinberg [11]). *Let (π_i, E_i) $(1 \leq i \leq s)$ be infinitesimally irreducible representations of G_K which are defined over $\mathbf{F}_p$, and $\mu_1, \ldots, \mu_s$ be distinct homomorphisms of k into K.*

(i) $\pi = \pi_1^{\mu_1} \otimes \ldots \otimes \pi_s^{\mu_s}$ is an irreducible representation of G_k.

(ii) Let the π_i's be non-trivial, let (σ_i, F_i), $(1 \leq i \leq t)$ be nontrivial infinitesimally irreducible representations of G_K defined over $\mathbf{F}_p$ and $\nu_1, \ldots, \nu_t$ distinct homomorphisms of k into K such that $\sigma = \sigma_1^{\nu_1} \otimes \ldots \otimes \sigma_t^{\nu_t}$ and π are equivalent representations of G_k. Then $s = t$, and, for a suitable permutation $i \mapsto \alpha(i)$ of the first s integers, $\mu_i = \nu_{\alpha(i)}$ and π_i is equivalent to $\sigma_{\alpha(i)}$, as a representation of G_K.

Let l_i be the highest weight of $\pi_i (1 \leq i \leq s)$. Let $l = \sum l_i$ and $E = E_1 \otimes \ldots \otimes E_s$. Clearly, a weight m of $\pi_0 = \pi_1 \otimes \ldots \otimes \pi_s$ is a sum $m_1 + \ldots + m_s$ $(m_i \in P(\pi_i))$. In particular, each m is of the form $m = l - \sum c_a \cdot a$ $(a \in \Delta, c_a \in \mathbf{N})$. We let again $d^0 m = \sum c_a$, and, for $d \in \mathbf{N}$, we put

$$(1) \qquad \hat{E}_d = \sum_{d^0 m = d} E_m, \qquad \hat{E}_{(d)} = \sum_{d' \geq d} \hat{E}_{d'}.$$

Then, taking 5.3 into account, we have

$$(2) \qquad E = \sum_d \hat{E}_d, \qquad \hat{E}_0 = E_{l_1} \otimes \ldots \otimes E_{l_s}.$$

An element $x \in \sum_{d' \leq d} \hat{E}_{d'}$, but not in $\sum_{d' < d} \hat{E}_{d'}$, will be said to have degree d.

In order to avoid confusion, we write the decomposition 5.13 of $\pi_i(x_a(t))$ as

$$(3) \qquad \pi_i(x_a(t)) = \sum_{j \geq 0} X_{a,j,i} \cdot t^j, \qquad (a \in \Phi; 1 \leq i \leq s).$$

Let then

$$X_{a,j}^{(i)} = 1 \otimes \ldots \otimes 1 \otimes X_{a,j,i} \otimes 1 \otimes \ldots \otimes 1,$$

$$(4)$$

$$X_a^{(i)} = X_{a,1}^{(i)}, \qquad (a \in \Phi, j \in \mathbf{N}, 1 \leq i \leq s).$$

Obviously, if $a > 0$ (resp. $a < 0$), $X_{a,j}^{(i)}$ decreases (resp. increases) the degree by $j \cdot d^0 a$ (resp. $j \cdot d^0(-a)$). Since π_i is infinitesimally irreducible we have $E_i = \bar{U}(\mathfrak{u}^-) \cdot E_{l_i}$, and consequently

$$(5) \qquad \hat{E}_d = \sum_{1 \leq i \leq s} \sum_{a > 0} X_{-a}^{(i)} \cdot \hat{E}_{d - d^0 a}.$$

We choose in E_i a basis rational over the prime field. (Since π_i is defined over $\mathbf{F}_p$, the space E_i is understood to have an $\mathbf{F}_p$-structure.) Then the transformations $X_{a,j,i}$ are represented by matrices with coefficients in the prime field. We have then, for $a \in \Phi$, $t \in k$:

$$\pi(x_a(t)) = \bigotimes_{1 \leq i \leq s} \mu_i \left(\sum_{j \geq 0} X_{a,j,i} \cdot t^j \right) = \bigotimes_{1 \leq i \leq s} \left(\sum_{j \geq 0} X_{a,j,i} \mu_i(t^j) \right),$$

and therefore, using the notation of (4),

$$(6) \qquad \pi(x_a(t)) = 1 + \sum_{1 \leq i \leq s} \mu_i(t) \cdot X_a^{(i)} + y_a(t),$$

where $y_a(t)$ is a sum of terms each of which lowers (resp. increases) the degree by at least $2 \cdot d^0 a$ (resp. $2 \cdot d^0(-a)$) if $a > 0$ (resp. $a < 0$).

The proof of (i) will be divided into three steps:

(a) To prove: *the space F annihilated by the $X_a^{(i)} (a > 0)$ is $\hat{E}_0$*. Fix an index i $(1 \leq i \leq s)$. Let $x \in F$. We can write $x = \sum_{j \geq 1} u_j \otimes v_j$ with $v_j \in E_i$ and u_j in the product of the $E_{i'}(i' \neq i)$. Moreover, we may assume the U_j's to be linearly independent. We have then

$$0 = X_a^{(i)} \cdot x = \sum_j u_j \otimes X_{a,1,i} v_j = 0,$$

whence $X_{a,1,i} \cdot v_j = 0$ $(a > 0)$, and, by 6.2, $v_j \in E_{l_i}$. This implies (a) *immediately.*

(b) To prove: *let $F \neq \{0\}$ be a subspace of E stable under G_k. Then $\hat{E}_0 \subset F$.* Let x be a non-zero element of F which has the least possible degree. Assume, contrary to what we have to prove, that this degree d is $\neq 0$. We can write $x = x_d + x_{d-1} + \ldots (x_{d'} \in \hat{E}_{d'}, 0 \leq d' \leq d)$. The element $\pi(x_a(t)) x - x$ belongs to F for every $a \in \Phi$, $t \in k$. However, by (6), we have

$$\pi(x_a(t)) \cdot x - x = \sum_i \mu_i(t) \cdot X_a^{(i)} \cdot x + y_a(t) \cdot x,$$

and, for $a > 0$, $X_a^{(i)} \cdot x$ has degree at most $d - d^0 a$, while $y_a(t)$ has degree at most $d - 2 \cdot d^0 a$. By our assumption, the right hand side is zero. Therefore its homogeneous component of degree $d - d^0 a$ is zero. But the latter is

$$\sum_{1 \leq i \leq s} \mu_i(t) \cdot X_a^{(i)} \cdot x_d.$$

By Dedekind's theorem, the homomorphisms $\mu_i: k \to K$ are linearly independent over K, hence we get $X_a^{(i)} \cdot x_d = 0$ for $a > 0$, and $i = 1, \ldots, s$. By (a), this gives $d = 0$, a contradiction.

104

(c) Let F be a non-zero subspace of E, invariant under G_k. We have to prove that $F = E$. Since $\hat{E}_{(d)} = 0$ for large enough d, it is enough to prove that $\hat{E}_d \subset F + \hat{E}_{(d+1)}$ for all $d \geq 0$. For $d = 0$, this follows from (b). Assume it has been established up to $d - 1$. By (5), our assertion amounts to

$$(7) \qquad X_{-a}^{(i)} \cdot \hat{E}_{d - d^\circ a} \subset F + \hat{E}_{(d+1)}, \qquad (a > 0, 1 \leq i \leq s) .$$

The induction assumption gives

$$\hat{E}_{d - d^\circ a} \subset F + \hat{E}_{(d - d^\circ a + 1)} ,$$

which can also be written

$$\hat{E}_{d - d^\circ a} \subset (F \cap \hat{E}_{(d - d^\circ a)}) + \hat{E}_{(d - d^\circ a + 1)} .$$

Since $X_{-a}^{(i)} \cdot \hat{E}_{(d')} \subset \hat{E}_{(d' + d^\circ a)}$ for any d' and any $a \in \Phi^+$, (7) will be a consequence of

$$(8) \qquad X_{-a}^{(i)} \cdot (F \cap \hat{E}_{(d - d^\circ a)}) \subset F + \hat{E}_{(d+1)}, \qquad (a > 0) .$$

Let $a \in \Phi^+$ and $x \in F \cap \hat{E}_{(d - d^\circ a)}$. By (6), we have

$$\pi(x_{-a}(t)) \cdot x = x + \sum_i \mu_i(t)\, X_{-a}^{(i)} x , \qquad (\bmod\ E_{d+1}) ,$$

hence

$$\sum_i \mu_i(t) \cdot X_{-a}^{(i)} \cdot x \in F + \hat{E}_{(d+1)}, \qquad (t \in k) .$$

By Dedekind's theorem, this yields $X_{-a}^{(i)} \cdot x \in F + \hat{E}_{(d+1)}$, and proves (8), and hence (i).

(ii) Write $l_i = \sum_{a \in \Delta} c_{i,a}\, l_a$. Since π_i is infinitesimally irreducible, we have $0 \leq c_{i,a} < p$, by 6.5, and, since π_i is not trivial, one of the $c_{i,a}$ at least is $\neq 0$. If e_i is a highest weight vector in E_i, we have (see 3.2(1)):

$$\pi_i(h_a(t)) \cdot e_i = t^{c_{i,a} \cdot} e_i, \qquad (t \in k^*, a \in \Delta) .$$

Since e_i may be taken rational over the prime field, we have then (7.1):

$$\pi_i(h_a(t)) \cdot e_i = \mu_i(t^{c_{i,a}}) \cdot e_i , \qquad (t \in k^*) .$$

For the vector $e_0 = e_1 \otimes \ldots \otimes e_s$, this yields

$$\pi(h_a(t)) \cdot e_0 = \left(\prod_i \mu_i(t^{c_{i,a}}) \right) \cdot e_0, \qquad (a \in \Delta; t \in k^*) .$$

The vector e_0 spans the unique line which is pointwise fixed under all $\pi(x_a(t))$, $(a > 0, t \in k)$, by (a) above. In order to prove (ii), it suffices therefore to show that the homomorphism

$$\text{(9)} \qquad\qquad t \mapsto \prod_i \mu_i(t^{c_{i,a}})$$

of k^* into K^* characterizes the μ_i and the $c_{i,a}$ uniquely. Changing the notation, we shall write t^{μ_i} for $\mu_i(t)$ and then $t^{\sum c_{i,a}\cdot\mu_i}$ for the right hand side of (9). Our assertion then follows from the following lemma:

Lemma [11]. *Let $\alpha_1, \ldots, \alpha_q$ be distinct homomorphisms of k into K. Then the p^q homomorphisms $t \mapsto t^{\sum m_i \cdot \alpha_i}$ of k^* into K^* $(0 \leq m_i < p; 1 \leq i \leq r)$ are linearly independent over K.*

Let C be the set of such homomorphisms. For an element $c = \sum m_i \cdot \alpha_i$ of C, let the degree $d^0 c$ be equal to the sum of the m_i's. Order the elements of C lexicographically. If the lemma is false, consider a relation in which the maximum degree is as small as possible, and among those, take one with the smallest possible number of terms. Write it

$$\text{(10)} \qquad\qquad t^c = \sum_{d<c} r_d \cdot t^d, \qquad (t \in k^*) .$$

We have $(t \cdot u)^c = \sum r_d(t \cdot u)^d$, hence $u^c = \sum r_d t^{d-c} \cdot u^d$. If $t^{d-c} \neq 1$ for some $t \in k^*$ and some d, then, subtracting from (10) we get a relation with strictly less terms. Hence (10) has the form $t^d = t^c$. If the degree of c is ≥ 2, then, by subtracting $t^c = d^d$ from $(t + u)^c = (t + u)^d$, we get readily a relation of strictly lower degree, whence $c = \alpha_i$ for some i. Since $c > d$, it follows that d involves only the a_j with $j > i$. Similarly, we seen then that $d = \alpha_j$ for some $j > i$, whence $\alpha_i = \alpha_j \, (i \neq j)$, a contradiction.

7.3. Corollary (Curtis [7]). *If (π, E) is infinitesimally irreducible, then its restriction to G_k is irreducible.*

The important point here is that k may be arbitrary. In particular it may be finite.

7.4. In the sequel, we let Fr be the Frobenius homomorphism $x \mapsto x^p$ of a field of characteristic p, and Fr^i be the i-th power of $\mathrm{Fr}\,(i \in \mathbf{N})$.

Let π be a representation of G_K defined over the prime field $\mathbf{F}_p$. Then we may assume $\pi(H)$ to be diagonal with respect to a base rational over $\mathbf{F}_p$. To obtain $\pi^{\mathrm{Fr}^i}(g)$ from $\pi(g)$, we raise the matrix coefficients to the p^i-st power. As a consequence of these two facts, we see that $m \mapsto p^i \cdot m$ is a bijection of $P(\pi)$ onto $P(\pi^{\mathrm{Fr}^i})$. In particular, if π is irreducible, with highest weight l, then π^{Fr^i} is irreducible, with highest weight $p^i \cdot l$. If $i > 0$, the differentials of the weights of π^{Fr^i} are then all zero; in fact, more generally, the differential of π^{Fr^i} is identically zero.

7.5. Theorem (Steinberg [11]). *Let $M(G)$ be as in 6.3.*

(i) *If $\pi_i \in M(G)$ $(0 \leq i \leq s)$, π_s not trivial, then $\pi_0 \otimes \pi_1^{\mathrm{Fr}} \otimes \ldots \otimes \pi_s^{\mathrm{Fr}^s}$ is an irreducible rational representation of G_K. Every rational irreducible representation is equivalent to one and only one of these.*

(ii) *If k is finite, with p^c elements, those representations in* (i) *for which $s < c$ are inequivalent irreducible representations of G_k.*

(iii) $M(G)$ *is the set of all classes of infinitesimally irreducible rational representations of G_K.*

(i) The representations π_i are infinitesimally irreducible by 6.4, hence the first assertion follows from 7.2.

Let now (π, E) be an irreducible representation of G_K, and $l = \sum_a c_a l_a$ its highest weight. Replacing each c_a by its p-adic expansion

$$c_a = \sum_i c_{a,i} \cdot p^i, \qquad (c_{a,i} \in \mathbf{N}, 0 \leqq c_{a,i} < p),$$

we get

$$l = \sum_{i \leqq 0} p^i \cdot l_i, \quad \text{with} \quad l_i = \sum_a c_{a,i} l_a.$$

Let π_i be an irreducible representation with highest weight l_i. Then its class $[\pi_i]$ belongs to $M(G)$, hence $\pi' = \otimes_i \pi_i^{\mathrm{Fr}^i}$ is irreducible. But it follows from 7.4 that its highest weight is l, hence π' is equivalent to π. It is also clear from the above discussion of the highest weight that two such representation, corresponding to different sequences $\{\pi_i\}$, have distinct highest weights. Thus (i) is proved.

(ii) If k has p^c elements, then the homomorphisms Fr^i $(0 \leqq i < c)$ are distinct homomorphisms of k into K, and our assertion follows from 7.2 and 6.4.

(iii) Let (π, E) be irreducible, $\pi \notin M(G)$. By (i), we can write $\pi = \pi_1 \otimes \pi_2^{\mathrm{Fr}}$, with $\pi_1 \in M(G)$ and π_2 a non-trivial representation of the type considered in (i). Let $E = E_1 \otimes E_2$ be the corresponding decomposition of E. For $x \in \mathfrak{g}$, we have

$$d\pi(x) = d\pi_1(x) \otimes 1 + 1 \otimes d\pi_2^{\mathrm{Fr}}(x).$$

But the second term is zero, therefore,

$$d\pi(x)\,(u \otimes v) = (d\pi_1(x) \cdot u) \otimes v, \qquad (u \in E_1, v \in E_2; x \in \mathfrak{o}),$$

whence

$$d\pi(\mathfrak{g})\,(u \otimes v) \subset E_1 \otimes v,$$

which shows that the representation $d\pi$ of $\mathfrak{g}$ is not irreducible. Since on the other hand the elements of $M(G)$ are infinitesimally irreducible (6.4), this proves (iii).

Remark. When k is finite, Steinberg has also shown in [11] that every irreducible representation of G_k is equivalent to one of the $p^{cr} = (\mathrm{Card}\,k)^r$ representations mentioned in (ii). A more general result will be proved by Curtis in Part B.

7.6. In this last section, we assume that either k is of characteristic two, and G of one of the types $\mathbf{B}_n$, $\mathbf{C}_n$, $\mathbf{F}_4$, or k is of characteristic three and G of type $\mathbf{G}_2$. In these cases G has roots of two different lengths, and we have a partition $\Delta = \Delta' \cup \Delta''$, where Δ' (resp. Δ'') is the set of short (resp. long) simple roots. Let $M'(G)$ (resp. $M''(G)$) be the set of $\pi \in M(G)$ whose, highest weight is a linear combination of the fundamental weights l_a ($a \in \Delta'$; resp. $a \in \Delta''$). Given $\pi \in M(G)$, there are unique elements $\pi' \in M'(G)$, $\pi'' \in M''(G)$ such that $l_\pi = l'_{\pi'} + l_{\pi''}$. Steinberg has shown that

$$(1) \qquad\qquad \pi = \pi' \otimes \pi''.$$

Thus, in the cases under consideration here, there is a further splitting of the elements of $M(G)$, and the irreducible representations of G can all be expressed as tensor products of elements in $M'(G) \cup M''(G)$.

The equality (1) was proved first in [11], and then, in a simpler way, in ([12], Cor. to Thm. 41, p. 218). We reproduce the latter argument.

In view of 5.3, it suffices to show that $\pi' \otimes \pi''$ is irreducible. Let Φ^* be the root system dual to $\Phi = \Phi(G)$, and $a \mapsto a^*$ the natural bijection of Φ onto Φ^*. Let G^* be a simply connected Chevalley group over k, with root system Φ^*. There exists a k-isogeny $\lambda \colon G^* \to G$ such that

$$(2) \qquad \begin{aligned} {}^t\lambda(a) &= p \cdot a^*, \quad (a \in \Delta''; p = \operatorname{char} k), \\ {}^t\lambda(a) &= a^*, \qquad (a \in \Delta'). \end{aligned}$$

(See [4; Exp. 23], or [12, § 10].) We have to show that $\pi' \otimes \pi''$ is irreducible. This is equivalent to proving that the representation $(\pi' \circ \lambda) \otimes (\pi'' \circ \lambda)$ of G^* is irreducible. But (2) implies that $\pi' \circ \lambda \in M'(G^*)$ and that the highest weight of $\pi' \circ \lambda$ is the product by p of the highest weight of an element of $M''(G^*)$. The irreducibility of $(\pi' \circ \lambda) \otimes (\pi'' \circ \lambda)$ then follows from 7.5 (i), applied to G^*.

References

1. Borel, A.: Linear algebraic groups, Benjamin Lecture Notes Series, New York 1969
2. Bourbaki, N.: Groupes et algèbres de Lie, Chapitre I, Algèbres de Lie, Act. Sci. Ind. 1285, Hermann Paris 1960
3. Chevalley, C.: Sur certains groupes simples, Tohoku Math. J. (2) 7 (1955), 14–66
4. Chevalley, C.: Classification des groupes de Lie algébriques, Notes polycopiées, Inst. H. Poincaré, Paris (1956-58)
5. Chevalley, C.: Certains schémas de groupes semi-simples. Sém. Bourbaki, 13è année, (1960-61), Exp. 219
6. Curtis, C. W.: Representations of Lie algebras of classical type with applications to linear groups, J. Math. Mech. 9 (1960), 307–326.
7. Curtis, C. W.: On projective representations of certain finite groups, Proc. A.M.S. 11 (1960), 852–860
8. Jacobson, N.: Lie algebras, Intersc. Tracts in pure and applied math. 10, Interscience Publ. New York 1962
9. Kostant, B.: Groups over **Z**, Algebraic groups and discontinuous subgroups. Proc. Symp. pure math. 9, A.M.S. Providence, R.I. (1966)
10. Serre, J.-P.: Algèbres de Lie semi-simples complexes, Benjamin, New York 1966
11. Steinberg, R.: Representations of algebraic groups, Nagoya M. J. 22 (1963), 33–56
12. Steinberg, R.: Lectures on Chevalley groups, Notes by J. Faulkner and R. Wilson, Yale University (1967)
13. Tits, J.: Algebraic and abstract simple groups, Annals of Math. (2) 80 (1964), 313–329

90.

(avec J-P. Serre)

Adjonction de coins aux espaces symétriques;
Applications à la cohomologie des groupes arithmétiques

C. R. Acad. Sci., Paris **271** (1970) 1156–1158

Soit G un groupe algébrique linéaire réductif connexe sur $\mathbf{Q}$, sans caractère non trivial, et soit $G(\mathbf{R})$ [resp. $G(\mathbf{Q})$] le groupe de ses points réels (resp. rationnels). On associe à G une *variété à coins* [1] $\overline{X}$ dont l'intérieur X est un espace homogène à droite de $G(\mathbf{R})$, isomorphe au quotient de $G(\mathbf{R})$ par un sous-groupe compact maximal. Tout sous-groupe arithmétique Γ de $G(\mathbf{Q})$ opère proprement sur $\overline{X}$, et le quotient $\overline{X}/\Gamma$ est compact. Si Γ est sans torsion, sa cohomologie s'identifie à celle de $\overline{X}/\Gamma$ et vérifie une certaine *formule de dualité*; en particulier, la dimension cohomologique de Γ est $d - l$, où $d = \dim (X)$ et $l = \mathrm{rg}_{\mathbf{Q}}(G)$.

1. ADJONCTION DE COINS A L'ESPACE SYMÉTRIQUE X. — Soient G et X comme ci-dessus et soit $\mathfrak{P}$ l'ensemble des sous-groupes paraboliques de G (définis sur $\mathbf{Q}$).

On peut plonger X dans une variété à coins $\overline{X}$ jouissant des propriétés suivantes :

(i) $\overline{X}$ est séparée, dénombrable à l'infini, contractile.

(ii) Si $\partial\overline{X}$ désigne le bord de $\overline{X}$, on a $X = \overline{X} - \partial\overline{X}$.

(iii) A tout $P \in \mathfrak{P}$ est attachée une partie e_P de $\overline{X}$, et $\overline{X}$ est réunion disjointe des e_P, $P \in \mathfrak{P}$. On a $e_G = X$.

(iv) Si $P \in \mathfrak{P}$, l'adhérence de e_P dans $\overline{X}$ est réunion des e_Q pour $Q \subset P$; c'est une sous-variété à coins de $\overline{X}$; elle est contractile.

(v) Si $P \in \mathfrak{P}$, la réunion $X(P)$ des e_Q, pour $Q \supset P$, est une sous-variété ouverte de $\overline{X}$.

La construction de $\overline{X}$ se fait en définissant directement les $X(P)$ et en les recollant suivant leurs intersections [2]; les propriétés (i) à (v) sont alors faciles à vérifier [le fait que $\overline{X}$ soit *séparée* résulte de la proposition 12.6 de [3]].

Indiquons brièvement comment on définit $X(P)$; pour simplifier, nous nous bornons au cas où P est un sous-groupe parabolique *minimal* (le cas général est analogue) :

Soient N le radical unipotent de P et S le tore déployé maximal de P/N. Soient R le système de racines de G par rapport à un relèvement de S dans P [4], et $\{\alpha_1, \ldots, \alpha_l\}$ la base de R définie par N. Le groupe de Lie réel $S(\mathbf{R})$ est isomorphe à $(\mathbf{R}^*)^l$; soit A sa composante neutre. Les α_i définissent un isomorphisme de A sur $(\mathbf{R}_+^*)^l$, ce qui permet de faire opérer A sur la variété à coins $\overline{A} = (\mathbf{R}_+)^l = (\mathbf{R}_+^* \cup \{o\})^l$. Le groupe A opère également sur X : si $x \in X$, il existe un unique sous-groupe de Lie A_x de $P(\mathbf{R})$ qui soit stable par l'involution de Cartan [5] associée au stabilisateur de x, et qui s'applique isomorphiquement sur A par la projection

109

$P(\mathbf{R}) \to P(\mathbf{R})/N(\mathbf{R})$; si $a \in A$, notons a_x l'élément correspondant de A_x; le transformé de x par a est défini comme $x.a_x$. L'action de A sur X ainsi obtenue est appelée *l'action géodésique* de A; elle commute à l'action naturelle de $P(\mathbf{R})$ sur X; de plus, elle fait de X un *espace fibré principal de groupe structural A*. L'espace fibré *associé* $X \times^A \overline{A}$, de fibre type $\overline{A}$, est la variété à coins $X(P)$ cherchée; en particulier, e_P s'identifie au quotient X/A de X par l'action géodésique de A.

THÉORÈME 1. — *Le bord $\partial\overline{X}$ de $\overline{X}$ a le même type d'homotopie que l'immeuble de Tits* $T(\mathfrak{P})$ *défini par* $\mathfrak{P}$.

[Rappelons ([6]) que $T(\mathfrak{P})$ est un complexe simplicial dont les faces σ_P correspondent bijectivement aux éléments P de $\mathfrak{P}$ distincts de G; on a $\sigma_P \supset \sigma_Q$ si et seulement si Q contient P. Le groupe $G(\mathbf{Q})$ opère simplicialement sur $T(\mathfrak{P})$, et le stabilisateur de σ_P est $P(\mathbf{Q})$.]

COROLLAIRE 1. — *La variété $\partial\overline{X}$ a même type d'homotopie qu'un bouquet de sphères de dimension $l-1$, où $l = \mathrm{rg}_\mathbf{Q} G$.*

Cela résulte d'un théorème de Solomon-Tits ([7]), appliqué à $T(\mathfrak{P})$.

COROLLAIRE 2. — *Soit $d = \dim(X)$. Les groupes de cohomologie à supports compacts $H_c^q(\overline{X}, \mathbf{Z})$ sont nuls pour $q \neq d-l$; le groupe $I = H_c^{d-l}(\overline{X}, \mathbf{Z})$ est libre sur* $\mathbf{Z}$.

Cela résulte du corollaire 1, combiné avec la dualité de Poincaré.

Remarque. — Si $g \in G(\mathbf{Q})$, l'application $x \mapsto xg$ se prolonge en un automorphisme de $\overline{X}$; on voit ainsi que $G(\mathbf{Q})$ *opère* sur $\overline{X}$, donc aussi sur $I = H_c^{d-l}(\overline{X}, \mathbf{Z})$; la représentation de $G(\mathbf{Q})$ obtenue de cette manière est analogue à la *représentation de Steinberg* des groupes finis munis d'un système de Tits ([7]).

2. GROUPES ARITHMÉTIQUES. — Soit Γ un sous-groupe arithmétique de $G(\mathbf{Q})$. Il opère sur $\overline{X}$.

THÉORÈME 2. — *L'action de Γ sur $\overline{X}$ est propre; le quotient $\overline{X}/\Gamma$ est compact.*

C'est essentiellement une reformulation de deux des principaux résultats de la « théorie de la réduction », *cf.* ([3]), th. 13.1 et th. 15.4.

Supposons désormais Γ *sans torsion*. Il opère alors *librement* sur $\overline{X}$ et le quotient $\overline{X}/\Gamma$ est une *variété à coins compacte*; comme $\overline{X}$ est contractile, la cohomologie de $\overline{X}/\Gamma$ s'identifie à celle de $\overline{X}$. Plus précisément, si M est un Γ-module, et $\tilde{M}$ le système local correspondant sur $\overline{X}/\Gamma$, les groupes $H^q(\overline{X}/\Gamma, \tilde{M})$ s'identifient aux groupes $H^q(\Gamma, M)$.

Or, du fait que $\overline{X}/\Gamma$ est compacte, on a une suite spectrale

$$H_*(\Gamma, H_c^*(\overline{X}, M)) \implies H^*(\overline{X}/\Gamma, \tilde{M}).$$

Vu le corollaire 2 au théorème 1, le groupe $H_c^q(\overline{X}, M)$ est isomorphe à $I \otimes M$ si $q = d - l$, et il est réduit à o si $q \neq d - l$. La suite spectrale ci-dessus dégénère donc en un isomorphisme

$$H_{d-l-q}(\Gamma, I \otimes M) \simeq H^q(\overline{X}/\Gamma, \tilde{M}) \simeq H^q(\Gamma, M).$$

D'où le *théorème de dualité* suivant :

Théorème 3. — *Si M est un Γ-module et q un entier, le groupe de cohomologie $H^q(\Gamma, M)$ est isomorphe au groupe d'homologie $H_{d-l-q}(\Gamma, I \otimes M)$, où* $I = H_c^{d-l}(\overline{X}, \mathbf{Z})$.

En particulier :

Corollaire 1. — *La dimension cohomologique $cd(\Gamma)$ de Γ est $d - l$.*

(Il s'agit de dimension cohomologique relativement à un anneau commutatif non nul quelconque.)

Exemple. — Si G est *déployé*, $cd(\Gamma)$ est égal au nombre de racines positives du système de racines de G.

Corollaire 2. — *On a $H^q(\Gamma, \mathbf{Z}[\Gamma]) = o$ pour $q \neq d - l$ et $H^{d-l}(\Gamma, \mathbf{Z}[\Gamma]) = I$.*

En particulier, Γ n'a qu'*un seul bout* si $d - l \geq 2$; si $d - l = 1$, Γ est un groupe libre.

Remarque. — Le théorème 3 et ses corollaires restent valables pour tous les sous-groupes arithmétiques sans torsion des groupes algébriques linéaires sur $\mathbf{Q}$, moyennant une définition convenable de I.

(*) Séance du 23 novembre 1970.

(¹) Pour tout ce qui concerne les *variétés à coins* (ou *variétés à bords anguleux*), *voir* les exposés de A. Douady dans le Séminaire H. Cartan, 1961,1962, W. A. Benjamin, New York, 1968.

(²) Dans le cas du groupe $\mathbf{SL}_n$, X(P) et $\overline{X}$ sont essentiellement les espaces définis par C. L. Siegel en ajoutant à X des points « frontières » et des points « idéaux » (cf. *Zur Reduktionstheorie der quadratischer Formen*, Publ. Math. Soc. Japan, 5, 1959, § 1 et 12; *Gesamm. Abh.*, III, p. 275-327). Dans le cas général, la construction de X(P) esquissée ci-dessous doit beaucoup à des remarques faites à l'un de nous par H. Garland.

(³) A. Borel, *Introduction aux groupes arithmétiques*, Hermann, Paris, 1969.

(⁴) *Cf.* A. Borel et J. Tits, *Groupes réductifs*, Publ. Math. I. H. E. S., 27, 1965, p. 55-151.

(⁵) Si H est un groupe algébrique réductif connexe sur $\mathbf{R}$ et K un sous-groupe compact maximal de $H(\mathbf{R})$, *l'involution de Cartan* associée à K est l'unique automorphisme involutif de H dont la restriction à $H(\mathbf{R})$ admet K pour ensemble de points fixes.

(⁶) *Cf.* J. Tits, *Structures et groupes de Weyl*, Séminaire N. Bourbaki, exposé 288 (fév. 1965), W. A. Benjamin, New York, 1966.

(⁷) *Cf.* L. Solomon, *The Steinberg character of a finite group with* BN-*pair*, Theory of Finite Groups (Symposium édité par R. Brauer et C.-H. Sah, W. A. Benjamin, New York, 1969, p. 213-221).

(*Institute for Advanced Study,*
Princeton, N. J. 08540,
États-Unis
et *Collège de France,*
place Marcelin-Berthelot,
75-Paris, 5ᵉ.)

91.

(avec J-P. Serre)

Cohomologie à supports compacts des immeubles
de Bruhat-Tits;
Applications à la cohomologie des groupes S-arithmétiques

C. R. Acad. Sci., Paris **272** (1971) 110–113

Soient K un corps local [1], G un groupe algébrique linéaire semi-simple sur K et X l'immeuble de Bruhat-Tits [2] associé à G. Soit $l = \mathrm{rg}_K G = \dim(X)$. Les groupes de cohomologie à supports compacts $H_c^q(X, \mathbf{Z})$ sont nuls pour $q \neq l$ et le groupe $H_c^l(X, \mathbf{Z})$ est libre sur $\mathbf{Z}$. Ce résultat, combiné avec ceux d'une Note précédente [3], permet de démontrer une formule de dualité pour la cohomologie des groupes S-arithmétiques, et de déterminer la dimension cohomologique de ces groupes.

1. UN COMPLEXE AUXILIAIRE. — Soient G et K comme ci-dessus. Munissons le groupe $G(K)$ d'une structure de système de Tits $(G(K), B, N, S)$ comme indiqué dans [4]. Le groupe de Weyl $W = N/(B \cap N)$ est un groupe de Coxeter fini, de rang égal au rang l de G sur K. Si J est une partie de S, notons P_J le groupe BW_JB correspondant [5]; c'est le groupe des K-points d'un sous-groupe parabolique de G. L'ensemble $G(K)/P_J$ a une structure naturelle de K-variété analytique compacte, quotient de la structure de variété de $G(K)$. Nous noterons F_J le groupe des fonctions localement constantes sur $G(K)/P_J$ à valeurs entières; un élément de F_J s'identifie à une fonction localement constante sur $G(K)$, à valeurs dans $\mathbf{Z}$, invariante à droite par P_J. Si $J \subset J'$, on a $F_{J'} \subset F_J$. Cela permet de définir un *complexe* $C = \sum_{n=0}^{n=l} C_n$, où les éléments de C_n sont les familles

$$\{ f_{s_1 \ldots s_n} \}_{s_i \in S} \quad (\text{avec } f_{s_1 \ldots s_n} \in F_{\{ s_1, \ldots, s_n \}}),$$

dépendant de façon alternée des s_i, et où l'opérateur bord $d : C_n \to C_{n-1}$ est donné par

$$(df)_{s_1 \ldots s_{n-1}} = \sum_{s \in S} f_{s s_1 \ldots s_{n-1}}.$$

THÉORÈME 1. — *Les groupes d'homologie* $H_p(C)$ *sont nuls pour* $p \geq 1$.

Le groupe $H_0(C)$ se détermine de la manière suivante : pour tout $w \in W$, posons $\varepsilon(w) = (-1)^{l(w)}$, où $l(w)$ est la longueur [6] de w, et choisissons un représentant $\overline{w}$ de w dans N. Si $f \in C_0$, f est une fonction localement constante sur $G(K)$, invariante à droite par B. Si $x \in G$, posons

$$\tilde{f}(x) = \sum_{w \in W} \varepsilon(w) f(x \overline{w}).$$

La fonction f ainsi définie ne dépend pas du choix des $\overline{w}$, et l'on a $\tilde{f} = 0$ si $f \in d(C_1)$; de plus, $\tilde{f}$ est invariante à droite par $T = B \cap N$. Notons f_1 la fonction sur B/T induite par $\tilde{f}$.

112

Théorème 2. — *L'application $f \mapsto f_1$ définit par passage au quotient un isomorphisme de $H_0(C)$ sur le groupe des fonctions localement constantes à support compact sur B/T.*

En particulier, $H_0(C)$ est un **Z**-module libre.

On peut démontrer des résultats analogues aux théorèmes 1 et 2 pour les groupes semi-simples *réels* : on définit F_J comme le groupe des fonctions continues réelles sur $G(\mathbf{R})/P_J$, et l'on trouve que $H_p(C) = 0$ pour $p \geqq 1$ et que $H_0(C)$ est isomorphe au groupe des fonctions continues réelles sur B/T qui tendent vers zéro à l'infini.

2. Cohomologie a supports compacts de l'immeuble de Bruhat-Tits.

— Soit d'abord Y l'immeuble ([7]) attaché au système de Tits $(G(K), B, N, S)$ du n° 1. Le groupe $G(K)$ opère sur Y. Si σ est un simplexe de Y de dimension maximale $l - 1$, l'application $G(K) \times \sigma \to Y$ est surjective. Nous munirons Y de la *topologie quotient* de celle de $G(K) \times \sigma$, étant entendu que $G(K)$ est muni de sa topologie naturelle de groupe localement compact, et σ de sa topologie naturelle de simplexe. L'espace Y^{top} ainsi défini est *compact*. Nous noterons $H^i(Y^{\mathrm{top}}, \mathbf{Z})$ ses groupes de cohomologie de Čech à valeurs dans $\mathbf{Z}$, et $\tilde{H}^i(Y^{\mathrm{top}}, \mathbf{Z})$ les groupes de cohomologie « réduits » correspondants :

$$\tilde{H}^i(Y^{\mathrm{top}}, \mathbf{Z}) = H^i(Y^{\mathrm{top}}, \mathbf{Z}) \quad \text{si } i \neq 0, -1,$$
$$\tilde{H}^0(Y^{\mathrm{top}}, \mathbf{Z}) = \mathrm{Coker}(\mathbf{Z} \to H^0(Y^{\mathrm{top}}, \mathbf{Z})),$$
$$\tilde{H}^{-1}(Y^{\mathrm{top}}, \mathbf{Z}) = \mathrm{Ker}(\mathbf{Z} \to H^0(Y^{\mathrm{top}}, \mathbf{Z})).$$

Ces groupes sont liés aux groupes d'homologie considérés au n° 1 :

Lemme. — *On a $\tilde{H}^i(Y^{\mathrm{top}}, \mathbf{Z}) \simeq H_{l-1-i}(C)$ pour tout $i \in \mathbf{Z}$.*

D'autre part, l'immeuble de Bruhat-Tits X attaché à G est un complexe polysimplicial localement fini (donc localement compact), de dimension l; on sait qu'il est contractile. On peut *compactifier* X en lui adjoignant l'espace Y^{top} défini ci-dessus, et l'espace $X \cup Y^{\mathrm{top}}$ ainsi obtenu est encore contractile. La suite exacte de cohomologie montre alors que la cohomologie à supports compacts de X est donnée par

$$H_c^i(X, \mathbf{Z}) \simeq \tilde{H}^{i-1}(Y^{\mathrm{top}}, \mathbf{Z}) \quad \text{pour tout } i \in \mathbf{Z}.$$

D'où, en combinant le lemme ci-dessus avec les théorèmes 1 et 2 :

Théorème 3. — *Les groupes $H_c^i(X, \mathbf{Z})$ sont nuls pour $i \neq l$. Le groupe $I = H_c^l(X, \mathbf{Z})$ est isomorphe au groupe des fonctions localement constantes à support compact sur B/T; c'est un **Z**-module libre.*

3. Groupes S-arithmétiques.

— Soient k un corps de nombres algébriques et S un ensemble fini de places de k contenant l'ensemble Σ des places archimédiennes; si $v \in S$, on note k_v le complété de k pour v. Soit G un groupe algébrique semi-simple sur k, et soit Γ un sous-groupe S-arithmétique de $G(k)$; c'est un sous-groupe discret du produit des $G(k_v)$, pour $v \in S$.

Soit X_∞ l'espace homogène des sous-groupes compacts maximaux du groupe de Lie réel $G_\infty = \prod_{v \in \Sigma} G(k_v)$, et soit $\overline{X}_\infty$ la *variété à coins* [3] d'intérieur X_∞ associée au groupe algébrique sur **Q** déduit de G par restriction des scalaires.

Si $v \in S - \Sigma$, notons X_v l'immeuble de Bruhat-Tits de G sur le corps local k_v. Posons :

$$X_S = \overline{X}_\infty \times \prod_{v \in S - \Sigma} X_v.$$

Le groupe $G(k)$ (donc *a fortiori* le groupe Γ) opère sur X_S.

Théorème 4. — *Le groupe Γ opère proprement sur X_S; le quotient X_S/Γ est compact.*

Notons l (resp. l_v) le rang de G sur k (resp. sur k_v). Posons

$$d = \dim(X_\infty) \quad \text{et} \quad m = \dim(X_S) - l = d - l + \sum_{v \in S - \Sigma} l_v.$$

Il résulte du corollaire 2 au théorème 1 de [3] et du théorème 3 que les groupes de cohomologie à supports compacts $H^i_c(X_S, \mathbf{Z})$ sont nuls pour $i \neq m$ et que le groupe $I_S = H^m_c(X_S, \mathbf{Z})$ est libre sur **Z**. On en déduit, comme dans [3] :

Théorème 5. — *Soient Γ un sous-groupe S-arithmétique sans torsion de $G(k)$, M un Γ-module et q un entier. Le groupe $H^q(\Gamma, M)$ est isomorphe au groupe d'homologie $H_{m-q}(\Gamma, I_S \otimes M)$.*

Corollaire. — *On a $cd(\Gamma) = m$. Le groupe $H^q(\Gamma, \mathbf{Z}[\Gamma])$ est nul pour $q \neq m$ et $H^m(\Gamma, \mathbf{Z}[\Gamma])$ est isomorphe à I_S.*

4. Sous-groupes discrets a quotient compact. — Soit L un groupe localement compact, produit direct d'un nombre fini de groupes L_α où L_α est, soit un groupe de Lie réel ayant un nombre fini de composantes connexes, soit le groupe des k_α-points d'un groupe semi-simple sur un corps local k_α. Soit X le produit des X_α, où X_α désigne, dans le premier cas, le quotient de L_α par un sous-groupe compact maximal, et, dans le second cas, l'immeuble de Bruhat-Tits de L_α. Posons $m = \dim(X)$. Vu le théorème 3, le groupe $I_L = H^m_c(X, \mathbf{Z})$ est libre sur **Z** et $H^i_c(X, \mathbf{Z}) = o$ pour $i \neq m$. Si Γ est un sous-groupe discret de L tel que L/Γ soit compact, Γ opère proprement sur X, et X/Γ est compact. On en déduit comme plus haut :

Théorème 6. — *Soient Γ un sous-groupe discret sans torsion de L tel que L/Γ soit compact, M un Γ-module et q un entier. Les groupes $H^q(\Gamma, M)$ et $H_{m-q}(\Gamma, I_L \otimes M)$ sont isomorphes.*

Corollaire. — *On a $cd(\Gamma) = m$. Le groupe $H^q(\Gamma, \mathbf{Z}[\Gamma])$ est nul pour tout $q \neq m$ et $H^m(\Gamma, \mathbf{Z}[\Gamma])$ est isomorphe à I_L.*

Remarque. — Le théorème 6 s'applique en particulier aux sous-groupes S-arithmétiques sans torsion d'un groupe semi-simple G sur un corps de fonctions d'une variable sur un corps fini, pourvu que le rang de G sur le corps en question soit *nul* ([8]).

(*) Séance du 4 janvier 1971.

([1]) Nous entendons par là un corps complet pour une valuation discrète à corps résiduel fini; un tel corps est localement compact.

([2]) *Cf.* F. Bruhat et J. Tits, *Proc. Conf. Local Fields*, Springer-Verlag, 1967. (*Voir aussi Comptes rendus*, 263, 1966, p. 598, 766, 822 et 867.)

([3]) *Comptes rendus*, 271, série A, 1970, p. 1156.

([4]) *Cf.* A. Borel et J. Tits, *Publ. Math. I. H. E. S.*, 16, 1963, p. 5-30, § 5.

([5]) *Cf.* N. Bourbaki, *Groupes et Algèbres de Lie*, chap. IV, § 2, Hermann, Paris, 1968.

([6]) *Cf.* N. Bourbaki, *Groupes et Algèbres de Lie*, chap. IV, § 1, Hermann, Paris, 1968.

([7]) *Cf.* J. Tits, *Structures et groupes de Weyl*, Séminaire N. Bourbaki, exposé 288 (février 1965), W. A. Benjamin, New York, 1966.

([8]) *Cf.* G. Harder, *Invent. Math.*, 7, 1969, p. 33-54, Kor. 2.2.7.

(*Institute for Advanced Study,
Princeton, N. J.,* 08540, *U. S. A.
et Collège de France,
place Marcelin-Berthelot,*
75-*Paris,* 5e.)

92.

(avec J. Tits)

Eléments unipotents et sous-groupes paraboliques de groupes réductifs I

Invent. Math. **12** (1971) 95–104

Introduction

Le premier but de cet article est d'associer à un sous-groupe unipotent k-fermé U d'un k-groupe connexe réductif G un sous-groupe parabolique k-fermé P de G, contenant le normalisateur $\mathcal{N}(U)$ de U, et dont le radical unipotent $R_u P$ contient U. Si U est défini sur k, mais k n'est pas parfait, alors P n'est pas nécessairement défini sur k. Toutefois, lorsque U est contenu dans un sous-groupe de Borel de G défini sur une clôture séparable k_s de k (nous disons alors que U est k_s-*plongeable*) nous lui associons aussi un k-sous-groupe parabolique $\mathcal{P}(U)$ contenant $\mathcal{N}(U) \cap G(k_s)$ et tel que $U \subset R_u \mathcal{P}(U)$. Si k est parfait, le groupe $P = \mathcal{P}(U)$ s'obtient ainsi: on définit deux suites N_i, U_i $(i = 1, 2, \ldots)$ de k-sous-groupes de G par

$$N_1 = \mathcal{N}(U), \quad U_1 = U \cdot R_u N_1, \quad N_i = \mathcal{N}(U_{i-1}), \quad U_i = U_{i-1} \cdot R_u N_i, \quad (i \geq 2).$$

Ces suites sont évidemment stationnaires, et $\mathcal{P}(U)$ est par définition égal à N_i pour i assez grand. Pour faire voir que $\mathcal{P}(U)$ est parabolique, on montre que si un sous-groupe connexe Q de G est le normalisateur dans G de son radical unipotent $R_u Q$, alors Q est parabolique (cf. 2.3). (En caractéristique 0, cela résulte aussi de [4]; la démonstration donnée ici dans le cas général est analogue à celle de la Proposition 1 de Platonov [5].) La construction de $\mathcal{P}(U)$ sur un corps quelconque est donnée au §2. On utilise notamment quelques propriétés des groupes résolubles déployés, dues à Rosenlicht [6], qui sont rappelées et complétées sur certains points au §1.

Le §3 est consacré à quelques applications, en particulier: (i) un sous-groupe fermé propre maximal de G est parabolique ou de composante neutre réductive (3.3) (en caractéristique 0, ce résultat est dû à Morozov [4]); (ii) le groupe dérivé $\mathcal{D}(G)$ de G est anisotrope sur k si et seulement si $G(k)$ ne contient aucun élément unipotent $\neq 1$ et k_s-plongeable[1].

---

[1] Ce résultat est bien connu en caractéristique zéro: cf. par exemple [2, §8]. Dans le cas général, il a été annoncé dans [9] (Théorème 1.1), où il est, à peu près comme ici, ramené à la possibilité d'associer de façon invariante à tout élément unipotent rationnel contenu

Disons que G est bon sur k si tout élément unipotent de $G(k)$ est k_s-plongeable. Il en est évidemment ainsi si k est parfait. Dans un travail ultérieur on donnera des exemples de G qui ne sont pas bons sur k et des conditions suffisantes pour que G soit bon sur k. Supposons G simplement connexe, absolument presque simple, et soit p la caractéristique de k. On verra notamment que si $p \neq 2$ et si G est de type classique, alors G est toujours bon sur k, mais que si $p = 2$ et si G est déployé (de type quelconque) sur k, alors G est bon si et seulement si $[k:k^2] \leq 2$ ou si G est de type $\mathbf{A}_n$ ou $\mathbf{C}_n$. Ces résultats et d'autres (relatifs notamment au cas $p = 3$) que nous ne résumons pas ici suggèrent la

Conjecture. *Supposons G simplement connexe, absolument presque simple sur k. Si $p = \mathrm{car}\, k$ n'est pas un nombre premier de torsion de G [7, p. 178] alors G est bon sur k. Si p est un nombre premier de torsion de G, si G est déployé sur k et $[k:k^p]$ suffisamment grand, alors G n'est pas bon sur k.*

§0. Notations. Terminologie

0.1. Dans ce travail, k désigne toujours un corps commutatif et p la caractéristique de k. Si L est un corps, on note $\bar{L}$ (resp. L_s), une clôture algébrique (resp. séparable) de L. L'adhérence de Zariski *absolue* d'une partie Y d'une variété algébrique est notée $\bar{Y}$. En général, les notations sont conformes à celles de [1].

0.2. Si X est un groupe et Y une partie de X, nous désignons par $\mathscr{Z}_X(Y)$ ou $\mathscr{Z}(Y)$ le centralisateur de Y dans X, par $\mathscr{N}_X(Y)$ ou $\mathscr{N}(Y)$ son normalisateur, et par $\mathscr{C}(X)$ le centre de X.

Si X est un groupe algébrique, X^0 (resp. $R_u X$) désigne la composante connexe de l'élément neutre de X (resp. le radical unipotent de X). Par définition, $R_u X = R_u X^0$.

0.3. Un groupe algébrique connexe résoluble est dit *k-déployé* ou *déployé sur k*, s'il est défini sur k et s'il possède une suite de composition définie sur k dont les quotients sont tous isomorphes sur k au groupe additif ou au groupe multiplicatif.

0.4. Soient G, G' des groupes réductifs, $\varphi: G \to G'$ un épimorphisme, T un tore maximal de G et $T' = \varphi(T)$ son image par φ. L'épimorphisme φ est dit *quasi-central* si le noyau de $\varphi(\bar{L}): G(\bar{L}) \to G'(\bar{L})$ (où L désigne un corps de définition de φ) est central dans $G(\bar{L})$, et il est dit *central* si

---

dans un k-sous-groupe de Borel un k-sous-groupe parabolique. Mais la démonstration d'existence de ce dernier, à laquelle il est fait allusion dans [9], consiste en une construction «géométrique» distincte pour les divers types de groupes simples, et que l'auteur n'a pu reconstituer pour E_7 et E_8. La construction beaucoup plus simple (fonction $\mathscr{P}$) donnée ici est, on l'a vu, basée sur un principe tout différent.

l'homomorphisme $X^*(T') \to X^*(T)$ induit par $\rho|_T$ applique l'ensemble $\Phi(T', G')$ des racines de G' sur l'ensemble $\Phi(G, T)$ des racines de G (cela signifie aussi que le «noyau schématique» de φ est central dans G).

Supposons G, G' et φ définis sur k et φ central. Alors, si R désigne un k-sous-groupe parabolique de G, $\varphi(P)$ est un k-sous-groupe parabolique de G' et φ induit un k-isomorphisme de $R_u(P)$ sur $R_u(\varphi(P))$; réciproquement, si P' est un k-sous-groupe parabolique de G', son image inverse $\varphi^{-1}(P')$ est un k-sous-groupe parabolique de G [2b].

0.5. Soit G un k-groupe connexe. Un élément ou un sous-ensemble de $G(\bar{k})$ sera dit k-*plongeable* ou *plongeable sur k* s'il est contenu dans le radical unipotent d'un k-sous-groupe parabolique de G.

Si $P \subset P'$ sont des sous-groupes paraboliques du groupe G, on a évidemment $R_u P' \subset R_u P$, donc si M est une partie k-plongeable de $G(\bar{k})$, il existe un k-sous-groupe parabolique minimal P de G tel que $M \subset R_u P$.

§1. Rappels et compléments sur les groupes résolubles déployés

Dans ce paragraphe, G désigne un k-groupe résoluble connexe.

1.1. Théorème (Rosenlicht). (i) *Si G est déployé sur k, alors $R_u(G)$ et tout k-tore de G le sont aussi.*

(ii) *G est déployé sur k si et seulement si $k[G]$ est k-isomorphe à une sous-algèbre d'une algèbre de la forme $k[X_1, \ldots, X_N, X_1^{-1}, \ldots, X_N^{-1}]$, où les X_i sont algébriquement indépendants sur k.*

Pour (i), voir [6], Lemma 1, p.103, et Theorem 4, p.110. Pour (ii), voir [6], Theorem 2, p.103.

1.2. Corollaire. *Soient H un k-groupe, $(G_i)_{i \in I}$ une famille de k-groupes résolubles déployés sur k, $\varphi_i \colon G_i \to H$ $(i \in I)$ un k-morphisme dont l'image contient 1, et X le sous-groupe de H engendré par les $\varphi_i(G_i)$. Alors, si X est résoluble, il est k-déployé.*

C'est une conséquence immédiate de 1.1(ii) et du fait qu'il existe une suite finie $i_1, \ldots, i_m$ d'éléments de I telle que l'application produit

$$\varphi_{i_1}(G_{i_1}) \times \cdots \times \varphi_{i_m}(G_{i_m}) \to X$$

soit surjective.

1.3. Corollaire. *Soient H un k-groupe et L un sous-groupe résoluble fermé. Alors, les sous-groupes résolubles k-déployés de H contenus dans L engendrent un groupe k-déployé. Autrement dit, L possède un «plus grand sous-groupe k-déployé».*

1.4. Corollaire. (i) *Si M et N sont des k-sous-groupes connexes déployés de G, alors (M, N) est déployé sur k. (ii) Si G est k-déployé, il en est de même de $\mathscr{D}^j G$ et $\mathscr{C}^j G$ $(j = 0, 1, \ldots)$.*

(i) est une conséquence immédiate de 1.2 (cf. aussi [6], Corollary to Theorem 2, p. 105). (ii) résulte de (i) par récurrence sur j.

1.5. Corollaire. *Supposons G nilpotent et déployé sur k et soit H un sous-groupe fermé propre de G. Alors, G possède un sous-groupe distingué G', déployé sur k et normalisant H, tel que* dim $G'H >$ dim H.

Il suffit de prendre pour G' le dernier terme de la série centrale descendante de G non contenu dans H; il est k-déployé en vertu de 1.4.

1.6. Lemme. *Supposons G déployé sur k et soit S un k-tore de G. Alors, $\mathscr{Z}(S)$ est déployé sur k.*

Soit T un k-tore maximal de G contenant S. Posons $U = R_u(G) \cap \mathscr{Z}(S)$. On a $\mathscr{Z}(S) = T \cdot U$ et U est connexe [1, 9.4]. Vu 1.1, T est déployé sur k, et il nous suffit donc de montrer qu'il en est de même de U. Si k est fini (ou, plus généralement, parfait), c'est évident. Si k est infini, il existe $s \in S(k)$ tel que $\mathscr{Z}(s) = \mathscr{Z}(S)$ [1, 1.10]; il résulte alors de [1, 9.3] que U est défini sur k et que $R_u(G)$ est isomorphe comme k-variété au produit direct de U et d'une autre k-variété; comme $R_u(G)$ est k-déployé (1.1), il en est de même de U, vu 1.1.

§ 2. Un plongement canonique des sous-groupes plongeables

Dans ce paragraphe, K désigne un corps séparablement clos et G un groupe connexe réductif défini sur K.

Remarquons, que, puisque K est séparablement clos, tout K-tore de G est déployé sur K [1, 8.11], donc G est déployé sur K.

2.1. Lemme. *Soient H un k-groupe, B un k-sous-groupe de Borel de H, L une partie de B et M un sous-groupe résoluble k-déployé de H normalisant L. Alors, $M \cdot L$ est contenu dans un k-sous-groupe de Borel de H.*

Soit $X = H/B$ et soit F la variété des points fixes de L dans X. Il s'agit de montrer que $M \cdot L$ possède un point fixe dans $X(k)$, ou encore que M possède un point fixe dans $X(k) \cap F$. Notons o l'image de B dans X et posons $V = M \cdot o$. Comme M normalise L, on a $V \subset F$, donc aussi $\overline{V} \subset F$. La variété complète $\overline{V}$ est définie sur k et stable par M; de plus, $o \in \overline{V}(k)$ donc $\overline{V}(k) \neq \emptyset$. Par conséquent, M a un point fixe dans $\overline{V}(k)$ [1, 15.2].

2.2. Lemme. *Soient H un K-groupe et L un sous-groupe résoluble fermé de H. Alors, il existe $h \in H(K)$ tel que $R_u(L \cap hLh^{-1}) \subset R_u(H)$.*

En effet, soit B un sous-groupe de Borel de H^0 contenant L^0. En appliquant [2, 4.12] à $H^0/R_u H$, on voit qu'il existe un ouvert non vide X de H tel que, pour tout $x \in X$, on ait $R_u(B \cap xBx^{-1}) = R_u(H)$. Le corps K

étant séparablement clos, $X \cap H(K)$ n'est pas vide et il est clair que tout point de cet ensemble possède la propriété requise.

2.3. Proposition. *Soient U un sous-groupe unipotent K-plongeable de $G(\overline{K})$, H l'adhérence de $\mathcal{N}(\overline{U}) \cap G(K)$ et R le plus grand sous-groupe K-déployé de $R_u(H)$. Alors H est défini sur K et $R \cdot \overline{U}$ est K-plongeable. De plus, si $\dim R \cdot \overline{U} = \dim \overline{U}$, on a $R \cdot \overline{U} = \overline{U}$ et $\overline{U}$ est le radical unipotent d'un K-sous-groupe parabolique de G.*

La première assertion résulte du fait que $H \cap G(K)$ est dense dans H [1, AG 14.4] et la deuxième du Lemme 2.1.

Supposons désormais que $\dim R \cdot \overline{U} = \dim \overline{U}$. Il est alors évident que $R \subset \overline{U}$, d'où $R \cdot \overline{U} = \overline{U}$. Soient B un K-sous-groupe de Borel de G contenant U et h un élément de $H(K)$ tel que $R_u(H \cap B \cap hBh^{-1}) \subset R_u(H)$ (cf. 2.2). Posons $B' = hBh^{-1}$. Comme $h \in H \subset \mathcal{N}(\overline{U})$, on a $\overline{U} \subset B \cap B'$. Tout sous-groupe unipotent K-déployé L de $B \cap B'$ normalisant $\overline{U}$ est contenu dans $R_u(H)$, donc dans R, donc aussi dans $\overline{U}$, de sorte que $\dim L \cdot \overline{U} = \dim \overline{U}$. Vu 1.5 et le fait que $B \cap B'$ est déployé [2, 4.7 et 3.18], cela implique que

$$R \cdot \overline{U} = \overline{U} = R_u(B \cap B').$$

Soient T un K-tore maximal de G contenu dans $B \cap B'$ [1, 14.7, p. 352], Δ la base du système de racines $\Phi = \Phi(T, G)$ contenue dans $\Phi^+ = \Phi(T, B)$, $\Psi = \Phi(T, \mathcal{N}(\overline{U}))$ l'ensemble des poids non nuls de T dans $\mathcal{N}(\overline{U})$ et Δ' l'ensemble des racines $a \in \Delta$ telles que $U_a \not\subset \overline{U}$. Pour $b \in \Delta$, notons B_b le sous-groupe de Borel de G engendré par T, U_{-b} et les U_c ($c \in \Phi^+ - \{b\}$). Pour $a \in \Delta'$, on a $U_a \not\subset B'$, donc $U_{-a} \subset B'$, de sorte que U_{-a} normalise $R_u(B \cap B_a)$ et B' donc aussi $R_u(B \cap B_a \cap B') = \overline{U}$. Ainsi $-a \in \Psi$. Comme $U_{-a} \not\subset R$, cela implique qu'on a aussi $a \in \Psi$. En conclusion $\Delta \cup (-\Delta') \subset \Psi$. En particulier, $B \subset \mathcal{N}(\overline{U})$, et $\mathcal{N}(\overline{U})$ est le sous-groupe parabolique «standard» $P_{\Delta'}$ [2, 4.2]. Comme B est un sous-groupe de Borel de $\mathcal{N}(\overline{U})$, il en est de même de B', et on a donc

$$R_u\big(\mathcal{N}(\overline{U})\big) \subset R_u(B \cap B') = \overline{U}.$$

Mais l'inclusion inverse est évidente. Par conséquent, $\overline{U} = R_u(P_{\Delta'})$, ce qui achève la démonstration.

2.4. *La fonction $\mathscr{P}$.* Pour tout sous-groupe unipotent K-plongeable U de $G(\overline{K})$, posons $\mathscr{L}_{G,K}(U) = R \cdot \overline{U}$ où R désigne le plus grand sous-groupe K-déployé de $R_u((\mathcal{N}(\overline{U}) \cap G(K))^-)$. Vu 2.1, l'opérateur $\mathscr{L}_{G,K}$ peut être itéré. Évidemment il existe $n \in \mathbf{N}$ tel que $\mathscr{L}_{G,K}^n(U) = \mathscr{L}_{G,K}^{n'}(U)$ pour tout $n' \geq n$; vu 2.3, le groupe $\mathscr{L}_{G,K}^n(U)$ correspondant est le radical unipotent d'un K-sous-groupe parabolique de G que nous noterons $\mathscr{P}_{G,K}(U)$, l'un des indices G et K, ou les deux, étant omis le cas échéant. Nous avons ainsi défini un K-sous-groupe parabolique attaché à U de façon invariante. Plus précisément:

2.5. Théorème. *Soit $U \subset G(\overline{K})$ un groupe unipotent K-plongeable.*

(i) *On a $U \subset R_u(\mathscr{P}(U))$.*

(ii) *Supposons que k soit un sous-corps de K, que G soit défini sur k, que U soit k-fermé et que K soit séparable sur une extension radicielle de k. Alors $\mathscr{P}(U)$ est k-fermé. Si K est séparable sur k, alors $\mathscr{P}(U)$ est défini sur k.*

(iii) *Soit $\varphi\colon G \to G'$ un K-épimorphisme quasi-central. Supposons que K soit algébriquement clos ou que φ soit central. Alors $\varphi(\mathscr{P}_{G,K}(U)) = \mathscr{P}_{G',K}(\varphi(U))$.*

Ces propriétés sont des conséquences immédiates de la définition et, pour (iii), du Lemme 2.7 ci-dessous.

2.6. Lemme. *Soient H un groupe algébrique, A une partie dense de H et U un sous-groupe fermé unipotent connexe de H. Supposons que U soit normalisé par $a\,U\,a^{-1}$ pour tout $a \in A$. Alors, $U \subset R_u(H)$.*

En effet, U est contenu dans le radical unipotent du groupe engendré par les $a\,U\,a^{-1}$, lequel est distingué dans H.

2.7. Lemme. *Soit $\varphi\colon G \to G'$ un K-épimorphisme quasi-central.*

(i) *Pour tout sous-groupe unipotent U de $G(\overline{K})$, on a $\mathscr{N}_{G(\overline{K})}(U) = \varphi^{-1}(\mathscr{N}_{G'(\overline{K})}(\varphi(U)))$.*

(ii) *Soient N un sous-groupe fermé de G et R (resp. R') le plus grand sous-groupe K-déployé de $R_u((N \cap G(K))^-)$ (resp. $R_u((\varphi(N) \cap G'(K))^-))$. Supposons que K soit algébriquement clos ou que φ soit central. Alors $\varphi(R) = R'$.*

(i) Soient Z le noyau de φ et n un élément de $\varphi^{-1}(\mathscr{N}(\varphi(U)))$. On a alors $n\,U\,n^{-1} \subset U \cdot Z$. Mais Z est central et formé d'éléments semi-simples, donc $n\,U\,n^{-1} \subset U$, d'où $\varphi^{-1}(\mathscr{N}(\varphi(U))) \subset \mathscr{N}(U)$. L'autre inclusion est évidente.

(ii) Pour tout sous-groupe unipotent fermé connexe U' de G', posons $\psi(U') = R_u(\varphi^{-1}(U'))$, de sorte que $\varphi^{-1}(U') = \psi(U') \times Z$. Si U' est K-déployé, il est contenu dans le radical unipotent d'un K-sous-groupe de Borel de G' et il résulte alors des propriétés rappelées en 0.4 que $\psi(U')$ est aussi K-déployé. Si $U' \subset \varphi(N)$, on a $\psi(U') \subset N$. Si U' est normalisé par $\varphi(N) \cap G'(K)$, alors $\psi(U')$ est normalisé par $N \cap G(K)$. De ces diverses remarques, il résulte que $\psi(R') \subset R$, donc que $R' \subset \varphi(R)$.

D'autre part, pour $a \in \varphi(N) \cap G'(K)$, le groupe $\psi(a \cdot \varphi(R) \cdot a^{-1}) \cap G(K)$, qui est contenu dans $N \cap G(K)$, normalise R; par conséquent, $\psi(a \cdot \varphi(R) \cdot a^{-1})$ normalise R, et $a \cdot \varphi(R) \cdot a^{-1}$ normalise $\varphi(R)$. Vu 2.6, ceci implique que $\varphi(R) \subset R_u((\varphi(N) \cap G'(K))^-)$ donc que $\varphi(R) \subset R'$.

2.8. *Remarques.* (1) Le groupe $\mathscr{P}_{G,K}(U)$ dépend effectivement de K en général. Par exemple, supposons $p=2$, $[k:k^2]>2$ et G simplement connexe, déployé, de type $\mathbf{D}_4$. On peut alors trouver dans $G(k)$ un élément unipotent k-plongeable u dont le centralisateur ne fait partie d'aucun k-sous-groupe parabolique propre de G. Soit U le sous-groupe engendré par u. Alors $\mathscr{P}_{G,k_s}(U)$ est défini sur k et ne contient pas $\mathscr{Z}_G(U)$ tandis que $\mathscr{P}_{G,\bar{k}}(U)$ est k-fermé, mais non défini sur k, et contient $\mathscr{Z}_G(U)$, donc $\mathscr{P}_{G,k_s}(U)\neq\mathscr{P}_{G,\bar{k}}(U)$. Cela montre aussi que dans 3.1(ii), on ne peut remplacer k_s par $\bar{k}$, et que dans 3.9, on ne peut affirmer que P est défini sur k, même si U l'est.

(2) On peut se demander s'il existe un procédé, différent de 2.5 en général, qui permette d'associer à tout k-sous-groupe unipotent k_s-plongeable U un k-sous-groupe parabolique P tel que $U\subset R_u P$ et $\mathscr{N}(U)\subset P$. L'exemple précédent montre aussi que la réponse est négative. Cela est cependant possible pour certains groupes, par exemple pour $\mathbf{SL}_n$ sur un corps quelconque.

§3. Applications. Sous-groupes maximaux non réductifs et sous-groupes unipotents maximaux

Dans ce paragraphe, G désigne un groupe connexe réductif défini sur k.

3.1. Proposition. *Soit $U\subset G$ un sous-groupe unipotent k-fermé et k_s-plongeable de G. Alors, il existe un k-sous-groupe parabolique P de G tel que:*

   (i) $U\subset R_u(P)$;

   (ii) $\mathscr{N}(U)\cap G(k_s)\subset P$;

   (iii) *tout k_s-automorphisme de G qui stabilise U stabilise aussi P.*

En particulier, U est k-plongeable.

Il suffit de poser $P=\mathscr{P}_{G,k_s}(U)$. Les propriétés (i) et (iii) résultent alors du Théorème 2.5 et (ii) est une conséquence immédiate de (iii), appliqué aux automorphismes intérieurs de G correspondant aux éléments de $\mathscr{N}(U)\cap G(k_s)$.

3.2. Corollaire. *Si un sous-groupe unipotent fermé d'un groupe algébrique connexe H est le radical unipotent de son normalisateur, ce dernier est un sous-groupe parabolique de H.*

En effet, soit U un sous-groupe unipotent fermé de H tel que $U=R_u(\mathscr{N}(U))$. Vu la Proposition 3.1 (appliquée au groupe réductif $H/R_u(H)$), il existe un sous-groupe parabolique P de H tel que $U\subset R_u(P)$ et $\mathscr{N}(U)\subset P$. Mais alors, on a $\mathscr{N}(U)\cap R_u(P)=U$, d'où $U=R_u(P)$ puisque, dans un groupe nilpotent, tout sous-groupe propre est distinct de son normalisateur.

3.3. Corollaire. *Un sous-groupe fermé propre maximal de G est parabolique ou de composante neutre réductive.*

Soit H un tel sous-groupe. Si H^0 n'est pas réductif, l'hypothèse de maximalité implique que $H = \mathcal{N}(R_u(H))$, et H est parabolique en vertu de 3.2.

3.4. *Remarque.* En caractéristique 0, ce résultat est dû à Morozov [4]; en fait, sa démonstration donne aussi le Corollaire 3.2 (toujours en caractéristique 0). (Pour $k = \mathbf{C}$, cf. aussi Morozov, Thèse, Kazan, 1943 et F.I. Karpelevič, Dokl. Akad. Nauk SSSR, N.S. **76**, 775 – 778 (1951).)

3.5. Corollaire. *Soient P un k-sous-groupe parabolique minimal de G et S un tore déployé sur k maximal de G contenu dans P (cf. [2, 4.16]). Alors, $R = S \cdot R_u(P)$ est un sous-groupe résoluble connexe déployé maximal de G et tout sous-groupe résoluble connexe k-déployé de G est conjugué par un élément de $G(k)$ à un sous-groupe de R.*

Il suffit évidemment d'établir la seconde assertion. Soient R' un sous-groupe résoluble connexe k-déployé de G et S' un k-tore maximal de R'. Vu 1.1 et 2.1, $R_u(R')$ est k_s-plongeable, donc (3.1) il existe un k-sous-groupe parabolique P' de G tel que $R_u(R') \subset R_u(P')$ et $R'(k_s) \subset P'$, d'où $R' \subset P'$. Mais alors, il existe $g \in G(k)$ tel que $g \cdot R_u(P') \cdot g^{-1} \subset R_u(P)$ et $g \cdot S' \cdot g^{-1} \subset S$ [2, 4.13 et 11.6], d'où $g R' g^{-1} \subset R$.

3.6. Proposition. *Un sous-groupe unipotent de $G(k)$ dont tous les éléments sont k_s-plongeables est k-plongeable.*

Raisonnons par récurrence sur $\dim G$. Soit U le sous-groupe de $G(k)$ en question. On peut évidemment supposer $U \neq \{1\}$. Soit $u \neq 1$ un élément du centre de U. Vu la Proposition 3.1 appliquée à l'adhérence du sous-groupe engendré par u, il existe un k-sous-groupe parabolique P de G tel que $u \in R_u(P)$ et $U \subset P(k)$. Notre assertion résulte à présent de l'hypothèse d'induction appliquée au groupe réductif $P/R_u(P)$ et à l'image de U dans ce groupe.

3.7. Corollaire. *Supposons k parfait et soit H un k-groupe connexe. Alors, tout sous-groupe unipotent de $H(k)$ (resp. tout k-sous-groupe unipotent de H) est contenu dans le radical unipotent d'un k-sous-groupe parabolique de H. En particulier, les sous-groupes unipotents maximaux de $H(k)$ (resp. les k-sous-groupes unipotents maximaux de H) sont les sous-groupes de la forme $R_u(P)(k)$ (resp. $R_u(P)$), où P désigne un k-sous-groupe parabolique minimal de H ; ils sont tous conjugués par des éléments de $H(k)$.*

Soient U un k-sous-groupe unipotent de H et U_1 son image canonique dans $H_1 = H/R_u(H)$. Tous les éléments de U_1 étant $\bar{k}$-plongeables, U_1 est $\bar{k}$-plongeable (3.6), donc k-plongeable (3.1), et U l'est aussi. Les autres assertions de l'énoncé sont alors évidentes.

3.8. Corollaire. *Le groupe dérivé de G est k-anisotrope si et seulement si 1 est le seul élément unipotent k_s-plongeable de G(k). Si k est parfait, 1 est alors le seul élément unipotent de G(k).*

3.9. Corollaire. *Soit $U \subset G$ un sous-groupe unipotent k-fermé. Alors il existe un sous-groupe parabolique k-fermé P de G tel que:*

   (i) $U \subset R_u(P)$;

   (ii) $\mathcal{N}(U) \subset P$;

   (iii) *tout $\bar{k}$-automorphisme de G qui stabilise U stabilise aussi P.*

Il suffit d'appliquer 3.1 en remplaçant k par son extension radicielle maximale k_i; le groupe U est plongeable sur $(k_i)_s = \bar{k}$ en vertu de 3.6.

3.10. *Remarque.* Des cas particuliers des Corollaires 3.5, 3.7 et 3.8 ont été obtenus dans [2, §8]. Les Propositions 8.6 et 8.7 de [2], que nous ne réénonçons pas, ainsi que leurs démonstrations, s'étendent aussi à présent, sans modification, à un corps parfait quelconque.

3.11. L'énoncé suivant, qui répond à une question de D. Gorenstein, concerne un sous-groupe H de $G(k)$. Nous supposons que k est parfait et que l'une des conditions suivantes est remplie:

   (i) H contient le groupe G^+ engendré par les éléments unipotents de $G(k)$;

   (ii) car $k = 3$ (resp. 2), l'endomorphisme de Frobenius de k possède une racine carrée σ, G est un groupe déployé de type $\mathbf{G}_2$ (resp. $\mathbf{B}_2$ ou $\mathbf{F}_4$) et H est un groupe de Ree (resp. de Suzuki ou de Ree), formé des points fixes d'un automorphisme α d'ordre 2 de $G(k)$ induit par une isogénie $G \to {}^\sigma G$ (où ${}^\sigma G$ désigne le groupe déduit de G par le changement de base σ).

Dans les deux cas, H possède une BN-paire naturelle (à conjugaison près) dont les sous-groupes paraboliques sont les sous-groupes de la forme $P \cap H$ où P est un k-sous-groupe parabolique de G quelconque dans le cas (i) et tel que $\alpha(P(k)) = P(k)$ dans le cas (ii). Pour un tel P, on pose, par définition, $R_u(P \cap H) = R_u(P) \cap H$, ce qui est légitime parce que P est entièrement déterminé par $P \cap H$.

Si U est un sous-groupe unipotent de H, posons

$$\mathcal{Q}_H(U) = Q = H \cap \mathcal{P}_{G,\bar{k}}(U), \tag{1}$$

ce qui a un sens puisque U est k-plongeable (3.7). Vu 2.5, $\mathcal{Q}_H(U)$ est un sous-groupe parabolique de H, et l'on a

$$U \subset R_u(Q), \quad \mathcal{N}_H(U) \subset Q, \tag{2}$$

d'où la proposition suivante:

3.12. Proposition. *Soit H comme en 3.11. Les éléments maximaux de l'ensemble des normalisateurs de sous-groupes unipotents non triviaux de H sont les sous-groupes paraboliques maximaux de H.*

3.13. *Remarque.* Soient γ un automorphisme de k, ${}^{\gamma}G$ le k-groupe déduit de G par changement de base et $\gamma_{*}: G(k) \to {}^{\varphi}G(k)$ l'isomorphisme canonique. Il est clair que l'on a

$$\mathscr{Q}_{\gamma_{*}(H)}\big(\gamma_{*}(U)\big) = \gamma_{*}\big(\mathscr{Q}_{H}(U)\big). \tag{1}$$

En tenant compte de [8] si k est fini, et, si k est infini, de [3; 10] dans le cas (i) de 3.11 et d'un résultat analogue, non publié, dans le cas (ii) de 3.11, on en déduit aisément que $\mathscr{Q}_{H}(U)$ *est invariant par tout automorphisme de H stabilisant U.*

Références

1. Borel, A.: Linear algebraic groups. Notes by H. Bass. New York: Benjamin 1969.
2. — Tits, J.: a) Groupes réductifs, Publ. math. I.H.E.S. **27**, 55–150 (1965). b) Compléments, ibid. (à paraître).
3. — — On "abstract" homomorphisms of simple algebraic groups. Proceedings of the Bombay Colloquium on Algebraic Geometry, 75–82 (1968).
4. Morozov, V.V.: Démonstration du théorème de régularité. Usp. M. Nauk **XI**, fasc. 5, 191–194 (1956).
5. Platonov, V. P.: Proof of the finiteness hypothesis for solvable subgroups of algebraic groups. Sibirskii M. J. **X**, 1084–1090 (1969).
6. Rosenlicht, M.: Questions of rationality for solvable algebraic groups over nonperfect fields. Annali di Mat. (IV) **61**, 97–120 (1963).
7. Springer, T.A., Steinberg, R.: Conjugacy classes, Seminar on algebraic groups and related finite groups, part E, 100 p., Springer Lecture Notes **131** (1969).
8. Steinberg, R.: Automorphisms of finite linear groups. Canadian J. M. **12**, 606–615 (1960).
9. Tits, J.: Groupes semi-simples isotropes. Coll. sur la théorie des groupes algébriques, Bruxelles 1962, 137–147.
10. — Homomorphismes et automorphismes "abstraits" de groupes algébriques et arithmétiques. Proceedings Int. Congress of Math. Nice 1970 (à paraître).

A. Borel
The Institute for Advanced Study
School of Mathematics
Princeton, N.J. 08540 USA

J. Tits
Mathematisches Institut der Universität
BRD-5300 Bonn, Wegelerstr. 10
Deutschland

(Reçu le 8 décembre 1970)

93.

Cohomologie réelle stable de groupes S-arithmétiques

C. R. Acad. Sci., Paris **274** (1972) 1700–1702

G désigne un groupe algébrique linéaire sur $\mathbf{Q}$, dont la composante neutre est presque simple sur $\mathbf{Q}$ et de $\mathbf{Q}$-rang [1] non nul. On énonce un théorème de comparaison, en basses dimensions, entre la cohomologie réelle d'un sous-groupe S-arithmétique de G et la cohomologie relative de l'algèbre de Lie L (G (**R**)) du groupe des points réels de G modulo l'algèbre de Lie L (K) d'un sous-groupe compact maximal K de G (**R**). On l'applique à la détermination de la cohomologie réelle stable de groupes S-arithmétiques classiques et du rang sur $\mathbf{Q}$ de certains groupes K_i de Quillen [2] et $_{\pm 1}L_i$ de M. Karoubi [3].

1. On note X l'espace riemannien symétrique des sous-groupes compacts maximaux de G (**R**), C_X le complexe des formes différentielles extérieures C^∞ sur X, et, si L est un sous-groupe de G (**R**), C_X^L le sous-complexe des éléments de C_X invariants par L. On écrira I_G pour C_X^L si $L = G (\mathbf{R})^0$ est la composante neutre en topologie ordinaire de G (**R**). On sait que I_G est de dimension finie, formé de formes harmoniques, s'identifie canoniquement à H* (L (G (**R**)), L (K)) ou à la cohomologie réelle de l'espace symétrique « dual » de X, quotient par K d'un sous-groupe compact maximal de G (**C**) contenant K, ou encore à la cohomologie d'Eilenberg-Mac Lane réelle de G (**R**)^0, calculée à l'aide de cochaînes C^∞ ou continues [4]. L'algèbre I_G est stable par G (**R**), qui y opère par l'intermédiaire d'un groupe fini. Si L est un sous-groupe discret de G (**R**), alors

$$\mathrm{H}^* (C_X^L) = \mathrm{H}^* (X/L) = \mathrm{H}^* (L) \quad [5].$$

L'inclusion $I_G^L \to C_X^L$ induit donc un homomorphisme $j^* : I_G^L \to \mathrm{H}^* (L)$ dont la restriction à l'espace $I_G^{q,L}$ des éléments homogènes de degré q ($q \geqq 0$) de I_G^L est notée j^q.

Vu l'hypothèse sur le $\mathbf{Q}$-rang, L (G (**R**)) est produit direct d'algèbres de Lie simples non-compactes. Les résultats de Y. Matsushima [6] entraînent l'existence d'un entier m (G) $\geqq 0$, défini à partir de la structure de L (G (**R**)), tel que si L est discret et G (**R**)/L compact, alors j^q est un isomorphisme pour $q \leqq m$ (G) [7].

2. Fixons un ordre sur l'ensemble des $\mathbf{Q}$-racines [1] de G. Notons r la demi-somme des $\mathbf{Q}$-racines positives, d_0 celle des $\mathbf{Q}$-racines simples. Soit m' (G) le plus grand entier q tel que $r - d_0 - b$ soit combinaison linéaire à coefficients *tous* > 0 des $\mathbf{Q}$-racines simples lorsque b parcourt les sommes de q $\mathbf{Q}$-racines positives, une $\mathbf{Q}$-racine pouvant intervenir au plus un nombre de fois égal à sa multiplicité.

THÉORÈME 1. — *On conserve les notations précédentes. Soit* S *un ensemble fini de places de* $\mathbf{Q}$ *contenant la place à l'infini. Soient* Γ *un sous-groupe arithmétique de* G *et* Γ_S *un sous-groupe* S-*arithmétique de* G *contenant* Γ.

Alors $j^q : \mathrm{I}_\mathrm{G}^{q,\Gamma} \to \mathrm{H}^q\,(\Gamma)$ *est un isomorphisme pour*

$$q \leqq q\,(\mathrm{G}) = \min\,(m\,(\mathrm{G}),\, m'\,(\mathrm{G}))\ (^9).$$

Il en est de même de res. : $\mathrm{H}^q\,(\Gamma_\mathrm{S}) \to \mathrm{H}^q\,(\Gamma)$ *si* $\mathrm{I}_\mathrm{G}^q = (\mathrm{I}_\mathrm{G}^q)^{\mathrm{G}(\mathbf{R})}$, *en particulier si* G *est connexe et simplement connexe.*

Soient maintenant $(\mathrm{G}_n,\, \Gamma_n,\, \Gamma_{n,\mathrm{S}},\, f_n)$ une suite de groupes tels que $(\mathrm{G},\, \Gamma,\, \Gamma_\mathrm{S})$ et de $\mathbf{Q}$-morphismes injectifs, où f_n envoie $(\mathrm{G}_n,\, \Gamma_n,\, \Gamma_{n,\mathrm{S}})$ dans $(\mathrm{G}_{n+1},\, \Gamma_{n+1},\, \Gamma_{n+1,\mathrm{S}})$ $(n = 1,\, 2,\, \ldots)$.

Corollaire. — *Supposons que* $q\,(\mathrm{G}_n) \to \infty$ *avec* n *et que* $\mathrm{I}_{\mathrm{G}_n}^q = (\mathrm{I}_{\mathrm{G}_n}^q)^{\mathrm{G}_n(\mathbf{R})}$ *pour* $q \leqq q\,(\mathrm{G}_n)$, $n \geqq 1$. *Alors* j^q *et res. induisent des isomorphismes*

$$(1) \qquad \varprojlim \mathrm{I}_{\mathrm{G}_n}^q \cong \varprojlim \mathrm{H}^q\,(\Gamma_n) \cong \varprojlim \mathrm{H}^q\,(\Gamma_{n,\mathrm{S}}), \qquad (q = 0,1,\ldots).$$

La première limite projective est prise par rapport à des homomorphismes f_n^* naturellement associés aux f_n. Soit $q \geqq 0$. Disons que la suite $\{\,\mathrm{I}_{\mathrm{G}_n}^q\,\}$ est stationnaire s'il existe $n\,(q)$ tel que f_n^* soit un isomorphisme en dimension q pour tout $n \geqq n\,(q)$. On a alors

$$(2) \qquad \mathrm{I}_{\mathrm{G}_{a(q)}}^q \cong \mathrm{H}^q\,(\Gamma_n) \cong \mathrm{H}^q\,(\Gamma_{n,\mathrm{S}}) \cong \mathrm{H}^q\,(\mathrm{G}_n\,(\mathbf{Q}))$$

pour n assez grand, et cela vaut encore, aussi en homologie, si l'on remplace Γ_n [resp. $\Gamma_{n,\mathrm{S}}$, resp. $\mathrm{G}_n\,(\mathbf{Q})$] par la limite inductive des Γ_n [resp. $\Gamma_{n,\mathrm{S}}$, resp. $\mathrm{G}_n\,(\mathbf{Q})$].

3. Applications. — Soient k un corps de nombres, Σ l'ensemble des places à l'infini de k, $\mathfrak{o}$ l'anneau des entiers de k, S un ensemble fini de places de k contenant Σ et $\mathfrak{o}_\mathrm{S}$ l'anneau des S-entiers de k. Étant donné un anneau A à élément unité, soit

$$(3) \quad \mathbf{SL}\,(\mathrm{A}) = \varinjlim \mathbf{SL}_n\,(\mathrm{A}), \qquad \mathbf{Sp}\,(\mathrm{A}) = \varinjlim \mathbf{Sp}_{2n}\,(\mathrm{A}), \qquad \mathbf{O}\,(\mathrm{A}) = \varinjlim \mathbf{O}_{n,n}\,(\mathrm{A}).$$

Notons $\mathrm{E}\,(\{\,x_i,\, d^0\,x_i\,\})$ [resp. $\mathrm{P}\,(\{\,x_i,\, d^0\,x_i\,\})$] l'algèbre extérieure (resp. de polynomes) sur $\mathbf{R}$, sur les générateurs x_i $(i \in \mathrm{I})$, $d^0\,x_i$ étant le degré de x_i. Le nº 2 entraîne :

Théorème 2. — *Soient* A *l'un des anneaux* k, $\mathfrak{o}$, $\mathfrak{o}_\mathrm{S}$ *et* L *l'un des groupes* $\mathbf{SL}\,(\mathrm{A})$, $\mathbf{Sp}\,(\mathrm{A})$, $\mathbf{O}\,(\mathrm{A})$. *Alors* $\mathrm{H}^*\,(\mathrm{L}) = \bigotimes\limits_{v \in \Sigma} \mathrm{I}_v$, *où les* I_v *sont donnés par le tableau suivant, dans lequel* $i = 1,\, 2,\, \ldots$:

$\mathbf{SL}\,(\mathrm{A})$:	$\mathrm{E}\,(\{\,x_i,\, 4\,i + 1\,\})$ (v réelle),	$\mathrm{E}\,(\{\,x_i,\, 2\,i + 1\,\})$ (v complexe),
$\mathbf{Sp}\,(\mathrm{A})$:	$\mathrm{P}\,(\{\,x_i,\, 4\,i - 2\,\})$ (v réelle),	$\mathrm{E}\,(\{\,x_i,\, 4\,i - 1\,\})$ (v complexe),
$\mathbf{O}\,(\mathrm{A})$:	$\mathrm{P}\,(\{\,x_i,\, 4\,i\,\})$ (v réelle),	$\mathrm{E}\,(\{\,x_i,\, 4\,i - 1\,\})$ (v complexe).

On retrouve en particulier la nullité de $\mathrm{H}^2\,(\mathbf{SL}_n\,(\mathfrak{o}))$ pour n assez grand, démontrée par H. Garland (^{8}). D'autre part, M. Karoubi (^{3}) a obtenu des résultats partiels sur l'homologie de $\mathbf{Sp}\,(\mathrm{A})$ ou $\mathbf{O}\,(\mathrm{A})$ pour des anneaux (et des coefficients) plus généraux qui, pour $\mathrm{A} = k$, mettent en évidence l'existence des facteurs I_v (v réelle).

Si $i \geq 2$, le groupe $K_i A$ de $(^2)$ est le $i^{\text{ième}}$ groupe d'homotopie d'un espace de lacets infinis ayant même homologie (à coefficients triviaux quelconques) que $\mathbf{SL}(A)$. De même, le groupe $_{-1}L_i(A)$ [resp. $_1L_i(A)$] défini dans $(^3)$, lorsque 2 est inversible dans A, est le $i^{\text{ième}}$ groupe d'homotopie d'un espace de lacets infinis ayant l'homologie de $\mathbf{Sp}(A)$ [resp. $\mathbf{O}(A)$].

COROLLAIRE. — *Pour $i \geq 2$, la dimension du $\mathbf{Q}$-espace vectoriel $K_i A \otimes \mathbf{Q}$ [resp. $_{-1}L_i(A) \otimes \mathbf{Q}$, resp. $_1L_i(A) \otimes \mathbf{Q}$] est périodique de période quatre et vaut 0, $r_1 + r_2$, 0, r_2 (resp. 0, 0, r_1, r_2, resp. r_1, 0, 0, r_2) suivant que i est congru à 0, 1, 2, 3 mod 4, où r_1 (resp. r_2) est le nombre de places réelles (resp. complexes) de k.*

4. PRINCIPE DE DÉMONSTRATION DU THÉORÈME 1. — Pour établir la première assertion il suffit, en vertu du paragraphe 3.4 de $(^8)$ et de résultats connus sur les formes harmoniques de carré intégrable, de prouver que l'on peut calculer la cohomologie de X/Γ à l'aide d'un sous-complexe de C_X^Γ contenant I_G^Γ et dont les éléments de degré $\leq m'(G)$ sont de carré intégrable. Pour cela on utilise le quotient $\overline{X}/\Gamma$, où $\overline{X}$ est la variété à coins attachée à X $(^{10})$, une description de la métrique dans un coin, et la théorie des faisceaux. On passe aux groupes S-arithmétiques en se servant, comme en $(^{11})$, du produit $X_S = \overline{X} \times X_S'$, où X_S' est le produit des immeubles de Bruhat-Tits attachés aux groupes $G(\mathbf{Q}_p)$ ($p \in S - \{\infty\}$), et en étudiant la suite spectrale de la projection $X_S/\Gamma_S \to X_S'/\Gamma_S$.

(*) Séance du 5 juin 1972.

$(^1)$ *Cf.* A. BOREL et J. TITS, *Publ. Math. I. H. E. S.*, 27, 1965, p. 55-150.

$(^2)$ *Cf.* D. QUILLEN, *Actes Congrès int. Math.*, Nice, 2, 1970, p. 47-51.

$(^3)$ *Cf.* M. KAROUBI, *Comptes rendus*, 273, série A, 1971, p. 1030.

$(^4)$ *Cf.* W. T. VAN EST, *Proc. Konink. Nederl. Ak. v. Wet.*, Series A, 58, 1955, p. 542-544; 61, 1958, p. 399-413; G. HOCHSCHILD et G. D. MOSTOW, *Illinois J. Math.*, 6, 1962, p. 367-401.

$(^5)$ Sauf mention expresse du contraire, l'homologie et la cohomologie sont à coefficients réels.

$(^6)$ *Cf.* Y. MATSUSHIMA, *Osaka Math. J.*, 14, 1962, p. 1-20.

$(^7)$ Il faut légèrement modifier les calculs des paragraphes 6 et 7 de $(^6)$ pour englober le cas où L (G $(\mathbf{R})$) n'est pas simple. Pour les valeurs de m (G) lorsque G $(\mathbf{R})$ est simple, *voir* $(^6)$ et S. KANEYUKI-T. NAGANO, *Osaka Math. J.*, 14, 1962, p. 241-252.

$(^8)$ *Cf.* H. GARLAND, *Annals of Math.*, (2), 94, 1971, p. 534-548.

$(^9)$ La surjectivité de j^q résulte aussi du paragraphe 3.4 de $(^8)$, compte tenu d'un résultat de H. GARLAND et W. C. HSIANG, *Proc. Nat. Acad. Sci. U. S. A.*, 59, 1968, p. 354-360, jusqu'à une constante pouvant être éventuellement plus grande que q (G).

$(^{10})$ *Cf.* A. BOREL et J.-P. SERRE, *Comptes rendus*, 271, série A, 1970, p. 1156.

$(^{11})$ *Cf.* A. BOREL et J.-P. SERRE, *Comptes rendus*, 272, série A, 1971, p. 110.

Institute for Advanced Study,
Princeton, N. J. 08540,
U. S. A.

94.

(avec J. Tits)

Compléments à l'article: ‹Groupes réductifs›

Publ. Math., Inst. Hautes Etud. Sci. **41** (1972) 253–276

Les compléments à [3] donnés ici portent principalement sur les isogénies centrales (§ 2), l'adhérence des doubles classes d'une décomposition de Bruhat (§ 3) et le groupe fondamental des sous-groupes déployés maximaux construits dans le § 7 de [3] (§ 4). Nous profitons aussi de cette occasion pour rectifier quelques erreurs de [3]; celles indiquées au § 5 nous ont été en partie signalées par J. Humphreys et S. P. Wang; nous les en remercions vivement.

Les notations utilisées sans explication sont celles de [3].

L'algèbre de Lie d'un groupe algébrique G, H, ... est notée L(G), L(H), ... et $L(f) : L(G) \to L(H)$ désigne le morphisme d'algèbres de Lie associé à un morphisme de groupes algébriques $f : G \to H$. C'est donc la différentielle de f en l'élément neutre.

Si A est une partie d'un groupe algébrique G, le centralisateur de A dans L(G) est par définition

$$\mathscr{Z}_{L(G)}(A) = \{X \in L(G) \mid (\mathrm{Ad}\, a)(X) = X, \quad (a \in A)\}.$$

Si X est un groupe ou une algèbre de Lie, $\mathscr{Z}(X)$ désigne son centre.

Si k est un corps commutatif, $_k\mathbf{Add}$ ou $\mathbf{Add}$ désigne le groupe additif sur k.

Dans toute la suite, k est un corps commutatif, p sa caractéristique, $\bar{k}$ une clôture algébrique de k, k_s la clôture séparable de k dans $\bar{k}$, et G un k-groupe connexe réductif.

1. Groupes orthogonaux.

(1.1) Dans [3; 4.24], on a supposé implicitement que le groupe orthogonal O(Q) est défini sur k. C'est toujours le cas si la caractéristique p de k est $\neq 2$. Lorsque $p = 2$, cela est vrai si Q est non dégénérée et de défaut ≤ 1 (cf. [10], Remarque p. 30, en tenant compte de la terminologie introduite au bas de la p. 21).

(1.2) Le nº 4.26 de [3] a pour but de fournir un exemple de deux k-groupes connexes réductifs G, G′ de k-rangs différents mais liés par une k-isogénie (qui ne peut être séparable vu [3; 4.25], ni même centrale vu (2.19) ci-dessous). Cet exemple est erroné, et le nº 4.26 de [3] doit être remplacé, à partir de la troisième ligne, par le texte suivant :

Supposons k de caractéristique 2 ; soit Q′ une forme quadratique à deux variables, non dégénérée et non défective, qui ne représente pas 1 (on suppose qu'une telle forme

129

existe, ce qui implique en particulier que k est non parfait); soient Q la forme à trois variables définie par $Q(x, y, z) = x^2 + Q'(y, z)$, et enfin $G = O(Q)$ le groupe orthogonal associé à Q. La forme Q ne représente pas zéro sur k, donc G est anisotrope sur k (4.24). D'autre part, la forme bilinéaire associée à Q admet la droite $y = z = o$ comme noyau. On en déduit aisément que les éléments de G sont les matrices de la forme

$$M = \begin{pmatrix} 1 & Q'(a, b)^{1/2} + \xi & Q'(c, d)^{1/2} + \eta \\ o & a & c \\ o & b & d \end{pmatrix}, \qquad (ad - bc = 1),$$

où $\xi = Q'(1, o)^{1/2}$, $\eta = Q'(o, 1)^{1/2}$, donc que $M \mapsto \begin{pmatrix} a & c \\ b & d \end{pmatrix}$ est une k-isogénie de G sur $\mathbf{SL}_2 = G'$. Cependant, $r_k(G) = o$ et $r_k(G') = 1$.

2. Isogénies centrales.

La proposition 4.25 de [3] montre que le k-rang est conservé par une isogénie séparable. Nous voulons prouver ici qu'il en est encore de même pour une isogénie centrale et que, plus généralement, une telle isogénie conserve les invariants sur k d'un k-groupe réductif étudiés dans [3; §§ 4, 5].

Définition (**2.1**). — Soit $f : H \to H'$ un morphisme de groupes algébriques. Nous disons que f est *quasi-central* si son noyau (ensembliste) est contenu dans le centre de H. En d'autres termes, f est quasi-central si le commutateur $H \times H \to H$ se factorise à travers $f(H) \times f(H)$ *en tant qu'application*, c'est-à-dire s'il existe une application $\kappa : f(H) \times f(H) \to H$ telle que $\kappa(f(x), f(y)) = xyx^{-1}y^{-1}$ pour $x, y \in H$.

Définition (**2.2**). — Un morphisme $f : H \to H'$ de groupes algébriques est dit *central* si le commutateur $H \times H \to H$ se factorise à travers $f(H) \times f(H)$ *en tant que morphisme*, c'est-à-dire s'il est quasi-central et si l'application κ de (2.1) (qui est évidemment unique) est un morphisme.

Remarque (**2.3**). — Soit k un corps de définition de f. Il est facile de voir que, avec les notations de [1; § 1], f est central si et seulement si $\operatorname{Ker} f(C) \subset \mathscr{Z}(H(C))$ pour toute k-algèbre C. Dans le langage des schémas en groupes, cela revient à dire que le noyau (schématique) de f est contenu dans le centre de H.

Proposition (**2.4**). — *Soit* $f : H \to H'$ *un morphisme de groupes algébriques. Les conditions suivantes sont équivalentes :*

(i) *f est quasi-central (resp. central);*

(ii) *pour tout* $h \in H$, *il existe une application (resp. un morphisme de variétés)* $\varphi_h : f(H) \to H$ *tel que* $\varphi_h(f(x)) = xhx^{-1}$ *pour* $x \in H$;

(iii) *l'action de* H *sur soi-même par les automorphismes intérieurs induit une action ensembliste* (resp. *algébrique*) *de* $f(\mathrm{H})$ *sur* H, *c'est-à-dire qu'il existe une application* (resp. *un morphisme de variétés*) $\iota : f(\mathrm{H}) \times \mathrm{H} \to \mathrm{H}$ *tel que* $\iota(f(x), y) = xyx^{-1}$ *pour* $x, y \in \mathrm{H}$.

Dans le cas quasi-central, la proposition est évidente. Pour établir l'implication (i) $\Rightarrow$ (ii) dans le cas central, il suffit de remarquer que si κ est défini comme en (2.1), le morphisme $\varphi_h : x' \mapsto \kappa(x', f(h)) . h$ (pour $x' \in f(\mathrm{H})$) possède la propriété requise. Enfin, les implications (ii) $\Rightarrow$ (iii) et (iii) $\Rightarrow$ (i) sont des conséquences immédiates du lemme suivant, où on pose successivement $\mathrm{X} = \mathrm{Z} = \mathrm{H}$, $\mathrm{X}' = f(\mathrm{H})$, $\mathrm{Y} = \mathrm{H}$, $\psi(x, y) = xyx^{-1}$ et $\mathrm{X} = \mathrm{Z} = \mathrm{H}$, $\mathrm{X}' = f(\mathrm{H})$, $\mathrm{Y} = f(\mathrm{H})$, $\psi(x, y') = x . \iota(y', x^{-1})$ (d'où $\psi' = \kappa$).

Lemme (**2.5**). — *Soient* X, X', Y *des ensembles algébriques* (*noethériens*), Z *un ensemble algébrique affine,* $f : \mathrm{X} \to \mathrm{X}'$ *un morphisme surjectif,* $\psi : \mathrm{X} \times \mathrm{Y} \to \mathrm{Z}$ *un morphisme et* $\psi' : \mathrm{X}' \times \mathrm{Y} \to \mathrm{Z}$ *une application telle que* $\psi = \psi' \circ (f \times \mathrm{id}_{\mathrm{Y}})$. *Alors, pour que* ψ' *soit un morphisme, il faut et il suffit que* $\psi'|_{\mathrm{X}' \times \{y\}}$ *en soit un pour tout* $y \in \mathrm{Y}$.

(N.B. — Nous ignorons si ce lemme reste valable pour un ensemble algébrique Z quelconque. P. Gabriel nous a fait remarquer qu'il l'est en tout cas si on suppose le morphisme f fidèlement plat. Cela permet d'étendre la proposition (2.4) à des groupes algébriques non affines.)

La condition est évidemment nécessaire. Montrons qu'elle est aussi suffisante. L'assertion est évidemment locale en X' et Y, que nous pouvons donc supposer affines. Si (U_i) est un recouvrement fini de X par des ouverts affines, il suffit de remplacer X par $\coprod_i \mathrm{U}_i$ et f par le composé de f et de l'application canonique $\coprod_i \mathrm{U}_i \to \mathrm{X}$ pour se ramener au cas où X est affine, ce que nous supposerons aussi.

Soient $\mathrm{K}[\mathrm{X}], \mathrm{K}[\mathrm{X}'], \ldots$ les algèbres affines de X, X', … sur un corps de définition K algébriquement clos. Nous devons faire voir que si $a \in \mathrm{K}[\mathrm{Z}]$, alors $\psi'^*(a) \in \mathrm{K}[\mathrm{X}'] \otimes \mathrm{K}[\mathrm{Y}]$. Comme f est surjectif, cela revient à prouver que $\psi^*(a) \in f^*(\mathrm{K}[\mathrm{X}']) \otimes \mathrm{K}[\mathrm{Y}]$. Posons

$$\psi^*(a) = \sum_{1 \le i \le r} b_i \otimes c_i, \qquad (b_i \in \mathrm{K}[\mathrm{X}], c_i \in \mathrm{K}[\mathrm{Y}], i = 1, \ldots, r),$$

les c_i étant supposés linéairement indépendants. Soient $y_1, \ldots, y_r$ des points de Y tels que la matrice $(c_i(y_j))$ soit inversible. Comme la restriction de ψ à $\mathrm{X} \times \{y_j\}$ est un morphisme $(j = 1, \ldots, r)$, on a $\sum_i c_i(y_j) . b_i \in f^*(\mathrm{K}[\mathrm{X}'])$ pour tout j, d'où $b_i \in f^*(\mathrm{K}[\mathrm{X}'])$, q.e.d.

Remarque (**2.6**). — Si H, H' sont des k-groupes et $f : \mathrm{H} \to \mathrm{H}'$ un k-morphisme central, les morphismes κ et ι définis en (2.1) et (2.4) (iii) sont définis sur k, et il en est de même de φ_h ((2.4) (ii)) si $h \in \mathrm{H}_k$: cela résulte de ce que si $\alpha : \mathrm{X} \to \mathrm{X}'$ est un k-morphisme surjectif de k-ensembles algébriques, un morphisme α' de X' dans un autre k-ensemble algébrique est défini sur k si et seulement si $\alpha' \circ \alpha$ l'est.

Proposition (**2.7**). — *Soient* H, H' *des* k-*groupes algébriques et* $f : \mathrm{H} \to \mathrm{H}'$ *un* k-*morphisme central surjectif. Alors,* $f(\mathrm{H}_k)$ *est un sous-groupe distingué de* H'_k *et* $\mathrm{H}'_k / f(\mathrm{H}_k)$ *est commutatif.*

(Cette propriété est bien connue : cf. par ex. [8].)

Le commutateur $H' \times H' \to H'$ coïncidant avec le composé $f \circ \kappa$ (où κ est défini comme en (2.1)), on a, vu la remarque (2.6), $(H'_k, H'_k) = f(\kappa(H'_k \times H'_k)) \subset f(H_k)$, d'où l'assertion.

Lemme (**2.8**). — *Soient* H, H$'$, H$''$ *des groupes algébriques*, $h : H'' \to H$ *un morphisme surjectif et* $f : H \to H'$ *un morphisme de noyau fini. Supposons* H$''$ *connexe.*

(i) *Si* $h' : H'' \to H$ *est un morphisme tel que* $f \circ h' = f \circ h$, *on a* $h' = h$.

(ii) *Tout automorphisme de* H$'$ *« se relève » d'au plus une façon en un automorphisme de* H.

(iii) *Supposons* H, H$'$, H$''$ *et* $f \circ h$ *définis sur* k, *et* f *et* h *définis sur une extension séparable* k' *de* k. *Alors, pour que* h *soit défini sur* k, *il faut et il suffit que* f *le soit.*

(i) Si $f \circ h' = f \circ h$, l'application $x \mapsto h(x) . h'(x)^{-1}$ est un morphisme de variétés de H$''$ dans Ker f, dont l'image contient 1. Comme H$''$ est connexe et Ker f fini, cela implique que $h(x) . h'(x)^{-1} = 1$ pour tout x.

(ii) est un cas particulier de (i) (poser H$'' = $H).

(iii) Nous pouvons évidemment supposer k' séparablement clos. Tout élément γ de Aut(k'/k) définit des automorphismes de H(k'), H$'(k')$, H$''(k')$, que nous désignerons aussi par γ. Par hypothèse, $f \circ h \circ \gamma = \gamma \circ f \circ h$ pour tout $\gamma \in$ Aut(k'/k). Si f est défini sur k, on a $f \circ \gamma = \gamma \circ f$, d'où $f \circ h \circ \gamma = f \circ \gamma \circ h$ et, vu (i), $h \circ \gamma = \gamma \circ h$. Réciproquement, si h est défini sur k, on a $h \circ \gamma = \gamma \circ h$, d'où $f \circ \gamma \circ h = \gamma \circ f \circ h$ et, comme h est surjectif, $f \circ \gamma = \gamma \circ f$.

Proposition (**2.9**). — *Soient* T *un tore maximal et* N *un sous-groupe distingué fermé de* G. *Alors, les propriétés suivantes sont équivalentes* : (i) N *est central*; (ii) N $\subset$ T; (iii) N *est contenu dans l'intersection des tores maximaux de* G; (iv) N *est formé d'éléments semi-simples*; (v) *pour tout* $a \in \Phi(T, G)$, *on a* $U_a \not\subset N$.

Les implications (i) $\Leftrightarrow$ (ii) $\Leftrightarrow$ (iii) $\Rightarrow$ (iv) $\Rightarrow$ (v) sont évidentes, compte tenu de la conjugaison des tores maximaux de G et du fait que leur intersection est le centre de G [1; 13.17, p. 316].

Montrons que (v) implique (ii). Le groupe N^0 étant normalisé par T, il est engendré par son intersection avec T et par les U_a qu'il contient [3; 3.4]. Si $U_a \not\subset N$ quel que soit $a \in \Phi(T, G)$, on a donc N$^0 \subset$ T et, quitte à remplacer G, T et N par G/N^0, T/N^0 et N/N^0, nous pouvons supposer que N est fini. Mais alors, N est central, donc contenu dans T, ce qui achève la démonstration.

Remarque. — La proposition est encore valable si on ne suppose pas N fermé.

Corollaire (**2.10**). — *Pour que* N *soit central, il faut et il suffit que* N^0 *le soit.*
On utilise la propriété (2.9) (v).

Proposition (**2.11**). — *Soient* G$'$ *un groupe algébrique*, $f : G \to G'$ *un morphisme et* T *un tore maximal de* G. *Alors, les conditions suivantes sont équivalentes* :

(i) f *est quasi-central*;

(ii) *la restriction de f à tout sous-groupe unipotent fermé et connexe de* G *est injective*;

(iii) *pour toute racine* $a \in \Phi(T, G)$, *il existe un entier n et une racine* $a' \in \Phi(f(T), f(G))$ *tels que* $a = n.(f|_T)^*(a')$;

(iv) *l'image réciproque par f d'un tore maximal de* G' *est un tore maximal de* G.

Lorsque ces conditions sont remplies, la restriction de f à tout sous-groupe unipotent de G *est injective, et l'entier n de* (iii) *est une puissance de l'exposant caractéristique de k.*

Si f est quasi-central, son noyau est formé d'éléments semi-simples, vu (2.9), donc la restriction de f à tout sous-groupe unipotent de G est injective; en particulier, (ii) est satisfaite.

Supposons que la restriction de f à tout sous-groupe unipotent fermé connexe soit injective. Si $a \in \Phi(T, G)$, le groupe $f(U_a)$ est un sous-groupe additif de $f(G)$ normalisé par $f(T)$; il existe donc $a' \in \Phi(f(T), f(G))$ tel que $f(U_a) = U_{a'}$. Comme f induit une isogénie purement inséparable de U_a sur $U_{a'}$, on a $a = n.(f|_T)^*(a')$, où n est une puissance de l'exposant caractéristique de k.

Enfin, (i) $\Leftrightarrow$ (iv) vu (2.9) et $[1 ; 11.14]$. Si (iii) est satisfaite, il est clair que $\operatorname{Ker} f$ ne contient aucun des U_a, pour $a \in \Phi(T, G)$, donc f est quasi-central, vu (2.9).

Corollaire $(\mathbf{2.12})$. — *Soient* G' *un k-groupe algébrique et* $f : G \to G'$ *un k-morphisme quasi-central. Alors* $\operatorname{Ker} f$ *est défini sur k.*

Comme G possède un tore maximal défini sur k $[1 ; 18.2]$, (2.9) et $[3 ; 1.6]$ entraînent que $\operatorname{Ker} f$ est défini sur une extension séparable de k. Comme il est k-fermé, l'assertion s'ensuit.

Proposition $(\mathbf{2.13})$. — *Soient* T *un tore maximal de* G *et* M *un idéal de* L(G) *stable par* G. *Alors les propriétés suivantes sont équivalentes* : (i) $M \subset \mathscr{L}_{L(G)}(G)$; (ii) $M \subset L(T)$; (iii) M *est contenu dans l'intersection* Z *des algèbres de Lie des tores maximaux de* G; (iv) M *est formé d'éléments semi-simples*; (v) M *ne contient aucune des algèbres* $L(U_a)$ *pour* $a \in \Phi(T, G)$. *On a* $\mathscr{L}(L(G)) = \mathscr{L}_{L(G)}(G) = Z$.

(i) $\Rightarrow$ (ii) parce que $\mathscr{L}_{L(G)}(T) = L(\mathscr{L}(T)) = L(T)$ (cf. $[1 ; (9.2)$, Cor. p. 229$]$).

(ii) $\Rightarrow$ (iii) vu la conjugaison des tores maximaux de G.

(iii) $\Rightarrow$ (i) puisque tout tore centralise son algèbre de Lie et que G est engendré par ses tores maximaux.

Les implications (iii) $\Rightarrow$ (iv) $\Rightarrow$ (v) sont évidentes.

L'idéal M est la somme de ses sous-espaces propres pour T, donc est engendré par son intersection avec $L(T)$ et par les $L(U_a)$ qu'il contient, d'où l'implication (v) $\Rightarrow$ (ii). L'équivalence des propriétés (i) à (v) est ainsi établie.

L'idéal $\mathscr{L}(L(G))$ est stable par tout automorphisme de L(G), donc par G, et il ne contient aucun des $L(U_a)$ car $L(U_a)$ n'est pas central dans l'algèbre de Lie du groupe isomorphe à SL_2 ou à PSL_2 engendré par U_a et U_{-a}. Par conséquent $\mathscr{L}(L(G)) \subset \mathscr{L}_{L(G)}(G)$. L'inclusion inverse étant évidente, la dernière assertion s'ensuit, vu l'équivalence de (i) et (iii).

Lemme (**2.14**). — *Soient* G' *un groupe algébrique,* $f : G \to G'$ *un morphisme quasi-central surjectif,* T *un tore maximal de* G, U^+ *et* U^- *deux sous-groupes unipotents maximaux de* G *normalisés par* T *et opposés, et* $\psi : G \times X \to X$ *une action algébrique de* G *sur un ensemble algébrique* X. *Alors, pour que* ψ *se factorise à travers* $G' \times X$, *c'est-à-dire pour qu'il existe un morphisme* $\psi' : G' \times X \to X$ *tel que* $\psi = \psi' \circ (f \times \mathrm{id}_X)$, *il faut et il suffit que les restrictions de* ψ *à* $U^+ \times X$, $U^- \times X$ *et* $T \times X$ *se factorisent respectivement à travers* $f(U^+) \times X$, $f(U^-) \times X$ *et* $f(T) \times X$.

La condition est évidemment nécessaire. Supposons donc qu'elle soit remplie et soient $\upsilon^\pm : f(U^\pm) \times X \to X$ et $\tau : f(T) \times X \to X$ les morphismes définis par

$$\psi|_{U^\pm \times X} = \upsilon^\pm \circ (f \times \mathrm{id}_X), \qquad \psi|_{T \times X} = \tau \circ (f \times \mathrm{id}_X).$$

L'existence de τ implique que le noyau de f opère trivialement sur X, donc qu'il existe une action ensembliste $\psi' : G' \times X \to X$ de G' sur X telle que $\psi = \psi' \circ (f \times \mathrm{id}_X)$. Il est clair que $\upsilon^\pm$ et τ sont les restrictions de ψ' à $f(U^\pm) \times X$ et à $f(T) \times X$ respectivement. Pour $u_+ \in f(U^+)$, $u_- \in f(U^-)$ et $t \in f(T)$, on a

$$\psi'(u_+ t u_-, x) = \psi'(u_+, \psi'(t, \psi'(u_-, x))) = \upsilon^+(u_+, \tau(t, \upsilon^-(u_-, x))).$$

Posant $V = f(U^+) . f(T) . f(U^-)$, on voit donc que la restriction de ψ' à $V \times X$ est un morphisme. D'autre part, pour tout $g \in G$, l'application $x \mapsto \psi(g, x) = \psi'(f(g), x)$ est un automorphisme de X, donc la restriction de ψ' à $(f(g) . V) \times X$ est aussi un morphisme. Notre assertion résulte à présent de ce que les ouverts $(f(g) . V) \times X$ recouvrent $G' \times X$.

Proposition (**2.15**). — *Soient* G' *un groupe algébrique,* $f : G \to G'$ *un morphisme et* T *un tore maximal de* G. *Alors les conditions suivantes sont équivalentes* :

(i) f *est central*;

(ii) f *est quasi-central et* $\operatorname{Ker} L(f) \subset \mathscr{Z}(L(G))$;

(iii)   *la restriction de* f *à tout sous-groupe unipotent fermé et connexe de* G *est une immersion*;

(iv) $(f_T)^* : X^*(f(T)) \to X^*(T)$ *applique les racines de* $f(G)$ *par rapport à* $f(T)$ *sur les racines de* G *par rapport à* T.

Lorsque ces conditions sont remplies, la restriction de f *à tout sous-groupe unipotent fermé de* G *est une immersion.*

Soient g un élément de G et $\gamma : G \to G$ l'application définie par $\gamma(x) = gxg^{-1}x^{-1}$. Si f est central, il existe un morphisme $\gamma' : f(G) \to G$ tel que $\gamma = \gamma' \circ f$, d'où $(d\gamma)_1 = \operatorname{Ad} g - 1 = (d\gamma')_1 \circ L(f)$. On a donc, vu (2.13),

$$\operatorname{Ker} L(f) \subset \operatorname{Ker}(\operatorname{Ad} g - 1) = \mathscr{Z}_{L(G)}(G) = \mathscr{Z}(L(G)),$$

ce qui établit l'implication (i) $\Rightarrow$ (ii).

Si (ii) est satisfaite, il résulte de (2.11) que la restriction de f à tout sous-groupe unipotent fermé de G est une injection; de plus, celle-ci est une immersion, et en particulier (iii) est satisfaite, puisque, vu (2.13), $\operatorname{Ker} L(f)$ est formé d'éléments semi-simples.

Pour faire voir que la condition (iii) implique (iv), il suffit de l'appliquer aux groupes U_a $(a \in \Phi(T, G))$.

Enfin, supposons (iv) satisfaite et montrons que f est central. Soient U^+ et U^-

deux sous-groupes unipotents maximaux de G normalisés par T et opposés. De (iv), il résulte aussitôt que l'homomorphisme $U^{\pm} \to f(U^{\pm})$ induit par f est un isomorphisme de groupes algébriques; son inverse sera noté $\alpha^{\pm}$. On a $(\mathrm{Ker}\,f) \cap U^+ = \{1\}$. Comme $\mathrm{Ker}\,f$ est distingué dans G, cela implique qu'il est formé d'éléments semi-simples, donc que f est quasi-central (2.9). Soit $\iota : f(G) \times G \to G$ l'application définie par $\iota(f(x), y) = xyx^{-1}$ (cf. (2.4) (iii)). Pour établir notre assertion, il suffit, vu (2.4), de montrer que ι est un morphisme, ou encore, en vertu de (2.14), de montrer que les restrictions $\upsilon^{\pm}$ et τ de ι à $f(U^{\pm}) \times G$ et à $f(T) \times G$ sont des morphismes. Mais $\upsilon^{\pm}$ n'est autre que le composé de $\alpha^{\pm} \times \mathrm{id}_G$ et du morphisme $(u, x) \mapsto uxu^{-1}$ de $U^{\pm} \times G$ dans G. Reste donc à faire voir que τ est un morphisme ou, ce qui revient au même vu (2.5), que pour tout $g \in G$, l'application $\tau_g : f(T) \to G$ définie par $\tau_g(t') = \tau(t', g)$ est un morphisme. Vu l'identité $\tau_{gg'}(t') = \tau_g(t') . \tau_{g'}(t')$, il suffit d'établir cette assertion pour g appartenant à un système générateur de G, par exemple pour $g \in T \cup U^+ \cup U^-$. Or, si $g \in T$, τ_g est l'application constante $f(T) \to \{g\}$ et si $g \in U^{\pm}$, on a $\tau_g(t') = \alpha^{\pm}(t'\, f(g) . t'^{-1})$, relation qui définit bien un morphisme.

Remarque (**2.16**). — Utilisons à nouveau les notations de [1]. En particulier, soit $\bar{k}[\delta]$ l'algèbre des nombres duaux sur $\bar{k}$. Il résulte de l'équivalence des propriétés (i) et (ii) de (2.15), compte tenu des propriétés de la suite exacte scindée

$$\{1\} \to L(G)(\bar{k}) \to G(\bar{k}[\delta]) \to G(\bar{k}) \to \{1\}$$

[1; I, (3.5)], que $f : G \to G'$ est central si et seulement si $\mathrm{Ker}\,f(\bar{k}[\delta]) \subset \mathscr{Z}(G(\bar{k}[\delta]))$. (On comparera avec (2.3), où H n'était pas supposé réductif.)

Corollaire (**2.17**). — *Supposons le morphisme* $f : G \to G'$ *central et surjectif. Alors,* $\mathrm{Im}\,L(f)$ *contient tous les éléments nilpotents de* $L(G')$. *Si* T' *est un tore maximal de* G', *on a* $L(G') = L(T') + \mathrm{Im}\,L(f)$. *Soit* H' *un k-sous-groupe de* G'. *Supposons que* H' *contienne* T', *ou soit contenu dans* T', *ou soit connexe et réductif; alors* $f^{-1}(H')$ *est défini sur k.*

Tout élément nilpotent de $L(G')$ est tangent à un sous-groupe unipotent connexe de G' [1; 14.17]. Si X est un tel sous-groupe, le radical unipotent de $f^{-1}(X)$ est appliqué surjectivement sur X par f. La première assertion résulte donc de (2.15) (iii). La seconde s'ensuit, puisque $L(G') = L(T') + L(U') + L(U'^-)$, où U' et U'^- désignent des sous-groupes unipotents maximaux opposés de G' normalisés par T'. Si $H' \supset T'$, on a $L(G') = L(H') + \mathrm{Im}\,f$, et $f^{-1}(H')$ est défini sur k, vu [2; 1.18]. Si H' est contenu dans T', il est aussi contenu dans un k-tore maximal T''; d'après ce qu'on vient de voir et (2.11), $f^{-1}(T'')$ est un k-tore de G, dont $f^{-1}(H')$ est un sous-groupe k-fermé donc défini sur k, vu [3; 1.6]. Enfin, si H' est connexe et réductif, il est engendré par ses k-tores [2; 7.10]; chacun d'eux est contenu dans un tore maximal de G', donc a son image réciproque définie sur k et il en est alors de même pour H'.

Corollaire (**2.18**). — *Soient* G' *et* G'' *des groupes algébriques,* $f : G \to G'$ *un morphisme surjectif et* $h : G' \to G''$ *un morphisme. Alors,* $h \circ f$ *est central (resp. quasi-central) si et seulement si* h *et* f *sont centraux (resp. quasi-centraux).*

Si $h \circ f$ est central (resp. quasi-central), le commutateur $G \times G \to G$ se factorise en tant que morphisme (resp. en tant qu'application) à travers $h(G') \times h(G')$. Cela implique qu'il se factorise aussi à travers $G' \times G'$, et que le commutateur $G' \times G' \to G'$ se factorise à travers $h(G') \times h(G')$; par conséquent, f et h sont centraux (resp. quasi-centraux). (N.B. Dans cette partie de la démonstration, le fait que G est réductif n'a pas été utilisé.)

Réciproquement, si f et h sont centraux (resp. quasi-centraux), ils possèdent la propriété (2.15) (iv) (resp. (2.11) (iii)) et il en est donc de même de $h \circ f$.

Corollaire (**2.19**). — *Un morphisme surjectif* $f : G \to G'$ *est central si et seulement s'il est quasi-central et possède une factorisation* $f = f_m \circ \ldots \circ f_0$ *où* $f_0 : G \to G_1 = G/\mathrm{Ker} f$ *est la projection canonique de* G *sur* $G/\mathrm{Ker} f$, $f_i : G_i \to G_{i+1} = G_i/\mathfrak{m}_i$ $(0 < i < m)$ *est la projection canonique du but* G_i *de* f_{i-1} *sur le quotient de* G_i *par un idéal* $\mathfrak{m}_i$ *de* $L(G_i)$ *centralisé par* G_i [1; § 17] *et* $f_m : G_m \to G'$ *est un isomorphisme.*

On sait que tout morphisme surjectif $f : G \to G'$ possède une factorisation $f = f_m \circ \ldots \circ f_0$ du type décrit dans l'énoncé, à cela près que $\mathfrak{m}_i$ n'est pas nécessairement centralisé par G_i. Vu (2.18), f est central si et seulement si les f_i le sont, c'est-à-dire si $\mathrm{Ker} f \subset \mathscr{Z}(G)$ et si $\mathfrak{m}_i$ est centralisé par G_i pour $0 < i < m$ (cf. (2.13) et (2.15) (ii)).

Théorème (**2.20**). — *Soient* G' *un k-groupe et* $f : G \to G'$ *un k-morphisme central surjectif.*

(i) *Les k-sous-groupes paraboliques de* G' (*resp.* G) *sont les images* (*resp. les images réciproques*) *par* f *des k-sous-groupes paraboliques de* G (*resp.* G').

(ii) *Les tores déployés sur* k *maximaux de* G' (*resp.* G) *sont les images* (*resp. les tores déployés maximaux des images réciproques*) *des tores déployés sur* k *maximaux de* G (*resp.* G').

(iii) *Soient* S *un tore déployé sur* k *maximal de* G *et* $f(S) = S'$. *Alors,* f *applique* $\mathscr{N}(S)$ (*resp.* $\mathscr{Z}(S)$) *sur* $\mathscr{N}(S')$ (*resp.* $\mathscr{Z}(S')$) *et induit un isomorphisme de* $_k W(G) = \mathscr{N}(S)/\mathscr{Z}(S)$ *sur* $_k W(G') = \mathscr{N}(S')/\mathscr{Z}(S')$. *L'homomorphisme* $(f|_S)^* : X^*(S') \to X^*(S)$ *applique* $\Phi(S', G')$ *isomorphiquement sur* $\Phi(S, G)$. *Si* $a' \in \Phi(S', G')$ *et* $a = (f|_S)^*(a')$, *alors* f *induit un isomorphisme de* $U_{(a)}$ *sur* $U'_{(a')}$.

Dans cet énoncé, $U_{(a)}$ est le groupe défini en [3; 5.2] et $U'_{(a')}$ est le sous-groupe analogue dans G'.

(i) Le seul point non évident est que l'image réciproque d'un k-sous-groupe parabolique est définie sur k. Mais cela résulte de (2.17).

(ii) Comme l'image d'un tore déployé sur k est déployée sur k, il suffit de montrer, compte tenu de (2.11), que tout tore X' déployé sur k maximal de G' est l'image d'un tore déployé sur k de G. Soit T' un k-tore maximal de G' contenant X' et soit T_d le tore déployé sur k maximal du tore $f^{-1}(T')$ (cf. (2.11)). Alors, vu [1; 8.15], $f(T_d)$ est le tore déployé sur k maximal de T', donc $f(T_d) = X'$.

(iii) On a $f(\mathscr{Z}(S)) = \mathscr{Z}(S')$ d'après [2; 4.6], donc aussi $\mathscr{Z}(S) = f^{-1}(\mathscr{Z}(S'))$. Comme les automorphismes intérieurs de G associés aux éléments de $\mathscr{N}(S)$ permutent transitivement les k-sous-groupes paraboliques minimaux contenant S ([3; 5.9]) il résulte de (i) que les automorphismes intérieurs de G' associés aux éléments de $f(\mathcal{N}(S))$ permutent transitivement les k-sous-groupes paraboliques minimaux de G' contenant S'. D'après [3; 5.9], cela implique que $f(\mathcal{N}(S)) = \mathcal{N}(S')$ et que f induit un isomorphisme de $_kW(G)$ sur $_kW(G')$.

Soient P, P^- deux k-sous-groupes paraboliques minimaux opposés de G d'intersection $\mathscr{Z}(S)$, et U, U^- leurs radicaux unipotents. Alors, $f(P)$ et $f(P^-)$ sont des k-sous-groupes paraboliques minimaux opposés de G', d'intersection $\mathscr{Z}(S')$ et ayant $f(U)$ et $f(U^-)$ pour radicaux unipotents. De plus, f induit un k-isomorphisme de U sur $f(U)$ et de U^- sur $f(U^-)$ ((2.15) (iii)). Les assertions restantes de l'énoncé résultent alors de ce que $\Phi(S, G)$ est la réunion des ensembles de poids de S dans U et U^-, que $U_{(a)}$ est le plus grand sous-groupe connexe X de G normalisé par S et tel que les poids de S dans X soient des multiples entiers positifs de a [3; 3.12], et qu'on a des caractérisations analogues pour $\Phi(S', G')$ et $U'_{(a')}$.

Corollaire (**2.21**). — *Si la composante neutre de* $\mathrm{Ker}\,f$ *est un tore anisotrope sur* k, *on a* $\mathrm{rg}_kG = \mathrm{rg}_kG'$.

En effet, $(S \cap \mathrm{Ker}\,f)^0$ est alors un k-tore à la fois déployé et anisotrope sur k, donc réduit à $\{1\}$, et $\dim S = \dim f(S)$.

(**2.22**) Soient M, N deux sous-groupes fermés connexes distingués d'intersection finie de G. Alors, M et N se centralisent mutuellement, donc $M \cap N$ et $L(M) \cap L(N)$ sont centralisés par M.N. Par suite, si G est produit presque direct de k-sous-groupes fermés connexes $G_1, \ldots, G_n$, l'application produit $G_1 \times \ldots \times G_n \to G$ est une k-isogénie *centrale*. Elle n'est pas nécessairement séparable, comme cela est implicitement admis dans la démonstration de [3; 4.27], mais cette démonstration subsiste si l'on remplace la référence à 4.25 par une référence à (2.20) ci-dessus. Le même raisonnement montre que si P est un k-sous-groupe parabolique de G, alors $P \cap G_i$ est un k-sous-groupe parabolique de G_i et P est produit presque direct des $P \cap G_i$. Réciproquement, si P_i est un k-sous-groupe parabolique de G_i $(1 \leqslant i \leqslant n)$, alors, le produit des P_i est un k-sous-groupe parabolique P de G. De plus, $R_u(P)$ est le produit direct des $R_u(P_i)$.

(**2.23**) Rappelons brièvement quelques faits concernant la classification des groupes semi-simples déployés et des isogénies entre tels groupes. Pour plus de détails, voir [5].

a) Soient Φ un système de racines réduit dans un espace vectoriel réel, $P(\Phi)$ le réseau des poids de Φ et $Q(\Phi)$ le réseau engendré par Φ. A tout réseau X intermédiaire entre $Q(\Phi)$ et $P(\Phi)$ correspond un et, à isomorphisme près, un seul système (H, T, φ) formé d'un k-groupe semi-simple H déployé sur k, d'un tore déployé maximal T de H et d'un isomorphisme $\varphi : X^*(T) \to X$ tel que $\varphi(\Phi(T, H)) = \Phi$. Le groupe H, et plus généralement tout k-groupe semi-simple isomorphe à H sur $\bar{k}$, est dit *simplement connexe* (resp. *adjoint*) si $X = P(\Phi)$ (resp. $Q(\Phi)$). Pour qu'un groupe réductif G soit semi-simple et adjoint, il faut et il suffit que sa représentation adjointe $\mathrm{Ad} : G \to GL(L(G))$ soit une immersion, ou encore que $\mathscr{Z}(G) = \{1\}$ et $\mathscr{Z}_{L(G)}(G) = \{0\}$. Un groupe semi-simple simplement connexe ou adjoint est le produit direct de ses facteurs presque simples.

b) Soient H_1, H_2 des k-groupes semi-simples déployés sur k, T_i $(i=1, 2)$ un tore déployé sur k maximal de H_i et $\varphi : X^*(T_2) \to X^*(T_1)$ un monomorphisme. Posons $\Phi_i = \Phi(T_i, H_i)$. Pour qu'il existe une k-isogénie $f : H_1 \to H_2$ telle que $f(T_1) = T_2$ et que $(f|_{T_1})^* = \varphi$, il faut et il suffit qu'il existe une bijection $\varphi' : \Phi_2 \to \Phi_1$ telle que, pour tout $a \in \Phi_2$, on ait $\varphi(a) = \lambda(a) . \varphi'(a)$, où $\lambda(a)$ est une puissance entière de l'exposant caractéristique de k. Lorsque cette condition est remplie, f est entièrement déterminé à la multiplication à droite près par un automorphisme de H_1 centralisant T_1. De plus, φ transforme le groupe de Weyl W_2 de Φ_2 en le groupe de Weyl W_1 de Φ_1, donc aussi toute forme bilinéaire sur $X^*(T_1)$ invariante par W_1 en une forme bilinéaire invariante par W_2. On en déduit aussitôt que φ, étendu par linéarité à $X^*(T_2) \otimes \mathbf{R}$, applique $P(\Phi_2)$ dans $P(\Phi_1)$.

Proposition (**2.24**). — (i) *Soient* $f : G' \to G$ *une k-isogénie centrale, G_1 un k-groupe semi-simple simplement connexe et* $h : G_1 \to G$ *un k-morphisme. Alors, il existe un morphisme* $h' : G_1 \to G'$ *et un seul tel que* $h = f \circ h'$; *il est défini sur k.*

(ii) *Si G est semi-simple, il existe une k-isogénie centrale* $\widetilde{G} \to G$ *telle que* $\widetilde{G}$ *soit un k-groupe simplement connexe. Cette isogénie est unique à G-isomorphisme près.*

L'assertion d'unicité signifie que si, pour $i=1, 2$, $f_i : \widetilde{G}_i \to G$ est une k-isogénie centrale telle que $\widetilde{G}_i$ soit un k-groupe simplement connexe, il existe un k-isomorphisme $\alpha : \widetilde{G}_1 \to \widetilde{G}_2$ tel que $f_1 = f_2 \circ \alpha$.

(i) Vu (2.17), le groupe $G'' = f^{-1}(h(G_1))^0$ est défini sur k. Quitte à remplacer G par $h(G_1)$, G' par G'' et G_1 par $G_1/(\operatorname{Ker} h)^0$, nous pouvons supposer que G et G' sont semi-simples et que h est une isogénie. Vu (2.23) *b)*, il existe alors un k_s-morphisme $h' : G_1 \to G'$ tel que $h = f \circ h'$. Ce morphisme est unique et défini sur k en vertu de (2.8).

(ii) A un k-isomorphisme près, le groupe G peut être obtenu en tordant un k-groupe semi-simple H déployé sur k par un cocycle ξ de $\operatorname{Aut}(k_s/k)$ à valeurs dans $\operatorname{Aut}_{k_s} H$. Soit $f : \widetilde{H} \to H$ une k-isogénie centrale telle que $\widetilde{H}$ soit un k-groupe simplement connexe. Une telle isogénie existe vu (2.23). De (i), il résulte que tout k_s-automorphisme de H « se relève » de façon unique en un k_s-automorphisme de $\widetilde{H}$. Cela étant, le groupe $\widetilde{G}$ et l'isogénie $\widetilde{G} \to G$ s'obtiennent en tordant $\widetilde{H}$ et f par ξ. L'unicité est une conséquence immédiate de (i) (et d'ailleurs aussi des principes généraux de la cohomologie galoisienne).

Définition (**2.25**). — Si G est semi-simple, une (k)-isogénie centrale $\widetilde{G} \to G$ telle que $\widetilde{G}$ soit simplement connexe est appelée un $(k$-$)$*revêtement universel* de G.

Proposition (**2.26**). — (i) *Soient* $f : G \to G'$ *une k-isogénie centrale, G_1 un k-groupe semi-simple adjoint et* $h : G \to G_1$ *un k-morphisme. Alors, il existe un morphisme* $h' : G' \to G_1$ *et un seul tel que* $h = h' \circ f$; *il est défini sur k.*

(ii) *Il existe un et, à G-isomorphisme près, un seul k-morphisme central surjectif de* G *sur un k-groupe semi-simple adjoint, à savoir, le morphisme canonique de* G *sur son image par la représentation adjointe* Ad : $G \to GL(L(G))$.

(i) Pour montrer l'existence de h', on peut, vu (2.19), se borner à considérer le cas où f est la projection canonique de G sur un quotient, soit par un sous-groupe fermé central, soit par un idéal de $L(G)$ centralisé par G. Mais alors, cette existence résulte de la propriété universelle du quotient. Le morphisme h' est évidemment unique, puisque f est surjectif. Enfin, h' est défini sur k en vertu de (2.8).

(ii) On sait que Ad G est un groupe adjoint (il suffit de le vérifier pour G déployé). Le noyau de $L(Ad)$ est le centre de $L(G)$, donc (2.15) Ad est un morphisme central. Enfin, l'unicité est une conséquence immédiate de (i).

3. Décomposition de Bruhat.

Dans ce paragraphe, nous donnons diverses propriétés des doubles classes de la décomposition de Bruhat et des produits de deux telles doubles classes. En particulier, nous décrivons l'adhérence dans G_k d'une double classe pour la topologie de Zariski et nous montrons que si k est un corps topologique non discret, le même résultat est valable pour la topologie associée à celle de k.

(3.1) On fixe un tore déployé sur k maximal S de G et un k-sous-groupe parabolique minimal P contenant S; on pose $\Phi = \Phi(S, G)$, $\Phi^+ = \Phi(S, P)$, $\Phi^- = -\Phi^+ = \Phi - \Phi^+$, $\Phi_0 = \{a \in \Phi \mid a/2 \notin \Phi\}$, $\Phi_0^\pm = \Phi^\pm \cap \Phi_0$, $W = \mathcal{N}(S)/\mathcal{Z}(S) = \mathcal{N}(S)_k/\mathcal{Z}(S)_k = W(S, G)$. Selon les circonstances, un élément de W sera considéré comme une classe latérale de $\mathcal{Z}(S)$ dans $\mathcal{N}(S)$, comme une classe latérale de $\mathcal{Z}(S)_k$ dans $\mathcal{N}(S)_k$ ou comme une transformation linéaire du dual de $X^*(S) \otimes \mathbf{R}$. Si X est une partie de G normalisée par $\mathcal{Z}(S)_k$, on pose $w(X) = nXn^{-1}$, où n désigne un représentant quelconque de w dans $\mathcal{N}(S)_k$. On note Δ la base de Φ contenue dans Φ^+ et $\sigma : \Delta \to W$ l'application canonique qui envoie une « racine simple » sur la réflexion correspondante. La projection canonique $G \to G/P$ est désignée par π. Pour $w \in W$, $\ell(w)$ représente la « longueur » de w, c'est-à-dire le plus petit entier q tel que w puisse s'écrire comme produit de q éléments de $\sigma(\Delta)$. Une suite $(s_1, \ldots, s_q)$, $(q = \ell(w))$ d'éléments de $\sigma(\Delta)$ telle que $w = s_1 \ldots s_q$ est appelée une *décomposition réduite* de w.

(3.2) Pour $w \in W$, nous désignons par Ψ_w l'ensemble $w(\Phi_0^-) \cap \Phi^+$. On a $w(\Phi_0^-) = \Psi_w \cup (\Phi_0^- \cap \complement(-\Psi_w))$, donc l'élément w de W est déterminé par l'ensemble Ψ_w, vu [4; 3.3, Th. 1]. Pour $w, w' \in W$, on a

(1) $$\Psi_w \cap w(\Psi_{w'}) = \varnothing,$$

(2) $$\Psi_{ww'} \subset \Psi_w \cup w(\Psi_{w'}),$$

(3) $$\mathrm{Card}\, \Psi_w = \ell(w).$$

Les relations (1), (2) sont immédiates et (3) résulte de [4; Cor. 2, p. 158].

Pour $w \in W$, nous posons $C(w) = PwP$ et $U_w = \prod_{a \in \Psi_w} U_{(a)}$ (dans [3], U_w est noté U'_w); ce dernier produit ne dépend pas de l'ordre des facteurs et, quel que soit cet ordre, l'application produit du produit direct des $U_{(a)}$ $(a \in \Psi_w)$ dans U_w est un k-isomorphisme de variétés [3; 3.11]. D'autre part, on déduit facilement de [3; 4.10] que si $w \in W$ et si $n \in N(S)_k$ est un représentant de w, alors l'application produit définit un k-isomorphisme de variétés

$$(4) \qquad (U_w . n) \times P \to U_w . w . P = C(w).$$

En particulier, $\pi(C(w))$ est k-isomorphe à l'espace affine U_w et

$$(5) \qquad C(w)_k = P_k . w . P_k = U_{w,k} . w . P_k.$$

Remarque (**3.3**). — Le groupe G_k est réunion disjointe des $C(w)_k$ [3; 5.16] mais G n'est réunion des $C(w)$ que s'il est déployé sur k (*i.e.*, si S est un tore maximal de G). Si G est quasi-déployé sur k, autrement dit si P est un sous-groupe de Borel de G, les $C(w)$ sont les doubles classes de P dans G qui sont définies sur k; en effet, $\mathscr{Z}(S) = T$ est alors un tore maximal de G, toute double classe de P dans G est de la forme $P . w' . P$ avec $w' \in \overline{W} = N(T)/T$, et l'on sait qu'un élément de $\overline{W}$ appartient à W si et seulement s'il est invariant par le groupe de Galois d'un corps normal de déploiement de G opérant sur $\overline{W}$ de la façon évidente (cf. [3; 6.11]). Enfin, dans tous les cas, les $C(w)$ sont les doubles classes de P dans G qui possèdent des points rationnels sur k, mais une double classe peut être définie sur k sans pour autant posséder de tels points.

Lemme (**3.4**). — *Les conditions*

$$(i) \qquad \ell(w . w') = \ell(w) + \ell(w')$$

$$(ii) \qquad \Psi_{ww'} = \Psi_w \cup w(\Psi_{w'}),$$

sont équivalentes. Lorsqu'elles sont remplies, on a

$$(iii) \qquad C(w) . C(w') = C(w . w')$$

$$(iv) \qquad C(w)_k . C(w')_k = C(w . w')_k.$$

Plus précisément, l'application produit $C(w) \times C(w') \to C(w . w')$ *identifie* $C(w . w')$ *au « produit fibré », quotient de* $C(w) \times C(w')$ *par* P *opérant par* $(g, g') . p = (g . p, p^{-1} . g')$ *pour* $(g \in C(w), g' \in C(w'), p \in P)$.

L'équivalence de (i) et (ii) résulte de (3.2) (1), (2), (3). Supposons-les remplies et soient n, n' des représentants de w, w' dans $N(S)_k$. Vu (ii) et (3.2), l'application produit

$$U_w . n \times U_{w'} . n' \to U_{ww'} . nn'$$

est un k-isomorphisme de variétés. Il en est alors de même de l'application produit

$$(U_w . n) \times C(w') \to C(w . w')$$

ce qui démontre (iii), (iv). D'autre part, l'application $P \times C(w') \to C(w') \times P$ qui applique (p, g') sur (pg', p) est un isomorphisme de variétés. Par conséquent, l'application $\alpha : C(w) \times C(w') \to C(ww') \times P$ définie par

$$\alpha(unp, g') = (unpg', p) \qquad (u \in U_w; \ p \in P; \ g' \in C(w'))$$

en est aussi un et la dernière assertion de l'énoncé résulte à présent de ce que l'application produit $C(w) \times C(w') \to C(ww')$ est composée de α et de la première projection du produit $C(ww') \times P$.

Remarque. — On verra plus loin (3.18) que (iii) est équivalente à (i), (ii).

(3.5) Soient $a \in \Delta$, $s = \sigma(a)$ et $w \in W$. Vu [3; 5.16] et [4, p. 23 et Th. 2, p. 25], on a

$$(1) \qquad C(w)_k . C(s)_k = C(w.s)_k \qquad \text{si} \quad \ell(w.s) > \ell(w),$$

$$(2) \qquad C(w)_k . C(s)_k = C(w.s)_k \cup C(w)_k \qquad \text{si} \quad \ell(w.s) < \ell(w).$$

On a évidemment $\ell(w.s) > \ell(w)$ si et seulement si $\ell(ws) = \ell(w) + \ell(s) = \ell(w) + 1$, donc (1) est aussi un cas particulier de (3.4) (iv) (et l'entraîne par une récurrence immédiate). De même, (3.4) (iii) entraîne

$$(3) \qquad C(w) . C(s) = C(w.s) \qquad \text{si} \quad \ell(ws) > \ell(w).$$

La double classe $C(s)$ est ouverte dans le sous-groupe parabolique standard $P_{\{s\}}$ de G (cf. [3; 5.12] ou (3.6) ci-dessous pour la notation). On a donc, en notant par une barre l'adhérence en topologie de Zariski,

$$(4) \qquad P_{\{s\}} = \overline{C(s)} = C(s) . C(s) \supset C(s) \cup P.$$

En multipliant les trois derniers membres de (4) à gauche par $C(ws)$ et en tenant compte de (3), on en déduit :

$$(5) \qquad \overline{C(w)} \supset C(w) . C(s) \supset C(w) \cup C(ws) \qquad \text{si} \quad \ell(ws) < \ell(w).$$

(3.6) Pour toute partie θ de Δ, nous notons W_θ le sous-groupe de W engendré par $\sigma(\theta)$, et P_θ ou $P_{\sigma(\theta)}$ le k-sous-groupe parabolique « standard » engendré par P et W_θ (*i.e.*, défini par $P_{\theta, k} = P_k . W_\theta . P_k$; on n'a pas, en général, $P_\theta = P . W_\theta . P$, cf. (3.3)), P_θ^- le sous-groupe parabolique opposé à P_θ et contenant $\mathscr{Z}(S)$, $\Phi_{0, \theta}$ l'ensemble des éléments de Φ_0 qui sont combinaisons linéaires d'éléments de θ, et nous posons

$$\Phi_{0, \theta}^{\pm} = \Phi_{0, \theta} \cap \Phi^{\pm},$$

$$\widetilde{\Phi}_{0, \theta}^{\pm} = \Phi_0 \cap \Phi(S, R_u P_\theta^{\pm}) = \Phi_0^{\pm} - \Phi_{0, \theta}^{\pm}.$$

Lemme (3.7). — *Soit* $n \in \mathbf{N}^*$. *Pour* $i \in \{1, \ldots, n\}$, *soient* θ_i *une partie de* Δ *et* w_i *un élément de* W_{θ_i}. *Posons* $w = w_1 \ldots w_n$. *Alors, il existe* $w_i' \in W_{\theta_i}$ $(1 \leqslant i \leqslant n)$ *tels que*

$$\ell(w_i') \leqslant \ell(w_i), \qquad w = w_1' \ldots w_n' \qquad \text{et} \qquad \ell(w) = \sum_{i=1}^{n} \ell(w_i').$$

C'est une conséquence immédiate de la « condition d'échange » caractérisant les groupes de Coxeter ([4; chap. IV, n° 1.5]).

Lemme **(3.8)**. — *Soient* P_1 *un k-sous-groupe parabolique minimal de* G *et* P_2 *un k-sous-groupe parabolique quelconque. Alors* $P_3 = (P_1 \cap P_2) . R_u(P_2)$ *est un k-sous-groupe parabolique minimal.*

On sait [3; 4.4, 4.7] que P_3 est un k-sous-groupe parabolique. Pour voir qu'il est minimal, il suffit d'observer que si S' est un tore déployé maximal contenu dans $P_1 \cap P_2$ [3; 4.18] et si $a \in \Phi(S', G)$, $U_{(a)}$ et $U_{(-a)}$ ne peuvent être contenus simultanément dans P_3.

Proposition **(3.9)**. — *Soient* $w \in W$ *et* $\theta \subset \Delta$. *Alors il existe une décomposition unique* $w = \widetilde{w}_\theta . w_\theta$ *possédant les propriétés suivantes et caractérisée par chacune d'elles :*

(i) $\widetilde{w}_\theta$ *est l'unique élément de longueur minimum dans* $w . W_\theta$;

(ii) $\Psi'_{\widetilde{w}_\theta} = w(\widetilde{\Phi}^-_{0,\theta}) \cap \Phi^+$;

(iii) $\Psi'_{w_\theta^{-1}} = w^{-1}(\Phi^-_0) \cap \Phi^+_\theta$;

(iv) $w_\theta^{-1}(P) = (w^{-1}(P) \cap P_\theta) . R_u(P_\theta)$.

On a $w_\theta \in W_\theta$ *et*

$$(1) \qquad \ell(w) = \ell(\widetilde{w}_\theta) + \ell(w_\theta).$$

(N.B. — L'existence d'un unique élément de longueur minimum dans $w . W_\theta$ est une propriété générale bien connue des systèmes de Coxeter (cf. p. ex. [4; chap. IV, ex. 3 du § 1]). Nous la redémontrons ici dans le cas qui nous intéresse.)

Le produit $(w^{-1}(P) \cap P_\theta) . R_u(P_\theta)$ étant un k-sous-groupe parabolique minimal (3.8), il existe un unique élément $w_\theta \in W$ satisfaisant à (iv), et cet élément appartient à W_θ puisque $w_\theta^{-1}(P) \subset P_\theta$.

La relation (iv) peut aussi s'écrire

$$w_\theta^{-1}(\Phi^+_0) = (w^{-1}(\Phi^+_0) \cap \Phi_\theta) \cup \widetilde{\Phi}^+_{0,\theta},$$

ou encore

$$(2) \qquad w_\theta^{-1}(\Phi^-_0) = (w^{-1}(\Phi^-_0) \cap \Phi_\theta) \cup \widetilde{\Phi}^-_{0,\theta},$$

d'où on déduit la relation (iii), qui caractérise aussi w_θ, vu (3.2).

Soit w'' un élément quelconque de W_θ. Posons $w' = w w''$. On a alors

$$\Psi'_{w'} = w w''(\Phi^-_{0,\theta} \cup \widetilde{\Phi}^-_{0,\theta}) \cap \Phi^+ = (w w''(\Phi^-_{0,\theta}) \cap \Phi^+) \cup (w(\widetilde{\Phi}^-_{0,\theta}) \cap \Phi^+).$$

Les deux termes du dernier membre sont disjoints, le second ne dépend pas de w'' et il résulte de (2) que le premier est vide si et seulement si $w'' = w_\theta^{-1}$. Les propriétés (i) et (ii) s'ensuivent. Enfin, (1) est une conséquence immédiate de (i) et (3.7) (ou, si l'on préfère, de (ii), (iii) et (3.2) (3), par un calcul facile).

Corollaire (**3.10**). — *Soient* θ, $\theta_0 = \varnothing$, θ_1, ..., θ_n *des parties de* Δ *et* $w \in W$.

(i) *Si* $w \in W_{\theta_1} \ldots W_{\theta_n}$, *alors* $\widetilde{w}_{\theta_n} \in W_{\theta_1} \ldots W_{\theta_{n-1}}$.

(ii) *Soient* $g \in P_k w P_k$ *et* $h \in G_k$. *Supposons que* $h^{-1}Ph = (g^{-1}Pg \cap P_\theta) . R_u P_\theta$. *Alors* $gh^{-1} \in P_k . \widetilde{w}_\theta . P_k$.

(i) résulte de (3.7) et (3.9) (i). Démontrons (ii). Posons $g' = gh^{-1}$ et soit $u \in W$ tel que $g' \in P_k u P_k$. Comme $hPh^{-1} \in P_\theta$, on a $h \in P_\theta$ [3; 5.18], par conséquent

$$g = g'h \in P_k u P_k W_\theta P_k \subset P_k u W_\theta P_k$$

(vu (3.5) (1) et (2) et [3; 5.20]), d'où $w \in u W_\theta$ et $\widetilde{w}_\theta = \widetilde{u}_\theta$. D'autre part, l'hypothèse peut s'écrire $P = (g'^{-1}Pg' \cap P_\theta) . R_u(P_\theta)$, ce qui veut dire que $u_\theta = 1$, vu (3.9) (iv). On a donc $u = \widetilde{u}_\theta = \widetilde{w}_\theta$, c.q.f.d.

Proposition (**3.11**). — *Soient* P_1, ..., P_n *des k-sous-groupes paraboliques de* G *contenant* P. *Alors* $(P_1 \ldots P_n)_k = P_{1,k} \ldots P_{n,k}$.

Soient T un tore maximal de P contenant S et B un sous-groupe de Borel de P contenant T. Posons $\overline{\Phi} = \Phi(T, G)$, $\overline{\Phi}^+ = \Phi(T, B)$ et $\overline{W} = {}_{\overline{k}}W = \mathcal{N}(T)/T$. Soit $\overline{\Delta}$ la base de $\overline{\Phi}$ contenue dans $\overline{\Phi}^+$. Pour $\theta \subset \overline{\Delta}$ et $w \in \overline{W}$, nous définissons P_θ, W_θ, $\widetilde{w}_\theta$, w_θ comme en (3.6) et (3.9). Soit $\theta_i \subset \overline{\Delta}$ tel que $P_i = P_{\theta_i}$ ($1 \leqslant i \leqslant n$). Nous allons montrer par récurrence sur n que si $g \in (P_1 \ldots P_n)_k$, alors $g \in P_{1,k} \ldots P_{n,k}$. Soit $w \in \overline{W}$ tel que $g \in BwB$. On a $w \in \overline{W}_{\theta_1} \ldots \overline{W}_{\theta_n}$, d'où $\widetilde{w}_{\theta_n} \in W_{\theta_1} \ldots W_{\theta_{n-1}}$ vu (3.10) (i). D'après (3.8), $P' = (g^{-1}Pg \cap P_n) . R_u P_n$ est un k-sous-groupe parabolique minimal de G contenu dans P_n, donc il existe $h \in P_{n,k}$ tel que $h^{-1}Ph = P'$. Mais alors $gh^{-1} \in B\widetilde{w}_{\theta_n}B \subset P_1 \ldots P_{n-1}$ vu (3.10) (ii) et notre assertion résulte de l'hypothèse de récurrence.

Lemme (**3.12**). — *Soient* Q *et* R *deux sous-groupes paraboliques et* M *une partie fermée de* G. *Supposons que* $Q \subset R$ *et que* $MQ = M$. *Alors* MR *est fermé*.

Comme G/Q est une variété complète, la projection canonique $G/Q \to G/R$ est propre, donc fermée. D'autre part, les projections canoniques de G sur G/Q et G/R sont ouvertes, donc une partie de G/Q ou G/R est fermée si et seulement si son image inverse dans G l'est ; d'où le lemme.

Théorème (**3.13**). — *Supposons k infini. Soit* $w \in W$ *et soit* $(s_1, \ldots, s_l)$ *une décomposition réduite de* w. *Alors, l'ensemble* $A_w = \{ s_{i_1} \ldots s_{i_m} \mid m \in \mathbf{N}, 1 \leqslant i_1 < \ldots < i_m \leqslant \ell \}$ *ne dépend que de* w *et non de la décomposition réduite choisie, et on a, en notant par une barre l'adhérence pour la topologie de Zariski,*

$$(1) \qquad \overline{(C(w))}_k = \overline{C(w)}_k \cap G_k = \bigcup_{w' \in A_w} C(w')_k.$$

Il suffit évidemment d'établir les relations (1).

Vu [3; 3.20], on a $\overline{U_{w,k}} = U_w$ et $\overline{P}_k = P$, donc, compte tenu de (3.2) (4), $\overline{C(w)}_k = \overline{C(w)}$, ce qui établit la première égalité de (1).

Pour $X \subset G_k$, notons $A(X) = \overline{X} \cap G_k$ l'adhérence relative de X dans G_k. C'est une propriété connue de la topologie de Zariski que $\overline{A.B} \supset \overline{A}.\overline{B}$ quels que soient $A, B \subset G$. Par conséquent, si $X_1, \ldots, X_s$ sont des parties de G_k, on a

$$(2) \qquad A(X_1 \ldots X_s) = A\left(\prod_{1 \le i \le s} A(X_i)\right).$$

Posons $P_i = P_{\{s_i\}}$ ($1 \le i \le m$). Il résulte de (3.12), par récurrence sur j, que le produit $P_1 \ldots P_j$ est fermé pour tout $j \le m$. On a donc, compte tenu de (3.11),

$$(3) \qquad A\left(\prod_{i=1}^{m} P_{i,k}\right) = \prod_{i=1}^{m} P_{i,k}.$$

D'autre part, vu la première égalité (1), déjà établie, on a aussi

$$(4) \qquad A(C(s_i)_k) = P_{i,k} = C(s_i)_k \cup P_k.$$

Utilisant (2) et (3.4), on en déduit que

$$A(C(w)_k) = A\left(\prod_{i=1}^{m} C(s_i)_k\right) = A\left(\prod_{i=1}^{m} A(C(s_i)_k)\right) = \prod_{i=1}^{m} P_{i,k} = \prod_{i=1}^{m} (C(s_i)_k \cup P_k);$$

or le dernier membre de cette égalité n'est autre que le dernier membre de (1), vu (3.5) (1), (2). Le théorème est démontré.

Proposition (**3.14**). — *Supposons G_k muni d'une topologie $\mathscr{E}$ ayant les propriétés suivantes :*

(i) *$\mathscr{E}$ est plus fine que la topologie induite par la topologie de Zariski.*

(ii) *L'application produit $G_k \times G_k \to G_k$ est continue, $G_k \times G_k$ étant muni de la topologie produit.*

(iii) *Pour tout $a \in \Delta$, le groupe P_k n'est pas ouvert dans $P_{\{a\},k}$.*

Alors l'adhérence de $C(w)_k$ pour $\mathscr{E}$ coïncide avec son adhérence relative $\overline{C(w)_k} \cap G_k$ pour la topologie de Zariski, décrite en (3.13).

Dans la démonstration de (3.13), la topologie de Zariski est intervenue uniquement par l'intermédiaire des propriétés (2), (3), (4). Il suffit donc de voir qu'elles sont satisfaites par la topologie $\mathscr{E}$. Les relations (2) et (3) pour $\mathscr{E}$ résultent évidemment de (ii) et (i) respectivement. Soient $a \in \Delta$ et $s = \sigma(a)$. Si P_k n'est pas ouvert dans $P_{\{a\},k}$ alors $A(C(s)_k) \cap P_k \ne \varnothing$ et l'on a, compte tenu de (ii),

$$A(C(s)_k) = A(P_k . C(s)_k) \supset P_k . A(C(s)_k) \supset P_k,$$

d'où (3.13) (4) pour $\mathscr{E}$.

Corollaire (**3.15**). — *Supposons k muni d'une topologie $\mathscr{S}$ non discrète (satisfaisant à l'axiome T_1) qui en fait un anneau topologique. Alors l'adhérence de $C(w)_k$ pour la topologie $\mathscr{E}$ sur G_k associée à $\mathscr{S}$ coïncide avec son adhérence relative pour la topologie de Zariski, décrite en (3.13).*

Rappelons que $\mathscr{E}$ est la topologie la moins fine telle que les fonctions $G_k \to k$ restrictions de fonctions k-régulières sur G soient continues. Elle satisfait évidemment à (3.14) (i), (ii). De plus, comme k n'est pas discret et $U_{(-a),k}$ est un espace affine sur k, l'ensemble $\{1\} = P_k \cap U_{(-a),k}$ n'est pas ouvert dans $U_{(-a),k}$, donc (3.14) (iii) est aussi satisfaite. On peut donc appliquer (3.14).

Remarque. — Pour k algébriquement clos, (3.13) a aussi été obtenu par C. Chevalley (manuscrit non publié). Pour $k = \mathbf{C}$ muni de sa topologie usuelle, (3.15) est démontré dans [9; p. 107].

Proposition **(3.16)**. — *On conserve les notations de* (3.6), (3.9) *et l'on pose* $\widetilde{W}_\theta = \{\widetilde{w}_\theta \mid w \in W\}$. *Soient* $\theta \subset \Delta$, $w \in W$, *et notons* π_θ *la projection canonique de* G *sur* G/P_θ.

(i) *Pour tout* $w \in W$, *on a* $PwP_\theta = P\widetilde{w}_\theta P_\theta$ *et* $\pi_\theta(C(w)) = \pi_\theta(C(\widetilde{w}_\theta))$.

(ii) *Les applications canoniques* $\pi(C(w)) \to \pi_\theta(C(\widetilde{w}_\theta))$ *et* $\pi(C(w))_k \to \pi_\theta(C(\widetilde{w}_\theta))_k$ *sont surjectives. Chacune d'elle est injective si et seulement si* $w \in \widetilde{W}_\theta$, *auquel cas* $\pi(C(w)) \to \pi_\theta(C(w))$ *est un k-isomorphisme de variétés.*

(iii) *On a* $(C(w).P_\theta)_k = C(w)_k.P_{0,k}$.

(iv) *Si* w', w'' *sont des éléments distincts de* $\widetilde{W}_\theta$, *on a* $\pi_\theta(C(w')) \cap \pi_\theta(C(w'')) = \varnothing$. *Les* $\pi_\theta(C(x))_k$, *pour* $x \in \widetilde{W}_\theta$, *forment une partition de* $(G/P_\theta)_k$.

Les relations (i) résultent de (3.9).

Soient n, $\widetilde{n}_\theta$, n_θ des représentants de w, $\widetilde{w}_\theta$, w_θ dans $\mathcal{N}(S)_k$. L'ensemble $X = \widetilde{n}_\theta.U_{w_\theta,k}.n_\theta$, qui est contenu dans $C(\widetilde{w}_\theta)_k.C(w_\theta)_k = C(w)_k$, est appliqué injectivement dans $\pi(C(w))_k$ par π, et on a $\pi_\theta(X) = \{\widetilde{n}_\theta.P_\theta\}$; donc si l'application $\pi(C(w))_k \to \pi_\theta(C(\widetilde{w}_\theta))_k$ est injective, on a $U_{w_\theta} = \{1\}$, d'où $w_\theta = \{1\}$ et $w \in \widetilde{W}_\theta$. Les autres assertions de (ii) et (iii) résultent immédiatement des relations

$$U_{\widetilde{w}_\theta}.n \subset C(w), \qquad U_{\widetilde{w}_\theta}.n.P_\theta = C(w).P_\theta,$$

et du fait que, comme $n^{-1}.U_{\widetilde{w}_\theta}.n \subset R_u(P_\theta^-)$ (vu (3.9) (ii)), l'application produit $(U_{\widetilde{w}_\theta}.n) \times P_\theta \to C(w).P_\theta$ est un k-isomorphisme de variétés, en vertu de [3; 4.10].

Soient w', $w'' \in \widetilde{W}_\theta$. Si $\pi_\theta(C(w')) \cap \pi_\theta(C(w'')) \neq \varnothing$, cela signifie que les doubles classes $C(w').P_\theta = Pw'P_\theta$ et $C(w'').P_\theta = Pw''P_\theta$ ont une intersection non vide, donc coïncident. Vu (iii), on a alors $P_k w' P_{\theta,k} = P_k w'' P_{\theta,k}$, d'où $w' \in w''W_\theta$ (cf. [3; 5.20]). Comme w' et w'' appartiennent à $\widetilde{W}_\theta$, cela implique qu'ils sont égaux. La dernière assertion de (iv) résulte à présent de (i), de la relation $G_k = P_k W P_k$ et de la surjectivité de $G_k \to (G/P_\theta)_k$ [3; 3.25].

Corollaire **(3.17)**. — *Pour* $w \in W$ *et* $s \in \sigma(\Delta)$, *on a* $(C(w).C(s))_k = C(w)_k.C(s)_k$.

Si $\ell(ws) = \ell(w) + 1$, cela résulte de (3.4). Sinon, on a, compte tenu de (3.16) (iii) et (3.5) (2),

$$(C(w).C(s))_k \subset (C(w).P_{\{s\}})_k = C(w)_k.P_{\{s\},k} = C(w)_k.(C(s)_k \cup P_k)$$
$$= C(w)_k.C(s)_k \cup C(w)_k = C(w)_k.C(s)_k \subset (C(w).C(s))_k.$$

Nous terminerons ce paragraphe par quelques remarques sur le produit de deux cellules $C(w)$, $C(w')$ lorsque $\ell(w.w')$ n'est pas nécessairement égal à $\ell(w) + \ell(w')$.

Proposition **(3.18)**. — *Soient* w, $w' \in W$, $\ell' = \ell(w')$ *et* $(s_1, \ldots, s_{l'})$ *une décomposition réduite de* w'. *Soit* $i_0, \ldots, i_m$ *la suite d'indices définie récursivement par les propriétés sui-*

vantes : $0=i_0<\ldots<i_m\leq\ell'$ *et, pour* $q\geq 1$, i_q *est le plus petit entier* $h\in[i_{q-1}+1,\ell']$ *tel que* $\ell(ws_{i_1}\ldots s_{i_{q-1}}s_h)>\ell(ws_{i_1}\ldots s_{i_{q-1}})$. *Posons* $w''=s_{i_1}\ldots s_{i_m}$. *Alors*

$$(1) \qquad\qquad \ell(w'')=m, \qquad \ell(w.w'')=\ell(w)+\ell(w'')$$

$$(2) \qquad\qquad C(ww'')\subset C(w).C(w')\subset\overline{C(w).C(w')}=\overline{C(ww'')},$$

et w'' *est le seul élément de* W *satisfaisant aux relations* (2). *On a* $w'=w''$ *si et seulement si* $\ell(ww')=\ell(w)+\ell(w')$.

Les relations (1) sont évidentes par construction, et (2) se déduit immédiatement de (3.5) (3), (5) en raisonnant par récurrence sur ℓ'. Pour établir l'unicité de w'', il suffit de remarquer que toute double classe PgP étant ouverte et dense dans son adhérence, deux doubles classes distinctes ont des adhérences distinctes. La dernière assertion est conséquence immédiate de la définition de w''.

Remarques (**3.19**). — *a)* Conservons les notations de (3.18). Soit J l'ensemble des suites $(j_0,\ldots,j_m)$ telles que $0=j_0<\ldots<j_m\leq\ell'$ et que, pour $q\geq 1$, j_q soit *inférieur* au plus petit entier $h\in[j_{q-1}+1,\ell']$ tel que $\ell(ws_{j_1}\ldots s_{j_{q-1}}s_h)>\ell(ws_{j_1}\ldots s_{j_{q-1}})$, et posons

$$X(w,w')=\{ws_{j_1}\ldots s_{j_m}\mid (0,j_1,\ldots,j_m)\in J\}.$$

Alors, il résulte de (3.5) (1) et (3.5) (2) que

$$(1) \qquad\qquad C(w)_k.C(w')_k=\bigcup_{x\in X(w,w')}C(x)_k.$$

En particulier, l'ensemble $X(w,w')$ ne dépend pas de la décomposition réduite de w' choisie — ce qui justifie la notation *a posteriori* — et l'on a

$$(2) \qquad\qquad X(w,w')^{-1}=X(w'^{-1},w^{-1}).$$

L'ensemble $X(w,w')$ possède un unique élément de longueur maximum, à savoir ww''.

b) On peut montrer que si w, w' sont des éléments d'un groupe de Coxeter quelconque, l'ensemble $X(w,w')$ défini comme en *a)* à partir d'une décomposition réduite de w' ne dépend pas de cette décomposition, possède un unique élément de longueur maximum et satisfait à la relation (2). De plus, la première assertion de (3.13) est valable dans ce cas, et l'on a $X(w,w')\subset A_x$, où x désigne l'élément de longueur maximum de $X(w,w')$.

4. Le groupe fondamental de certains sous-groupes semi-simples.

(**4.1**) *Poids fondamentaux spéciaux.* — Soient V un espace vectoriel de dimension finie sur **Q**, Φ un système de racines dans V et $\check{\Phi}$ le système de racines inverse de Φ dans le dual V' de V [4; VI, 1.1]; par définition, P(Φ) (resp. P($\check{\Phi}$)) s'identifie au dual de Q($\check{\Phi}$) (resp. Q(Φ)) [4; VI, 1.9] (pour les notations P( ), Q( ), cf. (2.23) *a)*); les groupes abéliens finis P(Φ)/Q(Φ) et P($\check{\Phi}$)/Q($\check{\Phi}$) sont en dualité sur **Q**/**Z**, donc isomorphes.

Supposons Φ irréductible et réduit et soient Δ une base de Φ, $d = \sum_{a \in \Delta} m_a . a$ la racine dominante correspondante, $\overline{\omega}_a$ $(a \in \Delta)$ le « poids fondamental » de $\check{\Phi}$ correspondant à a, c'est-à-dire l'élément de V' défini par

$$\langle \overline{\omega}_a, b \rangle = \delta_{ab} \qquad (a, b \in \Delta)$$

et J l'ensemble des éléments a de Δ tels que $m_a = 1$. On sait alors [4; VI, 2.4, corollaire] que $\{o\} \cup \{\overline{\omega}_a \,|\, a \in J\}$ est un système de représentants de $P(\check{\Phi})/Q(\check{\Phi})$ dans $P(\check{\Phi})$.

Lemme **(4.2)**. — *Si Φ est un système de racines irréductible non réduit, et si Φ^{nm} est le système de racines formé par les éléments non multipliables de Φ, on a*

$$P(\Phi^{\mathrm{nm}}) = P(\Phi) = Q(\Phi).$$

En effet, Φ et Φ^{nm} sont alors respectivement de type BC et de type C. On a donc $[P(\Phi^{\mathrm{nm}}) : Q(\Phi^{\mathrm{nm}})] = 2$. D'autre part, $Q(\Phi^{\mathrm{nm}}) \subsetneq Q(\Phi)$; en effet, une racine multipliable $a \in \Phi$ ne peut appartenir à Φ^{nm} puisque $2a$ est une racine et donc un élément primitif de $Q(\Phi^{\mathrm{nm}})$. Notre assertion résulte alors des inclusions évidentes

$$Q(\Phi) \subset P(\Phi) \subset P(\Phi^{\mathrm{nm}}).$$

Proposition **(4.3)**. — *Supposons G semi-simple. Soient G_1 un sous-groupe semi-simple de G, T un tore maximal de G contenant un tore maximal T_1 de G_1 et $r : X^*(T) \otimes \mathbf{Q} \to X^*(T_1) \otimes \mathbf{Q}$ l'application linéaire prolongeant l'épimorphisme de restriction $X^*(T) \to X^*(T_1)$. Posons $\Phi = \Phi(T, G)$ et $\Phi_1 = \Phi(T_1, G_1)$. Supposons que l'ensemble $r(\Phi)^{\times}$ des éléments non nuls de $r(\Phi)$ soit un système de racines dont Φ_1 est l'ensemble des éléments non multipliables, et que Φ_1 possède une base Δ telle que $\Delta_1 = r(\Delta) \cap r(\Phi)^{\times}$ soit une base de $r(\Phi)^{\times}$. Alors*

(i) $r(Q(\Phi)) = Q(r(\Phi)^{\times})$.

(ii) $r(P(\Phi)) = P(r(\Phi)^{\times}) = P(\Phi_1)$.

(iii)   *L'application r induit un épimorphisme de $P(\Phi)/X^*(T)$ sur $P(\Phi_1)/X^*(T_1)$.*

La relation (i) est évidente, et (iii) est conséquence immédiate de (ii). Démontrons donc (ii).

Il existe un groupe simplement connexe $\widetilde{G}$ et une isogénie centrale $\widetilde{G} \to G$. Quitte à remplacer G par $\widetilde{G}$ et G_1, T, T_1 par les composantes neutres de leurs images réciproques dans $\widetilde{G}$, nous pouvons supposer G simplement connexe, c'est-à-dire $P(\Phi) = X^*(T)$. On a alors

(1) $$r(P(\Phi)) = r(X^*(T)) = X^*(T_1) \subset P(\Phi_1).$$

L'égalité $P(\Phi_1) = P(r(\Phi)^x)$ étant conséquence de (4.2), il reste seulement à établir l'inclusion

(2) $$P(\Phi_1) \subset r(P(\Phi)).$$

Si $a, b \in \Delta$ appartiennent à un même facteur simple de Φ et si $r(a) \neq o \neq r(b)$, alors $r(a)$ et $r(b)$ appartiennent à un même facteur simple de $r(\Phi^{\times})$. En effet, il existe une suite $a=a_1, a_2, \ldots, a_n=b$ d'éléments de Δ distincts tels que a_i, a_{i+1} soient non orthogonaux pour $i=1, \ldots, n-1$; mais alors $\sum_{i=1}^{n} a_i \in \Phi$, donc $\sum_{i=1}^{n} r(a_i) \in r(\Phi)^\times$ ce qui implique notre assertion. De celle-ci, il résulte qu'on peut décomposer G en un produit direct $\prod_{j=1}^{m} G^{(j)}$ de telle façon que $T_1 = \prod_{j=1}^{m} (T_1 \cap G^{(j)})$ et que les restrictions de Φ_1 aux tores $T_1 \cap G^{(j)}$ soient les facteurs simples de Φ_1. Décomposant la relation à établir en ses composantes suivant les divers $G^{(j)}$, nous sommes ainsi ramenés au cas où Φ_1 est irréductible.

Soit $G=G' \times G''$ une décomposition de G en produit direct d'un facteur G' presque simple et d'un facteur G'' ne contenant pas G_1. Soient T', T'_1, G'_1 les images de T, T_1, G_1 par la première projection $G \to G'$. Celle-ci induit une injection $X^*(T') \otimes \mathbf{Q} \to X^*(T) \otimes \mathbf{Q}$ qui applique $P(\Phi(T', G'))$ dans $P(\Phi)$ et une bijection $X^*(T'_1) \otimes \mathbf{Q} \to X^*(T_1) \otimes \mathbf{Q}$ qui applique $\Phi(T'_1, G_1)$ sur Φ_1. On voit donc que, pour établir (2), il suffit de montrer l'inclusion analogue pour $\Phi(T', G')$ et $\Phi(T'_1, G'_1)$. Autrement dit, il nous est loisible de supposer G presque simple, ce que nous ferons désormais.

Si le système de racines $r(\Phi)^\times$ n'est pas réduit, (2) est une conséquence immédiate de (i) et de (4.2). Nous supposerons donc que $\Phi_1 = r(\Phi)^\times$.

Soient $d = \sum_{a \in \Delta} m_a . a$ la racine dominante de Φ et $d_1 = \sum_{b \in \Delta_1} m_b . b$ celle de Φ_1. On a évidemment $d_1 = r(d)$, d'où

$$(3) \qquad m_b = \sum_{\substack{a \in \Delta \\ r(a) = b}} m_a.$$

Soient V', V'_1 les duals des espaces vectoriels V, V_1. Identifions V'_1 à son image dans V' par l'adjoint de r. La relation (2) est alors équivalente à la suivante :

$$(4) \qquad Q(\check{\Phi}) \cap V'_1 \subset Q(\check{\Phi}_1).$$

Or on a, en vertu de (i) et (1),

$$(5) \qquad Q(\check{\Phi}) \cap V'_1 \subset P(\check{\Phi}) \cap V'_1 = P(\check{\Phi}_1)$$

$$(6) \qquad Q(\check{\Phi}_1) \subset Q(\check{\Phi}) \cap V'_1.$$

Vu les résultats rappelés en (4.1), il suffit donc — pour établir (4) — de montrer que pour tout $b \in \Delta_1$ tel que $m_b = 1$, le poids fondamental $\overline{\omega}_b$ (cf. (4.1)) n'appartient pas à $Q(\check{\Phi})$. Or il résulte de (3) que b est la restriction à T_1 d'un seul élément a de Δ, et qu'on a $m_a = 1$. Mais alors

$$\overline{\omega}_b = \overline{\omega}_a \notin Q(\check{\Phi})$$

(toujours en vertu de (4.1)), et la démonstration est terminée.

Corollaire **(4.4)**. — *Si* G *est simplement connexe, ou si le système de racines* $r(\Phi)^\times$ *est de type* BC, *alors* G_1 *est simplement connexe.*

Il suffit de montrer que $X^*(T_1) = P(\Phi_1)$ (cf. (2.23)). Si G est simplement connexe, alors $P(\Phi) = X^*(T)$ et cela résulte de (iii). Si $r(\Phi)^\times$ est de type BC, alors $P(\Phi_1) = P(r(\Phi)^\times) = Q(r(\Phi)^\times)$ vu (4.2), donc $P(\Phi_1) \subset X^*(T_1)$ vu (i).

Remarques (**4.5**). — *a*) Soient $\pi : \widetilde{G} \to G$ et $\pi_1 : \widetilde{G}_1 \to G_1$ des revêtements universels de G et G_1 respectivement (2.23). Vu (2.24), il existe un unique morphisme $j : \widetilde{G}_1 \to \widetilde{G}$ tel que $\pi \circ j = i \circ \pi_1$, où i est l'inclusion de G_1 dans G. Le corollaire (4.4) entraîne que j *est injectif*. En particulier, j applique injectivement le noyau (schématique) de π_1 dans celui de π. Si $k = \mathbf{C}$, ces noyaux s'identifient respectivement aux groupes fondamentaux de G_1 et G.

b) Rappelons que si S est un tore, alors $X_*(S)$ désigne le groupe des morphismes de $\mathbf{GL}_1$ dans S, groupe qui s'identifie canoniquement au dual sur $\mathbf{Z}$ de $X^*(S)$ ([1; 8.6], [3; § 1]). Le dual V_1' (resp. V') de V_1 (resp. V) s'identifie à $X_*(T_1) \otimes \mathbf{Q}$ (resp. $X_*(T) \otimes \mathbf{Q}$). Par dualité, (ii) et (iii) équivalent respectivement à :

(ii)′ $Q(\check{\Phi}) \cap V_1' = Q(\check{\Phi}_1)$;

(iii)′ *le monomorphisme* $r_* : X_*(T_1) \to X_*(T)$ *associé au morphisme d'inclusion* $T_1 \to T$ *induit un monomorphisme de* $C_1 = X_*(T_1)/Q(\check{\Phi}_1)$ *dans* $C = X_*(T)/Q(\check{\Phi})$.

En fait, c'est essentiellement sous cette forme qu'ils ont été démontrés.

Si $k = \mathbf{C}$, les groupes C_1 et C s'identifient aux groupes fondamentaux de G_1 et G respectivement, et l'on retrouve l'injectivité signalée ci-dessus.

(**4.6**) La proposition (4.3) et son corollaire nous ont été suggérés par le résultat suivant de J. Humphreys, qui en a donné une démonstration d'ailleurs très différente de la nôtre.

Corollaire (J. Humphreys). — *Supposons que le groupe G soit simplement connexe, ou bien qu'il soit presque simple et que son système de racines relatives $_k\Phi$ soit non réduit. Alors, le sous-groupe semi-simple déployé maximal F de G défini en* [3; 7.2] *est simplement connexe.*

Cela résulte de (4.4), vu [3; 6.8].

Corollaire (**4.7**). — *Supposons G semi-simple et $k = \mathbf{R}$. Soient S un tore $\mathbf{R}$-déployé maximal de G, $\varphi : \widetilde{G} \to G$ un revêtement universel de G, $\widetilde{S}$ la composante neutre de l'image réciproque de S dans $\widetilde{G}$ et $G_{\mathbf{R}}^0$ la composante neutre du groupe de Lie $G_{\mathbf{R}}$. Alors $\widetilde{S}$ est un tore déployé sur $\mathbf{R}$ maximal de $\widetilde{G}$; on a*

$$|G_{\mathbf{R}}/G_{\mathbf{R}}^0| = |(\mathrm{Ker}\ \varphi)_{\mathbf{R}}|/|(\mathrm{Ker}\ \varphi/(\widetilde{S} \cap \mathrm{Ker}\ \varphi))_{\mathbf{R}}|,$$

où $|Y|$ désigne l'ordre du groupe Y. En particulier, si G est simplement connexe, $G_{\mathbf{R}}$ est connexe.

Rappelons en outre que $G_{\mathbf{R}}/G_{\mathbf{R}}^0$ est un 2-groupe abélien élémentaire [3; 14.5]. Plus précisément [3; 14.4]

(1) $$G_{\mathbf{R}}/G_{\mathbf{R}}^0 \xrightarrow{\sim} S_{\mathbf{R}}/(S_{\mathbf{R}} \cap G_{\mathbf{R}}^0).$$

La première assertion résulte de (2.20). Pour tout groupe algébrique X défini sur $\mathbf{R}$, notons $X_{\mathbf{R}}^0$ la composante neutre du groupe de Lie $X_{\mathbf{R}}$. Soit F le sous-groupe déployé de G défini en [3; 7.2] et soit $\widetilde{F}$ la composante neutre de $\varphi^{-1}(F)$. Vu (4.6), $\widetilde{F}$ est simplement connexe, et on sait alors que le tore maximal $\widetilde{S}$ de $\widetilde{F}$ est produit direct de ses intersections avec les sous-groupes de type $\mathbf{SL_2}$ de $\widetilde{F}$ correspondant aux racines simples. On a donc $\widetilde{S}_{\mathbf{R}} \subset \widetilde{F}_{\mathbf{R}}^0 \subset \widetilde{G}_{\mathbf{R}}^0$ d'où $\widetilde{G}_{\mathbf{R}} = \widetilde{G}_{\mathbf{R}}^0$, vu [3; 14.4]. Par conséquent, $\varphi(\widetilde{G}_{\mathbf{R}}) = G_{\mathbf{R}}^0$ et

(2) $$S \cap G_{\mathbf{R}}^0 = \varphi((\widetilde{S}.\mathrm{Ker}\,\varphi)_{\mathbf{R}}) \cong (\widetilde{S}.\mathrm{Ker}\,\varphi)_{\mathbf{R}}/(\mathrm{Ker}\,\varphi)_{\mathbf{R}}.$$

D'autre part, il résulte du théorème 90 de Hilbert que l'application canonique

$$(\widetilde{S}.\mathrm{Ker}\,\varphi)_{\mathbf{R}} \to ((\widetilde{S}.\mathrm{Ker}\,\varphi)/\widetilde{S})_{\mathbf{R}} \cong (\mathrm{Ker}\,\varphi/(\widetilde{S} \cap \mathrm{Ker}\,\varphi))_{\mathbf{R}}$$

est surjective. Par conséquent, le nombre de composantes connexes de $(\widetilde{S}.\mathrm{Ker}\,\varphi)_{\mathbf{R}}$ est égal à $2^l.|(\mathrm{Ker}\,\varphi/(\widetilde{S} \cap \mathrm{Ker}\,\varphi))_{\mathbf{R}}|$, où $l = \dim S = \dim \widetilde{S}$. Comme $(\mathrm{Ker}\,\varphi)_{\mathbf{R}} \cap (\widetilde{S}.\mathrm{Ker}\,\varphi)_{\mathbf{R}}^0 = \{1\}$, il résulte alors de (2) que le nombre de composantes connexes de $S \cap G_{\mathbf{R}}^0$ est égal à

$$2^l.|(\mathrm{Ker}\,\varphi/(\widetilde{S} \cap \mathrm{Ker}\,\varphi))_{\mathbf{R}}|/|(\mathrm{Ker}\,\varphi)_{\mathbf{R}}|.$$

La relation à établir s'ensuit aussitôt, vu (1) et le fait que $S_{\mathbf{R}}$ possède 2^l composantes connexes.

(**4.8**) La dernière assertion du corollaire (4.7) est connue et remonte à E. Cartan. Elle est essentiellement équivalente au cas réel de la conjecture selon laquelle le groupe G_k des points rationnels d'un groupe presque simple, simplement connexe, non anisotrope G est engendré par les points rationnels des radicaux unipotents de ses k-sous-groupes paraboliques. La démonstration donnée ici est à peu près celle de V. P. Platonov [7].

5. Remarques et rectifications.

(**5.1**) La seconde égalité (2) de [3; 3.8] doit être remplacée par une inclusion :

$$G_{\tau}^{*(S')} \supset G_{\psi}^{*(S)}.$$

L'égalité n'a pas toujours lieu comme le montre l'exemple suivant. Supposons $G = G_1 \times G_2$, où G_i est un groupe déployé de type A_i. Posons $\Phi(T, G_1) = \{\pm a\}$ et $\Phi(T, G_2) = \{\pm a_i \mid i = 1, 2, 3\}$. On obtient le contre-exemple annoncé en faisant $S = (\mathrm{Ker}\,a \cap \mathrm{Ker}\,a_1)^0$, $S' = (\mathrm{Ker}(a-a_1))^0$ et $\psi = \{\pm a_2|_S\} = \{\pm a_3|_S\}$. En effet, on a alors $\tau = \{\pm a_i|_{S'} \mid i = 1, 2, 3\}$, $G_{\psi}^{*(S)} = G_2$ et $G_{\tau}^{*(S')} = G$.

(**5.2**) La proposition 5.7 de [3] est valable pour toute partie ψ de $_k\Phi(G)$, sans restriction. En effet, on se ramène aussitôt au cas considéré dans [3] en remplaçant G par G_Ψ (cf. [3; 3.8]) où Ψ est l'ensemble des racines relatives qui sont combinaisons linéaires à coefficients entiers d'éléments de ψ.

Dans la démonstration du corollaire 5.8 de [3], c'est cette généralisation de 5.7 — et non 5.7 elle-même — qu'il convient d'appliquer, en posant $\psi = \{a\}$.

(**5.3**) Notons une conséquence immédiate mais utile de [3; 6.4 (2)]. Supposons G semi-simple et soit Γ le groupe de Galois de k_s sur k. Alors, dim S est égal au nombre d'orbites de Γ dans $\Delta - \Delta^0$ (pour l'action $\gamma \mapsto {}_\Delta\gamma$ de Γ sur Δ), et la dimension du centre de $\mathscr{Z}(S)$ est $\mathrm{card}(\Delta - \Delta^0)$. Par conséquent, S est la composante neutre du centre de $\mathscr{Z}(S)$ si et seulement si Γ fixe tous les éléments de $\Delta - \Delta^0$.

(**5.4**) L'assertion (iv) de [3; 12.6] doit être complétée comme suit :

(iv) *Il existe une algèbre à division* C *centrale sur* k *et un* k-*morphisme*

$$G \to \mathbf{GL}_m(\mathrm{C}) \qquad (d(\xi) = m \,.\, c,\, c^2 = [\mathrm{C} : k])$$

équivalent sur $\bar{k}$ *à un élément de* ξ.

(Ici, $\mathbf{GL}_m(\mathrm{C})$ désigne le k-groupe algébrique réductif isomorphe sur $\bar{k}$ à $\mathbf{GL}_{mc}$, dont le groupe des points à valeurs dans une k-algèbre A est $\mathbf{GL}_m(\mathrm{C} \underset{k}{\otimes} \mathrm{A})$, au sens usuel de la notation.)

Pour plus de détails sur les questions traitées au § 12 de [3] et la généralisation des résultats de ce paragraphe à un corps k quelconque, cf. [11].

(**5.5**) *Autres corrections à* [3] :

P. 70, l. 3, lire G au lieu de G_k.

P. 72, l. 28, lire L_{i-1}/L_i au lieu de H/L_{i-1}.

P. 76, l. 23, lire $\Phi(\mathrm{S}, \mathrm{G})$ au lieu de $\Phi(\mathrm{S}, \mathrm{T})$.

P. 83, ll. 6, 7, 8, lire $\mathbf{G}_a$ au lieu de $\mathbf{G}_m$.

P. 83, l. 9, lire $\mathrm{H} \not\subset \mathrm{U}_\mu$ au lieu de $\mathrm{H} \subset \mathrm{U}_\mu$.

P. 84, l. 4 du bas, lire positifs au lieu de positive.

P. 91, l. 4, lire *de Levi de* P au lieu de *de Levi de* G.

P. 113, l. 28, lire $\mathscr{Z}(\mathrm{S})_\mathrm{K}$ au lieu de $\mathscr{Z}(\mathrm{T})_\mathrm{K}$.

P. 141, la lettre ρ, utilisée sans avoir été introduite (sinon en (6.3)), représente la restriction de j (définie en (12.12) mais non utilisée par la suite) à Δ.

P. 142, l. 11, lire $d\pi_e$ au lieu de ρ.

Les références à [6] doivent être remplacées par les suivantes :

P. 62, l. 2 du bas : [6; VI, 1.1].

P. 63, l. 9 : [6; VI, 1.3, Prop. 8].

P. 71, ll. 13, 14 : Prop. 19 de [6; VI, 1.6].

P. 74, l. 19 : [6; VI, 1.7, Prop. 22].

P. 83, l. 13 du bas : [6; V, 3.3, Prop. 1].

Pp. 85, 86, dernière ligne : [6; VI, 1.7, Prop. 20].

P. 97, l. 6, p. 98, l. 4 : [6; V, 3.3, Prop. 1].

P. 98, l. 10 : [6; 1.10, Prop. 27].

P. 100, l. 2 : [6; VI, 1.7, Prop. 20].

P. 107, l. 16 : [6; V, 3.3, Prop. 1].

P. 143, l. 6 : [6; V, 3.3, Th. 2].

P. 149, l. 10 : [6] N. Bourbaki, *Groupes et algèbres de Lie*, chap. IV, V, VI, *Act. Sci. Ind.*, 1337, Paris, Hermann, 1968.

The Institute for Advanced Study et Universität Bonn.

RÉFÉRENCES

[1] A. Borel, *Linear Algebraic Groups*, notes by H. Bass, New York, Benjamin, 1969.

[2] — and T. A. Springer, Rationality properties of linear algebraic groups, II, *Tôhoku Math. Jour.*, **20** (1968), 443-497.

[3] — et J. Tits, Groupes réductifs, *Publ. Math. I.H.E.S.*, **27** (1965), 55-151.

[4] N. Bourbaki, *Groupes et algèbres de Lie*, chap. IV, V, VI, *Act. Sci. Ind.*, 1337, Paris, Hermann, 1968.

[5] C. Chevalley, *Séminaire sur la classification des groupes de Lie algébriques*, 2 vol., notes polycopiées, Inst. H. Poincaré, Paris, 1958.

[6] M. Demazure et P. Gabriel, *Groupes algébriques*, t. I, Paris, Masson, 1970.

[7] V. P. Platonov, The problem of strong approximation and the conjecture of Kneser-Tits on algebraic groups, *Isvestia Ak. Nauk. USSR*, **33** (1969), 1211-1219.

[8] C. Riehm, The congruence subgroup problem over local fields, *Amer. Jour. Math.*, **92** (1970), 771-778.

[9] R. Steinberg, *Lectures on Chevalley groups*, notes by J. Faulkner and R. Wilson, Yale University, 1967.

[10] J. Tits, Formes quadratiques, groupes orthogonaux et algèbres de Clifford, *Inventiones Math.*, **5** (1968), 19-41.

[11] —, Représentations linéaires irréductibles d'un groupe réductif sur un corps quelconque, *J. Reine Angew. Math.*, **247** (1971), 196-220.

Manuscrit reçu le 15 octobre 1971.

95.

Some metric properties of arithmetic quotients of symmetric spaces and an extension theorem

J. Differ. Geom. **6** (1972) 543–560

This paper has two main objectives. One is to prove:

Theorem A. *Let D be the open unit disc $|z| < 1$ in C, and $D^* = D - \{0\}$. Let X be a bounded symmetric domain, and Γ an arithmetically defined torsion-free group of automorphisms of X. Let V^* be the complex analytic compactification of $V = X/\Gamma$ constructed in [3], a and b positive integers, and $f: D^{*a} \times D^b \to V$ a holomorphic map. Then f extends to a holomorphic map of D^{a+b} into V^*.*

In fact, a slightly more general result will be obtained (see Thm. 3.7). Together with some known facts, this implies that if S is an algebraic variety, $h: S \to V$ is a holomorphic map, and V is endowed with its natural structure of quasi-projective variety defined in [3, Thm. 3,10], then h is a morphism of algebraic varieties.

The proof of Theorem A makes use of an extension theorem of M. H. Kwack [12], or rather of a slight variant of it [9], and the main point is to check that its assumptions are satisfied in our case. Let d_0 be the Kobayashi invariant pseudo-distance [10] on X; since X is a bounded symmetric domain, it is a distance (cf. § 3.3). Let d'_0 be the associated distance on V defined by

$$(1) \qquad d'_0(\pi(x), \pi(y)) = \inf_{\gamma \in \Gamma} d_0(x, y \cdot \gamma) , \qquad (x, y \in X) ,$$

where $\pi: X \to V$ is the canonical projection. In view of some distance decreasing properties of f, we have essentially to prove the following result (where Γ may have torsion):

Theorem B. *Let $p, q \in V^* - V$, and let p_n, q_n $(n = 1, 2, \cdots)$ be sequences of points in V converging to p and q respectively. If $d'_0(p_n, q_n) \to 0$, then $p = q$.*

Theorem B will be derived in § 3.5 from properties of Siegel sets and arithmetic groups, whose discussion is the other purpose of this paper. Since they have some independent interest, they will be proved in greater generality and in a stronger form than is needed in § 3.5. Let then Γ be an arithmetic subgroup of a connected semi-simple Q-group $\mathscr{G}$, X the symmetric space of maximal compact subgroups of the group G of real points of $\mathscr{G}$, and d_X the

---

Received March 8, 1972.

153

distance function associated to a G-invariant Riemannian metric on X. Our main result, Theorem 2.3, is:

Theorem C. *Let $\mathfrak{S}$ be a Siegel set in X, and C a finite subset of $\mathcal{G}_{\mathbf{Q}}$. Then there exists a constant δ such that $d_X(x \cdot c, x' \cdot c' \cdot \gamma) \geq d_X(x, x') + \delta$ for all $x, x' \in \mathfrak{S},\ c, c' \in C$ and $\gamma \in \Gamma$.*

If $\pi: X \to X/\Gamma$ is the canonical projection, and d' the associated distance function on X/Γ, defined as in (1) above, Theorem C asserts in particular that the difference $d_X(x, x') - d'(\pi(x), \pi(x'))$ is bounded when x and x' vary through $\mathfrak{S}$.

In the case where $G = SL(n, \mathbf{R})$, $\Gamma = SL(n, \mathbf{Z})$, $C = \{e\}$, and x is fixed, Theorem C reduces to [16, Thm. 4]. The fact that x is also allowed to vary gives a positive answer in general to a question raised in that case at the end of [16, § 4].

Theorems A and 2.5 were proved in 1968, and Theorem A is stated in P. Griffiths' report [7, Thm. 6.6]. Since then, results closely related to Theorems A and B have been proved by P. Kiernan [9] and S. Kobayashi–S. Ochiai [11]. They have influenced the presentation given here, in particular by focussing attention on Theorem B, which had been essentially proved, but not made explicit, originally. The relations between these results are discussed in § 3.9.

Notation. In general, we use that of [5], [6], with one main exception: algebraic groups, which are always defined over $\mathbf{R}$ in this paper, are denoted by script letters $\mathcal{G}, \mathcal{H}, \cdots$, while the corresponding Roman capitals $G, H, \cdots$ stand for the group of real points of $\mathcal{G}, \mathcal{H}, \cdots$.

Let X be a differentiable manifold. The tangent space to X at x is denoted $T_x(X)$. Let Y be a differentiable manifold, $\mu: X \to Y$ an isomorphism, and g a Riemannian metric on Y. Then $\mu^*(g)$ denotes the induced Riemannian metric on X, i.e.,

$$\mu^*(g)(A, B) = g(d\mu_x(A), d\mu_x(B)) , \qquad (A, B \in T_x(X); x \in X) .$$

If u, v are complex valued functions on a set S and $|u - v|$ is bounded on S, we write $u \approx v$. Assume u and v to have real positive values. We write $u \succ v$ if there exists a constant $c > 0$ such that $u(s) \geq c \cdot v(s)$ for all $s \in S$, and $u \prec v$ (resp. $u \asymp v$) if $v \succ u$ (resp. $u \prec v$ and $v \prec u$).

If a, b are elements of a group H, then ${}^a b$ stands for $a \cdot b \cdot a^{-1}$. The value of a rational character α of an algebraic group $\mathcal{G}$ on an element $x \in \mathcal{G}$ is denoted $\alpha(x)$ or x^α.

Throughout this paper, $\mathcal{G}$ is a connected semi-simple $\mathbf{R}$-group, X the symmetric space of maximal compact subgroups of G, and K a maximal compact subgroup of G.

1. Right invariant metrics

1.1. Let B be the Killing form on the Lie algebra $L(G)$ of G, θ be the Cartan involution of $L(G)$ with respect to $L(K)$, and

$$(1) \qquad g_0(X, Y) = -B(X, \theta(Y)) , \qquad (X, Y \in L(G)) .$$

Then g_0 is a positive nondegenerate scalar product on $L(G)$, invariant under Ad K. Let dg^2 be the right invariant metric on G which is equal to g_0 on $L(G)$. Let d or d_G be the associated distance function on G, and

$$(2) \qquad |x| = d(e, x) , \qquad (x \in G) .$$

Since d is symmetric, right invariant under G, left invariant under K, and satisfies the triangular inequality, we have

$$(3) \qquad d(x, y) = |y \cdot x^{-1}| ,$$

$$(4) \qquad |x| = |x^{-1}| , \qquad |x \cdot y| \leq |x| + |y| , \qquad (x, y \in G) ,$$

and also, since $x = x \cdot y \cdot y^{-1}$,

$$(5) \qquad |x| \leq |x \cdot y| + |y| , \qquad (x, y \in G) ,$$

from which it follows immediately that if C is a compact subset of G, then

$$(6) \qquad d(u \cdot x, v \cdot y) \approx d(x, y) , \qquad |u \cdot x \cdot v| \approx |x| ,$$

as x, y vary in G and u, v in C.

1.2. Let $\mathfrak{p}$ be the orthogonal complement to $L(K)$ with respect to the Killing form (or to g_0). We have then

$$\theta(A + B) = A - B \qquad (A \in L(K), B \in \mathfrak{p}) .$$

Let $\sigma : G \to X = K \backslash G$ be the canonical projection, and $o = \sigma(K)$. For $C \in L(G)$ we write $o \cdot C$ for $d\sigma_e(C)$. The map $C \mapsto o \cdot C$ induces an isomorphism of $\mathfrak{p}$ onto $T_0(X)$, whence a scalar product on $T_0(X)$, defined by the restriction of g_0 to $\mathfrak{p}$, to be denoted also by g_0. For $Z \in L(G)$, we have

$$(1) \quad Z = Z_k + Z_p \qquad (Z_k = (Z + \theta(Z))/2 \in L(K); Z_p = (Z - \theta(Z))/2 \in \mathfrak{p}) ,$$

$$(2) \qquad g_0(o \cdot Z, o \cdot Z) = g_0(Z_p, Z_p) = B(Z_p, Z_p) = B(Z, Z_p) .$$

Let dx^2 be the G-invariant Riemannian metric on X which is equal to g_0 on $T_0(X)$, and d_X the associated distance function. It is elementary that $d_X(x, y) = d_G(\sigma^{-1}(x), \sigma^{-1}(y))(x, y \in X)$, whence

$$(3) \qquad d_G(x, y) \geq d_X(o \cdot x, o \cdot y) ,$$

$$(4) \qquad d_X(o \cdot x, o \cdot y) \approx d_G(x, y) \qquad (x, y \in G) .$$

1.3. By definition, a parabolic subgroup P of G is the group of real points of a parabolic $\mathbf{R}$-subgroup $\mathscr{P}$ of $\mathscr{G}$. Let L be a subfield of $\mathbf{R}$, and assume $\mathscr{G}$ and $\mathscr{P}$ to be defined over L. Let $\mathscr{S}$ be a maximal L-split torus of the radical of $\mathscr{P}$, and A the connected component of e in S, in the ordinary topology. After conjugation of K by some element of G we may (and shall) assume that $L(A) \subset \mathfrak{p}$.

The exponential map $\exp : L(A) \to A$ is an isomorphism of Lie groups, which carries onto one another the invariant metrics defined by g_0 onto $L(A)$ and A. In particular, if Q is a set of rational characters of $\mathscr{S}$ which form a basis of $X(\mathscr{S}) \otimes \mathbf{Q}$, then the invariant metric on A may be written

$$(1) \qquad da^2 = \sum_{\alpha, \beta \in Q} c_{\alpha\beta} \alpha^{-1} \beta^{-1} d\alpha d\beta ,$$

where the $c_{\alpha\beta}$ are constants such that

$$(2) \qquad g_0 = \sum c_{\alpha\beta} d\alpha d\beta .$$

Let d_A be the distance function on A associated to da^2. In view of (1) there exists a constant $c > 0$ such that

$$(3) \quad c^{-1} \cdot d_A(a, b) \le (\sum_{\alpha \in Q} \ln^2 (\alpha(a)/\alpha(b)))^{1/2} \le c \cdot d_A(a, b) , \qquad (a, b \in A) .$$

Moreover, A is a totally geodesic submanifold of G, hence

$$(4) \qquad d_A(a, b) = d_G(a, b) \qquad (a, b \in A) .$$

The group $\mathscr{P}$ is semi-direct product of its unipotent radical $\mathscr{U}$ by the centralizer $\mathscr{Z}(\mathscr{S})$ of $\mathscr{S}$. Let $\mathscr{M}$ be the intersections of the characters χ^2, where χ runs through $X(\mathscr{Z}(\mathscr{S}))$. Then $\mathscr{Z}(\mathscr{S}) = \mathscr{M} \cdot \mathscr{S}$, the intersection $\mathscr{M} \cap \mathscr{S}$ is finite, and $Z(S)$ is the direct product of M and A.

1.4. Lemma. (i) *The Lie algebra $L(M)$ of M is stable under θ and orthogonal to $L(A)$ with respect to B and g_0.* (ii) *We have*

$$(1) \qquad g_0(C, C) = (1/2)g_0(o \cdot C, o \cdot C) , \qquad (C \in L(U)) .$$

Let Φ be the set of roots of $\mathscr{G}$ with respect to $\mathscr{S}$. There exists an ordering on $X(\mathscr{S})$ such that the weights of $\mathscr{S}$ in $L(\mathscr{U})$ are the positive elements of Φ [6, § 3]. The restrictions to $L(A)$ of the differentials of the roots are the roots of $L(G)$ with respect to $L(A)$, in the sense of the theory of Lie algebras. For $\alpha \in \Phi$, let

$$(2) \qquad \mathfrak{g}_\alpha = \{C \in L(G) \,|\, [X, C] = d\alpha(X) \cdot C, X \in L(A)\} .$$

We have the decompositions

$$(3) \qquad L(G) = L(Z(S)) \oplus \bigoplus_{\alpha \in \Phi} \mathfrak{g}_\alpha \, , \qquad L(U) = \sum_{\alpha > 0} \mathfrak{g}_\alpha \, .$$

Let us put $\mathfrak{g}_0 = L(Z(S))$. We have then

$$(4) \qquad B(\mathfrak{g}_\alpha, \mathfrak{g}_\beta) = 0 \, , \qquad (\alpha, \beta \in \Phi \cup \{0\}; \alpha + \beta \neq 0) \, ,$$

which implies that the restriction of B to $\mathfrak{g}_\alpha + \mathfrak{g}_{-\alpha} (\alpha \in \Phi \cup \{0\})$ is nondegenerate. Let $C \in L(M)$. By definition of M, the trace of $\mathrm{ad}\, C$ in $\mathfrak{g}_\alpha (\alpha \in \Phi)$ is zero, whence $B(L(M), L(A)) = 0$. Since the restrictions of B to $L(A)$ and $L(\mathscr{Z}(S))$ are nondegenerate, it follows that $L(M)$ is the orthogonal complement of $L(A)$ in $\mathfrak{g}_0$, hence $L(M)$ is stable under θ.

The automorphism θ is $-Id.$ on $L(A)$, hence

$$(5) \qquad \theta(\mathfrak{g}_\alpha) = \mathfrak{g}_{-\alpha} \qquad (\alpha \in \Phi \cup \{0\}) \, ,$$

and consequently, using (4),

$$(6) \qquad B(\mathfrak{g}_\alpha + \mathfrak{g}_{-\alpha}, \mathfrak{g}_\beta + \mathfrak{g}_{-\beta}) = \mathfrak{g}_0(\mathfrak{g}_\alpha + \mathfrak{g}_{-\alpha}, \mathfrak{g}_\beta + \mathfrak{g}_{-\beta}) = 0$$
$$(\alpha, \beta > 0; \alpha \neq \beta) \, ,$$

$$(7) \qquad B(\mathfrak{g}_0, L(U)) = g_0(\mathfrak{g}_0, L(U)) = 0 \, .$$

Let now $C = \sum C_\alpha \, (C \in \mathfrak{g}_\alpha, \alpha > 0)$ be an element of $L(U)$. The C_α are mutually orthogonal, and so are the $C_{\alpha, p} = (1/2)(C_\alpha - \theta(C_\alpha)) \in \mathfrak{g}_\alpha + \mathfrak{g}_{-\alpha}$ by (6). To prove (ii), it suffices therefore to consider the case where $C = C_\alpha$ for some $\alpha > 0$. We have then, by (3) and § 1.2 (2):

$$g_0(o \cdot C, o \cdot C) = g_0(C, C_p) = (1/2)g_0(C, C - \theta(C)) \, ,$$
$$g_0(o \cdot C, o \cdot C) = (1/2)g_0(C, C) + (1/2)B(C, C) = (1/2) \cdot g_0(C, C) \, ,$$

which proves (ii).

1.5. We already noticed that $P = M \cdot A \cdot U$. More precisely, the map $A \times M \times U \to P$ defined by the product is an isomorphism of analytic manifolds. For $p \in P$, we shall denote by $a(p), m(p), u(p)$ the elements of A, M, U such that $p = a(p) \cdot m(p) \cdot u(p)$.

It is known that $G = K \cdot P = K \cdot M \cdot A \cdot U$. If

$$x = k \cdot a \cdot m \cdot u \, , \qquad (x \in G, k \in K, a \in A, m \in M, u \in U) \, ,$$

then a and u are uniquely determined by x, and are analytic functions of x. They will often be denoted $a(x)$ and $u(x)$. The elements k and m are determined up to the product by an element of $K \cap M$. The group $K \cap M$ is maximal compact in M or P. Let $Z = (K \cap M)\backslash M = (K \cap M^0)\backslash M^0$, and let $\tau : M \to Z$ be the canonical projection. It is known that the map $(a, m, u) \mapsto o \cdot a \cdot m \cdot u$ $(a \in A, m \in M, u \in U)$ induces an isomorphism of analytic manifolds

$$(1) \qquad \mu: Y = A \times Z \times U \xrightarrow{\sim} X .$$

The group P operates on Y by

$$(2) \quad (a, z, u) \cdot p = (a \cdot a(p), z \cdot m(p), a(p)^{-1} \cdot m(p)^{-1} \cdot u \cdot m(p) \cdot a(p) \cdot u(p)) ,$$

and we have

$$(3) \qquad \mu(y \cdot p) = \mu(y) \cdot p \qquad (y \in Y; p \in P) .$$

Let us identify $L(M)_{\mathfrak{p}} = L(M) \cap \mathfrak{p}$ with the tangent space to Z at the origin, and let dz^2 be the M-invariant Riemannian metric defined at the origin by g_0. Let further du^2 be the right-invariant Riemannian metric on U which is equal to the restriction of g_0 on $L(U)$. If φ is an automorphism of U, then $g' = \varphi^*(du^2)$ is also right-invariant and we have

$$(4) \quad g'(C, C') = g_0(d\varphi_e(C \cdot u^{-1}))d\varphi_e(C' \cdot u^{-1})) , \qquad (u \in U; C, C' \in T_u(U)) .$$

1.6. Proposition. *We keep the previous notation. Let $dy^2 = \mu^*(dx^2)$.*

(i) *For any $z \in Z, u \in U$, the metric induced by dy^2 on $A \times \{z\} \times \{u\}$ is da^2.*

(ii) *Let $y = (a, z, u) \in Y$ and $m \in \tau^{-1}(z)$. The tangent spaces at y to the submanifolds $A \times \{z\} \times \{u\}, \{a\} \times Z \times \{u\}$ and $\{a\} \times \{z\} \times U$ are orthogonal, and we have*

$$(1) \qquad (dy^2)_y = (da^2)_a + (dz^2)_z + (1/2) ((\text{Int } am)^*(du^2))_u .$$

It is well-known that the map $C \mapsto o \cdot \exp C$ induces an isomorphism of $L(A)$ onto a closed and flat totally geodesic submanifold of X, whence (i).

For part (ii), let $C \in T_y(Y)$. Write it in the form

$$(2) \quad C = C_1 \cdot a + C_2 \cdot m + C_3 \cdot u \qquad (C_1 \in L(A), C_2 \in L(M)_{\mathfrak{p}}, C_3 \in L(U)) .$$

It is clear that we have

$$(3) \qquad d\mu(C) = o \cdot (C_1 + C_2 + \text{Ad } am(C_3)) \cdot amu .$$

Let $\langle C, C \rangle$ denote the value of dy^2 on C, and let D be the projection of $C_1 + C_2 + \text{Ad } am(C_3)$ in $\mathfrak{p}$. By (3) and § 1.2 (2), we have

$$(4) \qquad \langle C, C \rangle = g_0(D, D) .$$

Since A and M normalize U, we have $\text{Ad } am(C_3) \in L(U)$. The elements C_1 and C_2 are in $\mathfrak{p}$. By Lemma 1.4 and § 1.4 (7), they are orthogonal to each other and to $L(U)$, which implies the first assertion of (ii). Moreover, by § 1.4 (1), we have

$$\langle C, C \rangle = g_0(C_1, C_1) + g_0(C_2, C_2) + \tfrac{1}{2}g_0(\text{Ad } am(C_3), \text{Ad } am(C_3)) ,$$

which, in view of § 1.5 (4), is just another way to write (1).

1.7. Corollary. *Let d_A, d_Z, d_X, d_Y be the distance functions associated to the Riemannian metrics da^2, dz^2, dx^2 and dy^2, and $y = (a, z, u), y' = (a', z', u')$ be two points of Y. Then*

$$(1) \qquad\qquad d_X(\mu(y), \mu(y')) \geq \max\,(d_A(a, a'), d_Z(z, z')) \;,$$

$$(2) \qquad d_G(g, h) \geq d_A(a(g), a(h)) = d_G(a(g), a(h)) \;, \qquad (g, h \in G) \;.$$

(1) follows from 1.6 by an obvious computation. (2) is a consequence of (1) and § 1.2 (3), § 1.3 (4).

1.8. The space Z is the Riemannian direct product of the symmetric spaces of maximal compact subgroups of the simple noncompact factors of M by a flat space (which has strictly positive dimension if and only if the center of M is not compact). Let F be a direct factor of Z in this decomposition, and F' the remaining factor. Then d_Z majorizes its restrictions $d_F, d_{F'}$ to F and F', which are distance functions associated to invariant Riemannian metrics. Let $\nu\colon X \to F$ be the composition of $\mu^{-1}\colon X \to Y$ by the projections $Y \to Z \to F$. It follows from Proposition 1.6 that we have

$$(1) \qquad\qquad d_F(\nu(o \cdot x), \nu(o \cdot y)) \leq d_X(o \cdot x, o \cdot y)) \qquad (x, y \in G) \;.$$

2. Siegel sets and invariant distances

2.1. From now on, $\mathscr{G}$ is defined over Q, and $\mathscr{P}$ is a minimal parabolic Q-subgroup of $\mathscr{G}$. We keep the notation of § 1.5 (with $L = Q$). Moreover, Φ is the set of roots of $\mathscr{G}$ with respect to $\mathscr{S}$, and Δ the set of simple roots for the ordering associated to $\mathscr{P}$ [5, § 11].

We recall that a Siegel set $\mathfrak{S}$ or $\mathfrak{S}_{t,\omega}$ (with respect to K, P, S, as will always be understood) is a set of the form $\mathfrak{S} = K \cdot A_t \cdot \omega$ where ω is a relatively compact subset of $M \cdot U$ and

$$A_t = \{a \in A \,|\, \alpha(a) \leq t, (a \in \Delta)\}$$

[5, § 12]. For $x \in \mathfrak{S}$, the decomposition of § 1.5 will sometimes be written

$$x = k_x \cdot a(x) \cdot m_x \qquad (k_x \in K, a_x \in A, m_x \in \omega) \;.$$

2.2. Lemma. *We keep the previous notation. The differences*

$$d_G(x \cdot u, x' \cdot u') - d_G(a(x) \cdot u, a(x') \cdot u') \;, \qquad d_G(x, x') - d_A(a(x), a(x'))$$

are bounded in absolute value as x, x' range through $\mathfrak{S}$ and u, u' through G.
We have

$$d_G(x \cdot u, x' \cdot u') = d_G(k_x \cdot {}^{a(x)}m_x \cdot a(x) \cdot u, k_{x'} \cdot {}^{a(x')}m_{x'} \cdot a(x') \cdot u') \;.$$

159

The elements $k_x, k_{x'}$ run through a compact set. By a fundamental property of Siegel sets [5, Lemma 12.2], so do $^{a(x)}m_x$ and $^{a(x')}m_{x'}$, hence (§ 1.1 (6), § 1.3 (4))

$$d_G(x \cdot u, x' \cdot u') \approx d_G(a(x) \cdot u, a(x') \cdot u') , \qquad (x, x' \in \mathfrak{S}; \ u, u' \in G) ,$$

$$d_G(x, x') \approx d_A(a(x), a(x')) \qquad\qquad (x, x' \in \mathfrak{S}) .$$

By definition, a Siegel set in X is the projection $\sigma(\mathfrak{S}) = o \cdot \mathfrak{S}$ of a Siegel set $\mathfrak{S}$ in G. Hence Theorem 2.3 below is Theorem C of the introduction.

2.3. Theorem. *Let $\mathfrak{S}$ be a Siegel set in G (with respect to K, P, S), C a finite subset of $\mathscr{G}_Q$, and Γ an arithmetic subgroup of G. Then there exists a constant δ such that*

$$(1) \qquad d_X(o \cdot x \cdot c, o \cdot x' \cdot c' \cdot \gamma) \geq d_X(x, x') + \delta ,$$

for all $x, x' \in \mathfrak{S}, c, c' \in C$ and $\gamma \in \Gamma$.

In view of Lemma 2.2, § 1.1 (3) and § 1.2 (4), our assertion is equivalent to the existence of a constant δ' such that

$$(2) \qquad |a(x') \cdot c' \cdot \gamma \cdot c^{-1} \cdot a(x)^{-1}| \geq |a(x') \cdot a(x)^{-1}| + \delta' ,$$

for all $x, x' \in \mathfrak{S}, c, c' \in C, \gamma \in \Gamma$.

Using the Bruhat decomposition in G_Q, we can write

$$(3) \qquad c' \cdot \gamma \cdot c^{-1} = u \cdot w \cdot t \cdot v \qquad (u \in U_w, v \in U, w \in \mathscr{N}(S)_Q, t \in S)$$

(see [5, § 11.4; 6, § 5]) where U_w is a certain subgroup of U, and w runs through a set of representatives of $\mathscr{N}(S)/\mathscr{Z}(S)$ in $\mathscr{N}(S)_Q$, chosen once and for all. Let

$$(4) \qquad z = a(x') \cdot c' \cdot \gamma \cdot c^{-1} \cdot a(x)^{-1} , \qquad q = w^{-1} \cdot z .$$

We have $|z| \approx |q|$, (§ 1.1), and $|q| \geq |a(q)|$, (Corollary 1.7). Therefore (2) will be proved if we show the existence of a constant δ'' such that

$$(5) \qquad |a(q)| \geq |a(x') \cdot a(x)^{-1}| + \delta'' ,$$

for all $x, x' \in \mathfrak{S}, c, c' \in C$ and $\gamma \in \Gamma$.

We note first that $q = w^{-1} \cdot a(x') \cdot u \cdot w \cdot t \cdot v \cdot a(x)^{-1}$, whence

$$(6) \qquad q = {}^{w^{-1} \cdot a(x')}u \cdot (^{w^{-1}}a(x')) \cdot t \cdot a(x)^{-1} \cdot {}^{a(x)}v .$$

For $\alpha \in \Delta$, let (π_α, V_α) be a strongly rational representation of G whose highest weight λ_α is orthogonal to $\Delta - \{\alpha\}$ (see [6, § 12]). Fix on $V_{\alpha,R}$ a euclidean norm $\| \ \|$ invariant under K, and with respect to which S is represented by self-adjoint operators (see e.g. [5, § 9]). Let e_0 be a unit vector in the (unique) line stable under P. We have $\pi(g)e_0 = \pm e_0$ for $g \in M \cdot U$, whence

$$(7) \qquad \|\pi_\alpha(x) \cdot e_0\| = a(x)^{\lambda_\alpha} \qquad (x \in G) \ .$$

By construction of U_w, the element $w^{-1} \cdot a(x') \cdot u \cdot a(x')^{-1} \cdot w$ belongs to the unipotent radical U^- of the group P^- opposed to P and containing $\mathscr{Z}(S)$. Now, if $g \in U^-$, then $\pi_\alpha(g) \cdot e_0 = e_0$ modulo the sum of the eigenspaces of S corresponding to lower weights, i.e., modulo the orthogonal complement of $\boldsymbol{R} \cdot e_0$. Therefore

$$(8) \qquad \|\pi_\alpha(g) \cdot \lambda \cdot e_0\| \geq |\lambda| \qquad (\lambda \in \boldsymbol{R}; g \in U^-) \ .$$

We have then, using (6) and (7):

$$
(9) \qquad
\begin{aligned}
a(q)^{\lambda_\alpha} = \|\pi(q) \cdot e_0\| &\geq \|\pi_\alpha(^{w^{-1}}a(x') \cdot t \cdot a(x)^{-1}) \cdot e_0\| \\
&= a(x')^{w(\lambda_\alpha)} \cdot t^{\lambda_\alpha} \cdot a(x)^{-\lambda_\alpha} \ .
\end{aligned}
$$

There is a matrix realization of $\mathscr{G}$ over $\boldsymbol{Q}$ in which Γ is represented by integral matrices [5, Cor. 7.13]. The elements of $C \cdot \Gamma \cdot C \cup C^{-1} \cdot \Gamma \cdot C^{-1}$ are then rational matrices whose entries have bounded denominators. This implies that $t^{\lambda_\alpha} \succ 1$ (see the proof of Cor. 15.3 in [5]), whence

$$(10) \qquad a(q)^{\lambda_\alpha} \succ a(x')^{w(\lambda_\alpha)} \cdot a(x)^{-\lambda_\alpha} \ .$$

$w(\lambda_\alpha)$ is a weight of π_α, therefore [6, § 12]

$$w(\lambda_\alpha) = \lambda_\alpha - \sum_{\beta \in \varDelta} c_\beta \cdot \beta \qquad (c_\beta \in \boldsymbol{Z}, c_\beta \geq 0) \ .$$

(10) can then be written

$$(11) \qquad a(q)^{\lambda_\alpha} \succ a(x')^{\lambda_\alpha} \cdot a(x)^{-\lambda_\alpha} \cdot a(x)^{-\Sigma c_\beta \beta} \ .$$

Since $c_\beta \geq 0$ and $a(x)^\beta \leq t$, the last factor is $\succ 1$, and we have proved the existence of a constant $\delta_1 > 0$ such that

$$(12) \qquad a(q)^{\lambda_\alpha} \geq \delta_1 \cdot a(x')^{\lambda_\alpha} \cdot a(x)^{-\lambda_\alpha} \ ,$$

for all $x, x' \in \mathfrak{S}, c, c' \in C, \gamma \in \Gamma$ and $\alpha \in \varDelta$, with q defined by (4).

For an element $a \in A$, let $n(a)$ be the positive square root of

$$(13) \qquad n(a)^2 = \sum_{\alpha \in \varDelta} \ln^2 a^{\lambda_\alpha} \ .$$

In order to prove (5), it suffices, by § 1.3 (3), to show the existence of a constant δ_2 such that

$$(14) \qquad n(a(q)) - n(a(x') \cdot a(x)^{-1}) \geq \delta_2 \ ,$$

for all $x, x' \in \mathfrak{S}, c, c' \in C, \gamma \in \Gamma$. We have

$$n(a(q))^2 - n(a(x') \cdot a(x)^{-1})^2$$

(15)
$$= \sum_{\alpha \in \varDelta} (\ln a(q)^{\lambda_\alpha} + \ln (a(x')^{\lambda_\alpha} \cdot a(x)^{-\lambda_\alpha})$$
$$\cdot (\ln a(q)^{\lambda_\alpha} - \ln (a(x')^{\lambda_\alpha} \cdot a(x)^{-\lambda_\alpha})) \ .$$

If $n(a(q)) = n(a(x') \cdot a(x)^{-1}) = 0$, there is nothing to prove. If not, it is clear that

$$(\ln a(q)^{\lambda_\alpha} + \ln (a(x')^{\lambda_\alpha} \cdot a(x)^{-\lambda_\alpha})/(n(a(q)) + n(a(x') \cdot a(x)^{-1})$$

is ≤ 1 in absolute value. On the other hand, (12) implies that

$$\ln a(q)^{\lambda_\alpha} - \ln (a(x')^{\lambda_\alpha} \cdot a(x)^{-\lambda_\alpha}) = \ln \lambda_\alpha(a(q) \cdot a(x')^{-1} \cdot a(x))$$

is bounded from below. It follows then that $n(a(q)) - n(a(x') \cdot a(x)^{-1})$ is bounded from below, which proves (14), and ends the proof of the theorem.

2.4. Let $J \subset \varDelta$. As usual, $\mathscr{P}_J$ denotes the standard parabolic subgroup generated by U and the centralizer of $\mathscr{S}_J$, where $\mathscr{S}_J$ is the identity component of $\bigcap_{\alpha \in J} \ker \alpha$. The group $\mathscr{P}_J$ is the semi-direct product over Q of its unipotent radical $\mathscr{U}_J$ by $\mathscr{Z}(\mathscr{S}_J)$. An element $g \in \mathscr{P}_J$ can be written uniquely as $g = r \cdot u$ $(r \in \mathscr{Z}(\mathscr{S}_J), u \in \mathscr{U}_J)$. The element r will be called the *reductive part of g.*

A sequence of elements $x_n \in \mathfrak{S}$ is said to be *of type J* if $a(x_n)^\alpha$ converges for all $\alpha \in \varDelta$ and if $\lim (a(x_n)^\alpha \neq 0$ if and only if $\alpha \in J$.

2.5. Theorem. *Let $J, J' \subset \varDelta$. Let $\{x_n\}, \{x'_n\} (n = 1, 2, \cdots)$ be sequences of elements in $\mathfrak{S}$, of types J and J' respectively. Assume there exist an element $c \in G_Q$ and a sequence of elements $\gamma_n \in \Gamma$ such that $d_G(x_n, x'_n \cdot c \cdot \gamma_n)$ remains bounded as $n \to \infty$. Then*

(1)
$$a(x_n)^\alpha \asymp a(x'_n)^\alpha \ , \qquad (\alpha \in \varDelta, n \geq 1) \ ,$$

in particular $J = J'$, and there exists $n_0 \geq 1$ such that $c \cdot \gamma_n \in P_J$ for $n \geq n_0$. Moreover, the set of reductive parts of the elements $c \cdot \gamma_n (n \geq n_0)$ is finite.

By Theorem 2.3, $d_G(x_n x'_n)$ is bounded as $n \to \infty$. Then so is $d_A(a(x_n), a(x'_n))$ by Corollary 1.7. In view of § 1.3 (3), this implies that $\ln a(x_n)^\alpha \cdot a(x'_n)^{-\alpha}$ is bounded in absolute value for every $\alpha \in \varDelta$, whence (1) and the equality $J = J'$.

We now revert to the proof of Theorem 2.3, let $x = x_n, x' = x'_n, \gamma = \gamma_n$ and write z_n, q_n, w_n instead of z, w, q. By assumption and Lemma 2.2, z_n is bounded; hence so are q_n and, by Corollary 1.7, $a(q_n)$. In view of (1), § 2.3 (11) yields

(2)
$$a(x_n)^{-\Sigma c_\beta \cdot \beta} \prec 1 \ , \qquad (n \geq 1) \ .$$

Assume that $c_\beta \neq 0$ for some $\beta \in \varDelta$. By standard properties of weights [6, § 12.14, Prop. 12.16] we have then $c_\alpha \neq 0$. Since $a(x_n)^\beta \leq t$ for all n's and

$\beta \in \varDelta$, the relation (2) then forces $\lim \alpha(x_n)^\alpha$ to be $\neq 0$, i.e., $\alpha \in J$. Otherwise said, if $\alpha \notin J$, then $w_n(\lambda_\alpha) = \lambda_\alpha$ for n big enough. But then we have $w_n \in P_J$, hence $c \cdot \gamma_n \in P_J$ for those values of n (see § 2.3 (3)), which proves the second assertion. Let now $n \geq n_0$. Then we can write

$$(3) \qquad c \cdot \gamma_n = r_n \cdot u_n , \qquad (r_n \in \mathscr{Z}(S_J), u_n \in U_J) .$$

There exists a matrix realization $\mathscr{G} \subset \boldsymbol{GL}(n, \boldsymbol{C})$ of $\mathscr{G}$ over $\boldsymbol{Q}$ such that $\varGamma$ is represented by elements of $\boldsymbol{GL}(n, \boldsymbol{Z})$. Then $\{c \cdot \gamma_n\}$ consists of rational matrices with bounded denominators. Since the decomposition (3) is over $\boldsymbol{Q}$, there exists then also a rational number f such that $f \cdot r_n \in M(n \cdot \boldsymbol{Z})$ for $n \geq n_0$. In particular the r_n's form a discrete set. In order to show that this set is finite, it therefore remains to show that $\{r_n\}_{n \geq n_0}$ is relatively compact.

Let A_J be the identity component of S_J, in the ordinary topology, and M_J the analogue of the group M in § 1.3 (for $P = P_J$). Then

$$Z(A_J) = Z(S_J) = M_J \times A_J , \qquad A = (M_J \cap A) \times A_J .$$

Write accordingly

$$a(x_n) = a_{1n} \cdot a_{2n} , \quad a(x'_n) = a'_{1n}, a'_{2n} , \quad (a_{1n}, a'_{1n} \in M_J ; a_{2n}, a'_{2n} \in A_J) ,$$
$$r_n = r_{1n} \cdot r_{2n} , \qquad\qquad\qquad (r_{1n} \in M_J, r_{2n} \in A_J) .$$

For $n \geq n_0$, we have $z_n = a(x'_n) \cdot r_n \cdot u_n \cdot a(x_n)^{-1} \in P_J$, and the reductive part of z_n is

$$a(x'_n) \cdot r_n \cdot a(x_n)^{-1} = (a'_{1n} \cdot r_{1n} \cdot a_{1n}^{-1}) \cdot (a'_{2n} \cdot a_{2n}^{-1} \cdot r_{2n}) .$$

z_n is bounded ($n \geq n_0$) hence so are its reductive part and the components

$$a'_{1n} \cdot r_{1n} \cdot a_{1n}^{-1} \in M_J , \qquad a'_{2n} \cdot a_{2n}^{-1} \cdot r_{2n} \in A_J$$

of the latter. Since any $\alpha \in J$ is trivial on A_J, we have

$$a(x_n)^\alpha = a_{1n}^\alpha , \quad a(x'_n)^\alpha = a_{1n}'^\alpha \quad (\alpha \in J, n \geq n_0) ,$$

hence $a_{1n}^\alpha, a_{1n}'^\alpha \asymp 1$ for $\alpha \in J$. But J is a set of coordinates on $M \cap A_J$; therefore a_{1n} and a'_{1n} are bounded, and then so is r_{1n}. The elements a_{1n} and a'_{1n} being bounded, we have

$$a_{2n}^\alpha \asymp a(x_n)^\alpha , \quad a_{2n}'^\alpha \asymp a(x'_n)^\alpha , \quad (\alpha \in \varDelta, n \geq n_0) ,$$

and therefore, by (1),

$$a_{2n}^\alpha \asymp a_{2n}'^\alpha \qquad (\alpha \in \varDelta, n \geq n_0) ,$$

which implies that $a'_{2n} \cdot a_{2n}^{-1}$ runs through a relatively compact subset of A_J. It follows then that r_{2n} is bounded, hence r_n is bounded, which ends the proof of the theorem.

Remark.   In view of § 1.2 (4), Theorem 2.5 and its proof remain valid if $d_G(x_n, x'_n \cdot c \cdot \gamma_n)$ is replaced by $d_X(o \cdot x_n, o \cdot x'_n \cdot c \cdot \gamma_n)$. In the case where $G = SL(n, R), \Gamma = SL(n, Z), c = e$, this theorem, thus formulated, yields [16, Thm. 1, p. 19].

3.   An extension theorem

From now on, and up to § 3.10 inclusive, X is assumed to be a bounded symmetric domain.

3.1.   We recall that a complex analytic space E is *hyperbolic* if its Kobayashi pseudo-distance d_0 is a distance [10]. The open unit disc D and the punctured unit disc $D^* = D - \{0\}$ are hyperbolic [10, Chap. IV, § 4].

If E and E' are hyperbolic, then $E \times E'$ is so, and for $x, y \in E, x', y' \in E'$ we have

$$(1)\quad \max (d_0(x, y), d_0(x', y')) \leq d_0(x, x'), (y, y')) \leq d_0(x, y) + d_0(x', y')$$

[10, Chap. IV, Prop. 2.6 and § 4].

Let E be a complex manifold endowed with a hermitian metric whose holomorphic curvature is bounded from above by a strictly negative constant, and d be the associated distance. Then [10, Chap. IV, Thm. 4.11] there exists a constant $c > 0$ such that

$$(2)\qquad\qquad d(x, y) \leq c \cdot d_0(x, y) , \qquad (x, y \in E) .$$

In particular, E is hyperbolic, and d_0 is a complete metric if d is so. We recall that a holomorphic mapping always decreases d_0.

3.2.   Assume now X to be irreducible. We claim that there exist constants $c, c' > 0$ such that

$$(1)\qquad c \cdot d_X(x, y) \leq d_0(x, y) \leq c' \cdot d_X(x, y) , \qquad (x, y \in X) .$$

Any two invariant Riemannian metrics on X are proportional, hence dx^2 is associated to a hermitian metric with holomorphic curvature bounded from above by a strictly negative constant, and § 3.1 (2) yields the first inequality in (1). X contains a totally geodesic polydisc $F = D^l$ (where l is the rank of X). Given $x, y \in X$ there exists an automorphism of X which brings x, y in F. In order to prove the second inequality in (2), we may therefore assume $x, y \in F$. The restriction of dx^2 to F is invariant under Aut F, hence majorizes a multiple of the Poincaré metric of F. But the inclusion map $F \to X$ is holomorphic, hence decreases the Kobayashi distance. Therefore it suffices to prove the second inequality of (1) in the case of a polydisc. But then it follows from [10, Example 1, p. 47].

3.3. Let $X = X_1 \times \cdots \times X_m$ be the canonical decomposition of X as a product of irreducible bounded symmetric domains. This decomposition is also a Riemannian product with respect to any invariant Riemannian metric. In particular if d_i is the restriction of d_X to X_i, then d_X is majorized by the sum of the d_i's. It follows further from § 3.1 (1) and § 3.2 (1) that there exist constants $c, c' > 0$ such that

$$(1) \qquad c \cdot \max_i \, (d_i(x_i, y_i)) \le d_0(x, y) \le c' \cdot \textstyle\sum_i d_i(x_i, y_i) \, ,$$

for all $x = (x_i), y = (y_i), (x_i, y_i \in X_i, 1 \le i \le s)$.

3.4. Assume now $\mathscr{G}$ to be simple over $\mathbf{Q}$. Let Γ be an arithmetic subgroup of $\mathscr{G}$, $V = X/\Gamma$, V^* be the complex analytic compactification of V constructed in [3], and $\pi \colon X \to V$ be the canonical projection. We shall now use the results and notation of [3], and recall here briefly those which are most pertinent for the sequel. The numbering of the elements of Δ will be the canoical one [3, § 1.2, Prop. 2.9].

V^* is by definition the quotient by Γ of a space X^* on which $G_{\mathbf{Q}}$ operates. X^* is the union of X, which is open in X^*, and of so-called rational boundary components. Those are the transforms under $G_{\mathbf{Q}}$ of the standard boundary components F_b ($1 \le b \le s = rk_{\mathbf{Q}}\mathscr{G}$). For each boundary component F there is a canonical holomorphic projection $\sigma_F \colon X \to F$, which is constructed as the map ν of § 1.8. In particular, § 1.8 (1) holds true for σ_F.

Let $\{x_n\}_{n \ge 1}$ be a sequence of elements in $\mathfrak{S}, d \in G_{\mathbf{Q}}$, and F be a rational boundary component. Then $x_n \cdot d$ tends to a point $u \in F$ in the topology of X^*, if b is the greatest index i such that $\lim a(x_n)^{\alpha i} = 0$, $F_b \cdot d = F$ and $\sigma_F(o \cdot x_n \cdot d) \to u$.

Conversely, let $\{p_n\}$ be a sequence of elements of V which tends to an element $p \in V^* - V$. Let $b \in \{1, \cdots, s\}$ and $c \in G_{\mathbf{Q}}$ be such that $p \in \pi(F_b \cdot c)$, and $u \in F = F_b \cdot c$ be such that $\pi(u) = p$. Then, after having replaced $\{p_n\}$ by an infinite subsequence, we may assume that there exist x_n and d as before, with $\pi(x_n \cdot d) = p_n$ for all n's and $F_b \cdot d = F$.

This follows from the description of the topology of X^* or V^* in terms of "truncated Siegel sets" [3, § 4.12, Lemma 4.13], if we take into account the fact that $\sigma_{F_b}(o) = o_b$.

3.5. *Proof of Theorem* B. We first reduce the proof to the case where $\mathscr{G}$ is $\mathbf{Q}$-simple. We may assume $\mathscr{G}$ to be adjoint type [3, Lemma 11.5]. It is then the direct product over $\mathbf{Q}$ of $\mathbf{Q}$-simple groups $\mathscr{G}_i$ ($1 \le i \le m$). The symmetric space X_i of maximal compact subgroups of G_i is a bounded symmetric domain, and X is (complex analytically) the product of the X_i's. Let $\Gamma_i = \Gamma \cap G_i$, and Γ' be the product of the Γ_i. It is a normal subgroup of Γ, which is arithmetic [5, § 7], hence of finite index in Γ. The canonical projection $V' = X/\Gamma' \to V$ extends to a holomorphic map $V'^* \to V^*$ with finite fibers, which sends boundary components onto boundary components. From this it is elementary that it suffices to prove Theorem B for V' and V'^*. But V' is the product of the quotients $V_i = X_i/\Gamma_i$, and then V'^* is the product of the complexifications V_i^* of the V_i's, whence the reduction to the $\boldsymbol{Q}$-simple case.

Let now p_n, q_n and p, q be as in the statement of Theorem B. Let F and F' be rational boundary components of X^* whose projections contain p and q respectively, and $u \in F, u' \in F'$ be such that $\pi(u) = p, \pi(u') = q$. We have to show the existence of $\gamma \in \Gamma$ such that $u' = u \cdot \gamma$. Let $b, b' \in \{1, \cdots, s\}$ and $c, c' \in G_{\boldsymbol{Q}}$ be such that $F = F_b \cdot c, F' = F_{b'} \cdot c'$. In view of § 3.4 we may assume (after having taken subsequences and renumbered), the existence of elements $x_n, x'_n \in \mathfrak{S}$ and $d, d' \in G_{\boldsymbol{Q}}$ such that b (resp. b') is the greatest index i for which $a(x_n)^{\alpha_i}$ (resp. $a(x'_n)^{\alpha_i}$) tends to zero, $F_b \cdot d = F, F_{b'} \cdot d' = F'$ and $\nu_F(o \cdot x_n \cdot d) \to u$ (resp. $\nu_{F'}(o \cdot x'_n \cdot d') \to u'$).

Let d'_γ be the distance function on X/Γ defined by

$$d'_\gamma(\pi(x), \pi(y)) = \operatorname*{Inf}_{\gamma \in \Gamma} d_X(x, y \cdot \gamma) , \qquad (x, y \in X) .$$

By § 3.3 and the assumption of Theorem B, $d'_\gamma(p_n, q_n) \to 0$; therefore we can find $\gamma_n \in \Gamma$ such that

$$d'_X(o \cdot x_n \cdot d, o \cdot x'_n \cdot d' \cdot \gamma_n) \to 0 .$$

By Theorem 2.5, we have then

$$a(x_n)^\alpha \asymp a(x'_n)^\alpha \qquad (\alpha \in \Delta) ,$$

which implies in particular that $b = b'$. Let $J = \Delta - \{b\}$. The group P_J is the normalizer of F_b. After having gone over to subsequences, we may assume $a(x_n)^\alpha$ and $a(x'_n)^\alpha$ to converge for all $\alpha \in \Delta$; x_n and x'_n are then of type J' for some $J' \supset J$; by Theorem 2.5, there exists then n_0 such that

$$(1) \qquad\qquad d' \cdot \gamma_n \cdot d^{-1} \in P_J \qquad (n \geq n_0) .$$

We have therefore

$$(2) \qquad F' \cdot \gamma_n = F_b \cdot d' \cdot \gamma_n \cdot d^{-1} \cdot d = F_b \cdot d = F , \qquad (n \geq n_0) ,$$

whence also

$$(3) \qquad\qquad \gamma_{n_0}^{-1} \cdot \gamma_n \in d^{-1} \cdot P_J \cdot d = \mathcal{N}(F) \qquad (n \geq n_0) .$$

By construction

$$(4) \qquad\qquad \lim \sigma_F(o \cdot x_n \cdot d) = u , \qquad \lim \sigma_{F'}(o \cdot x'_n \cdot d') = u' .$$

Since $F' \cdot \gamma_{n_0} = F$, the second relation implies

$$(5) \qquad\qquad \lim \sigma_F(o \cdot x'_n \cdot d' \cdot \gamma_{n_0}) = u' \cdot \gamma_{n_0} .$$

As pointed out at the end of § 3.4, σ_F is distance decreasing, F being endowed with a suitable invariant distance function d_F. We have then

$$(6) \qquad \lim d_F(\sigma_F(o \cdot x_n \cdot d), \sigma_F(o \cdot x'_n \cdot d' \cdot \gamma_n)) = 0 \ .$$

Together with (4), this yields

$$(7) \qquad \lim \sigma_F(o \cdot x'_n \cdot d' \cdot \gamma_n) = u \ .$$

We have $o \cdot x'_n \cdot d \cdot \gamma_n \in F$ and

$$o \cdot x'_n \cdot d \cdot \gamma_n = o \cdot x'_n \cdot d' \cdot \gamma_{n_0} \cdot \gamma_{n_0}^{-1} \cdot \gamma_n \ .$$

The map ν commutes with the action of $\mathcal{N}(F)$ on X and F. Since $\gamma_{n_0}^{-1} \cdot \gamma_n \in \mathcal{N}(F)$ we have

$$(8) \qquad \sigma_F(o \cdot x_n \cdot d' \cdot \gamma_n) = \sigma_F(o \cdot x'_n \cdot d' \cdot \gamma_{n_0}) \cdot \gamma_{n_0}^{-1} \cdot \gamma_n \ .$$

Let $\mathcal{Z}(F) = \{g \in \mathcal{N}(F) \,|\, f \cdot g = f, (f \in F)\}$ be the centralizer of F, and $G(F) = \mathcal{N}(F)/\mathcal{Z}(F)$. Since F is a rational boundary component, the image of $\mathcal{N}(F) \cap \Gamma$ in $G(F)$ is a discrete subgroup (in fact of arithmetic type, see [3, Thm. 3.7]). In particular, it acts in a properly discontinuous manner on F. In view of (4) and (8), there exists then $\gamma \in \mathcal{N}(F) \cap \Gamma$ such that

$$(9) \quad \sigma_F(o \cdot x'_n \cdot d' \cdot \gamma_{n_0}) \cdot \gamma = \sigma_F(o \cdot x'_n \cdot d' \cdot \gamma_{n_0}) \gamma_{n_0}^{-1} \cdot \gamma_n = \sigma_F(o \cdot x'_n \cdot d' \cdot \gamma_n)$$

for infinitely many n's. We have then, using (5) and (7):

$$u' \cdot \gamma_{n_0} \cdot \gamma = \lim \sigma_F(o \cdot x'_n \cdot d' \cdot \gamma_{n_0}) \cdot \gamma = \lim \sigma_F(o \cdot x'_n \cdot d' \cdot \gamma_n) = u \ ,$$

hence $u \in u' \cdot \Gamma$ and $p = \pi(u) = \pi(u') = q$.

3.6. Let Z be a complex space. A holomorphic map $f: Z \to V$ is said to be *locally liftable* if for every $z \in Z$ there exist a neighborhood U_z of z in Z and a holomorphic map f_z of U_z into X such that the restriction of f to U_z is equal to $\pi \circ f_z$. If Γ is torsion-free, then it operates freely on X, the projection π is a covering map, and every holomorphic map of Z into V is locally liftable. Since a product of open discs is a hyperbolic space, the following theorem contains Theorem A of the introduction as a special case.

3.7. Theorem. *Let Z be a normal hyperbolic space, a a positive integer and $f: D^{*a} \times Z \to V$ a locally liftable holomorphic map. Then f extends to a holomorphic map of $D^a \times Z$ into V^*.*

The space $E = D^{*a} \times Z$ is a normal hyperbolic analytic space, since both factors are so. Let d_E be the hyperbolic metric on E. Then [10, Prop. 6.1, p. 104]

$$(1) \qquad d'_0(f(u), f(v)) \leq d_E(u, v) \ , \qquad (u, v \in E) \ .$$

Since $D^a \times Z$ is normal, it suffices, by Riemann's extension theorem [1, Thm. 44.42, p. 420], to show that f extends to a continuous map of $D^a \times Z$ into V^*.

First let Z be reduced to a point. In view of (1) and § 3.5, our assertion is essentially contained in [9, Thm. 2]. For the sake of completenss, we sketch a proof. First let $a = 1$. Let C_n be a sequence of concentric circles in D^*, with center at the origin, whose radii r_n tend to zero. Then the hyperbolic length $L(C_n) = 2\pi(\ln(1/r_n))^{-1}$ (see [10, p. 81]) tends to zero. By (1), the length of $f(C_n)$ also tends to zero, and § 3.5 then shows that, for some infinite subsequence, the images $f(C_n)$ converge to a point of V^*. Hence M. H. Kwack's theorem [12, Thm. 3], [10, Thm. 3.1] obtains and yields our assertion.

Let $a \geq 2$, and $u = (u_i) \in D^a - D^{*a}$. Let r be a strictly positive number $< \max(|u_i|, 1 - |u_i|)$ if $u_i \neq 0$ $(1 \leq i \leq a)$. Let $\lambda = (\lambda_i)$ be a sequence of complex numbers of modulus one. Then $f_{u,\lambda} \colon z \mapsto (u_i + \lambda_i \cdot z)$ is a holomorphic embedding of $D_r^* = \{z \in C | 0 < |z| < r\}$ into D^{*a}; by the above, $f \circ f_{u,\lambda}$ extends to a holomorphic mapping of $D_r = \{z \in C | 0 \leq |z| < r\}$ into V^*. Let $v_{u,\lambda}$ be its value at the origin. In fact, $v_{u,\lambda}$ is independent of λ. This follows from Theorem 1 in [9], or also from § 3.1, § 3.5 and the fact that if $\lambda' = (\lambda_i')$ is another sequence of complex numbers of modulus one, and if $z_n \in D_r^*$ tends to 0, then the hyperbolic distance of $\lambda_i z_n$ and $\lambda_i' \cdot z_n$ also tends to zero. We then define an extension f' of f to D^a by putting $f'(u) = f_{u,\lambda}(0)$ for $u \in D^a - D^{*a}$. Let now z_n be a sequence of elements in D^{*a} which converges to u. By Theorem 1 of [9] applied to the sequence of maps f_{u,λ_n} where $\lambda_n = (z_{ni}/|z_{ni}|)$, we have

$$\lim f(z_n) = \lim f_{u,\lambda_n}(z_n) = \lim f_{u,\lambda_n}(0) = f'(u) ,$$

whence the continuity of f'.

Let now Z be not reduced to a point. For each $z \in Z$, there is a continuous extension $f_z \colon D^a \times \{z\} \to V^*$ of the restriction of f to $D^{*a} \times \{z\}$. Let then $f' \colon D^a \times Z \to V^*$ be the map whose restriction to $D^a \times \{z\}$ is equal to f_z for every $z \in Z$. It extends f, and there remains to show that it is continuous. To prove the continuity of f' it suffices to show that if $(y, z) \in D^a \times Z$ and (y_n, z_n) is a sequence of elements in E which converges to (y, z), then $f((y_n, z_n)) \to f_z(y, z)$. By § 3.1,

$$d_E((y_n, z_n), (y_n, z)) \leq d_Z(z_n, z) ,$$

hence the left hand side tends to zero. By (1) and § 3.5, it follows (after having taken subsequences convergent in V^*) that

$$\lim_{n \to \infty} f((y_n, z_n)) = \lim_{n \to \infty} f((y_n, z)) .$$

Since the right hand side is $f_z((y, z))$ by definition, this proves our contention.

3.8. Remark. If Γ has torsion, d_0' is not necessarily the hyperbolic distance on V, but majorizes it [10, p. 103]. Thus, strictly speaking, V is not necessarily hyperbolically imbedded in V^* in the sense of [9], [11], and the theorems of [9] do not apply directly to our situation. However, this is a harmless point: since we have the "hyperbolic embedding" condition for the modified pseudo-distance d_0', and the distance decreasing property § 3.7 (1), the arguments of [9] apply without change to our case (see [11] for similar remarks).

3.9. Let us denote by V^{**} the set V^* endowed with the topology defined by Piateckii-Sapiro [12], using "cylindrical sets" in realizations of X as "Siegel domains of the third kind". The identity map $\iota: V^* \to V^{**}$ is continuous [2]. Since V^* is compact and Hausdorff, ι is a homeomorphism if and only if V^{**} is Hausdorff. In [12], it is asserted that this follows from the results announced in [4], but the assertion is not really convincing to the author of the present paper.

Theorem B is obviously equivalent to the following assertion:

$(*)$ *Given $p \in V^* - V$ and a neighborhood U_1 of p in V^*, there exists a neighborhood U_2 of p in U_1 such that*

$$d_0'(U_2 \cap V, V - (U_1 \cap V)) > 0 .$$

This statement is proved in [11] for V^{**}. From this, the authors deduce Theorem 3.7 for $E = D^* \times D^b$ by using Theorem 6.1 of [10]. However, in the latter, all spaces under consideration are of course Hausdorff. Similarly [9] proves Theorem A for V^{**}, using $(*)$ above, and Theorem 1 of [9], where again the spaces are Hausdorff. It is also pointed out there that the proof of Theorem 2 is also valid for V^*, provided $(*)$ holds true in V^*.

All these distinctions will be happily superfluous and the situation more satisfactory once it is shown that V^{**} is Hausdorff. The author believes it follows from his joint work with J–P. Serre on corners, but prefers not to commit himself firmly on this point until everything is fully written up. The proof of $(*)$ in [11], unlike the one of Theorem B here, does not involve any reduction theory; this is maybe an indication that indeed some rather strong results in reduction theory are hidden behind the equivalence of the two topologies.

In the following theorem, which was pointed out by P. Deligne, V is endowed with its canonical structure of quasi-projective variety [3, § 10].

3.10. Theorem. *Let S be a complex algebraic variety, f a holomorphic map of S, viewed as an analytic space, into V. Then f is a morphism of algebraic varieties.*

This assertion is local in the Zariski topology of S, so we may assume S to be affine, and in particular to be quasi-projective. Furthermore, it suffices to show that the restriction of f to an open Zariski-dense subset is algebraic. So we may assume S to be smooth. By Hironaka's desingularization theorem [8], S may be embedded as a Zariski-open subset in a smooth projective variety $\bar{S}$

in such a way that $\bar{S} - S$ consists of smooth divisors with normal crossings; more precisely, around any point of $\bar{S} - S$ there are coordinates with respect to which S is of the form $D^{*a} \times D^b \, (a + b = n)$. By Theorem A, f extends to a holomorphic map f' of $\bar{S}$ into V^*. Since $\bar{S}$ and V^* are projective varieties, the map f' is a morphism of algebraic varieties by Chow's theorem [13, Prop. 15], whence the theorem.

3.11. Let us now drop the assumption that X carries an invariant complex structure. Theorem B is then also true for the embedding of $V = X/\Gamma$ into any Satake compactification [4, § 1.4]. The proof is essentially the same and relies on suitable analogues of the facts recalled in § 3.4. However, since there is no adequate reference for them in the literature, we shall not try to make this more precise.

References

[1] S. Abhyankar, *Local analytic geometry*, Academic Press, New York, 1964.
[2] W. L. Baily, Jr., *Fourier-Jacobi series, Algebraic groups and discontinuous subgroups*, Proc. Sympos. Pure Math. Vol. 9, Amer. Math. Soc. 1966, 296–300.
[3] W. L. Baily, Jr. & A. Borel, *Compactification of arithmetic quotients of bounded symmetric domains*, Ann. of Math. (2) **84** (1966) 442–528.
[4] A. Borel, *Ensembles fondamentaux pour les groupes arithmétiques*, Colloque sur la théorie des groupes algébriques, Bruxelles, 1962, 23–40.
[5] ——, *Introduction aux groupes arithmétiques*, Actualités Sci. Indust. No. 1341, Hermann, Paris, 1969.
[6] A. Borel & J. Tits, *Groupes réductifs*, Inst. Hautes Études Sci. Publ. No. 27 (1965) 55–150.
[7] P. Griffiths, *Periods of integrals on algebraic manifolds: summary of main results and discussion of open problems*, Bull. Amer. Math. Soc. **76** (1970) 228–296.
[8] H. Hironaka, *Resolution of singularities of an algebraic variety over a field of characteristic zero. I, II*, Ann. of Math. (2) **79** (1964) 109–326.
[9] P. Kiernan, *Extension of holomorphic maps* [Trans. A.M.S. **172** (1972), 347–355].
[10] S. Kobayashi, *Hyperbolic manifolds and holomorphic mappings*, Dekker, New York, 1970.
[11] S. Kobayashi & T. Ochiai, *Satake compactification and the great Picard theorem*, J. Math. Soc. Japan **23** (1971) 340–350.
[12] M. H. Kwack, *Generalization of the big Picard theorem*, Ann. of Math. (2) **90** (1969) 9–22.
[13] I. I. Piateckii-Sapiro, *Arithmetic groups in complex domains*, Uspehi Mat. Nauk **19** (1964) 93–120; Russian Math. Surveys **19** (1964) 83–109.
[14] I. Satake, *On compactifications of the quotient spaces for arithmetically defined discontinuous groups*, Ann. of Math. (2) **72** (1960) 555–580.
[15] J. P. Serre. *Géométrie algébrique et géométrie analytique*, Ann. Inst. Fourier (Grenoble) **6** (1955) 1–42.
[16] C. L. Siegel, *Zur Reduktionstheorie quadratischer Formen*, Publ. Math. Soc. Japan **5** (1959), Ges. Werke, Bd. 3, 274–327.

INSTITUTE FOR ADVANCED STUDY

97.

(avec J. Tits)

Homomorphismes ‹abstraits› de groupes algébriques simples

Ann. Math., (2) **97** (1973) 499–571

TABLE DES MATIÈRES

Introduction

I. Préliminaires

II. Compléments sur les groupes réductifs

III. Homomorphismes

Introduction

Le point de départ de ce travail est l'étude des automorphismes du groupe $G(k)$ des points rationnels d'un k-groupe connexe presque simple, lorsque k est un corps infini, et le k-rang de G est > 0. En fait, on élargira le problème en considérant plus généralement un homomorphisme $\alpha: H \to G'(k')$, où k' est un corps commutatif, G' un k'-groupe connexe semi-simple et H un sous-groupe de $G(k)$ contenant le sous-groupe G^+ de $G(k)$ engendré par les points rationnels des radicaux unipotents des k-sous-groupes paraboliques de G. Si $\varphi: k \to k'$ est un homomorphisme, désignons par $^\varphi G$ le k'-groupe obtenu à partir de G par changement de base (1.7), et par φ^0 l'homomorphisme canonique de $G(k)$ dans $^\varphi G(k')$. [Si G est donné comme groupe matriciel, $^\varphi G$ s'obtient en appliquant φ aux équations sur k définissant G, et, si $g = (g_{ij}) \in G(k)$, alors $\varphi^0(g) = (\varphi(g_{ij}))$.] On démontrera notamment, sous ces hypothèses, le théorème suivant:

* Ce travail a été en partie élaboré en 1971–1972, alors que le deuxième auteur bénéficiait de l'aide financière de la N.S.F. (contrat GP 7952X3).

171

(A) *Supposons G et G' absolument presque simples, G simplement connexe ou G' adjoint, $G' \neq \{e\}$ et $\alpha(G^+)$ dense dans G'. Alors il existe un homomorphisme $\varphi\colon k \to k'$, une k'-isogénie $\beta\colon {}^{\varphi}G \to G'$ de différentielle $d\beta$ non nulle et un homomorphisme γ de H dans le centre de $G'(k')$, tous trois uniques, tels que $\alpha(g) = \gamma(g) \cdot \beta(\varphi^0(g))$, $(g \in H)$.*

(En fait, γ est probablement toujours trivial sous ces hypothèses. On renvoie au §8 pour des énoncés plus complets et plus généraux. (A) est intermédiaire entre 8.1 et 8.16. Dans ce dernier, G' n'est pas nécessairement presque simple.)

On tire facilement de (A) la détermination des automorphismes de H, sans supposer G simplement connexe ou adjoint (8.14). Le résultat est analogue, avec cependant une légère complication pour les groupes de type $\mathbf{D}_{2n}$ non simplement connexes ou adjoints; de plus, si l'on n'est pas dans ce dernier cas et si β n'est pas un automorphisme, alors k est parfait de caractéristique deux (resp. trois), G est déployé de type $\mathbf{C}_2$, $\mathbf{F}_4$ (resp. $\mathbf{G}_2$) et β est une isogénie exceptionnelle (3.3).

Lorsque k et k' sont localement compacts non discrets, différents de $\mathbf{C}$, on verra au §9 que α est toujours continu pour les topologies de groupe localement compact déduites de celles de k et k', sous des hypothèses du reste plus générales quant à G, G' et H.

Si φ est un homomorphisme de k dans une clôture algébrique $\bar{k}'$ de k, et $\rho\colon G \to \mathbf{PGL}_n$ une représentation rationnelle projective de G, alors ${}^{\varphi}\rho\colon {}^{\varphi}G \to {}^{\varphi}\mathbf{PGL}_n \cong \mathbf{PGL}_n$ est une représentation rationnelle projective de même degré de ${}^{\varphi}G$, et ${}^{\varphi}\rho \circ \varphi^0$ un homomorphisme de $G(k)$ dans $\mathbf{PGL}_n(\bar{k}')$. On prouvera au §10 l'assertion suivante:

(B) *Soient k, k', G, H comme dans (A), et $\rho\colon H \to \mathbf{PGL}_n(k')$ un homomorphisme tel que $\rho(G^+)$ soit absolument irréductible. Alors il existe des représentations rationnelles projectives irréductibles π_i de G, et des homomorphismes distincts $\varphi_i\colon k \to \bar{k}'$ $(1 \leq i \leq m)$ tels que ρ soit la restriction à H du produit tensoriel des ${}^{\varphi_i}\pi_i \circ \varphi_i^0$.*

Si k est de caractéristique p non nulle, on peut prendre les π_i dans un ensemble fini $\mathfrak{M}'(G)$ de classes de représentations irréductibles de G, donné une fois pour toutes (10.2, 10.3). Si k est algébriquement clos, cela démontre en particulier une conjecture de R. Steinberg [28].

La démonstration de ces résultats occupe le troisième partie de ce travail, les deux premières étant consacrées à des rappels, préliminaires et compléments sur les groupes réductifs, motivés par les besoins de la troisième partie. Certains sont assez techniques, aussi, après avoir pris connaissance des notations et conventions générales, le lecteur intéressé à cette dernière aura-t-il avantage à sauter directement au § 6, quitte à revenir aux paragraphes antérieurs au fur et à mesure des besoins. En particulier, s'il fait abstraction des corps non parfaits de caractéristique deux, il n'aura pas à recourir aux §§ 4 et 5.

Nous passons maintenant à quelque indications sur le contenu des différents paragraphes.

Le § 1 rappelle un certain nombre de notations et de résultats sur les groupes réductifs et systèmes de racines, fixe quelques conventions et prouve deux résultats élémentaires sur les corps de définition de morphismes de variétés algébriques. Le § 2 établit quelques propriétés de morphismes de corps. Le § 3 est consacré principalement à un rappel et à des compléments sur les isogénies dites exceptionnelles, découvertes par C. Chevalley [14]. À partir de 3.18, on y suppose le corps de base k localement compact non discret et on établit quelques propriétés relatives à la topologie déduite de celle k de certains k-morphismes de groupes. Le § 4 discute les séries centrales descendantes du radical unipotent d'un sous-groupe parabolique. Il répond à des questions assez naturelles, qui sont traitées avec plus de généralité qu'il n'est strictement nécessaire ici. En particulier, on aura besoin de 4.7, 4.8 au § 5 seulement et 4.9, 4.10 et 4.11 ne seront pas utilisés plus loin.

Supposons k non parfait de caractéristique deux. Alors k^2 est d'indice infini dans k. Par suite, le sous-groupe trigonal supérieur strict $U(k)$ de $\mathbf{SL}_2(k)$ est réunion d'une infinité de sous-groupes, chacun réunion de e et d'une orbite du groupe $T(k)$ des matrices diagonales de $\mathbf{SL}_2(k)$, opérant par automorphismes intérieurs. Cela est à la source d'une difficulté assez sérieuse dans la démonstration du théorème 8.1, et le but du § 5 est d'établir un lemme (5.4) qui permettra de la circonvenir. En fait, on ne fait pas d'hypothèse sur k dans ce paragraphe, mais les démonstrations sont aisées si l'on n'est pas dans le cas précité et, du reste, 5.4 ne sera utilisé plus loin (dans 8.5) que sous ces hypothèses. Le § 6 est consacré à diverses caractérisations et à des propriétés générales du groupe G^+ introduit plus haut, qui joue un rôle essentiel dans ce travail.

Les homomorphismes "abstraits" sont l'objet de la troisième partie. Les démonstrations principales sont celles de 7.1, 7.2 et 8.1. Pour en donner les lignes directrices, nous esquissons ici la démonstration de (A) sous les hypothèses simplificatrices suivantes: $H = G^+$, G' est adjoint, et

(*) car $k \neq 2$ *et* G *possède des* k-*sous-groupes paraboliques propres opposés et conjugués* P, P^- *dont les radicaux unipotents sont commutatifs.*

La dernière hypothèse signifie que le système de racines relatif de G n'est pas de l'un des types $\mathbf{BC}$, $\mathbf{G_2}$, $\mathbf{F_4}$ ou $\mathbf{E_8}$; dans le cas général, nous avons seulement à notre disposition des sous-groupes paraboliques P, P^- dont les radicaux unipotents sont de classe deux (cf. 8.2), ce qui est l'une des causes des complications techniques rencontrées au §8.

Pour $M \subset G$, nous notons $\bar{\alpha}(M)$ l'adhérence de $\alpha(M \cap G^+)$. Posons $R_u(P) = U$, $R_u(P^-) = U^-$, $P \cap P^- = Z$, $\bar{\alpha}(U) = U'$, $\bar{\alpha}(U^-) = U'^-$ et $\bar{\alpha}(Z) = Z'$. On montre (cf. 4.8) que, vu (*), le centre connexe S de Z est un tore déployé sur k de dimension un possédant un seul poids a dans U, ce qui veut dire que U a une structure d'espace vectoriel sur lequel $s \in S$ opère par multiplication par $a(s)$; de plus, on vérifie que

$$(1) \qquad\qquad a\big(S(k) \cap G^+\big) \supset k^{*2} \,.$$

(Dans la démonstration que nous donnons au §8, il est procédé un peu différemment: on commence par choisir de façon appropriée le tore déployé unidimensionnel S et on s'en sert pour définir P, P^-, U, U^- et Z, ce qui évite le recours au §4.)

Pour tout sous-groupe X de $S(k)$ tel que $\alpha(X) \neq \{1\}$, le groupe $X \cdot U(k)$ est résoluble et a $U(k)$ pour groupe dérivé. Utilisant le fait que $U(k) \subset G^+$ et que l'intersection $S(k) \cap G^+$ est infinie, on en déduit aussitôt que U' est le groupe dérivé de tout sous-groupe d'indice fini du groupe résoluble $\bar{\alpha}(SU)$, donc que U' est unipotent connexe. Cela montre déjà que k et k' ont même caractéristique (pour un résultat plus général, voir 7.1).

Soient T' un tore maximal de Z' et V', V'^- des sous-groupes unipotents connexes maximaux de Z' normalisés par T' et tels que $V' \cdot T' \cdot V'^-$ soit dense dans Z'. Vu l'hypothèse de densité de $\alpha(G^+)$ dans G', on voit que $U' \cdot V' \cdot T' \cdot V'^- \cdot U'^-$ est dense dans G'. Comme $U'V'$ et $V'^-U'^-$ sont des groupes unipotents connexes, on en déduit aussitôt que ce sont des sous-groupes unipotents connexes maximaux, que $P' = Z'U' = \bar{\alpha}(P)$ et $P'^- = Z'U'^- = \bar{\alpha}(P^-)$ sont des k'-sous-groupes paraboliques opposés de G', que $Z' = P' \cap P'^- = \bar{\alpha}(Z)$ est un groupe réductif et enfin, compte tenu de (*) et du fait que G' est adjoint, que le centre S' de Z' est un tore déployé de dimension un qui a un seul poids a' dans U'. (Ici, nous utilisons à nouveau 4.8; dans la démonstration de 8.1, U' est extension d'un espace vectoriel par un autre et, dans chacun de ces espaces, S' peut posséder plusieurs poids qui, toutefois, ne diffèrent que par des facteurs puissances de l'exposant caractéristique de k.)

Les espaces $U(k)$ et $U'(k')$ sont des espaces vectoriels sur k et k'. Cela étant, l'homomorphisme de corps $\varphi\colon k \to k'$ s'obtient ainsi: pour $t \in k$, $\varphi(t)$ est l'élément de k' tel que

$$(2) \qquad \alpha(t \cdot u) = \varphi(t) \cdot \alpha(u) \qquad \qquad \big(u \in U(k)\big) \, .$$

L'unicité de $\varphi(t)$ est évidente. Son existence est claire lorsque $t \in a(S \cap G^+)$: si $s \in S \cap G^+$ est tel que $t = a(s)$, il suffit de poser $\varphi(t) = a'(\alpha(\beta))$. Mais vu (1) et l'hypothèse faite sur k, l'ensemble $a(S \cap G^+)$ engendre k additivement et l'existence de $\varphi(t)$ pour tout t s'en déduit aussitôt. (Si k est un corps non parfait de caractéristique 2, ce raisonnement fournit encore un homomorphisme de corps $k^2 \to k'$ dont on peut déduire $\varphi \colon k \to k'^{1/2}$ en extrayant des racines carrées, mais il faut alors montrer que (2) est aussi satisfaite pour $t \notin k^2$, ce pourquoi nous avons besoin du §5. D'autre part, si on abandonne les hypothèses (*), une autre complication surgit du fait que S' possède plusieurs poids dans U'; cela nous conduit à associer d'abord à chacun de ces poids un homomorphisme de k dans une extension algébriquement close de k' et à faire voir ensuite que tous ces homomorphismes sont produits de l'un d'entre eux, appelé φ, par des puissances de l'endomorphisme de Frobenius.)

Identifions désormais k à un sous-corps de k' par φ. Il résulte de (2) que $\alpha|_{U(k)}$ est alors la restriction à $U(k)$ d'une application linéaire $\beta_U \colon U \to U'$ définie sur k'. On montre de même facilement (par exemple en utilisant le fait que les poids de S et S' dans U^- et U'^- sont $-a$ et $-a'$) que $\alpha|_{U^-(k)}$ est la restriction à $U^-(k)$ d'une application linéaire $\beta_{U^-} \colon U^- \to U'^-$. L'opération par automorphismes intérieurs de Z et Z' sur U et U' définit des homomorphismes $\gamma \colon Z \to GL(U)$ et $\gamma' \colon Z' \to GL(U')$, et γ' est une immersion parce que G' est adjoint. Soient U_0 le noyau de β_U, $GL(U, U_0)$ le stabilisateur de U_0 dans $GL(U)$ et $\psi \colon GL(U, U_0) \to GL(U')$ le k'-morphisme induit par β_U. Il est facile de voir que, pour $z \in Z \cap G^+$, on a $\gamma(z) \subset GL(U, U_0)$ et $\gamma'(\alpha(z)) = \psi(\gamma(z))$. Comme $Z \cap G^+$ est dense dans Z (6.11), il s'ensuit que $\gamma(Z) \subset GL(U, U_0)$ et que $\psi(\gamma(Z)) \subset \gamma'(Z')$. Cela étant, $\beta_Z = \gamma'^{-1} \circ \psi \circ \gamma$ est un k'-morphisme de Z dans Z' qui coïncide avec α sur $Z \cap G^+$. Enfin, le produit de β_U, β_Z et β_{U^-} est un k'-morphisme de l'ouvert $\Omega = UZU^-$ dans G' qui coïncide avec α sur $\Omega \cap G^+$ et, par translations et recollement, on en déduit aisément le k'-morphisme $\beta \colon G \to G'$ de l'énoncé.

Dans le §9, on suppose k et k' localement compacts non discrets, différents de C. Comme tout homomorphisme de k dans k' est un isomorphisme (topologique) sur un sous-corps fermé (2.3), les résultats du §8 impliquent que α est continu pour les topologies de groupes localement compacts de $G(k)$ et $G'(k')$ déduites de celles de k et k' (9.8). Cela présuppose G isotrope. En fait, une généralisation convenable (9.1) d'un raisonnement de B. L. van der Waerden [36] nous permettra de nous affranchir de cette condition et de démontrer des théorèmes de continuité ou d'isomorphisme topologique sous des hypothèses plus générales, (9.11, 9.12, 9.13). Il résulte en particulier de 9.13 que si k et k' sont $\neq$ C, G semi-simple simplement connexe et $\alpha\colon G(k) \to G'(k')$ surjectif (resp. bijectif), alors α est continu (resp. est un isomorphisme de groupes topologiques).

Enfin, le §10 est consacré aux représentations projectives et obtient (B) comme conséquence de 8.16 et de propriétés connues des représentations rationnelles.

Un exemple (8.19) montre que l'hypothèse G' réductif est nécessaire, mais il suggère en même temps une reformulation de notre théorème principal admettant une généralisation assez naturelle qui pourrait peut être valoir sans hypothèse particulière sur G'. D'autre part, il ne semble pas exclu que les résultats du §8 restent en gros valables, pour $H = G(k)$, même si G est anisotrope sur k, comme semblent l'indiquer les résultats connus sur les groupes classiques.

Terminons par quelques remarques bibliographiques. La détermination des automorphismes de $G(k)$ a été faite tout d'abord pour le groupe projectif $\mathbf{PSL}_n(k)$ par O. Schreier et B. L. van der Waerden [25], et ensuite pour de nombreux groupes classiques, notamment par J. Dieudonné, L. K. Hua, C. E. Rickart, O. T. O'Meara (*voir* l'exposé d'ensemble de J. Dieudonné [18] et, pour des résultats plus récents, [21], [35]). Pour les groupes isotropes, et k infini, ces résultats sont contenus dans ceux du §8. D'un autre côté E. Cartan [12] et B. L. van der Waerden [36] montraient que tout homomorphisme d'un groupe de Lie compact connexe semi-simple dans un groupe de Lie compact est continu. Plus tard, H. Freudenthal [19] faisait voir que tout isomorphisme "abstrait" d'un groupe de Lie connexe, d'algèbre de Lie absolument simple, sur un groupe de Lie, est continu. Ces résultats sont retrouvés et étendus à d'autres corps localement compacts au §9.

Notons encore que les travaux précités sur les groupes classiques déterminent Aut $G(k)$ aussi lorsque k est fini. Sous cette hypothèse, ce groupe a été décrit pour tout G absolument simple par R. Steinberg [27].

Les résultats de ce travail ont été en partie annoncés dans [5] et [35].

I. PRÉLIMINAIRES

1. Notations. Généralités

Dans ce travail, nous utilisons les notations de [4], avec quelques modifications ou additions indiquées ci-dessous.

1.1. Soit H un groupe et soient $A_1, \cdots, A_q$ des parties de H. On note $\langle A_1, \cdots, A_q \rangle$ ou $\langle A_i | 1 \leq i \leq q \rangle$ le sous-groupe de H engendré par les A_i.

Le centre de H est désigné par $\mathcal{C}(H)$. Comme dans [4], le sous-groupe engendré par les commutateurs $(x, y) = x \cdot y \cdot x^{-1} \cdot y^{-1}$ avec $x \in A_1$, $y \in A_2$ est noté (A_1, A_2). On pose

$$\mathcal{C}^1 H = H \quad \text{et} \quad \mathcal{C}^i H = (H, \mathcal{C}^{i-1}H) \quad \text{pour} \ i \in \mathbf{N}, \ i \geq 2 \, ,$$

décalant ainsi d'une unité la numérotation adoptée dans [4] pour la série centrale descendante de H. Si H est nilpotent, sa *classe de nilpotence* est le plus petit entier j tel que $\mathcal{C}^{j+1}H = \{e\}$. On note $(\mathcal{C}_i H)$ la série centrale ascendante de H, définie par

$$\mathcal{C}_0 = \{e\} \quad \text{et} \quad \mathcal{C}_i H = \{h \in H \,|\, (h, x) \in \mathcal{C}_{i-1}H \ \text{pour tout} \ x \in H\} \ \text{si} \ i \geq 1 \, .$$

Si X, X', X'' sont des groupes tels que $X \supset X' \supset X''$ et si x est un élément de X normalisant X' et X'', nous notons $\mathrm{Int}_{X'} x$ (ou simplement $\mathrm{Int}\, x$) et $\mathrm{Int}_{X'/X''} x$ les permutations de X' et X'/X'' induites par l'automorphisme intérieur $y \mapsto xyx^{-1}$ de X.

1.2. Soit k un corps commutatif et soit p son exposant caractéristique. On note Fr^i l'endomorphisme $x \mapsto x^{p^i}$ de k ($i = 0, 1, \cdots$). Rappelons que k_s et $\bar{k}$ désignent une clôture séparable et une clôture algébrique de k.

1.3. Si V est une k-variété et k_1 une extension de k, on note $V(k_1)$ l'ensemble des points de V rationnels sur k_1 et $k_1[V]$ l'algèbre des k_1-fonctions régulières sur V.

En même temps qu'un corps k, k', $\cdots$, nous nous donnerons généralement une extension algébriquement close K, K', $\cdots$ de ce corps, jouant le rôle de "domaine universel". Cela signifie que toutes les extensions de k envisagées sont contenues dans K et que nous prenons le point de vue qui consiste à attribuer à toute k-variété V l'ensemble $V(K)$ — aussi noté V — comme "ensemble sous-jacent". Cette convention nous permet de noter K (resp. K^*) le groupe additif (resp. multiplicatif) "considéré comme groupe défini sur k". Lorsqu' aucun corps de définition n'est spécifié, un "morphisme" d'une k-variété dans une autre est un K-morphisme, et, s'appliquant à des k-variétés ou à des parties de telles variétés, les termes "fermé", "dense", "adhérence", $\cdots$ se rapportent à la K-topologie de Zariski ou à la topologie induite par celle-ci.

1.4. *Soient V, W des k-variétés et $\beta : V \to W$ un morphisme. S'il existe une partie A de $V(k)$ dense dans V telle que $\beta(A) \subset W(k)$, le morphisme β est défini sur k.*

L'assertion étant locale, nous pouvons supposer V et W affines (ce qui sera d'ailleurs toujours le cas dans les applications) et nous devons montrer alors que si $f \in k[W]$, on a $\beta^*(f) \in k[V]$. Posons $\beta^*(f) = \sum c_i f_i$ avec $c_i \in K$, $f_i \in k[V]$, les f_i étant linéairement indépendants sur k. Pour tout $x \in A$, on a $\sum c_i f_i(x) = \beta^*(f)(x) = f(\beta(x)) \in k$, donc $c_i \in k$ pour tout i, puisque les restrictions des f_i à A sont linéairement indépendantes.

1.5. *Soient V, W, W' des k-variétés, $\beta\colon V \to W$ un k-morphisme surjectif, et $\gamma\colon W \to W'$ un K-morphisme tel que le morphisme $\gamma \circ \beta\colon V \to W'$ soit défini sur k. Alors γ est défini sur k.*

Si k est séparablement clos, cela résulte de 1.4; dans ce cas, en effet, $V(k)$ est dense dans V [1: AG 13.3], donc $\beta(V(k))$ est dense dans W, et l'on a $\gamma(\beta V(k)) = (\gamma \circ \beta)(V(k)) \subset W'(k)$. Le cas général se ramène à celui-là par une descente galoisienne facile.

1.6. *Soient V, W des groupes algébriques et $\beta\colon V \to W$ un morphisme de variétés. S'il existe un sous-groupe dense A de V tel que $\beta|_A$ soit un homomorphisme de groupes, alors β est un morphisme de groupes algébriques.*

Il suffit en effet de remarquer que les deux morphismes $V \times V \to W$ déduits du diagramme

$$
\begin{array}{ccc}
V \times V & \xrightarrow{\ \text{prod.}\ } & V \\
{\scriptstyle \beta \times \beta}\big\downarrow & & \big\downarrow{\scriptstyle \beta} \\
W \times W & \xrightarrow[\ \text{prod.}\ ]{} & W
\end{array}
$$

coïncident sur $A \times A$, qui est dense dans $V \times V$, d'où il résulte que le diagramme est commutatif.

1.7. Soient k, k' des corps commutatifs, $\varphi\colon k \to k'$ un homomorphisme et V une k-variété. On note $_{k'}^{\varphi}V$ la k'-variété obtenue à partir de V par le "changement de base" φ. Si k'' est un sous-corps de k' contenant $\varphi(k)$, φ peut aussi être vu comme un homomorphisme de k dans k'' ce qui donne un sens à la notation $_{k''}^{\varphi}V$; comme $_{k'}^{\varphi}V$ est la k'-variété déduite de $_{k''}^{\varphi}V$ par extension du corps de base, nous pourrons généralement, sans grand inconvénient, omettre l'indice et écrire $^{\varphi}V$ au lieu de $_{k'}^{\varphi}V$. L'injection canonique $V(k) \to {}^{\varphi}V(k')$ est notée φ^0; elle est bijective si φ est un isomorphisme. On a $k'[^{\varphi}V] = k[V] \otimes_k k'$, où k' est muni de la structure de k-algèbre définie par φ, et l'application φ^0 est compatible avec l'homomorphisme $f \mapsto f \otimes 1$ de $k[V]$ dans $k'[^{\varphi}V]$; cela définit φ^0 lorsque V est affine. Si $\beta\colon V \to W$ est un k-morphisme de k-variétés, il lui est canoniquement associé un k'-morphisme $^{\varphi}\beta\colon {}^{\varphi}V \to {}^{\varphi}W$ et l'on a $^{\varphi}\beta \circ \varphi^0 = \varphi^0 \circ \beta$ sur $V(k)$. On en déduit aussitôt que si V est un k-groupe algébrique, alors $^{\varphi}V$ est un k'-groupe et φ^0 est un homomorphisme de groupes. Si $V = \mathbf{GL}_n$, alors V est le groupe $\mathbf{GL}_n$ "vu comme k'-groupe", et si $g = (g_{ij}) \in \mathbf{GL}_n(k)$, on a $\varphi^0(g) = (\varphi(g_{ij}))$.

Soit K comme en 1.3. Supposons que $Fr^i(k) \subset k' \subset K$ et que $\varphi = Fr^i$, et notons $\bar{\varphi}$ l'automorphisme Fr^i de K. Alors, $\bar{\varphi}^0 \colon V \to {}^\varphi V$ est un morphisme de variétés, défini sur le composé de k et k', que nous noterons encore Fr^i. On a $\varphi^0 = Fr^i|_{V(k)}$.

Si k_1 est une extension séparable de k et H un k_1-groupe, nous notons $R^0_{k_1/k}$ la bijection canonique $H(k_1) \to (R_{k_1/k}H)(k)$.

1.8. Soient k, K comme en 1.3. Tout espace vectoriel V de dimension finie sur K est, de façon naturelle, un groupe algébrique pour sa structure additive. Si on se donne en outre une "k-structure", c'est-à-dire un k-sous-espace vectoriel engendré sur k par une K-base de V, le groupe acquiert une structure de k-groupe algébrique, et nous dirons que l'espace vectoriel V "est défini sur k".

Soit $i \in \mathbf{N}$ et posons $\varphi = Fr^i$. Le groupe algébrique ${}^\varphi V$ est le groupe additif d'un espace vectoriel (sur K) caractérisé par la condition que le morphisme $Fr^i \colon V \to {}^\varphi V$ soit Fr^i-linéaire (c'est-à-dire que $Fr^i(tv) = Fr^i(t) \cdot Fr^i(v)$ pour $t \in K$ et $v \in V$). Le groupe $GL({}^\varphi V)$ s'identifie alors avec ${}^\varphi GL(V)$ de telle façon que les applications $Fr^i \colon V \to {}^\varphi V$ et $Fr^i \colon GL(V) \to {}^\varphi GL(V)$ soient compatibles. Toute bijection Fr^i-linéaire de V sur un autre espace vectoriel V' est composée de Fr^i et d'un isomorphisme ${}^\varphi V \to V'$; elle induit donc une isogénie de $GL(V)$ sur $GL(V')$.

1.9. L'algèbre de Lie d'un groupe algébrique X est désignée par $L(X)$. Si $\beta \colon G \to H$ est un homomorphisme de groupes algébriques, l'homomorphisme $(d\beta)_e \colon L(G) \to L(H)$ est aussi noté $L(\beta)$.

1.10. Soit Φ un système de racines dans un espace vectoriel V sur $\mathbf{Q}$ [11: VI, 1.1]. On suppose V muni d'un produit scalaire positif non-dégénéré invariant par le groupe de Weyl $W(\Phi)$ de Φ, on note $|a|$ la longueur d'un élément a de V, et l'on pose

$$n_{ab} = 2(a, b)(b, b)^{-1} \qquad \text{pour } a, b \in \Phi \, .$$

On sait [11: VI, 1.3] que

$$n_{ab} \in \mathbf{Z} \, , \quad |n_{ab}| \leq 3 \, .$$

Supposons Φ réduit et $(a, b) \neq 0$. Alors

$$|n_{ab}| = 1 \text{ si } |a| \leq |b| \, , \text{ et } |n_{ab}| = 2 \text{ ou } 3 \text{ si } |a| > |b| \, .$$

Supposons Φ irréductible et réduit; alors l'ensemble des $|a|$ $(a \in \Phi)$ a un ou deux éléments, et nous disons que a est *longue* (resp. *courte*) si $|a| \geq |b|$

(resp. $|a| \leqq |b|$) pour tout $b \in \Phi$. La racine dominante, par rapport à un ordre quelconque, est toujours longue.

1.11. Rappelons encore la signification de quelques notations de [4] qui seront souvent utilisées. Soient G un groupe réductif, S un tore de G et T un tore maximal contenant S. On désigne par $\Phi(S, G)$ (resp. $\Phi(T, G)$) l'ensemble des poids non nuls de S (resp. T) dans G. Pour $a \in \Phi(T, G)$, le "groupe à un paramètre" normalisé par T et sur lequel T opère par a est noté U_a. Si ψ est une partie *close* de $\Phi(S, G)$ (i.e. une partie telle que toute somme d'éléments de ψ appartient à ψ dès qu'elle appartient à $\Phi(S, G)$), on désigne par $G_\psi^{*(S)}$ le groupe engendré par les U_a pour $a|_S \in \psi$; le produit $\mathcal{Z}(S) \cdot G_\psi^{*(S)}$ est alors un groupe, noté $G_\psi^{(S)}$. Si les éléments de ψ sont tous positifs pour un ordre compatible avec la loi de groupe dans $X^*(S)$, le groupe $G_\psi^{*(S)}$ est unipotent et on le note aussi U_ψ; si, en particulier, ψ est l'ensemble des multiples entiers positifs d'un élément $a \in \Phi(S, G)$ qui sont contenus dans $\Phi(S, G)$, on pose encore $U_\psi = U_{(a)}$.

Supposons $S = T$ et soit ψ une partie close de $\Phi(T, G)$ dont les éléments sont positifs pour un ordre compatible avec la loi de groupe dans $X^*(T)$. Alors, l'application produit

$$\prod_{a \in \psi} U_a \longrightarrow U_\psi$$

(c'est-à-dire l'application $(u_a)_{a \in \psi} \mapsto \prod u_a$) est un isomorphisme de variétés, quel que soit l'ordre dans lequel sont rangés les facteurs du produit.

Dans la suite de ce travail, les corps sont commutatifs, k est un corps, K une extension algébriquement close de k ("domaine universel pour k"), k_s (resp. $\bar{k}$) la clôture séparable (resp. algébrique) de k dans K, p l'exposant caractéristique de k et G un k-groupe connexe réductif. Les groupes algébriques simples ou presque simples sont toujours supposés semi-simples.

2. Propriétés des homomorphismes de corps

2.1. Proposition. *Soient k' un corps et $\lambda_i \colon k \to k'$ $(1 \leqq i \leqq m)$ des homomorphismes tels que toute égalité $\lambda_i \circ Fr^s = \lambda_j \circ Fr^t$ $(s, t \in \mathbf{N}; 1 \leqq i, j \leqq m)$ entraîne $i = j$. Soit $F \in k'[T_1, \cdots, T_m]$, où les T_i sont des indéterminées. Si $F(\lambda_1(x), \cdots, \lambda_m(x)) = 0$ pour tout $x \in k$, on a $F = 0$.*

Supposons que la proposition soit fausse et que F soit un polynôme de degré minimal fournissant un contre-exemple. Le théorème de Dedekind [8: V, §7, n° 5, Thm. 3] montre qu'il existe alors des monômes distincts $\prod_{i=1}^m T_i^{c_i}$ et $\prod_{i=1}^m T_i^{d_i}$ intervenant dans F avec des coefficients non nuls et tels que

$$\prod_{i=1}^m \lambda_i(x)^{c_i} = \prod_{i=1}^m \lambda_i(x)^{d_i}$$

pour tout $x \in k$. Vu l'hypothèse de minimalité faite sur le degré de F, c_i et d_i ne peuvent être simultanément non nuls. Quitte à permuter les λ_i, nous pouvons donc supposer qu'il existe $q \in \{2, \cdots, m\}$ tel que

$$F = T_1^{c_1} \cdots T_{q-1}^{c_{q-1}} - T_q^{d_q} \cdots T_m^{d_m}$$

et que, de plus,

$$(1) \qquad c_1 \neq 0 \ , \quad d_q \neq 0 \ .$$

Fixons $y \in k^*$. Alors

$$G(T_1, \cdots, T_m) = F(T_1, \cdots, T_m) - F(T_1 + \lambda_1(y), \cdots, T_m + \lambda_m(y))$$

est de degré strictement plus petit que celui de F et

$$\begin{aligned} G\big(\lambda_1(x), \cdots, \lambda_m(x)\big) &= F\big(\lambda_1(x), \cdots, \lambda_m(x)\big) \\ &\quad - F\big(\lambda_1(x+y), \cdots, \lambda_m(x+y)\big) = 0 \end{aligned}$$

pour tout $x \in k$; donc $G = 0$, c'est-à-dire que

$$(2) \qquad F(T_1, \cdots, T_m) = F\big(T_1 + \lambda_1(y), \cdots, T_m + \lambda_m(y)\big) \ .$$

Comparant les coefficients de $T_1^{c_1}$, on en déduit, vu (1), que

$$T_2^{c_2} \cdots T_{q-1}^{c_{q-1}} = \big(T_2 + \lambda_2(y)\big)^{c_2} \cdots \big(T_{q-1} + \lambda_{q-1}(y)\big)^{c_{q-1}} \ ,$$

d'où $c_i = 0$ pour $2 \leqq i \leqq q - 1$. De même, comparant les coefficients de $T_q^{d_q}$, on trouve que $d_i = 0$ pour $q + 1 \leqq i \leqq m$. On a donc

$$F = T_1^{c_1} - T_q^{d_q} \ .$$

Utilisant à nouveau (2), on voit que c_1 et d_q doivent être des puissances de p, soient p^s et p^t, et que $\lambda_1(y)^{p^s} = \lambda_q(y)^{p^t}$ pour tout $y \in k$. Mais cette dernière relation peut s'écrire $\lambda_1 \circ Fr^s = \lambda_q \circ Fr^t$ et contredit l'hypothèse.

2.2. Corollaire. *Soient k' un corps, $\lambda_i \colon k \to k'$ $(1 \leqq i \leqq m)$ des homomorphismes et, n, n_i $(1 \leqq i \leqq m)$ des entiers positifs. Supposons que, pour tout $x \in k$, on ait*

$$(1) \qquad \lambda_1(x)^n = \prod_{i=1}^{m} \lambda_i(x)^{n_i} \ .$$

Alors, pour tout i tel que $n_i \neq 0$, il existe un entier positif q tel que $\lambda_1 = \lambda_i \circ Fr^q$ ou bien $\lambda_i = \lambda_1 \circ Fr^q$.

Raisonnons par récurrence sur m. Si $m = 1$, ou, plus généralement, si $n_h = 0$ pour tout $h \neq 1$, il n'y a rien à démontrer. S'il existe $h \neq 1$ tel que $n_h \neq 0$, la proposition 2.1 implique l'existence d'indices i, j distincts, avec $1 \leqq i, j \leqq m$, et d'entiers positifs s, t tels que $s \geqq t$ et $\lambda_i \circ Fr^s = \lambda_j \circ Fr^t$. Remplaçant $\lambda_j(x)$ par $\lambda_i(x)^{p^{s-t}}$ dans (1), on diminue d'une unité le nombre d'homomorphismes considérés, et il suffit alors d'appliquer l'hypothèse de récurrence.

2.3. La proposition suivante est bien connue. Nous en indiquons brièvement la démonstration pour la commodité du lecteur.

PROPOSITION. *Soit k' un corps non isomorphe à C. Supposons k et k' munis de topologies qui en font des corps localement compacts non discrets. Alors, tout homomorphisme de corps $\alpha\colon k \to k'$ est un isomorphisme de corps topologiques de k sur un sous-corps fermé de k'.*

Supposons $k' \cong \mathbf{R}$. Comme k est alors formellement réel, il est aussi isomorphe à $\mathbf{R}$ et nous pouvons supposer que $k = k' = \mathbf{R}$. Les éléments positifs de k sont des carrés donc sont appliqués par α sur des éléments positifs de k'. Il s'ensuit que α est un homomorphisme de corps ordonnés et, par conséquent, est continu. Comme $\alpha|_{\mathbf{Q}}$ est l'identité, α l'est aussi, par continuité.

Si $k \cong \mathbf{R}$, $\mathbf{C}$, alors $k(\sqrt{-1})$ est algébriquement clos, donc $k'(\sqrt{-1})$ contient un corps algébriquement clos, et $k' \cong \mathbf{R}$.

Nous supposerons donc désormais que k et k' sont des corps valués complets à corps résiduels finis. Notons ω, ω' leurs valuations et l, l' les caractéristiques résiduelles. Si $n \in \mathbf{N}$ est premier à l et $l - 1$ (resp. l' et $l' - 1$), l'application $x \mapsto x^n$ est une bijection du groupe des unités $\omega^{-1}(0)$ (resp. $\omega'^{-1}(0)$) sur lui-même. On en déduit aussitôt qu'un élément x de k (resp. k') appartient à $\omega^{-1}(0)$ (resp. $\omega'^{-1}(0)$) si et seulement si, pour tout $N \in \mathbf{N}$, il existe $n > N$ tel que $x \in k^{*n}$ (resp. $x \in k'^{*n}$). Par conséquent, $\alpha(\omega^{-1}(0)) \subset \omega'^{-1}(0)$. Soit π une uniformisante de k. Posons $\omega'(\alpha(\pi)) = c$. Vu ce qu'on vient de voir, on a $\omega'(\alpha(x)) = c\cdot\omega(x)$ pour tout $x \in k$. Notre assertion résultera manifestement de cette identité si nous montrons que c est strictement positif. Or, s'il en était autrement, le corps $\alpha(k)$, qui est l'anneau engendré par $\alpha(\omega^{-1}(0))$ et $\alpha(\pi^{-1})$, serait contenu dans l'anneau de valuation de k', ce qui est absurde.

2.4. COROLLAIRE. *Soit k_0 (resp. k_0') l'adhérence du corps premier dans k (resp. k'). Alors $\alpha(k_0) = k_0'$ et la restriction de α à k_0 est l'unique isomorphisme de corps de k_0 sur k_0'.*

La première assertion est une conséquence immédiate de la proposition. Si k_0 est fini (i.e. $p \neq 1$), la seconde assertion est évidente. Supposons k_0 et k_0' infinis et soit $\alpha_0\colon k_0 \to k_0'$ un isomorphisme. Vu la proposition, appliquée à k_0, k_0' et α_0, α_0 est continu. Comme les isomorphismes α_0 et $\alpha|_{k_0}$ coïncident sur le corps premier de k, ils sont identiques.

II. COMPLÉMENTS SUR LES GROUPES RÉDUCTIFS

3. Isogénies

3.1. Soient G' un K-groupe réductif, T un tore maximal de G, T' un tore maximal de G', $X = X^*(T)$ et $X' = X^*(T')$ les groupes de caractères de T et T', $X_* = X_*(T) = \mathrm{Hom}\,(K^*, T)$ et $X'_* = X_*(T') = \mathrm{Hom}\,(K^*, T')$ leurs duals et $\Phi = \Phi(T, G)$, $\Phi' = \Phi(T', G')$ les systèmes de racines de G, G' relatifs à T, T'. À tout élément a de Φ ou Φ' est associée une *coracine* $a^\vee \in X_*$ ou X'_* caractérisée par les relations $\langle a^\vee, a \rangle = 2$ et $a^\vee(K^*) \in \langle U_a, U_{-a} \rangle$. Les résultats rappelés dans ce n° ont été démontrés par C. Chevalley [14] pour les groupes semi-simples; leur extension à des groupes réductifs quelconques est un exercice facile.

Si $\beta: G \to G'$ est une isogénie telle que $\beta(T) = T'$, les homomorphismes $(\beta|_T)^*: X' \to X$ et $(\beta|_T)_*: X_* \to X'_*$, que nous noterons simplement β^*, β_* par la suite, sont injectifs, et leurs conoyaux sont finis, d'ordre égal au degré de β; de plus, il existe une bijection $\rho: \Phi \to \Phi'$ et une fonction $\lambda: \Phi \to \{p^n \,|\, n \in \mathbf{N}\}$ telle que

$$(1) \qquad \beta^*\big(\rho(a)\big) = \lambda(a)\cdot a \ \text{ et } \ \beta_*(a^\vee) = \lambda(a)\cdot\rho(a)^\vee \qquad\qquad \text{pour } a \in \Phi\,.$$

Réciproquement, si $\beta^*: X' \to X$ et $\beta_*: X_* \to X'_*$ sont des monomorphismes transposés l'un de l'autre (donc de conoyaux finis) et s'il existe une bijection $\rho: \Phi \to \Phi'$ et une fonction $\lambda: \Phi \to \{p^n \,|\, n \in \mathbf{N}\}$ satisfaisant aux relations (1), β^* et β_* sont induits par une isogénie $\beta: G \to G'$, unique à la multiplication près par un automorphisme intérieur associé à un élément de T. Si $x_a: K \to U_a$ est un isomorphisme du groupe additif sur le groupe U_a, pour $a \in \Phi$, alors $t^{p^{\lambda(a)}} \mapsto \beta(x_a(t))$ est un isomorphisme de K sur $U_{\rho(a)}$; autrement dit, la restriction de β à U_a est le composé de $Fr^{\lambda(a)}: U_a \to {}^{Fr^{\lambda(a)}}U_a$ et d'un isomorphisme ${}^{Fr^{\lambda(a)}}U_a \to U_{\rho(a)}$.

Remarquons encore que G' et l'isogénie β sont entièrement déterminés à K-isomorphisme près par la donnée du sous-réseau $Y = \beta^*(X')$ de X et de la fonction λ (cela résulte de ce que (G', T') est déterminé, à isomorphisme près, par X', $\Phi' \subset X'$ et $\Phi'^\vee$).

Dans la suite de ce paragraphe, β désignera toujours une isogénie de G sur un groupe réductif G', et Y, λ conserveront la signification qui leur a été donnée ci-dessus. Notons encore que si $\beta = Fr^i$, on a $Y = p^i\cdot X$ et $\lambda(\Phi) = \{p^i\}$.

3.2. Soient $\beta_1: G \to G'_1$ une autre isogénie de G sur un groupe réductif et Y_1, λ_1 le sous-réseau de X et la fonction sur Φ correspondants; alors β se factorise à travers β_1, c'est-à-dire est composée de β_1 et d'une isogénie $G'_1 \to G'$, si et seulement si $Y \subset Y_1$ et $\lambda_1 \leqq \lambda$.

3.3. L'isogénie β est dite *centrale* si $\lambda(\Phi) = \{1\}$ (cf. [7: 2.15]) et *spéciale* si $1 \in \lambda(\Psi)$ pour toute composante irréductible Ψ de Φ.

Supposons G absolument presque simple. Un coup d'oeil aux systèmes de racines montre alors que si $\beta\colon G \to G'$ est une isogénie spéciale non centrale, alors ρ envoie les racines longues (courtes) de G sur les racines courtes (longues) de G' et l'une des conditions suivantes est remplie:

$p = 2$, G est de type $\mathbf{B}_n$, $\mathbf{C}_n$ ou $\mathbf{F}_4$, G' du type "dual" $\mathbf{C}_n$, $\mathbf{B}_n$ ou $\mathbf{F}_4$, et $\lambda(a) = 1$ ou 2 selon que a est longue ou courte;

$p = 3$, G et G' sont de type $\mathbf{G}_2$ et $\lambda(a) = 1$ ou 3 selon que a est longue ou courte.

On en déduit aussitôt que le composé de deux isogénies de groupes absolument presque simples est spécial, si et seulement si l'une d'elles est centrale et l'autre est spéciale.

Nous dirons qu'on se trouve dans le *cas exceptionnel* si $p = 2$ et G est de type $\mathbf{B}_n$, $\mathbf{C}_n$, $\mathbf{F}_4$ on si $p = 3$ et G est de type $\mathbf{G}_2$.

3.4. Le noyau de l'isogénie β est le sous-groupe de T intersection des noyaux des éléments de Y. En particulier, on voit que les propriétés suivantes sont équivalentes:

(i) β est radicielle (i.e. Ker $\beta = \{e\}$);

(ii) $\beta|_T$ est injective;

(iii) il existe $i \in \mathbf{N}$ tel que $p^i \cdot X \subset Y$;

(iv) il existe $i \in \mathbf{N}$ et une isogénie $\beta'\colon G' \to {}^{Fr^i}G$ telle que $\beta' \circ \beta = Fr^i$.

3.5. On a $L(G) = L(T) \oplus \coprod_{a \in \Phi} L(U_a)$, et $L(T)$ peut être identifiée à $X_* \otimes K$. Cela étant, un calcul immédiat montre qui si $p \neq 1$ et si l'on pose $Z = \{x \in X_* \,|\, \langle x, Y \rangle \subset p \cdot \mathbf{Z}\}$, on a

$$\mathrm{Ker}\, L(\beta) = Z \otimes K + \sum_{\substack{a \in \Phi \\ \lambda(a) \neq 1}} L(U_a) \,.$$

Si de plus $p \cdot X \subset Y$ et $\lambda(\Phi) \subset \{1, p\}$, on a $G' = G/\mathrm{Ker}\, L(\beta)$ (cf. [1: § 17]); en effet, on vérifie aisément, utilisant 3.2, que toute isogénie de G annulant Ker β se factorise alors à travers β.

3.6. Les remarques suivantes sont dues à C. Riehm [23]. Pour $a \in \Phi$ et $u \in U_a - \{e\}$, on a

(1)
$$a^\vee \otimes 1 \in [L(U_a), L(U_{-a})]$$

et

(2)
$$a^\vee \otimes 1 \in L(U_{-a}) \oplus (\mathrm{Ad}\, u)(L(U_{-a})) \oplus L(U_a) \,.$$

(Cela se vérifie immédiatement pour le groupe $\mathbf{SL}_2$ et le cas général s'en déduit en considérant un morphisme central $\mathbf{SL}_2 \to G$ dont l'image est $\langle U_a, U_{-a} \rangle$.) De (1), on déduit aussitôt que *si G est simplement connexe, l'algèbre de Lie $L(G)$ est engendrée par les $L(U_a)$*; en effet, X_* est alors engendré par les $a^\vee$.

De (2), il résulte de même que *si G est simplement connexe, le seul sous-espace linéaire de $L(G)$ invariant par* Ad G *et contenant les $L(U_a)$ est $L(G)$ lui-même.*

3.7. Lemme. *Supposons G simplement connexe ou G' adjoint, et G presque simple. Alors, l'isogénie β est spéciale si et seulement si $L(\beta) \neq 0$.*

Il résulte immédiatement de la définition des isogénies spéciales que β est spéciale si et seulement s'il existe $a \in \Phi$ telle que $L(U_a) \not\subset \mathrm{Ker}\, L(\beta)$, ce qui implique $L(\beta) \neq 0$. Si G est simplement connexe, la réciproque résulte de 3.6. Supposons G' adjoint et β non spéciale. Alors, $L(\beta)(L(G)) = L(\beta)(L(T))$ est une sous-algèbre de $L(T')$ invariante par Ad $\beta(G) = \mathrm{Ad}\, G'$, donc est réduite à $\{0\}$, vu [7: 2.13, 2.23(a)].

3.8. Supposons $p \neq 1$ et G absolument presque simple, et soit Ψ l'ensemble des racines courtes ou l'ensemble de toutes les racines de G selon qu'on est ou non dans le cas exceptionnel. Posons

$$Y_0 = \{x \in X \,|\, \langle a^\vee, x \rangle \subset p\cdot\mathbf{Z} \ \text{ pour } \ a \in \Psi\}$$

$$\lambda_0(a) = \begin{cases} p \ \text{ si } \ a \in \Psi \,, \\ 1 \ \text{ sinon} \,. \end{cases}$$

Il résulte des assertions du n° 3.1 que (Y_0, λ_0) est le système (Y, λ) d'une isogénie spéciale $\beta_0\colon G \to G_0$. De plus,

$$\mathrm{Ker}\, \beta_0 = \sum_{a \in \Psi} \big(a^\vee \otimes K + L(U_a)\big)$$

est engendré (comme algèbre de Lie) par les $L(U_a)$, $(a \in \Psi)$, vu 3.6 (1), et on a $G_0 = G/\mathrm{Ker}\, L(\beta_0)$. Il résulte immédiatement de 3.2 et 3.3 que toute isogénie non centrale de G se factorise (de façon évidemment unique) à travers β_0.

Si, dans ce qui précède, on choisit T défini sur k, il est clair que $\mathrm{Ker}\, L(\beta_0)$ est une k-sous-algèbre de $L(G)$ et l'on peut prendre $G_0 = G/\mathrm{Ker}\, L(\beta_0)$ et $\beta_0\colon G \to G_0$ définis sur k. Il résulte alors de 1.5 que *toute k-isogénie non centrale $G \to G'$ se factorise sur k à travers β_0, de façon unique,* c'est-à-dire qu'il existe une k-isogénie $\beta''\colon G_0 \to G'$, bien déterminée, telle que $\beta = \beta'' \circ \beta_0$. On voit en particulier que la classe de k-isomorphisme (resp. de K-isomorphisme) de β_0 ne dépend que de G et non des autres choix que nous avons faits — de celui de T par exemple —; ses éléments seront appelés les *k-isogénies* (resp. les *isogénies*) *non centrales minimales de G.*

Soit $\gamma\colon G \to G'$ une isogénie centrale et $\beta_0'\colon G' \to G_0'$ une isogénie non centrale minimale de G'. Vu la propriété universelle de β_0, il existe une isogénie $\gamma_0\colon G_0 \to G_0'$ telle que $\gamma_0 \circ \beta_0 = \beta_0' \circ \gamma$. Cette isogénie γ_0 est centrale, comme on le voit facilement en utilisant le fait que les fonctions λ se multiplient lorsqu'on compose des isogénies, et elle est définie sur k si γ, β_0 et β_0' le sont.

3.9. Supposons $p = 2$ et G adjoint de type $\mathbf{B}_n$. Alors, G est, à k-isomorphisme près, le groupe orthogonal (réduit) d'une forme quadratique q non dégénérée de défaut 1 dans un espace vectoriel V de dimension impaire sur k (cf. [33: n° 3]). Soient f la forme bilinéaire associée à q, $V^\perp$ son noyau, $V_0 = V/V^\perp$ et f_0 la forme alternée dans V_0 "image de f". L'application β qui envoie tout $g \in G$ sur la transformation de $V_0 \otimes K$ induite par g est une k-isogénie de G sur le groupe symplectique de la forme f_0. Il est facile de voir que β est une k-isogénie non centrale minimale de G (comme le groupe symplectique est simplement connexe, il suffit d'ailleurs, vu 3.3, de vérifier que β est spéciale et non centrale). Compte tenu du dernier alinéa de 3.8 et de [7: 2.20], nous pouvons donc énoncer la

PROPOSITION. *Si $p = 2$ et si G est de type $\mathbf{B}_n$, l'image de G par toute k-isogénie non centrale est un groupe déployé sur k.*

3.10. COROLLAIRE. *Si $p = 2$ et si G est de type $\mathbf{C}_n$, alors ^{Fr}G est déployé sur k.*

En effet, soit $\beta_0 \colon G \to G_0$ une k-isogénie non centrale minimale de G. Vu la propriété universelle de β_0, il existe une isogénie $\beta' \colon G_0 \to {}^{Fr}G$ telle que $\beta' \circ \beta_0 = Fr$. Mais, vu 3.3, G_0 est de type $\mathbf{B}_n$ et β' n'est pas centrale. Notre assertion résulte donc de la proposition précédente appliquée à β'.

3.11. COROLLAIRE. *Si $p = 2$ et si G est un groupe isotrope de type $\mathbf{F}_4$, alors l'image de G par toute isogénie non centrale est déployée.*

Nous pouvons supposer G non déployé. Soient β une isogénie non centrale de G, S un k-tore déployé maximal de G, T un k-tore maximal contenant S et H le groupe dérivé de $\mathcal{Z}(S)$. Vu [32: p. 60], le groupe H est de type $\mathbf{B}_3$. Comme $\Phi(T, H)$ contient des racines courtes, l'isogénie $\beta|_H$ n'est pas centrale, donc $\beta(H)$ est déployé et $\beta(\mathcal{Z}(S))$ contient un tore maximal déployé, d'où le corollaire.

3.12. *Remarque.* Plus loin, nous aurons seulement à utiliser le fait que, sous les hypothèses de 3.11, ^{Fr}G est déployé, ce qui peut se déduire directement de 3.9 et 3.10 sans faire usage de la classification de [32].

3.13. COROLLAIRE. *Dans le cas exceptionnel de 3.3, G est anisotrope ou bien ^{Fr}G est déployé.*

Cela résulte de 3.9, 3.10, 3.11 et du fait que tout groupe de type $\mathbf{G}_2$ est anisotrope ou déployé [32].

3.14. *Remarque.* Lorsque $p = 2$ (resp. $p = 3$) et *G est de type* $\mathbf{F}_4$ (resp. $\mathbf{G}_2$), il peut arriver que G et ${}^{Fr}G$ soient tous deux anisotropes sur k.

3.15. Etant donné un entier $d \geq 0$, on note λ_d l'application de puissance d-ième: $x \mapsto x^d$ d'un groupe dans lui-même. Si H est un k-groupe, alors λ_d: $H \to H$ est un k-morphisme de variétés. En remarquant que λ_d est le composé de l'application diagonale $x \mapsto (x, \cdots, x)$ de H dans $H \times \cdots \times H$ (d facteurs) et de l'application produit $H \times \cdots \times H \to H$, on voit tout de suite que la différentielle $(d\lambda_d)_x$ de λ_d en $x \in H$ est donnée par la formule

$$(1) \qquad (d\lambda_d)_x(x \cdot X) = x^d \cdot \sum_{0 \leq i < d} \mathrm{Ad}\, x^{-i}(X) , \qquad (X \in L(G)) .$$

3.16. PROPOSITION. *Soient G' un k-groupe, β: $G \to G'$ une k-isogénie centrale, et d le degré de β. Alors il existe un unique k-morphisme de variétés μ: $G' \to G$ tel que $\mu \circ \beta$: $G \to G$ soit égal à λ_d. On a aussi $\beta \circ \mu = \lambda_d$: $G' \to G'$.*

Soient G'' un k-groupe et supposons que l'on ait une factorisation

$$\beta: G \xrightarrow{\ \beta''\ } G'' \xrightarrow{\ \beta'\ } G'$$

de β en deux k-isogénies centrales. Si μ'': $G'' \to G$ et μ': $G' \to G''$ sont des k-morphismes de variétés tels que $\mu' \circ \beta' = \lambda_{d'}$, $\mu'' \circ \beta'' = \lambda_{d''}$, où d' et d'' sont les degrés de β' et β'' alors, comme $d = d' \cdot d''$ et β'' commute à $\lambda_{d'}$, il est clair que $\mu = \mu'' \circ \mu'$ satisfait à notre condition. Vu [7: 2. 19] cela nous ramène au cas où β est la projection de G sur son quotient, soit par un k-sous-groupe central fini N, soit par une k-sous-algèbre $\mathfrak{n} \neq \{0\}$ de $L(G)$, centralisée par G. Dans le premier cas, β est séparable, d est d'ordre premier à p, et N est un groupe d'ordre d. Par suite $(x \cdot n)^d = x^d$ quels que soient $x \in G$ et $n \in N$, λ_d est constante sur les classes $x \cdot N$, et l'existence de μ résulte de la propriétés universelle du quotient par un k-sous-groupe [1: 6.1]. Dans le deuxième cas on a $p \neq 1$ et d est une puissance de p. Soient $x \in G$, $X \in \mathfrak{n}$. Comme X est centralisé par G, 3.15(1) montre que $(d\lambda_d)_x(x \cdot X) = x^d \cdot d \cdot X = 0$, donc $(d\lambda_d)_x$ est nulle sur $x \cdot \mathfrak{n}$ quel que soit $x \in G$ et l'existence de μ est à nouveau conséquence d'une propriété universelle [1: 17.2]. La dernière assertion de l'énoncé résulte de ce que β commute à λ_d.

3.17. COROLLAIRE. *Sous les hypothèses de 3.16, $\beta(G(k))$ est un sous-groupe distingué de $G'(k)$ et $G'(k)/\beta(G(k))$ est un groupe commutatif dont l'exposant divise d.*

Cela résulte de [7: 2.7] et du fait que, vu la dernière assertion de 3.16, $\lambda_d(G'(k)) \subset \beta(G(k))$.

3.18. Supposons k doté d'une topologie qui en fait un corps topologique localement compact non discret. Si V est une k-variété, alors $V(k)$ est muni canoniquement d'une topologie d'espace localement compact, et, si V est non-singulière, d'une structure de k-variété analytique [38: App. III], visiblement dénombrable à l'infini. Si H est un k-groupe, alors $H(k)$, muni de ces structures, est un k-groupe analytique, et en particulier un groupe topologique localement compact, dénombrable à l'infini. Supposons que H opère k-morphiquement sur V et soit $v \in V(k)$. Si l'application $\varphi_v \colon h \mapsto hv$ de H sur Hv est séparable, alors, pour tout $w \in (Hv)(k)$, l'application $h \mapsto hw$ de $H(k)$ dans $(Hv)(k)$ est une submersion de variétés k-analytiques; par conséquent, les orbites de $H(k)$ dans $(Hv)(k)$ sont ouvertes et fermées dans $(Hv)(k)$, donc localement fermées dans $V(k)$. Il s'ensuit que φ_v induit un homéomorphisme — donc un isomorphisme de variétés k-analytiques — de $H(k)/H(k)_v$ sur $H(k)v$, où $H(k)_v$ désigne le stabilisateur de v dans $H(k)$; en effet, on sait (cf. [9: VII, App. 1, Lemme 2]) que si un groupe topologique localement compact dénombrable à l'infini opère transitivement et continûment sur un espace localement compact, ce dernier s'identifie comme espace topologique au quotient du groupe par le stabilisateur d'un point.

Si H° est réductif, connexe, anisotrope sur k, alors $H(k)$ est compact: en caractéristique zéro, cf. [4: 9.4]; en caractéristique non nulle, ce résultat a été annoncé par F. Bruhat et l'un des auteurs (C.R. Acad. Sci. Paris **263** (1966), 822-825).

3.19. PROPOSITION. *Conservons l'hypothèse de 3.18 et soit* $\beta \colon G \to G'$ *un k-morphisme central surjectif.*

(i) *Le groupe* $\beta(G(k))$ *est fermé distingué dans* $G'(k)$ *et le quotient* $C = G'(k)/\beta(G(k))$ *est compact et commutatif. Soit q le degré de l'isogénie radicielle canonique* $\beta'' \colon G/\mathrm{Ker}\,\beta \to G'$. *Alors C contient un sous-groupe ouvert d'exposant divisant q, et qui est d'indice premier à p.*

(ii) *L'isomorphisme* $G(k)/(\mathrm{Ker}\,\beta)(k) \to \beta(G(k))$ *est un isomorphisme de groupes topologiques. Si* $(\mathrm{Ker}\,\beta)^\circ$ *est anisotrope sur k, la restriction de β à $G(k)$ est une application propre.*

Nous établirons tout d'abord la première assertion de (i).

Le groupe $\beta(G(k))$ est distingué dans $G'(k)$ et C est commutatif vu [7: 2.7]. Soient S un tore déployé sur k maximal de G et U^+, U^- les radicaux unipotents de deux k-sous-groupes paraboliques minimaux opposés de G contenant $\mathcal{Z}(S)$. Posons $Z = \mathcal{Z}(S)$, $S' = \beta(S)$, $U'^{\pm} = \beta(U^{\pm})$, $Z' = \mathcal{Z}(S') = \beta(Z)$, $\Omega = Z \cdot U^+ \cdot U^-$ et $\Omega' = Z' \cdot U'^+ \cdot U'^-$. On a $\beta^{-1}(\Omega') = \Omega$ et $\Omega(k) = Z(k) \cdot U^+(k) \cdot U^-(k)$. Par conséquent

$$\Omega'(k) \cap \beta(G(k)) = \beta(Z(k)) \cdot U'^+(k) \cdot U'^-(k) \, .$$

Comme l'application produit $Z' \times U'^+ \times U'^- \to \Omega'$ est un k-isomorphisme de variétés, on voit que $\Omega'(k) \cap \beta(G(k))$ est fermé dans l'ouvert $\Omega'(k)$, donc que $\beta(G(k))$ est fermé dans $G'(k)$, si et seulement si $\beta(Z(k))$ est fermé dans $Z'(k)$. D'autre part, comme le groupe $Z'(k) \cdot \beta(G(k))$ contient $Z'(k)$, $U'^+(k)$ et $U'^-(k)$, il coïncide avec $G'(k)$ et si $\beta(Z(k))$ est fermé dans $Z'(k)$ on a un isomorphisme de groupes topologiques:

$$(1) \qquad G'(k)/\beta(G(k)) \xrightarrow{\sim} Z'(k)/\beta(Z(k)) \ .$$

On voit donc que, quitte à remplacer G par Z et G' par Z', nous pouvons supposer S central dans G.

Sous cette hypothèse, le quotient G/S est un groupe anisotrope, donc $(G/S)(k)$ est compact (3.18). Comme S est déployé, l'homomorphisme $G(k) \to (G/S)(k)$ est surjectif [1: 15.7]. Il s'ensuit (3.18) que $G(k)/S(k)$ est topologiquement isomorphe à $(G/S)(k)$ donc est compact; en particulier, il existe une partie compacte M de $G(k)$ telle que $G(k) = M \cdot S(k)$. De même, $G'(k)/S'(k)$ est compact.

Si k est connexe, (i) est évident car β est séparable, donc $\beta(G(k))$ est ouvert dans $G'(k)$, qui n'a qu'un nombre fini de composantes connexes topologiques (cf. par exemple [4: §14]). Nous pouvons par conséquent supposer k muni d'une valuation discrète complète. Soient o l'anneau de valuation et π une uniformisante. Notons S^o (resp. S'^o) le groupe de tous les éléments s de $S(k)$ (resp. $S'(k)$) tels que $\chi(s) \in o$ pour tout caractère χ de S (resp. S'). Soit $\lambda \colon X_*(S) \to S(k)$ (resp. $\lambda' \colon X_*(S') \to S'(k)$) l'homomorphisme $\eta \mapsto \eta(\pi)$. Posons $X = \lambda(X_*(S))$ et $X' = \lambda'(X_*(S'))$. Alors, S^o et S'^o sont des groupes compacts et $S(k) = S^o \times X$, $S'(k) = S'^o \times X'$ (produits directs topologiques). On a $\beta(S^o) \subset S'^o$ et β applique X sur un sous-groupe d'indice fini de X'. Il s'ensuit immédiatement que $\beta(S(k))$ est fermé dans $S'(k)$ et que $S'(k)/\beta(S(k))$ est compact (c'est-à-dire notre assertion pour $G = S$). Par conséquent, $\beta(G(k)) = \beta(M) \cdot \beta(S(k))$ est fermé dans $G'(k)$ et $G'(k)/\beta(S(k))$, donc *a fortiori* $G'(k)/\beta(G(k))$, est compact, ce qui termine la démonstration de la première assertion de (i).

(ii) en résulte aussitôt vu 3.18.

Soit $\beta' \colon G \to G'' = G/\mathrm{Ker}\,\beta$ la projection canonique. C'est un k-morphisme central séparable et l'on a $\beta = \beta'' \circ \beta'$ [7: 2.19]. Compte tenu de ce qui a déjà été démontré, on a une suite exacte de groupes compacts commutatifs

$$(2) \qquad 1 \longrightarrow \beta''(G''(k))/\beta(G(k)) \longrightarrow C \longrightarrow G'(k)/\beta''(G''(k)) \longrightarrow 1 \ ;$$

comme β'' est injectif, il induit un isomorphisme

$$(3) \qquad G''(k)/\beta'(G(k)) \xrightarrow{\sim} \beta''(G''(k))/\beta(G(k)) \ .$$

L'exposant de $G'(k)/\beta''(G''(k))$ divise q vu 3.17. Comme β' est séparable, $L(\beta')$ est surjective, donc $\beta'(G(k))$ est un sous-groupe ouvert de $G''(k)$. Le quotient $G''(k)/\beta'(G(k))$, étant compact, est fini. Si β' est une isogénie (ce qui a lieu si et seulement si β en est une), alors le degré r de β' est premier à p, et (3.17) $G''(k)/\beta'(G(k))$ est d'exposant divisant r. Il en résulte que dans (2) le deuxième terme est un groupe fini, d'ordre premier à l'exposant du quatrième terme; C est donc le produit direct de ces deux groupes, ce qui termine la démonstration de la proposition lorsque β est une isogénie. Sinon soit G_1 un k-facteur de G tel que $G = (\mathrm{Ker}\,\beta)^\circ \cdot G_1$ et que $\mathrm{Ker}\,\beta \cap G_1$ soit fini. Alors β induit une k-isogénie centrale de G_1 sur G' et $G'(k)/\beta(G(k))$ est un quotient de $G'(k)/\beta(G_1(k))$, d'où la deuxième assertion de (i) aussi dans ce cas.

3.20. COROLLAIRE. *Soit* $\alpha\colon G \to G'$ *un k-morphisme séparable surjectif. Alors $\alpha(G(k))$ est un sous-groupe distingué ouvert d'indice fini de $G'(k)$ et le quotient $G'(k)/\alpha(G(k))$ est commutatif.*

Le groupe $N = (\mathrm{Ker}\,\alpha)^\circ$ est défini sur k [4: 3.13]. Comme $L(\alpha)$ est surjective, $\alpha(G(k))$ est un sous-groupe ouvert de $G'(k)$. Soit G_1 un k-facteur de G tel que $G = N \cdot G_1$ et que $N \cap G_1$ soit fini. La restriction de α à G_1 est une k-isogénie centrale de G_1 sur G'. Vu 3.19 $\alpha(G_1(k))$ est distingué fermé dans $G'(k)$ et le quotient $G'(k)/\alpha(G_1(k))$ est compact commutatif. Comme $\alpha(G(k))$ contient $\alpha(G_1(k))$ et est ouvert, le corollaire s'ensuit.

4. Suites centrales du radical unipotent d'un sous-groupe parabolique

Dans ce paragraphe, G est supposé absolument presque simple, sauf mention expresse du contraire.

4.1. *Notations.* On note T un tore maximal de G, $\Phi = \Phi(T, G)$ le système des racines de G par rapport à T, $W = W(T, G) = \mathfrak{N}(T)/\mathfrak{Z}(T)$ le groupe de Weyl, Δ une base de Φ, Φ^+ l'ensemble des racines qui sont combinaisons linéaires à coefficients positifs d'éléments de Δ et d la racine dominante. Pour $a \in \Phi$, on désigne par s_a la réflexion associée à a et on pose $a = \sum_{b \in \Delta} m_b(a) \cdot b$; les $m_b(a)$ sont donc des entiers tous positifs ou tous négatifs. Pour toute partie θ de Δ, W_θ désigne le sous-groupe de W engendré par les s_b $(b \in \theta)$, et P_θ le "sous-groupe parabolique standard" engendré par T et les U_b $(b \in \Delta \cup (-\theta))$, on pose $m_\theta(a) = \sum_{b \in \theta} m_b(a)$ (pour $a \in \Phi$) et $r_\theta = m_\theta(d)$. Sauf dans les n°s 4.10 et 4.11, le corps de définition n'interviendra pas, et on peut donc, si l'on veut, le supposer algébriquement clos. Pour les propriétés des systèmes de racines utilisées sans démonstration, nous renvoyons à [11: VI].

4.2. Pour $a, b \in \Phi$ avec $a + b \neq 0$, posons

$$\kappa(a, b) = \{ia + jb| \quad i, j \in \mathbf{N}^*; \quad ia + jb \in \Phi\} \, .$$

C'est une partie close de Φ et on a

$$(1) \qquad\qquad (U_a, U_b) \subset U_{\kappa(a,b)} \, .$$

Plus précisément, les formules de [13: p. 27], établies en caractéristique zéro mais valables en toute caractéristique vu les résultats de [14] et [15] (cf. aussi [2]) montrent qu'il existe des isomorphismes $x_a \colon K \to U_a$ $(a \in \Phi)$ du groupe additif sur les U_a et, pour un ordre fixé dans l'ensemble des paires (i, j) d'entiers positifs, des entiers $m_{a,b;i,j}$ $(a, b \in \Phi; i, j \in \mathbf{N}^*)$, dépendant seulement de Φ et non de p, tels que

$$\big(x_a(s), x_b(t)\big) = \prod_{\substack{i,j \in \mathbf{N}^* \\ ia+jb \in \Phi}} x_{ia+jb}(m_{a,b;i,j}s^i t^j) \quad (s, t \in K) \, ,$$

et que

$$m_{a,b,1,1} = \pm (q + 1)$$

où q est le plus grand entier tel que $b - q \cdot a \in \Phi$.

Etant normalisé par T, le groupe (U_a, U_b) est de la forme U_ψ, où $\psi \subset \kappa(a, b)$ est une partie quasi-close de Φ^+ (cf. [4: 3.8]). Nous dirons que la commutation de U_a et U_b est *irrégulière* si $a + b \in \Phi$ et $a + b \notin \psi$, c'est-à-dire si $q + 1$ est un multiple de la caractéristique de k, et qu'elle est *exceptionnelle* si $a + b \in \Phi$ et $\psi = \varnothing$ $\big($d'où $(U_a, U_b) = \{e\}\big)$.

4.3. L'examen des systèmes de racines de rang 2 montre que les cas de commutation irrégulière sont les suivants:

(i) $p = 2$, G est de type $\mathbf{B}_n$, $\mathbf{C}_n$, $\mathbf{F}_4$, les racines a, b sont courtes et orthogonales et $a + b \in \Phi$;

(ii) $p = 3$, G est de type $\mathbf{G}_2$ et a, b sont courtes et forment un angle de $60°$;

(iii) $p = 2$, G est de type $\mathbf{G}_2$ et a, b sont courtes et forment un angle de $120°$.

Dans les cas (i) et (ii), on a $\kappa(a, b) = \{a + b\}$, donc $\psi = \varnothing$ et il y a commutation exceptionnelle. Dans le cas (iii), on déduit des formules de [13: p. 27] que $\psi = \{2a + b, 2b + a\}$, et la commutation n'est pas exceptionnelle. En particulier, on voit qu'*il n'y a pas de commutation exceptionnelle entre U_a, U_b si $(a, b) < 0$*, et qu'*il n'y a pas de commutation irrégulière si a ou b est longue*.

4.4. LEMME. *Soient $a, b \in \Phi$ telles que $a + b \in \Phi$. Si a est longue, on a $(a, b) < 0$ et $a + b = s_a(b)$.*

Comme a est longue, on a $2(a + b, a)(a, a)^{-1} \leqq 1$, et $2(a, b)(a, a)^{-1} \geqq -1$. Par conséquent $2(a, b)(a, a)^{-1} = -1$, et le lemme s'ensuit.

4.5. LEMME. *Soit $\theta \subset \Delta$. La fonction m_θ définie en 4.1 est constante sur les orbites de $W_{\Delta - \theta}$. Si $a \in \Phi$ est tel que $m_{\Delta - \theta}(a)$ est le minimum (resp. le maximum) de $m_{\Delta - \theta}$ sur $W_{\Delta - \theta}(a)$, alors $(a, c) \leqq 0$ (resp. $(a, c) \geqq 0$) pour tout $c \in \Delta - \theta$.*

C'est une conséquence immédiate des relations suivantes, où b, $c \in \Delta$ et $b \neq c$:

$$m_b\big(s_c(a)\big) = m_b(a) \; ;$$

$$m_c\big(s_c(a)\big) = m_c(a) - 2(a, c)(c, c)^{-1} \; .$$

4.6. LEMME. *Soient $\theta \subset \Delta$ et $a \in \Phi^+$ tels que $m_\theta(a) < r_\theta$. Alors, il existe $b \in \Phi^+$ telle que $m_\theta(b) = 1$ et $a + b \in \Phi^+$; si a est longue, on peut choisir $b \in W_{\Delta - \theta}(\theta)$. Si $a_1 \in \Phi^+ - \{d\}$, il existe $b_1 \in \Delta$ telle que $a_1 + b_1 \in \Phi^+$.*

La dernière assertion (qui est cas particulier de la première, pour $\theta = \Delta$) résulte du lemme 12.15 de [4] appliqué à la représentation adjointe d'un groupe algébrique complexe ayant Φ pour système de racines.

Démontrons les deux premières assertions. Vu 4.5, on peut supposer que $m_{\Delta - \theta}(a)$ est le maximum de $m_{\Delta - \theta}$ sur $W_{\Delta - \theta}(a)$, et on a alors

$$(1) \qquad\qquad\qquad (a, c) \geqq 0 \quad \text{pour tout } c \in \Delta - \theta \; .$$

Supposons a longue. Alors $a \in W(d)$ et comme $a \neq d$, il existe $b \in \Delta$ telle que $(a, b) < 0$. Vu (1), on a $b \in \theta$ donc $m_\theta(b) = 1$, d'où le lemme sous l'hypothèse faite.

Reste à établir la première assertion lorsque a n'est pas longue. S'il existe $c \in \theta$ telle que $a + c \in \Phi$, $b = c$ satisfait à l'énoncé. Nous pouvons donc supposer que

$$(2) \qquad\qquad pour\ tout\ c \in \theta, \quad on\ a\ a + c \notin \Phi,\ donc\ (a, c) \geqq 0 \; .$$

Vu la dernière partie de l'énoncé, déjà établie, il existe $c \in \Delta$ telle que $a + c \in \Phi$. Vu (1) et (2), $a + c$ est une racine longue et on a $c \in \Delta - \theta$, d'où $m_\theta(a + c) = m_\theta(a)$. Comme le lemme est déjà démontré pour les racines longues, il existe $c' \in \Phi^+$ telle que $m_\theta(c') = 1$ et $a + c + c' \in \Phi^+$. Le lemme 4.4 implique alors que $(a + c, c') < 0$. Mais (1) et (2) entraînent que $(a, c') \geqq 0$. Par conséquent, $(c, c') < 0$ et $b = c + c'$ est une racine satisfaisant à nos conditions.

4.7. PROPOSITION. *Soit $\theta \subset \Delta$. Posons $r = r_\theta$. Pour $i \in \mathbf{N}^*$, soit ψ_i (resp. λ_i) l'ensemble des racines (resp. des racines longues) a telles que*

$m_\theta(a) \geqq i$. *Les ensembles* ψ_i, λ_i *sont clos. Posons* $U_i = U_{\psi_i}$, $U_{lg,i} = U_{\lambda_i}$, $U = U_1 = R_u(P_{\Delta-\theta})$ *et* $U_{lg} = U_{lg,1}$.

    (i) *On a* $\mathcal{C}^i U \subset U_i \subset \mathcal{C}_{r-i+1} U$ $(1 \leqq i \leqq r+1)$.

    (ii) *La classe de nilpotence de* U *est* $\geqq \mathrm{Card}\,\theta$ *et* $\leqq r$.

    (iii) *Si* G *n'a pas de commutation irrégulière* (4.3), *alors*

$$U_i = \mathcal{C}^i U = \mathcal{C}_{r-i+1} U \qquad\qquad (1 \leqq i \leqq r+1)\,.$$

En particulier, la classe de nilpotence de U *est égale à* r.

    (iv) *Si* θ *est formé de racines longues, on a*

$$U_{lg} \cap \mathcal{C}^i U = \mathcal{C}^i U_{lg} = \mathcal{C}_{r-i+1} U_{lg} = U_{lg} \cap \mathcal{C}_{r-i+1} U = U_{lg,i} \quad (1 \leqq i \leqq r+1)\,.$$

En particulier, les classes de nilpotence de U *et* U_{lg} *sont égales à* r.

Il est évident que ψ_i est clos, et il s'ensuit, vu 4.4, que λ_i l'est aussi.

(i) On a $\mathcal{C}^1 U = U = U_1$. Il résulte de 4.2 (1) que U_i est un sous-groupe distingué de U et que

$$(1) \qquad\qquad (U,\, U_{i-1}) \subset U_i\,.$$

On en déduit aussitôt, par récurrence sur i, que $\mathcal{C}^i U \subset U_i$. D'autre part, $U_{r+1} = \{e\} = \mathcal{C}_0 U$, et la deuxième inclusion de (i) en résulte aussitôt par récurrence descendante sur i, car si $U_{i+1} \subset \mathcal{C}_{r-i} U$, on a, toujours vu (1):

$$U_i \subset \{x \in U \mid (x,\, U) \subset U_{i+1}\} \subset \{x \in U \mid (x,\, U) \subset \mathcal{C}_{r-i} U\} = \mathcal{C}_{r-i+1} U\,.$$

(ii) Le fait que la classe de nilpotence de U est $\leqq r$ résulte immédiatement de (i). Posons $\Delta = \{a_1,\, \cdots,\, a_l\}$ (d'où $\mathrm{Card}\,\Delta = l$) et $\mathrm{Card}\,\theta = l'$, et supposons les a_i numérotés de telle façon que $\{a_1,\, \cdots,\, a_i\}$ soit connexe, donc que $b_i = a_1 + \cdots + a_i$ soit une racine [11: VI, 1.6, Cor. 3] pour $1 \leqq i \leqq l$. On a alors, pour $1 \leqq i < l$,

$$(b_i,\, a_{i+1}) < 0$$

et, vu 4.3,

$$(2) \qquad\qquad U_{b_{i+1}} \subset (U_{b_i},\, U_{a_{i+1}})$$

(le cas (iii) de 4.3 ne peut se présenter, car si G est de type $\mathbf{G}_2$, $i = 1$ et l'une des racines $b_1 = a_1$, a_2 est longue). Pour $i \leqq l$, soit $j(i)$ le nombre d'éléments de $\theta \cap \{a_1,\, \cdots,\, a_i\}$. Pour tout i, U_{a_i} normalise $U = R_u P_\theta$, donc aussi tous les $\mathcal{C}^h U$. Cela étant, on déduit de (2), par une récurrence immédiate sur i, que

$$U_{b_i} \subset \mathcal{C}^{j(i)} U\,.$$

En particulier $U_{b_l} \subset \mathcal{C}^{l'} U$.

(iii) Vu (i), nous avons seulement à prouver que si G n'a pas de commutation irrégulière, on a les relations

(3) $$U_i \subset (U, U_{i-1}) \quad \text{pour } 2 \leqq i \leqq r+1,$$

et

(4) $$U_i/U_{i+1} \supset \mathcal{C}(U/U_{i+1}) \quad \text{pour } 1 \leqq i \leqq r,$$

les inclusions $\mathcal{C}^i U \supset U_i \supset \mathcal{C}_{r-i+1} U$ s'en déduisant alors par récurrence sur i.

Pour démontrer (3), il suffit de faire voir que si $a \in \Phi^+$ est telle que $m_\theta(a) \geq 2$, il existe b, $c \in \Phi^+$ telles que $a = b + c$ et $m_\theta(c) = 1$. Vu 4.5, nous pouvons supposer que $m_{\Delta-\theta}(a)$ est la valeur minimum de $m_{\Delta-\theta}$ dans $W_{\Delta-\theta}(a)$, donc (toujours vu 4.5), que $(a, f) \leqq 0$ pour tout $f \in \Delta - \theta$. Mais alors, il existe $c \in \theta$ telle que $(a, c) > 0$, et on a $b = a - c \in \Phi^+$, d'où l'assertion.

Pour démontrer (4), il suffit de prouver que si $a \in \Phi^+$ est telle que $1 \leqq m_\theta(a) \leqq i$ et $(U, U_a) \subset U_{i+1}$, alors $m_\theta(a) = i$, ou encore que si $m_\theta(a) < i$, il existe $c \in \Phi^+$ satisfaisant à $m_\theta(c) = 1$ et $a + c \in \Phi^+$, or ceci est la première assertion de 4.6.

(iv) Supposons θ formé de racines longues. Soient $i \geq 2$ et a une racine longue telle que $m_\theta(a) \geqq i$. Montrons qu'il existe $b \in W_{\Delta-\theta}(\theta)$ telle que $a - b \in \Phi^+$. Vu 4.5, nous pouvons supposer que $m_{\Delta-\theta}(a)$ est le minimum des valeurs prises par $m_{\Delta-\theta}$ sur $W_{\Delta-\theta}(a)$, donc que $(a, c) \leq 0$ pour tout $c \in \Delta - \theta$, ce qui entraîne l'existence de $b \in \theta$ telle que $(a, b) > 0$, d'où $a - b \in \Phi^+$. Vu 4.4, $a - b = s_a(-b)$ est longue. Comme il n'y a pas de commutation irrégulière mettant en jeu une racine longue (4.3), on a

$$U_a \subset (U_{lg,i-1}, U_{lg})$$

d'où

$$U_{lg,i} \subset (U_{lg,i-1}, U_{lg}) .$$

On en tire $U_{lg,i} \subset \mathcal{C}^i U_{lg}$, par récurrence sur i, d'où, tenant compte de (i),

$$\mathcal{C}^i U_{lg} \subset \mathcal{C}^i U \cap U_{lg} \subset U_{lg,i} \subset \mathcal{C}^i U_{lg} ,$$

et finalement

$$\mathcal{C}^i U_{lg} = \mathcal{C}^i U \cap U_{lg} = U_{lg,i} .$$

Comme d est longue, cela montre en particulier, compte tenu de (ii), que les classes de nilpotence de U et U_{lg} sont égales à r.

Pour $1 \leqq i \leqq r + 1$, on a, en vertu de (i),

$$U_{lg,i} = U_{lg} \cap U_i \subset \mathcal{C}_{r-i+1} U \cap U_{lg} \subset \mathcal{C}_{r-i+1} U_{lg} .$$

Pour terminer la démonstration, il suffit donc de prouver que

$$\mathcal{C}_{r-i+1} U_{lg} \subset U_{lg,i} .$$

Pour $i = r + 1$, les deux membres sont réduits à $\{e\}$. Raisonnant par récurrence descendante sur i, on voit qu'il suffit de montrer que si a est une racine longue et si $(U_a, U_b) \subset U_{lg,i}$ quelle que soit la racine longue b telle que $m_\theta(b) = 1$, alors $m_\theta(a) \geq i - 1$. Mais cela résulte de 4.3 et 4.6.

4.8. Corollaire. *Supposons U commutatif. Alors* $\operatorname{Card} \theta = 1$. *Si on n'est pas dans le cas exceptionnel (3.3), θ est formé d'une racine longue ayant le coefficient 1 dans la racine dominante. Dans le cas exceptionnel, la même chose est vraie après transformation éventuelle de G et U par une isogénie spéciale radicielle.*

La première assertion résulte de 4.7 (ii). Soit a l'élément de θ. Si a est longue, on a $r = m_a(d) = 1$, d'après 4.7 (iv). Supposons donc que a n'est pas longue. Comme $m_a(d)$ est divisible par $(d, d)(a, a)^{-1}$ [11: Exerc. 13, p. 224], on a $m_a(d) > 1$ et 4.7 (iii) implique que G a de la commutation irrégulière. De plus, G ne peut être de type $\mathbf{G}_2$ sinon, désignant par b la deuxième racine simple, on en déduirait que U contient U_a et U_{a+b}, qui ne commutent pas entre eux (4.3). Vu 4.3, on est donc dans le cas exceptionnel et, d'après 3.8, il existe une isogénie spéciale radicielle $\beta \colon G \to G'$ telle que l'application $\rho \colon \Phi(T, G) \to \Phi(\beta(T), G')$ du n° 3.1 échange racines longues et racines courtes. Si l'on pose $a' = \rho(a)$ et $\Delta' = \rho(\Delta)$, on a $\beta(U) = R_u(P_{\Delta - \{a'\}})$, la racine a' est longue et son coefficient dans la racine dominante de G' est égal à 1, vu 4.7 (iv). Le corollaire est ainsi démontré.

4.9. Corollaire. *On suppose que $\theta = \{a\}$, où a est courte et $m_a(d) = 2$. Alors $\mathcal{C}(U) = U_2$ sauf dans les cas suivants, où U est commutatif:*

$p = 2$ *et G est de type $\mathbf{B}_n$;*

$p = 2$, *G est de type $\mathbf{C}_n$ et a est la première racine simple dans la numérotation de* [11: Pl. II, p. 252].

Vu 4.7, il suffit de montrer que $\mathcal{C}(U) \subset U_2$ et de considérer le cas où G a de la commutation irrégulière. Par ailleurs, G n'est pas de type $\mathbf{G}_2$ puisque, a étant courte, on aurait alors $m_a(d) = 3$. On est donc ramené au cas 4.3 (i); en particulier, $p = 2$.

Supposons qu'il existe $b \in \Phi$ tel que $m_a(b) = 1$ et $U_b \subset \mathcal{C}(U)$. Soit $\beta \colon G \to G'$ une isogénie non centrale minimale (3.8), et $\rho \colon \Phi \to \Phi' = \Phi(\beta(T), G')$ la bijection associée (3.1). Pour $b' \in \Phi'$, posons $b' = \sum_{a' \in \rho(\Delta)} m_{a'}(b') \cdot a'$. Comme $m_a(b)$ est divisible par $(b, b)(a, a)^{-1}$, la racine b est courte, donc $\rho(b)$ est longue et on a $m_{\rho(a)}(\rho(b)) = 1$. Mais $U_{\rho(b)} \subset \mathcal{C}(\beta(U))$. Comme $\rho(a)$ est longue, il suit alors de 4.7 (iv) que $\rho(a)$ a le coefficient un dans la racine dominante d' de G'

par rapport à $\rho(\Delta)$. Comme on le vérifie immédiatement à l'aide de la classification [11: p. 250–275], les relations $m_a(d) = 2$, $m_{\rho(a)}(d') = 1$ (et le fait que ρ est la bijection canonique de Φ sur le système de racines inverse) impliquent qu'on se trouve dans un des deux cas mentionnés à la fin de l'énoncé. Réciproquement, dans ces cas-là, on a effectivement $m_{\rho(a)}(d') = 1$, donc $\beta(U)$, et par conséquent aussi U, est commutatif.

4.10. Dans ce n°, nous abandonnons l'hypothèse selon laquelle G est absolument presque simple.

COROLLAIRE. *Soient S un k-tore déployé maximal de G et $c \in \Phi(S, G)$ une racine multipliable, i.e. telle que $2c \in \Phi(S, G)$. Alors, U_{2c} est le centre de $U_{(c)}$.*

Le groupe $G_1 = G^{*}_{\{\pm c, \pm 2c\}}$ engendré par $U_{(c)}$ et $U_{(-c)}$ est semi-simple [4: 3.8 (ii), 3.13] et de rang relatif 1, puisque $\Phi(S, G_1) = \{\pm c, \pm 2c\}$. Vu [4: 5.11], il est presque simple sur k. Quitte à remplacer G par G_1, nous pouvons donc supposer G presque simple sur k et de rang relatif 1.

À une isogénie centrale près, G est à présent le groupe obtenu par restriction des scalaires à partir d'un groupe absolument presque simple de rang relatif 1 sur une extension séparable de k [4: 6.21 (ii)]. Utilisant ce fait et [4: 6.19, 6.21 (i)], une réduction facile, dont nous laissons le détail au lecteur, permet de se ramener au cas où G est absolument presque simple, ce que nous supposerons désormais.

Supposons en outre, ce qui est loisible, que le tore maximal T contient S et que Δ est l'ensemble des racines simples pour un ordre dans $X^*(T)$ compatible avec l'ordre sur $X^*(S)$ pour lequel c est positive. Posons $\theta = \{b \in \Delta \mid b_{|S} = c\}$. Si Card $\theta \neq 1$, le diagramme de Dynkin de G possède un automorphisme non trivial [4: 6.4]; vu la classification, on en déduit que toutes les racines de G ont même longueur, donc (4.3) qu'il n'y a pas de commutation irrégulière, et notre assertion résulte de 4.7 (iii). Si $\theta = \{a\}$, les tables de [32] montrent que a est une racine courte et qu'on ne se trouve pas dans un des cas d'exception du corollaire 4.9; ce dernier implique alors l'égalité à établir. Le corollaire est ainsi démontré.

4.11. *Remarque.* Reprenons les notations et les hypothèses de la proposition 4.7 et supposons T et $P_{\Delta-\theta}$ définis sur k. Alors, les groupes U, U_i, $U_{lg,i}$ sont définis sur k, parce que k-fermés et définis sur k_s, Si k est infini, un argument de densité montre que toutes les assertions de la proposition restent valables lorsqu'on remplace U, U_i, $U_{lg,i}$ par leurs groupes de points rationnels sur k. En fait, on peut montrer que cela est encore vrai pour k fini; dans le cas particulier où G est déployé, il suffit d'ailleurs de remarquer que la démonstration donnée ici dans le cas "absolu" reste valable, *mutatis mutandis.*

5. Propriétés de certains sous-groupes paraboliques à radicaux unipotents commutatifs

5.1. PROPOSITION. *Supposons G absolument presque simple. Soient P^+, P^- des sous-groupes paraboliques opposés de G et $U^{\pm} = R_u(P^{\pm})$, $S = \mathcal{C}(P^+ \cap P^-)^{\circ}$. Supposons que U^+ soit commutatif et que P^+ et P^- soient conjugués. Alors, $\mathcal{Z}(S) = P^+ \cap P^-$ est d'indice deux dans $\mathfrak{N}(S)$. Posons $M = \mathfrak{N}(S) - \mathcal{Z}(S)$ et $\Omega = U^+ \cap U^- M U^-$.*

(i) Ω est un ouvert non vide de U^+.

(ii) Pour $u \in \Omega$, les éléments $m \in M$ et $u', u'' \in U^-$ tels que $u = u'mu''$ sont uniques et on a $u' = u'' = mu^{-1}m^{-1}$. Posons $m = \mu(u)$ et $u' = \nu(u)$.

(iii) Pour $s \in S$ et $u \in \Omega$, on a $^s u \in \Omega$ et $\mu(^s u) = s^2 \cdot \mu(u)$.

(iv) Soient $u \in \Omega$ et $X = {}^s u \cup \{e\}$. Alors $L = X \cdot S \cup X \cdot S \cdot \mu(u) \cdot X$ est un groupe isomorphe à $\mathbf{SL}_2$ ou à $\mathbf{PSL}_2$.

(v) Soient $u, u' \in \Omega$. Supposons que $\mu(u') \subset S \cdot \mathcal{C}(G) \cdot \mu(u)$. Alors $u' \in {}^s u$.

Soit T un tore maximal de G contenu dans $P^+ \cap P^-$. Nous reprenons les notations de 4.1 en supposant la base Δ choisie de telle façon que P^+ soit un sous-groupe parabolique standard [4:4.2]. Quitte à transformer éventuellement G par une isogénie spéciale radicielle, nous pouvons supposer que $P^+ = P_{\Delta - \{a\}}$ où a est telle que

$$(1) \qquad m_a(d) = 1$$

(cf. 4.8). Les groupes P^+ et P^- étant conjugués, il résulte de [4: 4.9] que

$$(2) \qquad a \text{ est invariante par l'involution d'opposition.}$$

Le groupe S est la composante neutre de l'intersection des noyaux des éléments de $\Delta - \{a\}$; c'est donc un tore à une dimension. En particulier, $[\mathfrak{N}(S): \mathcal{Z}(S)] \leqq 2$. Mais P^+ et P^- sont conjugués et contiennent T; il existe donc $g \in G$ tel que $^g P^- = P^+$ et $^g T = T$, d'où, vu [4: 3.13], $^g \mathcal{Z}(S) = \mathcal{Z}(S)$ et $^g S = S$ (car S est le centre connexe de $\mathcal{Z}(S)$). Comme $g \notin \mathcal{Z}(S)$, ceci montre que $[\mathfrak{N}(S): \mathcal{Z}(S)] = 2$ et que, pour $m \in M = \mathfrak{N}(S) - \mathcal{Z}(S)$, on a

$$^m P^- = P^+$$

d'où aussi

$$(3) \qquad {}^m P^+ = P^- \text{ et } {}^m U^{\pm} = U^{\mp} .$$

Notons encore que si a' est la restriction de a à S, on a $U^{\pm} = U_{\pm a'}^{(S)}$ (1.11); par conséquent [4: 3.17] U^+ (resp. U^-) a une structure naturelle d'espace vectoriel sur lequel tout $s \in S$ opère par l'homothétie de rapport $a'(s)$ (resp.

$a'(s)^{-1})$ et $\mathcal{Z}(S)$ opère linéairement. Ces remarques préliminaires étant faites, nous passons à la démonstration des assertions (i) à (v).

Vu [4: 4.10], l'ensemble $\mathcal{Z}(S)\cdot U^+\cdot U^-$ est un ouvert dense de G. Il en est donc de même de $M\cdot U^+\cdot U^- = U^-\cdot M\cdot U^-$. Par conséquent, Ω est ouvert dans U^+, et il n'est pas vide car l'ensemble

$$(M\cdot U^+\cdot U^-) \cap \big(\mathcal{Z}(S)\cdot U^+\cdot U^-\big)$$

est un ouvert non vide, réunion de doubles classes modulo $\mathcal{Z}(S)$ et U^-, dont chacune rencontre $M\cdot U^+\cdot U^- \cap U^+ = \Omega$, d'où (i).

Comme le tore S a une dimension, ses seuls automorphismes sont l'identité et $x \mapsto x^{-1}$. Pour $m \in M$ et $s \in S$, on a donc $m^{-1}sm = s^{-1}$, d'où

$$(4) \qquad\qquad {}^s m = s^2\cdot m \ .$$

Soient $u \in U^+$, u', u'_1, u'', $u''_1 \in U^-$ et $m, m_1 \in M$ tels que $u = u'mu'' = u'_1 m_1 u''_1$. On a

$$u'^{-1}_1 u' = m_1 u''_1 u''^{-1} m_1^{-1}\cdot m_1 m^{-1}$$

d'où, vu (3),

$$u'^{-1}_1 u' \in U^- \cap U^+\cdot \mathcal{Z}(S) = \{e\}$$

et $u'_1 = u'$. Comme $u''^{-1} m^{-1} u'^{-1} = u''^{-1}_1 m_1^{-1} u'^{-1}$, on voit de même que $u''_1 = u''$, d'où l'unicité de u', m, u''.

Si h est un élément de S tel que $a'(h) = -1$, on a, compte tenu de (4),

$$u'mu'' = u = {}^h u^{-1} = {}^h u''^{-1}\cdot {}^h m^{-1}\cdot {}^h u'^{-1} = u''\cdot h^2\cdot m^{-1}\cdot u' \ ,$$

d'où, vu l'unicité de la décomposition $u = u'mu''$, déjà établie,

$$(5) \qquad\qquad u' = u''$$

et

$$(6) \qquad\qquad m^2 = h^2 \in S \ .$$

Appliquant le résultat (5) à la décomposition $u''^{-1} = u^{-1}\cdot m\cdot m^{-1} u'm$, on voit encore que

$$(7) \qquad\qquad u' = mu^{-1}m^{-1} \ ,$$

ce qui achève la démonstration de (ii).

Vu (4), on a $\mu({}^s u) = {}^s\mu(u) = s^2\cdot\mu(u)$, c'est-à-dire (iii).

Soit $u = u'mu'$ comme ci-dessus. Posons $Y = {}^s u$, $Y' = {}^s u'$, $X = Y \cup \{e\}$ et $L = X\cdot S \cup X\cdot m\cdot S\cdot X = \{e, m\}\cdot X\cdot S \cup Y\cdot m\cdot S\cdot X$. Comme X est un sous-espace vectoriel de U^+ normalisé par S, on a

$$(8) \qquad\qquad S\cdot X\cdot L = S\cdot L = L \ .$$

D'autre part, il résulte de (4) que

$$Y \subset {}^{s}u' \cdot {}^{s}m \cdot {}^{s}u' = Y' \cdot S \cdot m \cdot Y' = Y' \cdot m \cdot S \cdot Y'$$

et de (7) et (6) que

$$Y' = m^{-1}Ym = mYm^{-1},$$

d'où

$$m \cdot Y \cdot m \cdot S \cdot X \subset m \cdot Y' \cdot m \cdot S \cdot Y' \cdot m \cdot S \cdot X = Y \cdot S \cdot Y' \cdot m \cdot S \cdot X$$
$$= Y \cdot Y' \cdot m \cdot S \cdot X = Y \cdot m \cdot Y \cdot S \cdot X = Y \cdot m \cdot S \cdot X.$$

Par conséquent,

$$(9) \qquad\qquad m \cdot L \subset L.$$

Comme $L^{-1} = L$, compte tenu de (6), il résulte de (8) et (9) que L est le sous-groupe de G engendré par $X \cdot S$ et m. Ce groupe est connexe, de dimension 3, et est son propre groupe dérivé, car (L, L) contient $(S, m) = S$ et $(S, X) = X$, donc ${}^{m}Y \cdot Y \cdot {}^{m}Y = Y' \cdot Y \cdot Y'$ donc aussi $m = u'^{-1} \cdot u \cdot u'^{-1}$. Par conséquent L est isomorphe à $\mathbf{SL}_2$ ou à $\mathbf{PSL}_2$, ce qui prouve (iv).

Il reste à établir l'assertion (v). Celle-ci étant invariante par isogénie centrale, nous pouvons supposer G simplement connexe. Soient $\rho\colon G \to GL(V)$ une représentation irréductible et $\bar{\omega}$ le poids dominant de ρ. Pour $v \in V$ et $g \in G$, nous posons $gv = \rho(g)(v)$. Tout poids de T dans V s'obtient en soustrayant de $\bar{\omega}$ une combinaison linéaire à coefficients entiers positifs d'éléments de Δ, donc les poids de S dans V sont de la forme $\bar{\omega}|_{s} - q \cdot a'$ ($q \in \mathbf{N}$, $a' = a|_{s}$). Pour $z \in \mathbf{Z}$, notons V_z l'espace propre de S dans V correspondant au poids $(1/2) za'$. Pour $x \in U^{+}$ (resp. U^{-}) et $v \in V_z$, on a

$$(10) \qquad\qquad xv - v \in \sum_{n \in \mathbf{N}^*} V_{z+2n} \quad (\text{resp.} \ \sum_{n \in \mathbf{N}^*} V_{z-2n}).$$

En effet, on sait que si $x \in U_b$ avec $b \in \Phi(T, G)$ et si v est un vecteur propre de T de poids λ, alors $xv - v$ est une somme de vecteurs propres de poids $\lambda + nb$ ($n \in \mathbf{N}^*$) [2: 5.2], et (10) s'ensuit par une réduction immédiate.

Comme les éléments de M transforment a en $-a$, on a $mV_z = V_{-z}$ pour tout $m \in M$.

Soient $u, u' \in \Omega$ tels que $\mu(u') \in S \cdot \mathcal{C}(G) \cdot \mu(u)$. Nous devons montrer que $u' \in {}^{s}u$. Les restrictions de $\rho(\mu(u))$ et $\rho(\mu(u'))$ à V_{-1} ne diffèrent que par une homothétie. Quitte à remplacer u par ${}^{s}u$, pour $s \in S$ convenablement choisi (ce qui a pour effet de remplacer $\mu(u)$ par $s^2 \cdot \mu(u)$, vu (iii), déjà établi), nous pouvons donc supposer que

$$\rho\big(\mu(u)\big)|_{V_{-1}} = \rho\big(\mu(u')\big)|_{V_{-1}},$$

et nous montrerons alors que $u = u'$. Posons $X_1 = \{v \in V_1 \,|\, uv = u'v = v\}$ et $X_{-1} = \{v \in V_{-1} \,|\, \nu(u)v = \nu(u')v = v\}$. Pour $v \in X_{-1}$, on a, vu (10),

$$uv = \nu(u)\mu(u)\nu(u)v = \nu(u)\mu(u)v \in \mu(u)v + \sum_{n \leq -1} V_n$$

et, d'autre part,

$$uv \in v + \sum_{n \geq 1} V_n \, ,$$

d'où

$$uv = v + \mu(u)v \, .$$

De même,

$$u'v = v + \mu(u')v = v + \mu(u)v \, .$$

On voit donc que $uv = u'v$ pour $v \in X_{-1}$; par suite

(11) $X_1 + X_{-1}$ *est contenu dans l'espace des points fixes de* $u^{-1}u'$ *dans* V.

La relation $mV_z = V_{-z}$ $(m \in M)$ montre que l'ensemble des poids de S dans V est symétrique. Par conséquent, si c désigne le coefficient de a dans $\bar{\omega}$ exprimé comme combinaison linéaire de racines simples, $2c$ est entier et les poids de S dans V sont de la forme qa' avec $q \equiv c(\mathrm{mod}\,1)$ et $-c \leq q \leq c$.

S'il est possible de choisir $\bar{\omega}$ de telle façon que c soit égal à $1/2$, on a alors $V = V_{-1} + V_1$. Vu (10), ceci implique que $X_{\pm 1} = V_{\pm 1}$, et (11) signifie que $u = u'$, ce qui achève le démonstration.

Il nous reste à considérer le cas où, pour tout poids dominant non nul $\bar{\omega}$, le coefficient c de a dans $\bar{\omega}$ est ≥ 1, ce qui, compte tenu de (1), (2) et de la classification [11: p. 250–275], implique que G est de type $\mathbf{E}_7$ et que $a = a_1$ (avec la numérotation des planches citées). Choisissons pour $\bar{\omega}$ le poids fondamental dominant correspondant à a_1.

On sait [14: Exp. 20, n° 2], que l'ensemble Π des poids de ρ est l'orbite $W\bar{\omega}$ de $\bar{\omega}$ sous l'action du groupe de Weyl, tous ces poids étant simples. Le poids dominant $\bar{\omega}$ étant orthogonal à tous les éléments de $\Delta - \{a_1\} = \{a_2, \cdots, a_7\}$, tout élément de $\Pi - \{\bar{\omega}\}$ est de la forme $\bar{\omega} - a_1 - \sum_{1 \leq i \leq 7} c_i a_i$ avec $c_i \geq 0$ ([4: 12.16], [2: 5.4]); le coefficient de a_1 dans $\bar{\omega}$ étant $3/2$, on voit donc que $\dim V_3 = \dim V_{-3} = 1$. Vu (10), on en déduit aussitôt que $\dim X_1 \geq \dim V_1 - 2$ et $\dim X_{-1} \geq \dim V_{-1} - 2$.

Pour $\bar{\omega}' \in \Pi$, notons $V_{\bar{\omega}'}$ l'espace propre correspondant. Montrons que

(12)   si $b \in \Phi$, $x \in U_b - \{e\}$, $\bar{\omega}' \in \Pi$, $(b, \bar{\omega}') < 0$ et $v \in V_{\bar{\omega}'} - \{0\}$,

    alors $xv - v$ est un élément non nul de $V_{\bar{\omega}'+b}$.

Comme W est transitif sur Π, on peut supposer $\bar{\omega}' = \bar{\omega}$. Nous avons déjà rappelé que $xv - v \in \sum_{n \in \mathbf{N}^*} V_{\bar{\omega}+nb}$. Mais comme tous les poids de ρ sont frontière, Π ne peut contenir à la fois $\bar{\omega}$, $\bar{\omega} + b$ et $\bar{\omega} + 2b$. Donc $xv - v \in V_{\bar{\omega}+b}$. De plus, $xv - v \neq 0$ sinon $V_{\bar{\omega}}$ serait stable par le groupe engendré par x et $P_{\Delta-\{a\}}$, c'est-à-dire par G, puisque $P_{\Delta-\{a\}}$ est un sous-groupe propre maximal de G. L'assertion (12) est ainsi démontrée.

On a Card $\Phi = 126$ et Card $\Pi = 56$. Les racines orthogonales à $\bar{\omega}$, qui forment un système de type E_6, sont au nombre de 72, donc il y a $(126{-}72)/2 = 27$ racines b telles que $(b, \bar{\omega}) < 0$. Compte tenu de la transitivité de W sur Φ et Π, on en déduit que, pour $b \in \Phi$, il existe $(56 \cdot 27)/126 = 12$ éléments $\bar{\omega}'$ de Π tels que $(b, \bar{\omega}') < 0$. Vu (12), cela implique que si $b \in \Phi$ et $x \in U_b - \{e\}$, le rang de $\rho(x) - $ Id. est égal à 12.

Supposons $u^{-1}u' \neq e$. Dans l'espace vectoriel U^+, l'adhérence Y de $^T(u^{-1}u')$ est une variété invariante par T, donc un cône (car $S \subset T$), dans l'image projective duquel T possède un point fixe [1: 10.4]. Par conséquent, il existe $b \in \Phi$ tel que $U_b \subset Y$. D'autre part, vu (11), le rang de $\rho(u^{-1}u') - $ Id. est inférieur à 5. Il en est donc de même du rang de $\rho(y) - $ Id. pour tout $y \in Y$, et en particulier pour $y \in U_b - \{e\}$, et cela contredit ce qu'on vient de voir. Ainsi, $u = u'$, et la proposition est démontrée.

5.2. *Remarques.* Dans les remarques (a) et (b) ci-après, on suppose que les hypothèses de la proposition 5.1 sont remplies et que $P = P_{\Delta - \{a\}}$ où a est une racine longue ayant le coefficient un dans la racine dominante (cf. le début de la démonstration de 5.1).

(a) Dans la démonstration de 5.1 (v), on a vu qu'il existe toujours un poids dominant $\bar{\omega}$ de G tel que $\bar{\omega}|_S$ soit un multiple impair de $a'/2$. On en déduit aussitôt que

> *si G est simplement connexe, le groupe L de 5.1 (v) est*
> *isomorphe à* $\mathbf{SL}_2$.

(b) Sous les hypothèses faites, U possède une structure naturelle d'algèbre de Jordan, au sens de K. McCrimmon [20] et T. A. Springer [26], et on obtient ainsi toutes les algèbres de Jordan simples de dimension finie (cf. [26]); la proposition 5.1 exprime donc en fait des propriétés de ces algèbres, que nous laissons au lecteur le soin de formuler s'il en a le goût.

5.3. CorOLLAIRE. *On conserve les notations de* 5.1. *Soient $u \in \Omega$ et $s \in S$. Posons $Q = U^+ \cap (U^- \cdot s^2 \cdot \mu(u) \cdot U^-)$. Alors,*

$$^s u \in Q \subset \{^s u, \, ^s u^{-1}\} \, ,$$

et si $\operatorname{car} k = 2$, $Q = \{^s u\}$.

Soit L défini comme en 5.1 (iv). On a alors

$$(1) \qquad Q = (U^+ \cap L) \cap \big((U^- \cap L) \cdot s^2 \cdot \mu(u) \cdot (U^- \cap L)\big) \, .$$

En effet, si u', $u'' \in U^-$ et $u_1 \in U^+$ sont tels que $u' \cdot s^2 \cdot \mu(u) \cdot u'' = u_1$, il résulte de 5.1 (v) que $u_1 \in L$, puis de 5.1 (ii) que $u' = u'' \in L$. Vu (1), nous pouvons remplacer G par L, donc supposer qu'il existe une isogénie centrale $\beta\colon \mathbf{SL}_2 \to$ G telle qu'on ait

$$(2) \qquad U^+ = \{u(t)\,|\,t \in K\}\,, \quad S = \{s(t)\,|\,t \in K^*\}\,, \quad u = u(1)\,,$$

en posant

$$(3) \qquad u(t) = \beta\begin{pmatrix} 1 & t \\ 0 & 1 \end{pmatrix} \quad \text{et} \quad s(t) = \beta\begin{pmatrix} t & 0 \\ 0 & t^{-1} \end{pmatrix}.$$

De l'identité

$$(4) \qquad \begin{pmatrix} 1 & t \\ 0 & 1 \end{pmatrix} = \begin{pmatrix} 1 & 0 \\ -t^{-1} & 1 \end{pmatrix}\cdot\begin{pmatrix} 0 & t \\ -t^{-1} & 0 \end{pmatrix}\cdot\begin{pmatrix} 1 & 0 \\ -t^{-1} & 1 \end{pmatrix} \qquad (t \in K^*)\,,$$

il résulte que, pour $t \in K^*$, on a

$$(5) \qquad \mu\big(u(t)\big) = \beta\begin{pmatrix} 0 & t \\ -t^{-1} & 0 \end{pmatrix} = s(t)\cdot\beta\begin{pmatrix} 0 & 1 \\ -1 & 0 \end{pmatrix} = s(t)\cdot\mu(u)\,.$$

Soit $t_0 \in K^*$ tel que $s = s(t_0)$. Pour $t \in K^*$, on a $^{s(t)}u = u(t^2) \in Q$ si et seulement si $\mu\big(u(t^2)\big) = s^2\cdot\mu(u)$, c'est-à-dire si $s(t)^2 = s(t_0)^2$, ce qui signifie que $t^2 = t_0^2$ ou que $t^2 = \pm\, t_0^2$ selon que $G = L \cong \mathbf{SL}_2$ ou $\mathbf{PSL}_2$, d'où le corollaire.

5.4. Lemme. *Soient P, P^- des k-sous-groupes paraboliques opposés de G, $Z = P \cap P^-$ leur intersection, V, V^- des sous-groupes fermés connexes centraux de $R_u(P)$ et $R_u(P^-)$ respectivement, S le centre connexe de Z et n un élément de $V^-(k)\cdot V(k)\cdot V^-(k)$. On suppose que $\operatorname{car} k = 2$, que $^nP = P^-$, $^nP^- = P$, $^nV = V^-$, $^nV^- = V$, et que l'ensemble VZV^- est dense dans le groupe H qu'il engendre. Alors*

(i) *$V(k) \cap \big(V^-(k)\cdot n\cdot Z(k)\cdot V^-(k)\big)$ est un ouvert dense de $V(k)$;*

(ii) *si $u \in V$ et $n_1 \in nZ$ sont tels que $u \in V^-n_1V^-$, on a, pour tout $s \in S$, $(V^-\cdot s^2\cdot n_1\cdot V^-) \cap V = \{^su\}$.*

Comme l'application produit $R_u(P) \times Z \times R_u(P^-) \to G$ est une k-immersion, il en est de même de $V \times Z \times V^- \to G$. Par conséquent, $V(k)\cdot Z(k)\cdot V^-(k)$ est un ouvert de $H(k)$, et il en est de même de $n\cdot V(k)\cdot Z(k)\cdot V^-(k) = V^-(k)\cdot n\cdot Z(k)\cdot V^-(k)$, d'où (i).

Démontrons (ii) par induction sur la dimension de G. Nous supposerons dorénavant G semi-simple et k algébriquement clos, ce qui est manifestement loisible. Pour établir l'inclusion $\{^su\} \subset (V^-\cdot s^2\cdot n_1\cdot V^-) \cap V$, on peut remplacer G par un revêtement simplement connexe et V, V^- par les composantes neutres de leurs images réciproques dans ce revêtement. De même, pour établir l'inclusion inverse, on peut remplacer G par son groupe adjoint et V, V^- par leurs images dans ce groupe. On voit donc qu'il suffit d'établir (ii) dans l'hypothèse où G est produit direct de ses facteurs presque simples et comme l'assertion à démontrer se décompose alors en les assertions analogues pour ces facteurs, nous pouvons dorénavant supposer G presque simple, vu l'hypothèse d'induction.

Soit T un tore maximal de Z. Si les racines de G n'ont pas toutes la même longueur, il existe $n' \in \mathfrak{N}(T)$ transformant chaque racine de G relative à T en son opposée. On a alors $^{n'}P = P^-$, $^{n'}P^- = P$, d'où $n'n \in P \cap P^- = Z$ et $^{n'}V = ^{n'n}V^- = V^-$, $^{n'}V^- = ^{n'n}V = V$. Le groupe H est donc normalisé par n', c'est-à-dire que $\Phi(T, H) = -\Phi(T, H)$ et H est réductif. Vu l'hypothèse d'induction, nous pouvons alors supposer que $G = H$.

Si, par contre, les racines de G ont toutes même longueur, G n'a pas de commutation irrégulière (4.3) et, comme $V \subset \mathcal{C}R_uP$, il résulte de 4.7 (ii) que S n'a qu'un seul poids dans V, soit d; toujours en vertu de l'hypothèse d'induction, nous pouvons donc supposer que $G = G_{\{\pm d\}}^{(S)}$ (cf. 1.11).

Dans l'un et l'autre cas, $R_u(P)$ est à présent commutatif, et (ii) résulte du corollaire 5.3.

6. Le groupe G^+

6.1. Si H est un k-groupe connexe, on note $H(k)^+$ ou H^+ le sous-groupe distingué de $H(k)$ engendré par les $U(k)$, où U parcourt l'ensemble des k-sous-groupes unipotents de H qui sont déployés sur k. L'image d'un groupe unipotent déployé par un k-morphisme est déployée [24: Prop. 6]; par conséquent, si $f: H \to H'$ est un k-morphisme de k-groupes connexes, on a $f(H^+) \subset H'^+$. De même, si $\varphi: k \to k'$ est un homomorphisme de corps, on a $\varphi^0(H^+) \subset (^\varphi H)^+$. D'autre part, si k est une extension séparable de degré fini d'un sous-corps k'', alors

$$R_{k/k'}^0\big(H(k)^+\big) = (R_{k/k'}H)(k'')^+ .$$

L'inclusion du premier membre dans le second résulte de ce que la restriction des scalaires transforme un groupe unipotent déployé en un groupe unipotent déployé. Pour établir l'inclusion inverse, il suffit de remarquer que si U'' est un k''-sous-groupe unipotent déployé de $R_{k/k'}H$, sa projection canonique U dans H est un k-groupe unipotent déployé et $U'' \subset R_{k/k'}U$.

6.2. PROPOSITION. *Soient G_i $(1 \leq i \leq r)$ les k-sous-groupes distingués connexes presque simples et isotropes sur k de G, et P, P^- deux k-sous-groupes paraboliques opposés de G ne contenant aucun des G_i.*

(i) G^+ est engendré par les éléments unipotents de $G(k)$ qui appartiennent à des k_s-sous-groupes de Borel de G. Si k est parfait, G^+ est engendré par les éléments unipotents de $G(k)$.

(ii) G^+ est engendré par les $R_u(Q)(k)$ où Q parcourt l'ensemble des k-sous-

groupes paraboliques minimaux de G.

(iii) G^+ *est le produit des G_i^+* $(1 \leq i \leq r)$.

(iv) G^+ *est engendré par les $R_u(Q)(k)$ où Q parcourt l'ensemble des k-sous-groupes paraboliques de G conjugués à P.*

(v) G^+ *est engendré par $R_u(P)(k)$ et $R_u(P^-)(k)$.*

(vi) G^+ *est engendré par les $X(k)$, où X parcourt l'ensemble des k-sous-groupes de G isomorphes sur k au groupe additif, ou bien $k \cong \mathbf{F}_2$ et l'un des G_i est non-déployé de type $\mathbf{A}_2$.*

Soient M_1 le groupe engendré par les $R_u(Q)(k)$, où Q parcourt l'ensemble' des k-sous-groupes paraboliques minimaux de G, et M_2 le groupe engendré par les éléments unipotents de $G(k)$ qui appartiennent à un k_s-sous-groupe de Borel. Les inclusions $M_2 \subset M_1$ et $G^+ \subset M_1$ résultent respectivement de la proposition 3.6 et du corollaire 3.5 de [6]; les inclusions inverses sont évidentes puisque, pour tout k-sous-groupe parabolique Q, $R_u(Q)$ est déployé sur k et contenu dans un k_s-sous-groupe de Borel de G. La deuxième assertion de (i) étant une conséquence immédiate de la première, vu [6: 3.7], cela démontre (i) et (ii).

On sait [7: 2.22] que si Q parcourt l'ensemble des k-sous-groupes paraboliques minimaux de G, alors $Q \cap G_i$ parcourt l'ensemble des k-sous-groupes paraboliques minimaux de G_i, et que pour un tel Q, $R_u(Q)$ est le produit direct (sur k) des $R_u(Q \cap G_i)$. Vu (ii), G^+ est donc engendré par les G_i^+, et (iii) en résulte puisque les G_i^+ sont distingués dans G^+.

Pour la démonstration des assertions restantes, nous pouvons, vu (iii), supposer que G est presque simple et isotrope sur k (on utilise à nouveau [7: 2.22] et, en ce qui concerne (vi), on note qu'il suffit de montrer que G^+ est contenu dans le groupe engendré par les $X(k)$, l'inclusion inverse résultant de [6: 3.5]).

Le groupe M_3 engendré par les $R_u(Q)(k)$, où Q parcourt l'ensemble des k-sous-groupes conjugués à P, est contenu dans G^+, distingué dans $G(k)$ et non-contenu dans le centre de G. Si on ne se trouve pas dans l'un des cas d'exception du théorème principal de [31], celui-ci permet d'en déduire que $G^+ = M_3$. Dans les cas d'exception en question, G est soit de rang relatif 1, soit déployé sur k et de rang 2. S'il est de k-rang 1, le k-sous-groupe parabolique propre P est minimal et (iv) résulte de (ii). Supposons donc G déployé et de rang 2, et soit T un k-tore maximal k-déployé contenu dans P. On vérifie aisément que, pour toute longueur de racine, il existe une racine $a \in \Phi(T, G)$ ayant cette longueur et telle que $U_a \subset R_u(P)$, d'où $U_a(k) \subset M_3$. Comme M_3 est distingué dans $G(k)$, on en déduit aussitôt que $R_u(B)(k) \subset M_3$ pour tout

k-sous-groupe de Borel B de G, donc, vu (ii), que $G^+ \subset M_3$, ce qui achève la démonstration de (iv).

Le groupe M_4 engendré par $R_u(P)(k)$ et $R_u(P^-)(k)$ est normalisé par $(P \cap P^-)(k)$, donc aussi par $G(k)$, puisque celui-ci est engendré par $P(k) = (P \cap P^-)(k) \cdot R_u(P)(k)$ et $R_u(P^-)(k)$ (cf. [4: 6.25]). Vu (iv), ceci implique que $G^+ \subset M_4$. L'inclusion inverse étant évidente, (v) s'ensuit.

Pour établir (vi), supposons que le groupe M_5 engendré par les $X(k)$, où X parcourt l'ensemble des k-sous-groupes de G isomorphes sur k au groupe additif, ne contienne pas G^+. Le groupe M_5 est distingué dans $G(k)$ et n'est pas contenu dans le centre de G. Par conséquent, on se trouve dans un des cas d'exception du théorème de [31]. Mais G n'est pas déployé, sinon M_5 contiendrait G^+ en vertu de (ii). Les seuls groupes non-déployés faisant exception au théorème de [31] étant les groupes de type $\mathbf{A}_2$ sur un corps à deux éléments, ceci termine la démonstration de (vi), et de la proposition.

Remarque. Supposons $k = \mathbf{F}_2$ et G non déployé de type $\mathbf{A}_2$. On peut alors montrer que G^+ est le produit semi-direct d'un sous-groupe distingué F d'ordre 27 ou 9 selon que G est simplement connexe ou adjoint et d'un groupe U d'ordre 8, à savoir le groupe des points rationnels du radical unipotent d'un k-sous-groupe de Borel. Le centre $C(U)$ de U est d'ordre 2 et tous les éléments non centraux de U sont d'ordre 4. Il s'ensuit que le groupe H engendré par les points rationnels des k-sous-groupes de G isomorphes au groupe additif est contenu dans $C(U) \cdot F$, puisque $C(U) \cdot F$ contient tous les éléments d'ordre 2 de G^+. (En fait, on a $H = C(U) \cdot F$, d'où $[G^+ : H] = 4$.)

6.3. Corollaire. *Si $f: G \to G'$ est un k-morphisme central de G dans un k-groupe réductif G', dont l'image $f(G)$ contient le groupe dérivé de G', on a $G'^+ = f(G^+)$.*

C'est une conséquence immédiate de 6.2 (ii) et [7: 2.20].

6.4. Corollaire. *Le groupe G^+ est son propre groupe dérivé, sauf si l'une des conditions suivantes est remplie:*

$k = \mathbf{F}_2$ et l'un des G_i est de type $\mathbf{A}_1$, $\mathbf{B}_2$, $\mathbf{G}_2$ ou non-déployé de type $\mathbf{A}_2$;

$k = \mathbf{F}_3$ et l'un de G_i est de type $\mathbf{A}_1$.

Vu 6.2 (iii), on peut supposer G presque-simple et non-anisotrope sur k. On a

$$(1) \qquad\qquad (G^+, G^+) \not\subset C(G) \ .$$

Pour le montrer, il suffit, vu 6.3, de considérer le cas où $C(G) = \{e\}$ et de faire voir que G^+ n'est pas commutatif, ce qui résulte par exemple du fait que le normalisateur de $R_u(P)(k)$ dans $G(k)$, qui n'est autre que $P(k)$ [4: 5.19], ne contient pas $R_u(P^-)(k)$. Si on ne se trouve pas dans un des cas d'exception énumérés, le corollaire résulte du théorème de [31], vu (1). Il reste donc à faire voir que si l'une des conditions de l'énoncé est remplie, on a $(G^+, G^+) \neq G^+$, or ceci est bien connu si G est adjoint (cf. par ex. [30: p. 214]) et le cas général s'en déduit en utilisant 6.3.

6.5. COROLLAIRE. *Supposons G semi-simple et soit $f: \widetilde{G} \to G$ un k-revêtement universel de G. Supposons que $\widetilde{G}^+ = \widetilde{G}(k)$. Alors $G^+ = f\bigl(\widetilde{G}(k)\bigr)$ et G^+ est le groupe dérivé de $G(k)$, sauf peut-être si l'une des conditions du corollaire 6.4 est remplie.*

La première assertion est un cas particulier de 6.3 et la seconde résulte aussitôt de 6.4 et du fait que $f\bigl(\widetilde{G}(k)\bigr)$ est un sous-groupe distingué de $G(k)$ et que le quotient $G(k)/f\bigl(\widetilde{G}(k)\bigr)$ est commutatif [4: 2.7].

6.6. *Remarque.* Il est conjecturé que $\widetilde{G}(k) = \widetilde{G}^+$ dès que G n'a pas de k-sous-groupe distingué anisotrope non trivial. Cela est connu lorsque G est déployé ou quasi-déployé (cf. par ex. [29: lemme 64, p. 183]).

6.7. COROLLAIRE. *Supposons k infini. Le groupe G^+ ne possède pas de sous-groupe propre d'indice fini. Si H est un sous-groupe de $G(k)$ et F une partie finie de $G(k)$ tels que $G^+ \subset H \cdot F$, alors $G^+ \subset H$.*

Pour établir la première assertion, on peut, vu 6.3, supposer G semi-simple et simplement connexe. Le groupe G^+ est alors le produit direct des G_i^+ et aucun de ceux-ci ne possède de sous-groupe propre d'indice fini, vu le théorème de [31].

La seconde assertion est une conséquence immédiate de la première. En effet, quitte à remplacer F par $(H \cdot G^+) \cap F$ puis à multiplier les éléments de F à gauche par des éléments appropriés de H, on peut supposer que $F \subset G^+$; on a alors

$$G^+ \subset (H \cdot F) \cap G^+ = (H \cap G^+) \cdot F$$

ce qui implique que $H \cap G^+$ est un sous-groupe d'indice fini de G^+.

6.8. COROLLAIRE. *Supposons G semi-simple et soient $f: \widetilde{G} \to G$ un k-revêtement universel de G, S un k-tore déployé de G et S' la composante neutre de $f^{-1}(S)$. Alors, $f(S'(k)) \subset G^+ \cap S$. Si m désigne l'exposant du conoyau de l'homomorphisme $(f|_{S'})^*: X^*(S) \to X^*(S')$ (exposant qui divise le degré de l'isogénie $S' \to S$, donc aussi le degré de f), G^+ contient x^m pour tout $x \in S(k)$. Si k est infini, $S \cap G^+$ est dense dans S.*

Remarquons tout d'abord que S' est un k-tore déployé de $\widetilde{G}$ (cf. [7: 2.17]). Cela étant, les deux dernières assertions sont des conséquences immédiates de la première. Pour établir celle-ci, il suffit, vu 6.1, de montrer que $S'(k) \subset \widetilde{G}^+$;

or cela résulte de 6.5, 6.6 et du fait que S' est contenu dans un sous-groupe déployé simplement connexe de G (cf. [7: 4.6, Cor.]).

Remarque. Il n'est pas vrai que, sous les hypothèses de 6.8, on ait toujours $f\big(S'(k)\big) = G^+ \cap S$.

6.9. Corollaire. *Supposons k infini. Alors l'adhérence $\mathcal{C}(G^+)$ de G^+ est le produit des G_i.*

Vu 6.2 (iii), on peut supposer G presque simple sur k et de k-rang > 0. De 6.2 (v), il résulte que $\mathcal{C}(G^+)$ est engendré par $R_u(P)$ et $R_u(P^-)$. Comme P ne contient manifestement aucun sous-groupe distingué presque simple de G, on déduit à nouveau de 6.2 (v), appliqué cette fois en remplaçant k par K, que $\mathcal{C}(G^+) = G(K)^+ = G$.

6.10. Lemme. *Soit H un sous-groupe fermé de G.*

(i) $\mathfrak{N}(H) \cap G(k) \subset \mathfrak{N}(H \cap G^+) \cap G(k)$.

(ii) *Si $H \cap G^+$ est dense dans H, alors $\mathfrak{N}(H) \cap G(k) = \mathfrak{N}(H \cap G^+) \cap G(k)$.*

L'assertion (i) est immédiate, puisque G^+ est distingué dans $G(k)$. D'autre part, tout élément de $\mathfrak{N}(H \cap G^+)$ normalise l'adhérence de $H \cap G^+$, d'où (ii).

6.11. Proposition. *Soient P, P^- deux k-sous-groupes paraboliques opposés et S un k-tore déployé de G.*

(i) *On a $G(k) = G^+ \cdot \mathcal{Z}(S)(k)$.*

(ii) *Il existe une partie finie C de G^+ telle que $G = C \cdot R_u(P^-) \cdot P$. On a alors $G(k) = C \cdot R_u(P^-)(k) \cdot P(k)$.*

(iii) *Si G^+ est dense dans G, $\mathcal{Z}(S) \cap G^+$ est dense dans $\mathcal{Z}(S)$.*

Nous supposerons, ce qui est évidemment loisible, que $\mathcal{Z}(S) = P \cap P^-$, d'où $P = \mathcal{Z}(S) \cdot R_u(P)$ et $P^- = \mathcal{Z}(S) \cdot R_u(P^-)$.

(i) Le groupe $G^+ \cdot \mathcal{Z}(S)(k)$ est normalisé par

$$\langle \mathcal{Z}(S)(k),\ R_u(P)(k),\ R_u(P^-)(k) \rangle = \langle P(k),\ P^-(k) \rangle = G(k)\ .$$

Comme il contient $P(k)$, cela implique qu'il coïncide avec $G(k)$ [4: 5.18].

(ii) Posons $\Omega = R_u(P^-) \cdot P$. C'est un ouvert de G. De [4: 6.25], il résulte que

$$G^+ \cdot \Omega \supset R_u(P)(k) \cdot \Omega = G\ ,$$

d'où la première assertion de (ii), puisque G est quasi-compact. La deuxième s'ensuit aussitôt, compte tenu de ce que l'application produit $R_u(P^-) \times P \to G$ est une k-immersion.

(iii) Comme G^+ contient $R_u(P)(k)$ et $R_u(P^-)(k)$, on a

$$G^+ \cap \Omega = R_u(P^-)(k) \cdot \big(G^+ \cap \mathcal{Z}(S)\big) \cdot R_u(P)(k)\ .$$

Si G^+ est dense dans G, $G^+ \cap \Omega$ est dense dans Ω, donc $G^+ \cap \mathfrak{Z}(S)(k)$ est dense dans $\mathfrak{Z}(S)$, puisque l'application produit $R_u(P^-) \times \mathfrak{Z}(S) \times R_u(P) \to \Omega$ est un isomorphisme de variétés algébriques. La proposition est ainsi démontrée.

6.12. COROLLAIRE. *Soient P' un k-sous-groupe parabolique minimal et S' un k-tore déployé maximal de G. Alors $R_u(P)$ est conjugué par un élément de G^+ à un sous-groupe de $R_u(P')$ et S est conjugué par un élément de G^+ à un sous-groupe de S'.*

Comme précédemment, nous supposons $P = \mathfrak{Z}(S) \cdot R_u(P)$. Vu [4: 4.13] (resp. [4: 4.21]), il existe $g \in G(k)$ tel que $^g R_u(P) \subset R_u(P')$ (resp. $^g S \subset S'$). Quitte à multiplier g par un élément de $\mathfrak{Z}(S)(k)$, on peut le choisir dans G^+ (6.11 (i)), q.e.d.

6.13. COROLLAIRE. *Si k est infini, on a*

$$\mathfrak{N}\big(\mathfrak{Z}(S) \cap G^+\big) \cap G^+ = \mathfrak{N}\big(\mathfrak{Z}(S)\big) \cap G^+ .$$

Quitte à remplacer G par l'adhérence de G^+, on peut supposer G^+ dense dans G. Alors, l'assertion résulte de 6.10 (ii) et 6.11 (iii).

6.14. PROPOSITION. *Supposons G semi-simple et k doté d'une topologie qui en fait un corps topologique localement compact non discret. Munissons $G(k)$ de la topologie localement compacte déduite de celle-là et soit q le degré d'inséparabilité d'un revêtement universel de G. Alors, G^+ est un sous-groupe distingué fermé de $G(k)$. Le quotient $G(k)/G^+$ est compact. Si G n'a pas de k-sous-groupe distingué connexe anisotrope $\neq \{e\}$, alors $G(k)/G^+$ possède un sous-groupe ouvert commutatif dont l'exposant divise q; si de plus $k = \mathbf{R}$, G^+ est la composante neutre topologique de $G(k)$.*

La dernière assertion est une conséquence immédiate des précédentes et du fait que G^+ n'a pas de sous-groupe propre d'indice fini.

Soit $(\widetilde{G}, \pi)$ un k-revêtement universel de G. Soient $\widetilde{G}_1$ (resp. $\widetilde{G}_2$) le produit des facteurs isotropes (resp. anisotropes) sur k de $\widetilde{G}$, et $G_i = \pi(\widetilde{G}_i)$ $(i = 1, 2)$. On a $(\widetilde{G}, \widetilde{G}_1) \subset \widetilde{G}_1$ et l'application de commutation $\widetilde{G} \times \widetilde{G}_1 \to \widetilde{G}_1$ se factorise à travers un k-morphisme $G \times G_1 \to \widetilde{G}_1$ (cf. [7: §2] et 1.5); il s'ensuit que $(G(k), G_1(k)) \subset \pi(\widetilde{G}_1(k))$. En particulier $\pi(\widetilde{G}_1(k))$ est distingué dans $G(k)$. D'après 3.19, $\pi(\widetilde{G}(k)) = \pi(\widetilde{G}_1(k)) \cdot \pi(\widetilde{G}_2(k))$ est distingué et fermé dans $G(k)$ et $G(k)/\pi(\widetilde{G}(k))$ est compact. Comme $\pi \colon \widetilde{G}(k) \to G(k)$ est propre et $\widetilde{G}_2(k)$ compact, on voit que $\pi(\widetilde{G}_1(k))$ est fermé dans $G(k)$ et que $G(k)/\pi(\widetilde{G}_1(k))$ est compact.

Le groupe G^+ est distingué dans $G(k)$ et l'on a $\widetilde{G}^+ = \widetilde{G}_1^+$ et $\pi(\widetilde{G}^+) = G^+$ (6.2, 6.3). Supposons avoir démontré que $\widetilde{G}_1^+$ est ouvert d'indice fini dans

$\widetilde{G}_1(k)$. Alors, vu 3.19, G^+ est ouvert d'indice fini dans $\pi(\widetilde{G}_1(k))$, donc aussi fermé dans $G(k)$ et la projection $G(k)/G^+ \to G(k)/\pi(\widetilde{G}_1(k))$ est un isomorphisme local de noyau fini. Les trois premières assertions de 6.14 résultent alors de 3.19, compte tenu du fait que G_2 est le plus grand k-sous-groupe connexe anisotrope distingué de G [7: 2.21]. Il reste à faire voir que $\widetilde{G}_1^+$ est ouvert d'indice fini dans $\widetilde{G}_1(k)$, ce qui revient à prouver que si G est simplement connexe, presque simple et isotrope sur k, alors G^+ est ouvert d'indice fini dans $G(k)$. Soient S un tore déployé sur k maximal de G, T un k-tore maximal de G contenant S et U le radical unipotent d'un k-sous-groupe parabolique minimal de G contenant $\mathcal{Z}(S)$. Toute racine $a \in \Phi(T, G)$ est transformée par un élément du groupe de Weyl $\mathfrak{N}(T)/T$ en un élément de $\Phi(T, U)$. Vu 3.6, il s'ensuit que $L(G)$ ne possède pas de sous-espace propre stable par Ad G et contenant $L(U)$. Comme $G(k)$ est dense dans G pour la topologie de Zariski, cela implique qu'il existe des éléments $g_1, \cdots, g_r$ de $G(k)$ tels que les $(\mathrm{Ad}\, g_i)(L(U))$ engendrent $L(G)$. Alors, la dérivée à l'élément neutre de l'application produit

$$\prod g_i U g_i^{-1} \longrightarrow G$$

est surjective et il s'ensuit que le produit des $g_i U(k) g_i^{-1}$ contient un voisinage de e, donc que G^+ est un sous-groupe ouvert de $G(k)$.

Soit Z le plus grand k-sous-groupe distingué connexe anisotrope de $\mathcal{Z}(S)$. Vu 3.19 (i) appliqué à l'isogénie produit $Z \times S \to \mathcal{Z}(S)$, il existe une partie compacte C_1 de $\mathcal{Z}(S)(k)$ telle que $\mathcal{Z}(S)(k) = C_1 \cdot Z(k) \cdot S(k)$. Comme $Z(k)$ est compact (voir 3.18), il en est de même de $C = C_1 \cdot Z(k)$. Vu 6.11 (i) et 6.8, on a $C \cdot G^+ = \mathcal{Z}(S)(k) \cdot G^+ = G(k)$; par conséquent, le quotient $G(k)/G^+$ est compact. Etant aussi discret, il est fini, c.q.f.d.

6.15. *Remarque.* Soit H un k-groupe presque simple isotrope sur k et simplement connexe. Alors on a en fait $H^+ = H(k)$: la démonstration donnée dans [22] en caractéristique zéro — démonstration légèrement incorrecte mais qu'il est facile de corriger — s'étend sans difficulté au cas général.

Si on admet ce résultat, 3.19 entraîne que $G(k)/G^+$ est *commutatif* lorsque G ne possède pas de facteur $\neq \{e\}$ anisotrope sur k.

III. HOMOMORPHISMES

Dans cette troisième partie, k' désigne un corps, K' un "domaine universel", extension algébriquement close de k' et G' un k'-groupe de dimension strictement positive. Si X est un groupe, Y, Z deux sous-groupes de X et $\alpha\colon Y \to G(k')$ un homomorphisme, on note $\bar{\alpha}(Z)$ l'adhérence de $\alpha(Y \cap Z)$ dans G'; c'est un k'-sous-groupe de G' [1: AG 14.4].

7. Images de sous-groupes unipotents et de sous-groupes paraboliques

7.1. PROPOSITION. *Soient X un k-groupe, S un k-sous-tore de X et U un k-sous-groupe unipotent connexe normalisé par S, tel que $\mathfrak{Z}(S) \cap U = \{e\}$. Soient H un sous-groupe de $X(k)$ contenant $U(k)$ et $\alpha \colon H \to G'(k')$ un homomorphisme. Supposons $H \cap S$ dense dans S.*

(i) $V = \bar{\alpha}(U(k))$ *est un k'-sous-groupe unipotent connexe de G'.*

(ii) *Si $U(k) \not\subset \operatorname{Ker} \alpha$, alors $\operatorname{car} k = \operatorname{car} k'$.*

(iii) *Si $x \in H \cap S$ et si $\mathfrak{Z}(x) \cap U = \{e\}$, on a $Z(\alpha(x)) \cap V = \{e\}$.*

Pour tout $s \in S$, le groupe $\mathfrak{Z}(s) \cap U$ est connexe [1: 9.3], donc [1: 9.5] l'ensemble A des $s \in S$ tels que $\mathfrak{Z}(s) \cap U = \{e\}$ contient un ouvert dense de S. Si $s \in A \cap H$, l'application $u \mapsto (s, u)$ est un k-isomorphisme de variétés de U sur U [1: 9.3]; on a donc $\{(s, u) \mid u \in U(k)\} = U(k)$, d'où

$$(1) \qquad \{(\alpha(s), v) \mid v \in \alpha(U(k))\} = \alpha(U(k)) \qquad (s \in A \cap H) \, .$$

Le groupe SU est résoluble. Il en est donc de même de $\alpha(SU \cap H)$, et par suite de $L = \bar{\alpha}(SU)$ [1: 2.4, Cor. 2]. Soit D un sous-groupe d'indice fini de $S \cap H$ tel que $\bar{\alpha}(D)$ soit connexe. Ce groupe est dense dans S, puisque $S \cap H$ l'est, donc $A \cap D \neq \varnothing$. Pour $s \in A \cap D$, on a $\alpha(s) \in L^\circ$; utilisant (1) à deux reprises, on en déduit successivement que

$$\alpha(U(k)) \subset (\alpha(s), L) \subset L^\circ$$

puis que

$$\alpha(U(k)) \subset (\alpha(s), L^\circ) \subset \mathfrak{D}L^\circ \, ,$$

d'où $V \subset \mathfrak{D}L^\circ$. Mais $\mathfrak{D}L \subset V$, car $\mathfrak{D}(SU) \subset U$. Par conséquent $V = \mathfrak{D}L^\circ$, ce qui démontre (i).

(ii) Supposons $U(k) \not\subset \operatorname{Ker} \alpha$ et soit p' l'exposant caractéristique de k'. Si $p = 1$ (i.e. $\operatorname{car} k = 0$) et $p' \neq 1$, alors $U(k)$ est q-divisible pour tout entier q et $\alpha(U(k))$ a un exposant fini (puissance de p') donc $\alpha(U(k)) = \{e\}$, contrairement à l'hypothèse. Si $p \neq 1$, $U(k)$ est un p-groupe, donc $\alpha(U(k))$ en est un aussi et $p = p'$, d'où (ii).

(iii) Soit x comme dans l'énoncé, c'est-à-dire $x \in A \cap H$. Supposons $\mathfrak{Z}(\alpha(x)) \cap V \neq \{e\}$. Soient x'_u et x'_s la composante unipotente et la composante semi-simple de $\alpha(x)$. L'intersection $V' = \mathfrak{Z}(x'_s) \cap V$ est un groupe unipotent connexe, de dimension strictement positive puisqu'il contient $\mathfrak{Z}(\alpha(x)) \cap V$. Considérons alors la suite de groupes $(V'_i)_{i \in \mathbf{N}}$ définie inductivement par $V'_0 = V'$ et $V'_{i+1} = (\{x'_u\}, V'_i)$. Comme x'_u normalise V, les V'_i sont des sous-groupes de V. Pour i assez grand, V'_{i-1} est réduit à l'élément neutre puisqu'il est contenu dans le i-ème terme de la série centrale descendante du groupe unipotent engendré par x'_u et V'. Le dernier terme nontrivial de la suite (V'_i) est centralisé par x'_u, donc par $\alpha(x)$, et on a $\dim\big(\mathfrak{Z}(\alpha(x)) \cap V\big) \geqq 1$. L'application $v \mapsto (\alpha(x), v)$ de V dans V est constante sur les classes $v \cdot (\mathfrak{Z}(\alpha(x)) \cap V)$ donc définit un morphisme de $V/(\mathfrak{Z}(\alpha(x)) \cap V)$ sur $W = \{(\alpha(x), v) \,|\, v \in V\}$. Par conséquent, $\dim \bar{W} < \dim V$ et W n'est pas dense dans V, ce qui contredit (1). La proposition est ainsi démontrée.

7.2. PROPOSITION. *Supposons k infini et G semi-simple. Soient P, P^- deux k-sous-groupes paraboliques opposés, U, U^- leurs radicaux unipotents respectifs, $Z = P \cap P^-$, H un sous-groupe de $G(k)$ contenant G^+ et $\alpha \colon H \to G'(k')$ un homomorphisme.*

(i) *Le groupe $\bar{\alpha}(U)$ est unipotent connexe, le groupe $\bar{\alpha}(G^+)$ est connexe et l'ensemble $\bar{\alpha}(U^-) \cdot \bar{\alpha}(Z) \cdot \bar{\alpha}(U)$ est ouvert dans $\bar{\alpha}(H)$. Si $\alpha(G^+) \neq \{e\}$, on a $\operatorname{car} k = \operatorname{car} k'$.*

(ii) *Supposons G' réductif et $\alpha(H)$ dense dans G'. Alors, $P' = \bar{\alpha}(P)$ et $P'^- = \bar{\alpha}(P^-)$ sont des k'-sous-groupes paraboliques opposés de G' et on a $\bar{\alpha}(U) = R_u(P')$, $\bar{\alpha}(U^-) = R_u(P'^-)$ et $\bar{\alpha}(Z) = P' \cap P'^-$.*

(i) Soit S un k-tore déployé sur k maximal de Z. On a $\mathfrak{Z}(S) \cap U = \{e\}$. Vu 6.8 $S \cap G^+$ est dense dans S. La première et la dernière assertions de (i) résultent donc de 7.1. Comme aucune hypothèse n'a été faite sur le k-sous-groupe parabolique P, il est ainsi établi que pour tout k-sous-groupe parabolique Q de G, $\bar{\alpha}\big(R_u(Q)\big)$ est connexe; il en est donc de même de $\bar{\alpha}(G^+)$, vu 6.2 (iv). Posons

$$(1) \qquad\qquad \Omega' = \bar{\alpha}(U^-) \cdot \bar{\alpha}(Z) \cdot \bar{\alpha}(U) \ .$$

D'après 6.11 (ii), il existe une partie finie C de G^+ telle que $G = C \cdot U^- \cdot Z \cdot U$, d'où $H = C \cdot U^-(k) \cdot (Z \cap H) \cdot U(k)$. Par conséquent, $\bar{\alpha}(H)$ est réunion d'un nombre fini de translatés de Ω'; donc Ω' contient un ouvert non vide de $\bar{\alpha}(H)$. Comme Ω' est une orbite du groupe $\bar{\alpha}(U^-) \cdot \bar{\alpha}(Z) \times \bar{\alpha}(U)$ opérant sur $\bar{\alpha}(H)$ par $(a \times b, c) \mapsto a \cdot c \cdot b^{-1}$ $(a \in \bar{\alpha}(U^-) \cdot \bar{\alpha}(Z), \ b \in \bar{\alpha}(U), \ c \in \bar{\alpha}(H))$, il s'ensuit que Ω' est ouvert dans $\bar{\alpha}(H)$.

(ii) Désignons respectivement par U'^-; Z', U' les trois facteurs de second membre de (1); soit T' un tore maximal de Z'° et soient V^-, V des sous-groupes unipotents connexes maximaux de Z'°, normalisés par T' et tels que $V^- \cdot T' \cdot V$ contienne un ouvert non vide de Z'° [1: 14.1, Cor. 1]. Vu (i) et l'hypothèse faite sur $\alpha(H)$, $U'^- \cdot V^- \cdot T' \cdot V \cdot U'$ contient un ouvert non vide de G' et $V^- \cdot U'^-$, $V \cdot U'$ sont des sous-groupes unipotents connexes normalisés par T'. Comme G' est réductif, cela entraîne que $B'^- = T' \cdot V^- \cdot U'^-$ et $B' = T' \cdot V \cdot U'$ sont deux sous-groupes de Borel opposés de G'. En particulier, $V^- \cap V = \{e\}$ et le groupe Z'° est réductif. Ce groupe normalise évidemment

U'^- et U'; il s'ensuit que $Q = Z'^\circ \cdot U'$ et $Q^- = Z'^\circ \cdot U'^-$, qui contiennent respectivement B' et B'^-, sont des sous-groupes paraboliques opposés ayant pour radicaux unipotents U' et U'^-. Comme Z' normalise aussi U' et U'^- on a $Z' \subset Q \cap Q^- = Z'^\circ$, d'où $Z' = Z'^\circ$, $Q = Z' \cdot U'$ et $Q^- = Z'^- \cdot U'^-$. D'autre part, $H \cap P = (H \cap Z) \cdot U(k)$ et $H \cap P^- = (H \cap Z) \cdot U^-(k)$. Par conséquent, $P' = \bar{\alpha}(P) = \bar{\alpha}(Z) \cdot \bar{\alpha}(U) = Z' \cdot U' = Q$, et de même $P'^- = Q^-$, ce qui démontre (ii).

8. Principaux résultats

8.1. Théorème. *Supposons G absolument presque simple, G' absolument simple de type adjoint et $\alpha\colon G^+ \to G'(k')$ un homomorphisme dont l'image est dense dans G'. Alors, il existe un unique homomorphisme $\varphi\colon k \to k'$ et une unique isogénie spéciale $\beta\colon {}^\varphi G \to G'$ tels que $\alpha = \beta \circ \varphi^0|_{G^+}$. L'isogénie β est définie sur k'. Si $\bar{\varphi}\colon k \to k'$ est un homomorphisme et $\bar{\beta}\colon {}^\varphi G \to G'$ une isogénie tels que $\alpha = \bar{\beta} \circ \bar{\varphi}^0|_{G^+}$, il existe un entier positif m tel que $\varphi(k) \subset Fr^m(k')$ et que $\varphi = Fr^m \circ \bar{\varphi}$ et $\bar{\beta} = \beta \circ Fr^m$.*

Remarquons que l'existence de α implique que G *est isotrope sur k* (i.e. $G^+ \neq \{e\}$) et que k *est infini*.

La démonstration du théorème 8.1 occupe les numéros 8.2 à 8.9.

8.2. *Notations et préliminaires.* Soient S_m un tore k-déployé maximal de G, a_m la racine dominante de S_m dans G pour un ordre donné dans $\Phi(S, G)$ et $a_m^\vee\colon \mathrm{Mult} \to S_m$ la "racine duale" de a_m, c'est-à-dire l'homomorphisme défini par la condition que, pour tout $\chi \in X^*(S_m)$, $\chi \circ a_m^\vee$ soit l'élévation à la puissance $2(a_m, \chi)/(a_m, a_m)$ (où la parenthèse représente un produit scalaire invariant par le groupe de Weyl relatif). Nous posons $S = a_m(\mathrm{Mult})$ et $a = a_{m|S}$, et nous désignons par L un k-sous-groupe presque simple de dimension 3 de G contenant S et tel que $L \cap U_{(a_m)}$ et $L \cap U_{(-a_m)}$ soient de dimension 1; l'existence d'un tel sous-groupe est assurée par le théorème 7.2 de [4], vu aussi [4: 5.11]. En vertu de 6.2 et 6.8, on a

$$(1) \qquad\qquad a^{-1}(k^{*2}) = L^+ \cap S \subset G^+ .$$

(Si b est un caractère d'un tore et X une partie du groupe multiplicatif d'un corps de définition de celui-ci, nous notons $b^{-1}(X)$ l'ensemble des éléments y du tore tels que $b(y) \subset X$; rappelons que les groupes de caractères sont notés additivement de sorte que l'inverse de b dans le groupe des caractères est noté $-b$, c'est-à-dire qu'on a $(-b)(y) = b(y)^{-1}$.) Le normalisateur $\mathfrak{N}(S) \cap L$ de S dans L est la réunion de S et d'une classe latérale M. On a $M(k) \neq \varnothing$ donc, vu 6.11 (i), $M \cap L^+ \neq \varnothing$. Choisissons un élément n dans $M \cap L^+$. L'automorphisme $\mathrm{Int}_S n$ induit sur $X^*(S)$ l'automorphisme $\chi \mapsto -\chi$.

Comme a_m est une racine de longueur maximum, on a, pour tout $\chi \in \Phi(S_m, G)$, $2|(a_m, \chi)/(a_m, a_m)| \leq 2$, d'où l'on déduit aussitôt que $\chi|_S \in \{0, \pm(1/2)a, \pm a\}$. Par conséquent

$$\Phi(S, G) = \{\pm a\} \quad \text{ou} \quad \left\{\pm \frac{1}{2}a, \pm a\right\} .$$

Nous posons, avec les notations de 1.11,

$$\Phi^+ = \Phi(S, G) \cap (\mathbf{R}_+ \cdot a) , \quad U = G_{\Phi^+}^{*(S)} , \quad U^- = G_{-\Phi^+}^{*(S)} ,$$

$$U_2 = G_{\{a\}}^{*(S)} , \quad U_2^- = G_{-\{a\}}^{*(S)} , \quad U_1 = U/U_2 , \quad Z = \mathcal{Z}(S);$$

les groupes ZU et ZU^- sont des k-sous-groupes paraboliques opposés de G. On a

$$(2) \qquad\qquad {}^nU = U^- , \quad {}^nU_2 = U_2^- , \quad {}^nZ = Z$$

et

$$(3) \qquad n \in (L \cap U_{(-a_m)}) \cdot (L \cap U_{(a_m)}) \cdot (L \cap U_{(-a_m)}) \subset U_2^- U_2 U_2^- .$$

Rappelons que, pour toute partie X de G, nous notons $\bar{\alpha}(X)$ l'adhérence de $\alpha(X \cap G^+)$ dans G'. Posons

$$U' = \bar{\alpha}(U) , \quad U'^- = \bar{\alpha}(U^-) , \quad U_2' = \bar{\alpha}(U_2) , \quad U_2'^- = \bar{\alpha}(U_2^-) ,$$

$$U_1' = U'/U_2' , \quad Z' = \bar{\alpha}(Z) , \quad S' = \bar{\alpha}(S)^\circ .$$

Vu 7.2 (ii), $Z'U'$ et $Z'U'^-$ sont des k'-sous-groupes paraboliques opposés de G' dont U' et U'^- sont les radicaux unipotents et Z' le sous-groupe de Levi commun. De plus, S' est un tore contenu dans le centre de Z'. Le sous-groupe U_2' de U' est stable par Z' donc connexe.

Le tore S (resp. S') opère sur U et U_2 (resp. U' et U_2') par automorphismes intérieurs, ce qui induit aussi une action de S (resp. S') sur U_1 (resp. U_1'). Pour $\varepsilon = 1, 2$, nous notons Ψ_ε l'ensemble des poids de S' dans U_ε'. Le groupe U_ε est commutatif et d'exposant p si $p \neq 1$; il en est donc de même de U_ε', qui, vu [3: 9.9], possède alors une décomposition unique

$$(4) \qquad\qquad U_\varepsilon' = \prod_{b \in \Psi_\varepsilon} U_{\varepsilon, b}' \qquad\qquad \text{(produit direct)} ,$$

où $U_{\varepsilon, b}'$ est un groupe stable par S' et possède une structure d'espace vectoriel tel que tout $s \in S'$ opère par l'homothétie de rapport $b(s)$. Désignant par Ψ l'ensemble des poids de S' dans U', on a $\Psi = \Psi_1 \cup \Psi_2$. De plus, $\Psi \cup (-\Psi) = \Phi(S', G')$.

Vu (1) et 7.2 (i) appliqué à $\alpha|_{L^+}$, le produit $\bar{\alpha}(L \cap U) \cdot \bar{\alpha}(a^{-1}(k^{*2})) \cdot \bar{\alpha}(L \cap U^-)$ est connexe. Comme ses facteurs sont contenus respectivement dans U', Z' et U'^-, et que l'application produit $U' \times Z' \times U'^- \to G'$ est une immersion, cela implique que $\bar{\alpha}(a^{-1}(k^{*2}))$ est connexe, donc contenu dans S'.

D'autre part, l'image de $S \cap G^+$ par l'application $x \mapsto x^2$ est contenue dans $a^{-1}(k^{*2})$, donc l'image de $S' = \bar{\alpha}(S \cap G^+)^\circ$ par $x \mapsto x^2$ est contenue dans $\bar{\alpha}(a^{-1}(k^{*2}))$, et on a

$$(5) \qquad S' = \bar{\alpha}(a^{-1}(k^{*2})) \;.$$

8.3. *L'homomorphisme φ.*

Soient (ε, b) un couple tel que $\varepsilon \in \{1, 2\}$ et $b \in \Psi_\varepsilon$. On a $(1/2)\varepsilon a \in \Phi(S, G) \subset X^*(S)$, car si $\varepsilon = 1$, alors $U_1' \neq \{e\}$, donc $U_1 \neq \{e\}$ et $(1/2)a \in \Phi(S, G)$. Quel que soit ε, on a $((1/2)\varepsilon a)^{-1}(k^{*\varepsilon}) \subset G^+$, vu 8.2 (1). Nous nous proposons de montrer qu'il existe un homomorphisme de corps $\varphi_{\varepsilon, b} \colon k \to K'$ et un seul tel que

$$(1) \qquad \{\varphi_{\varepsilon, b}(t)\} = b\Big(\alpha\Big(\Big(\tfrac{1}{2}\varepsilon a\Big)^{-1}(t)\Big)\Big) \quad \text{pour} \quad t \in k^{*\varepsilon} \;.$$

Remarquons tout d'abord que si $t \in k^{*\varepsilon}$, (1) définit effectivement un élément $\varphi_{\varepsilon, b}(t)$ de K'; en effet, si s, $s' \in ((1/2)\varepsilon a)^{-1}(t)$, on a $\mathrm{Int}_{U_\varepsilon} s = \mathrm{Int}_{U_\varepsilon} s'$, donc $\mathrm{Int}_{U'_\varepsilon} \alpha(s) = \mathrm{Int}_{U'_\varepsilon} \alpha(s')$, d'où $\mathrm{Int}_{U'_{\varepsilon, b}} \alpha(s) = \mathrm{Int}_{U'_{\varepsilon, b}} \alpha(s')$ et, finalement, $b(\alpha(s)) = b(\alpha(s'))$. Il est clair que l'application $\varphi_{\varepsilon, b}|_{k^{*\varepsilon}} \colon k^{*\varepsilon} \to K'^*$ définie par (1) est un homomorphisme de groupes multiplicatifs. D'autre part, si (t_i) est une famille finie d'éléments de $k^{*\varepsilon}$ de somme nulle, et si $s_i \in ((1/2)\varepsilon a)^{-1}(t_i)$ pour tout i, on a $\sum_i \mathrm{Int}_{U_\varepsilon} s_i = 0$, donc $\sum_i \mathrm{Int}_{U'_{\varepsilon, b}} \alpha(s_i) = 0$ et $\sum_i \varphi_{\varepsilon, b}(t_i) = 0$. Par conséquent, $\varphi_{\varepsilon, b}|_{k^{*\varepsilon}}$ s'étend, évidemment de façon unique, en un homomorphisme d'anneaux de l'anneau k_1 engendré par $k^{*\varepsilon}$ dans K'. Mais $k_1 = k$ sauf si k est un corps non parfait de caractéristique 2 et $\varepsilon = 2$, et dans ce dernier cas, tout homomorphisme ψ de $k_1 = k^2$ dans K' se prolonge à k de façon unique par $t \mapsto (\psi(t^2))^{1/2}$. L'existence et l'unicité de $\varphi_{\varepsilon, b}$ est ainsi établie.

Soit T' un tore maximal de G' contenant S'. Choisissons dans $\Phi(T', G')$ une base Δ' telle que les racines positives par rapport à cette base soient des poids de T' dans $Z' \cdot U'$. Les poids de T' dans Z' sont toutes les racines qui sont combinaisons linéaires des éléments d'une certaine partie propre de Δ'. Comme tous les éléments de Δ' interviennent avec un coefficient non nul dans la racine dominante, celle-ci est nécessairement un poids de T' dans U' et sa restriction à S', que nous notons d, appartient à Ψ. Soit $\eta \in \{1, 2\}$ tel que $d \in \Psi_\eta$. Dans $\Phi(T', G')$, la différence entre la racine dominante et une racine positive quelconque est une somme de racine positives [11: VI, 1.8, Prop. 25]; par restriction à S', on en déduit qu'il existe des éléments $b_1, \cdots, b_m$ de Ψ tels que

$$(2) \qquad d = b + b_1 + \cdots + b_m \;.$$

Pour $1 \leq i \leq m$, soit $\varepsilon_i \in \{1, 2\}$ tel que $b_i \in \Psi_{\varepsilon_i}$.

Soient $s \in a^{-1}(k^{*2})$ et $t \in k^*$ tels que $((1/2)a)(s) = t$ ou $a(s) = t^2$ selon que

$(1/2)a$ appartienne ou non à $X^*(S)$. Soient $\varepsilon' \in \{1, 2\}$ et $b' \in \Psi_{\varepsilon'}$. Vu (1), on a:

$$\text{si } \varepsilon' = 1, \quad \mathcal{P}_{1,b'}(t) = b'\Big(\alpha\Big(\Big(\tfrac{1}{2}a\Big)^{-1}(t)\Big)\Big) = b'(\alpha(s)) \ ;$$

$$\text{si } \varepsilon' = 2, \quad \mathcal{P}_{2,b'}(t)^2 = \mathcal{P}_{2,b'}(t^2) = b'(\alpha(a^{-1}(t^2))) = b'(\alpha(s)) \ .$$

Dans l'un et l'autre cas,

$$(3) \qquad\qquad\qquad \mathcal{P}_{\varepsilon',b'}(t)^{2z'} = b'(\alpha(s))^2 \ .$$

Vu (2), on voit donc que

$$\mathcal{P}_{\eta,d}(t)^{2\eta} = \mathcal{P}_{\varepsilon,b}(t)^{2\varepsilon} \cdot \prod_{i=1}^{m} \mathcal{P}_{\varepsilon_i,b_i}(t)^{2\varepsilon_i} \ .$$

Comme t peut prendre n'importe quelle valeur dans k^*, il s'ensuit, compte tenu de 2.2, que $\mathcal{P}_{\eta,d}$ et $\mathcal{P}_{\varepsilon,b}$ ne diffèrent que par une puissance de l'automorphisme de Frobenius de K'. Cela étant vrai quel que soit le couple (ε, b), les homomorphismes $\mathcal{P}_{\varepsilon,b}$ ne diffèrent entre eux que par des puissances de Fr et ils sont tous produits de l'un d'entre eux, soit $\mathcal{P}_{\varepsilon_0,b_0}$, par des puissances à exposants positifs de Fr. Nous poserons $\mathcal{P} = \mathcal{P}_{\varepsilon_0,b_0}$, $a' = 2\varepsilon_0^{-1} \cdot b_0$ et

$$(4) \qquad\qquad\qquad \mathcal{P}_{\varepsilon,b} = Fr^{r(\varepsilon,b)} \circ \mathcal{P} \qquad\qquad\qquad (r(\varepsilon, b) \in \mathbf{N}) \ .$$

En particulier, on a

$$(5) \qquad\qquad\qquad r(\varepsilon_0, b_0) = 0 \ .$$

Soient s et t comme ci-dessus. Compte tenu de (3) et (4), on a

$$(2\varepsilon_0 b)(\alpha(s)) = \mathcal{P}_{\varepsilon,b}(t)^{2\varepsilon\varepsilon_0} = \mathcal{P}_{\varepsilon_0,b_0}(t)^{2\varepsilon\varepsilon_0 \cdot p^{r(\varepsilon,b)}}$$
$$= (2\varepsilon p^{r(\varepsilon,b)} \cdot b_0)(\alpha(s)) = (\varepsilon\varepsilon_0 p^{r(\varepsilon,b)} \cdot a')(\alpha(s)) \ .$$

Comme $\alpha(a^{-1}(k^{*2}))$ est dense dans S' (cf. 8.2 (5)), cela implique, vu la connexité de S', que

$$(6) \qquad\qquad\qquad b = \frac{1}{2}\varepsilon p^{r(\varepsilon,b)} \cdot a' \ .$$

Le groupe G' étant semi-simple, $X^*(S') \otimes \mathbf{Q}$ est engendré linéairement par $\Phi(S', G')$, donc par Ψ, ce qui signifie, vu (6), que *le tore S' est de dimension un*. De plus, toujours en vertu de (6), la somme des poids de S' dans U', qui est un caractère de S' défini sur k', n'est pas nulle. Donc, S' *est déployé sur k'*, d'où l'on déduit encore que *les espaces vectoriels $U'_{\varepsilon,b}$ sont définis sur k'*.

Nous verrons plus loin (8.6 (5)) que $\mathcal{P}(k) \subset k'$; ce qui précède montre déjà que $\mathcal{P}(k_1) \subset k'$, mais nous ne ferons pas usage de cette remarque.

Nous supposerons jusqu' à la fin de 8.8 que $K = K'$ et que $\mathcal{P}$ est l'homomorphisme d'inclusion, ce qui est évidemment loisible: il suffit en effet, pour se ramener à ce cas, de remplacer k, K, G et α par $\varphi(k)$, K', $^{\varphi}G$ et $\alpha \circ (\varphi^0)^{-1}$. Moyennant cette convention, on déduit de (1), où l'on fait $\varepsilon = \varepsilon_0$ et $b = b_0$, que

$$(7) \qquad a'\big(\alpha(a^{-1}(t))\big) = \{t\} \quad \text{pour} \quad t \in k^{*2} .$$

Lorsque $a/2 \in \Phi(S, G)$ et $p \neq 2$, il résulte aussi de (1), (6) et (7) que $a'/2 \in X^*(S')$ et que

$$(8) \qquad \left(\frac{1}{2}a'\right)\left(\alpha\left(\left(\frac{1}{2}a\right)^{-1}(t)\right)\right) = \{t\} \quad \text{pour} \quad t \in k^* .$$

8.4. Nous interrompons le cours de la démonstration du théorème 8.1 pour établir un

LEMME. *Soient Y un k-tore déployé, χ un caractère non trivial de Y, V un k-groupe unipotent connexe sur lequel Y opère morphiquement et u un élément de $V(k) - \{e\}$. Supposons que tous les poids de Y dans V soient des multiples entiers strictement positifs de χ.*

(i) *Il existe un k-morphisme de variétés $\psi \colon K \to V$ et un seul tel que $\psi(1) = u$ et que*

$$y\big(\psi(t)\big) = \psi\big(\chi(y) \cdot t\big) \qquad\qquad (y \in Y; \ t \in K^*) .$$

On a $\psi(0) = e$.

(ii) *S'il existe un sous-groupe distingué fermé V' de V, stable par Y, ne contenant pas u et tel que χ soit le seul poids de Y dans V/V', alors ψ est un isomorphisme de variétés de K sur son image.*

(i) On sait [3: 9.12] qu'il existe un k-isomorphisme de variétés $V \to L(V)$ qui commute avec l'action de Y. Comme la loi de groupe de V n'intervient pas dans l'énoncé, nous pouvons substituer $L(V)$ à V et supposer que V est un espace vectoriel sur lequel Y opère linéairement. Alors, V est le produit direct des sous-espaces propres pour Y et il suffit de considérer le cas où Y a un seul poids dans V, soit $n \cdot \chi$. Mais lorsqu'il en est ainsi, il est clair que le morphisme

$$(1) \qquad\qquad t \longmapsto t^n \cdot u \qquad\qquad (t \in K)$$

possède les propriétés de l'énoncé. L'unicité est évidente.

(ii) La formule (1) montre que si χ est le seul poids de Y dans V et si $u \neq e$, alors ψ est un isomorphisme de K sur son image, c'est-à-dire que (ii) est vrai pour $V' = \{e\}$. Le cas général de (ii) s'en déduit en remarquant que si on substitue à V le quotient V/V' et à u son image canonique dans V/V', alors ψ est remplacé par son composé avec la projection canonique $V \to V/V'$; comme ce composé est un isomorphisme de K sur son image, il en est évidemment de même de ψ.

8.5. *Le morphisme β_u.*

Pour tout $u \in U(k) - \{e\}$, nous posons $X_u = {}^s u \cup \{e\}$. Vu 8.4, l'application $\iota_u \colon K \to X_u$ définie par les relations

$$\iota_u(0) = e$$

$$\iota_u(t) = {}^s u \quad \text{pour} \quad t \in K^* \quad \text{et} \quad \begin{cases} s \in a^{-1}(t) \ \ \text{si} \ \ u \in U_2 \\ s \in \left(\dfrac{1}{2}a\right)^{-1}(t) \ \ \text{si} \ \ u \notin U_2 \end{cases}$$

est un k-isomorphisme de variétés. La relation 8.3 (6) montre que les poids de S' dans U'_2 sont des multiples entiers positifs de a' et que les poids de S' dans U' sont des multiples entiers de $(1/2)a'$ si $(1/2)a' \in X^*(S')$ et de a' sinon; les relations suivantes définissent donc une application $\iota'_u \colon K \to U'$ et cette application est un k'-morphisme vu 8.4 (i):

$$\iota'_u(0) = e$$

$$\iota'_u(t) = {}^{s'}\alpha(u) \quad \text{pour} \quad t \in K^* \quad \text{et} \quad \begin{cases} s' \in a'^{-1}(t) \ \ \text{si} \ \ u \in U_2 \\ s' \in a'^{-1}(t^2) \ \ \text{si} \ \ u \notin U_2 \ \ \text{et} \ \ \dfrac{1}{2}a' \notin X^*(S') \\ s' \in \left(\dfrac{1}{2}a'\right)^{-1}(t) \ \ \text{si} \ \ u \notin U_2 \ \ \text{et} \ \ \dfrac{1}{2}a' \in X^*(S') \ . \end{cases}$$

(Dans le deuxième cas, on a $p = 2$ vu 8.3 (6), et ι'_u est le composé de Fr et de l'application ψ de 8.4 (i), où l'on fait $\chi = a'$.) Le composé $\beta_u = \iota'_u \circ \iota_u^{-1} \colon X_u \to U'$ est un morphisme de variétés défini sur le corps $k \cdot k'$ engendré par k et k'. Nous nous proposons de montrer qu'on a

$$(1) \qquad \beta_u|_{X_u(k)} = \alpha|_{X_u(k)} \ .$$

Remarquons d'emblée qu'il résulte de la définition de ι_u et ι'_u, et de 8.3 (7), 8.3 (8) que

$$(2) \qquad \beta_u \ et \ \alpha \ co\ddot{\imath}ncident \ sur \ l'ensemble \ \ Y_u = \{{}^s u \mid s \in a^{-1}(k^{*2})\} \cup \{e\} \ .$$

Cela démontre déjà (1) lorsque $u \notin U_2$, parce qu'on a alors $X_u(k) = \{{}^s u \mid s \in ((1/2)a)^{-1}(k^*)\} = Y_u$.

Supposons que $u \in U_2(k)$. Il résulte alors de 8.3 (6) que ι_u et ι'_u sont des morphismes de groupes algébriques. En particulier, $\beta_u|_{X_u(k)}$ est un homomorphisme, et il en est évidemment de même de $\alpha|_{X_u(k)}$. Si k n'est pas un corps non parfait de caractéristique 2, k^{*2} engendre additivement k, donc $Y_u = \iota_u(k^{*2})$ engendre $X_u(k) = \iota_u(k)$ et (2) implique (1).

Il nous reste à établir (1) dans le cas où $u \in U_2(k)$ et où k est un corps non parfait de caractéristique 2. Nous supposerons seulement que car $k = 2$.

Soit s un élément quelconque de $a^{-1}(k^*)$ et soit s' la racine carrée de l'élément $\alpha(s^2)$ (lequel appartient à S' vu 8.2 (5)) dans S'. Nous devons montrer que

$$(3) \qquad \qquad {}^{s'}\alpha(u) = \alpha({}^{s}u) \quad \text{pour tout } u \in U_2(k) .$$

Vu 7.2 (i) appliqué au groupe $G_{(\pm a)}^{(S)}$ (notations de 1.11) et à ses sous-groupes paraboliques opposés ZU_2 et ZU_2^-, le produit $U_2' Z' U_2'^-$ est ouvert, donc dense, dans le groupe qu'il engendre. De plus, $\operatorname{Int}\alpha(n)$ permute U_2' et $U_2'^-$, ainsi que les sous-groupes paraboliques $Z'U'$ et $Z'U'^-$. Les hypothèses du lemme 5.4 sont donc satisfaites si on y remplace G, P, P^-, V, n soit par G, $Z \cdot U$, $Z \cdot U^-$, U_2, n, soit par G', $Z' \cdot U'$, $Z' \cdot U'^-$, U_2', $\alpha(n)$. Vu l'assertion (i) de ce lemme, l'ensemble $\Omega = \big(U_2^-(k) \cdot n \cdot Z(k) \cdot U_2^-(k) \big) \cap U_2(k)$ engendre $U_2(k)$. Comme les deux membres de (3) sont multiplicatifs en u, il nous suffit donc d'établir cette relation pour $u \in \Omega$. Soit alors $n_1 \in n \cdot Z$ tel que $u \in U_2^-(k) \cdot n_1 \cdot U_2^-(k)$, d'où $n_1 \in G^+$ et $\alpha(u) \in U_2'^-(k') \cdot \alpha(n_1) \cdot U_2'^-(k')$. Appliquant à deux reprises l'assertion (ii) du lemme 5.4, on a

$$\alpha({}^{s}u) \in \alpha\big(\big(U_2^-(k) \cdot s^2 \cdot n_1 \cdot U_2^-(k) \big) \cap U_2(k) \big)$$
$$\subset \big(U_2'^-(k') \cdot s'^2 \cdot \alpha(n_1) \cdot U_2'^-(k') \big) \cap U_2'(k') \subset \{ {}^{s'}\alpha(u) \} ,$$

d'où (3). La relation (1) est ainsi démontrée dans tous les cas.

8.6. *Les morphismes* β_U, β_{U^-} *et* $\beta_{U,\varepsilon,b}$.

Soient $(u_i)_{1 \leq i \leq m_1} (m_1 = \dim U_1)$ une famille d'éléments de $U(k)$ se projetant sur une base de l'espace vectoriel U_1 et $(u_i)_{m_1+1 \leq i \leq m} (m = \dim U)$ une base de l'espace vectoriel $u_2(k)$. Utilisant les notations X_u, ι_u de 8.5, posons $X_i = X_{u_i}$ et $\iota_i = \iota_{u_i}$, et désignons par $\pi \colon U \to U_1$ la projection canonique. L'espace vectoriel U_1 (resp. U_2) est le produit direct de ses sous-espaces $\pi(X_1), \cdots,$ $\pi(X_{m_1})$ (resp. $X_{m_1+1}, \cdots, X_m$). Pour $1 \leq i \leq m_1$, $\pi \circ \iota_i \colon K \to U_1$ est une application linéaire, donc $\pi|_{X_i} \colon X_i \to \pi(X_i)$ est un k-isomorphisme de variétés; notons $\lambda_i \colon U \to X_i$ le k-morphisme composé de π, de la projection canonique de U_1 sur son facteur $\pi(X_i)$ et de l'inverse $\pi(X_i) \to X_i$ de $\pi|_{X_i}$. L'application

$$(1) \qquad \qquad u \longmapsto \big(\textstyle\prod_{i=1}^{m_1} \lambda_i(u) \big)^{-1} \cdot u \qquad \qquad (u \in U)$$

est un k-morphisme de U dans U_2; pour $m_1 + 1 \leq i \leq m$, soit $\lambda_i \colon U \to X_i$ le k-morphisme composé de (1) et de la projection canonique de U_2 sur son facteur X_i. On a alors

$$(2) \qquad \qquad u = \textstyle\prod_{i=1}^{m} \lambda_i(u) \quad \text{pour } u \in U .$$

Nous désignons par $\beta_U \colon U \to U'$ le $(k \cdot k')$-morphisme

$$u \longmapsto \textstyle\prod_{i=1}^{m} \big(\beta_{u_i}(\lambda_i(u)) \big) \qquad \qquad (u \in U)$$

et par $\beta_{U^-} \colon U^- \to U'^-$ le $(k \cdot k')$-morphisme $(\operatorname{Int}\alpha(n)) \circ \beta_U \circ (\operatorname{Int} n^{-1}|_{U^-})$. Vu

(2) et 8.5 (1), on a

$$(3) \qquad \beta_U|_{U(k)} = \alpha|_{U(k)} \quad \text{et} \quad \beta_{U^-}|_{U^-(k)} = \alpha|_{U^-(k)} \ ,$$

d'où il résulte (1.6) que β_U et β_{U^-} sont des morphismes de groupes algébriques.

Si $\varepsilon \in \{1, 2\}$, β_U induit un morphisme de U_ε sur U'_ε; pour tout $b \in \Psi_\varepsilon$, nous notons $\beta_{U,\varepsilon,b}$ le composé de ce morphisme et de la projection canonique $\pi_{\varepsilon,b}\colon U'_\varepsilon \to U'_{\varepsilon,b}$. Soient $u \in U_\varepsilon(k)$, $t \in k^{*2}$ et $s \in \big((1/2)\varepsilon a^{-1}\big)(t)$; on a, avec des conventions de notation évidentes, $tu = {}^s u$, d'où

$$\beta_{U,\varepsilon,b}(tu) = \pi_{\varepsilon,b}\big(\alpha({}^s u)\big) = \pi_{\varepsilon,b}\big({}^{\alpha(s)}\alpha(u)\big) = {}^{\alpha(s)}\pi_{\varepsilon,b}\big(\alpha(u)\big) = b\big(\alpha(s)\big)\cdot\beta_{U,\varepsilon,b}(u) \ ,$$

et enfin, vu les relations (6), (7) et (8) de 8.3,

$$\beta_{U,\varepsilon,b}(tu) = Fr^{r(\varepsilon,b)}(t)\cdot\beta_{U,\varepsilon,b}(u) \ .$$

Les deux membres de cette égalité conservent un sens pour $t \in K^*$ et, pour u fixé, ce sont des fonctions régulières en t. Comme elles coïncident sur k^{*2} qui est dense dans K, elles coïncident pour toute valeur de t; autrement dit:

$$(4) \qquad \textit{le morphisme } \beta_{U,\varepsilon,b} \textit{ est } Fr^{r(\varepsilon,b)}\textit{-linéaire} \ .$$

En particulier, il résulte de 8.3 (5) que le morphisme $\beta_{U,\varepsilon_0,b_0}$ est linéaire; comme il applique $U_{\varepsilon_0}(k)$ dans $U'_{\varepsilon_0,b_0}(k')$, ceci implique que $k \subset k'$ ou encore, en réécrivant ce résultat sous une forme indépendante de la convention faite à la fin de 8.3

$$(5) \qquad \varphi(k) \subset k' \ .$$

8.7. *L'homomorphisme β_Z.*

Soient ε, b comme ci-dessus. Pour $z \in Z \cap G^+$, on a

$$(1) \qquad \big(\mathrm{Int}_{U'_\varepsilon,b}\,\alpha(z)\big) \circ \beta_{U,\varepsilon,b} = \beta_{U,\varepsilon,b} \circ \big(\mathrm{Int}_{U_\varepsilon} z\big) \ ,$$

car il résulte de 8.6 (3) que les morphismes des deux membres coïncident sur $U_\varepsilon(k)$. Vu (1), $Z \cap G^+$ normalise le noyau $N_{\varepsilon,b}$ de $\beta_{U,\varepsilon,b}$ et il en est donc de même de son adhérence Z (cf. 6.11 (iii)). Notons $V_{\varepsilon,b}$ l'espace vectoriel $U_\varepsilon/N_{\varepsilon,b}$ et $\psi_{\varepsilon,b}(z)$, pour $z \in Z$, la transformation linéaire de $V_{\varepsilon,b}$ induite par $\mathrm{Int}_{U_\varepsilon} z$; l'application $\psi_{\varepsilon,b}\colon Z \to GL(V_{\varepsilon,b})$ est une représentation linéaire rationnelle de Z. En la composant avec le morphisme $GL(V_{\varepsilon,b}) \to GL(U'_{\varepsilon,b})$ induit par $\beta_{U,\varepsilon,b}$ (cf. 8.6 (4) et 1.8) on obtient un morphisme de groupes algébriques $\beta_{Z,\varepsilon,b}\colon Z \to GL(U'_{\varepsilon,b})$, et il résulte aussitôt de (1) que

$$(2) \qquad \beta_{Z,\varepsilon,b}(z) = \mathrm{Int}_{U'_\varepsilon,b}\big(\alpha(z)\big) \quad \text{pour} \ z \in Z \cap G^+ \ .$$

Soit $\theta\colon Z' \to \prod_{\varepsilon,b} GL(U'_{\varepsilon,b})$ le morphisme de groupes algébriques défini par

$$\theta(z') = \prod_{\varepsilon,b} \mathrm{Int}_{U'_\varepsilon,b} z' \ ,$$

le produits étant étendu à tous les couples (ε, b) tels que $\varepsilon \in \{1, 2\}$ et $b \in \Psi_\varepsilon$. Montrons que

(3) *la restriction de θ à tout tore T' de Z' est une immersion.*

Comme G' est de type adjoint, la restriction de sa représentation adjointe à T' est une immersion, c'est-à-dire que $\Phi(T', G')$ engendre $X^*(T')$. Pour établir (3), il suffit donc de faire voir que le groupe X engendré par les poids de T' dans $\prod_{\varepsilon,b} U'_{\varepsilon,b}$, c'est-à-dire par les poids de T' dans U', contient $\Phi(T', G')$. Or ceci résulte aussitôt de ce que le groupe $G^{(T')}_{X \cap \Phi(T',G')}$ (cf. 1.11) contient U' et U'^-, donc coïncide avec G', d'après 6.2 (v).

Vu (2), on a pour tout $z \in Z \cap G^+$,

(4) $$\theta(\alpha(z)) = (\textstyle\prod_{\varepsilon,b} \beta_{Z,\varepsilon,b})(z) \ .$$

Par conséquent

$$(\textstyle\prod_{\varepsilon,b} \beta_{Z,\varepsilon,b})\,(Z) \subset \{(\textstyle\prod_{\varepsilon,b} \beta_{Z,\varepsilon,b})\,(z)\,|\,z \in Z \cap G^+\}^-$$
$$= \{\theta(\alpha(z))\,|\,z \in Z \cap G^+\}^- = \theta(\alpha(Z \cap G^+))^- = \theta(Z') \ .$$

En vertu de (3) et 3.4 le morphisme θ est injectif; nous désignerons par θ^{-1} l'isomorphisme de groupes $\theta(Z') \to Z'$ inverse de θ. L'inclusion précédente nous permet alors de poser

$$\beta_Z = \theta^{-1} \circ (\textstyle\prod_{\varepsilon,b} \beta_{Z,\varepsilon,b})\colon Z \longrightarrow Z'$$

(où, selon la convention générale de 1.3, on écrit Z, Z' pour $Z(K)$, $Z'(K)$). La relation (4) peut à présent s'écrire:

(5) $$\beta_Z|_{Z \cap G^+} = \alpha|_{Z \cap G^+} \ .$$

Il résulte de (3) que

(6) *la restriction de β_Z à tout tore de Z est un morphisme.*

Soit j un entier positif tel que $Fr^j \circ \theta^{-1}$ soit un morphisme; l'existence d'un tel entier résulte de (3) et 3.4 (en fait, (3) implique même que $Fr \circ \theta^{-1}$ est un morphisme, mais nous n'aurons pas besoin de cette précision). Alors

(7) $$Fr^j \circ \beta_Z \colon Z \longrightarrow {}^{Fr^j}Z' \text{ est un morphisme.}$$

8.8. *L'isogénie β.*

Posons $X = UZU^-$ et, pour tout $g \in G^+$, soit $\beta^{(g)}\colon gX \to G'$ l'application définie par

$$\beta^{(g)}(guzu^-) = \alpha(g) \cdot \beta_U(u) \cdot \beta_Z(z) \cdot \beta_{U^-}(u^-) \qquad (u \in U,\ z \in Z,\ u^- \in U^-) \ .$$

Vu 8.6 (3) et 8.7 (5), on a

(1) $$\beta^{(g)}|_{gX \cap G^+} = \alpha|_{gX \cap G^+} \ .$$

Comme l'application produit $U \times Z \times U^- \to G$ est une immersion, il résulte de 8.7 (7) que $Fr^j \circ \beta^{(g)}\colon gX \to {}^{Fr^j}G'$ est un morphisme de variétés. Pour g, $g' \in G^+$, les morphismes $Fr^j \circ \beta^{(g)}$ et $Fr^j \circ \beta^{(g')}$ coïncident sur $gX \cap g'X \cap G^+$, vu (1), donc sur $gX \cap g'X$, "par continuité", de sorte que $\beta^{(g)}$ et $\beta^{(g')}$ coïncident aussi sur $gX \cap g'X$. Les applications $\beta^{(g)}(g \in G^+)$ "se recollent" donc en une application $\beta\colon G \to G'$ telle que

$$(2) \qquad\qquad \beta|_{G^+} = \alpha|_{G^+} \; .$$

Comme $Fr^j \circ \beta$ est un morphisme de variétés coïncidant avec $Fr^j \circ \alpha$ sur le sous-groupe dense G^+, c'est un morphisme de groupes algébriques (1.6) et β est un homomorphisme.

Soit T un tore maximal de G contenant S. Pour toute racine $c \in \Phi(T, G)$, il existe $h \in \mathfrak{N}(T)$ tel que $hU_c h^{-1} \subset U$; en effet, G étant presque simple, aucune orbite du groupe de Weyl dans $\Phi(T, G)$ ne peut être contenue dans le sous-espace propre de $X^*(T) \otimes \mathbf{R}$ engendré par $\Phi(T, Z)$. Mais $\beta|_U = \beta_U$ est un morphisme. Il en est donc de même de

$$\beta|_{U_c} = \big(\mathrm{Int}\, \beta(h)^{-1}\big) \circ \beta_U \circ \big(\mathrm{Int}\, h|_{U_c}\big) \; .$$

D'autre part, $\beta|_T$ est un morphisme en vertu de 8.7 (6). Comme l'application produit

$$T \times \prod_{c \,\in\, \Phi(T,G)} U_c \longrightarrow G$$

est une immersion ouverte pour un ordre convenable des facteurs, on voit que la restriction de β à un ouvert de G est un morphisme et il s'ensuit, par raison d'homogénéité, que β lui-même est un morphisme. Comme $\beta(G^+) = \alpha(G^+)$ est dense dans G', β est surjectif; c'est donc une isogénie puisque G est presque simple. Comme G^+ est dense dans G et $\alpha(G^+) \subset G(k')$, β est défini sur k' (cf. 1.4).

Il résulte immédiatement de la construction de β que les homomorphismes $\beta_{U,c,b}$ de 8.6 sont induits par β, en un sens évident. Comme β_{U,c_0,b_0} est une application linéaire non nulle (cf. 8.3 (5) et 8.6 (4)), sa différentielle à l'élément neutre n'est pas nulle, donc $(d\beta)_e \neq 0$ et l'isogénie β est spéciale (3.7).

8.9. *Relations entre $\bar\varphi$, $\bar\beta$ et φ, β. Unicité de φ, β.*

Soient $\bar\varphi$ et $\bar\beta$ comme dans l'énoncé du théorème. On a $\bar\beta\big(\bar\varphi^0(S \cap G^+)\big) = \alpha(S \cap G^+) \subset S'$, donc $\bar\beta(\bar\varphi S) = S'$. Soit $\bar a$ le caractère de S défini par $\bar\varphi \bar a = \bar\beta^*(a')$ et soient l, $\bar l$ des entiers tels que $\bar l \bar a = la$. Pour tout $t \in k^*$ on a, en notant s un élément de $a^{-1}(t^2)$,

$$\bar\varphi(t^{2l}) = \bar\varphi\big((la)(s)\big) = \bar\varphi\big((\bar l \bar a)(s)\big) = (\bar l a')\big(\alpha(s)\big) = \varphi(t^{2\bar l}) \; .$$

En vertu de 2.2, il existe $m \in \mathbf{N}$ tel que l'on ait ou bien $\varphi = Fr^m \circ \bar\varphi$ ou bien

$\bar{\varphi} = Fr^m \circ \varphi$. Dans le premier cas, on a

$$\alpha(g) = (\beta \circ \varphi^0)(g) = (\beta \circ Fr^m) \circ \bar{\varphi}^0(g) = (\beta \circ Fr^m)(\bar{\varphi}^0(g)) = \bar{\beta} \circ (\bar{\varphi}^0(g)) \quad (g \in G^+)$$

d'où

$$(1) \qquad\qquad\qquad \bar{\beta} = \beta \circ Fr^m \ ,$$

sur G^+, donc, par densité, sur G. De même, dans le deuxième cas:

$$(2) \qquad\qquad\qquad \beta = \bar{\beta} \circ Fr^m \ .$$

La différentielle $(dFr)_e$ de Fr est nulle. Comme $d\beta$ n'est pas nulle, on ne peut avoir (2) avec $m > 0$ et l'on est donc dans le cas (1). Si $\bar{\beta}$ est spéciale, (1) entraine alors $m = 0$, d'où $\beta = \bar{\beta}$ et $\varphi = \bar{\varphi}$.

8.10. Lemme. *Soient A un groupe, B un sous-groupe distingué de A, H un groupe algébrique et ν, ν' des homomorphismes de A dans H dont les restrictions à B coïncident et ont une image dense dans H. Alors, il existe un homomorphisme $\mu\colon A \to \mathcal{C}(H)$ tel que $\nu'(x) = \mu(x) \cdot \nu(x)$ pour tout $x \in A$.*

Posons $\mu(x) = \nu(x)^{-1} \cdot \nu'(x)$ pour $x \in A$. Il suffit manifestement de montrer que $\mu(A) \subset \mathcal{C}(H)$. Soient $x \in A$ et $y \in B$, d'où $xyx^{-1} \in B$. On a

$$\nu'(x)\nu(y)\nu'(x)^{-1} = \nu'(xyx^{-1}) = \nu(xyx^{-1}) = \nu(x)\nu(y)\nu(x)^{-1} \ ,$$

ce qui montre que $\mu(x)$ centralise $\nu(y)$. Comme $\nu(B)$ est dense dans H, l'assertion s'ensuit.

8.11. Théorème. *Supposons k, k' infinis et G, G' absolument presque simples et isotropes. Soient H et H' des groupes tels que $G^+ \subset H \subset G(k)$, $G'^+ \subset H' \subset G'(k')$ et soit $\alpha\colon H \to H'$ un homomorphisme surjectif. Notons $\mathrm{Ad}_G\colon G \to \mathrm{Ad}\, G$ et $\mathrm{Ad}_{G'}\colon G' \to \mathrm{Ad}\, G'$ les isogénies canoniques de G et G' sur leurs groupes adjoints respectifs, et soit $\pi\colon \tilde{G} \to G$ un revêtement universel de G. Supposons $\alpha(G^+) \neq \{e\}$.*

(i) Il existe un unique isomorphisme $\varphi\colon k \to k'$ et une unique k'-isogénie $\beta'\colon {}^\varphi G \to \mathrm{Ad}\, G'$ tels que

$$(1) \qquad\qquad\qquad \beta' \circ \varphi^0|_H = \mathrm{Ad}_{G'} \circ \alpha \,..$$

(ii) Si les isogénies Ad_G et $\mathrm{Ad}_{G'}$ ont le même degré, alors l'une des conditions suivantes est réalisée:

(a) l'isogénie β' se factorise à travers une isogénie $\beta\colon {}^\varphi G \to G'$ (i.e. $\beta' = \mathrm{Ad}_{G'} \circ \beta$) et il existe un homomorphisme $\mu\colon H \to \mathcal{C}(G')(k')$ tel que

$$\alpha(h) = \mu(h) \cdot \beta(\varphi^0(h))$$

pour tout $h \in H$;

(b) les groupes G et G' sont de type $\mathbf{D}_{2n}$ (pour un certain entier n), non simplement connexes et non adjoints, et il existe un unique monomorphisme

$\lambda\colon H \to \widetilde{G}(k)$ *et une unique isogénie centrale* $\pi'\colon {}^{\varphi}\widetilde{G} \to G'$ *tels que* $\pi \circ \lambda = \mathrm{id}_H$ *et* $\alpha = \pi' \circ \varphi^0 \circ \lambda$; *de plus,* π' *n'est pas la composée de* ${}^{\varphi}\pi$ *et d'un isomorphisme* ${}^{\varphi}G \to G'$. *On a* $\lambda(G^+) = \widetilde{G}^+$.

(iii) *Si l'isogénie* β' *n'est pas centrale ou, sous les conditions de* (ii) (a), *si* β *n'est pas un isomorphisme, alors* G *et* G' *sont déployés,* k *et* k' *sont parfaits et on se trouve dans le cas exceptionnel* (3.3).

Le groupe $\mathrm{Ad}_{G'}(\alpha(G^+))$ est normalisé par $\mathrm{Ad}_{G'}(H')$, donc par $\mathrm{Ad}_{G'}(G'^+) = (\mathrm{Ad}\,G')^+$ (cf. 6.3); par conséquent [31], il contient $(\mathrm{Ad}\,G')^+$, lequel est dense dans $\mathrm{Ad}\,G'$. D'après le théorème 8.1, il existe donc un unique homomorphisme $\varphi\colon k \to k'$ et une unique k'-isogénie spéciale $\beta'\colon {}^{\varphi}G \to \mathrm{Ad}\,G'$ tels que $\beta' \circ \varphi^0|_{G^+} = \mathrm{Ad}_{G'} \circ \alpha|_{G^+}$. Vu le lemme 8.10, on a alors $\beta' \circ \varphi^0|_H = \mathrm{Ad}_{G'} \circ \alpha$.

Comme $\mathrm{Ad}\,G'$ est adjoint, le noyau de β' est le centre de ${}^{\varphi}G$. Par conséquent, le noyau de $\mathrm{Ad}_{G'} \circ \alpha$ est $H \cap \mathcal{C}(G)$ et α induit un isomorphisme $\alpha_1\colon \mathrm{Ad}_G H \to \mathrm{Ad}_{G'} H'$. Appliquant ce qu'on vient de voir à α_1 et à son inverse, on en déduit qu'il existe des homomorphismes $\sigma\colon k \to k'$ et $\tau\colon k' \to k$ et des isogénies spéciales $\gamma\colon \mathrm{Ad}\,{}^{\sigma}G \to \mathrm{Ad}\,G'$ et $\delta\colon \mathrm{Ad}\,{}^{\tau}G' \to \mathrm{Ad}\,G$ tels que

$$\alpha_1 = \gamma \circ \sigma^0|_{\mathrm{Ad}_G H} \quad \text{et} \quad \alpha_1^{-1} = \delta \circ \tau^0|_{\mathrm{Ad}_{G'} H'}\;.$$

Vu l'unicité de β', φ, on a $\varphi = \sigma$ et $\beta' = \gamma \circ \mathrm{Ad}_{({}^{\sigma}G)}$. Posons $\varphi_0 = \tau \circ \sigma$ et $\beta_0 = \delta \circ {}^{\tau}\gamma$. Alors, la restriction de $\beta_0 \circ \varphi_0^0$ à $\mathrm{Ad}_G H$ est l'identité et, en vertu du théorème 8.1, il existe un entier positif m tel que

$$(2) \qquad\qquad \beta_0 = Fr^m$$

et

$$(3) \qquad\qquad Fr^m \circ \varphi_0 = \mathrm{id}\;.$$

De (3), il résulte que φ_0 est un automorphisme de k, donc $\varphi = \sigma$ et τ sont des isomorphismes. Cela démontre (i).

Supposons β' non centrale, ce qui implique qu'on se trouve dans le cas exceptionnel du n° 3.3. Alors, γ n'est pas centrale [7: 2.18], $d\gamma$ n'est pas injectif, $d\beta_0$ ne l'est pas non plus, et $m \geq 1$, vu (2). Cela étant, il résulte de (3) que k est parfait, et de (3) et 3.13 que $\mathrm{Ad}\,G$ est déployé (car $\mathrm{Ad}\,G = {}^{Fr^m}({}^{\varphi_0}\mathrm{Ad}\,G)$). Comme $\mathrm{Ad}\,G' = \gamma({}^{\varphi}\mathrm{Ad}\,G)$, le groupe $\mathrm{Ad}\,G'$ est aussi déployé. Si G et G' sont adjoints, (ii)(a) est satisfait pour $\beta = \beta'$. Sinon, il résulte de 3.3 que G et G' sont simplement connexes; il existe alors une unique isogénie $\beta\colon G \to G'$ telle que $\mathrm{Ad}_{G'} \circ \beta = \beta'$ (cf. [7: 2.24]) et comme $\mathrm{Ad}_{G'}$ est injectif, on déduit de (1) que $\beta \circ \varphi^0|_H = \alpha$. Cela achève la démonstration du théorème dans l'hypothèse où β' n'est pas centrale.

Désormais, nous identifions k et k' par φ, supposons que $K = K'$, que β' est central, et identifions $\mathrm{Ad}\,G$ et $\mathrm{Ad}\,G'$ de telle façon que $\beta' = \mathrm{Ad}_G$ (ce qui est permis, vu [7: 2.26]). Alors, $\tilde{G}$ est un revêtement universel de G'; plus exactement, il existe une k-isogénie centrale $\pi'\colon \tilde{G} \to G'$ telle que $\mathrm{Ad}_{\tilde{G}} = \mathrm{Ad}_{G'} \circ \pi'$. S'il existe un k-isomorphisme $\beta\colon G \to G'$ tel que

$$\text{(4)} \qquad\qquad \beta \circ \pi = \pi' \, ,$$

on a $\mathrm{Ad}_{G'} \circ \beta = \mathrm{Ad}_G = \beta' = \mathrm{Ad}_{G'} \circ \alpha$ (sur H) et l'application $\mu\colon h \mapsto \alpha(h) \cdot \beta(h)^{-1}$ est un homomorphisme de H dans $(\mathrm{Ker}\,\mathrm{Ad}_{G'})(k) = \mathcal{C}(G')(k)$, d'où (ii) (a). Supposons donc, pour finir la démonstration, qu'il n'existe pas d'isomorphisme β satisfaisant à (4) et que Ad_G et $\mathrm{Ad}_{G'}$ aient le même degré. Pour tout groupe presque simple qui n'est pas de type $\mathbf{D}_{2n}$ ($n \in \mathbf{N}$), le quotient du groupe des poids par le sous-groupe engendré par les racines est cyclique; d'après les résultats rappelés en 3.1, un tel groupe possède donc, à un isomorphisme canonique près, au plus une isogénie centrale de degré donné. Vu l'hypothèse de non-existence de l'isomorphisme β, G est donc de type $\mathbf{D}_{2n}$. De plus, π et π' sont des isogénies "différentes", de degré 2, et l'on en déduit aussitôt que $\pi \times \pi'\colon \tilde{G} \to G \times G'$ est un k-isomorphisme λ' de $\tilde{G}$ sur le groupe

$$\tilde{G}_1 = \{(g,\, g') \in G \times G' \mid \mathrm{Ad}_G\, g = \mathrm{Ad}_{G'}\, g'\} \, .$$

Soit $\lambda\colon H \to \tilde{G}(k)$ l'homomorphisme défini par $\lambda(h) = \lambda'^{-1}(h,\, \alpha(h))$. Il est clair que $\pi \circ \lambda = \mathrm{id}_H$ (donc que λ est injectif), que $\alpha = \pi' \circ \lambda$ et que ces relations caractérisent l'isogénie π' et l'homomorphisme λ. De plus, $\pi(\lambda(G^+)) = G^+ = \pi(\tilde{G}^+)$ et comme le noyau de π est d'ordre au plus égal à 2, cela implique que $[\tilde{G}^+\colon (\tilde{G}^+ \cap \lambda(G^+))] \leqq 2$. Mais $\tilde{G}^+$ ne possède pas de sous-groupe propre d'indice fini (6.7). Par conséquent $\tilde{G}^+ \subset \lambda(G^+)$, et, comme π applique $\lambda(G^+)$ injectivement sur $\pi(\tilde{G}^+)$, il y a égalité, ce qui termine la démonstration.

8.12. *Remarques.* (a) Il est probable que les conditions de (ii) (b) impliquent $p = 2$. Cela est vrai en tout cas lorsque $\tilde{G}^+ = \tilde{G}(k)$ (cf. 6.6), car alors $\lambda(G^+) = \tilde{G}^+$ contient le noyau de π, qui doit donc être réduit à l'élément neutre.

(b) Le théorème 8.11 reste vrai à un petit nombre d'exceptions près lorsque k et k' sont finis (cf. notamment [27] et les références de [30: 4.5]).

8.13. Corollaire. *Supposons k, k' infinis et G, G' absolument presque simples et isotropes. Soient H et H' des groupes tels que $G^+ \subset H \subset G(k)$ et $G'^+ \subset H' \subset G'(k')$ et soit $\alpha\colon H \to H'$ un isomorphisme. Supposons que G et G' sont tous deux adjoints, ou qu'ils sont tous deux simplement connexes et que $G^+ = H$ ou $G'^+ = H'$ (cf. 6.6). Alors, il existe un unique isomorphisme $\varphi\colon k \to k'$ et une unique k'-isogénie spéciale $\beta\colon {}^{\varphi}G \to G'$ tels que $\alpha = \beta \circ \varphi^0|_H$. Si β n'est pas un isomorphisme, G et G' sont déployés, k et k' sont parfaits et on se trouve dans le cas exceptionnel (3.3).*

Il n'y a quelque chose à démontrer que si G et G' sont simplement connexes. Vu 8.11 (i), leurs systèmes de racines sont isomorphes ou inverses. Il s'ensuit que Ad_G et $\mathrm{Ad}_{G'}$ ont le même degré, donc que (ii)(b) est applicable. Si $H = G^+$, on a $\mu(H) = \{e\}$ vu 6.4. Supposons donc que $H' = G'^+$. Ce qu'on vient de voir, appliqué à α^{-1}, montre l'existence d'un isomorphisme $\psi\colon k' \to k$ et d'une isogénie spéciale $\gamma\colon {}^\psi G' \to G$ tels que $\alpha^{-1} = \gamma \circ \psi^0|_{H'}$. Si γ est un isomorphisme, $\varphi = \psi^{-1}$ et $\beta = ({}^\varphi \gamma)^{-1}$ possèdent les propriétés requises. Sinon, G est déployé, vu 8.11 (iii), donc $H = G^+$ (6.6), cas déjà envisagé.

8.14. Corollaire. *Supposons k infini et G absolument presque simple et isotrope. Soient H un sous-groupe de $G(k)$ contenant G^+ et α un automorphisme de H. Alors, l'une des conditions suivantes est remplie:*

(i) *il existe un unique automorphisme φ de k, une unique k-isogénie spéciale $\beta\colon {}^\varphi G \to G$ et un unique homomorphisme $\mu\colon H \to \mathcal{C}(G)(k)$ tels que $\alpha(h) = \mu(h)\cdot\beta\big(\varphi^0(h)\big)$ pour tout $h \in H$;*

(ii) *G est isomorphe sur $\bar{k}$ à l'image d'un groupe Spin_{4n} par une représentation semi-spinorielle, et si $\pi\colon \widetilde{G} \to G$ désigne un k-revêtement universel de G, il existe un unique monomorphisme $\lambda\colon H \to \widetilde{G}$, un unique automorphisme φ de k et un unique isomorphisme $\widetilde{\beta}\colon {}^\varphi\widetilde{G} \to \widetilde{G}$ tels que $\pi\circ\lambda = \mathrm{id}_H$, $\widetilde{\beta}\big(\varphi^0(\lambda(H))\big) = \lambda(H)$ et $\alpha = \lambda^{-1}\circ\widetilde{\beta}\circ\varphi^0\circ\lambda$; de plus, il n'existe pas d'isomorphisme $\beta\colon {}^\varphi G \to G$ tel que $\pi\circ\widetilde{\beta} = \beta\circ{}^\varphi\pi$. On a $\lambda(G^+) = \widetilde{G}^+$.*

Si la condition (i) *est remplie et si β n'est pas un isomorphisme, alors k est parfait, G est déployé et ou bien $p = 2$ et G est de type $\mathbf{C}_2$, $\mathbf{F}_4$ ou bien $P = 3$ et G est de type $\mathbf{G}_2$.*

8.15. *Remarque.* Le cas (ii) du corollaire précédent correspond au cas (ii) (b) de 8.11. Comme nous l'avons dit en 8.12 (a), cette condition implique probablement que $p = 2$. Lorsque $p = 2$, elle peut évidemment être réalisée: il suffit par exemple de supposer $\widetilde{G}$ déployé, de faire $H = \pi(\widetilde{G}(k))$ et de prendre pour α l'automorphisme de H induit par un k-automorphisme de $\widetilde{G}$ qui ne provient pas d'un k-automorphisme de G (ce qui est le cas pour n'importe quel automorphisme extérieur de $\widetilde{G}$ si $n \neq 2$).

8.16. Théorème. *Supposons G absolument presque simple sur k et G' réductif. Soient H un sous-groupe de $G(k)$ contenant G^+ et $\alpha\colon H \to G'(k')$ un homomorphisme tel que $\alpha(G^+)$ soit dense dans G'. Supposons G simplement connexe ou G' adjoint. Soient G'_i $(1 \leq i \leq m)$ les k'-sous-groupes distingués presque simples sur k' de G'. Alors, il existe des extensions séparables finies k_i $(1 \leq i \leq m)$ de k', des homomorphismes de corps $\varphi_i\colon k \to k_i$, une k'-isogénie spéciale*

$$\beta : \prod_{i=1}^{m} R_{k_i/k'}({}^{\varphi_i}G) \longrightarrow G'$$

et un homomorphisme $\mu\colon H \to \mathcal{C}(G')(k')$ *tels que*

$$\beta\big(R_{k_i/k'}(^{\varphi_i}G)\big) = G'_i \qquad (1 \leqq i \leqq m)$$

et que

$$(1) \qquad \alpha(h) = \mu(h)\cdot\beta\big(\textstyle\prod_{i=1}^{m}\big(R^0_{k_i/k'}(\varphi_i^0(h))\big)\big) \qquad (h \in H)\ .$$

Ces k_i, φ_i, β et μ sont uniques au sens suivant: si k'_i, φ'_i, β' et μ' ont les mêmes propriétés, alors $\mu = \mu'$, il existe des k'-isomorphismes uniques $\kappa_i\colon k_i \to k'_i$ tels que $\varphi'_i = \kappa_i \circ \varphi_i$, et si on identifie les k_i aux k'_i au moyen des κ_i, on a $\beta = \beta'$.

Soient $\sigma_{i,j}$ $(1 \leqq i \leqq m;\ 1 \leqq j \leqq r_i)$ les k'-homomorphismes de k_i dans K'. Alors, une égalité

$$(2) \qquad Fr^n \circ \sigma_{i,j} \circ \varphi_i = Fr^{n'} \circ \sigma_{i',j'} \circ \varphi_{i'}\ ,$$

avec n, $n' \in \mathbf{N}$, n'a lieu que si $n = n'$, $i = i'$ et $j = j'$. Le corps k_i est engendré par $\varphi_i(k)$ et k'.

Si $\alpha(G^+) \supset G'^+$, alors $m = 1$ et φ_1 est un isomorphisme. Si de plus β n'est pas centrale, alors k et k' sont parfaits, G est déployé sur k et on se trouve dans le cas exceptionnel (3.3).

Remarquons que, comme en 8.1, l'hypothèse faite sur $\alpha(G^+)$ implique que k est infini et que G est isotrope. Vu 6.4, G' est nécessairement semi-simple.

Considérons d'abord le cas où G' est adjoint. Il est alors le produit direct des G'_i et nous pouvons supposer [4: 6.21 (ii)] que

$$G'_i = R_{k_i/k'}\, G''_i$$

où k_i est une extension séparable finie de k' et G''_i un k_i-groupe absolument simple adjoint. Soit π_i la projection canonique de G' sur G'_i. Vu le théorème 8.1, appliqué à la restriction à G^+ de l'homomorphisme $(R^0_{k_i/k'})^{-1}\circ \pi_i \circ \alpha$, il existe un homomorphisme $\varphi_i\colon k \to k_i$ et une k_i-isogénie spéciale $\beta_i\colon {}^{\varphi_i}G \to G''_i$ tels que

$$\pi_i \circ \alpha = R^0_{k_i/k} \circ \beta_i \circ \varphi_i^0\ \text{ sur }\ G^+\ .$$

Posant $\beta = \prod_{i=1}^{m}(R_{k_i/k'}\beta_i)$, on en déduit que

$$(3) \qquad \alpha(h) = \beta\big(\textstyle\prod_{i=1}^{m}\big(R^0_{k_i/k'}((\varphi_i^0(h)))\big)\big)\ ,$$

pour $h \in G^+$, et cette égalité reste vraie pour tout $h \in H$ en vertu du lemme 8.10. Posant $\mu(H) = \{e\}$, nous avons ainsi établi l'existence de k_i, φ_i, β et μ. Remettant à plus tard la preuve de leur unicité, nous passons à présent à la démonstration des autres assertions de l'énoncé (en conservant l'hypothèse que G' est adjoint).

Supposons k' algébriquement clos. Alors $r_i = 1$ et $\sigma_{i,1}$ est l'identité.

Supposons qu'il existe i, $i' \in \{1, \cdots, m\}$ et $n \in \mathbf{N}$ tels que $i \neq i'$ et

$$Fr^n \circ \varphi_i = \varphi_{i'} \, .$$

L'ensemble $\{(\varphi_i^0(g), \varphi_{i'}^0(g)) \,|\, g \in G(k)\} \subset {}^{\varphi_i}G \times {}^{\varphi_{i'}}G$ est alors contenu dans le graphe du morphisme $Fr^n \colon {}^{\varphi_i}G \to {}^{\varphi_{i'}}G$. Par conséquent, l'ensemble $\{\prod_{i=1}^m \varphi_i^0(g) \,|\, g \in G(k)\}$ est contenu dans une sous-variété fermée propre du produit des ${}^{\varphi_i}G$ et, vu (3), $\alpha(H)$ est contenu dans une sous-variété fermée propre de G', contrairement à l'hypothèse de densité. Cela démontre l'assertion d'indépendance des φ_i lorsque k' est algébriquement clos. Le cas général se ramène immédiatement à celui-là: il suffit d'appliquer ce qu'on vient de voir à l'homomorphisme $H \to G'(\bar{k}')$ composé de α et de l'inclusion $G'(k) \to G'(\bar{k}')$ (où $\bar{k}'$ désigne la clôture algébrique de k' dans K'), et de remarquer que les φ_i correspondants sont les produits $\sigma_{i,j} \circ \varphi_i$ $(1 \leqq i \leqq m;\ 1 \leqq j \leqq r_i)$. Du résultat ainsi établi, on déduit en particulier que deux k'-homomorphismes de k_i dans K' qui coïncident sur $\varphi_i(k)$ sont identiques, c'est-à-dire que k_i est engendré par k' et $\varphi_i(k)$.

Remarquons que l'existence de β_i entraîne que G_i' est isotrope. Supposons que $\alpha(H) \supset G'^+$. Alors, pour tout i, le groupe $\alpha^{-1}(G_i'^+)$ est distingué et non-central dans H, donc contient G^+ [31], et $\alpha(G^+) \subset G_i'^+$. Il s'ensuit que $m = 1$, donc G'^+ est simple [31]. Comme $\alpha(G^+)$ est normalisé par $\alpha(H)$, donc par G'^+, il s'ensuit que $\alpha(G^+) = G'^+$. Il résulte alors du théorème 8.11 appliqué à l'homomorphisme $(R_{k_1/k'}^0)^{-1} \circ \alpha|_{G^+}$ que φ_1 est un isomorphisme et que si de plus β n'est pas central (ce qui signifie que β_1 ne l'est pas non plus), alors k est parfait, G est déployé et on se trouve dans le cas exceptionnel du n° 3.3.

Il nous reste à démontrer l'assertion d'unicité. Soient k_i', φ_i', β' comme dans l'énoncé (μ' étant nécessairement l'homomorphisme trivial dans le cas qui nous occupe). L'isogénie β' induit une isogénie γ_i de $R_{k'_i/k'}{}^{\varphi_i}G$ sur $G_i' = R_{k_i/k'}G_i''$. Il résulte immédiatement de la définition du foncteur R (cf. par ex. [16]) qu'il existe un k'-isomorphisme $\kappa_i \colon k_i \to k_i'$ tel que, si l'on identifie k_i et k_i' par κ_i, alors γ_i est l'image d'une k_i-isogénie $\beta_i' \colon {}^{\varphi_i}G \to G_i''$ par le foncteur $R_{k_i/k'}$. Moyennant l'identification en question, on a

$$\pi_i \circ \alpha = \gamma_i \circ R_{k_i/k'}^0 \circ \varphi_i'^0|_H = R_{k_i/k}^0 \circ \beta_i' \circ \varphi_i'^0|_H$$

et il résulte de l'assertion d'unicité du théorème 8.1 qu'on a $\varphi_i' = \varphi_i$ et $\beta_i' = \beta_i$, d'où $\beta' = \beta$. Une fois k_i et k_i' "désidentifiés", l'égalité de φ_i' et φ_i devient la relation $\varphi_i' = \kappa_i \circ \varphi_i$, et celle-ci détermine le k'-isomorphisme κ_i puisque, comme on l'a vu plus haut, $\varphi_i(k)$ et k' engendrent k_i.

Cela termine la démonstration du théorème lorsque G' est adjoint. Supprimons cette hypothèse et supposons G simplement connexe. Ce qui vient d'être démontré, appliqué à l'homomorphisme $\mathrm{Ad}_{G'} \circ \alpha \colon H \to \mathrm{Ad}\, G'$ montre l'existence et l'unicité d'une décomposition

$$(4) \qquad \mathrm{Ad}_{G'} \circ \alpha = \beta_* \circ \prod_{i=1}^{m} (R^0_{k_i/k'} \circ \varphi_i^{\eta})|_H \ .$$

Comme G est simplement connexe, β_* se relève en une isogénie

$$\beta: \prod_{1 \le i \le m} (R_{k_i/k'} {}^{\varphi_i}G) \to G'$$

(cf. [7: 2.24]). Posons $\rho = \beta \circ \prod_{i=1}^{m} (R^0_{k_i/k'} \circ \varphi_i^{\eta})$ et, pour $h \in H$, $\mu(h) = \alpha(h)\rho(h)^{-1} \in G'(k')$. Vu (3), on a $\mathrm{Ad}_{G'} \circ \alpha = \mathrm{Ad}_{G'} \circ \rho|_H$, d'où $\mu(H) \subset \mathcal{C}(G') \cap G'(k') = \mathcal{C}(G')(k')$. Il est alors évident que $\mu: H \to \mathcal{C}(G')(k')$ est un homomorphisme. Compte tenu du fait que $\mathrm{Ad}_{G'}(G'^+) = (\mathrm{Ad}\, G')^+$ (6.3), les autres assertions de l'énoncé découlent immédiatement des assertions correspondantes pour G' adjoint.

8.17. *Remarques.* (a) Si $k = \mathbf{Q}$, on a $m = 1$, $\mathbf{Q}$ peut s'identifier par φ_i au corps premier de k', β est une k'-isogénie de G sur G' et la formule (1) s'écrit $\alpha(h) = \mu(h) \cdot \beta(h)$.

(b) Supposons que le "revêtement simplement connexe" $\tilde{G}$ de G satisfasse à la relation $\tilde{G}^+ = \tilde{G}(k)$ (cf. 6.6). Alors, on peut, dans l'énoncé de 8.16, remplacer l'hypothèse $\bar{\alpha}(G^+) = G'$ par l'hypothèse en apparence plus faible $\bar{\alpha}(H) = G'$, et supprimer le facteur $\mu(h)$ de la formule (1) (c'est-à-dire qu'on a $\mu(H) = \{e\}$). En effet, l'hypothèse faite sur $\tilde{G}$ implique que H/G^+ est commutatif [7: 2.7] donc que $\bar{\alpha}(G^+)$ est un sous-groupe distingué de $\bar{\alpha}(H)$ à quotient commutatif. Si G' est adjoint, donc semi-simple, la relation $\bar{\alpha}(H) = G'$ entraîne $\bar{\alpha}(G^+) = G'$. Si G est simplement connexe, on a $G^+ = H$, donc $\bar{\alpha}(G^+) = \bar{\alpha}(H)$, et μ est trivial vu 6.4.

(c) Soient G, G', H, m comme dans l'énoncé du théorème. Soient k_i des extensions finies séparables de k', $\varphi_i: k \to k_i$ des homomorphismes tels que toute égalité 8.16 (2) entraîne $n = n'$, $i = i'$ et $j = j'$, $\beta: \prod_{1 \le i \le m} (R_{k_i/k'} {}^{\varphi_i}G) \to G'$ une k'-isogénie et $\mu: H \to \mathcal{C}(G')(k')$ un homomorphisme. Définissons l'homomorphisme $\alpha: H \to G'(k')$ par la formule 8.16 (1). Alors

$$\alpha(G^+) \ \text{est dense dans } G' \ .$$

Pour le faire voir, nous pouvons évidemment supposer que G' est adjoint et, par le raisonnement déjà utilisé en 8.16 dans la démonstration de l'indépendance des φ_i, que k' est algébriquement clos. Soient $\psi: G(k) \to \prod_{1 \le i \le m} ({}^{\varphi_i}G)$ le produit direct des φ_i^0 et X un k-sous-groupe de G k-isomorphe au groupe additif. Le groupe $\psi(X(k))$ est dense dans l'espace vectoriel $\prod_{1 \le i \le m} ({}^{\varphi_i}X)$, car s'il ne l'était pas, il existerait un polynôme non identiquement nul dans cet espace qui s'annulerait sur $\psi(X(k))$ et, vu 2.1, cela impliquerait l'existence d'indices j, j' distincts et d'un entier m tels que $\varphi_{j'} = Fr^m \circ \varphi_j$, en contradiction avec l'hypothèse faite sur les φ_i. On voit donc que, pour tout i, ${}^{\varphi_i}X$

(identifié à un sous-groupe de $\prod_{1 \leq i \leq m} ({}^{\varphi_i}G)$) est contenu dans l'adhérence de $\psi(G^+)$. Il en est donc de même de ${}^{\varphi_i}G$, car G est engendré par ses k-sous-groupes k-isomorphes au groupe additif (6.9), c'est-à-dire que $\psi(G^+)$ est dense dans $\prod_{1 \leq i \leq m} ({}^{\varphi_i}G)$. Comme $\alpha(G^+) = \beta(\psi(G^+))$, notre assertion s'ensuit.

8.18. *Contre-exemples.* (a) Pour montrer ce qui peut se passer si on ne suppose pas G' adjoint ou G simplement connexe, prenons $k = k' = \mathbf{R}$, $G = \mathbf{PSL}_3$ et $G' = \mathbf{SL}_3$. L'homomorphisme canonique $\mathbf{SL}_3(\mathbf{R}) \to \mathbf{PSL}_3(\mathbf{R})$ est bijectif et possède donc un inverse $\alpha\colon G(\mathbf{R}) \to G'(\mathbf{R})$ qui n'est évidemment induit par aucune isogénie de G sur G'.

(b) L'hypothèse d'après laquelle G' est réductif n'est pas conséquence de la densité de $\alpha(G^+)$ comme le montre l'exemple suivant. Supposons $G \subset \mathbf{GL}_n$, d'où $L(G) \subset \mathbf{M}_n$ (algèbre des matrices d'ordre n), ces inclusions étant supposées définies sur k. Prenons $k = k'$ et $G' = L(G) \rtimes G$, produit semi-direct pour l'action de G sur $L(G)$ par la représentatif adjointe. Soit d une dérivation non nulle de k. Pour $g = (g_{ij}) \subset \mathbf{M}_n(k)$, posons $dg = ((dg_{ij}))$. Alors, on vérifie aisément que si $g \in G(k)$ on a $g^{-1} \cdot dg \in L(G)(k)$, que l'application $\alpha\colon g \mapsto (g^{-1} \cdot dg, g)$ de $G(k)$ dans $G'(k)$ est un homomorphisme, que si k est infini $\alpha(G(k))$ (qui est en quelque sorte un "sous-groupe de Levi abstrait" de $G'(k)$) est dense dans $G'(k)$, et que si, de plus, G est absolument presque simple et isotrope, alors $\alpha(G^+)$ est aussi dense dans $G'(k)$. Nous verrons ci-après une interprétation plus "conceptuelle" de cet exemple.

8.19. *Remarque.* Dans ce qui précède, nous sommes restés toujours dans la catégorie des groupes algébriques (schémas en groupes absolument réduits) sur des corps. En l'élargissant quelque peu, on peut énoncer le théorème 8.16 sous une forme plus simple. Nous en reprenons seulement l'assertion principale:

Sous les hypothèses de 8.16, il existe une k'-algèbre séparable commutative de dimension finie L, un homomorphisme $\varphi\colon k \to L$, une k'-isogénie $\beta\colon R_{L/k'}G \to G'$ et un homomorphisme $\mu\colon H \to \mathcal{C}(G')(k')$ tels que

$$\alpha(h) = \mu(h) \cdot \beta\big(R^0_{L/k'}(\varphi^0(h))\big) \qquad (h \in H) \,.$$

(Pour une définition très générale du foncteur de restriction des scalaires, cf. [16: I, §1, 6.6]; les notations φ^0 et R^0 introduites en 1.7 s'étendent de façon évidente aux changements de base et restrictions de scalaires quelconques.)

Il est probable que, sous cette forme plus "abstraite", le résultat reste valable — moyennant sans doute quelques restrictions supplémentaires imposées à G et k — sans aucune hypothèse sur le groupe algébrique G' et sur l'homomorphisme α, à condition de supprimer le qualificatif "séparable" et de remplacer "isogénie" par "morphisme". On reviendra ailleurs sur cette conjecture, qui a été vérifiée par exemple pour $p \neq 2$ et G simplement connexe déployé sur k ou encore pour $k = k' = \mathbf{R}$. L'exemple 8.18 (b) rentre aussi dans ce cadre: on prend pour L l'algèbre $k[\varepsilon]$ des nombres duaux et pour φ l'homomorphisme $x \mapsto x + \varepsilon \cdot dx$.

9. Cas des corps localement compacts

Dans ce paragraphe, on suppose k et k' munis de topologies qui en font des corps topologiques localement compacts non discrets et on suppose G semi-simple. Sauf mention explicite du contraire, les expressions "ouvert", "fermé", "application continue" etc. se rapportent aux topologies données sur k et k', aux topologies sur $G(k)$ et $G'(k')$ déduites de celles-là (cf. par ex. [38: App. III]) ou aux topologies induites par ces dernières sur des parties de $G(k)$ ou $G'(k')$.

9.1. La proposition suivante et, plus particulièrement, le corollaire 9.3, généralisent des résultats de E. Cartan [12], J. von Neumann et B. L. van der Waerden [36]. La démonstration que nous en donnons est essentiellement celle de van der Waerden (cf. aussi [10: exercices 9, p. 264, 36, p. 283]).

PROPOSITION. *Soient H un groupe k-analytique, L son algèbre de Lie et H_1 l'ensemble des $x \in H$ tels que $(\mathrm{Ad}\, x - 1)(L)$ ne soit contenu dans aucun sous-espace de L stable par $\mathrm{Ad}\, H$ et distinct de L.*

(i) Tout sous-groupe distingué de H rencontrant H_1 est ouvert.

(ii) Si l'adhérence de H_1 contient e, tout homomorphisme de H dans un groupe topologique compact est continu.

Soient d la dimension de H et $\alpha\colon H \to H'$ un homomorphisme de H dans un groupe topologique compact. Notons $H^{(d)}$ (resp. $H'^{(d)}$) le produit direct de d copies de H (resp. H') et désignons par $f\colon H \times H^{(d)} \times H^{(d)} \to H$ (resp. $f'\colon H' \times H'^{(d)} \times H'^{(d)} \to H'$) l'application

$$(h, \mathbf{h}, \mathbf{g}) \longmapsto \prod_{i=1}^{d} g_i \cdot (h, h_i) \cdot g_i^{-1} \text{ pour } \mathbf{h} = (h_1, \cdots, h_d)$$
$$\text{et } \mathbf{g} = (g_1, \cdots, g_d)\,.$$

Pour $h \in H$ et $\mathbf{g} = (g_1, \cdots, g_d) \in H^{(d)}$, la différentielle de l'application $(h, \mathbf{x}, \mathbf{g}) \mapsto f(h, \mathbf{x}, \mathbf{g})$ de $H^{(d)}$ dans H au point $(e, \cdots, e)$ est l'application linéaire

$$(1) \qquad (l_i) \longmapsto \sum_{1 \le i \le d} \mathrm{Ad}\, g_i \cdot (\mathrm{Ad}\, h - 1)(l_i)\,, \qquad (l_i \in L)\,.$$

L'élément h appartient à H_1 si et seulement s'il existe $\mathbf{g} \in H^{(d)}$ tel que l'application (1) soit surjective, auquel cas $f(h, H^{(d)}, \mathbf{g})$ est un voisinage de e dans H. On voit donc que

(2) $\qquad$ *pour* $h \in H_1, f(h, H^{(d)}, H^{(d)})$ *est un voisinage de e dans* H.

D'autre part, pour **g** donné, l'ensemble des h tels que l'application (1) soit surjective est manifestement ouvert; donc

(3) $\qquad\qquad\qquad\qquad H_1$ *est ouvert dans* H .

Tout sous-groupe distingué de H contenant h contient aussi $f(h, H^{(d)}, H^{(d)})$, donc (i) est une conséquence immédiate de (2).

Montrons (ii). Soit $(h_i')_{i \in \mathbf{N}}$ une suite d'éléments de H_1 tendant vers e. Quitte à la remplacer par une suite extraite, nous pouvons supposer que la suite $(\alpha(h_i'))_{i \in \mathbf{N}}$ converge dans H'. Pour tout $i \in \mathbf{N}$, soit $j(i) > i$ tel que $h_i' \cdot h_{j(i)}'^{-1} \in H_1$ (un tel entier $j(i)$ existe, vu (3)). Posons $h_i = h_i' \cdot h_{j(i)}'^{-1}$. On a

$$\lim_{i \to \infty} h_i = e \text{ et } \lim_{i \to \infty} \alpha(h_i) = e .$$

Nous devons montrer que, pour tout voisinage V de e dans H', il existe un voisinage de e dans H dont l'image par α est contenue dans V. Mais on a $f'(e, H'^{(d)}, H'^{(d)}) = \{e\}$. Compte tenu de la compacité de $H'^{(d)} \times H'^{(d)}$ et de la continuité de f', on en déduit ausitôt l'existence d'un entier positif i tel que

$$\alpha\big(f(h_i, H^{(d)}, H^{(d)})\big) \subset f'\big(\alpha(h_i), H'^{(d)}, H'^{(d)}\big) \subset V ,$$

et notre assertion en résulte, vu (2).

9.2. Soit $\pi \colon \widetilde{G} \to G$ un revêtement universel de G. Nous inspirant d'une terminologie de C. Riehm [23] nous dirons que *le groupe fondamental de* G *est réduit* si π est séparable, c'est-à-dire si G est le quotient d'un groupe simplement connexe par un sous-groupe central fini. Cela a toujours lieu si $p = 1$, ou si $p > 3$ et si G ne possède pas de sous-groupe distingué de type $\mathbf{A}_n$ avec $p \,|\, n + 1$.

Si G a un groupe fondamental réduit, il en est de même de tout sous-groupe distingué fermé connexe de G.

9.3. Corollaire. *Supposons le groupe fondamental de* G *réduit et soit* H *un groupe topologique (resp. k-analytique) localement isomorphe à* $G(k)$.

(i) *Si* X *est un sous-groupe distingué de* H, *alors* X *est fermé (resp. fermé et analytique) et il existe un k-sous-groupe distingué connexe* G_1 *de* G *tel que les couples* (H, X) *et* $\big(G(k), G_1(k)\big)$ *soient localement isomorphes au sens de la théorie des groupes topologiques (resp. analytiques).*

(ii) *Tout homomorphisme de* H *dans un groupe compact est continu.*

Si k est totalement discontinu, soit H' un sous-groupe ouvert de H isomorphe (comme groupe topologique ou analytique) à un sous-groupe ouvert de $G(k)$, et soit $\psi \colon H' \to H$ l'inclusion. Si k est connexe, soit $\psi \colon H' \to H^\circ$ un revêtement universel du groupe de Lie connexe H°, composante neutre de H. Dans les deux cas, ψ est un homomorphisme ouvert et il existe un homomorphisme de H' dans $G(k)$ qui est un isomorphisme local. Quitte à remplacer H et X par H' et $\psi^{-1}(X)$, nous sommes donc en droit de supposer qu'il existe un homomorphisme $H \to G(k)$ qui est un isomorphisme local. Soit φ un tel homomorphisme.

Soit Ω l'ensemble des éléments semi-simples réguliers de G. Si $g \in \Omega$ et $T = \mathcal{Z}_G(g)^\circ$, on a

$$(1) \qquad L(U_a) \subset (\operatorname{Ad} g - 1)\big(L(G)\big)$$

pour tout $a \in \Phi(T, G)$. Tout sous-espace de $L(G)$ stable par $\operatorname{Ad}\big(\varphi(H)\big)$ est stable par $\operatorname{Ad} G$, puisque $\varphi(H)$ est dense dans G pour la topologie de Zariski. Si un tel sous-espace contient $L(U_a)$ pour tout a, il coïncide avec $L(G)$, vu 3.6. Définissant l'ensemble H_1 comme en 9.1, on voit donc, compte tenu de (1), que $\varphi^{-1}(\Omega) \subset H_1$. Comme Ω est un ouvert de Zariski non vide de G, l'adhérence de $\Omega(k)$ (pour la topologie ordinaire) contient e; par conséquent, e appartient à l'adhérence de H_1 dans H et (ii) est un cas particulier de 9.1 (ii).

Soit X un sous-groupe distingué de H et soit G_1 la composante neutre de l'adhérence de $\varphi(X)$ pour la topologie de Zariski. C'est un k-sous-groupe distingué de G. Nous voulons montrer que les paires (H, X) et $\big(G(k), G_1(k)\big)$ sont localement isomorphes. Quitte à remplacer X par $\varphi^{-1}(G_1) \cap X$, nous pouvons supposer que $\varphi(X) \subset G_1$ et il suffit alors de faire voir que $\varphi(X)$ est ouvert dans G_1. Quitte à substituer G_1 à G, nous pouvons supposer que $\varphi(X)$ est dense dans G pour la topologie de Zariski. Mais alors, $\varphi(X) \cap \Omega$ n'est pas vide, donc $X \cap H_1 \neq \varnothing$ et X est ouvert, vu 9.1 (i).

9.4. COROLLAIRE. *Soient G et H comme en 9.3, soit X un sous-groupe distingué de H et soit G_1 comme en 9.3 (i).*

(i) Le quotient H/X est localement isomorphe à $(G/G_1)(k)$.

(ii) Si H/X possède une suite de composition dont chaque facteur est commutatif ou dénombrable, alors X est ouvert dans H.

(iii) Si G est presque simple sur k, X est ouvert ou discret.

(i) et (iii) sont conséquences immédiates de 9.3 (i) puisque $G(k)/G_1(k)$ s'identifie à un ouvert de $(G/G_1)(k)$ (cf. 3.18). Pour montrer (ii), on se ramène, par une induction évidente sur la longueur de la suite de composition, au cas où H/X lui-même est commutatif ou dénombrable, et on utilise (i).

9.5. COROLLAIRE. *Soient $\pi\colon \tilde{G} \to G$ un revêtement universel de G, q le degré d'inséparabilité de π (égal à la p-composante du degré de π) et H un sous-groupe de $G(k)$ tel que $H \cap \pi(\tilde{G}(k))$ soit ouvert dans $\pi(\tilde{G}(k))$. Alors, le groupe dérivé $\mathfrak{D}(H)$ de H est fermé dans H et $H/\mathfrak{D}(H)$ possède un sous-groupe* *ouvert d'exposant divisant q.*

Si $q = 1$, alors $\mathfrak{D}(H)$ est ouvert dans H (9.4 (ii)) et l'assertion est claire. Supposons donc $q \neq 1$. Vu [7: 2.7, 2.24], $\mathfrak{D}(H)$ est contenu dans $H \cap \pi(\widetilde{G}(k))$. Comme $\pi(\widetilde{G}(k))$ est localement isomorphe à $\widetilde{G}(k)$ (en tant que groupe topologique), il résulte de 9.4 (ii) que $\mathfrak{D}(H)$ est ouvert dans $\pi(\widetilde{G}(k))$. L'homomorphisme canonique $H/\mathfrak{D}(H) \to H/(\pi(\widetilde{G}(k)) \cap H)$ est donc un isomorphisme local. Comme $H/\mathfrak{D}(H)$ possède des sous-groupes ouverts arbitrairement petits, notre assertion résulte de 3.19 (i).

9.6. LEMME. *Soient* $\pi\colon \widetilde{G} \to G$ *un revêtement universel de* G, H *un groupe topologique localement isomorphe à un sous-groupe de* $G(k)$ *contenant* $\pi(\widetilde{G}(k))$, H' *et* H_1' *des groupes localement compacts,* $\alpha\colon H \to H'$ *un homomorphisme et* $\varphi\colon H' \to H_1'$ *un homomorphisme continu dont le noyau est central dans* H' *et qui est localement un isomorphisme. Si* π *n'est pas séparable, supposons que* $\mathrm{Ker}\,\varphi$ *ne possède pas d'élément d'ordre* p. *Supposons* $\varphi \circ \alpha$ *continu. Alors* α *est continu.*

Supposons d'abord k isomorphe à $\mathbf{R}$ ou à $\mathbf{C}$. Alors $\pi(\widetilde{G}(k))$ est ouvert dans $G(k)$, donc H est un groupe de Lie localement isomorphe à $G(k)$. Soit $\psi\colon \widetilde{H} \to H^\circ$ un revêtement universel de la composante neutre de H. L'homomorphisme ψ est ouvert donc, quitte à remplacer H et α par $\widetilde{H}$ et $\alpha \circ \psi$, nous pouvons supposer H connexe et simplement connexe. Mais alors, vu la propriété universelle des groupes simplement connexes, il existe un homomorphisme continu $\alpha'\colon H \to H'$ tel que $\varphi \circ \alpha' = \varphi \circ \alpha$. Comme $\mathrm{Ker}\,\varphi \subset \mathcal{C}(H')$, l'application $\alpha''\colon H \to \mathrm{Ker}\,\varphi$ définie par

$$(1) \qquad\qquad \alpha''(h) = \alpha(h) \cdot \alpha'(h)^{-1} \qquad\qquad (h \in H)$$

est un homomorphisme, dont le noyau est ouvert, vu 9.4 (ii). Par conséquent, α'' est continu, et il en est de même de α.

Supposons à présent k totalement discontinu. Quitte à remplacer H par un sous-groupe ouvert et α par sa restriction à ce sous-groupe, nous pouvons supposer que H est un sous-groupe de $G(k)$ dont l'adhérence est compacte et tel que $H \cap \pi(\widetilde{G}(k))$ est ouvert dans $\pi(\widetilde{G}(k))$. Le groupe H possède des sous-groupes distingués ouverts d'indice fini arbitrairement petits. Il en est donc de même de l'adhérence H_2' du groupe $\alpha(\varphi(H))$, et H_2' est un groupe compact totalement discontinu. Soient U un sous-groupe ouvert de H_2' et $\varphi'\colon U \to H'$ un homomorphisme continu tels que $\varphi \circ \varphi' = \mathrm{id}_U$ (l'existence de U, φ' résulte de ce que φ est un isomorphisme local). Quitte à remplacer H par $(\varphi \circ \alpha)^{-1}(U)$, nous pouvons supposer que $\alpha(\varphi(H)) \subset U$. Posons $\alpha' = \varphi' \circ (\varphi \circ \alpha)$. C'est un homomorphisme continu de H dans H' et on a $\varphi \circ \alpha' = \varphi \circ \alpha$. L'application

$\alpha''\colon H \to \operatorname{Ker}\varphi$ définie par (1) est à nouveau un homomorphisme dont le noyau est ouvert, vu cette fois 9.5 et l'hypothèse faite sur $\operatorname{Ker}\varphi$. Comme précédemment, on en déduit que α'' est continu, donc que α l'est aussi.

9.7. *Définition.* Soient h un corps, F un groupe algébrique connexe défini sur h et F_1 un h-sous-groupe fermé connexe $\neq \{e\}$ de F. Par abus de langage, nous dirons que $F_1(h)$ est un *facteur complexe* de $F(h)$ si F_1 est un facteur presque direct de F (c'est-à-dire s'il existe un h-sous-groupe fermé F_2 commutant avec F_1 et tel que $F = F_1 \cdot F_2$ et que $F_1 \cap F_2$ soit fini) et si $h \cong \mathbf{C}$ ou bien $h \cong \mathbf{R}$ et F_1 est isogène à un groupe de la forme $R_{\bar{h}/h}F_1'$, où $\bar{h}$ désigne une clôture algébrique de h.

9.8. PROPOSITION. *Supposons que G ne possède pas de sous-groupe distingué connexe anisotrope sur k non trivial, que G' soit réductif et que $G'(k')$ ne possède pas de facteur complexe. Soit k_0 (resp. k_0') l'adhérence du corps premier de k (resp. k') dans k (resp. k'). Soient H un sous-groupe de $G(k)$ contenant G^+ et $\alpha\colon H \to G'(k')$ un homomorphisme dont l'image est dense dans G' pour la topologie de Zariski. Alors α est continu et $k_0 \cong k_0'$. Si $G''(k)$ est un facteur complexe de $G(k)$, on a $\alpha(G''^+) = \{e\}$; en particulier, $k \not\cong \mathbf{C}$.*

Vu 6.14, $G(k)/G^+$ est un groupe d'exposant fini. Il s'ensuit que $\bar{\alpha}(H)/\bar{\alpha}(G^+) = G'/\bar{\alpha}(G^+)$ est un groupe réductif d'exposant fini, donc est réduit à l'élément neutre, c'est-à-dire que $\bar{\alpha}(G^+) = G'$. Comme G^+ est son propre groupe dérivé (6.4), il en est de même de G', qui est donc un groupe semi-simple. Vu 7.2 (i), on a car $k = $ car k'.

Comme $\operatorname{Ad}_{G'}(\alpha(H))$ est dense dans $\operatorname{Ad} G'$ pour la topologie de Zariski, son centre est réduit à l'élément neutre, donc $\operatorname{Ker}(\operatorname{Ad}_{G'} \circ \alpha)$ contient le centre de H et il existe $\alpha'\colon \operatorname{Ad}_G H \to (\operatorname{Ad} G')(k')$ tel que $\alpha'(\operatorname{Ad}_G h) = \operatorname{Ad}_{G'}(\alpha(h))$ pour tout $h \in H$. Comme l'exposant caractéristique de k' est égal à p, le noyau de $\operatorname{Ad}_{G'}$ ne possède pas d'élément d'ordre p distinct de e. Par conséquent, le lemme 9.6 s'applique et il suffit, pour établir la proposition, de montrer que α' est continu (car alors, $h \mapsto \alpha'(\operatorname{Ad}_G h)$ l'est aussi). Nous pouvons donc supposer que les groupes G et G' sont adjoints et (compte tenu de ce que α est alors le produit direct de ses composés avec les projections de G' sur les divers facteurs simples de celui-ci) que G' est simple sur k'.

Soit G_1 un facteur simple sur k de G tel que $\alpha(G_1^+) \neq \{e\}$ et soit G_2 le produit des autres facteurs simples de G. Comme G_1^+ est distingué dans H, on a $\bar{\alpha}(G_1^+) = G'$. Par conséquent, $\alpha(G_2(k) \cap H)$, qui centralise $\alpha(G_1^+)$, est réduit à l'élément neutre. Il s'ensuit que si on désigne par $pr_1\colon G \to G_1$ la projection canonique, il existe un homomorphisme $\alpha_1\colon pr_1 H \to G'(k')$ tel que $\alpha = \alpha_1 \circ pr_1|_H$. Quitte à remplacer G par G_1, α par α_1 et H par $pr_1 H$, nous pouvons supposer G simple sur k. Il existe alors une extension séparable l de k et un l-groupe absolument simple F tel que $G = R_{l/k}F$. Comme $R_{l/k}^{\circ}$ est un isomorphisme de groupes topologiques, on voit que, remplaçant k par l, G par F et H par son image canonique dans $F(l)$, on est ramené au cas où G est absolument simple sur k, et notre proposition est alors une conséquence immédiate du théorème 8.16 et de 2.3, 2.4.

9.9. LEMME. *Soit H un groupe k-analytique simple localement isomorphe à $G(k)$. Alors, H est isomorphe (comme groupe analytique) à un sous-groupe ouvert de $G(k)/\mathcal{C}(G)(k)$.*

Posons $G_1 = G(k)/\mathcal{C}(G)(k)$. Soient L l'algèbre de Lie de H, que nous identifions à $L(G)(k)$, et $\lambda: U \to U'$ un isomorphisme analytique d'un voisinage ouvert de e dans H sur un voisinage ouvert de e dans $G(k)$ tel que Ad_H et $\mathrm{Ad}_G \circ \lambda$ coïncident sur U. Pour $h \in H$, il existe des voisinages V_1 et V_2 de e dans U tels que $hV_1h^{-1} = V_2$, d'où $(\mathrm{Ad}\,h)(\mathrm{Ad}\,V_1)(\mathrm{Ad}\,h)^{-1} = \mathrm{Ad}\,V_2$. L'adhérence de $\mathrm{Ad}\,V_i$ ($i = 1, 2$) pour la topologie de Zariski étant $(\mathrm{Ad}\,G)(k)$, il en résulte que $(\mathrm{Ad}\,G)(k)$ est normalisé par $\mathrm{Ad}\,h$. Utilisant l'hypothèse de simplicité de H, on en déduit successivement que $\mathrm{Ad}\,H \subset (\mathrm{Ad}\,G)(k)$ puis, comme $\mathrm{Ad}(G(k))$ est distingué dans $(\mathrm{Ad}\,G)(k)$ et contient $\mathrm{Ad}_H U$, que $\mathrm{Ad}\,H \subset \mathrm{Ad}(G(k))$. L'homomorphisme Ad_G définit un isomorphisme de groupes topologiques ψ de G_1 sur $\mathrm{Ad}(G(k))$. Donc, $\psi' = \psi^{-1} \circ \mathrm{Ad}_H$ est un homomorphisme injectif continu de H dans G_1. Soient π la projection canonique de $G(k)$ sur G_1 et V un voisinage ouvert de e dans U' tel que $\pi|_V$ soit un isomorphisme de variétés analytiques de V sur $\pi(V)$ (qui est un voisinage ouvert de e dans G_1). Alors, ψ', qui coïncide avec $\pi \circ \lambda$ sur $\lambda^{-1}(V)$, induit un isomorphisme de variétés analytiques de $\lambda^{-1}(V)$ sur $\pi(V)$, donc ψ' est un isomorphisme de groupes analytiques de H sur un sous-groupe ouvert de G_1.

9.10. Pour établir la proposition 9.11 et le corollaire 9.12, nous aurons à utiliser le résultat suivant, inédit à notre connaissance (cf. cependant [34: 3.6] pour le cas des groupes de rang relatif 1):

²

(*) *si G est presque simple sur k, tout sous-groupe
propre ouvert de G^+ est compact.*

Le lecteur qui n'est pas prêt à l'admettre pourra s'en passer à condition de supposer en 9.11 que $\alpha(H)$ contient G'^+, et en 9.12 que $\varphi(H'')$ contient G'^+.

9.11. PROPOSITION. *Supposons que le groupe fondamental de G soit réduit et que $R_u(G')$ soit défini et déployé sur k'. Soient H un groupe k-analytique dénombrable à l'infini, localement isomorphe à $G(k)$, et $\alpha: H \to G'(k')$ un homomorphisme. Supposons $\alpha(H) \cap G'^+$ ouvert dans G'^+. Alors G' est réductif. De plus, α est continu ou bien les groupes $G(k)$ et $G'(k')$ possèdent tous* *deux un facteur complexe.*

Vu 9.3 (i), il existe un k-sous-groupe distingué connexe G_1 de G tel que $\alpha^{-1}(R_u(G'))$ soit localement isomorphe à $G_1(k)$. Il résulte alors de 9.4 (ii) que $\operatorname{Ker}\alpha$ est ouvert, donc d'indice au plus dénombrable, dans $\alpha^{-1}(R_u(G'))$. Par conséquent, $\alpha(H) \cap R_u(G')$ est un groupe d'ordre au plus dénombrable. Comme il est ouvert dans $R_u(G')(k')$, par hypothèse, on a $R_u(G') = \{e\}$ et G' est réductif. Nous supposerons dorénavant que l'un au moins des groupes $G(k)$ et $G'(k')$ ne possède pas de facteur complexe, et nous nous proposons de montrer que α est continu.

Quitte à remplacer H par $\mathcal{D}(H)$, qui est ouvert dans H vu 9.4 (ii), et G' par $\mathcal{D}(G')$ (qu'on peut évidemment supposer non réduit à l'élément neutre) nous pouvons supposer G' semi-simple. Vu 9.6, il suffit alors de faire voir que $\operatorname{Ad}_{G'} \circ \alpha$ est continu, c'est-à-dire que nous pouvons aussi supposer G' adjoint et même (en composant α avec les projections de G' sur ses divers facteurs) simple sur k'.

Si $\alpha(H)$ est relativement compact, la proposition résulte de 9.3 (ii). Supposons donc $\alpha(H)$ non relativement compact, donc G' isotrope vu 3.18. Le groupe $\alpha(H) \cap G'^+$ est distingué dans $\alpha(H)$ et le quotient $\alpha(H)/(a(H) \cap G'^+)$ est extension d'un groupe fini par un groupe commutatif (6.14). (En fait, il est commutatif: cf. 6.15.) Par suite, $\alpha^{-1}(a(H) \cap G'^+)$ est ouvert dans H (9.4 (ii)) et, quitte à remplacer H par ce sous-groupe, nous pouvons supposer que $\alpha(H) \subset G'^+$, donc $\alpha(H) = G'^+$ vu 9.10 (*). En vertu de 9.4 (i), $H/\operatorname{Ker}\alpha$ est aussi localement isomorphe au groupe des points rationnels d'un k-groupe semi-simple simplement connexe. Quitte à remplacer G par celui-ci et H par $H/\operatorname{Ker}\alpha$, nous pouvons donc supposer que α est un isomorphisme (de groupes "abstraits") de H sur G'^+. Vu le lemme 9.9 et [31], H peut être identifié à un sous-groupe ouvert de $G(k)/\mathcal{C}(G)(k)$, donc aussi (3.18) à un sous-groupe fermé de $(\operatorname{Ad} G)(k)$, dense pour la topologie de Zariski.

Si G ne possède pas de facteur complexe, nous pouvons appliquer la proposition 9.8 à l'isomorphisme $\alpha^{-1}: G'^+ \to H$ et il en résulte que α^{-1} est continu, donc est un isomorphisme de groupes topologiques.

Il reste à considérer le cas où G possède un facteur complexe. Comme H est simple, ce facteur est alors le groupe G lui-même et $H = (\operatorname{Ad} G)^+$. Cela étant, nous pouvons, sans nuire à la généralité, supposer que $k = \mathbf{C}$ et il résulte aussitôt du théorème 8.16 que G' est aussi un groupe complexe.

9.12. COROLLAIRE. *Supposons que G et G' remplissent les conditions de 9.11 et que l'un au moins des groupes $G(k)$ et $G'(k')$ ne possède pas de facteur complexe. Soient H un groupe k-analytique localement isomorphe à $G(k)$ et*

H' un groupe localement compact. Supposons que H et H' soient dénombrables à l'infini et qu'il existe un sous-groupe distingué ouvert H'' de H' et un homomorphisme $\varphi\colon H'' \to G'(k')$ qui soit un isomorphisme local dont le noyau est central dans H''. Soit α un homomorphisme surjectif de H sur H'. Alors α est continu. S'il est aussi injectif, c'est un isomorphisme de groupes topologiques.

Vu 3.18, il suffit de montrer que α est continu. Comme H'' est d'indice au plus dénombrable dans H', il résulte de 9.4 (ii) que $\alpha^{-1}(H'')$ est ouvert dans H. Quitte à remplacer H par $\alpha^{-1}(H'')$, nous pouvons donc supposer que $H' = H''$. Il résulte alors de la proposition 9.11 que $\varphi \circ \alpha$ est continu et il suffit d'appliquer le lemme 9.6 pour achever la démonstration.

9.13. Le corollaire suivant répond à une question qui nous a été posée par M. Goto et H. C. Wang.

COROLLAIRE. *Supposons que l'un au moins des groupes $G(k)$ et $G'(k')$ ne possède pas de facteur complexe, et que l'une des conditions suivantes soit remplie:*

(i) *le groupe fondamental de G est réduit;*

(ii) *G ne possède pas de k-sous-groupe distingué connexe anisotrope $\neq \{e\}$ et G' est réductif.*

Alors, tout homomorphisme surjectif de $G(k)$ sur $G'(k')$ est continu et tout isomorphisme de $G(k)$ sur $G'(k')$ est un isomorphisme de groupes topologiques.

Si (i) est satisfaite, c'est un cas particulier de 9.12. Sinon, car $k \neq 0$, donc car $k' \neq 0$ vu 7.2 (i); en particulier $G'(k')$ n'a pas de facteur complexe et on peut appliquer la proposition 9.8.

9.14. Le corollaire suivant généralise un théorème de H. Freudenthal [19].

COROLLAIRE. *Soient H et H' des groupes de Lie réels ayant au plus une infinité dénombrable de composantes connexes et L et L' leurs algèbres de Lie. Supposons que L soit semi-simple et que tous les facteurs directs simples de L ou de L' soient absolument simples. Alors tout homomorphisme surjectif de H sur H' est continu et tout isomorphisme de H sur H' est un isomorphisme de groupes de Lie.*

Soient $\alpha\colon H \to H'$ un homomorphisme surjectif et R le radical de la composante neutre de H'. Vu 9.3 (i) les sous-groupes distingués $\alpha^{-1}(R)$ et $\mathrm{Ker}\,\alpha$ de H sont fermés. Comme L est semi-simple, cela implique que le groupe résoluble $\alpha^{-1}(R)/\mathrm{Ker}\,\alpha$ est discret, donc dénombrable. Par conséquent $R = \{e\}$ et L' est semi-simple. Cela étant, le corollaire est un cas particulier de 9.12: il suffit de prendre pour G (resp. G') la composante neutre algébrique du groupe algébrique Aut L (resp. Aut L') et pour H'' la composante neutre topologique de H'.

9.15. *Remarques.* (a) Si k est totalement discontinu et si H est un groupe analytique localement isomorphe à $G(k)$, il peut arriver, même en caractéristique 0, que H possède des automorphismes non continus. Prenons par exemple $k = \mathbf{Q}_l$ (l un nombre premier) et $G = \mathbf{SL}_2$. Considérons le groupe localement compact F produit semi-direct de $\mathbf{SL}_2(\mathbf{Z}_l)$ et de son algèbre de Lie $L = \mathbf{M}_2^0(\mathbf{Q}_l)$ (nous désignons par $\mathbf{M}_2^0(X)$ le groupe des matrices d'ordre 2 et de trace nulle à coefficients dans l'anneau X). Le sous-groupe $L_1 = \mathbf{M}_2^0(\mathbf{Z}_l)$ de L est distingué dans F. Posons $H = F/L_1$. L'image de $\mathbf{SL}_2(\mathbf{Z}_l)$ dans H est un sous-groupe ouvert H_0 de H isomorphe à $\mathbf{SL}_2(\mathbf{Z}_l)$, donc H est localement isomorphe à $\mathbf{SL}_2(\mathbf{Q}_l)$. D'autre part, nous avons vu en 8.18 (b) qu'à toute dérivation non triviale d de $\mathbf{Q}_l$, on peut associer un "section de Levi" $\mathbf{SL}_2(\mathbf{Z}_l) \to F$ non continue, donc aussi un automorphisme non continu de F fixant tous les points de L. Cet automorphisme induit, par passage au quotient, un automorphisme α de H dont on va voir qu'il n'est pas continu. Il laisse fixes tous les points d'un sous-groupe dense de H, à savoir, le produit semi-direct de L/L_1 et de l'image canonique de $\mathbf{SL}_2(\mathbf{Q} \cap \mathbf{Z}_l)$ dans H_0. Supposons α continu, donc égal à l'identité. Cela implique par exemple que, pour toute unité z de $\mathbf{Z}_l$, on a $z^{-1}dz \in \mathbf{Z}_l$, d'où $dz \in \mathbf{Z}_l$, et par conséquent que $d\mathbf{Z}_l \subset \mathbf{Z}_l$. Mais alors, $d(l^n\mathbf{Z}_l) \subset l^n\mathbf{Z}_l$ pour tout entier n, et d est continu. Comme $d\mathbf{Q} = 0$, cela entraîne $d = 0$, contrairement à l'hypothèse.

Cet exemple montre que dans l'énoncé de 9.12, on ne peut pas remplacer l'hypothèse faite sur H' (existence de (H'', φ)) par celle, plus faible, que H' est localement isomorphe à $G'(k')$.

(b) Le corollaire 9.12 n'est pas valable si l'on ne suppose pas le groupe fondamental de G réduit. Prenons par exemple $k = k' = \mathbf{F}_2((t))$, $G = G' = \mathbf{PGL}_2$ et $H = G(k) \times (\mathbf{Z}/2\mathbf{Z})$. Le déterminant $\mathbf{GL}_2(k) \to k^*$ induit, par passage au quotient, un homomorphisme $\delta \colon G(k) \to k^*/k^{*2}$. Si ε désigne un homomorphisme non-continu de k^*/k^{*2} dans $\mathbf{Z}/2\mathbf{Z}$ (il est facile de voir qu'il en existe), l'automorphisme de H défini par

$$(x, y) \mapsto \big(x, \varepsilon(\delta(x)) + y\big) \qquad\qquad \big(x \in G(k),\ y \in \mathbf{Z}/2\mathbf{Z}\big)$$

n'est pas continu.

(c) En 9.13 (ii), on ne peut se passer de l'hypothèse que G' est réductif. Prenons en effet $k = k' = \mathbf{F}_2((t))$, $G = \mathbf{PGL}_2$ et $G' = \mathbf{Add}$. Le groupe $G(k)/\mathfrak{D}(G(k))$ est un groupe commutatif d'exposant 2 ayant la puissance du continu, et il en est de même de $G'(k')$. Ces deux groupes sont donc isomorphes, et il existe un homomorphisme surjectif $G(k) \to G'(k')$. Mais un tel homomorphisme ne peut être continu, car $G(k)/\mathcal{D}(G(k))$ est compact et $G'(k')$ ne l'est pas.

Nous ignorons si, toujours en 9.13 (ii), l'hypothèse faite sur G est indispensable.

(d) Nous ne savons pas non plus si le lemme 9.9, la proposition 9.11 et le corollaire 9.12 restent vrais lorsqu'on suppose seulement que H est un groupe *topologique* localement isomorphe à $G(k)$. Toutefois, il en est bien ainsi si car $k = 0$; en effet, on peut alors, moyennant une restriction de scalaires évidente, supposer que k est l'adhérence de son corps premier, auquel cas tout groupe topologique localement isomorphe à $G(k)$ est analytique. Remarquons aussi que si car $k = $ car $k' = 0$, on peut remplacer, dans 9.8, 9.11, 9.12 et 9.13, "continu" par "k_0-analytique" et "topologiques" par "k_0-analytiques", où k_0 est l'adhérence du corps premier de k dans k [10: §8, n°1].

10. Application aux représentations linéaires et projectives

10.1. Soient m et n des entiers strictement positifs. On désigne par $\mathbf{PGL}_n$ le groupe adjoint de $\mathbf{GL}_n$. Le groupe $\mathbf{PGL}_n(k)$ s'identifie au groupe des transformations projectives de l'espace projectif $\mathbf{P}_{n-1}(k)$ à $n - 1$ dimensions sur k. L'homomorphisme $\mathbf{GL}_m \times \mathbf{GL}_n \to \mathbf{GL}_{mn}$ donné par le produit tensoriel induit par passage au quotient un morphisme $\tau_{m,n} \colon \mathbf{PGL}_m \times \mathbf{PGL}_n \to \mathbf{PGL}_{mn}$, défini sur le corps premier.

Soit M un groupe. Si $\rho \colon M \to \mathbf{PGL}_m(k)$ et $\sigma \colon M \to \mathbf{PGL}_n(k)$ sont des représentations projectives, alors $\tau_{m,n} \circ (\rho \times \sigma)$ est une représentation projective que l'on appellera le *produit* ou *produit tensoriel* de ρ et σ. On appellera aussi produit de deux représentation ρ_i ($i = 1, 2$) de deux groupes M_i différents, la représentation de $M_1 \times M_2$ produit des représentations $\rho_i \circ pr_i$. On définit de même, en remplaçant $\tau_{m,n}$ par le produit tensoriel ordinaire, le produit (tensoriel) de deux représentations linéaires d'un groupe ou de deux groupes distincts. Les opérations de produit ainsi définies sont compatibles avec l'équivalence et passent donc aux classes d'équivalence de représentations linéaires ou projectives.

On note $[\pi]$ la classe d'équivalence d'une représentation linéaire ou projective π.

10.2. Soient F un corps algébriquement clos de caractéristique q et H un groupe algébrique semi-simple défini sur F. On note $\mathcal{R}(H)$ (resp. $\mathcal{R}'(H)$) l'ensemble des classes d'équivalence de représentations rationnelles linéaires (resp. projectives) irréductibles de H. On a une injection naturelle $\mathcal{R}(H) \to$

$\mathcal{R}'(H)$, qui est bijective si H est simplement connexe. l'ensemble $\mathcal{R}'(H)$ n'est pas affecté par une isogénie centrale. Si $H = H_1 \times H_2$, les opérations de produit décrites en 10.1 identifient $\mathcal{R}(H)$ et $\mathcal{R}'(H)$ respectivement à $\mathcal{R}(H_1) \times \mathcal{R}(H_2)$ et $\mathcal{R}'(H_1) \times \mathcal{R}'(H_2)$.

Si $q \neq 0$, notons $\mathfrak{M}'(H)$ l'ensemble des q^l ($l = rg\, H$) représentations projectives irréductibles dont le poids dominant est combinaison linéaire des poids dominants fondamentaux avec des coefficients compris entre 0 et $q - 1$. D'après un résultat de R. Steinberg ([28] ou [2]), tout élément $[\pi]$ de $\mathcal{R}'(H)$ peut s'écrire, de façon unique aux facteurs trivaux près, comme un produit fini de la forme $\prod_i ([\pi_i \circ Fr^i])$ avec $[\pi_i] \in \mathfrak{M}'(^{Fr^i}H)$. De plus, si H est presque simple et si π_0 n'est pas triviale, π est une isogénie spéciale de H sur son image.

Si $q = 0$, on pose $\mathfrak{M}'(H) = \mathcal{R}'(H)$

À isomorphisme canonique près, les ensembles $\mathcal{R}(H)$, $\mathcal{R}'(H)$, $\mathfrak{M}'(H)$ ne dépendent pas du corps de définition algébriquement clos choisi. Plus généralement, si $\varphi: F \to F'$ est un homomorphisme de corps algébriquement clos, l'application $\pi \mapsto {}^\varphi\pi$ est une bijection de $\mathcal{R}(H)$ sur $\mathcal{R}(^\varphi H)$, de $\mathcal{R}'(H)$ sur $\mathcal{R}'(^\varphi H)$ et de $\mathfrak{M}'(H)$ sur $\mathfrak{M}'(^\varphi H)$. En particulier, les ensembles $\mathfrak{M}'(^{Fr^i}H)$ intervenant dans notre énoncé du résultat de Steinberg peuvent tous être identifiés à $\mathfrak{M}'(H)$ (ce que Steinberg fait d'ailleurs).

Si H est simplement connexe, nous noterons $\mathfrak{M}(H)$ l'ensemble correspondant à $\mathfrak{M}'(H)$ par la bijection canonique $\mathcal{R}(H) \to \mathcal{R}'(H)$.

Le résultat suivant a été conjecturé par R. Steinberg [28] dans le cas d'un corps k algébriquement clos.

10.3. Théorème. *Supposons k infini, k' algébriquement clos et G absolument presque simple et isotrope sur k. Soient H un sous-groupe de $G(k)$ contenant G^+ et $\rho: H \to \mathbf{PGL}_n(k')$ un homomorphisme ($n \geq 2$). Supposons que $\rho(G^+)$ soit un groupe irréductible de transformations projectives de $\mathbf{P}_{n-1}(k')$. Alors, il existe des homomorphismes distincts $\varphi_i: k \to k'$, en nombre fini, et, pour chaque i, une représentation rationnelle projective non triviale π_i de $^{\varphi_i}G$, tels que $[\pi_i] \in \mathfrak{M}'(^{\varphi_i}G)$ et que, sur H, ρ soit équivalente au produit $\prod_i (\pi_i \circ \varphi_i^0)$; en particulier, car $k = $ car k'. Les couples $([\pi_i], \varphi_i)$ sont uniques à l'ordre près.*

Soit G' l'adhérence de $\rho(G^+)$. C'est un k'-groupe connexe $\neq \{e\}$ vu les hypothèses, et l'on a car $k = $ car k' (7.2). Comme G' est irréductible en tant que groupe de transformations projectives, il est réductif. Le centre de $L(G')$ est diagonalisable, donc trivial, sinon G' laisserait stable un sous-espace projectif propre non vide de $\mathbf{P}_{n-1}(k')$. Par conséquent, G' est semi-simple adjoint. Soient G'_j ses facteurs simples.

La représentation identique ι de G' peut s'écrire sous la forme d'un produit $\prod_j \lambda_j$, où λ_j désigne une représentation projective irréductible de G'_j. En vertu du théorème 8.16, il existe des homomorphismes $\psi_j \colon k \to k'$ et des isogénies spéciales $\beta_j \colon {}^{\psi_j}G \to G'_j$, tous uniques, tels qu'on ait

$$(1) \qquad \rho = \iota \circ \prod_j (\beta_j \circ \psi_j^0) = \prod_j (\lambda_j \circ \beta_j \circ \psi_j^0) \quad sur \ G^+ \ .$$

De plus

$$(2) \qquad \psi_j \circ Fr^a = \psi_{j'} \circ Fr^{a'} \qquad (a,\ a' \in \mathbf{N}) \Longrightarrow j = j' \ \text{et} \ a = a' \ .$$

Vu le résultat de Steinberg rappelé en 10.3, $\lambda_j \circ \beta_j$ peut s'écrire sous la forme d'un produit $\prod_h (\pi_{jh} \circ Fr^h)$ avec $[\pi_{jh}] \in \mathfrak{M}'({}^{\varphi_{jh}}G)$ non trivial, où l'on a posé $\varphi_{jh} = Fr^h \circ \psi_j$ et où h prend seulement la valeur 0 ou bien parcourt un nombre fini de valeurs entières positives distinctes selon que $p = 1$ ou $p \neq 1$. Désignons simplement par φ_i et π_i les φ_{jh} et π_{jh} et posons

$$\rho' = \prod_i (\pi_i \circ \varphi_i^0) \ .$$

Vu (2), les φ_i sont tous distincts, et vu (1), ρ et ρ' coïncident sur G^+. Mais alors, si $x \in H$, le quotient $\rho(x) \cdot \rho'(x)^{-1}$ centralise $\rho(G^+)$, donc est égal à e en vertu du lemme de Schur, et on a $\rho = \rho'|_H$.

Soit maintenant

$$(3) \qquad \rho = \prod_{i \in I}(\sigma_i \circ \eta_i^0)$$

une décomposition de ρ satisfaisant à nos conditions. Soit J la plus petite partie de I telle que tout η_i soit de la forme $Fr^h \circ \eta_j$ avec $h \in \mathbf{N}$ et $j \in J$. En particulier

$$(4) \qquad Fr^a \circ \eta_j = Fr^{a'} \circ \eta_{j'} \qquad (j,\ j' \in J;\ a,\ a' \in \mathbf{N}) \Longrightarrow j = j' \ \text{et} \ a = a' \ .$$

Pour $i \in I$, soient $j(i) \in J$ et $h(i) \in \mathbf{N}$ définis par $\eta_i = Fr^{h(i)} \circ \eta_{j(i)}$. On a alors, sur H,

$$(5) \qquad \rho = \prod_{j \in J} (\gamma_j \circ \eta_j^0)$$

avec

$$(6) \qquad \gamma_j = \prod_{j(i)=j} (\sigma_i \circ Fr^{h(i)}) \ .$$

Soit G''_j l'image de ${}^{\eta_j}G$ par γ_j. Vu (4) et la remarque 8.17 (c), $\rho(G^+)$ est dense dans $\prod G''_j$. Par conséquent, les G''_j ne sont autres que les G'_j. D'autre part, comme $h(j) = 0$, γ_j est une isogénie spéciale. Le théorème 8.16 implique alors que les (γ_j, η_j) coïncident à l'ordre près avec les (β_j, ψ_j) du début de la démonstration. L'assertion d'unicité s'ensuit en vertu de (6) et, pour $p \neq 1$, de l'unicité de la décomposition de Steinberg.

10.4. Corollaire. *Supposons G simplement connexe. Soit $\rho \colon G^+ \to$ $\mathbf{GL}_n(k')$ un homomorphisme $(n \geq 2)$ tel que $\rho(G^+)$ soit un groupe irréductible*

de transformations linéaires. Alors il existe des homomorphismes $\varphi_i\colon k \to k'$
en nombre fini, et pour chaque i une représentation linéaire non triviale π_i
de $^{\varphi_i}G$ *tels que* $[\pi_i] \in \mathfrak{M}(^{\varphi_i}G)$ *et que* ρ *soit équivalente au produit* $\prod_i (\pi_i \circ \varphi_i^0)$,
restreint à G^+.

Soit $\tau\colon \mathbf{GL}_n \to \mathbf{PGL}_n$ la projection canonique. Soient $\varphi_i\colon k \to k'$ et
$[\pi_i'] \in \mathfrak{M}'(G)$ tels que, sur G^+, $\tau \circ \rho'$ soit équivalente à $\prod_i (\pi_i' \circ \varphi_i^0)$ (cf. 10.3).
Comme $^{\varphi_i}G$ est simplement connexe, π_i' se relève en une représentation linéaire
π_i, et $[\pi_i] \in \mathfrak{M}(^{\varphi_i}G)$. On a alors

$$\rho(x) = \mu(x) \cdot \left(\prod_i (\pi_i \circ \varphi_i^0) \right)(x) , \qquad\qquad (x \in G^+)$$

où μ est un homomorphisme de G^+ dans le centre de $\mathbf{GL}_n$. Mais G^+ est égal
à son groupe dérivé (6.4), donc μ est trivial, d'où le corollaire.

Institute for Advanced Study, Princeton
Universität Bonn (Sonderforschungsbereich Theoretische Mathematik)

Bibliographie

[1] A. Borel, Linear Algebraic Groups, Notes by H. Bass, Benjamin, New York 1969.

[2] ————, *Properties of linear representations of Chevalley groups*, Seminar on algebraic groups and related finite groups, Springer Lecture Notes 131, Part A, 55 pages.

[3] ———— and T. A. Springer, *Rationality properties of linear algebraic groups* II, Tôhoku Math. J. **20** (1968), 443-497.

[4] ———— et J. Tits, *Groupes réductifs*, Publ. Math. I.H.E.S. n° 27 (1965), 55-151.

[5] ———— and ————, *On "abstract" homomorphisms of simple algebraic groups*, Proc. Bombay Colloquium on Algebraic Geometry, 1968, 75-82.

[6] ———— et ————, *Éléments unipotents et sous-groupes paraboliques des groupes réductifs* I, Inv. Mat. **12** (1971), 95-104.

[7] ———— et ————, *Compléments à l'article "Groupes réductifs"*, Publ. Math. I.H.E.S. n° 41 (1972), 253-276.

[8] N. Bourbaki, Algèbre, Chap. IV, V, Act. Sci. Ind. 1102, Hermann éd., Paris (1959).

[9] ————, Intégration, Chap. VII, VIII, Act. Sci. Ind. 1306, Hermann éd., Paris (1963).

[10] ————, Groupes et Algèbres de Lie, Chap. II, III, Act. Sci. Ind. 1349, Hermann éd., Paris (1972).

[11] ————, Groupes et Algèbres de Lie, Chap. IV, V, VI, Act, Sci. Ind. 1337, Hermann éd., Paris (1968).

[12] E. Cartan, *Sur les représentations linéaires des groupes clos*, Comm. Math. Helv., **2** (1930), 269-283.

[13] C. Chevalley, *Sur certains groupes simples*, Tôhoku Math. J. (2) **7** (1955), 14-66.

[14] ————, Classification des groupes de Lie algébriques, Notes polycopiées, Inst. H. Poincaré, Paris (1956-58).

[15] ————, *Certain schémas de groupes semi-simples*, Sém. Bourbaki 13è année, (1960-61), Exp. 219.

[16] M. Demazure et P. Gabriel, Groupes algébriques, T. 1, Masson, Paris (1970).

[17] ———— et A. Grothendieck, Schémas en groupes I, II, III, Springer Lecture Notes 151, 152, 153.

[18] J. Dieudonné, La géométrie des groupes classiques, 2 ième édition, Springer-Verlag, Berlin-Göttingen-Heidelberg, 1963.

[19] H. FREUDENTHAL, *Die Topologie der Lieschen Gruppen als algebraisches Phänomen* I, Ann. of Math. (2) **42** (1941), 1051-1074. *Erratum ibid.* **47** (1946), 829-830.

[20] K. MCCRIMMON, *A general theory of Jordan rings*, Proc. Nat. Ac. Sci. USA **56** (1966), 1072-1079.

[21] O. T. O'MEARA, *The automorphisms of the orthogonal groups* $\Omega_n(V)$ *over fields*, Amer. J. Math. **90** (1968), 1260-1306.

[22] V. P. PLATONOV, *Le problème d'approximation forte et la conjecture de Kneser-Tits*, Izvestiya Ak. Nauk. USSR **33** (1969), 1211-1219; *compléments, ibid.* **34** (1970), 775-777.

[23] C. RIEHM, *The congruence subgroup problem over local fields*, Amer. J. Math. XCII (1970), 771-778.

[24] M. ROSENLICHT, *Some rationality questions on algebraic groups*, Annali di Math. (IV), **43** (1957), 25-50.

[25] O. SCHREIER und B. L. VAN DER WAERDEN, *Die Automorphismen der projektiven Gruppen*, Abh. Math. Sem. Hamburg **6** (1928), 303-322.

[26] T. A. SPRINGER, Jordan algebras and algebraic groups (to appear).

[27] R. STEINBERG, *Automorphisms of finite linear groups*, Canadian J. Math. **12** (1960), 606-615.

[28] ————, *Representations of algebraic groups*, Nagoya Math. J. **22** (1963), 33-56.

[29] ————, Lectures on Chevalley groups, Yale University lecture notes, 1967.

[30] J. TITS, *Groupes simples et géométries associées*, Proc. Intern. Congress Math., Stockholm, 1962, 197-221.

[31] ————, *Algebraic and abstract simple groups*, Annals of Math. (2) **80** (1964), 313-329.

[32] ————, *Classification of semi-simple algebraic groups*, Algebraic groups and discontinuous subgroups, Proc. Symp. Pure Math. **9**, A.M.S., Providence, R. I. (1966), 33-62.

[33] ————, *Formes quadratiques, groupes orthogonaux et algèbres de Clifford*, Inventiones Math. **5** (1968), 19-41.

[34] ————, *Sur le groupe des automorphismes d'un arbre*, Essays in Topology and Related Topics, Mémoires dédiés à G. de Rham, Springer (1970), 188-211.

[35] ————, *Homomorphismes et automorphismes "abstraits" de groupes algébriques et arithmétiques*, Actes Congrès Int. Math. Nice, 1970, Vol. 2, 349-355.

[36] B. L. VAN DER WAERDEN, *Stetigkeitssätze für halb-einfache Liesche Gruppen*, Math. Zeit, **36** (1933), 780-786.

[37] A. WEIL, Adeles and algebraic groups (Notes by M. Demazure and T. Ono), The Institute for Advanced Study, Princeton, 1961.

[38] ————, *Foundations of algebraic geometry*, Amer. Math. Soc. Colloquium Publ., n° 29, 2nd éd., 1962.

(Received August 30, 1972)

98.

(with J-P. Serre)

Corners and arithmetic groups

Comment. Math. Helv. **48** (1973) 436–491

Introduction

Let Γ be a torsion-free arithmetic subgroup of a semi-simple **Q**-group G. It operates properly and freely on the symmetric space X of maximal compact subgroups of the group $G(\mathbf{R})$ of real points of G and the quotient X/Γ is a manifold. If the **Q**-rank $r_{\mathbf{Q}}(G)$ of G is not zero, X/Γ is not compact and the main purpose of this paper is to provide a suitable compactification $\bar{X}/\Gamma$ for it. Topologically $\bar{X}/\Gamma$ is a compact manifold with interior X/Γ, whose boundary has the homotopy type of the quotient by Γ of the Tits building of parabolic **Q**-subgroups of G. However, from the differential-geometric point of view, $\bar{X}/\Gamma$ comes naturally equipped with a structure of (real analytic) manifold "with corners" (with boundary if $r_{\mathbf{Q}}(G)=1$), which is of interest in its own right, and which we shall therefore not smooth out to a boundary in the general case. As the notation suggests, we shall in fact first enlarge X to a space $\bar{X}$, whose construction involves the **Q**-structure of G, but not Γ. It is a manifold with (countably many) corners, on which $G(\mathbf{Q})$ operates so that the action is proper for any arithmetic

244

subgroup of $G(\mathbf{Q})$. In the classical case where $G=\mathbf{SL}_2$, $\Gamma \subset \mathbf{SL}_2(\mathbf{Z})$ and X is the open unit disc, our $\bar{X}$ is the union of X with countably many lines, one for each cusp point on the unit circle, and $\bar{X}/\Gamma$ is a compact surface whose boundary consists of finitely many circles, one for each Γ-equivalence class of parabolic points. Thus, the cusp points, which are classically added to X, are here blown-up to lines, and this allows us to get a space on which Γ still acts *properly*. In this case, and more generally if $G=\mathbf{SL}_n$, our construction is equivalent to one of C. L. Siegel [27; §§1, 12].

The space $\bar{X}$ is set-theoretically the union of X and of euclidean spaces $e(P)$, one for each proper parabolic $\mathbf{Q}$-subgroup P of G (the codimension of $e(P)$ being equal to the parabolic rank $\mathrm{park}(P)$ of P). For a given P, the set of $e(Q)$'s ($Q \supset P$) is organized into a corner $X(P)$, isomorphic to $\mathbf{R}_+^* \times \mathbf{R}^{n-k}$, where $\mathbf{R}_+$ is the closed half-line of positive real numbers, $n=\dim X$ and $k=\mathrm{park}(P)$. By definition, the $X(P)$'s form an open cover of $\bar{X}$. Now we wish to consider the closure $\overline{e(P)}$ of $e(P)$ in $\bar{X}$ as a space obtained from $e(P)$ by a similar construction, using the parabolic $\mathbf{Q}$-subgroups of P. However $e(P)$ is a homogeneous space under the group $P(\mathbf{R})$, which is not semi-simple and the isotropy groups of $P(\mathbf{R})$ are bigger than the maximal compact subgroups of $P(\mathbf{R})$. This led us to enlarge our framework, drop the assumption that G is semi-simple, or even reductive, and replace X by a suitable generalization of the above symmetric space, which we call a space of type S or $S-\mathbf{Q}$. Our construction has then a hereditary character well-suited for proofs by induction on $\dim G$, besides allowing us to handle directly a general arithmetic group. The price to pay is the appearance of some technical complications, mainly in §§1, 2, 3; in first approximation, it may be best for the reader not to dwell too much on them, and to keep in mind the case of a semi-simple G.

The properties of $\bar{X}$ and $\bar{X}/\Gamma$ are applied to the cohomology of Γ. It is shown that $H^i(\Gamma, \mathbf{Z}[\Gamma])=0$ except in dimension $m=\dim X-r_\mathbf{Q}(G)$, where it is a free module I, and that we have an isomorphism

$$H^i(\Gamma; A) = H_{m-i}(\Gamma; I \otimes A), \quad (i \in \mathbf{Z}), \tag{1}$$

for any Γ-module A. In particular, the cohomological dimension of Γ is m. If X/Γ is compact, then $I \cong \mathbf{Z}$ and (1) is just Poincaré duality. If X/Γ is not compact, then the rank of I is infinite and I is in a natural way a $G(\mathbf{Q})$-module which is a direct analogue of the Steinberg module of a finite Chevalley group.

We now give some more details on the contents of the various paragraphs. Let G be an affine algebraic group over a subfield k of $\mathbf{R}$. §1 is technical. It introduces a normal k-subgroup ^0G of G which is more or less a supplement to a maximal k-split torus of the radical of G, and discusses Cartan involutions of reductive groups. In particular, it is shown that if G is semi-simple, K a maximal compact subgroup of $G(\mathbf{R})$, and P a parabolic $\mathbf{R}$-subgroup of G, then $P(\mathbf{R})$ has a unique Levi subgroup stable under the Cartan involution θ_K associated to K (see 1.9 for a more general statement).

§2 is devoted to spaces of type $S-k$, or more generally of type S, but in this introduction we limit ourselves to the former. The homogeneous space X of $G(\mathbf{R})$ is of type $S-k$ if: (i) the isotropy groups $H_x (x \in X)$ are of the form $K \cdot S(\mathbf{R})$, where S is a maximal k-split torus of the radical $R(G)$ of G and K a maximal compact subgroup of $G(\mathbf{R})$ normalizing $S(\mathbf{R})$; (ii) there is given a map $x \mapsto L_x$ of X to Levi subgroups of $G(\mathbf{R})$ such that $L_{x \cdot g} = g^{-1} \cdot L_x \cdot g$ and $L_x \supset H_x (x \in X; g \in G(\mathbf{R}))$. Condition (i) determines completely the homogeneous space structure of X (see 2.1), but there is a choice involved in (ii) (unless G is reductive). If P is a parabolic k-subgroup of G, then X is canonically of type $S-k$ under $P(\mathbf{R})$, the choice of the Levi subgroups being given by Corollary 1.9 mentioned above.

§3 introduces the notion of *geodesic action* on a space of type S. For simplicity, assume here G to be connected and semi-simple. Let P be a parabolic $\mathbf{R}$-subgroup of G, and Z the center of the quotient $P/R_u P$ of P by its unipotent radical. For $x \in X$, denote by $Z(\mathbf{R})_x$ the unique lifting of $Z(\mathbf{R})$ in the Levi subgroup of P associated to x as above. There is then an action of $Z(\mathbf{R})$ on X, which commutes with $P(\mathbf{R})$, and is given by $x \circ z = x \cdot z_x$ $(z \in Z(\mathbf{R}), x \in X)$. The orbits of $Z(\mathbf{R})$ are totally geodesic flat submanifolds of X. If A is the identity component of the group of real points of the biggest $\mathbf{R}$-split torus of Z, then X becomes a principal A-bundle under this action. For a simple example, see 3.5.

§4 reviews some facts on parabolic k-subgroups. If P is such a group, let A_P be the identity component of the group of real points of the greatest k-split torus of the center of $P/R_u P$. There is a canonical isomorphism $A_P \to (\mathbf{R}_+^*)^d$, where $d = \dim A_P$, which is provided by a suitable set of simple k-roots. We let $\bar{A}_P$ be the closure of A_P in $\mathbf{R}^d$. It is therefore isomorphic to the positive quadrant $(\mathbf{R}_+)^d$. In §§4.3, 4.5 we discuss various decompositions of A_P or $\bar{A}_P$.

§5 defines the *corner* $X(P)$ associated to P. Take the simplest case, where $k = \mathbf{R}$ and P is minimal. We have then the familiar Iwasawa decomposition $G(\mathbf{R}) = K \cdot A \cdot N$, where A is the identity component of the group of real points of a maximal $\mathbf{R}$-split torus of G stable under the Cartan involution θ_K. Then $R_u P(\mathbf{R}) = N$, $P(\mathbf{R})$ is the normalizer of $A \cdot N$ in $G(\mathbf{R})$, and the projection $P \to P/R_u P$ maps A isomorphically onto A_P. There is then a canonical isomorphism $X \to A \times N$ and $X(P)$ is defined to be $\bar{A} \times N$. More intrinsically, in the general case, $X(P)$ is the associated bundle $X \times^{A_P} \bar{A}_P$ with typical fibre $\bar{A}_P$ associated to X, viewed as a principal A_P-bundle under the geodesic action of A_P. The stratification of $\bar{A}_P$ in orbits of A_P yields a stratification of $X(P)$ into locally closed subspaces. In particular, $X \times^{A_P} \{o\} = X/A_P$ is the face $e(P)$ associated to P. If Q is a k-parabolic subgroup of G containing P, then $X(Q)$ may be identified to an open submanifold of $X(P)$. In §6, it is shown that if G is reductive and P minimal, the Siegel sets in X, with respect to P [3, §12] allow one to describe the topology of $X(P)$ around $e(P)$. More precisely any point $y \in e(P)$ has a fundamental set of neighborhoods which are the closures of a suitable family of Siegel sets (6.2). The space X, and hence also $X(P)$, is a trivial bundle, and to any $x \in X$ is associated a trivialization of $X(P)$, whose cross-sections are orbits of the group $^0P(\mathbf{R})$.

§7 defines $\bar{X}$, and shows that it is a Hausdorff manifold with corners, which is paracompact if k is countable (7.8). The main point is the Hausdorff property, which is derived from 6.5, itself an immediate consequence of a known property of Siegel sets [3, Prop. 12.6]. It is also shown that the closure of $e(P)$ in $\bar{X}$ can be identified to the space $\overline{e(P)}$ associated by this construction to $e(P)$, viewed as a space of type $S-k$ under P (7.3).

§8 describes the homotopy type of $\partial \bar{X}$ and the cohomology with compact supports of $\bar{X}$.

In the last three paragraphs, $k = \mathbf{Q}$, and Γ is an arithmetic subgroup of G. It is shown that Γ acts properly on $\bar{X}$ and that $\bar{X}/\Gamma$ is compact (9.3). The proof is mainly an appeal to the main theorems of reduction theory [3, §§13, 15]. In turn, the properties of $\bar{X}$ and $\bar{X}/\Gamma$ yield various strengthenings and generalizations of some results of reduction theory, in particular with regard to the "Siegel property" and related facts, which are discussed in §10. Finally §11 gives the applications to the cohomology of Γ already mentioned.

In what follows, the notion of "manifold with corners" is taken for granted. Although this notion has already occurred at various places, there was a lack of foundational material on it, and we are grateful to A. Douady and L. Hérault, who have been willing to supply it; their paper is included here as an appendix.

The main results of this work have been announced in a Comptes Rendus Note [7].

The first named author gave a set of lectures on this topic at the University of Utrecht, in the Spring of 1971. We thank very much Mr. van der Hout, who wrote them up and whose Notes were helpful to us in the preparation of the present paper.

§0. Notation and Conventions

0.1. Let G be a group. If L is a subgroup of G, then $\mathscr{Z}_G(L) = \{g \in G \mid g \cdot x \cdot g^{-1} = x \, (x \in L)\}$ is the centralizer of L in G and $\mathscr{N}_G(L) = \{g \in G \mid g \cdot L \cdot g^{-1} = L\}$ the normalizer of L in G. The center $\mathscr{Z}_G(G)$ of G is denoted $\mathscr{C}(G)$. If $x \in G$ and $A \subset G$, then $^xA = x \cdot A \cdot x^{-1}$ and $A^x = x^{-1} \cdot A \cdot x$. Let N, M be subgroups of G. Then $N \triangleleft G$ means that N is normal in G, and $G = M \ltimes N$ that N is normal in G and G is the semi-direct product of M and N.

0.2. Algebraic groups will be affine and defined over fields of characteristic zero (mostly subfields of $\mathbf{R}$); we follow the notation and conventions of [4]. If G is an $\mathbf{R}$-group, then $G(\mathbf{R})$, endowed with the topology associated to the one of $\mathbf{R}$, is a real Lie group. The symbol $L(\ )$ will denote the Lie algebra both for algebraic groups and real Lie groups, as the case may be. Similarly, G^0 will denote the connected component of the identity in the Zariski topology if G is an algebraic group or in the ordinary topology if G is a real Lie group.

By definition, a parabolic subgroup of an algebraic group G is one of G^0, i.e. a closed subgroup P of G^0 such that G^0/P is a projective variety [4]. If G is defined over $\mathbf{R}$, and H is an open subgroup of $G(\mathbf{R})$, then a parabolic subgroup of H is by definition the intersection of H with a parabolic subgroup of G defined over $\mathbf{R}$.

0.3. As usual, the radical (resp. unipotent radical) of a connected k-group G is denoted RG (resp. R_uG). Then the greatest connected k-split subgroup ([4], §15) of RG is normal in G (because $G(k)$ is Zariski dense in G, [4], §18) and is the semi-direct product of R_uG by any maximal k-split subtorus of RG. It will be denoted R_dG and called the *split* or *k-split radical of G*.

If G is not connected, then, by definition, its radical, unipotent radical and split radical are those of G^0; they are also denoted RG, R_uG and R_dG respectively.

0.4. Let k be a field of characteristic zero. A k-group H is said to be *reductive* if H^0 is so, i.e. if $R_uH = \{e\}$. Let G be a k-group. Any reductive k-subgroup of G is contained in a maximal one. The maximal ones are called the Levi k-subgroups of G and are conjugate under $R_uG(k)$. If L is one of them, then $G = L \ltimes R_uG$ ([22], [6]).

Sometimes, the set of k-points of a Levi k-subgroup of G will be called a Levi subgroup of $G(k)$.

Recall that if $k = \mathbf{R}$, every compact subgroup of $G(\mathbf{R})$ is the group of real points of a reductive k-group; hence every compact subgroup is contained in a Levi $\mathbf{R}$-subgroup.

0.5. If G is a $\mathbf{Q}$-group, an *arithmetic subgroup Γ of G* is a subgroup of $G(\mathbf{Q})$ which is commensurable with $\varrho(G) \cap \mathbf{GL}_n(\mathbf{Z})$ for any injective $\mathbf{Q}$-morphism $\varrho: G \to \mathbf{GL}_n$ ([3], §7).

In this paper, k is a subfield of $\mathbf{R}$, G a k-group, U the unipotent radical of G and $\mathfrak{P}$ or $\mathfrak{P}(G)$ the set of parabolic k-subgroups of G. From §9 on, we assume $k = \mathbf{Q}$.

I. CARTAN INVOLUTIONS. GEODESIC ACTION. PARABOLIC SUBGROUPS

§1. The Group ^0G. Cartan Involutions

1.1. Assume G to be connected, defined over k. We put

$$^0G = \bigcap_{a \in X(G)_k} \ker a^2, \tag{1}$$

where, as usual, $X(G)_k$ is the group of k-morphisms of G into $\mathbf{GL}_1$. The group ^0G is normal in G, defined over k. If $a \in X(G)_k$, its restriction to ^0G is of order $\leqslant 2$, hence is trivial on $(^0G)^0$, therefore

$$(^0G)^0 = \left(\bigcap_{|a \in X(G)_k} \ker a \right)^0 . \tag{2}$$

Any character is trivial on U, hence

$$^0G = {}^0L \ltimes U \tag{3}$$

for any Levi k-subgroup L of G.

1.2. PROPOSITION. *Assume G to be connected. Let S be a maximal k-split torus of RG and $A = S(\mathbf{R})^0$. Then $G(\mathbf{R}) = A \rtimes {}^0G(\mathbf{R})$. The group $^0G(\mathbf{R})$ contains every compact subgroup of $G(\mathbf{R})$ and also, if $k = \mathbf{Q}$, every arithmetic subgroup of G.*

Let $a \in X(G)_k$ and let M be a compact subgroup of $G(\mathbf{R})$. Then $a(M)$ is a compact subgroup of $\mathbf{R}^*$, hence contained in $\{\pm 1\}$. Similarly $a(M) \subset \{\pm 1\}$ if $k = \mathbf{Q}$ and M is arithmetic. In both cases $M \subset \ker a^2$, which yields the second assertion.

To prove the first one we may assume, by 1.1(3), that G is reductive. The group G is the almost direct product of S and of $(^0G)^0$, as follows from 1.1(2) and ([3], Prop. 10.7), hence $G(\mathbf{R})^0 \subset A \cdot {}^0G(\mathbf{R})$. Moreover $A \cap {}^0G(\mathbf{R})$ is finite, hence reduced to $\{e\}$ since A is torsion-free. On the other hand, $G(\mathbf{R})$ has finitely many connected components, hence is generated by $G(\mathbf{R})^0$ and a compact subgroup H [21]; since $H \subset {}^0G(\mathbf{R})$ by the above, this gives $G(\mathbf{R}) = A \cdot {}^0G(\mathbf{R})$, whence the proposition.

We shall on occasion use a slight variant of 1.2:

1.2'. PROPOSITION. *We keep the previous notation. Let S', S'' be k-tori in RG such that S' is k-split, $S' \cdot S''$ is a torus and $S' \cap S''$ is finite. There exists then a normal k-subgroup N of G containing S'' and ^0G such that $G(\mathbf{R}) = S'(\mathbf{R})^0 \ltimes N(\mathbf{R})$.*

Dividing out by U reduces us to the case where G is reductive. The tori S' and S'' belong to the center of G. Using the decomposition of a torus T in anisotropic and split factors T_a and T_d ([4], §8), we can write $\mathscr{C}(G)^0 = V \cdot S'$ where $V \supset \mathscr{C}(G)_a^0 \cdot S_d''$ and $V \cap S'$ is finite. Let Y be the set of elements in $X(G)_k$ which are trivial on V, and let

$$N = \bigcap_{|a \in Y} \ker a^2 .$$

Let $S = \mathscr{C}(G)_d^0$. It follows from ([3], 10.7) that the restriction map $X(G)_k \to X(S)$ is injective, with finite cokernel, and that $X(G) \to X(S')$ maps Y injectively onto a subgroup of finite index. From this we see that $N \supset {}^0G$, $G = N \cdot S'$ and $N \cap S'$ is finite; therefore $G(\mathbf{R})^0 = S'(\mathbf{R})^0 \ltimes N(\mathbf{R})^0$. Since $N(\mathbf{R})$ contains $^0G(\mathbf{R})$, it meets every connected component of $G(\mathbf{R})$ by 1.2, hence $G(\mathbf{R}) = S'(\mathbf{R})^0 \ltimes N(\mathbf{R})$.

1.3. LEMMA. *Let L, L' be two Levi k-subgroups of G, and N a normal k-subgroup of G containing U. Then $L \cap N$ is a Levi k-subgroup of N. The groups L and L' are conjugate by an element of $U(k) \cap \mathscr{Z}_G(L \cap L')$.*

We have $G = L \ltimes U$, hence $N = (N \cap L) \ltimes U$. Moreover $U \subset R_u N$, and since N is normal, $R_u N \subset U$, whence $R_u N = U$, and the first assertion.

Let $u \in U(k)$ be such that $^u L = L'$ (0.4) and let $x \in L \cap L'$. Then $u \cdot x \cdot u^{-1} = x \cdot v$ for some $v \in U$. Since $u \cdot x \cdot u^{-1}$ and x belong to L', we have $v \in L' \cap U = \{e\}$, hence u centralizes $L \cap L'$.

1.4. We recall now a few standard facts about maximal compact subgroups or consequences thereof.

Let H be a real Lie group with finitely many connected components. Then any compact subgroup of H is contained in a maximal one. If K is a maximal one, then H is diffeomorphic to the direct product of K with a euclidean space. Moreover $K/K^0 = H/H^0$. Any two maximal compact subgroups are conjugate by an element of H^0 [21].

If N is a closed normal subgroup of H, with finitely many connected components, then the maximal compact subgroups of N are the intersections of N with the maximal compact subgroups of H. If M is a closed subgroup of H with finitely many connected components such that all maximal compact subgroups of H are conjugate by elements of M (e.g. if $H = K \cdot M$), then similarly the maximal compact subgroups of M are the intersections of M with the maximal compact subgroups of H. (In both cases, by taking a maximal compact subgroup K of H containing a maximal one of M, we see that $M \cap K$ is compact maximal in M for at least one K. It is then so for all maximal compact subgroups of H by conjugacy.)

Let $H \to H'$ be a surjective morphism of Lie groups whose kernel N has finitely many connected components. Then the maximal compact subgroups of H' are the images of the maximal compact subgroups of H. (This is well-known if H and N are connected and the reduction to that case is immediate.)

1.5. PROPOSITION. *Let P be a parabolic k-subgroup of G, S a maximal k-split torus of $R_d G$ and $A = S(\mathbf{R})^0$. Let K be a maximal compact subgroup of $G(\mathbf{R})$. Then $K \cap P$ is a maximal compact subgroup of $P(\mathbf{R})$ and $G(\mathbf{R}) = K \cdot P(\mathbf{R}) = K \cdot A \cdot {}^0P(\mathbf{R})$. If $K \cdot a \cdot {}^0P(\mathbf{R}) = K \cdot a' \cdot {}^0P(\mathbf{R})$ $(a, a' \in A)$, then $a = a'$. The map which assigns to $g \in G(\mathbf{R})$ the element $a = a(g) \in A$ such that $g \in K \cdot a \cdot {}^0P(\mathbf{R})$ is real analytic.*

The equality $G(\mathbf{R}) = K \cdot P(\mathbf{R})$ is well-known and follows from the Iwasawa decomposition (see e.g. [8], 14.7). We have then $G(\mathbf{R}) = K \cdot A \cdot {}^0P(\mathbf{R})$ by 1.2 applied to P. The group $K \cap P$ is a maximal compact subgroup of $P(\mathbf{R})$ by 1.4 and is contained in ${}^0P(\mathbf{R})$ by 1.2, hence we can identify $(K \cap P) \backslash P(\mathbf{R})$ and $A \times (K \cap P) \backslash {}^0P(\mathbf{R})$. Composing the obvious maps

$$G(\mathbf{R}) \to K\backslash G(\mathbf{R}) \stackrel{\sim}{\to} (K \cap P)\backslash P(\mathbf{R}) \stackrel{\sim}{\to} A \times (K \cap P)\backslash^{0}P(\mathbf{R}) \stackrel{pr_1}{\to} A,$$

we get a real analytic map $f: G(\mathbf{R}) \to A$. It is clear that $f(kap) = a\,(k \in K, a \in A, p \in {}^{0}P(\mathbf{R}))$, which proves the uniqueness and analyticity of a.

1.6. PROPOSITION. *Let G be reductive and let K be a maximal compact subgroup of $G(\mathbf{R})$. There exists one and only one involutive automorphism θ_K of $G(\mathbf{R})$ whose fixed point set is K and which is "algebraic," i.e. the restriction to $G(\mathbf{R})$ of an involutive automorphism of algebraic groups of the Zariski-closure of $G(\mathbf{R})$ in G. Let $\mathfrak{p}$ be the (-1)-eigenspace of θ_K in $L(G(\mathbf{R}))$. Then $L(G(\mathbf{R})) = L(K) \oplus \mathfrak{p}$ and $(k, X) \mapsto K \cdot \exp X$ is an isomorphism of analytic manifolds of $K \times \mathfrak{p}$ onto $G(\mathbf{R})$. Let N be a normal $\mathbf{R}$-subgroup of G. Then $\theta_K(N(\mathbf{R})) = N(\mathbf{R})$.*

By a result of G. D. Mostow [22] (see also [5], §1), we may arrange that $G \subset \mathbf{GL}_n$, $K \subset \mathbf{O}(n, \mathbf{R})$, and $G(\mathbf{R})$ is stable under $\theta: g \mapsto \check{g} = {}^t g^{-1}$, and then the latter automorphism has all the properties required from θ_K. There remains to prove the uniqueness. Let then θ' be an involutive automorphism of $G(\mathbf{R})$ whose fixed point set is K and which is algebraic in the above sense. Since K meets every connected component of $G(\mathbf{R})$, it suffices to show that θ and θ' coincide on $G(\mathbf{R})^0$, and hence that they define the same automorphism of $L(G(\mathbf{R}))$. For this, it is enough to prove that the (-1)-eigenspaces $\mathfrak{p}$ and $\mathfrak{p}'$ of θ and θ' are equal, and we may also assume G to be connected.

The group G is then the almost direct product of its derived group G', which is semi-simple, and of the identity component S of its center, which is a torus. Both $G'(\mathbf{R})$ and $S(\mathbf{R})$ are stable under θ', θ hence

$$\mathfrak{p}' = L(G') \cap \mathfrak{p}' \oplus L(S) \cap \mathfrak{p}', \quad \mathfrak{p} = L(G') \cap \mathfrak{p} \oplus L(S) \cap \mathfrak{p}.$$

The group $G' \cap K$ is a maximal compact subgroup of $G'(\mathbf{R})$ (1.4), hence $L(G') \cap \mathfrak{p}'$ and $L(G') \cap \mathfrak{p}$ are both equal to the orthogonal complement of $L(K \cap G')$ in $L(G'(\mathbf{R}))$ with respect to the Killing form. The group S is the almost direct product of its greatest $\mathbf{R}$-anisotropic torus S_a and its greatest $\mathbf{R}$-split torus S_d and $S_a(\mathbf{R})^0$ is the greatest connected compact subgroup of $S(\mathbf{R})$ ([3], 10.8). We have then, using 1.4,

$$L(S(\mathbf{R})) = L(K \cap S) \oplus L(S_d(\mathbf{R})).$$

But θ, θ' are the restriction to $G(\mathbf{R})$ of an $\mathbf{R}$-automorphism of G, hence they must leave both $S_a(\mathbf{R})$ and $S_d(\mathbf{R})$ stable, whence

$$L(S) \cap \mathfrak{p} = L(S) \cap \mathfrak{p}' = L(S_d(\mathbf{R})).$$

Let now N be a normal $\mathbf{R}$-subgroup of G. It is then also reductive. By [22] (see also [5]) there exists a maximal compact subgroup K_1' of $G(\mathbf{R})$ such that the associated involution θ'_1 leaves $N(\mathbf{R})$ stable. There exists $g \in G(\mathbf{R})$ such that ${}^g K = K'_1$. But then θ'_1 is conjugate to θ_K under $\operatorname{Int} g$, whence the last assertion.

1.7. DEFINITION. The automorphism θ_K in 1.6 will be called the *Cartan involution* of $G(\mathbf{R})$ with respect to K.

If G is semi-simple, then θ_K is the usual Cartan involution, and the uniqueness is obvious.

1.8. PROPOSITION. *Let H be a $\mathbf{R}$-subgroup of G containing U. Assume that all maximal compact subgroups of $G(\mathbf{R})$ are conjugate under $H(\mathbf{R})$. Let K be a maximal compact subgroup of $G(\mathbf{R})$ and L a Levi subgroup of $G(\mathbf{R})$ containing K. Let θ_K be the Cartan involution of L with respect to K and $L_1 = (H \cap L) \cap \theta_K(H \cap L)$. Then L_1 is the unique Levi subgroup of $H(\mathbf{R})$ contained in L and stable under θ_K.*

The group L_1 is stable under θ_K, hence is reductive ([5], 1.5), and contains every subgroup of $H' = H \cap L$ stable under θ_K, whence the uniqueness assertion. Let now L' be a Levi subgroup of H'. Since $H(\mathbf{R}) = H' \cdot U(\mathbf{R})$, the group L' is a Levi subgroup of $H(\mathbf{R})$. By a result of Mostow ([22], see also 1.9 in [5]), L admits a Cartan involution θ' leaving L' stable. Let K' be its fixed point set. In view of the assumption on H there exists $h \in H(\mathbf{R})$ such that $K = {}^h K'$. Since ${}^h K' \subset L \cap {}^h L$, there exists by 1.3 an element $u \in H(\mathbf{R}) \cap \mathscr{L}_H(K)$ such that ${}^{uh}L = L$. We have therefore $K = {}^{uh}K'$ and, by the uniqueness of Cartan involutions (1.6),

$$\theta_K \circ \operatorname{Int} uh = (\operatorname{Int} uh) \circ \theta'.$$

As a consequence ${}^{uh}L'$ is a Levi subgroup of $H(\mathbf{R})$ stable under θ_K, hence contained in L_1, hence equal to L_1 since L_1 is reductive; thus L_1 has all the required properties.

1.9. COROLLARY. *Let P be a parabolic $\mathbf{R}$-subgroup of G, K a maximal compact subgroup of $G(\mathbf{R})$ and L a Levi subgroup of $G(\mathbf{R})$ containing K. Then $L \cap P$ contains one and only one Levi subgroup of $P(\mathbf{R})$ stable under θ_K.*

Since $G(\mathbf{R}) = K \cdot P(\mathbf{R})$ by 1.5, this is a special case of 1.8.

§2. Homogeneous Spaces of Type S

2.1. LEMMA. *Let R be a solvable connected normal $\mathbf{R}$-subgroup of G containing U.*

(i) *Let K be a maximal compact subgroup of $G(\mathbf{R})$. Then R has a maximal $\mathbf{R}$-torus normalized by K.*

(ii) *If S is a maximal $\mathbf{R}$-torus of R, then $\mathscr{N}_G(S)$ contains a maximal compact subgroup of $G(\mathbf{R})$.*

(iii) *The subgroups of $G(\mathbf{R})$ of the form $K \cdot S(\mathbf{R})$, where S is a maximal torus of R*

defined over **R** *and K a maximal compact subgroup of* $G(\mathbf{R}) \cap \mathcal{N}_G(S)$, *form one conjugacy class of subgroups of* $G(\mathbf{R})$. *Let* $H = K \cdot S(\mathbf{R})$ *be one of them and* $\bar{H}$ *the Zariski closure of H. Then* $\bar{H}$ *is reductive,* $H = \bar{H}(\mathbf{R})$, $S = R \cap \bar{H}$ *and* $S(\mathbf{R}) = R \cap H$.

(iv) *Let L be a Levi* **R**-*subgroup of G containing H, and* θ_K *the Cartan involution of* $L(\mathbf{R})$ *with respect to K. Then* $L \cap R = S$ *and* θ_K *leaves* $H(\mathbf{R})$ *and* $S(\mathbf{R})$ *stable.*

(i) Let L be a Levi **R**-subgroup of G containing K. Then $R = (R \cap L) \ltimes U$, and $R \cap L$ is a maximal torus of R which is normalized by L, hence by K.

(ii) In view of (i) this assertion is true for at least one maximal **R**-torus of R. Since such tori are conjugate under $R(\mathbf{R})$ ([8], 11.4), it is then true for all of them.

(iii) The first assertion follows from the conjugacy of maximal **R**-tori of R and of maximal compact subgroups (1.4).

The Zariski-closure $\bar{K}$ of K is reductive and normalizes S, hence $\bar{K}^0$ centralizes S. We have $\bar{H} = \bar{K} \cdot S$, $\bar{H}^0 = \bar{K}^0 \cdot S$, hence $\bar{H}^0$, and therefore $\bar{H}$, is reductive. Moreover, $K \cdot S(\mathbf{R})$ contains $\bar{H}(\mathbf{R})^0$; but K is a maximal compact subgroup of $G(\mathbf{R})$, hence *a fortiori* of $\bar{H}(\mathbf{R})$, and consequently intersects every connected component of $\bar{H}(\mathbf{R})$; therefore $\bar{H}(\mathbf{R}) = K \cdot \bar{H}(\mathbf{R})^0 = H$. We have $S(\mathbf{R}) \subset R \cap H$, hence $S \subset R \cap \bar{H}$. The group $R \cap \bar{H}$ is normal in $\bar{H}$, hence reductive. Since S is maximal reductive in R, it follows that $S = R \cap \bar{H}$, whence also

$$S(\mathbf{R}) \subset R \cap H \subset (R \cap \bar{H})(\mathbf{R}) = S(\mathbf{R}),$$

which ends the proof of (iii).

(iv) Any subspace of the Lie algebra of $L(\mathbf{R})$ which contains $L(K)$ is stable under θ_K, hence $L(H(\mathbf{R}))$ and $H(\mathbf{R})^0$ are stable under θ_K. Since $H(\mathbf{R})$ is generated by K and $H(\mathbf{R})^0$ (1.4), it is also stable under θ_K. The group $L \cap R$ is a normal **R**-subgroup of L, hence its group of real points is stable under θ_K by 1.6. It is reductive, contains $R \cap \bar{H} = S$, hence is equal to S.

2.2. *Remark.* Notation being as above, assume G to be connected. Then L is also connected, hence centralizes the torus $L \cap R$. It follows that, in 2.2(iii), K *centralizes S and K is the unique maximal compact subgroup of* $K \cdot S(\mathbf{R})$.

2.3. DEFINITION. *A space of type S for G or* $G(\mathbf{R})$ is a pair consisting of a right homogeneous space X under $G(\mathbf{R})$ and of a family $(L_x)_{x \in X}$ of Levi subgroups of $G(\mathbf{R})$ satisfying the two following conditions:

SI. There exists a connected normal solvable **R**-subgroup R_X of G containing U, such that the isotropy groups $H_x (x \in X)$ of $G(\mathbf{R})$ in X are of the form $K \cdot S(\mathbf{R})$, where S is a maximal **R**-torus of R_X and K is a maximal compact subgroup of $G(\mathbf{R})$ normalizing S (cf. 2.1).

SII. We have $H_x \subset L_x$ and $L_{x \cdot g} = (L_x)^g$ for all $x \in X$ and $g \in G(\mathbf{R})$.

We shall often say simply that X is a homogeneous space of type S under $G(\mathbf{R})$. Note that, by 2.1, R_X is completely determined by the action of $G(\mathbf{R})$ on X. From §4 on, we shall be concerned only with the case where $R_X = R_d G$ is the k-split radical of G, in which case we shall say that X is of type $S-k$. If moreover $G = {}^0G$, then the isotropy subgroups are just the maximal compact subgroups of $G(\mathbf{R})$. As explained in the introduction, this is our main case of interest.

2.4. *Remarks.* (1) The condition SI of 2.3 implies that X is diffeomorphic to a euclidean space, $G(\mathbf{R})^0$ is transitive on X and the isotropy groups are conjugate under $G(\mathbf{R})^0$.

(2) Let X be a homogeneous space under $G(\mathbf{R})$ for which the isotropy groups $H_x\,(x\in X)$ are reductive. Then it is always possible to find a family $\{L_x\}_{x\in X}$ satisfying SII. Indeed, choose $y\in X$, a Levi subgroup $L\supset H_y$, and put $L_{y\cdot g}=L^g$ for $g\in G(\mathbf{R})$. Since $L^g = L$ for $g\in H$, the group L^g depends only on $y\cdot g$, and SII is then satisfied.

2.5. EXAMPLES. (1) Let G be semi-simple. Then $R_X = \{e\}$, and the isotropy groups are the maximal compact subgroups of $G(\mathbf{R})$. Since they are equal to their normalizers, X may be identified with the symmetric space of maximal compact subgroups of $G(\mathbf{R})$.

(2) Let G be reductive and let $C = \mathscr{C}(G^0)^0$ be its radical. The group R_X is an $\mathbf{R}$-subtorus of G, normal in G. The isotropy groups are then the subgroups $K\cdot R_X(\mathbf{R})$, where K runs through the maximal compact subgroups of $G(\mathbf{R})$. The group $K\cap RG$ is maximal compact in $RG(\mathbf{R})$. By standard facts on tori, there exists an $\mathbf{R}$-split subtorus D of RG normal in G, such that $RG(\mathbf{R}) = (K\cdot R_X(\mathbf{R})\cap RG)\times A$, with $A = D(\mathbf{R})^0$. The group A operates properly and freely on X. On X/A, the isotropy groups of $G(\mathbf{R})$ are the groups $K\cdot RG(\mathbf{R})$, (K maximal compact), hence X/A may be identified with the space of maximal compact subgroups of $\mathscr{D}G^0(\mathbf{R})$. If G is connected, then A is central. Note that when G is reductive, SII is vacuously fulfilled since we must have $L_x = G(\mathbf{R})$ for all $x\in X$.

2.6. LEMMA. *Let X be a homogeneous space of type S under $G(\mathbf{R})$ and $\mathscr{H} = \{H_x\}_{x\in X}$ the set of isotropy groups of $G(\mathbf{R})$ on X. Let G' be an $\mathbf{R}$-subgroup of G containing U such that $G'(\mathbf{R})$ is transitive on X. Then X satisfies SI under G', the corresponding solvable group R_X' being equal to $(G'\cap R_X)^0\cdot R_u G'$.*

Let K be a maximal compact subgroup of G such that $K\cap G'$ is maximal compact in $G'(\mathbf{R})$ and let $x\in X$ be such that $K\subset H_x$; thus $H_x' = H_x\cap G'$ contains a maximal compact subgroup of $G'(\mathbf{R})$; by conjugacy, this is then true for all $x\in X$. The group $R' = (G'\cap R_X)^0\cdot R_u G'$ is a normal connected solvable $\mathbf{R}$-subgroup of G' and $R_u R' = R_u G'$. Let S be a maximal $\mathbf{R}$-torus of R_X. Then $R_X = S\ltimes U$, hence $R_X\cap G' = (S\cap G')\ltimes U$, and $R' = S'\ltimes R_u G'$, where $S' = (S\cap G')^0$ is a maximal $\mathbf{R}$-torus of R', and we have $S'=(S\cap R')^0$. Choose $x\in X$ such that $S(\mathbf{R})\subset H_x$. Then

$$S(\mathbf{R})\cap G' = S(\mathbf{R})\cap H_x\cap G' = S(\mathbf{R})\cap H'_x,$$

which shows that

$$S'(\mathbf{R})\lhd H'_x,\quad S'(\mathbf{R})\subset S(\mathbf{R})\cap H'_x,\quad S'(\mathbf{R})^0 = (S(\mathbf{R})\cap H'_x)^0.\tag{1}$$

Let K' be a maximal compact subgroup of H'_x. We have already seen that it is maximal compact in $G'(\mathbf{R})$; by (1), it normalizes $S'(\mathbf{R})$. We wish to show that $H'_x = K'\cdot S'(\mathbf{R})$, which will prove the lemma. For this, it is enough, by the last assertion of 1.4, to prove that $H'_x/S'(\mathbf{R})$ is compact.

Let $\bar H$ be the Zariski-closure of H_x in G. The group S being normal in $\bar H$, the set $M=(\bar H\cap G')\cdot S$ is an $\mathbf{R}$-subgroup of G. Since $H_x=\bar H(\mathbf{R})$ by 2.1, $H'_x\cdot S(\mathbf{R})$ is an open subgroup of finite index in $M(\mathbf{R})$. In particular, it is a closed subgroup of $G(\mathbf{R})$. As a consequence, $H'_x\cdot S(\mathbf{R})/S(\mathbf{R})$ may be identified with a closed subgroup of $H_x/S(\mathbf{R})$. Since the latter is compact, so is the former. But $H'_x/(H'_x\cap S(\mathbf{R}))$ is isomorphic (as a Lie group) to $H'_x\cdot S(\mathbf{R})/S(\mathbf{R})$, hence is compact, too. By (1), $H'_x/(H'_x\cap S(\mathbf{R}))$ and $H'_x/S'(\mathbf{R})$ have a common finite covering, hence $H'_x/S'(\mathbf{R})$ is compact.

2.7. *Restriction to subgroups. Examples.* Let X and G' be as in 2.6. Then X satisfies S I with respect to $G'(\mathbf{R})$, and, by 2.4(2), there is then at least one way to make X of type S under G'. We shall now indicate some cases in which this can be done in a canonical manner. In the sequel, it will always be understood that X will be viewed of type S under G' with the choice of the Levi subgroups of $G'(\mathbf{R})$ given below. These will often be denoted $L_{x,G'}(x\in X)$.

(1) G' *is normal in* G. We define $L_{x,G'}$ as $G'\cap L_x$. This applies in particular to G^0.

(2) G *is connected.* By 2.2, H_x has a unique maximal compact subgroup, say K_x. Since $G'(\mathbf{R})$ operates transitively on X, it follows that all maximal compact subgroups of $G(\mathbf{R})$ are conjugate under $G'(\mathbf{R})$. We then define $L_{x,G'}$ as the unique Levi subgroup of $G'\cap L_x$ which is stable under the Cartan involution of L_x with respect to K_x (cf. Prop. 1.8).

(3) $G'=P$ *is parabolic.* We apply (1) to G^0 and (2) to $P\subset G^0$.

Remark. Assume G' to satisfy one of the above three conditions. Let $g\in G(\mathbf{R})$ and $G''=G'^g$. Then G'' satisfies the same condition, and we have

$$L_{x\cdot g,\,G''} = (L_{x,\,G'})^g\quad (x\in X).\tag{4}$$

2.8. Let X be a space of type S under G. It is the total space of a principal fibration with structure group $U(\mathbf{R})$, where $U(\mathbf{R})$ operates as a subgroup of $G(\mathbf{R})$. Let V be a (necessarily connected) $\mathbf{R}$-subgroup of U normal in G. Let $\pi:G\to G'=G/V$ and

$\sigma: X \to X' = X/V(\mathbf{R})$ be the natural projections. Let $x, y \in X$ be such that $\sigma(x) = \sigma(y)$. Then $x \in y \cdot V(\mathbf{R})$, hence $L_y = L_x^v$ for some $v \in V(\mathbf{R})$ and $\pi(L_x) = \pi(L_y)$. It is then immediate that X' is of type S under G', with $R_{X'} = \pi(R_X)$, and $L_{\sigma(x)} = \pi(L_x)$ for all $x \in X$. Assume V to be defined over k. Then $\pi(R_dG) = R_dG'$, therefore X' is of type $S-k$ if X is so.

§3. Geodesic Action

In this section, X is a space of type S under $G(\mathbf{R})$; for $x \in X$, H_x is the isotropy subgroup of x and L_x the Levi subgroup of $G(\mathbf{R})$ associated to x.

3.1. LEMMA. *Let P be a parabolic $\mathbf{R}$-subgroup of G, Z the center of P/R_uP and $\pi: P \to R/R_uP$ the canonical projection. Let Y be the greatest compact subgroup of $Z(\mathbf{R})$. Then, for $x \in X$, $Z_0 = Z \cap \pi(H_x \cap P)$ is generated by $\pi(R_X(\mathbf{R}))$ and by Y. In particular it is independent of $x \in X$.*

We have $R_X \cap R_uP = U$, therefore π defines an injective homomorphism of R_X/U into P/R_uP, whose image is a torus which is normal, hence central; we have thus $\pi(R_X) \subset Z$. We have $\pi(R_X) = \pi(S)$ where S is any maximal torus of R_X, hence, by 2.1,

$$\pi(R_X)(\mathbf{R}) = \pi(H_x \cap R_X) \quad \text{for any } x \in X.$$

On the other hand, $H_x = K \cdot (H_x \cap R_X)$, where K is a suitable maximal compact subgroup of $G(\mathbf{R})$, hence

$$H_x \cap P = (K \cap P) \cdot (H_x \cap R_X)$$
$$Z_0 = Z \cap \pi(H_x \cap P) = (\pi(K \cap P) \cap Z) \cdot \pi(H_x \cap R_X).$$

The group $K \cap P$ is maximal compact in $P(\mathbf{R})$ (1.5), hence $\pi(K \cap P)$ is maximal compact in $(P/R_uP)(\mathbf{R})$ (1.4), and then $\pi(K \cap P) \cap Z$ is maximal compact in $Z(\mathbf{R})$ (1.4). Therefore $\pi(K \cap P) \cap Z = Y$, whence the lemma.

Remark. Let S_1 be the $\mathbf{R}$-torus of Z which is generated by the greatest $\mathbf{R}$-anisotropic torus of Z and by the image of R_X. We have then $S_1(\mathbf{R})^0 = (Z_0)^0$. Let S_2 be an $\mathbf{R}$-torus in Z such that Z^0 is the almost direct product of S_1 and S_2. Then $S_2(\mathbf{R})^0 \cap \cap Z_0 = \{e\}$ and $S_2(\mathbf{R})^0$ maps isomorphically onto $Z(\mathbf{R})/Z_0$ under the natural projection. The torus S_2 splits over $\mathbf{R}$, and it follows from 1.2' that there exists a normal $\mathbf{R}$-subgroup N of P containing R_X and all compact subgroups of $P(\mathbf{R})$ and such that $P(\mathbf{R}) = S_2(\mathbf{R})^0 \ltimes N(\mathbf{R})$.

3.2. *Definition of the geodesic action.* Let P, Z, Z_0 be as before. We shall define here an action of $Z(\mathbf{R})$ on X which commutes with $P(\mathbf{R})$, is trivial for Z_0 and defines a proper and free action of $Z(\mathbf{R})/Z_0$.

By 2.7(3), X is canonically of type S under P. For $x \in X$, let L'_x be the Levi subgroup of $P(\mathbf{R})$ associated to x; it is contained in L_x. Let $Z_x = \mathscr{C}(L'_x)$. It follows from 1.8 and the definition of L'_x (2.7) that Z_x is the unique lifting of $Z(\mathbf{R})$ in $P(\mathbf{R})$ which is stable under the Cartan involution of L_x with respect to a maximal compact subgroup of H_x. Given $z \in Z(\mathbf{R})$, let z_x be its lifting in Z_x. Fix $x \in X$. Let $y \in X$ and $g \in P(\mathbf{R})$ be such that $y = x \cdot g$. Let us put

$$y \, \mathbf{o}_x \, z = x \cdot z_x \cdot g. \tag{1}$$

Let $g' \in P(\mathbf{R})$ be such that $x \cdot g' = y$. Then $g' = h \cdot g$ for some $h \in H_x \cap P \subset L'_x$; the element h then commutes with z_x, hence $x \cdot z_x \cdot g' = x \cdot z_x \cdot g$. This shows that the right hand side of (1) depends only on x, y, z and justifies the notation of the left hand side.

LEMMA. *We have*

(i) $y \cdot p \, \mathbf{o}_x \, z = (y \, \mathbf{o}_x \, z) \cdot p, \quad (x, y \in X; \ p \in P(\mathbf{R}), \ z \in Z(\mathbf{R}))$.

(ii) $y \, \mathbf{o}_x \, z = y \, \mathbf{o}_{x'} \, z, \qquad (x, x', y \in X; \ z \in Z(\mathbf{R}))$.

Let $g \in P(\mathbf{R})$ be such that $y = x \cdot g$. For $p \in P(\mathbf{R})$, we have $y \cdot p = x \cdot g \cdot p$, hence

$$(y \cdot p) \, \mathbf{o}_x \, z = x \cdot z_x \cdot g \cdot p = (y \, \mathbf{o}_x \, z) \cdot p, \tag{2}$$

which gives (i).

Let now $h \in G(\mathbf{R})$ and set $P' = P^h$. Then $\operatorname{Int} h$ induces an $\mathbf{R}$-isomorphism of Z onto $Z' = \mathscr{C}(P'/R_u P')$. Let us also denote by z^h the image of $z \in Z$ under this map. It is clear, by "transport de structure," that we have

$$(y \, \mathbf{o}_x \, z) \cdot h = y \cdot h \, \mathbf{o}_{x \cdot h} \, z^h \quad (x, y \in X; \ z \in Z(\mathbf{R})). \tag{3}$$

Let $x' \in X$, and choose $h \in P(\mathbf{R})$ such that $x \cdot h = x'$. By (3), applied to $y \cdot h^{-1}$, we have, taking (i) into account:

$$y \, \mathbf{o}_x \, z = (y \cdot h^{-1} \, \mathbf{o}_x \, z) \cdot h = y \, \mathbf{o}_{x'} \, z^h;$$

but, since $h \in P(\mathbf{R})$, we have $z^h = z$, whence (ii).

In particular, this shows that the actions $\mathbf{o}_x$ and $\mathbf{o}_{x'}$ are the same. We may therefore omit the reference point, and get an action

$$\gamma : (x, z) \mapsto x \, \mathbf{o} \, z \quad (x \in X; \ z \in Z(\mathbf{R}))$$

of $Z(\mathbf{R})$ on X. By (1), with $g = e$:

$$x \, \mathbf{o} \, z = x \cdot z_x, \quad (x \in X; \ z \in Z(\mathbf{R})), \tag{4}$$

which shows that

$$(x \mathbf{o}\, z) \mathbf{o}\, z' = x \mathbf{o}\, (z \cdot z') \quad (x \in X;\, z,\, z' \in Z(\mathbf{R})).$$

We can now write (3) in the form

$$(x \mathbf{o}\, z) \cdot h = x \cdot h \mathbf{o}\, z^h, \quad (x \in X,\, z \in Z(\mathbf{R}),\, h \in G(\mathbf{R})), \tag{5}$$

where $\mathbf{o}$ refers to the geodesic action of $Z(\mathbf{R})$ on the left-hand side and of $^h Z(\mathbf{R})$ on the right-hand side.

3.3. DEFINITION. The action γ defined above is called the *geodesic action* of $Z(\mathbf{R})$ on X.

The reader will note that this action depends only on the structure of type S of X under P. If P is reductive, then Z is a subgroup of P, and the geodesic action is just the ordinary action.

3.4. PROPOSITION. *The geodesic action commutes with $P(\mathbf{R})$. The group Z_0 operates trivially and $Z(\mathbf{R})/Z_0$ operates freely.*

(In fact, the action of $Z(\mathbf{R})/Z_0$ is a principal bundle action, see 3.6.)

The first assertion follows from 3.2(i). If $z \in Z_0$ then, by 3.1, $z_x \in H_x$ for all $x \in X$, hence z acts trivially by 3.2(4).

It also follows from 3.1 that if $z \in Z(\mathbf{R})$, $z \notin Z_0$, then $z_x \notin H_x$ for any $x \in X$, hence $Z(\mathbf{R})/Z_0$ acts freely on X.

Remark. The group Z_0 contains the maximal compact subgroup of $Z(\mathbf{R})$. We may therefore write $Z(\mathbf{R}) = Z_0 \times A$, where A is the identity component of the group of real points of an $\mathbf{R}$-split torus of Z. By 3.4, and 3.2(4), the orbit $x \mathbf{o}\, Z(\mathbf{R})$ of $x \in X$ may be identified with $x \cdot A_x$, i.e. with the orbit of x under the ordinary action of the identity component of the group of real points of an $\mathbf{R}$-split torus. If G is reductive, and hence X is a symmetric space with negative curvature, then $x \cdot A_x$ is a totally geodesic flat submanifold, isometric to a euclidean space, and the orbits of 1-dimensional subgroups of A_x are geodesics, whence the terminology.

3.5. EXAMPLE. Let $G = \mathbf{SL}_2$ and X be the upper half-plane. Let $G(\mathbf{R})$ act on X by

$$z \cdot \begin{pmatrix} a & b \\ c & d \end{pmatrix} = (dz + b)(cz + a)^{-1}.$$

Let P be the group of upper triangular matrices in G. The group Z_0 has two elements ± 1. Let $A = Z(\mathbf{R})/Z_0$. The stability group of i is $\mathbf{SO}(2, \mathbf{R})$ and the Cartan involution associated to it is $g \mapsto {}^t g^{-1}$. For $x = i$, $L_x = Z_x$ is the group of diagonal matrices in $G(\mathbf{R})$. For $a \in A$, let $\alpha(a)$ be the square of the upper left entry of the lifting of a in A_i.

Then, for $z = x + iy$, we have

$$z \mathbf{o} \, a = x + i\alpha(a)\cdot y.$$

Hence the orbits of A are the vertical lines and on each such line, identified to $\mathbf{R}_+^*$ by the y coordinate, $z \mapsto z \mathbf{o} \, a$ is the multiplication by $\alpha(a)$.

If we take as model of X the open unit disc, then the choice of a parabolic $\mathbf{R}$-subgroup P of G corresponds to that of the fixed point P_0 of $P(\mathbf{R})$ on the unit circle. The orbits of $Z(\mathbf{R})/Z_0$ under the geodesic action are then the geodesics abutting to P_0.

3.6. *Bundles defined by the geodesic action.* Let P, Z, Z_0 be as before. Let T be an $\mathbf{R}$-split torus in $\mathscr{C}(P/R_u P)$ whose intersection with Z_0 is finite and A be the image of $T(\mathbf{R})^0$ into $Z(\mathbf{R})/Z_0$. By 1.2′, there exists a normal k-subgroup M of P containing R_X and all maximal compact subgroups of $P(\mathbf{R})$, such that $P(\mathbf{R}) = T(\mathbf{R})^0 \ltimes M(\mathbf{R})$. Since $P(\mathbf{R})$ commutes with A, the latter operating by geodesic action, we have an action of $A \times M(\mathbf{R})$ onto X defined by

$$x \mapsto (x \mathbf{o} \, a)\cdot m, \quad \left(a \in A,\ m \in M(\mathbf{R});\ x \in X\right).$$

3.7. PROPOSITION. *The above action of $A \times M(\mathbf{R})$ on X is transitive. For $x \in X$, $H_x \cap P \subset M(\mathbf{R})$ and the isotropy group of x in $A \times M(\mathbf{R})$ is $\{e\} \times (H_x \cap P)$.*

Let $x \in X$. Then $(x \mathbf{o} \, A)\cdot M(\mathbf{R}) = x \cdot A_x \cdot M(\mathbf{R}) = x \cdot P(\mathbf{R}) = X$, which proves the first assertion. The space X is of type S under $P(\mathbf{R})$ (2.7) and in particular the isotropy group $H_x \cap P$ of x under $P(\mathbf{R})$ is generated by a maximal compact subgroup K'_x of $P(\mathbf{R})$ and a subgroup of $R_X(\mathbf{R})$. Since both of these groups are contained in $M(\mathbf{R})$, we have $H_x \cap P \subset M(\mathbf{R})$ for all $x \in X$. Let now $a \in A$ and $m \in M(\mathbf{R})$ be such that $(x \mathbf{o} \, a)\cdot m = x$. We then have $a_x \cdot m \in H_x \cap P$ hence $a_x \cdot m \in M(\mathbf{R})$ and $a_x = e$, whence the proposition.

3.8. COROLLARY. *Let $x \in X$. The map $(a \cdot m) \mapsto (x \mathbf{o} \, a)\cdot m$ $(a \in A;\ m \in M(\mathbf{R}))$ induces an isomorphism $\mu_x: A \times (H_x \cap P)\backslash M(\mathbf{R}) \xrightarrow{\sim} X$ of $(A \times M(\mathbf{R}))$-homogeneous spaces. The space X is a trivial principal A-bundle, and the orbits of $M(\mathbf{R})$ are cross-sections of this fibration.*

By 3.7, μ_x is an analytic bijective map of $(A \times M(\mathbf{R}))$-spaces. Since these are homogeneous spaces, it is then an isomorphism. This proves the first assertion; the second one is an obvious consequence.

Remark. Since Z_0 contains the greatest compact subgroup of $Z(\mathbf{R})$, we may in particular, in 3.6, choose T such that $A = Z(\mathbf{R})/Z_0$; hence 3.8 applies to the geodesic action of $Z(\mathbf{R})/Z_0$ on X.

3.9. *Structure of type S for X/A.* We keep the previous notation and let $\sigma: X \rightarrow$

$\rightarrow X' = X/A$ be the canonical projection. In view of 3.4, $P(\mathbf{R})$ commutes with σ, whence a transitive action of $P(\mathbf{R})$ on X'. For $x \in X$, the fibre $F_x = \sigma^{-1}(\sigma(x))$ is $x \cdot A_x$ (3.2), hence the isotropy group of $\sigma(x)$ is $(H_x \cap P) \cdot A_x$. The condition SI is then fulfilled, with respect to P, if we let $R_{X'}$ be the subgroup of P generated by R_X and the inverse image of T. In particular if $R_X = R_d G$ and T is the greatest k-split torus of $\mathscr{C}(P/R_u P)$, then $R_{X'} = R_d P$.

For $x \in X$, let $L_{x,P}$ be the Levi subgroup of P assigned to x by the condition SII. If $y \in F_x$, then L_y is conjugate to L_x by an element $a \in A_x$. But $A_x \subset L_x$, hence $L_x = L_y$ and therefore, by construction (2.7), $L_{y,P} = L_{x,P}$. Thus $x \mapsto L_{x,P}$ is constant along the fibres of σ, and SII is clearly satisfied for the action of $P(\mathbf{R})$ on X' if we put $L_{x'} = L_{x,P}$ for any $x \in \sigma^{-1}(x')$. This choice will be understood in the sequel. Therefore X' is canonically of type S under P; by the end remark of the previous paragraph, it is also of type $S - k$ if X is so under G.

The group $Z(\mathbf{R})$ is commutative, therefore the geodesic action of $Z(\mathbf{R})$ goes over to an action on X', trivial on A, and commuting with $P(\mathbf{R})$. In view of the definition of $L_{x'}$ $(x' \in X')$, it is clearly the geodesic action of $Z(\mathbf{R})$ on X', for the structure of space of type S under P just defined.

3.10. Assume G to be connected, and let G' be as in 2.6. Then, by 2.7(2), X is canonically of type S under G' and for $x \in X$ the Levi subgroup L'_x of $G'(\mathbf{R})$ associated to x is contained in L_x. The group $\mathscr{C}(G/U)$ is reductive, therefore the canonical homomorphism $G'/U \rightarrow G'/R_u G'$ maps $\mathscr{C}(G/U) \cap (G'/U)$ isomorphically onto an $\mathbf{R}$-subgroup of $\mathscr{C}(G'/R_u G')$. Let $x \in X$. Since $L'_x \subset L_x$, an element $z \in \mathscr{C}(G/U)(\mathbf{R}) \cap (G'/U)$ and its image in $\mathscr{C}(G'/R_u G')$ have the same lifting associated to x. Consequently, the geodesic actions of z on X, with respect to the structures of type S under G and G', are the same.

If G' is parabolic, then G'/U is parabolic in G/U, hence contains $\mathscr{C}(G/U)$, which is then identified to a subgroup of $\mathscr{C}(G'/R_u G')$, and this identification is compatible with the geodesic actions on X under G and G'. Returning to the situation of 3.2, we may in particular apply this to two $\mathbf{R}$-parabolic subgroups $P \subset Q$ of G, which play the role of G' and G in the preceding discussion, and get:

3.11. PROPOSITION. *Let $P \subset Q$ be two parabolic $\mathbf{R}$-subgroups of G. Then $\mathscr{C}(Q/R_u Q)$ may be canonically identified with a subgroup of $\mathscr{C}(P/R_u P)$, and the geodesic action of $\mathscr{C}(Q/R_u Q)(\mathbf{R})$ on X is the restriction of the geodesic action of $\mathscr{C}(P/R_u P)(\mathbf{R})$ on X.*

II. CORNERS

§4. Parabolic k-Subgroups

4.1. In this section, we review some standard facts on parabolic subgroups and fix some notation. We recall that if H is a k-group, $\mathfrak{P}(H)$ is the set of parabolic k-subgroups of H. Let R be a connected normal solvable k-subgroup of G and $\pi: G \to G/R$ the canonical projection. Then $P \mapsto \pi(P)$ induces a bijection of $\mathfrak{P}(G)$ onto $\mathfrak{P}(G/R)$ whose inverse map is given by $Q \mapsto \pi^{-1}(Q)$. Assume now R to be k-split [4, §15], which is automatically the case if R is unipotent. Then $\pi(k): G(k) \to (G/R)(k)$ is surjective, hence the bijection $\mathfrak{P}(G) \to \mathfrak{P}(G/R)$ preserves conjugacy classes over k. In particular the classification of parabolic k-subgroups up to conjugacy over k is "the same" in G, G/U or $G/R_d G$; this reduces us to the case of reductive groups. Let S be a maximal k-split torus of G^0/U, $_k\Phi = \Phi(S, G/U)$ the set of k-roots of G^0/U with respect to S, and Δ a basis of $_k\Phi$ ([8], §5). By [8, §5], the conjugacy classes in $\mathfrak{P}(G)$ with respect to $G^0(k)$ are in $1-1$ correspondence with the subsets of Δ. The class corresponding to $J \subset \Delta$ is represented by the standard parabolic subgroup P_J: the image P_J/U of P_J in G/U is the semi-direct product of its unipotent radical U_J by the centralizer $Z(S_J)$ of S_J, where $S_J = (\bigcap_{\alpha \in J} \ker \alpha)^0$, and its split radical is $S_J \cdot U_J$. Given $P \in \mathfrak{P}$, the only I such that P is conjugate to P_I under $G^0(k)$ will be denoted $I(P)$ and called the type of P.

4.2. Let $P \in \mathfrak{P}(G)$. The quotient $S_P = R_d P/(R_u P \cdot R_d G)$ is a k-split torus, and is also the greatest k-split torus in $C_P = \mathscr{C}(P/(R_u P \cdot R_d G))$. We let A_P be the identity component of $S_P(\mathbf{R})$. Let $P' \in \mathfrak{P}(G)$ be conjugate to P under G^0, and let $x \in G^0$ be such that $^x P' = P$. Then $\operatorname{Int} x$ induces an isomorphism of $C_{P'}$ onto C_P. If $P'^y = P$ with $y \in G^0$, then, since P' is its own normalizer in G^0, $y \in x \cdot P'$. Clearly, $\operatorname{Int} p'$ $(p' \in P')$ induces the trivial automorphism of $C_{P'}$, hence $\operatorname{Int} y$ induces the same isomorphism of $C_{P'}$ onto C_P as $\operatorname{Int} x$. Since we may take $y \in G^0(k)$, this isomorphism is defined over k, hence defines a canonical isomorphism

$$\sigma_{P', P}: S_{P'} \xrightarrow{\sim} S_P. \tag{1}$$

Let in particular $P' = P_I$ be standard. Then $S_{P'} = S_I/S_\Delta$. The elements of $\Delta - I$ define a basis of $X^*(S_I/S_\Delta) \otimes \mathbf{Q}$, where $X^*(\ )$ denotes the group of rational characters [4], which is carried over onto a basis of $X^*(S_P) \otimes \mathbf{Q}$, to be denoted in the same way. We thus have a canonical isomorphism:

$$A_P \xrightarrow{\sim} (\mathbf{R}_+^*)^{\Delta - I}. \tag{2}$$

4.3. Let Q be a parabolic k-subgroup containing P and let $J = I(Q)$. The inclusion

$R_dQ \subset R_dP$ induces an injective morphism of S_Q into S_P which maps S_Q onto $(\bigcap_{\alpha \in J-I} \ker \alpha)^0$. Let $A_{P,Q} = (\bigcap_{\alpha \in \Delta-J} \ker \alpha) \cap A_P$. Then the product decomposition

$$A_P = A_{P,Q} \times A_Q \tag{3}$$

corresponds to the factorization

$$(\mathbf{R}_+^*)^{\Delta-I} = (\mathbf{R}_+^*)^{J-I} \times (\mathbf{R}_+^*)^{\Delta-J} ; \tag{4}$$

The group A_P associated to P, viewed as subgroup of Q, is A_P/A_Q, i.e. is the isomorphic image of $A_{P,Q}$ under the canonical projection. Thus, our A_P above can be written A_{P,G^0}, and then (3) takes the form

$$A_{P,G^0} = A_{P,Q} \times A_{Q,G^0} . \tag{5}$$

For a subset L of $\Delta - I$, let

$$A_{P(L)} = (\bigcap_{\alpha \in L} \ker \alpha) \cap A_P, \tag{6}$$

where $P(L)$ is the parabolic subgroup of type $I \cup L$ containing P. In particular

$$A_{P(\emptyset)} = A_P, \quad A_{P(\Delta-I)} = A_{G^0} = \{e\} .$$

4.4. The isomorphism 4.2(2) yields an open embedding of A_P into $\mathbf{R}^{\Delta-I}$. The closure of A_P is $\mathbf{R}_+^{\Delta-I}$ and will be denoted $\bar{A}_P$ (or $\bar{A}_{P,G}$ if we wish to emphasize the ambient group). The elements of $\Delta - I$ are then coordinates on $\bar{A}_P$, taking all positive values (zero included), and they identify $\bar{A}_P$ to the positive quadrant in $\mathbf{R}^{\Delta-I}$. The action of A_P on itself by means of translations extends to one on $\bar{A}_P$, given by coordinate-wise multiplication.

4.5. For every $L \subset \Delta - I$, let o_L be the point with coordinates

$$\alpha(o_L) = \begin{cases} 0 & \alpha \notin L \\ 1 & \alpha \in L . \end{cases}$$

In particular:

$$o_\emptyset = (0, 0, ..., 0), \quad o_{\Delta-I} = (1, 1, ..., 1) = e .$$

Then $A_P \cdot o_L = \bar{A}_P(L)$ is the face of $\bar{A}_P$ given by

$$\bar{A}_P(L) = \{x \in \bar{A}_P \mid \alpha(x) = 0 \,(\alpha \notin L), \, \alpha(x) \neq 0 \,(\alpha \in L)\} . \tag{1}$$

In particular:

$$A_P \cdot o_\emptyset = \bar{A}_P(\emptyset) = \{o_\emptyset\}, \quad \bar{A}_P(\Delta - I) = A_P \cdot o_{\Delta-I} \cong A_P,$$

and we have the orbit space decomposition:

$$\bar{A}_P = \coprod_{L \subset \Delta - I} \bar{A}_P(L). \tag{2}$$

The isotropy group of o_L is $A_{P(L)}$; we have

$$A_P \cdot o_L = A_{P, P(L)} \cdot o_L \cong A_{P, P(L)} \quad (L \subset \Delta - I), \tag{3}$$

and the orbit map $a \mapsto a \cdot o_L$ extends to a diffeomorphism:

$$\bar{A}_{P, P(L)} \xrightarrow{\sim} Cl(\bar{A}_P(L)). \tag{4}$$

In this formula, the right-hand side is the closure of $\bar{A}_P(L)$ in $\bar{A}_P$ and the left-hand side is defined as in 4.3, but with G replaced by $P(L)$.

§5. The Corner $X(P)$ associated to a Parabolic Subgroup

5.0. From now on, X is a homogeneous space of type $S - k$ for G (2.3). Thus we have $R_X = R_d G$. This latter condition determines uniquely the isotropy groups $H_x (x \in X)$ of $G(\mathbf{R})$ on X; it involves the k-structure of G, as the case of a torus already shows. Note that X is also of type $S - k$ under G^0 or under the group $^0(G^0)$ of 1.1. If $G^0 = {}^0(G^0)$, the isotropy groups in X are the maximal compact subgroups of $G(\mathbf{R})$. By 2.1 and 2.4(2), G always has a homogeneous space of type $S - k$.

5.1. We keep the previous notation. By 3.8 and the definition of A_P (4.2), X is a principal A_P-bundle under the geodesic action. By definition, the *corner* $X(P)$ *associated to* P is the total space of the associated bundle with typical fibre $\bar{A}_P$:

$$X(P) = X \times^{A_P} \bar{A}_P. \tag{1}$$

Thus $X(P)$ is the quotient of $X \times \bar{A}_P$ under the equivalence relation: $(x, z) \sim (x', z')$ if and only if there exists $a \in A_P$ such that $x = x' \mathbf{o} \, a$ and $z' = a \cdot z$. The space $X(P)$ is endowed with a natural (real analytic) structure of manifold with corners coming from that of the fibres (the components of the boundary being the $e(Q)$ described below).

In view of 4.5(2), we have

$$X(P) = \coprod_{L \subset \Delta - I} X(P, L), \tag{2}$$

where

$$X(P, L) = X \times^{A_P} A_P \cdot o_L. \tag{3}$$

By 4.5(3)

$$X(P, L) = X/A_{P(L)} \times^{A_{P, P(L)}} A_{P, P(L)} \simeq X/A_{P(L)}, \tag{4}$$

in particular

$$X(P, \emptyset) = X/A_P, \qquad X(P, \Delta - I) = X. \tag{5}$$

Let us put

$$e_X(Q) = e(Q) = X/A_Q, \qquad (Q \in \mathfrak{P}),$$

in particular $e(G^0) = X$. By 3.9, $e(Q)$ is canonically of type $S - k$ under Q. The equality (2) can be written

$$X(P) = \coprod_{Q \in \mathfrak{P}, Q \supset P} e(Q). \tag{6}$$

We have a principal fibration

$$X \to e(Q) \qquad \text{with structural group} \qquad A_Q. \tag{7}$$

Let J be a subset of L. Then $P(J) \subset P(L) = Q$. Replacing X, G, P by $e(Q), Q, P(J)$, we then have also a principal fibration

$$v_{P(J), Q} : e(Q) \to e(P(J)) \qquad \text{with structural group} \qquad A_{P(J), Q}. \tag{8}$$

We have the factorization $A_{P(J)} = A_{P(J), Q} \times A_Q$. The group $A_{P(J)}$ operates by geodesic action on the fibration (7), and the action induced on $e(Q)$ is the geodesic action of $A_{P(J), Q}$ which underlies (8).

The group $P(\mathbf{R})$ operates on X, and commutes with A_P. The action of $P(\mathbf{R})$ extends then to $X(P)$, leaving the faces $e(Q)$ stable. Since A_P is commutative, its action on X also extends to $X(P)$, leaving each $e(Q)$ stable, and it still commutes with $P(\mathbf{R})$. Moreover, $P(\mathbf{R})$ operates on the fibrations (7), (8).

5.2. Let V be a normal unipotent k-subgroup of G, and $\pi : G \to G' = G/V$ the canonical projection. The group $V(\mathbf{R})$ operates properly and freely on X, and $X' = X/V(\mathbf{R})$ is canonically of type $S - k$ under G' (2.8). Moreover, if $\sigma : X \to X'$ is the canonical projection, then $L_{\sigma(x)} = \pi(L_x)$ $(x \in X)$. If $P \in \mathfrak{P}$ and $P' = \pi(P)$, then $A_{P'}$ is canonically identified with A_P and one checks that the geodesic actions of A_P on X and X' commute with σ. As a consequence, $X(P)$ is a principal $V(\mathbf{R})$-bundle over $X'(P')$, and the projection $\tau : X(P) \to X'(P')$ extends σ. For every $Q \supset P$, $Q \in \mathfrak{P}$ the restriction of τ to $e_X(Q)$ is the projection of a principal $V(\mathbf{R})$-fibration with base $e_{X'}(Q)$.

5.3. PROPOSITION. *Let $P \subset Q$ be two parabolic k-subgroups of G and $I_0 = I(Q) - I(P)$. The inclusion $X(Q) \hookrightarrow X(P)$ is an isomorphism of manifolds with corners of $X(Q)$ onto an open subset of $X(P)$. We have $Cl_{X(P)} e(Q) = \coprod_{Q \supset R \supset P, R \in \mathfrak{P}} e(R) = e(Q)(P)$,* *where $e(Q)$ is viewed as a space of type $S-k$ under Q (5.1), and $e(Q)(P)$ is the corner of $e(Q)$ associated to the parabolic k-subgroup P of Q.*

The canonical factorization (4.3(3))

$$A_P = A_{P,Q} \times A_Q$$

yields an embedding

$$A_{PQ} \times \bar{A}_Q \to \bar{A}_P \tag{1}$$

which, for every $L \subset \Delta - I(Q)$, induces a homeomorphism

$$A_{PQ} \times \bar{A}_Q(L) \xrightarrow{\sim} \bar{A}_P(L \cup I_0), \quad (L \subset \Delta - I(Q)). \tag{2}$$

We have clearly

$$X \times^{A_Q} \bar{A}_Q = X \times^{A_{PQ} \times A_Q}(A_{PQ} \times \bar{A}_Q), \tag{3}$$

$$X \times^{A_Q} \bar{A}_Q(L) = X \times^{A_{PQ} \times A_Q}(A_{PQ} \times \bar{A}_Q(L)). \tag{4}$$

In view of (1), (2), this yields

$$X(Q) = X \times^{A_P}(A_{PQ} \times \bar{A}_Q), \tag{5}$$

$$e(Q(L)) = X(Q, L) = X \times^{A_P}(\bar{A}_P(L \cup I_0)), \quad (L \subset \Delta - I(Q)). \tag{6}$$

The inclusion $X(Q) \hookrightarrow X(P)$ is then defined by the inclusion (1) of the typical fibres in the right-hand sides of (5) and 5.1(1). By (2) and (6), its restriction to $X(Q, L)$ is an isomorphism of $X(Q, L)$ onto $X(P, L \cup I_0)$. Taking 5.1(4) into account, we get canonical isomorphisms

$$e(Q(L)) = X/A_{Q(L)} \cong X(Q, L) \to e(P(L \cup I(Q)) \cong X/A_{P(L \cup I_0)}$$
$$(L \subset \Delta - I(Q)). \tag{7}$$

Thus $X(Q, L)$ is endowed with structures of space of type S under both Q and P (5.1). It is immediate from the definitions that the latter structure can also be associated to P viewed as a subgroup of Q, hence the inclusion commutes with the geodesic action of A_P.

Let $J \subset \Delta - I(P)$. The closure of $\bar{A}_P(J)$ is the set of points of $\bar{A}_P$ on which the elements of $\Delta - I(P) - J$ are zero. Therefore

$$Cl(\bar{A}_P(J)) = \coprod_{L \subset J} \bar{A}_P(L). \tag{8}$$

But

$$Cl(e(P(J))) = X \times^{A_P} Cl(\bar{A}_P(J)) \tag{9}$$

whence

$$Cl\big(e(P(J))\big) = \coprod_{L \subset J} X(P, L) = \coprod_{P(J) \supset R \supset P, R \in \mathfrak{P}} e(R). \tag{10}$$

Let now $J = I_0$, i.e. $P(J) = Q$. We have $A_P = A_{PQ} \times A_Q$ and A_Q acts trivially on $Cl(\bar{A}_P(J))$ whence

$$Cl\big(e(Q)\big) = (X/A_Q) \times^{A_{PQ}} Cl\big(\bar{A}_P(I_0)\big). \tag{11}$$

In view of 4.5(4), this can be written

$$Cl\big(e(Q)\big) = (X/A_Q) \times^{A_{PQ}} \overline{A_{PQ}} \tag{12}$$

or, taking 4.3 into account

$$Cl\big(e(Q)\big) = e(Q)(P). \tag{13}$$

Together with (10) and 4.3, this proves the second assertion of 5.3.

5.4. *Canonical cross-sections.* Let $J \subset L$ be subsets of $\Delta - I(P)$. Put $Q = P(J)$, $R = P(L)$ and consider the fibration 5.1(8) with structural group $A_{Q,R}$

$$v_{Q,R} : X(P, L) \to X(P, J) \tag{1}$$

which can also be written

$$v_{Q,R} : e(R) \to e(Q). \tag{2}$$

The space $X(P, L)$ is of type S under Q, associated to $R_d R$, hence it is of type S under $^0Q(\mathbf{R})$, and the isotropy groups in $^0Q(\mathbf{R})$ are its maximal compact subgroups.

Let $y \in X(P, L)$ and M its isotropy group in $^0Q(\mathbf{R})$. By 3.6, the map $A_{Q,R} \times \times {}^0Q(\mathbf{R}) \to X(P, L)$ defined by $(a, q) \mapsto (y \circ a) \cdot q$ induces an isomorphism

$$\mu_y : A_{Q,R} \times X(P, J) \xrightarrow{\sim} X(P, L), \tag{3}$$

which commutes with $^0Q(\mathbf{R})$ acting in the usual way on $X(P, L)$ and $X(P, J)$. The images of the sets $\{a\} \times X(P, J)$ are the orbits of $^0Q(\mathbf{R})$ and will be called the canonical or standard cross-sections of the fibration (1).

Let now $x \in X$. For $Q \supset P$, denote x_Q its projection on $e(Q)$. The trivialization

$$\mu_x : A_P \times e(P) \xrightarrow{\sim} X \tag{4}$$

induces one of the associated bundle $X(P)$

$$\mu_x : \bar{A}_P \times e(P) \xrightarrow{\sim} X(P) \tag{5}$$

which commutes with A_P and $^0P(\mathbf{R})$. It is immediate from the definitions that, on the face $e(Q)$, this trivialization coincides with the trivialization given by (3) with $y=x_Q$, and $J=\emptyset$. If we replace G by Q, we get similarly an isomorphism

$$\mu_x: \bar{A}_{P,Q} \times e(P) \to Cl_{X(P)}\, e(Q) \cong e(Q)(P) \tag{6}$$

(cf. 5.3). We have then $X(P) \cong \bar{A}_{P,Q} \times \bar{A}_Q \times e(P)$, whence also an isomorphism

$$\mu_x: \bar{A}_Q \times e(Q)(P) \overset{\sim}{\to} X(P) \tag{7}$$

which commutes with A_Q and $^0P(\mathbf{R})$, acting in the obvious way. It is immediate from the definitions that the diagram

$$
\begin{array}{ccc}
\bar{A}_{P,Q} \times e(P) & \overset{\mu_{xQ}}{\underset{\sim}{\longrightarrow}} & e(Q)(P) \\
\downarrow & & \downarrow \\
\bar{A}_P \times e(P) & \overset{\mu_x}{\underset{\sim}{\longrightarrow}} & X(P),
\end{array}
\tag{6}
$$

where the vertical arrows are the canonical injections, is commutative, and that all maps commute with the natural actions of $^0P(\mathbf{R})$ and A_P.

Together with 5.3, this shows the commutativity of the following diagram, where the vertical arrows are inclusions:

$$
\begin{array}{ccc}
\bar{A}_R \times e(R)(Q) & \overset{\mu_x}{\to} & X(Q) \\
\downarrow & & \downarrow \\
\bar{A}_R \times e(R)(P) & \overset{\mu_x}{\to} & X(P)
\end{array}
\tag{7}
$$

5.5. PROPOSITION. *Let $Q, R \in \mathfrak{P}$ be such that $Q \cap R \in \mathfrak{P}$. Then the geodesic action of A_Q on X extends to a geodesic action on $e(R)$.*

Let $P = Q \cap R$. Then $e(Q)$ and $e(R)$ may be canonically identified to faces of $X(P)$ (5.3) and A_Q, A_R to subgroups of A_P. The action of A_Q is then defined by the extended geodesic action of A_P on $X(P)$.

Remark. Let $P' \in \mathfrak{P}$, $P' \subset P$. The canonical inclusions $X(P) \hookrightarrow X(P')$ and $A_Q \hookrightarrow A_P$ being compatible with the extended geodesic actions, it is clear that the above action of A_Q on $e(R)$ can also be defined using the corner $X(P')$.

5.6. Let $g \in G(k)$, $P \in \mathfrak{P}$, $P' = P^g$ and $I = I(P) = I(P')$. Then $\mathrm{Int}\, g^{-1}$ induces an isomorphism $\alpha: A_P \to A_{P'}$ compatible with the identifications of both groups with $(\mathbf{R}_+^*)^{\mathit{\Delta}-I}$ (4.2), hence it extends to an isomorphism of $\bar{A}_P$ onto $\bar{A}_{P'}$ also denoted $a \mapsto a^g$. We have

$$x \cdot p \cdot g = (x \cdot g) \cdot p^g \quad (x \mathbf{o}\, a) \cdot g = (x \cdot g)\, \mathbf{o}\, a^g \quad (a \in A_P,\, x \in X,\, p \in P(\mathbf{R})) \tag{1}$$

cf. 3.2(5); therefore $x \mapsto x \cdot g$ extends to an isomorphism of $X(P)$ onto $X(P')$, also denoted $x \mapsto x^g$, which also satisfies (1) with $x \in X(P)$. Let $Q \in \mathfrak{P}$, $Q \supset P$ and $Q' = Q^g$. Then $x \mapsto x^g$ induces an isomorphism $e(Q) \to e(Q^g)$ which is induced by passage to the quotient from the translation $x \mapsto x \cdot g$. Translation by g also gives rise to a commutative diagram of trivializations:

$$\begin{array}{ccc} \bar{A}_Q \times e(Q)(P) & \to & \bar{A}_{Q'} \times e(Q')(P') \\ \downarrow{\scriptstyle \mu_x} & & \downarrow{\scriptstyle \mu_{x \cdot g}} \\ X(P) & \to & X(P') \end{array} \qquad (2)$$

The map μ_x (resp. μ_{xg}) commutes with $A_Q \times {}^0P(\mathbf{R})$ (resp. $A_{Q'} \times {}^0P'(\mathbf{R})$) (5.4). In particular, if $g \in Q$, then $Q = Q'$, $a^g = a$ $(a \in A_Q)$, and all maps in (2) commute with A_Q.

§6. Topology of $X(P)$ and Siegel Sets

6.1. Let $P \in \mathfrak{P}$, and $I = I(P)$. For $t > 0$, we put

$$\begin{aligned} A_{P,t} &= \{ a \in A_P \mid \alpha(a) \leqq t, \quad (\alpha \in \Delta - I) \} \\ \bar{A}_{P,t} &= \{ a \in \bar{A}_P \mid \alpha(a) \leqq t, \quad (\alpha \in \Delta - I) \} . \end{aligned} \qquad (1)$$

Let $x \in X$. A *Siegel set* in X, with respect to P, x, is a set

$$\mathfrak{S} = \mathfrak{S}_{t, \omega} = (x \mathbf{o} A_{P,t}) \cdot \omega \qquad (2)$$

where ω is a relatively compact subset of ${}^0P(\mathbf{R})$.

Let x_P be the canonical projection of x on $e(P)$. Then, if $\mu_x : A_P \times e(P) \to X$ is the canonical isomorphism of 5.5, we have

$$\mu_x^{-1}(\mathfrak{S}_{t, \omega}) = A_{P,t} \times x_P \cdot \omega . \qquad (3)$$

In particular, every point $y \in X$ has a neighborhood of this form. If $P = G^0$, then $A_P = \{e\}$, and the Siegel sets with respect to P are just relatively compact subsets.

Let S' be a maximal torus of $R_d P$ stable under the Cartan involution of L_x with respect to a maximal compact subgroup of H_x. Let $A' = S'(\mathbf{R})^0$. Then $P(\mathbf{R}) = A' \ltimes {}^0P(\mathbf{R})$ and there is a canonical projection $\sigma : A' \to A_P$. Let $y \in X$. There exists $p \in P(\mathbf{R})$ such that $y = x \cdot p$. Write $p = a' \cdot q$ with $a' \in A'$ and $q \in {}^0P(\mathbf{R})$. Then $y = (x \mathbf{o} \sigma(a')) \cdot q$, and

$$(y \mathbf{o} A_{P,t}) \cdot \omega = (x \mathbf{o} \sigma(a') \cdot A_{P,t}) \cdot q \cdot \omega .$$

From this it is clear that any Siegel set with respect to x, P is contained into one with respect to y, P and conversely. Thus the choice of the origin matters little.

6.2. PROPOSITION. *For $Q \supset P$, $Q \in \mathfrak{P}$, let $J \subset \Delta - I$ such that $Q = P(J)$ and x_Q be* *the canonical projection of x onto $X(P, J)$. Let $y \in e(P)$ and $p \in {}^0P(\mathbf{R})$ be such that $y = x_P \cdot p$. Let $\mathfrak{S} = \mathfrak{S}_{t, \omega}$ be a Siegel set with respect to P, x.*

(i) *The closure $\overline{\mathfrak{S}}$ of $\mathfrak{S}$ in $X(P)$ is compact. We have*

$$\overline{\mathfrak{S}} \cap X(P, J) = (x_Q \mathbf{o} A_{P, Q, t}) \cdot \overline{\omega},$$

where $A_{P, Q, t} = A_{P, Q} \cap A_{P, t}$. In particular, the left-hand side is a Siegel set in $X(P, J)$, with respect to P and x_Q, and any such Siegel set can be obtained in this way.

(ii) *Let $t_i \to 0$ and ω_i be a fundamental decreasing sequence of relatively compact neighborhoods of e in ${}^0P(\mathbf{R})$, $(i = 1, 2, \ldots)$. The closures in $X(P)$ of the sets $(x \mathbf{o} A_{P, t_i}) \cdot$ $\cdot p \cdot \omega_i$ form a fundamental system of neighborhoods of y.*

The canonical isomorphism μ_x extends to $\bar{A}_P \times e(P) \xrightarrow{\sim} X(P)$ and 6.1(3) implies

$$\mu_x^{-1}(\overline{\mathfrak{S}}) = \bar{A}_{P, t} \times x_P \cdot \overline{\omega}, \tag{1}$$

which proves (ii) and the first part of (i). The second part of (i) follows from (1) and 5.4(6).

6.3. It is clear from the definitions and the equalities

$$G(\mathbf{R}) = K \cdot G^0(\mathbf{R}), \qquad G^0(\mathbf{R}) = (G^0 \cap H_x) \cdot {}^0(G^0)(\mathbf{R})$$

(1.2, 1.4) that Siegel sets do not change if we replace G by G^0 or ${}^0(G^0)$. Similarly, let V be a normal k-subgroup of $R_d G$ and let $X' = X/V(\mathbf{R})$, viewed as usual as a space of type $S - k$ for $G' = G/V$ (2.8). Then the image of a Siegel set in X under the canonical projection is a Siegel set in X', and any such set can be obtained in this way. In particular, in discussing properties of Siegel sets in X, we may always assume G to be connected, reductive and even to have no non-trivial central k-split torus. The most important ones will be deduced from reduction theory. For this, we have to relate the present Siegel sets to those considered in [3], which are subsets of reductive groups.

6.4. Assume then G to be reductive. Let P be a minimal parabolic k-subgroup of G and K a maximal compact subgroup of $G(\mathbf{R})$. Let L_P be the Levi subgroup of $P(\mathbf{R})$ stable under the Cartan involution θ_K (1.9), $S'_P = L_P \cap R_d(P)$ and $A'_P = (S'_P)^0$. The elements of Δ define a surjective homomorphism $A'_P \xrightarrow{\sim} (\mathbf{R}_+^*)^\Delta$ which goes over to the canonical isomorphism $A_P \xrightarrow{\sim} (\mathbf{R}_+^*)^\Delta$ under the natural projection $A'_P \to A_P$ and whose kernel is $A'_P \cap \mathscr{C}(G)$.

We now want to prove:

$$(H_x \cap R_d G)^0 = (H_x \cap A'_P)^0, \qquad (x \in X \text{ fixed under } K). \tag{1}$$

The group $H_x \cap R_d G$ is contained in P, stable under θ_K (2.1 (iv)), hence contained in

L_P (1.9), and we have

$$(H_x \cap R_d G)^0 \subset (H_x \cap L_P \cap R_d P)^0 = (H_x \cap A_P')^0 \subset (H_x \cap R_d P)^0. \tag{2}$$

The group $H_x \cap R_d G$ is contained in $R_d P$, and $H_x = K \cdot (H_x \cap R_d G)$, hence

$$H_x \cap R_d P = (K \cap R_d P) \cdot (H_x \cap R_d G),$$
$$(H_x \cap R_d P)^0 = (K \cap R_d P (\mathbf{R})^0)^0 \cdot (H_x \cap R_d G)^0.$$

But $R_d P(\mathbf{R})^0$ is the semi-direct product of A_P' and $R_u P(\mathbf{R})$, hence is contractible, and has no compact subgroup $\neq \{e\}$. Therefore

$$(H_x \cap R_d P)^0 = (H_x \cap R_d G)^0$$

which, together with (2), proves (1).

Define $A_{P,t}'$ in the same way as $A_{P,t}$. A Siegel set of $G(\mathbf{R})$ (with respect to K, P, S') is then a set of the form

$$\mathfrak{S}' = \mathfrak{S}_{t,\omega}' = K \cdot A_{P,t}' \cdot \omega, \tag{3}$$

where ω is a relatively compact subset of $^0 P(\mathbf{R})$. This is the definition of a standard normal Siegel set in [3, §12], except that we do not require S' to be defined over k, and ω to be a neighborhood of e. The maximal tori defined over $\mathbf{R}$ of $R_d P$ are conjugate under $P(\mathbf{R})$. Therefore, given P, we may always choose x so that S' is defined and split over k.

Let $x \in X$ be such that $K \subset H_x$. Then

$$x \cdot \mathfrak{S}_{t,\omega}' = (x \mathbf{0} \, A_{P,t}) \cdot \omega = \mathfrak{S}_{t,\omega} \tag{4}$$

is a Siegel set of X, with respect to P, x, as defined in 6.1. By definition (see 2.3 and (1)), $H_x = K \cdot (H_x \cap R_d G)^0 = K \cdot (H_x \cap A_P')^0$. Since $H_x \cap A_P'$ is the intersection of the kernels of the elements of $\varDelta$, we have

$$H_x \cdot \mathfrak{S}' = \mathfrak{S}', \qquad \mathfrak{S}' = \pi_x^{-1}(x \cdot \mathfrak{S}'), \tag{5}$$

where $\pi_x : G(\mathbf{R}) \to X$ is the orbital map $g \mapsto x \cdot g$. Thus the Siegel sets in $G(\mathbf{R})$, with respect to K, P are the inverse images of the Siegel sets in X with respect to P, x, where x is fixed under K.

6.5. PROPOSITION. *Let G be reductive. Let P be a parabolic k-subgroup of G, P_0 be a minimal parabolic k-subgroup of G contained in P, and $g \in G^0(k)$. Let $x \in e(P)$, $\{x_n\}$ $(n = 1, 2, \ldots)$ a sequence of points of X which tends to x in $X(P_0)$ and such that $\{x_n \cdot g\}_{n \geq 1}$ is relatively compact in $X(P_0)$. Then $g \in P(k)$.*

Fix a point $x_0 \in X$. By 6.2, there exists a Siegel set $\mathfrak{S} = \mathfrak{S}_{t,\omega}$ in X, with respect to x_0, P_0, which contains x_n and $x_n \cdot g$ for all n's. Let then $a_n \in A_{P,t}$ and $p_n \in \omega$ be such that $x_n = (x_0 \circ a_n) \cdot p_n$, $(n = 1, 2, \ldots)$. Let I be such that $P = P_{0I}$. Then $e(P) = X(P_0, I)$. The isomorphism

$$\mu = \mu_{x_0} : \bar{A}_{P_0} \times e(P_0) \overset{\sim}{\to} X(P_0)$$

of 5.4 maps $A_{P_0, P} \times e(P_0)$ onto $e(P)$. Thus, if we write $\mu^{-1}(x) = (a, b)$ with $a \in \bar{A}_{P_0}$, $b \in e(P_0)$, we have $\alpha(a) = 0$ for $\alpha \in \Delta - I$. As a consequence

$$\lim \alpha(a_n) = 0 \quad (\alpha \in \Delta - I). \tag{1}$$

We now fix a maximal compact subgroup K of the stability group of x_0, let π: $G(\mathbf{R}) \to X$ be the orbital map $g \mapsto x_0 \cdot g$ and $\mathfrak{S}' = \pi^{-1}(\mathfrak{S})$. Then $\mathfrak{S}'$ is a Siegel set in $G(\mathbf{R})$, with respect to K, P_0 (6.4), and we have $\mathfrak{S}' = K \cdot A'_{P,t} \cdot \omega$, in the notation of 6.4. Let then a'_n be an element of $A'_{P,t}$ which maps onto a_n under the natural projection. Our assumptions and (1) imply

$$a'_n \cdot p_n, \, a'_n \cdot p_n \cdot g \in \mathfrak{S}' \quad (n = 1, 2, \ldots),$$
$$\lim \alpha(a'_n) = 0 \quad (\alpha \in \Delta - I).$$

We have then $g \in P$ by Prop. 12.6 of [3].

§7. The Manifold with Corners $\bar{X}$

7.1. We shall denote by $\bar{X}$ or $\bar{X}(G)$ the disjoint union of the sets $e(P)$ $(P \in \mathfrak{P})$ (where, by definition, $e(G^0) = X$). For $P \in \mathfrak{P}$, we identify $X(P)$ with $\bigcup_{Q \supset P} e(Q)$ (see 5.1 (6)). We have then

$$X(P) \cap X(Q) = X(R), \quad (P, Q \in \mathfrak{P}), \tag{1}$$

where R is the smallest parabolic k-subgroup of G containing P and Q. By 5.3, the inclusion map $X(P') \to X(P)$ $(P \subset P' \in \mathfrak{P})$ is an isomorphism, of manifolds with corners, of $X(P')$ onto an open submanifold of $X(P)$. There exists therefore one and only one structure of manifold with corners on $\bar{X}$ such that the $X(P)$'s are open submanifolds with corners of X. The space $\bar{X}$ will always be endowed with that structure.

For every $P \in \mathfrak{P}$, the subspace $e(P)$ has an open neighborhood which meets only finitely many $e(Q)$'s $(Q \in \mathfrak{P})$, namely $X(P)$. Consequently *the $e(P)$'s $(P \in \mathfrak{P})$, or their closures in $\bar{X}$, form a locally finite cover of $\bar{X}$.*

By definition, we have

$$\bar{X} = \coprod_{P \in \mathfrak{P}} e(P) = \bigcup_{P \in \mathfrak{P}} X(P), \tag{2}$$

and the $X(P)$ $(P \in \mathfrak{P})$ form an open cover of $\bar{X}$. For $Q \in \mathfrak{P}$, we let

$$Y(Q) = \bigcup_{P \in \mathfrak{P}(Q)} X(P). \tag{3}$$

We have then also

$$Y(Q) = \coprod_{R \in \mathfrak{P}, R \cap Q \in \mathfrak{P}} e(R). \tag{4}$$

7.2. (i) The space X is canonically of type S under G^0, and $\mathfrak{P}(G) = \mathfrak{P}(G^0)$ by definition. Therefore $\bar{X}(G) = \bar{X}(G^0)$.

(ii) *Assume G to be reductive.* Then $R_d G(\mathbf{R})$ operates trivially on X, and X is of type S under $G/R_d G$, and also under $G^0/R_d G$. Since $\mathfrak{P}(G) = \mathfrak{P}(G/R_d G) = \mathfrak{P}(G^0/R_d G)$, we also have natural identifications

$$\bar{X}(G) = \bar{X}(G/R_d G) = \bar{X}(G^0/R_d G).$$

(iii) Let V be a connected unipotent normal k-subgroup of G and $\pi: G \to G' = G/V$, $\sigma: X \to X' = X/V(\mathbf{R})$ the canonical projections; the latter is the projection map of a principal fibration with structural group $V(\mathbf{R})$. By 5.2, for every $P \in \mathfrak{P}(G)$, this fibration extends to a principal fibration

$$X(P) \to X'(P/V) \quad \textit{with structural group } V(\mathbf{R})$$

commuting with A_P, which is therefore also compatible with the inclusions $X(Q) \hookrightarrow X(P)$ $(Q \supset P; Q \in \mathfrak{P})$. It follows that these principal fibrations match to give one for $\bar{X}$ over $\bar{X}'$.

7.3. PROPOSITION. (i) *The embedding $e(P) \to \bar{X}$ $(P \in \mathfrak{P})$ extends to an isomorphism of manifolds with corners of $\overline{e(P)}$ onto the closure of $e(P)$ in $\bar{X}$.*

(ii) *For $Q \in \mathfrak{P}$, the space $Y(Q) = \bigcup_{P \in \mathfrak{P}(Q)} X(P)$ is an open neighborhood of $\overline{e(Q)}$. For $x \in X$, the isomorphism $\mu_x: A_Q \times e(Q) \xrightarrow{\sim} X$ (see 5.4) extends to an isomorphism of $\bar{A}_Q \times \overline{e(Q)}$ onto $Y(Q)$, which commutes with A_Q, acting on $\bar{A}_Q \times \overline{e(Q)}$ via its natural action on $\bar{A}_Q$, and on each $e(Q')$, $(Q' \in \mathfrak{P}(Q))$, by geodesic action (5.5).*

(In (i), $\overline{e(P)}$ means the manifold with corners extending $e(P)$, where $e(P)$ is endowed with its canonical structure of space of type $S-k$ under P (5.1).)

(i) Let Z be the closure of $e(Q)$ in $\bar{X}$. Let $P \in \mathfrak{P}$. Since $X(P)$ is open, Z meets $e(P)$ only if $X(P) \cap e(Q) \neq \emptyset$, i.e. if $P \subset Q$. Therefore Z is the union of the spaces $X(P) \cap Z$ for $P \in \mathfrak{P}$, $P \subset Q$. By 5.3, $Z \cap X(P)$ may be canonically identified with $e(Q)(P)$, whence (i).

(ii) Since the $X(P)$'s are open in $\bar{X}$, the space $Y(Q)$ is an open neighborhood of $\overline{e(Q)}$. For $P \in \mathfrak{P}(Q)$, we have, by the above and (6), (7) of 5.4, an isomorphism

$$\mu_x: \bar{A}_Q \times e(Q)(P) \xrightarrow{\sim} X(P) \tag{1}$$

which commutes with A_Q, $^0P(\mathbf{R})$ and with the inclusions

$$e(Q)(P) \to e(Q)(P') \quad \text{and} \quad X(P) \to X(P') \quad (P' \subset P;\, P,\, P' \in \mathfrak{P}(Q)).$$

This proves that the maps μ_x, for $P \in \mathfrak{P}(Q)$, match and define an isomorphism $\mu_x: \bar{A}_Q \times \overline{e(Q)} \xrightarrow{\sim} Y(Q)$ commuting with A_Q, whence (ii). From now on, we identify $\overline{e(P)}$ with the closure of $e(P)$ in $\bar{X}(P \in \mathfrak{P})$.

7.4. COROLLARY. *Let $P,\, Q \in \mathfrak{P}$. Then $\overline{e(P)} \cap \overline{e(Q)}$ is equal to $\overline{e(P \cap Q)}$ if $P \cap Q \in \mathfrak{P}$ and is empty otherwise. In particular $\overline{e(P)} = \overline{e(Q)}$ if and only if $P = Q$.*
This follows from 7.3(i) and the definition.

7.5. COROLLARY. *Let $P,\, Q \in \mathfrak{P}$. Then $e(P) \cap \overline{e(Q)} \neq \emptyset \Leftrightarrow e(P) \subset \overline{e(Q)} \Leftrightarrow P \subset Q$.*
This follows again from the fact that, by 7.3(i), $\overline{e(Q)}$ is the union of the $e(Q')$ with $Q' \in \mathfrak{P}(Q)$.

7.6. PROPOSITION. *The action of $G(k) \cdot R_u G(\mathbf{R})$ on X extends to one on $\bar{X}$, which preserves the structure of manifold with corners of $\bar{X}$, and in particular permutes the faces $e(P)\ (P \in \mathfrak{P})$. For $g \in G(k) \cdot R_u G(\mathbf{R})$ and $P \in \mathfrak{P}$, we have $e(P) \cdot g = e(P^g)$.*
This is clear from 5.6, in particular 5.6(1) and 7.2, or by "transport de structure."

7.7. COROLLARY. *Let $P,\, Q \in \mathfrak{P}$.*
(1) $\{g \in G(k) \mid P^g = Q\} = \{g \in G(k) \mid e(P) \cdot g \cap e(Q) \neq \emptyset\}$
$\qquad\qquad\qquad = \{g \in G(k) \mid e(P) \cdot g = e(Q)\}.$
(2) $\{g \in G(k) \mid P^g \cap Q \in \mathfrak{P}\} = \{g \in G(k) \mid \overline{e(P)} \cdot g \cap \overline{e(Q)} \neq \emptyset\}.$
(3) $Q(k) = \{g \in G^0(k) \mid \overline{e(Q)} \cdot g \cap \overline{e(Q)} \neq \emptyset\}.$

(1) and (2) follow from 7.4, 7.6. By (2) the right-hand side of (3) is $\{g \in G^0(k) \mid Q^g \cap Q \in \mathfrak{P}\}$, which is known to be $Q(k)$.

7.8. THEOREM. *The manifold with corners $\bar{X}$ is Hausdorff. If k is countable, then $\bar{X}$ is countable at infinity.*
To prove the first assertion, we proceed by induction on $\dim G$. If $\dim G = 0$, then $\bar{X}$ is reduced to a point, so we assume our assertion to be true for every k-group G' of dimension $< \dim G$.
Let $y,\, y' \in \bar{X}$ and let $\{V_n\}$ (resp. $\{V_n'\}$) $(n = 1,\, 2,\, \ldots)$ be a fundamental sequence of neighborhoods of y (resp. y') such that $V_n \cap V_n' \neq \emptyset$ for all n. We have to prove that $y = y'$. Since a corner $X(P)\ (P \in \mathfrak{P})$ is open, by definition, and Hausdorff, it suffices to show that y and y' belong to one. Assume first $U = R_u G \neq \{e\}$. Let $X' = X/U(\mathbf{R})$.

By 7.2, the projection $X \to X'$ extends to one σ of $\bar{X}$ onto $\bar{X}'$, and we have

$$\sigma(X(P)) = X'(P/U), \quad \sigma^{-1}(X'(P/U)) = X(P), \quad (P \in \mathfrak{P}).$$

Since $\bar{X}'$ is Hausdorff by induction, we have $\sigma(y) = \sigma(y') \in X'(P/U)$ for some $P \in \mathfrak{P}$, whence $y, y' \in X(P)$.

This reduces us to the case where G is reductive. Let $P, P' \in \mathfrak{P}$ be the parabolic k-subgroups such that $y \in e(P)$ and $y' \in e(P')$. By 7.1(1), $X(P) \cap X(P')$ is the union of the $e(Q)$, with $Q \in \mathfrak{P}$, $Q \supset P, P'$. Since these are finite in number, there exists $Q \in \mathfrak{P}$ such that $e(Q) \cap V_n \cap V_n' \neq \emptyset$ for all n's. The points y and y' then belong to the closure of $e(Q)$, which may be identified with $\overline{e(Q)}$ by 7.3. If $Q \neq G^0$, the induction assumption, applied to Q and $\overline{e(Q)}$, shows that $y = y'$. So assume $Q = G^0$, i.e. $V_n \cap V_n' \cap X \neq \emptyset$ for all n's, and let $x_n \in V_n \cap V_n' \cap X$ $(n = 1, 2, \ldots)$. Thus both y and y' are limit points of the sequence $\{x_n\}$. Let P_0 be a minimal parabolic k-subgroup of G contained in P, and $g \in G^0(k)$ be such that $P_0 \subset P'^g$. We have then

$$x_n \to y \in e(P) \subset X(P_0), \quad x_n \cdot g \to y' \cdot g \in e(P'^g) \subset X(P_0).$$

We may assume $x_n, x_n \cdot g \in X(P_0)$ for all n's. The $x_n \cdot g$ then form a relatively compact subset of $X(P_0)$, and we have $g \in P$ by 6.5. The relation $P'^g \cap P \supset P_0$ then yields

$$P' \cap P = (P'^g \cap P)^{g^{-1}} \supset P_0^{g^{-1}}$$

whence $e(P'), e(P) \subset X(P_0^{g^{-1}})$; this shows that y and y' are contained in one corner, and finally that $y = y'$ since, as remarked above, each corner is Hausdorff.

Assume now k to be countable. Then so is $G(k)$, and also $\mathfrak{P}$, since the latter is the union of finitely many orbits of $G^0(k)$. Since each $e(P)$ is a countable union of compact subsets, the second assertion follows.

7.9. COROLLARY. *Let $P \in \mathfrak{P}$ and $x \in X$. Then the closure in $\bar{X}$ of a Siegel set $\mathfrak{S}$ with respect to x, P is contained in $X(P)$ and is compact.*

The closure $A = Cl_{X(P)}(\mathfrak{S})$ of $\mathfrak{S}$ in $X(P)$ is compact by 6.2. Since $\bar{X}$ is Hausdorff, this implies that A is compact and closed in $\bar{X}$, hence $A = Cl_{\bar{X}}(\mathfrak{S})$.

§8. Homotopy Type of $\partial \bar{X}$

8.1. *Retracts*

We recall some basic facts about *absolute retracts* (AR) and *absolute neighborhood retracts* (ANR) in the category of metric spaces. Proofs can be found in [24], [16], [17] and [19], App. II.

A metric space X is an AR if and only if the following equivalent conditions are satisfied:

(AR_1) X is a retract of any metric space which contains it as a closed subspace.

(AR_2) For any continuous map $f: A \to X$, where A is a closed subspace of a metric space Y, there exists a continuous map $F: Y \to X$ which extends f.

Similarly, the fact that X is an ANR can be characterized by the equivalent conditions:

(ANR_1) For any embedding of X as a closed subspace of a metric space Z, there is a neighborhood of X in Z of which X is a retract.

(ANR_2) For any continuous map $f: A \to X$, where A is a closed subspace of a metric space Y, there is a neighborhood U of A in Y, and a continuous map $F: U \to X$, such that F extends f.

The property of being ANR is *local* ([24], p. 8); every metrizable manifold (with boundary) is an ANR ([24], p. 3).

If X and Y are ANR's, every weak homotopy equivalence $f: X \to Y$ is a homotopy equivalence ([24], th. 15).

If X is an ANR, then (cf. [24], p. 5):

X is an AR $\Leftrightarrow X$ is contractible $\Leftrightarrow$ all $\pi_i(X)$ are 0.

8.1.1. LEMMA. *Let Y be a metric space, X a closed subspace of Y and $f: X \to Z$ a continuous map of X into a topological space Z. Assume X is an ANR and Z is contractible. Then f can be extended to a continuous map $F: Y \to Z$.*

By (ANR_1) we can choose a neighborhood U of X in Y of which X is a retract. The map f can be extended to a continuous map $f': U \to Z$; since Z is contractible, f' is homotopic to a constant map. Since a constant map can be extended to Y, the same is true for f, by the "homotopy extension theorem," cf. [12], p. 1–05.

8.2. *Nerves*

We need a variant of Weil's theorem ([30], p. 141) comparing a space with the nerve of one of its covers.

Let Y be a space, and $(Y_i)_{i \in I}$ a locally finite cover of Y by closed non-empty subsets. Let T be the *nerve* of that cover; it is a simplicial complex, whose set of vertices is I; a simplex s of T is a finite subset of I such that $Y_s = \bigcap_{i \in s} Y_i$ is non-empty. We denote by S the set of simplices of T, and by $|T|$ (resp. by $|s|$, for $s \in S$) the geometrical realization of T (resp. s); we put on $|T|$ the *weak topology*: a subset U of $|T|$ is open if and only if $U \cap |s|$ is open in $|s|$ for any $s \in S$ ([19], p. 41). We make the following assumptions:

(1) T has *finite dimension*, i.e. there exists an integer N such that $\mathrm{Card}(s) < N$ for all $s \in S$.

(2) All the Y_s, $s \in S$, are *absolute retracts*, cf. 8.1.

8.2.1. THEOREM. *The spaces Y and $|T|$ have the same homotopy type.*

We prove a more precise statement. Identify $|T|$ as usual with the subspace of $\mathbf{R}^{(I)}$ made of those $(x_i)_{i\in I}$ with $0 \leqslant x_i \leqslant 1$, $\sum x_i = 1$, and $\{i \mid x_i > 0\} \in S$. If $i \in I$, call $|T_i|$ the subspace of $|T|$ made of those (x_i) such that $x_i \geqslant x_j$ for all $j \in I$. If $s \in S$, we put $|T_s| = \bigcap_{i \in s} |T_i|$; it is the *star* of the barycenter of s in the first barycentric subdivision $|T^1|$ of $|T|$, see below; it is contractible. The refined form of th. 8.2.1. is

8.2.2. THEOREM. (i) *There exist continuous maps $f: Y \to |T|$ and $g: |T| \to Y$ such that $f(Y_i) \subset |T_i|$ and $g(|T_i|) \subset Y_i$ for every $i \in I$; they are unique, up to homotopy.*

(ii) *If f and g are chosen as in* (i), *$f \circ g$ and $g \circ f$ are homotopic to the identity.*

Proof of (i).

(i_1) *Construction of $f: Y \to |T|$.*

If $n > 0$, call $S(n)$ the set of $s \in S$ with $\mathrm{Card}\,(s) \geqslant n$, and put $Y_n = \bigcup_{s \in S(n)} Y_s$. We have

$$\emptyset = Y_N \subset Y_{N-1} \subset \cdots \subset Y_2 \subset Y_1 = Y.$$

We use decreasing induction on n to construct a continuous map

$$f_n : Y_n \to |T|$$

such that $f_n(Y_s) \subset |T_s|$ for all $s \in S(n)$. To get f_n from f_{n-1}, we have to define $f_{n,s}: Y_s \to |T_s|$ for every s with $\mathrm{Card}\,(s) = n$, and $f_{n,s}$ is known on all Y_t with $t \supset s$, $t \neq s$. Using assumption (2) together with Lemma 3.2 of [24], one sees that the union of those Y_t is an ANR, which is closed in Y_s; the existence of $f_{n,s}$ then follows from Lemma 8.1.1 since $|T_s|$ is contractible. This completes the induction process, hence the construction of $f_1 = f$.

The uniqueness of f (up to homotopy) is proved in a similar way; one uses the fact that, if Z is an AR (resp. an ANR), the same is true for $Z \times [0, 1]$.

(i_2) *Construction of $g: |T| \to Y$.*

Let T^1 be the first barycentric subdivision of T. The set of vertices of T^1 is S. A subset σ of S is a simplex of T^1 if and only if it is totally ordered by inclusion; we then denote by $s(\sigma)$ (resp. $t(\sigma)$) its smallest (resp. biggest) element. We identify the topological spaces $|T^1|$ and $|T|$ in the usual way; a vertex s of $|T^1|$ corresponds to the barycenter of the simplex $|s|$ of $|T|$; moreover, $|s|$ is the union of the simplices $|\sigma|$ with $t(\sigma) \subset s$. The star of s in $|T^1|$ is $|T_s| = \bigcap_{i \in s} |T_i| = \bigcup_{s(\sigma) \ni s} |\sigma|$.

If σ is a simplex of T^1, put $Y_\sigma = Y_{s(\sigma)}$. One checks easily that the condition $g(|T_i|) \subset Y_i$ for all $i \in I$ is equivalent to $g(|\sigma|) \subset Y_\sigma$ for all σ's. Since the Y_σ's are contractible, the existence of g follows from the "aspherical carrier theorem" ([19], p. 75–76); the same argument proves the uniqueness of g, up to homotopy.

Proof of (ii).

(ii_1) *The maps $g \circ f$, $\mathrm{Id}_Y: Y \to Y$ are homotopic.*

The proof is analogous to (i₁). Using decreasing induction on n, one constructs homotopies

$$F_n : Y_n \times [0, 1] \to Y$$

between $g \circ f$ and Id_Y, such that $F_n(Y_s \times [0, 1]) \subset Y_s$ for all $s \in S(n)$. To get F_n from F_{n+1}, we have to define $F_{n,s} : Y_s \times [0, 1] \to Y_s$ for every s with $\mathrm{Card}(s) = n$, and $F_{n,s}$ is known on the union of $Y_s \times \{0\}$, $Y_s \times \{1\}$ and all $Y_t \times [0, 1]$ with $t \supset s$, $t \neq s$; since Y_s is an AR, the corresponding extension problem is solvable by (AR₂).

(ii₂) *The maps $f \circ g$, $\mathrm{Id}_{|T|} : |T| \to |T|$ are homotopic.*

Both maps send each simplex $|\sigma|$ of $|T^1| = |T|$ into $|T_{s(\sigma)}|$, which is contractible. We then apply the aspherical carrier theorem, as above.

8.3. *The $\overline{e(P)}$'s.*

We go back to the hypotheses and notation of §7; we assume moreover that the ground field k is *countable*. The manifold with corners $\bar{X}$ is Hausdorff and countable at infinity (th. 7.8), hence metrizable ([24], th. 1); the $\overline{e(P)}$'s, for $P \in \mathfrak{P}$, make up a locally finite closed cover of $\bar{X}$ (7.1, 7.3).

8.3.1. LEMMA. *For every $P \in \mathfrak{P}, \overline{e(P)}$ is an absolute retract.*

Note first that, from the topological point of view, "corners" and "boundaries" are the same thing, hence $\overline{e(P)}$ is a metrizable manifold with boundary; by 8.1, it is an ANR. Moreover, it is known that a metrizable manifold with boundary has the same homotopy type as its "interior" (this follows for instance from the collar theorem of M. Brown [10] – in the present case, we may also use the differentiable structure of $\overline{e(P)}$ to get a differentiable collar, cf. the Appendix to the present paper). By 3.9, the interior $e(P)$ of $\overline{e(P)}$ is a homogeneous space of type $S - k$ under $P(\mathbf{R})$, hence is homeomorphic to some euclidean space. This shows that $\overline{e(P)}$ is contractible; by 8.1, it is an AR.

8.3.2. *Remark.* Instead of the (global) collar theorem, one may use a local deformation argument to prove that

$$\pi_i(e(P)) \to \pi_i(\overline{e(P)})$$

is surjective for all i, hence that all $\pi_i(\overline{e(P)})$ are 0. The fact that $\overline{e(P)}$ is an AR then follows from 8.1.

8.4. *Comparison between $\partial \bar{X}$ and the Tits building of G.*

Recall that G^0 is the biggest element of $\mathfrak{P}$, and that $X = e(G^0)$; all other $e(P)$'s are contained in the boundary $\partial \bar{X}$ of $\bar{X}$. We denote by I the set of maximal elements of $\mathfrak{P} - \{G^0\}$; by abuse of language, an element of I is called a *maximal parabolic subgroup* of G.

8.4.1. THEOREM. *The $\overline{e(P)}$'s, for $P \in I$, make up a locally finite closed cover of $\partial \bar{X}$. This cover has properties* (1) *and* (2) *of* 8.2. *Its nerve is the Tits building T of G.*

(Recall cf. [29], that the *Tits building* of G is the simplicial complex whose set of vertices is I, and whose simplices are the non-empty subsets s of I such that $P_s = \bigcap_{P \in s} P$ is a parabolic subgroup of G. It is canonically isomorphic to the building attached to the Tits system of $G^0(k)/RG(k)$ constructed in [8], cf. Bourbaki, LIE IV, §2, exerc. 10.)

The cover of $\partial \bar{X}$ given by the $\overline{e(P)}$ is locally finite (7.1). If s is a finite non-empty subset of I, we know (cf. 7.4) that $\bigcap_{P \in s} \overline{e(P)}$ is non-empty if and only if $P_s = \bigcap_{P \in s} P$ is parabolic, i.e. if and only if s is a simplex of the Tits building T. If this is the case, we have $\mathrm{Card}(s) \leqslant l$, where l is the rank of the corresponding BN-pair (i.e. the k-rank of the semi-simple group G^0/RG, or equivalently the semi-simple k-rank of G^0/R_uG, cf. [8], def. 4.23); moreover, by 7.4, the intersection of the $\overline{e(P)}$, for $P \in s$, is $\overline{e(P_s)}$, which is an AR by 8.3.1. All the assertions of 8.4.1 are now obvious.

8.4.2. COROLLARY. *The spaces $\partial \bar{X}$ and $|T|$ have the same homotopy type.*
This follows from 8.2. More precisely, 8.2.2 gives homotopy equivalences

$$f : \partial \bar{X} \to |T| \quad \text{and} \quad g : |T| \to \partial \bar{X}$$

which are canonical and inverse to each other, up to homotopy, and allow us to identify the homology groups

$$H_i(\partial \bar{X}, \mathbf{Z}) \quad \text{and} \quad H_i(T, \mathbf{Z}) \quad (i = 0, 1, \ldots)$$

of $\partial \bar{X}$ and T. By *transport de structure*, this identification is compatible with the action of $G(k)$ on both groups.

8.4.3. *Remark.* For each $x \in X$, the geodesic action (3.2) allows one to construct an *explicit* homotopy equivalence $g_x : |T| \to \partial \bar{X}$ of the type required in 8.2.2. We sketch the construction:

Let T^1 be the first barycentric subdivision of T and σ a simplex of T^1; let t be a point of $|\sigma|$. If s is a vertex of σ (hence a simplex of T), we denote by t_s the s-coordinate of t. Choose now a maximal simplex s_0 of T containing all the $s \in \sigma$, so that $P = P_{s_0}$ is a minimal parabolic subgroup of G, contained in all P_s for $s \in \sigma$. If Δ is the corresponding basis of the k-roots (cf. 4.1), the elements s of σ may be identified with subsets of Δ. For every $\alpha \in \Delta$, put

$$a_\alpha(t) = \sum_{\alpha \notin s} t_s,$$

where the sum extends to those $s \in \sigma$ which do not contain α. We have $a_\alpha(t) \in [0, 1]$, and one of them at least is 0. Let $a(t)$ be the element of $\bar{A}_P$ whose coordinates are $a_\alpha(t)$, cf. 4.3. Using the natural map $X \times \bar{A}_P \to \bar{X}$, we get a point $x \cdot a(t)$ of $\bar{X}$, which belongs to $\partial \bar{X}$ and does not depend on the choice of s_0. We now define $g_{x,\sigma}$ on $|\sigma|$ by the formula

$$g_{x,\sigma}(t) = x \cdot a(t).$$

One checks that the $g_{x,\sigma}$ are compatible with each other, and define a continuous map $g_x \colon |T| \to \partial \bar{X}$ having the required property (there is also a natural extension of g_x to a map $\bar{g}_x \colon C(|T|) \to \bar{X}$, where $C(|T|)$ is the cone on $|T|$).

It would be interesting to have a similar explicit construction for one of the maps $f \colon \partial \bar{X} \to |T|$.

8.5. *Homotopy type of* $|T|$.

We keep the notations of 8.3 and 8.4. In particular, l denotes the k-rank of the semi-simple group G^0/RG; the dimension of the Tits building T is $l-1$. The following result is known (cf. [28], [14]):

8.5.1. THEOREM. *The space* $|T|$ *has the homotopy type of a bouquet of* $(l-1)$-*dimensional spheres with the weak topology.*

(When $l=0$, this means that T is empty.)

We just outline the proof.

Assume $l \geqslant 1$, and choose an $(l-1)$-dimensional simplex s of T. Let Σ be the set of "apartments" of T containing s (see, e.g., Bourbaki LIE IV, §2, exerc. 10). It is known that any apartment A is isomorphic to the Coxeter complex of the Weyl group W of G, hence is a subdivision of an $(l-1)$-sphere. This allows us to identify $|A|$ with the sphere S_{l-1}. Now, form the bouquet

$$Bo_\Sigma = \bigvee_{A \in \Sigma} |A|,$$

of the spheres $|A|$, with $A \in \Sigma$, choosing for base-point a point of $|s|$. The inclusion maps $|A| \to |T|$ define a continuous map

$$i \colon Bo_\Sigma \to |T|$$

and the refined form of th. 8.5.1 is:

8.5.2. THEOREM. *The map* $i \colon Bo_\Sigma \to |T|$ *is a homotopy equivalence.*

This is proved by remarking first that each apartment A contains a unique $(l-1)$-simplex s_A which is *opposite* to s (the corresponding notion for parabolic subgroups being the one defined in [8], n° 4.8). Moreover, the $(l-1)$-simplices which are not opposite to s make up a *contractible* subcomplex T' of T, and T' contains all the faces of the s_A's. Pinching $|T'|$ to a point, one thus gets a map

$$j: |T| \to Bo_\Sigma$$

which is a homotopy inverse of i (for more details, see [14]).

8.5.3. *Remark.* Note that Bo_Σ, i and j all depend on the choice of s, i.e. of the choice of a minimal parabolic subgroup P of G. Hence, one can only assert that the homotopy equivalences i and j between $|T|$ and Bo_Σ are compatible (up to homotopy) with the action of $P(k)$ on both spaces. Note also that Σ can be identified with the set of maximal k-split tori in $P/R_u G$; in particular, $P(k)$ acts transitively on Σ. When moreover G is reductive, Σ may be identified with $P(k)/Z(S)(k)$, where S is a maximal k-split torus of P, and $Z(S)$ its centralizer in G^0; in particular, writing P as a semi-direct product $Z(S) \cdot R_u P$, one sees that $R_u(P)(k)$ acts *transitively and freely* on Σ; if $l \geqslant 1$, this implies that $\mathrm{Card}(\Sigma) = \aleph_0$.

8.6. *Homology and cohomology of $\bar{X}$ and $\partial\bar{X}$.*
Putting 8.4 and 8.5 together we get:

8.6.1. THEOREM. *The boundary $\partial\bar{X}$ of $\bar{X}$ is empty if $l=0$. If $l \geqslant 1$, it has the homotopy type of a bouquet of an infinite number of $(l-1)$-spheres.*
For $l=1$, this means that $\partial\bar{X}$ has an infinite number of components, and that each component is contractible.

8.6.2. COROLLARY. *The space $\partial\bar{X}$ is $(l-2)$-connected, i.e. $\pi_i(\partial\bar{X})=0$ for $i \leq l-2$.*
In particular, $\partial\bar{X}$ is connected if $l \geqslant 2$ and simply connected if $l \geqslant 3$. When $l=2$, $\pi_1(\partial\bar{X})$ is a free (non-abelian) group with an infinite basis.

Denote by $\tilde{H}_i(\partial\bar{X})$ the *reduced* homology groups of $\partial\bar{X}$, defined by:
$\tilde{H}_i(\partial\bar{X}) = H_i(\partial\bar{X}, \mathbf{Z})$ if $i \geqslant 1$
$\tilde{H}_0(\partial\bar{X}) = \mathrm{Ker}: H_0(\partial\bar{X}, \mathbf{Z}) \to \mathbf{Z}$.
Th. 8.6.1 implies:

8.6.3. COROLLARY. *If $l \geqslant 1$, the only non-zero $\tilde{H}_i(\partial\bar{X})$ is $\tilde{H}_{l-1}(\partial\bar{X})$; it is free abelian of infinite rank.*
On the other hand, Lemma 8.3.1, applied to $P=G^0$, gives:

8.6.4. LEMMA. *The space $\bar{X}$ is contractible. We have*

$$H_0(\bar{X}) = H^0(\bar{X}) = \mathbf{Z} \quad and \quad H_i(\bar{X}) = H^i(\bar{X}) = 0 \quad for \quad i \neq 0.$$

Denote now by $H_c^i(\bar{X})$ the i-th cohomology group of $\bar{X}$ *with compact carriers* [13], the coefficient group being $\mathbf{Z}$.

8.6.5. THEOREM. *The groups $H_c^i(\bar{X})$ are 0 for $i \neq d-l$, where $d = \dim X$. The group $H_c^{d-l}(\bar{X})$ is free abelian; its rank is 1 if $l=0$ and $\aleph_0$ if $l \geq 1$.*

Let $\Omega_X = H_c^d(X)$ be the *orientation group* of X; it is a free abelian group of rank 1 whose bases correspond to the two orientations of X. If $l=0$, $\bar{X}$ is equal to X, hence is an orientable manifold, and Poincaré duality gives a canonical isomorphism

$$H_c^i(\bar{X}) = H_{d-i}(\bar{X}) \otimes \Omega_X,$$

whence the required result since $H_j(\bar{X}) = 0$ for $j > 0$ and $H_0(\bar{X}) = \mathbf{Z}$.

Assume now $l \geq 1$. Since $\bar{X}$ is contractible, the homology exact sequence yields an isomorphism between $\tilde{H}_j(\partial \bar{X})$ and the relative homology group $H_{j+1}(\bar{X}, \partial \bar{X})$. On the other hand, Poincaré duality for manifolds with boundary (see below) gives isomorphisms

$$H_c^i(\bar{X}) = H_{d-i}(\bar{X}, \partial \bar{X}) \otimes \Omega_X.$$

By 8.6.3, this gives $H_c^i(\bar{X}) = 0$ for $i \neq d-l$, and

$$H_c^{d-l}(\bar{X}) = \tilde{H}_{l-1}(\partial \bar{X}) \otimes \Omega_X = \tilde{H}_{l-1}(T) \otimes \Omega_X, \tag{8.6.6}$$

which is free abelian of infinite rank, see above.

8.6.7. *Remark.* Poincaré duality for non-compact manifolds with boundary is well-known, but not easy to find in the literature. One can for instance prove it by the sheaf-theoretic method of Cartan's seminar ([13], p. 20–04 and 20–05). Another possibility is to apply Poincaré duality to the manifolds (with empty boundary) X and $\partial \bar{X}$ and to use the exact sequence

$$\cdots \to H_c^i(X) \to H_c^i(\bar{X}) \to H_c^i(\partial \bar{X}) \to H_c^{i+1}(X) \to \cdots$$

The details may be left to the reader.

8.6.8. *Remark.* The isomorphism

$$H_c^{d-l}(\bar{X}) = \tilde{H}_{l-1}(T) \otimes \Omega_X, \quad \text{valid for } l \geq 1, \tag{8.6.6}$$

is canonical, hence compatible with the natural action of $G(k)$ on both groups. Using 8.5.3, this gives information on the action of $P(k)$ on $H_c^{d-l}(\bar{X})$, where P is a minimal parabolic subgroup of G. Let us assume for simplicity that G is reductive, and put $B = P(k)$, $H = Z(S)(k)$, where S is a maximal k-split torus of P. One then finds that the $\mathbf{Z}[B]$-module $H_c^{d-1}(\bar{X})$ is isomorphic to the induced module $\mathbf{Z}[B]\otimes_{\mathbf{Z}[H]}\Omega_X$. In particular, if we write B as a semi-direct product $B=H\ltimes U$, we see that *the $\mathbf{Z}[U]$-module $H_c^{d-1}(\bar{X})$ is free of rank* 1; this is analogous to what happens in the *Steinberg representation* of a finite group endowed with a BN-pair, cf. [28].

III. THE QUOTIENT OF $\bar{X}$ BY AN ARITHMETIC SUBGROUP

From now on $k=\mathbf{Q}$ and Γ is an arithmetic subgroup of $G(\mathbf{Q})$ (0.5).

§9. The quotient $\bar{X}/\Gamma$

9.1. LEMMA. *Let Y be a locally compact space, Z a closed subspace with empty interior, L a discrete group which operates continuously on Y and leaves Z stable. Assume the following condition to be fulfilled:*

(∗) *For any compact subsets C, D of Y*

$$\{g\in L\mid C\cdot g\cap D\cap(Y-Z)\neq\emptyset\}\quad\text{is finite.}$$

Then L operates properly on Y.

Let C', D' be compact subsets of Y and C, D compact neighborhoods in Y of C' and D' respectively. Let $g\in L$ be such that $C'\cdot g\cap D'\neq\emptyset$. Then $C\cdot g\cap D$ is a neighborhood of some point in Y, hence it meets $Y-Z$, and we have

$$E=\{g\in L\mid C'\cdot g\cap D'\neq\emptyset\}\subset\{g\in L\mid C\cdot g\cap D\cap(Y-Z)\neq\emptyset\}.$$

Therefore E is finite by (∗), which proves the lemma.

9.2. In this section, we consider the following situation: V is a real Lie group, T a locally compact principal V-bundle, L a discrete group operating continuously on T. Let H be the group of homeomorphisms of T, $\sigma:L\to H$ the natural homomorphism and identify V with a subgroup of H. Assume that $\sigma(L)$ normalizes V and that σ is injective on $N=\sigma^{-1}(\sigma(L)\cap V)$. We identify N with a subgroup of V. Let $\pi:T\to T'= =T/V$ be the natural projection. It follows from our assumptions that L commutes with π and that the action of L on T induces one of $L'=L/N$ on T' by passage to the quotient.

LEMMA. *We keep the previous notation and assumptions.*

(i) *Assume L' to act properly on T' and N to be discrete in V. Then L acts properly on T.*

(ii) *Assume moreover V/N and T'/L' to be compact. Then T/L is compact.*

(i) Let C, D be compact subsets of T, and $E=\{g\in L\mid C\cdot g\cap D\neq\emptyset\}$. Then $\pi(E)$ is finite, hence E consists of finitely many subsets of the form $N \cdot g \cap E$, with $g \in E$. But, for $g \in L$:

$$N \cdot g \cap E = \{ n \cdot g \mid n \in N \quad \text{and} \quad C \cdot n \cap D \cdot g^{-1} \neq \emptyset \},$$

and the latter set is finite since V acts properly on T and N is discrete in V.

(ii) Our assumptions imply the existence of compact subsets $C \subset T$ and $D \subset N$ such that $T' = \pi(C) \cdot L'$ and $V = D \cdot N$. We have then

$$T = C \cdot V \cdot L = (C \cdot D) \cdot L$$

with $C \cdot D = \bigcup_{d \in D} (C \cdot d)$ compact since both C and D are.

We now prove one of the main results of this paper:

9.3. THEOREM. *The group Γ operates properly on $\bar{X}$. The quotient $\bar{X}/\Gamma$ is compact.*

Let Γ' be a subgroup of finite index of Γ. If our assertions are true for Γ', then they are true for Γ. We may therefore replace Γ by $\Gamma \cap G^0$. Moreover, X is canonically of type $S - Q$ under G^0, and $\bar{X}(G) = \bar{X}(G^0)$ (7.2). Thus we may (and do) assume G to be connected.

We prove the theorem by induction on $\dim G$. Assume first that $V = R_u P \neq \{e\}$. Let $\sigma : G \to G' = G/V$ and $\pi : X \to X' = X/V(\mathbf{R})$ be the canonical projections. The group $\Gamma' = \sigma(\Gamma)$ is arithmetic in G' [2], $\Gamma \cap V$ is arithmetic in V, $V(\mathbf{R})/(\Gamma \cap V(\mathbf{R}))$ is compact [3; 8.4], X' is canonically of type $S - Q$ under G' (2.8). The space $\bar{X}$ is a principal $V(\mathbf{R})$-bundle and $\bar{X}/V(\mathbf{R}) = \bar{X}'$ (7.2(iii)). By induction assumption Γ' operates properly on $\bar{X}'$ and $\bar{X}'/\Gamma'$ is compact. Our conclusion then follows from 9.2. This reduces us to the case where G is *connected* and *reductive*.

We now prove that Γ acts properly. In view of 9.1, applied to $Y = \bar{X}$, $Z = \partial \bar{X}$ and $L = \Gamma$, it suffices to show that if C, D are compact subsets of $\bar{X}$, then

$$E = \{ \gamma \in \Gamma \mid C \cdot \gamma \cap D \cap X \neq \emptyset \} \quad \text{is finite}. \tag{1}$$

Fix $x \in X$ and let $P \in \mathfrak{P}$. The closure $Cl_{\bar{X}}(\mathfrak{S})$ in $\bar{X}$ of a Siegel set $\mathfrak{S}$ with respect to P, x is compact (7.9) and every point in the corner $X(P)$ has a neighborhood of this form (6.2). Since the corners $X(P)$, where P runs through the set $\mathfrak{P}_\theta$ of minimal parabolic k-subgroups of G, form an open cover of $\bar{X}$ (7.1), it suffices to consider the case where $C = Cl_{\bar{X}}(\mathfrak{S})$, $D = Cl_{\bar{X}}(\mathfrak{S}')$, where $\mathfrak{S}$ (resp. $\mathfrak{S}'$) is a Siegel set with respect to x and a minimal parabolic k-subgroup P (resp. P'). There exists $g \in G(k)$ such that $P' = P^g$ (4.1). Then $\mathfrak{S}' \cdot g^{-1}$ is a Siegel set with respect to P, x. Since any two Siegel sets are contained in a bigger one (see 6.1), we may assume that $\mathfrak{S}' = \mathfrak{S} \cdot g$. We may also assume the set ω occurring in the definition of $\mathfrak{S}$ in 6.1 to be compact. Then $\mathfrak{S}$ is closed in X, hence equal to $Cl_{\bar{X}}(\mathfrak{S}) \cap X$. Under those conditions, (1) may therefore be written

$$E = \{ \gamma \in \Gamma \mid \mathfrak{S} \cdot \gamma \cap \mathfrak{S} \cdot g \neq \emptyset \} \quad \text{is finite}. \tag{2}$$

Let $\pi: G(\mathbf{R}) \to X$ be the map $g \mapsto x \cdot g$, and let $\mathfrak{S}' = \pi^{-1}(\mathfrak{S})$. By 6.4, $\mathfrak{S} = \pi(\mathfrak{S}')$ and $\mathfrak{S}'$ is a Siegel set in $G(\mathbf{R})$, with respect to a maximal compact subgroup K of H_x, P and a suitable maximal torus S' of $R_d P$. We have then $E = \{\gamma \in \Gamma \mid \mathfrak{S}'\gamma \cap \mathfrak{S}' \cdot g \neq \emptyset\}$.

It follows from the end remark in 6.1 that we may change at will the origin x used to define the Siegel sets. In particular (6.4), we may assume x to be such that S' is defined and split over k. Then $\mathfrak{S}'$ is a Siegel set in the sense of [3, §12], and the finiteness of E follows from Theorem 15.4 in [3].

In view of the relation between Siegel sets in X and in $G(\mathbf{R})$ (6.3). and of Theorem 13.1 in [3], there exists a Siegel set $\mathfrak{S}$ in X (with respect to some minimal parabolic $\mathbf{Q}$-subgroup P) and a finite subset C of $G(\mathbf{Q})$, such that $X = \mathfrak{S} \cdot C \cdot \Gamma$. By 7.9, the closure M of $\mathfrak{S} \cdot C$ in $\bar{X}$ is compact. Since Γ acts properly on $\bar{X}$, the family of sets $M \cdot \gamma$ $(\gamma \in \Gamma)$ is locally finite in $\bar{X}$, hence is closed in $\bar{X}$. On the other hand, it contains X, which is dense in $\bar{X}$. Therefore $M \cdot \Gamma = \bar{X}$ and M is mapped onto $\bar{X}/\Gamma$ under the natural projection. Hence $\bar{X}/\Gamma$ is compact.

9.4. PROPOSITION. *Let $\pi: \bar{X} \to \bar{X}/\Gamma$ be the natural projection. For $P \in \mathfrak{P}$, let $\Gamma_P = \mathscr{N}_G(P) \cap \Gamma$ and $e'(P) = \pi(e(P))$. Let D be a set of representatives for $\mathfrak{P}/\Gamma$.*
(i) We have $e'(P) = e(P)/\Gamma_P$ and, for $Q \in \mathfrak{P}$,

$$e'(P) \cap e'(Q) \neq \emptyset \Leftrightarrow e'(P) = e'(Q) \Leftrightarrow \exists \gamma \in \Gamma \quad such\ that \quad P^\gamma = Q. \tag{1}$$

The set D is finite and $\bar{X}/\Gamma = \coprod_{P \in D} e'(P)$.

(ii) $Cl_{\bar{X}/\Gamma}(e'(P)) = \pi(\overline{e(P)})$. $\tag{2}$

If $\Gamma \subset G^0$, then $\Gamma_P = \Gamma \cap P$ and

$$\overline{\pi(e(P))} \cong \overline{e(P)}/\Gamma_P = \coprod_{Q \in \mathfrak{P}(P)/\Gamma_P} e'(Q); \tag{3}$$

in particular $e'(Q)$ is in the closure of $e'(P)$ if and only if Q is conjugate under Γ to a subgroup of P.

(i) The first equality and (1) follow from 7.7 and imply the last equality of (i). Since the $e(P)$'s $(P \in \mathfrak{P})$ are permuted by Γ and form a locally finite family in $\bar{X}$ (7.1), it follows that the $e'(P)$ form a locally finite family in $\bar{X}/\Gamma$, parametrized by D. Since $\bar{X}/\Gamma$ is compact, this shows that D is finite. (The finiteness of $\mathfrak{P}/\Gamma$ also follows from [3; 15.6].)

(ii) We have $\overline{e(P)} \cdot \gamma = \overline{e(\gamma^{-1} \cdot P \cdot \gamma)}$ $(\gamma \in \Gamma)$, therefore (7.1) the $\overline{e(P)} \cdot \gamma$ $(\gamma \in \Gamma)$ form a locally finite family in $\bar{X}$, and $\overline{e(P)} \cdot \Gamma$ is closed in $\bar{X}$. It is the inverse image of $\pi(\overline{e(P)})$, hence the latter is closed, and contains the closure of $e'(P)$. On the other hand, $e(P)$ is dense in $\overline{e(P)}$, hence $e'(P)$ is dense in $\pi(\overline{e(P)})$, which proves (2). The first equality in (3) follows from 7.7, and the second one from (i), applied to $\overline{e(P)}$ and Γ_P.

9.5. Clearly, any compact subgroup of $G(\mathbf{R})$ has a fixed point in X. Since Γ acts properly (9.3), it acts freely if and only if it is torsion-free. Assume this to be the case. Then $\pi: \bar{X} \to \bar{X}/\Gamma$ is a local homeomorphism and X/Γ inherits the structure of manifold with corners of $\bar{X}$. Let $y \in e(P)$, $y' = \pi(y)$ and U' a sufficiently small neighborhood of y'. Then U' is isomorphic under π, as a manifold with corners, with a suitable neighborhood U of $y \in X(P)$. In particular the faces of the corner are the $e'(Q) \cap$ $\cap U' (Q \in \mathfrak{P}; Q \supset P)$.

1 **§10. Strong Separation Properties**

10.1. *Distinguished neighborhoods of* $\overline{e(P)}$. We keep the notation of §9. For $x \in X$, $P \in \mathfrak{P}$ and $t > 0$, we put

$$U_{x,P,t} = (x \mathbf{o} A_{P,t}) \cdot {}^0P(\mathbf{R}). \tag{1}$$

Since x is fixed under a maximal compact subgroup of ${}^0P(\mathbf{R})$, we also have

$$U_{x,P,t} = (x \mathbf{o} A_{P,t}) \cdot ({}^0P(\mathbf{R}))^0. \tag{2}$$

In the notation of 5.4, (1) may be written

$$U_{x,P,t} = \mu_x(A_{P,t} \times e(P)) \tag{3}$$

and $U_{x,P,t}$ is closed in X; in view of 7.3, we have therefore

$$\bar{U}_{x,P,t} = \mu_x(\bar{A}_{P,t} \times \overline{e(P)}), \qquad U_{x,P,t} = \bar{U}_{x,P,t} \cap X, \tag{4}$$

and $\bar{U}_{x,P,t}$ is a neighborhood of $\overline{e(P)}$ in $\bar{X}$. Any neighborhood of $\overline{e(P)}$ containing some $\bar{U}_{x,P,t}$ will be called *distinguished*.

10.2. LEMMA. *The neighborhood* $\bar{U}_{x,P,t}$ *is stable under* $A_{P,1} \times {}^0P(\mathbf{Q})$, *where the semi-group* $A_{P,1}$ *acts by geodesic action and* ${}^0P(\mathbf{Q})$ *by ordinary action, and* μ_x *commutes with* $A_{P,1} \times {}^0P(\mathbf{Q})$. *If* V *is a neighborhood of* $\overline{e(P)}$ *stable under* $\Gamma \cap P$, *then* V *is distinguished.*

We have clearly $A_{P,t} \cdot A_{P,1} = A_{P,t}$ hence also $\bar{A}_{P,t} \cdot A_{P,1} = \bar{A}_{P,t}$. Together with 7.3 and 7.6 this implies the first assertion.

Let V be a neighborhood of $\overline{e(P)}$. Let C be a compact subset of $\overline{e(P)}$. As t varies, the sets $\mu_x(\bar{A}_{P,t} \times C)$ form a family of compact sets whose intersection is C. Therefore V contains one of them. By 9.3, we may choose C so that $\overline{e(P)} = C \cdot (\Gamma \cap P)$. If $V \cdot (\Gamma \cap P) = V$, we have then for a suitable t, taking 7.3 into account:

$$V = V \cdot (\Gamma \cap P) \supset \mu_x(\bar{A}_{P,t} \times C) \cdot (\Gamma \cap P) = \mu_x(\bar{A}_{P,t} \times C \cdot (\Gamma \cap P)) = \bar{U}_{x,P,t}.$$

10.3. PROPOSITION. *We keep the previous notation. Let $\pi: \bar{X} \to \bar{X}/\Gamma$ be the natural projection and assume G to be connected. There exists $t > 0$ such that the equivalence relations defined on $\bar{U}_{x,P,t}$ by Γ and $\Gamma \cap P$ are the same. For any such t, the isomorphism μ_x induces an isomorphism*

$$\mu_x' : \bar{A}_{P,t} \times \overline{e(P)}/(\Gamma \cap P) \overset{\sim}{\to} \pi(\bar{U}_{x,P,t}), \tag{1}$$

such that the following diagram

$$
\begin{array}{ccc}
\bar{A}_{P,t} \times \overline{e(P)} & \overset{\mu_x}{\to} & \bar{U}_{x,P,t} \\
{\scriptstyle \mathrm{id.} \times \pi} \downarrow & & \downarrow {\scriptstyle \pi} \\
\bar{A}_{P,t} \times \overline{e(P)}/(\Gamma \cap P) & \overset{\mu_x'}{\to} & \pi(\bar{U}_{x,P,t})
\end{array}
\tag{2}
$$

is commutative. The geodesic action of $A_{P,1}$ on $\bar{U}_{x,P,t}$ commutes with π and induces an action on $\pi(\bar{U}_{x,P,t})$. All the maps in (2) commute with $A_{P,1}$.

By 9.4, the equivalence relations defined on $e(P)$ by Γ and $\Gamma \cap P$ are the same, and $\pi(\overline{e(P)}) = \overline{e(P)}/(\Gamma \cap P)$. By 9.3, there exists a compact subset C of $\overline{e(P)}$ such that $\overline{e(P)} = C \cdot (\Gamma \cap P)$. Since Γ operates properly (9.3), there is a neighborhood U of C in $\bar{X}$ such that, for any $\gamma \in \Gamma$, $U \cdot \gamma \cap U \neq \emptyset$ implies $C \cdot \gamma \cap C \neq \emptyset$, and hence $\gamma \in \Gamma \cap P$. Since C is compact, there exists $t > 0$ such that $\mu_x(\bar{A}_{P,t} \times C) \subset U$. We wish to show that any such t satisfies our conditions.

Let $a, b \in \bar{U}_{x,P,t}$ and $\gamma \in \Gamma$ be such that $a \cdot \gamma = b$. Since $\Gamma \cap P$ commutes with μ_x (10.2), there exist $a', b' \in \mu_x(\bar{A}_{P,t} \times C)$ and $\sigma, \tau \in \Gamma \cap P$ such that $a = a' \cdot \sigma$, $b = b' \cdot \tau$. We have then $b' = a' \cdot \sigma \cdot \gamma \cdot \tau^{-1}$, hence $\sigma \cdot \gamma \cdot \tau^{-1} \in \Gamma \cap P$ and $\gamma \in \Gamma \cap P$.

This proves the first assertion. The other assertions then follow immediately from 7.3 and 10.2.

10.4. PROPOSITION. *Let $P, Q \in \mathfrak{P}$, $x, y \in X$ and $g \in G(\mathbf{Q})$. Then the following four conditions are equivalent:*

    (i) $U_{x,P,t} \cdot g \cap U_{y,Q,t} \neq \emptyset$ *for all $t > 0$.*
    (ii) $\bar{U}_{x,P,t} \cdot g \cap \bar{U}_{y,Q,t} \neq \emptyset$ *for all $t > 0$.*
    (iii) $\overline{e(P)} \cdot g \cap \overline{e(Q)} \neq \emptyset$.
    (iv) $P^g \cap Q = R$ *is parabolic.*
If they are fulfilled, the sets $\bar{U}_{x,P,t} \cdot g \cap \bar{U}_{y,Q,t}$ $(t > 0)$ form a basis of ${}^0R(\mathbf{Q})$-invariant neighborhoods of $\overline{e(R)}$.

It follows from 10.2 that we may assume $x = y$. Moreover, replacing P by P^g, we may take $g = e$.

The equivalence of (iii) and (iv) follows from 7.7. Clearly, (iii) $\Rightarrow$ (ii). By 9.3, there exist compact subsets $C \subset \overline{e(P)}$ and $C' \subset \overline{e(Q)}$ such that

$$C \cdot (\Gamma \cap P) = \overline{e(P)} \quad \text{and} \quad C' \cdot (\Gamma \cap Q) = \overline{e(Q)}.$$

For $t>0$, let

$$D_t = \mu_x(\bar{A}_{P,t} \times C), \qquad D_t' = \mu_x(\bar{A}_{Q,t} \times C'). \tag{1}$$

We have then

$$\bar{U}_{x,P,t} = D_t \cdot (\Gamma \cap P), \qquad \bar{U}_{x,Q,t} = D_t' \cdot (\Gamma \cap Q). \tag{2}$$

The D_t (resp. D_t') are compact and their intersection is C (resp. C'). Since Γ acts properly on $\bar{X}$ (9.3), there exists $t>0$ such that

$$\{\gamma \in \Gamma \mid D_t \cdot \gamma \cap D_t' = \emptyset\} = \{\gamma \in \Gamma \mid C \cdot \gamma \cap C' \neq \emptyset\} \tag{3}$$

and the set of such γ's is finite. Let in particular $\{\gamma_i\}_{1 \leq i \leq m}$ be those $\gamma \in \Gamma$ which satisfy (3) and are contained in $(\Gamma \cap P) \cdot (\Gamma \cap Q)$; for each of them choose a decomposition

$$\gamma_i = \sigma_i \cdot \tau_i^{-1} \quad (\sigma_i \in \Gamma \cap P; \tau_i \in \Gamma \cap Q; i = 1, \ldots, m). \tag{4}$$

If now $\sigma \in \Gamma \cap P$ and $\tau \in \Gamma \cap Q$ are such that $D_t \cdot \sigma \cap D_t' \cdot \tau \neq \emptyset$, then, for some $i \leq m$, we have $\sigma \cdot \tau^{-1} = \sigma_i \cdot \tau_i^{-1}$ and hence

$$\sigma_i^{-1} \cdot \sigma = \tau_i^{-1} \cdot \tau \in \Gamma \cap P \cap Q.$$

This implies readily

$$\bar{U}_{x,P,t} \cap \bar{U}_{x,Q,t} = \bigcup_{1 \leq i \leq m} (D_t \cdot \sigma_i \cap D_t' \cdot \tau_i) \cdot (\Gamma \cap P \cap Q). \tag{5}$$

Assume now (ii). Then there exists $i\,(1 \leq i \leq m)$ such that $D_t \cdot \sigma_i \cap D_t' \cdot \tau_i \neq \emptyset$ for all $t>0$, hence such that $C \cdot \sigma_i \cap C' \cdot \tau_i \neq \emptyset$. This proves that (iii) holds. Clearly, (i) $\Rightarrow$ (ii). Assume again (ii) to hold. Then, (iii) holds and $\bar{U}_{x,P,t} \cdot g \cap \bar{U}_{y,Q,t}$ is a neighborhood in $\bar{X}$ of any point in $\overline{e(P)} \cdot g \cap \overline{e(Q)}$. Therefore

$$X \cap (\bar{U}_{x,P,t} \cdot g \cap \bar{U}_{x,Q,t}) \neq \emptyset;$$

by 10.1 (4) this is condition (i).

Assume R to be parabolic. Then $\overline{e(P)} \cap \overline{e(Q)} = \overline{e(R)}$ by 7.4, and the left-hand side of (5) is a closed neighborhood of $\overline{e(R)}$, which is stable under $^0R(\mathbf{Q}) \subset {}^0P(\mathbf{Q}) \cap {}^0Q(\mathbf{Q})$, hence distinguished (10.2). For each $i\,(1 \leq i \leq m)$ we have

$$\bigcup_{t>0} D_t \cdot \sigma_i \cap D_t' \cdot \tau_i = C \cdot \sigma_i \cap C' \tau_i,$$

hence, given $s>0$, there exists $t>0$ such that

$$D_t \cdot \sigma_i \cap D_t' \cdot \tau_i \subset \bar{U}_{x,R,s}, \quad (1 \leq i \leq m);$$

(5) then shows that the left-hand side of (5) is contained in $\bar{U}_{x,R,s}$, whence the last assertion.

10.5. COROLLARY. *There exists $t>0$ such that for any P, $Q\in\mathfrak{P}$, we have*

$$\bar{U}_{x,P,t}\cap\bar{U}_{y,Q,t}\neq\emptyset \quad \text{*if and only if*} \quad \overline{e(P)}\cap\overline{e(Q)}\neq\emptyset.$$

This follows from 10.4 and the finiteness of $\mathfrak{P}/\Gamma$.

10.6. PROPOSITION. *Let P, $Q\in\mathfrak{P}$, x, $y\in X$. For $t>0$, let*

$$E_t = \{\gamma\in\Gamma \,|\, \bar{U}_{x,P,t}\cdot\gamma\cap\bar{U}_{x,Q,t} \neq \emptyset\}.$$

(i) *E_t is the union of finitely many double cosets modulo $(\Gamma\cap P)$ and $(\Gamma\cap Q)$.*
(ii) *For t small enough,*

$$E_t = \{\gamma\in\Gamma \,|\, \overline{e(P)}\cdot\gamma\cap\overline{e(Q)} \neq \emptyset\} = \{\gamma\in\Gamma \,|\, P^{\gamma}\cap Q \text{ is parabolic}\}.$$

(i) Let D_t and D'_t be as in 10.4(1). Then $E_t=(\Gamma\cap P)\cdot F_t\cdot(\Gamma\cap Q)$, where

$$F_t = \{\gamma\in\Gamma \,|\, D_t\cdot\gamma \cap D'_t \neq \emptyset\},$$

and F_t is finite since D_t and D'_t are compact and Γ acts properly on $\bar{X}$ (9.3).

(ii) Since F_t is finite and decreasing as $t\to 0$, it is independent of t for t small enough. Our assertion then follows from 10.4.

10.7. Let $P\in\mathfrak{P}$. Let S' be a maximal split torus of $R_d P$ and A' the identity component of $S'(\mathbf{R})$. Then $P(\mathbf{R})=A'\ltimes{}^0P(\mathbf{R})$, and there is a natural projection $\sigma:A'\to A_P$. We let for $t>0$

$$P(t) = \sigma^{-1}(A_{P,t})\cdot{}^0P(\mathbf{R}). \tag{1}$$

In the notation of 6.1, we can also write this $P(t)=A'_{P,t}\cdot{}^0P(\mathbf{R})$, and (1) shows that $P(t)$ does not depend on the choice of A'.

10.8. PROPOSITION. *Let P, $Q\in\mathfrak{P}$ and K, K' be maximal compact subgroups of $G(\mathbf{R})$.*

(i) *Let $g\in G(\mathbf{Q})$ and assume that $K\cdot P(t)\cdot g\cap K'\cdot Q(t)\neq\emptyset$ for all $t>0$. Then $P^g\cap Q$ is parabolic.*

(ii) *Given $t>0$, the set of $\gamma\in\Gamma$ for which $K\cdot P(t)\cdot\gamma\cap K'\cdot Q(t)\neq\emptyset$ is the union of finitely many double cosets modulo $(\Gamma\cap P)$ and $(\Gamma\cap Q)$.*

(iii) *There exists $t>0$ such that $K\cdot P(t)\cdot\Gamma\cap K'\cdot Q(t)=\emptyset$ unless $P^{\gamma}\cap Q$ is parabolic for some $\gamma\in\Gamma$.*

Let x (resp. y) be a point of X fixed under K (resp. K'). Then $K\cdot P(t)$ (resp. $K'\cdot Q(t)$) is the inverse image of $U_{x,P,t}$ (resp. $U_{y,Q,t}$) under the orbital map $g\mapsto x\cdot g$ (resp. $g\mapsto y\cdot g$). Therefore (i) follows from 10.4, (ii) from 10.6 and (iii) from 10.4, 10.6.

10.9. COROLLARY. *Assume G to be connected and $K\cdot P(t)\cdot g\cap K'\cdot P(t)\neq\emptyset$ for all $t>0$. Then $g\in P(\mathbf{Q})$.*

Indeed, $P^g \cap P$ is parabolic by 10.8. But in a connected algebraic group, two conjugate parabolic subgroups whose intersection is parabolic are identical. Hence g normalizes P, and then $g \in P$.

10.10. *Remark.* In §10, the ground field k is the field of rational numbers. However, the definitions in 10.1, 10.7 and the statements 10.4, 10.5, 10.8(i), (ii) and 10.9 make sense in the context of §§6, 7 where k is any subfield of **R**. For P, Q minimal these assertions can be proved using representative functions as in [3, §§14, 15], but we do not know whether they are true in general.

§11. Cohomology of Arithmetic Groups

We keep the hypotheses and notations of §9: $k = \mathbf{Q}$ and Γ is an *arithmetic subgroup* of $G(\mathbf{Q})$. Moreover, we assume that Γ is *torsion-free*.

11.1. *Qualitative results*

By 9.3 and 9.5, $\bar{X}/\Gamma$ is a compact C^∞-manifold with corners, hence is homeomorphic to a compact C^∞-manifold with boundary (cf. Appendix), and can be triangulated ([23], §10). Moreover, $\bar{X}$ is contractible (8.3.1), hence is a universal covering of $\bar{X}/\Gamma$. These properties imply:

a) The group Γ is isomorphic to the fundamental group of $\bar{X}/\Gamma$, hence is *finitely presented*.

b) The space $\bar{X}/\Gamma$ is a $K(\Gamma, 1)$-*space*. Its cohomology (or homology) is isomorphic to the one of Γ. More precisely, if A is a Γ-module, and $\tilde{A}$ the corresponding local system on $\bar{X}/\Gamma$, there are canonical isomorphisms

$$H_q(\Gamma, A) \approx H_q(\bar{X}/\Gamma, \tilde{A}) \quad \text{and} \quad H^q(\Gamma, A) \approx H^q(\bar{X}/\Gamma, \tilde{A})$$

for any q.

c) The group Γ is *of type* (FL) in the sense of [26], p. 84. Indeed, a triangulation of $\bar{X}/\Gamma$ lifts to a Γ-invariant triangulation of $\bar{X}$ and the corresponding complex of simplicial chains

$$0 \to C_d \to C_{d-1} \to \cdots \to C_0 \to \mathbf{Z} \to 0 \quad (d = \dim X)$$

gives a $\mathbf{Z}[\Gamma]$-free resolution of finite type of the $\mathbf{Z}[\Gamma]$-module $\mathbf{Z}$.

Remark. The above results depend only on the *existence* of a compactification of X/Γ as a manifold with boundary (or even as a finite complex), and not on the structure of the compactification; in the semi-simple case, they are due to Raghunathan [25].

11.2. *Comparison between $\bar{X}/\Gamma$ and its boundary $\partial\bar{X}/\Gamma$*
Let l be the $\mathbf{Q}$-rank of G/RG.

PROPOSITION. *The inclusion map $\partial \bar{X}/\Gamma \to \bar{X}/\Gamma$ is an $(l-2)$-homotopy equivalence.*
(This means that the natural maps $\pi_i(\partial \bar{X}/\Gamma) \to \pi_i(\bar{X}/\Gamma)$ are bijective for $i \leqslant l-2$.)
Since $\pi_1(\bar{X}/\Gamma) = \Gamma$ and $\pi_i(\bar{X}/\Gamma) = 0$ for $i \neq 1$, we have to prove that $\pi_1(\partial \bar{X}/\Gamma) \to \Gamma$ is an isomorphism if $l \geqslant 3$, and that $\pi_i(\partial \bar{X}/\Gamma) = 0$ for $i \leqslant l-2$, $i \neq 1$. This in turn follows from the fact that $\partial \bar{X}$ has the homotopy type of a bouquet of $(l-1)$-spheres (8.5.1), hence is simply connected if $l \geqslant 3$, and $\pi_i(\partial \bar{X}) = 0$ for $i \leqslant l-2$.

11.3. Euler-Poincaré characteristics

If Y is a finite complex, we denote by $\chi(Y)$ its Euler-Poincaré characteristic; we put $\chi(\Gamma) = \chi(\bar{X}/\Gamma)$, cf. [26], p. 91, prop. 9.

PROPOSITION. (a) $\chi(\partial \bar{X}/\Gamma) = 0$.
(b) *If $d = \dim X$ is odd, or if $R_u G \neq \{e\}$, we have $\chi(\Gamma) = 0$.*
Assume first Γ to be "net" [3, §17], hence contained in G^0, let $V = R_u G(\mathbf{R})$. The space $\bar{X}$ has a natural structure of principal V-bundle, cf. 7.2, (iii). By [3, 17.3], $\Gamma/(\Gamma \cap V)$ is torsion-free, hence acts freely on $\bar{X}/V$. This implies that $\bar{X}/\Gamma$ has a fibering with typical fiber $N = V/(\Gamma \cap V)$. If $\dim V \geqslant 1$, it is well-known that $\chi(N) = 0$, hence $\chi(\Gamma) = \chi(\bar{X}/\Gamma) = 0$ which proves the second assertion of (b).

Now, if $P \in \mathfrak{P}$ is distinct from G^0, its unipotent radical is non-trivial. Hence, by the above, applied to P, we see that the image $\overline{e'(P)} = \overline{e(P)}/\Gamma_P$ of $\overline{e(P)}$ in $\bar{X}/\Gamma$ (cf. 9.4) is such that $\chi(\overline{e'(P)}) = 0$. But the $\overline{e'(P)}$ make up a finite cover of $\partial \bar{X}/\Gamma$, and their intersections are either empty or of the form $\overline{e'(Q)}$ for some $Q \neq G^0$, hence have zero Euler-Poincaré characteristic. By an easy combinatorial argument, this implies that $\chi(\partial \bar{X}/\Gamma) = 0$.

If $\dim X$ is odd, the duality of manifolds with boundary implies that $\chi(\bar{X}/\Gamma) = \frac{1}{2}\chi(\partial \bar{X}/\Gamma)$, which is 0 by (a). This concludes the proof when Γ is net; the general case follows by a covering argument, using [3, 17.4].

Remark. The fact that $\chi(\Gamma) = 0$ when d is odd can also be proved by the method of Harder [18].

11.4. Duality theorem

We keep the above notation. In particular $d = \dim X$ and l is the $\mathbf{Q}$-rank of G/RG.

11.4.1. THEOREM. *We have $H^i(\Gamma, \mathbf{Z}[\Gamma]) = 0$ for $i \neq d-l$ and the group $I = H^{d-l}(\Gamma, \mathbf{Z}[\Gamma])$ is free abelian of rank 1 if $l = 0$ and of infinite rank if $l \geqslant 1$.*

This follows from theorem 8.6.5 together with the elementary fact that $H^i(\Gamma, \mathbf{Z}[\Gamma]) = H^i_c(\bar{X}, \mathbf{Z})$ for all i (cf. for instance [1], n° 6.3).

Note that the right action of Γ on $\mathbf{Z}[\Gamma]$ defines on $H^{d-l}(\Gamma, \mathbf{Z}[\Gamma])$ a structure of Γ-module. This Γ-module is the *dualizing module* of Γ:

11.4.2. THEOREM. *There is a homology class $e \in H_{d-l}(\Gamma, I)$ such that, for every*

Γ-module A, and for every integer q, the cap-product by e defines an isomorphism

$$H^q(\Gamma, A) \to H_{d-l-q}(\Gamma, I \otimes A).$$

This follows from theorem 4.5 of [1], combined with theorem 11.4.1 and 11.1.

Remark. In the language of [1], Γ is a *duality group*. It is a "Poincaré duality group" if I is isomorphic to $\mathbf{Z}$, i.e. if $l=0$, or, equivalently, if $\partial \bar{X} = \emptyset$.

11.4.3. COROLLARY. *We have* $cd(\Gamma) = d - l$.
(Recall that $cd(\Gamma)$ is the *cohomological dimension* of Γ, cf. [26], p. 84.)
This is clear from theorem 11.4.2.

11.4.4. THEOREM. *Let Δ be any arithmetic subgroup of $G(\mathbf{Q})$ (which may have torsion). Then Δ is of type (WFL), $vcd(\Delta) = d - l$, $H^i(\Delta, \mathbf{Z}[\Delta]) = 0$ for $i \neq d - l$ and $H^{d-l}(\Delta, \mathbf{Z}[\Delta])$ is isomorphic to I.*
(For the definitions of "type (*WFL*)" and "*vcd*," see [26], n° 1.8.)
This follows from 11.1, 11.4.1, 11.4.3 applied to the torsion-free subgroups of finite index of Δ. (Notice that $H^i(\Delta, \mathbf{Z}[\Delta])$ is isomorphic to $H^i(\Gamma, \mathbf{Z}[\Gamma])$ if Γ is of finite index in Δ, cf. [1], prop. 3.1.)

EXAMPLES. $vcd(\mathbf{SL}_3(\mathbf{Z})) = 5 - 2 = 3$; $vcd(\mathbf{Sp}_4(\mathbf{Z})) = 6 - 2 = 4$.

11.5. *Duality in cohomology*
Let R be a commutative ring and Ω an injective R-module. If V is any R-module, we define V' as $\mathrm{Hom}_R(V, \Omega)$. We also define I' as $\mathrm{Hom}_{\mathbf{Z}}(I, \Omega)$.

11.5.1. THEOREM. *For every $R[\Gamma]$-module V and every integer q, $H^q(\Gamma, V)'$ is naturally isomorphic to $H^{d-l-q}(\Gamma, \mathrm{Hom}_R(V, I'))$.*
This follows from the isomorphisms

$$H^q(\Gamma, V)' \cong H_{d-l-q}(\Gamma, V \otimes I)' \qquad \text{(theorem 11.4.2)}$$
$$H_{d-l-q}(\Gamma, V \otimes I)' \cong H^{d-l-q}(\Gamma, (V \otimes I)') \quad \text{(elementary)}$$
$$(V \otimes I)' \cong \mathrm{Hom}_k(V, I') \qquad\qquad\qquad \text{(linear algebra)}.$$

EXAMPLES. a) R is a field, $\Omega = R$, and V' is the dual vector space of V.
b) $R = \mathbf{Z}$ and $\Omega = \mathbf{Q}/\mathbf{Z}$.

11.6. *Remark.* If K is a number field, L an affine K-group, and Δ an arithmetic subgroup of $L(K)$, the above results can be applied to Δ, viewed as an arithmetic subgroup of the $\mathbf{Q}$-group $G = R_{K/\mathbf{Q}}L$, cf. [3], n° 7.16. We leave the details to the reader; he will notice, in particular, that the $\mathbf{Q}$-rank l of G/RG is equal to the K-rank of L/RL [8; 6.21].

Appendice

Arrondissement des variétés à coins

par A. Douady et L. Hérault

1. Secteurs

On pose $S_k^n = \mathbf{R}_+^k \times \mathbf{R}^{n-k}$. C'est le *secteur* type d'indice k dans $\mathbf{R}^n$. Soient U un ouvert de S_k^n, F un espace vectoriel réel de dimension finie, f une application de U dans F et $x_0 \in U$. On dit que f admet $\lambda \in \mathscr{L}(\mathbf{R}^n, F)$ pour dérivée en x_0 si $f(x) = f(x_0) + \lambda(x - x_0) + \eta(x)$ pour $x \in U$, où $\eta(x)$ est $o(x - x_0)$. La dérivée, si elle existe, est unique. On définit comme d'ordinaire les applications de classe C^1, C^r, C^∞ de U dans F.

On dit que f est analytique ou de classe C^ω en un point $x \in U$ si f est induite au voisinage de x par une application analytique d'un ouvert de $\mathbf{R}^n$ dans F.

PROPOSITION. 1.1. *Si* $U = S_k^n \cap V$, *où* V *est un ouvert de* $\mathbf{R}^n$, *toute application de classe* C^∞ *de* U *dans* $\mathbf{R}$ *se prolonge en une application de classe* C^∞ *de* V *dans* $\mathbf{R}$.

C'est un cas particulier du théorème de prolongement de Whitney ([31], [20]). On pourrait trouver pour ce cas particulier une démonstration plus simple que celle du cas général.

Soient U et U' des ouverts de S_k^n et $S_{k'}^{n'}$ respectivement, et f une application de U dans U'. On dit que f est de classe C^∞ (resp. C^ω) si elle est de classe C^∞ (resp. C^ω) de U dans $\mathbf{R}^{n'}$. On dit que f est un difféomorphisme si elle est bijective et si f et f^{-1} sont C^∞; on a alors $n = n'$ si $U \neq \emptyset$.

2. Variétés à coins

En prenant pour modèles les ouverts des S_k^n, $k \leqslant n \in \mathbf{N}$, et pour changements de cartes les difféomorphismes C^∞ (resp. C^ω), on obtient une catégorie locale qui est celle des *variétés à coins* C^∞ (resp. C^ω). (On peut aussi définir une variété à coins comme un espace annelé localement isomorphe à un modèle muni du faisceau des fonctions C^∞ (resp. C^ω).)

Soient X une variété à coins et $x \in X$. Il existe une carte φ de X centrée en x i.e. telle que $\varphi(x) = 0$. Si cette carte est à valeurs dans un ouvert de S_k^n, on dit que n est la dimension de X au voisinage de x et k est l'*indice* de x. On définit comme d'ordinaire l'espace vectoriel tangent $T_x X$. L'image réciproque de S_k^n dans $T_x X$ par $T_x \varphi$ est le *secteur rentrant* $ST_x X$ (il ne dépend pas du choix de la carte); son intérieur est noté

$S\mathring{T}_x X$. On dit que $t \in T_x X$ est rentrant (resp. strictement rentrant, resp. sortant, resp. strictement sortant) si $t \in ST_x X$ (resp. $S\mathring{T}_x X$, resp. $-ST_x X$, resp. $-S\mathring{T}_x X$).

On note $X^{(k)}$ l'ensemble des points de X d'indice $\geqslant k$. C'est le *k-bord* de X. L'ensemble $X = X - X^{(1)}$ est l'*intérieur* de X, $X^{(1)}$ est le *bord* de X. On dit que X est une *variété à bord lisse* si $X^{(2)} = \emptyset$.

Soit V une variété (sans bord) et soit X un fermé de V. On dit que X est une *pièce à coins* de V si, pour tout $x_0 \in X$, il existe un voisinage U de x_0 dans V et des fonctions $u_1, \dots, u_k$ de classe C^∞ (resp. C^ω) sur U telles que $d_{x_0} u_1, \dots, d_{x_0} u_k$ soient linéairement indépendants et que $X \cap U$ soit l'ensemble des $x \in U$ tels que $u_1(x) \geqslant 0, \dots, u_k(x) \geqslant 0$.

On dit qu'un champ de vecteurs θ sur une variété à coins X est strictement rentrant (resp. strictement sortant) si $\theta(x) \in S\mathring{T}_x X$ (resp. $\theta(x) \in -S\mathring{T}_x X$) pour tout $x \in X$ (il suffit de le vérifier pour $x \in X^{(1)}$). Sur toute variété à coins paracompacte, il existe un champ de vecteurs de classe C^∞ strictement rentrant (resp. sortant), comme on le voit avec une partition C^∞ de l'unité.

Soient X une variété à coins séparée, θ un champ de vecteurs de classe C^∞ sur X, strictement rentrant. Pour $x \in X$, soit $\gamma_x : I_x \to X$ la courbe intégrale maximale de θ d'origine x. L'intervalle I_x contient l'origine et est ouvert à droite; si $x \in \mathring{X}$, l'intervalle I_x est un voisinage de 0. Soit $\varrho_\theta(x)$ la borne supérieure de I_x. La fonction $\varrho_\theta : X \to \overline{\mathbf{R}}_+$ est strictement positive et semi-continue inférieurement.

3. Plongement d'une variété à coins comme pièce à coins d'une variété

PROPOSITION 3.1. *Toute variété à coins paracompacte peut être plongée dans une variété sans bord comme pièce à coins C^∞.*

Démonstration. Soit X une variété à coins paracompacte, et choisissons sur X un champ de vecteurs θ_1 de classe C^∞ strictement rentrant. On construit, au moyen d'une partition de l'unité, une fonction $\eta : X \to \mathbf{R}_+$, de classe C^∞, strictement inférieure à ϱ_{θ_1}, et strictement positive sur $X^{(1)}$. Posons $\theta = \eta \theta_1$. Le champ de vecteurs θ est de classe C^∞, strictement rentrant, et $\varrho_\theta > 1$.

On voit alors que l'application $\exp \theta : X \to X$, qui, à $x \in X$ associe $\gamma_x(1)$, où γ_x est la courbe intégrale de θ d'origine x, est un difféomorphisme de X sur une pièce à coins de $\mathring{X}$, cqfd.

PROPOSITION 3.2. *Toute variété à coins $\mathbf{R}$-analytique paracompacte peut être plongée dans une variété $\mathbf{R}$-analytique sans bord comme pièce à coins de classe C^ω.*

La démonstration est analogue à celle donnée par Whitney et Bruhat ([32]) pour prouver que toute variété $\mathbf{R}$-analytique paracompacte admet une complexification.

COROLLAIRE. *Soient X une variété à coins $\mathbf{R}$-analytique paracompacte et $\mathscr{F}$ un faisceau analytique cohérent sur X. On a $H^q(X; \mathscr{F}) = 0$ pour tout $q > 0$.*

Démonstration. Plongeons X comme pièce à coins dans une variété **R**-analytique V et soit W une complexification de V. D'après un résultat de Grauert ([15], p. 470, 1.11), tout voisinage ouvert U de X dans V admet un système fondamental de voisinages de Stein dans W, donc X admet dans W un système fondamental de voisinages de Stein. On obtient alors le corollaire en appliquant un résultat de Cartan ([1], prop. 6), cqfd.

PROPOSITION 3.3. *Sur toute variété **R**-analytique à coins paracompacte, il existe un champ de vecteurs strictement rentrant de classe C^{ω}.*

Démonstration. Soit X une pièce à coins d'une variété **R**-analytique paracompacte V. D'après ([15], Th. 3), on peut plonger V dans $\mathbf{R}^k$ pour k assez grand. Le fibré T_V admet dans le fibré trivial $V \times \mathbf{R}^k$ un supplémentaire analytique, par exemple l'orthogonal. Il résulte alors de ([15], Prop. 8) que tout champ de vecteurs C^∞ sur V peut être approché par un champ de vecteurs analytique au sens suivant: si θ est un champ de vecteurs C^∞ et $\varepsilon: V \to \mathbf{R}$ une fonction continue strictement positive, il existe un champ de vecteurs θ' de classe C^ω sur V tel que pour tout $x \in V$ on ait $\|\theta'(x) - \theta(x)\| < \varepsilon(x)$. Alors, si θ est strictement rentrant sur X, on peut choisir ε de façon que ceci entraîne que θ' est strictement rentrant sur X, cqfd.

4. Champs de vecteurs strictement sortants

Soient X une pièce à coins d'une variété V, et θ un champ de vecteurs de classe C^∞ (resp. C^ω) sur V, strictement sortant de X. Pour $x \in V$, soit $\gamma_x :]-a'(x), a(x)[\to V$ la courbe intégrale maximale de θ d'origine x dans V. S'il existe un $t_0 \in]-a'(x), a(x)[$ tel que $\gamma_x(t_0) \in X^{(1)}$, on voit en prenant des coordonnées locales que la courbe γ_x sort de X en t_0, i.e. qu'il existe $\varepsilon > 0$ tel que $\gamma_x(]t_0 - \varepsilon, t_0]) \subset X$ et $\gamma_x(]t_0, t_0 + \varepsilon[) \subset V - X$. Il en résulte que $\gamma_x(]-a'(x), t_0]) \subset X$ et $\gamma_x(]t_0, a(x)[) \subset V - X$, et il existe au plus un tel t_0.

PROPOSITION 4.1. *Avec ces notations, il existe un voisinage M de $X^{(1)}$ dans V tel que:*

(a) *pour tout $x \in M$, il existe un $b(x) \in]-a'(x), a(x)[$ et un seul tel que $\gamma_x(b(x)) \in X^{(1)}$;*

(b) *l'application $b: M \to \mathbf{R}$ ainsi définie est continue.*

Démonstration. Soit $x_0 \in X^{(1)}$. Il existe un voisinage U de x_0 dans V, des fonctions $u_1, \ldots, u_k$ de classe C^∞ sur U et un nombre $m > 0$ tels que $X \cap U = \{x \in U : u_1(x) \geq 0, \ldots, u_k(x) \geq 0\}$, $u_i(x_0) = 0$ et $\langle \theta(x), d_x u_i \rangle \leq -m$ pour $x \in U$, $i = 1, \ldots, k$. Il existe alors un voisinage U' de x_0 dans U et un nombre $r > 0$ tels que, pour tout $x \in U'$, on ait $[-r, +r] \subset]-a'(x), a(x)[$ et $\gamma_x([-r, +r]) \subset U$. Soit U'' l'ensemble des $x \in U'$ tels que $u_i(x) < mr$ pour $i = 1, \ldots, k$. Pour $x \in U''$, on a $\gamma_x(-r) \in U \cap \mathring{X}$ et $\gamma_x(r) \in U - X$, donc il existe un $b(x) \in]-r, +r[$ tel que $\gamma_x(b(x)) \in X^{(1)}$.

Pour $x \in U''$, $t \in\,]-r, +r[$ et $i = 1, \ldots, k$, posons $h_i(x, t) = u_i(\gamma_x(t))$. On a $\partial h_i/\partial t < 0$ et $h_i(x, t)$ tend vers $u_i(\gamma_x(r)) < 0$ (resp. $u_i(\gamma_x(-r)) > 0$) quand t tend vers r (resp. $-r$), donc l'ensemble des couples (x, t) tels que $h_i(x, t) = 0$ est le graphe d'une fonction $b_i : U'' \to\,]-r, r[$ de classe C^∞, et $b = \inf_i(b_i)$ est continue. La proposition en résulte.

PROPOSITION 4.2. *Avec les mêmes notations, soit M comme dans la Prop. 4.1; l'application $\varphi : x \mapsto (\gamma_x(b(x)), b(x))$ est un homéomorphisme de M sur un voisinage W de $X^{(1)} \times \{0\}$ dans $X^{(1)} \times \mathbf{R}$.*

Démonstration. Il est clair que φ est continue. Son image est l'ensemble des $(x_1, t) \in X^{(1)} \times \mathbf{R}$ tels que $-t \in\,]-a'(x_1), a(x_1)[$ et $\varphi_{x_1}(-t) \in M$. Cet ensemble est un voisinage de $X^{(1)} \times \{0\}$ et l'application $(x_1, t) \mapsto \gamma_{x_1}(-t)$ de W dans M est continue. Or cette application est φ^{-1}, d'où la proposition.

PROPOSITION 4.3. *Avec les mêmes notations, supposons V séparée et M ouvert. La relation $R\{x, y\} : \exists t$ tel que $y = \gamma_x(t)$ est une relation d'équivalence entre éléments de M et le quotient M/R admet une structure de variété C^∞ (resp. C^ω) et une seule telle que l'application canonique $\chi : M \to M/R$ soit une submersion. Cette application induit un homéomorphisme de $X^{(1)}$ sur M/R.*

Démonstration. Il est clair que R est une relation d'équivalence. L'ensemble $\Omega \subset M \times \mathbf{R}$ des couples (x, t) tels que $t \in\,]-a'(x), a(x)[$ et $\gamma_x(t) \in M$ est ouvert dans $M \times \mathbf{R}$ et $\Phi : (x, t) \mapsto (x, \gamma_x(t))$ de Ω dans $M \times M$ est une immersion dont l'image est le graphe de R. Il suffit de voir que Φ est un homéomorphisme de Ω sur un fermé de $M \times M$, la première assertion résultera alors de ([9], 5.9.5). Le graphe de R est $\varphi^{-1}(\Omega \times_{X^{(1)}} \Omega)$, donc est fermé, et l'inverse de Φ est

$$\left(\varphi^{-1}(x_1, t),\, \varphi^{-1}(x_1, t')\right) \mapsto \left(\varphi^{-1}(x_1, t),\, t - t'\right),$$

qui est continue, d'où la première assertion.

L'application $\chi/X^{(1)}$ est continue, et son inverse est l'application déduite de $x \mapsto \gamma_x(b(x))$, donc est continue.

5. Fonctions tapissantes

DÉFINITION. Soient U un voisinage de 0 dans S_k^n et h une fonction $U \to \mathbf{R}$. On dit que h est tapissante en 0 si h se met au voisinage de 0 sous la forme $h(x) = f(x) \cdot x_1 \ldots x_k$, où f est C^∞ et > 0.

Remarque. Dans cette définition, si h est analytique, il en est de même de f.

PROPOSITION 5.1. *Soient U et U' deux ouverts de S_k^n contenant 0 et φ un dif-*

féomorphisme de U sur U' tel que $\varphi(0)=0$. Si h est une fonction sur U tapissante en 0, la fonction $\varphi_*(h)=h\circ\varphi^{-1}$ sur U' est tapissante en 0.

Démonstration. Soit u_i la fonction $x\mapsto x_i$. L'application φ permute les faces de S_k^n au voisinage de 0 suivant une permutation $\sigma\in\mathfrak{S}_k$. La fonction $\varphi_*(u_i)$ est nulle sur la $\sigma(i)$-ème face au voisinage de 0, et a une dérivée normale >0, donc est au voisinage de 0 de la forme $g_i u_{\sigma(i)}$, où g_i est C^∞ et >0. Alors $\varphi_*(h)=\varphi_*(f)\cdot g_1\ldots g_k\cdot u_1\ldots u_k$ au voisinage de 0.

DÉFINITION. Soient X une variété à coins, et h une fonction sur X. On dit que h est tapissante en un point x de X s'il existe une carte centrée en x telle que l'expression de h dans cette carte soit tapissante en 0.

Remarques

5.2. Si h est tapissante en x pour une carte, elle l'est pour toutes d'après la prop. 5.1.

5.3. Si h est tapissante en x, elle l'est au voisinage de x.

5.4. Si $x\in\mathring{X}$, h est tapissante en $x\Leftrightarrow h$ est C^∞ au voisinage de x et $h(x)>0$. Si x est d'indice $k\geqslant 1$, et si h est tapissante en x, on peut trouver une carte centrée en x telle que l'expression de h soit $u_1\ldots u_k$.

5.5. Un barycentre de fonctions tapissantes est tapissante. Sur toute variété à coins paracompacte, il existe une fonction tapissante de classe C^∞.

PROPOSITION 5.6. *Sur toute variété à coins $\mathbf{R}$-analytique paracompacte, il existe une fonction tapissante de classe C^ω.*

Démonstration. Soient X une variété à coins $\mathbf{R}$-analytique paracompacte, (U_i) un recouvrement de X par des domaines de cartes, et pour tout i, h_i une fonction tapissante analytique sur U_i. La fonction $g_{i,j}=\mathrm{Log}(h_j/h_i)$ se prolonge en une fonction analytique sur $U_i\cap U_j$, et on a $g_{i,k}=g_{i,j}+g_{j,k}$ sur $U_i\cap U_j\cap U_k$. D'après le Cor. de la Prop. 3.2, il existe des fonctions analytiques $c_i:U_i\to\mathbf{R}$ telles que $g_{i,j}=c_j-c_i$. Les fonctions $e^{-c_i}h_i$ se recollent alors en une fonction analytique tapissante sur X, cqfd.

PROPOSITION 5.7. *Soient X une variété à coins, h une fonction tapissante sur X et θ un champ de vecteurs strictement sortant sur X. Il existe un voisinage N de $X^{(1)}$ dans X tel que on ait $\langle\theta(x),d_x h\rangle<0$ pour tout $x\in N-X^{(2)}$.*

Démonstration. Soit x_0 un point de $X^{(1)}$ et soit $(u_1,\ldots,u_n)$ un système de coordonnées centré en x_0 tel que $h=u_1\ldots u_k$. Soient $\theta_1,\ldots,\theta_n$ les coordonnées de θ dans ce système, i.e. $\theta_i=\langle\theta,du_i\rangle$. On a $\theta_1,\ldots,\theta_k<0$ au voisinage de x_0, d'où $\langle\theta,dh\rangle$

$= \sum u_1 \ldots \hat{u}_i \ldots u_k \theta_i \leqslant 0$, et $\langle \theta, dh \rangle < 0$ en tout point où un au plus des u_i s'annule, car alors la somme possède un terme non nul.

6. Arrondissement des coins

PROPOSITION 6.1. *Soient X une variété à coins paracompacte et h une fonction tapissante sur X. Supposons que X soit une pièce à coins d'une variété V et soit θ un champ de vecteurs C^∞ sur V strictement sortant de X. Soit M un voisinage ouvert de $X^{(1)}$ dans V répondant aux conditions de la Prop. 4.1. Alors, il existe un voisinage N de $X^{(1)}$ dans $M \cap X$ tel que l'application $\psi : x \mapsto (\chi(x), h(x))$ soit un homéomorphisme de N sur un voisinage de $M/R \times \{0\}$ dans $M/R \times \mathbf{R}_+$.*

Démonstration. Avec les notations de la proposition 4.2, soit f une fonction continue strictement positive sur $X^{(1)}$ telle que l'ensemble $N_1 = \{(x, t) : 0 \leqslant t \leqslant f(x)\}$ soit contenu dans W. Posons $N = \varphi^{-1}(N_1)$. L'ensemble N est un voisinage de $X^{(1)}$ dans $M \cap X$, et, quitte à diminuer f, on peut supposer qu'il répond à la condition de la Prop. 5.7. On a alors le diagramme commutatif

$$
\begin{array}{ccccc}
N_1 & \xrightarrow{\;\varphi^{-1}\;} & N & \xrightarrow{\;\psi = (\chi,\, h)\;} & M/R \times \mathbf{R}_+ \\
\downarrow{\scriptstyle \pi} & & & & \downarrow \\
X^{(1)} & & \xrightarrow{\qquad\chi_1\qquad} & & M/R
\end{array}
$$

où π est propre, χ_1 est un homéomorphisme et, pour tout $x \in X^{(1)}$, la fonction $t \mapsto h(\varphi^{-1}(x, t))$ est strictement croissante sur $[0, f(x)]$ car sa dérivée est > 0 pour $t > 0$. La proposition en résulte.

THÉORÈME ET DÉFINITION 6.2. *Soient X une variété à coins paracompacte de classe C^∞ (resp. C^ω), h une fonction tapissante de classe C^∞ (resp. C^ω) et θ un champ de vecteurs strictement sortant de classe C^∞ (resp. C^ω) sur X. Soit N un voisinage ouvert de $X^{(1)}$ dans X répondant aux conditions de la Prop. 6.1. Il existe alors une variété à bord lisse $\tilde{X}$ de classe C^∞ (resp. C^ω) et une seule, ayant même espace topologique sous-jacent que X, telle que les structures de X et $\tilde{X}$ coïncident sur $X - X^{(2)}$, et que ψ soit un difféomorphisme (resp. un difféomorphisme C^ω) de N muni de la structure induite par $\tilde{X}$ sur un ouvert de $M/R \times \mathbf{R}_+$. On dit que $\tilde{X}$ est la variété obtenue en arrondissant X au moyen de h et θ.*

Démonstration. Il reste à voir que ψ induit un difféomorphisme (resp. un difféomorphisme $\mathbf{R}$-analytique) de $N - X^{(2)}$ sur un ouvert de $M/R \times \mathbf{R}_+$, mais cela résulte du théorème d'inversion locale, cqfd.

BIBLIOGRAPHY

[1] BIERI, R. and ECKMANN, B., *Groups with homological duality generalizing Poincaré duality*, Invent. Math. *20* (1973), 103–124.

[2] BOREL, A., *Density and maximality of arithmetic subgroups*, J. f. reine angew. Math. *224* (1966), 78–89.

[3] ——, *Introduction aux groupes arithmétiques*, Act. Sci. Ind. 1341, Hermann éd. Paris 1969.

[4] ——, *Linear algebraic groups* (Notes by H. Bass), Benjamin Lecture Notes Series, New York 1969.

[5] BOREL, A. and HARISH-CHANDRA, *Arithmetic subgroups of algebraic groups*, Ann. of Math. *75* (1962), 485–535.

[6] BOREL, A. et SERRE, J-P., *Théorèmes de finitude en cohomologie galoisienne*, Comment. Math. Helv. *39* (1964), 111–169.

[7] ——, *Adjonction de coins aux espaces symétriques. Applications à la cohomologie des groupes arithmétiques*, C. R. Acad. Sci. Paris *271* (1970), 1156–1158.

[8] BOREL, A. et TITS, J., *Groupes réductifs*, Publ. Math. I.H.E.S. *27* (1965), 55–151. (Compléments, *ibid. 41* (1972), 253–276.)

[9] BOURBAKI, N., *Variétés différentielles et analytiques*. Fascicule de résultats. Act. Sci. Ind. 1333, 1347, Paris 1967.

[10] BROWN, M., *Locally flat imbeddings of topological manifolds*, Ann. of Math. *75* (1962), 331–341.

[11] CARTAN, H., *Variétés analytiques réelles et variétés analytiques complexes*, Bull. Soc. Math. France *85* (1957), 77–99.

[12] ——, *Espaces fibrés et homotopie*, Séminaire ENS 1949/50, W. A. Benjamin, New York, 1967.

[13] ——, *Cohomologie des groupes, suite spectrale, faisceaux*, Séminaire ENS 1950/51, W. A. Benjamin, New York, 1967.

[14] GARLAND, H., *p-adic curvature and the cohomology of discrete subgroups of p-adic groups*, Ann. of Math. *98* (1973), 375–423.

[15] GRAUERT, H., *On Levi's problem and the imbedding of real analytic manifolds*, Ann. of Math. *68* (1958), 660–671.

[16] HANNER, O., *Some theorems on absolute neighborhood retracts*, Arkiv för Mat. *1* (1950), 389–408.

[17] ——, *Retraction and extension of mappings of metric and non-metric spaces*, Arkiv för Mat. *2* (1952), 315–360.

[18] HARDER, G., *A Gauss-Bonnet formula for discrete arithmetically defined groups*, Annales Sci. E.N.S. *4* (1971), 409–455.

[19] LUNDELL, A. and WEINGRAM, S., *The Topology of CW Complexes*, van Nostrand, New York, 1969.

[20] MALGRANGE, B., *Ideals of differentiable functions*, Bombay and Oxford Univ. Press (1966).

[21] MOSTOW, G. D., *Self-adjoint groups*, Ann. of Math. *62* (1955), 44–55.

[22] ——, *Fully reducible subgroups of algebraic groups*, Amer. J. Math. *78* (1956), 200–221.

[23] MUNKRES, J. R., *Elementary Differential Topology* (revised edition), Ann. of Math. Studies 54, 1966.

[24] PALAIS, R., *Homology theory of infinite dimensional manifolds*, Topology *5* (1966), 1–16.

[25] RAGHUNATHAN, M. S., *A note on quotients of real algebraic groups by arithmetic subgroups*, Invent. Math. *4* (1968), 318–335.

[26] SERRE, J-P., *Cohomologie des groupes discrets*, Ann. of Math. Studies *70* (1971), 77–169.

[27] SIEGEL, C. L., *Zur Reduktionstheorie der quadratischen Formen*, Publ. Math. Soc. Japan 5, 1959 (Gesam. Abh. 111, p. 275–333).

[28] SOLOMON, L., *The Steinberg character of a finite group with BN-pair*, Theory of Finite Groups (R. Brauer and C-H. Sah ed.), W. A. Benjamin, New York, 1969, 213–221.

[29] TITS, J., *Structures et groupes de Weyl*, Séminaire N. Bourbaki, exposé *288* (févr. 1965), W. A. Benjamin, New York, 1966.

[30] WEIL, A., *Sur les théorèmes de de Rham*, Comment. Math. Helv. *26* (1952), 119–145.

[31] WHITNEY, H., *Analytic extensions of differentiable functions defined in closed sets*, Trans. Amer. Math. Soc. *36* (1936), 63–89.
[32] WHITNEY, H. et BRUHAT, F., *Quelques propriétés fondamentales des ensembles analytiques réels*, Comm. Math. Helv. *33* (1959), 132–160.

The Institute for Advanced Study, Princeton, N.J. 08540, USA
Collège de France, 75231 Paris Cedex 05, France

Received June 4, 1973

99.

Cohomologie de certains groupes discrets et Laplacien p-adique
(d'après H. Garland)

Séminaire Bourbaki, Exp. 437 (1973/74),
Lect. Notes Math. **431** (1975) 12−35

Les résultats principaux de [7] concernent la nullité, en certaines dimensions, du r-ième groupe de cohomologe $H^r(\Gamma; M)$ d'un sous-groupe discret Γ d'un groupe simple p-adique $G(K)$, (K corps p-adique), qui est co-compact (i.e. $G(K)/\Gamma$ compact), pour la cohomologie à coefficients dans un Γ-module unitaire de dimension finie M, (par exemple la cohomologie réelle). Cette cohomologie s'identifie à celle d'un complexe cellulaire fini, le quotient par Γ de l'immeuble de Bruhat-Tits T [5] du revêtement simplement connexe $\tilde{G}$ de G, ou encore à un espace de formes harmoniques (au sens combinatoire), et les démonstrations se divisent en gros en deux parties:

(1) Réduction à un problème local. On montre que si certaines valeurs propres du laplacien combinatoire sur le «link» d'un simplexe quelconque de dimension donnée $< r$ de T sont suffisamment grandes, alors $H^r(\Gamma; M) = 0$, ($1 \leqq r < \dim T$).

(2) Estimation de la plus petite des valeurs propres entrant en ligne de compte.

La première partie peut en fait s'exposer pour un complexe simplicial fini presque quelconque, ce qui est fait au § 1. On en tire une première forme des théorèmes de nullité (1.5, 1.6), qui est généralisée au § 2 et appliquée à Γ au § 3. La deuxième partie est un problème sur des immeubles de systèmes de Tits finis, complètement résolu en dimension un, mais loin de l'être en général. Les §§ 4, 5 exposent les résultats obtenus par Garland sur cette question, qui permettent de montrer (6.1) l'existence d'un entier N, ne dépendant que du K-rang l de G, tel que $H^r(\Gamma; M) = 0$ pour $0 < r < l$ si le corps résiduel de K a au moins N éléments (hypothèse que l'on espère bien être superflue, suivant une conjecture de Serre qui a été le point de départ de [7]).

En revanche, le groupe $H^l(\Gamma; \mathbf{R})$ n'est pas nul en général (7.3). On verra (7.2) que si G est simplement connexe, sa dimension est égale à la multiplicité de la représentation spéciale de $G(K)$ dans $L^2(\Gamma\backslash G(K))$.

Ces résultats et, dans une certaine mesure, leurs démonstrations, présentent une analogie assez frappante avec ceux de Y. Matsushima [12] concernant la cohomologie réelle de sous-groupes discrets co-compacts d'un groupe de Lie réel simple non-compact H, (connexe, à centre fini pour simplifier). Matsushima associe à H un entier $m(H) \geqq 0$ tel que $H^*(\Gamma; \mathbf{R})$ s'identifie à la cohomologie relative de l'algèbre de Lie de H modulo celle d'un sous-groupe compact maximal L de H au moins jusqu'en dimension $m(H)$, quel que soit le sous-groupe discret co-compact Γ de H. La constante $m(H)$ se détermine à partir de la plus petite valeur propre d'un opérateur auto-adjoint associé à la courbure riemannienne de l'espace riemannien symétrique H/L. Les laplaciens locaux intervenant dans (1) ci-dessus sur l'immeuble T jouent donc un rôle analogue à celui de la courbure de H/L, ce qui a conduit Garland à les baptiser «courbure p-adique».

Le théorème de Matsushima peut aussi s'exprimer en disant que sur H/L, toute forme harmonique Γ-invariante de degré $\leq m(H)$ est H-invariante. Il en est en somme de même pour celui de Garland, car on voit facilement que sur T, toute forme harmonique $\tilde{G}(K)$-invariante de dimension $\neq 0$ est nulle (4.2 (1)).

2   Les théorèmes de nullité valent aussi pour Γ non nécessairement co-compact, à condition de remplacer $H^r(\Gamma; M)$ par un espace de formes harmoniques de carré intégrable convenable (cf. 2.2, 3.5, 6.1). Lorsqu'ils sont valables, ils montrent en particulier que toute r-forme harmonique de carré intégrable sur T est nulle si $r \neq 0, l$, ce qui devrait aussi être vrai sans hypothèse sur le corps résiduel.

§ 1. Formes harmoniques sur un complexe simplicial fini

1.1. Soit X un complexe *simplicial* fini, satisfaisant à la condition: C1. Tout simplexe de X est face d'un simplexe de dimension $l = \dim X$.

On note $X(r)$ l'ensemble des simplexes de dimension r de X et, pour $s \in X(r)$, on note $c_X(s)$ ou $c(s)$ le *nombre de simplexes de dimension l contenant s*. Il est élémentaire que l'on a

$$(1) \qquad (l - r) \cdot c(s) = \sum_{t \in X(r+1),\, t \supset s} c(t)\,.$$

L'espace $C^r(X; \mathbf{R})$ ou C^r des cochaines réelles de degré r de X sera muni du produit scalaire

$$(2) \qquad (f, g) = \sum_{s \in X(r)} c(s) \cdot f(s) \cdot g(s)\,, \qquad (f, g \in C^r)\,.$$

Soient d l'opérateur cobord sur X, $\{[t : s]\}$ le système de coefficients d'incidence le définissant, et ∂ l'adjoint de d par rapport au produit scalaire (2). On a donc, pour $f \in C^r, s \in X(r-1), u \in X(r+1)$

$$(3) \qquad df(u) = \sum_{t \in X(r),\, t \subset u} [u : t] \cdot f(t)\,,$$

$$(4) \qquad \partial f(s) = \sum_{t \in X(r),\, t \supset s} [t : s] \cdot (c(t)/c(s)) \cdot f(t)\,.$$

Soient $\Delta = d \cdot \partial + \partial \cdot d$ le *laplacien* de X et $\Delta^+ = \partial \cdot d$. Un élément $f \in C^r$ est *harmonique* si $\Delta f = 0$, ce qui équivaut à $df = \partial f = 0$. Soit $\mathbf{H}^r = \mathbf{H}^r(X; \mathbf{R})$ l'espace des formes harmoniques de degré r. On a la décomposition orthogonale «de Hodge»:

$$(5) \qquad C^r = \mathbf{H}^r \oplus dC^{r-1} \oplus \partial C^{r+1},$$

où

(6) $$\mathbf{H}^r \oplus dC^{r-1} = C^r \cap (\ker d) , \qquad \mathbf{H}^r + \partial C^{r+1} = C^r \cap (\ker \partial) ,$$

donc

(7) $$\mathbf{H}^r \cong \mathbf{H}^r(X; \mathbf{R}) .$$

La décomposition (5) est stable par Δ, qui est auto-adjoint $\geqq 0$, et > 0 sur $dC^{r-1} + \partial C^{r+1}$. On note $k^r(X)$ *la valeur propre minimale* de Δ sur ∂C^{r+1} ($0 \leqq r < l$). Notons que sur ∂C^{r+1}, on a $\Delta = \Delta^+$ et que $\ker \Delta^+ = \mathbf{H}^r + dC^{r-1}$. Par suite, $k^r(X)$ est aussi la plus petite valeur propre strictement positive de Δ^+ sur C^r. Evidemment, $k^r(X)$ est invariant par changement d'orientation des simplexes de X, i.e. par changement du système des coefficients d'incidence. Par convention, $k^r(X) = + \infty$ si $\partial C^{r+1} = \{0\}$.

1.2. Si s est un simplexe de X, on note St s *l'étoile* (fermée) de s, sous-complexe formé des simplexes de X contenant s, et Lk s le *«link»* de s, réunion des faces de St s ne rencontrant pas s. Ce dernier est un sous-complexe de dimension $l - \dim s - 1$, satisfaisant aussi à C1.

Si P est un sommet de X, soit ϱ_P la transformation linéaire de C définie par

$$(\varrho_P f)(s) = f(s) \text{ si } P \in s , \quad (\varrho_P f)(s) = 0 \text{ sinon}, \quad (s \in X(r)) .$$

Il est immédiat que

(1) $$\varrho_P^2 = \varrho_P, \quad (\varrho_P f, g) = (f, \varrho_P g) , \quad \sum_{P \in X(0)} \varrho_P = (r+1) \cdot \mathrm{Id.} \text{ sur } C^r .$$

On pose

$$d_P = d \circ \varrho_P, \quad \partial_P = \varrho_P \circ \partial , \quad \Delta_P = d_P \cdot \partial_P + \partial_P \cdot d_P , \quad \Delta_P^+ = \partial_P \cdot d_P .$$

1.3. Lemme. *Pour tout* $f \in C^r (1 \leqq r \leqq \dim X)$, *on a*

$$r \cdot (\Delta^+ f, f) = \sum_{P \in X(0)} (\Delta_P^+ f, f) - (l - r) \cdot (f, f) .$$

Soient $t \in X(r+1)$ et P un sommet de t. Par définition

$$d_P f(t) = \sum_{P \in s \subset t} [t : s] \cdot f(s) , \quad df(t) = d_P f(t) = d_P f(t) + [t : s_0] \cdot f(s_0) ,$$

où s_0 est la face de t de dimension r ne contenant pas P. On en tire

$$r \cdot df(t)^2 = \sum_{P \in t} d_P f(t)^2 - \sum_{s \subset t} f(s)^2 .$$

Comme $(\Delta^+ f, f) = (df, df) = \sum_t c(t) \, df(t)^2$, cela entraîne

$$r \cdot (\Delta^+ f, f) = \sum_{t, P, P \in t} c(t) \cdot d_P f(t)^2 - \sum_{t, s, s \subset t} c(t) \cdot f(s)^2 ,$$

$(t \in X(r+1), s \in X(r), P \in X(0))$. La première somme est égale à

$$\sum_{P \in X(0)} \sum_{t \in X(r+1),\, P \in t} c(t) \cdot d_P f(t)^2 = \sum_{P \in X(0)} (\varDelta_P^+ f, f),$$

et 1.1 (1) montre que la deuxième est égale à $(l - r) \cdot (f, f)$.

1.4. Lemme. *Soient* $r \geqq 1$, $f \in C^r \cap \ker \partial$, $P \in X(0)$. *On suppose que* $\tilde{H}^{r-1}(\mathrm{Lk}\,P; \mathbf{R}) = 0$. *Alors* $(\varDelta_P^+ f, f) \geqq k^{r-1}(\mathrm{Lk}\,P) \cdot (\varrho_P f, f)$.

[$\tilde{H}^i$ est le i-ème groupe de cohomologie réduit: $\tilde{H}^i = H^i$ si $i \geqq 1$, et $\tilde{H}^0 = \mathrm{coker}\,(H^0(p\,t) \to H^0)$.]

Pour $m \geqq 1$ et $h \in C^m(X)$, désignons par h_0 la cochaine de degré $m-1$ sur $\mathrm{Lk}\,P$ définie par

$$h_0(s) = [P * s; s]\, h\,(P * s) \qquad (s \in (\mathrm{Lk}\,P)\,(m-1); \; P * s = \text{joint de } P \text{ et } s)\,.$$

Si d_0 est l'opérateur cobord sur $\mathrm{Lk}\,P$, défini à l'aide des coefficients

$$[t:s]_0 = [P * t : P * s] \cdot [P * s : s] \cdot [P * t : t]$$

alors $d_0 h_0 = (d_P h)_0$. Evidemment, $h_0 = (\varrho_P h)_0$. D'autre part, $c_X(P * s) = c_{\mathrm{Lk}P}(s)$ pour $s \in (\mathrm{Lk}\,P)\,(m-1)$, donc, vu 1.2 (1)

$$(1) \qquad\qquad (\varrho_P h, h) = (\varrho_P h, \varrho_P h) = (h_0, h_0)\,,$$

le dernier produit scalaire étant aussi défini par 1.1 (2), mais sur $\mathrm{Lk}\,P$. Vu 1.1 (4), on a

$$(2) \qquad c_X(P) \cdot \partial h\,(P) = \sum_{s \in (\mathrm{Lk}P)\,(0)} c_{\mathrm{Lk}P}(s)\, h_0(s) = (h_0, 1), \; (h \in C^1(X))\,.$$

Soient ∂_0 l'adjoint de d_0, $\varDelta_0 = \partial_0 \cdot d_0 + d_0 \cdot \partial_0$ et $\varDelta_0^+ = \partial_0 \cdot d_0$. Vu (1) et $(d_P h)_0 = d_0 h_0$, on a, pour $h \in C^m$,

$$(3) \qquad (\partial_P h)_0 = (\partial h)_0 = \partial_0 h_0, \quad (\varDelta_P h)_0 = \varDelta_0 h_0, \qquad (m \geqq 2)\,,$$

$$(4) \qquad (\varDelta_P^+ h)_0 = \varDelta_0^+ \cdot h_0, \qquad (\varDelta_P^+ h, h) = (\varDelta_0^+ h_0, h_0)\,, \quad (m \geqq 1)\,.$$

Plaçons-nous maintenant sous les hypothèses du lemme. Montrons tout d'abord que $f_0 \in \partial_0 C^r(\mathrm{Lk}\,P)$. Si $r \geqq 2$, alors $\partial_0 f_0 = 0$ vu (3); mais $\ker \partial_0 \cap C^{r-1} = \partial_0 C^r$ puisque $H^{r-1}(\mathrm{Lk}\,P; \mathbf{R}) = 0$, (cf. 1.1 (5), (6), (7)). Si $r = 1$, (2) montre que f_0 est orthogonale aux cochaines constantes, qui forment $\mathbf{H}^0$ puisque $\tilde{H}^0(\mathrm{Lk}\,P) = 0$, d'où $f_0 \in \partial_0 C^1$ aussi dans ce cas. On a alors, par définition de $k^{r-1}(\mathrm{Lk}\,P)$:

$$(5) \qquad (\varDelta_0 h_0, h_0) = (\varDelta_0^+ h_0, h_0) \geqq k^{r-1}(\mathrm{Lk}\,P) \cdot (h_0, h_0)\,,$$

et le lemme résulte de (1), (4) et (5).

1.5. Théorème. *Soient* $r \geq 1$ *et* $m^{r-1}(X) = \min\limits_{P \in X(0)} k^{r-1}(\operatorname{Lk} P)$. *On suppose que*
$\tilde{H}^{r-1}(\operatorname{Lk} P; \mathbf{R}) = 0$ *pour tout sommet P de X. Alors on a*

(1) $\qquad r \cdot (\Delta^{+}\!f, f) \geq ((r+1) \cdot m^{r-1}(X) - (l-r)) \cdot (f, f), \qquad (f \in C^r \cap \ker \partial).$

En particulier, $r \cdot k^r(X) \geq (r+1) \cdot m^{r-1}(X) - (l-r)$ et si $m^{r-1}(X) > (l-r)/(r+1)$, alors $\mathbf{H}^r = 0$.

1.4 entraîne que $(\Delta_P^{+}\!f, f) \geq m^{r-1}(X) \cdot (\varrho_P f, f)$ pour tout $P \in X(0)$. (1) résulte alors de 1.3, compte tenu de 1.2 (1).

1.6. Le théorème précédent est, dans le présent contexte, le «fundamental estimate» de [7: § 5]. Si $r \geq 2$ et si les links des sommets de $\operatorname{Lk} P$, (autrement dit les links des 1-simplexes de X contenant P) sont acycliques en dimension $r - 2$, on peut appliquer le théorème à $\operatorname{Lk} P$ pour estimer $k^{r-1}(\operatorname{Lk} P)$ à l'aide de $m^{r-2}(\operatorname{Lk} P)$, donc, en définitive, pour estimer $k^r(X)$ à l'aide des $k^{r-2}(\operatorname{Lk} s)$, $(s \in X(1))$ et ainsi de suite. Par une récurrence sans difficulté, laissée au lecteur, on démontre ainsi:

Théorème. *Soient $j, r \in \mathbf{N}$, $(1 \leq j \leq r < l)$. On suppose $\tilde{H}^{r-m}(\operatorname{Lk} s) = 0$ pour tout $s \in X(m-1)$ et $1 \leq m \leq j$. Soit $m^{r,j}(X) = \min\limits_{s \in X(j-1)} k^{r-j}(\operatorname{Lk} s)$. Alors*

(1) $\qquad (r - j + 1) \cdot (\Delta^{+}\!f, f) \geq ((r+1)\, m^{r,j}(X) - j \cdot (l-r)) \cdot (f, f), \qquad (f \in C^r \cap \ker \partial$.

En particulier $(r - j + 1) \cdot k^r(X) \geq (r+1) \cdot m^{r,j}(X) - j(l-r)$ et si $m^{r,j}(X) > j(l-r)/(r+1)$, alors $\mathbf{H}^r = 0$. De plus $m^{r,j}(X) > j(l-r)/(r+1)$ entraîne $m^{r,i} > i(l-r)/(r+1)$ pour $1 \leq i \leq j$.

La plus forte de ces conditions correspond donc à $j = r$; elle porte sur des sous-complexes de dimension $l - r$ et s'écrit

(2) $\qquad\qquad\qquad \min\limits_{s \in X(r-1)} k^0(\operatorname{Lk} s) > r(l-r)/(r+1) .$

On peut aussi obtenir 1.6 directement, sans récurrence, par une généralisation convenable de 1.3, 1.4, où l'on remplace P par un simplexe de dimension $j - 1$.

§ 2. Généralisations

2.1. La première consiste à considérer la cohomologie de X à coefficients dans un système local $\tilde{M}$ associé à un $\pi_1(X)$-module unitaire M de dimension finie. Le système local est alors formé d'espaces vectoriels M_s munis canoniquement d'un produit scalaire positif, et l'on définit sur l'espace $C^r(X; \tilde{M})$ des cochaines à coefficients dans $\tilde{M}$ un produit scalaire par 1.1 (2), étant entendu que $f(s) \cdot g(s)$ désigne maintenant le produit scalaire dans M_s. Le système $\tilde{M}$ est trivial sur l'étoile d'un simplexe et en particulier $C^m(\operatorname{Lk} s; \tilde{M}) = C^m(\operatorname{Lk} s; \mathbf{R}) \otimes M$. Il s'ensuit que 1.3, 1.4 subsistent, avec la même constante $k^{r-1}(\operatorname{Lk} P)$, (indépendante de M), et il en est de même pour 1.5, 1.6.

2.2. Dans une autre direction, ne supposons plus X fini, mais de dimension finie, satisfaisant à C 1 du § 1 et à:

C 2. Il existe une constante C telle que Card St $s \leq C$ pour tout simplexe s (i.e. X est «uniformément localement fini».) On munit $C^r(X; \tilde{M})$ du produit scalaire 1.1 (2) et on note $L^r(X; \tilde{M})$ ou L^r l'espace des cochaines de degré r qui sont «de carré intégrable», c'est-à-dire telles que $(f,f) < \infty$. En utilisant C 2, on voit facilement que d est un opérateur *borné* de L^r à L^{r+1} $(r = 0, 1, \ldots)$. Par suite, ∂, Δ, Δ^+ sont aussi bornés, et un élément $f \in L^r$ est harmonique si et seulement si $df = \partial f = 0$. On note $\mathbf{H}^r$ l'espace des formes harmoniques de degré r de carré intégrable. On a une décomposition orthogonale

$$(1) \qquad\qquad L^r = \mathbf{H}^r \oplus \overline{dL^{r-1}} \oplus \overline{\partial L^{r+1}} \qquad (r \in \mathbf{N})\,,$$

où $\overline{}$ dénote l'adhérence dans L^r, mais il n'y a bien entendu plus de lien direct entre $H^r(X; \tilde{M})$ et $\mathbf{H}^r$. On définit $k^r(X)$ comme inf $(\Delta^+ f, f)$ sur la boule unité de $\overline{\partial L^{r+1}}$. Alors les résultats du § 1 restent valables, de même que leurs démonstrations, à quelques changements d'écriture près. En particulier, si $m^{r-1} > (l - r)/(r + 1)$, alors il existe $c > 0$ tel que $(df, df) \geq c \cdot (f,f)$ pour tout $f \in \partial L^{r+1}$, ce qui entraîne que dL^r est *fermé* dans L^{r+1}.

§ 3. Sous-groupes discrets de groupes simples p-adiques I

3.1. K désigne un corps valué complet non discret, localement compact, non archimédien (autrement dit, une extension de $\mathbf{Q}_p$ de degré fini ou un corps de séries formelles à une variable sur un corps fini), v sa valuation et k_v le corps résiduel correspondant; k_v est donc fini.

G est un groupe affine connexe, défini sur K, presque simple sur K, et $\tilde{G}$ son revêtement universel. Le groupe $G(K)$ des points de G à valeurs dans K est muni d'une structure de groupe topologique localement compact totalement discontinu (et même de groupe de Lie sur K). On note l le K-rang $\mathrm{rg}_K(G)$, (dimension des sous-tores algébriques déployés sur K maximaux de G, cf. [1: § 5]) ou de $\tilde{G}$, ce qui revient au même [2].

3.2. On suppose $l \geq 1$, ce qui équivaut à $G(K)$ non compact, et on note $T(\tilde{G})$ ou T *l'immeuble* de Bruhat-Tits de $\tilde{G}(K)$ [5]. Il a, entre autres, les propriétés suivantes:

(i) T est un complexe simplicial localement fini, de dimension l, satisfaisant à C 1 de 1.1.

(ii) T est contractile.

(iii) $\tilde{G}(K)$ opère proprement sur T. Quels que soient $g \in \tilde{G}(K)$ et le simplexe s de T, l'élément g induit l'identité sur $s \cap g \cdot s$. Les groupes d'isotropie des sommets de T sont les sous-groupes compacts maximaux de $\tilde{G}(K)$; ceux des simplexes de T sont les *sous-groupes parahoriques* de $\tilde{G}(K)$; ils forment un système de Tits de rang $l + 1$, à groupe de Weyl de type affine, et T est l'immeuble de Tits associé à ce système.

(iv) $\tilde{G}(K)$ est transitif sur $T(l)$.

(v) Si $s \in T(r)$, $(0 \leq r < l)$, alors Lk s est l'immeuble d'un système de Tits fini, de rang $l - r$.

Fixons un simplexe C_0 de dimension l de T (une «chambre»). Vu (iv), l'étoile d'un simplexe de T est transformée, par un automorphisme de T, de l'étoile d'une face de C_0. En particulier, T est uniformément localement fini, et il n'y a qu'un nombre fine de types d'étoiles. On pose

$$(1) \qquad m^{r-1}(T) = \min_{P \in C_0(0)} k^{r-1}(\mathrm{Lk}\ P) = \min_{P \in T(0)} k^{r-1}(\mathrm{Lk}\ P)$$

$$(2) \qquad m^{r,j}(T) = \min_{s \in C_0(j-1)} k^{r-j}(\mathrm{Lk}\ s) = \min_{s \in T(j-1)} k^{r-j}(\mathrm{Lk}\ s), \qquad (0 \leqq j \leqq r < l).$$

3.3 Théorème. *Soient Γ un sous-groupe discret co-compact de $G(K)$, M un Γ-module unitaire de dimension finie et $j, r \in \mathbf{N}$ ($1 \leqq j \leqq r < l$). Si $m^{r,j}(T) > j(l-r)/(r+1)$, alors $H^r(\Gamma; M) = 0$.*

(a) On suppose tout d'abord que $G = \tilde{G}$ et que Γ opère «très librement» sur T, ce qui signifie:

$$(1) \qquad \gamma(\mathrm{St}\ P) \cap \mathrm{St}\ P = \emptyset \text{ quels que soient } P \in T(0) \text{ et } \gamma \in \Gamma, \gamma \neq e.$$

Cette condition entraîne que $X = T/\Gamma$, muni de la structure cellulaire quotient de celle de T, est en fait un complexe *simplicial*. De plus la projection canonique $\pi: T \to X$ est un isomorphisme de St s (resp. Lk s) sur St $\pi(s)$ (resp. Lk $\pi(s)$) pour tout simplexe s de T. Vu 3.2 (iv) et l'hypothèse sur Γ, X est fini. D'après le théorème de Solomon-Tits (*voir* par exemple [7: Appendix II]), un immeuble de Tits à groupe de Weyl fini, de dimension n, est un bouquet de sphères de dimension n. Par conséquent, vu 3.2 (v), on a

$$(2) \qquad \tilde{H}^m(\mathrm{Lk}\ s) = 0, \qquad (s \in X(j-1), 0 \leqq m < l-j).$$

D'après 1.6 pour la cohomologie réelle, ou 2.1 dans le cas général, on a donc $H^r(X; \tilde{M}) = 0$. Mais Γ opère librement sur T, qui est contractile, donc X est un $K(\Gamma, 1)$, et $H^r(X; \tilde{M}) = H^r(\Gamma; M)$, d'où 3.3 sous l'hypothèse (a).

(b) Dans le cas général, on remarque tout d'abord que si Γ' est un sous-groupe distingué d'indice fini de Γ, alors $H^r(\Gamma; M)$ s'identifie au sous-espace des invariants de Γ/Γ' dans $H^r(\Gamma'; M)$. Pour se ramener à (a), il suffit donc de prouver le lemme suivant.

3.4 Lemme. *Soit $\sigma: \tilde{G} \to G$ l'isogénie centrale canonique. Alors il existe un sous-groupe discret co-compact Γ' de $\tilde{G}(K)$, qui opère très librement sur T, et tel que $\sigma(\Gamma') \cong \Gamma'$ soit distingué d'indice fini dans Γ.*

Tout automorphisme défini sur K de $\tilde{G}$ induit un automorphisme de $\tilde{G}(K)$, donc de T. On en déduit une opération, aussi propre, de $G(K)$ sur T, *a fortiori* transitive sur $T(l)$. Par suite, Γ opère proprement sur T, et T/Γ est compact. Comme T est connexe, cela entraîne tout d'abord que Γ est de type fini. Le groupe $H = \sigma(\tilde{G}(K))$ est fermé distingué dans $G(K)$, et $G(K)/H$ est compact commutatif, d'exposant fini [2: 3.19]; par conséquent, l'image de Γ dans $G(K)/H$ est un groupe fini et $\Gamma_1 = H \cap \Gamma$ est un sous-groupe distingué d'indice fini de Γ. Il est co-compact, donc $\Gamma_2 = \sigma^{-1}(\Gamma_1) \cap \tilde{G}(K)$ est co-compact dans $\tilde{G}(K)$. Comme ker σ est fini, Γ_2 est aussi discret. Il s'ensuit immédiatement que l'ensemble des $\gamma \in \Gamma_2$ pour lesquels il existe $P \in T(0)$ tel que $\gamma(\mathrm{St}\, P) \cap \mathrm{St}\, P \neq \emptyset$ est réunion d'un nombre fini de classes de conjugaisons de Γ_2. Mais Γ_2 s'identifie à un sous-groupe de type fini d'un $\mathbf{GL}_n(\mathbf{K})$, donc l'intersection de ses sous-groupes d'indice fini ou, ce qui revient au même, de ses sous-groupes distingués d'indice fini, est réduite à $\{e\}$ [7: 2.7]. Il existe donc un sous-groupe distingué d'indice fini Γ_3 de Γ_2 qui opère très librement sur T. Il est sans torsion, et σ l'applique isomorphiquement sur un sous-groupe d'indice fini de Γ. Il suffit alors de prendre pour Γ' l'image réciproque dans Γ_3 de l'intersection de conjugués de $\sigma(\Gamma_3)$ dans Γ.

3.5. En utilisant 2.2, on démontre comme en 3.3(a):

Proposition. *Soient j, r et $m^{r,j}(T)$ comme en 3.3. Soient Γ un sous-groupe discret de $\tilde{G}(K)$ qui opère très librement sur T et M un Γ-module unitaire de dimension finie. Alors l'espace $\mathbf{H}^r(T/\Gamma; \tilde{M})$ des formes harmoniques de carré intégrable sur T/Γ, à coefficients dans le système local associé à M, est nul.*

[En fait, la démonstration de 3.3 montre aussi que si Γ est un sous-groupe discret de $G(K)$, alors l'espace $\mathbf{H}^r(T; M)^\Gamma$ des formes harmoniques Γ-invariantes sur T, de carré intégrable modulo Γ, est nul. Cet espace s'identifie à $\mathbf{H}^r(T/\Gamma; \tilde{M})$ si Γ opère très librement. Pour pouvoir le dire dans le cas général, il faudrait étendre les considérations du § 1 et de 2.2 à des complexes non simpliciaux, ce dont le conférencier n'a guère envie.]

§ 4. Estimation de valeurs propres: réduction au rang deux

4.1. Dans ce paragraphe, G est un groupe muni d'un système de Tits (G, B, N, S), T l'immeuble associé et l la dimension de T. A partir du n° 4.3, G est fini; jusque-là, G et T peuvent aussi être les $\tilde{G}(K)$ et T du § 3. On suppose $l \geq 1$.

Soit C le simplexe de dimension l de T fixe par le groupe B du système de Tits donné, et notons $I_l = \{0, 1, \ldots, l\}$ l'ensemble de ses sommets. Tout simplexe $s \in T(r)$ $(0 \leq r \leq l)$ est transformé par G d'une et une seule face C_I de C, dont l'ensemble $I \subset I_l$ des sommets définit le type $I(s)$ de s. La réunion T_I des simplexes de type I est donc une orbite de G. Pour $i \in I_l$, on pose (cf. 1.1, 1.2)

$$\varrho_i = \sum_{P \in T_{\{i\}}} \varrho_P, \qquad \Delta_i^+ = \sum_{P \in T_{\{i\}}} \Delta_P^+.$$

4.2. La deuxième propriété de 3.2(iii) et 3.2(iv) entraînent que·si $t \in T(r)$ et si s est une face de t, alors $c(s)/c(t)$ est le nombre des simplexes de type $I(t)$ contenant s, et ceux-ci forment une orbite du groupe d'isotropie de s. Comme les coefficients d'incidence sont invariants par G, on voit que dans l'expression de $\partial f(s)$ donnée par 1.1(4), $(f \in C^r(T), s \in T(r-1))$, le coefficient de $f(t)$ ne dépend que du type de t, et est égal à $[t:s]$ si f est invariante par le groupe d'isotropie de s. Il s'ensuit que, sur l'ensemble $C^r(T)^G$ des r-cochaines invariantes par G, la restriction $f \mapsto f|u$ à un simplexe $u \in T(l)$, qui commute évidemment à d, *commute aussi à* ∂. Elle commute alors aussi à Δ ou Δ^+. Comme il n'y a pas de forme harmonique non nulle en dimension > 0 sur un simplexe, on en déduit tout d'abord

(1) $$f \in C^r(T)^G, (1 \leqq r \leqq l), \quad df = \partial f = 0 \Rightarrow f = 0.$$

Il est élémentaire que sur un simplexe de dimension l, les valeurs propres de Δ sur C^0 sont 0 avec la multiplicité un (sur l'espace des cochaines constantes) et $l + 1$ avec la multiplicité l, (sur l'espace des fonctions de moyenne nulle). Par conséquent, si $f \in C^0(T)^G$ est une fonction propre non nulle de Δ, de valeur propre $c \neq 0$, alors $c = l + 1$.

4.3. Rappelons que G est dorénavant fini. $\partial C^1(T)$ est le sous-espace de $C^0(T)$ orthogonal aux constantes (vu 1.1 (5) et le fait que T est connexe). Comme $c(s)$ ne dépend que de $I(s)$ on a, en posant $c(i) = c(P)$ pour $P \in T_{\{i\}}$

(1) $$f \in \partial C^1(T) \Leftrightarrow \sum_{0 \leqq i \leqq l} c(i) \sum_{P \in T_{\{i\}}} f(P) = 0.$$

Soit maintenant $f \in \partial C^1(T)$ fonction propre de Δ, de valeur propre c. La somme $\bar{f}$ des transformées de f par G est encore dans le même espace propre de Δ, donc, vu la dernière remarque de 4.2, ou bien $c = l + 1$ ou bien $\bar{f} = 0$; autrement dit

(2) $$f \in \partial C^1(T), \Delta f = c \cdot f, \quad c \neq l + 1 \Rightarrow \sum_{P \in T_{\{i\}}} f(P) = 0, \quad (0 \leqq i \leqq l).$$

4.4. Le but principal de ce paragraphe est de prouver le:

Lemme. *Supposons* $l \geqq 2$ *et soit* $m^0(T) = \min\limits_{P \in T(0)} k^0(\mathrm{Lk}\, P)$. *Alors ou bien* $k^0(T) = l + 1$ *ou bien* $(l - k^0(T))^2 \leqq l^2 \cdot (l - 1 - m^0(T))/m^0(T)$.

Il s'ensuit que, étant donné $a > 0$, il existe $b > 0$ tel que si $m^0(T) \geqq l - 1 - b$, alors $k^0(T) \geqq l - a$. Ce lemme est en substance le lemme 6.3 de [7], et la démonstration est essentiellement la même.

Soit $f \in \partial C^1(T)$ et supposons $\Delta f = c \cdot f$. Il s'agit d'évaluer $(\Delta f, f) = (df, df)$. On voudrait pour cela appliquer 1.4 à df. Mais ce dernier n'est pas en général annulé par ∂. La méthode de Garland consiste à multiplier f sur les sommets d'un type i fixé par une constante convenable, de manière à ce que ∂f soit nulle sur $T_{\{i\}}$, d'appliquer 1.4 à cette transformée de f, puis de sommer sur i. La démonstration sera donnée en 4.6, après avoir discuté en 4.5 quelques propriétés de la transformation à faire subir à f.

4.5. Soient $i \in I_l$ et $t \in \mathbf{R}$. Etant donnée $f \in C^0$, on définit $f_{i,t} \in C^0$ par

(1) $$f_{i,t}(P) = t \cdot f(P)\ (P \in T_{\{i\}}), \quad f_{i,t}(P) = f(P)\ (P \in T_{\{j\}}, j \neq i).$$

De la définition de ϱ_i (cf. 4.1), on déduit

(2) $$(1 - \varrho_i) \cdot f_{i,t} = (1 - \varrho_i) \cdot f; \quad (1 - \varrho_i)\, df_{i,t} = (1 - \varrho_i)\, df, \quad (f \in C^0(T)),$$

ce qui entraîne tout d'abord

(3) $$(\varrho_i\, df_{i,t}, df_{i,t}) = (df_{i,t}, df_{i,t}) - ((1 - \varrho_i)\, df, df).$$

Soient $P \in T_{\{i\}}$ et $s \in (\text{Lk } P)(1)$. Comme $d^2 = 0$, on a

$$d_P(df_{i,t})(P * s) = \pm\, df_{i,t}(s) = \pm\, df(s)\,,$$

d'où

$$(4) \qquad (\Delta_i^+ df_{i,t}, df_{i,t}) = \sum_{P \in T_{\{i\}},\, s \in (\text{Lk } P)(1)} c(P * s) \cdot df(s)^2.$$

Mais il est immédiat que si s est un simplexe dont le type ne contient pas i, alors

$$(5) \qquad c(s) = \sum_{P \in T_{\{i\}} \cap \text{Lk } s} c(P * s),$$

par conséquent

$$(6) \qquad (\Delta_i^+ df_{i,t}, df_{i,t}) = \sum_{s \in T(1),\, i \notin I(s)} c(s) \cdot df(s)^2 = ((1 - \varrho_i)\, df, df)\,.$$

Vu la définition de ∂ (cf. 1.1 (4)), de d, et 1.1 (1), on a, pour $P \in T_{\{i\}}$

$$(7) \qquad \partial df_{i,t}(P) = t \cdot l \cdot f(P) + B\,,$$

où B est combinaison linéaire des valeurs de f sur des sommets de type $\neq \{i\}$, et ne dépend pas de t. Supposons maintenant que $\Delta^+ f = c \cdot f$. On a alors $\partial df(P) = c \cdot f(P)$ et (7), pour $t = 1$, montre que $B = (c - l) \cdot f(P)$, d'où le

Lemme. *Soient* $f \in \partial C^1(T)$ *et supposons que* $\Delta f = c \cdot f$. *Si* $t = (l - c)/l$, *alors* $\partial df_{i,t}(P) = 0$ *pour tout* $P \in T_{\{i\}}$, *et tout* $i \in I_l$.

4.6. Nous passons maintenant à la démonstration de 4.4. Soit $f \in \partial C^1(T)$ une fonction propre non nulle de Δ et soit c sa valeur propre. On suppose $c \neq l + 1$. D'après 4.3 (2), on a alors $\sum_{P \in T_{\{j\}}} f(P) = 0$ pour tout $j \in I_l$. Cette condition est encore satisfaite par $f_{i,t}$, donc (4.3 (1)), $f_{i,t}$ *est aussi contenue dans* $\partial C^1(T)$, et l'on a

$$(1) \qquad (df_{i,t}, df_{i,t}) = (\Delta f_{i,t}, f_{i,t}) \geqq k^0(T) \cdot (f_{i,t}, f_{i,t})\,.$$

Soit $P \in T_{\{i\}}$. Nous reprenons les notations de 1.4, et considérons l'application $h \mapsto h_0$ de $C^1(T)$ dans $C^0(\text{Lk } P)$ de 1.4. Faisons maintenant $t = (l - c)/l$. Alors $\partial df_{i,t}(P) = 0$ (4.5, Lemme), donc $(df_{i,t})_0$ est orthogonale aux constantes, (1.4 (2)), i.e. dans $\partial C^1(\text{Lk } P)$, et on peut lui appliquer 1.4 (5) d'où, vu 1.4 (1)

$$(\Delta_P^+ df_{i,t}, df_{i,t}) \geqq k^0(\text{Lk } P) \cdot (\varrho_P df_{i,t}, df_{i,t}) \geqq m^0(T) \cdot (\varrho_P df_{i,t}, df_{i,t})\,.$$

En sommant sur $P \in T_{\{i\}}$ et en utilisant 4.5 (3), (6), on obtient

$$((1 - \varrho_i)\, df, df) \geqq m^0(T) \cdot ((df_{i,t}, df_{i,t}) - ((1 - \varrho_i)\, df, df))\,,$$

d'où, compte tenu de (1),

$$(1 + m^0(T)) \cdot ((1 - \varrho_i)\, df, df) \geqq m^0(T) \cdot k^0(T) \cdot (f_{i,t}, f_{i,t})\,.$$

On a $\sum_i (1 - \varrho_i) = (l + 1) \cdot Id - \sum_i \varrho_i$, donc, sur C^1, la somme des $(1 - \varrho_i)$ est égale à $(l - 1) \cdot Id$. D'autre part, on tire immédiatement de la définition de $f_{i,t}$ que $\sum_i (f_{i,t}, f_{i,t}) = (l + t^2) \cdot (f, f)$, d'où, vu que $(df, df) = c \cdot (f, f)$,

$$(1 + m^0(T)) \cdot (l - 1) \cdot c \cdot (f, f) \geqq m^0(T) \cdot k^0(T) \cdot (l + t^2) \cdot (f, f) \,.$$

Si $k^0(T) = l + 1$, il n'y a rien à démontrer. Sinon, on peut prendre $c = k^0(T)$; on obtient alors

$$(1 + m^0(T)) \cdot (l - 1) \geqq m^0(T) \cdot (l + t^2) \,,$$

et le lemme s'ensuit en remplaçant t par sa valeur $(l - k^0(T))/l$.

4.7. Proposition. *Soient l un entier > 1, et $a \in \mathbf{R}$. Il existe $b \in \mathbf{R}$ ayant la propriété suivante: si T est l'immeuble d'un système de Tits fini, de rang $l + 1$, et si $k^0(\mathrm{Lk}\, s) \geqq 1 - b$ pour tout $s \in T(l - 2)$, alors $k^0(T) \geqq l - a$.*

Cela résulte de 4.4, par récurrence sur le rang.

§ 5. Valeurs propres pour le rang deux

Les valeurs propres de $\varDelta^+$ sur $C^0(T)$, lorsque T est l'immeuble d'un système de Tits de rang deux, de type de Lie, sont déterminées dans [7: § 7], et l'avaient aussi été antérieurement par Feit et Higman [6]. Nous nous bornerons ici à résumer la discussion de [7].

5.1. Nous nous plaçons tout d'abord dans le cas général de 4.1, dont nous reprenons les notations. Pour $I \subset I_l$, notons B_I le groupe d'isotropie de C_I; en particulier, $B_I = B$ si $I = I_l$. L'ensemble T_I des simplexes de type I s'identifie à G/B_I donc, si l'on désigne par $C(Y)$ l'espace des fonctions réelles sur l'ensemble Y, et par $|I|$ le cardinal de I:

$$C^r(T) = \bigoplus_{I \subset I_l, |I| = r+1} C(G/B_I) \,.$$

On note f_I la composante dans $C(G/B_I)$ d'un élément $f \in C^r(T)$. Soit encore π_{IJ} la projection canonique de G/B_J sur G/B_I lorsque $I \subset J \subset I_l$. On a alors (pour un choix évident des coefficients d'incidence);

$$(1) \qquad (df)_I = \sum_{j \in I} (-1)^{j+1} f \circ \pi_{I(j), I} \,, \qquad (f \in C^r(T)) \,,$$

où $I(j)$ désigne $I = \{i_0, \ldots, i_{r+1}\}$ $(i_0 < i_1 < \ldots < i_{r+1})$ privée de son j-ième terme. Posons $c(I) = c(C_I)$. Vu les propriétés de transitivité de G sur T déjà utilisées, on a

$$(2) \qquad c(I) = [B_I : B] \,, \qquad c(t) = c(I(t)) \ (t \in T(r)) \,,$$

$$(3) \qquad (\partial f)(x \cdot B_I) = \sum_{j \notin I} [C_{I \cup j} : C_I] \cdot (c(I \cup j)/c(I)) \cdot \sum_{y \in x B_I / B_{I \cup j}} f(x \cdot y) \,,$$

$(f \in C^r(T); I \subset I_l, |I| = r) \,.$

5.2. On note $H(G, B)$ l'algèbre de Hecke des fonctions bi-invariantes par B sur G, à support contenu dans un nombre fini de doubles classes modulo B, pour le produit de convolution. Elle est engendrée par les fonctions caractéristiques P_i des doubles classes $Bw_i B$, où $\{w_i\}$ $(0 \leq i \leq l)$ désigne l'ensemble générateur canonique S d'involutions du groupe de Weyl W de (G, B, N, S) [4]. Posons $Q_i = 1 + P_i$. C'est la fonction caractéristique de $B + B \cdot w_i \cdot B = B_{I_i(i)}$, groupe que l'on notera B^i. 5.1 (3) donne alors, pour $r = l$:

$$(1) \qquad \partial f(x \cdot B^i) = c(I_l(i))^{-1}(-1)^{i+1} f * Q_i, \qquad (f \in C^l(T)),$$

où le $*$ à droite indique l'opération de $H(G, B)$ sur $C(G/B)$ par convolution. Posons encore $q(i) = c(I_l(i)) - 1$. On a donc

$$(2) \qquad q(i) = [Bw_i B : B] = [B : w_i Bw_i^{-1} \cap B], \qquad (0 \leq i \leq l).$$

5.3. Supposons maintenant $l = 2$ et G fini. W est un groupe dihédral, caractérisé par l'ordre m de $w_0 w_1$, et l'on ne s'intéresse qu'aux cas où $m = 2, 3, 4, 6$ (i.e. où W est le groupe de Weyl d'un groupe de Lie semi-simple de rang deux, les valeurs de m correspondant respectivement aux types $\mathbf{A}_1 \times \mathbf{A}_1$, $\mathbf{A}_2$, $\mathbf{B}_2$ et $\mathbf{G}_2$). On a $C^0(T) = C(G/B^0) \oplus C(G/B^1)$, et il résulte immédiatement de 5.1 (1) et 5.2 (1) que, par rapport à cette décomposition de $C^0(T)$, on a $\Delta(f_0, f_1) = (f_0, f_1) * A$, où

$$(1) \qquad A = \begin{pmatrix} 1 & -c(0)^{-1} \cdot Q_1 \\ -c(1)^{-1} \cdot Q_0 & 1 \end{pmatrix},$$

donc que $(\Delta - Id)^2 \cdot c(0) \cdot c(1)$ est la matrice diagonale ayant $Q_1 Q_0$ et $Q_0 Q_1$ comme termes diagonaux. Il reste donc à trouver les valeurs propres de ces derniers. C'est trivial pour $m = 2$, car Q_0 et Q_1 commutent; dans les autres cas, on donne explicitement un polynôme (de degré 2, 3, 4 suivant que $m = 3, 4, 6$) en $Q_i Q_j$ $(i \neq j, 0 \leq i, j \leq 1)$ nul dans $H(G, B)$ [7: p. 407−408]. On en tire que les valeurs de $k^0(T)$ pour $m = 2, 3, 4, 6$ sont respectivement

$$1, \; 1 - q(1)^{1/2} \cdot c(1)^{-1}, \quad 1 - (q(0) + q(1))^{1/2} \cdot (c(0) + c(1))^{-1/2}$$

$$(2) \qquad 1 - (q(0) + q(1) + q(0)^{1/2} q(1)^{1/2})^{1/2} \cdot (c(0) \cdot c(1))^{-1/2}.$$

Cela montre qu'elles sont arbitrairement proches de un si $q(0)$ et $q(1)$ sont assez grands. Si maintenant K, k_v, G et T sont comme en 3.1, alors les coefficients $q(i)$ pour T, ou, ce qui revient au même, pour les links des simplexes de dimension $\leq l - 2$ de T, sont des puissances du cardinal $|k_v|$ de k_v. Par conséquent, vu que Lk s $(s \in T(l - 2))$ est de l'un des types considérés ici:

5.4. Proposition. *Soit $b > 0$. Il existe un entier $n(b) > 0$ tel que si T est l'immeuble de Tits d'un K-groupe simple simplement connexe G de K-rang $l \geq 2$ et si $|k_v| \geq n(b)$, alors $k^0(\text{Lk } s) \geq 1 - b$ pour tout $s \in T(l - 2)$.*

§ 6. Sous-groupes discrets de groupes simples p-adiques II

6.1. Théorème. *Soit $l_0 \geq 1$. Il existe un entier $N(l_0)$ ayant la propriété suivante. Si K, k_v, G, l sont comme en 3.1, si $l \leq l_0$, $|k_v| \geq N(l_0)$, alors $H^r(\Gamma; M) = 0$ ($1 \leq r < l$) quels que soient le sous-groupe discret co-compact Γ de $G(K)$ et le Γ-module unitaire de dimension finie M.*

Soit $a > 0$ tel que $l - a > r(l - r)/(r + 1)$ pour $2 \leq l \leq l_0$ et $1 \leq r < l$. On choisit ensuite $b > 0$ pour que 4.7 soit satisfaite si $l \leq l_0$, et on prend $N(l_0)$ égal au $n(b)$ de 5.4. On a alors $k^0(Lk\ s) > r(l - r)/(r + 1)$ pour tout $s \in T(l - 2)$ et 6.1 résulte de 3.3, appliqué au cas $j = r$.

Remarque. Si l'on ne suppose pas Γ co-compact, la conclusion reste valable pour l'espace $\mathbf{H}^r(T; M)$ des r-formes harmoniques Γ-invariantes sur T, de carré intégrable modulo Γ, avec la même démonstration, compte tenu de 3.5. Pour G simplement connexe, c'est le Théor. 8.3 de [7].

6.2. Supposons $l = 2$, $r = 1$. La condition de 3.3 est alors $k^0(Lk\ s) > \frac{1}{2}$ pour tout $s \in T(1)$. Les valeurs données en 5.3 montrent qu'elle est toujours remplie, sauf pour $m = 4$, $|k_v| = 2$ et $m = 6$, $|k_v| = 2, 3$. Cependant, par des calculs directs (non publiés), H. Garland a montré que l'on a aussi $H^1(\Gamma; M) = 0$ dans ces cas-là. Il n'y a donc pas de restriction sur le corps résiduel si $l = 2$.

6.3. La nullité de $H^1(\Gamma; \mathbf{R})$, (action triviale) équivaut au fait que $(\Gamma; \Gamma)$ est d'indice fini dans Γ. Rappelons qu'elle a été démontrée par A. Kazhdan (cf. [10]) sans hypothèse sur $|k_v|$, pour $l \geq 2$.

6.4. Les démonstrations utilisent essentiellement l'existence d'un produit scalaire positif non-dégénéré sur M invariant par Γ. «On» espère évidemment que cette hypothèse n'est pas nécessaire. Remarquons qu'elle est tout de même souvent satisfaite lorsque Γ est construit «arithmétiquement». Rappelons brièvement cette construction, qui est en fait le seul procédé général connu pour obtenir des Γ discrets co-compacts. Soit F un corps de nombres totalement réel admettant une complétion $F_v \cong K$ et soit H un F-groupe simple tel que $H \otimes_F K = G$ et que $(H \otimes_R F_w)(\mathbf{R})$ soit compact pour toute place à l'infini w de F. On sait que, sous ces hypothèses, tout sous-groupe S-arithmétique de H, où $S = \{v\}$, vu comme sous-groupe de $H(K)$, est discret et co-compact. D'autre part, comme H est compact à l'infini, il est clair que si M est l'espace d'une représentation rationnelle définie sur F de G, alors M admet une structure unitaire respectée par $H(F)$, et en particulier par Γ.

6.5. Enfin, il paraît assez plausible que l'hypothèse de simplicité faite sur G peut être considérablement affaiblie. Plus précisément, soient K_i ($1 \leq i \leq s$) un corps p-adique, G_i un K_i-groupe, simple sur K_i, et de K_i-rang $l_i \geq 1$. Soit $L = \prod_{1 \leq i \leq s} G_i(K_i)$. On suppose la somme l des l_i au moins égale à deux. Il est alors naturel de conjecturer que l'on a $H^i(\Gamma; \mathbf{R}) = 0$ ($1 \leq i < l$) quel que soit le sous-groupe discret co-compact Γ irréductible (projection sur tout produit partiel des $G_i(K_i)$ non discrète) de L. En tout cas on peut montrer, au moins en caractéristique zéro, que, étant donné r ($1 \leq r \leq \max l_i$), si l'on a $H^j(\Gamma_i; \mathbf{R}) = 0$ pour $1 \leq j$

$\leq \min (r, l_i - 1)$ quel que soit le sous-groupe discret co-compact Γ_i de $G_i(K_i)$ $(1 \leq i \leq s)$, alors $H^i(\Gamma; \mathbf{R}) = 0$ pour $1 \leq i \leq r$ et Γ discret, co-compact, irréductible dans L. L'hypothèse faite sur le i-ième facteur est vide si $r = 1$ ou si $l_i = 1$. Cela montre en particulier la nullité de $H^1(\Gamma; \mathbf{R})$ lorsqu'il y a au moins deux facteurs.

§ 7. Cohomologie de dimension maximum et représentation spéciale

7.1. Dans ce paragraphe, on suppose G simplement connexe. Comme précédemment T est l'immeuble de Tits de $G(K)$ et B un sous-groupe parahorique minimal de $G(K)$. On note χ l'unique caractère de l'algèbre de Hecke $H(G, B)$ qui est égal à -1 sur les P_i (notations de 5.2). Il existe sur G/B une et une seule fonction (sphérique élémentaire) φ qui est fonction propre de $H(G, B)$ pour le caractère χ et vaut un sur B. Elle est de carré intégrable, et le plus petit sous-espace fermé H_σ de $L^2(G)$ la contenant est irréductible. C'est un élément de la série discrète, la représentation *spéciale* de $G(K)$, qui sera notée σ (cf. [11, 14]). L'espace $L^2(G(K)/B)$ s'identifie, comme $G(K)$-module, à l'espace $L^l(T; \mathbf{C})$ des l-formes complexes de carré intégrable sur T. Vu 5.2 (1), une telle cochaine f est harmonique si et seulement si $f * Q_i = 0$, i.e. $f * P_i = -f (0 \leq i \leq l)$, donc si et seulement si $f * Q = \chi(Q) \cdot f (Q \in H(G, B))$, puisque les P_i engendrent $H(G, B)$. En particulier $\varphi \in \mathbf{H}^l(T; \mathbf{R})$. On montre qu'en fait l'isomorphisme $L^2(G/B) \xrightarrow{\sim} L^l(T; \mathbf{C})$ applique H_σ sur $\mathbf{H}^l(T; \mathbf{R})$, mais nous n'avons pas besoin de ce fait ici.

7.2. Théorème. *On conserve les notations et hypothèses de 7.1. Soit Γ un sous-groupe discret co-compact de $G(K)$. Alors* $\dim H^l(\Gamma; \mathbf{R})$ *est égale à la multiplicité de la représentation spéciale dans* $L^2(\Gamma \backslash G(K))$.

(cf. [7: 10.4]; pour $G = \mathbf{SL}_2$, ce résultat est dû à Ihara [9].) Rappelons que, comme $\Gamma \backslash G(K)$ est compact, $L^2(\Gamma \backslash G(K))$ est somme discrète à multiplicités finies de représentations irréductibles de $G(K)$. Soit

$$(1) \qquad V_\Gamma = \{ f \in L^2(\Gamma \backslash G(K)) \,|\, f * Q = \chi(Q) \cdot f, \, (Q \in H(G, B)) \} \,.$$

Un raisonnement standard montre que $\dim V_\Gamma$ est la multiplicité de la représentation spéciale dans $L^2(\Gamma \backslash G(K))$. Rappelons-le brièvement. Tout d'abord si $H \subset L^2(\Gamma \backslash G(K))$ est une copie de la représentation spéciale, on a $\dim H \cap V_\Gamma = 1$, vu 7.1. Soit maintenant $f \in V_\Gamma$, de norme un. Alors le «coefficient» $c_{f,f}$ défini par $c_{f,f}(g) = (f, g \cdot f) (g \in G(K))$ est égal à φ, vu la caractérisation de cette dernière. L'application $u \mapsto c_{f,u}$ de $L^2(\Gamma \backslash G(K))$ dans $L^2(G)$ est alors un opérateur d'entrelacement, qui applique isomorphiquement le plus petit sous-espace fermé invariant par $G(K)$ contenant f sur H_σ.

Soit $f \in V_\Gamma$. On a $f * Q = f$ si Q est la fonction caractéristique de B, donc f est invariante à droite par B. Vu ce qui a été dit en 7.1, on a alors

$$(2) \qquad\qquad V_\Gamma \cong C^l(T; \mathbf{C})^\Gamma \cap \ker \partial \,.$$

Supposons maintenant que Γ opère très librement (cf. 3.3). Alors $C^l(T)^\Gamma$

$= C^l(\Gamma \backslash T)$, donc

$$(3) \qquad V_\Gamma \cong H^l(\Gamma \backslash T; \mathbf{C}) = H^l(\Gamma; \mathbf{C}) = H^l(\Gamma; \mathbf{R}) \otimes \mathbf{C},$$

d'où le théorème dans ce cas. En général, on peut trouver un sous-groupe distingué Γ' d'indice fini de Γ qui opère très librement (cf. dém. de 3.4). Il est classique que $H^l(\Gamma; \mathbf{R}) = H^l(\Gamma'; \mathbf{R})^{\Gamma/\Gamma'}$ et évident que $V_\Gamma = (V_{\Gamma'})^{\Gamma/\Gamma'}$. Il suffit alors de remarquer que les isomorphismes de (3), pour $\Gamma = \Gamma'$, commutent avec l'action naturelle de Γ/Γ'.

7.3. On suppose Γ sans torsion et l'on dénote $\chi(\Gamma)$ sa caractéristique d'Euler-Poincaré: $\chi(\Gamma) = \sum_i (-1)^i \dim H^i(\Gamma; \mathbf{R})$. D'après [13: § 3], il existe une mesure de Haar sur $G(K)$ telle que l'on ait

$$\chi(\Gamma) = (-1)^l \mathrm{vol}(\Gamma \backslash G(K)),$$

pour tout Γ (discret, co-compact, sans torsion, ou même avec torsion pour une définition convenable de $\chi(\Gamma)$). Si $H^i(\Gamma; \mathbf{R}) = 0$ pour $0 < i < l$, alors

$$\dim H^l(\Gamma, \mathbf{R}) = \mathrm{vol}(\Gamma \backslash G(K)) + (-1)^{l+1}.$$

En passant à des sous-groupes d'indice fini de Γ, on voit que l'on peut rendre le membre de gauche arbitrairement grand.

Bibliographie

1. Borel, A., Tits, J.: Groupes réductifs, Publ. Math. I.H.E.S., n°27 (1965), 55−151
2. Borel, A., Tits, J.: Compléments à l'article «Groupes réductifs», Ibid. n° **41** (1972), 253−276
3. Borel, A., Tits, J.: Homomorphismes «abstraits» des groupes algébriques simples, Annals of Math. (2) **97** (1973), 499−571
4. Bourbaki, N.: Groupes et algèbres de Lie, chap. IV, V, VI, Hermann, Paris (1968)
5. Bruhat, F., Tits, J.: Groupes réductifs sur un corps local I, Publ. Math. I.H.E.S., n° **41** (1972), 1−251
6. Feit, W., Higman, G.: The nonexistence of certain generalized polygons, J. of Algebra **1** (1964), 114−131
7. Garland, H.: p-adic curvature and the cohomology of discrete subgroups of p-adic groups, Annals of Math. (2) **97** (1973), 375−423
8. Garland, H.: On the cohomology of discrete subgroups of semi-simple Lie groups, Proc. Int. Colloquium Bombay 1973 [Oxford University Press 1975]
9. Ihara, Y.: On discrete subgroups of the two by two projective linear group over p-adic fields, J. Math. Soc. Japan **18** (1966), 219−235
10. Kazhdan, A.: Connection of the dual space of a group with the structure of its closed subgroups, Funkcionalnyi Analiz **1** (1967), 71−74
11. Matsumoto, H.: Fonctions sphériques sur un groupe semi-simple p-adique, C. R. Acad. Sci. Paris **269** (1969), 829−832
12. Matsushima, Y.: On Betti numbers of compact, locally symmetric Riemannian manifolds, Osaka Math. J. **14** (1962), 1−20
13. Serre, J.-P.: Cohomologie des groupes discrets, dans «Prospects in Mathematics», Annals of Math. Studies **70**, Princeton University Press, Princeton 1970
14. Shalika, J. A.: On the space of cusp forms of a p-adic Chevalley group, Annals of Math. (2) **92** (1970), 262−278

100.

Stable real cohomology of arithmetic groups

Ann. Sci. Ec. Norm. Super., (4) **7** (1974) 235–272

A Henri Cartan,
à l'occasion de son
70ᵉ anniversaire

Let Γ be an arithmetic subgroup of a semi-simple group G defined over the field of rational numbers $\mathbf{Q}$. The real cohomology H* (Γ) of Γ may be identified with the cohomology of the complex Ω_X^Γ of Γ-invariant smooth differential forms on the symmetric space X of maximal compact subgroups of the group G (R) of real points of G. Let I_G be the space of differential forms on X which are invariant under the identity component G (R)^0 of G (R). It is well-known to consist of closed (in fact harmonic) forms, whence a natural homomorphism j^* : $I_G^\Gamma \to$ H* (Γ). Our main result (7.5) gives a range, computable from the algebraic group structure of G, up to which j^* is an isomorphism. For a given group, it is rather small, about $n/4$ for $\mathbf{SL}_n$ (Z) for instance; however, we may consider sequence (G_n, Γ_n), where G_n and Γ_n are as G and Γ, for which our range tends to infinity, and of injective morphisms f_n : $(G_n, \Gamma_n) \to (G_{n+1}, \Gamma_{n+1})$. Since much is known about I_G, this allows us to determine the " stable real cohomology " H^q (Γ_n), n large, and H* (lim Γ_n) for sequences of classical groups (§ 11). For example, H* (lim$\,\mathbf{SL}_n$(Z)) [resp. H* (lim $\mathbf{Sp}_{2n}$ (Z))] is an exterior (resp. polynomial) algebra over infinitely many generators, one for each dimension of the form $4\,i+1$ (resp. $4\,i-2$) $(i = 1, 2, \ldots)$. The result for **SL** (Z) then implies that the dimension of K_i (Z) $\otimes$ $\mathbf{Q}$ is one if $i \equiv 1$ mod 4 and zero otherwise $(i \geqq 2)$.

The homomorphism j^* is defined for any discrete subgroup Γ of G (R). If G (R)/Γ is compact, then, as a simple consequence of Hodge theory, j^* is injective in all dimensions. The results of Matsushima [19], recalled in paragraph 3, show the existence of a constant m (G (R)), determined by the Lie algebra of G (R), up to which j^* is surjective. Our theorem is thus the analogue for arithmetic subgroups of Matsushima's result. H. Garland had already shown in [9] that if G (R)/Γ is only assumed to have finite invariant volume, and if every cohomology class in a given dimension $q \leqq m$ (G (R)) is representable by a square integrable form, then Matsushima's arguments carry over without change and show that j^* is surjective in dimension q. Moreover [10] gives, for Γ arithmetic, a range up to which this condition is fulfilled (a so-called " square integrability criterion "). Our main concern is therefore the injectivity of j^* in the arithmetic case, but our discussion will also include the surjectivity and the square integrability criterion.

315

As a first step, assuming only that G (**R**)/Γ has finite invariant volume, we shall see (3.6) that, given an integer m' (G), j^* is bijective at least up to min $(m$ (G (**R**)), m' (G)) if there exists a subcomplex C of Ω_X^Γ such that, up to m' (G) : (i) C $\rightarrow \Omega_X^\Gamma$ induces an isomorphism in cohomology; (ii) $I_G^\Gamma \subset$ C; (iii) C^q consists of square integrable forms. The surjectivity (3.5) is essentially Garland's argument, while injectivity follows from the fact that on a complete Riemannian manifold, a non-zero square integrable harmonic form is not the differential of a square integrable form (2.5).

There remains then to find such a C when Γ is arithmetic, and it suffices to do so when Γ is moreover torsion-free. Its definition (7.1), and the preliminaries in paragraphs 4 and 5, involve only X/Γ but are better understood if X/Γ is viewed as the interior of the compact manifold with corners $\overline{X}/\Gamma$ of [4], (*see* also § 6). From that point of view, the main use of paragraph 4 is to give a decomposition of the invariant Riemannian metric near a boundary point y, with respect to a frame adapted to the corner structure around y, and paragraph 5 gives a criterion for the square integrability of a differential form σ near y, in terms of a growth condition of the coefficients of σ with respect to that frame. C is then defined by means of such growth conditions near the boundary points, i. e., on Siegel sets, for forms and their differentials. It follows immediately from paragraph 5 that C satisfies (ii) in all dimensions, and (iii) at least up to a constant c (G) defined in terms of the **Q**-roots of G. To prove (i) we introduce a presheaf $\mathscr{F}$ on $\overline{X}/\Gamma$. For y on the boundary and U a neighborhood of y in $\overline{X}/\Gamma$, $\mathscr{F}$ (U) is the space of forms on U $\cap$ (X/Γ) satisfying the growth condition which is part of the definition of C. For $y \in$ X/Γ, $\mathscr{F}$ (U) is simply the space of differential forms defined in U. This presheaf is a sheaf whose space of sections on $\overline{X}/\Gamma$ and X/Γ are C and Ω_X^Γ respectively. Since the inclusion X/Γ $\hookrightarrow \overline{X}/\Gamma$ is a homotopy equivalence, it suffices then to prove that $\mathscr{F}$ is a fine resolution of **R** on $\overline{X}/\Gamma$. The " fineness " is easy, so that the main point is to check a local Poincaré lemma near a boundary point, i. e., on a Siegel set. This is done by showing that the usual homotopy operator used to prove the Poincaré lemma on euclidean space does not alter our growth condition.

Theorem 7.5 yields a square integrability criterion which is analogous to, although somewhat weaker than, that of [10]. It can also be obtained more simply by proving that the cohomology of $\overline{X}/\Gamma$ can be computed by means of differential forms which, locally around the boundary, are lifted from forms on the boundary. This is done in paragraph 8, but is not used in the rest of the paper.

Paragraph 9 contains some remarks on m (G (**R**)) and c (G). In particular, they are both $\geq (rk_\mathbf{Q}$ (G)$/4)-1$, hence tend to infinity with $rk_\mathbf{Q}$ G. Paragraph 10 reviews some known facts about the stable cohomology of classical compact symmetric spaces, which all occur in the Bott periodicity [7]. Modulo well-known isomorphisms, this gives $\varinjlim I_G^q$ when G runs through suitable sequences of classical groups, and allows us in paragraph 11 to describe explicitly H* ($\varinjlim \Gamma_n$) for sequences of classical arithmetic groups which occur in K-theory or hermitian K-theory. This yields the ranks of the corresponding K_i or $_\varepsilon L_i$ groups (§ 12).

The main results of this paper have been announced in [2]. In fact, [2] states them more generally for S-arithmetic groups. Those will be considered in another paper.

The growth condition used in paragraph 7 is not the one underlying [2], and yields somewhat sharper results. I thank G. Harder very much for having suggested it. I would also like to thank G. Prasad, who read the manuscript and pointed out a considerable number of misprints and corrections.

0. Notation and conventions

0.1. For $x \in \mathbf{R}_+ = \{ x \in \mathbf{R} \mid x > 0 \}$, we let $[x]'$ be the greatest integer strictly smaller than x.

0.2. Let u, v be functions on a set X, with values in $\mathbf{R}_+$. We write $u \prec v$ if there exists a constant $c > 0$ such that $u(x) \leq c\, v(x)$ for all $x \in X$, $u \succ v$ if $v \prec u$ and $u \asymp v$ if $u \prec v$ and $v \prec u$.

0.3. Manifolds are smooth, i.e., C^∞ and oriented. Let M be a manifold. $C^\infty(M)$ is the space of real valued smooth functions and Ω_M the space of smooth exterior differential forms on M. For $x \in M$, $T_x(M)$ is the tangent space to M at x. If A is a tensor field on M, then A_x is the value of A at $x \in M$. If M is a Riemannian manifold, (A_x, B_x) is the scalar product defined by the metric of two tensors of the same type at x, and $|A_x| = (A_x, A_x)^{1/2}$; moreover $(A, B)_M$ denotes the integral of (A_x, B_x) over M, and $L^2(M)$ or $L^2(M, dv)$ [resp. $L^1(M)$ or $L^1(M, dv)$] is the space of square integrable (resp. integrable) functions on M, with respect to the Riemannian volume element dv.

0.4. The connected component of the identity of an algebraic group (in the Zariski topology), or of a topological group G, H is denoted G^0, H^0, ... The Lie algebra of an algebraic group or of a real Lie group G, H, ... is denoted $L(G)$, $L(H)$, ...

0.5. Our algebraic groups are linear and may be identified with algebraic subgroups of some $\mathbf{GL}_n(\mathbf{C})$. If we wish to emphasize a field of definition k for a given algebraic group H, we shall write H/k.

If k' is a finite separable extension of a field k, then $R_{k'/k}$ is the functor of restriction of scalars from k' to k ([23], Chap. I).

0.6. If a group G operates on a set A, then A^G is the set of fixed points of G in A.

1. A Stokes formula for complete Riemannian manifolds

In this section, M is a connected Riemannian manifold, g the metric tensor, $d(\ ,\ )$ and dv the associated distance function and volume element.

Our main aim in paragraphs 1 and 2 is to prove 1.3, 1.4 and 2.3, which are essentially known, although not exactly stated in the form we need them. The arguments are slight variations of the known ones.

1.1. We recall first that if M is *complete*, there exists a proper smooth function $\mu : M \to [0, \infty)$ and a constant $c > 0$ such that $\left| \operatorname{grad} \mu (x) \right| \leq c$ for all $x \in M$ ([22], § 35).

Such a function can be obtained by regularization ([22], § 15) from the function $r : x \to d (o, x)$, where o is some chosen point of M. (The latter function is proper and satisfies our second condition wherever it is differentiable, i. e., almost everywhere.)

Although this is not needed in this paper, we remark that the converse is true : assume μ to exist. Let $\gamma : s \mapsto x (s)$ be a geodesic segment parametrized by arc length, where s varies over some bounded interval J of the real line. Assume the length of γ to be finite. Then, since $\left| d\gamma/ds \right| = 1$ identically, the variation of $\mu \circ \gamma$ on J is bounded and Im γ is contained in a bounded set (μ being proper). But then Im γ can be extended (at both ends) to a longer geodesic segment. Hence M is complete.

1.2. LEMMA. — *Assume* M *to be complete. There exist compact sets* $C_r \subset D_r \, (r \in \mathbf{R}^{*}_{+})$ *such that* C_r *contains the interior of* $C_{r'}$ *if* $r > r'$, *and* M *is the union of the* C_r, *a family of smooth functions* $\sigma_r \, (r > 0)$ *and a constant* d *such that* $0 \leq \sigma_r (x) \leq 1 \, (x \in M)$, $\sigma_r (x) = 1$ *for* $x \in C_r$, $\sigma_r (x) = 0$ *for* $x \notin D_r$, *and*

$$\left| \operatorname{grad} \sigma_r(x) \right| \leq d \, r^{-1}, \qquad (r > 0).$$

Let $m : [0, \infty) \to [0, 1]$ be a smooth function equal to 1 on $[0, 1]$ and 0 on $[2, \infty)$. Then put $\sigma_r (x) = m (\mu (x)/r)$, where μ is as in 1.1.

1.3. PROPOSITION. — *Assume* M *to be complete. Let* X *be a continuous vector field on* M. *Assume that* $\left| X_x \right|$ *is bounded on* M *and that* $L_X (dv) = 0$. *Let* E *be a finite dimensional Hilbert space and* f *be a* C^1 *E-valued function on* M *such that* f, Xf *are integrable. Then*

$$\int_M Xf \, dv = 0.$$

Since $L_X (dv) = 0$, we have $Xf \, dv = L_X (f \, dv)$ hence, by H. Cartan's formula

$$(1) \qquad Xf \, dv = d(i(X)f \, dv) + i(X) d(f \, dv) = d(i(X)f \, dv).$$

Assume first that f has compact support. Then if B is a bounded open subset of M containing the support of f and having a sufficiently regular boundary ∂B, we have, by Stokes' formula

$$(2) \qquad \int_M Xf \, dv = \int_B d(i(X)f \, dv) = \int_{\partial B} i(X)f \, dv = 0.$$

Assume now simply f, Xf to be integrable and let σ_r be as in 1.2. Then, by (2) :

$$(3) \qquad 0 = \int X(\sigma_r f) \, dv = \int f X \sigma_r \, dv + \int \sigma_r Xf \, dv.$$

Clearly

$$(4) \qquad \lim_{r \to \infty} \int_M \sigma_r Xf \, dv = \int_M Xf \, dv.$$

On the other hand

$$\left| X \sigma_r(x) \right| = \left| (X_x, \operatorname{grad} \sigma_r(x)) \right| \leq \left| X_x \right| d/r, \qquad (x \in M),$$

is bounded on M, hence

$$\left| \int_M f X \sigma_r \, dv \right| \leq (d/r) \operatorname{Max}_x \left| X_x \right| \int_M |f(x)|_E \, dv \to 0$$

as $r \to \infty$; together with (3) and (4), this proves the proposition.

1.4. COROLLARY. — *Let u, v be* E-*valued functions on* M *of class* C^1. *Assume that*

$$h(x) = (u(x), v(x))_E \qquad and \qquad x \mapsto (X u(x), v(x))_E, \ x \mapsto (u(x), X v(x))_E$$

are in $L^1(M, dv)$. *Then* $(X u, v)_M + (u, X v)_M = 0$.

Since

$$X h(x) = (X u(x), v(x))_E + (u(x), X v(x))_E,$$

the function h satisfies the conditions imposed on f in 1.3, hence $\displaystyle\int X h(x) \, dv = 0$,

which proves 1.4.

1.5. COROLLARY. — *Let u, v be* E-*valued functions on* M *of class* C^1 *such that u, v,
$X u$, $X v$ are all square integrable on* M. *Then*

$$(X u, v)_M + (u, X v)_M = 0.$$

All the scalar products mentioned in the assumptions of 1.4 are integrable, hence 1.5 is a special case of 1.4.

2. Square integrable forms

2.1. Let M be a connected oriented complete Riemannian manifold, n its dimension, Ω_M^p the space of smooth real-valued exterior differential p-forms on M, $\Omega^* = \Omega_M^*$ the direct sum of the Ω^p. As usual, we have the exterior differentiation $d : \Omega_M^p \to \Omega_M^{p+1}$, the star operator $\bigstar : \Omega_M^p \xrightarrow{\sim} \Omega_M^{n-p}$, the operator $\partial = (-1)^{n(p+1)+1} \bigstar d \bigstar : \Omega^p \to \Omega^{p-1}$ and the Laplace-Beltrami operator $\Delta = d \partial + \partial d$.

For $\alpha, \beta \in \Omega^p$, the scalar product $(\alpha, \beta) = (\alpha, \beta)_M$ is defined by

$$(a, \beta)_M = \int_M (\alpha_x, \beta_x) \, dv$$

and we put $\| \alpha \| = (\alpha, \alpha)^{1/2}$. This scalar product is extended to Ω_M^* by decreeing that Ω^p and Ω^q are orthogonal if $p \neq q$. We let $\Omega_{M,(2)}^* = \Omega_{(2)}^p$ be the space of square integrable p-forms, and $\Omega_{(2)}^* = \Omega_{M,(2)}^*$ the direct sum of the $\Omega_{(2)}^p$.

2.2. PROPOSITION. — *Let $\alpha \in \Omega_M^p$, $\beta \in \Omega_M^{p+1}$. Assume that the functions*

$$x \to |\alpha_x| \cdot |\beta_x|, \qquad x \mapsto ((d\alpha)_x, \beta_x) \qquad and \qquad x \mapsto (\alpha_x, (\partial\beta)_x)$$

are integrable.   Then

$$(\alpha, \partial\beta)_M = (d\alpha, \beta)_M.$$

If one of α, β has compact support, this is standard.   Therefore, with σ_r as in 1.2, we have

(1) $(\sigma_r\alpha, \partial\beta) = (d(\sigma_r\alpha), \beta).$

We have $d(\sigma_r\alpha) = d\sigma_r \wedge \alpha + \sigma_r\, d\alpha$.   The scalar products $(\sigma_r\alpha, \partial\beta)$ and $(\sigma_r\, d\alpha, \beta)$ tend respectively to $(\alpha, \partial\beta)$ and $(d\alpha, \beta)$ as $r \to \infty$.   It suffices therefore to prove

(2) $\lim_{r \to \infty} (d\sigma_r \wedge \alpha, \beta) = 0.$

By elementary algebra, there exists a constant $c > 0$ independent of r, σ, α such that $\left| (d\sigma_r \wedge \alpha)_x \right| \leq c \left| (d\sigma_r)_x \right| . \left| \alpha_x \right|$.   Taking 1.2 into account, we have then

$$\left| d\sigma_r(x) \wedge \alpha_x \right| \leq c \left| d\sigma_r(x) \right| . \left| \alpha_x \right| \leq c(d/r) \left| \alpha_x \right|$$

whence,

$$\left| (d\sigma_r \wedge \alpha, \beta) \right| \leq c(d/r) \int \left| \alpha_x \right| \left| \beta_x \right| dv$$

and (2) follows.

2.3.   COROLLARY. — *Assume that α, $d\alpha$, β, $\partial\beta$ are square integrable.   Then*

$$(d\alpha, \beta)_M = (\alpha, \partial\beta)_M.$$

By the Schwarz inequality, all the assumptions of 2.2 are fulfilled.

2.4.   A form ω is *harmonic* if $\Delta\omega = 0$.   We let $\mathscr{H}^p_{(2)}$ or $\mathscr{H}^p_{M,\,(2)}$ denote the space of square integrable harmonic p-forms.   Since M is complete, $\omega \in \mathscr{H}^p_{(2)}$ if and only if $\omega \in \Omega^p_{(2)}$ and $d\omega = \partial\omega = 0$, by a theorem of Andreotti-Vesentini (*see* e. g. [22], § 35).

We recall that, by a result of Kodaira, if $\omega \in \Omega^p_{(2)}$ and $d\omega = 0$, then there exists $\sigma \in \Omega^{p-1}$ and $H\,\omega \in \mathscr{H}^p_{(2)}$ such that $\omega = H\,\omega + d\sigma$ (in [22], this follows from Theorems 14 and 24).

In fact, the result is also valid in the non-complete case, with " harmonic " being defined by the conditions $d\omega = \partial\omega = 0$, *loc cit.*

Let us denote by $H^p_{(2)}(M)$ the space of real p-dimensional cohomology classes of M which may be represented by a closed square integrable p-form.   We have then natural maps

$$\mathscr{H}^p_{(2)} \overset{\mu}{\to} H^p_{(2)}(M) \overset{v}{\to} H^p(M).$$

By the result of Kodaira quoted above, μ is surjective.   If M is compact then μ and v are bijective by Hodge theory.   In general, μ need not be injective, nor v surjective.   There is however a weaker form of injectivity which will be useful later :

2.5.   PROPOSITION. – *Let $\omega \in \mathscr{H}^p_{(2)}$ and assume that $\omega = d\sigma$, for some $\sigma \in \Omega^{p-1}_{(2)}$.   Then* $\omega = 0.$

In fact, $\partial\omega = 0$ by 2.4. Hence ω, $\partial\omega$, σ, and $d\sigma = \omega$ are all square integrable. By 2.3, we have then

$$(d\sigma, d\sigma) = (\omega, d\sigma) = (\partial\omega, \sigma) = 0,$$

whence $d\sigma = 0$ and $\omega = 0$.

2.6. REMARKS. $-$ The proposition holds true if the assumption $\sigma \in \Omega_{(2)}^{p-1}$ is replaced by : the function $x \mapsto |\sigma_x| \cdot |\omega_x|$ is integrable. The proof is the same, except for the use of 2.2 instead of 2.3.

<h3 align="center">3. The map $j^q : I_G^\Gamma \to H^* (\Gamma)$</h3>

3.1. In this section, G is a real semi-simple Lie group, whose identity component has finite index and finite center, K is a maximal compact subgroup of G and $X = K\backslash G$ the space of maximal compact subgroups of G. Endowed with a G-invariant Riemannian metric, X is a complete symmetric Riemannian manifold with negative curvature, diffeomorphic to euclidean space.

For a subgroup H of G, we let Ω_X^H be the algebra of differential forms on X which are invariant under H. If $H = G^0$, we write I_G instead of Ω_X^H. By well-known facts, which go back to E. Cartan, I_G consists of harmonic forms, and we have

$$(1) \qquad\qquad I_G = H^*(L(G), L(K)),$$

where the second term denotes relative Lie algebra cohomology.

Let Γ be a discrete subgroup of G and $\pi : X \to X/\Gamma$ the natural projection. Γ acts properly on X, and

$$(2) \qquad\qquad H^* (\Gamma) = H^* (\Omega_X^\Gamma) = H^* (X/\Gamma).$$

This is well-known if Γ is torsion free, because it then acts freely on X, X/Γ is a $K (\Gamma, 1)$, and π induces an isomorphism : $\Omega_{X/\Gamma} \xrightarrow{\sim} \Omega_X^\Gamma$. In the general case, this follows since the isotropy groups of Γ on X are finite, and therefore have trivial real cohomology (for a more general result, *see* e. g. [12])

Similarly, if $\Gamma \to GL (E)$ is a finite dimensional real or complex linear representation of Γ, then

$$(3) \qquad\qquad H^*(\Gamma; E) = H^*((\Omega_X \otimes E)^\Gamma).$$

Since the elements of I_G are closed, the inclusion $I_G^\Gamma \hookrightarrow \Omega_X^\Gamma$ induces a homomorphism

$$(4) \qquad\qquad j^* : \quad I_G^\Gamma \to H^* (\Gamma)$$

whose restriction to I_G^q will be denoted j^q, or j_Γ^q if necessary.

If $G = G^0$, there is a canonical isomorphism

$$(5) \qquad\qquad H^*(\Gamma) = H^*(L(G), L(K); C^\infty (\Gamma\backslash G)) \quad ([20], \S\ 3).$$

In view of (5), we may then identify j^* with the homomorphism of Lie algebra cohomology induced by the inclusion $\mathbf{R} \to C^\infty\,(\Gamma\backslash G)$ of $\mathbf{R}$ onto the space of constant functions. If $\Gamma\backslash G$ is compact, then $\mathbf{R}$ has an *invariant* supplement, hence j^* is injective. This is then also true for G not connected by a standard argument (*see* proof of 3.4 below).

For other interpretations of j^*, at any rate when G is linear, we refer to 10.2.

3.2. Paragraphs 3.4 and 3.5 are devoted to some known results of Y. Matsushima and H. Garland. However, we shall state them in a slightly more general manner, and shall have accordingly to make some remarks on the proofs. To this end we need some notation and facts recalled here. The reader willing to take 3.4 and 3.5 for granted may skip this, since 3.2 will not be referred to elsewhere in this paper.

Let $L(G) = L(K) + \mathfrak{p}$ be the Cartan decomposition of $L(G)$ and

$$\theta : \quad (x + y \mapsto x - y) \qquad (x \in L(K),\ y \in \mathfrak{p})$$

the corresponding Cartan involution. On G we consider the right invariant metric equal to $g_0\,(X,\,Y) = -B\,(X,\,\theta\,(Y))$. $(X,\,Y \in L(G))$ on $L(G)$, where B is the Killing form, whence also a metric on G/Γ. Let $(X_i)_{1 \leq i \leq N}$ be an orthonormal basis of $\mathfrak{p}$ and $(X_a)_{N < a \leq \dim G}$ one of $L(K)$. Identify the $(X_j)_{1 \leq j \leq \dim G}$ to right invariant vector fields on G, hence to vector fields on G/Γ. Since the metric is right invariant, and G is unimodular, we have

$$(1) \qquad\qquad |X_{i,\,x}| \;= 1, \qquad (1 \leq i \leq \dim G;\ x \in G/\Gamma),$$

$$(2) \qquad\qquad L_{X_i}(dv) = 0, \qquad (1 \leq i \leq \dim G).$$

Let (ω^j) be the basis of Maurer-Cartan forms dual to (X_j). For a strictly increasing sequence $I = \{\,i_1,\,\ldots,\,i_q\,\}$ of indices between 1 and $\dim G$, let $\omega^I = \omega^{i_1} \wedge \ldots \wedge \omega^{i_q}$ and $|I| = \mathrm{Card}\ I$. For simplicity, we shall assume Γ torsion-free. Let

$$\pi : \quad G/\Gamma \to X/\Gamma = K\backslash G/\Gamma$$

be the canonical projection. Then $\sigma \mapsto \sigma_0 = \sigma \circ \pi$ defines an isomorphism of $\Omega_{X/\Gamma} = \Omega_X^\Gamma$ onto the complex Ω_0 of forms on G/Γ which are left invariant under K and annihilated by the interior products $i\,(X)\,(X \in L(K))$. Any such form can be written uniquely

$$(3) \qquad\qquad \sigma_0 = \sum_{|I|=q} \sigma_I \omega^I,$$

with $\sigma_I \in C^\infty\,(G/\Gamma)$, and where I runs through the strictly increasing subsequences of $\Lambda = \{\,1,\,2,\,\ldots,\,N\,\}$. There is a constant c such that for $\sigma,\,\tau \in \Omega_{X/\Gamma}$:

$$(4) \qquad\qquad (\sigma,\,\tau)_{X/\Gamma} = c \sum_I \int_{G/\Gamma} (\sigma_I,\,\tau_I)\,dv.$$

We have, for σ of degree ≥ 1 :

$$(5) \qquad\qquad (\Delta\sigma)_0 = -\sum_I C\,\sigma_I \omega^I,$$

where C is the Casimir operator for G.

3.3. Assume G to be connected, simple non-compact. In [19], Y. Matsushima has attached to G a constant, depending in fact only on L (G), to be denoted m (G). We sketch its definition. Let $\mathfrak{m}$ be the second symmetric power of $\mathfrak{p}$. On $\mathfrak{m}$ let (,) be the scalar product defined by the Killing form and let P (,) be the quadratic form

$$P(\xi, \eta) = \sum R_{iklj}\,\xi_{kl}\,\eta_{ij},$$

where R_{iklj} denotes the curvature tensor. Then

$$m(G) = \max q \,|\, (A/q)(\xi, \xi) + P(\xi, \xi) > 0 \qquad (\xi \in \mathfrak{m} - \{0\}),$$

where A is a certain constant, defined in terms of the restriction of the Killing form to L (K).

DEFINITION. – For G as in 3.1, let m (G) = min m (G$_i$), where G$_i$ runs through the simple connected non-compact normal subgroups of G^0.

3.4. THEOREM (Matsushima [19]). — *Let Γ be a discrete subgroup of G and assume G/Γ to be compact. Then*

(i) *j^q is injective for all q' s;*

(ii) *j^q is surjective for $q \leq m$ (G).*

In fact, in [19], Theorem 1, G is connected, simple non-compact, and Γ is torsion-free. We discuss briefly how to go from there to the general case. Γ is linear, finitely generated. By a well-known result of A. Selberg, it has a torsion-free subgroup Γ' of finite index, which may then be assumed to be normal, and contained in G^0. We have then

$$H^*(\Gamma) = H^*(\Gamma')^{\Gamma/\Gamma'}, \qquad I_G^\Gamma = (I_G^{\Gamma'})^{\Gamma/\Gamma'}, \qquad \Omega_X^\Gamma = (\Omega_X^{\Gamma'})^{\Gamma/\Gamma'},$$

from which it is clear that the result for Γ' implies the result for Γ. Thus we may assume that G is connected and Γ torsion-free.

We already saw (3.1) that j^* is injective. This also follows from Hodge theory, since I_G consists of harmonic forms.

By examining Matsushima's proof of (ii), it is not difficult to see that the argument is valid for a non-simple group, provided the constant A is replaced by the minimum of the corresponding constants assigned to the various non-compact factors. This, which would suffice for our purposes, may yield a constant m (G) somewhat smaller than the one defined above. That one can use m (G) as defined above was pointed out to me by H. Garland. I sketch his argument. Let G$_t$ $(1 \leq t \leq s)$ be the simple non-compact factors of G. Then X is the Riemannian product of the symmetric spaces X$_t$ corresponding to the G$'_t$ s, and Δ is the sum of the corresponding Laplacian Δ_t. A form η on X/Γ is harmonic if and only if $(\Delta\eta, \eta) = 0$. We have $(\Delta\eta, \eta) = \sum (\Delta_t \eta, \eta)$ and, since the Δ_t are positive,

(1) $$\Delta\eta = 0 \quad \Leftrightarrow \quad \Delta_t \eta = 0 \qquad (1 \leq t \leq s).$$

The proof of 3.2 (5) also yields

$$(\Delta_t \eta)_0 = \sum -C_t \eta_I \,\omega^I$$

hence

$$(2) \qquad \Delta\eta = 0 \iff -C_t\eta_I = 0 \qquad (1 \leq t \leq s, \, I \subset [1, N]).$$

Let η be a harmonic q-form and $\eta_0 = \sum \eta_I \omega^I$ the corresponding form on G/Γ. Let $q \leq m(G)$. We have to prove it has constant coefficients. For this it is enough to show

$$(3) \qquad X_j \eta_I = 0 \qquad (1 \leq j \leq N),$$

for all $I's$ in Λ. We may assume that $\mathfrak{p}_t = L(G_t) \cap \mathfrak{p}$, is spanned by some of the $X_i's$ $(1 \leq t \leq s)$. Let then J_t be the set of indices for which $X_i \in \mathfrak{p}_t$ and J'_t its complement. We can write

$$(4) \qquad \eta_0 = \sum_{J \subset J'_t} \tau_J \wedge \omega^J,$$

where τ_J is homogeneous of degree $d(J) = q - |J|$ and of the form

$$(5) \qquad \tau_J = \sum_{\substack{I \subset J_t \\ |I| = d(J)}} \eta_{J,I}\omega^I.$$

τ_J is a $d(J)$-form on G/Γ which is left-invariant under $K_t = K \cap G_t$, annihilated by the interior products $i(X)$ $(X \in L(K_t))$. Thus, if we put

$$(6) \qquad \varphi(x) = (1/2\,q) \sum_{\substack{j,k \in J_t \\ I \subset J_t}} ([X_j, X_k]\eta_{J,I})^2$$

and consider its integral over G/Γ, all the formal computations of Matsushima [19] go through and yield (3) for $j \in J_t$ since $d(J) = q - |J| \leq m(G) \leq m(G_t)$.

3.5. THEOREM (Garland [9]). — *Let Γ be a discrete torsion-free subgroup of G. Let $q \leq m(G)$. Assume that every class of $H^q(\Omega_0)$ is representable by a square integrable form. Then j^q is surjective.*

In fact, Theorem 3.5 of [9] is a special case of our 3.5, but the argument is general. We sketch a slight variation of it. Let $q \leq m(G)$. We have to prove that $\mathscr{H}^q_{(2)} \subset I_G^{q;\Gamma}$, or, equivalently

$$(1) \qquad (\mathscr{H}^q_{(2)})_0 \subset (I_G^{q;\Gamma})_0.$$

Fix a Haar measure dy on G. As usual, the convolution product $u \star v$ of two functions on G is defined by

$$(2) \qquad u \star v(x) = \int_G u(xy^{-1})v(y)\,dy, \qquad (x \in G).$$

If $u \in C_c^\infty(G)$ and $v \in L^2(G/\Gamma)$, then $u \star v \in C^\infty(G/\Gamma) \cap L^2(G/\Gamma)$, and for any element X in the universal enveloping algebra $U(L(G))$ of $L(G)$, we have

$$(3) \qquad X(u \star v) = (Xu) \star v \in C^\infty(G/\Gamma) \cap L^2(G/\Gamma).$$

Let $\alpha \in C_c^\infty(G)$ be K-invariant [i. e. $\alpha(kxk^{-1}) = \alpha(x)$ $(k \in K; \; x \in G)$]. Then $\alpha \bigstar$ commutes with the Casimir operator and K acting on G/Γ in the natural way. Therefore, if $\eta = \sum \eta_I \omega^I$ is in Ω_0 [resp. $(\mathcal{H}^q_{(2)})_0$], then so is

$$\eta_{(\alpha)} = \sum_I (\alpha \bigstar \eta_I) \omega^I.$$

Let now $\eta \in (\mathcal{H}^q_{(2)})_0$. Matsushima's proof of (1) in the case where G/Γ is compact starts with the function

(4) $$\varphi(x) = \sum_{a,b,I} ([X_a, X_b] \eta_I)^2,$$

$(1 \leqq a, b \leqq N)$, studies its integral over G/Γ, and uses repeatedly Stokes' formula (1.5), which makes no difficulty when G/Γ is compact. Replace now η by $\eta_{(\alpha)}$. Then for any $Y \in U(L(G))$, $Y(\alpha \bigstar \eta_i)$ is square integrable; therefore by 3.2 (1), (2) we can apply 1.5 to $X = X_i$ $(1 \leqq i \leqq \dim G)$ and $u = Y(\alpha \bigstar \eta_I)$, $v = Y'(\alpha \bigstar \eta_{I'})$ $(Y, Y' \in U(L(G))$. Then all the steps in Matsushima's argument go through. This shows that $\eta_{(\alpha)} \in (I_G^{q,\Gamma})_0$ for any K-invariant $\alpha \in C_c^\infty(G)$. But we may choose a sequence α_n of such functions forming a Dirac sequence ([14], §2), and then η_I is the L^2-limit of the $\alpha_n \bigstar \eta_I$. Therefore η belongs to the L^2-closure of the space of square integrable elements in $(I_G^{q,\Gamma})_0$. Since $I_G^{q,\Gamma}$ is finite-dimensional, the latter space is closed, whence (1).

Remark. $-$ If $\eta \in (I_G^{q,\Gamma})_0$, then its coefficients η_I are constant on G/Γ (and conversely). Thus, if the volume of G/Γ is infinite, no non-zero element of $I_G^{q,\Gamma}$ is square integrable, and the theorem asserts simply that no element of $H^q(\Gamma)$ is representable by a square integrable form for $q \leqq m(G)$.

3.6. PROPOSITION. $-$ *Let Γ be a discrete subgroup of G and Γ' a torsion-free normal subgroup of finite index of Γ. Assume that X/Γ has finite volume.*

(i) *Let C' be a subcomplex of $\Omega_X^{\Gamma'}$ and m' a positive constant such that*

(a) *the inclusion $C' \to \Omega_X^{\Gamma'}$ induces an isomorphism in cohomology up to dimension m';*

(b) *C'^q consists of square integrable forms for $q \leqq m'$;*

(c) *$I_G^{q,\Gamma'} \subset C'^q$ for $q \leqq m'$.*

Then $j^q : I_G^{q,\Gamma'} \to H^q(\Gamma')$ is injective for $q \leqq m'(G)$ and bijective for $q \leqq \min(m(G), m')$.

(ii) *Assume C' to be stable under the natural action of Γ/Γ' on $\Omega_X^{\Gamma'}$ and let $C = (C')^{\Gamma/\Gamma'}$. Then (i) is valid with C' and Γ' replaced by C and Γ.*

(i) Let $q \leqq m'(G)$ and $\omega \in I_G^{q,\Gamma'}$. Assume that $j^q(\omega) = 0$. Let $\sigma \in \Omega_X^{q-1,\Gamma'}$ be such that $\omega = d\sigma$. In view of (c) and (a), we may (and do) assume $\sigma \in C'^{q-1}$. But then σ is square integrable by (b), and $\omega = 0$ by 2.5. Thus j^q is injective. For $q \leqq \min(m', m(G))$ every class is representable by a square integrable form by (a), (b), and j^q is surjective (3.5).

(ii) This follows from (i) and the elementary relations

$$H^*(C) = H^*(C')^{\Gamma/\Gamma'}, \qquad H(\Omega_X^\Gamma) = H(\Omega_X^{\Gamma'})^{\Gamma/\Gamma'}, \qquad I_G^\Gamma = (I_G^{\Gamma'})^{\Gamma/\Gamma'}.$$

3.7. REMARK. — The proof of injectivity shows more generally that for $q \leqq m'$, the space $\mathscr{H}^q_{(2)} \cap C^q$ of harmonic forms contained in C^q injects into the cohomology. In particular, an element of $H^q(\Omega^\Gamma_X)$ has at most one harmonic representative in C^q.

4. Decomposition of the metric on X, associated to a parabolic subgroup

4.1. In this section, k is a subfield of $\mathbf{R}$ and G a connected semi-simple k-group. The groupe G $(\mathbf{R})$ of real points of G plays the role of our former G.

As before, K is a maximal compact subgroup of G $(\mathbf{R})$, $X = K \backslash G(\mathbf{R})$, θ the Cartan involution of L $(G(\mathbf{R}))$, or of G $(\mathbf{R})$, with respect to K, B $(\, , \,)$ the Killing form of L $(G(\mathbf{R}))$ and g_0 the scalar product on L $(G(\mathbf{R}))$ defined by

$$g_0(\xi, \eta) = -B(\xi, \theta(\eta)), \qquad [\xi, \eta \in L(G(\mathbf{R}))].$$

Let $\sigma : G(\mathbf{R}) \to X$ be the canonical projection and $o = \sigma(K)$. Then $d\sigma_e$ identifies the orthogonal complement p of L (K) with respect to B (or, equivalently, to g_0) with $T_0(X)$. The restriction of g_0 to p then defines a metric on $T_0(X)$, invariant under the natural action of K, and we let dx^2 be the corresponding G $(\mathbf{R})$-invariant metric on X.

4.2. Let P be a parabolic k-subgroup of G, M the Levi k-subgroup of P stable under θ ([4], 1.9), $S_P = M \cap R_d P$ the greatest torus of the split radical $R_d P$ of P stable under θ and $A = S_P(\mathbf{R})^0$. With ^0P defined as in ([4], 1.1) we have

(1)
$$\begin{cases} P(\mathbf{R}) = A \ltimes {}^0P(\mathbf{R}), & {}^0P = {}^0M \ltimes U, \\ K \cap P = K \cap {}^0M \quad \text{maximal compact in } P(\mathbf{R}). \end{cases}$$

We put

$$Z = (K \cap P) \backslash {}^0M(\mathbf{R})$$

and have then

(2)
$$(K \cap P) \backslash {}^0P(\mathbf{R}) \cong Z \times U(\mathbf{R}),$$

which should be kept in mind to make the transition from [4] to this paper. The map σ induces, by passage to the quotient, an isomorphism

(3)
$$\mu_0 : \quad Y = A \times Z \times U(\mathbf{R}) \xrightarrow{\sim} X.$$

The map commutes with P $(\mathbf{R})$, where the action of P $(\mathbf{R})$ on Y is defined by

(4)
$$\begin{cases} (b, z, u) p = (ab, zm, m^{-1} a^{-1} uamv), \\ [p = amv; \, a \in A; \, m \in {}^0M(\mathbf{R}); \, v \in U(\mathbf{R})] \quad ([3], 1.5). \end{cases}$$

If ω is a differential form on X or Y, we shall also denote by $\omega.p$ its transform under the action of p. As in [3], we let $dy^2 = \mu_0^*(dx^2)$ and da^2 (resp. du^2) be the right-invariant metric on A [resp. U $(\mathbf{R})$] which is equal to the restriction of g_0 on L (A) [resp. L (U $(\mathbf{R})$)]. We want to write da^2 and du^2 more explicitly.

Let Δ be a basis of the set of k-roots of G with respect to some maximal k-split torus S. The conjugacy classes over k of parabolic k-subgroups are parametrized by the subsets of Δ. Let $I = I(P)$ be the type of P ([4] 4.1). There is then a canonical isomorphism

$$(5) \qquad\qquad A \xrightarrow{\sim} (\mathbf{R}_+^*)^{\Delta - I}, \quad ([4], 4.2);$$

we can identify Δ-I with a basis of $X(S_P) \otimes \mathbf{Q}$ and write

$$(6) \qquad\qquad da^2 = \sum_{\alpha, \, \beta \in \Delta - I} c_{\alpha\beta}\, \alpha^{-1} \beta^{-1}\, d\alpha\, d\beta,$$

where $\sum c_{\alpha\beta}\, d\alpha\, d\beta$ is the restriction of the Killing form to L(A).

Let Φ_P be the set of roots of P with respect to S_P. For $\beta \in \Phi_P$, let

$$(7) \qquad\qquad u_\beta = \{ X \in L(G(\mathbf{R})) \,|\, \mathrm{Ad}\, a\, X = a^\beta X, \ (a \in A) \}.$$

Then

$$(8) \qquad\qquad L(U(\mathbf{R})) = \bigoplus_\beta u_\beta,$$

and each u_β is stable under $\mathrm{Ad}_{L(G)}(A\,{}^0M(\mathbf{R}))$. For $\beta \in \Phi_P$, let h_β be the right invariant scalar product on $U(\mathbf{R})$ which is zero on u_α ($\alpha \neq \beta$), and equal to g_0 on u_β. By [3], 1.4 (6), the spaces u_β ($\beta \in \Phi_P$) are mutually orthogonal. For $a \in A$, $m \in {}^0M(\mathbf{R})$, we have then

$$(9) \qquad\qquad (\mathrm{Int}\, am)^*(du^2) = \bigoplus_\beta a^{2\beta}\, (\mathrm{Int}\, m)^*\, h_\beta.$$

For $k \in K$, $\mathrm{Ad}\, k$ leaves g_0 invariant; therefore, if $k \in K \cap M$, then $\mathrm{Int}\, k$ leaves h_β invariant, hence $(\mathrm{Int}\, m)^*\, h_\beta$ depends only on the image z of m in Z under the natural projection. Let us then put

$$(10) \qquad\qquad (\mathrm{Int}\, m)^*\, h_\beta = h_\beta(z), \qquad [m \in {}^0M(\mathbf{R}); \, z = om].$$

4.3. PROPOSITION. – *Let* $y = (a, z, u) \in Y$. *The spaces* L(A) a, $T_z(Z)$, *and* $u_\beta\, u$ ($\beta \in \Phi_P$) *are mutually orthogonal, and we have*

$$(1) \qquad\qquad (dy^2)_y = (da^2)_a \oplus (dz^2)_z \oplus \bigoplus_\beta 2^{-1}\, a^{2\beta}\, h_\beta(z).$$

By [3], 1.6, we have (1) with the last sum replaced by $(\mathrm{Int}\, am)^*\, (du^2)_u$, where $m \in {}^0 M(\mathbf{R})$ is such that $om = z$. The proposition then follows from 4.2 (9), (10).

4.4. COROLLARY. — *Let* dv_Y, dv_A, dv_Z *and* dv_U *be the volume elements of the metrics* dy^2, da^2, dz^2 *and* du^2. *Then*

$$(1) \qquad \left\{ \begin{array}{l} 2^e\, dv_Y = a^{2\rho}\, dv_A \wedge dv_Z \wedge dv_U, \qquad [e = (\dim U)/2], \\[2ex] \qquad\qquad dv_A = c \wedge_{\alpha \in \Delta - I} \dfrac{d\alpha}{\alpha}, \end{array} \right.$$

where $2\rho = 2\rho_P = \sum_{\beta \in \Phi_P} \beta \dim u_\beta$ *and* $c = (\det c_{\alpha\beta})^{1/2}$.

By definition, any rational $\mathbf{Q}$-character of M is equal to ± 1 on ${}^0\mathrm{M}$, hence $\mathrm{Ad}\, m\, |\mathrm{L}\,(\mathrm{U}\,(\mathbf{R}))$ is unimodular, and $(\mathrm{Int}\, m)^* \, du^2$ and du^2 have the same volume element. The corollary is then an obvious consequence of 4.3 and 4.2 (6).

5. Differential forms on Siegel sets

5.1.   We keep the assumptions of paragraph 4 and first fix some convention and notation to describe differential forms on certain subsets of Y. Let $\Delta - \mathrm{I} = \{\alpha_1, \ldots, \alpha_s\}$. We let $m = \dim \mathrm{X}$, fix a moving frame $(\omega^i)_{1 \leq i \leq m}$ on Y, where ω^i is lifted, under one of the natural projections, from $d \log \alpha_i$ on A if $i \leq s$, from an orthonormal frame on Z if $s < i \leq t$, and from a set of right invariant one-forms on $\mathrm{U}\,(\mathbf{R})$ which, at the origin, span the various subspaces $\mathfrak{u}_\beta^* \,(\beta \in \Phi_\mathrm{P})$ if $t < i \leq m$.

Let $\mathrm{I}_m = \{1, \ldots, m\}$. For $i \in \mathrm{I}_m$, we put

$$(1) \qquad \alpha(i) = \begin{cases} 0 & \text{if } 1 \leq i \leq t, \\ \beta & \text{if } \omega_e^i \in \mathfrak{u}_\beta^*. \end{cases}$$

For a subset J of I_m, we let

$$(2) \qquad \omega^\mathrm{J} = \Lambda_{i \in \mathrm{J}}\, \omega^i, \qquad \alpha(\mathrm{J}) = \sum_{i \in \mathrm{J}} \alpha(i).$$

A differential form τ of degree q on Y, or on an open subset of Y, will then be written

$$\tau = \sum_\mathrm{J} f_\mathrm{J}\, \omega^\mathrm{J}$$

where J runs through the subsequences of I_m whose number of elements $|\mathrm{J}|$ equals q.

5.2.   X (A) will denote the group of continuous homomorphisms of A into $\mathbf{R}_+^*$. We have $\mathrm{X}\,(\mathrm{A}) \cong \mathrm{X}\,(\mathrm{S}_\mathrm{P}) \underset{\mathbf{Z}}{\otimes} \mathbf{R}$, and, by restriction, $\mathrm{X}\,(\mathrm{S}_\mathrm{P})$ may be identified to a lattice in X (A). Any $\lambda \in \mathrm{X}\,(\mathrm{A})$ is a linear combinaison of the $\alpha_i's$ and can be written

$$\lambda = \sum c_i \alpha_i.$$

We write $\lambda \gg 0$ if $c_i > 0$ for *all* $i's$, and $\lambda \geqslant 0$ if $c_i \geq 0$ for all $i's$.

5.3.   For $t > 0$, let $\{\mathrm{A}_t = a \in \mathrm{A} \mid \alpha^a \leq t\, (\alpha \in \Delta - \mathrm{I})\}$. A Siegel *set* $\mathfrak{S}_{t,\omega}$ (resp. a *cylindrical set* $\mathfrak{C}_{t,\omega}$) in X, with respect to P, o, is a set of the form

$$\mathfrak{S}_{t,\omega} = \mu_0\,(\mathrm{A}_t \times \omega), \qquad [\text{resp. } \mathfrak{C}_{t,\omega} = \mu_0\,(\mathrm{A}_t \times \omega \times \mathrm{U}\,(\mathbf{R}))],$$

where ω is relatively compact in $\mathrm{Z} \times \mathrm{U}\,(\mathbf{R})$ (resp. Z).

In pratice, it is understood that ω is such that the complement of the set of interior points has measure zero; by definition a (smooth) differential form of such a set U is the restriction of a (smooth) form defined in some open neighborhood of U.

5.4. LEMMA. — *Let $\mathfrak{S}$ be a Siegel set with respect to* $\mathbf{P}$, o *and* f *a positive measurable function on* $\mathfrak{S}$. *Let* $\lambda \in X(A)$; *assume that* $(f \circ \mu_0)(a, q) \prec a^\lambda$ $(a \in A_t,\ q \in \omega)$, *and that* $2\rho + \lambda \gg 0$. *Then* $f \in L^1(\mathfrak{S}, dv_X)$. *In particular,* $\mathfrak{S}$ *has finite volume.*

This amounts to showing that $a^\lambda \in L^1(A_t \times \omega, dv_Y)$. Write

$$2\rho + \lambda = \sum_{1 \leq i \leq s} m_i \alpha_i.$$

Then, by 4.4, there is a constant $d > 0$ such that

$$\int_{A_t \times \omega} a^\lambda\, dv_Y = d \prod_{i \leq s} \int_0^t \alpha_i^{m_i - 1}\, d\alpha_i.$$

Since $m_i > 0$ for all i by assumption, each factor on the right hand side is finite, whence the first assertion. Since $\rho \gg 0$, this shows that any constant function is integrable, whence the second assertion.

5.5. PROPOSITION. — *Let* $\mathfrak{S}$ *be a Siegel set with respect to* $\mathbf{P}$, o, σ *a continuous* q-*form on* $\mathfrak{S}$, *and* $\tau = \mu_0^*(\sigma) = \sum f_J \omega^J$ *Assume there exists* $\lambda \in X(A)$ *such that* $\rho + \lambda - \nu \gg 0$ *whenever* ν *is a weight of* A *in* $\underset{i \leq q}{\otimes} \Lambda^i L(U(\mathbf{R}))$, *and that* $|f_J| \prec a^\lambda$ *on* $\mu_0^{-1}(\mathfrak{S})$ *for all* J. *Then* σ *is square integrable on* $\mathfrak{S}$.

We have to show

$$(1) \qquad \int_{\mathfrak{S}} (\sigma_x, \sigma_x)\, dv_X < \infty,$$

where $(\ ,\ )$ denotes the scalar product on $\Lambda^q T^*(X)$ associated to dx^2. This amounts to proving

$$(2) \qquad \int_{A_t \times \mathfrak{v}} (\tau_y, \tau_y)\, dv_Y < \infty,$$

where $(\ ,\ )$ now denotes the scalar product on $\Lambda^q T^*(Y)$ associated to dy^2. With (ω^i) as in 5.1, write

$$(3) \qquad dy^2 = \sum_{1 \leq i,\, j \leq m} g_{ij} \omega^i \omega^j.$$

As usual, let (g^{ij}) be the inverse matrix to (g_{ij}). For two strictly increasing sequences $J = \{j_1, \dots, j_q\}$, $J' = \{j_1', \dots, j_q'\}$ in I_m, put

$$(4) \qquad g^{J, J'} = \det(g^{j_i\, j_k'})$$

and let

$$(5) \qquad f^J = \sum_{J'} g^{J, J'} f_{J'}.$$

Then

$$(6) \qquad (\tau_y, \tau_y) = \sum_{J, J'} g^{J, J'} f_J f_{J'} = \sum_J f^J f_J.$$

We have therefore to prove that $f^J f_J$ is integrable on $A_t \times \omega$.   Write

$$(7) \qquad I_m = I_{-1} \cup I_0 \cup \bigcup_\beta I_\beta,$$

where $I_{-1} = \{1, \ldots, s\}$, $I_0 = \{s+1, \ldots, t\}$ and $I_\beta = \{i \in I_m \mid \omega_e^i \in u_\beta^*\}$.   Let, using the notation of $4.2\,(10)$:

$$(8) \qquad h_{\beta,z} = \sum_{i,j \in I_\beta} h_{\beta,ij}(z)\,\omega^i \omega^j.$$

Let $h_\beta^{ij}(z)$ be the inverse matrix to $(h_{\beta,ij}(z))$ and (c^{ij}) the inverse matrix to (c_{ij}).   We have

$$(9) \qquad g^{ij} = \begin{cases} c^{ij} & (i,j \in I_{-1}), \\ \delta_{ij} & (i,j \in I_0), \\ 2\,a^{-2\beta} h_\beta^{ij}(z) & (i,j \in I_\beta), \\ 0 & \text{otherwise.} \end{cases}$$

It follows that $g^{J,\,J'} = 0$ unless J and J' have the same number of elements in each of I_{-1}, I_0, I_β $(\beta \in \Phi_P)$.   Assume this is the case.   Then

$$(10) \quad \alpha(J) = \alpha(J') \qquad and \qquad |g^{J,\,J'}(y)| \prec a^{-2\alpha(J)}, \quad [y \in A_t \times C \times U(\mathbf{R}),\ a = \mathrm{pr}_A\,y],$$

where C is relatively compact in Z.
It follows then from the assumption on the $f_J's$ and from (10) that we have

$$(11) \qquad |f^J| \le \sum_{J'} |g^{J,\,J'}| \cdot |f_{J'}| \prec a^{-2\alpha(J)+\lambda},$$

$$(12) \qquad |f^J \cdot f_J| \prec a^{-2\alpha(J)+2\lambda},$$

on $A_t \times C \times U(\mathbf{R})$.   By 5.4, $f^J f_J$ is integrable on $A_t \times \omega$ if

$$(13) \qquad 2\rho + 2\lambda - 2\alpha(J) \gg 0.$$

But $\alpha(J)$ is the sum of at most q elements of Φ_P, and an element β occurs at most $|I_\beta| = \dim u_\beta$ times.   Thus our assumption on λ implies that (13) is fulfilled for all J with q elements, and the proposition is proved.

5.6.   PROPOSITION. $-$ *Let* $\sigma = \sum_J f_J \omega^J$ *be a q-form on* Y *and* $b \in A$.   *Then*

$$(1) \qquad (\sigma b)_y = \sum_J f_J(yb)\,b^{-\alpha(J)}\,\omega^J \qquad (y \in Y).$$

In particular, if $\sigma b = \sigma$, *we have*

$$(2) \qquad f_J((a,q)b) = b^{\alpha(J)} f_J(a,q), \qquad [y = (a,q);\ a \in A;\ q \in Z \times U(\mathbf{R})].$$

We have

$$(\sigma b)_y = \sum_J f_J(yb)(\omega^J b).$$

It suffices therefore to prove, using the notation of 5.1 (1) :

$$(3) \qquad \omega^j b = b^{-\alpha(j)} \omega^j,$$

for any $j \, (1 \leq j \leq m)$. By 4.2 (4), we have

$$(a, z, u) b = (ab, z, b^{-1} ub), \qquad [a \in A; z \in Z; u \in U(\mathbf{R})].$$

If $j \leq s$, then $\omega^j = d\alpha_j/\alpha_j$, $\alpha_j(i) = 0$; and (3) follows from the invariance of $d\alpha_j/\alpha_j$ under translations on A. For $s < j \leq t$, $\omega^j b = \omega^j$ and $\alpha(j) = 0$. If now $j > t$ and $\alpha(j) = \beta \in \Phi_p$, then $\omega^j b$ is the right-invariant form on $U(\mathbf{R})$ whose value at e is equal to $\operatorname{Ad} b^{-1}(\omega_e^j)$. Since $\omega_e^j \in \mathfrak{u}_\beta^*$ and $\operatorname{Ad} b^{-1}$ is the dilatation by $b^{-\beta}$ on $\mathfrak{u}^*$, this proves (3) also in that case.

5.7. COROLLARY. — *Assume σ to be invariant under A, and let $a_0 \in A$. Then*

$$(1) \qquad f_{\mathsf{J}}(a, z, u) = a^{\alpha(\mathsf{J})} f_{\mathsf{J}}(1, z, aua^{-1}) \qquad [a \in A, z \in Z, u \in U(\mathbf{R})].$$

In particular, on $\mathfrak{S}' = A_t \times \omega \, [\omega$ relatively compact in $Z \times U(\mathbf{R})]$, we have

$$(2) \qquad |f_{\mathsf{J}}(a, q)| \prec a^{\alpha(\mathsf{J})} \qquad (a \in A_t; q \in \omega)$$

and σ is square integrable.

(1) is a special case of 5.6 (2). It is elementary and well known that if C is relatively compact in $U(\mathbf{R})$, then $\{ aua^{-1} \mid a \in A_t; u \in C \}$ is relatively compact. Hence (1) $\Rightarrow$ (2). By 5.5 (10) we have

$$|f^{\mathsf{J}}(a, q)| \leq \sum_{\mathsf{J}'} |g^{\mathsf{J}, \mathsf{J}'}| |f_{\mathsf{J}'}| \prec a^{-\alpha(\mathsf{J})},$$

whence $|f^{\mathsf{J}} f_{\mathsf{J}}| \prec 1$. The latter is then integrable by (5.4).

6. The manifold with corners $\overline{X}/\Gamma$

In this section, G is a connected semi-simple $\mathbf{Q}$-group and Γ a torsion-free arithmetic subgroup. We shall recall first some results of [4], and then introduce neighborhoods and partitions of unity which reflect the structure of manifold with corners of $\overline{X}/\Gamma$.

6.1. As in [4], $\mathfrak{P}$ is the set of parabolic $\mathbf{Q}$-subgroups of G and $\pi : X \to X/\Gamma$ the natural projection. Our former A will be denoted A_P. This is in slight conflict with [4], where A_P denotes a subgroup of $P/R_u P$. Our present A_P is then the unique lifting in $P(\mathbf{R})$ of the previous one which is stable under the Cartan involution associated to K. It depends then on K, or o; if more precision appears to be desirable we then shall also write $A_{o, P}$ instead of A_P. By convention, the geodesic action of $a \in A_{o, P}$ is the geodesic action, as defined in [4], paragraph 3, of the canonical image of a in the center in $P/R_u P$. Consequently, modulo the canonical identification of $A_{o, P}$ with the $A_P \subset P/R_u P$ of [4], the maps μ_0 of this paper and of ([4], 5.4) are the same, and it follows that the results of [4] involving A_P to be used here are valid with the present interpretation of A_P and μ_0.

Given $P \in \mathfrak{P}$, there exists, by [4], paragraph 10, a number $t(o, P) > 0$ such that if $t \leq t(o, P)$, the equivalence relations defined by Γ and $\Gamma_P = \Gamma \cap P$ on

$$U_{o, P, t} = \mu_0(A_{P, t} \times Z \times U(\mathbf{R}))$$

are the same and consequently $\mu'_0 = \pi \circ \mu_0$ is an isomorphism

$$\mu'_0 : \quad A_{P, t} \times (Z \times U(\mathbf{R}))/\Gamma_P = A_{o, P, t} \times (Z \times U(\mathbf{R}))/\Gamma_P \xrightarrow{\sim} \pi(U_{o, P, t}).$$

Moreover, the action of the semi-group $A_{o, P, 1}$ on A_t by translations gives rise *via* μ'_0 to an invariantly defined action stemming from the geodesic action ([4], 10.3). In our present set up it can be characterized as follows : given $x \in \pi(U_{o, P, t})$ let $y \in \pi^{-1}(x)$. Then $x \circ a = \pi(y \circ a)$. If $y' \in \pi^{-1}(x)$, then $y' = y \sigma$, with $\sigma \in \Gamma_P$, whence $y' \circ a = (y \circ a) \sigma$ and $\pi(y' \circ a) = \pi(y \circ a)$, as it should be. This description does not involve the origin o. We recall that $U_{o, P, t}$ does indeed depend on o, however not in a drastic manner. In particular, it follows from ([4], 6.1) that, given $o' \in X$ and $t \leq t(o, P)$, $t(o', P)$, there exists $t' \leq t$ such that

$$\pi(U_{o, P, t}) \cap \pi(U_{o', P, t}) \supset \pi(U_{o, P, t'}) \cup \pi(U_{o', P, t'}).$$

Let $p \in P(\mathbf{R})$ be such that $o' = o p$. Then $\operatorname{Int} p^{-1} : A_{o, P} \xrightarrow{\sim} A_{o', P}$ commutes with the roots. If $x \in \pi(U_{o, P, t'} \cap U_{o', P, t'})$, then we have $x \circ a = x \circ a^p$ for $a \in A_{o, P, 1}$. In the sequel, whenever we write $U_{o, P, t'}$ it is understood that $t \leq t(o, P)$. Let $\gamma \in \Gamma$ and $P' = P^\gamma$. Then there exists $t' \leq t$ such that

$$\pi(U_{o, P, t}) \cap \pi(U_{o, P', t}) \supset \pi(U_{o, P, t'}) \cup \pi(U_{o, P', t'})$$

and the geodesic actions of $A_{o, P, 1}$ and $A_{o, P', 1}$ are compatible with the canonical isomorphisms

$$A_{o, P, 1} \xrightarrow{\sim} A_{P, 1} \xrightarrow{\sim} A_{P', 1} \xrightarrow{\sim} A_{o, P', 1},$$

where $A_{P, 1}$ and $A_{P', 1}$ are as in [4]. This follows from the above and 10.3, 5.6 (2) of [4].

6.2. In [4], paragraph 7, X is enlarged to a manifold with corners $\overline{X}$, which is the disjoint union of faces $e(P)$ $(P \in \mathfrak{P})$, where $e(G) = X$. The face $e(P)$ may be identified with $Z \times U(\mathbf{R})$. The group Γ operates freely and properly on $\overline{X}$, and $\overline{X}/\Gamma$ is a compact manifold with corners, disjoint union of faces

$$e'(P) = (Z \times U(\mathbf{R}))/\Gamma_P \cong e(P)/\Gamma_P,$$

where P runs through a set of representatives of $\mathfrak{P}/\Gamma$ ([4], § 9). The above map μ'_0 extends to an isomorphism μ''_0 of manifolds with corners of $\overline{A_{P, t}} \times (\overline{e(P)/\Gamma_P})$ onto a neighborhood $\pi(\tilde{U}_{o, P, t})$ of $\overline{e'(P)}$. We let pr_P denote the projection of $\pi(\tilde{U}_{o, P, t})$ onto $\overline{e'(P)}$, carried over from the projection of $\overline{A_{P, t}} \times (\overline{e(P)/\Gamma_P})$ onto its second factor. Its fibers are the orbits of $A_{P, 1}$ under the geodesic action. If P is replaced by a conjugate $P' = P^\gamma$

under an element $\gamma \in \Gamma$, then, by the end remarks of 6.1, the induced geodesic actions of $A_{o, P, 1}$ and $A_{o, P', 1}$ are defined in neighborhoods of $\overline{e'}(P) = \overline{e'}(P')$, and are the same, modulo the canonical isomorphism $\operatorname{Int} \gamma^{-1} : A_{o, P, 1} \xrightarrow{\sim} A_{o, P', 1}$. Thus $\operatorname{pr}_P = \operatorname{pr}_{P'}$ sufficiently close to $\overline{e'}(P)$, and the germ of projection pr'_P defined by pr_P depends only on $\overline{e'}(P)$, and not on the choice of P in its Γ-conjugacy class.

Let $Q \in \mathfrak{P}$, containing P. We have a canonical inclusion $A_Q \subset A_P$ and in fact a canonical factorization $A_P = A_Q \times A_{P, Q}$, and this inclusion is compatible with geodesic action (*see* 3.11, 4.3, 5.5 in [4]). From this it follows that $\operatorname{pr}_Q \circ \mu'_0$ induces an isomorphism of $\overline{A}_{P, Q, t} \times \overline{e'}(P)$ onto a neighborhood of $\overline{e'}(P)$ in $\overline{e'}(Q)$; if $\operatorname{pr}_{P, Q}$ is the canonical projection of the latter on $\overline{e'}(P)$, then $\operatorname{pr}_P = \operatorname{pr}_{P, Q} \circ \operatorname{pr}'_Q$.

6.3. SPECIAL NEIGHBORHOODS. — It will be slightly more convenient here to deal with open neighborhoods, and to use subsets of A_P defined by strict inequalities. Let then, for $t > 0$, $P \in \mathfrak{P}$ and $I = I(P)$:

$$A_{P, (t)} = \{ a \in A_P \mid a^\alpha < t \ (\alpha \in \Delta - I) \},$$

$$\overline{A}_{P, (t)} = \{ a \in \overline{A}_P \mid a^\alpha < t \ (\alpha \in \Delta - I) \}.$$

Thus $A_{P, (t)}$ is open in the corner $\overline{A}_P$. If ω is open relatively compact in $e(P)$, then $\mu_0 (A_{P, (t)} \times \omega)$ is the interior of the Siegel set $\mathfrak{S}_{t, \omega}$ and will be called an open Siegel set.

Let $y \in e(P)$ and y' its image in $e'(P)$. Let ω be an open relatively compact neighborhood of y in $e(P)$ on which $e(P) \to e'(P)$ is injective. Then $\mu_0 (A_{P, (t)} \times \omega)$ is an open Siegel set in X on which π is injective, and μ'_0 extends to an isomorphism of manifolds with corners of $\overline{A}_{P, (t)} \times \omega$ onto an open neighborhood U of y' in $\overline{X}/\Gamma$, to be called a *special neighborhood;* U is the isomorphic image under π (extended to $\overline{X}$) of a neighborhood of y in the corner $X(P)$. Clearly, y' has a fundamental system of special neighborhoods. If $Q \in \mathfrak{P}$ is such that $e'(Q) \cap U \neq \varnothing$, then a Γ-conjugate of Q contains P, and $\operatorname{pr}_Q U$ contains a special neighborhood of y' in $\overline{e'}(Q)$. We note that the Γ-conjugacy class of P is completely determined by U; in fact if s is the maximum of $|\Delta - I(Q)|$, $(Q \in \mathfrak{P})$ for Q such that $e'(Q) \cap U \neq \varnothing$, then s is attained only on the Γ-conjugates of P. We shall say that P *is associated to* U.

If $P = G$, i. e. if $y' \in X/\Gamma$, a neighborhood of y' is special if it is the isomorphic image under π of an open relatively compact neighborhood of y in X.

6.4. SPECIAL COVERS AND PARTITIONS OF UNITY. — A special cover of $\overline{X}/\Gamma$ is a finite cover by special neighborhoods. Any cover of $\overline{X}/\Gamma$ has a special refinement.

LEMMA. — *Let $\mathcal{U} = (U_i)_{i \in L}$ be a finite cover of $\overline{X}/\Gamma$. There exists a smooth partition of unity (λ_i) subordinated to $\mathcal{U}$ with the following property : given $P \in \mathfrak{P}$, $x \in \overline{e'}(P)$, there exists a special neighborhood U of x in $\overline{X}/\Gamma$ such that λ_i is constant along the fibres of $\operatorname{pr}_P \mid U$ for all $i \in L$.*

We may assume $\mathscr{U}$ to be special. For $i \in L$, let P_i be associated to U_i. We have then an isomorphism

$$(1) \qquad \mu_0' : \overline{A}_{P_i,\,(t_i)} \times \omega_i \overset{\sim}{\to} U_i$$

with ω_i open relatively compact in $e'(P_i)$.

There exist covers $\mathscr{V} = (V_i)_{i \in L}$ and $\mathscr{W} = (W_i)_{i \in L}$ such that $\overline{W}_i \subset V,\ \overline{V}_i \subset U$ and V_i (resp. W_i) is isomorphic under $\mu_0'^{-1}$ to $\overline{A}_{P_i,\,(t_i')} \times \omega_i'$ (resp. $\overline{A}_{P_i,\,(t_i'')} \times \omega_i''$), where $(0 < t'' < t' < t,\ \overline{\omega}_i'' \subset \omega_i',\ \overline{\omega}_i' \subset \omega_i;\ i \in L)$. Let ρ_{i0} be a smooth function on $e'(P)$ with support in ω_i', equal to one on ω_i''. On $\overline{A}_{P_i}$ we can construct a smooth function ρ_{i1} with support in $\overline{A}_{P_i,\,(t_i')}$ equal to one on $\overline{A}_{P_i,\,(t_i'')}$ $(i \in L)$, which, for each face F of $\overline{A}_P$, is independent of the coordinates transversal to F sufficiently close to F. Otherwise said, if $Q \supset P$ is such that $\mu_0 : \overline{A}_P \times e(P) \overset{\sim}{\to} X(P)$ maps $F \times e(P)$ onto $e(Q)$ (*see* [4], 5.3, 5.4), then, sufficiently close to F, ρ_{i1} is constant along the sets $A_Q \times \{\,b\,\}$ $(b \in A_{P,\,Q})$. On U_i, identified with $\overline{A}_{P_i,\,(t_i)} \times \omega_i$, we then put $\tau_i = \rho_{i0}\,\rho_{i1}$, and let τ be the sum of the τ_i. We claim that $\lambda_i = \tau_i/\tau$ defines the sought for partition of unity.

Let $P \in \mathfrak{P}$ and $y \in e'(P)$. Take a special neighborhood U of y such that for any $i \in L$, $U \cap U_i \neq \varnothing$ implies $U \subset U_i$. Let $J \subset L$ be the set of $i \in L$ for which $U \subset U_i$. Then $\tau \mid U$ is the sum of the $\lambda_i'\,s$ with $i \in J$ and if λ_i is not identically zero on U, then $i \in J$. It suffices therefore to see that for $i \in J$, λ_i is constant along the fibres of pr_P on some special neighborhood V_1 of y in V. If $i \in J$, then P is Γ-conjugate to a subgroup P' containing P_i. The relation $\mathrm{pr}_{P_i} = \mathrm{pr}_{P_i\,P'} \circ \mathrm{pr}_{P'}$ (*see* 6.2) implies that λ_i is constant along the fiber of $\mathrm{pr}_{P'}$. But (6.2) $\mathrm{pr}_{P'}$ and pr_P coincide sufficiently close to $e'(P)$, whence our assertion.

7. A square integrability criterion and the injectivity of j^q

In this section, G is a semi-simple $\mathbf{Q}$-group and Γ an arithmetic subgroup of G.

We use the notation of paragraph 5, except that we write ρ_P and A_P instead of ρ and A.

7.1. For $q \in \mathbf{N}$, $\lambda \in X(A_P)$, we let $c(P, q, \lambda)$ denote the condition

$$(\bigstar) \qquad \rho_P + \lambda - \nu \geqslant 0,$$

for any weight ν of A_P in $\underset{i \leqslant q}{\otimes} \Lambda^i L(R_u P(\mathbf{R}))$ (*see* 5.2 for $\geqslant$).

Let $Q \in \mathfrak{P}$ and assume P to be conjugate to a subgroup of Q. There is then a canonical monomorphism $A_Q \to A_P$ (an inclusion if $Q \supset P$), whence a canonical homomorphism $r_{QP} : X(A_P) \to X(A_Q)$. We have

$$(1) \qquad r_{QP}(\rho_P) = \rho_Q, \qquad \Phi_Q \subset r_{QP}(\Phi_P) \subset \Phi_Q \cup \{0\},$$

$$(2) \qquad \Delta - I(Q) \subset r_{QP}(\Delta - I(P)) \subset (\Delta - I(Q)) \cup \{0\},$$

and, for $\beta \in \Phi_Q$:

$$(3) \qquad \dim u_\beta = \sum_{\delta \in \Phi_P \cap r_{QP}^{-1}(\beta)} \dim u_\delta,$$

therefore

(4)$$c(P, q, \lambda) \;\Rightarrow\; c(Q, q, r_{QP}(\lambda)) \qquad [\lambda \in X(A_P),\ q \in \mathbf{N}].$$

Let P be *minimal*, and $\lambda \in X(A_P)$. We let then

(5)$$c(G, \lambda) = \max q \,\big|\, c(P, q, \lambda) \; holds \; true.$$

By (4), $c(Q, q, r_{QP}(\lambda))$ is then true for any $Q \in \mathfrak{P}$ and any $q \leqq c(G, \lambda)$. By definition, we have $c(G, \lambda) = c(G^0, \lambda)$.

If G is anisotropic over $\mathbf{Q}$, then $P = G$, $A_P = \{e\}$, $\lambda = 0$, and we agree to put $c(G, \lambda) = \infty$.

Assume G to be an almost direct product of $\mathbf{Q}$-subgroups $G_i\, (1 \leqq i \leqq s)$. Then P and A_P are accordingly decomposed and λ is a sum of elements $\lambda_i \in X(A_P \cap G_i)$. It is then clear that $c(G, \lambda) = \min_i c(G_i, \lambda_i)$.

In the sequel, we write $c(G)$ for $c(G, 0)$.

7.2.   As in 5.5, let σ be a form defined on a Siegel set $\mathfrak{S}_{t, \omega}$ and

(1)$$\tau = \mu_0^*(\sigma) = \sum_J f_J \omega^J.$$

We shall say that σ (or τ) has *logarithmic growth* if there exists a polynomial P in s variables, with real coefficients, such that

(2)$$\big| f_J(a, q) \big| \prec \big| P(\log a^{\alpha_1}, \ldots, \log a^{\alpha_s}) \big|^J \qquad (a \in A_t,\ q \in \omega,\ \text{all } J).$$

We have then, for any $\varepsilon > 0$:

(3)$$\big| f_J(a, q) \big| \prec a^{-\varepsilon d}, \qquad \text{where} \quad d = \sum_{1 \leqq i \leqq s} \alpha_i.$$

A form σ' on $\pi(\mathfrak{S}_{t, \omega})$ is said to have logarithmic growth if $\sigma' \circ \pi$ has logarithmic growth on $\mathfrak{S}_{t, \omega}$.

Let V be an open subset of $\overline{X/\Gamma}$. A form σ defined on $V \cap X/\Gamma$ has *logarithmic growth near the boundary* if any $y \in (\partial \overline{X/\Gamma}) \cap V$ has a special neighborhood W in V such that σ has logarithmic growth on $W \cap (X/\Gamma)$.

7.3.   LEMMA. — *Let $P \in \mathfrak{P}$, $x \in e'(P)$, U be a special neighborhood of x and σ a form with logarithmic growth on $U \cap (X/\Gamma)$. Then σ has logarithmic growth near the boundary of U.*

Let $y \in (\partial \overline{X/\Gamma}) \cap U$. Then there exists $Q \in \mathfrak{P}$ containing P such that $y \in e'(Q)$. We have to show that σ has logarithmic growth on the intersection of X/Γ with some special neighborhood of y.

After renumbering, if needed, we may assume that, for some $s' < s$, we have

(1)$$A_Q = \{ a \in A_P \mid a^{\alpha_i} = 1\ (s' < i \leqq s)\}.$$

We let $A_Q^\perp$ be the connected closed subgroup of A_P whose Lie algebra is orthogonal to $L(A_Q)$ with respect to the restriction of the Killing form. It is equal to $A_P \cap {}^0Q$, and plays the role of A_P, if P is viewed as a subgroup of Q rather than G.

Let $U' = R_u Q$, ${}^0M'$ the analogue of ${}^0M$ in 4.1 (1) and $Z' = (K \cap Q)\backslash {}^0M'(\mathbf{R})$. We have the canonical isomorphisms

$$(2) \qquad Y = A_P \times Z \times U(\mathbf{R}) \xrightarrow{\mu_{oP}} X \xleftarrow{\mu_{oQ}} Y' = A_Q \times Z' \times U'(\mathbf{R});$$

moreover, a similar result, applied to P and Q, gives an isomorphism

$$(3) \qquad A_Q^\perp \times Z \times U(\mathbf{R}) \overset{\sim}{\to} Z' \times U'(\mathbf{R}).$$

From this it follows that the special neighborhoods of y are boxed between sets of the form $\overline{A}_{Q,t} \times \omega_1 \times \omega_2$ with ω_1 (resp. ω_2) open relatively compact in $A_Q^\perp$ [resp. $Z \times U(\mathbf{R})$]. It suffices therefore to study the behavior of τ on $A_{Q,t} \times \omega_1 \times \omega_2$.

There is a unique choice of constants b_{ij} $(1 \leq i \leq s' < j \leq s)$ such that

$$(4) \qquad \tau^i = d \log \alpha_i + \sum_j b_{ij}\, d \log \alpha_j$$

is zero on $L(A_Q^\perp)$. Let

$$(5) \qquad \omega'^i = \omega^i + \sum_j b_{ij}\omega^j, \qquad (1 \leq i \leq s' < j \leq s).$$

Let ω'^i be defined by (5) for $i \leq s'$, by $\omega'^i = \omega^i$ for $i > s'$. Write

$$(6) \qquad \tau = \sum_J h_J \omega'^J.$$

Let $\nu = \mu_{o,P}^{-1} \circ \mu_{0,Q}$. Then

$$(7) \qquad \mu_{o,Q}^*(\sigma) = \nu^*(\tau) = \sum_J (h_J \circ \nu)\, \nu^*(\omega'^J).$$

It is immediate from the above that $(\nu^*(\omega'^i))$ is a moving frame which can be used to study the behavior of $\mu_{0,Q}^*(\sigma)$ on special neighborhoods of points of $e'(Q)$. Going back to Y, we see that we have to study the $h_J's$ on $A_{Q,t} \times \omega_1 \times \omega_2$. But the $h_J's$ are linear combinations with constant coefficients of the $f_J's$. Since $\left| \log a^{\alpha_i} \right|$ is bounded on ω_1, there exists a polynomial P' in s' variables such that

$$(8) \qquad \left| P(\log a^{\alpha_1}, \ldots, \log a^{\alpha_s}) \right| \leq \left| P'(\log a^{\alpha_1}, \ldots, \log a^{\alpha_{s'}}) \right| \quad \text{on } A_{Q,t} \times \omega_1.$$

We have then

$$(9) \qquad \left| h_J(a, q) \right| \prec \left| P'(a) \right|, \qquad (a \in A_{Q,t}; \; q \in \omega_1 \times \omega_2).$$

    7.4. THEOREM. — *Let G be connected and Γ be a torsion-free arithmetic subgroup of G. Let C be the subcomplex of the elements in $\Omega_{X/\Gamma}^*$ which, together with their exterior derivatives, have logarithmic growth near the boundary of X/Γ. Then :*

    (a) *The inclusion map $C \to \Omega_{X/\Gamma}^*$ induces an isomorphism in cohomology.*

    (b) *For $q \leq c(G)$, C^q consists of square integrable forms.*

    (c) *$I_G^\Gamma \subset C$.*

Assume that the form σ in 7.2 is invariant under A_P. Then, by 5.7, the coefficients f_J are bounded in absolute value on $A_{P,\,t} \times \omega$, hence σ has logarithmic growth. Since $d\sigma = 0$ if $\sigma \in I_G^\Gamma$, this proves (c).

Let $q \leq c(G)$ and $\sigma \in C^q$. To prove that σ is square integrable it suffices to show that if $P \in \mathfrak{P}$, $y \in e'(P)$ and U is a special neighborhood of y in $\overline{X/\Gamma}$, then σ is square integrable on $\mathfrak{S} = U \cap (X/\Gamma)$. The latter is a Siegel set and we are in the situation of 7.2. By assumption and definition (7.1), the condition $c(P, q, 0)$ holds true. But then, we also have $c(P, q, -\varepsilon\,d)$ for a suitably small $\varepsilon > 0$. By 7.2 (3) and 5.5, it follows that σ is square integrable on $\mathfrak{S}$, whence (b).

There remains to prove (a). For an open set $V \subset \overline{X/\Gamma}$, let $\mathscr{F}(V)$ be the space of smooth forms σ on $V \cap (X/\Gamma)$ such that σ and $d\sigma$ have logarithmic growth near the boundary (7.2). Clearly if V' is open in V, the restriction $\Omega_V \to \Omega_{V'}$, maps $\mathscr{F}(V)$ into $\mathscr{F}(V')$; hence $\mathscr{F} : V \mapsto \mathscr{F}(V)$ is a presheaf on $\overline{X/\Gamma}$. Obviously, it is a sheaf, and in fact a differential sheaf of algebras, the differential being defined by the exterior differentiation of forms, and we have

$$(1) \qquad\qquad \mathscr{F}(\overline{X/\Gamma}) = C, \qquad \mathscr{F}(X/\Gamma) = \Omega^*_{X/\Gamma}.$$

We now want to prove that $\mathscr{F}$ is a fine resolution of $\mathbf{R}$ on $\overline{X/\Gamma}$.

$\mathscr{F}$ IS FINE. — Let f be a smooth function on $\overline{X/\Gamma}$. On $A_{P,\,t} \times \omega$ (notation of 7.2), express df as a linear combination of the ω^i $(i > s)$ and the $d\alpha_i$. Then the coefficients are bounded in absolute value since $\mu_0^*(df)$ extends smoothly to $\overline{A_{P,\,t}} \times \omega$. This is then *a fortiori* true if df is expressed as linear combination of the ω^i and $d \log \alpha_i$. Thus f and df have logarithmic growth near the boundary; then, so have $f.\sigma$ and $d(f.\sigma)$ for $\sigma \in \mathscr{F}(V)$, V open in $\overline{X/\Gamma}$.

Let $\mathscr{U} = (U_i)_{i \in I}$ be a finite open cover of $\overline{X/\Gamma}$. We choose a smooth partition of unity (λ_i) subordinated to $\mathscr{U}$. Let $\sigma \in C$. We have then $\sigma = \sum \lambda_i \sigma$, with $\mathrm{supp}(\lambda_i \sigma) \subset U_i$, and $\lambda_i \sigma \in C$ by the above, hence $\mathscr{F}$ is fine.

F IS A RESOLUTION OF $\mathbf{R}$. — Let $\mathscr{H}^*(\mathscr{F})$ be the derived sheaf of $\mathscr{F}$ and $\mathscr{H}_x^*(\mathscr{F})$ its stalk at $x \in \overline{X/\Gamma}$. Since C contains the constant functions, it is clear that

$$\mathscr{H}^0(\mathscr{F}) \cong (\overline{X/\Gamma}) \times \mathbf{R}.$$

There remains to show that $\mathscr{H}_x^q(\mathscr{F}) = 0$ for all $x \in \overline{X/\Gamma}$ and $q > 0$. If $x \in X/\Gamma$, this is just the Poincaré lemma. Let now $x \in \partial\overline{X/\Gamma}$ and $P \in \mathfrak{P}$ be such that $x \in e'(P)$. We have to prove that if V is an (arbitrarily small) special neighborhood of x and σ a closed form in $\mathscr{F}(V)$, then there exists a special neighborhood V' of x in V and $\varphi \in \mathscr{F}(V')$ such that $\sigma = d\varphi$ on V'.

We let $\mathfrak{S} = V \cap (X/\Gamma) = \mathfrak{S}_{t,\,\omega}$ and use the notation of 7.2. We may assume that ω is an open ball in a coordinate patch of $Z \times U(\mathbf{R})$, with local coordinates $(x^i)_{s < i \leq m}$ centered on a point q_0 mapping onto x. Fix t_0 $(0 < t_0 < t)$. On A_P we takes as coordinates

$$x^i = \log a^{\alpha_i} - \log t_0.$$

In these coordinates, $U_t = \mu_0^{-1}(\mathfrak{S})$ is then given by

$$U_t = \{ x \mid x^i < \log t/t_0, \ (1 \leq i \leq s); \ |x^j| < 1, \ (s < j \leq m) \}.$$

In particular, U_t is star-shaped with respect to the origin. The $dx^j \ (s < j \leq m)$ are linear combinations with bounded coefficients of the $\omega^j \ (s < j \leq m)$ on ω, and conversely; moreover $dx^i = \omega^i$ for $1 \leq i \leq s$. Therefore, if

$$(2) \qquad \varphi = \sum_J c_J \omega^J = \sum d_J \, dx^J$$

is a q-form on U_t, then φ has logarithmic growth if and only if there exists a polynomial $Q(x^1, \ldots, x^s)$ such that

$$(3) \qquad |d_J(x)| \prec |Q(x)| \qquad (x \in U_t; \text{ all } J).$$

Let now

$$(4) \qquad \mu_0^*(\sigma) = \tau = \sum f_J \omega^J = \sum h_J \, dx^J$$

be a closed q-form on U_t $(q > 0)$. In view of the remark just made and of 7.3, it suffices to prove that on some set $A_{P, \, t'} \times \omega'$, where $0 < t' < t$ and ω' is an open neighborhood of q_0 in ω, there exists a $(q-1)$-form

$$\varphi = \sum c_I \, dx^I,$$

whose coefficients satisfy a polynomial growth condition, such that $\tau = d\varphi$. To check this, it is enough to see that the usual homotopy operator A in euclidean space is compatible with polynomial growth conditions, which is immediate from its definition. We recall it briefly.

On the form $f \, dx^J$ $(J = \{ j_1, \ldots, j_q \})$, the operator A is defined by

$$A(f \, dx^J) = \sum_{I \subset J} c_I \, dx^I,$$

where, for I equal to J with j_i erased :

$$(5) \qquad c_I = (-1)^{i-1} x^{j_i} \int_0^1 f(xt) t^{q-1} \, dt.$$

It leaves Ω_U^* stable if U is open, star-shaped with respect to the origin, and satisfies $d\,A + A\,d = \mathrm{Id}$. Since τ is closed, we have then $\tau = d\varphi$, with

$$(6) \qquad \varphi = A\tau = \sum c_I \, dx^I,$$

$$(7) \qquad c_I(x) = \sum_{j \notin I} \pm x^j \int_0^1 h_{I \cup \{j\}}(xt) t^{q-1} \, dt,$$

which clearly satisfies a polynomial growth condition in $U_{t'}$ $(0 < t' < t_0)$ if the h_J do so.

This proves that $\mathscr{F}$ is a fine resolution of $\mathbf{R}$. By one of the main theorems of sheaf theory ([11], 4.6.1) we have then a canonical isomorphism

$$(8) \qquad \alpha : \ H^*(C) \to H^*(\overline{X/\Gamma}).$$

We now consider the diagram

(9)
$$\begin{array}{ccc} H^*(C) & \xrightarrow{\alpha} & H^*(\overline{X}/\Gamma) \\ \downarrow{i^*} & & \downarrow{r^*} \\ H^*(\Omega^*_{X/\Gamma}) & \xrightarrow{\beta} & H^*(X/\Gamma) \end{array}$$

where β is the isomorphism of the de Rham theorem, i^* is defined by inclusion and r^* by restriction. The isomorphisms of (1) carry the inclusion $C \to \Omega^*_{X/\Gamma}$ over to the restriction map of cross-sections. Therefore, the diagram (9) is commutative. The maps α and β are isomorphisms; since $X/\Gamma \hookrightarrow \overline{X}/\Gamma$ is a homotopy equivalence, so is r^*. Consequently i^* is an isomorphism, which ends the proof of (a) and of the theorem.

7.5. THEOREM. — *Let G be a semi-simple $\mathbf{Q}$-group and Γ an arithmetic subgroup of G. The homomorphism $j^q : I^\Gamma_G \to H^q(\Gamma)$ is injective for $q \leqq c(G)$, and surjective for $q \leqq \min(c(G), m(G(\mathbf{R})))$.*

Let Γ' be a normal torsion-free subgroup of finite index of Γ, contained in G^0. Let C' be the complex of elements in $\Omega^{\Gamma'}_X$ which have logarithmic growth at the boundary. Since Γ/Γ' acts as a group of automorphisms of $\overline{X}/\Gamma'$, it is clear that C' is stable under the natural action of Γ/Γ' on $\Omega^{\Gamma'}_X$. The theorem then follows from 7.4 and 3.6, and this also shows :

7.6. COROLLARY. — *For $q \leqq c(G)$, any cohomology class of $H^q(\Omega^\Gamma_X) = H^q(\Gamma)$ is representable by a closed q-form which is square integrable modulo Γ. More precisely the real cohomology of Γ is that of a subcomplex C of Ω^Γ_X all of whose elements of degree $q \leqq c(G)$ are square integrable modulo Γ.*

Remark. — The square integrability criterion given by 7.6 is quite analogous to, but slightly weaker than one announced by H. Garland-W. C. Hsiang [10]. For scalar valued forms, their bound is the maximum of q for which $\rho - \nu \gg 0$ when ν runs through *some* of the weights of A in $\otimes_{i \leqq q} \Lambda^i L(U(\mathbf{R}))$.

7.7. EXAMPLE. — Let k be an algebraic number field, $n = r_1 + 2r_2$ its degree, where r_1 (resp. r_2) is the number of real (resp. complex) places of k. Let $G = R_{k/\mathbf{Q}} \mathbf{SL}_2$, and P a minimal parabolic $\mathbf{Q}$-subgroup. Then Φ_P consists of one root α with multiplicity n, and $\rho_P = (r_2 + (r_1/2))\alpha$. We get therefore

$$c(G) = \begin{cases} [r_1/2 + r_2] & \text{if } r_1 \text{ is odd,} \\ (r_1/2) + r_2 - 1 & \text{if } r_1 \text{ is even.} \end{cases}$$

Comparison with the results of G. Harder [13] shows that this bound for square integrability is sharp if $r_1 \leqq 1$, but not otherwise.

8. Another subcomplex of Ω^Γ_X

8.1. We keep the assumptions of 7.4. Let V be an open subset of $\overline{X}/\Gamma$. A differential form σ on $V \cap (X/\Gamma)$ is said to be locally lifted from the boundary if for any $x \in V \cap (\partial\overline{X}/\Gamma)$

and $P \in \mathfrak{P}$ such that $x \in e'(P)$, there exists a special neighborhood U of x in V and a differential form σ' on $U' = \mathrm{pr}_P\, U$ such that $\sigma = (\mathrm{pr}_P)^*\, \sigma'$ on $U \cap (X/\Gamma)$. If, in the notation of 7.2, we write

$$\mu_0^*(\sigma) = \sum f_J \omega^J,$$

then $f_J\,(a, q)$ is independent of a, and is identically zero if J contains an index $i \leq s$. In particular, the coefficients f_J are bounded in absolute value, and σ has *a fortiori* logarithmic growth.

The forms on V of this type form a subalgebra closed under exterior differentiation, whose elements, together with their exterior differentials, have logarithmic growth near the boundary.

8.2. PROPOSITION. — *Let C_0 be the subcomplex of elements of Ω_X^Γ which are locally lifted from the boundary. Then*

(a) *The inclusion map $C_0 \to \Omega_X^\Gamma$ induces an isomorphism in cohomology.*

(b) C_0 *is contained in the complex C of 7.2. In particular C^q consists of square integrable forms* mod Γ *for $q \leq c\,(G)$.*

The first assertion of (b) follows from 8.1, and the second then from 7.4.

The proof of (a) is quite similar to, only a bit simpler than, that of 7.4 (a). For V open in $\overline{X/\Gamma}$, let $\mathscr{F}\,(V)$ be the space of forms on $V \cap (X/\Gamma)$ which are locally lifted from the boundary. The functor $V \mapsto \mathscr{F}\,(V)$ again defines a differential sheaf of algebras on $\overline{X/\Gamma}$, and we have

(1) $$\mathscr{F}\,(\overline{X/\Gamma}) = C_0, \qquad \mathscr{F}\,(X/\Gamma) = \Omega^*_{X/\Gamma}.$$

As in 7.4, it suffices then to show that $\mathscr{F}$ is a fine resolution of **R**.

$\mathscr{F}$ IS FINE. — Let $\mathscr{U} = (U)_{i \in I}$ be a finite open cover of $\overline{X/\Gamma}$ and $\sigma \in C_0$. We have to show that we can write

$$\sigma = \sum_{i \in I} \sigma_i \quad \text{with} \quad \sigma_i \in C_0 \quad \text{and} \quad \mathrm{supp}\,(\sigma_i) \subset U_i \quad (i \in I).$$

By 6.4 there exists a partition of unity $(\lambda_i)_{i \in I}$ which is subordinated to U and is *special*. But then $\lambda_i \in C_0^0$, and $\sigma_i = \lambda_i\, \sigma$ satisfies our conditions.

$\mathscr{F}$ IS A RESOLUTION OF **R**. — Let $\mathscr{H}^*_x\,(\mathscr{F})$ be the derived sheaf of $\mathscr{F}$ and $\mathscr{H}^*_x\,(\mathscr{F})$ its stalk at $x \in \overline{X/\Gamma}$. Since C_0 contains the constant functions, we have $\mathscr{H}^0\,(\mathscr{F}) \cong \overline{X/\Gamma} \times \mathbf{R}$. There remains to show that $H^q_x\,(\mathscr{F}) = 0$ for $x \in \overline{X/\Gamma}$ and $q > 0$. If $x \in X/\Gamma$, this is just Poincaré lemma. Let now x be on the boundary and $P \in \mathfrak{P}$ be such that $x \in e'(P)$. It suffices to show that if W is a special neighborhood of x, and σ is a closed q-form on W which can be written $\sigma = (\mathrm{pr}_P)^*\,(\sigma')$, with $\sigma' \in \Omega^q_{W'}\,(W' = \mathrm{pr}_P\, W)$, necessarily closed, then σ is the coboundary near x of a similar form. This follows from the Poincaré lemma on W'.

Remark. — The complex C_0 gives the same square integrability criterion as C, and is more natural from the differential geometric point of view. Its drawback for our purposes is that it does not contain I_G^Γ.

9. The constants c (G) and m (G (R))

We content ourselves with rough estimates, sufficient for the needs of paragraph 11. Only 9.5 (3) will be used there.

9.1.   Let Φ be a root system ([6], VI, § 1) in a real vector space V. Assume first V to be irreducible, and let Φ^0 be the subsystem consisting of the non-multipliable elements of Φ (i. e. $a \in \Phi$, $m\, a \in \Phi$, $m \in \mathbf{Z} \Rightarrow m = \pm 1$). It is an irreducible reduced root system. Fix an order on Φ and let Δ be the corresponding basis of simple roots of Φ. Any $v \in$ V can be written uniquely as a linear combination $v = \sum\limits_{\alpha \in \Delta} c_\alpha\, \alpha$. We put $v \gg 0$ if $c_\alpha > 0$ for all $\alpha \in \Delta$, $v \geq 0$ if $c_\alpha \geq 0$ for all $\alpha \in \Delta$, and $v \geq v'$ $(v' \in$ V) if $v - v' \geq 0$.

Let $2\,r$ be the sum of the positive roots of Φ and d_0 the highest root of Φ^0. Let

$$(1) \qquad\qquad c(\Phi) = \max p \,\big|\, r - p\, d_0 \gg 0.$$

If Φ is not irreducible, it is a sum of irreducible ones, and we put

$$(2) \qquad\qquad c(\Phi) = \min c(\Phi')$$

where Φ' runs through the irreducible factors of Φ.

The value of c (Φ) for the various irreducible root systems can be readily computed from the tables in [6], p. 250-270. The result is for the classical systems

$$(3) \qquad \begin{array}{cccccc} \mathbf{A}_n & \mathbf{B}_n & \mathbf{C}_n & \mathbf{D}_4 & \mathbf{D}_n \;\; (n \geq 5) & \mathbf{BC}_n \\[2pt] [n/2]' & n-2 & [n/2]' & 3 & n & [(n+1)/2]' \end{array}$$

while for the exceptional types $\mathbf{E}_6, \mathbf{E}_7, \mathbf{E}_8, \mathbf{G}_2, \mathbf{F}_4$, c (Φ) is equal to 4, 8, 14, 3, 1 respectively.

9.2.   Let k be a field, H a connected semi-simple k-group of strictly positive k-rank ([5], § 5), S a maximal k-split torus of H and P a minimal parabolic k-subgroup containing S. Replacing X (A_P) by X (S) $\otimes$ R, we define, in analogy with 7.1 :

$$(1) \qquad\qquad c(\mathrm{H}/k) = \max q \,\big|\, \rho_P - v \gg 0,$$

where v runs through the weights of S in $\bigotimes\limits_{i \leq q} \Lambda^i\, \mathrm{L}\, (\mathrm{R}_u\, (\mathrm{P}))$.

Replacing P by a parabolic k-subgroup Q, S by a maximal k-split torus of the radical of Q, we can also consider the condition c (Q, q, 0); again, the inequality

$$(2) \qquad\qquad c(\mathrm{H}/k) \leq \max q \,\big|\, c(\mathrm{Q}, q, 0)$$

holds true.

9.3.   The relative root system $_k\Phi$ (H) of H, also to be denoted Φ (H/k), decomposes into the direct sum of the relative root systems of the almost k-simple factors of H whose k-rank is > 0 and those are irreducible ([5], § 5). We claim

$$(1) \qquad\qquad c(\mathrm{H}/k) \geq c(\Phi(\mathrm{H}/k)).$$

In view of the definitions, it suffices to prove this for H almost simple over k. In this case $\rho - r \geq 0$ because 2ρ contains each term occurring in $2r$ with a coefficient ≥ 1, and, on the other hand, $p\,d_0 - v \geq 0$ for any sum v of p roots. More precisely, the coefficient α occurs in 2ρ a number of times equal to its multiplicity, hence if all the multiplicities are $\geq m$, we have

$$\rho - pd_0 \geq mr + pd_0 \geq m(r - pd_0),$$

and therefore

$$(2) \qquad c(H/k) \geq mc(\Phi(H/k)), \qquad (1 \leq m \leq \min_\alpha \text{ multiplicity } \alpha).$$

9.1 (3) and 9.3 (1) imply in particular that if H is almost simple over k, then

$$(3) \qquad c(H/k) \geq [rk_k(H)/2]'.$$

If k' is a finite separable extension and $H = R_{k'/k} H'$, where H' is an almost k'-simple k'-group, then

$$(4) \qquad c(H/k) \geq [k' : k]\, c(H'/k').$$

In fact, by [5], 6.21, the root systems $\Phi(H'/k')$ and $\varphi(H/k)$ are canonically isomorphic and the passage from the former to the latter multiplies the multiplicities by $[k' : k]$.

We note also that if k'' is any extension of k, then

$$(5) \qquad c(H/k) \geq c(H/k''),$$

because, if P is a minimal parabolic k-subgroup of H, we have, using 9.2 (2) :

$$c(H/k) = c(H/k, P) \geq c(H/k'').$$

9.4. Let now $k = \mathbf{R}$ and H be almost simple over $\mathbf{R}$, of strictly positive $\mathbf{R}$-rank. The tables of [16] and [19] then show immediately that the constant $m(H(\mathbf{R}))$ defined in 3.3 satisfies

$$(1) \qquad m(H(\mathbf{R})) \geq [(rk_{\mathbf{R}} H)/4]'.$$

9.5. Let now G be a connected almost $\mathbf{Q}$-simple $\mathbf{Q}$-group. Up to isogeny, which does not affect the constants in consideration here, $G = R_{k/\mathbf{Q}} G'$, where k is a finite extension of $\mathbf{Q}$ and G' is an absolutely simple k-group. We have then ([23], 1.3.1) :

$$(1) \qquad G/\mathbf{R} = \prod_{v \in S(k)} R_{k_v/\mathbf{R}}(G'/k_v),$$

where $S(k)$ denotes the set of infinite places of k, and k_v the completion of k at $v \in S(k)$, whence also

$$(2) \qquad G'(\mathbf{R}) = \prod_{v \in S(k)} G'(k_v) = \prod_{v \in S(k)} (R_{k_v/\mathbf{R}} G')(\mathbf{R}).$$

The k_v-rank of (G'/k_v) is at least equal to the k-rank of G', i.e. to the $\mathbf{Q}$-rank of G. Therefore, 9.3 (3) and 9.4 imply

$$(3) \qquad \min(c(G/\mathbf{Q}), m(G(\mathbf{R}))) \geq [(rk_m G)/4]'.$$

10. Some sequences of compact homogeneous spaces

In this section, we review some known facts about the projective limits of certain sequences of cohomology rings of compact symmetric spaces.

10.1. Given a sequence of graded modules A_n $(n \geq 1)$ over a ring R and of homomorphisms $f_n : A_{n+1} \to A_n$ preserving the degrees, we denote by $\varprojlim{}^* A_n$ the graded module $A = \otimes_j A^j$, where $A^j = \lim (A_n^j, f_n)$. If the A_n are graded algebras, and the $f_n' s$ are graded algebra homomorphisms, then A is an algebra in the obvious way. $\varprojlim{}^*$ is thus the projective limit in the graded category; the ordinary $\varprojlim$ would be the direct product of the $A^{j'} s$ rather than their direct sum. In all cases to be considered here, $R = \mathbf{R}, \mathbf{C}$ the $A_n' s$ are finite dimensional and, for fixed j, the sequence (A_n^j, f_n) is stationary.

10.2. Let G be a semi-simple algebraic group over $\mathbf{R}$, and K a maximal compact subgroup of $G(\mathbf{R})$. We have already mentioned the isomorphism

$$(1) \qquad I_G = I_{G(\mathbf{R})} \cong H^*(L(G(\mathbf{R})), L(K)) \quad (3.1).$$

Although we shall not need this, we also recall that

$$(2) \qquad H^*(L(G(\mathbf{R})), L(K)) \cong \tilde{H}^*(G(\mathbf{R})^0)$$

where $\tilde{H}$ denotes the Eilenberg-Mac Lane cohomology, computed with continuous or smooth real valued cochains ([8], [15]). From that point of view, j_Γ^* [Γ discrete in $G(\mathbf{R})^0$] is just the restriction map in Eilenberg-Mac Lane cohomology.

Let G_u be a maximal compact subgroup of $G(\mathbf{C})$ containing K and $X_u = K^0 \backslash G_u^0$. The space X_u is a simply connected compact symmetric space, the " compact dual " or " compact twin " to $X = K \backslash G(\mathbf{R})$. We also have, by E. Cartan :

$$(3) \qquad H^*(L(G(\mathbf{R})), L(K)) \cong H^*(L(G_u), L(K)) \cong H^*(X_u).$$

Since $G(\mathbf{R})/G(\mathbf{R})^0 \cong K/K^0$:

$$(4) \qquad (K\backslash G_u)^0 \cong (K \cap G_u^0)\backslash G_u^0,$$

hence

$$(5) \qquad H^*((K\backslash G_u)^0) = H^*((K \cap G_u^0)\backslash G_u^0) = H^*(X_u)^{(K \cap G_u^0)/K^0}.$$

10.3. Let G' be another semi-simple $\mathbf{R}$-group and $f : G \to G'$ an injective $\mathbf{R}$-morphism. There is then a canonical morphism

$$f^* : \quad I_{G'(\mathbf{R})} \to I_{G(\mathbf{R})}.$$

In fact, we can choose a maximal compact subgroup K' of $G'(\mathbf{R})$ containing $f(K)$. Then f induces an embedding of $K\backslash G(\mathbf{R})$ into $K'\backslash G'(\mathbf{R})$ (whose image is a totally geodesic closed submanifold), and f^* is just given by the restriction of differential forms. Since any two maximal compact subgroups of G' (**R**) are conjugate by an inner automosphism leaving their intersection pointwise fixed, f^* is independent, up to unique isomorphisms, of the choices made. The homomorphism can of course also be defined in a natural way from the other interpretations of I_G given in 10.2. For example, we may take a maximal compact subgroup G'_u of G' (**C**) containing $f(G_u)$, and a conjugate K'' of K', whence an embedding $f : X_u \hookrightarrow X'_u = (K'' \backslash G'_u)^0$, and f^* is the induced map on cohomology.

10.4. In this paragraph and the next, we fix some notation for the classical groups.

$\mathbf{SL}_n$ and $\mathbf{Sp}_{2n}$ are as usual the special linear group and the symplectic group.

$\mathbf{O}_n$ [resp. $\mathbf{O}_n$ (**C**)] is the automorphism group of $\mathbf{R}^n$ (resp. $\mathbf{C}^n$) endowed with the standard quadratic form $\sum x_i^2$.

$\mathbf{O}_{n,\,n}$ is the **Q**-group whose group of k points (k field of characteristic zero) is the automorphism group of k^{2n} endowed with the quadratic form $\displaystyle\sum_{1 \le i \le n} x_i \, x_{n+i}.$

$\mathbf{U}_n$ is the unitary group on $\mathbf{C}^n$, $\mathbf{U}_{n,\,n}$ the automorphism group of $\mathbf{C}^{2n}$ endowed with the hermitian form $\displaystyle\sum_{1 \le i \le n} (x_i \, \bar{x}_{n+1} + x_{n+i} \, \bar{x}_i).$

If G is one of the above orthogonal or unitary groups, SG is the subgroup of elements of determinant one in G.

10.5. We let **H** be the field of quaternion numbers over the reals, i, j, k the standard anticommuting units of square -1 of **H**.

$\alpha : x \mapsto \bar{x}$ is the conjugation, which maps i, j, k onto $-i, -j, -k$ respectively and $\beta : x \mapsto -k \, \bar{x} \, k$ is the involution which fixes i, j and sends k onto $-k$.

$\mathbf{USp}_n$ (resp. $\mathbf{Sp}_{n,\,n}$) is the automorphism group of $\mathbf{H}^n$ (resp. $\mathbf{H}^{2n}$) endowed with the hermitian form

$$\sum_{1 \le i \le n} x_i \bar{x}_i \quad \left[\text{resp.} \sum_{1 \le i \le n} (x_i \bar{x}_{n+i} + x_{n+i} \bar{x}_i) \right].$$

$\mathbf{SO}_{2n}^*$ is the group of real points of the **R**-form of $\mathbf{SO}_{2n}$ of type D III in E. Cartan's notation. *A priori*, since this classification is one of Lie algebras, this means that the maximal compact subgroups of the identity component of $\mathbf{SO}_{2n}^*$ are isomorphic to $\mathbf{U}_n$, but, in fact, $\mathbf{SO}_{2n}^*$ is connected. To see this, it is enough to check that the non-trivial involutive automorphism of $\mathbf{SO}_{2n}$ which fixes $\mathbf{U}_n$ has $\mathbf{U}_n$ as its full fixed point set, and this is immediate. $\mathbf{SO}_{2n}^*$ may be identified with the group of automorphisms of $\mathbf{H}^n$ endowed either with the β-hermitian form

$$(1) \qquad\qquad \Phi_0(x, y) = \sum_{1 \le i \le n} x_i \beta(y_i),$$

or with the anti-hermitian form

$$(2) \qquad\qquad \Phi'_0(x, y) = \sum_{1 \le i \le n} x_i k \bar{y}_i.$$

We recall that α is the unique involution μ of **H** such that $\mu(x) \in \mathbf{R} \, x^{-1}$ ($x \in \mathbf{H}$, $x \ne 0$), and that any other involution is conjugate to β by an inner automorphism. Moreover, any non-degenerate β-hermitian (resp. α-antihermitian) form on $\mathbf{H}^n$ is equivalent to Φ_0 (resp. Φ_0'). If $\varepsilon = \pm 1$, and Φ is an ε-α-hermitian form on $\mathbf{H}^n$, then $\Phi'(x, y) = \Phi(x, y).k$ is $-\varepsilon$-β-hermitian.

10.6. We now consider sequences of non-compact classical real Lie groups (H_n, f_n) where $f_n : \mathrm{H}_n \to \mathrm{H}_{n+1}$ is injective. In the following table we give the n-th term of the sequence, the corresponding compact twin to the symmetric space of maximal compact subgroups of H_n, and the value of $\lim^* \mathrm{I}_{\mathrm{H}_n}$. In all cases, f_n is the obvious map.

$\mathrm{E}(\{\, x_i, d^0 x_i \,\})$ [resp. $\mathrm{P}(\{\, x_i, d^0 x_i \,\})$] is the graded exterior (resp. polynomial) algebra over $\mathbf{R}$, with generators x_i $(i = 1, 2, \ldots)$, where $d^0 x_i$ is the degree of x_i.

H_n	$\mathrm{X}_{n,u}$	$\overset{\longleftarrow}{\lim^*}\ \mathrm{I}_{\mathrm{H}_n}$
$\mathrm{SL}_n(\mathbf{C})$	SU_n	$\mathrm{E}(\{\, x_i, 2i+1 \,\})$
$\mathrm{Sp}_{2n}(\mathbf{C})$	USp_n	$\mathrm{E}(\{\, x_i, 4i-1 \,\})$
$\mathrm{SO}_n(\mathbf{C})$	SO_n	$\mathrm{E}(\{\, x_i, 4i-1 \,\})$
$\mathrm{SL}_n(\mathbf{R})$	$\mathrm{SU}_n/\mathrm{SO}_n$	$\mathrm{E}(\{\, x_i, 4i+1 \,\})$
$\mathrm{SL}_n(\mathbf{H})$	$\mathrm{SU}_{2n}/\mathrm{USp}_n$	$\mathrm{E}(\{\, x_i, 4i+1 \,\})$
$\mathrm{Sp}_{2n}(\mathbf{R})$	$\mathrm{USp}_n/\mathrm{U}_n$	$\mathrm{P}(\{\, x_i, 4i-2 \,\})$
$\mathrm{SO}_{n,n}(\mathbf{R})$	$\mathrm{SO}_{2n}/((\mathrm{O}_n \times \mathrm{O}_n) \cap \mathrm{SO}_{2n})$	$\mathrm{P}(\{\, x_i, 4i \,\})$
$\mathrm{O}_{n,n}(\mathbf{R})$	$\mathrm{SO}_{2n}/((\mathrm{O}_n \times \mathrm{O}_n) \cap \mathrm{SO}_{2n})$	$\mathrm{P}(\{\, x_i, 4i \,\})$
$\mathrm{U}_{n,n}$	$\mathrm{U}_{2n}/\mathrm{U}_n \times \mathrm{U}_n$	$\mathrm{P}(\{\, x_i, 2i \,\})$
$\mathrm{SU}_{n,n}$	$\mathrm{SU}_{2n}/((\mathrm{U}_n \times \mathrm{U}_n) \cap \mathrm{SU}_{2n})$	$\mathrm{P}(\{\, x_i, 2i \,\})$
$\mathrm{Sp}_{n,n}$	$\mathrm{USp}_{2n}/\mathrm{USp}_n \times \mathrm{USp}_n$	$\mathrm{P}(\{\, x_i, 4i \,\})$
SO_{2n}^*	$\mathrm{SO}_{2n}/\mathrm{U}_n$	$\mathrm{P}(\{\, x_i, 4i-2 \,\})$

These spaces all occur in Bott periodicity, and the $\overset{\longleftarrow}{\lim^*}$ on the right hand side are computed in [7], in particular in Exp. 16, by H. Cartan. More precisely, for given series $(\mathrm{X}_{n,\,u})$, let $\mathrm{X} = \lim \mathrm{X}_{n,\,u}$ be the inductive limit of the $\mathrm{X}_{n,\,u}$, endowed with the inductive limit topology. Then

$$(1) \qquad \mathrm{H}^*(\mathrm{X}) \cong \overset{\longleftarrow}{\lim^*}\ \mathrm{H}^*(\mathrm{X}_{n,\,u}) \cong \overset{\longleftarrow}{\lim}{}^*\mathrm{I}_{\mathrm{H}_n}$$

is determined in [7]. In fact, it is shown that, in a given dimension, the sequence is stationary, i. e., given j, there exists $n(j)$ such that

$$(2) \qquad \mathrm{I}_{\mathrm{H}_m}^j = \mathrm{H}^j(\mathrm{X}_{m,\,u}) \cong \mathrm{H}^j(\mathrm{X}) = \overset{\longleftarrow}{\lim}{}^*\mathrm{I}_{\mathrm{H}_n}^j \qquad [m \geq n(j)].$$

[In particular, there is no problem about the first equality in (1).] The space X is a « weak » H-space ([7], Exp. 11), which explains why its real cohomology algebra is free anticommutative. The homotopy groups of X are finitely generated, periodic, of period dividing 8, and we have

$$(3) \qquad \pi_j(\mathrm{X}) \otimes \mathbf{R} \cong (\mathrm{Im}\,\pi_j(\mathrm{X})) \otimes \mathbf{R} \subset \mathrm{H}_j(\mathrm{X}),$$

under the Hurewicz map. The second term is the space $P_j(X)$ of primitive elements (for the diagonal map) in $H_j(X)$. It is the dual to the quotient of $H^j(X)$ by the space of decomposable elements, the space of indecomposable elements of degree j. As the above table shows, its dimension is periodic, of period 2 or 4.

11. Stable cohomology of classical arithmetic groups

11.1. THEOREM. — *Let G_n be a semi-simple $\mathbf{Q}$-group, Γ_n an arithmetic subgroup of G_n, and $f_n : G_n \to G_{n+1}$ an injective morphism which maps Γ_n into Γ_{n+1} ($i = 1, 2, \ldots$). Assume: (i) Given $q \in \mathbf{N}$, there exists $n(q)$ such that $(I_{G_n}^q)^{\Gamma_n} = I_{G_n}^q$ for $n \geq n(q)$; (ii) $m(G_n(\mathbf{R}))$ and $c(G_n)$ tend to infinity with n. Then*

$$(1) \qquad H^*(\varinjlim \Gamma_n) = \varprojlim{}^* H^*(\Gamma_n) = \varprojlim{}^* I_{G_n(\mathbf{R})}.$$

By 7.5, for every n :

$$j^q : \quad (I_{G_n}^q)^{\Gamma_n} \to H^q(\Gamma_n)$$

is an isomorphism for $q \leq \min(c(G_n), m(G_n(\mathbf{R})))$. Therefore, given q, j_n^q yields an isomorphism of $I_{G_n}^q$ onto $H^q(\Gamma_n)$ for every big enough n. Since the j_n^q are compatible with the homomorphisms f_n^* defining the two $\varprojlim^*$, this proves the second equality of (1).

We have obviously $H_*(\varinjlim \Gamma_n) = \varinjlim H_*(\Gamma_n)$. The first equality follows then from the fact that the dual of a countable inductive limit of finite dimensional vector spaces is the projective limit of the dual spaces. In fact, in all examples below, the sequences are stationary in each dimension and the $\varprojlim^*$ will be finite dimensional in each degree.

11.2. REMARKS. — The condition (i) is of course satisfied if $\Gamma_n \subset G_n(\mathbf{R})^0$. In that respect, we recall that $G_n(\mathbf{R})$ is connected (as a Lie group), if G_n is connected and simply connected as an algebraic group; in the present case, the latter condition is equivalent to : $G_n(\mathbf{C})$ or $G_{n,u}$ is connected and simply connected.

By 9.5, (ii) is fulfilled if $rk_{\mathbf{Q}}(G_n) \to \infty$.

11.3. Let k be a number field, S the set of archimedian places of k, $d = r_1 + 2r_2$ the degree of k, where r_1 (resp. r_2) is the number of real (resp. complex) places of k.

Let G'_n be an absolutely almost simple k-group, Γ'_n an arithmetic subgroup of G'_n, and $f'_n : G'_n \to G'_{n+1}$ an injective k-morphism which maps Γ'_n into Γ'_{n+1}. Assume that $rk_k G'_n \to \infty$. Then

$$(1) \qquad G_n = R_{k/\mathbf{Q}} G'_n, \qquad f_n = R_{k/\mathbf{Q}} f'_n,$$

and the image Γ_n of Γ'_n under the canonical isomorphism $G'_n(k) \xrightarrow{\sim} G_n(\mathbf{Q})$ satisfy the assumptions of 11.1 [except possibly for (i)].

For $v \in S$, let k_v be the completion of k at v. Thus $k_v = \mathbf{R}$ or $\mathbf{C}$ depending on whether v is real or complex. Let $G'_{n,v}$ be the k_v-group obtained from G'_n by extension of the ground-field $k \to k_v$. We have then ([23], Th. 1.3.1) :

$$(2) \qquad G_n(\mathbf{R}) = \prod_{v \in S} G'_{n,v}(k_v).$$

If v is complex, then $G'_{n,v}(k_v) = G'_n(\mathbf{C})$. If v is real, $G'_{n,v}(k_v)$ is a real form of G' which may depend on v. At any rate, the symmetric space of maximal compact subgroups of $G_n(\mathbf{R})$ is the product of the corresponding symmetric spaces for the groups $G'_{n,v}(k_v)$ and $f_n : G_n(\mathbf{R}) \to G_{n+1}(\mathbf{R})$ is also compatible with that decomposition. As a result

$$(3) \qquad \lim{}^* I_{G_n(\mathbf{R})} = \bigotimes_{v \in S} I_v,$$

where

$$(4) \qquad \begin{cases} I_v = \lim{}^* I_{G'_{n,v}(\mathbf{R})}, & (v \text{ real}), \\ I_v = \lim{}^* I_{G'_n(\mathbf{C})} & (v \text{ complex}). \end{cases}$$

If, in addition, (i) is fulfilled (and this is a rather harmless condition in practice), then we have

$$(5) \qquad H^*(\lim \Gamma'_n) = \lim{}^* H^*(\Gamma'_n) = \bigotimes_{v \in S} I_v.$$

We now proceed to give a series of examples. In all of them, the sequences $G'_{n,v}(k_v)$ are among those listed in 10.5, and, in a given degree, the corresponding sequence of cohomology spaces is always stationary. Thus we have more precisely, given $q \in \mathbf{N}$:

$$(6) \qquad H^q(\Gamma'_n) = \left(\bigotimes_{v \in S} I_v \right)^q \text{ for } n \text{ large enough.}$$

11.4. If $G'_n = \mathbf{SL}_n/k$, $\mathbf{Sp}_{2n}/k$ or $\mathbf{O}_{n,n}/k$, then I_v is given by the following table

G'_n	I_v (v real)	I_v (v complex)
$\mathbf{SL}_n/k$	$E(\{x_i, 4i+1\})$	$E(\{x_i, 2i+1\})$
$\mathbf{Sp}_{2n}/k$	$P(\{x_i, 4i-2\})$	$E(\{x_i, 4i-1\})$
$\mathbf{O}_{n,n}/k$	$P(\{x_i, 4i\})$	$E(\{x_i, 4i-1\})$

In the first two cases, $G_n(\mathbf{R})$ is connected (see 11.2), 11.1 (i) is always fulfilled, and 11.3 (5), (6) hold. As group Γ'_n, we may in particular take $\mathbf{SL}_n(\mathfrak{o})$ or $\mathbf{Sp}_{2n}(\mathfrak{o})$, where $\mathfrak{o}$ is the ring of integers of k.

Let now $G'_n = \mathbf{O}_{n,n}/k$. Then $G'_{n,v}(k_v) = \mathbf{O}_{n,n}(\mathbf{R})$, $\mathbf{O}_{n,n}(\mathbf{C})$, depending on whether v is real or complex. It is not connected, and in fact for v real, the $2n$-form corresponding to the Euler class, which is in $I_{\mathbf{SO}_{n,n}}$, is not invariant under $\mathbf{O}_{n,n}(\mathbf{R})$. However, it disappears in the stable range, so that (i) is also fulfilled for any choice of Γ'_n, and we again have 11.3 (5), (6).

11.5. Let D be a central division algebra over k, and s^2 its degree over k. Then $D \otimes\limits_k k_v$ is a matrix algebra over a division algebra L_v with center k_v. There are three possibilities

$$(1) \quad \begin{cases} (a) & k_v = \mathbf{C}, & L_v = \mathbf{C}, & D \otimes_k k_v = \mathbf{M}_s(\mathbf{C}); \\ (c) & k_v = \mathbf{R}, & L_v = \mathbf{R}, & D \otimes_k k_v = \mathbf{M}_s(\mathbf{R}); \\ (d) & k_v = \mathbf{R}, & L_v = \mathbf{H}, & D \otimes_k k_v = \mathbf{M}_{s/2}(\mathbf{H}). \end{cases}$$

Take $G'_n = \mathbf{SL}_n(D)$, and for Γ'_n the arithmetic subgroup consisting of elements with entries in some fixed order of D. We have then

$$(2) \qquad G'_{n,\,v}(k_v) = \mathbf{SL}_{ns}(\mathbf{C}), \quad \mathbf{SL}_{ns}(\mathbf{R}), \quad \mathbf{SL}_{ns/2}(\mathbf{H})$$

corresponding to the cases (a), (b) and (c). Since G'_n is simply connected, (i) is fulfilled and 11.5 obtains. Moreover, we see from 10.6 that the factors I_v associated to the cases (b) and (c) are the same. Hence

$$(3) \qquad H^*(\lim_{\longrightarrow} \Gamma'_n) = \overset{r_1}{\otimes} E(\{x_i,\, 4i+1\}) \otimes \overset{r_2}{\otimes} E(\{x_i,\, 2i+1\}).$$

In particular, we always have

$$(4) \qquad H^2(\lim_{\longrightarrow} \Gamma_n) = 0.$$

In fact, since $rk_\mathbf{Q} G_n = n-1$ in this case, 11.1 gives more precisely

$$H^2(\Gamma_n) = 0 \qquad \text{for} \quad n > 9.$$

This applies in particular to the case where $D = k$, $\Gamma_n = \mathbf{SL}_n(\mathfrak{o})$, and we get Garland's result [9].

11.6. The following examples will be associated to algebras with involution and we first fix some notation.

D is a division algebra over k, and j an involution of D over k. Let k' be the center of D. If j is of the first kind, i. e. is trivial on k', then we assume $k' = k$. It is known that D is then a quaternion algebra over k.

If j is of the second kind, i. e. non trivial on k', then k' is a quadratic extension of k, and j restricts to k' to the non-trivial automorphism of k' over k.

Let $\varepsilon = \pm 1$. On D^{2n}, we consider the ε-j-hermitian form

$$(1) \qquad {}_\varepsilon\Phi(x,\, y) = \sum_{1 \leq i \leq n} (x_i j(y_{n+i}) + \varepsilon\, x_{n+1} j(y_i)).$$

There is a k-group ${}_\varepsilon\mathbf{O}_{n,n,D}$ whose group of points over D is the automorphism group of D^{2n} endowed with the form (1). We let ${}_\varepsilon G'_n$ be the derived group of the identity component of ${}_\varepsilon\mathbf{O}_{n,n,D}$ and Γ_n be the arithmetic subgroup consisting of the elements of ${}_\varepsilon G'_n$ with coefficients in some order $\mathfrak{o}$ of D.

The groups $_\varepsilon G'_n$, together with those considered in 11.4, are k-forms of the classical groups. The isomorphism class of $_\varepsilon G'_{n,\,v}(k_v)$ depends only on that of $(D \underset{k}{\otimes} k_v, j)$. It will be described without further justification, but the latter can be easily extracted from [18], Chapter II, and from the elementary remarks on quaternions made in 10.5 and 11.7.

11.7. We now consider the case where D is a quaternion algebra with center k. Assume first that j is the " standard " involution, such that $x + j(x)$ and $x\, j(x)$ are respectively the reduced trace and the reduced norm of x $(x \in D)$. Let us denote by t' the involution of $\mathbf{M}_2(F)$, (F any field) which is given by

$$(1) \qquad\qquad t'(x) = s\, {}^t x\, s^{-1} \quad \text{with} \quad s = \begin{pmatrix} 0 & -1 \\ 1 & 0 \end{pmatrix}.$$

There are, up to isomorphism, three possibilities for $(D \underset{k}{\otimes} k_v, j)$:

$$(2) \qquad \begin{cases} (a) & v \text{ complex,} & (D \underset{k}{\otimes} k_v, j) \cong (\mathbf{M}_2(\mathbf{C}), t'); \\[2mm] (b) & v \text{ real} & (D \underset{k}{\otimes} k_v, j) \cong (\mathbf{M}_2(\mathbf{R}), t'); \\[2mm] (c) & v \text{ real} & (D \underset{k}{\otimes} k_v, j) \cong (\mathbf{H}, \alpha). \end{cases}$$

The following table gives the corresponding values of $_\varepsilon G'_{n,\,v}(k_v)$ and I_v :

	(a)	(b)	(c)
$_1G'_{n,\,v}(k_v)$.............	$\mathbf{Sp}_{4n}(\mathbf{C})$	$\mathbf{Sp}_{4n}(\mathbf{R})$	$\mathbf{Sp}_{n,\,n}$
I_v...................	$E(\{x_i, 4i-1\})$	$P(\{x_i, 4i-2\})$	$P(\{x_i, 4i\})$
$_{-1}G'_{n,\,v}(k_v)$...........	$\mathbf{SO}_{4n}(\mathbf{C})$	$\mathbf{SO}_{2n,\,2n}(\mathbf{R})$	$\mathbf{SO}^*_{4n}$
I_v...................	$E(\{x_i, 4i-1\})$	$P(\{x_i, 4i\})$	$P(\{x_i, 4i-2\})$

If j is not equivalent to the standard involution, then the three possibilities for $(D \underset{k}{\otimes} k_v, j)$ are respectively

$$(3) \qquad\qquad (\mathbf{M}_2(\mathbf{C}), t), \quad (\mathbf{M}_2(\mathbf{R}), t), \quad (\mathbf{H}, \beta)$$

and we get the table corresponding to the above one just by interchanging the cases $\varepsilon = 1$ and $\varepsilon = -1$.

The groups involved are connected, except for $\mathbf{SO}_{2n,\,2n}(\mathbf{R})$ which has two components (*see* 10.5 for $\mathbf{SO}^*_{2n}$). As in 11.4, the non-connectedness of $\mathbf{SO}_{2n,\,2n}(\mathbf{R})$ does not matter in the stable range, so that 11.1 (i) is fulfilled and 11.3 (5), (6) hold for any coherent choice of the Γ'_n.

11.8. Let now j be of the second kind, and denote $x \mapsto \bar{x}$ the non-trivial automorphism of k' over k. Let m^2 be the degree of D over k'. For $(D \underset{k}{\otimes} k_v, j)$ the possibilities are the following, up to isomorphism :

(*a*) v complex. Then $k' \underset{k}{\otimes} k_v = \mathbf{C} \oplus \mathbf{C}$, the algebra $D \underset{k}{\otimes} \mathbf{C}$ is isomorphic to $\mathbf{M}_m(\mathbf{C}) \oplus \mathbf{M}_m(\mathbf{C})$, and j is $(x, y) \mapsto ({}^t y, {}^t x)$.

(b) v real, extends to two real places of k'. Then $D \otimes_k k_v = \mathbf{M}_m(\mathbf{R}) \oplus \mathbf{M}_m(\mathbf{R})$ and j is as in (a).

(c) v real, but extends to a complex place of k'. Then $k' \otimes k_v = \mathbf{C}$, j induces the complex conjugation on $\mathbf{C}$, $D \otimes k_v \cong \mathbf{M}_m(\mathbf{C})$ and $j(x) = s\,{}^t\bar{x}\,s^{-1}$, with s hermitian. We have then, for both values of ε,

	(a)	(b)	(c)
${}_\varepsilon G'_{n,v}(k_v)$................	$\mathbf{SL}_{2nm}(\mathbf{C})$	$\mathbf{SL}_{2nm}(\mathbf{R})$	$\mathbf{SU}_{nm,nm}$
I_v.....................	$E(\{\,x_i, 2\,i+1\,\})$	$E(\{\,x_i, 4\,i+1\,\})$	$E(\{\,x_i, 2\,i\,\})$

Thus G'_n is simply connected, and 11.3 (5), (6) obtain.

11.9. We have considered only " hyperbolic forms ", for convenience, and because these cases are those of interest in K-theory. But 11.1 can be applied to other sequences. As a simple example, fix a non-degenerate quadratic form F_0 on k^m, say anisotropic, let G_n be the orthogonal group of

$$F_0 + \sum_{1 \le i \le n} x_i x_{n+i}$$

in k^{m+2n}, and $\Gamma_n = G_n(\mathfrak{o})$. We have then again 11.3 (5), (6), and in fact the values of I_v are *the same* as in the $\mathbf{O}_{n,n}$ case. This is clear for v complex. For v real, we have

$$I_v = \underleftarrow{\lim}{}^* H^*(\mathbf{O}_{m+2n}(\mathbf{R})/(\mathbf{O}_{m+n}(\mathbf{R}) \times \mathbf{O}_n(\mathbf{R})).$$

But the homogeneous space on the right hand side is as good an approximation to the classifying space of $\mathbf{O}_n(\mathbf{R})$ as $\mathbf{O}_{2n}(\mathbf{R})/(\mathbf{O}_n(\mathbf{R}) \times \mathbf{O}_n(\mathbf{R}))$ hence the $\underleftarrow{\lim}{}^*$ does not depend on m.

The same is true in the hermitian case.

11.10. REMARK. — In a sequel to this paper, we shall see that 11.1 is also true (with the same range) for S-arithmetic groups. It will follow the that $H^*(\underrightarrow{\lim}\, G_n(\mathbf{Q})) = H^*(\underrightarrow{\lim}\, \Gamma_n)$ or, in the set up of 11.3, that we also have

$$(1) \qquad\qquad H^*(\underrightarrow{\lim}\, G'_n(k)) = \underleftarrow{\lim}{}^* H^*(G'_n(k)) = \bigotimes_{v \in S} I_v.$$

12. Applications to K and L-theory

12.1. Let D be a division algebra over k and $\mathfrak{o}$ an order of D. For $i \ge 2$, Quillen's group $K_i\,\mathfrak{o}$, as defined in [21], is the i-th homotopy group of an H-space which has the same homology as $\underrightarrow{\lim}\, \mathbf{SL}_n\,\mathfrak{o}$. Therefore, by the fact recalled in 10.6, $\dim(K_i \otimes \mathbf{R})$ is equal to the dimension of the space of indecomposable elements in $\underleftarrow{\lim}{}^* H^*(\mathbf{SL}_n\,\mathfrak{o})$ and 11.5 (3) implies :

12.2. PROPOSITION. — *Let $\mathfrak{o}$ be an order in a central division algebra* D *over* k. *Then, for $i \geq 2$*, dim $(\mathrm{K}_i \, \mathfrak{o} \otimes \mathbf{R})$ *is periodic of period four and is equal to* 0, $r_1 + r_2$, 0, r_2 *depending on whether $i \equiv 0, 1, 2, 3 \bmod 4$, where r_1 (resp. r_2) is the number of real (resp. complex) places of* k.

12.3. We now revert to the notation of 11.6. A result of L. I. Wasserstein (*see* [1], Chap. II, § 5, p. 156) implies that $\mathrm{M} = \lim\limits_{\longrightarrow} {}_{\varepsilon}\mathrm{G}'_n(\mathfrak{o})$ and $\lim\limits_{\longrightarrow} {}_{\varepsilon}\mathbf{O}_{n,n,\mathrm{D}}(\mathfrak{o})$ have the same derived group, which is perfect (and generated by " elementary " matrices). Quillen's +-construction, applied to the classifying space $\mathrm{B_M}$ of M, yields a space $\mathrm{B_M^+}$ with fundamental group $\mathrm{M}/(\mathrm{M}, \mathrm{M})$ and the same homology as M, and we let ${}_{\varepsilon}\mathrm{L}_i \, \mathfrak{o} = \pi_i(\mathrm{B_M^+})$ $(i \geq 2)$. These are the groups introduced and so denoted by M. Karoubi [17] (for more general rings A, except for the blanket assumption that 2 is invertible in A, which plays no role here). The space $\mathrm{B_M^+}$ is also and H-space, and as in 10.6, the dimension if ${}_{\varepsilon}\mathrm{L}_i \otimes \mathbf{R}$ is equal to the dimension of the space of indecomposable elements in $\mathrm{H}^i(\mathrm{M})$. The results of paragraph 11 then also give the dimension of ${}_{\varepsilon}\mathrm{L}_i(\mathfrak{o}) \otimes \mathbf{R}$ $(i \geq 2)$ in the various cases considered there. It is again periodic of period 4. Corresponding to 11.4, 11.7 and 11.8, we have the following cases :

(1) $\mathrm{D} = k$, $j = \mathrm{Id}$, and $\mathfrak{o}$ is the ring of integers of k. Then the period giving dim $({}_{\varepsilon}\mathrm{L}_i(\mathfrak{o}) \otimes \mathbf{R})$ starting with $i \equiv 2 \bmod 4$ is

$$\varepsilon = 1 \quad : \quad 0 \;\; r_2 \;\; r_1 \;\; 0,$$

$$\varepsilon = -1 : \quad r_1 \;\; r_2 \;\; 0 \;\; 0.$$

(2) D is quaternion algebra and j the standard involution. We let r_b (resp. r_c) be the number of real places for which (*b*) [resp. (*c*)] of 11.7 holds. The period is then,

$$\varepsilon = 1 \quad : \quad r_b \;\; r_2 \;\; r_c \;\; 0,$$

$$\varepsilon = -1 : \quad r_c \;\; r_2 \;\; r_b \;\; 0.$$

If j is not the standard involution, then we have only to interchange the cases $\varepsilon = 1$ and $\varepsilon = -1$.

(3) In the case of 11.8, let again r_b (resp. r_c) be the number of real places for which (*b*) [resp. (*c*)] holds. Let $\mathfrak{o}$ be an order of D stable under j. The period for dim $({}_{\varepsilon}\mathrm{L}_i(\mathfrak{o}) \otimes \mathbf{R})$, always starting with $i \equiv 2$ (4), is now, for both values of ε :

$$r_c \;\; r_2 \;\; r_c \;\; r_b + r_2.$$

In view of [2], the above results are also valid if $\mathfrak{o}$ is replaced by $\mathfrak{o}_\mathrm{S}$ (S finite set of primes in k, stable under the involution) or by k. In particular, if $\mathrm{D} = k$, they can be compared with those of [17] paragraph 5, p. 96. This shows that the part of the homology explicity determined there in terms of the Witt group of k, corresponds of the one associated here to the real places of k.

REFERENCES

[1] H. Bass, *Unitary algebraic K-theory, Algebraic K-theory III* (*Springer Lecture Notes*, vol. 343, 1973, p. 57-209).

[2] A. Borel, *Cohomologie réelle stable de groupes S-arithmétiques* (*C. R. Acad. Sc.*, Paris, t. 274, série A, 1972, p. 1700-1702).

[3] A. Borel, *Some properties of arithmetic quotients of symmetric spaces and an extension theorem* (*J. Diff. Geometry*, vol. 6, 1972, p. 543-560).

[4] A. Borel and J.-P. Serre, *Corners and arithmetic groups* (*Comm. Math. Helv.*, vol. 48, 1974, p. 244-297).

[5] A. Borel and J. Tits, *Groupes réductifs* (*Publ. Math. I. H. E. S.*, vol. 27, 1965, p. 55-150).

[6] N. Bourbaki, *Groupes et algèbres de Lie*, chap. IV, V, VI (*Act. Sci. Ind.*, 1337, Hermann, Paris, 1968).

[7] H. Cartan, *Périodicité des groupes d'homotopie stables des groupes classiques, d'après Bott* (*Séminaire E. N. S.*, 12ᵉ année, Notes polycopiées, Institut H. Poincaré, Paris, 1961).

[8] W. T. Van Est, *On the algebraic cohomology concepts in Lie groups II* (*Proc. Konink. Nederl. Akad. v. Wet.*, Series A, vol. 58, 1955, p. 286-294).

[9] H. Garland, *A finiteness theorem for K_2 of a number field* (*Annals of Math.*, vol. 94, n° 2, 1971, p. 534-548).

[10] H. Garland and W. C. Hsiang, *A square integrability criterion for the cohomology of arithmetic groups* (*Proc. Nat. Acad. Sci. USA*, vol. 59, 1968, p. 354-360).

[11] R. Godement, *Théorie des faisceaux* (*Act. Sci. Ind.*, p. 1252, Hermann, Paris, 1958).

[12] A. Grothendieck, *Sur quelques points d'algèbre homologique* (*Tôhoku Math. J.*, vol. 9, 1957, p. 119-221).

[13] G. Harder, *On the cohomology of* SL (2, o) (preprint) [in "Lie groups and their representations", Hilger, London, 1975].

[14] Harish-Chandra, *Discrete series for semi-simple Lie groups II*, (*Acta Math.*, vol. 116, 1966, p. 1-111).

[15] G. Hochschild and G. D. Mostow, *Cohomology of Lie groups, III* (*Illinois J. Math.*, vol. 6, 1962, p. 367-401).

[16] S. Kaneyuki and T. Nagano, *Quadratic forms related to symmetric spaces* (*Osaka Math. J.*, vol. 14, 1962, p. 241-252).

[17] M. Karoubi, *Périodicité de la K-théorie hermitienne, Algebraic K-theory III* (*Springer Lecture Notes*, vol. 343, 1973, p. 301-411).

[18] M. Kneser, *Lectures on Galois cohomology of classical groups* (Notes by P. Jothilingam, Tata Institute of Fundamental Research, Bombay, 1969).

[19] Y. Matsushima, *On Betti numbers of compact, locally symmetric Riemannian manifolds* (*Osaka Math. J.*, vol. 14, 1962, p. 1-20).

[20] Y. Matsushima and S. Murakami, *On certain cohomology groups attached to hermitian symmetric spaces* (*Osaka J. of Math.*, vol. 2, 1965, p. 1-35).

[21] D. Quillen, *Cohomology of groups* (*Actes Congrès Int. Math. Nice*, vol. 2, 1970, p. 47-51).

[22] G. de Rham, *Variétés Différentiables* (*Act. Sci. Ind.*, 1222 b, 3ᵉ éd., Hermann, Paris, 1973).

[23] A. Weil, *Adeles and algebraic groups* (Notes by M. Demazure and T. Ono, The Institute for Advanced Study, Princeton, 1961).

A. Borel,
The Institute for Advanced Study,
Princeton, N. J. 08540,
U. S. A.

(Manuscrit reçu le 4 février 1974.)

101.

Cohomology of arithmetic groups

Proc. Int. Congr. of Mathematicians, Vancouver 1974, Vol. 1 (1975) 435–442

1   This paper is concerned with the Eilenberg-MacLane cohomology groups $H^i(\Gamma\,; E)$ of an arithmetic or S-arithmetic subgroup Γ of a reductive affine algebraic group G over a number field, with coefficients in a Γ-module E. For the sake of brevity we shall, however, unless otherwise stated, assume G to be connected, semisimple and the groundfield to be $\mathbf{Q}$. Then S is a finite set of rational primes. We recall that a subgroup Γ of the group $G(\mathbf{Q})$ of $\mathbf{Q}$-rational points of G is arithmetic (resp. S-arithmetic) if, given a $\mathbf{Q}$-embedding $\rho : G \to GL_n$, the group $\rho(\Gamma)$ is commensurable with $\rho(G) \cap GL_n(\mathbf{Z})$ (resp. $\rho(G) \cap GL_n(\mathbf{Z}_S)$, where $\mathbf{Z}_S$ is the ring of rational numbers whose denominator is a product of elements in S).

I. General Coefficients

1. *In this section S is fixed, Γ is S-arithmetic, torsion-free.* From G and S alone, one can construct a contractible locally compact $G(\mathbf{Q})$-space $\bar{X}_S$, whose ith integral cohomology group with compact supports $H_c^i(\bar{X}_S\,; \mathbf{Z})$ is zero except in one dimension $e = e(G, S)$, where it is a free module I_S, on which Γ operates properly and freely, so that the quotient $\bar{X}_S/\Gamma$ is compact, triangulable. It follows that $\mathbf{Z}$ admits a $\mathbf{Z}[\Gamma]$-free finitely generated resolution (as was shown first by Raghunathan [27] when Γ is arithmetic (i.e., S is empty) and by J.-P. Serre [28] in general); hence $H^*(\Gamma\,; E)$ is finitely generated if E is (as a $\mathbf{Z}$-module), the cohomological dimension $\mathrm{cd}(\Gamma)$ of Γ is finite, equal to $e(G, S)$, Γ is a duality group in the sense of Bieri-Eckmann [1], and we have a canonical isomorphism

$$(1) \qquad H^i(\Gamma\,; E) \xrightarrow{\sim} H_{e-i}(\Gamma\,; I_S \otimes E) \qquad (i \in \mathbf{Z})$$

[5], [7]. Let $l = \mathrm{rk}_{\mathbf{Q}}\, G$ be the $\mathbf{Q}$-rank of G (common dimension of its maximal $\mathbf{Q}$-split algebraic tori) and l_p the $\mathbf{Q}_p$-rank of G, viewed as a group over the field

Q_p of p-adic numbers ($p \in S$). The space $\bar{X}_S$ is the product of the manifold with corners $\bar{X}$ with interior the symmetric space X of maximal compact subgroups of $G(\mathbf{R})$, constructed in [6] (which is just X if $l = 0$), by the Bruhat-Tits buildings T_p of the groups $G(\mathbf{Q}_p)$. If $S = \varnothing$, then $\bar{X}_S/\Gamma = \bar{X}/\Gamma$ is a compact manifold with corners, hence triangulable. In the general case, triangulability follows from properties of the projection $\bar{X}_S/\Gamma \to T_S/\Gamma$ (where $T_S = \Pi_{p \in S} T_p$) and from results of F. E. A. Johnson [18].

REMARK. Even if G is not semisimple, any torsion-free arithmetic subgroup of G is a duality group [5], [7]. For further examples of discrete subgroups of products of reductive groups over local fields (not necessarily of characteristic zero) which have a homological duality, see [5], [7].

2. In the general case, Γ possesses torsion-free subgroups of finite index. Their common cohomological dimension is, by definition, the virtual cohomological dimension vcd(Γ) of Γ [28], which is then finite, equal to $e(G, S)$. The groups $H^i(\Gamma ; E)$ are also finitely generated if E is. The group Γ also acts properly on $\bar{X}_S$ with a compact quotient (presumably triangulable, too, but the author is not aware of any proof outside the torsion-free case), which implies that the finite subgroups of Γ form finitely many conjugacy classes. Further cohomological information on Γ can be extracted from those subgroups; we mention some examples.

2.1. The group Γ and its subgroups of finite index Γ' satisfy the conditions under which an Euler characteristic $\chi(\Gamma') \in \mathbf{Q}$, in the sense of C. T. C. Wall, can be defined. It is equal to $\bar{\chi}(\Gamma') = \sum(-1)^i \dim H^i(\Gamma' ; \mathbf{Q})$ if Γ' is torsion-free and satisfies the condition $\chi(\Gamma'')=[\Gamma' : \Gamma'']\chi(\Gamma')$ if $\Gamma'' \subset \Gamma' \subset \Gamma$ and $[\Gamma : \Gamma''] < \infty$. This already implies that $[\Gamma:\Gamma'] \cdot \chi(\Gamma) \in \mathbf{Z}$ if Γ' is torsion-free. K. Brown [8], however, has given an expression for $\chi(\Gamma) - \bar{\chi}(\Gamma)$, involving the lattice of finite subgroups of Γ, which implies in particular that $m \cdot \chi(\Gamma) \in \mathbf{Z}$ as soon as the order of any finite subgroup of Γ divides m. Now let k be a totally real number field, n its degree over Q, $\mathfrak{o}_T$ the ring of elements in k which are integral outside a given finite set T of finite primes of k and $\Gamma = \mathbf{SL}_2(\mathfrak{o}_T)$. Let further $\zeta_{k,T}$ be the function obtained from Dedekind's zeta-function ζ_k of k by omitting the local factors associated to the primes in T. Then, by a special case of a formula of G. Harder [14], if $S \neq \varnothing$, by [**28**, 3.7] in eneral, we have

$$(1) \qquad\qquad \chi(\mathbf{SL}_2(\mathfrak{o}_T)) = \zeta_{k,T}(-1).$$

The results mentioned above then allow one to get estimates of the denominator of the right-hand side using finite subgroups or subgroups of finite index of Γ. (See [**28**, 3.7], [**8**, § 4], where products of values $\zeta_{k,T}(1 - 2i)$ ($1 \leq i \leq n$) are also similarly related to the symplectic groups $\mathbf{Sp}_{2n}(\mathfrak{o}_T)$.)

2.2. Given a prime p, there is an algebra A_p over $\mathbf{Z}/p\mathbf{Z}$ constructed from the category of elementary commutative p-subgroups of Γ, and a homomorphism $H^*(\Gamma, \mathbf{Z}/p\mathbf{Z}) \to A_p$, whose kernel and cokernel are annihilated by some power of the pth power map [**26**, §§ 14, 15]. Thus, asymptotically, the cohomology mod p is to some extent determined by the commutative elementary p-subgroups of Γ.

2.3. Assume now G to be the algebraic group defined by the elements of norm one in a division algebra D over $\boldsymbol{Q}$, and Γ to be arithmetic. Then, via the regular representation, the finite subgroups of Γ (and even $G(\boldsymbol{Q})$) operate on $\boldsymbol{R}^n - 0$ ($n = [D:\boldsymbol{Q}]$) *freely* since D is a division algebra; hence their cohomology is periodic, of period dividing n. From this, using the spectral sequences of equivariant cohomology theory, B.B. Venkov [29] has shown the same to hold for Γ, from a certain dimension on. This is also true if Γ is S-arithmetic; the argument is the same, with $X \times T_S$ (notation of § 1) playing the role of X.

2.4. We note finally that an easy spectral sequence argument shows that if d is a common multiple of the orders of the finite subgroups of Γ, then it annihilates $H^i(\Gamma\,;E)$ for $i > \mathrm{vcd}(\Gamma)$.

II. Real or Complex Cohomology

3. From now on, E is the space of a finite-dimensional real or complex representation of $G(\boldsymbol{R})$, and, unless otherwise stated, Γ is arithmetic. (In fact, the results recalled in this section and their proofs are valid for any discrete subgroup of $G(\boldsymbol{R})$.) The group $G(\boldsymbol{R})$ operates in a natural way on the space $\Omega_X(E)$ of smooth E-valued differential forms on X, and $H^*(\Gamma\,;E)$ is canonically isomorphic to the cohomology of the complex $\Omega_X(E)^\Gamma$ of Γ-invariant elements in $\Omega_X(E)$. Fix a maximal compact subgroup K of $G(\boldsymbol{R})$. Using the canonical projection of $G(\boldsymbol{R})/\Gamma$ onto X/Γ, one identifies in a well-known way $\Omega_X(E)^\Gamma$ with a space of smooth vector-valued functions on $G(\boldsymbol{R})/\Gamma$, and this yields in fact a natural isomorphism of $\Omega_X(E)^\Gamma$ with the cochain complex $C^*(\mathfrak{g},\,\mathfrak{k}\,;\,\mathscr{F} \otimes E)$ of relative Lie algebra cohomology, where, if E is real (resp. complex), $\mathscr{F}$ is the space of smooth real (resp. complex) valued functions on $G(\boldsymbol{R})/\Gamma$, $\mathfrak{g}$ and $\mathfrak{k}$ are the Lie algebras (resp. the complexifications of the Lie algebras) of $G(\boldsymbol{R})$ and K, and $\mathscr{F} \otimes E$ is viewed as a $\mathfrak{g}$-module in the obvious way. Thus we get a canonical isomorphism

$$(1) \qquad\qquad H^*(\mathfrak{g},\,\mathfrak{k}\,;\,\mathscr{F} \otimes E) \overset{\sim}{\longrightarrow} H^*(\Gamma\,;E).$$

(For all this, see [23], [24]; the blanket assumption X/Γ compact made there is not used for these general remarks. Actually, (1) is correct as stated if $G(\boldsymbol{R})$ is connected, also a standing assumption in [23], [24], or if Γ is contained in the identity component of $G(\boldsymbol{R})$. Otherwise, the left-hand side has to be replaced by the invariants under a suitable finite group of automorphisms. We shall ignore this technicality.) Let now $\mathscr{V}$ be a subspace of $\mathscr{F}$ stable under K and $\mathfrak{g}$. The inclusion $\mathscr{V} \otimes E \to \mathscr{F} \otimes E$ then induces homomorphisms

$$(2) \qquad j^q : H^q(\mathfrak{g},\,\mathfrak{k}\,;\,\mathscr{V} \otimes E) \to H^q(\Gamma\,;E) \qquad (q=0,\,1,\,2,\cdots)$$

which we shall discuss in three cases (in §§ 4, 6, 7).

4. Stable cohomology. Take $E = \mathscr{V} = \boldsymbol{R}$, with the trivial action of $G(\boldsymbol{R})$, where $\mathscr{V}$ is the space of real constant functions on $G(\boldsymbol{R})/\Gamma$. We have then a natural homomorphism $j^q\colon H^q(\mathfrak{g},\,\mathfrak{k}\,;\boldsymbol{R}) \to H^q(\Gamma\,;\boldsymbol{R})$. If Γ is cocompact (i.e., $G(\boldsymbol{R})/\Gamma$ is compact), then $\mathscr{F}$ has a $G(\boldsymbol{R})$-invariant supplement to $\mathscr{V}$; hence j^q is injective for all q (as was remarked first, I think, by W. T. van Est [9, Theorem 7], from a different point of view) but this is not so in the noncocompact case. Moreover, again in the cocompact case, but without assuming Γ to be arithmetic, Matsushima has shown j^q to be surjective at least up to a constant $m(G)$ computable from $\mathfrak{g}$ [21]. Similarly, in the arithmetic case, we have the following:

THEOREM 4.1. *The homomorphism j^q is an isomorphism for $q < (\mathrm{rk}_Q G)/4$.*

(See [3], [4]; the surjectivity part had already been proved in substance by H. Garland [11, 3.5] or [4, 3.5], and used by him to show that $K_2\mathfrak{o}$ is a torsion group.)

Let G_u be a maximal compact subgroup of $G(C)$ containing K. Then $X_u = K^\circ \backslash G_u)$, where K° is the identity component of K, is a compact symmetric space, the "dual" space to X. As is well known, and goes back to E. Cartan, $H^*(\mathfrak{g}, \mathfrak{k}; R)$ is canonically isomorphic to $H^*(X_u ; R)$, whence a homomorphism $\alpha^* : H^q(X_u ; R) \to H^*(\Gamma ; E)$, which is an isomorphism at least in degrees $< (\mathrm{rk}_Q G)/\varphi$.

Let now $(G_n, \Gamma_n, X_n, X_{n,u})$ be objects similar to (G, Γ, X, X_u) and $f_n : (G_n, \Gamma_n) \to (G_{n+1}, \Gamma_{n+1})$ an injective Q-morphism $(n = 1, 2, \cdots)$. Let $X_u = \mathrm{inj}\ \lim X_{n,u}$. Assume that $\mathrm{rk}_Q G_n \to \infty$. Then, for many classical sequences of this type, in particular the one of the next section, there exists, given q, an integer $n(q)$ such that

(1) $H^q(X_u; R) \cong H^q(\mathrm{inj}\ \lim \Gamma_n ; R) = H^q(X_{n,u} ; R) = H^q(\Gamma_n ; R) \qquad (n \geq n(q))$.

(See [3], [4] for more precise statements, also pertaining to S-arithmetic groups.)

5. The cohomology of $SL(\mathfrak{o})$, higher regulators and values of ζ-functions. We consider the special case where $G_n = R_{k/Q}SL_n$, $\Gamma_n = SL_n\mathfrak{o}$ (where k is a number field, $\mathfrak{o}$ its ring of integers, $R_{k/Q}$ the restriction of scalars [30, Chapter 1]) and f_n comes from the natural inclusion $SL_n \to SL_{n+1}$. Passing to homology, we have then an isomorphism $\alpha_* : H_*(SL(\mathfrak{o}); R) \to H_*(X_u; R)$, where $SL(\mathfrak{o}) = \mathrm{inj}\ \lim SL_n(\mathfrak{o})$. For $m = 1, 2, \cdots$, let P_{2m+1} (resp. P'_{2m+1}) be the space of primitive elements in $H_{2m+1}(SL(\mathfrak{o}); R)$ (resp. $H_{2m+1}(X_u; R)$). Its dimension d_m is equal to r_2 (resp. $r_1 + r_2$) if $m \geq 1$ is odd (resp. even), where r_1 (resp. r_2) is the number of real (resp. complex) places of k; this also happens to be the order of the zero of $\zeta_k (s)$ at $s = -m$. There is a natural map of $\pi_{2m+1}(X_u)$ (resp. of Quillen's group $K_{2m+1}\mathfrak{o}$) into P'_{2m+1} (resp. P_{2m+1}), whose image is a lattice L'_{2m+1} (resp. L_{2m+1}). The mth *regulator* of k is then, by definition [20, § 4], the positive real number R_m such that the map $\Lambda^{d}\cdot P_{2m+1} \to \Lambda^{d}\cdot P'_{2m+1}$ defined by α_* sends $\Lambda^{d}\cdot L_{2m+1}$ onto $R_m \cdot (\Lambda^{d}\cdot L'_{2m+1})$.

Given two nonzero real numbers, write $a \sim b$ if a/b is rational.

THEOREM 5.1. *Let D_k be the discriminant of k over Q. For $m \geq 1$, we have*

$$R_m \sim D_k^{1/2} \cdot \pi^{-n(m+1)} \cdot \zeta_k(m + 1).$$

According to the functional equation for $\zeta_k(s)$, this is equivalent to

$$R_m \sim \pi^{-d\cdot} \cdot \lim_{s \to -m} \zeta_k(s)/(s + m)^{d\cdot}.$$

This, however, says nothing about the actual quotient of these two numbers. According to the conjectures in [20, § 4], slightly modified to take into account the factor π^{-d_n}, it should be closely related to the orders of $K_{2m}\mathfrak{o}$ and of the torsion subgroup of $K_{2m+1}\mathfrak{o}$.

REMARK. Applied to other sequences of classical groups, the results of § 6 also yield the ranks of Karoubi's groups $_\varepsilon L_i\mathfrak{o}$ or $_\varepsilon L_i k$ (see [3], [4]).

6. Cusp cohomology. We come back to the setting of § 3, assume E to be complex, endowed with a hermitian scalar product invariant under K (or more precisely admissible in the sense of [23]), and fix an invariant measure on $G(\mathbf{R})/\Gamma$. This allows one to define a scalar product on compactly supported elements of $C^*(\mathfrak{g}, \mathfrak{k}; \mathscr{F} \otimes E)$ (and then on various bigger spaces), an adjoint ∂ to d, and a Laplace operator $\partial d + d\partial$ [23], [24]. Let $^0L^2(G(\mathbf{R})/\Gamma)$ be the space of cusp forms in the space $L^2(G(\mathbf{R})/\Gamma)$ of complex-valued square-integrable functions on $G(\mathbf{R})$ [17], [19]. By a well-known result of Gel'fand and Piateckii-Shapiro [13], [17], it is a direct sum of closed irreducible (under $G(\mathbf{R})$) subspaces, with finite multiplicities. Take now for $\mathscr{V}$ the space $^0\mathscr{F}$ of C^∞-vectors (in the sense of infinite-dimensional representation theory) of $^0L^2(G(\mathbf{R})/\Gamma)$. Thus, $\mathscr{V}$ consists of the elements $f\in \mathscr{F}$ such that Xf is square-integrable for every $X \in U(\mathfrak{g})$ and

$$\int_{U(\mathbf{R})/(U\cap\Gamma)} f(g\cdot u)\,du = 0 \quad \text{for all } g \in G(\mathbf{R}),$$

where U is the unipotent radical of an arbitrary proper parabolic $\mathbf{Q}$ subgroup of G.

THEOREM 6.1. *The map* $^0j^* : H^*(\mathfrak{g}, \mathfrak{k}; {}^0\mathscr{F} \otimes E) \to H^*(\Gamma; E)$ *is injective.*

Its image will be denoted $H^*_{\text{cusp}}(\Gamma; E)$.

Let r be a finite-dimensional representation of K. For $c \in \mathbf{R}$, let $A(r, c)$ be the space of elements in $L^2(G(\mathbf{R})/\Gamma)$ which transform according to r with respect to left translations by K and are eigenfunctions of the Casimir operator with eigenvalue c. As is well known, these elements are in fact analytic. Let $^0A(r, c) = A(r, c) \cap {}^0L^2(G(\mathbf{R})/\Gamma)$.

THEOREM 6.2. *The space* $^0A(r, c)$ *is finite-dimensional, contained in the sum of finitely many closed irreducible subspaces of* $^0L^2(G(\mathbf{R})/\Gamma)$; *in particular, its elements are Z-finite (Z center of* $U(\mathfrak{g})$), *and are automorphic forms in the sense of* [17], [19]. *Given* r, *the set of* c *for which* $^0A(r, c)$ *is* $\neq \{0\}$ *is bounded from above and has no finite accumulation point.*

(6.2, the results below, and those of §7 are joint work of H. Garland and the author; they extend theorems proved by H. Garland for $\mathbf{Q}$-rank one groups [10].)

In view of Kuga's formula relating the Laplace operator and the Casimir operator [23], it follows immediately from 6.2 that $H^*(\mathfrak{g}, \mathfrak{k}; {}^0\mathscr{F} \otimes E)$ may be identified with the space of harmonic forms in $C^*(\mathfrak{g}, \mathfrak{k}; {}^0\mathscr{F} \otimes E)$. Thus $H^*_{\text{cusp}}(\Gamma; E)$ is isomorphic to the space of harmonic cusp forms. Assume now E to be irreducible. Let c be such that the Casimir operator on E is $c\cdot\text{Id}$. Write further $^0L^2(G(\mathbf{R})/\Gamma)$ as a Hilbert direct sum of irreducible subspaces H_i and let c_i be the eigenvalue of C in the space of differentiable vectors of H_i. Then we have

$$(1) \qquad H^q_{\text{cusp}}(\Gamma; E) = \bigoplus_{i,c_i=c} \text{Hom}_K(\Lambda^q(\mathfrak{g}/\mathfrak{k}) \otimes E, H_i).$$

If E is the trivial representation, then $c = 0$. In the cocompact case (where $^0\mathscr{F} = \mathscr{F}$) the formula overlaps with one of Matsushima's [22]. More generally, 6.2 implies that a number of arguments in the cocompact case (such as the generalized Eichler-Shimura isomorphism in [24, § 7]) or the vanishing theorems in [24, § 11] remain valid for cusp forms and cusp cohomology.

7. Square-integrable cohomology. Take now for $\mathscr{V}$ the space of C^∞-vectors in $L^2(G(\mathbf{R})/\Gamma)$. The image of j^* is then the space of cohomology classes which can be represented by square-integrable forms (hence also by square-integrable harmonic forms). It contains the image of the cohomology with compact supports, is equal to it if $G = R_{k/\mathbf{Q}}\,SL_2$ by [15], but is bigger in general. However, 6.2, induction and the results of § 7 in [19] allow one to prove that 6.2 remains true if $^0A(r, c)$ and $^0L^2$ are replaced by $A(r, c)$ and L^2 (this was shown to us by Langlands). In particular, a K-finite eigenfunction of C in $L^2(G(\mathbf{R})/\Gamma)$ is Z-finite, contained in the sum of finitely many elements of the discrete spectrum. This is the analogue of a result of Okamoto for $L^2(G(\mathbf{R}))$ [25]. It also follows that the space of square-integrable E-valued harmonic forms is finite-dimensional, isomorphic to $H^*(\mathfrak{g}, \mathfrak{k}; \mathscr{V} \otimes E)$ and given by a formula similar to (1), but where $^0L^2$ is replaced by the discrete spectrum in $L^2(G(\mathbf{R})/\Gamma)$, if E is irreducible.

8. The $\mathbf{Q}$-rank one case. Assume now $\mathrm{rk}_{\mathbf{Q}}\,G = 1$. The manifold with corners $V = \bar{X}/\Gamma$ is then in fact a manifold with boundary, and the connected components of its boundary ∂V correspond bijectively to the Γ-conjugacy classes of minimal parabolic $\mathbf{Q}$-subgroups of G [2, § 17]. Let $r: H^*(V; E) \to H^*(\partial V; E)$ be the restriction homomorphism. Using Langlands' theory of Eisenstein series [17], [19], G. Harder has shown the existence of a subspace $H^*_{\mathrm{inf}}(\Gamma; E)$ of $H^*(\Gamma; E) \cong H^*(V; E)$, which restricts isomorphically onto Im r, whose elements are obtained either by taking analytic continuation of suitable Eisenstein series, or residues of such at simple poles [16]. Thus, in this case, every element of $H^*(\Gamma; E)$ has a closed harmonic representative. If $G = R_{k/\mathbf{Q}}SL_2$, and $E = \mathbf{C}$, G. Harder has given a complete description of Im r; moreover, $H^*(\Gamma; \mathbf{C})$ is in this case the direct sum of H^*_{inf}, H^*_{cusp}, and the image of $H^*(\mathfrak{g}, \mathfrak{k}; \mathbf{C})$ [15].

9. No such results have yet been obtained in the higher rank case. Still, they point to extremely interesting relations between $H^*(\Gamma; E)$ and the theory of automorphic forms, and lead one to wonder whether (a) all cohomology classes are represented by closed harmonic forms; (b) there is a sum decomposition of $H^*(\Gamma; E)/H^*_{\mathrm{cusp}}(\Gamma; E)$, where each summand is naturally associated to Eisenstein series built from a class $\{P\}$ of associated proper parabolic $\mathbf{Q}$-subgroups, starting from harmonic cusp forms on pieces of the boundary of the manifold with corners $\bar{X}/\Gamma$ corresponding to the elements of $\{P\}$.

Finally, I would like to draw attention to two topics left out of this survey: vanishing theorems for subgroups of p-adic groups and related questions, for which we refer to H. Garland's article [12], and the use of cohomology of arithmetic groups in the discussion of zeta-functions or L-functions of certain algebraic varieties

(Eichler, Shimura and, more recently, I. I. Piateckii-Shapiro and R. P. Langlands (in *Modular functions of one variable*. II, Springer Lecture Notes in Mathematics, vol. 349) for modular curves; Shimura, Kuga-Shimura and Langlands for other quotients of bounded symmetric domains).

References

1. R. Bieri and B. Eckmann, *Groups with homological duality generalizing Poincaré duality*, Invent. Math. **20** (1973), 103–124.

2. A. Borel, *Introduction aux groupes arithmétiques*, Publ. Inst. Math. Univ. Strasbourg, 15, Actualités Sci. Indust., no. 1341, Hermann, Paris, 1969. MR **39** #5577.

3. ——, *Cohomologie réelle stable de groupes S-arithmétiques classiques*, C. R. Acad. Sci. Paris Sér. A-B 274 (1972), A1700–A1702. MR **46** #7400.

4. ——, *Stable real cohomology of arithmetic groups*, Ann. Sci. École Norm. Sup. (4) **7** (1974), 235–272.

5. A. Borel and J.-P. Serre, *Cohomologie à supports compacts des immeubles de Bruhat-Tits; applications à la cohomologie des groupes S-arithmétiques*, C. R. Acad Sci. Paris Sér. A-B **272** (1971), A110–A113. MR **43** #221.

6. ——, *Corners and arithmetic groups*, Comment. Math. Helv. **48** (1974), 244–297.

7. ——, *Cohomologie d'immeubles et de groupes S-arithmétiques* (in preparation).

8. K. Brown, *Euler-characteristics of discrete groups and G-spaces*, Invent. Math. **27** (1974), 229–264.

9. W. T. van Est, *A generalization of the Cartan-Leray spectral sequence*. I, II, Nederl. Akad. Wetensch. Proc. Ser. A **61** = Indag. Math. **20** (1958), 399–413. MR **21** #2236.

10. H. Garland, *The spectrum of noncompact G/Γ and the cohomology of arithmetic groups*, Bull. Amer. Math. Soc. **75** (1969), 807–811. MR **40** #269.

11. ——, *A finiteness theorem for K_2 of a number field*, Ann. of Math. (2) **94** (1971), 534–548. MR **45** #6785.

12. ——, *On the cohomology of arithmetic groups*, these PROCEEDINGS.

13. I. M. Gel'fand and I. I. Pjateckii-Šapiro, *Automorphic functions and representation theory*, Trudy Moskov. Mat. Obšč. **12** (1963), 389–412 = Trans. Moscow Math. Soc. **1963**, 438–464. MR **28** #3115.

14. G. Harder, *A Gauss-Bonnet formula for discrete arithmetically defined groups*, Ann. Sci. École Norm. Sup. **4** (1971), 409–455. MR **46** #8255.

15. ——, *On the cohomology of $SL(2, \mathfrak{o})$* (preprint).

16. ——, *On the cohomology of discrete arithmetically defined groups*, Proc. Internat. Colloquium Bombay 1972 (to appear).

17. Harish-Chandra, *Automorphic forms on semisimple Lie groups*, Lecture Notes in Math., no. 62, Springer-Verlag, Berlin and New York, 1968. MR **38** #1216.

18. F. E. A. Johnson, *On the triangulation of stratified sets and singular varieties* (to appear).

19. R. P. Langlands, *On the functional equations satisfied by Eisenstein series*, Proc. Sympos. Pure Math., vol. 9, Amer. Math. Soc., Providence, R.I., 1965, pp. 235–252. MR **40** #2784.

20. S. Lichtenbaum, *Values of zeta-functions, étale cohomology, and algebraic K-theory*, Algebraic *K*-Theory. II, Lecture Notes in Math., no. 342, Springer-Verlag, New York, 1973, pp. 489–499.

21. Y. Matsushima, *On Betti numbers of compact, locally symmetric Riemannian manifolds*, Osaka Math. J. **14** (1962), 1–20. MR **25** #4549.

22. ——, *A formula for the Betti numbers of compact locally symmetric Riemannian manifolds*, J. Differential Geometry **1** (1967), 99–109. MR **36** #5958.

23. Y. Matsushima and S. Murakami, *On vector bundle valued harmonic forms and automorphic forms on symmetric Riemannian manifolds*, Ann. of Math. (2) **78** (1963), 365–416. MR **27** #2997.

24. ———, *On certain cohomology groups attached to Hermitian symmetric spaces*, Osaka J. Math. **2** (1965), 1–35. MR **32** #1728.

25. K. Okamoto, *On induced representations*, Osaka J. Math. **4** (1967), 85–94. MR **37** #1519.

26. D. G. Quillen, *The spectrum of an equivariant cohomology ring*. I, II, Ann. of Math. (2) **94** (1971), 549–572, 573–602. MR **45** #7743.

27. M. S. Raghunathan, *A note on quotients of real algebraic groups by arithmetic subgroups*, Invent. Math. **4** (1967/68). 318–335. MR **37** #5894.

28. J.-P. Serre, *Cohomologie des groupes discrets*, Prospects in Mathematies, Ann. of Math. Studies, no. 70, Princeton Univ. Press, Princeton, N.J., 1971, pp. 77–169.

29. B. B. Venkov, *On homologies of unit groups in division algebras*, Trudy Mat. Inst. Steklov. **80** (1965), 66–89 = Proc. Steklov Inst. Math. **80** (1965), 73–100. MR **34** #1379.

30. A. Weil, *Adèles and algebraic groups*, Notes by M. Demazure and T. Ono, Institute for Advanced Study, Princeton, N.J., 1961.

INSTITUTE FOR ADVANCED STUDY
PRINCETON, NEW JERSEY 08540, U.S.A.

102.

Linear representations of semi-simple algebraic groups

Proc. Symp. in Pure Math. **29**, Amer. Math. Soc. (1975) 421−439

This paper is an outgrowth of two talks, whose aim was to give a survey of known results and of some open problems on rational representations of semi-simple groups over an algebraically closed ground field. I am grateful to A. Fauntleroy and W. Haboush, who made Notes from the talks which were very helpful to me in writing this up.

Throughout, k denotes an algebraically closed field, p the characteristic of k, and G a connected semi-simple group over k. All algebraic groups are affine and, unless otherwise stated, defined over k.

§ 1. Classification

1.1. We refer to [2] for the basic notions and facts on algebraic groups used here without further comment.

We recall that an (algebraic) torus T is a connected group isomorphic to a product of groups $\mathbf{GL}_1$. The character group $X(T) = \operatorname{Morph}(T, \mathbf{GL}_1)$ is a free abelian group of rank equal to dim T. Given $t \in T$, $a \in X(T)$, we shall often write t^a for the value of a on t. It is then understood that the composition law in $X(T)$, which is given by taking products of values, is written additively: $t^{a+b} = t^a \cdot t^b$.

1.2. Roots. The maximal tori of G are conjugate by inner automorphisms. Their common dimension is the rank rk G of G. Let T be one of them. It acts on the Lie algebra $L(G)$ of G (identified to the tangent space at the identity) via the adjoint representation. This representation is diagonalisable (as is any rational representation of a torus). For $a \in X(T)$, let

$$L(G)_a = \{X \in L(G) \,|\, \operatorname{Ad} t\, X = t^a \cdot X \ (t \in T)\}.$$

An element $a \in X(T)$ is a *root* of G with respect to T if $a \neq 0$ and $L(G)_a \neq \{0\}$. We let Φ or $\Phi(T, G)$ be the set of roots of G with respect to T. The space $L(G)$ is the direct sum of $L(T) = L(G)_0$ and of the $L(G)_a$ $(a \in \Phi)$. Moreover, $L(G)_a$ is one-dimensional for $a \in \Phi$.

Let $W = W(G)$ be the quotient of the normalizer $N_G(T)$ of T in G by T. Since T is its own centralizer in G, W operates faithfully, on T and $X(T)$ via inner automorphisms, on $L(T)$ by the adjoint representation.

361

1.3. Proposition. *The set Φ is a reduced root system in $X(T)_{\mathbf{R}} = X(T) \otimes \mathbf{R}$, with Weyl group W. Moreover, $X(T)$ contains the lattice $R(\Phi)$ spanned by Φ and is contained in the lattice $P(\Phi)$ of weights of Φ.*

Briefly, this means the following (for more details, see [6; 7; 22; 28]): the set Φ is symmetric with respect to the origin, contains a basis of $X(T)_{\mathbf{R}}$ and is invariant under W; let $(,)$ be a positive non-degenerate scalar product on $X(T)$ which is invariant under W. Then W is generated by the reflections through the hyperplanes orthogonal to the roots; we have $2(a,b)/(b,b) \in \mathbf{Z}$ for all $a, b \in \Phi$. Moreover, $a, ra \in \Phi$ ($a \in \Phi$, $r \in \mathbf{R}$) imply $r = \pm 1$. (This last condition accounts for the word "reduced.") Finally the weights are the elements $r \in X(T)_{\mathbf{R}}$ such that $2(r,a)/(a,a) \in \mathbf{Z}$ for all $a \in \Phi$. They form the lattice $P(\Phi)$.

For brevity, let us call *diagram* a couple (Φ, Γ) formed of a reduced root system in some finite dimensional vector space and a lattice between $R(\Phi)$ and $P(\Phi)$.

1.4. Theorem. *The map $\delta: G \mapsto (\Phi(T, G), X(T))$ induces a bijection between isomorphism classes of semi-simple groups over k and isomorphism classes of diagrams.*

If $X(T) = R(\Phi)$, then G is of "adjoint type," i.e. the natural morphism $G \to \operatorname{Ad} G \subset GL(L(G))$ is an isomorphism. If $X(T) = P(\Phi)$, then G is "simply connected." The latter notion may be defined in several ways; we give two: (a) any rational projective representation $G \to PGL(E)$ lifts to a rational linear representation $G \to GL(E)$; (b) if G' is a semi-simple group, $f: G' \to G$ an isogeny such that $\ker df: L(G') \to L(G)$ is pointwise fixed under G' (acting by the adjoint representation), then f is an isomorphism (The requirement on $\ker df$ characterizes "central isogenies"; see [4: § 2] for more details). Over $\mathbf{C}$, this is equivalent to simple-connectedness in the topological sense.

The theorem is in fact more functorial than stated. In particular, if G, G' correspond to (Φ, Γ) and (Φ, Γ') and if $\Gamma' \subset \Gamma$, then there is an (essentially unique) central isogeny $G \to G'$ whose degree is equal to the index of Γ' in Γ, and in fact whose kernel and infinitesimal kernel can be read off Γ/Γ' [3: 3.3 (2)]. Over $\mathbf{C}$, $P(\Phi)/\Gamma \xrightarrow{\sim} \pi_1(G)$, $\Gamma/R(\Phi) \xrightarrow{\sim} \operatorname{Center}(G)$. Thus the reduced root systems parametrize the central isogeny classes of semi-simple groups over k, or the simply connected semi-simple groups over k. In fact they also parametrize the isogeny classes up to one exception: in characteristic two, the types $\mathbf{B}_n$ and $\mathbf{C}_n$ ($n \geq 3$) are isogeneous, but not centrally isogeneous.

1.5. References. Over $\mathbf{C}$, 1.4 goes back to results of Killing, Weyl, Cartan, proved however in a different context. Briefly, it may be viewed as the conjunction of the following:

(a) Classification of complex semi-simple Lie algebras by reduced root systems.

(b) Classification of connected complex semi-simple Lie groups with a given Lie algebra $\mathfrak{g}$ with root system Φ by means of lattices between $R(\Phi)$ and $P(\Phi)$.

(c) A complex connected semi-simple Lie group has one and only one structure of affine algebraic group compatible with its complex analytic structure.

(a) is in essence due to Killing and Cartan, although the connection with root systems emerged gradually only later. It is now standard (cf. e.g. [22; 28]).

It is more difficult to give a direct reference for (b). Results of H. Weyl and E. Cartan, as reformulated later by E. Stiefel (C. M. H. 14 (1942), 350−380; see also

J. F. Adams, Lectures on Lie groups, Benjamin) show that diagrams also classify compact semi-simple Lie groups. One then uses the fact that the assignment: connected Lie group $\mapsto$ maximal compact subgroup induces a bijection between isomorphism classes of connected complex semi-simple Lie groups and of connected compact semi-simple Lie groups (see e.g. [19]). In the course of proving this, one also sees that a complex connected complex semi-simple Lie group always has a faithful finite dimensional representation [19: p. 200].

Finally, in view of this last fact, (c) amounts to showing that the **C**-algebra of holomorphic functions on G whose translates span a finite dimensional space (the "representative functions") is finitely generated. It is then the coordinate ring for the desired structure of algebraic group [20].

In positive characteristics, 1.4 is due to C. Chevalley. There are two parts to the proof:

(i) surjectivity of the map δ. More precisely, Chevalley associates to each diagram (Φ, Γ) a smooth group scheme G_0 over **Z** such that $G_0 \otimes_{\mathbf{Z}} k$ is the k-group with diagram (Φ, Γ), for every k. This construction is given first in [11]; see also [3] and, for a more general existence theorem over schemes [15; 16]. It will be sketched in § 5.

(ii) injectivity of δ. This is proved in [10; Vol. 2]; see also [15; 16].

1.6. Remarks. (1) In characteristic zero, the Lie algebra is an invariant of the isogeny class, but it is not necessarily so in positive characteristic. For example, in characteristic two, the Lie algebra of $\mathbf{SL}_2$ is nilpotent, with a one-dimensional center, namely $L(T)$, whereas the Lie algebra of $\mathbf{PSL}_2$ is solvable, not nilpotent, with a two-dimensional commutative ideal.

(2) With a slight extension of the notion of reduced root system, 1.4 can also be stated and proved for connected algebraic groups which are *reductive*, i.e. groups which are isogeneous to products of semi-simple groups and tori, or, equivalently, groups whose radical is a torus [15; 16].

§ 2. Irreducible representations

2.1. Dominant and Fundamental Weights. Let Δ be a basis of Φ. Thus, Δ is a basis of $X(T)_{\mathbf{R}}$ contained in Φ, such that every element of Φ is linear combination of elements in Δ with coefficients in **Z**, all of the same sign. The set of such positive linear combinations will be denoted Φ^+; it is the set of positive elements in Φ for some ordering.

An element $r \in P(\Phi)$ is *dominant* if $(r, a) \geqq 0$ for all $a \in \Delta$. For $a \in \Delta$, let d_a be the dominant weight such that $2(d_a, b)/(b, b) = \delta_{ab}$ (Kronecker symbol) for all $b \in \Delta$. The d_a's are the *fundamental dominant weights*. The fundamental dominant weights form a basis of $P(\Phi)$, and generate the semi-group of dominant weights. Any weight d can be written uniquely

$$d = \sum_{a \in \Delta} m_a(d) \, d_{\mathrm{a}}, \quad (m_a \in \mathbf{Z}),$$

and is dominant if and only if $m_a(d) \geqq 0$ for all $a \in \Delta$.

2.2. To Δ there is associated a maximal connected solvable subgroup B of G containing T. It is the semi-direct product of T by a normal unipotent subgroup U, whose Lie algebra is the direct sum of the $L(G)_\alpha$ ($\alpha \in \Phi^+$). This yields in fact a bijection between bases of Δ and maximal solvable subgroups containing T.

Every rational character of B is trivial on U, whence a natural bijection $X(B) \cong X(T)$.

2.3. Let E be a finite dimensional vector space over k and $\pi: G \to GL(E)$ a rational representation of G. Its restriction to T can be put in diagonal form. For $r \in X(T)$, let $E_r = \{e \in E, \pi(t) \cdot e = t^r e)$ and let $P(\pi) = \{r \mid E_r \neq \{0\}\}$. The elements of $P(\pi)$ are the *weights of* π. They form a finite set, invariant under W, since the latter obviously permutes the E_r's.

By the Lie-Kolchin theorem, $\pi(B)$ can be put in triangular form, in suitable coordinates, which implies in particular that there is a least one line $D \subset E$ stable under B. We then get a one-dimensional representation of B, which is described by a rational character of B (or T).

2.4. Theorem. *Let* $\pi: G \to GL(E)$ *be a rational irreducible representation of* G.

(i) *There is a unique line D stable under B. The weight d_π of T on D is dominant, with multiplicity one. Every other weight $r \in P(\pi)$ is of the form*

$$(1) \qquad\qquad r = d_\pi - \sum_{\alpha \in \Delta} m_\alpha(r)\,\alpha \quad \text{with} \quad m_\alpha(r) \geqq 0 \quad \text{integral.}$$

(ii) *The map $\pi \mapsto d_\pi$ induces a bijection between isomorphism classes of rational irreducible representations of G and dominant weights belonging to $X(T)$.*

The weight d_π will be called the *highest weight* of π. In general, $P(\pi)$ may contain several dominant weights, but only one corresponding to a line stable under B.

Thus the dominant weights contained in $X(T)$ parametrize the irreducible linear representations of G. If G is simply connected, then $X(T) = P(\Phi)$ and all dominant weights in $P(\Phi)$ occur. In the general case, we can also say that the dominant weights in $P(\Phi)$ parametrize the rational irreducible *projective* representations of G.

2.5. References. Over $\mathbf{C}$, this again goes back to E. Cartan and H. Weyl. In fact they proved these results in the complex analytic framework. See e.g. [22; 28]. The transition to algebraic groups is then automatic, if one takes into account the fact that a holomorphic representation of G is always rational for the associated algebraic group structure, which is clear from its description (see 1.5).

2.6. Remarks. (1) Like the classification, the parametrization of irreducible representations is independent of k. However some features of the representation with a given highest weight may depend on the characteristic of k; this is a source of problems, which are only partially solved and will be mentioned in § 5.

(2) In characteristic zero, all representations are fully reducible, but this is not so in positive characteristic (except for tori). In this connection we recall Mumford's conjecture*: "Let $\pi: G \to GL(E)$ be a rational representation of G and D a line sta-

---

* (Added October 30, 1974) This conjecture has recently been proved in the general case by W. Haboush – see Seshadri's article in this volume.

ble under G. Then there exists a homogeneous polynomial P on E which is invariant under G but not zero on D." If we had full reducibility, we could take for P a linear form defining an invariant hyperplane supplementary to D. In general, Mumford's conjecture calls for an invariant cone not containing D. In fact, this conjecture is usually stated more generally for representations of algebraic groups whose identity component is reductive (1.6(2)), the requirement on P being that the line $k \cdot P$ be stable under G. It would have many interesting consequences [27; 30]; so far, it has been proved only for groups whose identity component is isogeneous to a product of $\mathbf{S L}_2$ and tori [30].

2.7. The Highest Weight of the Contragredient Representation. There exists a unique element $w_0 \in W$ such that $w_0(\Phi^+) = \Phi^- = -\Phi^+$. We let i be the automorphism $r \mapsto -w_0(r)$ of $X(T)$. It leaves Φ^+, Δ and the set of fundamental dominant weights stable. Both w_0 and i have order ≤ 2. The latter is called the *opposition involution*. If $-\mathrm{Id}. \in W$, then it is equal to w_0 and i is the identity. This happens for the types $\mathbf{B}_n$, $\mathbf{C}_n$, $\mathbf{D}_n$ (n even), $\mathbf{G}_2$, $\mathbf{F}_4$, $\mathbf{E}_7$, $\mathbf{E}_8$.

Given a rational representation $\pi\colon G \to GL(E)$ we denote by $\check{\pi}$ the contragredient representation to π. Thus π operates on the dual space E^* to E and $\check{\pi}(g) = {}^t\pi(g^{-1})$ for all $g \in G$. Diagonalizing $\pi(T)$, we see immediately that $P(\check{\pi}) = -P(\pi)$. If π is irreducible, then so is $\check{\pi}$, and it follows from the above that

$$(1) \qquad\qquad d_{\check{\pi}} = i(d_\pi).$$

In particular, if $w_0 = -\mathrm{Id}.$, then every irreducible representation is selfcontragredient.

§ 3. Infinitesimally irreducible representations. The tensor product theorem

3.1. Let $\pi\colon G \to GL(E)$ be a rational representation. Its differential d_π at e defines a homomorphism of Lie algebras $L(G) \to \mathfrak{g}l(E)$, which is also a homomorphism of restricted Lie algebras if $p > 0$. We say that π is *infinitesimally irreducible* if d_π is irreducible. If π is infinitesimally irreducible, then it is irreducible. The converse is true in characteristic zero (and follows from the connection between Lie algebras and Lie groups given by the exponential mapping), but not in positive characteristic.

In this section, we assume $p > 0$, and let $M(G)$ be the set of rational infinitesimally irreducible representations of G. The results will be stated for simply connected groups and linear representations. Alternatively, we could drop the assumption G simply connected, but then consider projective representations.

3.2. Theorem. *Let G be simply connected. An irreducible representation π of G is infinitesimally irreducible if and only if its highest weight d_π is of the form*

$$d_\pi = \sum_{a \in \Delta} m_a(\pi)\, d_a \quad \text{with} \quad 0 \leq m_a(\pi) < p.$$

This shows in particular that $M(G)$ has p^r elements ($r = \mathrm{rk}\, G$). This theorem is due to C. Curtis and R. Steinberg. More precisely, C. Curtis determined (with some restrictions on the characteristic) all restricted irreducible representations of $L(G)$ [12], and Steinberg showed all these representations to be differentials of representations of G [34], see also [3].

Let Fr: $x \mapsto x^p$ be the Frobenius automorphism of k. Given a rational representation π of G on E, we get another rational representation by identifying $GL(E)$ to $\mathbf{GL}_n$ ($n = \dim E$) via a basis of E and applying Fr to the coefficients of $\pi(g)$ ($g \in G$). This representation will be denoted π^{Fr}; its equivalence class depends only on that of π. Clearly, $P(\pi^{\mathrm{Fr}}) = p \cdot P(\pi)$. If π is irreducible, so is π^{Fr} and its highest weight is $p \cdot d_\pi$. Thus, the representation with highest weight $p^i \cdot d_\pi$ is obtained by applying Fr^i to π and has the same degree as π (whereas this degree tends to infinity with i, in characteristic zero, unless $\pi = 1$).

3.3. Theorem. *Let G be simply connected. Let m be a positive integer, and $\pi_i \in M(G)$ ($0 \leq i \leq m$). Then the tensor product $\pi_0 \otimes \pi_1^{\mathrm{Fr}} \otimes \ldots \otimes \pi_m^{\mathrm{Fr}^m}$ is an irreducible representation, with highest weight $d = \sum\limits_{0 \leq i \leq m} p^i \cdot d_{\pi_i}$. Any non-trivial irreducible representation of G can be written uniquely in this way, for a suitable m, with $\pi_m \neq 1$.*

This theorem is due to R. Steinberg [34]. (See also [3; 35].)

Let k_0 be a finite subfield of k. Assume that G is obtained by extension of the ground field from a group over k_0, also to be denoted G, which is *split over k_0* (i.e. contains a T defined over k_0 and isomorphic over k_0 to product of $\mathbf{GL}_1$'s). The groups $G(k_0)$ obtained in this way are the *finite Chevalley groups.*

3.4. Theorem. *Let G be simply connected. Let $p^m = q = \mathrm{Card}\, k_0$. The restrictions to $G(k_0)$ of the q^r representations $\pi_0 \otimes \pi_1^{\mathrm{Fr}} \otimes \ldots \otimes \pi_{m-1}^{\mathrm{Fr}^{m-1}}$ ($\pi_i \in M(G)$) form a complete set of irreducible representations of $G(k_0)$ over k. In particular, the elements of $M(G)$ define irreducible representations of $G(k_0)$ and all irreducible representation of $G(k_0)$ over k are restrictions of rational irreducible representations of G.*

This is due to C. Curtis [13] and Steinberg [34]. For other accounts, see [3; 35]. In [34; 35], there are analogues for the so-called "twisted Chevalley groups": groups $G(k_0)$, where G is defined over k_0, but not necessarily split over k_0, or groups of Ree or Suzuki type. For a systematic discussion of those groups, see e.g. [7; 35]. A further extension to finite groups with "split BN pairs" was given by F. Richen and C. Curtis (see e.g. [14]).

The representations occurring in 3.4 are those with highest weight of the form $d = \sum\limits_{a \in \Delta} m_a \cdot d_a$ with $0 \leq m_a < q$. The one for which the m_a are all equal to $q - 1$ has degree p^{Nm}, where $N = \mathrm{Card}\, \Phi^+$, and any irreducible representation of $G(k_0)$ non-equivalent to it has a strictly smaller degree [34]. The number p^{Nm} is also the order of the group $U(k_0)$, which is a p-Sylow subgroup of $G(k_0)$. A similar representation exists over any field F, and is irreducible if the characteristic of F is zero, or prime to the index of $B(k_0)$ in $G(k_0)$ [33]. It is now called the Steinberg representation of $G(k_0)$. Again, it also exists for twisted Chevalley groups; it has been given a cohomological interpretation in terms of the Tits building of $G(k_0)$ by Solomon and Tits [31].

§ 4. Construction of irreducible representations

Our main purpose in this section is to outline the realization of irreducible representations via cross sections of line bundles on G/B (4.3; 4.4), which leads to the existence assertion in 2.4 (ii). However, since this is quite short, we shall also sketch the proof of the remaining assertions of 2.4. For more details, see [3; 10; 35].

4.1. Let $B^- = w_0 \cdot B \cdot w_0$, where w_0 is as in 2.7. The group B^- is the semi-direct product of T and of $U^- = w_0 \cdot U \cdot w_0$. The Lie algebra of U^- is spanned by the $L(G)_\alpha$ $(\alpha < 0)$. The set $U^- \cdot B = B^- \cdot B$ is Zariski-open in G. As an example, if $G = \mathbf{SL}_n$, we may take for B the group of upper triangular matrices. Then B^- is the lower triangular group, and $U^- \cdot B$ the set of $g = (g_{ij}) \in \mathbf{SL}_n$ for which the principal minors $\det (g_{ij})_{1 \leq i, j \leq m}$ are $\neq 0$ for $1 \leq m \leq n$.

For each $a \in \Phi$, there is an injective morphism x_a of the additive group of k into G, whose image is normalized by T, and which satisfies

$$(1) \qquad t \cdot x_a(z) \cdot t^{-1} = x_a(t^a \cdot z), \qquad (t \in T; z \in k).$$

The group U (resp. U^-) is generated by the images of the x_a for $a \in \Phi^+$ (resp. $a \in \Phi^-$).

4.2. Let now $\pi\colon G \to GL(E)$ be a rational representation of G and $r \in P(\pi)$. We claim first that

$$(1) \qquad \pi(x_a(z)) \cdot E_r \subset \sum_{i \geq 0} E_{r+i \cdot a}, \qquad (z \in k; a \in \Phi).$$

Let $y \in E_r$. Then we may write $x_a(z) \cdot y = \sum_s e_s(z)$, where $s \in P(\pi)$ and $z \mapsto e_s(z)$ is a regular function from k to E_s, hence an E_s-valued polynomial in z. We have then

$$(2) \qquad \pi(x_a(z)) \cdot y = \sum_{i \geq 0, s} z^i \cdot e_{s,i}, \qquad (e_{s,i} \in E_s).$$

Since $\pi(t) = t^s \cdot \mathrm{Id}$ on E_s for $t \in T$, we have

$$\pi(t \cdot x_a(z) \cdot t^{-1}) \cdot y = t^{-r} \sum_{s,i} z^i \cdot t^s \cdot e_{s,i},$$

and on the other hand

$$\pi(x_a(t^a \cdot z)) \cdot y = \sum_{s,i} z^i \cdot t^{ia} \cdot e_{s,i}.$$

Comparison of coefficients, and 4.1 (1), then show that $e_{s,i} \neq 0$ implies $s = r + i \cdot a$.

Let now D be a line stable under B and d the corresponding weight. Iterated application of (1) yields

$$(3) \qquad \pi(U^- \cdot B)(D) = \pi(U^-) \cdot D \subset D + \sum_s E_s,$$

where s runs through weights of the form

(4) $\qquad s = d - \sum_{a \in \Delta} m_a(s) \cdot a$, with the $m_a(s) \geq 0$, integral, not all zero.

Since $U^- \cdot B$ is Zariski dense in G, it follows that the Zariski closure of $G(D)$ is contained in the right-hand side of (3), hence so is the smallest G-invariant subspace containg D. In particular, if E is irreducible, then E is equal to the right-hand side, which proves that the weight d of T on D has multiplicity one and all other weights are of the form (4). On the other hand, if s_a is the reflection of the Weyl group associated to $a \in \Delta$, then $s_a(d) = d - 2(d, a) \cdot (a, a)^{-1} a$, whence $(d, a) \geq 0$; this shows that d is dominant and concludes the proof of 2.4 (i).

4.3. We shall denote by A the coordinate ring $k[G]$ of G. The group G acts on A by left and right translations. For $g \in G, f \in A$, let $l_g f$ (resp. $r_g f$) be defined by

(1) $\qquad\qquad l_g f(x) = f(g^{-1} \cdot x), \qquad (\text{resp. } r_g f(x) = f(x \cdot g)).$

We have then $l_{g \cdot g'} = l_g \cdot l_{g'}$ (resp. $r_{g \cdot g'} = r_g \cdot r_{g'}$). It is elementary that every $f \in A$ is contained in a finite dimensional subspace of A which is left and right invariant under G.

4.4. Definition. For $r \in X(T)$, let $A_r = \{f \in A \mid r_b f = b^r f, b \in B)\}$. This is a G-module with respect to left-translations. It can be canonically identified to the space of regular cross-sections of the line bundle ξ_{-r} on G/B associated to $-r$, on which G acts by left translations. (We recall that the total space of ξ_s ($s \in X(B)$) is the quotient of $G \times k$ by the equivalence relation $(g, x) \sim (g \cdot b, b^{-s} \cdot x)$ ($g \in G; x \in k)$.) Since G/B is complete, this space is finite-dimensional. However, this fact is not really needed in the sequel.

4.5. Proposition. *Assume $A_{i(r)} \neq \{0\}$. Then the space $A_{i(r)}$ contains a unique line D_r stable under B. The weight of B on D_r is r and is dominant. The smallest non-zero G-invariant subspace E_r of $A_{i(r)}$ contains D_r and is an irreducible module with highest weight r. Every irreducible G-module with highest weight r is isomorphic to E_r. If $p = 0$, then $A_{i(r)} = E_r$. If r is not dominant, then $A_r = \{0\}$.*

Let $D \subset A_{i(r)}$ be a line stable under B and s the corresponding weight. Let $f \in D$. We have then

$$f(b \cdot x \cdot b') = b^{-s} \cdot f(x) \cdot b'^{i(r)}.$$

Since $B \cdot w_0 \cdot B = w_0 B^- B$ is open in G (4.1), this shows that f is completely determined by its value at w_0. In particular $f \neq 0$ if and only if $f(w_0) \neq 0$. There exists therefore at most one B-stable line. Moreover

$$f(t \cdot w_0) = t^{-s} \cdot f(w_0) = f(w_0 \cdot w_0^{-1} \cdot t \cdot w_0) = f(w_0) \cdot (w_0^{-1} \cdot t \cdot w_0)^{i(r)} = f(w_0) \cdot t^{-r}.$$

whence $s = r$.

Since every non-zero G-stable subspace of $A_{i(r)}$ contains a line stable under B, it contains D, which proves the third assertion. The second and the last one then follow from 4.2. Assume $p = 0$. By full reducibility, every proper G-invariant submodule has a G-invariant complement. They cannot contain both D, hence

$$(1) \qquad E_r = A_{i(r)} = H^0(G/B; \xi_{-i(r)}) \quad \text{if} \quad p = 0.$$

Let now (π, E) be an irreducible representation with highest weight r. Given $e \in E, f \in E^*$, let $c_{e,e*}$ be the "coefficient" of π, given by $c_{e,e*}(g) = \langle e^*, g^{-1}(e) \rangle$ where $\langle , \rangle$ is the canonical bilinear form. We have $c_{g \cdot e, h \cdot e*} = l_g \cdot r_h \cdot c_{e,e*}$. In particular, for fixed e^*, the map $q_{e*}: e \mapsto c_{e,e*}$ sends G-equivariantly E into A (acted upon by left translations). Take now for e (resp. e^*) a highest weight vector of π (resp. $\check{\pi}$). The corresponding weight is then r (resp. $i(r)$), from which it follows that q_{e*} maps E injectively into $A_{i(r)}$; the uniqueness assertion of 2.4 (ii) follows.

In order to complete the proof of 2.4, there remains only to show the existence of an irreducible representation whose highest weight is a given dominant weight $r \in X(T)$. This amounts to proving that $A_r \neq \{0\}$ for every such r. This can be done directly (see e.g. [3: §5]) or follows by "reduction mod p" from a construction over $\mathbf{Z}$, which is sketched in §5 (see 5.4).

§ 5. Group schemes over Z.
Reduction mod p of irreducible representations

5.1. We shall first outline Chevalley's construction of a group scheme over $\mathbf{Z}$ already referred to in 1.5.

The starting points is, for $k = \mathbf{C}$, a suitable $\mathbf{Z}$-structure on $L(G)$, defined by means of what is now called a Chevalley basis [3; 7; 9; 22; 35]. It consists of elements $e_a \in L(G)_a$ $(a \in \Phi)$ and of a suitable basis of $L(T)$. (In fact, there may be several possibilities for the $\mathbf{Z}$-structure of $L(T)$: the space $L(T)$ may be identified to the dual of $X(T) \otimes \mathbf{C}$, and the dual lattice to any lattice between $R(\Phi)$ and $P(\Phi)$ is suitable.)

Let now (π, E) be a faithful finite dimensional complex representation of $L(G)$. It may also be viewed as a rational representation of G, assumed to be simply connected. A $\mathbf{Z}$-form $E_{\mathbf{Z}}$ of E is said to be *admissible* if

$$(1) \qquad \pi(e_a^j/j!)(E_{\mathbf{Z}}) \subset E_{\mathbf{Z}}, \quad \text{for all} \quad a \in \Phi \quad \text{and} \quad j \geq 0.$$

This implies that $E_{\mathbf{Z}}$ is the direct sum of its intersections with the weight spaces $E_s(s \in P(\pi))$ [3: 2.3]. Chevalley has shown that any E has an admissible $\mathbf{Z}$-form (This is stated in [11], proved in [3; 22; 35]). This really amounts to showing the existence of a nice $\mathbf{Z}$-form $U_{\mathbf{Z}}$ of the universal enveloping algebra U of $L(G)$ ([26]; see also [22; 35]).

Identify E to $\mathbf{C}^n$ $(n = \dim E)$ via a basis of $E_{\mathbf{Z}}$. We have then

$$\pi(x_a(t)) = \sum_{j \geq 0} t^j \, \pi(e_a^j/j!) \in \mathbf{SL}_n(\mathbf{Z}[t]),$$

$[\pi(e_a)$ is nilpotent, hence the exponential series is in fact a finite sum.] and this makes sense in fact for any indeterminate t. Let now F be a commutative field, and let

$$(2) \qquad G_{\pi,F} = \langle \pi(x_a(z)) \mid a \in \Phi, z \in F \rangle$$

be the subgroup of $\mathbf{SL}_n(F)$ generated by the elements $\pi(x_a(z))$ $(a \in \Phi, z \in F)$. The groups so constructed are the "Chevalley groups." If G is simple, then $G_{\pi,F}$ is simple modulo its center, except in a few cases where F has two or three elements [6; 9; 35]. This construction was first made by Chevalley when π is the adjoint representation [9] and yielded the first new finite simple groups since Dickson.

5.2. Theorem. *Let Γ_π be lattice of $X(T)_{\mathbf{R}}$ spanned by the weights of π. Then $G_{\pi,k}$ is a connected semi-simple k-group with diagram (Φ, Γ_π).*

More precisely, if $A_{\pi,\mathbf{Z}}$ is the ring of regular functions on $G_{\pi,\mathbf{C}}$ generated by the matrix coefficients, the statement is in fact that $A_{\pi,\mathbf{Z}}$ represents a smooth group scheme over $\mathbf{Z}$, and that $G_{\pi,k} = \mathrm{Hom}\,(A_{\pi,\mathbf{Z}}, k)$ [11; see also [3; 15; 16]]. There is also a functorial complement to it: if π' is another faithful representation of $L(G)$, and $\Gamma_{\pi'} \subset \Gamma_\pi$, then there is a canonical morphism $A_{\pi',\mathbf{Z}} \to A_{\pi,\mathbf{Z}}$ (an isomorphism if $\Gamma_{\pi'} = \Gamma_\pi$) such that the associated morphism

$$G_{\pi,k} = \mathrm{Hom}\,(A_{\pi,\mathbf{Z}}, k) \to \mathrm{Hom}\,(A_{\pi',\mathbf{Z}}, k) = G_{\pi',k}$$

is the central isogeny mentioned in 1.4.

5.3. The existence part of 1.4 (see 1.5 (i)) is then a consequence of the fact that any Γ between $R(\Phi)$ and $P(\Phi)$ is a Γ_π. In particular, if $\Gamma_\pi = P(\Phi)$, then $G_{\pi,k}$ is the simply connected group with root system Φ. In the sequel we denote it by G_k, and let $G/\mathbf{Z}$ be the corresponding group scheme over $\mathbf{Z}$. Then, for any π, we have a rational representation $\pi \otimes k: G_k \to G_{\pi,k}$ obtained by reduction mod k from a $\mathbf{Z}$-form of $\pi: G_{\mathbf{C}} \to GL(E)$. In particular, it has the same degree as π. We may also say that it has the "same character." In fact, Chevalley's construction of $G/\mathbf{Z}$ also yields schemes $T/\mathbf{Z}$ and $B/\mathbf{Z}$ which give our earlier $T, B \subset G_k$ by specialization. In particular, there are natural identifications

$$X(T/\mathbf{Z}) = X(T_{\mathbf{C}}) = X(T_k) = P(\Phi),$$

and the character of π or $\pi \otimes k$ may be written $\sum s \cdot \dim E_s$, $(s \in P(\pi) \subset P(\Phi))$. In this connection, we note that if $R_k(G_k)$ (resp. $R_{\mathbf{Z}}(G/\mathbf{Z})$) is the Grothendieck group of rational representations of G_k (resp. of $G/\mathbf{Z}$-rational modules, of finite type over $\mathbf{Z}$), then $R_k(G_k)$ and $R_{\mathbf{Z}}(G/\mathbf{Z})$ are canonically isomorphic [29: 3.6, 3.7]. In particular $R_k(G_k)$ and $R_{\mathbf{C}}(G_{\mathbf{C}})$ are isomorphic by an isomorphism which is compatible with the identification of characters.

5.4. Assume from now on $G_{\mathbf{C}}$ to be simple. Let d be a dominant weight, and (π_d, E) the corresponding irreducible representation of $G_{\mathbf{C}}$. If x_0 is a non-zero highest weight vector, then $E_{\mathbf{Z}} = U_{\mathbf{Z}}(x_0)$ is an admissible $\mathbf{Z}$-form of E and $(\pi \otimes k, E_k)$, where $E_k = E_{\mathbf{Z}} \otimes k$, is a rational representation of G_k. It is not necessarily irreducible (as is already clear from the end remark of 3.2, and 3.3); however, it has a biggest proper invariant subspace F and the representation $\pi_{d,k}$ of G_k in E_k/F is irreducible, with highest weight d. This then yields the existence part of 2.4 (ii), and also shows that $A_d \neq \{0\}$ if d is dominant (4.5). Furthermore we see that $P(\pi_{d,k}) \subset P(\pi_d)$.

In some cases, $\pi \otimes k$ is irreducible for all k. This happens if all weights of π are extremal (i.e. form only one orbit of W) [3: 5.12]. Examples are the spinor or half-spinor representations of **Spin**(n) or the exterior power representations of $\mathbf{SL}_n$. However, if $G_\mathbf{C}$ is also adjoint, no non-trivial representation of $G_\mathbf{C}$ has only extremal weights since every representation has the weight zero. For instance, the seven-dimensional representation of $\mathbf{G}_2$ has the zero weight with multiplicity one, besides extremal weights, and is in fact reducible in characteristic two.

There arises therefore the problem of knowing when $\pi \otimes k$ is irreducible, or more generally describing a composition series for $\pi \otimes k$ and finding a formula for the character and the degree of $\pi_{d,k}$, which would be the analogues of H. Weyl's formulae in characteristic zero [22; 28]. We note that the equivalence class of $\pi \otimes k$ may depend on the choice of an admissible $\mathbf{Z}$-form of π; however, the composition factors do not (and depend only on the characteristic of k); moreover, there is, up to dilation, a natural "minimal" one [21].

Given d, the representation $\pi \otimes k$ is irreducible if p is big enough [32], but p may depend on d, and not just on the isomorphism class of $G_\mathbf{C}$. There are a number of conjectures and results on this general problem, due notably to B. Braden, C. W. Curtis, R. Carter and G. Lusztig, J. E. Humphreys, J. C. Jantzen, D. Verma, W. J. Wong, for which we refer the reader to [8; 21; 23; 36].

§ 6. Cohomology of line bundles on G/B

6.1. To a rational character r of B there is associated a line bundle ξ_r on G/B (see 4.4), and hence coherent sheaf cohomology spaces $H^i(G/B; \xi_r)$; these are finite-dimensional (since G/B is complete) vector spaces over k, on which G acts rationally, the action stemming from left translations, and we are interested in identifying the representations thus obtained. We already saw (4.5) that $A_r \cong H^0(G/B; \xi_{-r})$ is zero if r is not dominant and contains the irreducible representation with highest weight $i(r)$ if r is dominant.

6.2. Take $k = \mathbf{C}$. If r is dominant, then (4.5 (1)), $H^0(G/B; \xi_{-r})$ is the irreducible representation of G with highest weight $i(r)$. (Theorem of Borel-Weil); moreover, it follows readily from Kodaira's vanishing theorem that $H^i(G/B; \xi_{-r}) = 0$ for $i \geq 1$.

Let ϱ be half the sum of the positive roots. Let r be a weight and assume $r + \varrho$ to be regular, i.e. not orthogonal to any root. There are then a unique $w \in W$ and a unique dominant weight s such that $r + \varrho = w(s + \varrho)$. Let $n(w) = \mathrm{Card}\,(w(\Phi^-) \cap \Phi^+)$. Then $H^i(G/B; \xi_{-r}) = 0$ except for $i = n(w)$, in which case it is an irreducible G-module with highest weight $i(s)$. If $r + \varrho$ is singular (i.e. not regular), then $H^i(G/B; \xi_{-r}) = 0$ for all i's. These results are due to R. Bott [5]; for other proofs, see [17; 25].

6.3. For a general k we want, as in 5.4, to compare the situations over k and over $\mathbf{C}$. We revert to the notation of 5.3, hence write G_k, B_k instead of G, B, and view these groups as obtained by change of basis from group schemes over $\mathbf{Z}$. Since r is defined over $\mathbf{Z}$, the line bundle ξ_r is also a specialization of a line bundle over $\mathbf{Z}$. Let

$$\chi_F(\xi_r) = \chi(G_F/B_F; \xi_r) = \sum_i (-1)^i \dim H^i(G_F/B_F; \xi_r), \qquad (F = k, \mathbf{C}).$$

It follows from general principles that $\chi_{\mathbf{C}}(\xi_r) = \chi_k(\xi_r)$. Now these two alternating sums can be viewed as elements of $X(T_{\mathbf{C}})$ and $X(T_k)$ respectively, or also as elements of the Grothendieck groups of rational representations of G_k and $G_{\mathbf{C}}$. It seems rather likely that, as such, they correspond to each other under the isomorphisms mentioned in 5.3.

A natural question is whether $H^i(G_k/B_k; \xi_r) = 0$ for $i \geq 1$ if $-r$ is dominant. If so, the invariance of the Euler-characteristic would show that $\dim H^0(G_k/B_k; \xi_r) = \dim H^0(G_{\mathbf{C}}/B_{\mathbf{C}}; \xi_r)$. So far, this has been proved for $G_k = \mathbf{SL}_n(k)$ in [23], and for G_k classical or isomorphic to $\mathbf{G}_2$ in [1]; moreover [18] shows that it would follow in general from a conjecture on admissible $\mathbf{Z}$-forms.

It seems therefore rather likely that one has in characteristic p rather straightforward analogues of the results in characteristic zero for r dominant. However, this is certainly not so otherwise: for $G = \mathbf{SL}_3$, $(p > 2)$, Mumford has given an example of an $r \in X(T)$ such that $r + \varrho$ is singular and $H^i(G_k/B_k; \xi_{-r})$ has dimension one for $i = 1, 2$; I am told that recently, Larry Griffith has found many examples in arbitrary characteristic, for $G = \mathbf{SL}_3(k)$, of $r \in X(T)$, with $r + \varrho$ either singular or regular, for which $H^i(G_k/B_k; \xi_r) \neq 0$ for two dimensions. I do not know whether the representations of G_k occurring in these groups have been determined. This might be helpful in trying to see whether there is a general pattern.

6.4. Remark. Let P be a parabolic subgroup of G, i.e. a closed subgroup containing a conjugate of B. Let r be an irreducible rational representation of P and ξ_r the associated line bundle on G/P. The cohomology spaces $H^i(G/P; \xi_r)$ are again finite dimensional G-modules. The results of [5; 25] give more generally the structure of these G-modules for any P if $k = \mathbf{C}$. The speaker does not know of results pertaining to the analogous problem when $p > 0$, $P \neq B$ and r is not one-dimensional.

References

1. Bai, L., Musili, C., Seshadri, C. S.: Cohomology of line bundles on G/B, Annales E. N. S. (4) 7 (1974), 89–138
2. Borel, A.: Linear algebraic groups, Benjamin, New York 1969
3. Borel, A.: "Properties and linear representations of Chevalley groups," Seminar in algebraic groups and related finite groups, Springer Lecture Notes **131** (1970), 1–55
4. Borel, A., Tits, J.: Compléments à l'article: Groupes réductifs, Publ. Math. IHES **41** (1972), 253–276
5. Bott, R.: Homogeneous vector bundles, Annals of Math. (2) **66** (1957), 203–248
6. Bourbaki, N.: Groupes et algèbres de Lie, Chap. IV, V, VI, Paris Hermann 1968
7. Carter, R.: Simple groups of Lie type, Pure and Applied Math. XXVIII, J. Wiley and Sons, New York (1972)
8. Carter, R., Lusztig, G.: On the modular representations of the general linear and symmetric groups, Math. Z. **136** (1974), 193–242

9. Chevalley, C.: Sur certains groupes simples, Tôhoku Math. J. (2) **7** (1955), 14−66
10. Chevalley, C.: Classification des groupes de Lie algébriques, Notes polycopiées, Inst. H. Poincaré, Paris (1956−58)
11. Chevalley, C.: Certain schémas de groupes semi-simples, Sém. Bourbaki, 13è année (1960−61), Exp. 219
12. Curtis, C. W.: Representations of Lie algebras of classical type with applications to linear groups, J. Math. Mech. **9** (1960), 307−326
13. Curtis, C. W.: On projective representations of certain finite groups, Proc. A.M.S. **11** (1960), 852−860
14. Curtis, C. W.: Modular representations of finite groups with split (B, N) pairs, Seminar on algebraic groups and related finite groups, Springer Lecture Notes **131** (1970), 58−95
15. Demazure, M.: Schémas en groupes réductifs, Bull. Soc. Math. France **93** (1965), 369−413
16. Demazure, M.: Schémas en groupes III, Springer Lecture Notes **153** (1970)
17. Demazure, M.: Une démonstration algébrique d'un théorème de Bott, Invent. Math. **5** (1968), 349−356
18. Demazure, M.: Désingularisation des variétés de Schubert généralisées, Annales E. N. S. (4) **7** (1974), 53−88
19. Hochschild, G.: The structure of Lie groups, Holden-Day Inc. 1965
20. Hochschild, G., Mostow, G. D.: On the algebra of representative functions on an analytic group, Amer. J. M. **83** (1961) 111−136
21. Humphreys, J. E.: Modular representations of classical Lie algebras and semi-simple groups, J. of Algebra **19** (1971), 51−79
22. Humphreys, J. E.: Introduction to Lie algebras and representation theory, Graduate Texts in Math. 9, Springer Verlag 1972
23. Jantzen, J. C.: Darstellungen halbeinfacher algebraischer Gruppen und zugeordnete Kontravariante Formen, Bonner math. Schriften Nr. 67, Universität Bonn, 1973
24. Kempf, G.: Schubert methods with an application to algebraic curves, Stichting mathematisch centrum, Amsterdam (1971)
25. Kostant, B.: Lie algebra cohomology and the generalized Borel-Weil theorem, Ann. of Math. (2) **74** (1961), 329−387
26. Kostant, B.: Groups over **Z**, Algebraic groups and discontinuous subgroups, Proc. Symp. pure math. **9**, A. M. S. Providence, R. I. (1966)
27. Nagata, M.: Invariants of a group in an affine ring, J. Math. Kyoto Univ. 3 (1964), 369−377
28. Serre, J.-P.: Algèbres de Lie semi-simples complexes, Benjamin 1966
29. Serre, J.-P.: Groupes de Grothendieck des schémas en groupes réductifs déployés, Publ. Math. I. H. E. S. **34** (1968), 37−52
30. Seshadri, C. S.: Mumford's conjecture for $GL(2)$ and applications, Proc. Bombay Colloquium on algebraic geometry, 1968, Oxford University Press 1969, 347−371
31. Solomon, L.: The Steinberg character of a finite group with BN-pair, Theory of finite groups, A Symposium, ed, by R. Brauer and C. H. Sah, Benjamin 1969, 213−221
32. Springer, T. A.: Weyl's character formula for algebraic groups, Invent. Math. **5** (1968), 85−105
33. Steinberg, R.: Prime power representations of finite linear groups II, Canadian J. M. **9** (1957), 347−351
34. Steinberg, R.: Representations of algebraic groups, Nagoya M. J. **22** (1963), 33−56
35. Steinberg, R.: Lectures on Chevalley groups, Notes by J. Faulkner and R. Wilson, Yale University (1967)
36. Verma, D.: Role of affine Weyl groups in the representation theory of algebraic Chevalley groups and their Lie algebras. Lie groups and their representations Budapest 1971, 151−217

The Institute for Advanced Study
Princeton, New Jersey 08540

103.

Formes automorphes et séries de Dirichlet
(d'après R. P. Langlands)

Séminaire Bourbaki, Exp. 466 (1974/75),
Lect. Notes Math. **514** (1976) 183−222

Cet exposé tente de donner un aperçu d'un ensemble de résultats, problèmes et conjectures qui établissent, en fait ou conjecturalement, des liens étroits entre formes automorphes sur des groupes réductifs, ou représentations de tels groupes, et une classe générale de produits eulériens comprenant beaucoup de ceux que l'on rencontre en arithmétique ou géométrie algébrique.

A l'heure actuelle, plusieurs des conjectures ou «questions» paraissent tout à fait inaccessibles sous leur forme générale. Elles définissent plutôt un vaste programme, élaboré par R. P. Langlands depuis environ 1967, souvent appelé la «philosophie de Langlands,» et déjà illustré de façon très frappante par les résultats classiques ou récents qui sont à sa source, et ceux qui ont été obtenus depuis.

Pour décrire ce programme, on est malheureusement condamné à des préliminaires techniques d'une nature et d'une longueur que beaucoup trouvent décourageantes. Pour y remédier ici dans la mesure du possible, on procèdera par approximations, en commençant par des énoncés vagues ou particuliers, que l'on cherchera ensuite à préciser ou généraliser, sans du reste prétendre atteindre la précision ou généralisation ultime.

Le problème à la base de ce programme est d'associer à une forme automorphe ou à certaines représentations de groupes adéliques (resp. à certaines représentations de groupes sur des corps locaux) un produit (resp. facteur) eulérien. On le discutera en quatre étapes. Cela conduit à un exposé qui se divise en quatre parties, dont les trois premières sont consacrées essentiellement à $\mathbf{GL}_n$.

(i) On associe à une forme automorphe sur $\mathbf{GL}_n(A)/\mathbf{GL}_n(\mathbf{Q})$, fonction propre d'opérateurs de Hecke, un produit eulérien portant sur presque tous les p (2.4). Pour $n = 1$ (resp. $n = 2$) ce procédé redonne les séries L associées par E. Hecke à un Grössencharakter (resp. une forme modulaire). Le § 3 donne une première idée des problèmes posés par Langlands [27] à propos de cette application et le § 4 indique quelques résultats connus pour $n = 1, 2$.

(ii) On remplace ensuite les formes automorphes par certaines représentations (irréductibles, admissibles) de $\mathbf{GL}_n(A)$, ou plutôt d'une algèbre de Hecke convenable, qui interviennent dans des espaces de formes automorphes (5.1). On veut donc leur associer un produit eulérien. Mais elles se décomposent naturellement en un produit infini de représentations (irréductibles, admissibles) de $\mathbf{GL}_n$ sur des

374

corps locaux, et le problème se subordonne à une question locale (cf. 5.2, et 5.3–5.6 pour quelques résultats).

(iii) Le § 6 donne une formulation du problème local qui fait directement intervenir les représentations de groupes de Galois ou de groupes de Weil (6.3) et passe ensuite aux conjectures et résultats le concernant (6.4, 6.5).

(iv) Enfin, les §§ 7 et 8 sont consacrés au cas général. Le point nouveau essentiel est l'introduction, par Langlands [25, 27, 30], d'un groupe associé à un groupe connexe réductif G sur un corps local, sur lequel sont définis les facteurs L des représentations de G; aussi, suivant une suggestion de H. Jacquet, nous l'appellerons le L-groupe de G et le noterons $^L G$. Si $G = \mathbf{GL}_n$, ce groupe est $\mathbf{GL}_n(\mathbf{C})$, ou plutôt le produit direct de $\mathbf{GL}_n(\mathbf{C})$ par un groupe de Galois ou de Weil, ce qui nous a permis d'esquiver cette construction dans les six premiers paragraphes.

§ 1. Formes automorphes sur G_A/G_F

1.1. La théorie générale concerne un groupe linéaire réductif G défini sur un corps global F, qui sera pour nous un corps de nombres, sauf mention expresse du contraire. Pour ne pas avoir à y revenir, nous introduisons déjà ici quelques notations générales. Cependant, les §§ 2 à 6 sont essentiellement consacrés au cas $G = \mathbf{GL}_n$ et $F = \mathbf{Q}$, et le lecteur peut sans dommage spécialiser les notations à ce cas (cf. 1.2). Nous y utilisons celles du cas général lorsqu'il s'agit de notions ou résultats qui s'y transposent sans modification importance, ce que nous supposerons du reste être fait dans les §§ 7, 8.

On note V (resp. V_∞, resp. V_f) l'ensemble des places (resp. places archimédiennes, resp. non-archimédiennes) de F, F_v la complétion de F en v, O_F (resp. O_v) l'anneau des entiers de F(resp. F_v), $|\cdot|_v$ la valuation normalisée de F_v, et Nv le nombre d'éléments du corps résiduel $k_v = O_v/\pi_v \cdot O_v$, où π_v est une uniformisante locale. On suppose G réalisé linéairement (ou muni d'une O_F-structure). Si C est une O_F-algèbre, G_C ou $G(C)$ est le groupe des points de G à valeurs dans C. On note A ou A_F l'anneau des adèles de F et G_A ou $G(A)$ le groupe des adèles de G. Ce dernier est le produit restreint des $G_v = G(F_v)$ par rapport aux groupes $G(O_v)$, autrement dit la limite inductive (topologique) des produits

$$\prod_{v \in S} G_v \times \prod_{v \notin S} G(O_v)\,,$$

où S parcourt l'ensemble des parties finies de V contenant V_∞. L'anneau A_f des adèles finies est le produit restreint des F_v par rapport aux $O_v(v \in V_f)$ et G_A est le produit direct de $G_\infty = \prod_{v \in V_\infty} G_v$ et du produit restreint $G(A_f)$ des $G_v(v \in V_f)$ par rapport aux $G(O_v)$. On choisit un sous-groupe compact maximal K_∞ de G_∞ et on note K_0 le produit de K_∞ par $K_{0f} = \prod_{v \in V_f} G(O_v)$. Si K est un sous-groupe compact ouvert de $G(A_f)$, il existe une partie finie S de V_f telle que K soit le produit des $G(O_v)$ $(v \in V_f - S)$ par un sous-groupe compact ouvert du produit des $G_v(v \in S)$. On identifie G_F au sous-groupe de G_A obtenu en plongeant G_F diagonalement dans le produit des G_v. C'est un sous-groupe discret.

Si $n = 1$, alors $G_A = A^*$ est le groupe des idèles de F, $G_F = F^*$ et $C_F = G_A/G_F$ est le groupe des classes d'idèles de F. On appellera *Grössencharakter de F* un caractère de C_F, i.e. un homomorphisme continu de C_F dans $\mathbf{C}^*$. Il est de la forme $a \mapsto \chi(a) \, \|a\|^r$ où $|\chi(a)| = 1$, $r \in \mathbf{R}$ et $\|a\| = \prod_v |a_v|_v \; (a = (a_v) \in A^*)$.

1.2. Sauf mention expresse du contraire, on suppose dans les §§ 2 à 6 que $G = \mathbf{GL}_n$. Le centre Z de G est donc isomorphe à $\mathbf{GL}_1$. On supposera $F = \mathbf{Q}$ lorsque l'on établit des liens avec la théorie classique (1.3, 1.4, 2.5, en particulier). Dans ce cas V_∞ est formé d'une place ∞, V_f s'identifie à l'ensemble des nombres premiers et si $v = p \in V_f$, alors $F_v = \mathbf{Q}_p$, $O_v = \mathbf{Z}_p$. On convient que $O_\infty = \mathbf{R}$.

1.3. Pour fixer les idées, nous rappelons brièvement la notion de *forme automorphe* sur G_A, quoiqu'en fait la définition précise n'interviendra guère ici, vu l'absence totale de démonstrations.

Une fonction continue f à valeurs complexes sur G_A est une *forme automorphe* si:

(i) elle est invariante à droite par G_F;

(ii) elle est K_0-finie, i.e. l'ensemble de ses translatés à gauche par K_0 engendre un espace vectoriel sur $\mathbf{C}$ de dimension finie;

(iii) il existe un Grössencharakter ψ tel que $f(z \cdot g) = \psi(z) \cdot f(g)$ $(z \in Z_A,$ $g \in G_A)$;

(iv) il existe un idéal I de codimension finie de l'algèbre $\mathscr{Z}$ des opérateurs différentiels biinvariants sur G_∞ qui, quel que soit $y \in G(A_f)$, annule la fonction $x \mapsto f(x \cdot y)$ sur G_∞;

(v) f satisfait à une condition de croissance convenable.

Enfin, f est *parabolique* ou *cuspidale* si $\int\limits_{U_A/U_F} f(u \cdot x) \, du = 0$ quels que soient $x \in G_A$ et le radical unipotent U d'un F-sous-groupe parabolique propre de G.

Il résulte en particulier de (ii), (iv) que $f(x \cdot y)$, vue comme fonction de $x \in G_\infty$ et $y \in G(A_f)$, est analytique en x et localement constante en y; elle est aussi invariante à gauche par un sous-groupe compact ouvert de $G(A_f)$.

1.4. Cette notion n'est pas essentiellement plus générale que celle de forme automorphe sur G_∞/Γ, où Γ est un sous-groupe de congruence de $G(O_F)$. Indiquons rapidement le lien entre les deux, au moins dans le cas qui nous intéresse ici. Soit $G_{\mathbf{R}}^+$ l'ensemble des éléments de $G_{\mathbf{R}} = \mathbf{GL}_n(\mathbf{R})$ de déterminant > 0. Soit K un sous-groupe compact ouvert de $G(A_f)$, produit direct de groupes $K_p \in G_p$ supposés tels que $x \mapsto \det x$ applique K_p sur O_p^* quel que soit p. Le groupe $\Gamma = G_{\mathbf{Q}} \cap (G_{\mathbf{R}} \times K)$ est un sous-groupe de congruence de $G_{\mathbf{Z}}$ et on obtient ainsi un système cofinal de sous-groupes de congruence. En utilisant la relation

$$(1) \qquad \mathbf{SL}_n(A) = \mathbf{SL}_n(\mathbf{Q}) \cdot (\mathbf{SL}_n(\mathbf{R}) \times (\mathbf{SL}_n(A_f) \cap K))$$

(approximation forte pour $\mathbf{SL}_n$) et l'égalité

$$(2) \qquad A^* = \mathbf{Q}^* \cdot \mathbf{R}^+ \cdot \prod_p O_p,$$

qui exprime le fait que le nombre de classes de $\mathbf{Z}$ est 1, on obtient aisément $G_A = (K_f \times G_{\mathbf{R}}^+) \cdot G_{\mathbf{Q}}$. On a donc une identification naturelle

$$(3) \qquad K \backslash G_A / G_{\mathbf{Q}} = G_{\mathbf{R}} / \Gamma, \quad \text{avec} \quad \Gamma = (G_{\mathbf{R}} \times K_f) \cap G_{\mathbf{Q}},$$

qui permet d'associer canoniquement à une fonction sur $G_{\mathbf{R}}/\Gamma$ une fonction sur $G_A/G_{\mathbf{Q}}$ invariante à gauche par K. Les conditions données plus haut se traduisent en leurs analogues sur $G_{\mathbf{R}}$ (cf. par exemple [13; 14]).

1.5. Formes modulaires. On suppose ici $n = 2$. Alors $H = \mathbf{GL}_2(\mathbf{R})/\mathbf{O}(2)$ est le demi-plan de Poincaré, sur lequel $G_{\mathbf{R}}^+$ (resp. $G_{\mathbf{R}}$) opère par transformations holomorphes (resp. holomorphes ou antiholomorphes). Soit Γ un sous-groupe de congruence de $\mathbf{SL}_2(\mathbf{Z})$. Rappelons qu'une *forme modulaire de poids k* sur Γ est une fonction holomorphe sur H, satisfaisant à

$$(1) \qquad f\left(\frac{a z + b}{c z + d}\right) = (c z + d)^k \cdot f(z), \quad \left(z \in H; \begin{pmatrix} a & b \\ c & d \end{pmatrix} \in \Gamma\right)$$

et à une condition de régularité aux pointes d'un domaine fondamental. En particulier, si d est le plus petit entier > 0 tel que $z \mapsto z + d$ soit dans Γ, alors

$$(2) \qquad f(z) = \sum_{j \geqq 0} a_j \cdot e^{2\pi i j / d}.$$

f est parabolique si $a_0 = 0$. Posons

$$(3) \qquad j(g, z) = (\det g)^{-1/2} \cdot (c z + d), \quad \left(g = \begin{pmatrix} a & b \\ c & d \end{pmatrix} \in G_{\mathbf{R}}^+\right).$$

Soit $\tilde{f}$ la fonction sur $\Gamma \backslash G_{\mathbf{R}}^+$ définie par

$$\tilde{f}(g) = f(g \cdot i) \cdot j(g, i)^{-k};$$

alors $\tilde{f}$ est une forme automorphe sur $\Gamma \backslash G_{\mathbf{R}}^+$, invariante par rapport à $Z_{\mathbf{R}}^+$ et qui satisfait à

$$(4) \qquad f(x \cdot g) = e^{-i\theta(x) \cdot k} \cdot f(g), \quad (g \in G_{\mathbf{R}}; x \in \mathbf{O}(2) \text{ rotation d'angle } \theta(x)).$$

Un des cas les plus importants est celui où Γ est un groupe de Hecke

$$\Gamma_0(N) = \left\{ \begin{pmatrix} a & b \\ c & d \end{pmatrix} \in \mathbf{SL}_2(\mathbf{Z}) \,\middle|\, c \equiv 0 \bmod N \right\}$$

$(N \geqq 1)$. Si ψ est un caractère de $(\mathbf{Z}/N \cdot \mathbf{Z})^*$, alors on considère plus généralement des formes modulaires de poids k et caractère ψ, i.e. qui satisfont à

$$(5) \qquad f(\gamma z) = f(z) \cdot (c z + d)^k \cdot \psi(d), \quad \left(\gamma = \begin{pmatrix} a & b \\ c & d \end{pmatrix} \in \Gamma\right).$$

Les formes correspondantes sur $\Gamma\backslash G_{\mathbf{R}}$ ou sur $G_{\mathbf{Q}}\backslash G_A$ satisfont alors à une relation $f(z\cdot g)=\tilde{\psi}(z)\cdot f(g)$, où $\tilde{\psi}$ est un caractère convenable du centre de G [13], (pour ces relations, voir aussi [3; 7; 41]).

§ 2. Produit eulérien attaché à une forme automorphe

Dans cette première étape, nous ne définissons les facteurs locaux des produits eulériens que pour presque toutes les places finies, en évitant toute place où se présentent des phénomènes de ramification en des sens variés. Nous rappelons tout d'abord deux exemples fondamentaux.

2.1. Soit $n=1$. Une forme automorphe sur G_A est donc un Grössencharakter. On peut écrire $f=\prod_v f_v$ où f_v est un caractère de F_v^*. Soit S le complément dans V de l'ensemble des $v\in V_f$ pour lesquels le noyau de f_v contient O_v^*. Il est fini. On associe à f le produit

$$(1) \qquad L(s,f)=\prod_{v\notin S}(1-f_p(\pi_v)(Nv)^{-s})^{-1}.$$

Ce produit converge dans le demi-plan $Rs>1$ et admet un prolongement analytique méromorphe. Complété par des facteurs convenables pour les $v\in S_\infty$, il est holomorphe si f est non trivial, a un (seul) pôle simple si $f\neq 1$, pour $s=1$, et satisfait à une équation fonctionnelle reliant $L(s,f)$ et $L(1-s,\bar{f})$, (voir par ex. [44; 48]).

2.2. Pour ce §, *voir* par ex. [32], [41: Chap. 3]. Soit f une forme modulaire de poids k pour $\mathbf{SL}_2(\mathbf{Z})$. On lui associe, dans les notations de 1.5 (2), avec $d=1$, la série de Dirichlet

$$(1) \qquad L(s,f)=\sum_{n\geq 1} a_n/n^s.$$

Supposons $a_0=0$, i.e. f parabolique, $a_1=1$, et f fonction propre des opérateurs de Hecke $T(p)$. Alors on a

$$(2) \qquad L(s,f)=\prod_p (1-a_p\cdot p^{-s}+p^{(k-1)-2s})^{-1}.$$

D'après Hecke, $L(s,f)$ converge dans le demi-plan $Rs>1+(k/2)$. La fonction $R(s,f)=(2\pi)^{-s}\cdot\Gamma(s)\cdot L(s,f)$ a un prolongement analytique holomorphe borné dans les bandes verticales et satisfait à l'équation fonctionnelle

$$(3) \qquad R(s,f)=i^k\cdot R(k-s,f).$$

Des résultats semblables valent aussi pour les formes modulaires pour $\Gamma_0(N)$, mentionnées en 1.5.

2.3. Pour passer à des cas plus généraux, on utilise une définition «locale» des opérateurs de Hecke. Soit $\mathbf{T}_n$ le tore maximal standard de $G=\mathbf{GL}_n$ et soit W le groupe de Weyl de $\mathbf{GL}_n$ par rapport à $\mathbf{T}_n$; c'est donc le groupe des automorphismes de $\mathbf{T}_n$ associés aux permutations des vecteurs base canoniques de F^n. Soit $v \in V_f$ et soit $H_{0,v} = H(G_v, K_{0v})$ l'algèbre de Hecke de G_v par rapport à $K_{0v} = \mathbf{GL}_n(O_v)$, i.e. l'algèbre de convolution des fonctions complexes biinvariantes par K_{0v} et à support compact. On sait [35] que $H_{0,v}$ s'identifie canoniquement à l'algèbre P des fonctions polynomiales sur $\mathrm{Hom}(M, \mathbf{C}^*)$, où $M = \mathbf{T}_n(F_v)/\mathbf{T}_n(O_v)$, qui sont invariantes par W. Or $\mathrm{Hom}(M, \mathbf{C}^*)$ s'identifie à $\mathbf{T}_n(\mathbf{C})$, donc P s'identifie à l'algèbre des fonctions régulières sur $\mathbf{T}_n(\mathbf{C})$ qui sont invariantes par W. Si $\chi: H_{0,v} \to \mathbf{C}$ est un caractère, il existe une et une seule orbite de W dans $\mathbf{T}_n(\mathbf{C})$ telle que $h \mapsto \chi(h)$ soit l'évaluation en un point de l'orbite, et cela établit une bijection entre $\mathrm{Hom}(H_{0,v}, \mathbf{C})$ et $\mathbf{T}_n(\mathbf{C})/W$. D'autre part les orbites de W dans $\mathbf{T}_n(\mathbf{C})$ représentent (exactement) les classes de conjugaison d'éléments semi-simples de $\mathbf{GL}_n(\mathbf{C})$. Il existe donc une *bijection canonique entre caractères de degré un de* $H_{0,v}$ *et classes de conjugaison semi-simples de* $\mathbf{GL}_n(\mathbf{C})$. Lorsque $\mathbf{GL}_n(\mathbf{C})$ intervient dans cette capacité, on le note $^L G_v^0$ et on l'envisage comme un sous-groupe du L-groupe de G_v (cf. § 7).

On étend cette terminologie aux places infinies. Si $G_v = \mathbf{GL}_n(\mathbf{R})$ (resp. $\mathbf{GL}_n(\mathbf{C})$), soit $K_{0v} = \mathbf{O}(n)$ (resp. $\mathbf{U}(n)$). L'algèbre $H_{0,v}$ est l'algèbre de convolution des mesures à support compact sur G_v biinvariantes par K_{0v}. L'assertion précédente reste valable, et l'on pose de nouveau $\mathbf{GL}_n(\mathbf{C}) = {}^L G_v^0$.

2.4. Soit f une forme automorphe sur G_A/G_F. En particulier, f est invariante à gauche par un sous-groupe compact ouvert K_1 de $G(A_f)$. Il existe une partie finie S de V contenant V_∞ et telle que K_1 ait K_{0v} comme facteur direct si $v \notin S$. *Supposons que f soit fonction propre de $H_{0,v}$ pour $v \notin S$*, l'algèbre $H_{0,v}$ opérant par convolution à gauche, et soit χ_v le caractère de $H_{0,v}$ tel que $h * f = \chi_v(h) \cdot f$, $(h \in H_{0,v})$. A χ_v correspond par 2.3 une classe de conjugaison semi-simple $\{g_v\}$ de $^L G_v = \mathbf{G}\,\mathbf{L}_n(\mathbf{C})$. On associe alors à f le produit

$$(1) \qquad L(s,f) = \prod_{v \notin S} L(s, \chi_v)\,, \quad \text{avec} \quad L(s, \chi_v) = (\det(1 - N\,v^{-s} \cdot g_v))^{-1}\,.$$

Plus généralement, si r est une représentation de dimension finie de $^L G^0$, on pose

$$(2) \qquad L(s,f,r) = \prod_{v \notin S} L(s, \chi_v, r)\,, \quad \text{avec} \quad L(s, \chi_v, r) = (\det(1 - N\,v^{-s} \cdot r(g_v)))^{-1}\,.$$

Cette construction est donnée dans [25] pour les groupes déployés, dans [27] pour le cas général (cf. 7.4). Le produit $L(s,f,r)$ converge absolument pour $R\,s$ assez grand [25; 27].

2.5. Nous supposons ici $n = 2$, et faisons le lien avec 2.2. Soit $T(p)$ (resp. $\tilde{T}(p)$ l'opérateur de Hecke associé à la double classe de $\begin{pmatrix} 1 & 0 \\ 0 & p \end{pmatrix}$ modulo $\mathbf{SL}_2(\mathbf{Z})$ (resp. $\mathbf{GL}_2(\mathbf{Z}_p)$). Soit f comme en 2.2, soit $\tilde{f}$ la forme automorphe correspondante sur $G_A/G_\mathbf{Q}$ (*via* 1.4, 1.5). Alors [13: p. 48] on a

$$(1) \qquad (f \mid T(p))^\sim = p^{(k/2)-1}(\tilde{T}(p) * \tilde{f})\,.$$

Par conséquent, si f est fonction propre de $T(p)$, de valeur propre a_p, alors $\tilde{f}$ est fonction propre de $\tilde{T}(p)$ de valeur propre $\tilde{a}_p = p^{1-k/2} \cdot a_p$. D'autre part, f est invariante par Z_A, et par $\mathbf{GL}_2(\mathbf{Z}_p)$ pour tout p. L'algèbre $H_{0,p}$ est engendrée par $T(p)$ et la double classe $T(p,p)$ de $p \cdot \mathrm{Id.}$, donc $\tilde{f}$ est fonction propre de $H_{0,p}$. En fait, le procédé de 2.4 associe à $\tilde{f}$ le produit

$$(2) \qquad L(s,\tilde{f}) = \prod_p (1 - p^{(1-k)/2} a_p \cdot p^{-s} + p^{-2s})^{-1}.$$

On a donc $L(s,\tilde{f}) = L(s + (k-1)/2, f)$.

§ 3. Problèmes globaux

Ils se rangent en gros en trois types:

3.1. Prolongement analytique, équation fonctionnelle. Les résultats de Hecke rappelés en 2.1, 2.2, suggèrent évidemment de se demander si les séries L de 2.4 admettent un prolongement analytique méromorphe, n'ayant qu'un nombre fini de pôles, et si la définition de $L(s, f, r)$ peut être complétée aux $v \in S$ de manière à ce que $L(s, f, r)$ et $L(1 - s, f, r^{\vee})$, où $r^{\vee}$ est la représentation contragrédiente de r, soient liés par une équation fonctionnelle.

On y reviendra dans les §§ 4, 5, 6, 8.

3.2. Surjectivité. Etant donné f comme dans 2.3, on lui a associé presque tout $v \in V_f$ une classe de conjugaison semi-simple dans $\mathbf{GL}_n(\mathbf{C})$, autrement dit un polynôme caractéristique complexe, de degré n, à racines non nulles. Or il y a plusieurs procédés bien connus pour faire cela, en particulier

(i) *Séries d'Artin.* Soit E une extension galoisienne de F et soit σ une représentation linéaire complexe de degré n du groupe de Galois $G(E/F)$ de E sur F. Pour v non ramifiée dans E, soit φ_v un élément de Frobenius en v; il est déterminé à conjugaison près. On peut alors former la série

$$(1) \qquad L(s, \sigma, E/F) = \prod_{v \text{ non ram.}} (\det(1 - \sigma(\varphi_v) N v^{-s}))^{-1}.$$

Elle admet un prolongement méromorphe au plan complexe. On sait comment multiplier cette série par des facteurs locaux correspondant aux $v \in S$ de manière à ce que le produit ainsi complété soit méromorphe et ait une équation fonctionnelle liant $L(s, \sigma)$ et $L(1 - s, \sigma^{\vee})$. La conjecture d'Artin est que cette fonction est holomorphe si σ est irréductible non triviale. Pour la forme de l'équation fonctionnelle, cf. [5].

(ii) Soit Z une variété projective lisse sur F. Pour tout $v \in V_f$ où Z admet une bonne réduction $Z_{(v)} = Z \times \bar{k}_v$, soit φ_v l'endomorphisme de Frobenius de $Z_{(v)}$. Fixons un entier $m \geqq 1$. Pour l premier et premier à $N v$, l'endomorphisme φ_v induit un endomorphisme $\varphi_{v,m}$ du m-ième groupe de cohomologie l-adique de $Z_{(v)}$. Soit $P_v(T) = \det(1 - \varphi_{v,m} \cdot T)$. Il est à coefficients entiers rationnels indépendants de l [9]. On considère alors le produit

$$(2) \qquad L_m(s, Z) = \prod_v P_v(N v^{-s})^{-1},$$

étendu aux v de bonne réduction. On conjecture qu'il admet un prolongement analytique méromorphe à nombre fini de pôles. Modulo d'autres conjectures, non toutes réglées par [9], Serre [38] a montré comment compléter cette définition aux $v \in V$ restants, et donné la forme d'une équation fonctionnelle conjecturale liant les valeurs du produit complet en s et $m + 1 - s$.

(iii) Comme dans (i), à partir d'un système strictement compatible de représentations rationnelles l-adiques de $G(E/F)$ [37], voire d'un groupe de Weil.

Le problème est ici de savoir si 2.4 (ou son extension à des groupes plus généraux), donne beaucoup des séries obtenues par l'un de ces procédés, ou des variantes. La «philosophie» espère qu'il en est bien ainsi. Comme le montrent les exemples connus, dont certains seront mentionnés plus loin, cela est d'un intérêt considérable pour les deux classes d'objets en question.

3.3. Comparaison. Soit D une algèbre à division centrale sur F, de rang n. Soit H le F-groupe algébrique associé à D: pour toute F-algèbre L, le groupe $H(L)$ est le groupe des éléments inversibles de $D \otimes_F L$. Pour presque tout v, on a $H_v = \mathbf{GL}_{n,v}$. La construction de 2.4 associe donc aussi des séries $L(s, f)$ aux formes automorphes sur H_A/H_F qui sont fonctions propres des $H_{0,v}$ pour presque tout v. Le problème est ici de voir si les séries L obtenues à partir de H proviennent aussi de G, ou en tout cas s'il y a des liens entre elles et par suite entre formes automorphes sur $H_A/H_\mathbf{Q}$ et sur $G_A/G_\mathbf{Q}$. Pour $n = 2$, une réponse affirmative est donnée dans [20] (cf. 8.6(b)).

3.4. Il y a en gros deux types de réponses ou de conjectures concernant le problème de surjectivité. Dans l'un on impose des conditions analytiques au produit eulérien considéré, et à d'autres qui en sont dérivés, suivant le modèle de la rèciproque de Weil au théorème de Hecke (4.2). Une autre manière de poser le problème se trouve dans [27], et ne se formule bien qu'une fois introduits représentations, groupe de Weil et L-groupe (cf. 8.3). A titre d'exemple, donnons déjà ici un cas particulier de la question 3 de [27].

Soit $M = \prod_1^m \mathbf{GL}_{n_i}$, $(n_1 + \ldots + n_m \leqq n)$. On pose $^L M_v^0 = \prod_i \mathbf{GL}_{n_i}(\mathbf{C})$. Soit E comme en 3.2. Soit $\mu: {}^L M_v^0 \times G(E/F) \to {}^L G^0$ un homomorphisme continu, injectif sur $^L M_v \times \{1\}$, indépendant de v. Soit f une forme automorphe sur M_A/M_F, fonction propre des opérateurs de Hecke ($H_{0,v}$ est ici le produit des algèbres $H_{0,v}$ des facteurs) pour presque tout v, et soit $\{m_v\}$ la classe de conjugaison semi-simple de M_v associée à f et à un tel v par 2.4, ou son extension évidente à un produit. La question est alors de savoir si le produit

$$\prod \det \left(1 - \mu((m_v, \varphi_v))(Nv)^{-s}\right)^{-1}$$

étendu aux v précédents qui sont de plus non ramifiés en E, provient par 2.4 d'une forme automorphe sur G_A/G_F.

§ 4. Quelques réponses pour $\mathbf{GL}_1$ et $\mathbf{GL}_2$

4.1. Le cas particulier non trivial le plus simple de 3.4 est celui où $M = \{e\}$, et $n = 1$. Le produit 3.4(1) est une série d'Artin abélienne et l'on demande si elle est associée à un Grössencharakter. La réponse (affirmative) est la loi de réciprocité d'Artin pour les extensions cycliques, qui, plus précisément, établit une bijection entre caractères de degré 1 du groupe de Galois $G(F_s/F)$ d'une clôture séparable de F et les Grössencharaktere d'ordre fini de F. Cela démontre aussi la conjecture d'Artin (cf. 3.2) pour σ de degré 1. Dans [46], Weil a introduit une autre classe de Grössencharaktere qui doivent avoir une interprétation arithmétique intéressante, les Grössencharaktere χ de «type A_0»: la restriction de χ à $F_\infty^* = \prod_{v \in V_\infty} F_v^*$, identifié au groupe des points réels du groupe algébrique $R_{F/Q}\,\mathbf{GL}_1$, est le produit d'un caractère de groupe algébrique de ce groupe par un caractère d'ordre fini. Les Grössencharaktere de type (A_0) qui ne sont pas «triviaux,» i.e. produits de $\| \ \|^m (m \in \mathbf{Z})$ par un Grössencharakter d'ordre fini, donnent lieu à des séries L susceptibles d'intervenir dans des fonctions zêta de variétés algébriques, et dont certaines au moins sont connues pour le faire [46].

4.2. Soit $n = 2$. Hecke a déjà donné une réciproque aux résultats de 2.2 pour $\Gamma = \mathbf{SL}_2(\mathbf{Z})$. Si une série de Dirichlet $\sum a_n \cdot n^{-s} = D(s)$ se prolonge analytiquement en une fonction holomorphe et si $R(s) = (2\pi)^{-s}\,\Gamma(s) \cdot D(s)$ satisfait à $R(s) = i^k \cdot R(k-s)$ et est bornée dans des bandes verticales, alors $D(s)$ est associée à une forme modulaire pour $\mathbf{SL}_2(\mathbf{Z})$. Weil [47] a étendu cela aux formes modulaires pour les sous-groupes $\Gamma_0(N)$: il faut imposer des conditions du type précédent non seulement à $D(s)$ mais à des séries $\sum a_n \chi(n)/n^{-s}$ obtenues en tordant $D(s)$ par certains caractères de Dirichlet, avec de plus une valeur prescrite à l'avance, dépendant de χ, de la constante intervenant dans l'équation fonctionnelle. Ce résultat est généralisé dans [22; 49] (cf. 5.3).

4.3. Considérons la série d'Artin de 3.2 (i) pour $F = \mathbf{Q}$ et supposons σ irréducible de degré deux. Supposons que, pour tout caractère de degré un de $G(E/\mathbf{Q})$, la fonction $L(s, \sigma \otimes \chi)$ satisfasse à la conjecture d'Artin. Alors les résultats mentionnés en 4.2 entraînent que $L(s, f)$ provient d'une forme modulaire de poids 1. C'est donc un analogue de la loi de réciprocité pour des extensions non abéliennes dont le groupe de Galois a une représentation fidèle de degré deux, *mais* prouvé modulo la conjecture d'Artin. On préférerait bien entendu le démontrer directement et en déduire la conjecture d'Artin, comme pour $n = 1$. Il y a tout de même au moins un nouveau cas où Tate, Atkin and Co ont construit directement une forme modulaire dont la série de Dirichlet associée est la série d'Artin donnée, et ainsi prouvé la conjecture d'Artin pour cette dernière. Dans ce cas, l'image de σ dans $\mathbf{PGL}_2(\mathbf{C})$ est le groupe alterné $\mathscr{A}_4$, et on ne peut se ramener au cas abélien car aucun multiple de σ n'est monomial.

Par ailleurs [10] fournit une réciproque (cf. 6.5 (b)).

4.4. Soit f une forme modulaire pour un $\Gamma_0(N)$ de poids $k \geqq 2$ qui soit primitive. Alors Deligne a montré que sa série de Dirichlet était justiciable de 3.2 (ii) et ainsi déduit la conjecture de Ramanujan-Petersson de sa démonstration des conjectures de Weil [9] (en fait, cette réduction n'est explicitée dans [4] que pour la fonction τ de Ramanujan).

4.5. Considérons le produit eulérien défini à partir du H^1 d'une courbe elliptique sur F, que les spécialistes savent écrire pour tout v. Le théorème de Weil montre que, si ce produit et les produits obtenus en le tordant par certains

Grössencharaktere satisfont à des conditions d'analyticité convenables, alors le produit provient d'une forme modulaire de poids 2. Si F est un corps de fonctions, ces conditions sont satisfaites et l'on a un vrai théorème. Si $F = \mathbf{Q}$, elles ont été vérifiées dans de nombreux cas, et cela conduit à se demander si une courbe elliptique sur $\mathbf{Q}$ est isogène à un facteur de la jacobienne de la compactification naturelle d'un quotient $H/\Gamma_0(N)$ [47]. En fait, l'évidence en faveur de cette conjecture et la possibilité d'associer des séries L à des formes automorphes sur des groupes semi-simples [25] ont rapidement amené Serre à suggérer que les séries L de 3.2 (ii) proviennent de formes automorphes.

On reviendra encore plusieurs fois à $\mathbf{GL}_2$ (cf. 5.3, 6.4, 6.5, 8.4 d) e) f), 8.6).

§ 5. Facteur ou produit eulérien à associer à une représentation

5.1. Soit ψ un Grössencharakter de F. On note $_0L^2_\psi(G_A/G_F)$ ou $_0L^2_\psi$ l'espace des fonctions mesurables f sur G_A/G_F satisfaisant à 1.3 (iii), telles que $x \mapsto f(x)$ $\cdot \,|\psi(\det x^{-1})|^{1/n}$ soit de carré intégrable sur $G_A/Z_A G_F$ et qui sont cuspidales (cf. 1.3, la condition étant imposée pour presque tout $x \in G_A$). C'est un G_A-module, somme directe discrète à multiplicités finies de G_A-modules unitaires irréductibles. Les formes paraboliques sont, en gros, les combinaisons linéaires finies des éléments K_0-finis des composants irréductibles de $_0L^2_\psi$ et en forment un sous-espace dense. Soit (π, M) un tel composant. Il existe alors pour tout $v \in V$ une représentation unitaire irréductible (π_v, M_v) de G_v qui, pour presque tout v, est «de classe 1,» (l'espace $(M_v)^{K_{0v}}$ des vecteurs fixes par K_{0v} est $\neq 0$, et alors nécessairement de dimension 1), de sorte que l'on puisse envisager M comme un produit tensoriel complété $M = \hat{\otimes}_v M_v$, et l'espace M^∞ des formes automorphes contenues dans M comme un produit tensoriel $\otimes_v M^\infty_v$, où M^∞_v est l'espace des éléments K_{0v}-finis de M_v. Par définition $\otimes_v M^\infty_v$ est engendré par des produits $\otimes_v x_v$ avec $x_v \in M^\infty_v$, et $x_v = e_v$ pour presque tout v, où e_v est un générateur de $(M_v)^{K_{0v}}$ choisi une fois pour toutes.

Pour pouvoir faire entrer dans ce cadre des espaces de formes automorphes plus généraux, il est devenu usuel d'algébriser la situation et de considérer directement des espaces du type $\otimes_v M^\infty_v$ plutôt que des complétions. Malheureusement, pour v archimédienne, M_v n'est pas un G_v-module. Mais c'est tout de même un module pour K_{0v} et pour l'algèbre enveloppante de l'algèbre de Lie de G_v. On peut exprimer cela en envisageant cet espace comme un module sur une algèbre $\mathscr{H}_v$, dite aussi de Hecke, algèbre de convolution engendrée par les fonctions C^∞ à support compact K_{0v}-finies (des deux côtés) et les coefficients des représentations de dimension finie de K_{0v}. Pour $v \in V_f$ on note $\mathscr{H}_v$ l'algèbre de convolution des fonctions à support compact sur G_v qui sont biinvariantes par un sous-groupe compact ouvert de G_v (dépendant de la fonction). Dans tous les cas, on définit une notion de $\mathscr{H}_v$-module admissible; c'est, entre autres, un K_{0v}-module dont tous les éléments sont K_{0v}-finis et dans lequel les représentations de K_{0v} sont de multiplicité finie. (Pour $v \in V_f$, cette notion équivaut en fait à celle de G_v-module admissible.) Si (π_v, N_v) est admissible irréductible, et de classe 1 pour presque tout v, alors on peut former $N = \otimes_v N_v$ comme plus haut. C'est un $\mathscr{H}$-module irréductible, admissible, où $\mathscr{H}$ est l'algèbre de Hecke globale; par définition $\mathscr{H} = \otimes_v \mathscr{H}_v$ est le produit tensoriel des $\mathscr{H}_v$, i.e. l'espace engendré par les produits $\otimes_v h_v$ où les h_v sont égaux à 1 pour presque tout v, avec le produit évident. Dans la suite on considère uniquement des représentations $\pi = \otimes_v \pi_v$ de ce type. Ce sont donc des $\mathscr{H}$-modules, mais on se permettra d'en parler comme des G_A-modules. Une telle représentation est dite *automorphe* si elle est réalisée dans un $\mathscr{H}$-module de formes automorphes, ou plus généralement comme un sous-quotient d'un $\mathscr{H}_v$-module de formes automorphes. On note $\mathscr{A}(G_v)$ ou $\mathscr{A}_v$ l'ensemble des classes d'équivalence de $\mathscr{H}_v$-modules admissibles irréductibles (pour tout cela, voir [17; 22]).

5.2. On veut maintenant associer à une représentation automorphe (irréductible admissible) un produit eulérien indexé par V tout entier. Ce problème global se subordonne à une question locale: associer à toute représentation admissible irréductible de $\mathscr{H}_v$ un «facteur eulérien» convenable, i.e. si $v \in V_f$, une fonction de la forme $(P((Nv)^{-s}))^{-1}$ où $P \in \mathbf{C}[T]$ et $P(0) = 1$; si $v \in V_\infty$, un produit de facteurs $\pi^{-s/2} \Gamma((s+m)/2)$ et $\pi^{-s} \Gamma(s+m)$ par une exponentielle e^{as+b}. De façon un peu moins vague, Langlands [27] a d'abord posé les problèmes suivants (dans le cas général, en fait, cf. § 8) qui, pour $n = 1$, sont traités dans la Thèse de Tate [44].

(i) Fixons un caractère additif non trivial ψ_v de F_v. Il faut assigner à $\pi_v \in \mathscr{A}_v$ et à une représentation holomorphe r de dimension finie de $^L G^0$ un facteur eulérien $L(s, \pi_v, r)$ et un facteur $\varepsilon(s, \pi_v, r, \psi_v)$ de la forme $c \times$ exponentielle. Si π_v est de classe 1, la représentation naturelle de $H_{0,v}$ sur l'espace des vecteurs fixes par K_{0v} définit un caractère χ_v de $H_{0,v}$; on veut que $L(s, \pi_v, r)$ soit égal au $L(s, \chi_v, r)$ de 2.4; de plus, on pose $\varepsilon(s, \pi_v, r, \psi_v) = 1$ si O_v est le plus grand idéal sur lequel ψ_v est trivial.

(ii) Supposons ψ_v associé à un caractère additif donné ψ de A/F quel que soit $v \in V$. On voudrait que la solution de (i) ait la propriété suivante:

Soit $(\pi, M) = \otimes_v (\pi_v, M_v)$ un $\mathscr{H}$-module automorphe irréductible admissible. Alors

$$(1) \qquad L(s, \pi, r) = \prod_v L(s, \pi_v, r), \quad \text{et} \quad L(s, \pi, r^{\vee})$$

admettent des prolongements analytiques méromorphes avec nombre fini de pôles, qui satisfont à une équation fonctionelle

$$(2) \qquad L(s, \pi, r) = \varepsilon(s, \pi, r) L(1 - s, \pi, r^{\vee}), \quad \text{où} \quad \varepsilon(s, \pi, r) = \prod_v \varepsilon(s, \pi_v, r, \psi_v),$$

et sont holomorphes si $M \subset {}_0 L^2$ et r est irréductible non triviale.

Remarquons que les facteurs L et ε sont en tout état de cause définis pour presque tout v, indépendamment de la résolution du problème local. La construction de 2.4 s'applique donc à tout $\mathscr{H}$-module irréductible admissible.

5.3. Ces problèmes sont résolus affirmativement lorsque $n = 2$ et que r est la représentation identique ([22], voir aussi [13; 17]). Le problème local est en fait traité par deux méthodes, qui fournissent (heureusement) les mêmes facteurs L et ε. L'une généralise assez directement celle de Tate [44]. On considère des fonctions zêta locales de la forme $\zeta(\Phi, c, s) = \int_{G_v} \Phi(x) c(x) |\det x|_v^{s+1/2} dx$, où Φ est une fonction de Schwartz-Bruhat sur $\mathbf{M}_2(F_v)$ et c est un coefficient de π. Cette intégrale converge pour $Rs \gg 0$, admet un prolongement analytique méromorphe. Il existe un facteur eulérien $L(\pi, s)$, combinaison linéaire de telles fonctions, tel que $L(\pi, s)^{-1} \cdot \zeta(\Phi, c, s)$ soit entière quels que soient Φ et c. C'est le facteur cherché. Le facteur ε s'obtient en considérant l'équation fonctionnelle qui lie $\zeta(\Phi, c, s)$ et $\zeta(\hat{\Phi}, \bar{c}, 1 - s)$, où $\hat{\Phi}$ est la transformée de Fourier de Φ. L'autre méthode utilise des intégrales sur F_v^* formées en considérant une réalisation de π dans un espace de fonctions sur G_v, le modèle de Whittaker de π.

Par exemple, si $k_v = \mathbf{R}$ et π est dans la série discrète, alors $L(s, \pi)$ est formé d'un facteur gamma; si v est archimédienne et π est dans la série principale, alors $L(s, \pi)$ est produit de 2 facteurs gamma (se réduisant parfois à 1 par la formule de multiplication). Si v est ultramétrique, $L(s, \pi)^{-1} = 1$ si π est supercuspidale, est de degré 1 si π est spéciale, de degré 2 sinon. Dans tous les cas, les fonctions $L(s, \pi \otimes \chi)$ et $\varepsilon(s, \pi \otimes \chi)$, où χ parcourt les caractères du centre, caractérisent π à équivalence près parmi les représentations dont la restriction au centre est donnée [22: 2.19, 5.18, 6.7].

De plus [22] établit une réciproque: soit $\pi = \otimes_v \pi_v$ admissible irréductible. On suppose qu'il existe un Grössencharakter ψ_v tel que $\pi_v(a \cdot \mathrm{Id}) = \psi_v(a) \cdot \mathrm{Id}$ quels que soient $v \in V$, $a \in F_v^*$ et surtout que, pour tout Grössencharakter, $L(s, \pi \otimes \chi)$ et $L(s, \pi^\vee \otimes \chi^{-1})$ admettent des prolongements analytiques holomorphes, bornés dans des bandes verticales, satisfaisant à l'équation 5.2(2). Alors π intervient dans ${}_0L_\psi^2$. [$\pi \otimes \chi$ est le produit tensoriel de π et de la représentation $x \mapsto \chi(\det x)$.]

Remarquons que, même dans ce cas, on sait très peu de choses lorsque r n'est pas la représentation identique. En fait, on n'a des renseignements que pour $r = \mathrm{Sym}^m$ avec $m = 2, 3$. Si $m = 2$, et π provient d'une forme modulaire parabolique, prolongements analytiques et équations fonctionnelles sont établis dans [34], et l'holomorphie dans [42; 43]. Si $m = 3$, l'existence d'un prolongement analytique méromorphe résulte de [25].

5.4. Pour $\mathbf{GL}_n$ (même sur une algèbre à division) une solution est donnée par Godement-Jacquet [18] et obtenue par une méthode généralisant celle esquissée ci-dessus.

5.5. Le rôle joué par les Grössencharaktere dans la réciproque de 5.3 a amené Gelfand et Kazdhan [15] à poser un problème différent pour $\mathbf{GL}_n$, mettant en jeu $\mathbf{GL}_n$ et $\mathbf{GL}_{n-1}$, (cf. 6.3(d)). Du point de vue de 5.3, cela conduit à la question suivante: associer à tout couple de représentations irréductibles admissibles π_v de G_v et π_v' de $G_v' = \mathbf{GL}_{n-1,v}$ des facteurs $L(s, \pi_v \times \pi_v')$ et $\varepsilon(s, \pi_v \times \pi_v', \psi_v)$, de manière à ce que, si $\pi = \otimes_v \pi_v$ et $\pi' = \otimes_v \pi_v'$ sont automorphes pour G_A et G_A' respectivement, alors $L(s, \pi \times \pi') = \prod_v L(s, \pi_v \times \pi_v')$ et $L(s, \pi^\vee \times \pi'^\vee)$ admettent des prolongements analytiques méromorphes, holomorphes si π est parabolique, satisfaisant à

$$L(s, \pi \times \pi') = \varepsilon(s, \pi \times \pi') \cdot L(1 - s, \pi^\vee \times \pi'^\vee), \quad \text{où} \quad \varepsilon(s, \pi \times \pi') = \prod_v \varepsilon(s, \pi_v, \pi_v', \psi_v).$$

Pour $n = 3$, ce programme est réalisé dans [23] (au moins si π' est aussi parabolique et F est un corps de fonctions). De plus [21] annonce une réciproque, aussi due à Jacquet et Shalika: supposons π irréductible admissible et satisfaisant aux conditions précédentes pour tout π' automorphe pour $\mathbf{GL}_2$ et pour tout Grössencharakter χ, alors π est automorphe parabolique.

5.6. Très récemment, Piateckiǐ-Šapiro a annoncé qu'il a démontré pour $\mathbf{GL}_3$ les analogues des théorèmes de Jacquet-Langlands. En particulier, il suffit de nouveau que $\pi \otimes \chi$ satisfasse à des conditions d'analyticité et équations fonctionnelles quel que soit le Grössencharakter χ pour que π soit automorphe parabolique. Il a également donné un exemple simple montrant que cette réciproque cesse d'être valable pour $n \geqq 4$. En fait, il construit une infinité continue de représentations irreductibles admissibles non équivalentes de $\mathbf{GL}_{4A}$, qui ont toutes les mêmes facteurs locaux L et ε, et dont l'une est automorphe parabolique. Ces représentations satisfont donc aux conditions de Jacquet-Langlands, mais ne peuvent être toutes automorphes puisque le spectre de $_0L^2(G_A/G_F)$ est dénombrable.

§ 6. Représentations de $\mathbf{GL}_n$ et représentations de groupes de Weil

Les caractères d'ordre fini de F_v^* correspondent canoniquement aux caractères du groupe de Galois $G(F_{v,ab}/F_v)$ de l'extension abélienne maximale de F_v, d'après la loi de réciprocité locale. Langlands conjecture pour $\mathbf{GL}_n$ des relations analogues entre représentations irréductibles admissibles de G_v et représentations complexes de dimension n du groupe de Galois d'une clôture séparable de F_v, qui se globaliseraient en une loi de réciprocité non abélienne liant séries d'Artin et représentations automorphes de G_A (en partie démontrée pour $n=2$, modulo la conjecture d'Artin). En fait, les conjectures mettent en jeu plus généralement des représentations de groupe de Weil et séries L d'Artin-Hecke (6.5); en principe, elles comprennent même les séries L obtenues à partir de «motifs,» mais le conférencier n'ira pas jusque là.

6.1. Groupes de Weil. On note $W_{E'/E}$ le groupe de Weil associé à un corps local ou global E et une extension galoisienne finie E' de E [1; 45]. Parmi ses nombreuses propriétés, on rappelle l'existence:

a) d'une suite exacte

$$(1) \qquad 1 \to W_{E'/E'} \to W_{E'/E} \to G(E'/E) \to 1 \,,$$

b) d'isomorphismes

$$(2) \qquad W_{E/E} = E^* \text{ si } E \text{ est local, } W_{E/E} = C_E \text{ si } E \text{ est un corps de nombres,}$$

c) d'un diagramme commutatif

$$(3) \qquad \begin{array}{ccccccccc} 1 & \to & W_{E'_v/E'_{v'}} & \to & W_{E'_v/E_v} & \to & G(E'_v/E_v) & \to & 1 \\ & & \downarrow & & \downarrow & & \downarrow & & \\ 1 & \to & W_{E'/E'} & \to & W_{E'/E} & \to & G(E'/E) & \to & 1 \end{array}$$

si E est global, $v \in V_E$, et $v' \in V_{E'}$ prolonge v, et enfin

d) d'un homomorphisme canonique surjectif $\alpha_{E',E''} : W_{E'/E} \to W_{E''/E}$ compatible avec $G(E'/E) \to G(E''/E)$ si E'' est intermédiaire entre E et E' et galoisien sur E, d'où en particulier, si E est local, un homomorphisme $a: W_{E'/E} \to \mathbf{Z}$ composé d'applications

$$(4) \qquad W_{E'/E} \xrightarrow{\alpha_{E'/E}} W_{E/E} = E^* \xrightarrow{\text{ord.}} \mathbf{Z} \,.$$

6.2. Une représentation continue complexe de dimension finie de $W_{E'/E}$ est dite *admissible* si son image est formée d'éléments semi-simples.

Dans [28] et un célèbre diplodocus non publié (mais Deligne a fourni des démonstrations dans [8]), Langlands a associé à toute paire E, E' (E local, E' galoisien de degré fini sur E) et toute représentation admissible σ de $W_{E'/E}$ un facteur $\varepsilon(s, \sigma, \psi)$ de manière à ce que les $\varepsilon(s, \sigma, \psi)$ et les facteurs $L(s, \sigma, E'/E)$ de Weil [45] satisfassent, entre autres, à la condition suivante; soit F' une extension galoisienne finie de F et soit σ une représentation admissible de $W_{F'/F}$. Pour $v \in V$, soit σ_v la classe de représentations de $W_{F'_v/F_v}$ obtenue à partir de σ à l'aide de 6.1 (3). Alors la série d'Artin-Hecke $L(s, \sigma) = \prod_v L(s, \sigma_v)$ satisfait à l'équation fonctionelle

$$(1) \qquad L(s, \sigma) = \varepsilon(s, \sigma) \cdot L(1 - s, \check{\sigma}) \,, \quad \text{avec} \quad \varepsilon(s, \sigma) = \prod_v e(s, \sigma_v, \psi_v) \,.$$

Suivant Artin et Weil, on conjecture que $L(s, \sigma)$ est holomorphe si σ est irréductible «non triviale,» i.e. non de la forme $w \to \| \alpha_{F'/F}(w) \|^t$ ($t \in \mathbf{R}$). Rappelons que

$$(2) \qquad L(s, \sigma, E'/E) = (\det(1 - \sigma^I(\varphi_v) \cdot Nv^{-s}))^{-1} \,,$$

où σ^I est la restriction de σ à l'espace des points fixes du groupe d'inertie local et φ_v un Frobenius.

6.3. Le problème local. Soit E un corps local. Si E est archimédien, le problème local est d'associer canoniquement à tout $\pi \in \mathscr{A}(\mathbf{GL}_n/E)$ une représentation admissible $\sigma = \sigma(\pi)$ de degré n de $W_{E/E}$, de manière à ce qu'en passant aux classes d'équivalence de telles représentations modulo automorphismes intérieurs de $\mathbf{GL}_n(\mathbf{C})$ on obtienne une bijection.

Si E est ultramétrique, le problème est en principe analogue, mais on a remarqué rapidement que, ainsi posé, il n'admettait pas une réponse affirmative déjà pour $\mathbf{GL}_2$. On veut évidemment que dans ce cas π et $\sigma(\pi)$ ait les mêmes facteurs L et ε (cf. 5.3, 6.2). Or, en caractéristique résiduelle impaire, cas où les deux ensembles à comparer sont connus, on constate l'existence d'une bijection naturelle satisfaisant à cette condition pour autant que l'on laisse de côté les représentations spéciales, i.e. celles dont le facteur L est de degré 1. Mais la théorie des courbes elliptiques mène à attribuer un tel facteur à une représentation l-adique dans le module de Tate d'une courbe alliptique d'invariant j non entier et il y a plusieurs raisons qui suggèrent d'associer cette représentation à la représentation spéciale de $\mathbf{GL}_2$ qui est triviale sur le centre [4: § 3]. Cela a conduit à raffiner le problème de manière à inclure les représentations l-adiques de groupes de Weil [8: § 8]. On peut cependant le formuler en restant sur les complexes. On considère des paires (σ, X) où σ est comme précédemment et X un élément nilpotent de l'algèbre de Lie de $\mathbf{GL}_n(\mathbf{C})$ qui satisfont à la condition:

$$(1) \qquad \text{Ad } \sigma(w) \cdot X = (Nv)^{a(w)} \cdot X, \qquad (w \in W_{E'/E};\ a(w) \text{ donné par } 6.1\,(4)).$$

Soit $\Phi(G/E)$ l'ensemble des classes d'équivalence de telles paires (ou de σ si $E = \mathbf{R}, \mathbf{C}$), modulo automorphismes intérieurs de $\mathbf{GL}_n(\mathbf{C})$ (opérant simultanément sur σ et X) et passage à des extensions plus grandes de E. On demande encore d'établir une bijection canonique entre $\mathscr{A}(G_E)$ et $\Phi(G/E)$. Dans les deux cas, on associe alors à $\pi \in \mathscr{A}(G_E)$ les facteurs L et ε de l'élément correspondant de $\Phi(G/E)$ mentionnés ci-dessus. Cette correspondance doit satisfaire à plusieurs conditions données *a priori*, notamment:

(a) la représentation $\pi(\sigma, X)$ associée à (σ, X) est de carré intégrable modulo le centre si et seulement s'il n'existe pas de sous-groupe de Levi d'un sous-groupe parabolique propre de ${}^L G^0$ qui contienne l'image de σ et dont l'algèbre de Lie contienne X;

(b) si π est de classe 1 et E' non ramifiée, on demande, pour être en accord avec 2.4, que σ soit triviale sur le groupe d'inertie et envoie un Frobenius sur un élément de la classe analogue au $\{g_v\}$ de 2.4, et on prend $X = 0$.

Une autre formulation est donnée dans [8: § 8]. On remplace $W_{E'/E}$ par un groupe de Weil «élargi» $W'_{E'/E}$ produit semi-direct d'un groupe distingué N isomorphe au groupe additif du corps par $W_{E'/E}$, l'élément $w \in W_{E'/E}$ opérant par l'homothétie de rapport $(Nv)^{a(w)}$. On considère alors des représentations complexes de dimension finie de $W'_{E'/E}$ qui sont admissibles sur $W_{E'/E}$ et unipotentes sur N, et $\Phi(G/E)$ est l'ensemble de leurs classes d'équivalence.

6.4. Conjectures et résultats locaux. (a) Si $E = \mathbf{R}, \mathbf{C}$, le problème de 6.3 est résolu affirmativement. Si $E = \mathbf{C}$, c'est contenu dans [30], pour tout n. Si $E = \mathbf{R}$ et $n = 2$, cf. [22]; pour $n \geq 3$, le résultat général de [30] donne une partition de $\mathscr{A}(G_E)$ en parties finies non vides indexées par les σ; mais il sera prouvé, Jacquet dixit, dans l'article donnant les démonstrations des résultats annoncés dans [21, 23], que ces parties de $\mathscr{A}(G_E)$ se réduisent à un élément. Cependant, pour $n \geq 3$, on ignore si les facteurs L et ε ainsi obtenus sont ceux de [18].

Les représentations de $W_{\mathbf{C}/\mathbf{R}}$ qui sont triviales sur $\mathbf{C}^*$, i.e. proviennent de représentations de $G(\mathbf{C}/\mathbf{R}) = \mathbf{Z}/2$, correspondent biunivoquement aux classes de conjugaison d'éléments d'ordre deux de ${}^L G$. Les π_v correspondants seront dits de type A_{00}. Par analogie avec [46], les π_v associés aux représentations de $W_{\mathbf{C}/\mathbf{R}}$ dont la restriction à $\mathbf{C}^*$ est somme de caractères de la forme $z \mapsto z^a \cdot \bar{z}^b$ $(a, b \in \mathbf{Z})$ seront dits de type A_0.

Supposons $n = 2$. Les représentations irréductibles de $W_{\mathbf{C}/\mathbf{R}}$ sont induites d'un caractère de $\mathbf{C}^*$ dont la restriction à $\mathbf{U}(1)$ est non-triviale. Elles correspondent à la série discrète. Pour $n > m > 0$ entiers, soit $\sigma_{n,m}$ la représentation induite de $W_{\mathbf{C}/\mathbf{R}}$ à partir du caractère $z \mapsto z^{-n}\bar{z}^{-m}$ de $\mathbf{C}^*$. L'élément $\pi_{n,m}$ de la série discrète lui correspondant est de caractère central $z \mapsto z^{1-n-m}$, et a même opérateur de Casimir que la représentation holomorphe $\tau_{n,m}$ de $\mathbf{GL}_2$ de degré $n - m$ et caractère central $z \mapsto z^{n+m-1}$ [29: p. 388]. Les représentations réductibles sont d'image commutative et sont sommes de deux caractères ω_1, ω_2 de $\mathbf{R}^*$. La représentation $\omega_1 \oplus \omega_2$ correspond à la représentation de la série principale $\pi(\omega_1, \omega_2)$ (notation de [22: 5.11], cela inclut les représentations de dimension finie). Il y en trois de type A_{00}, $\pi(1, \mathrm{sgn})$, $\pi(1, 1)$, $\pi(\mathrm{sgn}, \mathrm{sgn})$, correspondant aux classes de conjugaison $\begin{pmatrix} 1 & 0 \\ 0 & -1 \end{pmatrix}$, Id. et $-$Id., où sgn est le caractère $t \mapsto \mathrm{sgn}\, t$ de $\mathbf{R}^*$.

(b) Supposons E ultramétrique et $n = 2$. Si la caractéristique résiduelle est impaire, il y a correspondance biunivoque; cela résulte de [22] et de la formule de Plancherel (voir [4: § 3] pour une description de la correspondance et des références). Dans [22: § 12], on associe, par voie globale, un élément $\pi(\sigma) \in \mathscr{A}(G_E)$ à une représentation admissible σ de degré 2 d'un groupe de Weil $W_{E'/E}$, modulo la conjecture d'Artin, ce qui n'est donc pas une restriction si E est de caractéristique non nulle. En s'appuyant notamment sur ce fait et des résultats de Drinfeld [12], Deligne a établi la bijectivité pour E de caractéristique non nulle quelconque.

(c) Dans [15], Gelfand et Kazdhan associent à toute paire $\pi \in \mathscr{A}(\mathbf{GL}_n)$, $\pi' \in \mathscr{A}(\mathbf{GL}_{n-1})$, un nombre $\Gamma(\pi, \pi')$; ils conjecturent pour $j \geqq 1$ l'existence d'une injection $\varkappa_j\colon \hat{W}_j(E) \to \mathscr{A}(\mathbf{GL}_j)$, où $\hat{W}_j(E)$ est l'ensemble des classes d'équivalence de représentations de degré j de groupes de Weil sur E, de manière à ce que l'on ait

$$(1) \qquad \Gamma(\varkappa_n\, \sigma, \varkappa_{n-1}\, \tau) = L\left(\tfrac{1}{2}, \sigma \otimes \check{\tau}\right) \cdot \varepsilon\left(\tfrac{1}{2}, \sigma \otimes \check{\tau}\right)^{-1} \cdot L\left(\tfrac{1}{2}, \check{\sigma} \otimes \tau\right)^{-1},$$

quels que soient $\sigma \in \hat{W}_n(E)$, $\tau \in \hat{W}_{n-1}(E)$, où L et ε sont les fonctions de Langlands.

6.5. Conjectures et résultats globaux. (a) Soit F' une extension galoisienne finie de F et soit $\sigma\colon W_{F'/F} \to \mathbf{GL}_n(C)$ une représentation admissible. A l'aide de 6.1 (3), on déduit de σ une représentation admissible σ_v d'un groupe de Weil local aur F_v $(v \in V)$, dont la classe ne dépend que de σ et v. On conjecture que la série d'Artin-Hecke $L(s, \sigma)$ est associée à un espace irréductible admissible de formes automorphes, ou plus précisément, admettant 6.3 résolu, que $\otimes_v \pi(\sigma_v)$ est automorphe. De plus, ces formes doivent être paraboliques si σ est irréductible non triviale.

Modulo la conjecture d'Artin, cela résulte de [22] si $n = 2$.

(b) Disons qu'une représentation admissible irréductible globale $\pi = \otimes_v \pi_v$ de $\mathscr{H}$ (5.1) est de type A_{00}(resp. A_0) si π_v est de type A_{00} (resp. A_0) pour v archimédienne (6.4 (a)).

Si la représentation σ est en fait une représentation du groupe de Galois $G(E'/E$, donc $L(s, \sigma)$ est une série d'Artin, alors l'hypothétique π sera de type A_{00}. Réciproquement, on conjecture ou, disons plus prudemment on se demande, si tout π de type A_{00} dans l'espace des formes paraboliques est associé à une série d'Artin. Si $n = 1$, cela résulte de la loi de réciprocité d'Artin. Si $n = 2$, $F = \mathbf{Q}$, et si π est associé à $\begin{pmatrix} 1 & 0 \\ 0 & -1 \end{pmatrix}$ (cf. 6.4 (a)), cela est démontré par Deligne-Serre [10]. Ainsi [10] et [22] fournissent, modulo la conjecture d'Artin, une correspondance bijective canonique entre formes modulaires primitives paraboliques de poids un et représentations irréductibles de degré deux de $G(\bar{\mathbf{Q}}/\mathbf{Q})$ à déterminant impair. Le cas où π_∞ correspond à Id. ou $-$Id. (cf. 6.4) n'est pas encore traité cependant. Par ailleurs on sait, depuis Hecke, que les représentations réductibles correspondent aux séries d'Eisenstein.

(c) Les séries L associées à la cohomologie de variétés sur F, ou à des représentations l-adiques de groupes de Galois ou de Weil, doivent de même conduire à des π paraboliques de type A_0, voire à tous. Pour $n = 1$, [46] donne quelques exemples, mais cette question a été étudiée surtout pour $n = 2$. On a déjà mentionné quelques cas en 4.4, 4.5. Supposons $F = \mathbf{Q}$. Si K_f est un sous-groupe compact ouvert de $G(A_f)$ suffisamment petit, alors $M_{K_f} = (K_\infty K_f)\backslash G_A/G_{\mathbf{Q}}$ est réunion d'un nombre fini de courbes modulaires lisses, quotients du demi-plan de Poincaré par un sous-groupe de congruence de $\mathbf{SL}_2(\mathbf{Z})$. D'après Eichler-Shimura, [41], les fonctions zêta de telles courbes sont associées par 2.2 à des formes modulaires paraboliques. Cette relation est étudiée du point de vue représentations dans [29; 33]. Soient n, m, $\pi_{n,m}$ et $\sigma_{n,m}$ comme dans 6.4(a). Dans [29], Langlands associe à tout $\pi = \otimes_v \pi_v$ automorphe parabolique, tel que $\pi_\infty = \pi_{n,m}$, une représentation l-adique σ de degré 2 de $G(\bar{\mathbf{Q}}/\mathbf{Q})$, par considération de la cohomologie l-adique des courbes M_{K_f}, à coefficients dans un faisceau associé à la représentation $\sigma_{n,m}$; il conjecture que, pour tout $v \in V$, les facteurs L et ε de π_v sont ceux associés à σ et v. [29] le démontre pour les facteurs L et en partie pour les facteurs ε. Les cas restants ont été ensuite traités par Deligne, sauf pour $p = 2$ (lettre à Piateckiĭ-Šapiro). [33] est aussi consacré à cette question, dans le cas $n = 1$, $m = 0$, pour les p non ramifiés.

(d) Soient $n = 2$, $F = \mathbf{Q}$, et supposons que π provienne d'une forme modulaire parabolique sur $\mathbf{SL}_2(\mathbf{Z})$. La conjecture de Ramanujan-Petersson équivaut à demander que les valeurs propres des g_p soient de module 1, [13; 27; 36]. Une généralisation de cette condition est donc que les classes g_p rencontrent $\mathbf{U}_n$. Vu [27], cela aura lieu si $L(s, \pi, r)$ est holomorphe pour $Rs > 1$ quelle que soit r. Si de plus $L(s, \pi, r)$ est holomorphe pour $Rs \geqq 1$ et r irréductible non triviale, alors les g_v sont équidistribués pour la mesure de Haar normalisée de $\mathbf{U}_n$ [37: I, A.2].

§ 7. Le L-groupe d'un groupe réductif. Généralisation de 2.4

7.1. *Le L-groupe d'un groupe réductif.* Soient E un corps et G un E-groupe connexe réductif quasi-déployé. Soient B un sous-groupe résoluble connexe maximal de G défini sur E et T un tore maximal de B défini sur E. Soient $X^*(T) = \mathrm{Morph}(T, \mathbf{GL}_1)$ le groupe des caractères rationnels de T et $X_*(T) = \mathrm{Morph}(\mathbf{GL}_1, T)$ celui des sous-tores de dimension $\leqq 1$ de T. Ces deux derniers groupes sont commutatifs libres, de rang égal à $\dim T$, en dualité naturelle. Soient encore $\Phi = \Phi(T, G)$ le système des racines de G par rapport à T, Δ la base de Φ définie par B et $\Phi^\vee \subset X_*(T)$ l'ensemble des coracines. On a donc une bijection canonique $\alpha \mapsto \alpha^\vee$ de Φ sur $\Phi^\vee$ telle que $\langle \alpha, \alpha^\vee \rangle = 2$ et $\langle \alpha, \beta^\vee \rangle$ soit un entier (de Cartan). Il existe un groupe réductif complexe, déterminé à isomorphisme près, ayant un tore maximal $T^\vee$, un sous-groupe résoluble maximal $B^\vee$ contenant $T^\vee$ par rapport auxquels $X_*(T)$, $\Phi^\vee$, $\Delta^\vee$ jouent les rôles de $X^*(T)$, Φ, Δ. On le notera $^LG^0$.

Soit E' une extension galoisienne finie de E sur laquelle G et T sont déployés. Alors $G(E'/E)$ opère sur les données précédentes et on en déduit une représentation de $G(E'/E)$ comme groupe d'automorphismes de $^LG^0$ préservant un épinglage dont $T^\vee$ et $B^\vee$ font partie. On peut donc former le produit semi-direct (direct si

G est déployé sur E, auqel cas T l'est aussi)

$$(1) \qquad\qquad {}^L G = {}^L G^0 \rtimes G(E'/E) \,.$$

C'est une première forme, dite «galoisienne,» du L-groupe de G, notée $\hat{G}_{E'/E}$ dans
[27], ${}^c G$ dans [25], $G\check{\ }$ dans [30], (et appelée groupe associé). Si G_1 est aussi quasi-
déployé sur F et est une forme intérieure de G (i.e. correspond à un élément de
$H^1(G(E_s/E), \operatorname{Ad} G))$, alors ${}^L G$ et ${}^L G_1$ sont isomorphes [27; 30]. Enfin, si G n'est pas
quasi-déployé, il existe un groupe G' quasi-déployé sur F dont G est une forme
intérieure. On pose alors ${}^L G = {}^L G'$ par definition. Il y a un certain nombre de
vérifications à faire pour s'assurer que cette définition est suffisamment canonique,
pour lesquelles on renvoie à [27; 30].

Si $G = \mathbf{GL}_n$, alors ${}^L G^0 = \mathbf{GL}_n(\mathbf{C})$. Si $G = \mathbf{SL}_n$ (resp. $\mathbf{Sp}_{2n}$) alors ${}^L G^0 = \mathbf{PSL}_n(\mathbf{C})$
(resp. $\mathbf{SO}_{2n+1}(\mathbf{C})$).

7.2. Supposons E local ou global. Comme $W_{E'/E}$ s'envoie sur $G(E'/E)$, on peut
aussi former le produit semi-direct (toujours direct si G est déployé) ${}^L G^0 \times W_{E'/E}$.
C'est la «forme de Weil» du L-groupe [30], qui sera aussi notée ${}^L G$. Il y a donc une
forme galoisienne et une forme de Weil, et on utilise l'une ou l'autre suivant les
besoins du contexte. On se permet aussi d'agrandir E' à volonté. Le lecteur que
cette dernière indétermination gêne peut toujours passer à la limite et remplacer
$W_{E'/E}$ par un groupe de Weil absolu [8]. Suivant la littérature, ${}^L G^0$ est complexe,
mais rien n'empêche de considérer ses points dans d'autres corps, en particulier, il
devient utile de considérer ${}^L G^0(\mathbf{Q}_l)$.

Si $E = F$ est global, $v \in V_F$ et v' prolonge v, le groupe associé à $G \times F_v$ est
${}^L G^0 \rtimes W_{F_{v'}/F_v}$ ou ${}^L G_v^0 \rtimes G(F'_v/F_v)$. Il sera noté ${}^L G_v$.

7.3. Dorénavant, G est un F-groupe connexe réductif réalisé linéairement et F'
une extension galoisienne finie de F sur laquelle G est déployé. Pour preque tout v,
le groupe $G \times F_v$ est quasi-déployé sur F_v et $G(O_v)$ est un sous-groupe compact
maximal spécial de G_v. Modulo Bruhat-Tits, cela implique que l'algèbre de Hecke
$H_{0,v} = H(G_v, G(O_v))$ est commutative et justiciable de [35]. Supposons encore v
non ramifiée dans F' et soit $\varphi_v \in G(F'_{v'}/F_v)$ le Frobenius. Alors [35] fournit *une
bijection canonique entre caractères de $H_{0,v}$ et classes de conjugaison, par rapport à
${}^L G^0$, d'éléments de la forme $(g, \varphi_v) \in {}^L G$ avec g semi-simple* (cf. [27]).

Cela étant, si l'on se donne de plus une représentation r holomorphe de
dimension finie de ${}^L G^0$, on peut répéter *verbatim* la construction de 2.4, ou son
analogue en termes de représentations. Par exemple, si $(\pi, M) = \otimes_v (\pi_v, M_v)$ est
automorphe, alors, pour presque tout v, l'espace des vecteurs fixes de $G(O_v)$ dans
M_v est de dimension un, définit un caractère χ_v de $H_{0,v}$, d'où une classe (g_v, φ_v)
$\in {}^L G$, et l'on peut former le produit

$$(1) \qquad L(s, \pi, r) = \prod_v L(s, \pi_v, r), \quad \text{avec} \quad L(s, \pi_v, r) = (\det(1 - r(g_v) \cdot Nv^{-s}))^{-1}.$$

Il converge pour $R s$ assez grand [27].

7.4. Sous-groupes paraboliques de ${}^L G$. Par définition [30], ce sont les normalisateurs dans ${}^L G$ des sous-groupes paraboliques de ${}^L G^0$, qui de plus rencontrent toute

classe de ^{L}G modulo $^{L}G^{0}$. Un sous-groupe de Levi M d'un sous-groupe parabolique P est, par définition, un sous-groupe fermé contenant un sous-groupe de Levi au sens usuel de $P \cap {}^{L}G^{0}$ et telque P soit produit semi-direct de M et du radical unipotent de $P \cap {}^{L}G^{0}$. L'ensemble $\mathscr{P}(^{L}G)$ des classes de conjugaison par rapport à $^{L}G^{0}$ de tels sous-groupes est en correspondance bijective naturelle avec l'ensemble des parties de $\check{\varDelta}$ qui sont stables par $G(E'/E)$. Si G est quasi-déployé la bijection $\varDelta \mapsto \check{\varDelta}$ établit donc une bijection entre $\mathscr{P}(^{L}G)$ et l'ensemble $\mathscr{P}(G)$ des classes de conjugaison de F-sous-groupes paraboliques de G.

Si G n'est pas quasi-déployé, il y a encore une injection canonique $\mathscr{P}(G) \rightarrow \mathscr{P}(^{L}G)$ dont l'image sera notée $^{L}\mathscr{P}(^{L}G)$. Si G est anisotrope, cette image se réduit à $\{^{L}G\}$.

§ 8. Problèmes et résultats

Nous passons aux problèmes posés par [27], à cela près que nous tenons compte de la formulation plus récente et plus précise du problème local à l'aide de groupes de Weil.

8.1. Le problème local. Soient E un corps local, G un E-groupe connexe réductif et E' une extension galoisienne finie de E sur laquelle G est déployé. On a la suite exacte

$$(1) \qquad\qquad 1 \rightarrow {}^{L}G^{0} \rightarrow {}^{L}G \overset{v}{\rightarrow} G(E'/E) \rightarrow 1.$$

On considère des homomorphismes continus $\varphi\colon W_{E'/E} \rightarrow {}^{L}G$ ou aussi, si E est ultramétrique, des paires (φ, X), où X est un élément nilpotent de l'algèbre de Lie de $^{L}G^{0}$, et φ, X satisfont à 6.3(1). L'homomorphisme Φ, ou la paire (φ, X) est admissible si les conditions suivantes sont satisfaites:

(i) l'image de φ est formée d'éléments semi-simples;

(ii) $v \circ \varphi$ est la projection canonique;

(iii) si un sous-groupe parabolique propre P de ^{L}G (7.4) possède un sous-groupe de Levi M tel que $\operatorname{Im} \varphi \subset M$ et que X soit dans l'algèbre de Lie de $M \cap {}^{L}G^{0}$, alors $P \in {}^{L}\mathscr{P}(^{L}G)$ (cf. 7.4).

On note $\Phi(G/E)$ l'ensemble des classes d'équivalence de φ (resp. (φ, X)) admissibles modulo automorphismes intérieurs par éléments de $^{L}G^{0}$ et passage à des extensions plus grandes. On peut aussi considérer la forme de Weil de ^{L}G, i.e. remplacer $G(E'/E)$ par $W_{E'/E}$ dans (1). A ce moment, φ sera un scindage de (1), admissible s'il satifait à (i), (iii), et $\Phi(G/E)$ sera l'ensemble des classes d'équivalence de scindages admissibles [30].

Le problème est alors de définir une partition de $\mathscr{A}(G_{E})$ en parties finies non vides Π_{σ} ou $\Pi_{(\sigma, X)}$ indexées par $\Phi(G/E)$, satisfaisant à un certain nombre de conditions données *a priori*, en particulier:

les éléments de $\Pi_{(\sigma, X)}$ ou Π_{σ} sont simultanément de carré intégrable modulo le centre ou non; ils le sont si et seulement si il n'existe pas de sous-groupe de Levi M d'un sous-groupe parabolique propre $P \in {}^{L}\mathscr{P}(^{L}G)$ qui contienne l'image de σ et tel que l'algèbre de Lie de $M \cap {}^{L}G^{0}$ contienne X (7.4).

Il s'ensuit par exemple que, si G est anisotrope et si G' est une forme intérieure quasi-déployée de G, alors $\Phi(G/E)$ s'identifie à la partie de $\Phi(G'/E)$ qui doit définir une partition de la série discrète de $G'(E)$ qui doit définir une partition de la série discrète de $G'(E)$.

On exige également l'analogue de 6.3 (b), et des compatibilités avec l'induction de représentations. Contrairement à ce qui se passe, ou est conjecturé, pour $\mathbf{GL}_n$, on ne peut espérer ici que les Π_σ ou $\Pi_{(\sigma, X)}$ soient réduits à un élément. Par exemple, des éléments de la série discrète ayant même caractère infinitésimal sont dans le même paquet. Les éléments d'un tel ensemble sont dits L-indiscernables.

Si $E = \mathbf{R}, \mathbf{C}$, ce problème est résolu dans [30]. Si $E = \mathbf{C}$, les Π_σ sont réduits à un élément. Si G est un tore, alors ${}^LG^0$ est le tore dual, X est automatiquement nul, et il y a une bijection canonique entre $\mathscr{A}(G)$, i.e. les homomorphismes continus de $G(E)$ dans $\mathbf{C}^*$, et les représentations admises de $W_{E'/E}$ dans ${}^LG$ [26]. Si E est ultramétrique on n'a à part cela de résultats que pour $\mathbf{GL}_2$ (6.4). Il y a tout de même des signes encourageants. Par exemple, les résultats annoncés par Lusztig sur la conjecture de Macdonald (C.R. **280** (1975), 317−320) et démontrés dans un article de Deligne et Lusztig à paraître aux Annals of Mathematics fournissent, par induction, des représentations supercuspidales (étudiées par P. Gérardin) que Langlands veut associer à des σ modérément ramifiés.

8.2. Prolongement analytique. Equation fonctionnelle. On se place dorénavant sous les hypothèses de 7.3. On demande tout d'abord si la série L de 7.3 (1) admet un prolongement analytique méromorphe. [25] donne un certain nombre de paires (G, r) où G est simple déployé, pour lesquelles cela est vrai pour tout $\pi = \otimes_v \pi_v$ automorphe parabolique dont les facteurs locaux π_v appartiennent presque tous à la série principale (voir aussi [16]). La méthode consiste à envisager G, ou un groupe isogène, comme un sous-groupe semi-simple maximal d'un sous-groupe parabolique propre maximal d'un sous-groupe semi-simple déployé G' et à se ramener aux résultats de [24] (voir aussi [19]) sur le prolongement analytique et équations fonctionnelles des séries d'Eisenstein de G'.

Si l'on admet 8.1 résolu pour G/F_v ($v \in V$), le problème plus précis est le suivant: soit $\pi = \otimes_v \pi_v$ une représentation automorphe (5.1). A π_v est associé $\varphi_v \in \Phi(G/F_v)$ donc des facteurs $L(s, r \circ \Phi_v)$ et $\varepsilon(s, r \circ \varphi_v, \psi_v)$ d'après 6.2. On demande alors si les produits globaux correspondants admettent un prolongement analytique méromorphe à nombre fini de pôles et satisfont à une équation fonctionnelle comme en 5.2 (1)

Si $G = \mathbf{GL}_2 \times \mathbf{GL}_2$ et si π, π' sont automorphes paraboliques pour $\mathbf{GL}_2$ cela est démontré pour $\pi \times \pi'$ dans [20]. A l'origine de ce résultat est le fait que si $\sum a_n\, n^{-s}$ et $\sum b_n\, n^{-s}$ sont associées à des formes modulaires paraboliques (cf. 1.5, 2.2) alors $\sum a_n b_n\, n^{-s}$ admet un prolongement analytique avec équation fonctionnelle [34].

8.3. Surjectivité. Soient H un F-groupe réductif connexe et F'' une extension galoisienne finie de E' sur laquelle H est déployé. Soit $\alpha: {}^LH \to {}^LG$ un homomorphisme continu, holomorphe sur ${}^LH^0$, tel que le diagramme

(1)
$$
\begin{array}{ccc}
{}^LH & \longrightarrow & G(F''/F) \\
\downarrow{\scriptstyle\alpha} & & \downarrow{\scriptstyle\beta} \\
{}^LG & \longrightarrow & G(F'/F)
\end{array}
$$

soit commutatif, β étant l'homomorphisme canonique. Par restriction, on en déduit pour $v'' \in V_{F''}$ prolongeant $v' \in V_{F'}$, un diagramme commutatif

$$(2) \qquad \begin{array}{ccc} {}^L H_v & \longrightarrow & G\left(F''_{v''}/F_v\right) \\ {\scriptstyle \alpha_v}\downarrow & & \downarrow{\scriptstyle \beta_v} \\ {}^L G_v & \longrightarrow & G\left(F'_{v'}/F_v\right) . \end{array}$$

Soit $\pi' = \otimes_v \Pi'_v$ une représentation automorphe pour H. Pour presque tout v, il lui est associé une classe (h_v, Fr_v), avec h_v semi-simple dans ${}^L H^0$ (7.3). On demande alors si $\{\alpha_v(h_v, Fr_v)\}$ est associé de la même manière à une représentation automorphe sur G [27: Question 3].

On peut de nouveau préciser si l'on admet 8.1. A π'_v est associé un élément φ_v (ou (φ_v, X_v) si v est ultramétrique) de $\Phi(H/F_v)$. Supposons que $\alpha_v \circ \varphi_v$ (ou $(\alpha_v \circ \varphi_v, \alpha_v(X_v))$ soit dans $\Phi(G/F_v)$ (la condition (iii) de 8.1 empêche de dire que c'est automatique). On demande alors si l'on peut choisir $\pi_v \in \mathscr{A}_v$ assigné à l'élément donné de $\Phi(G/F_v)$ de manière à ce que $\pi = \otimes_v \pi_v$ soit automorphe pour G.

Ce problème qui maintenant englobe celui de comparaison (cf. 8.6) apparait en somme comme le problème de base de la théorie globale.

8.4. Exemples. (a) Supposons $H = \{e\}$. On a déjà remarqué qui si $G = \mathbf{GL}_1$, une réponse positive à cette question est fournie par la loi de réciprocité d'Artin, (4.1) et que si $G = \mathbf{GL}_n$, elle donnerait une «loi de réciprocité non abélienne.»

(b) Si $H = \{e\}$ et G est un tore, le réponse est aussi affirmative, grâce à la dualité de Tate-Nakayama [26]. En fait, les représentations automorphes sont ici les caractères de G_A/G_F et [26] fournit une surjection canonique, de noyau fini, du groupe des représentations admises de $W_{F'/F}$ (F' extension galoisienne finie de F) dans ${}^L G$ sur le groupe des caractères de G_A/G_F.

(c) Si H est un sous-groupe de Levi d'un F-sous-groupe parabolique propre de G, alors ${}^L H$ est un sous-groupe de Levi (cf. 8.1) d'un sous-groupe parabolique de ${}^L G$ appartenant à ${}^L \mathscr{P}({}^L G)$ (7.4, 8.1), et la théorie des séries d'Eisenstein [24] conduit aussi à des exemples.

(d) Soit E une extension quadratique séparable de F et soient $H = R_{E/F}\mathbf{GL}_1$, $G = \mathbf{GL}_2$. Alors H s'identifie à un F-tore maximal de G. On a

$$ {}^L H = (\mathbf{GL}_1 \times \mathbf{GL}_1) \rtimes G(E/F) , $$

l'élément non trivial de $G(E/F)$ permutant les deux facteurs $\mathbf{GL}_1$. Soit μ_1 un isomorphisme de ${}^L H$ sur le normalisateur d'un tore maximal de ${}^L G^0 = \mathbf{GL}_2(\mathbf{C})$. L'homomorphisme

$$ \mu\colon {}^L H \to {}^L G = \mathbf{GL}_2(\mathbf{C}) \times G(E/F) , $$

est alors défini par $\mu(h) = (\mu_1(h), \mu_2(h))$ où $\mu_2\colon {}^L H \to G(E/F)$ est la projection canonique. Une forme automorphe pour H est un Grössencharakter χ de E. On demande donc de lui associer une forme automorphe sur $\mathbf{GL}_2$ dont la série L est la série L de Hecke de χ. Si $F = \mathbf{Q}$ et E est imaginaire quadratique, Hecke a ainsi obtenu des formes modulaires paraboliques. Du point de vue représentations, on obtient des représentations automorphes paraboliques $\pi = \otimes_p \pi_p$ pour lesquelles π_∞

appartient à la série discrète. Si $F = \mathbf{Q}$ mais par contre E est réel quadratique, alors la série L a deux facteurs Γ et l'hypothétique π_∞ doit appartenir à la série principale. Du point de vue formes automorphes, les premiers exemples sont dûs à H. Maass, qui a ainsi construit des formes paraboliques non-holomorphes sur le demi-plan de Poincaré, et a jeté les bases de la théorie des formes automorphes non-holomorphes précisément à cette occasion (Math. Annalen **121** (1949), $121-143$). Cela a été généralisé dans [22; 39] (voir aussi [13]).

On peut de même partir d'une extension galoisienne E de degré n et d'un Grössencharakter χ de E. On est alors amené à chercher s'il existe une représentation ou forme automorphe sur $\mathbf{GL}_n$ dont la série L est celle de χ. Le résultat de Piateckiĭ-Šapiro mentionné en 5.6 devrait entraîner que c'est bien le cas pour $n = 3$ et ainsi permettre (enfin) d'exhiber des formes paraboliques pour $\mathbf{GL}_3$.

(e) Supposons que G soit le groupe multiplicatif d'une algèbre de quaternions D sur F. Une extension quadratique E de F contenue dans D définit un F-tore maximal H de G. Les formes automorphes de H sont les Grössencharakter de E. On est donc conduit à vouloir leur associer des formes automorphes sur G, à série L prescrite, ce qui est fait dans [39] (voir aussi [13]), pour des Grössencharaktere convenables.

(f) Soit F' une extension quadratique séparable de F. Prenons $H = \mathbf{GL}_2$ et $G = R_{F'/F}H$. Le groupe $G(F)$ (resp. $G(A_F)$) s'identifie donc canoniquement à $H(F')$ (resp. $H(A_{F'})$). La forme galoisienne la plus simple de ${}^L G$ est

$$
{}^L G = (\mathbf{GL}_2(\mathbf{C}) \times \mathbf{GL}_2(\mathbf{C})) \rtimes G(F'/F),
$$

l'élément non trivial de $G(F'/F)$ échangeant les deux facteurs $\mathbf{GL}_2(\mathbf{C})$. Soit

$$
\mu \colon {}^L H = \mathbf{GL}_2(\mathbf{C}) \times G(F'/F) \to {}^L G,
$$

l'homomorphisme qui applique $\mathbf{GL}_2(\mathbf{C})$ sur la diagonale de ${}^L G^0$ et est l'identité sur $G(F'F)$. Le problème est donc ici d'associer à une forme ou représentation automorphe de $\mathbf{GL}_{2,A_F}$ une forme ou représentation automorphe de $\mathbf{GL}_{2,A_{F'}}$. Cela a été fait dans certains cas par Doi-Naganuma [11] et H. Jacquet [20: Thm 20.6].

8.5. Fonctions zêta de variétés de Shimura. Supposons que le quotient G_∞/K_∞ soit un domaine borné symétrique; soient K_f un sous-groupe compact ouvert de $G(A_f)$ et $K = K_\infty K_f$. Le quotient $K \backslash G_A / G_F$ admet alors une structure de variété quasi-projective. Dans beaucoup de cas, où ce quotient paramètre certaines familles de variétés abéliennes, Shimura en a donné un corps de définition naturel; [6] énonce une conjecture générale «de Shimura» sur ce corps. En prenant K_f suffisamment petit, on peut toujours faire en sorte que cette variété soit lisse. Suivant une voie inaugurée par Eichler, Shimura et Shimura-Kuga ont dans certains cas construit une forme automorphe parabolique dont la série L associée est essentiellement la fonction zêta de Hasse-Weil (le produit alterné des $L_m(s, Z)$ de 3.2 (2)), (cf. [2]). Dans un certain nombre de cas où le quotient est compact, Langlands a un procédé conjectural pour décrire cette fonction, ou la série L associée à la cohomologie à coefficients dans un faisceau défini à partir d'une représentation de G, à l'aide de séries L de représentations automorphes, et a démontré l'égalité en question pour certains d'entre eux, en particulier lorsque G est le groupe multiplicatif d'une algèbre de quaternions sur un corps totalement réel non ramifiée à l'infini (non publié, cf. [31] pour quelques indications sur ce problème).

8.6. Comparaison (locale et globale). Ce problème, qui était distinct de la surjectivité dans le cadre du § 4, en est maintenant en fait un cas particulier, celui où H est *une forme intérieure* de G.

(a) Supposons H et G définis sur un corps local E, et G quasi-déployé. Alors $^LH = {}^LG$ et $^L\mathscr{P}(^LH)$ s'identifie à une partie de $^L\mathscr{P}(^LG)$. Il s'ensuit que $\Phi(H/E)$ s'identifie à une partie de $\Phi(G/E)$. Une solution positive à 8.1 entraîne donc l'existence d'une injection qui associe à tout paquet de représentations L-indiscernables de $\mathscr{A}(H)$ une partie L-indiscernable de $\mathscr{A}(G)$. Si E est archimédien, cela est contenu dans [30]. Si H est le groupe multiplicatif d'une algèbre de quaternions sur E, $G = \mathbf{GL}_2$, c'est un des résultats frappants de [22]. On a alors en fait une bijection entre éléments de $\mathscr{A}(H)$ et les éléments de la série discrète dans $\mathscr{A}(G)$ telle que les caractères de deux éléments prennent, au signe près, les «mêmes valeurs» sur les éléments elliptiques réguliers. D'autre part, les résultats de R. Howe sur les représentations du groupe multiplicatif d'une algèbre à division sur un corps local plaident en faveur d'une telle correspondance.

(b) Supposons maintenant H et G définis sur F. Alors ^LH_v est isomorphe à ^LG_v pour tout v et $H \times F_v$ est isomorphe à $G \times F_v$, et quasi-déployé sur F_v, pour presque tout v. Supposons de nouveau G quasi-déployé. On a donc $\Phi(H/F_v) \hookrightarrow \Phi(G/F_v)$ pour tout v et $\Phi(H/F_v) = \Phi(G/F_v)$ pour presque tout v. Le problème de surjectivité prend la forme suivante: supposons que $\pi' = \otimes_v \pi'_v$ soit automorphe pour H. Peut-on trouver $\pi_v \in \mathscr{A}(G_v)$ correspondant au même élément de $\Phi(G/F_v) \supset \Phi(H/F_v)$ que π'_v, et égal à π'_v pour presque tout v, de manière à ce que $\pi = \otimes_v \pi_v$ soit automorphe pour G?

Supposons que H soit le groupe multiplicatif d'une algèbre de quaternions D sur F et que $G = \mathbf{GL}_2$. Une réponse positive est alors fournie par [22], (voir aussi [13]) et généralise des résultats antérieurs de Shimizu [40]. Plus précisément, soit S l'ensemble des places de F en lesquelles D est ramifiée. Alors [22] établit une bijection entre représentations automorphes de H de degré >1 et les représentations automorphes paraboliques $\pi = \otimes_v \pi_v$ de G telles que π_v soit dans la série discrète quel que soit $v \in S$.

Remarquons enfin que comme $H \times F_v$ et $G \times F_v$ sont isomorphes pour presque tout v, on peut déjà poser un problème global dans le cadre de 7.3, indépendamment de 8.1, comme cela est fait dans [27]: à savoir si, pour r donné, les séries L de 7.3 (1) associées aux représentations automorphes de H proviennent aussi de représentations automorphes de G.

[Texte revisé en septembre 1975]

Bibliographie

1. Artin, E., Tate, J.: Class Field Theory, Benjamin, N.Y., 1967
2. Borel, A.: Opérateurs de Hecke et Fonctions Zêta, Sém. Bourbaki, 18e année, 1965/66, Exp. 307, Benjamin, N.Y.

3. Casselman, W.: On representations of $\mathbf{GL}_2$ and the arithmetic of modular curves, Modular Functions of One Variable II, Springer Lecture Notes **349** (1973), 107–142
4. Deligne, P.: Formes modulaires et représentations l-adiques, Sém. Bourbaki, 1968/69, Exp. 355, Springer Lecture Notes **179**, 139–186
5. Deligne, P.: Les Constantes des Equations Fonctionnelles, Sém. Delange-Pisot-Poitou, 11e année, 1969/70, n° 19 bis
6. Deligne, P.: Travaux de Shimura, Sém. Bourbaki, 1970/71, Exp. 389, Springer Lecture Notes **244**, 123–165
7. Deligne, P.: Formes modulaires et représentations de $\mathbf{GL}_2$, Modular Functions of One Variable, Springer Lecture Notes **349** (1973), 55–106
8. Deligne, P.: Les constantes des équations fonctionnelles des fonction L, ibid., 501–597
9. Deligne, P.: La conjecture de Weil I, Publ. Math. I.H.E.S., vol. **43** (1974), 273–307
10. Deligne, P., Serre, J.-P.: Formes modulaires de poids 1, Annales E.N.S., (4), t. **7** (1974), 507–530
11. Doi, K., Naganuma, H.: On the functional equation of certain Dirichlet series, Inv. Math. **9** (1969), 1–14
12. Drinfeld, V. G.: Modules elliptiques, Math. Sbornik **94** (1974), 594–627
13. Gelbart, S. S.: Automorphic functions on adele groups, Annals of Math. Studies **83** (1975), Princeton University Press
14. Gelfand, I. M., Graev, M. I., Piateckiĭ-Šapiro, I. I.: Representation theory and automorphic functions, Saunders, Philadelphia 1969
15. Gelfand, I. M., Kazdhan, D. A.: Representations of the group $\mathbf{GL}(n, K)$ where K is a local field, Inst. for Applied Mathematics, Moscow 1971
16. Godement, R.: Fonctions automorphes et produits eulériens, Sém. Bourbaki, 1968/69, Exp. 349, Springer Lecture Notes **179**, 37–53
17. Godement, R.: Notes on Jacquet-Langlands Theory, Institute for Advanced Study, Princeton, N.J., 1970
18. Godement, R., Jacquet, H.: Zeta-Functions of Simple Algebras, Springer Lecture Notes **260**, 1972
19. Harish-Chandra: Automorphic Forms on a Semi-Simple Group, Springer Lecture Notes **62**, 1968
20. Jacquet, H.: Automorphic Forms on $\mathbf{GL}(2)$, Part II, Springer Lecture Notes **278**, 1972
21. Jacquet, H.: Euler products and automorphic forms, Proc. I.C.M. Vancouver
22. Jacquet, H., Langlands, R. P.: Automorphic Forms on $\mathbf{GL}(2)$, Springer Lecture Notes **114**, 1970
23. Jacquet, H., Shalika, J. A.: Hecke theory for $\mathbf{GL}(3)$, Comp. Math. [**29** (1974), 75–87]
24. Langlands, R. P.: On the functional equations satisfied by Eisenstein series, preprint, Yale University [Springer Lecture Notes **544**]
25. Langlands, R. P.: Euler Products, Yale University Press, 1967
26. Langlands, R. P.: Representations of abelian algebraic groups, preprint, Yale University, 1968
27. Langlands, R. P.: Problems in the theory of automatic forms, in Lectures in Modern Analysis and Applications, Springer Lecture Notes **170** (1970), 18–86
28. Langlands, R. P.: On Artin's L-functions, in Complex Analysis, Rice Univ. Studies **56** (1970), 23–28
29. Langlands, R. P.: Modular forms and l-adic representations, in Modular Functions of One Variable II, Springer Lecture Notes **349** (1973), 361–500
30. Langlands, R. P.: On the classification of irreducible representations of real algebraic groups, preprint
31. Langlands, R. P.: Some contemporary problems with origins in the Jugendtraum, [Proc. Symp. Pure Math. **29** (1976), 401–428, AMS Providence R. I.]
32. Ogg, A.: Modular Forms and Dirichlet Series, Benjamin Lecture Notes, N.Y., 1968
33. Piateckiĭ-Šapiro, I. I.: Zeta-functions of modular curves, Modular Functions of One Variable II, Springer Lecture Notes **349**, 1973, 317–360
34. Rankin, R. A., Contributions to the theory of Ramanujan's function $\tau(n)$ and similar arithmetical function, I, II, Proc. Cambridge Phil. Soc. **35** (1936), 351–372
35. Satake, I.: Theory of spherical functions on reductive algebraic groups, Publ. Math. I.H.E.S., **18** (1963), 5–59

36. Satake, I.: Spherical functions and Ramanujan conjecture, in Proc. Symp. Pure Math., **9** (1966), 258−264, A.M.S., Providence, R. I.
37. Serre, J.-P.: Abelian *l*-Adic Representations, Benjamin Lecture Notes, 1968
38. Serre, J.-P.: Facteurs Locaux des Fonctions Zêta des Variétés Algébriques (Définitions et Conjectures), Sém. Delange-Pisot-Poitou, 11e année, 1969/70, Exp. 19
39. Shalika, J. A., Tanaka, S.: On an explicit construction of a certain class of automorphic forms, Amer. J. Math., **91** (1969), 1049−1076
40. Shimizu, H.: On zeta functions of quaternion algebras, Annals of Math., (2), **81** (1965), 166−193
41. Shimura, G.: Introduction to the Arithmetic Theory of Automorphic Functions, Iwanami Shoten and Princeton Univ. Press, 1971
42. Shimura, G.: Modular forms of half-integral weight, in Modular Functions of One Variable I, Springer Lecture Notes **320** (1973), 57−74
43. Shimura, G.: On the holomorphy of certain Dirichlet series, Proc. London Math. Soc. (3), **31** (1975), 79−98
44. Tate, J.: Fourier analysis in number fields and Hecke's zeta-functions, in Algebraic Number Theory, Cassels and Fröhlich, ed., Thompson Book Co., 1967, pp. 305−347
45. Weil, A.: Sur la théorie du corps de classes, Jour. Math. Soc. Japan, **3** (1951), 1−35
46. Weil, A.: On a certain type of characters of the idèle-class group of an algebraic number field, Proc. Int. Colloq. on Alg. Number Theory, Tokyo-Nikko, 1955, 1−7
47. Weil, A.: Über die Bestimmung Dirichletscher Reihen durch Funktionalgleichungen, Math. Annalen **168** (1967), 149−167
48. Weil, A.: Basic Number Theory, Grundl. der Math. Wiss. **144,** Springer Verlag, 1967
49. Weil, A.: Dirichlet Series and Automorphic Forms, Springer Lecture Notes **189,** 1971

104.

Cohomologie de sous-groupes discrets
et représentations de groupes semi-simples

Astérisque **32–33** (1976) 73–112

Cet exposé est consacré à quelques résultats et problèmes concernant

la cohomologie de certains sous-groupes discrets de groupes semi-simples,

à coefficients dans un espace vectoriel complexe. On y insiste principalement

sur le cas de sous-groupes cocompacts et les liens existant entre la

cohomologie d'Eilenberg-MacLane du sous-groupe discret Γ, la cohomologie

d'Eilenberg-MacLane continue du groupe ambiant G et la décomposition de

$L^2(G/\Gamma)$ en G-modules irréductibles. On considère successivement trois

cas, dont le dernier englobe les deux premiers: G semi-simple réel (§§1, 2),

G semi-simple p-adique (3.1, 3.2), et G produit de groupes d'un de ces

deux types (3.3 à 3.9). L'exemple le plus important de la situation du §3,

et en fait presque le cas général vu [29], est celui de sous-groupes S-

arithmétiques d'un groupe semi-simple défini sur un corps de nombres k

et anisotrope sur k; il fait l'objet du §4. En ce qui concerne les groupes

arithmétiques ou S-arithmétiques en général, on s'est borné à quelques re-

marques dans le §5, surtout pour signaler quelques problèmes naturellement

suggérés par les résultats du §4. Cet article est ainsi en large partie com-

plémentaire de [4]. En cela, il diffère assez sensiblement de l'exposé oral,

de titre "Cohomologie réelle des groupes arithmétiques", dans lequel on

s'était borné aux groupes réels et on avait aussi passé en revue des résultats

sur les sous-groupes arithmétiques non cocompacts de groupes semi-simples
résumés dans [4] et des problèmes ouverts les concernant.

Deux exposés sur la cohomologie continue faits par G. Zuckerman
à l'IAS au printemps 1975, et des discussions avec J.-P. Serre m'ont été
très utiles pour la préparation de cet article. Je les en remercie vivement.

§0. Notations.

0.1. Si un groupe G opère sur un ensemble A, alors A^G est
l'ensemble des points fixes de G.

0.2. Les variétés réelles sont C^∞. Si X est une variété réelle,
et V un espace vectoriel complexe de dimension finie, alors $C^\infty(X)$
(resp. $C^\infty(X; V)$) est l'espace des fonctions C^∞ sur G, à valeurs complexes (resp. dans V) et Ω_X (resp. $\Omega_X(V)$) l'espace des formes
différentielles C^∞ sur X à valeurs complexes (resp. dans V). On a donc
des isomorphismes canoniques

$$C^\infty(X) \otimes V = C^\infty(X; V) \qquad \Omega_X \otimes V = \Omega_X(V) \ .$$

0.3. Si X est un espace localement compact totalement discontinu
et V est comme ci-dessus, alors $C^\infty(X)$ (resp. $C^\infty(X; V)$) est l'espace
des fonctions localement constantes sur X à valeurs complexes (resp. dans V).

0.4. Si G est un groupe localement compact unimodulaire, alors $\hat{G}$
est l'ensemble des classes d'équivalence de représentations unitaires irréductibles de G muni de la topologie de Fell (cf. p. ex. [14; 38]).

Si π est une représentation unitaire, M_π désigne l'espace de π et

400

$[\pi] \in \hat{G}$ sa classe. On ne distinguera pas toujours soigneusement entre une représentation et sa classe et si $\pi \in \hat{G}$, on écrira aussi M_π pour l'espace d'un élément de π; on procèdera de même pour d'autres notions qui ne dépendent que de la classe d'équivalence, comme le caractère infinitésimal si G est de Lie.

0.5. Soit G un groupe de Lie réel semi-simple, dont la composante neutre G^o est d'indice fini dans G et de centre fini. Si (π, E) est une représentation continue de G dans un espace vectoriel topologique complexe, on note E^∞ l'espace des vecteurs différentiables de E. Muni d'une topologie convenable, c'est un G-module différentiable et l'injection $E^\infty \longrightarrow E$ est continue. C'est aussi un module sur l'algèbre enveloppante $U(\underline{g})$ de l'algèbre de Lie $\underline{g}$ de G. Supposons que G opère trivialement sur le centre $Z(\underline{g})$ de $U(\underline{g})$, via la représentation adjointe. Si π est unitaire et irréductible, il existe alors un caractère χ_π de $Z(\underline{g})$ tel que $\pi(z) = \chi_\pi(z).\mathrm{Id}$ sur E^∞ si $z \in Z(\underline{g})$, appelé le caractère infinitésimal de π. L'hypothèse faite sur G est satisfaite si l'image de G dans $\mathrm{Aut}(\underline{g} \otimes \mathbb{C})$ par la représentation adjointe est contenue dans $\mathrm{Ad}(\underline{g} \otimes \mathbb{C})$; cela a lieu en particulier si G est d'indice fini dans le groupe des points réels d'un groupe algébrique connexe (en topologie de Zariski) défini sur $\mathbb{R}$, qui est le cas principalement en vue ici.

0.6. Soit G un groupe localement compact totalement discontinu, dénombrable à l'infini, et soit (π, E) une représentation complexe continue de G. On note E^∞ l'ensemble des vecteurs fixes par un sous-groupe ouvert de G. Muni de la topologie discrète, c'est un G-module continu. La représentation (π, E) est <u>admissible</u> si E^U est de dimension finie pour tout sous-groupe ouvert U de G, et E^∞ est dense dans E.

0.7. Soient G un groupe localement compact unimodulaire et Γ un sous-groupe discret. On dit que Γ est <u>cocompact</u> (resp. de covolume fini) si G/Γ est compact (resp. de volume fini par rapport à toute mesure invariante).

0.8. Soient k un corps et $\underline{G}$ un groupe algébrique semi-simple défini sur k. On rappelle que les tores (algébriques) définis et déployés sur k maximaux de $\underline{G}$ sont conjugués par des éléments du groupe $\underline{G}(k)$ des points rationnels sur k de $\underline{G}$. Leur dimension commune est le k-rang $rg_k\underline{G}$ de $\underline{G}$ [6: §§4, 5].

§1. <u>Remarques générales</u>.

<u>Dans ce</u> §, <u>sauf en</u> 1.7 , G <u>est un groupe de Lie réel semi-simple, ayant un nombre fini de composantes connexes, dont la composante neutre</u> G^o <u>a un centre fini</u>, K <u>est un sous-groupe compact maximal de</u> G, $X = K\backslash G$ <u>et</u> Γ <u>est un sous-groupe discret de</u> G.

1.1. Soit (r, V) une représentation complexe de dimension finie de G. Nous nous intéresserons à l'espace de cohomologie d'Eilenberg-MacLane $H^*(\Gamma; V)$ de Γ à coefficients dans le Γ-module V. Il est susceptible d'une définition purement algébrique, mais qui est en général assez peu utile. Aussi commencerons-nous par passer en revue d'autres interprétations de

402

cet espace.

1.2. Le groupe Γ opère proprement sur X, qui est contractile.
Supposons tout d'abord Γ sans torsion. Alors il opère librement et X/Γ
est une variété. Le comorphisme $\sigma^o : \Omega_{X/\Gamma} \longrightarrow \Omega_X$ associé à la projection
$\sigma : X \longrightarrow X/\Gamma$ identifie $\Omega_{X/\Gamma}$ à Ω_X^Γ. De plus (r, V) définit un système
local (faisceau localement constant) $\widetilde{V}$ sur X/Γ, et σ^o induit un
isomorphisme de l'espace $\Omega_{X/\Gamma}(\widetilde{V})$ des formes à valeurs dans $\widetilde{V}$ sur
l'espace $\Omega_X(V)^\Gamma$. Vu que X est contractile, le théorème de de Rham
entraîne les isomorphismes

$$(1) \qquad H^*(\Gamma; V) = H^*((\Omega_X(V)^\Gamma) = H^*(X/\Gamma; \widetilde{V}) \ .$$

Si Γ a de la torsion, alors le quotient X/Γ est une "V-variété" et $\widetilde{V}$ est
un faisceau non nécessairement localement constant, mais ces égalités
restent vraies.

1.3. Soit $(X_j)_{1 \le j \le N}$, où $N = \dim G$, une base de l'espace des champs
de vecteurs invariants à droite sur G. On suppose que, à l'origine, les
X_j $(n = \dim X < j \le N)$ (resp. $1 \le j \le n$) forment une base de l'algèbre
de Lie $\underline{k}$ de K (resp. du complément orthogonal $\underline{p}$ de $\underline{k}$ dans l'algèbre
de Lie $\underline{g}$ de G par rapport à la forme de Killing). Soit (ω^j) la base duale
de l'espace des 1-formes invariantes àdroite sur G. On envisage aussi les
X_j et ω^j comme des champs de tenseurs sur G/Γ. Soit $\pi : G/\Gamma \longrightarrow X/\Gamma$
la projection naturelle. A $\omega \in \Omega_X^d(V)^\Gamma$ on associe la forme ω^o sur G définie

par

$$(1) \qquad x \longmapsto r(x^{-1})\pi^o(\omega)(x) , \qquad\qquad (x \in G) .$$

On peut écrire

$$(2) \qquad \omega^o = \Sigma_I \tau_I^o \omega^I ,$$

où $I = \{i_1, \ldots, i_d\}$, $(1 \leq i_1 < \ldots < i_d \leq n)$ et

$$(3) \qquad \omega^I = \omega^{i_1} \wedge \ldots \wedge \omega^{i_d} , \ \tau_I^o \in C^\infty(G/\Gamma, V) .$$

Soit $A^o(G, \Gamma, V)$ l'espace des formes ainsi obtenues.

On envisage $C^\infty(G/\Gamma; V) = C^\infty(G/\Gamma) \otimes V$ comme un G-module via translations à gauche et r. Soit $C^*(\underline{g}, \underline{k}; C^\infty(G/\Gamma; V))$ l'espace des cochaînes relatives d'algèbre de Lie de $\underline{g}$ mod $\underline{k}$, à coefficients dans $C^\infty(G/\Gamma; V)$. Il s'identifie à l'espace des éléments annulés par $\underline{k}$ dans $\mathrm{Hom}_{K^o}(\Lambda \underline{p}, C^\infty(G/\Gamma; V))$. Le groupe K opère naturellement sur cet espace, d'où une action naturelle de K/K^o sur ce complexe et sa cohomologie. On pose

$$(4) \qquad C^*(\underline{g}, K; C^\infty(G/\Gamma; V)) = C^*(\underline{g}, \underline{k}; C^\infty(G/\Gamma; V))^{K/K^o} ,$$

$$(5) \qquad H^*(\underline{g}, K; C^\infty(G/\Gamma; V)) = H^*(\underline{g}, \underline{k}; C^\infty(G/\Gamma; V))^{K/K^o} .$$

Des calculs simples (cf. [33: §3]) montrent que $A^o(G, \Gamma, V)$ s'identifie canoniquement à $C^*(\underline{g}, K; C^\infty(G/\Gamma; V))$, d'où un isomorphisme

$$(6) \qquad H^*(\Gamma; V) = H^*(\underline{g}, K; C^\infty(G/\Gamma \otimes V)) \ .$$

1.4. Supposons V muni d'une structure d'espace de Hilbert "admissible" i.e. invariante par K et telle que $dr(Y)$ soit hermitien pour $Y \in \underline{p}$, et X de la métrique riemannienne invariante qui, sur $\underline{p}$, est égale à la forme de Killing; on suppose aussi que les X_j $(1 \le j \le n)$ forment une base orthonormale de $\underline{p}$ à l'origine. Alors $A_o(G, \Gamma, V)$ est muni d'un produit scalaire (partiellement défini) donné par

$$(1) \qquad (\sigma, \tau) = \int_{G/\Gamma} \Sigma_I (\sigma_I, \tau_I) dx \ , \qquad (\sigma, \tau \in A_o^d(G, \Gamma, V) \ ,$$

où, dans les notations de 1.3(2), (3), on pose

$$(2) \qquad \sigma = \Sigma_I \sigma_I \omega^I \ , \qquad \tau = \Sigma_I \tau_I \omega^I \ .$$

Soient δ l'adjoint de la différentiation extérieure d par rapport à ce produit scalaire et $\Delta = \delta d + d\delta$ le laplacien. D'après une formule de Kuga [32: §6] on a

$$(3) \qquad (\Delta\omega)_I = (-C + dr(C)).\omega_I \ ,$$

où C est l'opérateur de Casimir et où $-C + dr(C)$ opère sur $C^\infty(G/\Gamma) \otimes V$ par

$$(-C + dr(C))(f \otimes v) = -C.f \otimes v + f \otimes dr(C).v) \qquad (f \in C^\infty(G/\Gamma); v \in V)$$

1.5. Soit E un G-module différentiable. Par là on entend ici un espace vectoriel topologique localement convexe, séparé, quasi complet, sur lequel G opère continûment et dont tout élément est un vecteur différentiable. C'est donc aussi un g-module. Soit $H_d^*(G; E)$ (resp. $H_{ct}^*(G; E)$) l'espace de cohomologie d'Eilenberg-MacLane de G à valeurs dans E, calculé à l'aide de cochaînes différentiables (resp. continues). D'après [20], ces deux espaces sont canoniquement isomorphes et d'après [17]:

$$(1) \qquad H^*(\underline{g}, \underline{k}, E) = H_d^*(G^\circ, E) \ .$$

Par conséquent, on a aussi

$$(2) \qquad H^*(\Gamma; V) = H_{ct}(G; C^\infty(G/\Gamma, V)) = H_d(G; C^\infty(G/\Gamma, V)) \ ,$$

au moins si G est connexe, mais en fait cet isomorphisme peut aussi s'établir directement dans le cadre de la cohomologie continue [13; 20] par un lemme de Shapiro convenable, et cela sans supposer G connexe.

1.6. On sait que $H^*(\underline{g}, \underline{k}; V) = 0$ si V est irréductible non trivial. Supposons le trivial. L'inclusion $\mathbb{C} \hookrightarrow C^\infty(G/\Gamma)$ qui associe à $c \in \mathbb{C}$ la fonction constante égale à c induit donc un homomorphisme

$$(1) \qquad H_d^*(G; \mathbb{C}) = H^*(\underline{g}, \underline{k}; \mathbb{C}) \longrightarrow H^*(\Gamma; \mathbb{C}) \ .$$

La construction de [16] de l'isomorphisme de van Est montre que le diagramme

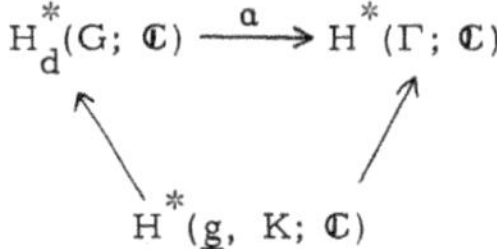

où α est l'homomorphisme de restriction, est commutatif.

1.7. <u>Cas des groupes localement compacts.</u> Une partie de 1.5, 1.6 admet une variante en termes de fonctions continues de portée plus générale, qui englobe aussi les cas considérés dans les §§3, 4, aussi l'indiquons-nous brièvement ici.

Dans ce n^o, on suppose que G est un groupe localement compact dénombrable à l'infini et Γ un sous-groupe discret de G. Admettons pour commencer que (r, V) soit seulement un Γ-module (complexe, de dimension finie). Soit $C(G; V)_\Gamma$ l'ensemble des fonctions continues $f : G \longrightarrow V$ telles que

$$(1) \qquad f(x. \gamma) = r(\gamma^{-1}). f(x) , \qquad (x \in G; \gamma \in \Gamma) .$$

Les Prop. 3, 4 de [13] entraînent l'égalité

$$(2) \qquad H^m(\Gamma; V) = H^m_{ct}(G; C(G; V)_\Gamma) , \qquad (m \in \mathbf{Z}) .$$

Supposons que (r, V) soit en fait un G-module. Alors

$$(3) \qquad C(G; V)_\Gamma \cong C(G/\Gamma; V) ,$$

où le membre de droite désigne l'ensemble des fonctions continues sur G/Γ, à valeurs dans V. On retrouve donc 1.5(2), à cela près que C remplace

C^∞. L'application $j_o : V \longrightarrow C(G/\Gamma; V)$ qui associe à v la fonction con-

stante sur G/Γ de valeur v induit donc un homomorphisme

$$(4) \qquad\qquad j^m : H^m_{ct}(G; V) \longrightarrow H^m(\Gamma; V) \ , \qquad\qquad (m \in \mathbb{Z})$$

qui généralise 1.6(1).

Si G/Γ est compact, j^m est injectif. En effet, dans ce cas G

est unimodulaire et G/Γ possède une mesure positive invariante dx;

l'ensemble des $f \in C(G/\Gamma; V)$ d'intégrale nulle par rapport à dx est alors

un sous-G-module fermé, supplémentaire de $j_o(V)$, et notre assertion

résulte de (2).

§2. Sous-groupes discrets cocompacts de groupes réels.

On conserves les hypothèses du §1. On suppose de plus que G opère

trivialement sur le centre $Z(\underline{g})$ de l'algèbre enveloppante de $\underline{g}$ (cf. 0.5) et

que Γ est cocompact, (0.7).

2.1. Soit $L^2(G/\Gamma)$ l'espace des fonctions de carré sommable (pour

une mesure de Haar) sur G/Γ. C'est un G-module unitaire et

$L^2(G/\Gamma)^\infty = C^\infty(G/\Gamma)$. On sait que l'on a une décomposition en somme directe

hilbertienne à multiplicités finies

$$(1) \qquad\qquad L^2(G/\Gamma) = \widetilde{\bigoplus}_{\pi \in \hat{G}} \ m(\pi, \ \Gamma) M_\pi \ ,$$

où $m(\pi, \Gamma)$ désigne donc la multiplicité de π dans $L^2(G/\Gamma)$, d'où aussi une décomposition en somme topologique

$$(2) \qquad C^\infty(G/\Gamma) = \widetilde{\bigoplus}_{\pi \epsilon \hat{G}} m(\pi, \Gamma) M_\pi^\infty \ .$$

2.2. PROPOSITION. <u>Soit</u> V <u>un</u> G-<u>module de dimension finie. On a</u>

$$(1) \qquad H^*(\Gamma; V) = \bigoplus_{\pi \epsilon \hat{G}} m(\pi, \Gamma) \ H^*(\underline{g}, K; M_\pi^\infty \otimes V) \ .$$

<u>En particulier, l'injection</u> $v \longmapsto 1 \otimes v$ <u>dans</u> V <u>dans</u> $C^\infty(G/\Gamma) \otimes V$ <u>induit un homomorphisme injectif de</u> $H^*(\underline{g}, K; V)$ <u>dans</u> $H^*(\Gamma; V)$.

Vu 1.4 et 2.1(2), le membre de gauche est la cohomologie d'un complexe qui s'écrit comme une somme topologique

$$C^*(\underline{g}, K; C^\infty(G/\Gamma) \otimes V) = \widetilde{\bigoplus}_{\pi \epsilon \hat{G}} C^*(\underline{g}, K; M_\pi^\infty \otimes V) \ ,$$

et le seul point non évident est le passage de la somme topologique à la somme directe algébrique. Pour cela, il suffit de savoir que l'image de l'opérateur cobord $d : C^{m-1} \longrightarrow (\ker d) \cap C^m$ est fermée. Or d est continu, et son conoyau $H^m(\Gamma : V)$ est de dimension finie, puisque G/Γ est compact; le fait que Im d est fermée résulte alors du Cor. 2 à la Prop. 4, p. 68 de [8: §4]. Pour cet argument, dans le cadre de la cohomologie continue, cf. [13: Prop. 6]. Une autre démonstration sera donnée plus bas (2.5(1)).

2.3. Pour tout $m \in \mathbb{N}$, l'espace $H^m(\underline{g}, K; M_\pi \otimes V)$ s'identifie à un espace de formes harmoniques, i.e. de zéros du laplacien. Supposons V irréductible. Alors, vu 1.5

(1) $\mathrm{Hom}_K(\Lambda^m \underline{p}; M_\pi^\infty \otimes V) = C^m(\underline{g}, K; M_\pi^\infty \otimes V) = H^m(\underline{g}, K; M_\pi^\infty \otimes V)$, si $\chi_r(C) = \chi_\pi(C)$

où C est l'opérateur de Casimir, et

(2) $$H^m(\underline{g}, K; M_\pi^\infty \otimes V) = 0 , \quad \text{si } \chi_\pi(C) \neq \chi_r(C) .$$

Joint à 2.2, cela donne la formule suivante, due à Matsushima [30] lorsque G est connexe et $V = \mathbb{C}$:

(3) $$H^m(\Gamma; V) = \bigoplus_{\pi \in \hat{G}} m(\pi, \Gamma) \ \mathrm{Hom}_K(\Lambda^m \underline{p}; M_\pi^\infty \otimes V) ,$$

la somme étant étendue aux $\pi \in \hat{G}$ tels que

(4) $$\chi_\pi(C) = \chi_r(C) .$$

En fait, d'après D. Wigner (non publié) la somme ne porte que sur les $\pi \in \hat{G}$ tels que

(5) $$\chi_\pi(z) = \chi_{\check{r}}(z) , \qquad \text{pour tout } z \in Z(\underline{g}) ,$$

où $\check{r}$ est la représentation contragrédiente de r.

La démonstration de Wigner (que je connais grâce à un exposé de G. Zuckerman à l'IAS) se plaçant dans le cadre de l'algèbre homologique, il s'impose de l'esquisser ici.

Vu 1.6(2), on peut aussi écrire 2.2 sous la forme

$$(6) \qquad H^m(\Gamma; V) = \oplus_{\pi \in \hat{G}} m(\pi,\ \Gamma)\ H_d^m(G;\ M_\pi^\infty \otimes V)\ .$$

Notre assertion résulte donc de la proposition plus générale suivante

2.4. PROPOSITION (D. Wigner). <u>Soient</u> $\pi \in \hat{G}$ <u>et</u> $(r,\ V)$ <u>un</u> G-<u>module irréductible de dimension finie tels que</u> $\chi_{\underset{r}{\vee}} \neq \chi_\pi$. <u>Alors</u> $H_d^m(G;\ M_\pi^\infty \otimes V) = 0$ <u>pour tout</u> $m \in \mathbb{Z}$. <u>En particulier, on a soit</u> $\operatorname{Hom}_K(\Lambda^m \mathfrak{p};\ M_\pi^\infty \otimes V) = 0$ <u>soit</u> $\chi_{\underset{r}{\vee}}(C) \neq \chi_\pi(C)$.

On a

$$H_d^m(G;\ M_\pi^\infty \otimes V) = \operatorname{Ext}_G^m(\mathbb{C},\ M_\pi^\infty \otimes V)\ ,$$

où Ext_G^* est calculé dans la catégorie des G-modules différentiables, pour une notion convenable de suite exacte [20]. Vu les égalités

$$\operatorname{Hom}_G(\mathbb{C},\ M_\pi^\infty \otimes V) = (M_\pi^\infty \otimes V)^G = \operatorname{Hom}_G(V^*,\ M_\pi^\infty)\ ,$$

où V^* est l'espace dual de V, on a aussi

$$H_d^m(G;\ M_\pi^\infty \otimes V) = \operatorname{Ext}_G^m(V^*,\ M_\pi^\infty))\ .$$

D'après Yoneda (cf. [28: Chap. III]), ce dernier module est l'ensemble des classes d'équivalence de suites exactes admises

$$(8) \quad 0 \longrightarrow V^* \longrightarrow E_0 \longrightarrow E_1 \longrightarrow \ldots \longrightarrow E_{m-1} \longrightarrow M_\pi^\infty \longrightarrow 0 \ ,$$

pour une relation d'équivalence convenable. Si maintenant $\chi_{\check{r}} \neq \chi_\pi$, il existe alors $z \in Z(\underline{g})$ tel que $\chi_{\check{r}}(z) = 0$, $\chi_\pi(z) = 1$; cet élément définit un endo-morphisme de (8), qui est zéro sur V^*, l'identité sur M_π^∞, ce qui entraîne que cette suite exacte représente l'élément zéro de $\text{Ext}_G^m(V^*, M_\pi^\infty)$.

2.5. <u>Remarques</u>. (1) Supposons Γ sans torsion. La somme topologique M' des $M_\pi^\infty \otimes V$, où π parcourt les éléments de $\hat{G}$ tels que $\chi_\pi(C) \neq \chi_r(C)$ est aussi la somme topologique des espaces propres du laplacien Δ correspondant aux valeurs propres non nulles. Comme X/Γ est une variété compacte, ces valeurs propres sont de multiplicité finie et n'ont pas de point d'accumulation fini. Par suite Δ^{-1} est un opérateur continu sur M'. Il s'ensuit alors que $H^m(\underline{g}, K; M') = 0$ ce qui établit (6) à partir de 1.4 et 2.1(2) sans utiliser 2.2, au moins si Γ est sans torsion; sinon, on se ramène à ce cas en prenant un sous-groupe distingué d'indice fini dont l'image dans le groupe adjoint est sans torsion.

(2) Si l'on ne suppose pas que G opère trivialement sur $Z(\underline{g})$, alors le caractère infinitésimal d'une représentation unitaire irréductible est défini sur $Z(\underline{g})^{G/G^o}$; cette algèbre contient toujours l'opérateur de Casimir C, et ce que précède reste valable à condition de remplacer $Z(\underline{g})$ par $Z(\underline{g})^{G/G^o}$.

2.6. La relation entre représentations et cohomologie fournie par 2.2, 2.3 a été notamment exploitée dans l'étude de $H^1(\Gamma; \mathbb{C})$. Supposons G

412

connexe et simple en tant que groupe de Lie. Soit ℓ le rang de l'espace symétrique X, autrement dit la dimension des sous-algèbres maximales de $\underline{p}$. D'après Kajdan [24; 14] on a $H^1(\Gamma; \mathbb{C})$ si la représentation triviale 1 est un point isolé de $\hat{G}$ et cette condition est satisfaite lorsque $\ell \geq 2$ [24; 38] ou encore si $\ell = 1$ et G est de type $Sp(n, 1)$, ou $\mathbb{F}_4$ avec sous-groupe compact maximal $Spin\ 9$ [25]. Cela implique déjà que dans ce cas $L^2(G/\Gamma)$ ne peut contenir d'élément π par rapport auquel le 1er groupe de cohomologie continue de G est non nul. En fait, on a plus généralement et plus précisément:

2.7. PROPOSITION. (i) <u>Soit</u> $\pi \in \hat{G}$. <u>Alors</u> $H^1_d(G, M^\infty_\pi) = 0$ <u>si</u> π <u>est séparée de la représentation triviale dans</u> $\hat{G}$.

(ii) <u>Si</u> G <u>est connexe de type</u> $SO(n, 1)$ (<u>resp.</u> $SU(n, 1)$), $\hat{G}$ <u>contient exactement</u> 1 <u>élément (resp.</u> 2 <u>éléments)</u> π <u>tels que</u> $H^1_d(G, M^\infty_\pi) \neq 0$. <u>Dans ces cas,</u> $H^1_d(G, M^\infty_\pi)$ <u>est de dimension</u> 1. <u>De plus</u> π <u>est non tempérée si</u> $n \geq 4$ (<u>resp.</u> $n \geq 2$).

(i) est dû à P. Delorme [15: Th. 3]; vu ce qui a été rappelé plus haut, il s'applique à tout $\pi \in \hat{G}$ si G est connexe simple non compact et non de l'un des types considérés dans (ii). Pour (ii), <u>voir</u> Hotta-Wallach [23] et, en ce qui concerne la dernière assertion, un article de K. Johnson et N. Wallach à paraître aux Trans. A.M.S.; l'article [15] établit aussi (ii) pour $SO(n, 1)^o$ et l'annonce en partie pour $SU(n, 1)$.

2.8. Cette proposition laisse donc ouverte la possibilité pour G de type $\text{SO}(n, 1)$ ou $\text{SU}(n, 1)$ de posséder un sous-groupe Γ cocompact de 1er nombre de Betti non nul. Si G de type $\text{SO}(n, 1)$ ce nombre est la multiplicité dans $L^2(G/\Gamma)$ de l'élément π mentionné dans 2.7(ii). Si $n = 2$, on rencontre parmi ces groupes les groupes fondamentaux des surfaces de Riemann de genre ≥ 2, donc des Γ de 1er nombre de Betti non nul. Pour $n \leq 3 \leq 5$, les premiers exemples ont été donnés par E. B. Vinberg [36]. Plus récemment J. Millson [34] et W. Thurston (non publié) ont construit de tels groupes arithmétiques de congruence pour tout $n \geq 3$. En fait, Thurston conjecture dans ce cas que tout quotient X/Γ admet un revêtement fini ayant un 1er nombre de Betti non nul. Millson et Thurston obtiennent plus précisément des Γ avec 1er nombre de Betti arbitrairement grand; mais cela résulte aussi d'un principe général simple (4.2), une fois obtenu un groupe de congruence de 1er nombre de Betti non nul.

Si G est de type $\text{SU}(n, 1)$, alors X s'identifie à un domaine borné symétrique, l'intérieur de la boule unité dans $\mathbb{C}^n$, et X/Γ est une variété projective compacte. Les multiplicités dans $L^2(G/\Gamma)$ des deux éléments mentionnés dans 2.7(ii) sont égales aux dimensions de $H^{1,0}(X/\Gamma)$ et $H^{0,1}(X/\Gamma)$, donc sont égales entre elles [23]. Mais, pour $n \geq 2$, on ne connait aucun exemple de Γ pour lequel elles sont non nulles. C'est aussi un des cas où l'on ne sait pas si G contient des sous-groupes discrets co-compacts ou de covolume fini non définissables arithmétiquement.

Les résultats précédents et la conjecture de Thurston suggèrent la question suivante:

(*) <u>Soit</u> $\pi \in \hat{G}$ <u>non séparée de la représentation triviale.</u> Existe-t-il <u>un sous-groupe d'indice fini</u> Γ' <u>de</u> Γ <u>tel que</u> $m(\pi, \Gamma') \neq 0$?

Une formulation adélique pour les sous-groupes arithmétiques de congruence sera indiquée en 4.3.

2.9. Supposons G connexe et $\pi \in \hat{G}$ dans la série discrète, ce qui implique que G et K ont même rang, donc aussi que la dimension de G/K est paire. De la conjecture de Blattner, établie dans [19] et 1.5(2), 2.3(1), (2) on tire que

$$(1) \qquad H^m_d(G, M^\infty_\pi \otimes V) = 0 \quad \text{si } m \neq (\dim X)/2 \text{ ou si } \chi_\pi \neq \chi_{\check{r}} \, ,$$

et que

$$(2) \qquad H^m_d(G, M^\infty_\pi \otimes V) = \mathbb{C} \quad \text{si } \chi_\pi = \chi_{\check{r}} \text{ et } m = (\dim X)/2 \, .$$

4 D'autre part, on sait (mais le rédacteur ne connait pas de référence) que si le poids dominant de r est suffisamment régulier, et si π est une représentation unitaire irréductible de caractère infinitésimal égal χ_r, alors π est dans la série discrète. Combinant cela avec 2.3(6), on voit que si le poids dominant de r est suffisamment régulier, alors $H^m(\Gamma; V) = 0$ si $m \neq \dim X/2$ et $\dim H^q(\Gamma; V)$ est la somme des multiplicités des représentations de la série discrète ayant $\chi_{\check{r}}$ pour caractère infinitésimal.

Je dois cette remarque à G. Zuckerman. Lorsque X est un domaine borné symétrique, ce résultat m'a été communiqué il y a plusieurs années par R. P. Langlands.

2.10. Du point de vue représentations, 2.2 montre que l'étude de $H^m(\Gamma; V)$ se subordonne naturellement à deux problèmes plus généraux:

1) <u>Déterminer les</u> $\pi \in \hat{G}$ <u>tels que</u>

$$\text{Hom}_K(\Lambda^m \underline{p}, M_\pi^\infty \otimes V) \neq 0 , \quad \chi_\pi = \chi_{\underset{r}{v}} .$$

D'après un théorème général de Harish-Chandra, on sait que ces π sont en nombre fini. Mais, sauf dans les cas sus-mentionnés, et $G = SU(2, 1)$ [37] ils n'ont pas été déterminés complètement; [23] donne aussi des résultats lorsque X est hermitien symétrique.

2) <u>Déterminer les multiplicités</u> $m(\pi, \Gamma)$.

Une méthode générale pour les étudier est la formule des traces de Selberg. Supposons Γ sans torsion, G connexe, et soit $\pi \in \hat{G}$ un élément intégrable de la série discrète. Soient d_π le degré formel de π et $\nu(G/\Gamma)$ le volume de G/Γ, calculés par rapport à une même mesure de Haar sur G. Alors on a [26]

$$(1) \qquad\qquad m(\pi, \ \Gamma) = \nu(G/\Gamma).d_\pi .$$

Le membre de gauche peut aussi s'interpréter comme la dimension d'un groupe de cohomologie convenable au moins si π est associée à un poids dominant assez régulier: dans ce cas, si T est un tore maximal de K, le quotient G/T admet une (en fait un nombre fini) structure complexe invariante par G, et $m(\pi, \Gamma)$ est la dimension d'un des groupes de cohomologie de $\Gamma \backslash G/T$ à coefficients dans le faisceau des germes de sections holomorphes d'un fibré holomorphe, les autres groupes étant nuls [35].

Cette égalité n'est pas nécessairement vraie si π est non intégrable, comme l'a tout d'abord signalé Langlands. En effet, si $G = SL_2(\mathbb{R})$ et si π est la représentation de plus petit degré formel de la série discrète holomorphe, alors $m(\pi, \Gamma)$ est le genre g de la courbe X/Γ, mais le membre de droite de (1) est égal à $g-1$ (cf. [37]). Cependant, pour d'autres groupes, il peut y avoir une infinité d'éléments non intégrables de la série discrète pour lesquels (1) est valable; cela se voit en reliant $m(\pi, \Gamma)$ à la dimension d'autres groupes de cohomologie [21, 22].

§3. <u>Groupes</u> p-adiques et groupes mixtes.

3.1. Soit k un corps local ultramétrique de corps résiduel fini. Soient $\underline{L}$ un k-groupe connexe semi-simple et $L = \underline{L}(k)$. Ce dernier est un groupe de Lie sur k, et un groupe topologique localement compact totalement discontinu dénombrable à l'infini pour la topologie associée à celle de k. Soit $\ell = \ell_k(\underline{L})$ le k-rang de $\underline{L}$ (0.8). On sait que L est compact si et seulement si $\ell = 0$, auquel cas $\underline{L}$ est dit anisotrope sur k; cela peut se présenter seulement si le revêtement universel de $\underline{L}$ est isomorphe (sur une extension de k), à un produit de groupes SL_n [10].

Supposons $\underline{L}$ presque simple sur k et $\ell > 0$. Soient $\underline{\tilde{L}}$ le revêtement universel de L et $\sigma : \underline{\tilde{L}} \longrightarrow \underline{L}$ l'isogénie centrale canonique. Tenant compte de [7: 6.4, 6.14] et du fait que la conjecture de Kneser-Tits

417

est vraie sur un corps local, on voit que $L^o = \sigma(\widetilde{L}(k))$ est le groupe dérivé de L, est fermé (ouvert d'indice fini en caractéristique zéro) et que L/L^o est compact commutatif. On utilisera aussi le fait, démontré par J. Tits (non publié), que tout sous-groupe ouvert propre de L^o est compact.

3.2. Soient Γ un sous-groupe discret de L et $(r; V)$ une représentation complexe de dimension finie de Γ. On note $C^\infty(L; V)$ l'ensemble des fonctions localement constantes sur L, à valeurs dans V et on pose

$$(1) \qquad C^\infty(L; V)_\Gamma = \{f \in C^\infty(L; V) \mid f(x.\gamma) = r(\gamma)^{-1}.f(x) , \quad (x \in L)\} \ .$$

Si r est la restriction d'une représentation complexe de L, on a donc un isomorphisme canonique

$$(2) \qquad C^\infty(L/\Gamma; V) = C^\infty(L; V)_\Gamma \ .$$

On a de nouveau [11]

$$(3) \qquad H^m(\Gamma; V) = H^m_{ct}(L; C^\infty(L; V)_\Gamma) \ .$$

Supposons dorénavant Γ cocompact et r unitaire. Alors si $f : L \longrightarrow V$ satisfait à la condition de (1), la fonction $x \longmapsto \|f(x)\|$ est invariante à droite par Γ. On note $L^2(L; V)_\Gamma$ l'ensemble de ces éléments qui sont de carré intégrable sur L/Γ. C'est un L-module unitaire admissible et on a comme dans 2.1,

$$(4) \qquad L^2(L, V)_\Gamma = \widetilde{\bigoplus}_{\pi \in \hat{L}} m(\pi, \Gamma)M_\pi \ ;$$

ce qui, en passant aux éléments différentiables, donne lieu à une somme
directe algébrique

$$(5) \qquad L^{\infty}(L, \ V)_{\Gamma} = \oplus_{\pi \in \hat{L}} \ m(\pi, \ \Gamma) M_{\pi}^{\infty}$$

de L-modules admissibles (munis de la topologie discrète) d'où, vu (3),

$$(6) \qquad H^{m}(\Gamma; \ V) = \oplus_{\pi \in \hat{L}} \ m(\pi, \ \Gamma) \ H_{ct}^{m}(L; \ M_{\pi}^{\infty}) \ .$$

Supposons $\underline{L}$ presque simple et simplement connexe. D'après W. Casselman
[11] on a $H_{ct}^{m}(L; \ M_{\pi}^{\infty}) = 0$ sauf si $m = 0$ et π est triviale, ou $\ell > 0$, $m = \ell$
et π est la représentation spéciale, auxquels cas cet espace est de dimension
1.

Si $\underline{L}$ n'est pas simplement connexe, soient $\underline{\tilde{L}}$, σ et L^{o} comme en
3.1. La suite spectrale de Hochschild-Serre en cohomologie continue
[13: Prop. 5] entraîne donc

$$H_{ct}^{m}(L; \ M_{\pi}^{\infty}) = (H_{ct}^{m}(L^{o}; \ M_{\pi}^{\infty}))^{L/L^{o}} \ .$$

Il s'ensuit, lorsque $\underline{L}$ est presque simple, que ce groupe est nul si $m \neq 0, \ell$,
de dimension 1 si $m = 0$ et π est triviale et ne peut être non nul pour
$m \neq 0$ que si $m = \ell > 0$ et la restriction de π à L^{o} est somme (nécessairement finie) de représentations spéciales.

On déduit alors de (5) que $H^{m}(\Gamma; \ V) = 0$ si $m \neq 0, \ell$, que dim $H^{o}(\Gamma; \ V)$
est la multiplicité de la représentation triviale dans V, que dim $H^{\ell}(\Gamma; \ V)$
est majorée par la multiplicité de la représentation spéciale de L^{o} dans

$C^{\infty}(L; V)_{\Gamma}$, et lui est égale si $\widetilde{\underline{L}} = \underline{L}$. On a un résultat semblable si $\underline{L}$ n'est pas simple [11], qui est formellement contenu dans 3.7 ci-dessous.

3.3. Nous considérons maintenant un cas qui englobe ceux de 3.2 et du §2. Soit S un ensemble fini. On suppose donné pour chaque $i \in S$ soit un groupe G_i satisfaisant aux hypothèses imposés à G dans les §§1,2 (cas archimédien), soit un corps local non archimédien k_i de caractéristique zéro, un groupe connexe semi-simple et presque simple $\underline{G}_i$ défini sur k_i et un sous-groupe ouvert G_i du groupe $\underline{G}_i(k_i)$ des points rationnels de $\underline{G}_i$ qui contient le sous-groupe dérivé G_i° de $\underline{G}_i(k_i)$ (cf. 3.1) (cas ultramétrique). On désigne par S_{∞} (resp. S_f) l'ensemble des $i \in S$ correspondants aux cas archimédiens (resp. ultramétriques). Si $S' \subset S$, on note $G_{S'}$ le produit des G_i pour $i \in S'$. On pose de plus

$$(1) \qquad G_{\infty} = \prod_{i \in S_{\infty}} G_i \, , \quad G_f = \prod_{i \in S_f} G_i \, , \quad G = G_S = G_{\infty} \times G_f \, .$$

On note K_{∞} un sous-groupe compact maximal de G_{∞}, X_{∞} le quotient $K_{\infty} \backslash G_{\infty}$ et $\underline{g}_{\infty}$ (resp. $\underline{k}_{\infty}$) l'algèbre de Lie de G_{∞} (resp. K_{∞}). On suppose G non compact. Soient encore

$$(2) \qquad \ell_i = \mathrm{rg}_{k_i} \underline{G}_i \, , \quad (i \in S_f) \, , \quad \text{et} \quad \ell = \Sigma_{i \in S_f} \ell_i \, .$$

Si $S' \subset S$ et $\pi \in \hat{G}_{S'}$, alors π se factorise de façon unique en un produit hilbertien

(3) $$\pi = \widetilde{\bigotimes}_{i \in S'} \pi_i \ , \qquad\qquad\qquad (\pi_i \in \hat{G}_i) \ .$$

Lorsque $S' = S$, on écrira aussi

(4) $$\pi = \pi_\infty \widetilde{\otimes} \pi_f \qquad\qquad (\pi_\infty = \widetilde{\bigotimes}_{i \in S_\infty} \pi_i; \ \pi_f = \widetilde{\bigotimes}_{i \in S_f} \pi_i) \ .$$

Si $S' \subset S_f$ et $G_i = G_i^o$ lorsque $\ell_i > 0$ $(i \in S')$, alors $\pi \in G_{S'}$ est dite

spéciale si π_i est triviale ou spéciale suivant que G_i est compact ou non,

i. e. suivant que ℓ_i est nul ou non.

    3. 4. Soit Γ un sous-groupe discret cocompact de G qui est

"irréductible", i. e. tel que la projection $\mathrm{pr}_{S'} : G \longrightarrow G_{S'}$ soit injective, et

d'image non discrète si $G_{S'}$ n'est pas cocompact, quel que soit $S' \subset S$,

$S' \neq \emptyset, \ S$.

    LEMME. <u>Si</u> $S' \subset S$, <u>soient</u> $\Gamma_{S'} = \mathrm{pr}_{S'} \Gamma$ <u>et</u> $\overline{\Gamma}_{S'}$ <u>l'adhérence de</u> $\Gamma_{S'}$

<u>dans</u> $G_{S'}$. <u>Soit</u> $S' \subset S$ <u>tel que</u> $G_{S'}$ <u>et</u> $G/G_{S'}$ <u>soient non compacts. Si</u>

$S' = \{i\} \subset S_f$, <u>alors</u> $\overline{\Gamma}_{S'}$ <u>contient</u> G_i^o. <u>Si</u> $S' \subset S_\infty$, <u>alors</u> $\overline{\Gamma}_{S'}$ <u>contient le</u>

<u>produit des facteurs simples non compacts de</u> $G_{S'}^o$. <u>En particulier,</u> $\overline{\Gamma}_{S'}$ <u>est</u>

<u>cocompact dans</u> $G_{S'}$.

    Supposons $i \in S_f$. Par restriction des scalaires, on se ramène au cas

où $k_i = \mathbb{Q}_p$, où p est la caractéristique résiduelle. Le groupe $\overline{\Gamma}_i$ est alors

un sous-groupe de Lie sur $\mathbb{Q}_p$ de G_i [9: §8, n° 2, Thm. 2], non discret,

dont l'algèbre de Lie $L(\overline{\Gamma}_i)$ est stable par Γ_i, opérant par la représentation

adjointe. Mais Γ_i est Zariski-dense dans $\underline{G}_i(k)$ [39], donc $L(G_i) = L(\overline{\Gamma}_i)$

et $\overline{\Gamma}_i$ est un sous-groupe ouvert de G_i. La projection pr_i induit une application surjective de G/Γ sur $G_i/\overline{\Gamma}_i$, donc $\overline{\Gamma}_i$ n'est pas compact. Notre assertion résulte alors du résultat de Tits mentionné en 3.1.

Si $S' \subset S_\infty$ alors le théorème de densité de [1], appliqué au quotient de $G_{S'}$ par le plus grand sous-groupe invariant compact connexe de $G_{S'}^o$ montre que l'image de $\overline{\Gamma}_{S'}$ dans ce quotient en contient la composante neutre, d'où notre assertion dans ce cas.

Remarque. Le lemme admet une version plus générale dans laquelle Γ est de covolume fini et S' une partie quelconque de S telle que $G_{S'}$ et $G/G_{S'}$ soient non compacts. Sous cette forme, il entraîne immédiatement le théorème d'approximation forte pour les groupes presque simples et simplement connexes sur un corps de nombres qui ne sont pas de type $\mathbf{A}_n$.

3.5. Exactement comme en 2.2, on a de nouveau une décomposition en somme discrète à multiplicités finies

$$(1) \qquad L^2(G/\Gamma) = \widetilde{\bigoplus}_{\pi \in \hat{G}} m(\pi, \Gamma) M_\pi \ ,$$

où de plus π admet les décompositions 3.3(3), (4).

LEMME. Supposons G_∞ et G_f non compacts. Soit $\pi \in \hat{G}$ tel que $M(\pi, \Gamma) \neq 0$ et que π_f soit de dimension finie. Alors π_∞ est de dimension finie.

Nous avons à montrer que $\ker \pi_\infty$ est cocompact, et notre hypothèse équivaut au fait que $N_f = \ker \pi_f$ est cocompact. Le groupe $\Gamma' = \Gamma \cap (G_\infty \times N_f)$ est cocompact dans G, et d'indice fini dans Γ. On a évidemment

$$(\mathrm{pr}_{S_\infty} \Gamma')^- \times N_f = (\Gamma' . N_f)^-.$$

M_π est réalisé comme un espace de fonctions sur G, invariantes à droite par Γ, sur lesquelles G opère par translations à gauche. Le groupe N_f opère trivialement à gauche, donc aussi à droite, puisqu'il est normal dans G. Par suite, $\overline{(\Gamma' . N_f)}$ opère trivialement à droite sur M_π et, vu ce qui précède, il en est de même pour $\overline{(\mathrm{pr}\, \Gamma')}$. Mais, d'après 3.4, ce dernier groupe contient un sous-groupe H de G^o, normal dans G et cocompact dans G_∞, à savoir le produit des facteurs simples non compacts de G_∞^o. Comme H est distingué, il opère alors aussi trivialement à gauche sur M_π, donc est contenu dans $\ker \pi_\infty$.

Remarque. Un raisonnement semblable, utilisant la remarque de 3.4, montre plus précisément que si $\pi = \widetilde{\otimes}_i \pi_i$ apparait dans $L^2(G/\Gamma)$, et si $i, j \in S$ sont tels que G_i et G_j ne soient pas compacts, alors π_i est de dimension finie si et seulement si π_j l'est.

3.6. On dira qu'un sous-groupe Γ' de Γ est de S-congruence s'il contient un sous-groupe d'indice fini de la forme $\Gamma_U = \Gamma \cap (G \times U)$, où U est un sous-groupe compact ouvert de G_f. La projection pr_∞ identifie Γ' à un sous-groupe discret cocompact de G_∞. La proposition suivante a été obtenue en collaboration avec J-P. Serre.

PROPOSITION. <u>On suppose</u> G_∞ <u>et</u> G_f <u>non compacts. Soit</u> Γ_o <u>un sous-groupe de</u> S-<u>congruence de</u> Γ <u>et soit</u> $\sigma \in \hat{G}_\infty$ <u>de dimension infinie qui intervienne dans</u> $L^2(G_\infty/\Gamma_o)$. <u>Soit</u> N <u>un entier. Alors il existe un sous-groupe de</u> S-<u>congruence</u> Γ_U <u>de</u> Γ_o <u>tel que la multiplicité de</u> σ <u>dans</u> $L^2(G_\infty/\Gamma_U)$ <u>soit</u> $\geq$ N.

Si Γ' est d'indice fini dans Γ_o alors l'homomorphisme canonique $L^2(G_\infty/\Gamma_o) \longrightarrow L^2(G_\infty/\Gamma')$ est injectif. Quitte à remplacer Γ_o par un sous-groupe de S-congruence, et G_i par un sous-groupe d'indice fini si $i \in S_f$, on peut supposer que l'on a

$$(1) \qquad \Gamma_o = \Gamma \cap (G_\infty \times U_o) , \quad \text{avec} \quad U_o = \textstyle\prod_{i \in S_f} U_{oi} ,$$

où U_{oi} est compact ouvert dans G_i et

$$(2) \qquad U_{oi} = G_i \ \text{si} \ G_i \ \text{est compact}, \quad G_i = G_i^o \ \text{sinon} \qquad (i \in S_f) .$$

Dans la suite, on considère uniquement des sous-groupes compacts ouverts U de G_f de la forme

$$(3) \qquad U = \textstyle\prod_{i \in S_f} U_i , \quad \text{où} \ U_i \subset G_i \ (i \in S_f) , \ U_i = G_i \ \text{si} \ G_i \ \text{est compact}.$$

Vu la remarque à 3.4, $\Gamma_f = \mathrm{pr}_f \Gamma$ est dense dans G_f, donc $\Gamma_f . U = G_f$ pour tout sous-groupe compact ouvert U de G_f, d'où

$$(4) \qquad \Gamma . (G_\infty \times U) = G \ \text{et} \ G/\Gamma \cong (G_\infty \times U)/\Gamma_U .$$

La projection $(G_\infty \times U)/\Gamma_U = G_\infty/\Gamma_U$ induit donc un isomorphisme

$$(5) \qquad L^2(G_\infty/\Gamma_U) \cong L^2(G/\Gamma)^U.$$

pour tout U du type (3). Vu 3.5, on a donc un isomorphisme de G_∞-modules

$$(6) \qquad L^2(G_\infty/\Gamma_U) = \widetilde{\bigoplus}_{\pi \in \hat{G}_\infty} m(\pi, \Gamma) \ M_{\pi_\infty} \otimes M_{\pi_f}^U .$$

De plus, vu 3.3(3)

$$(7) \qquad M_{\pi_f} = \widetilde{\bigotimes}_{i \in S_f} M_{\pi_i} \quad (\pi_i \in \hat{G}_i) \quad \text{donc} \quad M_{\pi_f}^U = \bigotimes_{i \in S_f} M_{\pi_i}^{U_i} .$$

Appliquons cela en particulier à U_o. Il existe donc $\pi_o \in \hat{G}$ tel que

$$(8) \qquad m(\pi_o, \Gamma) \neq 0, \ \pi_{o\infty} = \sigma \ \text{et} \ M_{\pi_{oi}}^{U_{oi}} \neq 0 \ \text{pour tout } i \in S_f .$$

Comme σ est de dimension infinie, il en est de même de π_f (3.5); il existe donc $j \in S_f$ tel que G_j soit non compact et π_j de dimension infinie. On peut alors trouver un sous-groupe ouvert V de U_{oj} tel que

$$\dim (M_{\pi_j})^V \geq N .$$

Si $U = \prod_{i \in S_f} U_i$, avec $U_i = U_{oi}$ pour $i \neq j$ et $U_j = V$, alors (6), (7), (8) montrent que la multiplicité de σ dans $L^2(G_\infty/\Gamma_U)$ est $\geq N$.

3.7. THÉORÈME. <u>On conserve les notations et hypothèses de 3.3, 3.4 et on suppose de plus que</u> $G_i = G_i^o$ <u>si</u> $i \in S_f$ <u>et</u> G_i <u>n'est pas compact. On a alors</u>

(1) $\qquad H^m(\Gamma; \mathbb{C}) = H^m_{ct}(G_\infty; \mathbb{C}) \oplus \oplus' \; \mathrm{Hom}_{K_\infty} (\Lambda^{m-\ell}(\underline{g}_\infty/\underline{k}_\infty), M^\infty_{\pi_\infty})$, $\quad (m \in \mathbb{Z})$,

où la somme est étendue aux $\pi \in \hat{G}$, $\pi \neq 1$ tels que χ_{π_∞} soit trivial et que π_f soit spéciale (3.3).

Pour $i \in S_f$ soit X_i l'immeuble de Bruhat-Tits de $\underline{G}_i$. Soient X_f le produit des X_i et $X = X_\infty \times X_f$. Ce dernier est un espace contractile sur lequel Γ opère proprement, donc $H^m(\Gamma; \mathbb{C}) = H^m(X/\Gamma; \mathbb{C})$, pour tout $m \in \mathbb{Z}$.

Le théorème coincide avec 2.2 si $G = G_\infty$, et s'y ramène immédiatement si G_f est compact. Si G_i est compact sauf pour un élément de S_f, notre assertion résulte des résultats de Casselman rappelés en 3.2. A partir de là on raisonne par récurrence sur Card S_f, en utilisant la suite spectrale (E_r) de la projection $\mathrm{pr}_j : X/\Gamma \longrightarrow X_j/\Gamma_j$, où $j \in S_f$ est tel que G_j soit non compact. Soit C une chambre de X_j; si F est une face de C, notons B_F son fixateur dans G_j et soit $\Gamma_F = (G_{S(j)} \times B_F) \cap \Gamma$, où $S(j) = S - \{j\}$. Par projection Γ_F s'identifie à un sous-groupe discret de $G_{S(j)}$ et la paire $(G_{S(j)}, \Gamma_F)$ satisfait aux conditions imposées à G et Γ. Le groupe Γ_j est dense dans G_j (3.1), donc X_j/Γ_j s'identifie à C. Il s'ensuit que la terme E_2 de la suite spectrale (E_r) de $\mathrm{pr}_j : X/\Gamma \longrightarrow X_j/\Gamma_j$ est la cohomologie de C relativement à un faisceau simplicial $F \longmapsto H^*(\Gamma_F; \mathbb{C})$, avec les homomorphismes de restriction évidents. En utilisant 3.2 et l'hypothèse de récurrence on trouve que la somme des termes de degré total m de E_2 est égale au membre de droite de (1) et que $E_2^{p, q} = 0$ si $p \neq 0, \ell_j$. En tenant compte de 1.7, on constate que $E_2^{o, q} = E_\infty^{o, q}$ $(q \in \mathbb{Z})$, d'où $E_2 = E_\infty$ et le théorème.

3.8. COROLLAIRE. <u>On supprime l'hypothèse que $G_i = G_i^o$ pour G_i</u> <u>non compact et</u> $i \in S_f$. <u>Supposons que</u> $m < \ell$. <u>Alors</u> $H^m(\Gamma; \mathbb{C}) = H_{ct}^m(G_\infty; \mathbb{C})$.

Soit $G_i' = G_i^o$ si $i \in S_f$ et G_i est non compact, $G_i' = G_i$ sinon; soient $G' = \prod_i G_i'$ et $\Gamma' = \Gamma \cap G'$. Le groupe Γ' est un sous-groupe d'indice fini de Γ, donc $H^m(\Gamma; \mathbb{C}) \longrightarrow H^m(\Gamma'; \mathbb{C})$ est injectif. Notre assertion résulte alors du théorème appliqué à (G', Γ').

3.9. <u>Relations avec la cohomologie continue</u>. Plaçons-nous de nouveau dans la situation de 3.7 et soit

$$Q^m = \varinjlim_U H^m(\Gamma_U; \mathbb{C}) \ ,$$

où U parcourt les sous-groupes compacts ouverts de G_j, la limite étant prise par rapport aux homomorphismes de restriction. Le groupe G_j, opérant par automorphismes intérieurs, permute les groupes Γ_U, donc opère sur Q^m. En fait, Q^m est un G_j-module admissible, et

$$(1) \qquad\qquad (Q^m)^U = H^m(\Gamma_U; \mathbb{C}) \ .$$

Vu [13], le terme E_2 décrit dans 3.7 n'est autre que la cohomologie continue de G_j dans la somme des Q^m. Plus précisément

$$(2) \qquad\qquad E_2^{r, s} = H_{ct}^r(G_j; Q^s) \ ;$$

l'égalité $E_2 = E_\infty$ entraîne donc:

$$(3) \qquad H^*(\Gamma; \mathbb{C}) = H^*_{ct}(G_j; Q^*) \ ,$$

où Q^* est la somme directe des Q^m.

Sans distinguer un des facteurs ultramétriques de G, on peut interpréter 3.7 directement en termes de cohomologie continue. Tout d'abord, on a de façon générale (1.7)

$$(4) \qquad H^m(\Gamma; \mathbb{C}) = H^m_{ct}(G; C(G/\Gamma)) \ ,$$

où $C(G/\Gamma)$ est l'espace des fonctions continues sur G/Γ. Admettons que l'on peut remplacer $C(G/\Gamma)$ par $L^2(G/\Gamma)$. On obtient alors

$$H^m(\Gamma; \mathbb{C}) = \bigoplus_{\pi \epsilon G} m(\pi, \ \Gamma) \ H^m_{ct}(G; M_\pi) \ ,$$

et on passe de là à 3.7 en admettant que l'on peut calculer $H^m_{ct}(G; M_\pi)$ par une règle de Künneth et en utilisant 3.2. Il est bien probable que cela peut être justifié (en remplaçant $C(G/\Gamma)$ par un sous-espace de vecteurs différentiables dans un sens convenable plutôt que par $L^2(G/\Gamma)$), mais il ne semble cependant pas à l'auteur que l'on puisse le faire sans ajouter aux "fondements" de la cohomologie continue publiés jusqu'à présent.

Remarquons que l'on peut aussi remplacer j par une partie quelconque S' de S_f dans (2), (3). Cela se voit en considérant la suite spectrale de la projection $X/\Gamma \longrightarrow X_{S'}/\Gamma_{S'}$, où $X_{S'}$ est le produit des X_i avec $i \epsilon S'$.

§4. <u>Groupes</u> S-<u>arithmétiques (cas anisotrope)</u>.

4.1. Dans ce paragraphe, on spécialise la situation précédente au cas
(le plus important) des groupes S-arithmétiques sur un corps de nombres k.
Soit $\underline{G}$ un k-groupe connexe absolument presque simple sur k, de k-rang
nul. S désigne maintenant un ensemble fini de places de k qui contient
l'ensemble S_∞ des places archimédiennes, k_i la complétion de k en i ϵ S,
$\underline{G}_i$ le groupe $\underline{G}$ vu comme k_i-groupe et $G_i = \underline{G}(k_i)$. On est alors dans la
situation du paragraphe précédent en prenant G égal au produit des G_i (i ϵ S)
et Γ égal à un sous-groupe S-arithmétique de $\underline{G}(k)$; autrement dit, si l'on
fixe un plongement $\underline{G} \hookrightarrow \mathbb{GL}_n$ défini sur k, Γ est un sous-groupe de $\underline{G}(k)$
commensurable à $\underline{G}(o_S)$, où o_S est l'anneau des éléments de k
entiers en dehors de S.

Supposons de plus $\underline{G}$ simplement connexe. Alors toutes les hypothèses
de 3.7 sont satisfaites et 3.7(1) est valable. La formule ainsi obtenue était
aussi connue de W. Casselman.

Remarquons encore que dans ce cas, si G_i n'est pas compact il est
simple modulo son centre et ne contient aucun sous-groupe ouvert propre
d'indice fini, donc n'a pas de représentation unitaire irréductible de
dimension finie à part la représentation triviale. En particulier, si les G_i
sont tous non compacts, la somme $\oplus'$ ne porte que sur des $\pi = \hat{\otimes}\, \pi_i$ avec
π_i de dimension infinie pour tout i.

D'autre part, $\underline{G}(k)$ est la limite inductive de groupes S-arithmétiques,

lorsque S parcourt une suite croissante d'ensembles de places de k dont
la réunion est l'ensemble de toutes les places de k. Le corollaire 3.8
entraîne donc,

$$(1) \qquad H^m_{ct}(G_\infty; \mathbb{C}) = H^m(\Gamma; \mathbb{C}) = H^m(\underline{G}(k); \mathbb{C}) \ , \qquad (m \in \mathbb{Z}) \ ,$$

si Γ est S-arithmétique et S assez grand, résultat d'abord obtenu en
collaboration avec H. Garland par une méthode différente, annoncé dans [18],
et aussi remarqué par Casselman.

4.2. Soit maintenant Γ_o un sous-groupe arithmétique de congruence
de $\underline{G}$. Cela signifie que si l'on fixe une représentation matricielle sur k de
$\underline{G}$ et désigne par Γ_S le groupe S-arithmétique des éléments de $\underline{G}(k)$ dont
les coefficients sont entiers en dehors de S_f, alors $\Gamma_o \cap \Gamma_S$ est un sous-groupe
de S-congruence de Γ_S, au sens de 3.6, pour un S convenable. Soit $\pi \in \hat{G}_\infty$
de dimension infinie qui apparaisse dans $L^2(G_\infty/\Gamma_o)$. Alors 3.6 entraîne
l'existence d'un sous-groupe d'indice fini Γ' de Γ_o tel que π intervienne
dans $L^2(G_\infty/\Gamma')$ avec une multiplicité arbitrairement grande. Supposons en
particulier que $\underline{G}(k_v)$ soit isomorphe à $SO(n, 1)$ ou $SU(n, 1)$ pour une place
archimédienne réelle et soit compact pour toute autre place archimédienne de k.
Alors Γ_o s'identifie à un sous-groupe discret cocompact de $SO(n, 1)$ ou
$SU(n, 1)$. Vu 2.8, ce qui précède montre que si $H^1(\Gamma_o; \mathbb{C}) \neq 0$, alors Γ_o
possède des sous-groupes d'indice fini de premier nombre de Betti arbitraire-
ment grand.

4.3. La formule 3.6(6) permet dans le cas présent de décrire le spectre d'un groupe arithmétique de congruence à partir celui d'un groupe S-arithmétique. On peut de manière similaire passer du spectre de $G(A)/G(k)$, où A est l'anneau des adèles de k, à celui d'un groupe arithmétique ou S-arithmétique. Supposons G simplement connexe (et anisotrope sur k comme précédemment). Alors $G(A) = G_\infty \times G(A_f)$, où A_f est l'anneau des adèles finies. Le groupe $G(k)$ s'identifie à un sous-groupe discret cocompact de $G(A)$. On peut écrire

$$(1) \qquad L^2(G(A)/G(k)) = \widetilde{\bigoplus}_{\pi \epsilon G(A)^\wedge} m(\pi)\, M_\pi \ ,$$

et

$$(2) \qquad M_\pi = M_{\pi_\infty} \widetilde{\otimes} M_{\pi_f} \qquad (\pi_\infty \epsilon \hat{G}_\infty ; \pi_f \epsilon G(A_f)^\wedge) \ .$$

Supposons G presque simple sur k et G_∞ non compact. Alors, par approximation forte, on a $G(A) = G(k).(G_\infty \times U)$ quel que soit le sous-groupe compact ouvert U de $G(A_f)$, d'où l'on déduit comme en 3.6

$$(3) \ L^2(G_\infty/\Gamma_U) = \widetilde{\bigoplus}_{\pi \epsilon G(A)^\wedge} m(\pi)\, M_{\pi_\infty} \otimes M_{\pi_f}^U \ , \qquad (\Gamma_U = G(k) \cap (G_\infty \times U)) \ .$$

Pour un groupe arithmétique de congruence (i.e. un sous-groupe de $G(k)$ contenant un sous-groupe tel que Γ_U comme sous-groupe d'indice fini), la question de 2.8 prend alors la forme suivante:

Soit $\sigma \epsilon \hat{G}_\infty$ non séparée de la représentation triviale. Existe-t-il $\pi \epsilon G(A)^\wedge$ tel que $m(\pi) \neq 0$ et $\pi_\infty = \sigma$?

4.4. Soient U un sous-groupe compact ouvert de G_f et $\Gamma_U = \underline{G}(o_S) \cap (G_\infty \times U)$. On déduit de 2.2 et 3.6(6),

$$(1) \qquad H^m(\Gamma_U; \mathbb{C}) = H^m_{ct}(G_\infty; \mathbb{C}) \oplus \oplus' \operatorname{Hom}(\Lambda^m \underline{g}_\infty/\underline{k}_\infty, M^\infty_{\pi_\infty}) \otimes M^U_{\pi_f} \ ,$$

la somme étant étendue aux $\pi \in \hat{G}$ non triviaux tels que χ_{π_∞} soit trivial. On a par conséquent

$$(2) \qquad Q^m = \varinjlim H^m(\Gamma_U; \mathbb{C}) = H^m_{ct}(G_\infty; \mathbb{C}) \oplus \oplus' \operatorname{Hom}(\Lambda^m \underline{g}/\underline{k}_\infty, M^\infty_{\pi_\infty}) \otimes M_{\pi_f} \ ,$$

la somme portant sur les mêmes π que précédemment; cette égalité est aussi valable pour les structures naturelles de G_f-modules, étant entendu que G_f opère trivialement sur le premier terme de droite. Supposons encore G_∞ et les G_i ($i \in S_f$) non compacts. Alors les M_{π_f} sont de dimension infinie (3.5), donc en particulier

$$(3) \qquad\qquad (Q^m)^{G_f} = H^m_{ct}(G_\infty; \mathbb{C}) \ ,$$

ce qui, dans le cas présent où $\underline{G}$ est anisotrope sur k, répond affirmativement à une question posée par Serre.

§5. <u>Groupes S-arithmétiques (cas général)</u>.

<u>Nous conservons les hypothèses de 4.1, excepté que le k-rang ℓ de $\underline{G}$ est supposé non nul, sauf mention expresse du contraire.</u>

5.1. Le quotient G/Γ est maintenant non compact, mais de mesure invariante finie, et l'homomorphisme de restriction
$$j^m : H^m_{ct}(G_\infty ; \mathbb{C}) \longrightarrow H^m(\Gamma; \mathbb{C})$$ n'est pas nécessairement injectif. Il existe une constante $c(G)$, calculable à partir de la structure de groupe algébrique de G et de l'algèbre de Lie de G_∞, telle que j^m soit bijectif pour $m \leq c(G)$. Cela est établi dans [3] lorsque Γ est arithmétique et $G = G_\infty$; le cas général (annoncé dans [2]), s'en déduit à l'aide de la suite spectrale de 3.7. C'est l'analogue d'un théorème de Matsushima valable dans la situation du §2 [30]. On en déduit que $H^m_{ct}(G_\infty ; \mathbb{C}) \longrightarrow H^m(\underline{G}(k); \mathbb{C})$ est un isomorphisme pour $m \leq c(G)$. Nous avons vu que si $\ell = 0$, c'est un isomorphisme pour tout m. Pour $\ell > 0$ et m quelconque, il n'est pas nécessairement injectif, mais je ne sais pas s'il est surjectif.

5.2. Supposons $k = \mathbb{Q}$, $\underline{G}$ presque simple sur k et simplement connexe comme précédemment, mais pas nécessairement absolument presque simple, et S_f réduit à une place $\{p\}$. Serre a posé la question d'étudier l'homomorphisme de restriction

$$(1) \qquad H^m_{ct}(G_p ; \mathbb{Q}_p) \longrightarrow H^m(\Gamma; \mathbb{Q}_p) \ ,$$

où Γ est S-arithmétique ou arithmétique. D'après [13], le membre de gauche est la cohomologie de l'algèbre de Lie de G_p. Si $\underline{G}$ est le groupe symplectique $\mathbb{S}p_{2n}$, les résultats de [3] entraînent que cet homomorphisme est trivial. Mais le cas en principe le plus intéressant est celui où $G = \mathbb{S}\mathbb{L}_n$

(ou plus généralement où $G = R_{k/\mathbb{Q}} \, \mathbb{SL}_n$). En basses dimensions, i.e. en dessous de la constante $c(G)$ ci-dessus, l'isomorphisme j^m donne lieu (mais cela est loin d'être canonique) à un isomorphisme

$$(2) \qquad H^m(\Gamma; \mathbb{Q}_p) \xrightarrow{\ \sim\ } H^m(\underline{g}_\infty, \underline{k}_\infty; \mathbb{Q}_p) \ ,$$

où $\underline{g}_\infty$, $\underline{k}_\infty$ sont envisagés comme des algèbres de Lie sur $\mathbb{Q}$ ou $\mathbb{Q}_p$. Cela laisse ouverte la possibilité que l'homomorphisme de (1) soit surjectif dans ces dimensions, de noyau l'algèbre de Lie de $\mathbb{SO}_n$. Si c'était le cas, on pourrait songer à s'en servir pour définir, au moins à une unité p-adique près, un s-ième régulateur p-adique lorsque $m = 2s+1$, d'une manière qui présenterait une certaine analogie formelle avec celle du s-ième régulateur à partir de la cohomologie complexe [4; 27].

5.3. La construction faite dans 3.7 vaut sans changement lorsque $\ell > 0$ et montre que $H^*(\Gamma; \mathbb{C})$ est l'aboutissement d'une suite spectrale (E_r) dans laquelle E_2 est la cohomologie de la chambre C par rapport au faisceau simplicial de 3.7. Mais j'ignore si $E_2 = E_\infty$. L'interprétation de E_2 en terms de cohomologie continue fournie par 3.9(2) est encore valable. De plus, [5: 4.12] montre que Q^s est somme directe d'un G_j-module à cohomologie continue nulle et du sous-G_j-module engendré par $H^s(\Gamma_C; \mathbb{C})$. Par définition Γ_C est le fixateur de C dans Γ, et est une généralisation naturelle du groupe de Hecke $\Gamma_0(p)$ (que l'on retrouve dans le cas où $G = \mathbb{SL}_2$, $\Gamma = \mathbb{SL}_2(\mathbb{Z}[p^{-1}])$ et $S_f = \{p\}$). Pour aller plus loin dans cette voie, il faudrait obtenir des renseignements sur les constituants du G_j-module Q^s. Vu [5: 4.10], cela équivaut à étudier les constituants de $H^m(\Gamma_C; \mathbb{C})$, vu comme module sur l'algèbre de Hecke $H(G_j, B_C)$, où B_C est le fixateur de C dans G_j, i.e. un sous-groupe d'Iwahori de G_j. Une première question dans cet ordre d'idées est de savoir si 4.4(3) est valable en général.

The Institute for Advanced Study, Princeton, N. J., U.S.A.

BIBLIOGRAPHIE

[1] A. Borel, "Density properties of certain subgroups of semi-simple
     groups," Annals of Math. (2) 72 (1960), 179-188.

[2] ————,"Cohomologie réelle stable de groupes S-arithmétiques,"
     C. R. Acad. Sci. Paris 274 (1972), 1700-1702.

[3] ————, "Stable real cohomology of arithmetic groups," Annales
     Sci. E. N. S. (4) 7 (1974), 237-272.

[4] ————, "Cohomology of arithmetic groups," Proc. Int. Cong. Math.
     Vancouver 1974.

[5] ————, "Admissible representations of a semi-simple group over a
     local field with vectors fixed under an Iwahori subgroup" (to appear).

[6] ———— et J. Tits, "Groupes réductifs," Publ. Math. I. H. E. S. 27
     (1965), 55-150.

[7] ———— et ————, "Homomorphismes 'abstraits' de groupes algébriques
     simples," Annals of Math. (2) 97 (1973), 499-571.

[8] N. Bourbaki, Espaces vectoriels topologiques, Chap. 1 et 2, Act. Sci.
     Ind. 1189, Paris 1966, Hermann éd.

[9] ————, Groupes et algèbres de Lie, Chap. 2, 3, Act. Sci. Ind.
     1349, Paris, Hermann éd. 1972.

[10] F. Bruhat et J. Tits, "Groupes algébriques simples sur un corps local:
     cohomologie galoisienne, décompositions d'Iwasawa et de Cartan,"
     C. R. Acad. Sci. Paris 263 (1966), 867-869.

[11] W. Casselman, "On a p-adic vanishing theorem of Garland,"
     Bull. A. M. S. 80 (1974), 1001-1004.

[12] ————, "Introduction to the theory of admissible representations
     of p-adic reductive groups," (to appear).

[13] ———— and D. Wigner, "Continuous cohomology and a conjecture
     of Serre's," Inv. Mat. 25 (1974), 199-211.

[14] C. Delaroche et A. Kirillov, "Sur les relations entre l'espace dual d'un
     groupe et la structure de ses sous-groupes fermés (d'après
     D. A. Kajdan)," Sém. Bourbaki 20e année (1967/68), Exp. 343,
     Benjamin New York.

[15] P. Delorme, "Sur la 1-cohomologie des représentations des groupes de
Lie semi-simples," C.R. Acad. Sci. Paris 280 (1975), 1101-1103.

[16] J. L. Dupont, "Simplicial de Rham cohomology and characteristic classes
of flat bundles," Preprint series 1974/75 no. 29, Matematisk
Institut, Aarhus Universitet, 36 p.

[17] V. T. van Est, "On the algebraic cohomology concepts in Lie groups II,"
Proc. Konink. Nederl. Akad. v. Wet. Series A, 58 (1955), 286-294.

[18] H. Garland, "On the cohomology of discrete subgroups of p-adic groups,"
Proc. Int. Congr. Math. Vancouver 1974.

[19] H. Hecht and W. Schmid, "A proof of Blattner's Conjecture,"
[Inventiones math. 31 (1978), 129−184].

[20] G. Hochschild and G. D. Mostow, "Cohomology of Lie groups," Ill. Jour.
Math. 6 (1962), 367-401.

[21] R. Hotta and R. Parthasarathy, "A geometric meaning of the multiplicity
of integrable discrete classes in $L^2(\Gamma \backslash G)$, Osaka J. Math. 10 (1973),
211-234.

[22] ——— and ———————, "Multiplicity formulae for discrete series,'
Inv. Mat. 26 (1974), 133-178.

[23] ——— and N. Wallach, "On Matsushima's formula for the Betti numbers
of locally symmetric spaces" (to appear).

[24] D. A. Kajdan, "On the connection of the dual space of a group with the
structure of its closed subgroups," Funct. Anal. and appl. 1 (1967),
63-65.

[25] B. Kostant, "On the existence and irreducibility of certain series of
representations," Bull. A.M.S. 75 (1969), 627-642.

[26] R. P. Langlands, "Dimension of spaces of automorphic forms," Proc.
Symp. pure math. 9 (1966), 235-252, A. M.S., Providence, R.I.

[27] S. Lichtenbaum, "Values of zeta-functions, étale cohomology, and
algebraic K-theory II, Springer Lecture Notes 342 (1973), 489-499.

[28] S. MacLane, Homology, Grund. math. Wiss. 144, Springer Verlag 1963.

[29] G. A. Margoulis, "Discrete groups of motions of manifolds of non-positive
curvature," Proc. Int. Congress Math. Vancouver, 1974.

[30] Y. Matsushima, "On Betti numbers of compact, locally symmetric Riemannian manifolds," Osaka Math. J. 14 (1962), 1-20.

[31] ——————————, "A formula for the Betti numbers of compact locally symmetric Riemannian manifolds," Jour. Diff. Geom. 1 (1967), 99-109.

[32] —————————— and S. Murakami, "On vector bundle valued harmonic forms and automorphic forms on symmetric Riemannian manifolds,' Annals of Math. (2) 78 (1963), 365-416.

[33] —————————— and ——————————, "On certain cohomology groups attached to hermitian symmetric spaces," Osaka Jour. Math 2 (1965), 1-35.

[34] J. Millson, "On the first Betti number of a compact constant negatively curved manifold," [Ann. Math. 104 (1976), 235—247].

[35] W. Schmid, "On a conjecture of Langlands," Annals of Math. (2) 93 (1971), 1-42.

[36] E. B. Vinberg, "Some examples of crystallographic groups on Lobacevskii spaces," Mat. Sbornik 7 (1969), 617-622.

[37] N. Wallach, "On the Selberg trace formula in the case of compact quotient," [Bull. Amer. math. Soc. 87 (1976), 171—195].

[38] S. P. Wang, "The dual space of semi-simple Lie groups," Amer J. Math. XCI (1969), 921-937.

[39] ——————————, "On density properties of S-subgroups of locally compact groups," Annals of Math. (2) 94 (1971), 325-329.

Armand BOREL
Institute for Advanced Study
PRINCETON, N.J. 08540, USA

105.
(avec J-P. Serre)

Cohomologie d'immeubles et de groupes S-arithmétiques

Topology **15** (1976) 211–232

INTRODUCTION

CE TRAVAIL se divise en deux parties, consacrées l'une (§§1 à 5) à la cohomologie de deux immeubles associés à un groupe réductif sur un corps local, et l'autre (§6) à la cohomologie des groupes S-arithmétiques, ou, plus généralement, de certains sous-groupes discrets de produits de groupes définis sur des corps locaux. Nous montrons que de tels groupes, lorsqu'ils sont sans torsion, ont une dualité homologique qui généralise celle des groupes arithmétiques[5: §11].

Soient k un corps local non archimédien, G un k-groupe réductif et X l'immeuble de Bruhat–Tits de G (§4). L'espace X est de dimension égale au k-rang $l = rg_k G$ de G. Notre principal objectif, dans la première partie, est d'en déterminer les groupes de cohomologie à supports compacts $H_c^i(X; \mathbf{Z})$: on trouve que $H_c^i(X; \mathbf{Z})$ est nul si $i \neq l$, et que c'est un groupe libre $\neq 0$ si $i = l$ (5.6, 5.9). Pour la démonstration, on se ramène tout de suite au cas où G est semi-simple et simplement connexe. On considère d'abord l'immeuble Y des k-sous-groupes paraboliques de G (§1). Il est simplicial, de dimension $l - 1$, mais non localement fini (à moins que $l = 0$, auquel cas il est vide); on sait qu'il a le type d'homotopie d'un bouquet de sphères; cependant, ce n'est pas sa topologie usuelle qui nous intéresse ici, mais une autre topologie, provenant de celle de k, qui en fait un espace compact et métrisable Y_t (§1). Soient P un k-sous-groupe parabolique minimal de G, et U son radical unipotent; utilisant une filtration de Y_t déduite de la décomposition de Bruhat de $G(k)/P(k)$, nous montrons au §2 que le i-ième groupe de cohomologie[1] réduite $\tilde{H}^i(Y_t; \mathbf{Z})$ de Y_t est nul pour $i \neq l - 1$, et que, pour $i = l - 1$, ce groupe est isomorphe au groupe $C_c^\infty(U(k); \mathbf{Z})$ des fonctions localement constantes, à valeurs dans $\mathbf{Z}$, et à support compact, sur $U(k)$ (2.6). On montre ensuite (§5) que l'on peut compactifier X par Y_t, de manière à obtenir un espace compact contractile Z_t. Le théorème 5.6 résulte alors de 2.6 par la suite exacte de cohomologie de Z_t mod Y_t.

Soient maintenant F un corps de nombres, S un ensemble fini de places de F contenant l'ensemble S^∞ des places archimédiennes de F, et $S' = S - S^\infty$. Si $v \in S$, on note F_v le complété de F en v. Soit G un F-groupe réductif. Notons $\bar{X}$ la variété à coins associée dans [5: §9] au $\mathbf{Q}$-groupe $R_{F/\mathbf{Q}}G$ obtenu à partir de G par restriction des scalaires de F à $\mathbf{Q}$. Si $v \in S'$, soit X_v l'immeuble de G sur F_v [17: 2.2], autrement dit le produit de l'immeuble de Bruhat–Tits du groupe dérivé $\mathscr{D}G$ de G par un espace euclidien de dimension égale à celle du plus grand F_v-tore déployé central de G. Soient enfin $\bar{X}_S = \bar{X} \times \Pi_{v \in S'} X_v$ et Γ un sous-groupe S-arithmétique de G. Alors Γ opère proprement sur $\bar{X}_S$ et $\bar{X}_S/\Gamma$ est compact (6.10). Soit $d = \dim X + \Sigma_{v \in S'} \dim X_v - rg_F G$. Il résulte de 5.9 et [5: §9] que $H_c^i(\bar{X}_S; \mathbf{Z})$ est nul pour $i \neq d$ et libre pour $i = d$. Si Γ est sans torsion, cela entraîne que Γ est un groupe à dualité au sens de [2], de module dualisant $I_S = H_c^d(\bar{X}_S; \mathbf{Z})$. On a en particulier un isomorphisme canonique

$$H^i(\Gamma; M) \xrightarrow{\sim} H_{d-i}(\Gamma; I_S \otimes M), \tag{1}$$

quels que soient $i \in \mathbf{Z}$ et le Γ-module M (6.2). On obtient également un résultat semblable pour les sous-groupes cocompacts sans torsion des produits finis $\Pi_v L_v(k_v)$, où k_v est un corps local et L_v un k_v-groupe réductif.

Les principaux résultats de ce travail ont été annoncés dans[4].

Notations et conventions. Si A est un groupe et B une partie de A, on note $\mathscr{Z}(B)$ ou $\mathscr{Z}_A(B)$ le

---

[1] Il s'agit ici de la cohomologie "d'Alexander–Spanier", i.e. de celle de la théorie des faisceaux, et non de la cohomologie singulière.

439

centralisateur de B dans A et $\mathcal{N}(B)$ ou $\mathcal{N}_A(B)$ le normalisateur de B dans A. On a donc

$$\mathscr{Z}_A(B) = \{a \in A \,|\, ab = ba\,(b \in B)\}, \qquad \mathcal{N}_A(B) = \{a \in A \,|\, a \cdot B = B \cdot a\};$$

le centre $\mathscr{Z}_A(A)$ de A est aussi noté $C(A)$. Si $\{B_i\}_{i \in I}$ est une famille de parties de A, on note $\langle(B_i)_{i \in I}\rangle$ le sous-groupe engendré par les B_i.

Si G est un groupe algébrique, G^0 désigne la composante connexe de e dans G, que l'on appelle aussi composante neutre de G, et $\mathscr{D}G$ le groupe dérivé de G. Si k est un corps de définition pour G et si G^0 est réductif, le k-rang $rg_k G$ de G est par définition celui de G^0, au sens de [6: §5].

Un *corps local* est, dans ce travail, $\mathbf{R}$, $\mathbf{C}$ ou un corps complet pour une valuation discrète à corps résiduel fini; un tel corps est localement compact.

Dans les §§1 à 4, k est un corps local non archimédien, G un k-groupe connexe semi-simple, $\tilde{G}$ le revêtement universel de G et l le k-rang $rg_k G$ de G.

§1. L'immeuble topologisé des sous-groupes paraboliques

1.1. On note $\mathfrak{P}$ l'ensemble des k-sous-groupes paraboliques de G, T un tore k-déployé maximal de G, et Φ le système des k-racines de G par rapport à T. Le quotient $W = \mathcal{N}(T)/\mathscr{Z}(T)$ est le groupe de Weyl de Φ. On a $\Phi \neq \emptyset$ et $W \neq \{1\}$ si et seulement si $l \geq 1$.

On note P un k-sous-groupe parabolique minimal contenant T, et Δ la base de Φ associée à P. Les racines de P par rapport à T sont donc les combinaisons linéaires à coefficients positifs d'éléments de Δ. Pour $I \subset \Delta$, soient $T_I = (\cap_{a \in I} \ker a)^0$ et P_I le k-sous-groupe parabolique engendré par P et $\mathscr{Z}(T_I)$. On sait que $I \mapsto P_I$ est une bijection entre l'ensemble des parties de Δ et celui des éléments de $\mathfrak{P}$ contenant P, que tout $Q \in \mathfrak{P}$ est conjugué à un et un seul P_I et que cette conjugaison peut être effectuée à l'aide d'un élément de $G(k)$ [6, §§4, 5].

On sait que $G(k)/P(k)$, muni de sa structure d'espace homogène du groupe topologique $G(k)$, est compact. Plus précisément, $G(k)/P(k)$ s'identifie à $(G/P)(k)$, donc à l'ensemble des k-points d'une k-variété projective lisse.

On note π_I la projection canonique de G/P sur G/P_I et $\pi_{I,J}\,(I \subset J)$ celle de G/P_I sur G/P_J.

1.2. L'immeuble Y

Si $l = 0$, on pose $Y = Y_t = \emptyset$. Supposons $l \geq 1$. Les sous-groupes $Q(k)$, $(Q \in \mathfrak{P})$ sont les sous-groupes paraboliques d'un système de Tits $\mathbf{T} = (G(k), B, N, S)$ de $G(k)$, dans lequel $B = P(k)$, $N = \mathcal{N}(T)(k)$ et $S \subset W$ est l'ensemble des réflexions associées aux éléments de Δ. Dans la suite on identifie Δ à S en associant à un élément $a \in \Delta$ la réflexion s_a qui, vue comme automorphisme de T, fixe $T_{\{a\}}$. On a $B \cap N = \mathscr{Z}(T)(k)$ et W s'identifie au groupe de Weyl de $\mathbf{T}$.

On note Y l'immeuble de Tits de $\mathbf{T}$. C'est un complexe simplicial de dimension $l - 1$, sur lequel $G(k)$ opère. Le groupe d'isotropie G_σ d'une face σ de Y est le groupe $P_\sigma(k)$ des points rationnels d'un k-sous-groupe parabolique $\neq G$ et $\sigma \mapsto P_\sigma$ est une bijection entre l'ensemble des faces de Y et $\mathfrak{P} - \{G\}$. De plus, $P_\sigma(k)$ laisse fixe tout point de σ. Pour $g \in G(k)$, on a $P_{g(\sigma)} = g \cdot P_\sigma \cdot g^{-1}$. Soit C le simplexe fixe par $P(k)$. Pour $I \subset S$, $I \neq S$, on note C_I la face de C formée des points fixés par $P_I(k)$. Avec cette convention, les faces de codimension 1 de C sont les $C_{\{s\}}(s \in S)$, et $C_\emptyset = C$. On a:

$$C_I \subset \partial C_J \Leftrightarrow J \subsetneqq I \Leftrightarrow P_J \subsetneqq P_I. \tag{1}$$

Soit $\mathring{C}_I = C_I - \cup_{J \supsetneqq I} C_J$ l'intérieur de C_I. On a une partition

$$Y = \amalg_{I \subsetneqq S} Y_I, \quad \text{où} \quad Y_I = G(k)/P_I(k) \times \mathring{C}_I. \tag{2}$$

Soient

$$G(k) \times C \xrightarrow{\ \lambda\ } Y \xrightarrow{\ \tau\ } C \cong Y/G(k), \tag{3}$$

les applications définies respectivement par l'action de $G(k)$ sur Y et par la projection de Y_I sur son deuxième facteur $(I \subsetneqq S)$. Comme $P(k)$ fixe C, λ se factorise en

$$G(k) \times C \xrightarrow{\ \pi \times Id.\ } (G(k)/P(k)) \times C \xrightarrow{\ \lambda'\ } Y, \tag{4}$$

où π est la projection canonique de $G(k)$ sur $G(k)/P(k)$.

1.3. Topologie sur Y. On munit $(G(k)/P(k)) \times C$ de la topologie produit et on note Y_t l'immeuble Y muni de la topologie quotient de $(G(k)/P(k)) \times C$ par la relation d'équivalence définie par λ'.

Comme C est un domaine fondamental pour $G(k)$, il résulte de 1.2(1) que la relation d'équivalence définie par λ' a un graphe fermé, donc Y_t est séparé. Il est compact puisque $(G(k)/P(k)) \times C$ l'est. Montrons que $G(k)$ opère continûment sur Y_t. L'action de $G(k)$ sur Y est définie par une application $\nu : G(k) \times Y \to Y$ qui s'insère dans le diagramme commutatif

$$G(k) \times (G(k)/P(k)) \times C \xrightarrow{\;Id. \times \lambda'\;} G(k) \times Y$$

$$\downarrow{\scriptstyle \rho \times Id.} \qquad\qquad\qquad\qquad \downarrow{\scriptstyle \nu}$$

$$(G(k)/P(k)) \times C \xrightarrow{\quad\lambda'\quad} Y \tag{1}$$

où ρ est l'application canonique de $G(k) \times (G(k)/P(k))$ dans $G(k)/P(k)$. L'application λ' est propre. Il en est donc de même de $Id. \times \lambda'$ [8: I.73, prop. 4] et $G(k) \times Y$ a la topologie quotient de $G(k) \times (G(k)/P(k)) \times C$ vu [8: I.74, cor. 4]. La continuité de ν résulte alors de celle des autres applications du diagramme (1). De fait que $P_l(k)$ opère trivialement sur C_l, on déduit immédiatement que la bijection naturelle $Y_l = (G(k)/P_l(k)) \times \overset{\circ}{C}_l \xrightarrow{\;\sim\;} \tau^{-1}(\overset{\circ}{C}_l)$, est un homéomorphisme sur un sous-espace localement fermé.

Si $l = 0$, on a $Y_t = \emptyset$; si $l = 1$, C est un point, $Y = G(k)/P(k)$ et Y_t est simplement $G(k)/P(k)$ muni de sa topologie naturelle.

1.4. *Remarque.* L'espace Y_t peut aussi se définir lorsque k, au lieu d'être ultramétrique, est le corps **R** des nombres réels. Soit K un sous-groupe compact maximal de $G(\mathbf{R})$ dont l'algèbre de Lie est orthogonale à celle de $A = T(\mathbf{R})^0$. On a $Y = K \cdot C$ et l'on peut identifier le K-espace Y_t à la *sphère unité* de l'espace tangent à G/K à i'origine, et cela de telle sorte que C soit l'image des vecteurs tangents t à A de longueur 1 tels que $a(t) \geqq 0$ pour tout $a \in \Delta$.

§2. Cohomologie de Y_t

2.1. Soient X un espace localement compact et M un **Z**-module. On désigne par $H^i(X; M)$ (resp. $H_c^i(X; M)$) le i-ième groupe de cohomologie d'Alexander–Spanier à supports fermés (resp. compacts) de X, à coefficients dans M. On a en particulier $H^0(X; M) = C^\infty(X; M)$ et $H_c^0(X; M) = C_c^\infty(X; M)$, où $C^\infty(X; M)$ (resp. $C_c^\infty(X; M)$) désigne le groupe des fonctions localement constantes (resp. localement constantes à support compact) sur X, à valeurs dans M.

Si X est compact non vide, $\tilde{H}^i(X; M)$ dénote le i-ième groupe de cohomologie *réduit* de X à coefficients dans M, i.e.

$$\tilde{H}^i(X; M) = H^i(X; M) \qquad \text{pour} \quad i \neq 0$$
$$\tilde{H}^0(X; M) = \operatorname{coker}\{M = H^0(\mathrm{pt}; M) \to H^0(X; M)\}.$$

Le groupe $\tilde{H}^0(X; M)$ s'identifie au quotient de $C^\infty(X; M)$ par le sous-groupe formé des fonctions constantes.

Si X est compact et contractile, on a $\tilde{H}^i(X; M) = 0$ pour tout $i \in \mathbf{Z}$.

Si Y est un sous-espace fermé non vide de X, la suite exacte de cohomologie à supports compacts de X mod Y peut s'écrire:

$$\cdots \to H_c^i(X - Y; M) \to \tilde{H}^i(X; M) \to \tilde{H}^i(Y; M) \to H_c^{i+1}(X - Y; M) \to \cdots \tag{1}$$

2.2. Lemme. *Soient X un espace localement compact, métrisable, totalement discontinu, et M un* **Z***-module. Alors $C_c^\infty(X; \mathbf{Z})$ est un* **Z***-module libre et $C_c^\infty(X; M) = C_c^\infty(X; \mathbf{Z}) \otimes_{\mathbf{Z}} M$.*

On peut écrire X comme réunion disjointe d'ouverts compacts $X_\alpha (\alpha \in J)$, d'où $C_c^\infty(X; M) = \bigoplus_{\alpha \in J} C^\infty(X_\alpha; M)$, ce qui nous ramène au cas où X est compact. L'espace X est alors limite projective dénombrable d'ensembles finis

$$X_1 \xleftarrow{\;f_1\;} X_2 \xleftarrow{\;f_2\;} X_3 \xleftarrow{\quad} \cdots \xleftarrow{\;f_{n-1}\;} X_n \xleftarrow{\;f_n\;} \cdots$$

où les applications f_i sont surjectives. On a donc $C^\infty(X; M) = \lim \cdot \operatorname{ind} \cdot M^{X_n}$ où M^{X_n} désigne l'ensemble des applications de X_n dans M. Evidemment, $M^{X_n} = \mathbf{Z}^{X_n} \otimes_{\mathbf{Z}} M$, d'où la deuxième assertion. Comme f_n est surjective, l'application canonique $M^{X_n} \to M^{X_{n+1}}$ a comme image un facteur direct; si l'on note S_n un supplémentaire, on a $C^\infty(X; M) = M^{X_1} \oplus \overset{\infty}{\underset{n=1}{\oplus}} S_n$, d'où la première assertion.

2.3. On reprend les hypothèses et notations du §1; on écarte le cas trivial $l = 0$. On désigne par $l(w)$ la longueur d'un élément $w \in W$ par rapport à S [9: IV, §1]. Soient $(w_i)_{1 \leqslant i \leqslant N}$ les éléments du groupe de Weyl W, numérotés de manière à ce que l'on ait

$$l(w_i) \leqq l(w_j) \quad \text{si} \quad i < j \quad (1 \leqslant i, j \leqslant N = |W|). \tag{1}$$

En particulier, $w_1 = 1$. Pour $m \in [1, N]$ entier, soit

$$I_m = \{s \in S \,|\, l(w_m \cdot s) > l(w_m)\}. \tag{2}$$

Il résulte de [9: IV, §1, Exer. 3, p. 37] que

$$I_1 = S, \quad I_N = \emptyset \quad \text{et} \quad I_m \neq \emptyset, S \text{ pour } 1 < m < N. \tag{3}$$

On a la décomposition de Bruhat

$$G(k) = \amalg_{w \in W} P(k) \cdot w \cdot P(k), \quad \text{cf. [6: §5]}, \tag{4}$$

d'où aussi

$$G(k)/P(k) = \amalg_{w \in W} C(w), \quad \text{avec} \quad C(w) = (P(k) \cdot w \cdot P(k))/P(k). \tag{5}$$

De plus, il existe un k-sous-groupe connexe U_w du radical unipotent U de P tel que $C(w)$ soit un espace homogène principal sous $U_w(k)$, donc homéomorphe à $k^{d(w)}$, où $d(w) = \dim U_w$. En particulier, si $w = w_N$, alors $U_w = U$, donc

$$C(w_N) \approx U(k) = (P/\mathscr{Z}(T))(k) \approx k^d, \quad (d = \dim U). \tag{6}$$

Si $1 \leqq m \leqq N$, posons

$$E_m = \cup_{1 \leqq i \leqq m} C(w_i), \qquad Y_m = \lambda'(E_m \times C) \subset Y_t.$$

D'après [7: 3.15], $C(w_m)$ est ouvert dans E_m et E_m est fermé dans $G(k)/P(k)$. Les Y_m forment donc une suite croissante de sous-espaces fermés de Y_t allant de $Y_1 = C$ à $Y_N = Y_t$.

On va utiliser cette filtration pour déterminer la cohomologie de Y_t.

2.4. LEMME. *Soient $m \in \mathbf{N}$ $(1 < m \leqslant N)$ et $I \subset S$. Alors:*

(i) *Si $I \subset I_m$, la projection $\pi_I \colon G/P \to G/P_I$ définit un homéomorphisme de $C(w_m)$ sur son image dans G/P_I.*

(ii) *Si $I \not\subset I_m$, alors $\pi_I(E_m) = \pi_I(E_{m-1})$.*

(iii) *$Y_m - Y_{m-1}$ est homéomorphe à $C(w_m) \times L_m$ où $L_m = \cup_{I \subset I_m} \mathring{C}_I$.*

Les assertions (i) et (ii) résultent de [7: 3.16]. La relation 1.2(2) et (ii) montrent que l'on a

$$Y_m - Y_{m-1} = \amalg_{I \subset I_m} \pi_I(C(w_m)) \times \mathring{C}_I, \tag{1}$$

ce qui, vu (i), entraîne (iii).

2.5. Les appartements de Y sont les sous-ensembles de points fixes des conjugués de $\mathscr{Z}(T)(k)$. Un appartement est un sous-complexe fini, isomorphe au complexe de Coxeter A de W, et en particulier est homéomorphe à la sphère S^{l-1}. L'espace Y est réunion de l'ensemble $\mathscr{A}_P$ des appartements contenant C. Si A_0 est l'appartement fixé par $\mathscr{Z}(T)(k)$, alors $g \mapsto g \cdot A_0$ induit une bijection de $P(k)/\mathscr{Z}(T)(k)$ sur $\mathscr{A}_P$, ce qui permet de munir $\mathscr{A}_P$ d'une structure de variété k-analytique.

La base S définit une chambre C de A et les chambres de A sont les transformés $w \cdot C$ de C par W. Identifions S à $\{1, \dots, l\}$. Cela fixe une orientation sur C, donc aussi sur A. On note $[A]$ la classe d'homologie dans $H_{l-1}(A\,;\mathbf{Z})$ du cycle $\Sigma(-1)^{l(w)} w \cdot C$. Si $l \geqq 2$, c'est la classe fondamentale de la sphère orientée A. Si $l = 1$, cette classe appartient au sous-groupe d'homologie réduit $\tilde{H}_0(A\,;\mathbf{Z}) = \mathrm{Ker}\{H_0(A\,;\mathbf{Z}) \to H_0(\mathrm{pt}.\mathbf{Z}) = \mathbf{Z}\}$. Si $A \in \mathscr{A}_P$, il existe un et un seul isomorphisme $\nu_A \colon A \to A$ qui prolonge l'isomorphisme canonique de C sur C. On note $[A] \in H_{l-1}(A\,;\mathbf{Z})$ l'image de $[A]$ par l'isomorphisme associé en homologie. Si $l \geqq 2$, c'est donc la classe fondamentale de A pour l'orientation qui, sur C, coincide avec l'orientation donnée par l'identification de S à $\{1, \dots, l\}$. Si $l = 1$, on a $[A] \in \tilde{H}_0(A\,;\mathbf{Z})$.

Soit M un $\mathbf{Z}$-module. Etant donnés $h \in \tilde{H}^{l-1}(Y_t\,;M)$, et $A \in \mathscr{A}_P$, on note $\tilde{h}(A)$ la valeur sur $[A]$ de la restriction de h à $\tilde{H}^{l-1}(A\,;\mathbf{Z})$. Cela a un sens puisque, pour $l = 1$, la classe $[A]$ est dans le 0-ème groupe d'homologie réduit de A. A tout $h \in \tilde{H}^{l-1}(Y_t\,;M)$ est ainsi associée une application $\tilde{h} \colon \mathscr{A}_P \to M$.

2.6. Théorème. (i) *On a* $H^i(Y_t; M) = 0$ *pour* $i \neq l - 1$. (ii) *L'application* $h \mapsto \bar{h}$ *de 2.5 définit un isomorphisme de* $\bar{H}^{l-1}(Y_t; M)$ *sur* $C_c^\infty(\mathscr{A}_P; M)$. *Le* $\mathbf{Z}$-*module* $\bar{H}^{l-1}(Y_t; \mathbf{Z})$ *est libre.*

Supposons le théorème démontré pour $M = \mathbf{Z}$. Alors $H^i(Y_t; M) = H^i(Y_t; \mathbf{Z}) \otimes M$ d'après la formule des coefficients universels; comme $C_c^\infty(\mathscr{A}_P; M) = C_c^\infty(\mathscr{A}_P; \mathbf{Z}) \otimes M$ vu 2.2, cela entraîne le théorème pour M quelconque. Il suffit donc de considérer le cas où $M = \mathbf{Z}$.

Pour $1 < m < N$, soit L'_m le complément de L_m dans C, i.e. la réunion des faces de codimension 1 $C_{\{s\}}(s \in S - I_m)$ de C. Comme $I_m \neq \emptyset, S$, le sous-espace L'_m est contractile, donc de cohomologie réduite nulle en toute dimension. La suite exacte de cohomologie de C mod L'_m montre alors que $H_c^*(L_m; \mathbf{Z}) = 0$; vu 2.4 (iii) et la formule de Künneth, il s'ensuit que $H_c^*(Y_m - Y_{m-1}; \mathbf{Z}) = 0$. L'espace Y_1 est égal à C, donc contractile; sa cohomologie réduite est nulle. En utilisant la suite exacte 2.1(1) pour $(X, Y) = (Y_m, Y_{m-1})$ on en tire tout d'abord, par récurrence sur m, que $H^*(Y_m; \mathbf{Z}) = 0$ si $m < N$ et ensuite que

$$\bar{H}^*(Y_t; \mathbf{Z}) = H_c^*(Y_t - Y_{N-1}; \mathbf{Z}). \tag{1}$$

Mais on a, vu 2.3(6), 2.4(iii) et 2.5

$$Y_t - Y_{N-1} \cong \mathring{C} \times C(w_N) \cong \mathring{C} \times P(k)/\mathscr{Z}(T)(k) \cong \mathring{C} \times \mathscr{A}_P. \tag{2}$$

L'espace $\mathring{C}$ est homéomorphe à $\mathbf{R}^{l-1}$, et orienté grâce à l'indexation de S; on a donc

$$H_c^i(\mathring{C}; \mathbf{Z}) = 0 \quad \text{si} \quad i \neq l - 1 \quad \text{et} \quad H_c^{l-1}(\mathring{C}; \mathbf{Z}) = \mathbf{Z}.$$

Appliquant la formule de Künneth, on obtient:

$$H_c^i(Y_t - Y_{N-1}; \mathbf{Z}) = H_c^{l-1}(\mathring{C}; \mathbf{Z}) \otimes H_c^{i-l+1}(\mathscr{A}_P; \mathbf{Z}) = \mathbf{Z} \otimes H_c^{i-l+1}(\mathscr{A}_P; \mathbf{Z}). \tag{3}$$

L'assertion (i) résulte alors de (1) et du fait que $\mathscr{A}_P$ est de dimension 0, donc de cohomologie nulle en dimensions $\neq 0$. De plus, pour $i = l - 1$, (1) et (3) donnent un isomorphisme

$$\bar{H}^{l-1}(Y_t; \mathbf{Z}) \xrightarrow{\sim} \mathbf{Z} \otimes H_c^0(\mathscr{A}_P; \mathbf{Z}) = C_c^\infty(\mathscr{A}_P; \mathbf{Z}). \tag{4}$$

Nous allons montrer que cet isomorphisme est défini par l'application $h \mapsto \bar{h}$ de (2.5), au signe près. Soit $e \in H_c^{l-1}(\mathring{C}; \mathbf{Z})$ le générateur défini par l'orientation choisie en 2.5 multipliée par $(-1)^N$. Soit $f \in C_c^\infty(\mathscr{A}_P; \mathbf{Z})$. Vu (1) et (4), l'application $f \mapsto e \otimes f$ induit un isomorphisme de $C_c^\infty(\mathscr{A}_P; \mathbf{Z})$ sur $\bar{H}^{l-1}(Y_t; \mathbf{Z})$. Il reste à voir que cet isomorphisme est l'inverse de l'application $h \mapsto \bar{h}$ de 2.5, i.e. que l'on a

$$(e \otimes f)[A] = f(A) \text{ quel que soit } A \in \mathscr{A}_P. \tag{5}$$

On a $A \cap (Y - Y_{N-1}) = \mathring{C}_A$, où C_A désigne la chambre de A "opposée" à C, celle qui correspond à $w_N(C)$ par l'isomorphisme $\nu_A: A \to A$ de 2.5. D'autre part, la restriction de $e \otimes f$ à $\mathring{C} \times \{A\}$ est $f(A) \cdot e \in H_c^{l-1}(\mathring{C}; \mathbf{Z})$. Considérons le diagramme commutatif

$$H_c^{l-1}(\mathring{C}; \mathbf{Z}) \xrightarrow{\alpha} H_c^{l-1}(\mathring{C}_A; \mathbf{Z}) \xrightarrow{\beta} H^{l-1}(A; \mathbf{Z})$$

$$\uparrow_\gamma \qquad\qquad\qquad \uparrow_{\gamma'}$$

$$H_c^{l-1}(Y_t - Y_{N-1}; \mathbf{Z}) \xrightarrow{\beta'} H^{l-1}(Y_t; \mathbf{Z})$$

où β, β' sont définis par inclusion, γ, γ' par restriction, et α par l'isomorphisme transporté de $w_N: C \to w_N(C)$ par $\nu_A: A \to A$. Alors $\beta \circ \alpha$ applique e sur la classe duale de $[A]$, i.e. $\beta \circ \alpha(e)[A] = 1$. On a déjà vu que $\alpha^{-1} \circ \gamma$ envoie $e \otimes f$ sur $f(A) \cdot e$, d'où (5), ce qui achève de démontrer (ii).

Remarque. Si $l = 1$, on a $Y_t = G(k)/P(k)$, $\mathscr{A}_P = Y_t - \{P\}$ et $\bar{H}^0(Y_t; \mathbf{Z})$ s'identifie au quotient de $C_c^\infty(Y_t; \mathbf{Z})$ par le sous-espace des fonctions constantes. Etant donné $h \in \bar{H}^0(Y_t; \mathbf{Z})$, la fonction $\bar{h}$ correspondante associe au point Q la différence $h(Q) - h(P)$. On a donc $\bar{h} \in C_c^\infty(\mathscr{A}_P; \mathbf{Z})$. Le théorème est évident dans ce cas.

§3. Un complexe auxiliaire

Les hypothèses et notations sont celles du §2; on suppose $l \geq 1$. Nous nous proposons de décrire un complexe dont la cohomologie est celle de l'espace Y_t.

Les résultats de ce paragraphe ne seront pas utilisés par la suite (mis à part 5.10).

3.1. Le complexe $\bar{L}(M)$

Soit M un **Z**-module et soit (E_r) la suite spectrale de Leray de l'application $\tau: Y_t \to C = Y/G(k)$, cf. 1.2(3), relativement au faisceau constant M. Cette suite spectrale aboutit à $H^*(Y_t; M)$ et son terme E_2 est donné par $E_2^{p,q} = H^p(C; F^q)$, où F^q est le faisceau $R^q\tau M$ associé au préfaisceau $U \mapsto H^q(\tau^{-1}(U); M)$.

La fibre F_x^q de F^q en $x \in C$ est égale à $H^q(\tau^{-1}(x); M)$. Si $x \in \mathring{C}_I$, on a $\tau^{-1}(x) \simeq G(k)/P_I(k)$, d'où:

$$F_x^q = H^q(G(k)/P_I(k); M) = \begin{cases} 0 & \text{si} \quad q \neq 0 \\ C^\infty(G(k)/P_I(k); M) & \text{si} \quad q = 0. \end{cases}$$

Les faisceaux F^q sont nuls pour $q \neq 0$. Si l'on pose $F^0 = F$, on a donc

$$H^p(Y_t; M) = H^p(C; F). \tag{1}$$

De plus, le faisceau F est constant sur chaque face ouverte $\mathring{C}_I$, où $I \subset S = \{1, \ldots, l\}$, $I \neq S$. On peut le considérer comme un "faisceau simplicial" sur le simplexe C: il associe à toute face C_I le groupe

$$F_I = C^\infty(G(k)/P_I(k); M), \tag{2}$$

et à toute inclusion $C_J \subset C_I$ $(J \supset I)$ l'homomorphisme

$$\pi_{I,J}^0: F_J \to F_I \tag{3}$$

défini par $f \mapsto f \circ \pi_{I,J}$, cf. 1.1.

On sait que la cohomologie d'un tel faisceau peut se calculer simplicialement (cela résulte, par exemple, de la suite spectrale associée à la filtration de C par ses squelettes). Vu (1), on en déduit que $H^*(Y_t; M)$ est la cohomologie d'un complexe $L(M) = \oplus L^m(M)$, où:

$$L^m(M) = \oplus F_I, \quad I \subset S, \quad I \neq S, \quad \text{Card}(I) + m = l - 1, \tag{4}$$

le cobord df d'un élément $f \in F_I$ étant donné par

$$df = \sum_{1 \leq i \leq n} (-1)^i \pi_{I-j_i,I}^0 f \in \bigoplus_{i=1}^n F_{I-j_i}, \tag{5}$$

où $I = \{j_1, \ldots, j_n\}$, $j_1 < j_2 < \cdots < j_n$, et $I - j$ désigne $I - \{j\}$.

Le complexe $L(M)$ est nul en degrés < 0. Il est commode de "l'augmenter", i.e. de définir un complexe $\bar{L}(M)$ tel que

$$\begin{cases} \bar{L}^q(M) = L^q(M) & \text{pour} \quad q \neq -1 \\ \bar{L}^{-1}(M) = F_S = C^\infty(G(k)/P_S(k); M) = M, \end{cases} \tag{6}$$

l'opérateur d étant encore donné par la formule (5).

La cohomologie de $\bar{L}(M)$ s'identifie à la cohomologie réduite $\tilde{H}^*(Y_t; M)$ de Y_t. Vu 2.6, on en déduit:

3.2. Théorème. *On a* $H^q(\bar{L}(M)) = 0$ *pour* $q \neq l - 1$, *et* $H^{l-1}(\bar{L}(M))$ *est canoniquement isomorphe à* $\tilde{H}^{l-1}(Y_t; M)$.

Pour la structure de $\tilde{H}^{l-1}(Y_t; M)$ et $H^{l-1}(\bar{L}(M))$, voir 2.6, ainsi que 3.5 et 3.8 ci-après.

3.3. Corollaire. *On a une suite exacte de* $G(k)$-*modules*

$$0 \to F_S \to \bigoplus_i F_{S-i} \to \cdots \to \bigoplus_i F_{\{i\}} \to F_\emptyset \to \tilde{H}^{l-1}(Y_t; M) \to 0,$$

où $F_I = C^\infty(G(k)/P_I(k); M)$, *le groupe* $G(k)$ *opérant sur* F_I *grâce à son action sur l'espace homogène* $G(k)/P_I(k)$.

Cela résulte de 3.2 et du fait que l'isomorphisme $H^{l-1}(\bar{L}(M)) \simeq \tilde{H}^{l-1}(Y_t; M)$ commute par construction à l'action de $G(k)$.

(Noter qu'on obtient ainsi une *résolution* du $G(k)$-module $\tilde{H}^{l-1}(Y_t; M)$ au moyen des F_I, cf. 5.10.)

3.4. Explicitation de $H^{l-1}(\bar{L}(M))$. La composante de degré $l - 1$ du complexe $\bar{L}(M)$ est

$F_\emptyset = C^\infty(G(k)/P(k); M)$; il en résulte (cf. 3.3) que $H^{l-1}(\bar{L}(M))$ est le quotient de $F_\emptyset$ par la somme des images des $F_{(i)}$, $1 \le i \le l$.

L'espace $G(k)/P(k)$ est réunion disjointe des $C(w)$, 2.3 (5). En particulier, la "grosse cellule" $C(w_N)$ est ouverte dans $G(k)/P(k)$. Nous noterons R l'ensemble des fonctions $f \in F_\emptyset$ qui s'annulent sur les $C(w_j)$, $j < N$, i.e. dont le support est contenu dans $C(w_N)$; on a $R = C_c^\infty(C(w_N); M)$.

3.5. PROPOSITION. *L'injection $R \to F_\emptyset$ définit par passage au quotient un isomorphisme de R sur $H^{l-1}(\bar{L}(M)) = F_\emptyset/(\Sigma \operatorname{Im}(F_{(i)}))$.*

D'après 2.6 (1), la cohomologie du sous-espace Y_{N-1} de Y_t est triviale, et l'on a $\bar{H}^{l-1}(Y_t; M) \simeq H_c^{l-1}(Y_t - Y_{N-1}; M)$. Or la cohomologie à supports compacts de $Y_t - Y_{N-1}$ peut se calculer, comme celle de Y_t, au moyen de la suite spectrale de Leray de la projection $Y_t - Y_{N-1} \to C$; cette suite spectrale est ici particulièrement simple, vu que $Y_t - Y_{N-1} \simeq \mathring{C} \times C(w_N)$, cf. 2.6 (2). On en déduit un isomorphisme $H^{l-1}(Y_t - Y_{N-1}; M) \simeq H_c^{l-1}(\mathring{C}, R)$, et le diagramme

$$\bar{H}^{l-1}(Y_t; M) \xleftarrow{\sim} H_c^{l-1}(Y_t - Y_{N-1}; M)$$

$$\wr \qquad\qquad\qquad \wr$$

$$\bar{H}^{l-1}(C; F) \xleftarrow{\sim} H_c^{l-1}(\mathring{C}; R)$$

est commutatif (car provenant d'un homomorphisme de suites spectrales).

On déduit de là que l'injection de R dans le faisceau F définit un isomorphisme de $H_c^{l-1}(\mathring{C}; R) = R$ sur $\bar{H}^{l-1}(C; F) = H^{l-1}(\bar{L}(M))$, d'où la proposition.

3.6. *Remarque.* On peut également démontrer 3.5 par le procédé suivant: on définit des sous-complexes L_j ($1 \le j \le N$) de $\bar{L}(M)$ par la condition qu'un élément f d'un F_l appartient à L_j si et seulement si f, considéré comme fonction sur $G(k)$ invariante à droite par $P_l(k)$, s'annule sur tous les $P(k)w_iP(k)$, pour $i < j$. On a $\bar{L}(M) = L_1 \supset L_2 \supset \cdots \supset L_N$, et le complexe L_N est égal à 0 en toute dimension $\ne l - 1$, et à R en dimension $l - 1$. La structure des quotients successifs L_j/L_{j+1}, $1 \le j < N$, se détermine facilement, grâce à 2.4; on trouve que la cohomologie de ces complexes est nulle en toute dimension. Il en résulte que l'injection de L_N dans $\bar{L}(M)$ induit un isomorphisme de $H^*(L_N)$ sur $H^*(\bar{L}(M))$, ce qui démontre à la fois 3.2 et 3.5.

3.7. Projection de $H^{l-1}(\bar{L}(M))$ sur R. On a défini dans 3.5 un isomorphisme de $R = C_c^\infty(C(w_N); M)$ sur $H^{l-1}(\bar{L}(M))$. Nous nous proposons d'expliciter l'isomorphisme réciproque.

Soit $f \in F_\emptyset = C^\infty(G(k)/P(k); M)$. Associons à f la fonction $\tilde{f}$ sur $G(k)$ définie par

$$\tilde{f}(x) = \sum_{w \in W} (-1)^{l(w)} f(x \cdot w), \qquad x \in G(k). \tag{1}$$

C'est une fonction localement constante sur $G(k)$, invariante à droite par $\mathscr{Z}(T)(k)$.

LEMME. *S'il existe $s \in S$ tel que f soit invariante à droite par $P_{\{s\}}(k)$, on a $\tilde{f} = 0$.*
Soit $W_s = \{w \in W | l(ws) > l(w)\}$. Alors W est réunion disjointe de W_s et $W_s \cdot s$ [9: IV, §1, Exer. 3, p. 37]. Comme

$$f(x \cdot w \cdot s) = f(x \cdot w) \quad \text{et} \quad l(w \cdot s) = l(w) + 1 \quad (x \in G(k), w \in W_s),$$

les différents termes de la somme (1) se détruisent deux à deux, et l'on a bien $\tilde{f}(x) = 0$.

Le lemme ci-dessus montre que $\tilde{f} = 0$ si $f \in \Sigma \operatorname{Im}(F_{(i)})$. Par passage au quotient, l'application $f \mapsto \tilde{f}$ définit donc un homomorphisme θ de $H^{l-1}(\bar{L}(M)) = F_\emptyset/(\Sigma \operatorname{Im}(F_{(i)}))$ dans le groupe $C^\infty(G(k)/\mathscr{Z}(T)(k); M)$; par restriction à $P(k)$, on en déduit un homomorphisme

$$\theta': H^{l-1}(\bar{L}(M)) \to C^\infty(P(k)/\mathscr{Z}(T)(k); M).$$

D'autre part, l'application $p \mapsto p \cdot w_N$ définit par passage au quotient un isomorphisme de $P(k)/\mathscr{Z}(T)(k)$ sur $C(w_N)$, d'où un isomorphisme

$$\iota: C^\infty(P(k)/\mathscr{Z}(T)(k); M) \to C^\infty(C(w_N); M).$$

3.8. PROPOSITION. *L'homomorphisme*

$$(-1)^N \iota \circ \theta': H^{l-1}(\bar{L}(M)) \to C^\infty(C(w_N); M)$$

est un isomorphisme de $H^{l-1}(\tilde{L}(M))$ *sur le sous-espace R de* $C^{\infty}(C(w_N); M)$ *formé des fonctions à support compact. Son inverse est l'isomorphisme* $R \to H^{l-1}(\tilde{L}(M))$ *de* 3.5.

Vu 3.5, il suffit de montrer que le composé de $R \to H^{l-1}(\tilde{L}(M))$ et de $(-1)^N \iota \circ \theta'$ est l'application identique de R sur R. Or c'est immédiat: si $f \in R$ et $x \in P(k)$, on a $f(x \cdot w) = 0$ pour $w \neq w_N$, d'où $\tilde{f}(x) = (-1)^N f(x \cdot w_N)$; si f' désigne l'image de f dans $H^{l-1}(\tilde{L}(M))$, cela signifie bien que $(-1)^N \iota \circ \theta'(f') = f$.

3.9. Corollaire. *L'application*

$$\theta: H^{l-1}(\tilde{L}(M)) \to C^{\infty}(G(k)/\mathscr{Z}(T)(k); M)$$

est injective.

C'est immédiat.

Remarque. Il serait intéressant de déterminer explicitement l'image de θ.

3.10. *Remarque.* Les résultats de ce § ont un analogue dans le cas où le corps de base k, au lieu d'être ultramétrique, est le corps $\mathbf{R}$ des nombres réels, le module M étant pris égal à $\mathbf{R}$.

Le complexe $\tilde{L} = \tilde{L}(\mathbf{R})$ est défini comme dans 3.1, en convenant que $C^{\infty}(G(k)/P_I(k); M)$ désigne l'espace des fonctions *continues*[2] réelles sur l'espace homogène $G(k)/P_I(k)$ (qui est ici une variété compacte). On trouve que la cohomologie de $\tilde{L}$ est la suivante:

(a) $H^q(\tilde{L}) = 0$ pour $q \neq l - 1$, où $l = rg_k G$;

(b) $H^{l-1}(\tilde{L})$ s'identifie à l'espace des fonctions continues sur la grosse cellule $C(w_N)$ qui tendent vers 0 à l'infini (i.e. qui se prolongent en une fonction continue sur $G(k)/P(k)$ nulle sur les $C(w_i)$ pour $i < N$).

Cela se démontre par des arguments analogues à ceux du cas ultramétrique, utilisant la filtration de $G(k)/P(k)$ par les $C(w_i)$. De plus, 3.5 et 3.8 restent valables, à condition, ici encore, de remplacer les fonctions localement constantes à support compact par les fonctions continues tendant vers 0 à l'infini.

La cohomologie de $\tilde{L}$ peut aussi s'interpréter comme la cohomologie réduite de l'espace Y_t (1.4) à valeurs dans le faisceau des fonctions continues réelles dont les restrictions à chaque simplexe sont localement constantes; cela résulte, comme dans 3.1, de la suite spectrale associée à $\tau: Y_t \to C$.

§4. L'immeuble de Bruhat–Tits des sous-groupes parahoriques

4.1. Rappelons que G est semi-simple et que $\tilde{G}$ est le revêtement universel de G. Soit $\pi: \tilde{G} \to G$ l'isogénie centrale canonique. Si X est un k-groupe connexe réductif, notons $\mathfrak{P}(X)$ l'ensemble de ses k-sous-groupes paraboliques. On sait que $Q \mapsto \pi(Q)$ et $Q \mapsto \pi^{-1}(Q)$ sont des bijections réciproques de $\mathfrak{P}(\tilde{G})$ et $\mathfrak{P}(G)$ [7: §2]. D'autre part (*loc. cit.*) $\tilde{T} = \pi^{-1}(T)$ est un k-*tore* déployé maximal de $\tilde{G}$, π induit un isomorphisme de $\mathcal{N}(\tilde{T})/\mathscr{Z}(\tilde{T})$ sur $\mathcal{N}(T)/\mathscr{Z}(T)$ et $\Phi(\tilde{T}, \tilde{G})$ s'identifie à Φ. Posons $\tilde{P} = \pi^{-1}(P)$ et $\tilde{N} = \mathcal{N}(\tilde{T})(k)$. Alors $\tilde{T} = (\tilde{G}(k), \tilde{P}(k), \tilde{N}, S)$ est un système de Tits, et l'immeuble correspondant s'identifie à l'immeuble Y de T.

Le groupe $\tilde{G}$ s'écrit de manière essentiellement unique comme produit direct de k-groupes presque simples simplement connexes G_j de k-rang > 0 $(1 \leq j \leq m)$ et d'un k-groupe simplement connexe G_0 de k-rang nul. Le groupe $\tilde{T}$ est le produit de ses intersections T_j avec les $G_j (1 \leq j \leq m)$ et $\tilde{P}$ (resp. $\mathcal{N}(\tilde{T})$) est le produit de G_0 et de ses intersections P_j (resp. N_j) avec les $G_j (1 \leq j \leq m)$. Le système de racines Φ est la somme directe des $\Phi(T_j, G_j)$, W le produit des $W_j = N_j/\mathscr{Z}_{G_j}(T_j)$ et S la réunion disjointe des $S_j = S \cap W_j (1 \leq j \leq m)$. Un simplexe σ de Y est une famille $\{\sigma_j\}(1 \leq j \leq m)$, où σ_j est, soit vide, soit un simplexe de l'immeuble Y_j des k-sous-groupes paraboliques de G_j, et au moins un σ_j est non vide. Il s'ensuit que Y est le joint des immeubles Y_j et que C est le joint des C_j, où C_j est l'ensemble des points fixes de $P_j(k)$ dans $Y_j (1 \leq j \leq m)$.

4.2. Dans ce qui suit, nous *admettrons* l'existence du système de Tits $\tilde{T}' = (\tilde{G}(k), B, N', S')$ des sous-groupes "parahoriques" de $\tilde{G}(k)$; cette existence est annoncée dans [10] et démontrée dans [11: §10] pour les groupes classiques. On note X l'immeuble correspondant [11: §2]; c'est un complexe polysimplicial dont les sommets (resp. les faces) correspondent aux sous-groupes compacts maximaux (resp. aux sous-groupes parahoriques) de $\tilde{G}(k)$.

---

1   [2]A la place des fonctions continues, on pourrait prendre les fonctions analytiques (ou différentiables); nous ignorons quelle serait alors la cohomologie du complexe $\tilde{L}$ correspondant.

Les données B, N', S' sont supposées compatibles avec les choix déjà faits de $\tilde{T}$, $\tilde{P}$, Φ, S; en particulier, on a un *double système de Tits*, au sens de [11: §5] et $N' = \tilde{N} = \mathcal{N}(\tilde{T})(k)$; le groupe de Weyl W' de T' s'identifie au quotient de N' par le plus grand sous-groupe compact H de $\mathcal{Z}(\tilde{T})(k)$, et est donc extension de W par le sous-groupe $\mathcal{Z}(\tilde{T})(k)/H$, qui est commutatif libre de rang l. On a $B \cap N' = H$. Le groupe B est le sous-groupe parahorique minimal standard ("sous-groupe d'Iwahori"); c'est le produit direct de $G_0(k)$ et des $B_j = B \cap G_j$, $1 \leqslant j \leqslant m$. On a de même des décompositions

$$S' = \coprod_{j=1}^{j=m} S'_j, \qquad W' = \prod_{j=1}^{j=m} W'_j, \qquad X = \prod_{j=1}^{j=m} X_j,$$

où X_j est l'immeuble associé au système de Tits $T_j = (G_j(k), B_j, N_j, S'_j)$, et W'_j le groupe de Weyl de $T_j (1 \leqq j \leqq m)$. Les sous-groupes parahoriques de $\tilde{G}(k)$ sont les produits $G_0(k) \times \Pi_1{}^m M_j$, où M_j parcourt les sous-groupes parahoriques de $G_j(k)$. En particulier, les sous-groupes parahoriques standard sont les sous-groupes $B_I = B \cdot W'_I \cdot B$ où I parcourt l'ensemble des parties de S' *qui ne contiennent aucun des* S'_j, et W'_I est le sous-groupe de W' engendré par I. Soit C'_j l'ensemble des points fixes de B_j dans $X_j (1 \leqq j \leqq m)$. C'est un simplexe de dimension égale au k-rang l_j de G_j, et l'ensemble C' des points fixes de B dans X est un polysimplexe de dimension l, le produit des C'_j.

On note A l'appartement standard de X; son fixateur est H. On suppose (ce qui est loisible, quitte à modifier le choix de B) qu'il existe un *point spécial* [11: 1.3.7, p. 22] $o = (o_1, \ldots, o_m)$, $o_j \in X_j$, de X ayant les propriétés suivantes (qui le caractérisent):

(a) o est un sommet de C';

(b) le quartier [11: 1.3.11, p. 23] de A de sommet o contenant l'intérieur de C' définit le germe de quartier correspondant à $\tilde{P}(k)$ [11: 5.1.33, p. 96].

On note s'_j l'élément de S'_j correspondant à la réflexion par rapport à la face de C'_j opposée à o_j. L'ensemble $S'_j - \{s'_j\}$ s'identifie à S_j, par la projection $W'_j \rightarrow W_j$. L'ensemble S' est réunion disjointe de S et de $\{s'_1, \ldots, s'_m\}$. Le fixateur de o_j dans $G_j(k)$ est $B_{S_j} = B \cdot W'_{S_j} \cdot B$. Le fixateur de o dans $\tilde{G}(k)$ est

$$B_S = B \cdot W \cdot B = G_0(k) \times \prod_{j=1}^{j=m} B_{S_j},$$

qui est un sous-groupe compact maximal spécial de $\tilde{G}(k)$ [11: 3.3.5, p. 66]; on écrira souvent $\tilde{K}$ pour B_S. On a

$$\tilde{G}(k) = \tilde{K} \cdot \tilde{P}(k) \quad \text{et} \quad G_j(k) = B_{S_j} \cdot P_j(k) \quad \text{pour} \quad 1 \leqslant j \leqslant m. \tag{1}$$

Pour toute partie I de S' ne contenant aucun S'_j, on note C'_I l'ensemble des points fixes de B_I dans X; on obtient ainsi, une fois et une seule, chaque face du polysimplexe C'.

4.3. Appartements. Les appartements de X sont les transformés de l'appartement standard A par $\tilde{G}(k)$. Etant donnés deux appartements A' et A'', il existe un élément de $\tilde{G}(k)$ qui transforme A' en A'' et induit l'identité sur $A' \cap A''$ [11: 2.5.8, p. 45]. Par suite, deux parties E, E' de A' transformées l'une de l'autre par un élément $g \in \tilde{G}(k)$ le sont aussi par un élément qui stabilise A' et coincide avec g sur E. Deux faces quelconques de X sont contenues dans un appartement [11: 2.3.1, p. 37] donc X est réunion des appartements contenant C', et ces derniers sont les transformés de A par B. Le groupe $N' = \mathcal{N}(\tilde{T})(k)$ stabilise A. Comme H opère trivialement sur A, on obtient ainsi une représentation de W' comme groupe d'automorphismes de A. L'application structurale j de [11: 2.2, p. 35] permet de munir A d'une structure d'espace vectoriel de dimension l, d'origine o, telle que Φ s'identifie à un système de racines dans le dual de A, de groupe de Weyl W, de manière à ce que la réflexion $s_a (a \in \Phi)$ soit la réflexion par rapport à l'hyperplan $a = 0$, et que W' soit un groupe de Weyl affine de partie linéaire W (cf. 4.5).

Montrons encore que N' est le normalisateur de H dans $\tilde{G}(k)$ ainsi que le stabilisateur de A dans $\tilde{G}(k)$, et que A est le seul appartement de X fixé par H. Le groupe H est Zariski-dense dans $\mathcal{Z}(\tilde{T})$, et $\tilde{T}$ est le plus grand tore déployé sur k central de $\mathcal{Z}(\tilde{T})$, donc $g \in \tilde{G}(k)$ normalise H si et seulement s'il normalise $\tilde{T}$, d'où l'égalité $N' = \mathcal{N}_{\tilde{G}(k)}(H)$. Par construction, N' stabilise A. D'autre part, comme X est associé à un système de Tits de $\tilde{G}(k)$, le groupe $\tilde{G}(k)$ y opère par automorphismes spéciaux; en particulier, si $g \in \tilde{G}(k)$ stabilise A, il y induit une transformation du groupe de Weyl de A, donc il existe $n \in N'$ tel que $n \cdot g$ fixe A, i.e. $n \cdot g \in H$, d'où $g \in N'$. Enfin, soit A' un appartement fixé par H. Il existe $g \in \tilde{G}(k)$ tel que

$g \cdot A = A'$. Comme $g \cdot H \cdot g^{-1}$ est le fixateur de A', on a $H \subset g \cdot H \cdot g^{-1}$, d'où, en passant aux adhérences de Zariski, $\mathscr{Z}(\tilde{T}) \subset g \cdot \mathscr{Z}(\tilde{T}) \cdot g^{-1}$ et par suite $\mathscr{Z}(\tilde{T}) = g \cdot \mathscr{Z}(\tilde{T}) \cdot g^{-1}$, d'où $g \in N'$ et $A' = A$. En associant à un appartement A' le plus grand tore déployé sur k du centre de l'adhérence de Zariski du fixateur de A', on établit donc une bijection entre appartements de X et tores déployés sur k maximaux de $\tilde{G}$. Remarquons que H peut avoir des points fixes en dehors de A (par exemple si $\tilde{G} = SL_2$ et $k = \mathbf{Q}_2$).

4.4. Quartiers fermés. Nous appellerons *quatiers fermés* les adhérences des quartiers de X au sens de [11: 7.1.4, p. 157]. Tout quartier fermé Δ est l'adhérence d'un unique quartier $\mathring{\Delta}$ qui est l'intérieur de Δ dans tout appartement de X contenant Δ. Le *sommet* (resp. la *direction*) de Δ est, par définition, le sommet (resp. la direction) de $\mathring{\Delta}$, *loc. cit.*

On note D le quartier fermé de A de sommet o contenant C'. C'est le plus petit cône de A de sommet o contenant C'. C'est un domaine fondamental pour W dans A; utilisant 4.3, on en déduit que c'est aussi un domaine fondamental pour $\tilde{K} = B \cdot W \cdot B$ dans X.

4.5. Racines affines. Les *murs* de A sont les hyperplans de points fixes des réflexions de W' et les *racines affines* sont les demi-espaces fermés de A limités par les murs [11: 1.3.3, p. 20]. On note Σ l'ensemble des racines affines. Il existe dans le dual de A un et un seul système de racines réduit ${}^v\Sigma$ ayant les propriétés suivantes:

(i) étant donné $\alpha \in \Sigma$ il existe un unique ${}^v\alpha \in {}^v\Sigma$ et un entier $r(\alpha)$ tels que α soit l'ensemble de points donnés par ${}^v\alpha + r(\alpha) \geqslant 0$;

(ii) tous les hyperplans $\beta + m = 0$ ($\beta \in {}^v\Sigma$, $m \in \mathbf{Z}$) sont des murs.

De plus, il existe une unique surjection $\nu: \Phi \to {}^v\Sigma$ commutant à W telle que $a \in \mathbf{R}^*_+ \cdot \nu(a)$ pour tout $a \in \Phi$. Le groupe W' est le groupe de Weyl affine de ${}^v\Sigma$; sa partie linéaire est W.

Une partie de X est un *demi-appartement* si elle est transformée d'une racine affine par un élément de $\tilde{G}(k)$.

4.6. Fixateurs. Pour $a \in \Phi$, on note $U_{(a)}$ le plus grand k-sous-groupe unipotent de $\tilde{G}$ normalisé par $\tilde{T}$ et dans lequel les poids de $\tilde{T}$ sont les multiples positifs de a contenus dans Φ [6: §5]. Le radical unipotent U (resp. U^-) de $\tilde{P}$ (resp. du sous-groupe parabolique $\tilde{P}^-$ opposé à $\tilde{P}$ et contenant $\mathscr{Z}(\tilde{T})$) est donc engendré par les $U_{(a)}$ avec $a > 0$ (resp. avec $a < 0$). Etant donné $r \in \mathbf{R}$, on note $U_{(a),r}$ le sous-groupe de $U_{(a)}$ qui fixe le demi-espace fermé $\nu(a) + r \geqq 0$. Les sous-groupes $U_{(a),r}$ sont compacts ouverts dans $U_{(a)}(k)$, d'intersection réduite à $\{e\}$, de réunion égale à $U_{(a)}(k)$, et définissent une filtration décroissante de $U_{(a)}(k)$, qui est en fait discrète, n'ayant de sauts que pour certaines valeurs entières de r. (Si la racine a n'est ni multipliable ni divisible, chaque entier correspond à un tel saut. Sinon, il y a plusieurs possibilités pour a et $2a$, qui sont décrites par un *échelonnage* de Φ par Σ [11: 1.4.1, p. 25].) Si α est une racine affine, on pose $U_\alpha = U_{(a),r}$, où $\nu(a) = {}^v\alpha$ et $r = r(\alpha)$.

Si $M \subset X$, et si E est un groupe opérant sur X, alors E_M dénote le fixateur de M dans E.

Si M est une partie de A, alors $\tilde{G}(k)_M$ est engendré par H et par les groupes U_α associés aux racines affines α contenant M [11: 5.2.28, p. 104].

4.7. LEMME. *Soient* $I \subset S$ *et* Φ_I *l'ensemble des éléments de* Φ *qui sont combinaisons linéaires d'éléments de* I. *Posons* $G_I = \mathscr{Z}(\tilde{T}_I)(k)$, $N_I = N' \cap G_I$ *et* $K_I = \tilde{K} \cap G_I$. *Soit* $U_0{}^I$ *le groupe engendré par les* $U_{(a),o}(a \in \Phi_I)$. *Alors*

$$G_I = U_0{}^I \cdot N_I \cdot U_0{}^I, \qquad K_I = U_0{}^I \cdot W_I \cdot H \cdot U_0{}^I, \tag{1}$$

et K_I *est un sous-groupe compact maximal de* G_I.

Le sous-groupe G_I de $\tilde{G}(k)$ satisfait aux conditions imposées au groupe noté G_1 dans [11: 7.6, p. 184] (avec $\Phi_1 = \Phi_I$, $T_1 = \mathscr{Z}(\tilde{T})(k)$). Notre groupe $U_0{}^I$ n'est autre que le $U^1_{\pi(x)}$ de *loc. cit.*, p. 185, pour $x = o$. La première égalité de (1) résulte alors de [11: 7.3.4, p. 165], comme cela est indiqué dans [11: p. 185, l. 21].

Le groupe K_I contient évidemment l'ensemble $U_0{}^I \cdot H \cdot W_I \cdot U_0{}^I$; tout revient donc à montrer qu'un sous-groupe de G_I contenant strictement $U_0{}^I \cdot H \cdot W_I \cdot U_0{}^I$ n'est pas compact. Soit M un tel sous-groupe, et soit $Q = M \cap N_I$. Vu la première égalité de (1), on a $M = U_0{}^I \cdot Q \cdot U_0{}^I$. Le groupe Q contient $K_I \cap N_I$; d'autre part, N_I/H est produit semi-direct de W_I et du groupe de translations $\mathscr{Z}(\tilde{T})(k)/H$. Supposons $M \neq K_I$. Alors $Q \neq W_I \cdot H$, donc Q contient un élément g dont l'image dans N_I/H est une translation non nulle. Par suite, g engendre un sous-groupe discret infini de M, et M n'est pas compact.

4.8. Action de $(\mathrm{Aut}\ \tilde{G})(k)$ et de $G(k)$ sur l'immeuble X. Nous admettrons que π est B–N-adapté [11: 1.2.13, p. 18] et de type connexe [11: 4.1.3, p. 72].

Tout élément de $\mathrm{Aut}\ \tilde{G}(k)$ laisse stable G_0, permute les G_j $(1 \le j \le m)$ et les sous-groupes parahoriques de $\tilde{G}(k)$, et laisse stable la classe de conjugaison de H. On en déduit une action de $(\mathrm{Aut}\ \tilde{G})(k)$ sur X qui en préserve la structure polysimpliciale et les appartements; elle est continue et propre. Par la représentation adjointe, G s'envoie dans $\mathrm{Ad}\ G = \mathrm{Ad}\ \tilde{G}$, d'où une action continue de $G(k)$ sur $\tilde{G}(k)$ qui commute à π (le groupe $G(k)$ opérant sur lui-même par automorphismes intérieurs), ainsi qu'une action, également continue et propre, de $G(k)$ sur X. On note $g \cdot x$ le transformé par $g \in G(k)$ d'un élément x de X. Comme X est localement fini, les fixateurs des faces de X dans $G(k)$ ou $\mathrm{Aut}\ \tilde{G}(k)$ sont des sous-groupes ouverts. Comme π commute à $G(k)$, le groupe $\pi(\tilde{G}(k))$ est distingué dans $G(k)$.

Le groupe N est le stabilisateur de A. Son image canonique W'' dans $\mathrm{Aut}\ A$ contient W' comme sous-groupe distingué. Comme π est de type connexe, les fixateurs W'_x et W''_x d'un point $x \in A$ dans W' et W'' sont les mêmes [11: 4.1.3, p. 72]; en particulier $W''_0 = W'_0 = W$.

Le noyau de π est contenu dans H, donc dans tous les conjugués de H, et opère trivialement sur X. Si $M \subset X$ et $g \in \tilde{G}(k)$, on a

$$g \in \tilde{G}(k)_M \Leftrightarrow \pi(g) \in G(k)_M. \tag{1}$$

4.9. LEMME. *Soit K (resp. L) le fixateur de o (resp. A) dans $G(k)$.*
(i) *Le groupe K est compact ouvert dans $G(k)$ et $L = K \cap \mathcal{Z}(T)(k)$.*
(iii) *On a $K = L \cdot \pi(\tilde{K})$, $G(k) = K \cdot P(k)$ et $K \cap P_I = L \cdot \pi(\tilde{K} \cap \tilde{P}_I)$ pour tout $I \subset S$.*
(iii) *L'ensemble D (resp. C) est un domaine fondamental pour K dans X (resp. Y).*

(i) On a déjà remarqué que K est ouvert dans $G(k)$; il est évidemment compact, et contient L. D'autre part, L normalise $\pi(H)$ puisque H est le fixateur de A dans $\tilde{G}(k)$; il normalise donc l'adhérence de $\pi(H)$ pour la topologie de Zariski, qui est $\mathcal{Z}(T)$, donc aussi T qui est le plus grand tore déployé sur k du centre de $\mathcal{Z}(T)$, et l'on a $L \subset N$. Le fait que L fixe A montre en outre que son image dans $W = N/\mathcal{Z}(T)(k)$ est triviale, d'où $L \subset \mathcal{Z}(T)(k)$, et par suite $L \subset K \cap \mathcal{Z}(T)(k)$. Inversement, si $g \in K \cap \mathcal{Z}(T)(k)$, l'élément g stabilise A et induit sur A une translation; comme celle-ci fixe o, c'est l'identité, et l'on a bien $g \in L$.

(ii) Soit $x \in K$. Le groupe $\tilde{K}$ est transitif sur l'ensemble des appartements contenant o, et W'_0 est transitif sur l'ensemble des quartiers fermés de sommet o contenus dans A. Il existe donc $y \in \pi(\tilde{K})$ tel que $y \cdot x$ stabilise o, D et A, donc fixe A puisque π est de type connexe; on a donc $y \cdot x \in L$, d'où $K = L \cdot \pi(\tilde{K})$.

Le groupe $\tilde{G}(k)$ permutant transitivement les appartements de X, on a $G(k) = N \cdot \pi(\tilde{G}(k))$. D'autre part, $\tilde{K} = B \cdot W \cdot B$. Comme N est engendré par W et $\mathcal{Z}(T)(k)$ et que $\pi(\tilde{G}(k))$ est distingué dans $G(k)$, on a

$$G(k) = N \cdot \pi(\tilde{G}(k)) \subset \pi(\tilde{K})(\mathcal{Z}(T)(k)) \cdot \pi(\tilde{G}(k)) = \pi(\tilde{K}) \cdot \pi(\tilde{G}(k)) \cdot (\mathcal{Z}(T)(k)),$$

d'où, vu 4.2(1),

$$G(k) \subset \pi(\tilde{K}) \cdot \pi(\tilde{K}) \cdot \pi(\tilde{P}(k)) \cdot (\mathcal{Z}(T)(k)) \subset K \cdot P(k).$$

Vu (i), on a $L \cdot \pi(\tilde{K} \cap \tilde{P}_I) \subset K \cap P_I$. Soit $x \in K \cap P_I$. Comme $K = L \cdot \pi(\tilde{K})$, on peut écrire x sous la forme $y \cdot \pi(z)$, avec $y \in L$ et $z \in \tilde{K}$. On a $y \in \mathcal{Z}(T) \subset P_I$, d'où $\pi(z) \in P_I$, ce qui entraîne $z \in \tilde{P}_I$, d'où $x \in L \cdot \pi(\tilde{K} \cap \tilde{P}_I)$.

(iii) La chambre C est un domaine fondamental pour $G(k)$ dans Y, fixé par $P(k)$. C'est donc un domaine fondamental pour K. La relation $X = \tilde{K} \cdot D$ (cf. 4.4) entraîne $X = K \cdot D$. Soient d, $d' \in D$ transformés l'un de l'autre par un élément de K. Vu 4.3, ils sont aussi transformés l'un de l'autre par un élément du stabilisateur de A dans K, i.e. par un élément de $W = W'_0 = W''_0$ (4.8). On a donc $d = d'$, ce qui montre que D est un domaine fondamental pour K dans X.

Remarque. Soient $\mu_X \colon K \times D \to X$ et $\mu_Y \colon K \times C \to Y$ les applications induites par l'action de K sur X et Y. D'après (iii), μ_X et μ_Y sont *surjectives*. De plus, ce sont des applications *propres*, puisque K est compact, cf. [8: III.28, prop. 2]. Vu [8: I.74, cor. 4], il en résulte que la topologie de X (resp. de Y) s'identifie à la *topologie quotient* de celle de $K \times D$ (resp. de $K \times C$) par la relation d'équivalence définie par μ_X (resp. par μ_Y).

4.10. Fixateurs de parties de D. Soit $I \subset S$. On note D_I l'intersection de D et des murs $\nu(a) = 0$ ($a \in I$). On a $D_S = \{o\}$ et $D_\emptyset = D$. Une partie M de D_I sera dite *intérieure à D_I* si $M \cap D_J = \emptyset$ quel que soit $J \subset S$ contenant I proprement. Par abus de langage, une demi-droite E issue de o, contenue dans D_I, sera dite intérieure à D_I si tout point de E différent de o est intérieur à D_I.

PROPOSITION. *Soit $I \subset S$, $I \neq S$ et soit E une demi-droite d'origine o intérieure à D_I. Alors*

$$G(k)_E = G(k)_{D_I} = K \cap P_I. \tag{1}$$

Nous reprenons les notations de 4.7, 4.9 et désignons par Ψ_I l'ensemble des racines >0 non contenues dans Φ_I, autrement dit l'ensemble des poids de $\tilde{T}$ dans le radical unipotent U_I de $\tilde{P}_I$.

Comme $G(k)_{D_I}$ est contenu dans $G(k)_E$, tout revient à prouver:

$$G(k)_E \subset K \cap P_I \subset G(k)_{D_I}. \tag{2}$$

Les groupes figurant dans (2) contiennent L et font partie de K. Vu 4.8(1) et 4.9(ii), il suffit donc de montrer:

$$\tilde{G}(k)_E \subset \tilde{K} \cap \tilde{P}_I \subset \tilde{G}(k)_{D_I}. \tag{3}$$

Le groupe $\tilde{G}(k)_E$ est engendré par H et les groupes U_α, où α parcourt l'ensemble des racines affines α qui contiennent E, cf. 4.6. Si α est une telle racine, et si $a \in \Phi$ est tel que $\nu(a) = {}^\nu\alpha$ (cf. 4.5), on a $a \in \Phi_I \cup \Psi_I$, d'où $U_\alpha \subset \tilde{P}_I$, ce qui démontre la première inclusion de (3).

Le groupe $\tilde{K} \cap \tilde{P}_I$ contient $K_I = \tilde{K} \cap G_I$. Son image dans G_I par la projection canonique de $\tilde{P}_I(k)$ sur $\tilde{P}_I(k)/U_I(k) = G_I$ contient donc aussi K_I. Comme ce dernier est compact maximal dans G_I (4.7), on a

$$\tilde{K} \cap \tilde{P}_I = K_I \cdot (\tilde{K} \cap U_I). \tag{4}$$

Pour établir la deuxième inclusion de (3), il nous suffit donc de montrer que K_I et $\tilde{K} \cap U_I$ fixent D_I.

On a $K_I = U_0^I \cdot W_I \cdot H \cdot U_0^I$ (4.7). Le groupe $W_I \cdot H$ fixe évidemment D_I. Par définition, le groupe U_0^I est engendré par les fixateurs $U_{(a),o}$ des racines affines $\nu(a) \geq 0$ ($a \in \Psi_I$). Comme ces dernières contiennent D_I on a $U_0^I \subset \tilde{G}(k)_{D_I}$, d'où $K_I \subset \tilde{G}(k)_{D_I}$.

Soit $u \in U_I(k)$. C'est un produit d'éléments $u_a \in U_{(a)}(k)$ où a parcourt Ψ_I. Soit E' une demi-droite d'origine o et intérieure à D_I. Si x est un point de E' tendant vers l'infini, alors $\nu(a)(x)$ tend vers l'infini, donc u_a fixe une demi-droite convenable de E'. Il en est alors de même de u. Soit y un point de cette demi-droite. Si u est de plus dans $\tilde{K}$, il fixe o, donc le segment joignant o et y, et finalement tout point de E'. Cela valant pour E' intérieure à D_I quelconque, et l'ensemble de ces demi-droites étant dense dans D_I, il s'ensuit que $\tilde{K} \cap U_I$ fixe D_I, ce qui achève la démonstration.

§5. Compactification de X par Y_I et cohomologie à supports compacts de X

Soit Z l'ensemble somme de X et de Y. Le but de ce § est de munir Z d'une topologie qui en fasse un espace compact contractile Z_t, et qui induise sur X (resp. Y) la topologie naturelle de X (resp. celle de Y_t), cf. 5.4; la détermination de la cohomologie à supports compacts de X en résultera (5.6).

5.1. Compactification d'un espace affine réel. Soient A un espace affine réel, et V_A l'espace vectoriel des translations de A; on suppose que la dimension l de V_A est finie. Soit S_A la sphère des demi-droites de V_A issues de l'origine, et soit $\bar{A} = A \amalg S_A$ la somme disjointe de A et de S_A. Il existe sur $\bar{A}$ une topologie naturelle qui en fait une *compactification* de A (compactification "par les directions de demi-droites"); cette topologie peut être caractérisée par les propriétés suivantes:

(a) elle induit sur A (resp. sur S_A) la topologie naturelle de A (resp. de S_A);

(b) A est ouvert dans $\bar{A}$;

(c) soient $o \in A$ et $d \in S_A$; soit (x_i) une suite de points de A; pour que x_i tende vers d dans $\bar{A}$, il faut et il suffit que x_i tende vers l'infini dans A et que la demi-droite d_i d'origine o contenant x_i tende vers d dans S_A.

L'espace $\bar{A}$ est homéomorphe à une boule fermée de dimension l. Si σ est un automorphisme affine de A, et σ_{lin} l'automorphisme correspondant de V_A (donc de S_A), le couple $(\sigma, \sigma_{\mathrm{lin}})$ définit un homéomorphisme de $\bar{A} = A \amalg S_A$ sur lui-même.

450

5.2. Appartements compactifiés

5.2.1. Compactification de l'appartement standard. Soit A l'appartement standard de X, et soit o le point spécial de A choisi au n° 4.2. Du fait que A est muni d'une structure naturelle d'espace affine, sa compactification $\bar{A} = A \amalg S_A$ est définie (5.1). Nous allons voir que l'on peut *identifier S_A à l'appartement standard $W \cdot C$ de l'immeuble Y.*

Reprenons les notations de 4.1, 4.2. Notons C'_{j,s'_j} la face de codimension 1 de C'_j opposée à o_j, i.e. le "link" de o_j dans C'_j ($1 \leq j \leq m$). Soit L_0 l'enveloppe convexe des C'_{j,s_j}. C'est un simplexe de dimension $l-1$, dont les sommets sont les sommets des C'_{j,s_j}. Il est contenu dans $D - \{o\}$, cf. 4.4, et toute demi-droite de D issue de o le rencontre en exactement un point. Soit λ_j l'unique application simpliciale de C_j sur C'_{j,s_j} qui, pour tout $I \subset S_j$, applique $C_{j,I}$ sur $C'_{j,I \cup \{s_j\}}$. C'est un isomorphisme. Il existe un unique isomorphisme $\lambda_C : C \to L_0$ qui prolonge les λ_j. Avec les notations de 1.2 et 4.10, on a $\lambda_C(C_I) = D_I \cap L_0$ pour tout $I \subset S$, $I \neq S$. Soit $A_\infty = W \cdot C$ l'appartement standard de Y, et soit $W \cdot L_0$ le sous-complexe de A réunion des $w \cdot L_0$, $w \in W$. On vérifie immédiatement que λ_C se prolonge en un unique isomorphisme simplicial $\lambda : A_\infty \to W \cdot L_0$ qui commute à l'action de W. D'autre part, toute demi-droite d de A issue de o rencontre $W \cdot L_0$ en un point $\mu(d)$ et un seul; on en déduit un homéomorphisme μ de S_A sur $W \cdot L_0$. D'où, en composant μ et λ^{-1}, un homéomorphisme ρ_A de S_A sur A_∞, par lequel nous *identifierons* ces deux espaces. On obtient ainsi une *topologie* sur $\bar{A} = A \amalg A_\infty$, et les éléments de A_∞ s'interprètent comme des *directions de demi-droites* dans A.

L'adhérence $\bar{D}$ de D dans $\bar{A}$ est $D \amalg C$, et $C = \bar{D} \cap A_\infty$ est l'espace des directions de demi-droites contenues dans D.

5.2.2. Invariance. Soit σ un automorphisme de $\tilde{G}$; alors σ opère de façon naturelle sur X (4.8) et sur Y. Supposons de plus que σ stabilise A (ou A_∞, ou H, ou $Z(T)$, cela revient au même); alors σ opère sur A, sur S_A et sur A_∞, et l'identification $\rho_A : S_A \to A_\infty = W \cdot C$ *commute à σ*; en effet, cela revient à dire que la partie linéaire σ_{lin} de $\sigma|A$ commute à l'homéomorphisme λ de $W \cdot C$ sur $W \cdot L_0$, ce qui est immédiat.

5.2.3. Compactification des appartements. Soit E un appartement de X, et soit E_∞ l'appartement correspondant de Y (cf. [11: 5.1.33]), i.e. l'ensemble des points de Y fixés par le fixateur de E dans $G(k)$. Soit S_E la sphère des directions de demi-droites de l'espace affine E. Il résulte de 5.2.2 qu'il existe une bijection $\rho_E : S_E \to E_\infty$ et une seule telle que, pour tout $\sigma \in (\text{Aut } \tilde{G})(k)$ tel que $\sigma A = E$, le diagramme

$$
\begin{array}{ccc}
S_A & \xrightarrow{\ \sigma\ } & S_E \\
{\scriptstyle \rho_A}\downarrow & & \downarrow{\scriptstyle \rho_E} \\
A_\infty & \xrightarrow{\ \sigma\ } & E_\infty
\end{array}
$$

soit commutatif. Dans ce qui suit, on identifiera S_E et E_∞ au moyen de ρ_E; vu 5.1, on en déduit une topologie (dite *naturelle*) sur $\bar{E} = E \amalg E_\infty$, qui fait de $\bar{E}$ un espace compact, contenu dans $Z = X \amalg Y$, et appelé le *compactifié* de E. Les topologies ainsi construites sont *compatibles* au sens suivant:

5.2.4. LEMME. *Soit M une partie de X, contenue dans deux appartements E_1 et E_2, et soit $\bar{M}_1$ (resp. $\bar{M}_2$) l'adhérence de M dans l'appartement compactifié $\bar{E}_1$ (resp. $\bar{E}_2$). On a alors $\bar{M}_1 = \bar{M}_2$, et les topologies induites sur cet ensemble par les topologies naturelles de $\bar{E}_1$ et $\bar{E}_2$ coïncident.*

On peut supposer que E_1 est l'appartement standard A. Choisissons un élément $g \in \tilde{G}(k)$ tel que $gA = E_2$ et que g fixe M, cf. 4.3. Vu 5.2.3, l'élément g définit un homéomorphisme de $\bar{A} = \bar{E}_1$ sur $\bar{E}_2$, et cet homéomorphisme applique $\bar{M}_1$ sur $\bar{M}_2$. Il reste donc à montrer que cet homéomorphisme est l'identité sur $\bar{M}_1$, autrement dit que g fixe $M_\infty = \bar{M}_1 - M$. D'après 4.6, le fixateur de M dans $\tilde{G}(k)$ est engendré par H et par les sous-groupes U_α associés aux racines affines α contenant M. On peut donc supposer que g appartient, soit à H, soit à l'un des U_α en question. Si $g \in H$, g fixe $\bar{A}$ donc aussi M_∞. Si $g \in U_\alpha$, et si $a \in \Phi$ est tel que $\nu(a) = {}^v\alpha$ (4.6), on a $g \in U_{(a)}$, et g fixe le demi-appartement de A_∞ défini par la racine a (i.e. l'ensemble des faces de Y associées aux k-sous-groupes paraboliques de $\tilde{G}$ contenant $\mathscr{Z}(\tilde{T}) \cdot U_{(a)}$); il résulte de 5.2 que ce demi-appartement est l'ensemble $\alpha_\infty = \bar{\alpha} - \alpha$ des directions de demi-droites contenues dans α; comme M est contenu dans α, on a $M_\infty \subset \alpha_\infty$, ce qui montre bien que g fixe M_∞, et achève la démonstration.

5.3. Soit M une partie de X qui soit contenue dans au moins un appartement. Le lemme précédent permet de définir la *compactification* $\bar{M}$ de M: on choisit un appartement E contenant M, et l'on prend pour $\bar{M}$ l'adhérence de M dans $\bar{E}$; on a $\bar{M} \subset Z$; vu 5.2.4, $\bar{M}$ ne dépend pas (topologie comprise) du choix de E. Ainsi, on peut parler du compactifié d'un *quartier fermé*, d'un *demi-appartement* (4.5), etc. Si σ est un automorphisme de $\tilde{G}$, l'action de σ sur Z définit un homéomorphisme de $\bar{M}$ sur $\overline{\sigma(M)}$: cela résulte de 5.2.3.

Nous allons maintenant "recoller" ces différentes topologies:

5.4. Théorème. (a) *Il existe une topologie $\mathcal{T}$ et une seule sur $Z = X \amalg Y$ ayant les propriétés suivantes*:

(a₁) *$\mathcal{T}$ est séparée*;

(a₂) *$\mathcal{T}$ induit sur chaque appartement compactifié (5.2.3) sa topologie naturelle*;

(a₃) *$\mathcal{T}$ est invariante par $G(k)$, et l'action de $G(k)$ sur l'espace topologique $Z_t = (Z, \mathcal{T})$ est continue.*

De plus:

(b) *X (resp. Y) est ouvert (resp. fermé) dans Z_t, et la topologie induite par $\mathcal{T}$ sur X (resp. Y) est la topologie naturelle de X (resp. celle de Y_t).*

(c) *L'espace Z_t est compact et contractile.*

(d) *La topologie $\mathcal{T}$ ne dépend que de $\tilde{G}$.*

(e) *$\mathcal{T}$ est invariante par $(\mathrm{Aut}\ \tilde{G})(k)$, et l'action de $(\mathrm{Aut}\ \tilde{G})(k)$ sur Z_t est continue.*

La démonstration occupe les n⁰ˢ 5.4.1 à 5.4.9 ci-après.

5.4.1. Soit K le fixateur de o dans $G(k)$, cf. 4.9, et soit $\bar{D} = D \amalg C$ la compactification du quartier fermé D (5.2.1). D'après 4.9, l'application $\mu: K \times \bar{D} \to Z$ définie par l'action de K sur Z est *surjective*. Si $\mathcal{T}$ satisfait à (a₂) et (a₃), μ est *continue*; si de plus (a₁) est satisfaite, le fait que $K \times \bar{D}$ soit compact entraîne que μ est propre [8: I.76, cor. 2]; d'après [8: I.74, cor. 4], on en déduit que μ identifie l'espace $Z_t = (Z, \mathcal{T})$ au *quotient* de $K \times \bar{D}$ par la relation d'équivalence R_μ définie par μ. L'assertion d'*unicité* de 5.4 en résulte. Il reste donc à montrer que, si l'on définit $Z_t = (Z, \mathcal{T})$ comme quotient de $K \times \bar{D}$ par R_μ, l'espace ainsi obtenu possède les propriétés (a₁), (a₂), (a₃), (b), (c), (d), (e).

5.4.2. Vérification de (b). Comme $K \times D$ (resp. $K \times C$) est ouvert (resp. fermé) dans $K \times \bar{D}$, et saturé pour R_μ, son image X (resp. Y) dans Z_t est ouverte (resp. fermée) et a pour topologie la topologie quotient de celle de $K \times D$ (resp. de $K \times C$), cf. [8: I.23, cor. 1], i.e. la topologie naturelle de X (resp. celle de Y_t), cf. 4.9, Remarque.

5.4.3. Vérification de (a₁). Il s'agit de montrer que Z_t est séparé. Comme $K \times \bar{D}$ est compact, il suffit pour cela de vérifier que le graphe de R_μ dans $(K \times \bar{D}) \times (K \times \bar{D})$ est fermé [8: I.78, prop. 8]. Cela revient à prouver l'assertion suivante:

(*) *Soient (k_n), (k_n') (resp. (d_n), (d_n')) ($n = 1, 2, \ldots$) deux suites d'éléments de K (resp. D) tendant vers des limites g, g' (resp. d, d'), et telles que $k_n \cdot d_n = k_n' \cdot d_n'$ pour tout n. Alors $g \cdot d = g' \cdot d'$.*

Pour n donné on a évidemment $d_n \in D$ (resp. $d_n \in C$) si et seulement si $d_n' \in D$ (resp. $d_n' \in C$). Si $d_n \in C$ pour une infinité de valeurs de n, notre assertion résulte du fait que K opère continûment sur Y_t. Cela nous ramène au cas où d_n, $d_n' \in D$ pour tout n. Comme D est un domaine fondamental pour K (4.9), on a $d_n = d_n'$ pour tout n, d'où $d = d'$. Posons

$$m_n = k_n^{-1} \cdot k_n' \qquad \text{et} \qquad m = g^{-1} \cdot g' = \lim_{n \to \infty} m_n.$$

On a $m_n \cdot d_n = d_n$, et l'on doit montrer que $m \cdot d = d$. Si $d \in D$, cela résulte de la continuité de l'action de K sur X. Supposons donc $d \in C$, i.e. que la distance euclidienne (dans A) $d(o, d_n)$ de o à d_n tende vers l'infini. Soit E la demi-droite de D d'origine o et de direction d. Pour $t > 0$, soit e_t le point de E à distance t de o. Il existe $n(t)$ tel que $d(o, d_n) \geq t$ pour tout $n \geq n(t)$; on a $e_t = \lim e_{n,t}$, où $e_{n,t}$ désigne le point du segment $[o, d_n]$ à distance t de o. Comme m_n fixe o et d_n, il fixe aussi l'unique segment géodésique $[o, d_n]$ joignant o à d_n, et en particulier il fixe $e_{n,t}$ ($n \geq n(t)$). Du fait que K opère continûment sur X, on a

$$m \cdot e_t = (\lim m_n) \cdot (\lim e_{n,t}) = \lim (m_n \cdot e_{n,t}) = \lim e_{n,t} = e_t.$$

Ainsi, m fixe la demi-droite E. Vu 5.3, m stabilise la compactification $\bar{E}$ de E, donc aussi $\{d\} = \bar{E} \cap Y$, ce qui montre bien que m fixe d (ce qui pourrait aussi se déduire de 4.10).

5.4.4. Continuité de l'action de K sur Z_t.

D'après 5.4.3, Z_t est séparé. Comme $K \times \bar{D}$ est compact, il en résulte que Z_t est *compact*, et que l'application $\mu: K \times \bar{D} \to Z_t$ est *propre*, cf. [8: I.78, prop. 8]. Par conséquent, si F est un espace topologique, l'application $\mathrm{Id}. \times \mu: F \times X \times \bar{D} \to F \times Z_t$ est aussi propre, et $F \times Z_t$ s'identifie au quotient de $F \times K \times \bar{D}$ par la relation d'équivalence définie par $\mathrm{Id}. \times \mu$ [8: I.74, cor. 4]. En particulier, si $\nu: K \times K \to K$ est l'application produit, l'unique application σ qui rend commutatif le diagramme

$$
\begin{array}{ccc}
K \times K \times \bar{D} & \xrightarrow{\ \nu \times \mathrm{Id}.\ } & K \times \bar{D} \\
{\scriptstyle \mathrm{Id}. \times \mu} \downarrow & & \downarrow {\scriptstyle \mu} \\
K \times Z_t & \xrightarrow{\ \ \sigma\ \ } & Z_t
\end{array}
$$

est continue, ce qui montre que K *opère continûment sur* Z_t.

5.4.5. Vérification de (c). On a vu que Z_t est compact; il s'agit de montrer qu'il est *contractile*. Remarquons d'abord que $\bar{D}$ est contractile puisqu'homéomorphe à un simplexe. On peut choisir une contraction $r: \bar{D} \times [0, 1] \to \bar{D}$ de $\bar{D}$ sur son sommet o telle que, si l'on pose $x_t = r(x, t)$ ($x \in \bar{D}, t \in [0, 1]$), les propriétés suivantes soient satisfaites:

(i) $x_0 = x$, $x_1 = o$;

(ii) $x_t \in [o, x]$ si $x \in D$;

(iii) si $x \in C$ et $t > 0$, le point x_t appartient à la demi-droite ox d'origine o et de direction x.

(Voici un choix possible de r: pour $x \in D$, on prend pour x_t le point de $[o, x]$ tel que $d(o, x_t) = (1 - t)d(o, x)/(1 + td(o, x))$, et pour $x \in C$, $t > 0$, on prend pour x_t le point de ox tel que $d(o, x_t) = (1 - t)/t$.)

Si K_x désigne le fixateur de x dans K, on a:

(iv) $K_x \subset K_{x_t}$ pour tout $x \in \bar{D}$ et tout $t \in [0, 1]$.

Si $x \in D$, cela résulte du fait que tout élément de $G(k)$ qui fixe o et x fixe aussi le segment $[o, x]$. Si $x \in C$, soit $I \subset S$ tel que $x \in \mathring{C}_I$; pour $t \neq 0$, on a $x_t \in D_I$ (4.10); comme $K_x = K \cap P_I$, (iv) résulte de la proposition 4.10.

La relation (iv) entraîne que le composé r' des applications continues

$$
K \times \bar{D} \times [0, 1] \xrightarrow{\ \mathrm{Id}. \times r\ } K \times \bar{D} \xrightarrow{\ \mu\ } Z_t
$$

est constant sur les ensembles de la forme $\mu^{-1}(x) \times \{t\}$ ($x \in Z, t \in [0, 1]$); donc r' se factorise en $r' = r_Z \circ (\mu \times \mathrm{Id}.)$, où r_Z est une application de $Z_t \times [0, 1]$ sur Z_t. Mais il résulte de 5.4.4 que $Z_t \times [0, 1]$ est le quotient de $K \times \bar{D} \times [0, 1]$ par la relation d'équivalence définie par $\mu \times \mathrm{Id}.$; par conséquent r_Z est continue, et définit une contraction de Z_t sur o, ce qui montre que Z_t est contractile.

5.4.6. Vérification de (a_2). Il s'agit de montrer que, pour tout appartement E, la topologie induite par $\mathcal{T}$ sur l'appartement compactifié $\bar{E}$ (5.2.3) est la topologie naturelle de $\bar{E}$. Il suffira pour cela de prouver que $\bar{E}$ est réunion finie de parties fermées F_i telles que les injections $F_i \to Z_t$ soient continues; en effet, cela prouvera que $\bar{E} \to Z_t$ est continue [8: I.19, prop. 4], et, comme $\bar{E}$ est compact et Z_t séparé (5.4.3), il en résultera que la topologie de $\bar{E}$ est induite par $\mathcal{T}$ [8: I.63, cor. 3]. Distinguons alors trois cas:

(1) $E = A$

On prend pour parties fermées F_i les transformés $w \cdot \bar{D}$ de $\bar{D}$ par les éléments w de W. Les injections $w \cdot \bar{D} \to Z_t$ sont continues: on les obtient en composant l'injection $\bar{D} \to Z_t$, continue par construction, avec les automorphismes de Z_t définis par les éléments de W (5.4.4).

(2) E *contient* o

Il existe alors $g \in K$ tel que $gE = A$ (4.3); comme g préserve $\mathcal{T}$ (5.4.4), on est ramené au cas précédent.

(3) *Cas général*

Soit $m = m(E)$ le plus petit entier $\geqslant 0$ tel qu'il existe une galerie minimale

$$
\gamma = (C' = C_0, C_1, \ldots, C_m)
$$

reliant la chambre C' de A (4.2) à une chambre C_m de l'appartement E. Si $m = 0$, on a $o \in E$, cas déjà traité. Supposons donc $m \geq 1$, et raisonnons par récurrence sur m. Soit γ comme ci-dessus, et soit F la cloison $C_{m-1} \cap C_m$; notons α_1 et α_2 les deux demi-appartements de E dont le mur commun contient F. D'après [11: 5.1.9, p. 85], il existe un appartement E_i contenant $C_{m-1} \cup \alpha_i$ $(i = 1, 2)$. Comme $C_{m-1} \subset E_i$, on a $m(E_i) \leq m - 1$, et l'hypothèse de récurrence entraîne que les injections $\bar{E}_i \to Z_t$ sont continues; *a fortiori*, il en est de même des injections $\bar{\alpha}_i \to Z_t$, d'où le résultat cherché, puisque $\bar{E}$ est réunion de $\bar{\alpha}_1$ et $\bar{\alpha}_2$.

5.4.7. Vérification de (a_3). Montrons d'abord que, si $g \in G(k)$, la bijection $z \mapsto g \cdot z$ de Z_t sur lui-même est *continue*. Du fait que K est ouvert dans $G(k)$, il existe un voisinage K_1 de l'élément neutre dans K tel que $g \cdot K_1 \cdot g^{-1} \subset K$; comme K est totalement discontinu, on peut en outre supposer que K_1 est un sous-groupe ouvert de K [8: III.36, cor. 1]. Choisissons une famille finie $(m_j)_{j \in J}$ d'éléments de K telle que K soit réunion des $K_1 m_j$, $j \in J$, et soit $\bar{D}_1$ le sous-espace de Z_t réunion des $m_j \cdot \bar{D}$; c'est une partie compacte de Z_t. L'application $\mu_1 \colon K_1 \times \bar{D}_1 \to Z_t$ définie par l'action de K_1 sur Z_t est continue d'après 5.4.4, et surjective. Comme $K_1 \times \bar{D}_1$ est compact, et Z_t séparé (5.4.3), cette application est propre [8: I.76, cor. 2], et identifie Z_t au quotient de $K_1 \times \bar{D}_1$ par la relation d'équivalence définie par μ_1, cf. [8: I.74, cor., 4]. Tout revient donc à prouver la continuité de l'application $(k_1, z) \mapsto g \cdot k_1 \cdot z$ de $K_1 \times \bar{D}_1$ dans Z_t. Or on a $g \cdot k_1 \cdot z = g k_1 g^{-1} \cdot g \cdot z$; vu (5.4.4) il suffit de démontrer la continuité des applications $k_1 \mapsto g k_1 g^{-1}$ de K_1 dans K et $z \mapsto g \cdot z$ de $\bar{D}_1$ dans Z_t. La première est claire. Pour la seconde, on remarque que $\bar{D}_1$ est réunion finie des parties fermées $m_j \cdot \bar{D}$, et qu'il suffit donc de prouver que la restriction de $z \mapsto g \cdot z$ à une telle partie est continue; comme D est contenu dans un appartement, cela résulte de 5.4.6 combiné avec la dernière assertion de 5.3.

Ainsi, $\mathcal{T}$ est invariante par $G(k)$. Comme K est ouvert dans $G(k)$ et opère continûment sur Z_t (5.4.4), il en résulte que $G(k)$ opère continûment sur Z_t.

5.4.8. Vérification de (d). Soit $\tilde{\mathcal{T}}$ l'unique topologie sur Z satisfaisant aux conditions (a_1), (a_2), (a_3) relativement à l'action de $\tilde{G}(k)$. Du fait que $\tilde{G}(k) \to G(k)$ est continue, la topologie $\mathcal{T}$ satisfait aux conditions en question. On a donc $\mathcal{T} = \tilde{\mathcal{T}}$, ce qui prouve que $\mathcal{T}$ ne dépend que de $\tilde{G}$ (et pas du choix du groupe G "intermédiaire" entre $\tilde{G}$ et le groupe adjoint $\mathrm{Ad}\,\tilde{G}$).

5.4.9. Vérification de (e). Si $\sigma \in (\mathrm{Aut}\,\tilde{G})(k)$, il résulte de (5.3) que $\sigma(\mathcal{T})$ satisfait aux conditions (a_1), (a_2), (a_3) relativement à l'action de $\tilde{G}(k)$. D'après (5.4.8), on a donc $\sigma(\mathcal{T}) = \mathcal{T}$, ce qui montre que $\mathcal{T}$ est invariante par $(\mathrm{Aut}\,\tilde{G})(k)$. D'autre part, le groupe $(\mathrm{Ad}\,\tilde{G})(k)$ est ouvert dans $(\mathrm{Aut}\,\tilde{G})(k)$ et opère continûment sur Z_t d'après (5.4.7) appliqué au groupe $\mathrm{Ad}\,\tilde{G}$ (ce qui ne change pas $\mathcal{T}$, grâce à (5.4.8)). On en déduit que $(\mathrm{Aut}\,\tilde{G})(k)$ opère continûment sur Z_t, ce qui achève la démonstration de (5.4).

5.5. Remarques. (1) Si $l = 1$, X est un arbre, et la compactification Z_t de X définie ci-dessus n'est autre que celle donnée par la *théorie des bouts* de H. Freudenthal.

(2) La compactification Z_t est isomorphe à celle définie dans un cadre plus général par J. Tits (cours au Collège de France, 1974). Nous en résumons brièvement la construction, dans le cas envisagé ici:

Soit $\bar{A}$ la compactification naturelle de A (5.2.1). Choisissons une chambre C_0 de A. Si F appartient à l'ensemble $\mathscr{C}$ des chambres de X, notons φ_F^0 l'isomorphisme canonique de F sur C_0; les φ_F^0 commutent aux opérations de $\tilde{G}(k)$. Pour tout $F \in \mathscr{C}$, l'isomorphisme φ_F^0 se prolonge de façon unique en un morphisme polysimplicial $\varphi_F \colon X \to A$ qui, sur chaque appartement A' contenant F, est l'unique isomorphisme $A' \to A$ prolongeant φ_F^0. Les φ_F, $F \in \mathscr{C}$, définissent une application

$$\varphi \colon X \to V = \prod_{F \in \mathscr{C}} \bar{A},$$

où V est le produit topologique de copies de $\bar{A}$ indexées par $\mathscr{C}$. L'application φ est injective; l'espace V est compact; la compactification de X en question est, par définition, *l'adhérence de* $\varphi(X)$ dans V.

5.6. THÉORÈME. *Soient R un anneau et M un R-module. Soient X l'immeuble de Bruhat–Tits de $\tilde{G}(k)$ (4.2) et Y_t l'immeuble topologisé des k-sous-groupes paraboliques de G (1.3). Alors $H_c^i(X; M) = 0$ pour $i \neq l$ et $H_c^l(X; M)$ est canoniquement isomorphe, comme $G(k)$-R-module, à $H^{l-1}(Y_t; M)$ si $l \geq 1$ et à M si $l = 0$. En particulier, $H_c^l(X; M)$ est un R-module libre si M l'est.*

Comme Z_t est contractile (5.4), la suite exacte de cohomologie de Z_t modulo Y_t définit, pour tout i, un isomorphisme $\tilde{H}^i(Y_t; M) \to H_c^{i+1}(X; M)$, qui commute évidemment à tout homéomorphisme de la paire (Z_t, Y_t). Le théorème résulte alors de 2.6 si $l \geq 1$; le cas $l = 0$ est trivial, X étant réduit à un point.

5.7. Extension aux groupes dont la composante neutre est réductive. Soit L un k-groupe dont la composante neutre L^0 est réductive. Nous allons associer à L un immeuble X, produit de l'immeuble X_0 du groupe dérivé $\mathcal{D}L^0$ de L^0 par un espace euclidien.

Soit $M = L/\mathcal{D}L^0$. La composante neutre M^0 de M est un tore. Par conséquent $M^0(k)$ possède un unique plus grand sous-groupe compact, soit Q. Posons $E = M(k)/Q$. On a une suite exacte

$$1 \to E' \to E \to E'' \to 1, \tag{1}$$

où $E' = M^0(k)/Q$ est commutatif libre, de rang égal au k-rang de M^0, i.e. à la dimension du plus grand k-sous-tore déployé de M^0, et où $E'' = M(k)/M^0(k)$ est fini. L'action de E'' sur E' se prolonge canoniquement à $E'_\mathbf{R} = \mathbf{R} \otimes_\mathbf{Z} E'$, donc (1) se plonge dans une suite exacte de groupes

$$1 \to E'_\mathbf{R} \to E_\mathbf{R} \to E'' \to 1, \tag{2}$$

qui est en fait une suite exacte de groupes de Lie réels, et est scindée puisque $H^2(E''; E'_\mathbf{R}) = 0$. Choisissons un *scindage* de (2), i.e. un sous-groupe fini Ψ de $E_\mathbf{R}$ tel que $E_\mathbf{R}$ soit produit semi-direct de Ψ par $E'_\mathbf{R}$. Faisons opérer Ψ (resp. $E'_\mathbf{R}$) sur l'espace euclidien $X_1 = E'_\mathbf{R}$ par l'action donnée par l'isomorphisme naturel $\Psi \simeq E''$ (resp. par translations); ces opérations définissent une opération de $E_\mathbf{R} = \Psi \cdot E'_\mathbf{R}$, et les homomorphismes naturels $L(k) \to M(k) \to E \to E_\mathbf{R}$ permettent de faire opérer $L(k)$ sur X_1 (l'image de $L(k)$ dans Aut (X_1) étant un "groupe de Bieberbach").

D'autre part, $L(k)$ opère naturellement sur l'immeuble X_0 de $\mathcal{D}L^0$.

L'immeuble X de L sur k est alors, par définition, le produit $X = X_0 \times X_1$, muni de l'action produit de $L(k)$. Soit Z la composante neutre du centre de L^0. Alors $L^0 = Z \cdot \mathcal{D}L^0$ et la restriction à Z de $L^0 \to L^0/\mathcal{D}L^0$ est une isogénie centrale. Il résulte alors de [7: §2] que $rg_k L^0 = rg_k Z + rg_k \mathcal{D}L^0$, et $rg_k Z = rg_k M^0$, par suite dim $X = rg_k L^0$.

5.8. Remarque. L'espace X_1 admet une triangulation différentiable stable par $M(k)$, donc par $L(k)$. En effet, E opère différentiablement sur X_1, et E' est un groupe de translations, donc X_1/E' est une variété différentielle (un produit de cercles) sur laquelle E/E' opère différentiablement. Comme E/E' est fini, il résulte de [19] que $X/E = (X/E')/(E/E')$ est triangulable, d'où notre assertion.

Le produit d'une telle structure et de la structure polysimpliciale canonique de X_1 définit donc une structure polysimpliciale sur X qui est invariante par $L(k)$. Il est clair que les stabilisateurs des faces sont des sous-groupes compacts ouverts et que $X/L(k)$ admet une structure de complexe cellulaire fini.

Si L est connexe, toute base du groupe des k-caractères de $L/\mathcal{D}L$ définit de manière évidente une structure polysimpliciale sur X_1 invariante par $L(k)$, d'où, par produit, une structure polysimpliciale sur X invariante par $L(k)$; c'est celle définie dans [17: §2].

5.9. THÉORÈME. *Soient L un k-groupe dont la composante neutre L^0 est réductive, et X l'immeuble de L sur k défini en 5.7. Alors $H_c^i(X; \mathbf{Z})$ est nul si $i \neq rg_k L^0$ et libre si $i = rg_k L^0$.*

C'est clair si L^0 est un tore puisque X est alors un espace euclidien de dimension $rg_k L^0$, et cela résulte de 5.6 si L^0 est semi-simple. Le cas général s'en déduit par la formule de Künneth $H_c^l(X; \mathbf{Z}) \simeq H_c^u(X_1; \mathbf{Z}) \otimes H_c^v(X_0; \mathbf{Z})$, où $u = rg_k Z$, $v = rg_k \mathcal{D}L^0$, $l = u + v = rg_k L^0$. On a en outre $H_c^u(X_1; \mathbf{Z}) \simeq \Lambda^u X(L^0)$, où $X(L^0) = \mathrm{Hom}_k(L^0, \mathbf{G}_m)$. Ces isomorphismes sont compatibles avec l'action de $L(k)$.

5.10. La représentation spéciale. Nous terminons ce paragraphe en décrivant, sans démonstration, une interprétation du $G(k)$-module $E = H_c^l(X; \mathbf{C}) \simeq \tilde{H}^{l-1}(Y_t; \mathbf{C})$.

Avec les hypothèses et notations des §§2, 3, supposons que M soit un corps de caractéristique zéro. Notons $\sigma_I(I \subset S)$ la représentation de $G(k)$ dans $C^\infty(G(k)/P_I(k); M)$ définie par translations à gauche, et St la représentation de $G(k)$ dans $\tilde{H}^{l-1}(Y_t; M)$. On voit facilement que σ_I est *admissible*, au sens de Jacquet–Langlands. Vu 2.6 et 3.3, il en est de même de St, qui est quotient de σ_θ, et l'on a l'égalité

$$[St] = \sum_{I \subset S} (-1)^{\operatorname{Card} I} [\sigma_I] \tag{1}$$

dans le groupe de Grothendieck des représentations admissibles de $G(k)$. Ainsi, St est l'analogue de la *représentation de Steinberg* des groupes de Chevalley finis.

Supposons maintenant que $M = \mathbf{C}$. Il y a un homomorphisme canonique ν de $H_c^l(X; \mathbf{C})$ dans l'espace de Hilbert $\mathscr{H}$ des formes harmoniques sur X (au sens simplicial) de dimension l qui sont de carré intégrable (pour la norme $\|f\|^2 = \Sigma |f(C)|^2$, où C parcourt l'ensemble des chambres de X). L'espace $\mathscr{H}$ est un $G(k)$-module unitaire, qui fournit une réalisation de la *représentation spéciale* de $G(k)$ définie par H. Matsumoto[16] et J. Shalika[18]. On peut montrer que ν est injectif et applique E sur l'espace des vecteurs "différentiables" de $\mathscr{H}$[3]. En particulier, la représentation St de $G(k)$ est irréductible (algébriquement) et préunitaire. Une autre démonstration de ces résultats a été obtenue par W. Casselman[12; 13].

§6. Applications à la cohomologie de certains groupes discrets

6.1. Hypothèses et notations. Soit $L = \Pi_{v \in S} L_v$ un produit fini de groupes localement compacts L_v. On suppose que chacun des L_v est de l'un des deux types suivants [17: 2.3]:

(i) un groupe de Lie réel ayant un nombre fini de composantes connexes;

(ii) le groupe $M_v(F_v)$ des F_v-points d'un groupe algébrique M_v, de composante neutre réductive, sur un corps local non archimédien F_v.

Soit S_∞ (resp. S_f) l'ensemble des $v \in S$ pour lesquels L_v est de type (i) (resp. (ii)). On note $d(L_v)$ la dimension du quotient de L_v par un sous-groupe compact maximal si $v \in S_\infty$, et le rang sur F_v de M_v si $v \in S_f$; on pose

$$d(L) = \sum_{v \in S} d(L_v).$$

On considère d'autre part un sous-groupe discret Γ de L. On se place dans l'un des deux cas suivants:

(a) "cas cocompact": le quotient L/Γ est compact;

(b) "cas S-arithmétique": l'ensemble S est un ensemble fini de places d'un corps de nombres F et S_∞ celui des places archimédiennes de F; pour tout $v \in S$, le corps F_v est le complété de F en v et il existe un F-groupe G, de composante neutre réductive, tel que $L_v = G(F_v)$; en particulier, si $v \in S_f$, le groupe M_v de (ii) est le groupe obtenu par extension du corps de base à partir de G. Enfin, Γ est un sous-groupe S-arithmétique de $G(F)$, plongé dans L de façon "diagonale" [17: 2.4].

6.2. Théorème. *On conserve les hypothèses et notations de 6.1.*

(i) *Le groupe Γ est de présentation finie. Ses sous-groupes d'ordre fini forment un nombre fini de classes de conjugaison.*

Supposons Γ sans torsion. Alors:

(ii) *Γ est de type (FL) [17: §1].*

(iii) *Γ est un groupe à dualité, au sens de [2].*

(iv) *La dimension cohomologique $cd(\Gamma)$ de Γ est égale à $d(L)$ dans le cas cocompact et à $d(L) - rg_F G$ dans le cas S-arithmétique.*

Compte tenu de (i), (ii), l'assertion (iii) équivaut à l'existence d'un entier d tel que

$$H^i(\Gamma; \mathbf{Z}[\Gamma]) = 0 \quad (i \neq d), \qquad H^d(\Gamma; \mathbf{Z}[\Gamma]) = I \text{ est libre, cf. [2].} \tag{1}$$

Elle entraîne que, si e est la "classe fondamentale" de $H_d(\Gamma; I)$, le cap-produit par e définit, pour tout Γ-module M et tout entier q, un isomorphisme

$$H^q(\Gamma; M) \overset{\sim}{\to} H_{d-q}(\Gamma; I \otimes M); \tag{2}$$

en particulier, on a $d = cd(\Gamma)$. Une fois (iii) établie, l'assertion (iv) revient donc à dire que l'entier d intervenant dans (1) est égal à $d(L)$ ou à $d(L) - rg_F G$ suivant que l'on est dans le cas cocompact ou dans le cas S-arithmétique.

La démonstration de 6.2 sera donnée en 6.6 et 6.11.

6.3. *Remarques.* (1) On peut se demander dans quel cas Γ est un groupe à dualité "de Poincaré," i.e. I est isomorphe à **Z**. La description de I donnée plus loin (6.6(2), 6.11(2)) montre que cela se produit: dans le cas cocompact si et seulement si le F_v-rang de $\mathscr{D}M_v^0$ est nul pour tout $v \in S_f$, dans le cas S-arithmétique si et seulement si $rg_F(\mathscr{D}G^0) = 0$ et $rg_{F_v}(\mathscr{D}G^0) = 0$ pour tout $v \in S_f$; sinon I est de rang infini. De plus, l'action naturelle de Γ sur I se prolonge en une action de $G(F)$ dans le cas S-arithmétique et de L dans le cas cocompact.

(2) L'hypothèse "Γ sans torsion" est peu gênante. En effet, si elle n'est pas satisfaite, mais si Γ est *séparé* pour la topologie des sous-groupes d'indice fini, alors, vu 6.2(i), Γ contient un sous-groupe d'indice fini sans torsion, auquel on peut appliquer le théorème. On a alors

$$vcd(\Gamma) = d(L) \qquad (\text{resp. } vcd(\Gamma) = d(L) - rg_F G), \tag{1}$$

dans le cas cocompact (resp. S-arithmétique), et aussi [5: 11.4.4]

$$H^i(\Gamma; \mathbf{Z}(\Gamma)) = 0 \quad (i \neq d), \qquad H^d(\Gamma; \mathbf{Z}(\Gamma)) = I, \tag{2}$$

avec $d = d(L)$ dans le cas cocompact, $d = d(L) - rg_F G$ dans le cas S-arithmétique, et I comme dans 6.2. L'hypothèse de séparation est notamment satisfaite lorsque Γ est plongeable dans un groupe linéaire, ce qui est en particulier vrai dans le cas S-arithmétique. Les relations (1), (2), avec I donné comme précédemment par 6.6(2), 6.11(2), sont donc valables si Γ est S-arithmétique. Par exemple, si $p_1, \ldots, p_n$ sont des nombres premiers distincts, on a

$$vcd(\mathbf{SL}_3(\mathbf{Z}[1/p_1, \ldots, 1/p_n])) = 3 + 2n,$$

$$vcd(\mathbf{Sp}_4(\mathbf{Z}[1/p_1, \ldots, 1/p_n])) = 4 + 2n, \text{ etc.} \tag{3}$$

(3) Le théorème 6.2 fournit une condition nécessaire pour qu'un groupe discret Δ soit isomorphe à un sous-groupe S-arithmétique d'un groupe à composante neutre réductive: il faut que tous les groupes $H^i(\Delta; \mathbf{Z}[\Delta])$ soient nuls, à l'exception d'un seul, et que ce dernier soit libre.

(4) Le cas S-arithmétique est cocompact si et seulement si le F-rang de G est nul. Il existe bien entendu des cas cocompacts, avec des corps F_v de caractéristique zéro, qui ne sont pas S-arithmétiques. Cependant, les résultats annoncés récemment par A. G. Margulis[15] montrent qu'en fait on est à peu de chose près dans le cas S-arithmétique si les groupes L_v sont semi-simples, sans facteurs anisotropes, si la somme de leurs rangs relatifs est $\geqslant 2$, et si Γ est irréductible.

6.4. LEMME. *Soit X un espace localement compact sur lequel un groupe discret Γ opère proprement et librement. On suppose que X/Γ est compact et que $H^0(X; \mathbf{Z}) = \mathbf{Z}$ et $H^i(X; \mathbf{Z}) = 0$ pour $i \geqq 1$. Alors les groupes de cohomologie $H^q(\Gamma; \mathbf{Z}[\Gamma])$ s'identifient aux groupes de cohomologie à supports compacts $H_c^q(X; \mathbf{Z})$ ($q \geqq 0$).*

Démonstration.[3] La projection canonique $\pi: X \to X/\Gamma$ est un revêtement, donc la suite spectrale de Leray de π dégénère en un isomorphisme

$$H_c^q(X; \mathbf{Z}) \cong H^q(X/\Gamma; M), \qquad (q \geqq 0),$$

où M est le faisceau localement constant sur X/Γ formé par les 0-ièmes groupes de cohomologie à supports compacts des fibres de π. On vérifie tout de suite que M est associé au Γ-module $\mathbf{Z}[\Gamma]$. Vu nos hypothèses, la cohomologie de X/Γ s'identifie à celle de Γ, d'où le lemme.

6.5. LEMME. *Soit X un espace localement compact sur lequel un groupe discret Γ opère proprement. On suppose que tout sous-groupe fini de Γ admet un point fixe dans X et que X/Γ est compact. Alors les sous-groupes d'ordre fini de Γ forment un nombre fini de classes de conjugaison.*

Soit C un sous-ensemble compact de X tel que $X = \Gamma \cdot C$. L'ensemble Γ_C des $\gamma \in \Gamma$ tels que $\gamma \cdot C \cap C \neq \emptyset$ est fini. Soit M un sous-groupe fini de Γ. Par hypothèse il existe un point $x \in X$ fixe par M. Si $\gamma \in \Gamma$ est tel que $\gamma \cdot x \in C$, alors $\gamma \cdot M \cdot \gamma^{-1} \subset \Gamma_C$, d'où le lemme.

6.6. Démonstration de 6.2 dans le cas cocompact. Pour $v \in S$, notons X_v le quotient de L_v par un sous-groupe compact maximal si $v \in S_\infty$, et l'immeuble de M_v sur F_v (5.7) si $v \in S_f$. Soit X (resp. X_∞, resp. X_f) le produit des X_v pour $v \in S$ (resp. $v \in S_\infty$, resp. $v \in S_f$).

---

[3]Cette démonstration a été suggérée à l'un de nous par B. Eckmann.

Il est immédiat que L, donc aussi Γ, opère continûment et proprement sur X, et que X/Γ est compact. Comme X est localement et globalement contractile, cela entraîne que Γ est de présentation finie [1: Satz 2]. D'autre part, tout sous-groupe compact de L_v laisse fixe un point de X_v: si $v \in S_\infty$, cela résulte du fait que tout sous-groupe compact de L_v est contenu dans un sous-groupe compact maximal et que les sous-groupes compacts maximaux de L_v sont conjugués; si $v \in S_f$, vu la définition de X_v (5.7), cela résulte de [11: 3.2.4] et du fait que tout groupe compact de transformations affines d'un espace euclidien admet un point fixe. Il s'ensuit que tout sous-groupe fini de Γ laisse fixe un point de X, et la deuxième assertion de 6.2(i) résulte donc de 6.5.

Supposons maintenant Γ sans torsion. C'est alors le groupe fondamental de X/Γ. Si $S = S_\infty$, alors X/Γ est une variété différentielle compacte, donc Γ est de type (FL). Sinon, en utilisant la remarque 5.8, on montre par récurrence sur Card S_f, exactement comme dans [17: Théor. 3, p. 121] que Γ est de type (FL).

Pour tout $v \in S$, le groupe $H_c^i(X_v; \mathbf{Z})$ est libre si $i = d(L_v)$, nul sinon; c'est clair si $v \in S_\infty$, puisqu'alors X_v est homéomorphe à un espace euclidien de dimension $d(L_v)$, et cela résulte de 5.9 si $v \in S_f$. La formule de Künneth entraîne donc que

$$H_c^i(X; \mathbf{Z}) = 0 \qquad \text{si} \qquad i \neq d(L), \tag{1}$$

et que

$$I = H_c^{d(L)}(X; \mathbf{Z}) \cong \otimes_{v \in S} H_c^{d(L_v)}(X_v; \mathbf{Z}) \tag{2}$$

est un $\mathbf{Z}$-module libre. D'autre part, comme X est globalement et localement contractile, $H^i(X; \mathbf{Z})$ est égal à $\mathbf{Z}$ pour $i = 0$, et à 0 pour $i \neq 0$. Vu 6.4 et (1), (2), Γ satisfait à 6.2(1) pour $d = d(L)$, ce qui entraîne (iii) et (iv) de 6.2 dans le cas cocompact.

6.7. *Remarque.* En utilisant[14], on peut montrer que X/Γ est triangulable si Γ est sans torsion (d'où une autre démonstration de 6.2(i), (ii)). Nous indiquons ici comment on se ramène aux hypothèses de [14]. Soit L_∞ (resp. L_f) le produit des L_v pour $v \in S_\infty$ (resp. $v \in S_f$). On fait opérer L sur X_f via la projection $L \to L_f$. Vu 5.8, X_f admet une structure polysimpliciale $\mathscr{S}$ stable par L_f, donc par L. Quitte à remplacer $\mathscr{S}$ par une subdivision suffisamment fine, on peut faire en sorte que $\mathscr{S}$ soit simpliciale, passe au quotient par L, donc par Γ, et que le fixateur L_σ dans L d'un simplexe quelconque σ de $\mathscr{S}$ soit égal au stabilisateur de σ dans L. De plus, $L_\sigma = L_\infty \times L_{f,\sigma}$, où $L_{f,\sigma} = L_f \cap L_\sigma$ est compact ouvert dans L_f. Les orbites de L_σ dans L/Γ sont ouvertes, donc fermées, donc compactes. En particulier $L_\sigma/(L_\sigma \cap \Gamma)$ est compact. Par suite, la projection de $\Gamma_\sigma = L_\sigma \cap \Gamma$ dans L_∞ est un sous-groupe discret, sans torsion, cocompact, et X_∞/Γ_σ est une variété différentielle compacte. Si τ est une face de σ, alors Γ_σ est d'indice fini dans Γ_τ, d'où une projection naturelle $X_\infty/\Gamma_\sigma \to X_\infty/\Gamma_\tau$ qui est un revêtement fini. Soit $\pi_f: X/\Gamma \to X_f/\Gamma$ la projection canonique. Si σ est un simplexe de X_f, alors X_∞/Γ_σ s'identifie à la fibre de π_f sur un point intérieur de l'image de σ dans X_f/Γ. Par suite, X/Γ est un ensemble "compact stratifié" [14: Théor. 4.1] donc triangulable [14: Théor. 3.2].

Il est vraisemblable que X/Γ est triangulable même si Γ a de la torsion, mais nous n'en connaissons pas de démonstration.

6.8. Lemme. *Soient* Y, Z *des espaces localement compacts et* Γ *un groupe discret opérant continûment sur* Y *et* Z. *On fait opérer* Γ *sur* $Y \times Z$ *par l'action produit. On suppose*:

(i) Z *est un complexe simplicial localement fini. L'action de* Γ *sur* Z *est simpliciale. Si* σ *est une face de* Z, *le stabilisateur* Γ_σ *de* σ *dans* Γ *opère proprement sur* Y.

Alors Γ *opère proprement sur* $Y \times Z$.
Si de plus:

(ii) *les faces de* Z *sont en nombre fini modulo* Γ, *et* Y/Γ_σ *est compact quelle que soit la face* σ *de* Z.

Alors $(Y \times Z)/\Gamma$ *est compact.*

Soit $\mathscr{C}$ l'ensemble des faces de Z. Si $\sigma \in \mathscr{C}$, notons b_σ le barycentre de σ, et $V(\sigma)$ l'étoile ouverte de b_σ dans la subdivision barycentrique de Z. Les $V(\sigma)$ $(\sigma \in \mathscr{C})$ forment un recouvrement ouvert de l'espace Z. On a $\gamma \cdot V(\sigma) = V(\gamma \cdot \sigma)$ pour tout $\gamma \in \Gamma$. De plus:

(1) $V(\sigma) \cap V(\tau) = \emptyset$ si $\sigma, \tau \in \mathscr{C}$, dim $\sigma = $ dim τ, et $\sigma \neq \tau$,

d'où

(2) $\gamma \cdot V(\sigma) \cap V(\sigma) \neq \emptyset \Rightarrow \gamma \in \Gamma_\sigma$, si $\gamma \in \Gamma$ et $\sigma \in \mathscr{C}$.

Soient $y \in Y, z \in Z$, et soit $\sigma \in \mathscr{C}$ tel que $z \in V(\sigma)$. Vu (i), il existe un voisinage U de y dans Y qui ne rencontre qu'un nombre fini de ses transformés par $\Gamma\sigma$. Vu (2), $U \times V(\sigma)$ est un voisinage de (y, z) dans $Y \times Z$ qui ne rencontre qu'un nombre fini de ses transformés par Γ. Il en résulte que Γ opère proprement sur $Y \times Z$.

Supposons maintenant que (ii) soit aussi satisfaite. Pour $\sigma \in \mathscr{C}$, soit K_σ un compact de Y tel que $Y = \Gamma_\sigma \cdot K_\sigma$. Si J est un système de représentants de $\mathscr{C}/\Gamma$, alors

$$Y \times Z = \cup_{\sigma \in J} \Gamma \cdot (K_\sigma \times \sigma).$$

En effet, si $y \in Y, z \in Z$, il existe $\gamma \in \Gamma, \sigma \in J$ et $\gamma' \in \Gamma_\sigma$ tels que $\gamma \cdot z \in \sigma$ et $\gamma' \cdot \gamma \cdot y \in K_\sigma$ d'où $\gamma' \cdot \gamma(y, z) \in (K_\sigma \times \sigma)$. Comme J est fini par hypothèse, et que $(Y \times Z)/\Gamma$ est séparé puisque Γ opère proprement [8: III.29, Prop. 3], cela termine la démonstration du lemme.

6.9. Nous nous plaçons maintenant dans le cas S-arithmétique (6.1). Si $v \in S_f$ on note X_v l'immeuble de G sur F_v. Soit d'autre part G' le $\mathbf{Q}$-groupe algébrique obtenu à partir de G par restriction des scalaires de F à $\mathbf{Q}$ et soit $\bar{X}_\infty$ la variété à coins associée à G' dans [5]. On pose

$$X_f = \Pi_{v \in S_f} X_v, \qquad \bar{X}_S = \bar{X}_\infty \times X_f. \tag{1}$$

Le groupe $G(F) \cong G'(\mathbf{Q})$ opère sur $\bar{X}_S$. Il en est *a fortiori* de même du groupe S-arithmétique Γ.

Montrons que

$$\dim \bar{X}_\infty - rg_{\mathbf{Q}}(\mathscr{D}(G'^0)) = \sum_{v \in S_\infty} d(G(F_v)) - rg_F(G). \tag{2}$$

Soient

$$a = rg_F(C(G^0)), \qquad b = rg_F(\mathscr{D}(G^0)). \tag{3}$$

On a donc $rg_F(G) = a + b$ vu [7: §2]. D'autre part, le rang relatif se conserve par restriction des scalaires, donc

$$a = rg_{\mathbf{Q}}(C(G'^0)), \qquad b = rg_{\mathbf{Q}}(\mathscr{D}(G'^0)) \quad \text{et} \quad a + b = rg_{\mathbf{Q}}(G'). \tag{4}$$

Si K est un sous-groupe compact maximal de $G'(\mathbf{R})$, on a, par construction de $\bar{X}_\infty$,

$$\dim \bar{X}_\infty = \dim (G'(\mathbf{R})/K) - a, \quad \text{cf. [5: §7].} \tag{5}$$

Mais $G'(\mathbf{R})/K$ est le produit direct des quotients $G(F_v)/K_v$, où K_v est un sous-groupe compact maximal de $G(F_v)$, et v parcourt S_∞; d'où (2). On a donc aussi

$$d(L) - rg_F(G) = d(L_\infty) + \Sigma_{v \in S_f} d(L_v), \quad \text{avec} \quad d(L_\infty) = \dim \bar{X}_\infty - rg_{\mathbf{Q}}(\mathscr{D}G'^0). \tag{6}$$

6.10. PROPOSITION. *Conservons les hypothèses et notations de 6.9. Le groupe Γ opère proprement sur $\bar{X}_S$ et le quotient $\bar{X}_S/\Gamma$ est compact.*

Si $S_f = \emptyset$, cela résulte de [5: 9.3]. Dans le cas général on raisonne par récurrence sur Card S_f. Soient $v \in S_f$ et posons $T_f = S_f - \{v\}$, $T = S - \{v\}$. Si M est un sous-groupe ouvert compact de $G(F_v)$, soit Γ_M l'ensemble des éléments de Γ se projetant dans M. C'est un sous-groupe T-arithmétique de G et l'on sait que $\Gamma \backslash G(F_v)/M$ est fini [17: p. 126]. Vu l'hypothèse de récurrence, Γ_M opère proprement sur $\bar{X}_T$ et $\bar{X}_T/\Gamma_M$ est compact. Notre assertion résulte alors de 6.8, appliqué à Γ et à $Y = \bar{X}_\infty \times (\Pi_{w \in T_f} X_w)$, $Z = X_v$.

Remarque. Supposons Γ sans torsion. Il résulte aussi de [14] que $\bar{X}_S/\Gamma$ est triangulable. Cela se voit comme en 6.7, en remplaçant X par $\bar{X}_S$ et X_∞ par $\bar{X}_\infty$. La seule différence est que le quotient $\bar{X}_\infty/\Gamma_\sigma$ est maintenant une variété différentielle compacte *à coins*, mais cela est permis dans [14]. Ici encore, il est vraisemblable que $\bar{X}_S/\Gamma$ est triangulable même lorsque Γ a de la torsion.

6.11. Démonstration de 6.2 dans le cas S-arithmétique. Comme on va le voir, cette démonstration est tout à fait analogue à celle donnée dans le cas cocompact, $\bar{X}_S$ et $\bar{X}_\infty$ jouant les rôles de X et X_∞ respectivement.

Les espaces $\bar{X}_\infty$ et X_v ($v \in S_f$) sont globalement et localement contractiles. Il en est donc de même de $\bar{X}_S$ et il résulte alors de 6.10 et de [1: Satz 2] que Γ est de présentation finie.

L'espace $\bar{X}_\infty$ de 6.9 est obtenu en ajoutant des coins au quotient X_∞ de $G'(\mathbf{R})$ par un sous-groupe H contenant un sous-groupe compact maximal de $G'(\mathbf{R})$, donc tout sous-groupe compact de $G'(\mathbf{R})$ admet un point fixe dans X_∞; en particulier tout sous-groupe fini de $G'(\mathbf{Q})$

admet un point fixe dans X_∞, et *a fortiori* dans $\bar{X}_\infty$. D'autre part (cf. 6.6), tout sous-groupe compact de $G(F_v)$ admet un point fixe dans X_v si $v \in S_f$. Il s'ensuit que tout sous-groupe fini de Γ (ou de $G(F)$), admet un point fixe dans $\bar{X}_S$, et la deuxième partie de 6.2(i) est conséquence de 6.5, 6.10.

Supposons maintenant Γ sans torsion. Si $S_f = \emptyset$, on a $S = S_\infty$ et $\bar{X}_S/\Gamma = \bar{X}_\infty/\Gamma$ est une variété différentielle compacte à coins, donc triangulable, ce qui entraîne que Γ est de type (FL); si $S_f \neq \emptyset$, on peut, soit raisonner par récurrence sur Card S_f, soit utiliser le fait que $\bar{X}_S/\Gamma$ est triangulable (Remarque à 6.10).

La variété $\bar{X}_\infty$ est contractile [5: 8.6.4], le groupe $H_c^i(\bar{X}_\infty; \mathbf{Z})$ est libre si $i = d(L_\infty)$ (notation de 6.9(6)), nul sinon [5: 8.6.5]. Comme précédemment $H_c^i(X_v; \mathbf{Z})$ est nul pour $i \neq d(G(F_v))$ et libre pour $i = d(G(F_v))$ si $v \in S_f$. La formule de Künneth et 6.9(6) montrent donc que

$$H_c^i(\bar{X}_S; \mathbf{Z}) = 0 \quad \text{si} \quad i \neq d(L) - l, \quad \text{où} \quad l = rg_F G, \tag{1}$$

et que

$$I = H_c^{d(L)-l}(\bar{X}_S; \mathbf{Z}) = H_c^{d(L_\infty)}(\bar{X}_\infty; \mathbf{Z}) \otimes \bigotimes_{v \in S_f} H_c^{d(L_v)}(X_v; \mathbf{Z}) \tag{2}$$

est un $\mathbf{Z}$-module libre. Comme $\bar{X}_S$ est localement et globalement contractile, $H^i(\bar{X}_S; \mathbf{Z})$ est nul pour $i \neq 0$, égal à $\mathbf{Z}$ pour $i = 0$; le lemme 6.4 et (1), (2) entraînent donc que Γ satisfait à 6.1(1), avec $d = d(L) - l$ (et I donné par (2)), d'où le théorème.

RÉFÉRENCES

1. H. BEHR: Ueber die endliche Definierbarkeit von Gruppen, *Jour. f. reine u. ang. Math.* **211** (1962), 116–122.
2. R. BIERI and B. ECKMANN: Groups with homological duality generalizing Poincaré duality, *Inv. Math.* **20** (1973), 103–124.
3. A. BOREL: Admissible representations of a reductive p-adic group having vectors fixed under an Iwahori subgroup, (to appear in *Inv. Math.*).
4. A. BOREL et J.-P. SERRE: Cohomologie à supports compacts des immeubles de Bruhat-Tits; application à la cohomologie des groupes S-arithmétiques, *C.R. Acad. Sci., Paris* **272** (1971), 110–113.
5. A. BOREL and J.-P. SERRE: Corners and arithmetic groups, *Comm. Math. Helv.* **48** (1974), 244–297.
6. A. BOREL et J. TITS: Groupes réductifs, *Publ. Math. I.H.E.S.* **27** (1965), 55–150.
7. A. BOREL et J. TITS: Compléments à l'article: "*Groupes réductifs*", *Ibid.* **41** (1972), 253–276.
8. N. BOURBAKI: Topologie générale, Chapitres 1 à 4, Hermann, Paris (1971).
9. N. BOURBAKI: Groupes et algèbres de Lie, Chapitres IV, V, VI, Act. Sci. Ind. 1337, Hermann, Paris (1968).
10. F. BRUHAT et J. TITS: Groupes algébriques simples sur un corps local, *C.R. Acad. Sci., Paris* **263** (1966), 822–825.
11. F. BRUHAT et J. TITS: Groupes réductifs sur un corps local, Chap. I, *Publ. math. I.H.E.S.* **41** (1972), 1–251.
12. W. CASSELMAN: The Steinberg character as a true character, *Harmonic analysis on homogeneous spaces, Proc. Symp. Pur. math. A.M.S.* **XXVI**, (1973), 413–417.
13. W. CASSELMAN: Introduction to the theory of admissible representations of p-adic reductive groups, to appear.
14. F. E. A. JOHNSON: On the triangulation of stratified sets and singular varieties, Trans. A.M.S. **172** (1972), 347–355.
15. A. G. MARGULIS: Discrete groups of motions of manifolds of non-positive curvature, *Proc. Intern. Congr. Math.* Vancouver (1974), Vol. II, 21–34.
16. H. MATSUMOTO: Fonctions sphériques sur un groupe semi-simple p-adique, *C.R. Acad. Sci., Paris* **269** (1969), 829–832.
17. J.-P. SERRE: Cohomologie des groupes discrets, *Prospects in Mathematics, Ann. Math. Stud.* **70**, Princeton U. Press (1971), 77–169.
18. J. SHALIKA: On the space of cusp forms of a p-adic Chevalley group, *Ann. Math.* **92**(2) (1970), 262–278.
19. C. T. YANG: The triangulability of the orbit space of a differentiable transformation group, *Bull. Am. Math. Soc.* **69** (1963), 405–408.

The Institute for Advanced Study, Princeton, N.J. 08540 U.S.A.

Collège de France, 75231 Paris Cedex 05, France

106.

Admissible representations of a semi-simple group over a local field with vectors fixed under an Iwahori subgroup

Invent. Math. **35** (1976) 233–259

To Jean-Pierre Serre

Let G be the group of rational points of a connected semi-simple algebraic group $\mathcal{G}$ over a locally compact nonarchimedian local field k, and let (r, V) be an admissible representation of G in a vector space V over a field R of characteristic zero. If B is a compact open subgroup of G, then the fixed point set V^B of B in V is a finite dimensional space, acted upon in a natural way by the Hecke algebra $H(G, B)$ of compactly supported B-biinvariant R-valued functions on G. We shall be concerned here with the case where B is an Iwahori subgroup of G and prove that, in that case, every finite dimensional $H(G, B)$-module E occurs in this way. In fact, given E, there are two natural algebraic constructions of a smooth G-space of which the fixed point set of B is isomorphic to E as an $H(G, B)$-module: the analogues, in the context of smooth representations, of the induced and produced modules of [10], also to be denoted $I(E)$ and $P(E)$ (see 2.3)[1]. We shall prove more precisely that $I(E)$ and $P(E)$ are admissible (4.4), canonically isomorphic, irreducible if and only if E is, and that $E \mapsto P(E)$ defines an exact functor from finite dimensional $H(G, B)$-modules to admissible G-spaces (4.10). The proof of admissibility uses, besides standard facts on $H(G, B)$, mainly a lemma on buildings (4.1), due to F. Bruhat, which expresses a strong transitivity property of compact open subgroups of G, while that of 4.10 depends on results of [7] and on a lemma of Casselman (4.8).

In §5 we consider the case where E is one-dimensional, i.e., when r is a character of $H(G, B)$, and $R = \mathbf{C}$, and determine in which case $P(E)$ belongs to the discrete series (or rather, is the space of smooth vectors of an element of the discrete series). There is always the special representation of Matsumoto [13] and Shalika [16]. In our setup, it corresponds to the special character σ, which assigns -1 to the standard generators e_s of $H(G, B)$ (see 3.5). However, if G is almost simple over k, of k-rank $l \geqq 2$ and $H(G, B)$ has more than two characters of degree one (see 3.4), then there is at least one other. Their list is derived in 5.8 from a criterion for a character σ to give rise to a square integrable representation (5.2) and results of MacDonald [12]. These representations are not cuspidal.

The special representation may be viewed as the natural representation of G in the space H_2^l of square integrable harmonic l-forms on the Bruhat-Tits building

---

[1] However, we shall use the word *coinduced* instead of *produced*.

X of G, where $l = rk_k \mathcal{G} = \dim X$, (6.1). There is a natural G-morphism η of the l-th cohomology group $H_c^l(X)$ of X with compact supports and complex coefficients into H_2^l. We shall prove that η is an isomorphism of $H_c^l(X)$ onto the space of smooth vectors in H_2^l, (6.2). One of the main results of [3] implies that the G-space $H_c^l(X)$ is admissible and that its character is the "Steinberg character": the alternating sum of induced characters from the trivial representations of the groups of rational points of the parabolic k-subgroups containing a given minimal one (see 6.3(2)). It follows then that this character is also the one of the special representation (6.4).

The above summarizes the contents of §§ 4, 5, 6, which contain the main results of this paper. §1 recalls some notions on admissible representations, Hecke algebras and convolution, mainly to fix some notation. §2 introduces induced and coinduced modules from a finite dimensional module over a Hecke algebra $H(G, B)$, where G is a locally compact totally disconnected group, B a compact open subgroup, and describes some of their properties. This is quite analogous to [10], where this is carried out for algebras, with some minor changes due to the fact that we work with smooth representations.

In § 3, G and B are specialized as above, and we recall some standard properties of $H(G, B)$. Finally § 7 is devoted to the proof of 4.1.

The fact that the Steinberg character is an irreducible unitary character was also announced by Casselman in [6] and proved in [7]. His argument is quite different from ours. In [14], Matsumoto also states, in a somewhat more general framework than ours, that every finite dimensional $H(G, B)$-module occurs as the fixed point set of B in an admissible representation of G.

In the course of this work, I have received precious help from several people, to whom I am glad to express my thanks. The starting point of this paper was a question raised to me by Harish-Chandra in December, 1970, namely whether the Steinberg character (6.3(2)) was an irreducible unitary character. The joint work with J-P. Serre summarized in [2] showed readily that this character was effective, and that it would be the character of the special representation if and only if the canonical map $\eta: H_c^l(X) \to H_2^l$ of 6.2 were injective. That was proved shortly afterwards. At that time, I benefited from a correspondance with J-P. Serre, where he notably emphasized the role of the Hecke algebra $H(G, B)$. This led me naturally to the more general questions studied in this paper. The results of § 5 were proved in Fall, 1971, with the help of J. Tits, the admissibility of the coinduced modules $P(E)$ was realized a bit later and discussed in a seminar at the I.H.E.S. at Bures, in Fall, 1973; finally, the other results of § 4 were obtained in Spring, 1974, with the help of P. Cartier, who suggested to use $I(E)$ and $P(E)$ concurrently, and of W. Casselman, who supplied Lemma 4.8.

I. Generalities

R denotes a field of characteristic zero[2], *G a locally compact totally disconnected unimodular group. Vector spaces are over R, and modules over groups or algebras are vector spaces.*

---

[2]   In fact, without substantial change, the characteristic p of R can be allowed to be >0, provided it is assumed that G has no compact pro-p-subgroup $\neq \{1\}$.

§1. Convolution. Smooth and Admissible Representations

We recall here a few definitions and facts about smooth representations, in a form convenient for the sequel, mainly to fix some conventions and notation. See [7] for a general discussion.

1.1. Let E be a vector space, and X a set. Then $C(X, E)$ is the vector space of maps of X into E. If X is a locally compact totally disconnected space, then $C_c(X, E)$ is the space of elements of $C(X, E)$ with compact support, and $C^\infty(X, E)$ (resp. $C_c^\infty(X, E)$) is the space of smooth (i.e., locally constant) functions in $C(X, E)$, (resp. $C_c(X, E)$). If $E = R$, we write $C(X)$, $C_c(X)$, etc.

If $X = G$, and $f \in C(G, E)$, then $\check{f}$ denotes the function defined by $\check{f}(x) = f(x^{-1})$, $(x \in G)$. Fix a Haar measure on G. Assume E to be an algebra over R. Then $C_c^\infty(G, E)$ is an algebra with respect to the convolution product

$$(u * v)(x) = \int_G u(x \cdot y^{-1}) \cdot v(y)\, dy = \int_G u(y) \cdot v(y^{-1} \cdot x)\, dy, \tag{1}$$

which is in fact a finite sum. We have

$$(u * v)^\vee = \check{v} * \check{u}. \tag{2}$$

More generally, if E', E'' and E are vector spaces and $(e', e'') \mapsto e' \cdot e''$ is a bilinear map from $E' \times E''$ to E, then (1) defines a pairing $C_c^\infty(G, E') \times C_c^\infty(G, E'') \to C_c^\infty(G, E)$. These definitions are valid if R is only a commutative ring, and E', E'', E modules over R.

The right hand side already makes sense if one of the two factors has compact support. In particular, it allows one to give $C^\infty(G, E)$ a bimodule structure over $C_c^\infty(G)$.

Let B be a compact open subgroup of G, and assume the total measure of B to be 1. Then the space $C_c(B \backslash G/B)$ of compactly supported B-biinvariant R-valued functions on G is an algebra under convolution, the "Hecke algebra" of G with respect to B to be denoted $H_R(G, B)$, or $H(G, B)$ or simply H. For $g \in G$, we let e_g be the characteristic function of BgB. We have

$$\check{e}_g = e_{g^{-1}}.$$

The elements e_w, where w runs through a set of representatives of $B \backslash G/B$, form a vector space basis of $H(G, B)$, and the characteristic function e_1 of B is a left and right identity.

If E is a vector space, then $C^\infty(G, E)$ and $C_c^\infty(G, E)$ are H-bimodules under convolution, and $* e_1$ is a projector of $C^\infty(G, E)$ onto $C(G/B, E)$. Thus we may also view $C(G/B, E)$ and $C_c(G/B, E)$ as H-modules under right convolution. We have

$$(f * u)(x) = \int_G f(xy) \cdot u(y^{-1})\, dy = \sum_{y \in G/B} f(xy) \cdot u(y^{-1}), \tag{3}$$

$(f \in C(G/B, E), u \in H(G, B))$. In particular,

$$(f * \check{e}_w)(x) = \sum_{y \in BwB/B} f(x \cdot y), \quad (x, w \in G; f \in C(G/B, E)). \tag{4}$$

Let us put

$$q_w = \mathrm{Card}(BwB/B) = \int_{BwB} dx. \tag{5}$$

We have then

$$(f * \check{e}_w)(1) = q_w \cdot f(w), \quad (f \in C(B \setminus G/B, E)), \tag{6}$$

$$(\check{e}_w * f)(1) = q_w \cdot f(w), \quad (f \in C(B \setminus G/B, E)). \tag{7}$$

Note that $q_w = q_{w^{-1}}$ since G is unimodular.

Clearly, $H(G, B)$ commutes with G acting on $C(G/B, E)$ by left translations. In fact, it is easily seen that $H(G, B)$ is the full commuting algebra of G on $C_c(G/B, E)$.

1.2. A representation of G will be denoted either by the vector space V acted upon, or by the homomorphism $\pi: G \to GL(V)$, or by the pair (π, V). Let (π, V) be one. An element $v \in V$ is smooth if its isotropy group is open in G.

The set V^∞ of smooth vectors of V is a vector subspace (since the open subgroups of G form a fundamental system of neighborhoods of the identity) stable under G. The representation π is *smooth* if $V = V^\infty$, *admissible* if, in addition, V^U is finite dimensional for every open subgroup U of G. Every G-submodule or quotient module of a smooth (resp. admissible) representation is smooth (resp. admissible). If (π, V) is smooth, then it defines a representation of the convolution algebra $C_c^\infty(G)$ characterised by

$$\pi(f) \cdot v = \int_G f(x) \cdot \pi(x) \cdot v \, dx, \quad (f \in C_c^\infty(G)). \tag{1}$$

If B and dx are as above, $V = C(G/B, E)$ and π is given by translations, then it follows from the definitions that $\pi(f) \cdot v = f * v$.

1.3. Assume π to be smooth and G to be *compact*. Then $G \cdot v$ is finite for all $v \in V$, hence V is union of finite dimensional semi-simple G-modules. Therefore, V is semi-simple and if

$$0 \to V' \to V \to V'' \to 0 \tag{2}$$

is an exact sequence of smooth modules, then the sequence of isotypic submodules of a given type is exact; in particular,

$$0 \to V'^G \to V^G \to V''^G \to 0 \tag{3}$$

is exact. If dx has total measure one, then $v \mapsto \int_G \pi(x) \cdot v \, dx$ is a projector of V onto V^G, whose kernel is the subspace $V(G)$ of V generated by the elements of the form $v - \pi(g) \cdot v$ $(g \in G; v \in V)$ [7: 3.2.1].

1.4. *Contragredient Representation.* Assume (π, V) to be smooth. The natural representation π of G in the space $\check{V} = V'^\infty$ of smooth elements in the dual V' of V is the contragredient representation $\tilde{\pi}$ of π. It is admissible if and only if π is. For every open subgroup K of G, the space $\check{V}^K$ is the orthogonal subspace to $V(K)$, and may be canonically identified to the dual of V^K [7: 2.1.9]. We have

$$\langle v, \tilde{\pi}(g) \cdot \tilde{v} \rangle = \langle \pi(g^{-1}) \cdot v, \tilde{v} \rangle; \quad \langle v, \tilde{\pi}(f) \cdot \tilde{v} \rangle = \langle \pi(\check{f}) \cdot v, \tilde{v} \rangle, \tag{1}$$

$(v \in V; \tilde{v} \in \check{V}; g \in G; f \in C_c^\infty(G))$. In particular

$${}^t\pi(f) = \tilde{\pi}(\check{f}), \quad (f \in C_c^\infty(G)). \tag{2}$$

1.5. *Coefficients.* Given $v\in V$, $\tilde{v}\in\tilde{V}$, we denote by $c_{v,\tilde{v}}$ the function on G defined by

$$c_{v,\tilde{v}}(g)=\langle v,\tilde{\pi}(g)\cdot\tilde{v}\rangle=\langle\pi(g^{-1})\cdot v,\tilde{v}\rangle,\qquad(g\in G). \tag{1}$$

We let l_g (resp. r_g) denote the effect of the left (resp. right) translation by $g\in G$ on functions on G. Thus

$$l_g f(x)=f(g^{-1}\cdot x),\qquad r_g f(x)=f(x\cdot g),\qquad(x\in G), \tag{2}$$

where f is a map of G in some set. Straightforward computations yield the formulae

$$c_{\pi(x)\cdot v,\tilde{\pi}(y)\cdot\tilde{v}}=l_x\cdot r_y\cdot c_{v,\tilde{v}},\qquad(v\in V,\tilde{v}\in\tilde{V};x,y\in G), \tag{3}$$

$$c_{\pi(f)\cdot v,\tilde{\pi}(h)\cdot\tilde{v}}=f*c_{v,\tilde{v}}*\check{h},\qquad(v\in V,\tilde{v}\in\tilde{V};f,h\in C_c^\infty(G)) \tag{4}$$

(of which (3) differs slightly from [7:2.5.1] since our $c_{v,\tilde{v}}$ is the V-transform of the $c_{v,\tilde{v}}$ there).

§ 2. Induced and Coinduced Modules from a Representation of a Hecke Algebra

In this section, B is a compact open subgroup of G, and convolution is taken with respect to a Haar measure dx giving B total measure one.

2.1. Let E be a vector space. The group G acts by left translations on $C(G/B, E)$ and $C_c(G/B, E)$. Clearly, the map

$$C_c(G/B)\otimes_R E\to C_c(G/B, E), \tag{1}$$

which assigns to $f\otimes e$ the function $x\mapsto f(x)\cdot e\ (x\in G)$ is an isomorphism. Since the intersection of finitely many conjugates of B is open in G, the representation of G on $C_c(G/B, E)$ is smooth. On the other hand, $C(G/B, E)$ is not smooth in general. The canonical bilinear form $\langle\ \rangle$ on $E'\times E$ allows one to define a pairing of $C(G/B, E')$ and $C_c(G/B, E)$ by

$$\langle\varphi,\psi\rangle=\sum_{x\in G}\langle\varphi(x),\psi(x)\rangle,\qquad(\varphi\in C(G/B, E'),\psi\in C_c(G/B, E)), \tag{2}$$

or, equivalently, by

$$\langle\varphi,\psi\rangle=(\varphi*\check{\psi})(1). \tag{3}$$

The map $j\colon C(G/B, E')\to C_c(G/B, E)'$ is readily seen to be an isomorphism, and therefore yields an isomorphism

$$C(G/B, E')^\infty\xrightarrow{\ \sim\ } C_c(G/B, E)^\sim. \tag{4}$$

2.2. In this paper a representation (r, E) of $H(G, B)$ in a vector space E is a homomorphism $r\colon H(G, B)\to\operatorname{End}(E)$ which maps e_1 onto the identity. By definition, the contragredient representation to r is the representation $\tilde{r}$ in the dual space E' to E characterised by

$$^t r(u)=\tilde{r}(\check{u}),\qquad(u\in H(G, B)). \tag{1}$$

We have therefore

$$\langle r(u)\cdot e, e'\rangle=\langle e,\tilde{r}(\check{u})\cdot e'\rangle,\qquad(e\in E, e'\in E';u\in H(G, B)). \tag{2}$$

If V is a smooth G-module, then V^B is stable under $H(G, B)$ acting via the natural representation of $C_c^\infty(G)$, and is an $H(G, B)$-module. If V is admissible, then $\tilde{V}^B$ is the contragredient H-module to V^B. If W is a smooth G-module, the restriction to the fixed points under B yields a natural map

$$\operatorname{Hom}_G(V, W) \to \operatorname{Hom}_{H(G, B)}(V^B, W^B). \tag{3}$$

2.3. Let (r, E) be a representation of $H(G\ B)$. We let

$$I_{B,G}(E) = I_{B,G}(r) = C_c(G/B) \otimes_H E, \tag{1}$$

$$P_{B,G}^0(E) = P_{B,G}^0(r) = \{f \in C(G/B, E) \mid f * u = r(u) \cdot f, (u \in H)\}. \tag{2}$$

Both $I_{B,G}(r)$ and $P_{B,G}^0(r)$ will be viewed as G-modules via left translations. The former is smooth, but the latter is not in general, and we let

$$P_{B,G}(E) = P_{B,G}(r) = (P_{B,G}^0(r))^\infty. \tag{3}$$

We shall sometimes suppress the indices B, G if they are clear from the context.

We note that $I(E)$ is the quotient of $C_c(G/B) \otimes_R E$ by the subspace M spanned by elements of the form

$$(f * h) \otimes e - f \otimes r(h) \cdot e \quad (f \in C_c(G/B); e \in E; h \in H),$$

and that the canonical projection $C_c(G/B) \otimes_R E \to I(E)$ commutes with G.

Remark. In view of the elementary relation

$$C(G/B, E) = \operatorname{Hom}(C_c(G/B), E), \tag{4}$$

and of the fact that e_1 is a projector of $C_c^\infty(G, E)$ onto $C_c(G/B, E)$, we have

$$I_{B,G}(E) = C_c^\infty(G) \otimes_H E, \qquad P_{B,G}^0(E) = \operatorname{Hom}_H(C_c^\infty(G), E), \tag{5}$$

where

$$\operatorname{Hom}_H(C_c^\infty(G), E)$$
$$= \{f \in \operatorname{Hom}(C_c^\infty(G), E) \mid f(\varphi * u) = r(u) \cdot f(\varphi), (u \in H; \varphi \in C_c^\infty(G))\}. \tag{6}$$

Therefore $I_{B,G}(E)$ (resp. $P_{B,G}^0(E)$) is the induced (resp. produced) module from H to $C_c^\infty(G)$, as defined by G. D. Higman in [10].

2.4. Proposition. *We keep the notation of 2.3.*

(i) *The map $\mu_0: e \mapsto e_1 \otimes e$ induces an H-isomorphism of E onto $I(E)^B$.*

(ii) *The map $v_0: f \mapsto f(1)$ is an H-morphism of $P(E)$ onto E, which maps $P(E)^B$ isomorphically onto E. For $e \in E$, the element $f_e \in P(E)^B$ mapped onto e by v_0 is the function*

$$f_e = \sum_{w \in B \setminus G/B} e_w \cdot q_w^{-1} \cdot r(\check{e}_w) \cdot e,$$

which assigns $q_w^{-1} \cdot r(\check{e}_w) \cdot e$ to $x \in BwB$.

(iii) *Every non-zero G-submodule of $P(E)$ contains a non-zero element fixed under B. The G-module $I(E)$ is generated by $I(E)^B$.*

(i) Left convolution by e_1 on $C_c(G/B)$ defines projectors of $I(E)$ onto $I(E)^B$ and of $C_c(G/B)$ onto $H = C_c(G/B)^B$. Therefore

$$I(E)^B = H \otimes_H E = E,$$

and (i) follows.

(ii) In the notation of 1.1(4)(5), we have

$$q_w \cdot f(x) = r(\check{e}_w) \cdot f(1), \qquad (f \in P(E)^B; w \in B \smallsetminus G/B; x \in BwB). \tag{1}$$

In fact, $f(x) = f(w)$; by 1.1(6), the left-hand side is equal to $(f * \check{e}_w)(1)$, which equals $r(\check{e}_w) \cdot f(1)$ by definition of $P(E)$. This implies that $v_0 \colon P(E)^B \to E$ is injective. Given $e \in E$, let us now define $f \in C(G/B, E)$ by (1) and the condition $f(1) = e$. This function is left-invariant under B; by 1.1(6), it satisfies

$$(f * e_w)(1) = r(e_w) \cdot f(1), \qquad (w \in B \smallsetminus G/B). \tag{2}$$

Since the e_w ($w \in B \smallsetminus G/B$) span H, this implies by linearity

$$(f * u)(1) = r(u) \cdot f(1), \qquad \text{for all } u \in H. \tag{3}$$

Together with 1.1(6), it yields, for $x \in BwB$:

$$q_w \cdot (f * u)(x) = (f * u * \check{e}_w)(1) = r(u) \cdot r(\check{e}_w) \cdot f(1) = r(u) \cdot (f * \check{e}_w)(1),$$

$$q_w(f * u)(x) = r(u) \cdot q_w \cdot f(w) = r(u) \cdot q_w \cdot f(x).$$

This shows that $f \in P(E)^B$, hence that v_0 is surjective. Taking (3) and 1.1(7) into account, we have, for $f \in P(E)^B$ and $w \in G$,

$$v_0(\check{e}_w * f) = (\check{e}_w * f)(1) = q_w \cdot f(w) = r(\check{e}_w) \cdot f(1) = r(\check{e}_w) \cdot v_0(f).$$

Since the e_w's span H, it follows that v_0 induces an H-isomorphism of $P(E)^B$ onto E.

Let M be the kernel of $e_1 *$ on $P(E)$. Then $P(E)$ is the direct sum of M and $P(E)^B$ and M is also annihilated by all elements of H. For any $f \in C(G/B, E)$, we have

$$f(1) = (e_1 * f)(1), \tag{4}$$

hence $M = \ker v_0$, which completes the proof of the first part of (ii). The second one then follows from (1).

(iii) Let V be a non-zero G-submodule of $P(E)$. Being invariant under left translations, it contains an element f such that $f(1) \neq 0$. But then $(e_1 * f)(1) \neq 0$, hence $V^B \neq 0$.

For $x \in G$, let ε_x be the characteristic function of xB on G/B. The tensor product $C_c(G/B) \otimes_R E$ is spanned by the subspaces

$$\varepsilon_x \otimes E = l_x^{-1}(e_1 \otimes E), \qquad (x \in G),$$

hence $C_c(G/B)\otimes_R E$ is generated, as a G-module, by $e_1\otimes E$. Since $I(E)$ is a quotient of $C_c(G/B)\otimes_R E$, this is *a fortiori* true for $I(E)$, whence the second assertion of (iii).

In the sequel, we shall often identify E with $I(E)^B$ or $P(E)^B$ by means of μ_0 or ν_0.

2.5. Proposition. *Let (r, E) be a finite dimensional H-module, and (π, V) a smooth G-module. Then the natural restriction maps (2.2(3))*

$$\rho_P\colon \mathrm{Hom}_G(V, P(E))\to\mathrm{Hom}_H(V^B, E) \quad and \quad \rho_I\colon \mathrm{Hom}_G(I(E), V)\to\mathrm{Hom}_H(E, V^B)$$

are bijective.

Given $\alpha\in\mathrm{Hom}_G(V, P(E))$, let us denote by α_0 the restriction of α to V^B. If $\alpha_0=\beta_0$, then $\alpha-\beta$ is a G-morphism which is zero on V^B. Since $P(E)$ has no non-zero G-module F with $F^B=0$ (2.4), we have $\alpha=\beta$, hence ρ_P is injective. Let β_0 be an H-morphism of V^B into E. We define a map $\beta\colon V\to C(G/B, E)$ by the rule

$$\beta(v)(x)=\beta_0(e_1(x^{-1}\cdot v)), \quad (v\in V; x\in G). \tag{1}$$

Let (v_i) be a basis of V^B, and (v^i) be the dual basis of $\tilde{V}^B$ (see 1.4). Then, in the notation of 1.5, we have

$$\beta(v)=\sum c_{v,v^i}\cdot\beta_0(v_i). \tag{2}$$

The relation 1.5(3) shows that β is a G-morphism of V into $C(G/B, E)$. Let us show that $\beta(V)\subset P(E)$. Let $h\in H$. The transformation $\pi(h)$ leaves V^B stable and we may write uniquely

$$\pi(h)\cdot v_i=\sum_j a_i^j\cdot v_j, \quad (a_i^j\in R). \tag{3}$$

By 1.4(2), $'\pi(h)=\pi(\check{h})$, hence

$$\tilde{\pi}(\check{h})\cdot v^j=\sum_i a_i^j\cdot v^i. \tag{4}$$

Using 1.5(4), we get

$$\beta(v)*h=\sum_i(c_{v,v^i}*h)\cdot\beta_0(v_i)=\sum_i c_{v,\tilde{\pi}(\check{h})v^i}\cdot\beta_0(v_i),$$

$$\beta(v)*h=\sum_{i,j}c_{v,v^j}\cdot a_j^i\cdot\beta_0(v_i).$$

Since β_0 is an H-isomorphism, we also have

$$r(h)\cdot\beta_0(v_i)=\beta_0(\pi(h)\cdot v_i))=\sum_j a_i^j\cdot\beta_0(v_j),$$

whence

$$\beta(v)*h=\sum_i c_{v,v^i}r(h)\cdot\beta_0(v_i)=r(h)\cdot\beta(v), \tag{5}$$

which shows that $\beta(v)\in P(E)$. If $v\in V^B$, then $\beta(v)\in P(E)^B$ and moreover $\nu_0(\beta(v))=\beta(v)(1)=\sum\langle v, v^i\rangle\beta_0(v_i)=\beta_0(v)$, hence $\beta_0=\rho_P(\beta)$.

Since $I(E)$ is generated by E as a G-module, the map ρ_I is injective. Let $\alpha_0\colon E\to V^B$ be an H-morphism. It extends uniquely to an R-linear map

$$\alpha_1\colon C_c(G/B)\otimes_R E\to V$$

given by

$$\alpha_1(f\otimes e)=\pi(f)\cdot\alpha_0(e),\qquad(f\in C_c(G/B);\ e\in E).$$

If $m\in C_c^\infty(G)$, then

$$\alpha_1((m*f)\otimes e)=\pi(m*f)\cdot\alpha_0(e)=\pi(m)\cdot\pi(f)\cdot\alpha_0(e)=\pi(m)\cdot\alpha_1(f\otimes e),$$

hence α_1 is $C_c^\infty(G)$-equivariant, and therefore also G-equivariant [7:2.2.1]. If $h\in H$, then

$$\alpha_1((f*h)\otimes e)=\pi(f*h)\cdot\alpha_0(e)=\pi(f)\cdot\pi(h)\cdot\alpha_0(e)=\pi(f)\cdot\alpha_0(r(h)\cdot e)$$
$$=\alpha_1(f\otimes r(h)\cdot e),$$

hence α_1 annihilates the kernel of the canonical projection $C_c(G/B)\otimes_R E\to I(E)$ (see 2.3). The map $\alpha\colon I(E)\to V$ obtained from α_1 by going over to the quotient is then a G-morphism which extends α_0.

2.6. Proposition. *Let (r,E) be an H-module. The canonical isomorphism*

$$C(G/B,E')^\infty\xrightarrow{\ \sim\ }(C_c(G/B,E))^\sim$$

of 2.1(4) induces an isomorphism of $P(E')$ onto $I(E)^\sim$.

By definition, $P(E')$ is a G-submodule of $C(G/B,E')$. It suffices to show that it is the orthogonal subspace to the kernel M of the projection

$$q\colon\ C_c(G/B)\otimes_R E=C_c(G/B,E)\to C_c(G/B)\otimes_H E=I(E).$$

Let $f\in C(G/B,E')$, $\varphi\in C_c(G,B)$, $e\in E$, $h\in H$. From 2.1, we get

$$\langle f,(\varphi*h)\otimes e\rangle=\langle f*(\varphi*h)^\vee(1),e\rangle=\langle f*\check{h}*\check{\varphi})(1),e\rangle,$$

$$\langle f,(\varphi*h)\otimes e\rangle=\langle f*\check{h},\varphi\otimes e\rangle.\tag{1}$$

$$\langle f,\varphi\otimes r(h)e\rangle=\langle(f*\check{\varphi})(1),r(h)\cdot e\rangle=\langle\tilde{r}(\check{h})(f*\check{\varphi})(1),e\rangle,$$

$$\langle f,\varphi\otimes r(h)e\rangle=\langle\tilde{r}(\check{h})f,\varphi\otimes e\rangle.\tag{2}$$

The proposition follows immediately from (1) and (2) and 2.3.

II. Semi-Simple Groups Over a Local Field

From now on, k is a non-archimedean local field with a finite residue field, q the number of elements of the residue field, $\mathcal{G}$ a connected semi-simple k-group, $\tilde{\mathcal{G}}$ its universal covering, $\omega\colon\tilde{\mathcal{G}}\to\mathcal{G}$ the canonical central isogeny, and l the k-rank of $\mathcal{G}$. We assume $l\geq 1$.

§ 3. The Hecke Algebra with Respect to an Iwahori Subgroup

3.1. Algebraic groups over k will usually be denoted by script letters, and the groups of their k-rational points by the corresponding Roman letters. In particular

$$G=\mathcal{G}(k),\qquad\tilde{G}=\tilde{\mathcal{G}}(k).$$

We let $\tilde{T}=(\tilde{G}, \tilde{B}, \tilde{N}, S)$ be the Tits system of $\tilde{G}$ considered in [5], where $\tilde{B}$ is an Iwahori subgroup of $\tilde{G}$, and X the associated building. X is a polysimplicial chamber complex of dimension l. We let C_0 be the chamber fixed by B and A_0 the apartment fixed by $\tilde{B} \cap \tilde{N}$. Therefore $A_0 = \tilde{N} \cdot C_0$. The Weyl group $\tilde{N}/(\tilde{N} \cap \tilde{B})$ of $\tilde{T}$ will be denoted $\tilde{W}$. The apartment A_0 has a canonical structure of affine euclidean space with respect to which $\tilde{W}$ acts faithfully as a euclidean reflection group. S is the set of reflections to the hyperplanes containing the faces of co-dimension 1 of C_0.

The group G maps naturally into $(\operatorname{Aut}\tilde{G})(k)$, hence operates on $\tilde{G}$ and on X. The latter action is also continuous and proper. We let B be the subgroup of G fixing C_0 (pointwise), N be the stabilizer of A_0 in G, and G_0 (resp. N_0) the biggest subgroup of G (resp. N) which acts on X by special automorphisms, i.e., automorphisms which preserve the type of a face [5:2.1.1]. Let L be the intersection of N with the normalizer $\mathcal{N}_G(B)$ of B in G. Then $G = L \cdot G_0$, the group G_0 is normal, of finite index, in G. The group $G/G_0 = L/(L \cap G_0)$ may be identified with a finite commutative subgroup ψ of the group $\operatorname{Aut}\operatorname{Cox} S$ of automorphisms of the Coxeter diagram of (W, S) which leaves stable every connected component of the latter. The isogeny ω induces an isomorphism

$$\tilde{W} \xrightarrow{\sim} N_0/(N_0 \cap B), \tag{1}$$

and bijections

$$\tilde{G}/\tilde{B} = G_0/B, \quad \tilde{B} \smallsetminus \tilde{G}/\tilde{B} = B \smallsetminus G_0/B. \tag{2}$$

We have semi-direct product decompositions

$$G = \Psi \ltimes G_0, \quad W = N/(B \cap N) = \Psi \ltimes \tilde{W}, \tag{3}$$

and $T_0 = (G_0, B, N, S)$, $T = (G, B, N)$ are respectively a Tits system [4: IV, §2] and a generalized Tits system [4: IV, §2, Exer. 8] or [11]. We have the decompositions

$$G_0 = \coprod_{w \in \tilde{W}} BwB, \quad G = \coprod_{w \in W} BwB. \tag{4}$$

Moreover

$$BwBw'B = Bww'B, \quad \text{if } w, w' \in \tilde{W} \quad \text{and} \quad l(ww') = l(w) + l(w'), \tag{5}$$

where $l(\;)$ denotes the length in $\tilde{W}$ with respect to S [4: IV, §1]. Let

$$q_w = \operatorname{Card}(BwB/B), \quad (w \in W). \tag{6}$$

Then

$$q_w = \operatorname{Card}(\tilde{B}w\tilde{B}/\tilde{B}), \quad \text{if } w \in \tilde{W}, \tag{7}$$

$$q_{ww'} = q_w \cdot q_{w'}, \quad \text{if } w, w' \in \tilde{W} \quad \text{and} \quad l(ww') = l(w) + l(w'). \tag{8}$$

We have

$$L \cap G_0 = B, \tag{9}$$

so that the elements of Ψ can be viewed as cosets modulo B, whose elements normalize B. We have

$$B\psi B = \psi B = B\psi, \quad B\psi B\psi' B = B\psi\psi' B, \quad (\psi, \psi' \in \Psi), \tag{10}$$

$$\psi BwB = B\psi wB, \quad q_{\psi w} = q_w, \quad (w \in \tilde{W}; \psi \in \Psi). \tag{11}$$

For all this and the facts recalled in 3.2 below, see [4: IV, §2, Exer. 8], or §3 of [11], where the underlying groups are split but the arguments are general.

3.2. The Hecke algebra $H(G_0, B)$ will often be denoted by H_0. It follows from 3.1(2) that ω yields an isomorphism of $H(\tilde{G}, \tilde{B})$ with H_0. The algebra H_0 is spanned over R by the characteristic functions e_w of the cosets BwB ($w \in \tilde{W}$). The e_w's have the following properties:

1) $e_w \cdot e_{w'} = e_{w \cdot w'}$, if $w, w' \in \tilde{W}$ and $l(w \cdot w') = l(w) + l(w')$

2) $e_s^2 = (q_s - 1) \cdot e_s + q_s \cdot e_1$, $(s \in S)$.

3) Let $s, s' \in S$ be distinct and let $m = m(s, s')$ be the order of $s \cdot s'$. Then m is finite, except if $|S| = 2$, and we have

$$(e_s \cdot e_{s'})^r \cdot e_s = e_{s'} \cdot (e_s \cdot e_{s'})^r, \quad \text{if } m(s, s') = 2r+1,$$

$$(e_s \cdot e_{s'})^r = (e_{s'} \cdot e_s)^r, \quad \text{if } m(s, s)' = 2r.$$

4) Let $w \in \tilde{W}$, $s \in S$ and assume that $l(s \cdot w) < l(w)$. Then

$$e_s \cdot e_w = (q_s - 1) e_w + q_s \cdot e_{sw}.$$

5) The algebra H_0 is generated, as an R-algebra, by 1 and the elements e_s $(s \in S)$. The relations (2), (3) form a presentation of H_0.

Let $\psi \in \Psi$. It defines a permutation of $(e_s)_{s \in S}$ given by

$$e_s \mapsto \psi \cdot e_s = e_{\psi(s)}.$$

By the above, this permutation extends to an automorphism of H_0. Let $R[\psi]$ be the group algebra of Ψ over R and let $R[\Psi] \tilde{\otimes} H_0$ be the ordinary tensor product $R[\Psi] \otimes H_0$ of modules, endowed with the product defined by

6) $(\psi \otimes h) \cdot (\psi' \otimes h') = \psi \cdot \psi' \otimes (\psi'^{-1} \cdot h) \cdot h'$.

Then [11: §3] or [4: IV, §2, Ex. 25]

7) $H \cong R[\Psi] \tilde{\otimes} H_0$.

3.3. *Characters of Degree One.* Two elements s, s' of S are conjugate in $\tilde{W}$ if and only if there exists a chain of elements of S: $s = s_0, s_1, \ldots, s_q = s'$ such that $s_i \cdot s_{i+1}$ is of finite odd order for $i = 0, 1, \ldots, q-1$ [4: IV, 1.3, Prop. 3]. The intersections S_i of S with the conjugacy classes of $\tilde{W}$ are therefore the connected components of the graph obtained by erasing the multiple edges in the Coxeter graph Cox S of S. The number m of such classes in equal to

    1 if Cox S is of type $\mathbf{A}_n$ $(n \geq 2)$, $\mathbf{D}_n$ $(n \geq 3)$, $\mathbf{E}_i$ $(i = 6, 7, 8)$,

    2 if Cox S is of type $\mathbf{A}_1$, $\mathbf{B}_n$ $(n \geq 2)$, $\mathbf{G}_2$, $\mathbf{F}_4$,

    3 if Cox S is of type $\mathbf{C}_n$ $(n \geq 2)$.

3.4. Proposition. *Let K be a commutative algebra over R. Let S_i $(1 \leq i \leq m)$ be the intersections of S with the conjugacy classes in $\tilde{W}$. A map $(e_s)_{s \in S} \to K$ is the restriction of a homomorphism $H_0 = H(G_0, B) \to K$, if and only if it is constant on each S_i and equal to -1 or q_s $(s \in S_i)$ on S_i. In particular, $H(G_0, B)$ has 2^m characters of degree one.*

The necessity of the condition follows from 3.2 (2), (3) and its sufficiency from 3.2 (5).

3.5. Proposition. *Let $\tau \in \hat{\Psi}$ be a character of Ψ, and σ a character of H_0. Then $\chi = \sigma \otimes \tau$ is a character of H if and only if σ is constant on the orbits of Ψ in S.*

This is obvious.

Special Cases. a) The algebra H_0 has a unique character σ_0 which is equal to -1 on every e_s, to be called *the special character* of H_0. If $\tau \in \hat{\Psi}$, then $\tau \otimes \sigma_0$ is a character of H, also to be called special.

b) Clearly, $q_s = q_{s'}$ if s and s' are conjugate in $\tilde{W}$. The map $e_s \mapsto q_s$ extends therefore to a character σ_1 of degree 1 of H_0; it satisfies $\sigma_1(e_w) = q_w$ for all $w \in \tilde{W}$. Given $\psi \in \hat{\Psi}$, the map $s \mapsto \psi(s)$ extends to an automorphism of G_0; therefore q_s is constant on the orbits of Ψ, and, for any character τ of $\hat{\Psi}$, the product $\tau \otimes \sigma_1$ is a character of H.

3.6. Proposition. *Let (r, E) be a finite dimensional representation of $H(G, B)$. Then the elements $r(e_g)$ $(g \in G)$ are invertible.*

In view of 3.1 (3), (4) it suffices to show this for $g = \psi \cdot w$ $(\psi \in \Psi; w \in \tilde{W})$. From 3.1 (10), it follows that $\psi \mapsto r(e_\psi)$ is a representation of Ψ, hence $r(\psi)$ is invertible. If $w \in \tilde{W}$ and $w = s_1 \ldots s_q$ $(q = l(w))$ is a reduced decomposition of w, then $r(e_w) = r(e_{s_1}) \ldots r(e_{s_q})$ by 3.2 (1), hence there remains only to show that $r(e_s)$ is invertible for $s \in S$. But this follows from 3.2 (2), which shows more precisely that the only possible eigenvalues of $r(e_s)$ are q_s and -1.

§ 4. Admissible Representations

The following lemma and its proof are due to F. Bruhat. The proof will be given in §7.

4.1. Lemma (F. Bruhat). *Let U be a compact open subgroup of G_0 (cf. 3.1). There exists a number $d_0 > 0$ with the following property: given a chamber C of the building X of $\tilde{G}$ such that $d_s(C, C_0) > d_0$, there exists a chamber D adjacent to C satisfying the two following conditions:*

(i) $d_s(D, C_0) = d_s(C, C_0) - 1$.

(ii) the group U is transitive on the set (C, D) of chambers C' such that $C' \cap D = C \cap D$.

If C_1 is a chamber of X, then $d_s(C_1, C_0)$ is the integer d such that the minimal galleries connecting C_1 and C_0 have $d+1$ elements. We recall that two distinct chambers C_1, C_2 are said to be adjacent if their intersection is a face of codimension one. If so, we let (C_1, C_2) be the set of chambers C such that $C \cap C_2 = C_1 \cap C_2$.

4.2. To use 4.1, we give first an interpretation in X of the convolution formula

$$(f * e_s)(x) = \sum_{y \in BsB/B} f(x \cdot y), \qquad (x \in G_0, \ s \in S, \ f \in C(G_0/B, E)),\tag{1}$$

which is a special case of 1.1(4) since $e_s = \check{e}_s$. Let C_{0s} be the face of C_0 which is fixed (pointwise) under the reflection s. As is well-known, the isotropy group of C_{0s} is the group $B_{\{s\}} = BsB \cup B$ and is transitive on the set of chambers of X containing C_{0s}.

Therefore, if

$$BsB = \coprod_{i \leq i \leq q_s} x_i B,\tag{2}$$

then $x_i \cdot C_0$ runs through all chambers of X which contain C_{0s} and are $\neq C_0$, i.e.

$$\bigcup_i x_i \cdot C_0 = \{C' \mid C' \cap C_0 = C_{0s}\},\tag{3}$$

and if $x \in G$ and $x \cdot C_0 = C_1$, then

$$\bigcup_i x \cdot x_i C_0 = \{C' \mid C' \cap C_1 = x \cdot C_{0s}\};\tag{4}$$

with the notation introduced at the end of 4.1, (4) can also be written

$$\bigcup_i x x_i \cdot C_0 = (C, C_1),\tag{5}$$

if C is any chamber such that $C \cap C_1 = x \cdot C_{0s}$. We recall further that if $x, x' \in G_0$ map C_0 onto C_1, they define the same isomorphism of C_0 onto C (since G_0 consists by definition of special automorphisms): hence $x \cdot C_{0s} = x' \cdot C_{0s}$. This is the face of type s of $x \cdot C_0$.

In view of this, (1) translates into the following: let C, D be adjacent chambers of X and s the type of $C \cap D$. Then

$$(f * e_s)(D) = \sum_{C' \in (C, D)} f(C'), \qquad (f \in C(G_0/B, E)).\tag{6}$$

$$\mathrm{Card}\,(C, D) = q_s.\tag{7}$$

4.3. Proposition. *Let U be a compact open subgroup of G_0. Then $C_c(U \smallsetminus G_0/B)$ is an H_0-module of finite type, with respect to convolution on the right.*

Let d be a positive number, and $D(d) = \{C \mid d_s(C, C_0) \leq d\}$. The latter is a finite set. We may replace U by a smaller open subgroup, hence assume $U \subset B$. Then $D_0(d)$ is stable under U.

Let L_d be the H_0-submodule of $C_c(U \smallsetminus G_0/B)$ generated by the elements of $C_c(U \smallsetminus G_0/B)$ with support in $D(d)$. Since $D(d)$ consists of finitely many chambers, L_d is finitely generated. It suffices therefore to show that if d is equal to the d_0 of Lemma 4.1, then $L_d = C_c(U \smallsetminus G_0/B)$.

By 2.1, the G_0-module $C(G_0/B)^\infty$ is the contragredient module to $C_c(G_0/B)$, the pairing being given by

$$\langle \varphi, \psi \rangle = (\varphi * \check{\psi})(1) = \sum_{x \in G_0/B} \varphi(x) \psi(x) \qquad (\varphi \in C(G_0/B)^\infty, \ \psi \in C_c(G_0/B)).\tag{1}$$

We have then also

$$\langle \varphi * h, \psi \rangle = \langle \varphi, \psi * \check{h} \rangle \quad (\varphi \in C(G_0/B)^\infty, \ \psi \in C_c(G_0/B), \ h \in H_0). \tag{2}$$

Under the pairing (1), the space

$$C(U \smallsetminus G_0/B) = C(G_0/B)^U$$

is the dual to

$$C_c(U \smallsetminus G_0/B) = C_c(G_0/B)^U.$$

To prove our assertion, it is enough therefore to show that if $\lambda \in C(U \smallsetminus G_0/B)$ is zero on L_{d_0} then $\lambda = 0$.

We prove first that $\lambda(C) = 0$ if $C \in D(d_0)$. Let χ_C be the characteristic function of $U \cdot C$ in G_0/B. Then (1) shows that

$$c \cdot \lambda(C) = \langle \lambda, \chi_C \rangle, \quad \text{where } c = \operatorname{Card} U \cdot C; \tag{3}$$

but, since $U \cdot D(d_0) = D(d_0)$, the function χ_C belongs to L_{d_0} hence $\lambda(C) = 0$.

Let now $C \notin D(d_0)$. Let D be as in 4.1 and s be the type of $C \cap D$ (4.2). Then U is transitive on (C, D), hence λ is constant on (C, D) and 4.2(6)(7) yield

$$(\lambda * e_s)(D) = \sum_{C' \in (C, D)} \lambda(C') = q_s \cdot \lambda(C). \tag{4}$$

Using induction on $d_s(C, C_0)$ and 3.1(8), we see that, given a chamber C not in $D(d_0)$, there exist $w \in \check{W}$ and a chamber $C' \in D(d_0)$ such that

$$(\lambda * e_w)(C') = q_w \cdot \lambda(C). \tag{5}$$

Let $c' = \operatorname{Card} U \cdot C'$. By (2), (3) and (5)

$$\langle \lambda, \chi_{C'} * \check{e}_w \rangle = \langle \lambda * e_w, \chi_{C'} \rangle = c' \cdot (\lambda * e_w)(C') = c' \cdot q_w \cdot \lambda(C).$$

Since $\chi_{C'} * \check{e}_w \in L_{d_0}$ this shows that $\lambda(C) = 0$.

4.4. Theorem. *Let (r, E) (resp. (r_0, E_0)) be a finite dimensional $H = H(G, B)$-module (resp. $H_0 = H(G_0, B)$-module). Then $P_{B, G}(E)$ and $I_{B, G}(E)$ (resp. $P_{B, G_0}(E_0)$ and $I_{B, G_0}(E_0)$) are admissible G-modules (resp. G_0-modules).*

By 2.6, $P_{B, G}(E)$ and $P_{B, G_0}(E)$ are the contragredient modules to $I_{B, G}(E')$ and $I_{B, G_0}(E_0')$ respectively. It suffices therefore to prove our statement for the induced modules $I_{B, G}(E)$ and $I_{B, G_0}(E_0)$. We next reduce the proof to the case of G_0.

We have $G = L \cdot G_0$, where L normalizes B (3.1). Since B is its own normalizer in G_0, it follows that each orbit of G_0 on G/B contains exactly one fixed point under B. If x is such a point, then $g \mapsto g \cdot x$ provides an isomorphism of G_0/B onto $G_0 \cdot x$, whence a canonical G_0-equivariant bijection of G/B onto the disjoint union of $m = [G : G_0]$ copies of G_0/B. This yields canonical G_0-equivariant isomorphisms

$$C_c(G/B, E) \xrightarrow{\sim} C_c(G_0/B, E) \oplus \cdots \oplus C_c(G_0/B, E), \quad (m \text{ summands}), \tag{1}$$

$$C(G/B, E) \xrightarrow{\sim} C(G_0/B, E) \oplus \cdots \oplus C(G_0/B, E), \quad (m \text{ summands}). \tag{2}$$

These isomorphisms are furthermore H_0-equivariant, H_0 being identified to a subalgebra of $H \cong R[\Psi] \tilde{\otimes} H_0$ (see 3.2(7)) by the map $h \mapsto 1 \otimes h$ $(h \in H_0)$.

Since G_0 is open, normal, of finite index in G, a G-module is smooth (resp. admissible) if and only it is so with respect to G_0. Thus we have to show that $I_{B,G}(E)$ is admissible as a G_0-module. Since H_0 is a subalgebra of H, the module $I(E) = C_c(G/B) \otimes_H E$ is a quotient of $C_c(G/B) \otimes_{H_0} E$. But, by (1), the latter is the direct sum of m copies of $I_{B,G_0}(E)$, hence it is admissible if and only if $I_{B,G_0}(E)$ is.

Let U be a compact open subgroup of G_0. Then

$$I_{B,G_0}(E_0)^U = C_c(U \smallsetminus G_0/B) \otimes_{H_0} E_0.$$

Since $C_c(U \smallsetminus G_0/B)$ is a H_0-module of finite type (4.3), this shows that the left-hand side is finite dimensional, hence $I_{B,G_0}(E_0)$ is admissible.

4.5. _Remark._ Theorem 4.4 can be proved directly for the coinduced modules without using 2.6. In fact this was my original argument. P. Cartier suggested that it would also yield 4.2 and that I consider the modules $I(E)$ as well.

We outline here the argument for the coinduced modules. First, it is easily seen that $P_{B,G}(E)$ is a G_0-invariant subspace of the direct sum of m copies of $P_{B,G_0}(E)$, whence the reduction to G_0.

Let U be a compact open subgroup of G_0, and d_0 be as in Lemma 4.1. To show that $P_{B,G_0}(E_0)^U$ is finite dimensional, it suffices to prove that if $f \in P_{B,G_0}(E_0)^U$ is zero on $D(d_0)$, then it is identically zero. Let C be a chamber such that $d_s(C,C_0) > d_0$. Let D be as in 4.1, and s the type of $C \cap D$. We have then again, using 4.2(6),(7):

$$(f * e_s)(D) = q_s \cdot f(C), \tag{1}$$

whence, by induction, the existence of $w \in \tilde{W}$ and $C' \in D(d_0)$ such that

$$(f * e_w)(C') = q_w \cdot f(C); \tag{2}$$

this yields

$$q_w \cdot f(C) = r_0(e_w) \cdot f(C'), \tag{3}$$

and our assertion.

4.6. Let $\mathscr{P}; \mathscr{P}^-$ be two opposite minimal parabolic k-subgroups of $\mathscr{G}$, $\mathscr{M} = \mathscr{P} \cap \mathscr{P}^-$ their common Levi subgroup and $\mathscr{N}, \mathscr{N}^-$ their respective unipotent radicals. We may (and do) choose them in such a way that B admits the "Iwahori factorization" [7: 1.4.4]:

$$B = N_0^- \cdot M_0 \cdot N_0, \quad (N_0^- = N^- \cap B; \; M_0 = M \cap B; \; N_0 = N \cap B), \tag{1}$$

and that M_0 is the greatest compact subgroup of $G_0 \cap M$. In particular, M_0 is normal in M.

We shall use some notation and results of [7]. If V is a smooth N-module, V_N denotes the greatest quotient of V on which N acts trivially [7: 3.2]. If V is a P-module, V_N is viewed as an M- or P-module in the obvious way and the projection $V \to V_N$ commutes with P. The assignment $V \mapsto V_N$ defines an exact functor from smooth N-modules to vector spaces [7: 3.2.3].

The following lemma is a slight strengthening of Theorem 3.3.3 in [7], valid for Iwahori subgroups:

4.7. Lemma. *Let (π, V) be an admissible G-module. Then the canonical projection $V \to V_N$ induces an isomorphism of V^B onto $(V_N)^{M_0}$.*

Let α be the restriction of $V \to V_N$ to V^B. Obviously $\alpha(V^B) \subset (V_N)^{M_0}$.
Recall that for $a \in G$, $e_a \in H(G, B)$ is the characteristic function of the double coset BaB and q_a its volume (1.1). It follows from the definitions that we have:

$$\pi(e_a) \cdot v = q_a \cdot \pi(e_1) \cdot \pi(a) \cdot v, \quad (v \in V^B). \tag{1}$$

By [7: 4.1.4], there exists $a \in G$ such that moreover:

$$\alpha \text{ induces an isomorphism of } \pi(e_1) \cdot \pi(a) \cdot V^B \text{ onto } (V_N)^{M_0}. \tag{2}$$

However, the restriction of $\pi(e_a)$ to V^B is an invertible automorphism (3.6), hence, by (1), $\pi(e_1) \cdot \pi(a) \cdot V^B = V^B$.

4.8. Lemma (Casselman). *Let E be an admissible G-module. Assume that E is generated by E^B as a G-module and that, if E' is a non-zero G-submodule of E, then $E'^B \neq 0$. Then*

(i) *Any exact sequence of admissible G-modules*

$$0 \to E \to F \to F' \to 0$$

where $F'^B = 0$, splits.

(ii) *if E' is a G-submodule of E, then E' is generated by E'^B as a G-module.*

Proof. By 4.6, 4.7:

$$(F'_N)^{M_0} = 0, \quad (E_N)^{M_0} = (F_N)^{M_0} \quad \text{and} \quad \alpha \colon E^B \to (E_N)^{M_0} \text{ is an isomorphism.} \tag{1}$$

The exact sequence [7: 3.2.3]

$$0 \to E_N \to F_N \to F'_N \to 0 \tag{2}$$

gives then rise to a direct sum decomposition

$$F_N = (E_N)^{M_0} \oplus L \tag{3}$$

where L is the sum of the isotypic subspaces of M_0 in F_N for the irreducible representations of M_0 different from the trivial representation. Since M_0 is normal in M, both spaces are stable under M. Thus (3) is also a direct sum decomposition of M- or P-modules.
Let V be the space of the unnormalized induced representation $\mathbf{Ind}((E_N)^{M_0}|P, G)$ [7:2.2]. By Frobenius reciprocity [7: Thm 2.4.1], the projection $\beta_N \colon F_N \to (E_N)^{M_0}$ is the map canonically associated to a G-morphism $\beta \colon F \to V$. We want to prove:

$$E \cap \ker \beta = 0, \quad \beta(E) = \beta(F). \tag{4}$$

β_N is an isomorphism of $E_N = (F_N)^{M_0}$ onto itself. Hence, by 4.7, the homomorphism $\beta \colon F^B = E^B \to V^B$ is injective. Our second assumption on E then implies the first part of (4).

The G-module $\beta(F)/\beta(E)$ is isomorphic to a decomposition factor of V. The space $(\beta(F)/\beta(E))^B$ is a quotient of $(F/E)^B$ hence is zero. That $\beta(F)=\beta(E)$ then follows from the following known fact (see 4.9 below)

($*$) *Let σ be a finite dimensional representation of M trivial on M_0, and L a non-zero decomposition factor of* $\mathbf{Ind}\,(\sigma\,|\,P, G)$. *Then* $L^B \neq 0$.

This proves (4). It follows that $\ker\beta$ is isomorphic to F' as a G-module and is a supplement to E whence (i).

Let now E' be a G-submodule of E and E'' the G-submodule generated by E'^B. By (i), the exact sequence

$$0 \to E'' \to E' \to E'/E'' \to 0,$$

splits, hence E'/E'' may be identified with a G-submodule of E'. However $(E'/E'')^B = E'^B/E''^B = 0$, whence $E' = E''$.

4.9. The assertion ($*$) above is essentially proved in [7]. A similar one is also announced in [14: p. 19]. We indicate briefly how to derive it from [7]. To refer freely to [7] we remark first that, since the modulus δ_P of P is trivial on the compact group M_0 and since, by definition [7: 3.1] the normalized induced representation $\mathrm{Ind}\,(\sigma\,|\,P, G)$ is $\mathbf{Ind}\,(\sigma\,\delta_P^{\frac{1}{2}}\,|\,P, G)$, we may shift from one to the other.

By [7: Thm. 3.3.3], it suffices to prove that $L_N \neq 0$. Since the functor $\sigma \mapsto \mathrm{Ind}\,\sigma$ is exact, as follows from its definition, we may, by an easy induction on the length of a Jordan-Hölder series of σ, assume that σ is irreducible. The G-module L has finite length [7: 6.3.8], so we may assume it to be irreducible. But then L embeds in an induced module $(\mathrm{Ind}\,(\tau\,|\,P, G)$, where τ is a finite dimensional irreducible representation of M trivial on M_0 [7: 6.3.9] and $L_N \neq 0$ by [7: 3.2.5].

Here, B need not be an Iwahori subgroup. It suffices that it be a compact open subgroup of G admitting an Iwahori factorization (4.6(1)).

4.10. Theorem. *Let (r, E) be a finite dimensional $H(G, B)$-module. Then the canonical morphism $j\colon I_{B, G}(E) \to P_{B, G}(E)$ is an isomorphism. The G-module $I_{B, G}(E)$ is irreducible if and only if E is an irreducible $H(G, B)$-module. The assignment $E \mapsto I(E)$ (resp. $E \mapsto P(E)$) is an exact functor from finite dimensional H-modules to admissible G-modules. The H-module $C_c(G/B)$ is flat.*

Let M be the G-submodule of $P(E)$ spanned by $G \cdot P(E)^B$, and $V = P(E)/M$. Every non-zero G-submodule of $P(E)$ contains a non-zero B-fixed vector (2.4). On the other hand, $V^B = P(E)^B/M^B$ is zero. By 4.8, the exact sequence of G-modules

$$0 \to M \to P(E) \to V \to 0,$$

splits. Thus V may be viewed as a G-submodule of $P(E)$ and the equality $V^B = 0$ implies $V = 0$ by 2.4. Therefore $P(E)$ is generated by E as a G-module and j is surjective. This also applies to $P(E')$. Since $P(E')$ is the contragredient to $I(E)$ (2.6), it follows that every non-zero G-submodule of $I(E)$ has non-zero B-fixed vectors [7: 2.2.3], hence j is injective; furthermore, if V is a G-submodule of $I(E)$, then V is spanned by $G \cdot V^B$ (4.8); if E' is an H-submodule of E, then the smallest G-submodule F of $I(E)$ containing E' is a quotient of $I(E')$, hence (2.4) satisfies

$F^B = E'$. These two remarks imply the irreducibility statement. Let

$$E' \xrightarrow{\ \alpha\ } E \xrightarrow{\ \beta\ } E'', \tag{1}$$

be an exact sequence of finite dimensional H-modules, and

$$I(E') \xrightarrow{\ I(\alpha)\ } I(E) \xrightarrow{\ I(\beta)\ } I(E''), \tag{2}$$

the corresponding sequence of induced modules. Let $F = \ker I(\beta)$. By assumption and 2.5, $F^B = \ker \beta = \operatorname{Im} \alpha$. Since, by the above, F is generated by F^B as a G-module, we have $F = \operatorname{Im} I(\alpha)$. This proves the third assertion for induced modules. In view of the first one, it also follows for coinduced modules. By the definition of induced modules, the last one is just a reformulation of the third one (for induced modules).

4.11. Corollary. *Let V be an admissible G-module and assume that V is generated by $G \cdot V^B$. Then the canonical morphisms $I(V^B) \xrightarrow{\ \mu\ } V \xrightarrow{\ \nu\ } P(V^B)$ are isomorphisms. Every G-submodule of V is generated as a G-module by its B-fixed vectors.*

The composition $\nu \circ \mu$ is the morphism j of 4.10 hence is an isomorphism. Moreover, μ is surjective since V is generated by $G \cdot V^B$. This proves the first assertion. The second one follows from 2.4 and 4.8.

4.12. Corollary. *Let V be an admissible G-module. Let M be the G-submodule generated by V^B. Then V is the direct sum of M and of a G-submodule N such that $N^B = 0$.*

By 4.11, $M = I(V^B)$ and every G-submodule of M is generated by its B-fixed vectors. Since $(V/M)^B = V^B/V^B = 0$, 4.12 follows from 4.8.

§ 5. Square Integrable Representations Induced from Characters of Degree One

In this section, R is the field $\mathbf{C}$ of complex numbers; G' stands for G or G_0 (3.1).

5.1. Let $L^2(G'/B)$ be the space of elements $f \in C(G'/B)$ such that

$$\|f\|^2 = \sum_{x \in G'/B} |f(x)|^2 < \infty.$$

Endowed with the scalar product

$$(f, g) = \sum_{x \in G'/B} f(x) \cdot \overline{g(x)}, \qquad (f, g \in L^2(G'/B)),$$

it is a Hilbert space on which G' acts by unitary transformations via left translations. It contains $C_c(G'/B)$ as a dense subspace. It follows from 4.4(2) that $L^2(G/B)$, viewed as a G_0-module, is the direct sum of finitely many copies of $L^2(G_0/B)$. Since B is compact, $L^2(G'/B)$ may be identified with a closed G'-submodule of $L^2(G'/B)$, hence the closed non-zero irreducible G'-submodules of $L^2(G'/B)$ are members of the discrete series of G'. We want to determine those which are of the form $P_{B,G'}(\chi)$, where χ is a character of degree one of the Hecke algebra $H(G', B)$.

5.2. Proposition. *Let σ be a character of degree 1 of H_0 and τ a character of Ψ such that $\chi = \sigma \circ \tau$ is a character of H (see 3.5). Let $P_\chi = P_{B,G}(\chi)$, $L_\chi = L^2(G/B) \cap P_\chi$, $P_\sigma = P_{B,G_0}(\sigma)$ and $L_\sigma = L^2(G_0/B) \cap P_\sigma$. Then*

(i) $L_\chi \neq 0 \Leftrightarrow L_\chi^B \neq 0 \Leftrightarrow P_\chi = L_\chi$,

(ii) $L_\sigma \neq 0 \Leftrightarrow L_\sigma^B \neq 0 \Leftrightarrow P_\sigma = L_\sigma$,

and these conditions are equivalent to

(iii) $C_\sigma = \sum_{w \in \tilde{W}} q_w^{-1} \cdot \sigma(w)^2 < \infty$.

The convolution on the right by an element of $H(G', B)$ obviously preserves $L^2(G'/B)$ and is continuous. Hence L_σ and L_χ are closed subspaces of $L^2(G_0/B)$ and $L^2(G/B)$ respectively.

Let M be a closed invariant subspace of $L^2(G'/B)$ and pr_M the orthogonal projection of $L^2(G'/B)$ onto M. Then $\mathrm{pr}_M(C_c(G'/B))$ is dense in M. Since $C_c(G'/B)$ is the space spanned by the G'-transforms of e_1, it follows that $\mathrm{pr}_M(e_1) \neq 0$ if $M \neq 0$. This yields the first equivalence in (i), (ii); the second one follows from the irreducibility of P_σ and P_χ (4.10). We have $e_s = \check{e}_s$, hence $\sigma(e_w) = \sigma(\check{e}_w)$, since the $\sigma(e_s)$'s commute. Therefore, by 2.4(ii), the space $(P_\sigma)^B$ is spanned by

$$f_\sigma = \sum_{w \in \tilde{W}} q_w^{-1} \sigma(e_w) e_w, \tag{1}$$

and in view of 3.1(3)(4)(10)(11), $(P_\chi)^B$ is spanned by

$$f_\chi = \left(\sum_{\psi \in \Psi} \tau(\psi) \right) \cdot f_\sigma. \tag{2}$$

We have

$$\| f_\chi \|^2 = | \Psi | \cdot \| f_\sigma \|^2, \quad (|\Psi| = \mathrm{Card}\ \Psi). \tag{3}$$

Since G_0 is the disjoint union of the BwB ($w \in \tilde{W}$), we have

$$\| f_\sigma \|^2 = \sum_{w \in \tilde{W}} \sum_{x \in BwB/B} f_\sigma(x)^2; \tag{4}$$

f_σ being constant on BwB ($w \in \tilde{W}$), this gives

$$\| f_\sigma \|^2 = \sum_{w \in \tilde{W}} q_w \cdot |f_\sigma(w)|^2 = \sum_{w \in \tilde{W}} q_w^{-1} \cdot \sigma(w)^2 = C_\sigma. \tag{5}$$

Therefore

$$L_\sigma^B \neq 0 \Leftrightarrow \| f_\sigma \|^2 < \infty \Leftrightarrow C_\sigma < \infty \Leftrightarrow \| f_\chi \|^2 < \infty \Leftrightarrow L_\chi^B \neq 0,$$

whence the equivalence of (iii) with the conditions in (i), (ii).

5.3. Proposition. *We keep the notation of 5.2 and assume that $L_\sigma \neq 0$. Let d_σ (resp. d_χ) be the formal degree of the representation of G_0 in L_σ (resp. G in L_χ), computed with respect to the Haar measure dx for which B has volume 1. Then*

$$d_\chi = |\Psi|^{-1} d_\sigma, \quad d_\sigma = C_\sigma^{-1}.$$

The orthogonal projection of e_1 onto L_σ is non-zero (see proof of 5.2), invariant under B, hence is a non-zero multiple of f_σ. We may therefore write:

$$f_\sigma = c \cdot e_1 + u$$

with $u \in L^2(G_0/B)$ orthogonal to L_σ, whence

$$C_\sigma = (f_\sigma, f_\sigma) = c \cdot (f_\sigma, e_1) = c \cdot f_\sigma(1) = c,$$

$$f_\sigma = C_\sigma \cdot e_1 + u \qquad (u \in L^2(G_0/B), \ u \perp L_\sigma). \tag{1}$$

By definition of the formal degree, we have

$$\int_{G_0} |(f_\sigma, l_x \cdot f_\sigma)|^2 \, dx = d_\sigma^{-1} \cdot \|f_\sigma\|^4. \tag{2}$$

From (1) we get

$$(f_\sigma, l_x \cdot f_\sigma) = C_\sigma(f_\sigma, l_x \cdot e_1) = C_\sigma^2 \cdot f_\sigma(x^{-1}), \qquad (x \in G_0),$$

hence

$$(f_\sigma, l_x \cdot f_\sigma) = C_\sigma \cdot f_\sigma(w) = C_\sigma \cdot q_w^{-1} \cdot \sigma(w), \qquad (w \in \tilde{W}; \ x \in B \cdot w^{-1} \cdot B), \tag{3}$$

and, by (2)

$$\int_{G_0} |(f_\sigma, l_x \cdot f_\sigma)|^2 \, dx = C_\sigma^2 \cdot \|f_\sigma\|^2 = C_\sigma^3 = d_\sigma^{-1} \|f_\sigma\|^4 = d_\sigma^{-1} \cdot C_\sigma,$$

which yields our assertion for d_σ. The proof for d_χ is the same.

5.4. We now discuss the finiteness of C_σ, assuming that $\mathscr{G}$ is almost simple over k. As in 3.3, we let S_i $(1 \leq i \leq m)$ be the intersections of S with the conjugacy classes in $\tilde{W}$ and put $q_i = q_s$ for $s \in S_i$ (3.5). For $w \in \tilde{W}$, the number of elements of S_i occuring in a reduced decomposition of w depends only on w (as follows from Prop. 5, p. 16 of [4]) and will be denoted $l_i(w)$. Thus

$$l(w) = \sum_{1 \leq i \leq m} l_i(w). \tag{1}$$

We let t_i be an indeterminate,

$$\mathbf{l}(w) = (l_1(w), \ldots, l_m(w)), \qquad \mathbf{t} = (t_1, \ldots, t_m),$$

$$\mathbf{t}^{\mathbf{l}(w)} = \prod_{1 \leq i \leq m} t_i^{l_i(w)} \tag{2}$$

and consider the formal power series

$$W(\{t_i\}) = \sum_{w \in \tilde{W}} \mathbf{t}^{\mathbf{l}(w)}. \tag{3}$$

Given a character σ of degree 1 of H_0, let us put

$$\varepsilon_i = \begin{cases} 1 & \text{if } \sigma(e_s) = q_s \ \text{ for } s \in S_i, \\ -1 & \text{if } \sigma(e_s) = -1 \ \text{ for } s \in S_i. \end{cases} \tag{4}$$

We have

$$\sigma(e_w)^2 = \prod_i q_i^{(1 + \varepsilon_i) l_i(w)}, \tag{5}$$

$$\sigma(e_w)^2 / q_w = \prod_i q_i^{\varepsilon_i l_i(w)}, \tag{6}$$

whence the

5.5. Lemma. *The sum C_σ is finite if and only if $W(\{t_i\})$ converges for $t_i = q_i^{\varepsilon_i}$, and then we have*

$$C_\sigma = W(\{q_i^{\varepsilon_i}\}).$$

5.6. The series $W(\mathbf{t})$ represents a rational function [15: Prop. 26], which, according to [12] can be written in the form

$$W(\mathbf{t}) = P(\mathbf{t}) \cdot \left(\prod_1^l Q_j(\mathbf{t}) \right)^{-1},$$

where $P(\mathbf{t})$ is a polynomial with positive coefficients and $Q_j(\mathbf{t})$ is of the form

$$Q_j(\mathbf{t}) = 1 - t_1^{q_{j1}} \dots t_m^{q_{jm}}, \qquad (q_{ji} \geq 0, \text{ integral}).$$

Therefore

$$C_\sigma < \infty \Leftrightarrow \prod_{i=1}^m q_i^{\varepsilon_i q_{ji}} < 1 \qquad \text{for } j = 1, \dots, l. \tag{1}$$

We have $q_i = q^{a_i}$ with $q \geq 2$, and $a_i \geq 1$, integral, hence (1) may be written

$$C_\sigma < \infty \Leftrightarrow \sum_{i=1}^m a_i \cdot \varepsilon_i q_{ji} < 0, \qquad (j = 1, \dots, l). \tag{2}$$

5.7. If σ is the special character, then $\varepsilon_i = -1$ for all i and the conditions of 5.6(2) are fulfilled. The representations L_σ or L_χ are the special representations constructed by Matsumoto [13] and Shalika [16], to which we shall come back in § 6. In fact, in this case

$$C_\sigma = \sum q_w^{-1}. \tag{1}$$

If $m = 1$, this is then the only square integrable representation we obtain in this way.

5.8. We now consider the cases where $m \geq 2$. By [12], the polynomials $Q_j(\mathbf{t})$ $(1 \leq j \leq l)$ are

$(1 - t_1 t_2)$	for the type $\mathbf{A}_1$
$(1 - t_1^{l-2+j} \cdot t_2)$	for the type $\mathbf{B}_l$ $(l \geq 3)$
$(1 - t_1^{l-2+j} t_2 t_3)$	for the type $\mathbf{C}_l$ $(l \geq 2)$
$(1 - t_1^3 t_2^2),\ (1 - t_1^4 t_2^3),\ (1 - t_1^5 t_2^3),\ (1 - t_1^6 t_2^5)$	for the type $\mathbf{F}_4$
$(1 - t_1^2 t_2),\ (1 - t_1^3 t_2^2),$	for the type $\mathbf{G}_2$

with the following assignment of the t_i's to the conjugacy classes in S

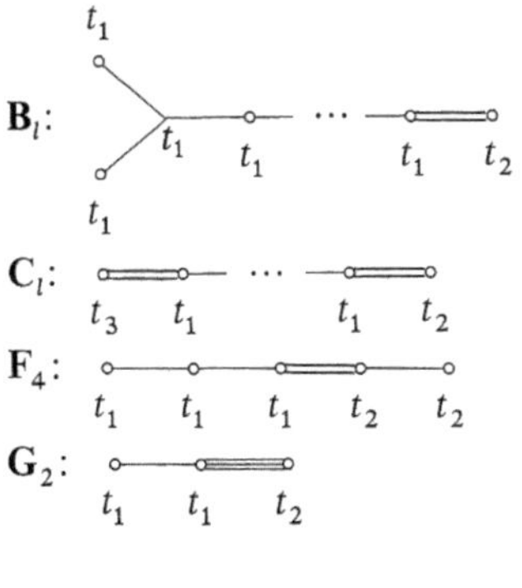

$$\mathbf{B}_l: \quad \substack{t_1 \\ t_1} \rangle\!\!-\!\!\circ\!-\cdots-\circ\!=\!=\!\circ \quad \substack{t_1 \quad t_1 \qquad t_1 \quad t_2}$$

$$\mathbf{C}_l: \quad \circ\!=\!=\!\circ\!-\!-\cdots-\circ\!=\!=\!\circ \quad \substack{t_3 \quad t_1 \qquad t_1 \quad t_2}$$

$$\mathbf{F}_4: \quad \circ\!-\!-\circ\!-\!-\circ\!=\!=\!\circ\!-\!-\circ \quad \substack{t_1 \quad t_1 \quad t_1 \quad t_2 \quad t_2}$$

$$\mathbf{G}_2: \quad \circ\!-\!-\circ\!\equiv\!\!\equiv\!\circ \quad \substack{t_1 \quad t_1 \quad t_2}$$

In the sequel we denote a character by the sequence of the ε_i defined by 5.4(4). For each type, and possible set of values of the q_i, we shall indicate the σ such that L_σ belongs to the discrete series, as given by application of 5.6(2) and 5.8.

(i) *Assume first the q_i's to be equal.* This is the case if G is split over k or, more generally, if G is quasi-split over k and splits over a totally ramified extension of k. The characters different from the special one which yield a square integrable representation are then

$$(-1, -1, 1),\ (-1, 1, -1),\ (-1, 1, 1) \quad \text{for the types } \mathbf{C}_l\ (l \geq 4),$$
$$(-1, -1, 1),\ (-1, 1, -1) \quad\qquad\qquad \text{for the types } \mathbf{C}_2, \mathbf{C}_3,$$
$$(-1, 1) \quad\qquad\qquad\qquad\qquad\qquad\quad \text{for the types } \mathbf{B}_l\ (l \geq 3),\ \mathbf{F}_4,\ \mathbf{G}_2.$$

(ii) *Assume the q_i's are not equal.* If $l=1$, and say, $q_1 < q_2$, then $(-1, 1)$ gives rise to a square integrable representation, whereas $(1, -1)$ does not.

Let now $l \geq 2$. The following table gives for each type, the possible values of the q_i, which were communicated to me by J. Tits, and the characters other than the special one which yield a square integrable representation.

Type of $\hat{W}$	(q_1, q_2)	Characters	Type of $\hat{W}$	(q_1, q_2)	Characters
$\mathbf{B}_l$ $(l \geq 4)$	(q, q^2)	$(-1, 1)$	$\mathbf{F}_4$	(q, q^2)	$(1, -1)$
$\mathbf{B}_l$ $(l \geq 2)$	(q^2, q)	$(-1, 1)$	$\mathbf{F}_4$	(q^2, q)	$(-1, 1)$
$\mathbf{B}_l$ $(l \geq 3)$	(q^2, q^3)	$(-1, 1)$	$\mathbf{G}_2$	(q, q^3)	$(1, -1)$
			$\mathbf{G}_2$	(q^3, q)	$(-1, 1)$

For the type $\mathbf{C}_l$, the list is as follows.

(q_1, q_2, q_3)	l	Characters	(q_1, q_2, q_3)	l	Characters
(q, q, q^2)	2	$(1, -1, -1), (-1, 1, -1)$	(q^2, q, q^4)	2	$(1, -1, -1)$
	≥ 3	$(-1, -1, 1), (-1, 1, -1)$		≥ 2	$(-1, -1, 1), (-1, 1, -1)$
	≥ 5	$(-1, 1, 1)$		≥ 4	$(-1, 1, 1)$
(q, q^2, q^2)	2	$(1, -1, -1)$	(q^2, q^2, q^3)	2	$(1, -1, -1)$
	≥ 2	$(-1, -1, 1), (-1, 1, -1)$		≥ 2	$(-1, -1, 1), (-1, 1, -1)$
	≥ 6	$(-1, 1, 1)$		≥ 4	$(-1, 1, 1)$
(q^2, q, q)	≥ 2	$(-1, -1, 1), (-1, 1, -1)$	(q^2, q^3, q^3)	2	$(1, -1, -1)$
	≥ 3	$(-1, 1, 1)$		≥ 2	$(-1, -1, 1), (-1, 1, -1)$
(q^2, q, q^2)	≥ 2	$(-1, -1, 1), (-1, 1, -1)$		≥ 5	$(-1, 1, 1)$
	≥ 3	$(-1, 1, 1)$			
(q^2, q, q^3)	≥ 2	$(-1, 1, -1)$	(q^2, q^3, q^4)	2	$(1, -1, -1)$
	≥ 3	$(-1, -1, 1)$		≥ 2	$(-1, -1, 1), (-1, 1, -1)$
	≥ 4	$(-1, 1, 1)$		≥ 5	$(-1, 1, 1)$

5.9. Remarks. (1) Let σ be a character of H_0 and π the representation of G_0 in L_σ. We use the notation of 4.6, 4.7. By 4.7 and 4.7(1), the space $(L_\sigma)_N$ is one-dimensional and we have

$$\pi(a) \cdot v = \sigma(e_a) \cdot q_a^{-1} \cdot v, \quad (v \in (L_\sigma)_N;\ a \in A^-), \tag{1}$$

where A^- is a suitable negative Weyl chamber in a k-split torus of M. Moreover,

$$q_a = |\delta_P(a)|^{-1}, \qquad (a \in A^-). \tag{2}$$

It follows that (π, L_σ) embeds in the induced representation $\mathrm{Ind}\,(\chi | P, G)$ where χ is the character of P which is equal to $\sigma(a)\,\delta_P(a)$ on A^-.

(2) [7:6.5.1] gives a criterion for the square integrability of an irreducible submodule of an induced representation. In view of the previous remark it yields in our case

$$|\sigma(a)| < (q_a)^{\frac{1}{2}}, \qquad (a \in A^-). \tag{3}$$

It can be checked directly that this is equivalent to 5.6(2). In fact, the first determination of the square integrable L_σ, obtained with the help of J. Tits, was based on a criterion quite similar to (3), proved by some geometric considerations on euclidean reflection groups.

(3) I understand that the results stated without proof or reference in 5.8 will eventually be found somewhere in the sequel to [5].

5.10. *Remarks on the formal degree.* It follows from 5.3 and 5.5, 5.6 that d_σ is a rational number. If σ_0 is the special character and $\sigma \neq \sigma_0$, then $|\sigma(w)| \geq |\sigma_0(w)|$ for all $w \in \tilde{W}$, and $|\sigma(w)| \neq |\sigma_0(w)|$ for at least one w, hence $d_\sigma < d_{\sigma_0}$. Examples show that in general neither of d_σ and d_{σ_0} is an integral multiple of the other. We note that the elements of the discrete series constructed here are not cuspidal, since the coefficient $x \mapsto (f_\sigma, l_x \cdot f_\sigma)$ is not compactly supported in view of 5.2(1).

5.11. The algebra $H = H_{\mathbf{C}}(G, B)$ is in a natural way an involutive Lie algebra, the involution $h \mapsto h^*$ being defined by $f \mapsto f^*$ where $f^*(x) = \overline{f(x^{-1})}$ ($f \in C_c(B \smallsetminus G/B)$, $x \in G$). In particular $e_w^* = e_{w^{-1}}$ ($w \in G$). Let E be a finite dimensional Hilbert space. A representation r of H into $\mathrm{End}\,E$ is self-adjoint if, for every $h \in H$, $r(h^*)$ is the adjoint of $r(h)$. In particular, every one-dimensional representation is so. Clearly, if (π, V) is an admissible unitary representation of G, then the associated representation of H into V^B is self-adjoint. It is not known whether conversely, if (r, E) is self-adjoint, the representation $P(E)$ is unitary. If $l=1$, however this is the case according to [14: p. 20].

§ 6. The Special Representation and Cohomology of Buildings

In this section $\mathcal{G}$ is simply connected and almost simple over k.

6.1. As before, X is the Bruhat-Tits buildung of G. It is a locally finite simplicial complex of dimension l, union of its simplices of dimension l, which are called the chambers. We let $C^j(X)$ (resp. $C_c^j(X)$) be the space of j-dimensional complex valued cochains with arbitrary (resp. compact) supports on X. In particular $C^l(X) = C(G/B)$ and $C_c^l = C_c(G/B)$. We let $d: C^j \to C^{j+1}$ be the coboundary operator and $\delta: C^j(X) \to C^{j-1}(X)$ its adjoint with respect to a suitable scalar product [1;9] which, for $j=l$, coincides with the one introduced on $C(G/B)$ in § 5. A element c of $C^j(X)$ is *harmonic* if it is annihilated by d and δ. If c is square integrable, this is equivalent to being annihilated by the Laplace operator $\Delta = d\delta + \delta d$ [1;9].

$H^j_c(X)$ (resp. $H_j(X)$) is the j-th cohomology group *with compact supports* (resp. j-th homology group *with arbitrary supports*) of X with complex coefficients. In particular

$$H^l_c(X) = C^l_c(X)/dC^{l-1}_c, \qquad H_l(X) = C^l(X) \cap \ker \delta = Z_l. \tag{1}$$

$C^j_2(X)$ is the space of square integrable j-cochains, and H^j_2 the space of square integrable harmonic cochains. In particular, $C^l_2 \cong L^2(G/B)$; there is an orthogonal decomposition

$$L^2(G/B) = C^l_2 = H^l_2 \oplus \overline{dC^{l-1}_2} \tag{2}$$

where $\overline{}$ denotes closure in the Hilbert space $L^2(G/B)$ [1; 9]. An element $u \in C^l(X)$ is a cycle if and only if, viewed as an element of $C^l(G/B)$, it satisfies the relations

$$f * e_s = -f, \qquad (s \in S). \tag{3}$$

It follows that f is a cycle if and only if

$$f * h = \sigma_0(h) \cdot f, \qquad \text{for all } h \in H(G, B), \tag{4}$$

where σ_0 is the special character (3.5). By 5.2, the identification $C^l(X) = C(G/B)$ then provides isomorphisms of G-modules

$$H_l(X) = P^0(\sigma_0), \qquad H^l_2 = P^0(\sigma_0) \cap L^2(G/B), \tag{5}$$

where $P^0(\sigma_0)$ is as in 2.3 and

$$H_l(X)^\infty = P(\sigma_0) = L_{\sigma_0}. \tag{6}$$

Let pr_h be the orthogonal projection of C^l_2 onto H^l_2. Since $C^l_c(X) \subset C^l_2$, $\ker \mathrm{pr}_h$ contains dC^{l-1}_2 (see (2)) and this projection induces a homomorphism of G-modules

$$\eta\colon H^l_c(X) \to H^l_2 = L_{\sigma_0}. \tag{7}$$

6.2. Theorem. *The homomorphism $\eta\colon H^l_c(X) \to H^l_2$ is injective. Its image is the space of smooth elements of H^l_2.*

Since $H^l_c(X)$ is a quotient of $C^l_c(X) = C_c(G/B)$, it is generated by $G \cdot e_1$ and is smooth, hence $\mathrm{Im}\,\eta$ is generated by $G \cdot \eta(e_1)$ and contained in L_{σ_0}. As remarked in the proof of 5.2, $\eta(e_1) \neq 0$, hence $\mathrm{Im}\,\eta$ is a non-zero G-submodule of L_{σ_0}. Since the latter is irreducible (4.10; it also follows from the irreducibility of the special representation and [7: 2.1.5]), $\mathrm{Im}\,\eta$ is equal to L_{σ_0}. There remains to show that η is injective.

The scalar product on $C^l(X)$ defines a pairing between $C^l(X)$ and $C^l_c(X)$. It is well-known, and elementary, that $Z_l = C^l \cap \ker \delta$ is the annihilator of dC^{l-1}_c, so that we may identify $H_l(X)$ with the dual of $H^l_c(X)$ in such a way that the canonical pairing is defined by our scalar product. By [3: 5.6, 5.10] the G-space $H^l_c(X)$ is admissible. Therefore $H_l(X)^\infty$ is admissible and is the contragredient to $H^l_c(X)$ [7: 2.1.9] in particular, if U is a compact open subgroup of G, then $H_l(X)^U$ is canonically isomorphic to the dual of $H^l_c(X)^U$ and we have

$$\dim H_l(X)^U = \dim H^l_c(X)^U. \tag{1}$$

On the other hand, since η is surjective, $L^U_{\sigma_0}$ is a quotient of $H^l_c(X)^U$.

However, by 6.1(6), $L_{\sigma_0} = H_l(X)^\infty$, hence by (1), $\dim L_{\sigma_0}^U = \dim H_c^l(X)^U$ and η is injective on $H_c^l(X)^U$. Since $H_c^l(X)$ is the union of such subspaces, our assertion follows.

6.3. Let $\mathscr{P}$ be a minimal parabolic k-subgroup of G, Φ the set of roots of $\mathscr{G}$ with respect to a maximal k-split torus contained in $\mathscr{P}$ and Δ the set of simple roots associated to $\mathscr{P}$. Let $\mathscr{P}_I$ $(I \subset \Delta)$ be the parabolic k-subgroups of $\mathscr{G}$ containing $\mathscr{P}$, and $C^\infty(G/P_I)$ the space of locally constant complex valued functions on G/P_I. It is an admissible G-module (with respect to left translations), the unnormalized induced representation $\mathbf{Ind}(1|P_I, G)$. Let θ_I be its character. In the Grothendieck group of admissible G-modules, we have, by [3: 5.6, 5.10], the equality

$$H_c^l(X) = \sum_{I \subset \Delta} (-1)^{|I|}\, C^\infty(G/P_I), \tag{1}$$

where $|I| = \mathrm{Card}\, I$. The character of the admissible G-module $H_c^l(X)$ is then the alternating sum

$$\sum_{I \subset \Delta} (-1)^{|I|}\, \theta_I, \tag{2}$$

the "Steinberg character" of G. In view of 6.2, we have

6.4. Corollary. Let θ_I be the character of $\mathbf{Ind}(1|P_I, G)$ $(I \subset \Delta)$. Then the character of the special representation is equal to

$$\sum_{I \subset \Delta} (-1)^{|I|}\, \theta_I.$$

6.5. *Remark.* As we saw, the fact that in 6.2 $\eta(H_c^l(X)) = (H_2^l)^\infty$ is rather elementary, so that the main point is the injectivity of η, which is equivalent to $H_c^l(X)$ being an irreducible G-module. If the isomorphism 6.3(1) is granted, this is in turn equivalent to the right hand side of 6.3(1) representing the class of an irreducible admissible G-module. This is included in a more general Theorem announced in [8].

§7. Proof of Lemma 4.1

7.1. In this section, we use freely some terminology and known facts on buildings, to be found in [5]. Much of what we need is also reviewed in [3: §§4, 5].

As in 3.1, C_0 is the chamber fixed under $\tilde{B}$ and A_0 the apartment of X fixed under $\tilde{H} = \tilde{B} \cap \tilde{N}$. We recall that an apartment (resp. a wall) in X is the transform by some element of $\tilde{G}$ of A_0 (resp. of the fixed point set of a reflection in $\tilde{W}$). Any two chambers are contained in an apartment [5: 2.3.1]. A half-apartment is a closed half-space in an apartment whose boundary is a wall. If α is one, then $\partial\alpha$ denotes the wall which is its boundary (in any apartment containing it) and U_α the greatest unipotent subgroup of $\tilde{G}$ which fixes it. Thus U_α is normal in the fixer $\tilde{G}_\alpha$ of α in $\tilde{G}$, and $\tilde{G}_\alpha$ is the semidirect product of $\tilde{H}$ and U_α.

We let $d(\,,)$ be the canonical metric on X [5: §2]. It is invariant under $\tilde{G}$ and induces a euclidean metric on each apartment. If M is closed and N compact in X, then

$$d(M, N) = \min_{x \in M, y \in N} d(x, y).$$

The ball of radius a and center $x \in X$ is the set of $y \in X$ such that $d(x, y) \leq a$.

7.2. Lemma. (i) *Let C be a chamber of X. Let A be an apartment containing C and C_0 and α a half-apartment of A such that $C \not\subset \alpha$ and $C \cap \partial \alpha$ is a face of co-dimension 1 of C. Let D be the chamber contained in α such that $D \cap C = C \cap \partial \alpha$. Then U_α is transitive on the set of chambers C' of X such that $C' \cap D = C \cap D$.*

(ii) *There exists a constant $\lambda > 0$ with the following property: if a is a half-apartment and $x \in \alpha$, then U_α fixes the ball of radius $\lambda \cdot d(x, \partial \alpha)$ and center x.*

(i) follows from 5.1.10, p. 86, and (ii) from 7.4.33, p. 179 in [5]. We note that it suffices to prove the existence of λ for one given half-apartment, since the half-apartments form finitely many orbits under $\tilde{G}$.

7.3. Lemma. *Let a_0 be a strictly positive real number. There exists a number $a_1 > 0$ with the following property: if C is a chamber of X, $d_s(C, C_0) \geq a_1$ and A is an apartment containing C, C_0, then there exists a half-apartment $\alpha \subset A$ such that $d(C_0, \partial \alpha) \geq a_0$, $C_0 \subset \alpha$, $C \subset (A - \alpha) \cup \partial \alpha$ and $C \cap \partial \alpha$ is a face of codimension one of C.*

Since $\tilde{G}$ is transitive on the set of pairs consisting of an apartment A' and a chamber $C' \subset A'$ [5: 2.26, p. 36], it suffices to prove this in a given apartment, say A_0. It follows there by an elementary argument in euclidean geometry, using the fact that the chambers are all congruent polysimplices and hence that the set of angles of two faces of codimension one of all chambers is finite.

7.4. Lemma. *Let U be a compact open subgroup of $\tilde{G}$. There exists a constant b_0 with the following property: let C be a chamber of X such that $d_s(C, C_0) \geq b_0$. Then there exists a minimal gallery $\gamma = \{C_0, C_1, \ldots, C_m = C\}$, $(m = d_s(C, C_0) + 1)$, connecting C_0 and C such that U is transitive on the set (C, C_{m-1}) of chambers C' in X such that $C' \cap C_{m-1} = C \cap C_{m-1}$.*

We may replace U by a smaller group, hence assume that $U \subset B$ fixes C_0. Let D be a ball in X, with center a point $x_0 \in C_0$, big enough so that the subgroup of $\tilde{G}$ which fixes it is contained in U. Let r be the radius of D, $a_0 = r/\lambda$, and choose a_1 as in 7.3. Let C be a chamber such that $d_s(C, C_0) \geq a_1$. Let A and α be as in 7.3. Then the wall $\partial \alpha$ contains a face of codimension 1 of C and separates C and C_0. There exists therefore in A a minimal gallery $\gamma = (C_0, C_1, \ldots, C_m = C)$ such that

$$C_i \subset \alpha \, (0 \leq i < m), \qquad C \cap C_{m-1} = C \cap \partial \alpha = C_{m-1} \cap \partial \alpha \tag{1}$$

[4: IV, §1, Ex. 16]. The gallery γ, being minimal in A, is also minimal in X [5: 2.3.6, p. 38] hence

$$m = d_s(C, C_0) + 1, \qquad d_s(C_{m-1}, C_0) = d_s(C, C_0) - 1. \tag{2}$$

By 7.2(ii), the group U_α fixes the ball of radius $\lambda \cdot d(x_0, \partial \alpha)$. Since

$$d(x_0, \partial \alpha) \geq d(C_0, \partial \alpha) \geq a_0 = r/\lambda,$$

the group U_α fixes the ball D chosen above hence,

$$U_\alpha \subset U. \tag{3}$$

Finally, the chamber C_m, C_{m-1} satisfy, with respect to A and α, the assumptions imposed on C, D in 7.2(i), therefore U_α is transitive on the set of (C, C_{m-1}). This proves 7.4, with $b_0 = a_1$.

7.5. *Proof of Lemma* 4.1. Since the isogeny $\omega \colon \tilde{G} \to G$ induces a homomorphism $\tilde{G} \to G_0$ which commutes with the actions of $\tilde{G}$ and G_0 on X, it suffices to consider the case where $G = \tilde{G}$, hence $G_0 = \tilde{G}$.

Let U be a compact open subgroup of $\tilde{G}$. Choose b_0 as in 7.4. Then, by 7.4, all our conditions are fulfilled if we take $d_0 = b_0$ and choose for D the element C_{m-1} of the gallery γ (see 7.4(1), (2)).

References

1. Borel, A.: Cohomologie de certains groupes discrets et laplacien p-adique. Sém. Bourbaki 26ème année (1973/74), Exp. 437, 24 p
2. Borel, A., Serre, J-P.: Cohomologie à supports compacts des immeubles de Bruhat-Tits; application à la cohomologie des groupes S-arithmétiques. C.R. Acad. Sci. Paris **272**, 110–113 (1971)
3. Borel, A., Serre, J-P.: Cohomologie d'immeubles et de groupes S-arithmétiques. A paraître dans Topology
4. Bourbaki, N.: Groupes et algèbres de Lie, Chap. IV, V, VI, Act. Sci. Ind. 1337. Paris: Hermann 1968
5. Bruhat, F., Tits, J.: Groupes réductifs sur un corps local. Chap. I. Publ. Math. I.H.E.S. **41**, 1–251 (1972)
6. Casselman, W.: The Steinberg character as a true character. Harmonic analysis on homogeneous spaces. Proc. Symp. pur. math. **26**, 413–417 (1974)
7. Casselman, W.: Introduction to the theory of admissible representations of $\mathfrak{P}$-adic reductive groups. To appear
8. Casselman, W.: On a p-adic vanishing theorem of Garland. Bull. A.M.S. **80**, 1001–1004 (1974)
9. Garland, H.: p-adic curvature and the cohomology of discrete subgroups of p-adic groups. Annals of Math. **97** (2), 375–423 (1973)
10. Higman, D.G.: Induced and produced modules. Canadian J.M. **VII**, 490–508 (1955)
11. Iwahori, N., Matsumoto, H.: On some Bruhat decompositions and the structure of the Hecke rings of p-adic Chevalley groups. Publ. Math. I.H.E.S. **25**, 5–48 (1965)
12. MacDonald, I.G.: The Poincaré series of a Coxeter group. Math. Annalen **199**, 161–174 (1972)
13. Matsumoto, H.: Fonctions sphériques sur un groupe semi-simple p-adique. C.R. Acad. Sci. Paris **269**, 829–832 (1969)
14. Matsumoto, H.: Analyse harmonique dans certains systèmes de Coxeter et de Tits. Analyse sur les harmonique et groupes de Lie. Lecture Notes in Math. **497**, 257–276. Berlin-Heidelberg-New York: Springer 1975
15. Serre, J-P.: Cohomologie des groupes discrets. Prospects in Mathematics. Annals of Math. Studies 70. Princeton: Princeton University Press 1970
16. Shalika, J.A.: On the space of cusp forms of a p-adic Chevalley group. Annals of Math. **92** (2), 262–278 (1970)

Armand Borel
The Institute for Advanced Study
School of Mathematics
Princeton, N.J. 08540
USA

Received December 23, 1975

107.

(with B. M. Schreiber)

p-adic linear groups with ergodic automorphisms

Isr. J. Math. **24** (1976) 199–205

ABSTRACT

Let k be a locally compact, totally disconnected, nondiscrete field, and let G be a Lie group over k satisfying suitable conditions which depend on the characteristic of k. It is shown that G is compact if it admits a bicontinuous automorphism which is ergodic with respect to Haar measure.

Let G be a locally compact group and T a bicontinuous automorphism of G. The automorphism T is called ergodic if every Borel subset A of G such that $T(A) = A$ must either be null or have null complement with respect to Haar measure on G. It has been conjectured that only compact groups can possess ergodic automorphisms, and there is considerable supporting evidence for this conjecture. See, for example, [7] and the earlier references listed there. In particular, it was shown in [7] that if G is a Zariski-connected linear algebraic group defined over the locally compact, totally disconnected, nondiscrete field k such that the unipotent radical of G is defined over k and T is an automorphism of G also defined over k, then the group $G(k)$ of rational points of G over k is compact with respect to the usual locally compact topology if the restriction of T to $G(k)$ is ergodic. In this paper we shall extend this last result in several ways. Firstly, we shall show that $G(k)$ as above must be compact without the assumption that the ergodic automorphism T of $G(k)$ be the restriction of a k-automorphism of G. Moreover, $G(k)$ need not be connected and may be replaced by a sufficiently large subgroup of $G(k)$ (Theorem 2). In fact, when k is of characteristic zero, we may substitute for $G(k)$ any Lie group over k with a finite-dimensional, continuous representation over k whose kernel is, say, compact or solvable (Theorem 1).

† Research supported in part by NSF Grant GP 20150 for the second-named author.
Received September 11, 1975

488

1. Preliminaries

Throughout this paper k is a nondiscrete, locally compact, totally disconnected field, $\bar{k}$ an algebraic closure of k, and $|\cdot|$ a multiplicative, discrete valuation on k. We also denote by $|\cdot|$ the canonical extension of this valuation to any finite extension of k.

Suppose G is a Zariski-connected linear algebraic group defined over k whose unipotent radical R_uG is also defined over k. We shall denote by G^* the subgroup of $G(k)$ generated by the subgroups $V(k)$, as V runs through the unipotent radicals of parabolic k-subgroups of G. The group G^* is normal in $G(k)$, and $G^* = \{e\}$ if G is reductive and anisotropic over k. When G is reductive, we shall also write G^+ for G^* to conform with the notation in [4].

For standard notation and results on Lie groups and linear algebraic groups see [6] and [2], [3], respectively.

We need the following two results, which appear as corollaries 1.3 and 1.7 in [7].

LEMMA 1. *Let G be a locally compact group which possesses an ergodic automorphism. If G is the union of an increasing sequence of compact open subgroups, then G is compact.*

LEMMA 2. *If the locally compact group G has a finite normal series of closed subgroups whose successive quotients are all unions of increasing sequences of compact open subgroups, then G is also the union of such an increasing sequence.*

LEMMA 3. *Let G be a σ-compact, locally compact, totally disconnected group with a closed, solvable, normal subgroup N. If the set of elements of G which lie in a compact subgroup is dense in G, then N is union of an increasing sequence of compact open subgroups.*

PROOF. Write

$$N = N^0 \supset N^1 \supset \cdots \supset N^n = \{e\},$$

where each N^j is a closed normal subgroup of G such that N^{j-1}/N^j is abelian, $1 \leq j \leq n$. Our lemma will follow from Lemma 2 if we show that each N^{j-1}/N^j is the union of an increasing sequence of compact open subgroups. Since N^{j-1}/N^j is abelian, totally disconnected and σ-compact, this is equivalent to showing that every element of N^{j-1}/N^j lies in a compact subgroup.

Fixing j and passing to the quotient group G/N^j we are reduced to considering the case where N is abelian, which we now assume. Let L be the set of elements of N which lie in a compact subgroup. Then L is an open subgroup of N which is normal in G. Thus $N' = N/L$ is a discrete, abelian, torsion-free, normal subgroup of $G' = G/L$. Let $x \in N'$. There exists a compact open subgroup H of G' such that $H \cap N' = \{e\}$ and every element of H centralizes x, and our hypothesis on G implies that for some $h \in H$, the product xh lies in a compact subgroup of G'. But then there exists a positive integer n such that $x^n h^n = (xh)^n \in H$, whence $x^n \in H \cap N' = \{e\}$. Thus $x = e$, and we have shown that $L = N$ as desired.

LEMMA 4. *Let $x \in GL_n(k)$. Then x is contained in a compact subgroup of $GL_n(k)$ if and only if all of its eigenvalues have valuation one.*

PROOF. If K is a finite extension of k, then $GL_n(k)$ is a closed subgroup of $GL_n(K)$. Hence we may replace k by a finite extension over which x may be put in upper triangular form, so let us assume x is upper triangular. Let B be the group of upper triangular matrices in $GL_n(k)$, let T and U be the subgroups of B consisting of all diagonal and of all unipotent upper triangular matrices, respectively, and let T_0 be the subgroup of T of all elements whose diagonal entries are units in the valuation ring of k. Then T_0 is compact, and U has a normal series of closed subgroups whose successive quotients are isomorphic to the additive group of k. Thus $T_0 \cdot U$ is the union of an increasing sequence of compact open subgroups by Lemma 2. If all of the eigenvalues of x have valuation one, then $x \in T_0 \cdot U$, so x lies in a compact subgroup of B. Conversely, T_0 is the largest compact subgroup of $T = B/U$. Thus if x is contained in a compact subgroup of B, then $x \in T_0 \cdot U$.

LEMMA 5. *The set of elements of $GL_n(k)$ which are contained in a compact subgroup is open and closed in $GL_n(k)$.*

PROOF. Let $x \in GL_n(k)$ and let λ be an eigenvalue of x. Then λ belongs to an extension K of degree n of k, so $|\lambda| = |N_{K/k}(\lambda)|^{1/n}$. Thus the set of valuations of eigenvalues of elements of $GL_n(k)$ is discrete. Furthermore, it is standard that if $P \in k[X]$ and Q is another element of $k[X]$ all of whose coefficients are sufficiently close in k to those of P, then the valuations of the roots of P and of Q are the same [1, ch. 6, §3]. It follows that the set of elements of $GL_n(k)$ whose eigenvalues have any prescribed valuations is open, and therefore also closed. Our lemma now follows from Lemma 4.

2. Ergodic automorphisms (k of characteristic zero)

Suppose that k is of characteristic zero, so that k is a finite separable extension of the field Q_p of p-adic numbers for some prime p. There is then a functor R_{k/Q_p} of "restriction of scalars" which maps the category of Lie groups over k to the category of p-adic Lie groups [5, 5.14], [6, p. 99]. If G is a Lie group over k, then $R_{k/Q_p}G$ is topologically isomorphic to G. Thus if G, G' are Lie groups over k, any continuous homomorphism $\rho: G \to G'$ may be viewed as a continuous homomorphism of $R_{k/Q_p}G$ into $R_{k/Q_p}G'$. If $G' = GL_n(k)$, then $R_{k/Q_p}G'$ may be canonically identified to a closed subgroup of $GL_{dn}(Q_p)$, where $d = [k : Q_p]$. Therefore if ρ is a continuous finite-dimensional representation of G over k, then it yields one of $R_{k/Q_p}G$ over Q_p. We recall that if $k = Q_p$, then every continuous homomorphism of Lie groups over k is analytic [6, III, §8, no. 1, th. 1] and every closed subgroup of a Lie group over k is a Lie group over k (loc. cit. no. 2, th. 2).

LEMMA 6. *Let k be of characteristic zero. Let G be a σ-compact Lie group over k and $\rho: G \to GL_n(k)$ a faithful continuous representation of G over k. Suppose that the set of elements of G which are contained in a compact subgroup is dense in G. Then G is the union of an increasing sequence of compact open subgroups.*

PROOF. The above remarks show that the restriction of scalars reduces us to the case where $k = Q_p$, and then ρ is a morphism of Lie groups. We let $L(G)$ denote the Lie algebra of G, and we distinguish three cases:

a) $L(G) = \{0\}$. Then G is a countable, discrete, periodic group, and it suffices to show G is locally finite. But since ρ is faithful this follows from a classical result of Schur [8] (see also [9, cor. 4.9]).

b) $L(G)$ *is semi-simple*. Let M be the unique connected algebraic subgroup of $GL_n(\bar{k})$ with Lie algebra $d\rho(L(G)) \otimes_k \bar{k}$. Then M is normalized by $\rho(G)$ and the group $M(k)$ of k-rational points of M is a closed subgroup of $GL_n(k)$ which is a Lie group over k with Lie algebra $d\rho(L(G))$. If H is a compact open subgroup of G, then $\rho(H)$ is a Lie subgroup of $GL_n(k)$ with Lie algebra $d\rho(L(G))$ (recall $k = Q_p$), so $\rho(H) \cap M(k)$ is open in $M(k)$ [6, III, §7, no. 1, th. 2], whence so is $\rho(G) \cap M(k)$. Thus $\rho(G) \cap M(k)$ is closed in $M(k)$, hence in $GL_n(k)$, the group $N = \rho^{-1}(M(k))$ is an open, normal subgroup of G and $\rho \mid N$ is a topological isomorphism of N onto $\rho(G) \cap M(k)$.

Let $\tilde{M}$ denote the normalizer of M in $GL_n(\bar{k})$. There is a k-rational linear representation α of $\tilde{M}$ with kernel M [2, th. 5.6]. Since $\rho(G) \subset \tilde{M}$, the map $\alpha \circ \rho$ yields a faithful representation of G/N, and it follows from a) that G/N is the union of an increasing sequence of finite subgroups. Thus, to complete the proof in case b), it suffices to show that N, or equivalently $\rho(G) \cap M(k)$, is compact. Let $M_1, \cdots, M_t$ be the isotropic simple quotients of M over k and $\pi_i: M \to M_i$ the canonical projection ($1 \le i \le t$). The group $L_i = \pi_i(\rho(N))$ is open ($1 \le i \le t$); let us show it is compact. By Lemma 5 every element of $\rho(N)$, hence of L_i, lies in a compact group. The group M_i^+ is closed in $M_i(k)$ [6, 6.14], hence $L_i \cap M_i^+$ is an open subgroup of M_i^+ which is the union of its compact subgroups. It is then a proper open subgroup of M_i^+, since the latter contains the k-rational points of a k-split torus and hence an element which generates an infinite discrete subgroup of M_i^+. But it is known that every proper open subgroup of M_i^+ is compact [4, 9.10], hence L_i is compact. It follows that the image of $\rho(N)$ under the mapping

$$\pi = \pi_1 \times \cdots \times \pi_t : M \to M_1 \times \cdots \times M_t$$

is compact. The kernel Q of π is anisotropic over k, hence $Q(k)$ is compact and so is $\rho(N) \cap Q(k)$. Consequently $\rho(N)$ is compact, which completes the proof of the lemma in the case b).

c) $L(G)$ *has a nonzero radical* $\mathfrak{r}$. Let $\mathfrak{a}$ be the algebraic hull of $d\rho(\mathfrak{r})$ and A the unique connected algebraic subgroup of $\boldsymbol{GL_n}(\bar{k})$ with Lie algebra $\mathfrak{a}$. Then A is solvable and normalized by $\rho(G)$, hence $N = \rho^{-1}(A(k))$ is a closed, normal, solvable subgroup of G. Since $\mathfrak{a} \cap d\rho(L(G)) \supset d\rho(\mathfrak{r})$, we have $(d\rho)^{-1}(\mathfrak{a}) \supset \mathfrak{r}$, whence $L(N) = \mathfrak{r}$. Thus G/N is a semi-simple Lie group over k with Lie algebra $L(G)/\mathfrak{r}$. Clearly G/N is σ-compact and the union of its compact subgroups is dense. There exists a k-rational linear representation β of the normalizer B of A in $\boldsymbol{GL_n}(\bar{k})$ with kernel A [2, th. 5.6.]. Then $\beta \circ \rho$ is a faithful, continuous representation of G/N over $\boldsymbol{Q_p}$. Thus G/N satisfies the hypotheses of our lemma, so by b) it is the union of an increasing sequence of compact open subgroups. This is then also true for G by Lemmas 2 and 3.

THEOREM 1. *Let k be of characteristic zero. Let G be a σ-compact Lie group over k with a finite-dimensional, continuous representation ρ over k such that* ker ρ *is either the union of an increasing sequence of compact open subgroups or solvable. If G admits an ergodic automorphism, then G is compact.*

PROOF. Let T be an ergodic automorphism of G. If H is a compact open subgroup of G, then $\bigcup_{n \in \mathbb{z}} T^n(H)$ is open, non-empty and invariant under T, hence dense in G. It follows that $G/\ker \rho$ satisfies the hypotheses of Lemma 6, hence $G/\ker \rho$ is the union of an increasing sequence of compact open subgroups. Moreover, if ker ρ is solvable, it is also such a union by Lemma 3. The theorem now follows from Lemmas 1 and 2.

3. Ergodic automorphisms (k of arbitrary characteristic)

LEMMA 7. *Let G be a reductive, connected linear algebraic group defined over k, and suppose that the derived group $\mathscr{D}G$ of G has k-rank zero. Then $G(k)$ has a unique maximal compact subgroup.*

PROOF. Let S be the maximal k-split torus in the radical R of G, and let D be the group generated by $\mathcal{D}G$ and the maximal anisotropic k-torus in R. Then $G = S \cdot D$, and the canonical mapping $S \times D \to G$ is a central isogeny. Thus $S(k)D(k)$ is a closed cocompact subgroup of $G(k)$ [4, prop. 3.19], and D is anisotropic, so $D(k)$ is compact. Hence $G(k)/S(k)$ is compact.

Let $\chi_1, \cdots, \chi_d \in X(G)_k$ be chosen so that their restrictions to S form a basis of $X(S) \otimes Q$ ($d = \dim S$). This can be done because the group of restrictions of elements of $X(G)_k$ to S has finite index in $X(S)$. Let $\boldsymbol{R}_+^*$ denote the multiplicative group of strictly positive real numbers, and define $\rho : G(k) \to (\boldsymbol{R}_+^*)^d$ by

$$\rho(x) = (|\chi_1(x)|, \cdots, |\chi_d(x)|).$$

The map ρ is a continuous homomorphism whose image is discrete and isomorphic to Z^d. The group $K = \ker \rho$ is an open subgroup of $G(k)$ containing all compact subgroups of $G(k)$. The quotient $K/(S(k) \cap K)$ is then an open, hence compact, subgroup of $G(k)/S(k)$. Since $S(k) \cap K$ is compact, we see that K is compact, which proves the lemma.

THEOREM 2. *Let G be a linear algebraic group defined over k, G^0 its (Zariski) identity component, and suppose that $R_u G^0$ is also defined over k. Let H be a closed subgroup of $G(k)$ which contains $(G^0)^*$ and admits an ergodic automorphism. Then H is compact.*

PROOF. We shall again show that H is the union of an increasing sequence of compact open subgroups. Since $H \cap G^0$ has finite index in H, we may assume G to be connected (Lemma 2). Let $\pi : G \to G' = G/R_u G$ be the canonical homomorphism. Since $(R_u G)(k) \subset H$ and $\pi(G^*) = G'^+$, we see that $\pi(H)$ is a closed subgroup of $G'(k)$ containing G'^+. By Lemma 5 every element of H lies in a compact subgroup. Then G'^+ contains no elements generating infinite discrete subgroups, hence $G'^+ = \{e\}$. Thus the k-rank of the derived group of G' is zero, so by Lemma 7 $G'(k)$ has a unique maximal compact subgroup K. It follows that $\pi(H) \subset K$. Now, $R_u G$ is trigonalizable over k, so $(R_u G)(k)$ is the union of an increasing sequence of compact open subgroups. The theorem is then proved upon applying Lemmas 1 and 2.

<h2 style="text-align:center">REFERENCES</h2>

1. E. Artin, *Theory of Algebraic Numbers* (notes by Gerhard Würges), George Striker Pub., Göttingen, 1959.

2. A. Borel, *Linear Algebraic Groups*, W. A. Benjamin, New York, 1969.

3. A. Borel and J. Tits, *Groupes réductifs*, Inst. Hautes Études Sci. Publ. Math. no. **27** (1965), 55–151; Compléments, Ibid., no. **41** (1972), 253–276.

4. A. Borel and J. Tits, *Homomorphismes "abstraits" de groupes algébriques simples*, Ann. of Math. (2) **97** (1973), 499–571.

5. N. Bourbaki, *Variétés différentielles et analytiques*, Actualités Sci. Indust., no. 1333 (1967).

6. N. Bourbaki, *Groupes et algèbres de Lie, Chapt. II et III*, Actualités Sci. Indust., no. 1349 (1972).

7. M. Rajagopalan and B. Schreiber, *Ergodic automorphisms and affine transformations of locally compact groups*, Pacific J. Math. **38** (1971), 167–176.

8. I. Schur, *Über Gruppen periodischer Substitutionen*, Sitzungsb. Preuss. Akad. Wiss. (1911), 619–627.

9. B. A. F. Wehrfritz, *Infinite Linear Groups*, Ergebnisse der Math. und ihrer Grenzgeb., Band 76, Springer-Verlag, New York, 1973.

SCHOOL OF MATHEMATICS, THE INSTITUTE FOR ADVANCED STUDY
  PRINCETON, NEW JERSEY 08540 USA

DEPARTMENT OF MATHEMATICS, WAYNE STATE UNIVERSITY
  DETROIT, MICHIGAN 48202 USA

108.

Cohomologie de SL$_n$ et valeurs de fonctions zeta aux points entiers

Ann. Sc. Norm. Super. Pisa, Cl. Sci., (4) **4** (1977) 613–636
Correction, ibid. **7** (1980) 373

dédié à Jean Leray

Cet article est un suite à [4] et est consacré principalement à l'étude d'un problème concernant l'homomorphisme canonique

$$j_\Gamma\colon H^*(X_u;\, \boldsymbol{R}) \to H^*(\Gamma;\, \boldsymbol{R})\,,$$

où Γ est un sous-groupe arithmétique d'un $\boldsymbol{Q}$-groupe semi-simple G et X_u l'espace symétrique « dual » de l'espace X des sous-groupe compacts maximaux du groupe $G(\boldsymbol{R})$ des points réels de G. Cet homomorphisme est défini à l'aide de formes différentielles invariantes (cf. 3.4) et est un isomorphisme en basses dimensions [4]. Ces deux espaces vectoriels réels ont des $\boldsymbol{Q}$-structures rationnelles naturelles, définies par la cohomologie rationnelle, et notre but est d'étudier le comportement de j_Γ par rapport à elles. Le résultat principal de cet article concerne le cas où $G = R_{k/\boldsymbol{Q}}\boldsymbol{SL}_n$ est obtenu par restriction des scalaires du corps de nombres k à $\boldsymbol{Q}$, et met en jeu les valeurs de la fonction zêta de Dedekind $\zeta_k(s)$ de k aux entiers positifs. Plus précisément, fixons un entier positif m et soit n un entier impair suffisamment grand. Alors $H^*(X_u;\, \boldsymbol{Q})$ est une algèbre extérieure et la dimension de l'espace $I^{2m+1}(X_u;\, \boldsymbol{Q})$ des éléments indécomposables de degré $2m+1$ de $H^*(X_u;\, \boldsymbol{Q})$ se trouve être égale à l'ordre d_m du zéro de $\zeta_k(s)$ au point $s = -m$. Soient $L_m(X_u;\, \boldsymbol{Q}) = \Lambda^{d_m} I^{2m+1}(X_u;\, \boldsymbol{Q})$ et définissons de même $L_m(\Gamma;\, \boldsymbol{Q})$ à partir de $H^*(\Gamma;\, \boldsymbol{Q})$. Alors j_Γ induit un isomorphisme de $L_m(X_u;\, \boldsymbol{Q}) \otimes_{\boldsymbol{Q}} \boldsymbol{R}$ sur $L_m(\Gamma;\, \boldsymbol{Q}) \otimes_{\boldsymbol{Q}} \boldsymbol{R}$ et l'on montrera que l'on a

$$(1) \qquad j_\Gamma\big(L_m(X_u;\, \boldsymbol{Q})\big) = R'_m \cdot L_m(\Gamma;\, \boldsymbol{Q})\,,$$

(*) The Institute for Advanced Study, Princeton, New Jersey.
Pervenuto alla Redazione l'11 Novembre 1976.

495

avec

$$(2) \qquad R'_m = |D|^{\frac{1}{2}} \zeta_k(m+1) \cdot \pi^{-d(m+1)},$$

où D est le discriminant et d le degré de k sur $\boldsymbol{Q}$ (6.2). On a supposé ici implicitement $d_m \neq 0$, ce qui est le cas sauf si m est impair et k totalement réel. En fait, il n'y a pas lieu d'imposer cette condition dans la démonstration, qui redonne donc aussi le fait bien connu que le membre de droite de (2) est un nombre rationnel lorsque $\zeta_k(-m) \neq 0$ (6.3).

Multiplié par un facteur rationnel convenable, le nombre R'_m est le m-ième régulateur R_m de k qui intervient dans les conjectures de S. Lichtenbaum en K-théorie algébrique [12]. L'égalité (2) montre que si l'on divise par π^{dm} le membre de gauche de l'égalité conjecturale de [12: § 4], qui est par définition $\lim\limits_{s \to -m} \zeta_k(s)(s+m)^{-d_m}$, alors les deux membres de cette égalité sont commensurables (6.4).

Nous résumons maintenant la marche de la démonstration. Soient K (resp. G_u) le sous-groupe compact maximal standard de $G(\boldsymbol{R})$ (resp. $G(\boldsymbol{C})$). On a donc

$$K = (\boldsymbol{SO}(n))^{r_1} \times (\boldsymbol{SU}(n))^{r_2}, \qquad G_u = (\boldsymbol{SU}(n))^d,$$

où r_1 (resp. r_2) est le nombre de places réelles (resp. complexes) de k. Comme n est supposé impair, K est totalement non homologue à zéro dans G_u et dans $G(\boldsymbol{R})/\Gamma$. Cela permet de remplacer l'étude de j_Γ par celle de l'homomorphisme

$$\mu^* \colon H^*(G_u; \boldsymbol{C}) \to H^*(G(\boldsymbol{R})/\Gamma; \boldsymbol{C}),$$

défini par restriction de formes différentielles invariantes (3.4). Plus précisément, on peut identifier naturellement l'algèbre de cohomologie $H^*(\mathfrak{g}; \boldsymbol{C})$ de l'algèbre de Lie $\mathfrak{g}$ de G à l'algèbre des formes différentielles biinvariantes sur $G(\boldsymbol{C})$. Alors $\mu^* = \beta^* \circ \alpha^{*-1}$, où

$$\alpha^* \colon H^*(\mathfrak{g}; \boldsymbol{C}) \to H^*(G_u; \boldsymbol{C}) \quad \text{et} \quad \beta^* \colon H^*(\mathfrak{g}; \boldsymbol{C}) \to H^*(G(\boldsymbol{R})/\Gamma; \boldsymbol{C}),$$

sont définis par restriction de formes différentielles. Cela nous ramène à étudier l'effet de α^* et β^* sur $H^*(\mathfrak{g}(\boldsymbol{Q}); \boldsymbol{Q})$. C'est aisé dans le cas de α^*, et conduit à la puissance de π et au facteur $|D|^{\frac{1}{2}}$ de (2), (cf. 5.4). Soient $R\omega_{2j+1}$ une forme biinvariante sur $G(\boldsymbol{C})$ qui représente un élément de la d-ième puissance extérieure de l'espace des éléments indécomposables de degré $2j+1$ de $H^*(\mathfrak{g}(\boldsymbol{Q}); \boldsymbol{Q})$, et η_m le produit des $R\omega_{2j+1}$ $(1 \leq j \leq m)$. Pour décrire l'effet de β^* sur la cohomologie rationnelle, on se ramène à prouver que l'on a

$$(3) \qquad \beta_*(\eta_m) \in i^{r_2 \cdot m} \Big(\prod_{1 \leq j \leq m} \zeta_k(j+1) \Big) \cdot H^*(\Gamma; \boldsymbol{Q})$$

et, pour cela, à construire une variété compacte Z_m dans $G(\boldsymbol{R})/\Gamma$, de dimension réelle $d(m^2 + 2m)$, sur laquelle l'intégrale de η_m est, à un facteur rationnel non nul près, le coefficient numérique dans la relation (3). Dans ce but, on part d'une algèbre à division D centrale sur k, de rang $(m+1)^2$, triviale à l'infini, on plonge le k-groupe algébrique H des éléments de norme réduite 1 de $D \otimes_k \boldsymbol{C}$ dans G par la représentation régulière, et on prend pour Z_m le quotient $H(\boldsymbol{R})/(\Gamma \cap H)$. C'est une variété (si Γ est sans torsion, ce que l'on peut supposer sans inconvénient), qui est compacte (théor. de Fr. Hey), sur laquelle η_m définit une forme rationnelle invariante non nulle de degré maximum. L'assertion sur son intégrale (cf. 5.5) résulte alors du fait que le nombre de Tamagawa de H est un nombre rationnel (2.4).

Pour pouvoir poser notre problème, nous avons admis que j_Γ^* est un isomorphisme en basses dimensions. Mais en fait la construction précédente fournit une autre démonstration de l'injectivité de j_Γ^* qui, à l'encontre de celle de [4], ne fait pas intervenir la compactification de X/Γ en une variété à coins (cf. 5.6).

Donnons pour terminer quelques indications sur les différents paragraphes. Le § 1 fixe quelques notations et conventions, le § 2 est consacré aux algèbres à division. Les homomorphismes $\alpha^*, \beta^*, j_\Gamma^*$ sont définis au § 3 dans le cas général; le § 4 considère α^* pour un groupe déployé. Les §§ 5, 6 sont consacrés au cas où $G = R_{k/\boldsymbol{Q}} \boldsymbol{SL}_n$; le premier étudie α^* et β^*, le deuxième démontre (1), (2) et discute leurs relations avec la K-théorie algébrique. Enfin, le § 7 contient quelques remarques générales sur l'homomorphisme j_Γ^*.

Le principal résultat de ce travail a été annoncé dans [5: § 5].

1. – Notations et conventions.

Nous suivons en général celles de [4], avec une déviation notée en 1.2, et adoptons en outre les suivantes.

1.1. – Si $a, b \in \boldsymbol{C}^*$, alors $a \sim b$ signifie que $a/b \in \boldsymbol{Q}^*$.

1.2. – Si M est une variété différentielle (de classe C^∞), on note $A^p(M)$ (resp. $A^p(M, \boldsymbol{C})$) l'espace des formes différentielles extérieures de degré p sur M, à valeurs réelles (resp. complexes) et $A^*(M)$ (resp. $A^*(M, \boldsymbol{C})$) la somme directe de ces espaces. Soient M_1, M_2 deux variétés différentielles,

π_i la projection de $M = M_1 \times M_2$ sur M_i et $\alpha_i \in A^*(M, C)$ $(i = 1, 2)$. Alors la forme différentielle $\pi_1^*\alpha \wedge \pi_2^*\alpha$ sera simplement notée $\alpha_1 \wedge \alpha_2$. On adopte une convention semblable pour un nombre quelconque de facteurs.

1.3. – Soit A une algèbre sur un corps F graduée par des sous-espaces A^p $(p \in N)$. Par abus de langage, on appelle espace des *éléments indécomposables* de degré p de A le quotient de A^p par le sous-espace des éléments décomposables, i.e. le sous-espace engendré par les produits d'éléments de degrés $< p$.

1.4. – On note k un corps de nombres, $\mathfrak{o}(k)$ l'anneau des entiers de k, d le degré et D ou $D_{k/Q}$ le discriminant de k sur Q, V ou V_k l'ensemble des places de k, V_∞ (resp. V_f) celui des places archimédiennes (resp. ultramétriques) de k. Si $v \in V$, alors k_v est le complété de k en v, et, pour $v \in V_f$, $\mathfrak{o}_v$ est l'anneau des entiers de k_v. Comme d'habitude A_k, ou simplement A, est l'anneau des adèles de k et A_f est l'anneau des adèles finies de k, i.e. le produit restreint des $k_v(v \in V_f)$.

L'ensemble des isomorphismes de k dans C est noté Σ et, suivant l'usage, r_1 (resp. r_2) est le nombre des places réelles (resp. complexes) de k. On a donc $d = r_1 + 2r_2$, et r_1 est le nombre des $\sigma \in \Sigma$ tels que $\sigma(k) \subset R$.

Pour $m \in N$, d_m désigne l'ordre du zéro en $-m$ de la fonction zêta de Dedekind $\zeta_k(s)$ de k. Il résulte de l'équation fonctionnelle que $d_0 = r_1 + r_2 - 1$, $d_m = r_1 + r_2$ si m est pair non nul et $d_m = r_2$ si m est impair.

1.5. – Soient G' un k-groupe connexe, n la dimension de G et $G = R_{k/Q}G'$ le Q-groupe obtenu par restriction des scalaires à partir de G' [15: I]. Il existe un isomorphisme canonique

$$\nu \colon G \xrightarrow{\sim} \prod_{\sigma \in \Sigma} {}^\sigma G',$$

où ${}^\sigma G'$ $(\sigma \in \Sigma)$ est le $\sigma(k)$-groupe obtenu à partir de G par le changement de base $k \to \sigma(k)$. Soit η une q-forme différentielle algébrique sur G', invariante à droite et définie sur k. Il lui correspond canoniquement par σ une forme ${}^\sigma\eta$ sur ${}^\sigma G'$ invariante à droite et définie sur $\sigma(k)$; le produit extérieur des ${}^\sigma\eta$ définit donc une forme invariante à droite sur le produit des groupes ${}^\sigma G'$. On pose

$$R\eta = |D|^{-q/2}\nu^{*-1}(\Lambda_{\sigma \in \Sigma}\sigma_\eta) \, .$$

C'est une forme de degré dq sur G invariante à droite et définie sur Q, qui est aussi invariante à gauche si η l'est.

1.6. – Si $v \in V$, on pose $G'_v = G'(k_v)$. C'est un groupe de Lie sur k_v. L'isomorphisme ν induit un isomorphisme

$$\nu_\infty \colon G(\boldsymbol{R}) \xrightarrow{\sim} \prod_{v \in V_\infty} G'_v .$$

Soit maintenant ω une n-forme non nulle sur G', invariante à droite et définie sur k. Elle définit canoniquement une mesure de Haar à droite ω_v sur G'_v $(v \in V)$ [15 : II]. Posons

$$(1) \qquad \omega_\infty = |D|^{-n/2} \prod_{v \in V_\infty} \omega_v .$$

Il résulte des définitions que l'on a

$$(2) \qquad \nu_\infty^{*-1}(\omega_\infty) = i^{r_2} \cdot R\omega|_{G(\boldsymbol{R})} .$$

Supposons G' semi-simple. Alors le produit de ω_∞ par les ω_v $(v \in V_f)$, est absolument convergent et définit la mesure de Tamagawa ω_A de $G(A_k)$ (loc. cit.). (En fait, la convergence de ce produit est démontrée dans [15] pour les groupes classiques, ce qui nous suffit. Pour le cas général, *voir* T. Ono, Annals of Math. (2) **82** (1965), 88-111.)

2. – Algèbres à division.

2.1. – On note Br_K le groupe de Brauer d'un corps K. Si K est local, on a un homomorphisme injectif canonique $\mathrm{inv}_K \colon \mathrm{Br}_K \to \boldsymbol{Q}/\boldsymbol{Z}$, qui est un isomorphisme si K est ultramétrique, et dont l'image est triviale si $K = \boldsymbol{C}$, d'ordre deux si $K = \boldsymbol{R}$. Si $K = k_v$, on écrira inv_v pour inv_K. Etant donné $a \in \mathrm{Br}_k$, on note a_v l'image canonique de a dans Br_v et on pose $\mathrm{inv}_v a = \mathrm{inv}_v a_v$. On sait que $\mathrm{inv}_v a$ est nul pour presque tout v et que la somme des $\mathrm{inv}_v a$ est nulle. Réciproquement, si l'on se donne pour tout $v \in V$ un élément $b_v \in \mathrm{Im}\,(\mathrm{inv}_v)$, nul pour presque tout v, et si la somme des b_v est nulle, alors il existe $a \in \mathrm{Br}_k$ tel que $b_v = \mathrm{inv}_v\,a$ pour tout v (*voir* par exemple [16 : XIII, Th. 4, p. 264]).

Le lemme suivant est bien connu. Ne pouvant fournir de référence, nous en donnons brièvement la démonstration

2.2. – LEMME. *Soit n un entier ≥ 1. Alors il existe une algèbre centrale à division D sur k de dimension n^2 sur k, qui est triviale à l'infini.*

Soit S un ensemble de n places ultramétriques de k. Vu 2.1, il existe $a \in \mathrm{Br}_k$ tel que

$$(1) \qquad \mathrm{inv}_v\,a = 0 \quad \text{si } v \notin S \quad \text{et} \quad \mathrm{inv}_v\,a = 1/n \quad \text{si } v \in S .$$

Comme $a_v = 0$ si $v \in V_\infty$, il suffit de montrer que a peut être représenté par une algèbre à division de rang n^2 sur k.

Il existe une extension L de k, cyclique de degré n, telle que $[L_w : k_v] = n$ quels que soient $v \in S$ et $w \in V_L$ prolongeant v [1: p. 105]. Pour $v \in S$, l'homomorphisme canonique $\mathrm{Br}_v \to \mathrm{Br}_{L_w}$ se traduit par la multiplication par n dans $\boldsymbol{Q}/\boldsymbol{Z}$ [14: p. 202], donc annule a_v. Comme a_v est nul pour $v \notin S$ par hypothèse, il s'ensuit que a est décomposé par L_w pour tout $w \in V_L$, donc a est aussi décomposé par L [16: XI, Th. 2, p. 206]. Cela entraîne que A peut être représenté par une algèbre A de dimension n^2 sur k [14: p. 106]. Il existe alors m divisant n et une algèbre à division centrale D sur k, de rang $d^2 = (n/m)^2$ sur k, tels que $A \equiv \boldsymbol{M}_m(D)$. Mais cela entraîne que $d \cdot \mathrm{inv}_v a = 0$ pour tout v, d'où, vu (1), $d = n$ et $m = 1$.

2.3. – Soit D satisfaisant au conditions de 2.2. Soit H le k-groupe algébrique tel que $H(L)$ soit le groupe des éléments de norme réduite 1 de $D \otimes_k L$ quelle que soit l'extension L de k. Le quotient $H(A_k)/H(k)$ est compact [15: 3.1.1].

Pour $v \notin S$, le groupe H est évidemment isomorphe à $\boldsymbol{SL}_n$ sur k_v. Cependant, nous avons besoin d'une assertion plus précise. H est une forme intérieure de $\boldsymbol{SL}_n$, définie par un élément α du premier ensemble de cohomologie galoisienne $H^1(k, \boldsymbol{PSL}_n)$. Par suite, si l'on fixe une k-réalisation linéaire $H \to \boldsymbol{GL}_m$ de H, on peut trouver une partie finie S' de V_f contenant S telle que, pour $v \notin S'$, il existe un k_v-isomorphisme f_v de $\boldsymbol{SL}_n$ sur H, qui applique $\boldsymbol{SL}_n \mathfrak{o}_v$ sur $H(\mathfrak{o}_v) = H \cap \boldsymbol{GL}_m \mathfrak{o}_v$. [Pour le voir, on remarque que α définit dans l'ensemble des $\bar{k}$-isomorphismes de $\boldsymbol{SL}_n$ sur H une k-variété irréductible lisse X, qui est un k-espace homogène principal sous $\boldsymbol{SL}_n$, et que X possède un $\mathfrak{o}_v$-point pour presque tout v.] Il s'ensuit en particulier que si U_v est un sous-groupe compact ouvert de H_v $(v \in V_f)$, égal à $f_v(\boldsymbol{SL}_n \mathfrak{o}_v)$ pour $v \notin S''$, avec $S'' \subset V_f$ fini, contenant S', alors le produit U des U_v est un sous-groupe compact *ouvert* de $H(A_f)$.

2.4. – PROPOSITION. *Soit ω_∞ la mesure sur H_∞ associée à une forme différentielle rationnelle ω non nulle invariante de degré maximum $n^2 - 1$ définie sur k (cf. 1.6). Soit Γ un sous-groupe arithmétique de H. Alors*

$$m_\infty(\Gamma) = \int_{H_\infty/\Gamma} \omega_\infty \sim \prod_{1 \leqq j < n} \zeta_k(j+1) \, .$$

Soient U_v $(v \in V_v)$ et U comme ci-dessus. Le groupe $\Gamma' = (H_\infty \times U) \cap H(F)$ est un sous-groupe arithmétique de H, donc est commensurable à Γ, et il suffit de montrer notre assertion pour Γ'. En prenant U assez petit, on peut faire en sorte que Γ' soit sans torsion. Par approximation forte on a $H(A_f) = H(F) \cdot U$. On en déduit que $H(A_k)/H(F)$ est fibré principal de groupe structural U, base H_∞/Γ'. Par suite le nombre de Tamagawa $\tau(H)$ de H, i.e. le volume de $H(A)/H(F)$ pour la mesure de Tamagawa, est produit de $m_\infty(\Gamma')$ et du volume $m(U)$ de U pour la mesure produit des ω_f ($v \in V_f$). Comme $\tau(H) = 1$ ([15: Thm. 3.3.1)]), il suffirait du reste ici de savoir que $\tau(H)$ est rationnel), il reste à montrer:

$$(1) \qquad m(U)^{-1} \sim \prod_{1 \leqq j < n} \zeta_k(j+1) \,.$$

Or

$$m(U) = \prod_{v \in V_f} m_v(U_v)\,, \quad \text{avec} \quad m_v(U_v) = \int_{U_v} \omega_v \,.$$

Les nombres $m_v(U_v)$ sont tous rationnels [13: 4.2.5]. Vu la remarque finale de 2.3, on est ramené à montrer (1) pour le groupe $\boldsymbol{SL}_n$, avec $U_v = \boldsymbol{SL}_n(\mathfrak{o}_v)$ pour tout $v \in V_f$. Mais on a alors

$$m_v(U_v) = Nv^{-(n^2-1)} \cdot \mathrm{Card}(\boldsymbol{SL}_n(F_v)) \,,$$

où F_v est le corps résiduel de $\mathfrak{o}_v$ et Nv le nombre d'éléments de F_v [15: 2.2.5], et d'autre part

$$\mathrm{Card}\,\boldsymbol{SL}_n(F_v) = Nv^{(n^2-n)/2} \prod_{1 \leqq j < n} (Nv^{j+1} - 1)\,,$$

d'où le résultat.

3. – Définition de quelques homomorphismes.

Le but de ce paragraphe est de définir les homomorphismes α^*, β^* et j_Γ^* de l'introduction. Auparavant, nous rappelons quelques notions et faits connus concernant la cohomologie des algèbres de Lie semi-simples et les formes différentielles invariantes.

3.1. – *Cohomologie d'algèbres de Lie et formes invariantes.* Soient F un corps de caractéristique zéro, F' une extension de F, et $\mathfrak{g}$ une algèbre de Lie sur F. On note $C^*(\mathfrak{g}; F')$ le complexe des cochaînes de $\mathfrak{g}$, à coefficients dans F' et $H^*(\mathfrak{g}; F')$ l'espace de cohomologie correspondant [8; 9]. On a donc, $C^*(\mathfrak{g}; F') = \mathrm{Hom}_F(\Lambda\mathfrak{g}, F')$. Si $\mathfrak{g}$ est semi-simple, alors $H^*(\mathfrak{g}; F')$ s'identifie à l'algèbre des invariants (au sens infinitésimal) de $\mathfrak{g}$, opérant dans $C^*(\mathfrak{g}; F')$ par la représentation adjointe, et est une algèbre extérieure [9: Thm. 10.2].

Si $F = \boldsymbol{R}$ or $\boldsymbol{C}$ et $F' \subset \boldsymbol{C}$ et si G est un groupe de Lie connexe sur F d'algèbre de Lie $\mathfrak{g}$ alors, en associant à $\alpha \in C^*(\mathfrak{g}; F')$ la forme invariante à droite sur G égale à α en l'élément neutre, on définit un isomorphisme de $C^*(\mathfrak{g}; F')$ sur l'espace des formes invariantes à droite sur G. Rappelons que si $F = \boldsymbol{R}$ et si G est compact, connexe, cet isomorphisme induit un isomorphisme de $H^*(\mathfrak{g}; F')$ sur $H^*(G; F')$; ce dernier espace s'identifie donc aussi dans ce cas à l'espace des formes biinvariantes sur G; de plus, l'espace des éléments indécomposables de degré $2j + 1$ $(j \in \boldsymbol{N})$ de $H^*(G; F')$ a un représentant canonique dans $H^{2j+1}(G; F')$, celui des éléments primitifs de degré $2j + 1$ de $H^*(G; F')$ pour sa structure d'algèbre de Hopf naturelle (Théorème de Samelson, cf. *loc. cit.*). On le notera $P^{2j+1}(G; F')$ et on désignera aussi par $P^{2j+1}(\mathfrak{g}; F')$ le sous-espace correspondant de $H^*(\mathfrak{g}; F')$. Tout cela vaut aussi sur $\boldsymbol{C}$ pour un groupe connexe réductif complexe et son algèbre de Lie.

3.2. – *Cohomologie relative.* Soit $\mathfrak{u}$ une sous-algèbre de $\mathfrak{g}$. On note $C^*(\mathfrak{g}; \mathfrak{u}; F')$ le complexe des cochaines relatives de $\mathfrak{g}$ modulo $\mathfrak{u}$, à coefficients dans F' et $H^*(\mathfrak{g}, \mathfrak{u}; F')$ l'espace de cohomologie correspondant [8, 9]. En particulier, $C^*(\mathfrak{g}, \mathfrak{u}; F')$ est un sous-espace de $C^*(\mathfrak{g}; F')$, qui est canoniquement isomorphe à l'espace des invariants de $\mathfrak{u}$, opérant par la représentation adjointe, dans $\mathrm{Hom}_F(\Lambda(\mathfrak{g}/\mathfrak{u}), F')$.

Si $F = \boldsymbol{R}, \boldsymbol{C}$ et $F' \subset \boldsymbol{C}$, soit G comme plus haut et supposons que $\mathfrak{u}$ soit l'algèbre de Lie d'un sous-groupe fermé connexe U de G. On identifie $\mathfrak{g}/\mathfrak{u}$ à l'espace tangent à l'origine à $U \backslash G$. Etant donné $\alpha \in C^*(\mathfrak{g}, \mathfrak{u}; F')$, il existe alors une et une seule forme invariante sur $U \backslash G$ égale à α à l'origine et cette correspondance définit un isomorphisme de $C^*(\mathfrak{g}, \mathfrak{u}; F')$ sur l'espace $A(U \backslash G; F')^G$ des formes différentielles invariantes sur $U \backslash G$, à valeurs dans F', d'où un isomorphisme canonique de $H^*(\mathfrak{g}, \mathfrak{u}; F')$ sur $H^*(U \backslash G; F')$.

On s'intéressera surtout au cas où $F = \boldsymbol{R}$, $\mathfrak{g}$ est semi-simple, $\mathfrak{u}$ est l'ensemble des points fixes d'un automorphisme involutif de $\mathfrak{g}$ et ou bien $\mathfrak{g}$ est compacte, ou bien $\mathfrak{u}$ est compacte maximale dans $\mathfrak{g}$, ou encore à la complexification d'une telle situation. On sait qu'alors la différentielle extérieure est identiquement nulle sur $C^*(\mathfrak{g}, \mathfrak{u}; F')$ [9: p. 101], donc $C^*(\mathfrak{g}, \mathfrak{u}; F') = {} = H^*(\mathfrak{g}, \mathfrak{u}; F')$. Si $F = \boldsymbol{C}$, ou si $F = \boldsymbol{R}$ et G est compact, alors on a de même, compte tenu du théorème de de Rham:

$$H^*(U \backslash G; F') = A(U \backslash G; F')^G = C^*(\mathfrak{g}; \mathfrak{u}; F') .$$

3.3. – *Notations.* Dans la suite de ce paragraphe, G est un groupe semi-simple connexe défini sur $\boldsymbol{Q}$, $\mathfrak{g}$ son algèbre de Lie, K un sous-groupe compact maximal de $G(\boldsymbol{R})$, $\mathfrak{k}$ son algèbre de Lie et G_u un sous-groupe compact maximal de $G(\boldsymbol{C})$ contenant K. On suppose, ce qui est loisible, que les algèbres de Lie $\mathfrak{g}(\boldsymbol{R})$ et $\mathfrak{g}_u$ de $G(\boldsymbol{R})$ et G_u admettent les décompositions de Cartan

$$(1) \qquad \mathfrak{g}(\boldsymbol{R}) = \mathfrak{k} \oplus \mathfrak{p}\,, \qquad \mathfrak{g}_u = \mathfrak{k} \oplus \mathfrak{p}_u\,,$$

avec

$$(2) \qquad \mathfrak{p}_u = i \cdot \mathfrak{p}\,.$$

En particulier, $\mathfrak{k}$ est l'ensemble des points fixes d'un automorphisme involutif de $\mathfrak{g}(\boldsymbol{R})$ (resp. $\mathfrak{g}_u$) qui laisse $\mathfrak{p}$ (resp. $\mathfrak{p}_u$) stable et est donc égal à -Id. sur cet espace. Soient encore

$$(3) \qquad X = K\backslash G\,, \qquad X_u = K\backslash G_u\,,$$

$\sigma: G \to X$, $\sigma': G_u \to X_u$ les projections naturelles, et Γ un sous-groupe discret de $G(\boldsymbol{R})$.

3.4. – *Les homomorphismes* α^*, β^*, $\bar\alpha^*$, $\bar\beta^*$, j_Γ^*. Si $\omega \in C^*(\mathfrak{g}; \boldsymbol{C})$, on note $\alpha(\omega)$ la forme invariante à droite sur G_u égale à ω en l'élément neutre (3.1). De même on associe à ω une forme invariante à droite sur $G(\boldsymbol{R})$, donc aussi une forme $\beta(\omega)$ sur $G(\boldsymbol{R})/\Gamma$, et à $\eta \in C^*(\mathfrak{g}, \mathfrak{k}_c; \boldsymbol{C})$, où $\mathfrak{k}_c = \mathfrak{k} \otimes_{\boldsymbol{R}} \boldsymbol{C}$, on fait correspondre une forme invariante $\bar\alpha(\eta)$ sur G_u et une forme invariante sur X, donc une forme $\bar\beta(\eta)$ sur X/Γ. En passant à la cohomologie, on obtient un diagramme commutatif

$$(1) \qquad \begin{array}{ccccc} H^*(G_u; \boldsymbol{C}) & \xleftarrow{\ \alpha^*\ } & H^*(\mathfrak{g}; \boldsymbol{C}) & \xrightarrow{\ \beta^*\ } & H(G(\boldsymbol{R})/\Gamma; \boldsymbol{C}) \\ \big\uparrow{\scriptstyle\sigma^*} & & \big\uparrow{\scriptstyle\lambda^*} & & \big\uparrow{\scriptstyle\sigma'^*} \\ H^*(X_u; \boldsymbol{C}) & \xleftarrow{\ \bar\alpha^*\ } & H^*(\mathfrak{g}, \mathfrak{k}_c; \boldsymbol{C}) & \xrightarrow{\ \bar\beta^*\ } & H^*(X/\Gamma; \boldsymbol{C})\,, \end{array}$$

où λ^* est associé à l'inclusion $C^*(\mathfrak{g}, \mathfrak{k}_c; \boldsymbol{C}) \to C^*(\mathfrak{g}; \boldsymbol{C})$. Comme on l'a rappelé ci-dessus (3.1, 3.2), α^* et $\bar\alpha^*$ sont des isomorphismes et $H^*(G_u; \boldsymbol{C})$ (resp. $H^*(X_u; \boldsymbol{C})$) s'identifie à l'espace des formes biinvariantes (resp. invariantes) sur G_u (resp. X_u).

On peut aussi définir directement

$$(2) \qquad \mu = \tau \circ \alpha^{-1}: A(G_u; \boldsymbol{C})^{G_u} \to A(G(\boldsymbol{R})/\Gamma; \boldsymbol{C})\,,$$

$$(3) \qquad \bar\mu = \bar\beta \circ \alpha^{-1}: A(X_u; \boldsymbol{C})^{G_u} \to A(X; \boldsymbol{C})^{G(\boldsymbol{R})} \to A(X/\Gamma; \boldsymbol{C}) = A(X; \boldsymbol{C})^\Gamma\,,$$

en associant à une forme invariante à droite ω sur G_u (resp. une forme invariante η sur X_u) la forme invariante sur $G(\boldsymbol{R})/\Gamma$ (resp. X/Γ) qui, à l'origine, a même extension que ω (resp. η) à l'espace tangent complexe. L'homomorphisme $\bar{\mu}^*$ associé en cohomologie à $\bar{\mu}$ est essentiellement celui qui nous intéresse, mais il a l'inconvénient de ne pas envoyer la cohomologie réelle en la cohomologie réelle, aussi voulons-nous en modifier légèrement la définition. Vu 3.3 (2), il est clair que si une forme multilinéaire η de degré m sur $\mathfrak{p} \otimes \boldsymbol{C}$ a une restriction réelle à $\mathfrak{p}_u$, alors $i^m \cdot \eta$ a une restriction réelle sur $\mathfrak{p}$. En posant

$$(4) \qquad j_\Gamma(\eta) = i^m \cdot \bar{\mu}(\eta), \qquad \left(m \in \boldsymbol{N}; \eta \in A^m(X_u; \boldsymbol{R})^{G_u} \right),$$

on définit donc un isomorphisme de $A^m(X_u; \boldsymbol{R})^{G_u}$ sur $A^m(X; \boldsymbol{R})^{G(\boldsymbol{R})} = I^m_{G(\boldsymbol{R})}$, d'où l'homomorphisme canonique

$$(5) \qquad j_\Gamma^*: H^*(X_u; \boldsymbol{R}) \to H^*(X/\Gamma; \boldsymbol{R}) = H^*(\Gamma; \boldsymbol{R}).$$

3.5. – Nous voulons encore décrire une compatibilité qui sera utile plus loin. On suppose Γ sans torsion. Soient (E_r) et (E'_r) les suites spectrales des fibrations définies par σ et σ', et (E''_r) la suite spectrale de cohomologie relative de $\mathfrak{g} \bmod \mathfrak{k}_c$ [8, 9]. Cette dernière est définie à partir d'une filtration de $C^*(\mathfrak{g}; \boldsymbol{C})$. On peut aussi définir (E'_r) et (E_r) à partir d'une filtration de l'algèbre des formes différentielles (voir par exemple [3]). Il est immédiat que α et β sont compatibles avec ces filtrations, donc induisent des homomorphismes de suites spectrales. En particulier

$$(1) \quad \alpha_2^*: E''_2 = H^*(\mathfrak{g}, \mathfrak{k}_c; \boldsymbol{C}) \otimes H^*(\mathfrak{k}_c; \boldsymbol{C}) \to E_2 = H^*(X_u; \boldsymbol{C}) \otimes H^*(K_c; \boldsymbol{C}),$$

$$(2) \qquad \beta_2^*: E''_2 \to E'_2 = H^*(X/\Gamma; \boldsymbol{C}) \otimes H^*(K; \boldsymbol{C}),$$

satisfont à

$$(3) \qquad \alpha_2^* = \bar{\alpha}_2^* \otimes \iota^*, \qquad \beta_2^* = \bar{\beta}_2^* \otimes \iota^*,$$

où $\iota^*: H^*(\mathfrak{k}; \boldsymbol{C}) \xrightarrow{\sim} H^*(\mathfrak{k}_c; \boldsymbol{C}) \xrightarrow{\sim} H^*(K_c; \boldsymbol{C})$ est l'isomorphisme canonique de 3.1.

4. – L'isomorphisme $H^*(\mathfrak{g}; \boldsymbol{C}) \to H^*(G_u; \boldsymbol{C})$ dans le cas déployé.

4.1. - Soient V un espace vectoriel réel de dimension finie, L un réseau de V et $T = V/L$ le tore quotient. Alors $H^*(T; \boldsymbol{R})$ s'identifie canoniquement à l'algèbre des formes différentielles biinvariantes sur T, donc à ΛV^*.

Etant donné $c \in V^*$, soit $\tilde{c}$ l'élément de $H^1(T; \boldsymbol{R})$ correspondant. Il est élémentaire qu'il existe un isomorphisme naturel $x \mapsto \tilde{x}$ de L sur $H_1(T; \boldsymbol{Z})$ tel que

$$\langle \tilde{c}, \tilde{x} \rangle = c(x), \qquad (c \in V^*, x \in L).$$

4.2. – Soit G un groupe semi-simple simplement connexe défini et déployé sur $\boldsymbol{R}$. On fixe un épinglage de l'algèbre de Lie $\mathfrak{g}$ de G [7]. Soient $\mathfrak{h}$ la sous-algèbre de Cartan donnée de $\mathfrak{g}$, R le système de racines de $\mathfrak{g}$ par rapport à $\mathfrak{h}$, S une base de R, h_s $(s \in S)$, x_a $(a \in R)$ une base de Chevalley de $\mathfrak{g}$ associée à l'épinglage donné et $\mathfrak{g}_{\boldsymbol{Z}}$ la forme entière de $\mathfrak{g}$ définie par cette base. Si F est un sous-corps de $\boldsymbol{C}$, on pose $\mathfrak{g}(F) = \mathfrak{g}_{\boldsymbol{Z}} \otimes_{\boldsymbol{Z}} F$.

Soit $\mathfrak{g}_u$ la forme compacte standard de $\mathfrak{g}$ associée aux données précédentes. Elle est engendrée par $\mathfrak{t} = i \cdot \mathfrak{h}_{\boldsymbol{R}}$ et les éléments $u \cdot x_a + \overline{u} \cdot x_{-a}$ $(u \in \boldsymbol{C}; a \in R)$ et possède une forme entière $\mathfrak{g}_{u\boldsymbol{Z}}$ sous-tendue par $i \cdot \mathfrak{h}_{\boldsymbol{Z}}$, $(x_a + x_{-a})$, $i(x_a - x_{-a})$ $(a \in R)$. On note G_u et T les sous-groupes intégraux de $G(\boldsymbol{C})$ (vu comme groupe de Lie réel), d'algèbres de Lie $\mathfrak{g}_u$ et $\mathfrak{t}$.

L'algèbre $\mathfrak{k} = \mathfrak{g}_u \cap \mathfrak{g}(\boldsymbol{R})$ est engendrée par les $x_a + x_{-a}$ $(a \in R)$. On a les décompositions de Cartan

$$(1) \qquad \mathfrak{g}(\boldsymbol{R}) = \mathfrak{k} \oplus \mathfrak{p}, \mathfrak{g}_u = \mathfrak{k} \oplus i \cdot \mathfrak{p} \qquad \text{avec} \qquad \mathfrak{p} = \mathfrak{h}(\boldsymbol{R}) + \sum_{a \in R} \boldsymbol{R}(x_a - x_{-a}).$$

Les espaces $H^*(\mathfrak{g}(\boldsymbol{C}); \boldsymbol{C})$ et $H^*(G_u; \boldsymbol{C})$ ont des $\boldsymbol{Q}$-structures naturelles définies par $H^*(\mathfrak{g}(\boldsymbol{Q}); \boldsymbol{Q})$ et $H^*(G_u; \boldsymbol{Q})$. Nous nous proposons de déterminer le comportement de l'isomorphisme $\alpha^* : H^*(\mathfrak{g}(\boldsymbol{C}); \boldsymbol{C}) \to H^*(G_u; \boldsymbol{C})$ de 3.3 par rapport à elles. Chacune de ces algèbres est une algèbre extérieure sur un espace d'éléments primitifs (3.1). Il suffit donc de connaître l'effet de α^* sur ces derniers.

4.3. – PROPOSITION. *Soient* $m \in \boldsymbol{N}$ et $P^{2m+1}(\mathfrak{g}(\boldsymbol{Q}); \boldsymbol{Q})$ $\big(resp.\ P^{2m+1}(\mathfrak{g}_u(\boldsymbol{Q}); \boldsymbol{Q}),$ *resp.* $P^{2m+1}(G_u; \boldsymbol{Q})\big)$ *l'espace des éléments primitifs de degré* $2m + 1$ *de* $H^*(\mathfrak{g}(\boldsymbol{Q}); \boldsymbol{Q})$ $\big(resp.\ H^*(\mathfrak{g}_u(\boldsymbol{Q}); \boldsymbol{Q}),$ *resp.* $H^*(G_u; \boldsymbol{Q})\big)$. *Alors* α^* *applique* $P^{2m+1}(\mathfrak{g}(\boldsymbol{Q}); \boldsymbol{Q})$ *isomorphiquement sur* $(\pi i)^{m+1} \cdot P^{2m+1}(G_u; \boldsymbol{Q})$. *L'injection canonique* $H^*(\mathfrak{g}_u(\boldsymbol{Q}); \boldsymbol{Q}) \to H^*(G_u; \boldsymbol{C})$ *applique* $P^{2m+1}(\mathfrak{g}_{u\boldsymbol{Q}}; \boldsymbol{Q})$ *isomorphiquement sur* $\pi^{m+1} \cdot P^{2m+1}(G_u; \boldsymbol{Q})$.

Si $c \in \mathfrak{h}^*$, on note w_c la forme invariante à droite nulle sur les x_a et induisant c sur $\mathfrak{h}$. Soit w_a la forme invariante à droite duale de x_a, nulle sur $\mathfrak{h}$.

On identifie $\mathfrak{t}$ au revêtement universel de T par l'application exponentielle. Le groupe fondamental de T s'identifie alors à $2\pi i \cdot \mathfrak{h}_{\boldsymbol{Z}}$. Soit P le groupe des poids. C'est donc la $\boldsymbol{Z}$-forme de $\mathfrak{h}^*$ duale de $\mathfrak{h}_{\boldsymbol{Z}}$. L'isomorphisme de 4.1 induit un isomorphisme γ de $(2\pi i)^{-1} \cdot P$ sur $H^1(T; \boldsymbol{Z})$. On a (cf. [6: Part I,

§ 14.5]) :

$$(1) \qquad dw_c = -\sum_{a>0} c(h_a) \cdot w_a \wedge w_{-a}, \qquad (c \in \mathfrak{h}^*),$$

et dw_c est un cocycle de $C^2(\mathfrak{g}, \mathfrak{h}; \boldsymbol{C})$. L'application $\tau \colon H^1(\mathfrak{h}; \varLambda) \to H^2(\mathfrak{g}, \mathfrak{h}; \boldsymbol{C})$ ainsi définie est la transgression [9]. On a de même la transgression $H^1(T; \boldsymbol{C}) \to H^2(T \backslash G_u; \boldsymbol{C})$ (cf. [2]). L'isomorphisme α^* commute évidemment à la transgression donc, puisque γ définit un isomorphisme de $H^1(\mathfrak{h}(\boldsymbol{Q}); \boldsymbol{Q})$ sur $(\pi i)H^1(T; \boldsymbol{Q})$ on a

$$(2) \qquad \alpha^*\big(\tau(P \otimes_{\boldsymbol{Z}} \boldsymbol{Q})\big) = \pi i \cdot H^2(T \backslash G_u; \boldsymbol{Q}).$$

En particulier, α^* définit un isomorphisme

$$(3) \qquad \alpha^* \colon H^2(\mathfrak{g}(\boldsymbol{Q}), \mathfrak{h}(\boldsymbol{Q}); \boldsymbol{Q}) \xrightarrow{\sim} (\pi i) \cdot H^2(T \backslash G_u; \boldsymbol{Q}).$$

Comme $H^*(\mathfrak{g}, \mathfrak{h}; \boldsymbol{C})$ et $H^*(T \backslash G_u; \boldsymbol{C})$ sont isomorphes et que $H^*(T \backslash G_u; \boldsymbol{C})$ est engendrée par ses éléments de degré deux [2: § 26; 11], on voit que α^* induit des isomorphismes

$$(4) \qquad \alpha^* \colon H^{2m}(\mathfrak{g}(\boldsymbol{Q}), \mathfrak{h}(\boldsymbol{Q}); \boldsymbol{Q}) \to (\pi i)^m \cdot H^{2m}(T \backslash G_u; \boldsymbol{Q}), \qquad (m \in \boldsymbol{N}).$$

De même, γ et α^* définissent des isomorphismes

$$(5) \qquad \gamma \colon H^1(\mathfrak{h}(\boldsymbol{Q}); \boldsymbol{Q}) \xrightarrow{\sim} i \cdot H^1(\mathfrak{t}(\boldsymbol{Q}); \boldsymbol{Q}),$$

$$(6) \qquad \alpha^* \colon H^{2m}(\mathfrak{g}(\boldsymbol{Q}), \mathfrak{h}(\boldsymbol{Q}); \boldsymbol{Q}) \xrightarrow{\sim} i^m \cdot H^{2m}(\mathfrak{g}_u(\boldsymbol{Q}), \mathfrak{t}(\boldsymbol{Q}); \boldsymbol{Q}), \qquad (m \in \boldsymbol{N}).$$

On considère les suites spectrales de $\mathfrak{g} \bmod \mathfrak{h}$, $\mathfrak{g}_u \bmod \mathfrak{t}$ et $G_u \bmod T$. Elles sont isomorphes. En particulier, au niveau des termes E_2 on a un isomorphisme

$$(7) \qquad \alpha^* \otimes \gamma \colon H^{2j}(\mathfrak{g}(\boldsymbol{Q}), \mathfrak{h}(\boldsymbol{Q}); \boldsymbol{Q}) \otimes H^k(\mathfrak{h}(\boldsymbol{Q}); \boldsymbol{Q}) \xrightarrow{\sim}$$
$$\xrightarrow{\sim} (\pi i)^{j+k} H^{2j}(T \backslash G_u; \boldsymbol{Q}) \otimes H^k(T; \boldsymbol{Q}),$$

$(j, k \in \boldsymbol{N})$ et aussi, vu (5), (6)

$$(8) \qquad H^{2j}(\mathfrak{g}(\boldsymbol{Q}), \mathfrak{h}(\boldsymbol{Q}); \boldsymbol{Q}) \otimes H^k(\mathfrak{h}(\boldsymbol{Q}); \boldsymbol{Q}) \xrightarrow{\sim}$$
$$\xrightarrow{\sim} i^{j+k} \cdot H^{2j}(\mathfrak{g}_u(\boldsymbol{Q}), \mathfrak{t}(\boldsymbol{Q}); \boldsymbol{Q}) \otimes H^k(\mathfrak{t}; \boldsymbol{Q}).$$

Soit (E_r) la suite spectrale de $G_u \bmod T$. D'après J. Leray [11 : p. 110], elle se termine à E_3, et l'isomorphisme $\operatorname{Gr} H^*(G_u; \mathbf{Q}) = E_\infty \xrightarrow{\sim} E_3$ applique les primitifs dans $E_3^{*,1}$. On a alors une assertion similaire pour les deux autres suites spectrales considérées ici, et la proposition résulte de (7), (8).

4.4. – *Normalisation de mesures invariantes sur* $T\backslash G_u$ *et* G_u. On sait que $T\backslash G_u$ admet un nombre fini de structures complexes invariantes, chacune associée à un ordre sur les racines [6 : Part I, § 10]. Supposons $T\backslash G_u$ muni de la structure complexe donnée par les racines positives. Alors les classes de Chern entières c_j de $T\backslash G_u$ sont les fonctions symétriques élémentaires en les $-\tau((2\pi i)^{-1}w_a)$ $(a > 0)$ [6 : Part I, § 10.8], donc c_j est représentée par

$$(2\pi i)^{-j} \cdot \sigma_j\left(\left\{\sum_{b>0} a(h_b)\,w_b \wedge \overline{w}_b\right\}_{a>0}\right),$$

où σ_j désigne la j-ème fonction symétrique élémentaire. On sait que $c_1/2$ est aussi une classe entière et que $(c_1/2)^m$ $(m = (\dim T/G_u)/2)$ est $(m!)\cdot F$, où F est la classe fondamentale [6 : Part II, § 14.3]. Si r est la somme des racines positives, alors c_1 est représentée par la forme

$$(2\pi i)^{-1} \sum_{a>0} r(a)(w_a \wedge \overline{w}_a),$$

donc

$$F = (2\pi i)^{-m} \cdot D \cdot \prod_{a>0} (w_a \wedge \overline{w}_a), \quad \text{avec} \quad D = \prod_{a>0} \left(r(a)/2\right).$$

D'autre part, vu 4.1, $\prod_{s\in S} w_s$ est la classe fondamentale de T multipliée par $\pm (2\pi i)^{-l}$, où l est la dimension de T. Par suite, à l'orientation près, la classe fondamentale de G_u est représentée par la forme

$$(2\pi i)^{-m-l} \cdot D \cdot \left(\prod_{a>0} w_a \wedge w_{-a}\right) \wedge \left(\prod_{s\in S} w_s\right).$$

Cette forme définit donc la mesure de Haar de G_u de volume total 1.

5. – **Les homomorphismes** α^* **et** β^* **pour** SL_n.

5.1. – Soient $n \in \mathbf{N}$, on pose $G'_n = SL_n$ et on note $\mathfrak{g}'_n$ l'algèbre de Lie de G'_n. On fixe un corps de nombres k et on pose $G_n = R_{k/\mathbf{Q}} G'_n$. On utilise les notations du § 1. On envisage ici G'_n comme un k-groupe, mais comme il est déjà défini sur Q, on a évidemment $^\sigma G'_n \cong G'_n$ pour tout $\sigma \in \Sigma$, et en particulier

$$(1) \qquad G'_{nv} = SL_n(R) \, (\text{resp. } G'_{nv} = SL_n(C))$$

$$\text{si } v \in V_\infty \text{ est réelle (resp. complexe)}.$$

On pose

$$(2) \quad K_{nv} = SO(n) \,, \quad G_{nv,u} = SU(n) \,, \quad \text{si } v \in V_\infty \text{ est réelle} \,,$$

$$(3) \quad K_{nv} = SU(n) \,, \quad G'_{nv,u} = SU(n) \times SU(n) \quad \text{si } v \in V_\infty \text{ est complexe} \,,$$

et, dans les deux cas,

$$(4) \qquad X_{nv} = K_{nv} \backslash G'_{nv} \,, \quad X_{nv,u} = K_{nv} \backslash G'_{nv,u} \,.$$

Si v est complexe, K_{nv} est identifié à la diagonale de $G_{nv,u} = K_{nv} \times K_{nv}$ et X est donc isomorphe à K_{nv} en tant qu'espace homogène. Soient encore

$$(5) \qquad K_n = \prod_v K_{nv} = SO(n)^{r_1} \times SU(n)^{r_2} \,,$$

$$(6) \qquad G_{nu} = \prod_v G'_{nv,u} = (SU(n))^d \,,$$

$$(7) \qquad X_n = \prod_v X_{nv} = (SO(n) \backslash SU(n))^{r_1} \times (SU(n) \backslash SL_n(C))^{r_2}$$

$$(8) \qquad X_{n,u} = \prod_v X_{nv,u} = (SO(n) \backslash SU(n))^{r_1} \times SU(n)^{r_2} \,,$$

où les produits sont pris sur l'ensemble V_∞ des places archimédiennes de k.

Enfin, Γ_n désigne un sous-groupe arithmétique de G_n. C'est donc un sous-groupe de $G_n(Q)$ commensurable à l'image de $SL_n \mathfrak{o}_k$ par l'isomorphisme canonique $G'_n(k) \xrightarrow{\sim} G_n(Q)$ défini par la restriction des scalaires.

5.2. – Si v est complexe, K_{nv} est évidemment totalement non homologue à zéro dans G'_{nv}. Si n est impair, il en est de même pour v réelle [2: § 30], donc K_{nv} est totalement non homologue à zéro dans $G_{nv,u}$, et la suite spectrale (E_r) de $G_{nu} \to X_{nu}$ dégénère. Il en est alors de même pour la suite spectrale (E''_r) de $\mathfrak{g}_n$ modulo $\mathfrak{k}_{nc}$ (cf. 3.5), qui lui est isomorphe, et pour celle de $G_{nr} \to X_{nr}/\Gamma_n$ (Γ_n sans torsion), vu 3.5.

Le fait que (E_r) dégénère signifie qu'il existe un isomorphisme canonique d'espaces vectoriels gradués entre E_2, gradué par le degré total, et $\text{Gr} \, H^*(G_{nu}; C)$, où Gr signifie l'espace gradué associé à une filtration convenable de $H^*(G_{nu}; \boldsymbol{C})$. Mais ici on a en fait, toujours pour n impair, un isomorphisme naturel d'algèbres graduées. En effet ces trois algèbres sont des algèbres extérieures; $H^*(G_{nu}; \boldsymbol{C})$ est une algèbre extérieure sur un espace d'éléments primitifs P. De plus P est somme directe de deux sous-espaces P_1, P_2 tels que σ^* applique $H^*(X_{nu}; \boldsymbol{C})$ isomorphiquement sur ΛP_1 et que la restriction $H^*(G_{nu}; \boldsymbol{C}) \to H^*(K_n; \boldsymbol{C})$ applique ΛP_2 isomorphiquement sur $H^*(K_n; \boldsymbol{C})$ (l'assertion relative à σ^* est un théorème de H. Samelson, cf. par exemple [8: Théorème 13.2]). On a donc un diagramme commutatif

$$(1) \qquad \begin{array}{ccc} H^*(G_{nu}; \boldsymbol{C}) & \xleftarrow{\quad\alpha^*\quad} & H^*(\mathfrak{g}_n; \boldsymbol{C}) \\ \big\uparrow{\scriptstyle\iota} & & \big\uparrow{\scriptstyle\iota} \\ H^*(X_{nu}; \boldsymbol{C}) \otimes H^*(K_n; \boldsymbol{C}) & \xleftarrow{\alpha^*\otimes\iota} & H^*(\mathfrak{g}_n, \mathfrak{k}_{nv}; \boldsymbol{C}) \otimes H^*(\mathfrak{k}_{nv}; \boldsymbol{C}) \end{array}$$

où les flèches verticales sont des isomorphismes. Si l'on se limite aux degrés $< (n-1)/4$, pour lesquels $\bar\beta^*$ est un isomorphisme [4], alors β^* est un isomorphisme et on a de même un diagramme commutatif

$$(2) \qquad \begin{array}{ccc} H^*(\mathfrak{g}_n; \boldsymbol{C}) & \xrightarrow{\quad\beta^*\quad} & H^*(G_n(\boldsymbol{R})/\Gamma_n; \boldsymbol{C}) \\ \big\uparrow{\scriptstyle\iota} & & \big\uparrow{\scriptstyle\iota} \\ H^*(\mathfrak{g}_n, \mathfrak{k}_n; \boldsymbol{C}) \otimes H^*(\mathfrak{k}_{nv}; \boldsymbol{C}) & \xrightarrow{\bar\beta^*\otimes\iota} & H^*(X_n/\Gamma_n; \boldsymbol{C}) \otimes H^*(K_n; \boldsymbol{C}) \,. \end{array}$$

5.3. – L'algèbre de cohomologie $H^*(\mathfrak{g}_n'(\boldsymbol{Q}); \boldsymbol{Q})$ est une algèbre extérieure sur un espace d'éléments primitifs P', et P'^{2m+1} est de dimension 1 si $1 \leq m < n$, zéro sinon. On fixe un générateur ω_{2m+1} de P'^{2m+1} et on note aussi de la même manière la forme rationnelle biinvariante sur $\boldsymbol{SL}_n(\boldsymbol{C})$ égale à ω_{2m+1} à l'origine. De même, pour $\sigma \in \Sigma$, $^\sigma\omega_{2m+1}$ dénote la forme sur $^\sigma\mathfrak{g}_n' \cong \mathfrak{g}_n'(\boldsymbol{Q}) \otimes_{\boldsymbol{Q}} \sigma(\boldsymbol{F})$ définie par $\omega_{2m}\omega_1$ ou la forme biinvariante sur $^\sigma G_n'$ associée.

L'algèbre $H^*(\mathfrak{g}_n; \boldsymbol{C})$ est le produit tensoriel des algèbres $H^*(^\sigma\mathfrak{g}_n; \boldsymbol{C})$, donc de d copies de $H^*(\mathfrak{g}_n', \boldsymbol{C})$. Par suite l'espace P^{2m+1} des éléments primitifs de $H^{2m+1}(\mathfrak{g}_n; \boldsymbol{C})$ est de dimension d et admet les $^\sigma\omega_{2m+1}$ $(\sigma \in \Sigma)$ comme base. La forme

$$R\omega_{2m+1} = |D|^{-(2m+1)/2} v_\infty^{*-1}(\Lambda_{\sigma\in\Sigma}{}^\sigma\omega_{2m+1})\,,$$

(cf. § 1 pour les notations) est une forme biinvariante sur G_n, définie sur $\boldsymbol{Q}$, dont la valeur à l'origine représente un générateur de

$$L_m(\mathfrak{g}_n, \boldsymbol{Q}) = \Lambda^d P^{2m+1}\,.$$

Soit $L_m(G_{nu}; \boldsymbol{Q}) = \Lambda^d P^{2m+1}(G_{nu}; \boldsymbol{Q})$.

5.4. – Proposition. *On a*

$$\alpha^*(L_m(\mathfrak{g}_n; \boldsymbol{Q})) = (\pi i)^{d(m+1)} |D|^{-(2m+1)/2} L_m(G_{nu}; \boldsymbol{Q}).$$

En effet, α^* est le produit d'isomorphismes

$$\alpha_\sigma^*: H^*(\mathfrak{g}_n'; \boldsymbol{C}) \xrightarrow{\sim} H^*(({}^\sigma G_n')_u; \boldsymbol{C}) \qquad (\sigma \in \Sigma),$$

donc en fait d'isomorphismes $H^*(\mathfrak{g}_n'; \boldsymbol{C}) \to H^*(G_{nu}'; \boldsymbol{C})$, auxquels on peut appliquer 4.3, ce qui entraîne la proposition.

5.5. – Théorème. *Soit $m \in \boldsymbol{N}$ et supposons $n - 1 > 4d(2m + 1)$. Posons $Y_n = G_n(\boldsymbol{R})/\Gamma_n$. Soit $L_m(\Gamma_n; \boldsymbol{Q})$ la d-ième puissance extérieure de l'espace des éléments indécomposables de $H^{2m+1}(Y_n; \boldsymbol{Q})$. Alors*

$$(1) \qquad \beta^*(L_m(\mathfrak{g}_n; \boldsymbol{Q})) = i^{r_2} \cdot \zeta_k(m + 1) \cdot L_m(Y_n; \boldsymbol{Q}).$$

Si (1) est vraie pour un sous-groupe arithmétique de G_n, elle l'est pour tous. D'autre part, soit q un entier $\geq n$. Le groupe G_n est naturellement plongé dans G_q. Si l'on choisit Γ_n et Γ_q tels que $\Gamma_n = \Gamma_q \cap G_n$, alors on a un diagramme commutatif

$$(2) \qquad \begin{array}{ccc} H^*(\mathfrak{g}_q; \boldsymbol{C}) & \xrightarrow{\ \beta^*\ } & H^*(G_q(\boldsymbol{R})/\Gamma_q; \boldsymbol{C}) \\ \downarrow{\scriptstyle \varrho} & & \downarrow{\scriptstyle \varrho'} \\ H^*(\mathfrak{g}_n; \boldsymbol{C}) & \xrightarrow{\ \beta^*\ } & H^*(G_n(\boldsymbol{R})/\Gamma_n; \boldsymbol{C}), \end{array}$$

où ϱ, ϱ' sont les homomorphismes associés à la restriction, et sont donc compatibles avec le structures rationnelles. De plus $\mathfrak{g}_n$ est totalement non homologue à zéro dans $\mathfrak{g}_q$ donc ϱ est surjective. Il en est alors de même pour ϱ' en dimensions $< (n-1)/4$. L'assertion (1) pour q implique donc (1) pour n. Par conséquent, il suffit d'établir (1) pour n arbitrairement grand, et un choix quelconque de Γ_n. On supposera $n > 4d(m + 1)^2$ impair et Γ_n sans torsion.

Soit s_j un nombre complexe tel que

$$(3) \qquad \beta^*(L_j(\mathfrak{g}_n; \boldsymbol{Q})) = s_j \cdot L_j(Y_n; \boldsymbol{Q}).$$

Il est déterminé à un facteur rationnel près et nous devons montrer que l'on a

$$(4) \qquad s_j \sim i^{r_2} \cdot \zeta_k(j + 1), \qquad (1 \leq j \leq m).$$

Soient ω_{2j+1}, $R\omega_{2j+1}$ comme en 4.2 et

$$(5) \qquad \eta_j = \Lambda_{1 \leq s \leq j} R\omega_{2s+1}, \qquad (1 \leq j \leq m).$$

Les nombres s_j sont définis en calculant modulo les éléments décomposables, i.e., en considérant des quotients de la cohomologie. Mais nous voulons nous placer dans la cohomologie elle-même; il faut alors avoir quelques renseignements sur les éléments décomposables ainsi introduits. Notons $\omega'_{2j+1,\sigma}$ des éléments de $H^{2j+1}(Y_n; \boldsymbol{Q})$ qui forment une base d'un supplémentaire dans $H^{2j+1}(Y_n; \boldsymbol{Q})$ de l'espace des éléments décomposables, donc qui, par passage au quotient, définissent une base de l'espace des éléments indécomposables de degré $2j+1$. On peut écrire

$$(6) \qquad \beta^*(^{\sigma}\omega_{2j+1}) = \sum_{\tau \in \Sigma} c^{\tau}_{j,\sigma} \cdot \omega'_{2j+1,\tau} + d_{j,\sigma}; \qquad (c^{\tau}_{j,\sigma} \in \boldsymbol{C}),$$

où $d_{j,\sigma}$ est dans la sous-algèbre de $H^*(Y_n; \boldsymbol{C})$ engendrée par les $\omega'_{2s+1,\sigma}$ avec $s < j$, $\sigma \in \Sigma$. Par suite, si l'on pose

$$(7) \qquad \omega''_j = \prod_{\sigma \in \Sigma} \omega'_{2j+1,\sigma},$$

on a

$$\beta^*(R\omega_{2j+1}) = s'_j \cdot \omega''_j + e_j, \qquad \text{avec} \qquad s'_j = |D|^{j+(1/2)} \cdot \det(c^{\tau}_{j\sigma}),$$

et e_j dans l'idéal I_j engendré par les $\omega'_{2s+1,\sigma}$ avec $s < j$. En passant au quotient par les éléments décomposables on voit que

$$(9) \qquad s'_j \sim s_j.$$

D'autre part I_j est aussi l'idéal engendré par les $\beta^*(^{\sigma}\omega_{2s+1})$. Il est donc annulé par $\beta_*(\eta_{j-1})$, d'où

$$\beta^*(\eta_j) = s'_1 \ldots s'_j \cdot \omega''_1 \wedge \ldots \wedge \omega''_j,$$

ou encore

$$(10) \qquad \beta^*(\eta_j) \in s_1 \ldots s_j \cdot H^{d(j^2+2j)}(Y_n; \boldsymbol{Q}).$$

Comme $i^{r_2(2j+1)} \sim i^{r_2}$, il suffit, pour établir (4) et le théorème, de montrer l'existence d'une sous-variété compacte Z_j de Y_n, de dimension $j^2 + 2j$, telle que

$$(11) \qquad \int_{Z_j} \beta^*(\eta_j) \sim i^{r_2(j^2+2j)} \cdot \prod_{1 \leq s \leq j} \zeta_k(s+1), \qquad (1 \leq j \leq m).$$

Fixons $j \leq m$. Soit D une algèbre à division centrale sur k, de rang $(j+1)^2$ et soit H' le groupe algébrique semi-simple correspondant (2.3). On plonge D dans l'algèbre des matrices carrées d'ordre n par la représentation régulière, d'où des plongements de H' dans $G'_n = \boldsymbol{SL}_n$ et de $H = R_{k/Q}H'$ dans G_n. Le groupe $H \cap \Gamma_n$ est un sous-groupe arithmétique de H sans torsion et le quotient

$$Z_j = H(\boldsymbol{R})/(H(\boldsymbol{R}) \cap \Gamma_n),$$

est donc une sous-variété compacte connexe orientable de Y_n. La forme $\beta(R\eta_j)$ est une forme rationnelle sur G_n, biinvariante et définie sur $\boldsymbol{Q}$. Sa restriction à H est une forme rationnelle biinvariante définie sur $\boldsymbol{Q}$ de degré maximum $j^2 + 2j$. Montrons qu'elle est *non nulle*. Avec nos conventions, $\beta(R\eta_j)$ est une forme sur Y_n, mais cette assertion est indépendante de Γ_n. Convenons de désigner aussi par β l'application qui associe à une forme sur l'algèbre de Lie la forme invariante à droite sur le groupe qui lui est égale à l'origine. Nous devons donc montrer que la restriction de $\beta(R\eta_j)$ à $H(\boldsymbol{R})$, ou, ce qui revient au même, à $H(\boldsymbol{C})$, est non nulle. Comme $H(\boldsymbol{C})$ est le produit des $^\sigma H'$ et que $v^\sigma_\infty(R\eta_j)$ est, à un facteur numérique près, le produit des $^\sigma\eta_j$ $(\sigma \in \Sigma)$, il suffit de faire voir que la restriction de η_j à H'_j est non nulle.

On a $H(\boldsymbol{C}) = \boldsymbol{SL}_{j+1}(\boldsymbol{C})$. Comme l'inclusion $H' \to G'_n$ provient de la représentation régulière de D on peut supposer, après conjugaison, que le plongement de $H(\boldsymbol{C})$ dans $\boldsymbol{SL}_n(\boldsymbol{C})$ est le composé de l'application diagonale de $H'(\boldsymbol{C})$ dans le produit de $j+1$ copies H_s $(1 \leq s \leq j+1)$ de $\boldsymbol{SL}_{j+1}(\boldsymbol{C})$ et de l'injection standard de ce produit dans $\boldsymbol{SL}_n(\boldsymbol{C})$. On a des isomorphismes naturels $v_s \colon H' \to H_s$ $(1 \leq s \leq j+1)$ et les H_s sont conjugués au sous-groupe standard $\boldsymbol{SL}_{j+1}(\boldsymbol{C})$ de $\boldsymbol{SL}_n(\boldsymbol{C})$ défini par les $j+1$ premières coordonnées. Les restrictions τ_s de $\beta(\eta_j)$ aux H_s sont des formes non nulles qui se correspondent par les isomorphismes $v_t \circ v_s^{-1}$ $(1 \leq s,\, t \leq j+1)$. La restriction de $\beta(\eta_j)$ à H' est donc égale à $(j+1)v_1^*(\tau_1)$.

Ainsi $\beta(R\eta_j)$ a une restriction non nulle à H. L'égalité (11) résulte alors de 1.6 (2) et 2.4.

5.6. – *Remarque.* Nous avons utilisé le fait que β^* est un isomorphisme dans les dimensions considérées, par exemple pour pouvoir établir 5.5 (10). Mais la démonstration du fait que $\beta(R\eta_j)$ a une intégrale non nulle sur Z_j en est évidemment indépendante. Cela montre donc en particulier, sans utiliser [4], que $\beta^*(R\eta_m) \neq 0$. Vu la définition de η_m, cela entraîne que β^* est injectif en degrés $\leq 2m + 1$. Or on a le diagramme commutatif

$$\begin{array}{ccc} H^*(\mathfrak{g}_n'; \boldsymbol{C}) & \xrightarrow{\ \beta^*\ } & H^*(Y_n; \boldsymbol{C}) \\ \Big\uparrow{\scriptstyle \lambda^*} & & \Big\uparrow \\ H^*(\mathfrak{g}_n', k_n'; \boldsymbol{C}) = I_{G_n'(\boldsymbol{R})} & \xrightarrow{\ j_\Gamma^*\ } & H^*(\Gamma_n; \boldsymbol{C})\,, \end{array}$$

et l'application λ^* est injective si n est impair (cf. 5.2). Par suite j_Γ^* est injectif en dimensions $\leq 2m + 1$. Cela donne donc une démonstration de l'injectivité différente de celle de [4], tout au moins jusqu'à une dimension qui est plus petite que $(n-1)/4$, mais au moins égale à $(dn)^{\frac{1}{2}}$, et qui tend donc aussi vers l'infini avec n, ce qui suffit pour les applications à la cohomologie stable.

6. – Les régulateurs supérieurs de k.

6.1. – On conserve les notations de 5.1, 5.2. Soient m, $n \in \boldsymbol{N}$ et supposons $(n-1) > 4d(2m+1)$. Soit $I^{2m+1}(X_{nu}; \boldsymbol{Q})$ (resp. $I^{2m+1}(\Gamma_n; \boldsymbol{Q})$) l'espace des éléments indécomposables de degré $2m+1$ de $H^*(X_{nu}; \boldsymbol{Q})$ (resp. $H^*(\Gamma_n; \boldsymbol{Q})$). Il est donc de dimension d_m (cf. [4: 11.4] et 1.4). Soit

$$(1) \qquad L_m(X_{nu}; \boldsymbol{Q}) = \Lambda^{d_m} I^{2m+1}(X_{nu}; \boldsymbol{Q})\,, \qquad L_m(\Gamma_n; \boldsymbol{Q}) = \Lambda^{d_m} I^{2m+1}(\Gamma_n; \boldsymbol{Q})\,.$$

6.2. – THÉORÈME. *L'homomorphisme canonique* j_Γ^* *(cf. 314) applique* $L_m(X_{nu}; \boldsymbol{Q})$ *sur* $R_m' \cdot L_m(\Gamma_n; \boldsymbol{Q})$, *avec* $R_m' = |D|^{\frac{1}{2}} \cdot \zeta_k(m+1) \cdot \pi^{-d(m+1)}$.

Les homomorphismes considérés ici sont compatibles avec la restriction $\boldsymbol{SL}_n \to \boldsymbol{SL}_q$ $(n \geq q > 4d(2m+1)+1)$. Notre assertion pour n l'entraîne donc pour q. Il suffit par suite de la démontrer pour n arbitrairement grand. En particulier, on peut supposer n impair. Dans ce cas, 5.2 (1), (2) donnent un diagramme commutatif

$$(1) \quad \begin{array}{ccc} H^*(G_{nu}; \boldsymbol{C}) & \xrightarrow{\qquad\ \mu^*\ \qquad} & H^*(Y_n; \boldsymbol{C}) \\ \Big\uparrow{\scriptstyle \iota} & & \Big\uparrow{\scriptstyle \iota} \\ H^*(X_{nu}; \boldsymbol{C}) \otimes H^*(K_n; \boldsymbol{C}) & \xrightarrow{\ \bar\mu^* \otimes \mathrm{id.}\ } & H^*(\Gamma_n; \boldsymbol{C}) \otimes H^*(K_n; \boldsymbol{C}) \end{array}$$

avec

$$(2) \qquad\qquad \mu^* = \beta^* \circ \alpha^{*-1}\,, \qquad \bar\mu^* = \bar\beta^* \circ \bar\alpha^{*-1}\,.$$

Soit s_m (resp. t_m) un nombre complexe tel que

$$(3) \qquad \mu^*(L_m(G_{nu}; \boldsymbol{Q})) = s_m \cdot L_m(Y_n; \boldsymbol{Q}) (\text{resp. } \bar{\mu}^*(L_m(X_{nu}; \boldsymbol{Q})) = t_m \cdot L_m(\Gamma_n; \boldsymbol{Q})).$$

Soit $L_m(K_n, \boldsymbol{Q})$ la $(d - d_m)$-ième puissance extérieure de l'espace des éléments primitifs de degré $2m + 1$ de $H^*(K_n; \boldsymbol{Q})$ (qui est de dimension $d - d_m$). On a des isomorphismes canoniques

$$(4) \qquad \begin{cases} L_m(G_{nu}; \boldsymbol{Q}) = L_m(X_{nu}; \boldsymbol{Q}) \otimes L_m(K_n; \boldsymbol{Q}), \\ L_m(Y_n; \boldsymbol{Q}) = L_m(\Gamma_n; \boldsymbol{Q}) \ \otimes L_m(K_n; \boldsymbol{Q}) \end{cases}$$

donc, vu (1), $s_m \sim t_m$.

D'après 5.4 et 5.5 on peut écrire

$$(5) \qquad s_m \sim (\pi i)^{-d(m+1)} \cdot |D|^{\frac{1}{2}} \cdot i^{r_2} \cdot \zeta_k(m + 1).$$

Mais, en dimension s, on a $j_\Gamma^* = i^s \cdot \bar{\mu}^*$ vu 3.4 (4), donc

$$R_m' \sim i^{d_m(2m+1)} \cdot s_m \sim i^{d_m(2m+1)} \cdot t_m \sim \pi^{-d(m+1)} \cdot |D|^{\frac{1}{2}} \cdot \zeta_k(m + 1) \cdot i^{r_2 + d(m+1) + d_m}.$$

Comme $d_m = r_1 + r_2$ si m est pair non nul et $d_m = r_2$ si m est impair, l'exposant de i est toujours pair, ce qui établit le théorème.

6.3. – REMARQUE. Supposons k totalement réel et m impair. Alors $d_m = 0$ donc (cf. 6.2 (4)), on a des isomorphismes

$$L_m(G_{nu}; \boldsymbol{Q}) \cong L_m(K_n; \boldsymbol{Q}) \cong L_m(Y_n; \boldsymbol{Q}),$$

et, vu les définitions, μ^* devient l'identité de $L_m(K_n; \boldsymbol{Q})$. On a alors $s_m = 1$, et 6.2 fournit donc une démonstration du fait bien connu que $|D|^{\frac{1}{2}} \cdot \zeta_k(m + 1) \cdot \pi^{-d(m+1)}$ est un nombre rationnel sous ces hypothèses.

6.4. – *Les régulateurs de k.* Soient

$$X_u = \varinjlim_n X_{nu}, \qquad \boldsymbol{SL}\mathfrak{o} = \varinjlim_n \boldsymbol{SL}_n \mathfrak{o},$$

(cf. [4: §§ 10, 11]). Définissons $L_m(X_u; \boldsymbol{Q})$ et $L_m(\boldsymbol{SL}\mathfrak{o}; \boldsymbol{Q})$ comme $L_m(X_{nu}; \boldsymbol{Q})$ et $L_m(\Gamma_n; \boldsymbol{Q})$. On a aussi, vu [4: § 11])

$$(1) \qquad j_\Gamma(L_m(X_u; \boldsymbol{Q})) = R_m' \cdot L_m(\boldsymbol{SL}\mathfrak{o}; \boldsymbol{Q}), \qquad (m = 1, 2, \ldots).$$

Le nombre R'_m est défini à un facteur rationnel près. Nous voulons en fixer une normalisation. On sait que $I^{2m+1}(X_u; \mathbf{Q})$ s'identifie au dual de $\pi_{2m+1}(X_u) \otimes_{\mathbf{Z}} \mathbf{Q}$ et que de même $I^{2m+1}(SL\mathfrak{o}; \mathbf{Q})$ s'identifie au dual de $K_{2m+1}\mathfrak{o} \otimes_{\mathbf{Z}} \mathbf{Q}$ (cf. [4: 10.6, 12.1]). Par suite $L_m(X_u; \mathbf{Q})$ (resp. $L_m(SL\mathfrak{o}; \mathbf{Q})$) s'identifie au dual de

$$\Lambda^{d_m}\big(\pi_{2m+1}(X_u) \otimes_{\mathbf{Z}} \mathbf{Q}\big) \ (\text{resp.} \ (\Lambda^{d_m} K_{2m+1}\mathfrak{o} \otimes_{\mathbf{Z}} \mathbf{Q})) \ .$$

Soit x_m (resp. y_m) un générateur du dual de

$$\Lambda^{d_m}\big(\pi_{2m+1}(X_u) \otimes 1\big) \ (\text{resp.} \ \Lambda^{d_m}(K_{2m+1}\mathfrak{o} \otimes 1)) \ .$$

On définit alors le nombre R_m par l'égalité

$$(2) \qquad\qquad j_\Gamma^*(x_m) = R_m \cdot y_m \, ,$$

en supposant de plus $R_m > 0$, ce qui est loisible, quitte à remplacer x_m par $-x_m$. On obtient ainsi le m-ième régulateur de k ($m \geq 1$) défini dans [12: § 4]. La conjecture de [12] impliquerait en particulier $R_m \sim c_m$, avec

$$(3) \qquad\qquad c_m > 0 \, , \quad c_m = \pm \lim \zeta_k(s) \cdot (s + m)^{-d_m} \, .$$

Mais il résulte de l'équation fonctionnelle (cf. [15: VII, Thm. 3, p. 129]) que l'on a

$$(4) \qquad\qquad c_m \sim \pi^{-d(m+1) + d_m} \cdot |D|^{\frac{1}{2}} \cdot \zeta_k(m + 1) \, .$$

Le théorème 6.2 montre que si l'on remplace dans cette égalité conjecturale le membre de gauche par $c'_m = c_m \cdot \pi^{-d_m}$, autrement dit par

$$(5) \qquad\qquad c'_m = \pm \lim_{s \to -m} \zeta_k(s) \cdot \big(\pi(s + m)\big)^{-d_m} \, , \quad c'_m > 0 \, ,$$

alors le quotient des deux membres est en tout cas un nombre rationnel.

7. – Remarques générales sur l'homomorphisme j_Γ^*.

7.1. – Nous revenons à la situation générale de 3.3 et notons B_K un espace classifiant pour K [2]. La fibration principale $\sigma: G_u \to X_u$ est image réciproque de la fibration universelle $E_K \to B_K$ par une application classifiante $\varrho: X_u \to B_K$, définie à homotopie près. Supposons Γ sans torsion. Alors

$\sigma': G(\boldsymbol{R})/\Gamma \to X/\omega$ est aussi une K-fibration principale. Soit $\varrho': X/\Gamma \to B_K$ une application classifiante.

7.2. PROPOSITION. *Le diagramme*

$$H^*(X_u; \boldsymbol{R}) \xrightarrow{\ j_\Gamma^*\ } H^*(\Gamma; \boldsymbol{R})$$
$$\varrho^* \nwarrow \qquad \nearrow \varrho'^*$$
$$H^*(B_K; \boldsymbol{R})$$

est commutatif.

Cela se voit facilement en utilisant l'interprétation de ϱ^* et ϱ'^* à l'aide de l'homomorphisme de Chern-Weil. Indiquons brièvement comment. Les décompositions de Cartan 3.3 (1) définissent des connexions sur G_u et $G(\boldsymbol{R})/\Gamma$ respectivement, dont les tenseurs de courbure sont donnés par la relation $R(X, Y) = ad[X, Y]$ (X, Y dans $\mathfrak{p}$, ou dans $\mathfrak{p}_u$). Soit d'autre part I_K l'algèbre des polynomes sur $\mathfrak{k}$ invariants par la représentation adjointe de K. Il existe un isomorphisme canonique $\delta: I_K \xrightarrow{\sim} H^*(B_K; \boldsymbol{R})$ tel que si $P \in I_K^m$, alors $\varrho^*(\delta(P))$ (resp. $\varrho'^*(\delta(P))$) est représentée par la classe de cohomologie de la forme fermée invariante par G_u (resp. G) donnée par $(X_1 \ldots X_{2m}) \mapsto$ $\mapsto P(R(X_1, X_2), \ldots, R(X_{2m-1}, X_{2m}))$. Cela étant, la commutativité de (1) résulte de la définition même de j_Γ^*.

7.3. – COROLLAIRE. *Soit* $x \in H^*(X_u; \boldsymbol{Q})$. *Si* x *est contenu dans l'image de* ϱ^*, *alors* $j(x) \in H^*(\Gamma; \boldsymbol{Q})$. *Si* G *et* K *ont même rang, alors* $j^*(H^*(X_u; \boldsymbol{Q})) \subset$ $\subset H^*(\Gamma; \boldsymbol{Q})$.

Les homomorphismes ϱ^* et ϱ'^* proviennent d'applications continues, donc envoient la cohomologie rationnelle dans la cohomologie rationnelle. Vu 7.2, cela établit la première assertion. La deuxième résulte alors du fait que ϱ^* est surjectif si G_u et K sont de même rang [2: § 26].

7.4. – Supposons G connexe. D'après [10], $H^*(X_u; \boldsymbol{Q}) = A \otimes B$, où $A = \operatorname{Im} \varrho^*$ est la sous-algèbre caractéristique et est formée d'éléments de degrés pairs et B est une algèbre extérieure appliquée isomorphiquement par σ^* sur $\operatorname{Im} \sigma^*$. Cette dernière est une algèbre extérieure engendrée par des éléments primitifs, comme on l'a déjà rappelé (5.2). D'après 7.3, $j^*(A) \subset$ $\subset H^*(\Gamma; \boldsymbol{Q})$. On a, pour m entier, $I^{2m+1}(X_u; \boldsymbol{Q}) = I^{2m+1}(B)$, donc aussi $L_m(X_u; \boldsymbol{Q}) = L_m(B)$. Le problème général dont nous avons discuté un cas particulier plus haut est donc d'étudier les nombres réels s_m tels que

$$j_\Gamma^*(L_m(X_u; \boldsymbol{Q})) = s_m \cdot L_m(\Gamma; \boldsymbol{Q}),$$

lorsque j_r^* est un isomorphisme en dimensions $\leq 2m + 1$. Dans certains cas classiques, on retrouve en fait certains des régulateurs du § 6. Par exemple soit $G = G_{k/\boldsymbol{Q}}\,\boldsymbol{Sp}_{2n}$. Le plongement $\boldsymbol{Sp}_{2n} \to \boldsymbol{SL}_{2n}$ donne lieu a un diagramme commutatif

$$
\begin{array}{ccc}
H^*(X_{2n,u};\,\boldsymbol{R}) & \xrightarrow{\;j_r^*\;} & H^*(\boldsymbol{SL}_{2n}\mathfrak{o};\,\boldsymbol{R}) \\
{\scriptstyle v^*}\big\downarrow & & \big\downarrow{\scriptstyle v'^*} \\
H^*(X_u;\,\boldsymbol{R}) & \xrightarrow{\;j_r^*\;} & H^*(\boldsymbol{Sp}_{2n}\mathfrak{o};\,\boldsymbol{R}),
\end{array}
$$

où les flèches verticales sont les homomorphismes de restriction. La forme compacte standard de $\boldsymbol{Sp}_{2n}(\boldsymbol{C})$ est le groupe unitaire quaternionnien $\boldsymbol{USp}_n$ à n variables quaternionniennes. Sa cohomologie est une algèbre extérieure sur des générateurs de degrés $2m + 1$, avec m impair $\leq n$. L'espace dual de $\boldsymbol{U}(n)\backslash\boldsymbol{Sp}_{2n}(\boldsymbol{R})$ est $\boldsymbol{U}(n)\backslash\boldsymbol{USp}_n$; sa cohomologie réelle est concentrée en degrés pairs, et est égale à sa sous-algèbre caractéristique, puisque $\boldsymbol{U}(n)$ et $\boldsymbol{USp}_n$ ont même rang. On a, comme dans 3.4, une décomposition

$$
X_u = \left(\boldsymbol{U}(n)\backslash\boldsymbol{USp}_n\right)^{r_1} \times \left(\boldsymbol{USp}_n\right)^{r_2},
$$

donc,

$$
A = H^*\left(\boldsymbol{U}(m)\backslash\boldsymbol{USp}_n\right)^{r_1}, \qquad B = H^*\left((\boldsymbol{USp}_n)^{r_2};\,\boldsymbol{Q}\right),
$$

et $I^{2m+1}(X_u;\,\boldsymbol{Q})$ est de dimension r_2 pour m impair, nulle pour m pair > 0. Par conséquent, on n'a à considérer s_m que pour m impair. Mais $\boldsymbol{USp}_n$ est totalement non homologue à zéro dans $\boldsymbol{SU}_{2n}$ [2: § 30]. On en déduit immédiatement que v^* induit un isomorphisme de $L_m(X_{2n,u};\,\boldsymbol{Q})$ sur $L_m(X_u;\,\boldsymbol{Q})$, d'où $s_m \sim R_m$ pour m impair. On a un résultat semblable si $G = R_{k/\boldsymbol{Q}}\boldsymbol{SO}_{n,n}$.

BIBLIOGRAPHIE

[1] E. Artin - J. Tate, *Class field theory*, Benjamin, New York, 1967.

[2] A. Borel, *Sur la cohomologie des espaces fibrés principaux et des espaces homogènes de groupes de Lie compacts*, Annals of Math. (2), **57** (1953), pp. 115-207.

[3] A. Borel, *A spectral sequence for complex analytic bundles*, Appendix 2 in F. Hirzebruch, *Topological methods in algebraic geometry*, 3rd ed., Springer (1966), pp. 202-217.

[4] A. Borel, *Stable real cohomological of arithmetic groups*, Annales E.N.S. (4), **7** (1974), pp. 235-272.

[5] A. Borel, *Cohomology of arithmetic groups*, Proc. Int. Congress Math. Vancouver 1974, vol. 1, pp. 435-442.

[6] A. Borel - F. Hirzebruch, *Characteristic classes and homogeneous spaces - I*, Amer. Jour. Math., **80** (1958), pp. 458-536; *II, ibid.*, **81** (1959), pp. 315-382.

[7] N. Bourbaki, *Groupes et Algèbres de Lie*, Chap. 7, 8, Act. Sci. Ind. 1364, Hermann éd., Paris, 1975.

[8] G. Hochschild - J.-P. Serre, *Cohomology of Lie algebras*, Annals of Math. (2), **57** (1953), pp. 591-603.

[9] J.-L. Koszul, *Homologie et cohomologie des algèbres de Lie*, Bull. Soc. Math. France, **78** (1950), pp. 65-127.

[10] J.-L. Koszul, *Sur un type d'algèbres différentielles en rapport avec la transgression*, Coll. Topologie algébrique, Bruxelles 1950, G. Thone, Liège, 1951, pp. 73-81.

[11] J. Leray, *Sur l'homologie des groupes de Lie, des espaces homogènes et des espaces fibrés principaux*, Coll. Topologie algébrique, Bruxelles 1950, G. Thone, Liège 1951, pp. 101-115.

[12] S. Lichtenbaum, *Values of zeta-functions, étale cohomology, and algebraic K-theory*, Algebraic K-theory - II, Springer Lecture Notes in Mathematics, **342** (1973), pp. 489-501.

[13] T. Ono, *Algebraic groups and discontinuous groups*, Nagoya Math. Jour., **27** (1966), pp. 279-322.

[14] J.-P. Serre, *Corps locaux*, Act. Sci. Ind. 1296, Hermann, Paris 1966.

[15] A. Weil, *Adeles and algebraic groups*, Notes by M. Demazure and T. Ono, Institute for Advanced Study, Princeton, N.J., 1961.

[16] A. Weil, *Basic number theory*, Grund. Math. Wiss., **144**, Springer 1973.

Errata - Corrige

Ser. IV, vol. IV (1977), pp. 613-636.

Ces corrections sont dues au fait que dans la définition de $R\eta$, p. 616, il faut remplacer la valeur absolue $|D|$ du discriminant par le discriminant lui-même. Cela change les puissances de i dans un certain nombre de formules.

1. Il faut tout d'abord définir $D^{\frac{1}{2}}$. Pour cela, suivant l'usage, on fixe une base $(\alpha_i)_{1 \le i \le d}$ de $o(k)$ sur $\mathbf{Z}$, un ordre $\Sigma \equiv \{\sigma_1, ..., \sigma_d\}$ sur Σ, et l'on pose

$$D^{\frac{1}{2}} \equiv \det (\alpha_{\alpha_i}^{\sigma_j}) \ .$$

Rappelons que $(-1)^{r_1} D^2 > 0$.

Les changements à faire sont alors les suivants.

2. Remplacer $|D|$ par D:

    p. 616, ligne 3 du bas,

    p. 627, ligne 5 du bas,

    p. 628, ligne 2.

Dans les deux premiers cas, il est entendu que l'ordre sur Σ choisi pour définir $D^{\frac{1}{2}}$ est le même que dans le produit extérieur indexé par Σ.

3. Supprimer les puissances de i dans les relations:

    5.5(1), 5.5(4), p. 628,

    6.2(5), p. 632.

Pervenuto alla Redazione il 24 Maggio 1980.

109.

(with G. Harder)

**Existence of discrete cocompact subgroups
of reductive groups over local fields**

J. Reine Angew. Math. **298** (1978) 53–64

The main purpose of this paper is to prove:

Theorem A. *Let L be a non-discrete locally compact field of characteristic zero and H a reductive group over L. Then the group $H(L)$ of rational points of H contains a discrete cocompact subgroup.*

Here $H(L)$ is viewed as a topological group, for the topology associated to that of L, and Γ cocompact in $H(L)$ means that the quotient space $H(L)/\Gamma$ is compact. If $L=\mathbb{R}$, $\mathbb{C}$, the result was known [1] (In fact, it was stated there for semi-simple, non-necessarily algebraic groups, but the extension to the reductive case is trivial.), so that our main case of interest is when L is non-archimedean, i.e., is a finite extension of the field $\mathbb{Q}_p$ of p-adic numbers, for some prime p. However, the proof will cover both cases (see 2. 2). It is easily reduced to the case where H is absolutely simple, and we assume this here until the discussion of § 3.

As was to be expected, Γ is constructed by the arithmetic method. Let k be a number field, V_∞ the set of its archimedean places, v a place of k and $S=V_\infty \cup \{v\}$. We can always find k such that $S \neq \{v\}$, the completion k_v of k at v is isomorphic to L, and w is real for $w \in V_\infty$, $w \neq v$, (1. 13). The main point is then to show the existence of a simple k-group G which is anisotropic over k_w, for $w \in S-\{v\}$ and is isomorphic to H over k_v. We then take for Γ an S-arithmetic group if $S \neq V_\infty$, an arithmetic group if $S=V_\infty$ (see 2. 2).

The fact that this procedure would yield cocompact discrete subgroups of p-adic groups of classical type, or of type G_2, F_4, had already been pointed out in [24].

The existence of G is in fact a special case of the following theorem in Galois cohomology (see § 1 for the notation), proved in § 1. 11:

Theorem B. *Let k be a number field, S a finite set of places of k and G an absolutely almost simple k-group which is either simply connected or of adjoint type. Then the canonical map*

$$(1) \qquad \omega_S : H^1(k, \operatorname{Aut} G) \to \prod_{v \in S} H^1(k_v, \operatorname{Aut} G),$$

is surjective.

520

In other words, given a k_v-form $G_{(v)}$ of G/k_v, for $v \in S$, there always exists a k-form G' of G such that G' is isomorphic to $G_{(v)}$ over k_v for all $v \in S$. This theorem will itself be a simple consequence of two further statements which assert: (a) that G' can be chosen to be quasi-split if the $G_{(v)}$ are so (1. 10); (b) that the map ω_S of (1) above is also surjective if $\mathrm{Aut}\, G$ is replaced by a connected semi-simple k-group, (1. 7).

These results are proved in §§ 1, 2. In § 3 we assume that L is a local field of positive characteristic p, and show similarly that if the derived group of H is isogeneous to a product of groups of type A, then $H(L)$ has a discrete cocompact subgroup (3. 3). It may be that if H is of another type, then $H(L)$ does not contain any discrete cocompact subgroup (see 3. 6).

As is known, the existence of discrete cocompact subgroups in real semi-simple groups can be used to prove results on infinite dimensional unitary representations. Our interest in the analogous problem in the p-adic case was in fact prompted by a remark of P. Sally, who pointed out to one of us that the existence of such a discrete cocompact group would lead to another proof of the known and basic fact that the formal degrees of the cuspidal representations are all integers for a suitable Haar measure. § 4 discusses briefly this implication.

1. Galois cohomology and classification

1. 1. Let k be a field, k_s a separable closure of k and H an algebraic k-group. Then $H^i(k, H)$ denotes the i-th cohomology set of the Galois group $\mathrm{Gal}(k_s/k)$ of k_s over k, with coefficients in $H(k_s)$ $(i = 0, 1)$ and, if H is commutative, the i-th cohomology group of $\mathrm{Gal}(k_s/k)$ in $H(k_s)$ for all $i \in \mathbb{N}$ [3]; [16]; [23]. If k' is a Galois extension of k, then, similarly, $H^i(k'/k, H)$ stands for $H^i(\mathrm{Gal}(k'/k), H(k'))$.

1. 2. A *local field* is in this paper a non-discrete field, complete under a discrete valuation $|\cdot|$, with finite residue field.

If k is a global field, then V or V_k is the set of places of k, V_∞ (resp. V_f) the set of archimedean (resp. ultrametric) places of k, and k_v the completion of k at $v \in V$. If H is a k-group, and $v \in V$, then

$$(1) \qquad \omega_{v,i} : H^i(k, H) \to H^i(k_v, H),$$

is the canonical map associated to the inclusion $k \to k_v$. If $S \subset V$, then $\omega_{S,i}$ is the product of the $\omega_{v,i}$ $(v \in S)$. The most important case for us is when $i = 1$, and we shall then usually drop the index i. The group H is said to satisfy the Hasse principle in dimension i if $\omega_{V,i}$ is injective.

1. 3. Let L be a local field, and T a torus over L. Let $X_*(T) = \mathrm{Morph}(GL_1, T)$ be the group of morphisms of GL_1 into T (the group of one-parameter subgroups of T). It is a finitely generated free abelian group, of rank equal to $\dim T$, and a $\mathrm{Gal}(L_s/L)$-module. By Tate-Nakayama duality, there is a canonical isomorphism

$$(1) \qquad H^{r-2}(L, X_*(T)) \xrightarrow{\sim} H^r(L, T),$$

for $r \geq 2$, where, if $r = 2$ and $L = \mathbb{R}$ the left-hand side is the "modified" 0-th cohomology group, namely the quotient of the invariants by the norms, and if $r = 2$ and L is non-archimedean, the left-hand side is the limit of such modified groups. This follows by passage to the limit over finite Galois extensions of L from the corresponding statement for those [22], IX, § 8; the latter is in fact valid for all $r \in \mathbb{Z}$ for the Tate cohomology groups.

It follows first from this that if T is anisotropic over L, then $H^2(L, T) = 0$, since then there are no non-zero invariants of the Galois group in $X_*(T)$.

1. 4. Let now $L = \mathbb{R}$ and H a semi-simple group over L. Let us say that a maximal torus T defined over $\mathbb{R}$ of H is *nice* if it is of the form $B \cap {}^\sigma B$, where B is a Borel subgroup of G and σ is complex conjugation. This is the case if and only if there is a Weyl chamber in $X_*(T) \otimes \mathbb{R}$ transformed into its opposite by σ. Such tori always exist. This follows for instance from [11], A. 4. 4, or from the fact that a maximal compact subgroup of $H(\mathbb{R})$ always contains regular elements.

Lemma. *If H is simply connected, and T is a nice maximal $\mathbb{R}$-torus of H, then $H^2(L, T) = 0$.*

Let $\Phi^v \subset X_*(T)$ be the set of coroots (i.e. the inverse of the root system of H with respect to T) and Δ^v the base of Φ^v associated to a Borel subgroup B such that $T = B \cap {}^\sigma B$. Then $\sigma(\Delta^v) = -\Delta^v$. Therefore, given $\alpha \in \Delta^v$, then either $\sigma(\alpha) = -\alpha$ or there exists $\beta \in \Delta^v$, $\beta \neq \alpha$ such that $\sigma(\alpha) = -\beta$. It follows readily that the (modified) 0-th cohomology group $H^0(L, X_*(T)) = 0$, and our assertion again follows from 1. 3 (1).

1. 5. Lemma. *Let k be a number field, S a finite set of places of k and G a simply connected semi-simple group over k. Then there exists a maximal k-torus T of G which satisfies the Hasse principle in dimension 2 (1. 2) and such that $H^2(k_v, T) = 0$ for $v \in S$.*

Enlarging S if necessary, we may assume that it contains at least one finite place. If v is finite, then G/k_v contains a maximal torus $T_{(v)}$ defined and anisotropic over k [15], § 15. If $k_v = \mathbb{R}$, let $T_{(v)}$ be a nice maximal torus of G/k_v (1. 4), and if $k_v = \mathbb{C}$, let $T_{(v)}$ be any maximal torus of G/k_v. Let $t_{(v)}$ be the Lie algebra of $T(k_v)$, viewed as a Lie group over k_v. The set of conjugates under $G(k_v)$ of the regular elements in $t_{(v)}$ is an open set U_v of the Lie algebra of $G(k_v)$. By strong approximation in k, there exists $X \in g(k)$, where g is the Lie algebra of G, which is contained in U_v for all $v \in S$. Let T be the centralizer of X in H. Then T is a maximal k-torus and, for $v \in S$, T is conjugate to $T_{(v)}$ by an element of $G(k_v)$. In particular, T is isomorphic to $T_{(v)}$ over k_v $(v \in S)$. By 1. 3, 1. 4, we have therefore $H^2(k_v, T) = 0$ for $v \in S$. Furthermore, by our construction, T is anisotropic at at least one place. Then it satisfies the Hasse principle in dimension 2 [16], Theorem 7, p. 58.

1. 6. Proposition. *Let k be a number field, S a finite set of places of k and M a finite commutative module over $\mathrm{Gal}(k_s/k)$. Then the map*

$$\omega_{S,2} : H^2(k, M) \to \prod_{v \in S} H^2(k_v, M),$$

is surjective.

The following proof was suggested to us by J. Tate.

Let m be a multiple of the order of M, μ_m the group of m-th roots of unity, and $M' = \mathrm{Hom}(M, \mu_m)$ the dual of M. The modules

$$(1) \qquad P^0(k, M') = \prod_{v \in V} H^0(k_v, M') \quad \text{and} \quad P^2(k, M) = \bigoplus_{v \in V} H^2(k_v, M),$$

where, for v real, $H^0(k_v, M')$ is again the modified 0-th cohomology group (1. 3), are

Pontrjagin dual of each other [23], II-48, 6.3. For any $a \in H^2(k, M)$, the element $\omega_{v,2}(a)$ is zero for almost all $v \in V$, so that $\omega_{V,2}$ also defines a map

$$j_2 : H^2(k, M) \to P^2(k, M),$$

[23], II-45. By a theorem of Poitou and Tate, the image of j_2 is the orthogonal complement to the image of

$$j_0 = \omega_{V,0} : H^0(k, M') \to P^0(k, M'),$$

in the above Pontrjagin duality [20], Exp. XV, Prop. 5b, p. 274.

Let w be a finite place of k not in S. In order to prove our proposition it suffices to show that given $\alpha = (\alpha_v) \in P^2(k, M)$, we can modify α_w so as to get an element in the image of j_2. By the duality just mentioned, α defines a character χ of $P^0(k, M')$ hence, by restriction, a character χ' of $H^0(k, M')$. On the other hand, the map

$$\omega_{w,0} : H^0(k, M') = M'^{\mathrm{Gal}(k_s/k)} \to H^0(k_w, M') = M'^{\mathrm{Gal}(k_{ws}/k_w)},$$

is an inclusion and every character of $H^0(k, M')$ extends to one of $H^0(k_w, M')$. Choose an extension of χ' and let β_w be the element of $H^2(k_w, M)$ corresponding to it by duality. Replacing in α the component α_w by $\alpha_w - \beta_w$, we get then an element orthogonal to $\mathrm{Im}\, j_0$ and therefore belonging to $\mathrm{Im}\, j_2$.

1.7. Theorem. *Let k be a number field and S a finite set of places of k. Let G be a connected semi-simple k-group. Then the canonical map*

$$\omega_S : H^1(k, G) \to \prod_{v \in S} H^1(k_v, G),$$

is surjective.

Let $\tilde{G}$ be the universal covering of G, $\pi : \tilde{G} \to G$ the canonical isogeny and $F = \ker \pi$. We have then the commutative diagram with exact rows

$$(1) \quad \begin{array}{ccccc}
H^1(k, \tilde{G}) & \xrightarrow{\mu} & H^1(k, G) & \xrightarrow{\delta} & H^2(k, F) \\
\downarrow{\scriptstyle \tilde{\omega}_S} & & \downarrow{\scriptstyle \omega_S} & & \downarrow{\scriptstyle \omega_S'} \\
\prod_{v \in S} H^1(k_v, \tilde{G}) & \xrightarrow{\mu'} & \prod_{v \in S} H^1(k_v, G) & \xrightarrow{\delta'} & \prod_{v \in S} H^2(k_v, F).
\end{array}$$

By 1.6, ω_S' is surjective. On the other hand, δ is also surjective. A proof is outlined in [16], V, 5.2. For the sake of completeness, we repeat briefly the argument. Given $c \in H^2(k, F)$, let S be the set of places for which $\omega_{v,2}(c) \neq 0$. It is finite (see 1.6). Let $\tilde{T}$ be a maximal k-torus of $\tilde{G}$ satisfying the conditions of 1.5 for this S, and $T = \pi(\tilde{T})$. Consider the following commutative diagram, where the rows are exact

$$(2) \quad \begin{array}{ccccc}
H^1(k, T) & \xrightarrow{\alpha} & H^2(k, F) & \xrightarrow{\beta} & H^2(k, \tilde{T}) \\
\downarrow{\scriptstyle \omega_{V,1}} & & \downarrow{\scriptstyle \sigma} & & \downarrow{\scriptstyle \omega_{V,2}} \\
\prod_{v \in V} H^1(k_v, T) & \xrightarrow{\alpha'} & \prod_{v \in V} H^2(k_v, F) & \xrightarrow{\beta'} & \prod_{v \in V} H^2(k_v, \tilde{T}).
\end{array}$$

In view of our construction, we have $\beta'(\sigma(c)) = 0$, hence $\beta(c) \in \ker \omega_{V,2}$, and therefore $\beta(c) = 0$, i.e. $c \in \mathrm{Im}\, \alpha$. A fortiori, $c \in \mathrm{Im}\, \delta$.

Let now $a = (a_v)_{v \in S}$ with $a_v \in H^1(k_v, G)$. Since δ and ω_S are surjective, we can find $b \in H^1(k, G)$ such that $\delta'(\omega_S(b)) = \delta'(a)$. If we twist G by a cocycle belonging to b ([3], § 1, [16], 1.4), then we are reduced to the case where $b = 0$, i.e. where $\delta'(a) = 0$, i.e. $a \in \operatorname{Im} \mu'$. But $H^1(k_v, \tilde{G}) = 0$ if v is finite [15], and

$$\omega_{V_\infty} : H^1(k, \tilde{G}) \to \prod_{v \in V_\infty} H^1(k_v, \tilde{G})$$

is surjective (see e.g. Theorem 1b of [16], V, whose proof does not use the standing assumption made there that G is classical). Hence $\tilde{\omega}_S$ is surjective. From this our assertion follows.

1.8. Let G be a semi-simple group of adjoint type over some field k. We may identify G with its group of inner automorphisms $\operatorname{Int} G$, which is the identity component of the group $\operatorname{Aut} G$ of automorphisms of G. The quotient group $A = (\operatorname{Aut} G)/G$ is finite and may be identified with the group $\operatorname{Aut} D$ of automorphisms of the Dynkin diagram $D = D(G)$ of the root system of G [10], Exp. XXIV. We assume D endowed with the structure of $\operatorname{Gal}(k_s/k)$-module defined in [5], 6.2 by consideration of the action of that group of the conjugacy classes of parabolic subgroups of G, whence also a Galois module structure on $\operatorname{Aut} D$. The natural isomorphism $A \xrightarrow{\sim} \operatorname{Aut} D$ is then one of Galois modules. The exact sequence

$$(1) \qquad\qquad 1 \to G \to \operatorname{Aut} G \to \operatorname{Aut} D \to 1$$

also yields an exact sequence of $\operatorname{Gal}(k_s/k)$-modules

$$(2) \qquad\qquad 1 \to G(k_s) \to (\operatorname{Aut} G)(k_s) \to \operatorname{Aut} D \to 1.$$

Let now G_0 be moreover split over k and let D_0 be its Dynkin diagram. Fix a maximal k-split torus T of G_0 and a Borel k-subgroup B of G containing T. In that case D_0 and $\operatorname{Aut} D_0$ are trivial $\operatorname{Gal}(k_s/k)$-modules. Moreover, as shown by Chevalley, there is a splitting $\operatorname{Aut} D_0 \to \operatorname{Aut} G_0$ of (1), whose image consists of automorphisms of G defined over k leaving B and T stable. In particular, we have a semi-direct product decomposition

$$(3) \qquad\qquad \operatorname{Aut} G_0 = G_0 \rtimes \operatorname{Aut} D,$$

which is compatible with the action of $\operatorname{Gal}(k_s/k)$ on the k_s-points. This is proved in [10], Exp. XXIV. If k is of characteristic zero, the only case of interest here, then these statements are equivalent to similar assertions for the automorphism group of the Lie algebra of G_0, for which we can refer to § 5, n° 3, p. 109 of [8], VIII. (For this discussion, see also [12].) The splitting (3) then yields an injective map

$$(4) \qquad\qquad \iota : H^1(k, \operatorname{Aut} D_0) \to H^1(k, \operatorname{Aut} G_0).$$

Note that, since $\operatorname{Aut} D_0$ is a trivial Galois module, we have

$$(5) \qquad H^1(k, \operatorname{Aut} D_0) = \operatorname{Hom}(\operatorname{Gal}(k_s/k), \operatorname{Aut} D_0)/\operatorname{Int}(\operatorname{Aut} D_0).$$

1.9. Lemma. *We keep the notation of 1.8, assume G to be absolutely simple, and k to be a number field. Let S be a finite set of places of k. Then the canonical map*

$$H^1(k, \operatorname{Aut} D) \to \prod_{v \in S} H^1(k_v, \operatorname{Aut} D),$$

is surjective.

Since G is k_s-isomorphic to a split group we may, using twisting, reduce the proof to the case where G is split over k. Let D_v be the decomposition group at $v \in S$. In view of 1. 8 (5), our assertion amounts to showing that the product of restriction mappings

$$(1) \qquad \mathrm{Hom}\,(\mathrm{Gal}\,(k_s/k),\,\mathrm{Aut}\,D) \to \prod_{v \in S} \mathrm{Hom}\,(D_v,\,\mathrm{Aut}\,D)/\mathrm{Int}\,(\mathrm{Aut}\,D),$$

is surjective.

Let $a_v \in \mathrm{Hom}\,(D_v,\,\mathrm{Aut}\,D)$ $(v \in S)$ and $H_v = a_v(D_v)$. There is nothing to prove if all the H_v are reduced to the identity, in particular if $\mathrm{Aut}\,D = \{1\}$. If $\mathrm{Aut}\,D$ is of order two, this amounts to showing the existence of a quadratic extension of k with prescribed decomposition groups of order ≤ 2 at finitely many places, which is well-known and can be proved by an elementary construction similar to, but simpler than, the one outlined below.

In view of the classification [7], the only remaining case is when $\mathrm{Aut}\,D$ is isomorphic to the group S_3 of permutations of three objects and G is of type D_4. Of course $D_v = \mathrm{Gal}\,(k_{v,s}/k_v)$. Let $a_v \in \mathrm{Hom}\,(D_v,\,\mathrm{Aut}\,D)$ $(v \in S)$ and let k'_v be the Galois extension of k_v defined by a_v. Its degree is 1, 2, 3 or 6. There exists then a monic cubic polynomial $P_v(X) \in k_v[X]$ whose splitting field is k'_v. In view of Krasner's theorem, we may further choose $P_v(X)$ so that any monic cubic polynomial in $k_v[X]$ with coefficients sufficiently close to those of $P_v(X)$ also has k'_v as splitting field. Using approximation, we can then find a monic cubic polynomial $P(X) \in k[X]$ whose coefficients are arbitrarily close with respect to v to those of $P_v(X)$ for all $v \in S$. Then we get for each $v \in S$ a commutative diagram

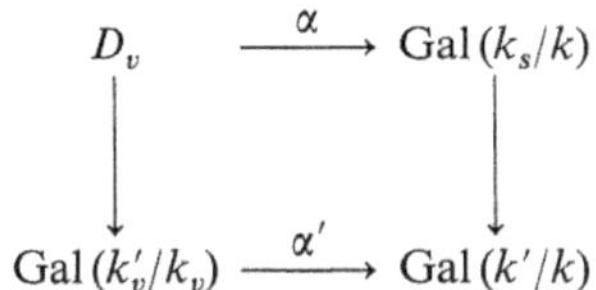

where α and α' are embeddings unique up to conjugation. We know that we have an embedding $\mathrm{Gal}\,(k'/k) \to S_3$ and giving the a_v's amounts to give embeddings $a_v \colon \mathrm{Gal}\,(k'_v/k_v) \to S_3$. Since the subgroups of S_3 of a given order form one conjugacy class, we may assume that the image of a_v is contained in $\mathrm{Gal}(k'/k)$. This ends the proof of our assertion.

1. 10. Remark. In the notation of 1. 8, let $a \in \mathrm{Im}\, \imath$. It is represented by a cocycle with values in $T(k_s)$, hence the k-form G' of G_0 associated to it contains a Borel subgroup defined over k, i.e. is quasi-split. It is wellknown that this map yields a bijection between $H^1(k,\,\mathrm{Aut}\,D_0)$ and k-isomorphism classes of quasi-split k-forms of G_0 [10], XXIV. Therefore 1. 9 is equivalent to (a) of the introduction, namely to the

Proposition. *Let k be a number field, S a finite set of places of k and G an absolutely almost simple k-group. Let $G_{(v)}$ be a quasi-split k_v-form of G $(v \in S)$. Then there exists a quasi-split k-form G' of G which is k_v-isomorphic to $G_{(v)}$ for every $v \in S$.*

1. 11. *Proof of Theorem* B (cf. introduction). A simply connected semi-simple group and its adjoint group have the same group of automorphisms. We may therefore assume G to be of adjoint type. Since G is k_s-isomorphic to a k-split group, we may assume further

G to be split over k. Let $a = (a_v)_{v \in S}$, where $a_v \in H^1(k_v, \operatorname{Aut} G)$. We consider the commutative diagram

$$
(1) \quad
\begin{array}{ccccc}
H^1(k, G) & \xrightarrow{\;\alpha\;} & H^1(k, \operatorname{Aut} G) & \xrightarrow{\;\beta\;} & H^1(k, \operatorname{Aut} D) \\
\big\downarrow{\omega_S} & & \big\downarrow{\omega'_S} & & \big\downarrow{\omega''_S} \\
\displaystyle\prod_{v \in S} H^1(k_v, G) & \xrightarrow{\;\alpha'\;} & \displaystyle\prod_{v \in S} H^1(k_v, \operatorname{Aut} G) & \xrightarrow{\;\beta'\;} & \displaystyle\prod_{v \in S} H^1(k_v, \operatorname{Aut} D),
\end{array}
$$

where the rows are associated to 1. 8 (2) and are exact. In view of the splitting 1. 8 (3), the map β is surjective. By 1. 9, there exists therefore $b \in H^1(k, \operatorname{Aut} G)$, in fact $b \in \operatorname{Im} \iota$, such that $\beta' \cdot \omega'_S(b) = \beta'(a)$. If we now twist G by a cocycle of b, then we arrive at a situation where $a_v \in H^1(k_v, G)$ for $v \in S$. Then we can use 1. 7.

1. 12. Corollary. *Let L be a local field of characteristic zero, k a number field, $v \in V_k$, and $S = V_\infty \cup \{v\}$. Assume that $k_v \cong L$ and $k_w \cong \mathbb{R}$ if $w \in S - \{v\}$. Let H be an absolutely almost simple L-group which is either simply connected or of adjoint type. Then there exists an absolutely almost simple k-group G which is isomorphic to H over k_v and is anisotropic over k_w for $w \in S - \{v\}$.*

Let G_0 be a split group over k, of which H is a k_v-form. Let $a_v \in H^1(k_v, \operatorname{Aut} G_0)$ be the corresponding cohomology class. Let $w \in S - \{v\}$. Since it is a real place, there exists an element $a_w \in H^1(k_w, \operatorname{Aut} G_0)$ which represents a compact form of G_0. We then apply Theorem B to $a = (a_w)_{w \in S}$.

1. 13. Remark. Given the local field L, it is easy to construct a number field k satisfying the conditions of 1. 12, and for which moreover $S \neq \{v\}$. In fact, if $L = \mathbb{R}, \mathbb{C}$, we have just to take a number field of degree $> [L : \mathbb{R}]$ with one completion at infinity isomorphic to L and all the others isomorphic to $\mathbb{R}$. Let now L be non-archimedean. We can write $L = \mathbb{Q}_p(\alpha)$ for some prime p, where α is algebraic over $\mathbb{Q}_p$. Let $P_0(X) \in \mathbb{Q}_p[X]$ be an irreducible polynomial having α as one of its roots, and let n be its degree. Let $P_1(X) = \prod_{1 \leq i \leq n} (X - i)$. Then if $P(X) \in \mathbb{Q}[X]$ is a polynomial of degree n whose coefficients are sufficiently close to those of $P_0(X)$ and $P_1(X)$, the field $\mathbb{Q}(\beta)$ generated over $\mathbb{Q}$ by a root of $P(X)$ satisfies our conditions.

2. Discrete cocompact subgroups

2. 1. We shall use the fact that if H, H' are algebraic groups over a local field L of characteristic zero, and if $f: H \to H'$ is a surjective morphism, then the image of the associated homomorphism $f(L): H(L) \to H'(L)$ is an open normal subgroup of finite index (for the L-topology). In fact, since we are in characteristic zero, f is submersive and hence $f(H(L))$ is open. That it is of finite index follows e.g. from 6. 4, p. 155 in [3]. If moreover $\ker f$ is finite, then $f(L)$ is proper.

2. 2. *Proof of Theorem* A. Since the identity component of H in the Zariski topology has finite index in H, we may assume H to be a connected reductive algebraic group. Let Z be the connected component of the identity of the center of H and H' the derived group of H. Then $H = Z \cdot H'$. Let $\tilde{H}'$ be the universal covering group of H'. There is a natural L-isogeny $\mu : Z \times \tilde{H}' \to H$. The natural homomorphism

$$\mu(L) : Z(L) \times \tilde{H}'(L) \to H(L),$$

has finite kernel, and its image is open, of finite index, (2. 1). It suffices therefore to prove the theorem for Z and $\tilde{H}'$, i.e. H may be assumed to be either a torus or a simply-connected semi-simple group.

If H is a torus, then $H(L)$ has a greatest compact subgroup K, the intersection of the kernels of the maps $h \mapsto |\chi(h)|$, where χ runs through the characters of H which are defined over k, and the quotient is a finitely generated free abelian group. The extension $H(L)$ of K therefore splits and any splitting gives a discrete cocompact subgroup. Let now H be semi-simple and simply connected. It is then a direct product of almost simple simply connected groups over L, which reduces us to the case where H is almost simple over L and simply connected. There exists then a finite extension L' of L and an absolutely almost simple simply connected L'-group H' such that $H = R_{L'/L} H'$ is obtained by restriction of scalars from H' [5], 6. 21 (ii). We have then an isomorphism of topological groups $H(L) = H'(L')$ [26], 1. 3. We may therefore assume H to be also absolutely almost simple over L.

Let k be a number field satisfying the conditions of 1. 13, and then choose G as in 1. 12. If $w \in S - \{v\}$, then w is real and G is anisotropic over k_w. Since $S \neq \{v\}$, this means that G is anisotropic at at least one place, hence *a fortiori* that G is anisotropic over k. Let $G_S = \prod_{w \in S} G(k_v)$ and let Γ be an S-arithmetic subgroup (or an arithmetic subgroup if v is also infinite). Then Γ, diagonally embedded in G_S, is discrete and, since G is anisotropic, it is cocompact [2], § 8. For $w \in S - \{v\}$, G is anisotropic over k_w, hence $G(k_w)$ is compact. The projection of Γ in $G(k_v) = H(L)$ is therefore discrete and cocompact. In other words Γ, viewed as a subgroup of $H(L)$, has the required properties.

2. 3. Remark. We recall that a discrete cocompact subgroup of $H(L)$ is always finitely generated. It suffices to see this when H is semi-simple, in which case we may refer e.g. to [4], 6. 3. Since such a group is linear, it follows moreover that it contains a torsion-free subgroup of finite index and that the intersection of its subgroups of finite index is reduced to the identity. This also obtains when L has positive characteristic (see H. Garland, Annals of Math. (2) **97** (1973), 375—423, § 2. 7).

3. The equal characteristic case

In this section, L is a local field (see 1. 2) and k a global field of characteristic $p > 0$.

3. 1. *Cohomology.* Our first objective is to prove Theorem B for a group of type $\mathbb{A}_n$ over k. The main step in the above proof is Theorem 1. 7 in the case where G is of adjoint type. But now the isogeny $\pi : \tilde{G} \to G$ of 1. 7 is not necessarily separable, though still central. Its schematic kernel is then not always a Galois module, but a commutative finite group scheme of multiplicative type. This makes it necessary to enlarge our cohomological framework. We can use Amitsur cohomology, in which the same general results are available. For a brief introduction to the latter, see [13], § 1. It corresponds to the cohomology for the flat topology in [21], where however, only commutative group schemes are considered. We note that if G is a reduced k-group, then Amitsur and Galois cohomology with coefficients in G coincide (see [13]; for G commutative, see [21], Theorem 43).

In view of Satz B of [13], we can reduce the proof of 1.7 directly to 1.6, without using 1.5. [We remark, however, that 1.5 and its proof go over in positive characteristic; the only change needed is to replace approximation in a vector space by weak approximation in a semi-simple k-group.] Proposition 1.6, however, has to be modified since we must allow M to have an infinitesimal part, and we cannot apply the results of Poitou-Tate proved in [20]. Over local fields, we could refer to [21], VI, § 6, but we do not know of a proof of global Poitou-Tate duality in the function field case. We shall content ourselves with proving 1.6 for the M's of interest to us here.

3.2. Let then M be the schematic center of an absolutely almost simple simply connected group of type $\mathbb{A}_n$ over k, and S a finite set of places of k. We want to prove the surjectivity of

$$(1) \qquad \omega_{S,2} : H^2(k, M) \to \prod_{v \in S} H^2(k_v, M),$$

where now H^2 refers to Amitsur cohomology. Assume first that G is an inner form of type $\mathbb{A}_n$. Then M is isomorphic to the scheme μ_{n+1} of $(n+1)$-st roots of unity, and $H^2(k, \mu_{n+1})$ is the group of elements of order dividing $n+1$ in the Brauer group of k. Our assertion then follows from class field theory [25], 9.6.

Assume now that G is an outer form of $\mathbb{A}_n$. There exists then a unique separable quadratic extension E of k over which G becomes an inner form of $\mathbb{A}_n$. Moreover, M is isomorphic to the kernel of the map $v : T \to T$ given by $x \mapsto x^{n+1}$, where T is a one-dimensional anisotropic torus over k which splits over E. For any finite normal extension F of k containing E, we have, by Hilbert's theorem 90:

$$(2) \qquad H^1(F, T) = 0, \quad H^1(k, T) = H^1(E/k, T) = k^*/N_{E/k}(E^*).$$

Let I_F (resp. C_F) denote the idele (resp. idele class) group of F. We have an exact sequence

$$(3) \qquad 1 \to T_F \to T_{A_F} \to T_{C_F} \to 1.$$

We deduce from (2):

$$(4) \qquad H^1(F/k, T_{A_F}) = I_k/N_{E/k}(I_E), \quad H^1(F/k, T_{C_E}) = C_k/N_{E/k}(C_E).$$

It follows that

$$(5) \qquad H^1(F/k, T_{A_F}) \to H^1(F/k, T_{C_F})$$

is surjective. On the other hand, we get

$$(6) \qquad H^2(F/k, T_{C_F}) = 0$$

by application of [22], IX, Theorem 14 (cf. [11], A 1). Using the exact cohomology sequence associated to (3), we see that

$$(7) \qquad j : H^2(k, T) \to H^2(k, T_A)$$

is an isomorphism. Consider now the commutative diagram, where the rows are segments of exact cohomology sequences:

$$(8) \quad \begin{array}{ccccccc}
H^1(k, T) & \xrightarrow{v_1^*} & H^1(k, T) & \longrightarrow & H^2(k, M) & \longrightarrow & H^2(k, T) \\
\downarrow & & \downarrow & & \downarrow{\scriptstyle i} & & \downarrow{\scriptstyle j} \\
H^1(k, T_A) & \xrightarrow{v_{1,A}^*} & H^1(k, T_A) & \xrightarrow{\delta} & H^2(k, M_A) & \longrightarrow & H^2(k, T_A).
\end{array}$$

Since j is an isomorphism, we see that, given $\xi \in H^2(k, M_A)$, there exists $\eta \in H^2(k, M)$ such that

$$(9) \qquad \xi - i(\eta) \in \operatorname{Im} \delta.$$

The assertion (1) now follows because the map

$$\omega_{S,1} : H^1(k, T) \to \prod_{v \in S} H^1(k_v, T),$$

which, by (2), amounts to

$$k^* / N_{E/k}(E^*) \to \prod_{v \in S} k_v^* / N_{E_{v'}/k_v}(E_{v'}^*),$$

where v' is a place of E lying over v, is surjective by class field theory [25], 9. 6.

3. 3. Theorem. *Let H be a group over L whose identity component is reductive and has a derived group isogeneous to a product of groups of type $\mathbb{A}$. Then $H(L)$ contains a cocompact discrete subgroup.*

Proof. As in 2. 2, we may assume H to be connected. Let Z, H', $\tilde{H}'$ and

$$\mu : Z \times \tilde{H}' \to H,$$

be as in 2. 2. Then μ is a central isogeny, therefore the image of

$$\mu(L) : Z(L) \times \tilde{H}'(L) \to H(L),$$

is a closed normal cocompact subgroup of $H(L)$ [6], 3. 17. This reduces us again to the cases where H is a torus, for which the proof is the same as in 2. 2, or is almost absolutely simple, simply connected, of type $\mathbb{A}_n$ for some n. We then proceed as in 2. 2, choose a global field k such that $k_v \xrightarrow{\sim} L$ for some $v \in V_k$, a place $w \neq v$ of k, and a k-group G of type $\mathbb{A}_n$ which is isomorphic to H over k_v and is anisotropic over k_w (3. 1, 3. 2). The group G is then anisotropic over k, hence G_A/G_k is compact (see [26], 4. 1. 2 if $p \neq 2$, [14] in general). Let Γ be the subgroup of elements of G_k which are integral outside $S = \{w, v\}$. Then G_S/Γ may be identified to an open and closed subset of G_A/G_k, hence is compact too. Since $G(k_w)$ is compact, the projection of Γ in $G(k_v)$ is discrete and cocompact.

3. 4. Our discrete cocompact subgroups have been obtained by the "arithmetic method" and are in fact "S-arithmetic groups". This construction relies therefore on the existence of anisotropic forms. By a theorem of Bruhat-Tits [9] and the results of [13], among simple groups, only those of type $\mathbb{A}$ have anisotropic forms over local or global fields of positive characteristic p. Therefore, our method cannot work in other cases. In characteristic zero, if say H is simple of L-rank ≥ 2, then the results of Margoulis [19] show that any discrete cocompact subgroup has to be commensurable to an S-arithmetic one. It does not seem to be known whether they are valid in characteristic p. If so, it would indeed follow that, in positive characteristic, 3. 3 would give all the cases in which there are discrete cocompact subgroups.

4. Remark on the formal degree of the cuspidal representations of $h(L)$

In this section, H is a semi-simple group over the local field L, Γ a torsion-free discrete cocompact subgroup of H(L) (see 2. 3) and dh a Haar measure on H(L).

4. 1. Assume first $L = \mathbb{R}$ and π to be integrable. Then, using the trace formula R. P. Langlands has shown that the multiplicity of π in $L^2(H(L)/\Gamma)$ is equal to $v(H(L)/\Gamma) \cdot d_\pi$, where $v(H(L)/\Gamma)$ is the volume of $(H(L)/\Gamma)$ and d_π the formal degree of π, both computed with respect to dh [18].

Let now L be non-archimedean, of characteristic zero. Recall that π is said to be cuspidal (or also supercuspidal) if its coefficients are compactly supported. Then, the remark of P. Sally alluded to in the introduction is that Langlands' result and its proof should remain valid and prove that d_π is an integrable multiple of $v(H(L)/\Gamma)^{-1}$. In particular, this implies the existence of a constant $c > 0$ such that $d_\sigma > 0$ for all cuspidal representations σ of $H(L)$. This last result, conjectured by Harish-Chandra, was in fact proved by R. Howe for $\mathbb{GL}_n$ (in arbitrary characteristic) [17] and then by Harish-Chandra in general in characteristic zero. But the method of [18] would provide another approach to it.

Langlands' proof in [18] hinges on the vanishing of the orbital integrals

$$I(\gamma) = \int\limits_{G/G_\gamma} f(g \cdot \gamma \cdot g^{-1}) \, dg^*,$$

where $\gamma \in \Gamma$, $\gamma \neq e$, $G = H(L)$, G_γ is the centralizer of γ in G, dg^* is a positive invariant measure on G/G_γ, and f a coefficient of π, proved by Harish-Chandra. Since Harish-Chandra has recently also established the analogous result in the p-adic case (characteristic zero), Langlands' argument does go over without change. However, the vanishing of $I(\gamma)$ has not yet been proved in characteristic p, and this proof cannot be applied to the groups considered in 3. 3.

References

[1] *A. Borel*, Compact Clifford-Klein forms of symmetric spaces, Topology **2** (1963), 111—122.

[2] *A. Borel*, Some finiteness properties of adele groups over number fields, Publ. Math. I.H.E.S. **16** (1963), 101—126.

[3] *A. Borel* et *J.-P. Serre*, Théorèmes de finitude en cohomologie galoisienne, Comm. Math. Helv. **39** (1964), 111—164.

[4] *A. Borel* et *J.-P. Serre*, Cohomologie d'immeubles et de groupes S-arithmétiques, Topology **15** (1976), 211—232.

[5] *A. Borel* et *J. Tits*, Groupes réductifs, Publ. Math. I.H.E.S. **27** (1965), 55—150.

[6] *A. Borel* et *J. Tits*, Homomorphismes "abstraits" de groupes algébriques simples, Annals of Math. (2) **97** (1973), 499—571.

[7] *N. Bourbaki*, Groupes et algèbres de Lie, Chap. 4, 5, 6, Act. Sci. Ind. **1337**, Hermann ed. Paris 1968.

[8] *N. Bourbaki*, Groupes et algèbres de Lie, Chap. 7, 8, Act. Sci. Ind. **1364**, Hermann ed. Paris 1975.

[9] *F. Bruhat* et *J. Tits*, Groupes algébriques simples sur un corps local: cohomologie galoisienne, décompositions d'Iwasawa et de Cartan, C.R. Acad. Sci. Paris **263** (1966), 867—869.

[10] *M. Demazure* et *A. Grothendieck*, Schémas en groupes. III, Springer Lecture Notes in Mathematics **153**, Berlin 1970.

[11] *G. Harder*, Über die Galoiskohomologie halbeinfacher Matrizengruppen. II, Math. Zeitschrift **92** (1966), 396—415.

[12] *G. Harder*, Bericht über neuere Resultate der Galoiskohomologie halbeinfacher Gruppen, Jahresbericht D. M. V. **70** (1967/68), 182—216.

[13] *G. Harder*, Über die Galoiskohomologie halbeinfacher algebraischer Gruppen. III, J. reine angew. Math. **274/275** (1975), 125—138.

[14] *G. Harder*, Minkowskische Reduktionstheorie über Funktionenkörper, Inv. Math. **7** (1969), 33—54.

[15] *M. Kneser*, Galoiskohomologie halbeinfacher algebraischer Gruppen über p-adischen Körpern. II, Math. Zeitschrift **89** (1965), 250—272.

[16] *M. Kneser*, Lectures on Galois cohomology of classical groups, Notes by P. Jothilingam, Tata Institute of Fundamental Research, Bombay 1969.

[17] *R. Howe*, The Fourier transform and germs of characters (case of GL_n over a p-adic field), Math. Annalen **208** (1974), 305—322.

[18] *R. P. Langlands*, Dimension of spaces of automorphic forms, Algebraic groups and discontinuous groups, Proc. Symp. pure math. IX, A.M.S. Providence, R.I. 1966.

[19] *G. A. Margoulis*, Discrete groups of isometries of manifolds with negative curvature, Proc. I.C.M. Vancouver 1974, Vol. **2**, 21—34.

[20] *G. Poitou*, Cohomologie galoisienne des modules finis, Séminaire de l'Institut mathématique de Lille, Dunod Paris 1967.

[21] *S. Schatz*, Profinite groups, arithmetic and geometry, Annals of Math. Sudy **67**, Princeton 1967.

[22] *J.-P. Serre*, Corps locaux, Act. Sci. Ind. **1296**, Hermann ed. Paris 1962.

[23] *J.-P. Serre*, Cohomologie galoisienne, Springer Lecture Notes in Mathematics **5**, Berlin-Heidelberg-New York 1965.

[24] *T. Tamagawa*, On discrete subgroups of p-adic algebraic groups, In: Arithmetical algebraic geometry, O. F. G. Schilling ed., Harper and Row, New York 1965.

[25] *J. Tate*, Global class field theory, in Algebraic number theory, ed. by J. W. S. Cassels and A. Fröhlich, Thompson Co. Washington, D.C. 1967.

[26] *A. Weil*, Adeles and algebraic groups, Notes by M. Demazure and T. Ono, The Institute for Advanced Study, Princeton, N.J., 1961.

Institute for Advanced Study, Princeton, N.J. 08540, USA,
Mathematisches Institut der Universität, Wegelerstr. 10, D-5300 Bonn

Eingegangen 16. März 1977

110.

(avec J. Tits)

Théorèmes de structure et de conjugaison pour les groupes algébriques linéaires

C. R. Acad. Sci., Paris **287** (1978) 55–57

La plupart des résultats faisant l'objet de cette Note généralisent à un groupe algébrique linéaire connexe G quelconque sur un corps k des théorèmes connus pour les groupes réductifs. On définit des sous-groupes « pseudo-paraboliques » qui jouent ici le rôle des sous-groupes paraboliques dans le cas réductif. Les k-sous-groupes pseudo-paraboliques minimaux de G sont tous conjugués sur k et leurs radicaux déployés (plus grands k-sous-groupes résolubles déployés distingués) sont les k-sous-groupes résolubles déployés maximaux de G. Si le radical unipotent de G ne contient pas de k-sous-groupe déployé non trivial, les poids non nuls d'un k-tore déployé maximal de G dans l'algèbre de Lie de G forment un système de racines. Enfin, on exhibe un système de Tits dans $G(k)$.

Most results presented in this Note generalize to an arbitrary connected linear algebraic group G over a field k known theorems on reductive groups. We define "pseudo-parabolic subgroups" which play the role of the parabolic subgroups in the classical situation. All minimal pseudo-parabolic k-subgroups are conjugate over k and their k-split radicals (i. e. their maximal normal k-split solvable subgroups) are the maximal k-split solvable subgroups of G. If the unipotent radical of G contains no nontrivial split k-subgroup, the nonzero weights of a maximal k-split torus of G in the Lie algebra of G form a root system. A Tits' system is exhibited in G(k).

1. NOTATIONS. — Nous utilisons la terminologie et les notations de notre article « Groupes réductifs » ([1]), à ceci près que le groupe des points rationnels d'un groupe algébrique X sur un corps l est noté $X(l)$ au lieu de X_l. Dans toute la Note, k désigne un corps, k_s une clôture séparable de k et G un groupe algébrique linéaire connexe défini sur k (c'est-à-dire un k-schéma en groupes affine, lisse et connexe). L'algèbre de Lie d'un groupe algébrique X est notée $L(X)$.

2. SOUS-GROUPES STABLES SOUS L'ACTION D'UN TORE. — Soient S un k-tore et $\psi \subset X^*(S)$ une partie du groupe des caractères de S. Si V est un k-espace vectoriel sur lequel S opère linéairement, on note V_ψ le plus grand sous-espace vectoriel V' de V stable par S et tel que tous les poids de S dans V' appartiennent à ψ (V' n'est éventuellement défini que sur une extension séparable de k). Pour tout groupe H sur lequel S opère, on désigne par H_ψ le sous-groupe de H engendré par tous les sous-groupes connexes M tels que $L(M)=L(M)_\psi$. L'ensemble ψ est dit *clos* s'il est fermé pour l'addition, et *clos dans* H si $L(H_\psi)=L(H)_\psi$.

THÉORÈME 1. — *Si ψ est défini sur k, le groupe H_ψ l'est aussi. Si ψ est clos, il est clos dans* H. *S'il existe un cocaractère $\lambda \in X_*(S)$ tel que $\langle \psi, \lambda \rangle \subset \mathbf{N}^*$, alors* H *est unipotent.*

Ce théorème précise certains résultats du paragraphe 3 de « Groupes réductifs » ([1]).

3. EXEMPLES DE GROUPES DONT LE RADICAL UNIPOTENT N'EST PAS DÉFINI SUR LE CORPS DE BASE. — Tous les résultats de la suite généralisent des théorèmes connus pour les groupes réductifs [([1], ([2])] et s'y ramènent facilement lorsque le radical unipotent $R_u(G)$ est défini sur k; en particulier, ils n'apportent rien de nouveau lorsque k est parfait. Rappelons en guise d'illustration deux exemples simples de groupes ayant un radical unipotent non défini sur k. Soient K une extension radicielle propre de k et H un K-groupe connexe non unipotent; alors, la « restriction des scalaires » [ou « restriction de Weil » selon ([3])] $R_{K/k} H$ a un radical unipotent non défini sur k (dans un certain sens — que nous ne préciserons pas ici —, cet exemple est « universel » si car $k \neq 2, 3$). Lorsque car $k=2$, un autre exemple est fourni par le plus grand k-sous-groupe du groupe orthogonal d'une forme quadratique de défaut $\geqq 2$ à forme bilinéaire associée non nulle.

4. Sous-groupes pseudo-paraboliques. — Si H est un k-groupe algébrique linéaire, on appelle *radical déployé* (resp. *radical unipotent déployé*) de H sur k et l'on note R_d(H) (resp. R_{ud}(H)) le plus grand k-sous-groupe résoluble (resp. unipotent) déployé contenu dans le radical R (H) ([4]). C'est un sous-groupe distingué de H. Un sous-groupe P de H est dit *pseudo-parabolique (relativement à k)* s'il existe un tore $S \subset H$ et un cocaractère $\lambda \in X_*(S)$ tels que, posant $\psi(\lambda) = \{ \chi \in X^*(S) \mid \langle \lambda, \chi \rangle \geqq 0 \}$, on ait $P = H_{\psi(\lambda)} \cdot R_{ud}$(H). Si l'on peut choisir S maximal dans H et λ ne s'annulant sur aucun poids non nul de S dans L (H), alors P est un sous-groupe résoluble connexe de H que l'on dit *pseudo-maximal* (relativement à k); les sous-groupes résolubles connexes pseudo-maximaux sont donc les sous-groupes pseudo-paraboliques minimaux. Toutes les notions ainsi définies, à l'exception du radical déployé, sont stables par extension *séparable* de k. Le radical déployé, le radical unipotent déployé, les sous-groupes pseudo-paraboliques et les sous-groupes résolubles connexes pseudo-maximaux jouent ici, relativement à k, des rôles analogues à ceux du radical, du radical unipotent, des sous-groupes paraboliques et des sous-groupes de Borel dans le cas d'un corps de base algébriquement clos. Si R (G) est défini sur k (et en particulier si k est parfait), les sous-groupes pseudo-paraboliques de G ne sont autres que ses sous-groupes paraboliques. Les assertions suivantes, où P désigne un k-sous-groupe pseudo-parabolique de G, se démontrent à peu près comme dans le cas réductif :

La fibration G $\rightarrow$ G/P possède une section rationnelle définie sur k et l'application $G(k') \rightarrow (G/P)(k')$ est surjective pour toute extension k' de k;

P est son propre normalisateur dans G (non seulement au sens ensembliste mais aussi au sens schématique); en particulier, G/P s'identifie à la variété des sous-groupes conjugués à P. De ce dernier résultat on déduit, par descente galoisienne, que la variété des sous-groupes résolubles connexes pseudo-maximaux de G est définie sur k.

5. Sous-groupes résolubles déployés. Théorèmes de conjugaison. — La proposition suivante est un « succédané relatif » du fait que le quotient d'un groupe par un sous-groupe parabolique est une variété complète.

Proposition 1. — *Si P est un k-sous-groupe pseudo-parabolique de G, l'application canonique* (G/P) $(k[[t]]) \rightarrow$ (G/P) $(k((t)))$ *est bijective.*

On en déduit aussitôt, de la façon habituelle, que tout k-groupe résoluble déployé opérant (k-morphiquement) sur G/P possède un point fixe dans (G/P) (k). Utilisant ce résultat et le fait (cas particulier du corollaire au théorème 4 ci-dessous) que l'intersection de deux k_s-sous-groupes résolubles connexes pseudo-maximaux contient un tore maximal, on peut imiter les raisonnements de ([2]) et prouver le :

Théorème 2. — (i) *Un k-sous-groupe P de G est pseudo-parabolique relativement à k si et seulement s'il est le plus grand k-sous-groupe du normalisateur de son radical unipotent déployé* R_{ud}(P).

(ii) *Les k-sous-groupes résolubles déployés maximaux de G sont les radicaux déployés des k-sous-groupes pseudo-paraboliques minimaux de G.*

Corollaire. — *Les k-sous-groupes résolubles déployés maximaux (resp. les k-tores déployés maximaux; resp. les k-sous-groupes unipotents déployés maximaux; resp. les k-sous-groupes pseudo-paraboliques minimaux) de G sont conjugués entre eux par des éléments de* G (k).

Du théorème 1, on déduit aussi l'équivalence des conditions suivantes (définissant la « k-isotropie ») : (i) G possède un k-tore déployé non contenu dans R (G); (ii) G possède un k-sous-groupe unipotent déployé non contenu dans R (G); (iii) G possède un k-sous-groupe pseudo-parabolique propre. D'autre part, on montre comme dans $(^2)$ que pour tout k-sous-groupe unipotent déployé U de G, il existe un k-sous-groupe pseudo-parabolique P contenant le normalisateur de U dans $G(k_s)$ (donc aussi le plus grand k_s-sous-groupe du normalisateur de U) et tel que $U \subset R_{ud}(P)$.

6. SYSTÈME DE RACINES ET SYSTÈME DE TITS. — Dans ce paragraphe, on choisit un k-tore déployé maximal S dans G et l'on désigne par Z (resp. N) son centralisateur (resp. son normalisateur) dans G. On sait que Z et N sont des k-sous-groupes de G. Si H est un sous-groupe de G normalisé par S, on note $\Phi(S, G/H)$ l'ensemble des poids non nuls de S dans $L(G)/L(H)$. Enfin, on pose $\Phi = \Phi(S, G/R_{ud}(G))$, $W = N(k)/Z(k)$ et, pour toute partie ψ de $X^*(S)$, $\psi_{nm} = \{ \chi \in \psi \mid (N^* \cdot \chi) \cap \psi = \{ \chi \} \}$.

THÉORÈME 3. — *L'ensemble Φ est un système de racines dans le sous-espace vectoriel de $X^*(S) \otimes \mathbf{R}$ qu'il engendre. Le groupe de Weyl de ce système est le groupe W (opérant sur $X^*(S)$ de façon évidente). On a $\Phi_{nm} = \Phi(S, G/R_u(G))_{nm}$; de plus $\Phi = \Phi(S, G/R_u(G))$ si car $k \neq 2$.*

Par contre, si k est un corps non parfait de caractéristique 2, on peut construire pour tout entier $n \geq 1$ un groupe G de dimension $4n(n+1)$ possédant un k-sous-groupe H isomorphe (sur k) à $\mathbf{Sp}_{2n}$, tel que $R_{ud}(G) = \{ 1 \}$ et $G = H \cdot R_u(G)$ [d'où $\Phi(S, G/R_u(G)) = \Phi(S, H)$, un système de racines de type $\mathbf{C}_n$] et tel que $\Phi = \Phi(S, G)$ soit un système de racines non réduit, de type $\mathbf{BC}_n$.

Soit λ un cocaractère de S ne s'annulant sur aucun élément de Φ et posons $P = G_{\psi(\lambda)} \cdot R_{ud}(G)$, où $\psi(\lambda)$ est défini comme au paragraphe 4.

THÉORÈME 4. — *Le système $(G(k), P(k), N(k))$ est un système de Tits de groupe de Weyl W $(^5)$. Les sous-groupes paraboliques de ce système sont les groupes de points rationnels des k-sous-groupes pseudo-paraboliques de G. Deux k-sous-groupes pseudo-paraboliques sont conjugués dans G (resp. sont égaux) si et seulement si leurs groupes de points rationnels sont conjugués dans $G(k)$ (resp. sont égaux).*

COROLLAIRE. — *L'intersection de deux k-sous-groupes pseudo-paraboliques de G est définie sur k et contient le centralisateur d'un k-tore déployé maximal de G.*

(*) Séance du 5 juin 1978.
$(^1)$ *Publ. Math. I.H.E.S.*, 27, 1965, p. 55-151.
$(^2)$ *Inventiones Math.*, 12, 1974, p. 95-104.
$(^3)$ M. DEMAZURE et P. GABRIEL, *Groupes algébriques*, Masson, Paris, 1970, p. 30.
$(^4)$ *Voir* $(^2)$, corollaire 2.3.
$(^5)$ N. BOURBAKI, *Groupes et Algèbres de Lie*, chapitres IV, V, VI, p. 22.

Institute for Advanced Study, Princeton, N. J. 08540, États-Unis.
Collège de France, 11, place Marcelin-Berthelot, 75231 Paris Cedex 05.

111.

On the development of Lie group theory

Proc. of the Bicentennial Congr. of the Dutch Math. Soc., Math. Centre Tract
100/101 (1979) 25–37

In his letter of invitation to give this lecture, Professor Freudenthal
pointed out that two choices had been made, first of a field, and then of
a mathematician having worked in it. In the present case, the field
is Lie groups. Accordingly, I shall devote this lecture to the award win-
ning field, so to say, and take as my main theme some aspects of the development and role in mathematics of Lie groups. In doing so, I shall view
"Lie groups" in a rather broad sense, including not only the classical real
or complex Lie groups, but linear algebraic groups and p-adic Lie groups
as well.

The theory of "finite and continuous groups", later called Lie groups,
was built from about 1873 on by the Norwegian mathematician Sophus Lie. It
arose out of his work on differential equations and contact transformations,
and he had a main goal in mind, namely to develop a Galois theory of differential equations, in which these groups would play the role of the Galois
group of an algebraic equation. It seems to me that, from that point of
view, Lie groups offer a, by no means unique, example of a theory created
for a certain purpose, but not fulfilling it up fully. However, it then
went off into many directions, so much so that, one hundred years later,
J. Dieudonné was led to write: "Les groupes de Lie sont devenus le centre
des mathématiques; on ne peut rien faire de sérieux sans eux" [1]. Such a
drastic statement was of course quickly challenged and, even to me, seems
too sweeping. But I do share the view that Lie groups, as understood here,
play a major, even central role in an important and ever spreading part of
mathematics.

In only one hour, I cannot present in a comprehensive manner all the
evidence to back up such a claim. Without trying to be systematic, I shall

---

[1] J. Dieudonné, "Orientation générale des mathématiques pures en 1973",
Gazette des Mathématiciens, Oct. 1974, p. 73-79, Soc. Math. France.

concentrate on some developments which seem to me to have been crucial in increasing the scope of the theory and of its connections with other parts of mathematics. In a topic with such a rich history, even the choice of those is not necessarily unique, and may depend upon one's own perspective. From mine, I would like to single out three:

(1) Global Lie groups and Riemannian symmetric spaces.

(2) Linear algebraic groups.

(3) Buildings and p-adic symmetric spaces.

Before taking them up, I should make some remarks on the first fifty years of the theory of Lie groups.

1.   It is a simple matter nowadays to define a real or complex Lie group, as a real or complex analytic manifold G endowed with a group structure such that the map $G \times G \to G$ given by $(x,y) \mapsto x \cdot y^{-1}$ is analytic. But S. Lie could not say that, and this definition differs from his in two respects. First, he considered only transformation groups. The notion of abstract group was not familiar at the time, and even later, when it had become more widespread, F. Klein had some misgivings about putting it in the foreground (Entwicklung der Mathematik I, 335-336). But S. Lie had a notion of isomorphism, called "Gleichzusammensetzung", which focussed attention on the law of composition, and he knew that each isomorphism class could be represented by the group acting on itself by left or right translations. So the difference here is more one of terminology than of substance. The second one is more important: his groups were local; G was in fact a neighborhood of the origin in C^n (mostly, occasionally R^n), the law of composition was defined for elements sufficiently close to the origin, and given by convergent power series. As you know, the thrust of the general theory was to reduce problems on such local groups to algebraic ones on what is now called the Lie algebra of G, that is, the vector space of left invariant vector fields, endowed with the bracket operation of infinitesimal transformations. Lie's approach was analytic, and much influenced by his work on contact transformations. For instance, it seems that an important step for him was the interpretation of the Poisson bracket of two functions as the bracket of two infinitesimal contact transformations. The purely algebraic problems to which his theory led were later solved mostly be other mathematicians, notably by W. Killing and, above all, by E. Cartan.

During this first period, the theory was not purely local, however. Lie and his contemporaries were familiar with the classical groups; namely the special or general linear group $SL_n(C)$ or $GL_n(C)$, the orthogonal and

symplectic groups $O(n,C)$ and $Sp_{2n}(C)$, except that they viewed them usually
as groups of projective transformations rather than of homogeneous linear
ones. Furthermore, a number of investigations were frankly global in character, as for instance Hurwitz's construction of invariants for $SL_n(C)$ or
$SO(n,C)$ by integration over the compact groups $SU(n)$ and $SO(n)$ (1897). Also,
I should not miss this opportunity to allude to the work of L.E.J. Brouwer
on Lie groups and on Brouwer's correspondence with F. Engel (1909-1911),
in which Brouwer clearly has a global picture of Lie transformation groups.
2.   In spite of this, I would still say that the global theory really got
off the ground with H. Weyl's famous Math. Zeitschrift papers (1925-26).
There, H. Weyl combined two approaches to representation theory which until
then had progressed unaware of one another: the infinitesimal one of
E. Cartan and the Frobenius-Schur theory of characters of finite groups,
which had been transposed to some classical compact groups by I. Schur,
using Hurwitz's integration trick. Particularly striking was the fact that
a topological result, the finiteness of the fundamental group of a compact
semi-simple group, played a key role in the proof of an algebraic theorem,
namely the full reducibility of the finite dimensional representations of
the complex semi-simple Lie algebras. This work led a bit later to the
Peter-Weyl theorem, and also paved the way for the applications of Lie group
representations to physics, which have steadily gained in importance since
then.

The work of H. Weyl also had a considerable and almost immediate impact
on E. Cartan. At that time, the latter was working on a seemingly unrelated
problem, the classification of Riemannian manifolds in which the curvature
tensor is invariant under paralleltransport, and had noticed a strange
relationship with the classification of simple real Lie algebras he had
carried out ten years earlier. His approach had been local. Under the
influence of H. Weyl, he recast the question in global terms, explored it
further and built up a beautiful theory of semi-simple Lie groups and
symmetric spaces in which both were inextricably linked. Apart from flat
factors, these spaces are homogeneous spaces of semi-simple groups, and are
products of quotients G/K, where either G is compact semi-simple, K the
fixed point set of an involution of G and G/K has positive curvature, or G
is simple non-compact with finite center, K is a maximal compact subgroup,
and G/K has negative curvature. A beautiful illustration of the interplay
between groups and differential geometry is the proof of the conjugacy of
the maximal compact subgroups of a semisimple group via a fixed point

theorem asserting that any compact group of isometries of a complete simply
connected Riemannian manifold with negative curvature has a fixed point.
For about thirty years, this was the only one.

For G compact, E. Cartan also extended the results of H. Weyl on $L^2(G)$
to $L^2(G/K)$, and began the study of the real homology of G/K via invariant
differential forms. This is the origin of the "de Rham theorems", which he
conjectured on that occasiom, and of Lie algebra cohomology. Later on (from
the late forties on), the topology of compact Lie groups, of the
Grassmannians, and representation theory also became prominent in fibre
bundle theory, in particular in the study of characteristic classes.

At this point, we can see that Lie group theory, which started as a
chapter of analysis and differential equations, then turned to algebra, had
become linked with topology and differential geometry. But the significance
of this development went beyond this, already because certain symmetric spaces
or homogeneous spaces occur in many parts of mathematics. On the compact
side, these spaces include in particular such well known ones as the spheres,
the projective spaces and the Grassmannians over the real, complex or
quaternionic numbers, the Cayley projective plane, and many projective
varieties, for instance the flag manifolds (see below), which play a basic
role in Schubert's enumerative calculus. Among the non-compact symmetric
spaces are to be found the Poincaré upper half-plane and its manifold generalizations: the n-dimensional hyperbolic space, the space $SL_n(R)/SO(n)$ of
positive non-degenerate quadratic forms of determinant one on R^n, the space
$O(p,q)/(O(p) \times O(q))$ of Hermite minimal majorizing forms of an indefinite
quadratic form of index (p,q) on R^{p+q}, the Siegel upper-half plane
$Sp_{2n}(R)/U(n)$. This, and the classification of the bounded symmetric domains
carried out a bit later by E. Cartan, made it clear that semi-simple groups
and symmetric spaces offered a natural framework to study reduction theory
with respect to arithmetic groups and automorphic forms in several variables,
developed notably by C.L. Siegel in the thirties and the forties. A number of
quotients of bounded symmetric domains by arithmetic groups were seen later
to parametrize families of abelian varieties, generalizing the relationship
between the upper half-plane, the modular group and elliptic curves. This
link between semi-simple groups and moduli problems in algebraic geometry
was reinforced much later by P. Griffiths' theory of the period mappings
for families of smooth projective varieties, whose target spaces are quotients of certain homogeneous spaces of semi-simple groups by arithmetic
groups.

The forties also saw the beginning of the study of infinite dimensional
unitary representations of semi-simple groups, a topic which grew into a
fullfledged harmonic analysis on semi-simple groups, including in particular
the spectral decomposition of $L^2(G)$ (the Plancherel formula) by Harish-
Chandra. The Weyl-Cartan theory was again an indispensable preliminary to
this. In particular a basic step towards the Plancherel formula was the
solution of the analogous problem for $L^2(G/K)$, where K is the maximal compact subgroup of G, i.e. the study of the so-called spherical functions on
non-compact symmetric spaces.

I could still go on a long way describing further outgrowths of the work
of Cartan and Weyl discussed above, but I would now like to pass to a quite
different set of ideas, originating in algebraic geometry.

3.   Let K be an algebraically closed field. A subgroup G of the general
linear group $GL_n(K)$ is said to be algebraic if there exists a set of polynomials in n^2 indeterminates

$$(1) \qquad P_\alpha \in K[(X_{ij})_{1 \le i,j \le n}], \qquad (\alpha \in I),$$

with coefficients in K, such that

$$(2) \qquad G = \{g = (g_{ij}) \in GL_n(K) \mid P_\alpha(g_{11}, g_{12}, \dots, g_{nn}) = 0, \ (\alpha \in I)\}.$$

In analogy with the above definition of a Lie group, one can also proceed
more intrinsically and define an affine algebraic group over K as an affine
variety over K which is endowed with a group structure such that the map
$(x,y) \mapsto x \cdot y^{-1}$ is a morphism of affine varieties. There is no substantial
difference between the two notions. If K = C, these groups are complex Lie
groups. As such they were already studied in the 19th century, chiefly by
E. Picard and L. Maurer. The former had in mind to build up a Galois theory
of linear differential equations; this led to the Picard-Vessiot theory.
Maurer's motivations were different, in part invariant theory, in part the
investigation of Lie groups in which the composition was given by algebraic
functions (in suitable coordinates), rather than just by analytic functions.
After that, and a Comptes Rendus Note by E. Cartan (1894), the topic seems
to have fallen into oblivion for about fifty years. To me this is a mildly
curious fact, since the Weyl-Cartan theory pertained first of all to semi-
simple groups, which are essentially algebraic. Interest in them was re-
vived in the early forties, first by C. Chevalley, then by E. Kolchin. At

that time, the emphasis was on groups over fields as general as possible.
Chevalley used an analogue of the exponential mappings of the classical
theory to go from the Lie algebra to the group, and was limited to characteristic zero. The first significant results valid in arbitrary characteristic
were obtained by E. Kolchin (1948). Interestingly enough, this important
step was also motivated by the wish to develop a Galois theory of differential equations, more precisely, a generalization to algebraic differential
equations in characteristic p of the Picard-Vessiot theory. Kolchin proved
an analogue of Lie's theorem on solvable linear groups now known as the
Lie-Kolchin theorem: "every connected solvable algebraic subgroup of $GL_n(K)$
can be put in triangular form", as well as several other results which contained implicitely a rather complete structure theory for solvable groups.
However, in the early fifties, the growing importance of certain algebraic
homogeneous spaces and the development of abstract algebraic geometry made
the need of a more general theory of linear algebraic groups rather widely
felt. To go further, some different methods had to be found. They were developed from 1955 on, and led to results and points of view which were new
even over C. To try to give an idea of the flavor of those, I would like to
sketch the starting point of this theory. For this we may assume that our
groundfield K is just C. Let then G be a linear algebraic group which operates on an algebraic variety V, the action being described by a morphism
$G \times V \to V$ of algebraic varieties. To make matters simple, assume that
$V \subset P_N(C)$ is a projective variety on which G operates by projective transformations, but it could be any variety. Then the "closed orbit lemma"
asserts in this case the existence of an orbit $G \cdot x$ ($x \in V$) which is itself
a projective variety. In fact, if $v \in V$, then the orbit map $g \mapsto g \cdot v$ is a
rational map of G into V. By a well-known fact, the complement of the image
$G \cdot v$ of this map in the smallest algebraic variety $Cl(G \cdot v)$ containing it
(its Zariski-closure) is contained in an algebraic variety of strictly
smaller dimension. Therefore, if we take $x \in V$ such that $Cl(G \cdot v)$ has the
smallest possible dimension, there is no room left and $G \cdot v$ is equal to its
closure. (For a general V, this argument shows the existence of a Zariski-
closed orbit). As a counterpart to that lemma, think of an irrational line
G in a two-dimensional torus T. It acts by translations on T and all the
orbits are dense. In fact, this example pertains to real analytic groups,
but it can easily be made complex, showing that there is no closed orbit
lemma in the complex analytic case. This lemma thus pinpoints a difference
between algebraic and analytic groups and, as simple as it is, is the

cornerstone of the theory of linear algebraic groups. Assume now G to be
connected, commutative. Let x be as above and H the subgroup of G fixing x.
Then G/H is on one hand a linear algebraic group, hence an affine variety,
and on the other hand a connected projective variety. It is then reduced
to a point, hence G has a fixed point on V. Using induction on dimension, one
then deduces that any connected solvable group G acting on a projective
variety has a fixed point. If $G \subset GL_n(K)$, we may then in particular take
for V the variety F_n of full flags in K^n, (i.e. the projective variety
whose points are increasing sequences of subspaces $\{V_1 \subset \ldots \subset V_{n-1}\}$ in K_n,
where dim V_i = i), and deduce that G leaves one flag stable. This is equivalent to saying that G can be put in triangular form, and proves anew the
Lie-Kolchin theorem. In this sketch, I have slurred over some technical
points, but the proof is in essence valid as is over any algebraically
closed field.

I hope this gives some idea of the global arguments which replaced
Lie algebra considerations. The theory was rather quickly developed in this
framework. A major achievement was the classification of simple algebraic
groups by C. Chevalley, which he proved to be independent of K:
The simple groups over K are classified by the root systems and lattices,
in the same way as over C. It is well-known that the classical groups can
be written over Z, i.e. by equations such as (2) above with integral coefficients. C. Chevalley also showed that every type of complex simple group
could be so described by conditions over Z, which made sense and defined
the corresponding simple group over any K.

4.   The next step was to extend the theory to non-algebraically closed
groundfields. If k is a subfield of K, let us say that the linear algebraic
group $G \subset GL_n(K)$ given by (2) above is defined over k if the ideal of all
polynomials $P \in K[(X_{ij})]$ which vanish on G is generated, as an ideal, by
elements with coefficients in k. If so, set $G(k) = G \cap GL_n(k)$. If k = R,
K = C, then G(k) is a (special kind of) real Lie group. If k is finite, then
G(k) is a finite group. In fact, Chevalley's construction over Z yields,
for every K, a group defined over the prime field of K, hence over any field.
In an earlier paper (1955), Chevalley had performed such an explicit construction starting from complex simple algebras and had shown that the
groups G(k) thus obtained were simple, as abstract groups. For k finite,
this yielded some new series of finite simple groups, attached to the
exceptional simple Lie algebras, the first new finite simple groups since
E. Dickson. Some variations of his construction (by R. Steinberg, R. Ree,

M. Suzuki, J. Tits) led to some others. As a result, a solid connection between algebraic groups and finite simple groups was established. As you
know, it is rather generally conjectured by the experts on finite simple
groups, that apart from those groups and the alternating groups there are
only finitely many simple non-commutative finite groups, the so-called
sporadic groups. An important part of this classification program is played
by the properties and various characterizations of the above simple groups,
often called of Lie type or of Chevalley type.

The general study of the groups G(k) was developed from about 1957 on.
It led notably to a structure theory of semi-simple, (or, slightly more
generally, reductive) groups over arbitrary fields by J. Tits and myself
in the early sixties. The notions of Cartan subgroups, roots, Weyl groups,
Bruhat decomposition of the classical theory have suitable analogues, but
I shall not attempt to describe it more precisely.

5.   Tits and I came to it for different reasons. My main motivation was
to find in G(k), when k is a number field, subgroups which would allow one
to generalize to fundamental domains of arbitrary arithmetic groups the
familiar notion of cusp of a fundamental domain of a fuchsian group; J. Tits
had been led to this theory by his investigations of various geometries and
by the role of the Bruhat decomposition in Chevalley's of 1955 referred to
above. Heobtained a far reaching axiomatization which allowed him to give
a unified treatment of many known geometries as well as to construct new
ones. As you know, according to Klein, a geometry on a space is the study
of properties invariant under a given group of transformations. From that
point of view, the geometry is governed by the properties of, and the
relations between, the isotropy groups of the objects under consideration.
Tits turned this around by defining a geometry starting from an aggregate
P of subgroups of a group G with suitable properties. The key notion is
that of a (B,N)-pair T, now generally called a Tits system. It consists
of two subgroups B,N of G satisfying some rather simple conditions (in
fact surprisingly simple, considering the amount of information which can
be extracted from them). In particular, $H = B \cap N$ is normal in N and
$W = N/H$, called the Weyl group of T, is a Coxeter group. The elements of
P, the "parabolic subgroups" of T, are then those which contain a conjugate
of B. To T there is associated a simplicial complex X, now called a
building. The vertices of X are the maximal elements of P different from
G, and $s + 1$ vertices $P_0, \ldots, P_s$ span an s-simplex if and only if their
intersection is in P. The group G operates on the building X by

conjugation and the isotropy groups of the various simplices are just the elements of P. This is the geometry associated to T. If $G = GL_n(k)$, B is the group of upper triangular matrices, N the group of monomial matrices, then P consists of the stability groups of the flags in k^n (strictly increasing sequences of subspaces of dimension $\neq 0,n$). If we replace any $P \in P$ by the flag which it stabilizes, then we may view X as the "building of flags in k^n": the vertices are the non-zero proper subspaces of k^n, and vertices $\{V_0, \ldots, V_s\}$ span a simplex if and only if, after renumbering, they form a flag. This complex thus describes the incidence relations among the subspaces of k^n, or of the projective space $P_{n-1}(k)$; by the fundamental theorem of projective geometry, it allows one to recover essentially projective geometry, at any rate for $n \geq 3$.

The Tits system just described has analogues in the groups G(k), where G is semi-simple defined over k. The geometries associated to them include many classical ones, as well as new ones attached to the exceptional groups. This now related algebraic groups with "abstract" geometry, or rather with many geometries such as projective or polar geometry over arbitrary fields. In these first applications of Tits systems and buildings, the Weyl group W had been in most cases a finite euclidean reflection group. It could be in principle also infinite. However, this possibility remained little exploited, until it was noticed that another type of Tits systems was the key to a problem which had been rather baffling until then, the study of maximal compact subgroups of p-adic semi-simple groups.

6.   To explain the background to this problem, I have to backtrack a little. I already pointed out that the symmetric spaces offered a natural framework for a general theory of automorphic forms with respect to arithmetic groups, But I was then implicitly referring mainly to its analytic aspects. However, it also has deep arithmetic ones, as exemplified by the work of E. Hecke or C.L. Siegel, for instance. Now in algebraic number theory we are taught to treat as symmetrically as possible all the completions of a number field; a convenient tool for this is the formalism of adeles and ideles, which was extended to algebraic groups in the late fifties. This led to consider, besides real or complex groups, also groups of the form G(k), where k is a non-archimedean local field, say the field Q_p of p-adic numbers, p prime, in particular for semi-simple G.

To study these groups, the first idea was naturally enough to see whether some properties of real groups would carry over, *mutatis mutandis*. After the development of the theory of algebraic groups, it was realized

that Cartan's results on the structure of real semi-simple groups were
really of two kinds: some, suitably reformulated, could be viewed as special
cases of theorems on algebraic groups while the others, such as the conjugacy of maximal compact subgroups and the properties of symmetric spaces,
were topological or differential geometric in nature. In other words, some
depended on the Zariski topology and the others on the ordinary topology
and the C^∞-structures associated to R. The former were now available over
non-archimedean fields, and there remained to see whether the others also
had some counterpart. With respect to the structures inherited from k, these
groups can be viewed as locally compact totally disconnected groups or as
Lie groups over k (the definition being the same as over R or C, based on
the notion of an analytic manifold over a non-discrete complete valued field).
A priori, as a topological group, such a group seems quite different from
a real or complex group. Still some simple examples led one to hope for the
existence of some significant similarities. The first item of business was
the determination of the maximal compact subgroups. They were shown to
exist in general, and were first determined effectively in some classical
cases. They were not necessarily conjugate, but at any rate formed finitely
many conjugacy classes. For instance, if $G(k) = SL_n(Q_p)$ these conjugacy
classes are represented by the stability groups of the n lattices in Q_p^n
spanned over the ring Z_p of p-adic integers by the vectors

$$e_1, \ldots, e_i, \ pe_{i+1}, \ldots, pe_n, \qquad (i = 1, 2, \ldots, n).$$

It was difficult at first to see any general principle behind this. Furthermore, these subgroups are open, so that the quotient G/K of G by one of
them is a countable discrete space, hardly a candidate to play the role of
a Riemannian symmetric space. The breakthrough came here with the discovery,
by N. Iwahori and H. Matsumoto, that in certain cases the maximal compact
subgroups could be described as the maximal elements in the set P of "parabolic subgroups" associated to a suitable Tits system T in G(k). This was
then shown to be a general phenomenon by F. Bruhat and J. Tits. Moreoever, the
building X associated to this Tits system supplied a space which turned out
to be amazingly analogous to the symmetric spaces of the real theory. In this
case, the Weyl group of the Tits system is an infinite reflection group in
some affine euclidean space A. The building X is a union of spaces naturally
isomorphic to A, endowed with the tesselation defined by the reflection
hyperplanes of W. It is contractible, the metrics on the copies of A combine

to define a complete metric on T, such that any two points are joined by a
unique shortest geodesic segment. There is a metric inequality for geodesic
triangles which, in the Riemannian symmetric case, follows from negative
curvature, and can be used again to prove that any compact group of isometries has a fixed point. In fact, the proof is valid without change both
in the p-adic and the real case. The fixed point theorem is used in an
essential way to insure that the maximal elements of P are exactly the maximal compact subgroups. As in the real case, a geometric fixed point theorem
is the key to the description of the maximal compact subgroups.

This theory is one striking evidence of a close analogy one keeps dis-
covering between real and p-adic semi-simple (or reductive) groups. At
first, the real case supplied the clues to the p-adic case, but this has
since become a two-way street, almost a shuttle, and many results on real
groups have been suggested by properties of p-adic groups. This analogy is
not formal, but is more in the form of some sort of dictionary, built little
by little and still growing. Some remarkable illustrations of this principle
are provided by harmonic analysis (for instance the Plancherel theorem or
the classification of irreducible admissible representations can be given
an essentially common formulation) or by the cohomology of discrete cocompact subgroups.

7.   The theory of Lie groups and algebraic groups has now attained a considerable degree of completeness and has found many applications. Its use-
fulness is nowhere more in evidence than in the present study of automorphic
forms and of their connections with arithmetic and algebraic geometry. In
the last twelve years or so, Lie groups, algebraic groups, arithmetic groups,
have become the framework of a vast program, often referred to as "Langlands' philosophy", in which infinite dimensional representations are
brought to bear on the study of Artin L-functions, Hasse-Weil zeta functions
of projective varieties, and non-abelian extensions of local or global
fields. The full realization of this program does not appear to be in sight,
but the old and recent results which illustrate it are so striking and pro-
mising that I could not resist alluding to it at this point.

8.   For lack of time, I have now to stop this survey, as incomplete as it
may be. Still, the entity "Lie groups-algebraic groups" has been related to
many parts of mathematics: analysis, differential equations, algebra, topology, differential geometry, fibre bundle theory, arithmetic groups, automorphic forms, algebraic geometry, moduli, finite simple groups, geometry,
L-functions, to name the main ones we have met. This may give the impression

that I am trying surrepticiously to prove that, after all, Lie groups are
the center of mathematics. But this is not my intention at all. Clearly,
some other topics could give rise to a rather similar picture. In fact, I
would rather schematize the structure of mathematics by a complicated graph,
where the vertices are the various parts of mathematics and the edges describe the connections between them. These connections go sometimes one
way, sometimes both ways, and the vertices can act both as sources and
sinks. The development of the individual topics is of course the life and
blood of mathematics, but, in the same way as a graph is more than the
union of its vertices, mathematics is much more than the sum of its parts.
It is the presence of those numerous, sometimes unexpected edges, which
makes mathematics a coherent body of knowledge, and testifies to its fundamental unity, in spite of its being too vast to be comprehended by one
single mind. I hope this lecture has made plausible my belief that Lie
group theory is a topic of great vitality and interest both in its own
right and by the remarkably important role it has played and is playing in
the ongoing process of expansion and unification of mathematics so well
described by D. Hilbert in his 1900 Paris address, as a meeting and testing
ground for many disciplines and as a starting point, tool, and framework
for many incursions into other parts of mathematics.

BIBLIOGRAPHY

A comprehensive bibliography would be out of proportion with the general character of this lecture. I limit myself to a few surveys in which
the reader will find many further references.

N. BOURBAKI, *Groupes et algèbres de Lie,* Chap. 2,3, note historique;
        Act. Sci. Ind. 1349 (1973), Hermann, Paris.

E. CARTAN, *La théorie des groupes finis et continus et l'analysis situs;*
        Memorial Sci. Math. XLII, 1930, Gauthier-Villars, Paris
        (Oeuvres complètes, t. 1_2, 1165-1225).

D.V. ALEKSEEVSKI, *Lie groups and homogeneous spaces,* Journal of Soviet
        Mathematics $\underline{4}$ (1975), 483-539, Plenum Publ. Corp., New York
        (translated from Itogi Nauk i Tekhniki $\underline{11}$ (1974), 37-123).

N. BOURBAKI, *Groupes et algèbres de Lie,* Chap. 4,5,6, note historique;
        Act. Sci. Ind 1337 (1968), Hermann, Paris.

J. TITS, *Groupes simples et géométries associées;* Proc. Int. Congress
        Mathematicians Stockholm 1962, Vol. 1, 197-221.

A. BOREL, *Arithmetic properties of linear algebraic groups;* ibid., 10-22.

V.P. PLATONOV, *Algebraic groups,* Journal of Soviet Mathematics $\underline{4}$ (1975),
        463-482, Plenum Publ. Corp., New York (translated from Itogi
        Nauk i Tekhniki $\underline{11}$ (1974), 5-36).

J. TITS, *On buildings and their applications;* Proc. Int. Congress Mathemati-
        cians Vancouver, 1974, Vol. 1, 209-220.

A. BOREL, *Formes automorphes et séries de Dirichlet, (d'après R.P. Langlands)*
        Sem. Bourbaki 1974/75, Exp. 466, 40 p., Lecture Notes in
        Mathematics $\underline{514}$ (1976), 183-222, Springer.

The Institute for Advanced Study
Princeton, N.J. 08540, U.S.A.

112.

(with H. Jacquet)

Automorphic forms and automorphic representations

Proc. Symp. Pure Math. **33**, Part 1, Amer. Math. Soc. (1979) 189–202

Originally, the theory of automorphic forms was concerned only with holomorphic automorphic forms on the upper half-plane or certain bounded symmetric domains. In the fifties, it was noticed (first by Gelfand and Fomin) that these automorphic forms could be viewed as smooth vectors in certain representations of the ambient group G, on spaces of functions on G invariant under the given discrete group Γ. This led to the more general notion of automorphic forms on real semisimple groups, with respect to arithmetic subgroups, on adelic groups, and finally to the direct consideration of the underlying representations. The main purpose of this paper is to discuss the notions of automorphic forms on real or adelic reductive groups, of automorphic representations of adelic groups, and the relations between the two. We leave out completely the passage from automorphic forms on bounded symmetric domains to automorphic forms on groups, which has been discussed in several places (see, e.g., [2], or also [5], [6], [15] for modular forms).

1. Automorphic forms on a real reductive group.

1.1. Let G be a connected reductive group over Q, Z the greatest Q-split torus of the center of G and K a maximal compact subgroup of $G(R)$. Let $\mathfrak{g}$ be the Lie algebra of $G(R)$, $U(\mathfrak{g})$ its universal enveloping algebra over C and $Z(\mathfrak{g})$ the center of $U(\mathfrak{g})$. We let $\mathcal{H}$ or $\mathcal{H}(G(R), K)$ be the convolution algebra of distributions on $G(R)$ with support in K [4] and A_K the algebra of finite measures on K. We recall that $\mathcal{H}$ is isomorphic to $U(\mathfrak{g}) \otimes_{U(\mathfrak{t})} A_K$. An idempotent in $\mathcal{H}$ is, by definition, one of A_K, i.e., a finite sum of measures of the form $d(\sigma)^{-1} \cdot \chi_\sigma \cdot dk$, where σ is an irreducible finite dimensional representation of K, $d(\sigma)$ its degree, χ_σ its character, and dk the normalized Haar measure on K. The algebra $\mathcal{H}$ is called the Hecke algebra of $G(R)$ and K.

1.2. A norm $\| \ \|$ on $G(R)$ is a function of the form $\|g\| = (\mathrm{tr}\, \sigma(g)^* \cdot \sigma(g))^{1/2}$, where $\sigma: G(R) \to \mathrm{GL}(E)$ is a finite dimensional complex representation with finite kernel and image closed in the space $\mathrm{End}(E)$ of endomorphisms of E and $*$ denotes the adjoint with respect to a Hilbert space structure on E invariant under K. It is easily seen that if τ is another such representation, then there exist a constant $C > 0$ and a positive integer n such that

---

AMS (MOS) subject classifications (1970). Primary 10D20; Secondary 20G35, 22E55.

548

(1) $\|x\|_\sigma \leq C \cdot \|x\|_\tau^n, \quad$ for all $x \in G(R)$.

A function f on $G(R)$ is said to be *slowly increasing* if there exist a norm $\| \ \|$ on $G(R)$, a constant C and a positive integer n such that

(2) $|f(x)| \leq C \cdot \|x\|^n, \quad$ for all $x \in G(R)$.

In view of (1) this condition does not depend on the norm (but n does).

REMARK. If $\sigma: G \to \mathrm{GL}(E)$ has finite kernel, but does not have a closed image in End E, then we can either add one coordinate and det $\sigma(g)^{-1}$ as a new entry, or consider the sum of σ and $\sigma^*: g \mapsto \sigma(g^{-1})^*$. The associated square norms are then $|\det \sigma(g)^{-1}|^2 + \mathrm{tr}(\sigma(g)^* \sigma(g))$ and $\mathrm{tr}(\sigma(g)^* \sigma(g)) + \mathrm{tr}(\sigma(g^{-1})\sigma(g^{-1})^*)$.

1.3. Let Γ be an arithmetic subgroup of $G(Q)$. A smooth complex valued function f on $G(R)$ is an *automorphic form* for (Γ, K), or simply for Γ, if it satisfies the following conditions:

(a) $f(\gamma \cdot x) = f(x)$ $(x \in G(R); \gamma \in \Gamma)$.

(b) There exists an idempotent $\xi \in \mathscr{H}$(cf. 1.1) such that $f * \xi = f$.

(c) There exists an ideal J of finite codimension of $Z(\mathfrak{g})$ which annihilates f: $f * X = 0$ $(X \in J)$.

(d) f is slowly increasing (1.2).

An automorphic form satisfying those conditions is said to be of type (ξ, J). We let $\mathscr{A}(\Gamma, \xi, J, K)$ be the space of all automorphic forms for (Γ, K) of type (ξ, J).

1.4. EXAMPLE. Let $G = \mathrm{GL}_2$, $\Gamma = \mathrm{GL}_2(Z)$, $K = O(2, R)$, $\xi = 1$, J the ideal of $Z(\mathfrak{g})$ generated by $(C - \lambda)$, where C is the Casimir operator and $\lambda \in C$. Then f is an eigenfunction of the Casimir operator which is left invariant under Γ, right invariant under K, and invariant under the center Z of $G(R)$. The quotient $G(R)/Z \cdot K$ may be identified with the Poincaré upper half-plane H. The function f may then be viewed as a Γ-invariant eigenfunction of the Laplace-Beltrami operator on X. It is a "wave-form," in the sense of Maass. See [2], [5], [6] for a similar interpretation of modular forms.

1.5. REMARKS. (1) Condition 1.3(b) is equivalent to: f is K-finite on the right, i.e., the right translates of f by elements of K span a finite dimensional space of functions. These functions clearly satisfy 1.3(a), (c), (d) if f does.

(2) Let $r : K \to \mathrm{GL}(V)$ be a finite dimensional unitary representation of K. One can similarly define the notion of a V-valued automorphic form: a smooth function $\varphi: G \to V$ satisfying (a), (c), (d), where $| \ |$ now refers to the norm in V, and

(b') $f(x \cdot k) = r(k^{-1}) \cdot f(x) \qquad (x \in G(R); k \in K)$.

For semisimple groups, this is Harish-Chandra's definition (cf. [2], [11]). For $v \in V$, the functions $x \mapsto (\varphi(x), v)$ are then scalar automorphic forms. Conversely, a finite dimensional space E of scalar automorphic forms stable under K yields an E-valued automorphic form.

(3) A similar definition can be given if $G(R)$ is replaced by a finite covering H of an open subgroup of $G(R)$ and Γ by a discrete subgroup of H whose image in $G(R)$ is arithmetic. For instance, modular forms of rational weight can be viewed as automorphic forms on finite coverings of $\mathrm{SL}_2(R)$.

1.6. *Growth condition.* (1) Let A be the identity component of a maximal Q-split torus S of G, and $_Q\Phi$ the system of Q-roots of G with respect to S. Fix an ordering on $_Q\Phi$ and let Δ be the set of simple roots. Given $t > 0$ let

$$(1) \qquad A_t = \{a \in A : |\alpha(a)| \geq t, (a \in \Delta)\}.$$

Let f be a function satisfying 1.3(a), (b), (c). Then the growth condition (d) is equivalent to:

(d') Given a compact set $R \subset G(\mathbf{R})$, and $t > 0$, there exist a constant $C > 0$ and a positive integer m such that

$$(2) \qquad |f(x \cdot a)| \leq C \cdot |\alpha(a)|^m, \quad \text{for all } a \in A_t, \alpha \in \Delta, x \in R.$$

This follows from reduction theory [11, §2]. More precisely, let G' be the derived group of G. Then A is the direct product of $Z(\mathbf{R})^\circ$ and $A' = A \cap G'(\mathbf{R})$. For a function satisfying 1.3(a), (b), the growth condition (d') is equivalent to (d) for $a \in A_t'$; but says nothing for $a \in Z(\mathbf{R})^\circ$. However condition (c) implies that f depends polynomially on $z \in Z(\mathbf{R})$, and this takes care of the growth condition on $Z(\mathbf{R})$.

(2) Assume f satisfies 1.3(a), (b), (c) and

$$(3) \qquad f(z \cdot x) = \chi(z) f(x) \qquad (z \in Z(\mathbf{R}), \ x \in G(\mathbf{R}))$$

where χ is a character of $Z(\mathbf{R})/(Z(\mathbf{R}) \cap \Gamma)$. Then $|f|$ is a function on $Z(\mathbf{R}) \cdot \Gamma \backslash G(\mathbf{R})$. If $|f| \in L^p(Z(\mathbf{R})\Gamma \backslash G(\mathbf{R}))$ for some $p \geq 1$, then f is slowly increasing, hence is an automorphic form. In view of the fact that $Z(\mathbf{R})\Gamma \backslash G(\mathbf{R})$ has finite invariant volume, it suffices to prove this for $p = 1$. In that case, it follows from the corollary to Lemma 9 in [11], and from the existence of a K-invariant function $\alpha \in C_c^\infty(G(\mathbf{R}))$ such that $f = f * \alpha$ (a well-known property of K-finite and $Z(\mathfrak{g})$-finite elements in a differentiable representation of $G(\mathbf{R})$, which follows from 2.1 below).

1.7. THEOREM [11, THEOREM 1]. *The space* $\mathcal{A}(\Gamma, \xi, J, K)$ *is finite dimensional.*

This theorem is due to Harish-Chandra. Actually the proof given in [11] is for semisimple groups, but the extension to reductive groups si easy. In fact, it is implicitly done in the induction argument of [11] to prove the theorem. For another proof, see [13, Lemma 3.5]. At any rate, it is customary to fix a quasi-character χ of $Z(\mathbf{R})/(\Gamma \cap Z(\mathbf{R}))$ and consider the space $\mathcal{A}(\Gamma, \xi, J, K)_\chi$ of elements in $\mathcal{A}(\Gamma, \xi, J, K)$ which satisfy 1.6(3). For those, the reduction to the semisimple case is immediate. Note that since the identity component $Z(\mathbf{R})^\circ$ of $Z(\mathbf{R})$ (sometimes called the split component of $G(\mathbf{R})$) has finite index in $Z(\mathbf{R})$ and $Z(\mathbf{R})^\circ \cap \Gamma = \{1\}$, it is substantially equivalent to require 1.6(3) for an arbitrary quasi-character of $Z(\mathbf{R})^\circ$.

The space $\mathcal{A}(\Gamma, \xi, J, K)$ is acted upon by the center $C(G(\mathbf{R}))$ of $G(\mathbf{R})$, by left or right translations. Since it is finite dimensional, we see that *any automorphic form is* $C(G(\mathbf{R}))$-*finite.*

1.8. *Cusp forms.* A continuous (resp. measurable) function on $G(\mathbf{R})$ is cuspidal if

$$(1) \qquad \int_{(\Gamma \cap N(\mathbf{R})) \backslash N(\mathbf{R})} f(n \cdot x)\, dn = 0,$$

for all (resp. almost all) x in $G(\mathbf{R})$, where N is the unipotent radical of any proper parabolic Q-subgroup of G. It suffices in fact to require this for any proper maximal parabolic Q-subgroup [11, Lemma 3].

A *cusp form* is a cuspidal automorphic form. We let $°\mathscr{A}(\Gamma, \xi, J, K)$ be the space of cusp forms in $\mathscr{A}(\Gamma, \xi, J, K)$.

Let f be a smooth function on $G(R)$ satisfying the conditions (a), (b), (c) of 1.3. Assume that f is cuspidal and that there exists a character χ of $Z(R)$ such that 1.6(3) is satisfied. Then the following conditions are equivalent:

(i) f is slowly increasing, i.e., f is a cusp form;

(ii) f is bounded;

(iii) $|f|$ is square-integrable modulo $Z(R) \cdot \Gamma$

(cf. [11, §4]). In fact, one has much more: $|f|$ decreases very fast to zero at infinity on $Z(R)\Gamma\backslash G(R)$, so that if g is any automorphic form satisfying 1.6(3), then $|f \cdot g|$ is integrable on $Z(R) \cdot \Gamma\backslash G(R)$ (loc. cit.).

The space $°\mathscr{A}(\Gamma, \xi, J, K)_\chi$ of the functions in $°\mathscr{A}(\Gamma, \xi, J, K)$ satisfying 1.6(3) may then be viewed as a closed subspace of bounded functions in the space $L^2(\Gamma\backslash G(R))_\chi$ of functions on $\Gamma\backslash G(R)$ satisfying 1.6(3), whose absolute value is square-integrable on $Z(R)\Gamma\backslash G(R)$. Since $Z(R)\Gamma\backslash G(R)$ has finite measure, this space is finite dimensional by a well-known lemma of Godement [11, Lemma 17]. This proves 1.7 for $\mathscr{A}(\Gamma, \xi, J, K)_\chi$ when $Z(R)\Gamma\backslash G(R)$ is compact, and is the first step of the proof of 1.7 in general.

1.9. Let $a \in G(Q)$. Then $^a\Gamma = a \cdot \Gamma \cdot a^{-1}$ is an arithmetic subgroup of $G(Q)$, and the left translation l_a by a induces an isomorphism of $\mathscr{A}(\Gamma, \xi, J, K)$ onto $\mathscr{A}(^a\Gamma, \xi, J, K)$. Let Σ be a family of arithmetic subgroups of $G(Q)$, closed under finite intersection, whose intersection is reduced to $\{1\}$. The union $\mathscr{A}(\Sigma, \xi, J, K)$ of the spaces $\mathscr{A}(\Gamma, \xi, J, K)$ $(\Gamma \in \Sigma)$ may be identified to the inductive limit of those spaces:

(1) $$\mathscr{A}(\Sigma, \xi, J, K) = \text{ind lim}_{\Gamma \in \Sigma} \mathscr{A}(\Gamma, \xi, J, K),$$

where the inductive limit is taken with respect to the inclusions

(2) $$j_{\Gamma''\Gamma'}: \mathscr{A}(\Gamma', \xi, J, K) \to \mathscr{A}(\Gamma'', \xi, J, K) \qquad (\Gamma'' \subset \Gamma')$$

associated to the projections $\Gamma''\backslash G(R) \to \Gamma'\backslash G(R)$.

Assume Σ to be stable under conjugation by $G(Q)$. Then $G(Q)$ operates on $\mathscr{A}(\Sigma, \xi, J, K)$ by left translations. Let us topologize $G(Q)$ by taking the elements of Σ as a basis of open neighborhoods of 1. Then this representation is admissible (every element is fixed under an open subgroup, and the fixed point set of every open subgroup is finite dimensional). By continuity, it extends to a continuous admissible representation of the completion $G(Q)_\Sigma$ of $G(Q)$ for the topology just defined. For suitable Σ, the passage to $\mathscr{A}(\Sigma, \xi, J, K)$ amounts essentially to considering all adelic automorphic forms whose type at infinity is prescribed by ξ, J, K; the group $G(Q)_\Sigma$ may be identified to the closure of $G(Q)$ in $G(A_f)$ and its action comes from one of $G(A_f)$. See 4.7.

1.10. Finally, we may let ξ and J vary and consider the space $\mathscr{A}(\Sigma, J, K)$ spanned by the $\mathscr{A}(\Sigma, \xi, J, K)$ and the space $\mathscr{A}(\Sigma, K)$ spanned by the $\mathscr{A}(\Sigma, J, K)$. They are $G(Q)_\Sigma$-modules and $(\mathfrak{g}, K)$-modules, and these actions commute. Again, this has a natural adelic interpretation (4.8).

1.11. *Hecke operators.* Let $\mathscr{H}(G(Q), \Gamma)$ be the Hecke algebra, over C, of $G(Q)$

mod Γ. It is the space of complex valued functions on $G(Q)$ which are bi-invariant under Γ and have support in a finite union of double cosets mod Γ. The product may be defined directly in terms of double cosets (see, e.g., [17]) or of convolution (see below). This algebra operates on $\mathscr{A}(\Gamma, \xi, J, K)$. The effect of $\Gamma a \Gamma$ $(a \in G(Q))$ is given by $f \mapsto \sum_{b \in (\Gamma \cap a\Gamma) \backslash \Gamma} l_b f$. More generally, let $\mathscr{H}(G(Q), \Sigma)$ be the Hecke algebra spanned by the characteristic functions of the double cosets $\Gamma' a \Gamma''$ $(\Gamma', \Gamma'' \in \Sigma,$ $a \in G(Q))$ [17, Chapter 3]. It may be identified with the Hecke algebra $\mathscr{H}(G(Q)_\Sigma)$ of locally constant compactly supported functions on $G(Q)_\Sigma$. This identification carries $\mathscr{H}(G(Q), \Gamma)$ onto $\mathscr{H}(G(Q)_\Sigma, \bar{\Gamma})$, where $\bar{\Gamma}$ is the closure of Γ in $G(Q)_\Sigma$ [12]. The product here is ordinary convolution (which amounts to finite sums in this case). Since $\mathscr{A}(\Sigma, \xi, J, K)$ is an admissible module for $G(Q)_\Sigma$, the action of $G(Q)_\Sigma$ extends in the standard way to one of $\mathscr{H}(G(Q)_\Sigma)$. The space $\mathscr{A}(\Gamma, \xi, J, K)$ is the fixed point set of $\bar{\Gamma}$, and the previous operation of $\mathscr{H}(G(Q), \Gamma)$ on this space may be viewed as that of $\mathscr{H}(G(Q)_\Sigma, \bar{\Gamma})$. For an adelic interpretation, see 4.8.

2. Automorphic forms and representations of $G(R)$. The notion of automorphic form has a simple interpretation in terms of representations (which in fact suggested its present form). To give it, we need the following known lemma (cf. [18] for the terminology).

2.1. LEMMA. *Let (π, V) be a differentiable representation of $G(R)$. Let $v \in V$ be K-finite and $Z(\mathfrak{g})$-finite. Then the smallest $(\mathfrak{g}, K)$-submodule of V containing v is admissible.*

Indeed, $\mathscr{H} \cdot v$ is a finite sum of spaces $\mathscr{H}^\circ \cdot w$, where $\mathscr{H}^\circ$ is the Hecke algebra of the identity component $G(R)^\circ$ of $G(R)$ and $K^\circ = K \cap G(R)^\circ$, and w is K°-finite and $Z(\mathfrak{g})$-finite. It suffices therefore to show that $\mathscr{H}^\circ \cdot v$ is an admissible $(\mathfrak{g}, K^\circ)$-module. By assumption, there exist an ideal R of finite codimension of the enveloping algebra $U(\mathfrak{k})$ of the Lie algebra $\mathfrak{k}$ of K and an ideal J of finite codimension of $Z(\mathfrak{g})$ which annihilate v and moreover $U(\mathfrak{k})/R$ is a semisimple $\mathfrak{k}$-module. Then $\mathscr{H}^\circ \cdot v$ may be identified with $U(\mathfrak{g})/U(\mathfrak{g}) \cdot R \cdot J$. By a theorem of Harish-Chandra (see [19, 2.2.1.1]), $U(\mathfrak{g})/U(\mathfrak{g}) \cdot R$ is $\mathfrak{k}$-semisimple and its $\mathfrak{k}$-isotypic submodules are finitely generated $Z(\mathfrak{g})$-modules. Hence their quotients by J are finite dimensional.

2.2. We apply this to $C^\infty(\Gamma \backslash G(R))$, acted upon by $G(R)$ via right translations. Therefore, if f is automorphic form, then $f * \mathscr{H}$ is an admissible $\mathscr{H}$- or $(\mathfrak{g}, K)$-module. This module consists of automorphic forms. In fact, 1.3(a) is clear, and 1.3(b) follows from 2.1; its elements are annihilated by the same ideal of $Z(\mathfrak{g})$ as v, whence (d). Finally, there exists $\alpha \in C_c^\infty(G)$ such that $f * \alpha = f$ so that $f * X$ satisfies 1.2(c) (with the same exponent as f) for all $X \in U(\mathfrak{g})$ [11, Lemma 14]. Thus the spaces

$$\mathscr{A}(\Gamma, J, K) = \Sigma_\xi \mathscr{A}(\Gamma, \xi, J, K), \qquad \mathscr{A}(\Gamma, K) = \Sigma_J \mathscr{A}(\Gamma, J, K),$$

are $(\mathfrak{g}, K)$-modules and unions of admissible $(\mathfrak{g}, K)$-modules.

If f is a cusp form, then $f * \mathscr{H}$ consists of cusp forms. Thus the subspace $^\circ\mathscr{A}(\Gamma, K)$ of cusp forms is also an $\mathscr{H}$-module. If χ is a quasi-character of Z, then the space $^\circ\mathscr{A}(\Gamma, K)_\chi$ of eigenfunctions for Z with character χ is a direct sum of irreducible admissible $(\mathfrak{g}, K)$-modules, with finite multiplicities. In fact, after a twist by $|\chi|^{-1}$, we may assume χ to be unitary, and we are reduced to the Gelfand-Piatetski-Shapiro theorem ([7], see also [11, Theorem 2], [13, pp. 41–42]) once we identify $^{\circ}\mathscr{A}(\Gamma, K)_{\chi}$ to the space of K-finite and $Z(\mathfrak{g})$-finite elements in the space $^{\circ}L^2(\Gamma\backslash G(\mathbf{R}))_{\chi}$ of cuspidal functions in $L^2(\Gamma\backslash G(\mathbf{R}))_{\chi}$ (see 1.8 for the latter).

3. Some notation. We fix some notation and conventions for the rest of this paper.

3.1. F is a global field, O_F the ring of integers of F, V or V_F (resp. V_∞, resp. V_f) the set of places (resp. archimedean places, resp. nonarchimedean places) of F, F_v the completion of F at $v \in V$, O_v the ring of integers of F_v if $v \in V_f$. As usual, A or A_F (resp. A_f) is the ring of adèles (resp. finite adèles) of F.

3.2. G is a connected reductive group over F, Z the greatest F-split torus of the center of G, $\mathscr{H}_v$ the Hecke algebra of $G_v = G(F_v)$ ($v \in V$) [4]. Thus $\mathscr{H}_v$ is of the type considered in §1 if $v \in V_\infty$ and is the convolution algebra of locally constant compactly supported functions on $G(F_v)$ if $v \in V_f$. We set

$$(1) \qquad \mathscr{H}_\infty = \bigotimes_{v\in V_\infty} \mathscr{H}_v, \qquad \mathscr{H}_f = \bigotimes_{v\in V_f} \mathscr{H}_v, \qquad \mathscr{H} = \mathscr{H}_\infty \otimes \mathscr{H}_f,$$

where the second tensor product is the restricted tensor product with respect to a suitable family of idempotents [4]. Thus $\mathscr{H}$ is the global Hecke algebra of $G(A)$ [4]. If F is a function field, then V_∞ is empty and $\mathscr{H} = \mathscr{H}_f$.

If L is a compact open subgroup of $G(A_f)$, we denote by ξ_L the associated idempotent, i.e., the characteristic function of L divided by the volume of L (relative to the Haar measure underlying the definition of $\mathscr{H}_f$). Thus $f * \xi_L = f$ if and only if f is right invariant under L.

The right translation by $x \in G(A)$ on $G(A)$, or on functions on $G(A)$, is denoted r_x or $r(x)$.

3.3. A continuous (resp. measurable) function on $G(A)$ is cuspidal if

$$\int_{N(F)\backslash N(A)} f(nx)\, dn = 0,$$

for all (resp. almost all) $x \in G(A)$, where N is the unipotent radical of any proper parabolic F-subgroup P of G. It suffices to chcek this condition when P runs through a set of representatives of the conjugacy classes of proper maximal parabolic F-subgroups.

4. Groups over number fields.

4.1. In this section, F is a number field. An element $\xi \in \mathscr{H}$ is said to be *simple* if it is of the form

$$(1) \qquad\qquad \xi = \xi_\infty \otimes \xi_f, \qquad \xi_f \in \mathscr{H}_f, \xi_\infty \text{ idempotent in } \mathscr{H}_\infty.$$

We let $G_\infty = \prod_{v\in V_\infty} G_v$ and $\mathfrak{g}_\infty$ be the Lie algebra of G_∞, viewed as a real Lie group. We recall that G_∞ may be viewed canonically as the group of real points $H(\mathbf{R})$ of a connected reductive group H, namely the group $H = R_{F/\mathbf{Q}}\, G$ obtained from G by restriction of scalars from F to $\mathbf{Q}$. This identification is understood when we apply the results and definitions of §§1, 2 to G_∞.

The group $G(A)$ is the direct product of G_∞ by $G(A_f)$. A complex valued function on $G(A)$ is *smooth* if it is continuous and, if viewed as a function of two arguments $x \in G_\infty$, $y \in G(A_f)$, it is C^∞ in x (resp. locally constant in y) for fixed y (resp. x).

4.2. *Automorphic forms.* Fix a maximal compact subgroup K_∞ of G_∞. A smooth function f on $G(A)$ is a K_∞-automorphic form on $G(A)$ if it satisfies the following conditions:

(a) $f(\gamma \cdot x) = f(x)$ $(\gamma \in G(F),\ x \in G(A))$.

(b) There is a simple element $\xi \in \mathcal{H}$ such that $f * \xi = f$.

(c) There is an ideal J of finite codimension of $Z(\mathfrak{g}_\infty)$ which annihilates f.

(d) For each $y \in G(A_f)$, the function $x \mapsto f(x \cdot y)$ on G_∞ is slowly increasing.

We shall sometimes say that f is then of type (ξ, J, K_∞). We let $\mathcal{A}(\xi, J, K_\infty)$ be the space of automorphic forms of type (ξ, J, K_∞).

There exists a compact open subgroup L of $G(A_f)$ such that $\xi_f * \xi_L = \xi_f$. Then $\mathcal{A}(\xi, J, K_\infty) \subset \mathcal{A}(\xi_\infty * \xi_L, J, K_\infty)$. We could therefore assume $\xi_f = \xi_L$ for some L without any real loss of generality.

4.3. We want now to relate these automorphic forms to automorphic forms on G_∞. For this we may (and do) assume $\xi = \xi_\infty \otimes \xi_L$ for some compact open subgroup L of $G(A_f)$. There exists a finite set $C \subset G(A)$ such that $G(A) = G(F) \cdot C \cdot G_\infty \cdot L$ [1]. We assume that C is a set of representatives of such cosets and is contained in $G(A_f)$. Then the sets $G(F) \cdot c \cdot G_\infty \cdot L$ form a partition of $G(A)$ into open sets. For $c \in C$, let

$$(1) \qquad \Gamma_c = G(F) \cap (G_\infty \times c \cdot L \cdot c^{-1}).$$

It is an arithmetic subgroup of G_∞.

Given a function f on $G(A)$ and $c \in C$, let f_c be the function $x \mapsto f(c \cdot x)$ on G_∞. Suppose f is right invariant under L. Then it is immediately checked that f is left invariant under $G(F)$ if and only if f_c is left invariant under Γ_c for every $c \in C$. More precisely, the map $f \mapsto (f_c)_{c \in C}$ yields a bijection between the spaces of functions on $G(F)\backslash G(A)/L$ and on $\coprod_c (\Gamma_c \backslash G_\infty)$. It then follows from the definitions that it also induces an isomorphism

$$(2) \qquad \mathcal{A}(\xi_\infty \otimes \xi_L, J, K_\infty) \xrightarrow{\ \sim\ } \bigoplus_{c \in C} \mathcal{A}(\Gamma_c, \xi_\infty, J, K_\infty),$$

so that the results mentioned in §1 transcribe immediately to properties of adelic automorphic forms. In particular, 1.6, 1.7 imply:

(i) The space $\mathcal{A}(\xi, J, K_\infty)$ is finite dimensional.

(ii) A smooth function f on $G(A)$ satisfying 4.2(a), (b), (c), and

$$(3) \qquad f(z \cdot x) = \chi(z) \cdot f(x) \qquad (z \in Z(A),\ x \in G(A)),$$

for some character χ of $Z(A)/Z(F)$, such that $|f| \in L^p(Z(A)G(F)\backslash G(A))$ for some $p \geq 1$, is slowly increasing.

(iii) For a smooth function satisfying 4.2(a), (b), (c), the growth condition 4.2(d) is equivalent to 1.6(d′), where R is now a compact subset of $G(A)$.

(iv) Any automorphic form is $C(G(A))$-finite, where $C(G(A))$ is the center of $G(A)$.

We note also that one can also define directly slowly increasing functions on $G(A)$, as in 1.2, using adelic norms: given an F-morphism $G \to \mathbf{GL}_n$ with finite kernel define, for $x \in G(A)$,

$$(4) \qquad \|x\| = \sup_{v \in V}\ \max_{ij}\left(|\sigma(g)_{ij}|_v,\ |\sigma(g^{-1})_{ij}|_v\right)$$

(or simply $\max |\sigma(g_{ij})|_v$ if $\sigma(G)$ is closed as a subset of the space of $n \times n$ matrices). For continuous functions satisfying 4.2(a), (b), this is equivalent to 4.2(d).

4.4. A *cusp form* is a cuspidal automorphic form. We let $°\mathscr{A}(\xi, J, K_\infty)$ be the space of cusp forms of type (ξ, J, K_∞).

The group N is unipotent, hence satisfies strong approximation, i.e., we have for *any* compact open subgroup Q of $N(A_f)$

$$(2) \qquad N(A) = N(F) \cdot N_\infty \cdot Q; \quad \text{hence } N(F)\backslash N(A) \cong (N(F) \cap (N_\infty \times Q))\backslash N_\infty$$

[1]. Let now f be a continuous function on $G(A)$ which is left invariant under $G(F)$ and right invariant under L. From (2) it follows by elementary computations that f is cuspidal if and only if the f_c's (notation of 4.3) are cuspidal on G_∞. Hence, the isomorphism of 4.3(2) induces an isomorphism

$$(3) \qquad °\mathscr{A}(\xi_\infty \otimes \xi_L, J, K_\infty) \overset{\sim}{\longrightarrow} \underset{c \in C}{\oplus} °\mathscr{A}(\Gamma_c, \xi_\infty, J, K_\infty),$$

so that the results of 1.8 extend to adelic cusp forms. In particular, assume that f satisfies 4.2(a), (b), (c), and also 4.3(3) for a character χ of $Z(A)/Z(F)$. Then the following conditions are equivalent:

(i) f is slowly increasing, i.e., f is a cusp form;

(ii) f is bounded;

(iii) $|f|$ is square-integrable modulo $Z(A) \cdot G(F)$.

REMARK. In 4.3, 4.4, we have reduced statements on adelic automorphic forms to the corresponding ones for automorphic forms on $G(R)$, chiefly for the convenience of references. However, it is also possible to prove them directly in the adelic framework, and then deduce the results at infinity as corollaries via 4.3(2), 4.4(3). In particular, as in 1.8, one proves using (ii) above and Godement's lemma that $°\mathscr{A}(\xi, J, K_\infty)$ is finite dimensional.

4.5. PROPOSITION. *Let f be a smooth function on $G(A)$ satisfying 4.2(a), (b), (d). Then the following conditions are equivalent:*

(1) *f is an automorphic form.*

(2) *For each infinite place v, the space $f * \mathscr{H}_v$ is an admissible $\mathscr{H}_v$-module.*

(3) *For each place $v \in V$, the space $f * \mathscr{H}_v$ is an admissible $\mathscr{H}_v$-module.*

(4) *The space $f * \mathscr{H}$ is an admissible $\mathscr{H}$-module.*

PROOF. The implications $(4) \Rightarrow (3) \Rightarrow (2) \Rightarrow (1)$ are obvious; $(1) \Rightarrow (4)$ follows from 4.3(i).

4.6. *Automorphic representations.* An irreducible representation of $\mathscr{H}$ is *automorphic* (resp. *cuspidal*) if it is isomorphic to a subquotient of a representation of $\mathscr{H}$ in the space of automorphic (resp. cusp) forms on $G(A)$. It follows from 4.5 that such a representation is always admissible. It will also be called an automorphic representation of G or $G(A)$, although, strictly speaking, it is not a $G(A)$-module. However, it is always a $G(A_f)$-module. More generally, a topological $G(A)$-module E will be said to be *automorphic* if the subspace of admissible vectors in E is an automorphic representation of $\mathscr{H}$. In particular, if χ is a character of $Z(A)/Z(F)$, any G-invariant irreducible closed subspace of

$$L^2(G(F)\backslash G(A))_\chi$$
$$= \{f \in L^2(G(F) \cdot Z(A)\backslash G(A))|, f(z \cdot x) = \chi(z) \cdot f(x), z \in Z(A), x \in G(A))\}$$

is automorphic in this sense, in view of [**4**, Theorem 4]. By a theorem of Gelfand and

Piatetski-Shapiro [7] (see also [8]), the subspace $^\circ L^2(G(F)\backslash G(A))_\chi$ of cuspidal functions of $L^2(G(F)\backslash G(A))_\chi$ is a discrete sum with finite multiplicities of closed irreducible invariant subspaces. Those give then, up to isomorphisms, all cuspidal automorphic representations in which $Z(A)$ has character χ. The admissible vectors in those subspaces are all the cusp forms satisfying 4.3(3).

4.7. Let $L \supset L'$ be compact open subgroups of $G(A_f)$. Then $\xi_L * \xi_{L'} = \xi_L$; hence $\mathscr{A}(\xi_\infty \otimes \xi_L; J, K_\infty) \subset \mathscr{A}(\xi_\infty \otimes \xi_{L'}, J, K_\infty)$. The space $\mathscr{A}(\xi_\infty, J, K_\infty)$ spanned by all automorphic forms of types $(\xi_\infty \otimes \xi_f, J, K_\infty)$, with ξ_f arbitrary in $\mathscr{H}_f$, may then be identified to the inductive limit

$$(1) \qquad \mathscr{A}(\xi_\infty, J, K_\infty) = \text{ind} \lim_L \mathscr{A}(\xi_\infty \otimes \xi_L, J, K_\infty),$$

where L runs through the compact open subgroups of $G(A_f)$, the inductive limit being taken with respect to the above inclusions. The group $G(A_f)$ operates on $\mathscr{H}_f$ by inner automorphisms. Let us denote by $^x\xi$ the transform of $\xi \in \mathscr{H}_f$ by Int x. We have in particular

$$(2) \qquad ^x(\xi_L) = \xi_{(^xL)} \qquad (x \in G(A_f), \xi_L \text{ as in } 4.1);$$

if f is a continuous (or measurable) function on $G(A)$, then

$$(3) \qquad (r_x f) * \xi = r_x(f * {}^{x^{-1}}\xi) \qquad (x \in G(A_f), \xi \in \mathscr{H}_f).$$

Therefore, $G(A_f)$ operates on $\mathscr{A}(\xi_\infty, J, K_\infty)$ by right translations. It follows from 4.3(i) that this representation is admissible.

In view of 4.3(3), $\mathscr{A}(\xi_\infty, J, K_\infty)$ is the adelic analogue of the space $\mathscr{A}(\Sigma, \xi_\infty, J, K_\infty)$ of 1.9, where Σ is the family of arithmetic subgroups of $G(F)$ of the form $\Gamma_L = G(F) \cap (G_\infty \times L)$, where L is a compact open subgroup of $G(A_f)$. These are the *congruence arithmetic subgroups* of $G(F)$, i.e., those subgroups which, for an embedding $G \hookrightarrow \mathbf{GL}_n$ over F, contain a congruence subgroup of $G \cap \mathbf{GL}_n(O_F)$. This analogy can be made more precise when G satisfies strong approximation, which is the case in particular when G is semisimple, simply connected, almost simple over F, and G_∞ is noncompact [16]. In that case, as recalled in 4.4(2), we have $G(A) = G(F) \cdot G_\infty \cdot L$ for any compact open subgroup of $G(A_f)$, so that we may take $C = \{1\}$ in 4.3. Then 4.3(2) provides an isomorphism

$$(4) \qquad \mathscr{A}(\xi_\infty \otimes \xi_L, J, K_\infty) \xrightarrow{\sim} \mathscr{A}(\Gamma_L, \xi_\infty, J, K_\infty),$$

for any L, whence

$$(5) \qquad \mathscr{A}(\xi_\infty, J, K_\infty) \cong \mathscr{A}(\Sigma, \xi_\infty, J, K_\infty),$$

where Σ is the set of congruence arithmetic subgroups of $G(F)$. Moreover, the projection of $G(F)$ in $G(A_f)$ is dense in $G(A_f)$ and $G(A_f)$ may be identified to the completion $G(F)_\Sigma$ of $G(F)$ with respect to the topology defined by the subgroups Γ_L. It is easily seen that the isomorphism (5) commutes with $G(Q)$, where, on the left-hand side $x \in G(Q)$ acts as in 1.9, via left translations, and, on the right-hand side, x acts as an element of $G(A_f)$ by right translations. It follows that the isomorphism (5) commutes with the actions of $G(F)_\Sigma = G(A_f)$ defined here and in 1.9. Also, the isomorphism $G(F)_\Sigma \xrightarrow{\sim} G(A_f)$ induces one of the Hecke algebra $\mathscr{H}(G(F)_\Sigma)$ (see 1.10) onto $\mathscr{H}_f$ and, again, (5) is compatible with the actions defined here and in 1.10. Note also that Γ_L is dense in L, by strong approximation, so that this isomorphism of Hecke algebras induces one of $\mathscr{H}(G(F)_\Sigma, \bar{\Gamma}_L)$, which is equal to $\mathscr{H}(G(F), \Gamma_L)$, onto $\mathscr{H}(G(A_f), L)$. [Strictly speaking, this applies at first for $F = Q$, but we can reduce the general case to that one, if we replace G by $R_{F/Q}G$ (4.1).]

In the general case, the isomorphisms 4.3(2) for various ξ_L are compatible with the action of $G(F)$ defined here and in 1.9 respectively, and this extends by continuity to the closure in $G(A_f)$ of the projection of $G(F)$.

4.8. Let $\mathscr{A}(J, K_\infty)$ be the span of the spaces $\mathscr{A}(\xi_\infty, J, K_\infty)$ and $\mathscr{A}(K_\infty)$ the span of the $\mathscr{A}(J, K_\infty)$. These spaces are $\mathscr{H}$-modules, and union of admissible $\mathscr{H}$-submodules. When 4.6(5) holds, they are isomorphic to the spaces $\mathscr{A}(\Sigma, J, K_\infty)$ and $\mathscr{A}(\Sigma, K_\infty)$ of automorphic forms on $G(R)$ defined in 1.10. Otherwise, the relationship is more complex, and would have to be expressed by means of the isomorphisms 4.3(2).

5. Groups over function fields. In this section, F is a function field of one variable over a finite field. A function on $G(A)$ is said to be smooth if it is locally constant.

5.1. Let χ be a quasi-character of $Z(A)/Z(F)$ and K an open subgroup of $G(A)$. We let $^\circ\mathscr{V}(\chi, K)$ be the space of complex valued functions on $G(A)$ which are right invariant under K, left invariant under $G(F)$, satisfy

$$(1) \qquad f(z \cdot x) = \chi(z) \cdot f(x) \qquad (z \in Z(A), x \in G(A)),$$

and are cuspidal (3.3). [These functions are cusp forms, in a sense to be defined below (5.7), but the latter notion is slightly more general.] We need the following:

5.2. PROPOSITION (G. HARDER). *Let K and χ be given. Then there exists a compact subset C of $G(A)$ such that every element of $^\circ\mathscr{V}(\chi, K)$ has support in $Z(A) \cdot G(F) \cdot C$. In particular, $^\circ\mathscr{V}(\chi, K)$ is finite dimensional.*

This follows from Corollary 1.2.3 in [10] when G is split over F and semisimple (the latter restriction because (1) is not the condition imposed in [10] with respect to the center). However, since $G(A)$ can be covered by finitely many Siegel sets, the argument is general (see [9, p. 142] for $\mathbf{GL}_n$).

5.3. COROLLARY. *Let X be a finite set of quasi-characters of $Z(A)/Z(F)$ and m a positive integer. Then the space $^\circ\mathscr{V}(X, m, K)$ of cuspidal functions which are right invariant under K and satisfy the condition*

$$(1) \qquad \prod_{\chi \in X} (r(z) - \chi(z))^m \cdot f = 0$$

is finite dimensional.

PROOF. The space $^\circ\mathscr{V}(X, m, K)$ is the direct sum of the spaces $^\circ\mathscr{V}(\{\chi\}, m, K)$; hence we may assume that X consists of one quasi-character χ. By 5.3, $^\circ\mathscr{V}(X, 1, K)$ is finite dimensional. We then proceed by induction and assume that $^\circ\mathscr{V}(X, s, K)$ is finite dimensional for some $s \geq 1$.

The group $Z(A)/Z(F) \cdot K'$, where $K' = Z(A) \cap K$, is finitely generated. Let $(z_j)_{1 \leq j \leq q}$ be a generating set. Set $A_{s+1,0} = ^\circ\mathscr{V}(X, s + 1, K)$ and, for $t = 1, ..., q$:

$$A_{s+1,t} = \{f \in A_{s+1,0} | (r(z_j) - \chi(z_j))^s \cdot f = 0 \ (j = 1, ..., t)\}.$$

Then

$$A_{s+1,0} \supset A_{s+1,1} \supset \ldots \supset A_{s+1,t} \supset \ldots \supset A_{s+1,q} = {}^{\circ}\mathscr{V}(X, s, K).$$

Fix t $(0 \leqq t < q)$. Then $f \mapsto (r(a_{t+1}) - \chi(a_{t+1})) \cdot f$ maps $A_{s+1,t}$ into $A_{s+1,t+1}$ and its kernel is contained in $A_{s+1,t+1}$. It follows, by descending induction on t, that $A_{s+1,0}$ is finite dimensional.

5.4. REMARK. If H is a commutative group, let $C[H]$ be its group algebra over C. Every complex representation (π, W) of H extends canonically to one of $C[H]$. An element $w \in W$ is H-finite if and only if it is annihilated by some ideal I of finite codimension of $C[H]$. If I is such an ideal, there exist a finite set X of quasi-characters of H and a positive integer m such that

$$(1) \qquad \prod_{\chi \in X}(\pi(h) - \chi(h))^m \cdot w = 0, \quad \text{for all } h \in H \text{ and all } w \text{ annihilated by } I.$$

If H is finitely generated, then, conversely, all elements $w \in W$ satisfying (1) for all $h \in H$ are H-finite, and annihilated by some ideal of finite codimension of $C[H]$. Therefore, 5.3 implies that the space ${}^{\circ}\mathscr{V}(I, K)$ of cuspidal functions on $G(F)\backslash G(A)/K$ which are annihilated by some ideal I of finite codimension of $C[Z(A)/Z(F)]$ is finite dimensional. In fact, since $Z(A)/Z(F) \cdot (K \cap Z(A))$ is finitely generated, these two statements are equivalent.

5.5. Let E be a local field, R a connected reductive group over E, $\mathscr{H}_R$ the Hecke algebra of $R(E)$. A left ideal J of $\mathscr{H}_R$ is said to be *admissible* if the natural representation of $\mathscr{H}_R$ on $\mathscr{H}_R/J$ is admissible.

LEMMA. *Let J be an admissible ideal of $\mathscr{H}_R$ and K a compact open subgroup of $R(E)$. Let P be a parabolic E-subgroup of H, N the unipotent radical of P and M a Levi E-subgroup of P. There is an admissible ideal J_M in the Hecke algebra $\mathscr{H}_M$ of $M(E)$ with the following property: if π is a smooth representation of $R(E)$ on a space W, and $w \in W$ is K-fixed, annihilated by J, then the image $\bar{w}$ of w in the space W_N (cf. [3]) is annihilated by J_M.*

PROOF. Let $\varphi_0 \in \mathscr{H}_R$ be the characteristic function of K, v_0 its image in $\mathscr{H}_R/J$ and $\bar{v}_0$ the image of v_0 in $(\mathscr{H}_R/J)_N$. The representation of $R(E)$ on $\mathscr{H}_R/J$ is admissible by assumption; therefore the representation of $M(E)$ on $(\mathscr{H}_R/J)_N$ is admissible [3], and the annihilator J_M of $\bar{v}_0$ in $\mathscr{H}_M$ is admissible. We claim that it has the required properties. In fact, if π and w are as in the lemma, then there is a unique $R(E)$-morphism $\mathscr{H}_R \to W$ taking φ_0 to w. It maps $\xi \in \mathscr{H}_R$ onto a scalar multiple of $\pi(\xi) \cdot w$. Therefore, it factors through an $R(E)$-morphism $\mathscr{H}_R/J \to W$, mapping v_0 onto w, whence an $M(E)$-morphism $(\mathscr{H}_R/J)_N \to W_N$ mapping $\bar{v}_0$ onto $\bar{w}$. It follows that $\bar{w}$ is annihilated by J_M.

5.6. THEOREM. *Let K be a compact open subgroup of $G(A)$. Let $v \in V$ and J an admissible ideal of $\mathscr{H}_v$. Then the space $\mathscr{V}(G, v, J, K)$ of complex valued functions f on $G(A)$ which are left invariant under $G(F)$, right invariant under K and annihilated by J is finite dimensional.*

PROOF. Since the representation of G_v on $\mathscr{H}_v/J$ is admissible (and finitely generated), there exists an ideal I_v of finite codimension of $C[Z(F_v)]$ which annihilates $\mathscr{H}_v/J$.

Let $f \in \mathscr{V}(G, v, J, K)$. Then $h \mapsto f * \check{h}$ ($h \mapsto \check{h}$ being the canonical involution of $\mathscr{H}_v$) gives a G_v-morphism $\mathscr{H}_v/J \to f * \mathscr{H}_v$. Therefore $\mathscr{V}(G, v, J, K)$ is annihilated by I_v. Let $Z' = Z(A) \cap K$ and regard $Z(F_v)$ as a subgroup of $Z(A)$. Then $Z(F) \cdot Z(F_v) \cdot Z'$ has *finite index* in $Z(A)$. As a consequence, there exists an ideal I of finite codimension of $Z(A)/Z(F)$ which annihilates $\mathscr{V}(G, v, J, K)$. The space $^\circ\mathscr{V}$ of cuspidal elements in $\mathscr{V}(G, v, J, K)$ is then contained in $^\circ\mathscr{V}(I, K)$ (notation of 5.4), hence is finite dimensional.

We now prove the theorem by induction on the F-rank $\mathrm{rk}_F G'$ of the derived group G' of G. If $\mathrm{rk}_F G' = 0$, then $\mathscr{V}(G, v, J, K) = {}^\circ\mathscr{V}$, and our assertion is already proved. So assume $\mathrm{rk}_F G' \geqq 1$ and the theorem proved for groups of strictly smaller semisimple F-rank. Let now $P = M \cdot N$ vary through a set $\mathscr{P}$ of representatives of the conjugacy classes of proper parabolic F-subgroups of G, where M is a Levi F-subgroup and N the unipotent radical of P. For each such P, let C_P be a set of representatives of $P(A) \backslash G(A)/K$. It is finite. The intersection of the kernels of the maps $f \mapsto f_{P,c}$, where $f_{P,c}(m) = \int_{N(F)\backslash N(A)} f(n \cdot m \cdot c)\, dn$ ($c \in C_P$, $P \in \mathscr{P}$) ($f \in \mathscr{V}(G, v, J, K)$) is then $^\circ\mathscr{V}$, hence is finite dimensional. It suffices therefore to show that, for given $P \in \mathscr{P}$, $c \in C_P$, the functions $f_{P,c}$ vary in a fixed finite dimensional space. After having replaced J and K by conjugates, we may assume that $c = 1$. We write f_P for $f_{P,1}$. Let now U_G (resp. U_M) be the space of functions on $G(F)\backslash G(A)$ (resp. $M(F)\backslash M(A)$) which are right invariant under some compact open subgroup (depending on the function). The representation r of $G(A)$ by right translations on U_G is smooth. If $x \in N(A)$ then $f_P = (r_x f)_P$; hence $\mu_P \colon f \mapsto f_P$ factors through $(U_G)_{N(F_v)}$. It follows then from 5.5 that the elements f_P ($f \in \mathscr{V}(G, v, J, K)$) are all annihilated by some admissible ideal J' of the Hecke algebra of $M(F_v)$. Since these elements are right invariant under $K' = K \cap M(A)$, it follows that μ_P maps $\mathscr{V}(G, v, J, K)$ into $\mathscr{V}(M, v, J', K')$. Since this last space is finite dimensional by our induction assumption, the proof is now complete.

5.7. Corollary. *Let f be a function on $G(A)$ which is left invariant under $G(F)$ and right invariant under some compact open subgroup of $G(A)$. Then the following conditions are equivalent:*

(1) There is a place v of F such that the representation of G_v on the $G(F_v)$-invariant subspace generated by $r(G_v) \cdot f$ is admissible.

(2) For each place v of F, the representation of G_v on the space generated by $r(G_v) \cdot f$ is admissible.

(3) The representation of $G(A)$ on the space spanned by $r(G(A)) \cdot f$ is admissible.

Proof. Clearly (3) $\Rightarrow$ (2) $\Rightarrow$ (1). Assume (1). Let v be as in (1). Then f is annihilated by an admissible ideal J of $\mathscr{H}_v$. Let $U = f * \mathscr{H}$. We have to prove that U^K is finite dimensional for any compact open subgroup K of $G(A)$. There is no harm in replacing K by a smaller group, so we may assume that K fixes f. We may also assume that $K = K_v \times K^v$, where K_v is compact open in G_v and K^v is compact open in the subgroup G^v of elements in $G(A)$ with v-component equal to 1. We also have $\mathscr{H} = \mathscr{H}_v \otimes \mathscr{H}^v$ where $\mathscr{H}^v$ is the Hecke algebra of G^v. Let ξ_v (resp. ξ^v) be the idempotents associated to K_v (resp. K^v) (3.3). Then $\xi_v \otimes \xi^v = \xi_K$ is the idempotent associated to K. Any element g in U is a finite linear combination of elements of the form $f * \alpha * \beta$ ($\alpha \in \mathscr{H}^v$, $\beta \in \mathscr{H}_v$). If such an element is fixed under K, then $g * \xi_K =$

g; hence we may assume that each summand is fixed under K, and that $\alpha * \xi^v = \alpha$, $\beta * \xi_v = \beta$. Since f is fixed under K, it follows that $f * \alpha$ is fixed under K. The elements $f * \alpha$ then belong to the space $\mathscr{V}(G, v, J, K)$, which is finite dimensional by the theorem. For each such element $f * \alpha * \beta$ is contained in the space of K_v-fixed vectors in the admissible $\mathscr{H}_v$-module $f * \alpha * \mathscr{H}_v = f * \mathscr{H}_v * \alpha$, whence our assertion.

5.8. DEFINITIONS. An *automorphic form* on $G(A)$ is a function which is left invariant under $G(F)$, right invariant under some compact open subgroup, and satisfies the equivalent conditions of 5.7. A *cusp form* is a cuspidal automorphic form. Any automorphic form is $Z(A)$-finite (as follows from 5.7(3)).

An *irreducible representation of $G(A)$ is automorphic* if it is isomorphic to a subquotient of the $G(A)$-module $\mathscr{A}$ of all automorphic forms on $G(F)\backslash G(A)$. It follows from 5.7 that it is always admissible.

More generally, a topologically irreducible continuous representation of $G(A)$ in a topological vector space is automorphic if the submodule of smooth vectors is automorphic.

As in 4.6, it follows from [**4**, Theorem 4] that if χ is a character of $Z(A)/Z(F)$, then any G-invariant closed irreducible subspace of $L^2(G(F)\backslash G(A))_\chi$ is automorphic.

5.9. PROPOSITION. *Let f be a function on $G(F)\backslash G(A)$. Then the following conditions are equivalent*:

(1) *f is a cusp form.*

(2) *f is $Z(A)$-finite, cuspidal (3.3), and right invariant under some compact open subgroup of $G(A)$.*

PROOF. That (1) $\Rightarrow$ (2) is clear. Assume (2). Then f is annihilated by an ideal I of finite codimension of $C[Z(A)/Z(F)]$. Let U be the space of functions spanned by $r(G(A)) \cdot f$. Every element of U is cuspidal, annihilated by I and right invariant under some compact open subgroup. If L is any compact open subgroup, then U^L is contained in the space $°\mathscr{V}(I, L)$ (notation of 5.4), hence is finite dimensional. Therefore U is an admissible $G(A)$-module and (1) holds.

5.10. It also follows in the same way that the space $°\mathscr{A}(I)$ (resp. $°\mathscr{A}(X, m)$) of all cusp forms which are annihilated by an ideal I of finite codimension of $C[Z(A)/Z(F)]$ (resp. which satisfy 5.3(1)) is an admissible $G(A)$-module. Moreover, if X consists of one element χ, and if $m = 1$, in which case we put $\mathscr{A}(X, m) = °\mathscr{A}_\chi$, then this space is a direct sum of irreducible admissible $G(A)$-modules, with finite multiplicities. To see this we may, after twisting with $|\chi|^{-1}$, assume that χ is unitary. Then, since $°\mathscr{A}_\chi$ consists of elements with compact support modulo $Z(A)$,

$$(f, g) = \int_{Z(A)G(F)\backslash G(A)} f(x) \cdot \overline{g(x)} \, dx$$

defines a nondegenerate positive invariant hermitian form on $°\mathscr{A}_\chi$. Our assertion follows from this and admissibility. This is the counterpart over function fields of the Gelfand-Piatetski-Shapiro theorem (4.6).

We note that, by [**14**], every automorphic representation transforming according to χ is a constituent of a representation induced from a cuspidal automorphic representation of a Levi subgroup of some parabolic F-subgroup, for any global field F.

REFERENCES

1. A. Borel, *Some finiteness theorems for adele groups over number fields*, Inst. Hautes Etudes Sci. Publ. Math. **16** (1963), 101–126.

2. ——, *Introduction to automorphic forms*, Proc. Sympos. Pure Math., vol. 9, Amer. Math. Soc., Providence, R.I., 1966, pp. 199–210.

3. P. Cartier, *Representations of reductive $\mathfrak{p}$-adic groups: A survey*, these PROCEEDINGS, part 1, pp. 111–155.

4. D. Flath, *Decomposition of representations into tensor products*, these PROCEEDINGS, part 1, pp. 179–183.

5. S. Gelbart, *Automorphic forms on adele groups*, Ann. of Math. Studies, no. 83, Princeton Univ. Press, Princeton, N.J., 1975.

6. I. M. Gelfand, M. I. Graev and I. I. Piatetski-Shapiro, *Representation theory and automorphic functions*, Saunders, 1969.

7. I. M. Gelfand and I. Piatetski-Shapiro, *Automorphic functions and representation theory*, Trudy Moscov. Mat. Obšč. **12** (1963), 389–412.

8. R. Godement, *The spectral decomposition of cusp forms*, Proc. Sympos. Pure Math., vol. 9, Amer. Math. Soc., Providence, R. I., 1966, pp. 225–234.

9. R. Godement and H. Jacquet, *Zeta functions of simple algebras*, Lecture Notes in Math., vol. 260, Springer-Verlag, New York, 1972.

10. G. Harder, *Chevalley groups over function fields and automorphic forms*, Ann. of Math. (2) **100** (1974), 249–306.

11. Harish-Chandra, *Automorphic forms on semi-simple Lie groups*, Notes by J. G. M. Mars, Lecture Notes in Math., vol. 68, Springer-Verlag, New York, 1968.

12. M. Kuga, *Fiber varieties over a symmetric space whose fibers are abelian varieties*, Mimeographed Notes, Univ. of Chicago, 1963–64.

13. R. P. Langlands, *On the functional equations satisfied by Eisenstein series*, Lecture Notes in Math., vol. 544, Springer, New York, 1976.

14. ——, *On the notion of an automorphic representation*, these PROCEEDINGS, part 1, pp. 203–207.

15. I. Piatetski-Shapiro, *Classical and adelic automorphic forms. An introduction*, these PROCEEDINGS, part 1, pp.185–188.

16. V. I. Platonov, *The problem of strong approximation and the Kneser-Tits conjecture*, Izv. Akad. Nauk SSSR Ser. Mat. **3** (1969), 1139–1147; Addendum, ibid. **4** (1970), 784–786.

17. G. Shimura, *Introduction to the arithmetic theory of automorphic forms*, Publ. Math. Soc. Japan **11**, I. Shoten and Princeton Univ. Press, Princeton, N.J., 1971.

18. N. Wallach, *Representations of reductive Lie groups*, these PROCEEDINGS, part 1, pp. 71–86.

19. G. Warner, *Harmonic analysis on semi-simple Lie groups*. I, Grundlehren der Math. Wissenschaften, Band 188, Springer, Berlin, 1972.

INSTITUTE FOR ADVANCED STUDY

COLUMBIA UNIVERSITY

113.

Automorphic *L*-functions

Proc. Symp. Pure Math. **33**, Part 2, Amer. Math. Soc. (1979) 27–61

This paper is mainly devoted to the L-functions attached by Langlands [**35**] to an irreducible admissible automorphic representation π of a reductive group G over a global field k and to local and global problems pertaining to them. In the context of this Institute, it is meant to be complementary to various seminars, in particular to the $\mathbf{GL}_2$-seminars, and to stress the general case. We shall therefore start directly with the latter, and refer for background and motivation to other seminars, or to some expository articles on this topic in general [**3**] or on some aspects of it [**7**], [**14**], [**15**], [**23**].

The representation π is a tensor product $\pi = \otimes_v \pi_v$ over the places of k, where π_v is an irreducible admissible representation of $G(k_v)$ [**11**]. Accordingly the L-functions associated to π will be Euler products of local factors associated to the π_v's. The definition of those uses the notion of the L-group $^L G$ of, or associated to, G. This is the subject matter of Chapter I, whose presentation has been much influenced by a letter of Deligne to the author. The L-function will then be an Euler product $L(s, \pi, r)$ assigned to π and to a finite dimensional representation r of $^L G$. (If $G = \mathbf{GL}_n$, then the L-group is essentially $\mathbf{GL}_n(C)$, and we may tacitly take for r the standard representation r_n of $\mathbf{GL}_n(C)$, so that the discussion of $\mathbf{GL}_n$ can be carried out without any explicit mention of the L-group, as is done in the first six sections of [**3**].) The local L- and ε-factors are defined at all places where G and π are "unramified" in a suitable sense, a condition which excludes at most finitely many places. Chapter II is devoted to this case. The main point is to express the Satake isomorphism in terms of certain semisimple conjugacy classes in $^L G$ (7.1). At this time, the definition of the local factors at the ramified places is not known in general. For $\mathbf{GL}_n$ and r_n, however, there is a direct definition [**19**], [**25**]. In the general case, the most ambitious scheme is to associate canonically to an irreducible admissible representation of a reductive group H over a local field E a representation of the Weil-Deligne group W'_E of E into $^L H$, and then use L- and ε-factors associated to representations of W'_E [**60**]. This problem is the main topic of Chapter III.

The L-function $L(s, \pi, r)$ associated to π and r as above is introduced in §13. In fact, it is defined in general as a product of local factors indexed by almost all places of k. It converges absolutely in some right half-plane (13.3; 14.2). Some of the main

---

AMS (MOS) subject classifications (1970). Primary 10D20; Secondary 12A70, 12B30, 14G10, 22E50.

562

conjectural analytic properties (meromorphic continuation, functional equation), and the evidence known so far, are discussed in §14.

From the point of view of [35], a great many problems on automorphic representations and their L-functions are special cases of one, the so-called lifting problem or problem of functoriality with respect to L-groups. It is discussed in Chapter V. It is closely connected with Artin's conjecture (see §17 and the base-change seminar [17]). In §18 brief mention is made of some known or conjectured relations between automorphic L-functions and the Hasse-Weil zeta-function of certain varieties, to be discussed in more detail in the seminars on Shimura varieties [8], [40].

Thanks are due to H. Jacquet and R. P. Langlands for various very helpful comments on an earlier draft of this paper.

Contents

CHAPTER I. *L*-GROUPS.

k is a field, $\bar{k}$ an algebraic closure of k, k_s the separable closure of k in $\bar{k}$, and Γ_k the Galois group of k_s over k. G is a connected reductive group, over $\bar{k}$ in 1.1, 1.2, 2.1, 2.2, over k otherwise.

§§1, 2 will be used throughout, §3 from Chapter III on. The reader willing to take on faith various statements about restriction of scalars need not read §§4, 5.

1. Classification. We recall first some facts discussed in [58].

1.1. There is a canonical bijection between isomorphism classes of connected reductive $\bar{k}$-groups and isomorphism classes of root systems. It is defined by associating to G the root datum $\psi(G) = (X^*(T), \varphi, X_*(T), \varphi^v)$ where T is a maximal torus of G, $X^*(T)$ $(X_*(T))$ the group of characters (1-parameter subgroups) of T and Φ (Φ^v) the set of roots (coroots) of G with respect to T.

1.2. The choice of a Borel subgroup $B \supset T$ is equivalent to that of a basis $\varDelta$ of $\Phi(G, T)$. The previous bijection yields one between isomorphism classes of triples (G, B, T) and isomorphism classes of based root data $\psi_0(G) = (X^*(T), \varDelta, X_*(T), \varDelta^v)$. There is a split exact sequence

$$(1) \qquad (1) \longrightarrow \operatorname{Int} G \longrightarrow \operatorname{Aut} G \longrightarrow \operatorname{Aut} \psi_0(G) \longrightarrow (1).$$

To get a splitting, we may choose $x_\alpha \in G_\alpha$ $(\alpha \in \varDelta)$ and then have a canonical bijection

$$(2) \qquad \operatorname{Aut} \psi_0(G) \xrightarrow{\ \sim\ } \operatorname{Aut} (G, B, T, \{x_\alpha\}_{\alpha \in \varDelta}).$$

Two such splittings differ by an inner automorphism $\operatorname{Int} t$ $(t \in T)$.

1.3. Given $\gamma \in \Gamma_k$ there is $g \in G(k_s)$ such that $g \cdot {}^\gamma T \cdot g^{-1} = T$, $g \cdot {}^\gamma B \cdot g^{-1} = B$, whence an automorphism of $\psi_0(G)$, which depends only on γ. We let $\mu_G \colon \Gamma_k \to \operatorname{Aut} \psi_0(G)$ be the homomorphism so defined. If G' is a k-group which is isomorphic to G over $\bar{k}$ (hence over k_s), then $\mu_G = \mu_{G'} \Leftrightarrow G, G'$ are inner forms of each other.

1.4. Let $f \colon G \to G'$ be a k-morphism, whose image is a normal subgroup. Then f induces a map $\psi(f) \colon \psi(G) \to \psi(G')$ (contravariant (resp. covariant) in the first (last) two arguments). Given $B, T \subset G$ as above, there exists a Borel subgroup B' (resp. a maximal torus T') of G' such that $f(B) \subset B'$, $f(T) \subset T'$, whence also a map $\psi_0(f) \colon \psi_0(G) \to \psi_0(G')$.

2. Definition of the *L*-group.

2.1. The inverse system $\Psi_0^\vee$ to the based root datum $\Psi_0 = (M, \varDelta, M^*, \varDelta^\vee)$ is $\psi_0^\vee = (M^*, \varDelta^\vee, M, \varDelta)$. To the $\bar{k}$-group G we first associate the group ${}^L G^\circ$ over C such that $\psi_0({}^L G^\circ) = \psi_0(G)^\vee$. We let ${}^L T^\circ$, ${}^L B^\circ$ be the maximal torus and Borel subgroup defined by $\psi_0^\vee$, and say they define the canonical splitting of ${}^L G^\circ$.

Let f be as in 1.4. Then f also induces a map $\psi_0^\vee(f) : \psi_0(G')^\vee \to \psi_0(G)^\vee$. An algebraic group morphism of ${}^L G'^\circ$ into ${}^L G^\circ$ associated to it will be denoted ${}^L f^\circ$. Given one, any other is of the form $\operatorname{Int} t \circ {}^L f^\circ \circ \operatorname{Int} t'$ $(t \in {}^L T^\circ, t' \in {}^L T'^\circ)$, and maps ${}^L T'^\circ$ (resp. ${}^L B'^\circ$) into ${}^L T^\circ$ (resp. ${}^L B^\circ$).

2.2. EXAMPLES. (1) Let $G = \mathbf{GL}_n$. Then ${}^L G^\circ = \mathbf{GL}_n$. In fact, let $M = \mathbf{Z}^n$ with $\{x_i\}$ its canonical basis. Let $\{e_i\}$ be the dual basis of $M^* = \mathbf{Z}^n$. Then $\Psi_0(\mathbf{GL}_n) = (M, \varDelta, M^*, \varDelta^\vee)$ with $\varDelta = \{(x_i - x_{i+1}), 1 \leq i < n\}$, $\varDelta^\vee = \{(e_i - e_{i+1}), 1 \leq i < n\}$, hence $\psi_0 = \psi_0^\vee$.

(2) Let G be semisimple and $\Psi_0(G) = (M, \Phi, M^*, \Phi^\vee)$. As usual, let $P(\Phi) \subset M \otimes \mathbf{Q}$ be the lattice of weights of Φ and $Q(\Phi)$ the group generated by Φ in M. Define $P(\Phi^\vee)$ and $Q(\Phi^\vee)$ similarly.

As is known G is simply connected (resp. of adjoint type) if and only if $P(\phi) = M$ (resp. $Q(\phi) = M$). Moreover

$$P(\Phi) = \{\lambda \in M \otimes \mathbf{Q} \mid \langle \lambda, \Phi^\vee \rangle \subset \mathbf{Z}\}, \qquad P(\Phi^\vee) = \{\lambda \in M^x \otimes \mathbf{Q} \mid \langle \lambda, \Phi \rangle \in \mathbf{Z}\}.$$

Therefore:

G simply connected $\Leftrightarrow$ $^LG^\circ$ of adjoint type;

G of adjoint type $\Leftrightarrow$ $^LG^\circ$ simply connected.

(3) Let G be simple. Up to central isogeny, it is characterized by one of the types A_n, ..., G_2 of the Killing-Cartan classification. It is well known that the map $\psi_0(G) \to \psi_0(G)^v$ permutes B_n and C_n and leaves all other types stable. Thus if $G = \mathbf{Sp}_{2n}$ (resp. $G = \mathbf{PSp}_{2n}$), then $^LG^\circ = \mathbf{SO}_{2n+1}$ (resp. $^LG^\circ = \mathbf{Spin}_{2n+1}$). In all other cases, $G \mapsto {}^LG^\circ$ preserves the type (but goes from simply connected group to adjoint group, and vice versa).

(4) Let again G be reductive and let $f : G \to G'$ be a central isogeny. Let

$$N = \operatorname{coker} f_* : X_*(T) \longrightarrow X_*(T') \qquad (T' = f(T)),$$
$$N' = \operatorname{coker} f^* : X^*(T') \longrightarrow X^*(T).$$

Then N and N' are isomorphic and $\ker {}^Lf^\circ = \operatorname{Hom}(N, C^*) \xrightarrow{\sim} N$. In particular, $^Lf^\circ$ is an isomorphism if and only if f is one.

(5) Let $f : G \to G'$ be a central surjective morphism, $Q = \ker f$, and Q° the identity component of Q. Then $\ker {}^Lf^\circ \simeq Q/Q^\circ$.

If Q is connected, then $T'' = T \cap Q$ is a maximal torus of Q, and the injectivity of $^Lf^\circ$ follows from the fact that the exact sequence $1 \to T'' \to T \to T' \to 1$ necessarily splits. If Q is not connected, then $r : H = G/Q^\circ \to G'$ is a nontrivial separable isogeny, with kernel Q/Q°. $^Lf^\circ$ factors through $^Lr^\circ$ and, by the first part and (4), $\ker {}^Lf^\circ = \ker {}^Lr^\circ \cong Q/Q^\circ$. In particular, if we apply this to the case where $G' = G_{\mathrm{ad}}$ is the adjoint group of G, and use (2), we see that the derived group of $^LG^\circ$ is simply connected if and only if the center of G is connected. As an example, let $G = \mathbf{GSp}_{2n}$ be the group of symplectic similitudes on a $2n$-dimensional space. Then the derived group of $^LG^\circ$ is isomorphic to $\mathbf{Spin}_{2n+1}$. In fact, we have $^LG^\circ = (\mathbf{GL}_1 \times \mathbf{Spin}_{2n+1})/A$ where $A = \{1, a\}$ and $a = (a_1, a_2)$, with a_1 of order two in $\mathbf{GL}_1$ and a_2 the nontrivial central element of $\mathbf{Spin}_{2n+1}$. If $n = 2$, then $\mathbf{Spin}_{2n+1} = \mathbf{Sp}_{2n}$. It follows that if $G = \mathbf{GSp}_4$, then $^LG^\circ = \mathbf{GSp}_4(C)$.

2.3. We have canonically $\operatorname{Aut} \Psi_0 = \operatorname{Aut} \Psi_0^\vee$. Therefore we may view μ_G as a homomorphism of Γ_k into $\operatorname{Aut} \psi_0^\vee$. Choose a monomorphism

$$(1) \qquad \operatorname{Aut} \psi_0^\vee \longrightarrow \operatorname{Aut} ({}^LG^\circ, {}^LB^\circ, {}^LT^\circ)$$

as in 1.2(2). We have then a homomorphism

$$\mu_G' : \Gamma_k \longrightarrow \operatorname{Aut} ({}^LG^\circ, {}^LB^\circ, {}^LT^\circ).$$

The associated group to, or L-group of, G is then by definition the semidirect product

$$(2) \qquad {}^L(G/k) = {}^LG = {}^LG^\circ \rtimes \Gamma_k,$$

with respect to μ_G'. We note that μ_G' is well defined up to an inner automorphism by an element of $^LT^\circ$. The group ^LG is viewed as a topological group in the obvious way. The canonical splitting of $^LG^\circ$ (2.1) is stable under Γ_k.

We have a canonical projection $^LG \to \Gamma_k$ with kernel $^LG^\circ$. The splittings of the exact sequence

$$(3) \qquad 1 \longrightarrow {}^LG^\circ \longrightarrow {}^LG \xrightarrow{\;\nu_G\;} \Gamma_k \longrightarrow 1$$

defined as in 1.2 via an isomorphism Aut $\Psi_0^\vee \xrightarrow{\sim}$ Aut($^LG^\circ$, $^LB^\circ$, $^LT^\circ$, $\{x_\alpha\}$) are called admissible. They differ by inner automorphisms Int t ($t \in {}^LT^\circ$). Note that if G splits over k, then Γ_k acts trivially on $^LG^\circ$ and ^LG is simply the direct product of $^LG^\circ$ and Γ_k.

2.4. REMARKS. (1) So far, we can in this definition take $^LG^\circ$ over any field. We have chosen C since this is the most important case at present, but it is occasionally useful to use other local fields.

(2) There are various variants of this notion, which may be more convenient in certain contexts. For instance we can divide Γ_k by a closed normal subgroup which acts trivially on $\Psi^\vee$, hence on $^LG^\circ$, e.g., by $\Gamma_{k'}$ if k' is a Galois extension of k over which G splits. Then Γ_k is replaced by $\mathrm{Gal}(k'/k)$, and ^LG is a complex reductive Lie group.

We can also define a semidirect product $^LG^\circ \rtimes \Sigma$, for any group Σ endowed with a homomorphism into Γ_k, e.g., the Weil group of k, if k is a local or global field. In that case, we get the "Weil form" of ^LG.

(3) Let G' be a k-group which is isomorphic to G over $\bar{k}$. Then G and G' are inner forms of each other if and only if ^LG is isomorphic to $^LG'$ over Γ_k. In fact, the first condition is equivalent to $\mu_G = \mu_{G'}$, and the latter is easily seen to be equivalent to the second condition. In particular, since two quasi-split groups over k which are inner forms of each other are isomorphic over k, it follows that if G, G' are quasi-split and $^LG \xrightarrow{\sim} {}^LG'$ over Γ_k then G and G' are k-isomorphic.

2.5. *Functoriality.* Let $f: G \to G'$ be a k-morphism whose image is a normal subgroup. Then $f_{\psi_0}: \psi_0(G) \to \psi_0(G')$ clearly commutes with Γ_k, hence so does $f_{\psi_0^\vee} : \psi_0(G')^v \to \psi_0(G)^v$ and $^Lf^\circ: {}^LG'^\circ \to {}^LG^\circ$. We get therefore a continuous homomorphism $^Lf: {}^LG' \to {}^LG$ such that

$$\begin{array}{ccc} {}^LG' & \xrightarrow{\ {}^Lf\ } & {}^LG \\ {}_{\nu_{G'}}\searrow & & \swarrow{}_{\nu_G} \\ & \Gamma_k & \end{array}$$

is commutative, which extends $^Lf^\circ$.

2.6. *Representations.* For brevity, by *representation* of ^LG we shall mean a continuous homomorphism $r: {}^LG \to \mathbf{GL}_m(C)$ whose restriction to $^LG^\circ$ is a morphism of complex Lie groups.

Clearly, ker r always contains an open subgroup of Γ_k, hence r factors through $^LG^\circ \rtimes \Gamma_{k'/k}$, where k' is a finite Galois extension of k over which G splits. The group $^LG^\circ \rtimes \Gamma_{k'/k}$ is canonically a complex algebraic group and r is a morphism of complex algebraic groups.

3. Parabolic subgroups.

3.1. *Notation.* We let $\mathscr{P}(G/k)$ denote the set of parabolic k-subgroups of G, and write $\mathscr{P}(G)$ for $\mathscr{P}(G/\bar{k})$. Let $p(G/k)$ be the set of conjugacy classes (with respect to $G(\bar{k})$ or $G(k)$, it is the same) of parabolic k-subgroups, and $p(G) = p(G/\bar{k})$. Let $p(G)_k$ be the set of conjugacy classes of parabolic subgroups which are defined over k (i.e., if $P \in \sigma \in p(G)_k$, then $^\gamma P \in \sigma$ for all $\gamma \in \Gamma_k$). In particular $p(G/k) \hookrightarrow p(G)_k$. There is equality if G is quasi-split/k.

3.2. We recall there is a canonical bijection between $p(G)$ and the subsets of Δ.

Then $p(G)_k$ corresponds to the Γ_k-stable subsets of Δ and $p(G/k)$ to those Γ_k-stable subsets which contain the set Δ_0 of simple roots of a Levi subgroup of a minimal parabolic k-group. In particular we have $p(G/k) = p(G)_k$ if G is quasi-split over k. Given $P \in \mathscr{P}(G)$, we let $J(P)$ be the subset of Δ assigned to the class of P

Since two conjugate parabolic subgroups whose intersection is a parabolic subgroup are identical, we see in particular that if P is defined over k, $P' \supset P$, and the class of P' is defined over k, then P' is defined over k.

3.3. *Parabolic subgroups of ^LG.* A closed subgroup P of ^LG is parabolic if $\gamma_G(P) = \Gamma_k$ and $P^\circ = {}^LG^\circ \cap P$ is a parabolic subgroup of $^LG^\circ$. Then $P = N_{^LG}(P^\circ)$. In other words, a parabolic subgroup is the normalizer of a parabolic subgroup P° of $^LG^\circ$, provided the normalizer meets every class modulo $^LG^\circ$. We say P is standard if it contains ^LB. The standard parabolic subgroups are the subgroups

$$\text{(1)} \qquad\qquad\qquad {}^LP^\circ \rtimes \Gamma_k,$$

where $^LP^\circ$ runs through the standard parabolic subgroup of $^LG^\circ$ such that $J(^LP^\circ) \subset \Delta^\vee$ is stable under Γ_k.

Every parabolic subgroup of ^LG is conjugate (under ^LG or, equivalently, $^LG^\circ$) to one and only one standard parabolic subgroup.

We let $\mathscr{P}(^LG)$ be the set of parabolic subgroups of ^LG and $p(^LG)$ the set of their conjugacy classes.

The given bijection $\Delta \leftrightarrow \Delta^\vee$ yields then, in view of 3.2, a bijection

$$\text{(2)} \qquad\qquad\qquad p(G)_k \leftrightarrow p(^LG).$$

We shall say that a parabolic subgroup of ^LG is *relevant* if its class corresponds to one of $p(G/k)$ under this map. We let $^L\mathscr{P}(^LG)$ be the set of relevant parabolic subgroups and $^Lp(^LG)$ the set of their conjugacy classes, the *relevant conjugacy classes* of parabolic subgroups. Thus, by definition

$$\text{(3)} \qquad\qquad\qquad p(G/k) \leftrightarrow {}^Lp(^LG).$$

Thus, if G and G' are inner forms of each other, $p(^LG)$ and $p(^LG')$ are the same, but $^Lp(^LG)$ and $^Lp(^LG')$ are not. If G' is quasi-split, then $^Lp(^LG') \doteq p(^LG')$; hence we have an injection

$$\text{(4)} \qquad\qquad {}^Lp(^LG) \subset {}^Lp(^LG') = p(^LG').$$

If $\mathscr{D}G$ is anisotropic over k, then $^Lp(^LG)$ consists of G alone.

3.4. *Levi subgroups.* Let P be a parabolic subgroup of ^LG. The unipotent radical N of P° is normal in P and will also be called the unipotent radical of P. Then $P/N \xrightarrow{\sim} P^\circ/N \rtimes \Gamma_k$. In fact, it follows from (1) that P is a split extension of N, and is the semidirect product of N by the normalizer in P of any Levi subgroup M° of P°. Those normalizers will be called the Levi subgroups of P.

Let $P \in \mathscr{P}(G/k)$, M a Levi k-subgroup of P. Let ^LP be the standard parabolic subgroup in the class associated to that of P (see (3)). Then ^LM may be identified to a Levi subgroup of ^LP. In fact if M corresponds to $(X^*(T), J, X_*(T), J^v)$, then $^LM^\circ$ corresponds to $(X_*(T), J^v, X^*(T), J)$ and $^LM^\circ \rtimes \Gamma_k$ is equal to ^LM by definition and is a Levi subgroup of ^LP, as defined above.

A Levi subgroup of a parabolic subgroup P of ^LG is *relevant* if P is.

For the sake of brevity, we shall sometimes say "Levi subgroup in G" for "Levi subgroup of a parabolic subgroup of G." Similarly for $^L G$.

3.5. LEMMA. *The proper Levi subgroups in $^L G$ are the centralizers in $^L G$ of tori in $\mathscr{D}(^L G^\circ)$, which project onto Γ_k.*

Let M be a proper Levi subgroup in $^L G$. It is conjugate to a subgroup $\mathscr{L}(S)^\circ \rtimes \Gamma_k$, where $S \subset {}^L T^\circ$ is the identity component of the kernel of a subset $J \subsetneqq \varDelta^v$ stable under Γ_k. Let then S' be the one-dimensional subtorus of $S \cap \mathscr{D}(^L G^\circ)$ on which the remaining simple roots are all equal. It is clear that $\mathscr{L}(S')^\circ = \mathscr{L}(S)$, and that S' is pointwise fixed under Γ_k. We have then $M = \mathscr{L}(S')$.

Let now S be a nontrivial torus in $\mathscr{D}(^L G^\circ)$ such that $\mathscr{L}(S)$ meets every connected component of $^L G$. Fix an ordering on $X^*(S)$. There is a proper parabolic subgroup P° of $^L G^\circ$ of the form $\mathscr{L}(S)^\circ \cdot U$, such that the weights of S in the unipotent radical U of P° are the roots of $^L G^\circ$ with respect to S which are positive for this ordering. U is normalized by $\mathscr{L}(S)$; hence $\mathscr{L}(S) \cdot U$ is a proper parabolic subgroup P of $^L G$, and then $\mathscr{L}(S)$ is a Levi subgroup of P.

3.6. PROPOSITION. *Let H be a subgroup of $^L G$ whose projection on Γ_k is dense in Γ_k. Then the Levi subgroups in $^L G$ which contain H minimally form one conjugacy class with respect to the centralizer of H in $^L G^\circ$.*

Let C be the identity component of the centralizer of H in $\mathscr{D}(^L G^\circ)$, and D a maximal torus of H. If $D = \{1\}$, then, by 3.5, H is not contained in any proper Levi subgroup in $^L G$, and there is nothing to prove. So assume $D \neq \{1\}$.

Let Γ' be a normal open subgroup of Γ_k which acts trivially on $^L G^\circ$. It is then normal in $^L G$, and $H \cdot \Gamma'$ projects onto Γ_k. Since $\mathscr{L}(D)$ contains $H \cdot \Gamma'$, it projects onto Γ_k, hence is a proper Levi subgroup by 3.5. Let M be a Levi subgroup containing H. By 3.5, $M = Z(S)$, where S is a torus in $\mathscr{D}(^L G^\circ)$. Then $S \subset C$, there exists $c \in C$ such that $c \cdot S \cdot c^{-1} \subset D$, hence $c \cdot M \cdot c^{-1} = \mathscr{L}(S') \supset \mathscr{L}(D)$.

4. Remarks on induced groups. (To be used mainly to discuss restriction of scalars in §5 and 6.4.)

4.1. Let A be a group, A' a subgroup of finite index of A and E a group on which A' operates by automorphisms. Then we let

$$(1) \qquad \operatorname{Ind}_{A'}^A(E) = I_{A'}^A(E) = \{f : A \longrightarrow E \,|\, f(a'a) = a' \cdot f(a) \ (a \in A; \ a' \in A')\}.$$

It is a group (composition being defined by taking products of values). It is viewed as an A-group by right translations:

$$(2) \qquad\qquad\qquad r_a f(x) = f(xa) \qquad (x, a \in A).$$

For $s \in A' \backslash A$, let

$$(3) \qquad\qquad\qquad E_s = \{f \in I_{A'}^A(E) \,|\, f(a) = 0 \text{ if } a \notin s\}.$$

Then E_s is a subgroup, $I_{A'}^A(E)$ is the direct product of the E_s's $(s \in A' \backslash A)$, and these subgroups are permuted by A. The subgroup $E_{\bar e}$ is stable under A' and is isomorphic to E as an A' module under the map $f \mapsto f(e)$. The product of the E_s's $(s \in A' \backslash A, \ s \neq e)$ is also stable under A'. We have therefore canonical homomorphisms

(4) $$E \rtimes A' \longrightarrow I(E) \rtimes A' \longrightarrow E \rtimes A'$$

whose composition is the identity.

4.2. Let B be a group, $\mu: B \to A$ a homomorphism. Let $B' = \mu^{-1}(A')$ and assume that μ induces a bijection: $B'\backslash B \xrightarrow{\sim} A'\backslash A$. Let E be a group on which A' operates by automorphisms, also viewed as a B'-group via μ. Then $f \mapsto \mu \circ f$ induces an isomorphism

(1) $$\mu': I^A_{A'}(E) \xrightarrow{\ \sim\ } I^B_{B'}(E),$$

whose inverse is μ-equivariant.

This follows immediately from the definitions.

4.3. Let A, E be as before, C a group and $\nu: C \to A$ a homomorphism. Let $\varphi: C \to E \rtimes A$ be a homomorphism over A. The map $\psi: C \to E$ such that $\varphi(c) = (\psi(c), \nu(c))$ $(c \in C)$ is a 1-cocycle of C in E and $\varphi \mapsto \psi$ induces a bijection

$$H^1(C; E) \xrightarrow{\ \sim\ } \varphi_A(C, E),$$

where, by definition, $\varphi_A(C, E)$ denotes the set of homomorphisms $\varphi: C \to E \rtimes A$ over A, modulo inner automorphisms by elements of E. `

4.4. Let A, A', B, B' and E be as in 4.2. We have a commutative diagram with exact rows

(1)
$$
\begin{array}{ccccccccc}
1 & \longrightarrow & I^A_{A'}(E) & \longrightarrow & I^A_{A'}(E) \rtimes A & \longrightarrow & A & \longrightarrow & 1 \\
& & \uparrow & & \uparrow & & \uparrow & & \\
1 & \longrightarrow & E & \longrightarrow & E \rtimes A' & \longrightarrow & A' & \longrightarrow & 1
\end{array}
$$

where the vertical maps are natural inclusions (4.1).

Let $\varphi: B \to I^A_{A'}(E) \rtimes A$ be a homomorphism over A. Using 4.1(4), we get by restriction a homomorphsim $\bar\varphi: B' \to E \rtimes A'$ over A'.

4.5. LEMMA. *The map $\varphi \mapsto \bar\varphi$ of 4.4 induces a bijection $\varphi_A(B, I^A_{A'}(E)) \xrightarrow{\sim} \varphi_{A'}(B', E)$.*

We have, using 4.2, 4.3:

(1) $$\varphi_A(B, I^A_{A'}(E)) = H^1(B; I^A_{A'}(E)) = H^1(B; I^B_{B'}(E)),$$

(2) $$\Phi_{A'}(B', E) = H^1(B'; E).$$

By a variant of Shapiro's lemma, contained, e.g., in [**4**, 1.29]:

$$H^1(B; I^B_{B'}(E)) \xrightarrow{\ \sim\ } H^1(B'; E),$$

and it is clear that the isomorphisms (1), (2) carry this isomorphism over to $\varphi \mapsto \bar\varphi$.

5. Restriction of scalars. *In this section, k' is a finite extension of k in k_s, G' is a connected k'-group, and $G = R_{k'/k} G'$.*

5.1. The Galois group $\Gamma_{k'}$ of k_s over k' is an open subgroup (of finite index) of Γ_k and $\Sigma_{k',k} = \Gamma_{k'}\backslash\Gamma_k$ may be identified with the set of k-monomorphisms of k' into k_s. We have, in the notation of 4.1 (with $A = \Gamma_k$, $A' = \Gamma_{k'}$)

(1) $$G(\bar k) = I^{\Gamma_k}_{\Gamma_{k'}}(G'(\bar k)) = \prod_{\sigma \in \Gamma_{k'}\backslash\Gamma_k} {}^\sigma G'(\bar k).$$

Assume G' to be reductive. Then we see easily that $\psi(G) = (M, \varphi, M^*, \varphi^\vee)$ is related to $\psi(G') = (M', \varphi', M'^*, \varphi'^\vee)$ by

$$(2) \qquad\qquad M = I^{\Gamma_k}_{\Gamma_{k'}}(M'), \qquad \varphi = \bigcup_{a \in A' \backslash A} \varphi' \cdot a.$$

Similarly, if Δ' is a basis of φ', then

$$(3) \qquad\qquad \Delta = \bigcup_a \Delta' \cdot a$$

is one for φ.

From this it follows that we have a natural isomorphism

$$(4) \qquad\qquad {}^L G^\circ \overset{\sim}{\longrightarrow} I^{\Gamma_k}_{\Gamma_{k'}}({}^L G'^\circ).$$

We have then a commutative diagram

$$(5) \quad \begin{array}{ccccccccc} 1 & \longrightarrow & {}^L G'^\circ & \longrightarrow & {}^L G' = {}^L G'^\circ \rtimes \Gamma_{k'} & \longrightarrow & \Gamma_{k'} & \longrightarrow & 1 \\ & & \downarrow & & \downarrow & & \downarrow & & \\ 1 & \longrightarrow & {}^L G^\circ = I^{\Gamma_k}_{\Gamma_{k'}}({}^L G'^\circ) & \longrightarrow & {}^L G = {}^L G^\circ \rtimes \Gamma_k & \longrightarrow & \Gamma_k & \longrightarrow & 1 \end{array}$$

5.2. The map $P' \mapsto R_{k'/k} P'$ induces a bijection between $\mathscr{P}(G'/k')$ and $\mathscr{P}(G/k)$. Moreover P' is a Borel subgroup of G' if and only if $R_{k'/k} P'$ is one of G. Hence G' is quasi-split$/k'$ if and only if G is quasi-split over k (see [**5**, §6]).

Since $G(k) \overset{\sim}{\to} G'(k')$, we also get a bijection $p(G'/k') \overset{\sim}{\to} p(G/k)$.

If $J' \subset \Delta'$ is stable under $\Gamma_{k'}$ then $J = \bigcup_{a \in A' \backslash A} r_a(J')$ is stable under Γ_k. This map is easily seen to yield a bijection between $\Gamma_{k'}$-stable subsets of Δ' and Γ_k-stable subsets of Δ, whence also canonical bijections

$$p({}^L G') \overset{\sim}{\longrightarrow} p({}^L G), \qquad {}^L p({}^L G') \overset{\sim}{\longrightarrow} {}^L p({}^L G).$$

CHAPTER II. QUASI-SPLIT GROUPS. THE UNRAMIFIED CASE.

In this chapter, G is a connected reductive quasi-split k-group. From 6.2 on, G is assumed to split over a cyclic extension k' of k, and σ denotes a generator of $\mathrm{Gal}(k'/k)$.

6. Semisimple conjugacy classes in ${}^L G$.

6.1. Assume B and T to be defined over k. Then the action of Γ_k on $X^*(T)$ or $X_*(T)$ given by μ_G coincides with the ordinary action. The greatest k-split subtorus T_d of T is maximal k-split in G, and its centralizer is T; in particular, T_d contains regular elements of G. Hence any element $w \in W$ which leaves T_d stable is completely determined by its restriction to T_d. It follows that ${}_k W$ may be identified with the subgroup of the elements of W which leave T_d stable or, equivalently, with the fixed-point set of Γ_k in W. If we go over to the L-group and identify canonically W with $W({}^L G^\circ, {}^L T^\circ)$, then ${}_k W$ is also the fixed-point set of Γ_k in W, and it operates on the greatest subtorus S of ${}^L T^\circ$ which is pointwise fixed under Γ_k. The group S always contains regular elements; hence any element of ${}_k W$ is determined by its restriction to S. We let ${}_k N$ be the inverse image of ${}_k W$ in the normalizer N of ${}^L T^\circ$ in ${}^L G^\circ$.

6.2. LEMMA. *Every element $w \in {}_k W$ has a representative in ${}_k N$ which is fixed under σ.*

Write $\varDelta^v = D_1 \cup \cdots \cup D_m$, where the D_i's are the distinct orbits of $\varGamma(k'/k)$ in $\varDelta^v$. Let δ_i be the common restriction to S of the elements of D_i and S_i the identity component of the kernel of δ_i. Then $_kW$, viewed as a group of automorphisms of S, is generated by the reflections s_i to the S_i ($1 \leq i \leq m$), and it suffices to prove the lemma for $w = s_i$ ($1 \leq i \leq m$). [The "reflection" s_i is the unique element $\neq 1$ of W which leaves S stable, fixes S_i pointwise and is of order two.]

We let $\mathrm{Lie}(M)$ denote the Lie algebra of the complex Lie group M. For $\check\alpha \in \varDelta^v$, let, as usual

$$(1) \qquad \mathfrak{g}_{\check\alpha} = \{X \in \mathrm{Lie}(^LG^\circ) | \mathrm{Ad}\, t(X) = \check\alpha(t) \cdot X \, (t \in {}^LT^\circ)\}.$$

It is one-dimensional. Fix i between 1 and m. By construction of ^LG, we can find nonzero elements $e_{\check\alpha} \in \mathfrak{g}_{\check\alpha}$ (resp. $e_{-\check\alpha} \in \mathfrak{g}_{-\check\alpha}$ ($\check\alpha \in D_i$)) which are permuted by σ. We have then

$$(2) \qquad [e_{\check\alpha}, e_{-\check\alpha}] = c \cdot \alpha,$$

where c is $\neq 0$, and independent of $\alpha \in D_i$ since $\varGamma(k'/k)$ is transitive on D_i. Here $X_*(^LT^\circ) \otimes C$ is identified with $\mathrm{Lie}(^LT^\circ)$, and $\check\alpha$ with $\check\alpha \otimes 1$. The element

$$(3) \qquad f_{\pm i} = \sum_{\check\alpha \in D_i} e_{\pm\check\alpha},$$

is fixed under σ. Moreover, since the difference of two simple roots is not a root, we have

$$(4) \qquad h_i = [f_i, f_{-i}] = \sum_{\check\alpha \in D_i} [e_{\check\alpha}, e_{-\check\alpha}] = c \cdot \sum_{\check\alpha \in D_i} \alpha.$$

Using (3) and (4), we get

$$(5) \qquad [h_i, f_{\pm i}] = c \sum_{\check\alpha, \check\beta \in D_i} \langle \alpha, \beta \rangle \, e_{\pm\beta}.$$

By the transitivity of $\mathrm{Gal}(k'/k)$ on D_i, the number

$$(6) \qquad d = \sum_{\check\alpha \in D_i} \langle \alpha, \check\beta \rangle$$

is also independent of $\check\beta \in D_i$; therefore

$$(7) \qquad [h, f_{\pm i}] = c \cdot d \cdot f_{\pm i}.$$

We claim that $d \neq 0$, in fact that $d = 1, 2$. The irreducible components of D_i are permuted transitively by $\mathrm{Gal}(k'/k)$ and have a transitive group of automorphisms. Therefore they are of type A_1 or A_2. Then, accordingly, $d = 2$ or $d = 1$. It follows that h_i, f_i and f_{-i} span a three-dimensional simple algebra pointwise fixed under σ. Then so is the corresponding analytic subgroup G_i of $^LG^\circ$. The group G_i centralizes S_i and $S \cap G_i$ is a maximal torus of G_i, with Lie algebra spanned by h_i. Then the nontrivial element of $W(G_i, S \cap G_i)$ is the required element.

REMARK. An equivalent statement is proved, in a different manner, in [35, pp. 19–22].

6.3. We let $Y = {}^L(T_d)^\circ$. The group $X_*(T_d)$ may be identified to the fixed-point set of $\varGamma_k$ in $X_*(T)$. The inclusion of $X_*(T_d) = X^*(Y)$ into $X_*(T) = X^*(^LT^\circ)$ induces a surjective morphism $^LT^\circ \to Y$, to be denoted ν.

The map $A : t \mapsto t^{-1} \cdot {}^\sigma t$ is an endomorphism of $^LT^\circ$, whose differential dA at 1 is $(d\sigma - \mathrm{Id})$. Let

(1) $$U = (\ker A)^\circ, \qquad V = \operatorname{im} A.$$

Then U is pointwise fixed under σ, the Lie algebra of U (resp. V) is the kernel (resp. image) of dA. Since dA is semisimple, they are transversal to each other; hence

(2) $${}^L T^\circ = U \cdot V, \quad \text{and } U \cap V \text{ is finite.}$$

Moreover,

(3) $$V = \ker \nu, \qquad \nu(U) = Y.$$

In the rest of this chapter, we let ${}^L G$ stand for the "finite Galois form" ${}^L G^\circ \rtimes \operatorname{Gal}(k'/k)$ of the *L*-group. We now want to discuss the semisimple conjugacy classes in ${}^L G^\circ \rtimes \sigma$ with respect to ${}^L G^\circ$. We have

(4) $$g^{-1} \cdot (h \rtimes \sigma) \cdot g = g^{-1} \cdot h \cdot {}^\sigma g \rtimes \sigma \qquad (g, h \in {}^L G^\circ);$$

therefore ${}^L G^\circ$-conjugacy in ${}^L G^\circ \rtimes \sigma$ is equivalent to σ-conjugacy in ${}^L G^\circ$.

6.4. LEMMA. *Let* $\nu' : {}^L T^\circ \rtimes \sigma \to Y$ *be defined by* $\nu'(t \times \sigma) = \nu(t)$ $(t \in {}^L T^\circ)$. *Then* ν' *induces a bijection*

(1) $$\bar{\nu} : ({}^L T^\circ \rtimes \sigma)/\operatorname{Int} {}_k N \xrightarrow{\ \sim\ } Y/{}_k W.$$

Let $n \in {}_k N$. By 6.2, we may write $n = w \cdot s$ with $w = {}^\sigma w$ and $s \in {}^L T^\circ$. Then the ${}^L T^\circ$-component of $n^{-1}(t \rtimes \sigma)\, n$ is

$$s^{-1} \cdot w^{-1} \cdot t \cdot w^\sigma s = s^{-1} \cdot {}^\sigma s \cdot (w^{-1} \cdot t \cdot w) \in V \cdot w^{-1} \cdot t \cdot w;$$

hence

$$\nu'(n^{-1} \cdot (t \rtimes \sigma) \cdot n) = \nu(w^{-1} \cdot t \cdot w) = w^{-1} \cdot \nu(t) \cdot w = w^{-1} \cdot \nu'(t \rtimes \sigma) \cdot w.$$

Thus ν' is equivariant with respect to the projection ${}_k N \to {}_k W$ and therefore induces a map of the left-hand side of (1) into the right-hand side of (1), which is obviously surjective. Let $t, t' \in {}^L T^\circ$ and assume that $\nu'(t \rtimes \sigma) = w^{-1} \cdot \nu'(t' \rtimes \sigma) \cdot w$ for some $w \in {}_k W$. Then we have $\nu(t) = \nu(w^{-1} \cdot t' \cdot w)$, where w is a representative of w fixed under σ, whence $t = v \cdot (w^{-1} \cdot t' \cdot w)$, with $v \in V$. We can write $v = s^{-1} \cdot {}^\sigma s$ for some $s \in {}^L T^\circ$, and get $t \times \sigma = n^{-1}(t' \times \sigma)n$, with $n = ws$.

6.5. LEMMA. *Let* $({}^L G^\circ \rtimes \sigma)_{\mathrm{ss}}$ *be the set of semisimple elements in* ${}^L G^\circ \rtimes \sigma$. *Then the map*

$$\bar{\mu} : ({}^L T^\circ \rtimes \sigma)/\operatorname{Int}_k N \longrightarrow ({}^L G^\circ \times \sigma)_{\mathrm{ss}}/\operatorname{Int} {}^L G^\circ,$$

induced by inclusion is a bijection.

By results of F. Gantmacher [12, Theorem 14], $\bar{\mu}$ is surjective. Let now $s, t \in {}^L T^\circ$ and $g \in {}^L G^\circ$ be such that $g^{-1} \cdot (s \rtimes \sigma) \cdot g = t \rtimes \sigma$, i.e., such that $g^{-1} \cdot s \cdot {}^\sigma g = t$. Using the Bruhat decomposition of ${}^L G^\circ$ with respect to ${}^L B^\circ$, we can write uniquely $g = u \cdot n \cdot v$, with u, v in the unipotent radical of ${}^L B^\circ$ and n in the normalizer N of ${}^L T^\circ$. These groups are stable under σ, and normalized by ${}^L T^\circ$. We have then

$$s \cdot {}^\sigma u \cdot {}^\sigma n \cdot {}^\sigma v = u \cdot n \cdot v \cdot t, \qquad (s \cdot {}^\sigma n \cdot s^{-1}) \cdot s \cdot {}^\sigma n \cdot {}^\sigma v = u \cdot n \cdot t \cdot (t^{-1} \cdot v \cdot t);$$

hence ${}^\sigma n \cdot s^{-1} = n \cdot t$. Therefore the connected component of n in N is stable under

σ, i.e., n represents an element of $_kW$; hence $n \in {_kN}$, and $(t \rtimes \sigma)$ and $(s \rtimes \sigma)$ are conjugate under $_kN$.

REMARK. This proof was suggested to me by T. Springer.

6.6. If M is a complex affine variety, we let $C[M]$ denote its coordinate algebra. The algebra $C[Y]$ may be identified with the group algebra of $X^*(Y) = X_*(T_d)$. The quotient $Y/_kW$ is also an affine variety (in fact isomorphic to an affine space) and $C[Y/_kW] = C[Y]^{_kW}$.

Let $\text{Rep}(^LG) \subset C[^LG]$ be the subalgebra generated by the characters of finite dimensional holomorphic representations. Its elements are constant on conjugacy classes. In particular, they define by restriction functions on $(^LG^\circ \rtimes \sigma)_{\text{ss}}/\text{Int}\ ^LG^\circ$.

6.7. PROPOSITION. *The map*

$$\alpha = \bar\mu \circ \bar\nu^{-1} : Y/_kW \longrightarrow (^LG^\circ \rtimes \sigma)_{\text{ss}}/\text{Int}\ ^LG^\circ$$

is a bijection, which induces an isomorphism of $C[Y/_kW]$ onto the algebra A of restrictions of elements of $\text{Rep}(^LG)$.

REMARK. We shall use 6.7 only when k is a nonarchimedean local field. In that case 6.7 is proved in [**35**, pp. 18–24].

PROOF. That α is bijective follows from 6.4, 6.5. We prove the second assertion as in [**35**]. Let ρ be a finite dimensional holomorphic representation of ^LG and f_ρ the function on $^LT^\circ$ defined by $f_\rho(t) = \text{tr}\ \rho(t \rtimes \sigma)$. It can be written as a finite linear combination $f = \sum c_\lambda \lambda$ of characters $\lambda \in X^*(^LT^\circ)$. Since $\text{tr}\ \rho$ is a class function on ^LG, we have $f_\rho(s^{-1} \cdot t \cdot {}^\sigma s) = f_\rho(t)$ for all s, $t \in {^LT^\circ}$. By the linear independence of characters, it follows that if $c_\lambda \neq 0$, then λ is trivial on V (cf. 6.3(1)), hence is fixed under σ, i.e., may be identified to an element of $X^*(Y)$. Thus we may view f_ρ as an element of $C[Y]$. But invariance by conjugation and 6.4 imply that $f \in C[Y/_kW]$, whence a map $\beta : A \to C[Y/_kW]$, which is obviously induced by α. There remains to see that β is surjective. Note that $C[Y/_kW]$ is spanned, as a vector space, by the functions

$$(1) \qquad\qquad \varphi_\lambda = \sum_{w \in {_kW}} w \cdot \lambda,$$

where λ runs through a fundamental domain C of $_kW$ on $X_*(T_d)$. But it is standard that we may take for C the intersection of $X_*(T_d)$ with the Weyl chamber of W in $X_*(T)$ defined by B. Therefore every $\lambda \in C$ is a dominant weight for $^LG^\circ$ with respect to $^LT^\circ$. It is then the highest weight of an irreducible representation π_λ of $^LG^\circ$. Since it is fixed under σ, the representation $^\sigma\pi_\lambda : g \mapsto \pi_\lambda(^\sigma g)$ is equivalent to π_λ. From this it is elementary that π_λ extends to an irreducible representation $\tilde\pi_\lambda$ of ^LG of the same degree as π_λ. The highest weight space is one-dimensional, stable under σ. Let c be the eigenvalue of σ on it. Then the trace gives rise to a function equal to $c \cdot \varphi_\lambda$ modulo a linear combination of functions φ_μ, with $\mu < \lambda$, in the usual ordering. That im β contains φ_λ ($\lambda \in C$) is then proved by induction on the ordering.

7. The Satake isomorphism and the L-group. Local factors.

7.1. We keep the previous notation and conventions. We assume moreover k to be a nonarchimedean local field, k' to be unramified over k, and σ to be the image of a Frobenius element Fr in Γ_k.

Let Q be a special maximal compact subgroup of $G(k)$ [**61**]. We assume $Q \cap T$ is the greatest compact subgroup of $T(k)$ and Q contains representatives of $_kW$. Let

$H(G(k), Q)$ be the Hecke algebra of locally constant, Q-bi-invariant, and compactly supported complex valued functions on $G(k)$. The Satake isomorphism provides a canonical identification $H \xrightarrow{\sim} C[Y/_k W]$, hence also one of $Y/_k W$ with the characters of H [6].

By 6.7, we have now a canonical isomorphism of H with the algebra A of restrictions of characters of finite dimensional representations of $^L G$ to semisimple $^L G^\circ$-conjugacy classes in $(^L G^\circ \rtimes \sigma)$, hence also a canonical bijection between characters of $H(G(k), Q)$ and semisimple classes in $^L G^\circ \rtimes \sigma$. Furthermore, each such class can be represented by an element of the form (t, σ), with $t \in {}^L T^\circ$ fixed under σ (and is determined modulo the finite group $U \cap V$, in the notation of 6.3).

7.2. *Local factors.* Assume now that U is *hyperspecial* [61]. Let ψ be an additive character of k. Let (π, U_π) be an irreducible admissible representation of $G(k)$ of class 1 for Q and r a representation of $^L G$. Then the space of fixed vectors of Q in U_π is one-dimensional, acted upon by H via a character χ_π. To the latter is assigned by 7.1 a semisimple class S_χ in $^L G^\circ \rtimes \sigma$. We then put

$$(1) \qquad L(s, \pi, r) = \det(1 - r((g \rtimes \sigma)), q^{-s})^{-1}, \qquad \varepsilon(s, \pi, r, \psi) = 1,$$

where q is the order of the residue field, and (g, σ) any element of S_χ.

CHAPTER III. WEIL GROUPS AND REPRESENTATIONS. LOCAL FACTORS.

In this section, k is a local field, W_k (resp. W'_k) the absolute Weil group (resp. Weil-Deligne group) of k. If H is a reductive k-group, then $\Pi(H(k))$ is the set of infinitesimal equivalence classes of irreducible admissible representations of $H(k)$.

G denotes a connected reductive k-group.

The main local problem is to define a *partition* of $\Pi(G(k))$ into finite sets $\Pi_{\varphi, G}$ or Π_φ indexed by the set $\Phi(G)$ of admissible homomorphisms of W'_k into $^L G$, modulo inner automorphisms (see §8 for $\Phi(G)$), and satisfying a certain number of conditions. So far, this has been carried out for any G if $k = R, C$ [37], for tori over any k [34] and (essentially) for $G = \mathbf{GL}_2$ (cf. 12.2). §9 recalls the results for tori; §10 describes some of the conditions to be imposed on this parametrization; §11 summarizes the construction over R or C. Such a parametrization would allow one to assign canonically local L- and ε-factors to any $\pi \in \Pi(G(k))$ and any complex representation of $^L G$. Two elements π, π' in the same set Π_φ would always have the same local factors, and are hence called *L-indistinguishable*. In the case of $\mathbf{GL}_n$ however, local factors have been defined in an a priori quite different way, so that the parametrization problem becomes subordinated to one concerning L- and ε-factors. This is discussed in §12.

8. Definition of $\Phi(G)$.

8.1. *Jordan decomposition in W'_k.* If $k = R, C$, then $W'_k = W_k$ and, by definition, every element of W_k is semisimple.

Let k be nonarchimedean. Then $x \in W'_k$ is said to be unipotent if and only if it belongs to G_a; the element x is semisimple if either $\varepsilon(x) \neq 0$ or x is in the inertia group. Here $\varepsilon: W'_k \to Z$ is the canonical homomorphism $W'_k \to W_k \to k^* \to Z$. Every element $x \in W'_k$ admits a unique Jordan decomposition $x = x_s \cdot x_u$ with x_s semisimple, x_u unipotent and $x_s x_u = x_u x_s$ [60].

8.2. *The set $\Phi(G)$.* We consider homomorphisms $\alpha : W'_k \to {}^LG$ over Γ_k, i.e., such that the diagram

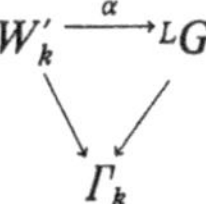

is commutative, and which satisfy moreover the following conditions:

(i) α is continuous, $\alpha(G_a)$ is unipotent, in ${}^LG^\circ$, and α maps semisimple elements into semisimple elements (in ${}^LG$: $x = (u, \gamma)$ is said to be semisimple if its image under any representation (2.6) is so).

(ii) If $\alpha(W'_k)$ is contained in a Levi subgroup of a parabolic subgroup P of ${}^LG$, then P is relevant (3.3).

Such α's are called admissible. We let $\Phi(G)$ be the set of their equivalence classes modulo inner automorphisms by elements of ${}^LG^\circ$.

If we write $\alpha(w) = (a(w), \nu(w))$ with $a(w) \in {}^LG^\circ$ then $w \mapsto a(w)$ is a 1-cocycle of W'_k (acting on ${}^LG^\circ$ via $W'_k \to \Gamma_k$) in ${}^LG^\circ$. It follows that

$$(1) \qquad \Phi(G) \subset H^1(W'_k; {}^LG^\circ).$$

Let H be a subgroup of W'_k. Then $\alpha: W'_k \to {}^LG$ is said to be trivial on H if $\nu(H)$ acts trivially on ${}^LG^\circ$ and $\alpha(H) = \{1\}$. Note that if $\nu(H)$ acts trivially on ${}^LG^\circ$, then $a|_H$ is a homomorphism.

8.3. Assume G' is an inner quasi-split form of G. Then

$$(1) \qquad \Phi(G) \subset \Phi(G').$$

In fact ${}^LG \cong {}^LG'$ and ${}^Lp({}^LG') \supset {}^Lp({}^LG)$; therefore $\alpha \in \Phi(G) \Rightarrow \alpha \in \Phi(G')$.

8.4. PROPOSITION. *Let k' be a finite separable extension of k; let G' be a connected reductive k-group and $G = R_{k'/k}G'$. Then there is a canonical bijection $\Phi(G) \xrightarrow{\sim} \Phi(G')$.*

We consider the situation of 5.2, 5.4 with $A = \Gamma_k$, $A' = \Gamma_{k'}$, $B = W'_k$, $B' = W'_{k'}$, $E = {}^LG'^\circ$. We have the injections (8.2):

$$\Phi(G) \subset H^1(W'_k; {}^LG^\circ), \qquad \Phi(G') \subset H^1(W'_{k'}; {}^LG'^\circ).$$

Moreover ${}^LG^\circ = I^{\Gamma_k}_{\Gamma_{k'}}({}^LG'^\circ)$ (see 5.1); whence, by Shapiro's lemma and 5.2:

$$H^1(W'_k; {}^LG^\circ) \xrightarrow{\sim} H^1(W'_{k'}; {}^LG'^\circ).$$

But it is clear that this isomorphism maps $\Phi(G)$ onto $\Phi(G')$.

8.5. Let $Z_L = C({}^LG^\circ)$. If $a: W_k \to Z_L$ and $b: W'_k \to {}^LG^\circ$ are 1-cocycles, then $ab: w \mapsto a(w)b(w)$ is again a 1-cocycle of W'_k in ${}^LG^\circ$. If a is continuous and b corresponds to $\varphi \in \Phi(G)$, then ab corresponds to an element of $\Phi(G)$. We get therefore maps

$$(1) \qquad H^1(W_k; Z_L) \times H^1(W'_k; {}^LG^\circ) \longrightarrow H^1(W'_k; {}^LG^\circ),$$

$$(2) \qquad H^1(W_k; Z_L) \times \Phi(G) \longrightarrow \Phi(G),$$

which define actions of the group $H^1(W_k; Z_L)$ on the sets $H^1(W'_k; {}^LG^\circ)$ and $\Phi(G)$.

8.6. PROPOSITION. *Let* $\varphi: W'_k \to {}^L G$ *be an admissible homomorphism. Then the Levi subgroups in* ${}^L G$ *which contain* $\varphi(W'_k)$ *minimally form one conjugacy class with respect to the centralizer of* $\varphi(W'_k)$ *in* ${}^L G^\circ$.

Since $\varphi(W'_k)$ projects onto a dense subgroup of Γ_k by definition, this follows from 3.6.

REMARK. Formally, this also applies to the archimedean case, but the proof in that case is simpler [**37**, pp. 78–79]. In fact, the argument there applies in all cases to admissible homomorphisms of the Weil (rather than Weil-Deligne) group because $\varphi(W_k)$ is always fully reducible. In this case, the Levi subgroups which contain $\varphi(W_k)$ minimally are those of the parabolic subgroups which contain $\varphi(W_k)$ minimally. Those parabolic subgroups form therefore one class of associated groups.

9. The correspondence for tori.

9.1. Let T be a complex torus. A continuous homomorphism $\varphi: T \to C^*$ is described by a pair of elements $\lambda, \mu \in X^*(T) \otimes C$ such that $\lambda - \mu \in X^*(T)$, by the rule $\varphi(t) = t^\lambda \bar{t}^\mu$.

Similarly, a continuous homomorphism $\varphi: C^* \to T$ is given by $\mu, \nu \in X_*(T) \otimes C$ such that $\mu - \nu \in X_*(T)$; we have $\varphi(z) = z^\mu \bar{z}^\nu$, meaning that, for any $\lambda \in X^*(T)$, $\lambda \circ \varphi: C^* \to C^*$ is given by

$$\lambda(\varphi(z)) = z^{\langle \lambda, \mu \rangle} \bar{z}^{\langle \lambda, \nu \rangle}.$$

This can also be interpreted in the following way: identify $X_*(T) \otimes C$ with the Lie algebra $\mathrm{Lie}(T(C))$. Then the exponential map yields an isomorphism $(X_*(T) \otimes C)/2\pi i X_*(T) = T(C)$. Then $\mu, \nu \in \mathrm{Lie}(T(C))$ are such that $\varphi(e^h) = e^{h \cdot \mu + \bar{h} \cdot \mu}$ $(h \in C)$.

9.2. Let $G = T$ be a k-torus, and $l = \dim T$.

Any $\varphi \in \Phi(G)$ is trivial on G_a; hence

$$(1) \qquad \Phi(G) = H^1_{\mathrm{ct}}(W_k; {}^L T^\circ) = H^1_{\mathrm{ct}}(W_k; X^*(T) \otimes C^*),$$

where H^1_{ct} refers to continuous cocycles.

On the other hand

$$(2) \qquad \Pi(G) = \mathrm{Hom}((X_*(T) \otimes k^*_s)^{\Gamma_k}, C^*).$$

We have canonically [**34**, Theorem 1]

$$(3) \qquad \Pi(G) = \Phi(G).$$

In fact, ${}^L T$ and W_k are replaced in [**34**] by a finite Galois form ${}^L T^\circ \rtimes \Gamma_{k'/k}$ and a relative Weil group $W_{k'/k}$, where k' is a finite Galois extension of k whose Galois group acts trivially on ${}^L T^\circ$; this is easily seen not to change $\Phi(G)$. The proof then consists in showing first that the transfer from $W_{k'/k}$ to k'^* yields an isomorphism

$$(4) \qquad H_1(W_{k'/k}; X_*(T)) \xrightarrow{\ \sim\ } H_1(k'^*; X_*(T))^{\Gamma_{k'/k}} = (k'^* \otimes X_*(T))^{\Gamma_{k'/k}},$$

and second that the pairing

$$(5) \qquad H^1_{\mathrm{ct}}(W_{k'/k}; {}^L T^\circ) \times H_1(W_{k'/k}; X_*(T)) \longrightarrow C^*,$$

associated to the evaluation map $(t, \lambda) \mapsto \lambda(t)$ $(t \in {}^L T^\circ; \lambda \in X_*(T))$ yields an isomorphism of the first group onto the group of characters of the second group, which is then (3) by definition.

For illustrations, we discuss some simple cases.

9.3. $k = C$. Then $W_k = C^*$ and $\Phi(G) = \mathrm{Hom}(C^*, {}^L T^\circ)$. The correspondence follows from 9.1 since both $\mathrm{Hom}(C^*, {}^L T^\circ)$ and $\mathrm{Hom}(T, C^*)$ are canonically identified with $\{(\lambda, \mu)|\lambda, \mu \in X^*(T) \otimes C, \lambda - \mu \in X^*(T)\}$.

9.4. $k = R$. We have

$$(1) \qquad W_R = C^* \rtimes \{\tau\} \quad \text{with } \tau^2 = -1,\ \tau \cdot z \cdot \tau^{-1} = \bar{z}\ (z \in C^*).$$

Put $C^* = S \times R^+$, with $S = \{z \in C^*, z \cdot \bar{z} = 1\}$. Then $\mathrm{Int}\ \tau$ is the identity on R^+, the inversion on S.

Write $\varphi(\tau) = (a, \sigma)$, where a is determined modulo σ-conjugacy, hence may be assumed to be fixed under σ (6.3(3)). We have then $\varphi(-1) = a^2$. Let μ, ν be the elements of $X^*(T) \otimes C$ such that

$$(2) \qquad \varphi(z) = z^\mu \cdot \bar{z}^\nu \qquad (z \in C^*),\ \mu - \nu \in X^*(T),$$

(see 9.1). We have $\varphi(\bar{z}) = \sigma(\varphi(z))\ (z \in C^*)$; hence $\nu = \sigma(\mu)$. Fix $h \in X^*(T) \otimes C$ such that $a = \exp 2\pi i h$. Then the character π associated to φ is given by

$$(3) \qquad \pi(e^x) = \exp(\langle h, x - \sigma \cdot \bar{x}\rangle) \cdot \exp(\langle \mu, x + \sigma \cdot x\rangle)/2 \quad (x \in X_*(T) \otimes C)$$

[37, p. 27]. Here $^-$ denotes the complex conjugation of $\mathrm{Lie}(T(C)) = X_*(T) \otimes C$ with respect to $X_*(T) \otimes R$; hence $x \mapsto \sigma \cdot \bar{x}$ is the complex conjugation with respect to $\mathrm{Lie}(T(R))$. It follows that $e^x \in T(R)$ if and only if $x - \sigma \cdot \bar{x} \in 2\pi i \cdot X_*(T)$.

EXAMPLES. (a) Let T be anisotropic over R. Then $\sigma = -1$ and we may assume $a = 1, h = 0$. We have $e^x \in T(R)$ if x is purely imaginary and then (3) yields $\pi = \mu$. The fact that $\varphi(-1) = 1$ shows that $\mu \in X^*(T)$, confirming that $\Pi(T(R)) = X^*(T)$.

(b) Let T be split over R. Then $\sigma = 1$, $\mu = \nu$, $\varphi(z) = (z \cdot \bar{z})^\mu$, $a^2 = 1$ and $h \in X^*(T)/2$. We have $e^x \in T(R)$ if and only if $x - \bar{x} \in 2\pi i \cdot X^*(T)$. It is then easily checked that π is given by μ on the connected identity component of $T(R)$, while its restriction to the torsion subgroup of $T(R)$ is the character naturally defined by h.

9.5. *The unramified case.* Let k be nonarchimedean, and assume T to split over an unramified extension k' of k. A character χ of $T(k)$ is said to be unramified if it is trivial on the greatest compact subgroup ${}^0 T(k)$ of $T(k)$. On the other hand, $\varphi \in \Phi\ (T)$ is unramified if it is trivial (see 6.2) on the inertia group. The bijection $\Phi(T) \xrightarrow{\sim} \Pi(T)$ induces a bijection between the sets $\Phi_{\mathrm{unr}}(T)$ and $\Pi_{\mathrm{unr}}(T)$ of unramified elements [34]. In view of its importance, we describe it in more detail.

Given $t \in T(k')$, let $v(t) \in \mathrm{Hom}(X^*(T), Z)$ be defined by $v(t)(m) = \mathrm{ord}\ m(t)$ $(m \in X^*(T))$. It is well known, and easily deduced from Hilbert's Theorem 90, that $H^1(\Gamma_{k'/k}; o_{k'}^*) = 0$, where $o_{k'}^*$ is the group of units in the ring $o_{k'}$ of integers of k'. Since T splits over k', it follows that $H^1(\Gamma_{k'/k}; {}^0 T(k')) = 0$. By Galois cohomology, this implies that $(T(k')/{}^0 T(k'))^{\Gamma_k} = T(k)/{}^0 T(k)$, therefore $t \mapsto v(t)$ yields a bijection

$$(1) \qquad T(k)/{}^0 T(k) \xrightarrow{\ \sim\ } \mathrm{Hom}(X^*(T), Z)^{\Gamma_k} = X_*(T)^{\Gamma_k} = X_*(T_d),$$

where T_d is the greatest k-split torus of T (this can also be expressed by saying that the inclusion $T_d \subset T$ induces an isomorphism $T_d(k)/{}^0T_d(k) \xrightarrow{\sim} T(k)/{}^0T(k)$, cf. [6]). We have then

$$(2) \qquad \Pi_{\mathrm{unr}}(T(k)) = \mathrm{Hom}(X_*(T)^{\Gamma_k}, C^*) = Y.$$

The group Γ_k operates on ${}^LT^\circ$ via the cyclic group $\Gamma_{k'/k}$ which is generated by the image σ of a Frobenius element Fr. An unramified φ is completely determined by $\varphi(\mathrm{Fr})$, which can be written $\varphi(\mathrm{Fr}) = (t, \mathrm{Fr})$, where $t \in {}^LT^\circ$ is determined up to conjugacy by ${}^LT^\circ$. Thus $\Phi_{\mathrm{unr}}(T) = ({}^LT^\circ \rtimes \sigma)/\mathrm{Int}\,{}^LT^\circ$; and elementary special case of 6.4 provides a canonical isomorphsim of the latter set onto Y, whence the desired isomorphism.

10. Desiderata. In order to formulate them, we need two preliminary constructions.

10.1. *The character χ_φ of $C(G)$ associated to $\varphi \in \Phi(G)$* (*cf.* [37, pp. 20–34]). We want to associate canonically to $\varphi \in \Phi(G)$ a character of the center $C(G)$ of G. Let $\mathfrak{G}_{\mathrm{rad}}$ be the greatest central torus of $\mathfrak{G}$. Then $G_{\mathrm{rad}} \to G$ yields a surjective homomorphism ${}^LG \to {}^LG_{\mathrm{rad}}$, whence a map $\Phi(G) \to \Phi(G_{\mathrm{rad}})$. In view of 7.2, this allows us to associate to $\varphi \in \Phi(G)$ a character χ_φ of G_{rad}. Thus, if $C(G(k)) \subset G_{\mathrm{rad}}(k)$, our problem is solved.

In the general case, G is enlarged to a bigger connected reductive G_1 generated by G and a central torus, whose center is a torus. One shows that $\Phi(G_1) \to \Phi(G)$ is surjective. Using the previous step, we get a character of $C(G_1)$, hence one of $C(G)$ by restriction. It is shown to be independent of the choice of G_1 (loc. cit.), and is χ_φ by definition.

The map $\varphi \mapsto \chi_\varphi$ is compatible with restriction of scalars [37, 2.11].

10.2. *The character π_α associated to $\alpha \in H^1(W'_k; Z_L)$* [37, pp. 34–36]. We recall that Z_L denotes the center of ${}^LG^\circ$ (8.5). We can always find a k-torus D such that $H^1(\Gamma_k; D) = 0$, and a k-group $\tilde{G}$ isogeneous to $G \times D$ such that there is an exact sequence

$$(1) \qquad 1 \longrightarrow D \longrightarrow \tilde{G} \longrightarrow G \longrightarrow 1.$$

Since $H^1(\Gamma_k; D) = 0$, the map $\mu\colon \tilde{G}(k) \to G(k)$ is surjective. Let G_{sc} be the universal covering of the derived group $\mathscr{D}G$ of G. We have a commutative diagram

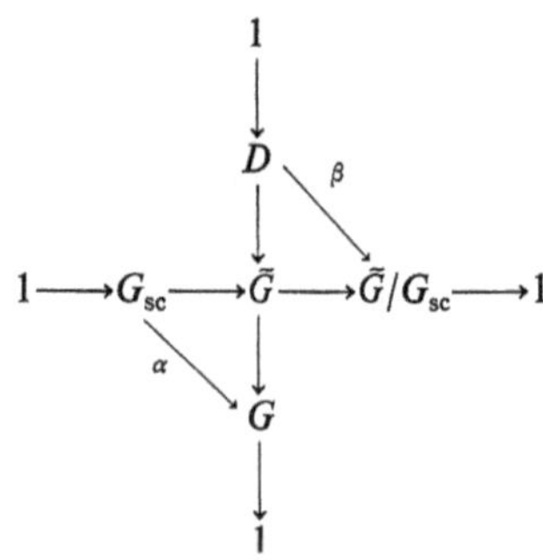

44   A. BOREL

Going over to L-groups, we get

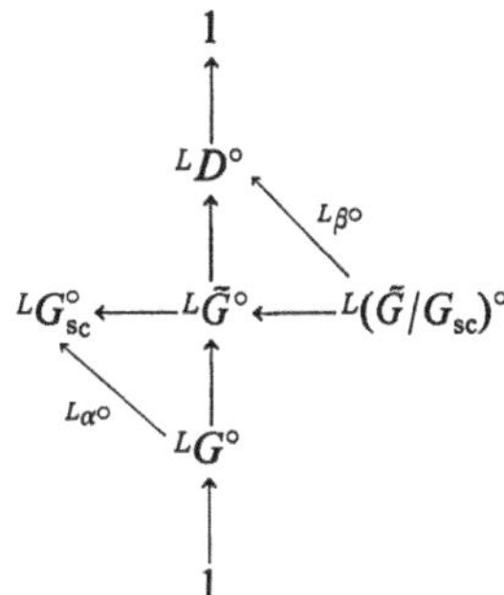

Since $^LG^{\circ}_{\mathrm{sc}}$ is of adjoint type (2.2), we see that $Z_L = \ker{}^L\alpha^{\circ}$. Moreover, it is easily seen that

$$(2) \qquad\qquad\qquad Z_L \cong \ker{}^L\beta^{\circ}.$$

This yields a map

$$(3) \qquad H^1(W'_k; Z_L) \longrightarrow \ker\{H^1(W'_k; {}^L(\tilde{G}/G_{\mathrm{sc}})^{\circ}) \longrightarrow H^1(W'_k; {}^LD^{\circ})\}.$$

This allows us to associate to $\alpha \in H^1(W'_k; Z_L)$ a character σ_α of $(\tilde{G}/G_{\mathrm{sc}})(k)$ which is trivial on $D(k)$, hence a character π_α of $G(k) = \tilde{G}(k)/D(k)$. It can be shown to be independent of the choice of D. The map $\alpha \mapsto \pi_\alpha$ is compatible with restriction of scalars [**37**, 2.12] and satisfies:

$$(4) \qquad\qquad \chi_{\alpha\varphi} = \pi_\alpha \cdot \chi_\varphi \qquad (\alpha \in H^1(W'_k; Z_L),\ \varphi \in \Phi(G)).$$

10.3. *Conditions on the sets Π_φ.* (1) If $\pi \in \Pi_\varphi$, then $\pi(z) = \chi_\varphi(z) \cdot \mathrm{Id}$ $(z \in C(G))$.

(2) If $\varphi' = \alpha \cdot \varphi$ $(\varphi, \varphi' \in \Phi(G), \alpha \in H^1(W'_k; Z_L))$ (see 6.5), then $\Pi_{\varphi'} = \{\pi_\alpha \otimes \pi | \pi \in \Pi_\varphi\}$.

(3) The following conditions on a set Π_φ are equivalent:

    (i) One element of Π_φ is square-integrable modulo $C(G)$.

    (ii) All elements of Π_φ are square-integrable modulo $C(G)$.

    (iii) $\varphi(W'_k)$ is not contained in any proper Levi subgroup in ^LG.

(4) Assume $\varphi(G_a) = \{1\}$. The following conditions on a set Π_φ are equivalent:

    (i) One element of Π_φ is tempered.

    (ii) All elements of Π_φ are tempered.

    (iii) $\varphi(W_k)$ is bounded.

(5) Let H be a connected reductive k-group and $\eta: H \to G$ a k-morphism with commutative kernel and cokernel. Let $\varphi \in \Phi(G)$ and $\varphi' = {}^L\eta \circ \varphi$. Then any $\pi \in \Pi_\varphi$, viewed as an $H(k)$-module, is the direct sum of finitely many irreducible admissible representations belonging to $\Pi_{\varphi'}$.

10.4. *The unramified case.* We say that $\varphi \in \Phi(G)$ is unramified if it is trivial, in the sense of 6.2, on G_a and on the inertia group I. If so, Im φ may be assumed to be in ^LT. Therefore, if $\Phi(G)$ contains an unramified element, then G is quasi-split (see 8.2 (ii)).

Assume now G to be quasi-split, to split over an unramified Galois extension

k' of k, and let $\varphi \in \Phi(G)$ be unramified. There exists $t \in ({}^{L}T^{\circ})^{\Gamma_{k}}$ such that

$$(1) \qquad\qquad \varphi(\text{Fr}) = (t, \text{Fr}),$$

(9.5) and we have

$$(2) \qquad\qquad \varphi(w) = (t, \text{Fr})^{\varepsilon(w)} \qquad (w \in W'_{k}),$$

where $\varepsilon: W'_{k} \to \mathbf{Z}$ is the canonical homomorphism. The element t defines an unramified character χ of a maximal k-torus T of a Borel k-subgroup B of G (9.5). It is then required that Π_{φ} consists of the constituents of the unramified normalized principal series $\text{PS}(\chi)$ which have a nonzero vector fixed under some hyperspecial maximal compact subgroup. Conversely let (π, V) be an irreducible admissible representation with a nonzero vector fixed under some hyperspecial maximal compact subgroup. There exists then an unramified character χ of T such that (π, V) is a constituent of $\text{PS}(\chi)$ (and χ is determined modulo the relative Weyl group). We have then $(\pi, V) \in \Pi_{\varphi}$, for the unramified φ which maps Fr to (t, Fr), where t represents χ (9.5). Note that if U is a special maximal compact subgroup of $G(k)$, then $G(k) = B(k) \cdot U$; hence the fixed-point set of U in $\text{PS}(\chi)$ is at most one-dimensional. It follows that $\text{PS}(\chi)$ has at most one irreducible constituent with nonzero fixed vectors under U.

This assignment is consistent with 7.2. Namely, if $\pi \in \Pi_{\varphi}$, then the semisimple class S_{χ} in ${}^{L}G^{\circ} \rtimes \sigma$ corresponding to the character of the Hecke algebra defined by π is indeed represented by $t \rtimes \sigma$. This follows from [6].

REMARK. Originally, it was thought that Π_{φ} should consist of those constituents of $\text{PS}(\chi)$ which had a nonzero fixed vector under some special maximal compact subgroup. However it was pointed out during the Institute by I. Macdonald that such representations may belong to the discrete series. If so, this condition would contradict 10.3(3). Upon a suggestion of J. Tits, this has led to the restriction to hyperspecial maximal compact subgroups made above. Those cannot belong to the discrete series, so that 10.3(3) and 10.4 are consistent.

10.5. EXAMPLE. Assume that $k = \mathbf{R}$ and that G is semisimple, possesses a Cartan subgroup T which is anisotropic over $\mathbf{R}$, and is an inner form of a split group. Then ${}^{L}G$ is the direct product of ${}^{L}G^{\circ}$ and Γ_{k}, the Weyl group W contains $-\,\text{Id}$ and $G(\mathbf{R})$ has a discrete series. We want to describe the parametrization of the latter in terms of $\Phi(G)$. As the notation implies, we shall view ${}^{L}T$ as the L-group of T. Let $\varphi \in \Phi(G)$. It is given by a continuous homomorphism $\varphi': W_{\mathbf{R}} \to {}^{L}G^{\circ}$. We may assume that $\text{Im } \varphi'$ is contained in the normalizer of ${}^{L}T^{\circ}$. Let $n = \varphi(\tau)$ and let $w \in W$ be the element of W represented by n. Then $w^{2} = 1$. Let $\mu, \nu \in X^{*}(T) \otimes \mathbf{C}$ be such that

$$(1) \qquad\qquad \varphi(z) = z^{\mu} \cdot \bar{z}^{\nu} \qquad (z \in \mathbf{C}^{*}), \ \mu - \nu \in X^{*}(T)$$

(see 9.1). We have

$$(2) \qquad\qquad \varphi(\bar{z}) = n \cdot \varphi(z) \cdot n^{-1} = z^{w \cdot \mu} \cdot \bar{z}^{w \cdot \nu};$$

hence $\mu = w \cdot \nu$, $\nu = w \cdot \mu$. Assume now that $\text{Im } \varphi$ is not contained in any proper Levi subgroup in ${}^{L}G$ or, equivalently, that $\text{Im } \varphi'$ is not contained in any proper Levi subgroup in ${}^{L}G^{\circ}$. Then $w = -\,\text{Id}$ and μ is regular: in fact, the proper Levi subgroups in ${}^{L}G^{\circ}$ are the centralizers of nontrivial tori. This implies first that w does not fix pointwise any nontrivial torus in $^LT°$, hence $w = -\,\mathrm{Id}$; if now μ were singular, then the centralizer of $\mu(\boldsymbol{C}^*)$ would contain a semisimple subgroup $H \neq \{1\}$ stable under $\mathrm{Int}\ n$, the latter would leave pointwise fixed a torus $S \neq \{1\}$ of H, and $\mathrm{Im}\ \varphi'$ would be contained in the centralizer $Z(S)$ of S, a contradiction. Since $\nu = w \cdot \mu = -\,\mu$, we have

$$(3) \qquad\qquad \lambda(\varphi(-\,1)) = (-\,1)^{\langle 2\mu,\,\lambda\rangle}, \quad \text{for all } \lambda \in X_*(T).$$

Let δ be half the sum of the roots α of G with respect to T such that $\langle \mu, \check\alpha \rangle > 0$. Then, Lemma 3.2 of [37] implies in particular

$$(4) \qquad\qquad \lambda(\varphi'(-\,1)) = (-\,1)^{\langle 2\delta,\,\lambda\rangle}, \quad \text{for all } \lambda \in X_*(T).$$

It follows that

$$(5) \qquad\qquad\qquad \mu \in \delta + X^*(T).$$

Therefore μ is among the elements of $X^*(T) \otimes \boldsymbol{Q}$ which parametrize the discrete series in Harish-Chandra's theorem. We then let $\mathit{\Pi}_\varphi$ be the set of discrete series representations of $G(\boldsymbol{R})$ with infinitesimal character χ_μ. If $G(\boldsymbol{R})$ is compact, then $\mathit{\Pi}_\varphi$ consists of the irreducible finite dimensional representation with dominant weight $\mu - \delta$. In that case, no proper parabolic subgroup of ^LG is relevant; hence $\Phi(G)$ consists of the φ considered here.

10.6. *Let* $G = \mathbf{GL}_n$, k *nonarchimedean.* Let ψ be an admissible representation of W'_k. If it is irreducible, then $\psi(G_a) = 1$. If it is indecomposable, then it is a tensor product $\rho \otimes \mathrm{sp}(m)$, where m divides n, ρ is irreducible of degree n/m, and $\mathrm{sp}(m)$ is m-dimensional, trivial on 1, maps a generator of the Lie algebra of G_a onto the nilpotent matrix with ones above the diagonal, zero elsewhere, and $w \in W'_k$ onto the diagonal matrix with entries $a(w)^i$ $(0 \le i < n)$ [9, 3.1.3]. If χ is a character of W_k (hence of k^*), and $\varphi = \chi \otimes \mathrm{sp}(n)$, then $\mathit{\Pi}_\varphi$ consists of the special representation with central character determined by χ. In fact, the Weil-Deligne group came up for the first time precisely to fit the special representations of $\mathbf{GL}_2$ into the general scheme (see [9]).

11. Outline of the construction over R, C. We sketch here the various steps which yield the sets $\mathit{\Pi}_\varphi$ when $k = \boldsymbol{R}$. For the proofs see [37].

We note first that we may always assume $\varphi(W_k) \subset N(^LT°)$, and we can write

$$(9.1)$$

$$\varphi(z) = z^\mu \bar z^\nu \qquad (z \in \boldsymbol{C}^*;\ \mu, \nu \in X^*(T) \otimes \boldsymbol{C},\ \mu - \nu \in X^*(T)).$$

11.1. LEMMA. *Let* $\varphi \in \Phi(G)$. *Assume* $\varphi(W_{\boldsymbol R})$ *is not contained in any proper Levi subgroup in* ^LG. *Then*

(i) G *has a Cartan k-subgroup C such that $(\mathscr{D}G \cap S)(\boldsymbol{R})$ is compact* [28, 3.1].

(ii) μ *is regular;* $\varphi(\boldsymbol{C}^*)$ *contains regular elements* [37, 3.3].

The group $^LC°$ may be viewed as a maximal torus of $^LG°$; hence there is an isomorphism $^LC \to^\sim {}^LT$ defined modulo an element of W. Therefore φ defines an orbit of W in $\Phi(C)$, hence, by 9.2, an orbit X_φ of W in $X(C(\boldsymbol{R}))$. [Note that W, which is defined in $G(\boldsymbol{C})$, operates on $C(\boldsymbol{R})$, since $C(\boldsymbol{R}) \cap \mathscr{D}G$ is compact, hence on $X(C(\boldsymbol{R}))$.]

11.2. Let $G_0 = C(\boldsymbol{R})((\mathscr{D}G)(\boldsymbol{R}))°$. Let A_0 be the set of representations of G_0 which are square-integrable modulo the center, and have infinitesimal character χ_λ ($\lambda \in X_\varphi$). The induced representations $\pi = I_{C_0(\mathbf{R})}^{G(\mathbf{R})}(\pi_0)$ ($\pi_0 \in A_0$) are irreducible [37, p. 50]. By definition, $I\!I_\varphi$ is the set of equivalence classes of these representations [37, p. 54].

11.3. Let $\varphi \in \Phi(G)$. Let $^L M$ be a minimal relevant Levi subgroup containing Im φ. It is essentially unique (8.6). We assume $^L M \neq {}^L G$; we may view φ as an element of $\Phi(M)$. By 11.2, there is associated to it a finite set of $I\!I_{\varphi,M}$ of discrete series representations of M.

We may assume $^L M$ to be a Levi subgroup of a relevant parabolic subgroup $^L P$ corresponding to $P \in \mathscr{P}(G/k)$. Then $U = X^*(T) \otimes \mathbf{R} = X_*(^L T^\circ) \otimes \mathbf{R}$. Let V be the subspace of elements of U which are orthogonal to roots of $^L M$, and fixed under Γ_k. It may be identified with the dual $\mathfrak{a}_p^*$ of the Lie algebra of a split component A of P.

Let ξ be the character of $C(M)$ defined by the elements of $I\!I_{\varphi,M}$. We may assume that $|\xi| \in \mathrm{Cl}(\mathfrak{a}_p^{*+})$. Let P_1 be the smallest parabolic k-subgroup containing P such that $|\xi|$, when restricted to $\mathfrak{a}_{P_1}$, is an element of the Weyl chamber $\mathfrak{a}_{P_1}^{*+}$. Let $M_1 = z(\mathfrak{a}_{P_1})$ and $P' = P \cap M_1$. Then P' is a parabolic subgroup of M_1. Moreover the restriction of $|\xi|$ to the split component $M_1 \cap A_P$ of P' is one; therefore, for each $\rho \in I\!I_{\varphi,M}$, the induced representation $\mathrm{Ind}_{P'}^{M_1}(\rho)$ is tempered. Let $I\!I'_\varphi$ be the set of all constituents of such representations. Then by definition, $I\!I_\varphi$ is the set of Langlands quotients $J(P_1, \sigma)$ with $\sigma \in I\!I'_\varphi$ (cf. [37, p. 82]).

11.4. *Complex groups.* Assume now $k = \mathbf{C}$. Then $W_k = \mathbf{C}^*$, and $\Phi(G)$ may be identified to the set of homomorphisms of $\mathbf{C}^*$ into $^L T^\circ$, modulo the Weyl group W, i.e., to

(1) $$\{(\lambda, \mu), \text{ where } \lambda, \mu \in X^*(T) \otimes \mathbf{C}, \lambda - \mu \in X^*(T)\}$$

modulo the (diagonal) action of W. In this case Im φ is in the Levi subgroup $^L T$ of $^L B$, which is the $^L P$ of 11.3. The set $I\!I_{\varphi,M}$ consists of one character of T (cf. 9.1). Choose P_1, M, as in 11.3. Since the unitary principal series of a complex group are irreducible (N. Wallach), the set $I\!I'_\varphi$ consists of one element. Hence so does $I\!I_\varphi$. Thus each $I\!I_\varphi$ is a singleton. The classification thus obtained is equivalent to that of Zelovenko.

11.5. Let $G = \mathbf{GL}_n$, $k = \mathbf{R}$. In this case, it is also true that the tempered representations induced from discrete series are irreducible [22]; therefore each set $I\!I'_\varphi$ (cf. 9.3) consists of only one element, hence so does $I\!I_\varphi$ and we get a bijection between $\Phi(G)$ and $I\!I G(\mathbf{R})$.

Let $n = 2$. If φ is reducible, then Im φ is commutative; hence φ factors through $(W_{\mathbf{R}})^{\mathrm{ab}} = \mathbf{R}^*$ and is described by two characters μ, ν of $\mathbf{R}^*$. Then $I\!I_\varphi$ consists of a principal series representation $\pi(\mu, \nu)$ (including finite dimensional representations, as usual). In particular there are three φ's with kernel $\mathbf{C}^*$, to which correspond respectively $\pi(1, 1)$, $\pi(\mathrm{sgn}, \mathrm{sgn})$ and $\pi(1, \mathrm{sgn})$, where sgn is the sign character. If φ is irreducible, then $\varphi(\tau)$ may be assumed to be equal to (s_0, τ), where s_0 is a fixed element of the normalizer of $^L T^\circ$ inducing the inversion on it. $\varphi(\mathbf{R}^+)$ belongs to the center of $^L G\circ$, and $\varphi(S)$ is sum of two characters, described by two integers. Then $I\!I_\varphi$ consists of a discrete series representation, twisted by a one-dimensional representation.

11.6. As is clear from these two examples, the main point to get explicit knowledge of the sets Π_φ is the decomposition of representations induced from tempered representations of parabolic subgroups. This last problem has been solved by A. Knapp and G. Zuckerman [**29**], [**30**].

11.7. *Remark on the nonarchimedean case.* Langlands' classification [**37**] is also valid over p-adic fields [**57**]. In view of 8.6, it is then clear that the last step (11.3) of the previous construction can also be carried out in the nonarchimedean case. Thus, besides the decomposition of tempered representations, the main unsolved problem in the p-adic case is the construction and parametrization of the discrete series.

12. Local factors.

12.1. Let $\pi \in \Pi(G(k))$ and r be a representation of ^LG (2.6). Assume that $\pi \in \Pi_\varphi$ for some $\varphi \in \Phi(G)$. For a nontrivial additive character ψ of k, we let

$$(1) \qquad L(s, \pi, r) = L(s, r \circ \varphi), \qquad \varepsilon(s, \pi, r) = \varepsilon(s, \pi, r, \psi) = \varepsilon(s, r \circ \varphi, \psi),$$

where on the right-hand sides we have the L- and ε-factors assigned to the representation $r \circ \varphi$ of W'_k [**60**]. In the unramified situation of 10.4, this coincides with the definition given in 7.2.

In view of what has been recalled so far, these local factors are defined if k is archimedean, or if k is nonarchimedean in the unramified case, or if G is a torus.

12.2. Let now $G = \mathbf{GL}_n$. In this case there are associated to $\pi \in \Pi(G(k))$ local factors $L(s, \pi)$ and $\varepsilon(s, \pi, \psi)$ defined by a generalization of Tate's method, in [**25**] for $n = 2$, in [**19**] for any n, which play a considerable role in the parametrization problem and in the local lifting. A natural question is then whether these factors can be viewed as special cases of 12.1, where $r = r_n$ is the standard representation of $\mathbf{GL}_n$, i.e., whether we have equalities

$$(1) \qquad L(s, \pi) = L(s, \pi, r_n), \qquad \varepsilon(s, \pi, \psi) = \varepsilon(s, \pi, r_n, \psi),$$

with the right-hand side defined by the rule of 12.1.

(a) Let $n = 2$. It has been shown in [**25**] that the equivalence class of π is characterized by the functions $L(s, \pi \otimes \chi)$, $\varepsilon(s, \pi \otimes \chi, \psi)$, where χ varies through the characters of k^*. In this case, the parametrization problem and the proof of (1) are part of the following problem:

(∗) Given $\sigma \in \Phi(G)$, find $\pi = \pi(\sigma)$ such that

$$(2) \qquad L(s, \sigma \otimes \chi) = L(s, \pi \otimes \chi), \qquad \varepsilon(s, \sigma \otimes \chi, \psi) = \varepsilon(s, \pi \otimes \chi, \psi)$$

for all χ's, and prove that $\sigma \mapsto \pi(\sigma)$ establishes a bijection between $\Phi(G)$ and $\Pi(G(k))$.

This problem was stated and partially solved in [**25**]. The most recent and most complete results in preprint form are in [**62**]; they still leave out some cases of even residual characteristic, although some arguments sketched by Deligne might take care of them (see [**63**] for a survey).

As stated, the problem is local, but, except at infinity, progress was achieved first mostly by global methods: one uses a global field E whose completion at some place v is k, a reductive E-group H isomorphic to G over k, an element $\rho \in \Phi(H/k)$ whose restriction to $^L(H/k_v) = {}^LG$ is σ, chosen so that there exists an automorphic representation $\pi(\rho)$ with the L-series $L(s, \rho)$ (see §14 for the latter). This construction relies, among other things, on Artin's conjecture in some cases, and [**38**]. In fact, it was already shown in [**25**] that (∗) for odd residual characteristics follows from Artin's conjecture, leading to a proof in the equal characteristic case. At present, there are in principle purely local proofs in the odd residue characteristic case [**63**]. Note also that the injectivity assertion is a statement on two-dimensional admissible representations of W'_k, namely, whether such a representation σ is determined, up to equivalence, by the factors $L(s, \sigma \circ \chi)$ and $\varepsilon(s, \sigma \circ \chi, \psi)$. But, so far, the known proofs all use admissible representations of reductive groups [**63**].

(b) For arbitrary n, (1) has been proved in the unramified case, for special representations, and by H. Jacquet for $k = \mathbf{R}, \mathbf{C}$ [**24**].

(c) Local L- and ε-factors are also introduced for $G = \mathbf{GL}_2 \times \mathbf{GL}_2$ in [**21**], at any rate for products $\pi \times \pi'$ of infinite dimensional irreducible representations. Partial extensions of this to $\mathbf{GL}_m \times \mathbf{GL}_n$ for other values of m, n are known to experts.

(d) For $n = 3$, $\pi \in \Pi(G(k))$ is again characterized uniquely by the factors $L(s, \pi \otimes \chi)$ and $\varepsilon(s, \pi \otimes \chi, \psi)$ [**27**], [**46**]. For $n \geq 4$ on, this is false [**46**]. However, it may be there are still such characterizations if χ is allowed to run through suitable elements of $\Pi(\mathbf{GL}_{n-1}(k))$ or maybe just $\Pi(\mathbf{GL}_{n-2}(k))$.

12.3. Local factors have also been defined directly for some other classical groups, in particular for $\mathbf{GSp}_4$ by F. Rodier [**48**], extending earlier work of M. E. Novodvorsky and I. Piatetskii-Shapiro, for split orthogonal groups, in an odd number $2n + 1$ of variables by M. E. Novodvorsky [**41**]. In the latter case $^L G^\circ = \mathbf{Sp}_{2n}$, and in the unramified case, the local factors coincide (up to a translation in s) with those associated by 7.2 to the standard $2n$-dimensional representation of the L-group. See also [**42**].

CHAPTER IV. THE L-FUNCTION OF AN AUTOMORPHIC REPRESENTATION.

From now on, k is a global field, $\mathfrak{o} = \mathfrak{o}_k$ the ring of integers of k, A_k or A the ring of adeles of k, V (resp. V_∞, resp. V_f) the set of places (resp. infinite places, resp. finite places) of V. For $v \in V$, k_v, $\mathfrak{o}_v$ and Nv have the usual meaning. Unless otherwise stated, G is a connected reductive k-group.

13. The L-function of an irreducible admissible representation of G_A.

13.1. Let π be an irreducible admissible representation of G_A and r a representation of $^L G$. There exists a finite Galois extension k' of k over which G splits and such that r factors through $^L G^\circ \rtimes \Gamma_{k'/k}$. We want to associate to π and r infinite Euler products $L(s, \pi, r)$ and $\varepsilon(s, \pi, r)$, whose factors are defined (at least) for almost all places of k.

Let $v \in V$. By restriction, r defines a representation r_v of $L(G/k_v) = {}^L G^\circ \rtimes \Gamma_k$. On the other hand, $\pi = \otimes_v \pi_v$, with $\pi_v \in \Pi(G(k_v))$ [**11**]. Assume the parametrization problem of Chapter III solved. Then there is a unique $\varphi_v \in \Phi(G/k_v)$ such that $\pi_v \in \Pi_{\varphi_v}$. Then we let

$$(1) \qquad\qquad L(s, \pi, r) = \Pi_v L(s, \pi_v, r_v),$$

$$(2) \qquad\qquad \varepsilon(s, \pi, r) = \Pi_v \varepsilon(s, \pi_v, r_v, \psi_v),$$

where ψ_v is an additive character of k_v associated to a given nontrivial additive character of k, and the factors on the right are given by 12.1(1).

The local problem is solved for archimedean v's, and for almost all finite v's (see below) so that the factors on the right are defined except for at most finitely many $v \in V_f$. For questions of convergence or meromorphic analytic continuation this does not matter, and we shall also denote such partial products by $L(s, \pi, r)$.

By 10.4, φ_v is well defined if the following conditions are fulfilled: G is quasi-split over k_v, $G(o_v)$ is a very special maximal compact subgroup of $G(k_v)$, k' is unramified over k, and π_v is of class one with respect to $G(o_v)$. All but finitely many $v \in V_f$ satisfy those conditions [61].

13.2. THEOREM [35]. *Let π be an irreducible admissible unitarizable representation of G_A and r be a representation of $^L G$ (2.6). Then $L(s, \pi, r)$ converges absolutely for* Re s *sufficiently large.*

We may and do view r as a complex analytic representation of $^L G° \rtimes \Gamma_{k'/k}$, where k' is a finite Galois extension of k over which G splits (2.7). We let V_1 be the set of $v \in V_f$ satisfying the conditions listed at the end of 13.1. We have to show that

$$(1) \qquad\qquad L' = \prod_{v \in V_1} L(s, \pi_v, r_v),$$

converges in some right half-plane.

Let Fr_v be the Frobenius element of $\Gamma_{k'_{v'}/k_v}$, where $v' \in V_{k'}$ lies over $v \in V_1$. We have

$$(2) \qquad\qquad \varphi_v(\mathrm{Fr}_v) = (t_v, \mathrm{Fr}_v), \quad \text{with } t_v \in {}^L T°$$

and

$$(3) \qquad\qquad L(s, \pi_v, r_v) = (\det(1 - r((t_v, \mathrm{Fr}_v))N_v^{-s}))^{-1}.$$

To prove the theorem, it suffices therefore to show the existence of a constant $a > 0$ such that

$$(4) \qquad |\mu| \leqq (Nv)^a \quad \text{for every } v \in V_1 \text{ and eigenvalue } \mu \text{ of } r((t_v, \mathrm{Fr}_v)).$$

Let $n = [k' : k]$. Since we may assume t_v fixed under Γ_{k_v} (6.3), we have $t_v^n = (t_v, \mathrm{Fr}_v)^n$; hence it is equivalent to show (4) for all eigenvalues μ of $r(t_v)$. These are of the form t_v^λ, where λ runs through the set P_r of weights of r, restricted to $^L G°$. Thus we have to show the existence of $a > 0$ such that

$$(5) \qquad\qquad |t_v|^{\mathrm{Re}\,\lambda} \leqq (Nv)^a \quad \text{for all } v \in V_1 \text{ and } \lambda \in P_r.$$

Let G' be a quasi-split inner k-form of G. Then $^L G = {}^L G'$, and G is isomorphic to G' over k_v for all $v \in V_1$. We may therefore replace G by G'; changing the notation slightly, we may (and do) assume G to be quasi-split over k. We then fix a Borel k-subgroup B of G and view $^L T$ as the L-group of a maximal k-torus T of G.

For a cyclic subgroup D of $\Gamma_{k'/k}$, let V_D be the set of $v \in V_1$ for which Γ_{k_v} is equal to the inverse image of D in Γ_k. The group $U = X_*(T)^D$ is then the group of one-parameter subgroups of a subtorus S of T such that S/k_v is a maximal k_v-split torus of G/k_v for all $v \in V_D$. The group

$$(6) \qquad Y = \mathrm{Hom}(U, C^*) = \mathrm{Hom}(X_*(T)^D, C^*) \qquad (v \in V_D),$$

is independent of v, and is the Y of §6 for G/k_v. The root datum $\psi(G/k_v)$, which is determined by the action of D, is also independent of $v \in V_D$.

Given $y \in Y$, let y_0 be a "logarithm" of y, i.e., an element of $\text{Hom}(X_*(T)^D, C)$ such that

$$(7) \qquad y(u) = Nv^{y_0(u)} = Nv^{\langle y_0, u \rangle}, \quad \text{for } u \in X_*(T)^D.$$

This element is determined modulo a lattice, but its real part $\text{Re } y_0 \in \text{Hom}(U, R)$, defined by

$$(8) \qquad y(u) = Nv^{\langle \text{Re } y_0, u \rangle}$$

is well defined. If y has values in R_+^*, then we choose y_0 to be equal to its real part. The space $\mathfrak{a}^*$ is the dual of $\mathfrak{a} = U \otimes R$ (the so-called real Lie algebra of S/k_v), and is acted upon canonically by ${}_kW$ as a reflection group. We let $\mathfrak{a}^{*+}$ be the positive Weyl chamber defined by B.

Let ρ_v be the unramified character of $T(k_v)$, given by $t \mapsto |\delta(t)|_v$, where $|\ |_v$ is the normalized valuation at v and δ half the sum of the positive roots. Then its real logarithm ρ_0 *is independent of* $v \in V_D$. In fact, it is a positive integral power of Nv whose exponent is determined by the k_v-roots, their multiplicities, and the indices q_α of the Bruhat-Tits theory [61]. But those are determined by the previous data and the action of Γ_{k_v} on the completed Dynkin diagram [61], which is also independent of $v \in V_D$. We write ρ_0 instead of $\rho_{v,0}$. We have $\rho_0 \in \mathfrak{a}^{*+}$.

The representation π_v is a constituent of an unramified principal series $\text{PS}(\chi_v)$, where χ_v is an unramified character of $T(k_v)$, or, equivalently, of $S(k_v)$, determined up to a transformation by an element of ${}_kW$. Thus we may assume $\chi_{v,0}$ to be contained in the closure $\mathscr{Cl}(\mathfrak{a}^{*+})$ of $\mathfrak{a}^{*+}$. Since π_v is unitary, the associated spherical function is bounded, and hence $\text{Re } \chi_{v,0}$ is contained in the convex hull of ${}_kW(\rho_0)$, i.e., we have

$$(9) \qquad \langle \rho_0 - \chi_{v,0}, \lambda \rangle \geq 0, \quad \text{for all } \lambda \in \mathfrak{a}^{*+}.$$

(See remark following the proof.)

For $\lambda \in X^*({}^LT^\circ)$, let λ' be the restriction of λ to $X_*(T^D)$. In view of 10.4 and our conventions, we have then

$$(10) \qquad \left| \lambda(t_v) \right| = Nv^{\langle \text{Re } \chi_{v,0}, \lambda' \rangle}.$$

Let $\bar{\lambda} = {}_kW(\lambda') \cap \mathscr{Cl}(\mathfrak{a}^{*+})$. Since $\text{Re } \chi_{v,0} \in \mathscr{Cl}(\mathfrak{a}^{*+})$, we have

$$(11) \qquad Nv^{\langle \text{Re } \chi_{v,0}, \lambda' \rangle} \leq Nv^{\langle \text{Re } \chi_{v,0}, \bar{\lambda} \rangle}.$$

Combined with (9), this implies

$$(12) \qquad \left| \lambda(t_v) \right| \leq Nv^{\langle \rho_0, \bar{\lambda} \rangle}.$$

If now λ runs through P_r, there are only finitely many possibilities for $\bar{\lambda}$, whence (4), with $a = \sup\langle \rho_0, \bar{\lambda} \rangle$ $(\lambda \in P_r)$, for $v \in V_D$. Since V_1 is a finite union of such sets, this proves (4).

REMARK. The relation (9) is proved in [35, pp. 27–29] for the split case. For a general semisimple simply connected group, see I. Macdonald, *Spherical functions on a group of p-adic type*, Publ. Ramanujan Institute 2, Madras, Theorem 4.7.1, or H. Matsumoto, Lecture Notes in Math., vol. 590, Springer-Verlag, Berlin and New York, Proposition 4.4.11. In fact, we have used it for a general connected reductive group but the reduction to the case of simply connected semisimple groups is easily carried out by going over to the universal covering of the derived group.

13.3. COROLLARY. *Let P be a parabolic k-subgroup of G, $P = M \cdot N$ a Levi decomposition over k of P. Assume that π is a constituent of a representation $\mathrm{Ind}\,{}^{G_A}_{P_A}(\sigma)$ induced from a unitarizable irreducible admissible representation σ of M_A, viewed as a representation of P_A trivial on N_A. Then $L(s, \pi, r)$ is absolutely convergent in some right half-plane.*

We view ${}^L M$ as a subgroup of ${}^L G$ (3.3). Let r' be the restriction of r to ${}^L M$.

Let $v \in V_f$ be such that the conditions listed at the end of 13.1 are satisfied by M, G, σ_v and π_v. Then, by the transitivity of induction, it follows that there exists χ_v as in the above proof such that σ_v (resp. π_v) is the constituent of class 1 with respect to $M(o_v)$ (resp. $G(o_v)$) of the principal series $\mathrm{PS}(\chi_v)$ for $M(k_v)$ (resp. $G(k_v)$). Then $L(s, \pi_v, r) = L(s, \sigma_v, r')$ (7.2, 10.4). This being true for almost all v's, we are reduced to 13.2.

14. The L-function of an automorphic representation.

14.1. A smooth representation of G_A is *automorphic* if it is a subquotient of the regular representation of G_A in $G_k \backslash G_A$. It is cuspidal if it consists of cusp forms. If so, it is unitary modulo the center. We let $\mathfrak{A}(G/k)$ denote the set of equivalence classes of irreducible admissible automorphic representations of G_A. By Proposition 2 of [39], every $\pi \in \mathfrak{A}(G/k)$ is a constituent of a representation induced from some cuspidal $\sigma \in \mathfrak{A}(M/k)$, where M is a Levi k-subgroup of a parabolic k-subgroup of G. Combined with 13.3 this yields the

14.2. THEOREM (LANGLANDS). *Let $\pi \in \mathfrak{A}(G/k)$ and r be a representation of ${}^L G$. Then $L(s, \pi, r)$ is absolutely convergent in some right half-plane.*

The L-function of an irreducible admissible automorphic representation will also be called an automorphic L-function.

14.3. There are several conjectures on the analytic character of $L(s, \pi, r)$ for automorphic π, all checked in some special cases, going back to the work of Hecke on L-series attached to Grössencharaktere and to modular forms.

(a) If $\pi \in \mathfrak{A}(G/k)$, then $L(s, \pi, r)$ admits a meromorphic continuation to the whole complex plane.

(b) Assume that π and G are such that the local solution to the local problem yields factors L and ε at all places. It is then conjectured that there is a functional equation $L(s, \pi, r) = \varepsilon(s, \pi, r) \cdot L(1 - s, \tilde{\pi}, r)$, where $\tilde{\pi}$ is the contragredient representation to π.

(c) In a number of cases, it has been shown that:

$(*)$ If π is cuspidal, r irreducible nontrivial, then $L(s, \pi, r)$ is entire.

Here and there, conjectures to the effect that this should be a general phenomenon have been stated. However, there are counterexamples. Heuristically, one sees this is likely to happen if π is lifted from a cuspidal representation of a reductive group H (in the sense of V below) and the restriction of r to ${}^L H$ contains the trivial representation.

14.4. (a) Let $G = \mathbf{GL}_n$ and $r = r_n$ be the standard representation of $\mathbf{GL}_n(C)$. Then 14.3(b), (c) are proved in [25] for $n = 2$, in [19] for $n \geq 2$, if L and ε are defined to be the products of the L- and ε-factors mentioned in 12.4. As recalled in 12.4, these are the same as those considered here at almost all places, and for $n = 2$, at all places.

(b) If $G = \mathbf{GL}_2 \times \mathbf{GL}_2$ and $r = r_2 \otimes r_2$, similar results are established by Jacquet in [**21**].

(c) Let $G = \mathbf{GL}_2$. If $r \colon \mathbf{GL}_2(C) \to \mathbf{GL}_3(C)$ is the adjoint representation, then 14.3(b), (c) are announced in [**16**]. This extends results of Shimura [**54**]. If $r = \mathrm{Sym}^3(r_2)$, $\mathrm{Sym}^4(r_2)$, then 14.3(b) is stated in [**15**], in the context of the global lifting (see V); for $\mathrm{Sym}^3(r_2)$, it is also proved in [**51**], in the framework of 14.5 below.

(d) Let k be a function field, $G = \mathbf{GL}_m \times \mathbf{GL}_n$ and $r = r_m \otimes r_n$. Let π (resp. π') be a cuspidal automorphic representation of the first (resp. second) factor. By the methods of [**19**], [**26**], [**27**], one can define L and ε, and (Jacquet dixit) show 14.3(b), and also the holomorphy, except when $m = n$ and π is contragredient to π'. These methods also yield further examples for other groups and for other representations. It is expected that similar results hold over number fields.

(e) 14.3(a) has also been checked when $G = \mathbf{PSp}(4)$ in some cases in [**1**], and, in general, in [**42**]. A functional equation is also established. 14.3(a), (b) are announced in [**41**] for orthogonal groups in an odd number of variables over functional fields, for the local factors mentioned in 12.3. For a survey and earlier references, see [**43**]. See also [**44**].

14.5. We describe some cases in which 14.3(a) has been verified in [**33**] (see also [**18**] for a survey). Let C be a split k-group, of adjoint type, endowed with its canonical $\mathfrak{o}$-structure. Fix a Borel subgroup B of C and a maximal torus T of B defined over $\mathfrak{o}$. Let P be a maximal proper standard parabolic subgroup and $P = M \cdot N$ its standard Levi decomposition. Since C is adjoint, it is easily seen that $C(M)$ is a torus. The group $M/C(M)$ is semisimple, split over k, of adjoint type, of rank equal to $\mathrm{rk}(C) - 1$. We let $G = M/C(M)$. The group ${}^L G^\circ$ is simply connected (2.2(2)). We have a natural inclusion ${}^L G \to {}^L M$, and ${}^L M$ is the Levi subgroup of a standard parabolic subgroup ${}^L P = {}^L M \cdot U$ with unipotent radical U (3.3). Let A be the split component of P in T, and ${}^L A^\circ$ the split component of ${}^L P^\circ$ in ${}^L T^\circ$. The group ${}^L A^\circ$ acts on the Lie algebra $\mathfrak{u}$ of U and its eigenspaces are irreducible ${}^L G^\circ$-modules. We let F_P denote the set of contragredient representations to these ${}^L G^\circ$-modules. The L-functions considered in [**33**] are of the form $L(s, \pi, r)$ with $r \in F_P$ and π an irreducible cuspidal automorphic representation of G. A number of examples are given in which $L(s, \pi, r)$ admits a meromorphic continuation. This is deduced from the results of [**32**]: let m be the length of a composition series of $\mathfrak{u}$ with respect to M. Then, for suitable numbering of the elements of F_P and strictly positive integers a_i, there is a relation

$$(1) \qquad M(s) = \prod_{1 \leq i \leq m} L(a_i s, \pi, r_i) \cdot L(s a_i + 1, \pi, r_i)^{-1},$$

where $M(s)$ is the intertwining operator occurring in the theory of Eisenstein series with respect to P, and is known to have a meromorphic continuation to the complex plane [**32**]. If $r = 1$, this and 13.2 yield the meromorphic continuation. In general, if we have the analytic continuation for all r_i's except one, (1) gives it for the remaining one.

14.6. The *converse problem* is to what extent automorphic representations can be characterized by analytic properties of their L-functions, or to give analytic conditions on a given L-function which will insure that it is automorphic. The first main result was Hecke's characterization of the Mellin transform of a parabolic modular form. Then came Weil's extension of this theorem to congruence subgroups [64], [65], its generalization in the context of representations in [25], and the extension to $\mathbf{GL}_3$ [46], [27]. In those results, conditions are imposed on the L-functions of π and of the twists $\pi \otimes \chi$ of π by characters. However, the analogous statement is false from $n = 4$ on [46]. It may remain true if one imposes conditions on the twist $\pi \otimes \rho$ of π by representations of $\mathbf{GL}_{n-1}$ or only of $\mathbf{GL}_{n-2}$. For results in that direction, over function fields, see [45].

Note however that in the general problem outlined here, one wishes rather to turn things around and deduce the analytical properties of some given L-series by showing directly that it is automorphic (see the seminars on base change and on zeta-functions of Shimura varieties [17], [8], [40]).

14.7. *Other problems.* (1) One "representation theoretic" form of "Ramanujan's conjecture" is the following: if $\pi = \bigotimes \pi_v$ is an irreducible nontrivial admissible cuspidal automorphic representation (and G is simple), then each π_v is tempered. It is now well known to be false for certain orthogonal or unitary groups, and even for one split group [20].

(2) Let π be a unitary irreducible representation of G_A. If $G = \mathbf{GL}_2$, then its multiplicity in the space of cusp forms ${}^0L_2(G(k)\backslash G(A))$ is at most one, "multiplicity one theorem" [25]. In fact there is even a "strong multiplicity one theorem" [38]: given π_v for almost all v's, there is at most one constituent π of the space of cusp-forms with those local factors.

The multiplicity one theorem has been proved for $\mathbf{GL}_n$ [52] and the strong form for $\mathbf{GL}_3$ [28]. It is unknown whether it is true for $\mathbf{SL}_2$. On the other hand, there are counterexamples for some inner forms of $\mathbf{SL}_2$ [31].

CHAPTER V. LIFTING PROBLEMS.

Although the problems on automorphic L-functions discussed in §14 are only partially solved, the solutions provide practically all cases in which an L-series (automorphic or not) has been proved to have meromorphic or holomorphic analytic continuation with functional equation. This suggests trying, given an L-series and a reductive group G, to see whether G has an automorphic representation with the given L-series. Many instances of such questions can be viewed more precisely as special cases of the "lifting problem" or of the "problem of functoriality with respect to morphisms of L-groups." There is also a local version. For the sake of exposition, we shall start with the latter, but it should be borne in mind that the motivation and requirements stem from the global one, and that local and global are at present inextricably linked in many proofs. These questions were raised by Langlands in [35].

15. L-homomorphisms of L-groups.

15.1. Let E be a field and H, G connected reductive E-groups. A homomorphism $u: {}^LH \to {}^LG$ over Γ_k is said to be an L-homomorphism if it is continuous and if its restriction to ${}^LH^\circ$ is a complex analytic homomorphism of ${}^LH^\circ$ into ${}^LG^\circ$. Let E be local and G quasi-split. If $\varphi \in \Phi(H)$, then $u \circ \varphi \in \Phi(G)$. In fact, condition 8.2(i) is clearly satisfied, by $u \circ \varphi$, and so is 8.2(ii) because every parabolic subgroup of ${}^LG$

is relevant, G being assumed to be quasi-split. Therefore $\varphi \mapsto u \circ \varphi$ defines a map $\Phi(H) \to \Phi(G)$, to be denoted $\Phi(u)$.

15.2. Let $E = k$ be a global field. For $v \in V$, the Galois group Γ_{k_v} is a subgroup of Γ_k; hence the L-group of G viewed as a k_v-group, to be denoted $^L(G/k_v)$, is a subgroup of $^LG = {}^L(G/k)$. Thus, in particular, the L-homomorphism u of 15.1 defines by restriction an L-homomorphism $u_v: {}^L(H/k_v) \to {}^L(G/k_v)$, hence also a map $\Phi(u_v): \Phi(H/k_v) \to \Phi(G/k_v)$ $(v \in V)$.

The "lifting problem" is, roughly speaking, whether such maps are mirrored by maps of representations in the local case, or of automorphic representations in the global case.

15.3. EXAMPLE: BASE CHANGE. Let H be a split over E, F a finite Galois extension of E, and $G = R_{F/E}H$. Then $^LG^\circ$ is a product of copies of $^LH^\circ$, indexed and permuted by $\Gamma_{F/E}$ (5.1). There is then a natural L-homomorphism u which is the identity on Γ_E and the diagonal map on $^LH^\circ$. If E is a local field, then W'_F is an open normal subgroup of W'_E, and the map $\Phi(u)$ may be viewed as given by the restriction to W'_F.

16. Local lifting.

16.1. Let $k = E$ be a local field, G quasi-split over E, H a connected reductive E-group and $u: {}^LH \to {}^LG$ an L-homomorphism. The problem of local lifting is, roughly, to establish a correspondence $\Pi(u): \Pi(H(k)) \to \Pi(G(k))$ which preserves L- and ε-factors. If the local parametrization problem of III is solved, then $\Pi(u)$ is the map between indistinguishable classes which assigns $\Pi_{u \circ \varphi, G}$ to $\Pi_{\varphi, H}$ $(\varphi \in \Phi(H))$. The element $\Pi \in \Pi(G(k))$ is said to be *a lift of* $\pi \in \Pi(H(k))$ if $\Pi \in \Pi_{u \circ \varphi, G}$, where $\varphi \in \Phi(H)$ is such that $\pi \in \Pi_{\varphi, H}$. We have then

$$(1) \qquad L(s, \Pi, r) = L(s, \pi, r \circ \varphi), \qquad \varepsilon(s, \Pi, r, \psi) = \varepsilon(s, \pi, r \circ u, \psi).$$

for every representation r of ^LG.

16.2. The local lifting is thus viewed as a map between classes of L-indistinguishable representations rather than one between representations. However it is possible to single out one lifting under assumptions which, in the global case, are satisfied almost everywhere: assume H, G to be quasi-split, split over an unramified extension F of E, endowed with an o_E-structure such that $H(o_E)$ and $G(o_E)$ are very special maximal compact subgroups, and π of class one with respect to $H(o_E)$. Then φ such that $\pi \in \Pi_{\varphi, H}$, and the set $\Pi_{u \circ \varphi, G}$ are well defined. Moreover, $\Pi_{u \circ \varphi, G}$ contains exactly one element of class one (with respect to $G(o_E)$), to be called the *natural lift* of π.

16.3. A full solution of the local parametrization problem does not seem to be in sight, and it is conceivable that it may require proving at the same time global results such as Artin's conjecture. Meanwhile, one wants to settle some approximations to it, notably to be able to prove some cases of Artin's conjecture. Note that if $G = \mathbf{GL}_n$, then the sets $\Pi_{\varphi, G}$ are either known or conjectured to consist of one element (12.2, 12.3). Such a lifting problem can then be stated as one of constructing a map $u_*: \Pi(H(k)) = \Pi(G(k))$ satisfying certain conditions. So far, there are two examples:

(a) Base change (cf. 15.3) when $H = \mathbf{GL}_2$ and F is cyclic of prime degree over E [17], [38], [49], [56]. Besides some naturality conditions and 16.3, the main requirements relate the characters of π and of the hypothetical $u_*(\pi)$. The results also describe the fibres and the image of u_*. [Note that the results of [38] on this problem are used in [62], so that we cannot invoke the solution of the local parametrization problem (12.4) for $\mathbf{GL}_2$ just to use the map $\varPi(u)$ of 16.1. If we could, then the local questions [38] would be mainly to relate the characters of π and $\varPi(u)(\pi)$.]

(b) $H = \mathbf{GL}_2$, $G = \mathbf{GL}_3$, and

$$(1) \qquad u: {}^{L}H^{\circ} = \mathbf{GL}_2(C) \longrightarrow {}^{L}G^{\circ} = \mathbf{GL}_3(C)$$

is given by the adjoint representation of ${}^{L}H^{\circ}$ (see [16]).

In this case, $\varPi = u_*(\pi)$ must be trivial on the center of ${}^{L}G^{\circ}$ and be such that the L- and ε-factors of $u_*(\pi) \otimes \chi$ (χ character of E^*) are certain given functions. There is at most one such $\varPi$ (12.4(d)). In [16], $\varPi$ is stated to exist, except possibly if E has even residual characteristic and π is "extraordinary."

16.4. In 16.3(a), the lifting problem was connected with the existence of relations between characters. This is a direct connection between $\varPi(H)$ and $\varPi(G)$, which is of great importance for the use of the trace formula in proving or using the local or global lifting. We now mention two other examples of such relations. Assume that G is a quasi-split inner form of H. There is then an isomorphism $u: {}^{L}H \xrightarrow{\sim} {}^{L}G$ and an embedding $\varPhi(u): \varPhi(H) \subset \varPhi(G)$. If $f: H \to G$ is a k_s-isomorphism such that $f^{-1} \cdot {}^{\gamma}f$ is an inner automorphism of G for every $\gamma \in \varGamma_k$, then f establishes a bijection between conjugacy classes which are stable under $\varGamma_k$. Using results of Steinberg [59], one then sees easily that maximal k-tori in H are isomorphic over k to maximal k-tori in G. This allows one in some cases to assign regular semisimple classes in $G(k)$ to such classes in $H(k)$, so that it makes sense to compare values of characters of $H(k)$ and of $G(k)$ on such classes.

(a) Let k be either $\mathbf{R}$ or nonarchimedean with odd residual characteristic. Let $G = \mathbf{GL}_2$ and H be the group of invertible elements in the quaternion algebra over k. The sets $\varPi_\varphi$ are singletons, $\varPhi(u)$ assigns to a (finite dimensional) irreducible representation π of $H(k)$ a discrete series representation π' of $G(k)$. In this case, the semisimple classes of $H(k)$ correspond to the elliptic classes in $G(k)$. It is proved in [25] that the characters of π and π' differ only by a sign on those classes.

(b) Let $k = \mathbf{R}$. For $\varphi \in \varPhi(H)$, $\varPhi(G)$, let χ_φ be the sum of the characters of the elements in $\varPi_\varphi$. Choose $\varphi \in \varPhi(H)$ such that $\varPi_\varphi$ consists of tempered representations. Then χ_φ and $\chi_{u \circ \varphi}$ are equal on the regular semisimple classes of $H(k)$, up to a sign depending only on H and G [53, 6.3].

16.5. We could also take the Weil forms of the L-groups. In that case an L-homomorphism, restricted to W_E, is assumed to satisfy the obvious analogue of 8.2(i). Take in particular the case where $H = \{1\}$. Then u is just an element of $\varPhi(G)$. The lifting problem in this case is part of the local problem of III.

17. Global lifting.

17.1. Assume G to be quasi-split. Let H be a reductive k-group and $u: {}^{L}H \to {}^{L}G$ an L-homomorphism. Let $u_v: {}^{L}(H/k_v) \to {}^{L}(G/k_v)$ and $\varPhi(u_v): \varPhi(H/k_v) \to \varPhi(G/k_v)$ be the associated maps ($v \in V$) (see 15.1).

Let $\pi = \otimes_v \pi_v$ (resp. $\varPi = \otimes_v \varPi_v$) be an irreducible admissible representation of H_A (resp. G_A). Then $\varPi$ is said to be a lift of π if $\varPi_v$ is one of π_v for every $v \in V$ (16.1). If that is the case, then, for every representation r of ${}^{L}G$, we have

$$(1) \qquad L(s, \varPi, r) = L(s, \pi, r \circ u), \qquad \varepsilon(s, \varPi, r) = \varepsilon(s, \pi, r \circ u).$$

It is also usually requested that Π_v be the natural lift (16.2) of π_v for almost all v's. The question is then whether every automorphic π has a lift, which is automorphic, or, somewhat more ambitiously, whether there is a map $u_*: \mathfrak{A}(H/k) \to \mathfrak{A}(G/k)$ with reasonable properties, which sends $\pi \in \mathfrak{A}(H/k)$ onto a lift of π. One also wants to describe the fibres and the image of u_*.

In that degree of generality, the problem appears to be inaccessible at present. However, there are many results, old and recent, which are very striking illustrations of this principle, some of which will be extensively discussed in various seminars. Here, for orientation, and to give an idea of the scope of the problem, I shall list briefly some special cases, referring to the literature or to other seminars for more details.

REMARK. Let r be a representation of ^LH of degree n. Then it defines an L-homomorphism $u : {}^LH \to {}^L\mathbf{GL}_n = \mathbf{GL}_n(C) \times \Gamma_k$ in the obvious way. A positive answer to the lifting problem would imply in particular that if π is an automorphic representation of H, then $L(s, \pi, r) = L(s, \Pi, r_n)$ where Π is an automorphic representation of $\mathbf{GL}_n$ and r_n the standard representation. This would therefore to a large extent reduce the study of automorphic L-functions to those of $\mathbf{GL}_n$, with respect to the standard representation.

17.2. Let $H = \{1\}$, $G = \mathbf{GL}_n$. Then an L-homomorphism u is just a continuous complex n-dimensional representation of Γ_k. The question is then whether the Artin L-series $L(s, u)$ is an automorphic L-series of $\mathbf{GL}_n$ (with respect to the standard representation of $\mathbf{GL}_n(C)$), which should be cuspidal if u is irreducible. In view of known results on $\mathbf{GL}_n$ (cf. 14.4) this would imply Artin's conjecture.

For $n = 1$, a positive answer is given by class-field theory. For $n = 2, 3$, a positive answer is equivalent to Artin's conjecture, since there are converses to Hecke theory [25], [65], [27], [46]. For $n = 2$, it has been proved for dihedral or tetrahedral representations of Γ_k, and for some others over Q (see [38], [17], [15]).

17.3. Let k' be a Galois extension of k, n the degree of k' over k. Take $H = R_{k'/k}\mathbf{GL}_1$, $G = \mathbf{GL}_n$. There is a natural homomorphism $f: {}^LH^\circ \rtimes \Gamma_{k'/k}$ into the normalizer of a maximal torus $^LT^\circ$ of $^LG^\circ$. Since the former group is a quotient of ^LH, and $^LG = {}^LG^\circ \times \Gamma_k$, we can define an L-homomorphism $u: {}^LH \to {}^LG$ by $u(h, \gamma) = (f(h), \gamma)$ $(h \in {}^LH^\circ, \gamma \in \Gamma_k)$. An automorphic representation of H is a Grössencharakter χ of k'. The problem is then whether the Artin L-series $L(s, \chi)$ is the L-series of an automorphic representation of G.

If $n = 2$, $k = Q$, and k' is imaginary, this was proved by Hecke; π is associated to a cuspidal holomorphic automorphic form. If $n = 2$, $k = Q$, and k' is real quadratic, this was established by H. Maass. π is then associated to a nonholomorphic automorphic form.

For $n = 3$, this is proved in [26], [27].

17.4. *Base change.* This is the global counterpart to 16.3(a). Let k' be a finite Galois extension of k. Assume H to be k-split and $G = R_{k'/k}H$. There is again an L-homomorphism $u: {}^LH \to {}^LG$ whose restriction to $^LH^\circ$ is a diagonal map. In this case $G(A)$ and $G(k)$ are canonically isomorphic to $H(A_{k'})$ and $H(k')$; therefore the problem is to associate an automorphic representation of $H(A_{k'})$ to an automorphic representation of $H(A_k)$. Again, it should be a counterpart to the restriction to $W_{k'}$ of homomorphisms $W_k \to {}^LH^\circ$.

If $H = \mathbf{GL}_2$ and k' is cyclic of prime degree, the lifting map u_* for representations is constructed in [38], which also gives a description of its image and fibres.

This extends work of Doi-Naganuma, Jacquet [21] (on the quadratic case) and of Saito [49], Shintani [55], [56] (cf. [17]).

17.5. Let G be quasi-split, and H an inner form of G. Then $^LH = {}^LG$ and $\Phi(H/k_v) \subset \Phi(G/k_v)$ for all v's (8.3). Moreover, for almost all v's, H and G are isomorphic over k_v; hence $\Phi(H/k_v) = \Phi(G/k_v)$ and $\Pi(H(k_v)) = \Pi(G(k_v))$. The question is then, given $\pi = \otimes_v \pi_v$, is there an automorphic representation $\Pi = \otimes_v \Pi_v$ of G such that $\Pi_v = \pi_v$ for almost all v's?

If $G = \mathbf{GL}_2$ and H is the group of invertible elements of a quaternion algebra D over k, a positive answer is given by Jacquet-Langlands [25]. Note that, in that case, because of the "strong multiplicity one theorem," at most one Π may be associated to a given π in this way. The possible Π's are in fact the cuspidal automorphic representations for which Π_v belongs to the discrete series for all v's over which D does not split (loc. cit.).

17.6. If $G = \mathbf{GL}_2$, $G = \mathbf{GL}_3$ and u is given by the adjoint representation, as in 13.4, the global lifting problem has been solved by Gelbart-Jacquet [16], the "local lifting" being the one of 16.2(b).

17.7. Let M be a Levi k-subgroup of a parabolic k-subgroup P of G. Then ^LM imbeds naturally into ^LG (3.3), whence an L-homomorphism $u: {}^LM \to {}^LG$. If π is cuspidal, then the analytic continuation and residues of Eisenstein series [32] are known to yield a unitary $u_*(\pi)$ in many cases, and, conjecturally, in general.

18. Relations with other types of L-functions.

18.1. In 17.2, the lifting problem amounts to identifying an Artin L-function with an automorphic L-function on $\mathbf{GL}_n$. One can also include in this problem more general representations of Weil groups if one passes to the Weil form of the L-groups. For simplicity, let us limit ourselves to relative Weil groups $W_{k'/k}$, where k' is a finite Galois extension of k over which H and G split. An L-homomorphism $u: {}^LH^\circ \rtimes W_{k'/k} \to {}^LG^\circ \rtimes W_{k'/k}$ is then a continuous homomorphism compatible with the projections on $W_{k'/k}$, whose restriction to $^LH^\circ$ is a complex analytic homomorphism into $^LG^\circ$, and such that, for $w \in W_{k'/k}$, $u(w) = (u'(w), w)$ with $u(w)$ semi-simple (cf. 8.2(i)).

If $H = \{1\}$, an L-homomorphism is said to be an admissible homomorphism of $W_{k/k}$ into ^LG. In analogy with the definition of $\Phi(G)$ in the local case, we can consider the set $\Phi_{k'/k}(G)$ of equivalence classes of such homomorphisms, modulo inner automorphisms of $^LG^\circ$, and then pass to a suitable limit $\Phi(G)$ over k'.

The lifting problem asks in this case to associate to any $\varphi \in \Phi(G)$ an automorphic representation π, such that, for any representation r of ^LG, $L(s, \pi, r)$ is equal to the Artin-Hecke L-series of $r \circ u$. In particular, is every Artin-Hecke L-series that of an automorphic representation of $\mathbf{GL}_n$, with respect to the standard representation?

If G is a torus, then [34] provides a positive answer. In fact, in this case the irreducible admissible automorphic representations of G are the characters of $G(k)\backslash G(A)$, and [34] gives a homomorphism with finite kernel of $\Phi_{k'/k}(G)$ onto the set of such characters.

18.2. In the same vein, it is natural to ask whether Hasse-Weil zeta-functions (or even L-functions of compatible systems of l-adic representations of Galois groups) can be expressed in terms of automorphic L-functions. For elliptic curves over function fields, it is a theorem. That it should be the case for elliptic curves over

Q is the Taniyama-Weil conjecture; it has been checked in a number of special cases (see [2], [14] for surveys from the classical and representation theoretic points of view respectively). Apart from that, this problem has been pursued mostly for Shimura curves and certain Shimura varieties; we refer to the corresponding seminars for a description of the present state of affairs.

Finally, one may ask whether it is possible to characterize a priori those automorphic representations whose *L*-series have an arithmetic or algebraico-geometric significance. A necessary condition if k is a number field is that for an infinite place v, π_v should be associated to a representation σ_v of W_{k_v} whose restriction to C^* is rational, C^* being viewed as real algebraic group, i.e., be of type A_0 in [3, 6.5]. If the *L*-series of π is to be an Artin *L*-series, then π should even be of type A_{00} (loc. cit.), i.e., σ_v should be trivial on C^*. Let $k = Q$. Then there are three possibilities for π_∞ (11.5). If $\pi_\infty = \pi(1, \text{sgn})$, then π corresponds to 2-dimensional representations of Γ_Q with odd determinant by the theorem of Deligne-Serre [10], [50]. Modulo the Artin conjecture for such representations, the correspondence is bijective. However, I am not aware of any result for the other two possible values of π_∞. A positive answer would involve nonholomorphic automorphic forms. In [36], it is shown in many cases for $\mathbf{GL}_2$ over Q that the *L*-series of a representation of type A_0 is that of a compatible system of *l*-adic representations of Γ_Q. Over a function field, there is no condition such as A_0. In fact, for $\mathbf{GL}_2$, Drinfeld has shown that all irreducible admissible automorphic representations are associated to *l*-adic representations (see the lectures on his work by G. Harder and D. Kazhdan).

REFERENCES

1. A. N. Andrianov, *Dirichlet series with Euler product in the theory of Siegel modular forms of genus two*, Trudy Mat. Inst. Steklov **112** (1971), 73–94.

2. B. J. Birch and H. P. F. Swinnerton-Dyer, *Elliptic curves and modular functions of one variable*. IV, Lecture Notes in Math., vol. 476, Springer, New York, 1975, pp. 2–32.

3. A. Borel, *Formes automorphes et séries de Dirichlet (d'après R. P. Langlands)*, Sém. Bourbaki, Exposé 466, (1974/75); Lecture Notes in Math., vol. 514, Springer, New York, pp. 189–222.

4. A. Borel and J.- P. Serre, *Théorèmes de finitude en cohomologie galoisienne*, Comment. Math. Helv. **39** (1964), 111–164.

5. A. Borel et J. Tits, *Groupes réductifs*, Inst. Hautes Etudes Sci. Publ. Math. **27** (1965), 55–151.

6. P. Cartier, *Representations of reductive p-adic groups*, these PROCEEDINGS, part 1, pp. 111–155.

7. W. Casselman, $\mathbf{GL}_n$, Proc. Durham Sympos. Algebraic Number Fields, London, Academic Press, New York, 1977.

8. ———, *The Hasse-Weil ζ-function of some moduli varieties of dimension greater than one*, these PROCEEDINGS, part 2, pp. 141–163.

9. P. Deligne, *Formes modulaires et représentations de $\mathbf{GL}_2$*, Modular Functions of One Variable, Lecture Notes in Math., vol. 349, Springer, New York, 1973, pp. 55–106.

10. P. Deligne and J.-P. Serre, *Formes modulaires de poids 1*, Ann. Sci. École Norm. Sup. (4) **7**(1974), 507–530.

11. D. Flath, *Decomposition of representations into tensor products*, these PROCEEDINGS, part 1, pp. 179–183.

12. F. Gantmacher, *Canonical representations of automorphisms of a complex semi-simple group*, Mat. Sb. **5** (1939), 101–142. (in English)

13. S. Gelbart, *Automorphic functions on adele groups*, Ann. of Math. Studies, no. 83, Princeton Univ. Press, Princeton, N. J., 1975.

14. ———, *Elliptic curves and automorphic representations*, Advances in Math. **21** (1976), 235–292.

15. S. Gelbart, *Automorphic forms and Artin's conjecture*, Lecture Notes in Math., vol. 627, Springer-Verlag, Berlin and New York, 1977, pp. 241–276.

16. S. Gelbart and H. Jacquet, *A relation between automorphic forms on* GL_2 *and* GL_3, Proc. Nat. Acad. Sci. U.S.A. **73** (1976), 3348–3350.

17. P. Gérardin and J.-P. Labesse, *The solution of a base change problem for* GL_2 *(following Langlands, Saito, Shintani)*, these PROCEEDINGS, part 2, pp. 115–133.

18. R. Godement, *Fonctions automorphes et produits eulériens*, Sém. Bourbaki (1968/69), Exposé 349; Lecture Notes in Math., vol. 179, New York, 1970, pp. 37–53.

19. R. Godement and H. Jacquet, *Zeta-functions of simple algebras*, Lecture Notes in Math., vol. 260, Springer, New York, 1972.

20. R. Howe and I. I. Piatetski-Shapiro, *A counterexample to the 'generalized Ramanujan conjecture' for (quasi-) split groups*, these PROCEEDINGS, part 1, pp. 315–322.

21. H. Jacquet, *Automorphic Forms on* $GL(2)$. II, Lecture Notes in Math., vol. 278, Springer, New York, 1972.

22. ———, *Generic representations*, Non-Commutative Harmonic Analysis, Lecture Notes in Math., vol. 587, Springer, New York, 1977, pp. 91–101.

23. ———, *From* GL_2 *to* GL_n, U.S.-Japan Seminar on Number Theory (Ann Arbor, Mich., 1975).

24. ———, *Principal L-functions of the linear group*, these PROCEEDINGS, part 2, pp. 63–86.

25. H. Jacquet and R. P. Langlands, *Automorphic forms on* $GL(2)$, Lecture Notes in Math., vol. 114, Springer, New York, 1970.

26. H. Jacquet, I. Piatetskii-Shapiro and J. Shalika, *Construction of cusp forms on* GL_3, Univ. of Maryland Lecture Notes, no. 16, 1975.

27. ———, *Hecke theory for* GL_3 [Ann. of Math. **109** (1979), 169–258].

28. H. Jacquet and J. Shalika, *Comparaison des représentations automorphes du groupe linéaire*, C. R. Acad. Sci. Paris Ser. A **284** (1977), 741–744.

29. A. Knapp and G. Zuckerman, *Classification of irreducible tempered representations of semi-simple Lie groups*, Proc. Nat. Acad. Sci. U.S.A. **73** (1976), 2178–2180.

30. ———, *Normalizing factors, tempered representations and L-groups*, these PROCEEDINGS, part 1, pp. 93–105.

31. J. P. Labesse and R. P. Langlands, *L-indistinguishability for* SL_2, Institute for Advanced Study, Princeton, N. J. (preprint) [Canadian J. M. **31** (1979), 726–785].

32. R. P. Langlands, *On the functional equations satisfied by Eisenstein series*, Lecture Notes in Math., vol. 544, Springer, New York.

33. ———, *Euler products*, Yale Univ. Press, 1967.

34. ———, *Representations of abelian algebraic groups*, Yale Univ., 1968 (preprint).

35. ———, *Problems in the theory of automorphic forms*, Lectures in Modern Analysis and Applications, Lecture Notes in Math., vol. 170, Springer, New York, 1970, pp. 18–86.

36. ———, *Modular forms and l-adic representations*, Modular Functions of One Variable. II, Lecture Notes in Math., vol. 349, Springer, New York, 1973, pp. 361–500.

37. ———, *On the classification of irreducible representations of real algebraic groups* (preprint).

38. ———, *Base change for* GL_2: *The theory of Saito-Shintani with applications*, Notes, Inst. Advanced Study, Princeton, N. J., 1975 [Annals of Math. Studies 96].

39. ———, *On the notion of an automorphic representation*, these PROCEEDINGS, part 1, pp. 203–207.

40. ———, *Automorphic representations, Shimura varieties, and motives. Ein Märchen*, these PROCEEDINGS, part 2, pp. 205–246.

41. M. E. Novodvorsky, *Théorie de Hecke pour les groupes orthogonaux*, C. R. Acad. Sci. Paris Sér. A, **285** (1975), 93–94.

42. ———, *Automorphic L-functions for the symplectic group* GSp(4), these PROCEEDINGS, part 2, pp. 87–95.

43. M. E. Novodvorsky and I. I. Piatetskii-Shapiro, *Rankin-Selberg method in the theory of automorphic forms*, Proc. Sympos. Pure Math., vol. 30, no. 2, Amer. Math. Soc., Providence, R. I., 1977, pp. 297–301.

44. I. Piatetski-Shapiro, *Euler subgroups*, Lie Groups and Their Representations, Proc. Summer School, Budapest, 1971.

45. ——, *Zeta-functions of* GL$_n$, Notes, Univ. of Maryland, College Park, Md.

46. ——, *Converse theorem for* GL$_3$, Notes, Univ. of Maryland, College Park, Md.

47. ——, *Multiplicity one theorems*, these PROCEEDINGS, part 1, pp. 209–212.

48. F. Rodier, *Les representations de* GSp(4, *k*), *où k est un corps local*, C. R. Acad. Sci. Paris **283** (1976), 429–431.

49. H. Saito, *Automorphic forms and algebraic extensions of number fields*, Lecture Notes in Math, vol. 8, Kinokuniya Book Store Co., Ltd., Tokyo, Japan, 1975.

50. J.-P. Serre, *Modular forms of weight one and Galois representations* (prepared in collaboration with C. J. Bushnell), Proc. Durham Sympos. Algebraic Number Fields (London), Academic Press, New York, 1977, pp. 193–268.

51. F. Shahidi, *Functional equations satisfied by certain L-functions*, Compositio Math. **37** (1978), 171–207.

52. J. A. Shalika, *The multiplicity one theorem for* GL$_n$, Ann. of Math. (2) **100** (1974), 171–193.

53. D. Shelstad, *Characters and inner forms of a quasi-split group over R*. Comp. Math. **39** (1979), 11–45.

54. G. Shimura, *On the holomorphy of certain Dirichlet series*, Proc. London Math. Soc. (3) **31** (1975), 79–98.

55. T. Shintani, *On liftings of holomorphic automorphic forms*, U.S.-Japan Seminar on Number Theory, (Ann Arbor, Mich., 1975).

56. ——, *On liftings of holomorphic cusp forms*, these PROCEEDINGS, part 2, pp. 97–110.

57. A Silberger, *The Langlands classification for reductive p-adic groups*, Math. Annalen **236** (1978), 95–104.

58. T. A. Springer, *Reductive groups*, these PROCEEDINGS, part 1, pp. 3–27.

59. R. Steinberg, *Regular elements of semi-simple algebraic groups*, Inst. Hautes Etudes Sci. Publ. Math. **25** (1965), 49–80.

60. J. Tate, *Number theoretic background*, these PROCEEDINGS, part 2, pp. 3–26.

61. J. Tits, *Reductive groups over local fields*, these PROCEEDINGS, part 1, pp. 29–69.

62. J. B. Tunnell, *On the local Langlands conjecture for* GL(2), Inv. Math. **46** (1978), 179–200.

63. ——, *Report on the local Langlands conjecture for* GL$_2$, these PROCEEDINGS, part 2, pp. 135–138.

64. A. Weil, *Ueber die Bestimmung Dirichletscher Reihen durch Funktionalgleichungen*, Math. Ann. **168** (1967), 149–167.

65. ——, *Dirichlet series and automorphic forms*, Lecture Notes in Math., vol. 189, Springer, New York, 1971.

THE INSTITUTE FOR ADVANCED STUDY

114.

Symmetric compact complex spaces

Arch. Math. **33** (1979) 49–56

A complex space X will be said to be *symmetric at a point* x if x is an isolated fixed point of some involutory automorphism of X and *symmetric* if it is so at all points. If X is a connected hermitian manifold and the symmetries can be chosen so as to be isometric, then X is a hermitian symmetric space in the sense of E. Cartan, and in particular is a homogeneous space of a Lie group. The purpose of this Note is to study to what extent this remains true when X is compact, without any metric assumption[1]). We shall see that a symmetric compact complex space X is always the union of finitely many orbits of a connected complex Lie group of automorphisms (2.2), but it is not necessarily homogeneous: examples are provided by the n-dimensional projective space with one point blown-up or a quadric surface with four suitable points blown-up (§ 3). On the other hand, if X is connected, homogeneous, then it is hermitian symmetric, isomorphic to the product of a complex torus by a hermitian symmetric projective rational variety (2.4).

§ 1. Automorphisms of complex spaces.

1.1. Complex spaces will always be assumed to be *reduced*. Let X be a *compact* one. Then the group $G(X)$ of automorphisms of X, endowed with the compact open topology, admits a structure of complex Lie group such that the map $G(X) \times X \to X$ defining the action of $G(X)$ on X is holomorphic: see [1] if X is a manifold, [7] for the present case (and [8: Satz 7] for non-necessarily reduced spaces). Furthermore $G(X)$ is countable at infinity. To see this, it suffices to show that the space $C_c(X, X)$ of continuous maps of X into itself, endowed with the compact open topology, has a countable basis for its open sets. Since X does so, this follows from Remarque 2), X.27 in [4]. In particular, the group $G(X)$ has (at most) countably many connected components.

1.2. Let X be a complex space and $x \in X$. An involution of X having x as an isolated fixed point will be called *a symmetry at* x. If X is a connected manifold. and g an automorphism of X, then the following conditions are equivalent: (1) g is

---

[1]) This question was raised over twenty-five years ago by S. Bochner in a conversation with the author.

597

a symmetry at x; (2) g is of order two and the differential $\mathrm{d}g_x$ of g at x is $-\,\mathrm{Id.}$; (3) $\mathrm{d}g_x = -\,\mathrm{Id.}$ and g belongs to a compact group of automorphisms of X. (These equivalences follow from the fact that if K is a compact group of automorphisms of X leaving x fixed, then K is linear in suitable local coordinates around x [5].) There may be more than one symmetry at x.

The fixed point set $F(g)$ of g is a closed analytic subset. Therefore, if X is compact the set of isolated fixed points of g is (at most) finite.

1.3. Given a complex Lie group of transformations of a complex space Y, and a positive integer k, the union of the orbits of G in Y, of dimension $< k$, is an analytic subset [6: Satz 5].

§ 2. Symmetric compact complex spaces.

In this section, X denotes a compact complex space.

2.0. We recall some facts about maximal compact subgroups of Lie groups to be used below. Let G be a Lie group with finitely many connected components. Then every compact subgroup is contained in a maximal one. The maximal ones are conjugate under inner automorphisms. If K is one of them, then K meets every connected component of G and its intersection with the identity component of G^0 of G is the identity component of K and is a maximal compact subgroup of G^0. (For G connected, this is the well-known Cartan-Malcev-Iwasawa theorem. For the case considered here, see [10].)

More generally, it follows that the intersection of K with any closed normal subgroup N of G with finitely many connected components is a maximal compact subgroup of N.

Let now C be a compact group of automorphisms of G. By applying the foregoing to the semi-direct product of C and G, we see that G contains a maximal compact subgroup stable under C.

2.1. Lemma. *Let X be irreducible and H be a complex Lie group, countable at infinity, which operates holomorphically on X. Let S be the set of isolated fixed points of symmetries of X contained in H. Then S is the union of at most countably many orbits of the identity component H^0 of H. If S has an interior point, then H^0 has an open orbit in Y.*

Let N be the subgroup of H acting trivially on X and $H' = H/N$. Then H' acts effectively, and the orbits of its identity component are those of H^0. We may therefore assume H to be effective (i.e. $N = (1)$).

To prove the first assertion it suffices to show that if M is a connected component of H, then the set S_M of isolated fixed points of symmetries contained in M is the union of finitely many orbits of H^0.

We need only consider the case where M contains at least one symmetry, say s_0. It is of order two, hence $H_1 = M \cup H^0$ is a group. Let K_1 be a maximal compact subgroup of H_1 containing s_0 and T a maximal torus in the centralizer of s_0 in $K_0 = K_1 \cap H^0$. The subgroup T_1 generated by s_0 and T is the direct product of T

and of a group of order two. By [11: p. 57], every element of K_1 is conjugate under K_0 to an element of T_1. Since every element of order two of H_1 is conjugate to an element of K_1 and T_1 has finitely many elements of order two, it follows that the symmetries in M form finitely conjugacy classes under H_1, hence under H^0. Let $(s_i)_{1 \leq i \leq m}$ be a set of representatives. Then any isolated fixed point of a symmetry $s \in M$ is the transform under H^0 of an isolated fixed point of one of the s_i's, whence the first assertion.

If H^0 has no open orbit in S, then S is the union of countably many closed sets without interior point, hence has no interior point by Baire's theorem [4: IX, § 5, n⁰. 3]. The second assertion follows.

2.2. Theorem. *Let X be symmetric. Then X is the union of finitely many orbits of $G(X)^0$.*

Since $G(X)$ is a complex Lie group countable at infinity (1.1), this theorem follows from the slightly move general:

2.3. Theorem. *Let H be a complex Lie group, countable at infinity, which acts holomorphically on X. Assume that for every $x \in X$, the group H contains a symmetry of X at x. Then X is the union of finitely many orbits of H^0.*

If $\dim X = 0$, then X consists of finitely many points, and there is nothing to prove. Let then $\dim X = n > 0$, and assume the theorem proved for spaces of strictly smaller dimension.

Let X_i $(1 \leq i \leq m)$ be the irreducible components of X of dimension n and Y the union of the irreducible components of X of strictly smaller dimension. The group H permutes the X_i's and leaves Y stable. Let H_i be the subgroup of H leaving X_i stable. The group H_i is closed, and its identity component is H^0.

Let U_i be the set of points of X_i not contained in any other irreducible component $(1 \leq i \leq m)$. It is Zariski-open, non-empty. If s is a symmetry of X at $x \in U_i$, then s must leave X_i stable, hence belongs to H_i. Since X is symmetric, the second assertion of 2.1 shows that the identity component of H_i', hence also H^0, has an open orbit V_i in X_i. Then V_i is the unique open orbit of H^0 in X_i: in fact, the complement of the orbits of dimension n is analytic (1.3) and proper since there is at least one open orbit. But the complement of a proper analytic set in an irreducible complex space is connected (cf. e.g. [2: 3.3]), whence our assertion. Then $Z_i = X_i - V_i$ is a closed analytic set, and the union Z of the Z_i's and of Y is a closed complex subspace of X, which is stable under H, of dimension $< n$. If $z \in Z$ and s is a symmetry of X at z contained in H, then s leaves Z invariant, and the restriction of s to Z is necessarily a symmetry of Z at z. We may apply the induction assumption to the action on Z of H. It follows that Z is the union of finitely many orbits of H^0, whence the theorem.

2.4. Theorem. *Let X be irreducible, symmetric, homogeneous. Then X is hermitian symmetric, and is the direct product of a complex torus by a hermitian symmetric projective rational variety.*

We let G denote the identity component of the group of automorphisms of X. Fix $x_0 \in X$ and let H be the isotropy group of x_0 in G. Then $X = G/H$.

Assume first that the fundamental group $\pi_1(X)$ is finite. Let M be a connected component of Aut X containing a symmetry s_0 at x_0. Then $G_1 = M \cup G$ is a group. Let K_1 be a maximal compact subgroup of G_1 containing s_0 (2.0). Then $K_0 = K_1 \cap G$ is a maximal compact subgroup of G, hence is transitive on X [9]. It follows that K_1 contains for each $x \in X$ a symmetry at x. Since K_1 is compact, X admits a hermitian metric invariant under K_1. Therefore X is hermitian symmetric.

We now drop the assumption on $\pi_1(X)$. Let N be the normalizer in G of the identity component H^0 of H. Then N is connected, contains the radical of G and $Y = G/N$ is a simply connected projective rational homogeneous variety [3: Thm. 7']. The projection $\pi: X \to Y$ defines a holomorphic fibration, whose typical fiber N/H is isomorphic to the quotient of the complex Lie group N/H^0 by the discrete co-compact subgroup H/H^0.

Let $y_0 = \pi(x_0)$. Let j_0 be the automorphism $g \mapsto s_0 \cdot g \cdot s_0$ of G. It leaves H, H^0 and hence N stable and we have $s_0(g \cdot x_0) = j_0(g) \cdot x_0$. Hence s_0 defines an automorphism t_0 of Y of order ≤ 2, leaving y_0 fixed, whose differential at y_0 is $-\mathrm{Id}$. It is then a symmetry of Y at y_0. Therefore, by homogeneity, Y is symmetric. Since it is simply connected, it is then hermitian symmetric by the first part of the proof.

The automorphism j_0 defines an automorphism of the connected group N/H^0. The tangent space of N/H^0 at the origin may be identified to a subspace of the tangent space of X at x_0. Therefore the differential at the origin of j_0, viewed as automorphism of N/H^0, is also $-\mathrm{Id}$. Using canonical coordinates, we see that $j_0: N/H^0 \to N/H^0$ is the map $g \mapsto g^{-1}$, therefore N/H^0 is commutative. As a consequence, H is normal in N and $A = N/H$ is a complex torus. To conclude the proof, it is then enough to show that X is isomorphic to the product of Y and A. The group A acts freely by right translations on X, and π is the projection of a principal holomorphic fibration with structural group A. It suffices therefore to prove that the fibration π admits a holomorphic cross-section.

Let R be the radical of G. By the Levi-Malcev theorem, if L is a maximal connected complex semi-simple subgroup of G, then $G = L \cdot R$ (and $L \cap R$ is finite). We may choose L to be stable under j_0: in fact, let G_1 be the subgroup of Aut X generated by G and j_0. It consists of one or two connected components, depending on whether s_0 is in G or not. Let K_1 be a maximal compact subgroup of G_1 containing s_0. We may then take for L the smallest complex subgroup containing the derived group of $K_0 = K_1 \cap G$.

Since N contains R, we have $Y = L/P$, where $P = L \cap N$ is the isotropy group of y_0 in L. We want to prove:

$$(1) \qquad P = L \cap H.$$

Assume this. Then $L(x_0)$ is a compact submanifold of X, and π obviously defines a bijective submersion of $L(x_0)$ onto Y, hence an isomorphism of $L(x_0)$ onto Y. Therefore $L(x_0)$ is the sought for cross-section. There remains then to prove (1).

Let L_i $(1 \leq i \leq s)$ be the simple factors of L which do not act trivially on Y, and M the greatest connected normal subgroup of L contained in N. The group $M \cdot R$ is normal in G and fixes y_0, hence acts trivially on Y. The group P is a parabolic subgroup of L. Hence

$$(2) \qquad P = P_1 \dots P_s \cdot M , \quad \text{where} \quad P_i = P \cap L_i \quad \text{is parabolic in } L_i ,$$

and

$$(3) \qquad Y = \prod_{1 \leq i \leq s} Y_i , \quad \text{with} \quad Y_i = L_i / P_i .$$

The spaces Y_i are then also hermitian symmetric; they are stable under s_0, since their tangent spaces at y_0 are so. Therefore j_0 leaves L_i, hence $P_i = L_i \cap N$ stable $(1 \leq i \leq s)$.

The group N/H^0 is commutative, hence H^0 contains the derived group of any subgroup of N, in particular of P. By standard facts on hermitian symmetric spaces, the group P_i is proper maximal in L_i, hence its "parabolic rank" is one, i.e. P_i is the semi-direct product of its unipotent radical U_i by a reductive subgroup Q_i with one-dimensional center. The derived group $\mathscr{D}P_i$ of P_i is the semi-direct product of U_i by the derived group $\mathscr{D}Q_i$ of Q_i, and has codimension one in P_i. It is then contained in H^0. We may choose Q_i to be the complexification of a maximal compact subgroup of P_i which is stable under j_0 (2.0), hence assume that j_0 leaves Q_i, $\mathscr{D}Q_i$ and A_i stable.

We now want to prove that $A_i \subset H^0$. We know that j_0 leaves P_i, A_i stable. It then also leaves U_i invariant. Since A_i is one-dimensional the automorphism j_0 of A_i is either the identity or the inversion $a \mapsto a^{-1}$. However the rational characters of A_i occurring in the Lie algebra of U_i are all strictly positive for a suitable ordering on the group of rational characters of A_i (which is infinite cyclic). This rules out the second possibility, hence j_0 leaves A_i pointwise fixed. Since the differential of j_0 at the origin of G/H is the same as the differential of s_0 at x_0, it follows that the Lie algebra of A_i is contained in that of H^0, and our assertion follows.

We have now $P_i \subset H^0$ for all i's, hence $P \subset L \cap H$. But any subgroup of L containing P properly must contain one of the L_i's. Since those do not act trivially on Y and are normal in L, they do not fix y_0, hence, a fortiori, do not fix x_0. This proves (1).

§ 3. Non-homogeneous symmetric compact complex manifolds.

3.1. Let D be the projective line. We let z be the coordinate on D, $(z \in \mathbb{C} \cup \infty)$. Let N be the semi-direct product of $\mathbb{C}^*$ by a group of order two, generated by an element w such that $w \cdot t \cdot w = t^{-1}$ $(t \in \mathbb{C}^*)$. We view N as a group of automorphisms of D in the obvious way: to $t \in \mathbb{C}^*$ we associate the map

$$A_t \colon z \mapsto tz \, (z \in \mathbb{C}) , \qquad A_t(\infty) = \infty ,$$

and to w the automorphism A_w permuting 0 and ∞ and mapping $z \in \mathbb{C}^*$ to z^{-1}. Then, for every $z \in D$, the group N contains a symmetry at z, namely, A_{-1} if $z = 0$, ∞, and $A_w \cdot A_t$ if $z \neq 0$, ∞ and $t = z^{-2}$.

3.2. Let now $X_0 = D_1 \times D_2$, where D_1, D_2 are two projective lines and let z_i be the coordinate on D_i $(i = 1, 2)$. Let

$$Z_0 = (0, 0) \cup (0, \infty) \cup (\infty, 0) \cup (\infty, \infty),$$

and let X be the complex surface obtained from X_0 by blowing up the four points of Z_0. Let $\pi\colon X \to X_0$ be the canonical projection and $Z = \pi^{-1}(Z_0)$. The divisor Z is the union of four projective lines and π is an isomorphism of $X - Z$ onto $X_0 - Z_0$.

3.3. Proposition. *The surface X defined in* 3.2 *is symmetric, but not homogeneous.*

The four components of Z are exceptional curves (self-intersection -1), and are all the exceptional curves on X, since X_0 has none. Therefore Z is stable under any automorphism of X, hence X is not homogeneous. There remains to see that it is symmetric.

Let N_1 and N_2 be copies of the group N of 3.1, acting on D_1 and D_2 as N on D, and G_1 be the direct product of N_1 and N_2. It is a group of automorphisms of X_0 which leaves Z_0 stable, and is transitive on Z_0. Let G be the semi-direct product of G_1 by a group of order two, generated by an element A which permutes N_1 and N_2. To A we associate the automorphism $(z_1, z_2) \mapsto (z_2, z_1)$ of X_0. It also leaves Z_0 stable. We have therefore now a representation of G as a group of automorphisms of X_0 leaving Z_0 stable, and transitive on Z_0. It then extends canonically to a group of automorphisms of X leaving Z stable, and transitive on the set of components of Z. We claim that for every $x \in X$, the group G contains a symmetry at x.

If $x \notin Z$, this follows readily from 3.1 and from the fact that π commutes with G and defines an isomorphism of $X - Z$ onto $X_0 - Z_0$. Since furthermore G permutes transitively the components of Z, it suffices to consider the case where

$$x \in Y = \pi^{-1}((0, 0)).$$

The identity component T of G is the product of two copies of $\mathbb{C}^*$, exchanged by A. The stability group of Y in G is the subgroup H generated by T and A. We can view (z_1, z_2) as homogeneous coordinates on Y. The elements of T multiply those coordinates by arbitrary non-zero complex numbers and A exchanges them. Therefore H yields the same transformations of the projective line as N in 3.1. Although it is not effective, it is easily seen that H contains an element h_x of order two whose restriction to Y is a symmetry at x: if x has homogeneous coordinates $(1, 0)$ or $(0, 1)$, take $h_x = (1, -1) \in T$. If x has homogeneous coordinates $(1, z)$, take $h_x = A \cdot (z, z^{-1})$. Let V be the (one-dimensional) quotient of the tangent space $T(X)_x$ to X at x by the tangent space $T(Y)_x$ to Y at x. Then the eigenvalue of h_x on V is ± 1. Let $t_0 = (-1, -1) \in T$. The differential at $(0, 0)$ of t_0, viewed as automorphism of X_0, is $-\mathrm{Id}$. As an automorphism of X, the element t_0 acts trivially on Y. Since π maps V injectively into $T(X_0)_{(0, 0)}$ and commutes with H, the eigenvalue of t_0 in V is -1. The element t_0 is in the center of H, hence $t \cdot h_x$ is also of order two. It follows that one of h_x, $t_0 \cdot h_x$ is a symmetry at x.

3.4. Let X_0 be the n-dimensional complex projective space $\mathbb{P}_n(\mathbb{C})$, $(n \geq 2)$. Let $(x_0, \ldots, x_n)$ be the homogeneous coordinates on X_0. We identify the group of automorphisms of X_0 with $\mathbb{SL}_{n+1}(\mathbb{C})$ modulo its center. Let H_0 be the stability group of $P_0 = (1, 0, \ldots, 0)$, B the group of upper triangular matrices and T the group of diagonal matrices. The symmetry at P_0 is represented by an element s_0 of finite order in T: if n is odd, it is the diagonal matrix with first entry 1 and all others -1. If n is even, it is the product of that matrix by $\lambda \cdot \mathrm{Id}$. where $\lambda^{n+1} = -1$. By Bruhat decomposition, which is elementary in the present case, X_0 is the union of finitely many orbits of B, each of which contains a fixed point of T which is the transform of P_0 by an element in the normalizer of T. Consequently, B contains a symmetry for every $x \in X_0$. A fortiori, the same is true for H_0.

3.5. Proposition. *Let X be the manifold obtained from $\mathbb{P}_n(\mathbb{C})$ by blowing up one point $(n \geq 2)$. Then X is symmetric, but not homogeneous.*

The argument is quite similar to that of 3.3.

By homogeneity, we may assume the point to be P_0. Let $\pi: X \to X_0$ be the canonical projection and $Y = \pi^{-1}(P_0)$. Then $Y \cong \mathbb{P}_{n-1}(\mathbb{C})$, and π is an isomorphism of $X - Y$ onto $X_0 - P_0$. The submanifold Y is the only irreducible divisor whose fundamental class, viewed as an element of the integral homology group $H_*(X; \mathbb{Z})$ of X, maps to zero under the homomorphism $\pi_*: H_*(X; \mathbb{Z}) \to H_*(X_0; \mathbb{Z})$; therefore Y is stable under any automorphism of X homotopic to the identity, and X is not homogeneous. There remains to check that X is symmetric. The group H_0 extends canonically to a group of automorphisms of X. If $x \notin Y$, then H_0 contains a symmetry to x by 3.4. The group H_0 is the semi-direct product of $\mathbb{GL}_n(\mathbb{C})$ by $\mathbb{C}^n$. The map which assigns to $g \in H_0$ its differential at P_0 induces an isomorphism of $\mathbb{GL}_n(\mathbb{C})$ onto the general linear group of the tangent space $T(X_0)_0$ to X_0 at P_0. Consequently given any automorphism σ of Y, there exists $g \in \mathbb{GL}_n(\mathbb{C})$ which induces σ on Y. In particular, by 3.4 we see that $\mathbb{GL}_n(\mathbb{C})$ contains an element of finite order h_x whose restriction to Y is the symmetry at x. Let $-\varrho$ be its eigenvalue on $V = T(X)_x/T(Y)_x$. Let $t = \varrho^{-1} \cdot \mathrm{Id}. \in \mathbb{GL}_n(\mathbb{C})$. Then t acts trivially on Y and its differential on $T(X_0)_0$ is $\varrho^{-1} \cdot \mathrm{Id}$. Since π commutes with H_0 and maps V isomorphically onto a subspace of $T(X_0)_0$, the eigenvalue of t on V is also ϱ^{-1}. Therefore $t \cdot h_x$ is an automorphism of finite order whose differential at x is $-\mathrm{Id}$., hence a symmetry at x (1.2).

References

[1] S. Bochner and D. Montgomery, Groups on analytic manifolds. Annals of Math. (2) **48**, 659—669 (1947).

[2] A. Borel et A. Haefliger, La classe fondamentale d'un espace analytique. Bull. Soc. Math. France **89**, 461—513 (1961).

[3] A. Borel und R. Remmert, Über kompakte homogene Kählersche Mannigfaltigkeiten. Math. Ann. **145**, 429—439 (1962).

[4] N. Bourbaki, Topologie générale. Chap. 5 à 10. Paris 1974.

[5] H. Cartan, Sur les groupes de transformations analytiques. Act. Sci. Inc. **198**, Paris 1935.

[6] H. Holman, Komplexe Räume mit komplexen Transformationsgruppen. Math. Ann. **150**, 327—360 (1963).

[7] H. Kerner, Über die Automorphismengruppen kompakter komplexer Räume. Arch. Math. **11**, 282—288 (1960).

[8] W. Kaup, Infinitesimale Transformationsgruppen komplexer Räume. Math. Ann. **190**, 72—92 (1965).

[9] D. Montgomery, Simply connected homogeneous spaces. Proc. Amer. Math. Soc. **1**, 467 —469 (1950).

[10] G. D. Mostow, Self-adjoint group. Ann. of Math. (2) **62**, 44—55 (1955).

[11] J. de Siebenthal, Sur les groupes de Lie compacts non connexes. Comm. Math. Helv. **31**, 41—89 (1956).

Eingegangen am 1. 6. 1979

Anschrift des Autors:

A. Borel
School of Mathematics
The Institute for Advanced Study
Princeton, New Jersey 08540 USA

115.

(with N. Wallach)

Continuous cohomology, discrete subgroups and representations of reductive groups

Ann. Math. Stud. **94** (1980)

INTRODUCTION

1. This monograph is mainly concerned with two types of cohomology spaces pertaining to a reductive Lie group G (real, p-adic, or product of such groups) and a discrete cocompact subgroup Γ of G. The first one is the Eilenberg-MacLane cohomology space $H^*(\Gamma; E)$ of Γ with coefficients in a finite dimensional unitary Γ-module (or a finite dimensional G-module if G is real). The second one is attached to G, or its Lie algebra $\mathfrak{g}$ and a maximal compact subgroup K if G is real, and a representation V of G, usually infinite dimensional, and appears in various guises: continuous, smooth, or also (for G real) relative Lie algebra cohomology. Our initial interest was in the former one. However, its study may be reduced in part to the latter one (see VII, XIII), where G is the ambiant group and V runs through the irreducible subspaces of $L^2(\Gamma \setminus G)$. The determination of this cohomology is then a first step towards that of $H^*(\Gamma; E)$. But, as this work developed, we were led to emphasize it more and more, and to treat it as our main topic rather than as an auxiliary one. In fact, ten out of thirteen chapters are devoted to it, or directly motivated by it.

The material presented here divides naturally into two parts, one devoted mainly to real Lie groups (I to IX), the other to locally compact totally disconnected groups (in short t.d. groups), in particular reductive p-adic groups, or products of real Lie groups and t.d. groups (X to XIII). Each part in turn contains roughly three main items: general results on the cohomology used, specific ones for cohomology and representations of reductive groups and applications to discrete cocompact subgroups.

We now give some indications on the contents of the various chapters.

605

2. In Chapters I to VIII, G is a real Lie group with finitely many connected components, and the underlying cohomology is the relative Lie algebra cohomology $H^*(\mathfrak{g},\mathfrak{k};V)$ or rather, to allow for non-connected G's, a slight modification of it denoted $H^*(\mathfrak{g},K;V)$. Chapter I is devoted to foundational material on that cohomology. In §§1 to 4, $\mathfrak{g}$ is a finite dimensional Lie algebra over a field of characteristic zero and $\mathfrak{k}$ a subalgebra. §1 recalls the direct definition of $H^*(\mathfrak{g},\mathfrak{k};V)$, §2 discusses more generally the derived functors of $\mathrm{Hom}_\mathfrak{g}$ in the category $\mathcal{C}_{\mathfrak{g},\mathfrak{k}}$ of $(\mathfrak{g},\mathfrak{k})$-modules, i.e., $\mathfrak{g}$-modules which are locally finite and semi-simple with respect to $\mathfrak{k}$. This approach differs only in minor details from that of G. Hochschild, in the framework of relative homological algebra. The translation in the formalism of Yoneda's long extensions is briefly recalled in §3. In §4, we give two proofs of a useful vanishing theorem of D. Wigner. From §5 on, $F = \mathbf{R}$, $\mathfrak{g}$ is the Lie algebra of G and $\mathfrak{k}$ that of a maximal compact subgroup K of G. In §5, we transpose the previous considerations to the category of $(\mathfrak{g},K)$-modules. In §6 we introduce a slightly different category $\mathcal{C}_{\mathfrak{g},\mathfrak{k},L}$, solely as a tool to prove the existence of a Hochschild-Serre spectral sequence for $(\mathfrak{g},K)$-modules. Also included are two results of Casselman (5.5) and of D. Vogan (2.8) on finitely generated or admissible modules, and a Poincaré duality theorem of D. Vogan when G is semi-simple and V irreducible admissible (§7).

Chapter II is devoted to the case where $\mathfrak{g}$ is semi-simple (or reductive) and the coefficient module is the tensor product of a finite dimensional G-module E by a unitary G-module V. The cochain complex for relative Lie algebra cohomology admits then a natural scalar product. Various constructions and results of Matsushima, Matsushima-Murakami, Kuga, originating in differential geometry and Hodge theory and discussed by them in the context of discrete cocompact subgroups, are adapted to our setting in §§1 to 4, and §8; in a similar vein, §§6, 7 prove some vanishing theorems by use of spinors, suggested by results of Hotta and Parthasarathy on discrete subgroups. In §5, we consider the case where V belongs to the discrete series and show, using the characterization of the minimal K-type in V, that $H^q(\mathfrak{g}, K; E \otimes V)$ vanishes unless $2q = \dim G/K$ and V has the same infinitesimal character as the contragredient representation E^* to E.

The main topic of Chapter III is the cohomology with respect to a principal series representation. The computation uses an analogue of Shapiro's lemma (2.5), a description of K-finite vectors in induced representations (2.4), results of B. Kostant on the cohomology of nilpotent radicals of parabolic subalgebras and the Hochschild-Serre spectral sequence (§3). The results are applied in §4 to the determination of the cohomology with respect to tempered representations: in particular, it can be non-zero only in a small interval around the middle dimension and if the underlying parabolic subgroup is fundamental. These results have also been proved independently by G. Zuckerman, and those of §3 for complex semi-simple Lie algebras by P. Delorme. The last paragraph of III contains some general remarks on C^∞-vectors of induced representations, proving in particular that these are smooth functions in the cases of interest to us.

The next step is the investigation of the cohomology with respect to non-tempered representations. It is based on the Langlands classification of irreducible admissible $(\mathfrak{g}, K)$-modules and on two complements to it: some information on the Langlands parameters of the constituents of the kernel of the intertwining operators used by Langlands, and a necessary condition for unitarizability (in fact, for uniform boundedness) in terms of the Langlands parameters. The latter sharpens a result of R. Howe stating that the coefficients of a unitary representation with compact kernel vanish at infinity. These results are proved in Chapter IV (see 4.13, 5.2), which also contains a proof of the Langlands classification (4.11).

The uniform boundedness condition singles out a subset denoted $\Pi_\infty(G)$ of the set $\Pi(G)$ of infinitesimal equivalence classes of irreducible admissible $(\mathfrak{g}, K)$-modules (V, §2). It contains the unitary representations with compact kernel. Chapters V and VI are devoted to the cohomology with coefficients in $\Pi_\infty(G)$, or also in $V \otimes E$, where V represents an element of $\Pi_\infty(G)$ and E is finite dimensional, irreducible. We prove first that $H^q(\mathfrak{g}, K; V \otimes E)$ vanishes for $q < \mathrm{rk}_{\mathbf{R}}\, G$ (3.3), a result also obtained independently by G. Zuckerman. For E trivial, this bound is sharp in $\Pi_\infty(G)$ (but not always in the unitary dual $\hat{G}$ of G, see (II, 8.7)): in §4, it is shown that the constituents of (an analogue of) the Steinberg representation are all in $\Pi_\infty(G)$, and that $H^q(\mathfrak{g}, K; V) \neq 0$ if $q = \mathrm{rk}_{\mathbf{R}}\, G$ for at least one of them. §5 reproves some results of P. Delorme on the relation between H^1 and the topology of $\hat{G}$.

Chapter VI gives some further information on the cohomology with respect to a Langlands quotient $J_{P,\sigma,\nu}$. We need only consider the $J_{P,\sigma,\nu}$ with the same infinitesimal character as the trivial representation. The criterion IV, 5.2 gives an upper bound for ν. The general pattern which emerges is that, roughly, the bigger ν (in a suitable order relation), the lower the first non-vanishing cohomology group. Since the cohomology with respect to tempered representations is non-zero only close to the middle dimension, this suggests proceeding by increasing induction on ν. Without attempting to do this in general, we illustrate this relationship in Chapter VI by some general results when ν is minimal (§§1, 2) or $\mathrm{rk}_{\mathbf{R}}\, G = 1$ (§3), and by a complete determination of the cohomology when $G = SO(n, 1)$, $SU(n, 1)$ in §4.

Chapter VII is devoted to the cohomology of discrete subgroups. First if Γ is a discrete subgroup of the Lie group G, and E is a G-module, then we have the (well-known) formula

$$(1) \qquad H^*(\Gamma; E) = H^*(\mathfrak{g}, K; C^\infty(\Gamma \backslash G) \otimes E)$$

(2.7). If now Γ is cocompact, then $L^2(\Gamma \backslash G)$ admits a Hilbert discrete sum decomposition with finite multiplicities

$$(2) \qquad L^2(\Gamma \backslash G) = \oplus_{\pi \in \hat{G}}\, m(\pi, \Gamma) H_\pi\ ,$$

and (1) transforms to

$$(3) \qquad H^*(\Gamma; E) = \oplus_{\pi \in \hat{G}}\, m(\pi, \Gamma) H^*(\mathfrak{g}, K; H_\pi \otimes E)$$

(5.2). There is also a counterpart to that formula when E is a unitary Γ-module, involving the decomposition of the unitarily induced representation $I^G_{\Gamma,2}(E)$ (3.2). Various consequences of the results of the previous chapters are drawn in §§4, 6.

Chapter VIII is concerned with cohomology at the **R**-rank q when $G = SU(p, q)\,(p \geq q)$. Let F_ℓ be the irreducible G-module whose highest weight is ℓ times the highest weight of the standard representation of $SU(p, q)$ in $\mathbf{C}^{p+q}$. For each $\ell \geq q$ there is a unitary irreducible representation H_ℓ of G such that $H^q(\mathfrak{g}, K; H_\ell \otimes F_{\ell-q}) \neq 0$ (2.13). It is then shown that certain cocompact arithmetically defined subgroups of G have subgroups of finite index Γ' such that H_ℓ occurs in $L^2(\Gamma \backslash G)$, whence in particular $H^q(\Gamma'; F_{\ell-q}) \neq 0$. This extends a result of Kazhdan concerning the case where $q = 1$, which gave the first examples of discrete cocompact subgroups of $SU(n, 1)$ with non-vanishing first Betti number for arbitrary n. The proof uses the metaplectic representation and the duality theorem, and is quite similar to that of Kazhdan, although the context is a bit different, since Kazhdan worked with adelic groups.

3. Chapters IX to XII are devoted to continuous and smooth cohomology. §§1 to 4 of Chapter IX contain some basic material concerning derived functors in the category $\mathcal{C}_G$ of continuous G-modules (always assumed to be locally convex Hausdorff topological vector spaces over **C**), when G is a locally compact group (countable at infinity). The approach is the one of Hochschild-Mostow, based on the use of injective modules relative to G-morphisms which are strong (i.e. split for the underlying structure of topological vector spaces). After that, we are concerned with real groups (IX, §§5, 6), t.d. (totally disconnected) groups, in particular p-adic groups (X, XI), and products of such groups (XII). The formal analogies between these three cases are emphasized. In each, besides $\mathcal{C}_G$, we consider the categories $\mathcal{C}^\infty_G$ of smooth topological G-modules and $\mathcal{C}^f_G$ of non-degenerate modules over a suitable Hecke algebra. The last one (introduced in substance by Jacquet-Langlands) is abelian and the modules in it are just complex vector spaces. The Hecke algebras occurring here have no unit in general, a situation not considered in standard texts on homological algebra. However they are idempotented and this allows one to extend some standard constructions to our case (XII, §0). In particular, $\mathcal{C}_G^f$ has enough injectives. There are natural functors

$$\mathcal{C}_G \xleftarrow[\alpha]{\gamma} \mathcal{C}_G^\infty \xrightarrow{\beta} \mathcal{C}_G^f \, ,$$

where α (resp. β) is the passage to smooth (resp. K-finite vectors) and γ is the inclusion. γ preserves derived functors and β cohomology for quasi-complete spaces, α preserves derived functors for quasi-complete spaces in the t.d. case, and cohomology for Fréchet spaces in the other two cases.

In the real case, $\mathcal{C}_G^\infty$ consists of the usual differentiable modules, with the C^∞-topology, while, up to Chapter IX, $\mathcal{C}_G^f$ is just the category of $(\mathfrak{g}, K)$-modules. But, as is known, it may also be viewed as the category of non-degenerate modules over the Hecke algebra $\mathcal{H}(\mathfrak{g}, K)$ of bi-K-finite distributions on G with support in K. This point of view is more convenient to treat the mixed case, and is introduced later (XII, §2). The above conservation theorems for derived functors in the real case (due to Hochschild-Mostow, W. v. Est, P. Blanc) are proved in IX, §§5, 6.

If G is a t.d. group (X, §1), then a topological G-module V is smooth if every $v \in V$ is fixed under an open subgroup and V is, topologically, the inductive limit of the subspaces V^L of fixed points under compact open subgroups L of G. The Hecke algebra underlying the definition of $\mathcal{C}_G^f$ is the convolution algebra of locally constant compactly supported functions. The main case of interest is when $G = \mathcal{G}(k)$, where k is a non-archimedean local field and $\mathcal{G}$ a connected reductive k-group. If $V \in \mathcal{C}_G$, then the V-valued cochains of the Bruhat-Tits building of G provide an s-injective resolution of V (X, §2). In §4 of X we prove the results of W. Casselman which give a complete description of the cohomology of G with respect to an irreducible admissible

G-module. §5 is devoted to $\mathcal{C}_G^f$, and the passage to $\mathcal{C}_G^f$ is used in §6 to prove some Künneth rules.

Chapter XI is a p-adic counterpart of IV. It discusses the analogue of the Langlands classification, and of the uniform boundedness condition. The latter is used to show that the only irreducible admissible representations with compact kernel, with respect to which G has non-vanishing cohomology in some dimension $q \neq 0$, $\mathrm{rk}_k \mathcal{G}$, are non-unitarizable (a result due to W. Casselman).

Let now $G = G_1 \times G_2$ be the product of a real Lie group G_1 and a t.d. group G_2. A topological continuous G-module V is said to be smooth if it is smooth with respect to G_1 and G_2 and if it is the topological inductive limit of the subspaces V^L, where L runs through the compact open subgroups of G_2.

There are also intermediate categories of continuous G-modules smooth with respect to one of the factors. The relations between the corresponding derived functors are discussed in §1. In §2, we fix a maximal compact subgroup K_1 of G_1 and pass to the $(K_1 \times L)$-finite vectors, where L is a compact open subgroup of G_2, which brings us to the non-degenerate modules over the Hecke algebra $\mathcal{H}(G) = \mathcal{H}(\mathfrak{g}_1, K_1) \otimes \mathcal{H}(G_2)$. §3 is devoted to some Künneth rules and to applications to the cohomology of products of reductive groups or of adelic groups.

In Chapter XIII, we consider the cohomology space $H^*(\Gamma; E)$, where Γ is a discrete cocompact subgroup of G and E a finite dimensional unitary Γ-module, first in general (§1), then when G is a product of reductive groups G_s $(s \in S)$. In the latter case, we have a formula quite similar to (3), except that $L^2(\Gamma \backslash G)$ is replaced by the unitarily induced representation from E. Furthermore, since the G_s's are of type I, each $\pi \in \hat{G}$ is a Hilbert tensor product $\pi = \hat{\otimes}_s \pi_s$ $(\pi_s \in \hat{G}_s)$, and the Künneth rule gives

$$(4) \qquad H_{ct}^*(G; H_\pi) = \otimes_s H_{ct}(G_s; H_{\pi_s}) .$$

This allows us to apply the earlier results on continuous cohomology of

real or p-adic groups. We then pass to some applications. We prove the
Casselman vanishing theorem (2.6) and extend it to the case where Γ is
irreducible (3.1) in a product of semi-simple groups over non-archimedean
fields (3.6). Following a suggestion of G. Prasad, we also show it to be
valid when E is a finite dimensional vector space over an arbitrary field
of characteristic zero, and G has rank ≥ 2, using a theorem of Margulis
(3.7). Finally, we prove that if $G = \mathcal{G}(A)$ is the adele group of a semi-
simple anisotropic group $\mathcal{G}$ over a global field, then $H^*(\mathcal{G}(k); \mathbf{R})$ reduces
to the continuous cohomology of the archimedean factor of $\mathcal{G}(A)$ (3.9).

A survey of some of the main results on vanishing and non-vanishing
cohomology is given at the end of the book.

4. This monograph is an outgrowth of a seminar on the "Cohomology
of discrete subgroups of semi-simple Lie groups" held at The Institute
for Advanced Study in 1976-77. A first set of Notes was written and dis-
tributed at that time. Most of the material of these Notes is incorporated
in Chapters I to IX, except for some results which were rendered somewhat
obsolete by others found in the course of the seminar. There was also
some discussion of the p-adic case in the seminar but it was not written
up then. In the first version, we kept track of who did what and each
chapter was accordingly authored or coauthored. It would have been quite
awkward to do so in the present version, which represents a considerable
reorganization and expansion of the first one. Rather, we prefer to take
joint responsibility for the results and mistakes in this book, except however that the first (resp. second) named author wishes to leave credit for
Chapters IV, VIII, XI (resp. III (§§3 to 6), IX, XII, XIII) to the second (resp.
first) named author.

The transition from the first to the final version was a rather painful
process, involving a long series of changes, additions, amplifications,
corrections upon corrections, reshuffling and renumbering. We are very
grateful to the secretaries of the School of Mathematics, and in particular
to Peggy Murray, who had by far the greatest load, to have taken care so
skillfully and so speedily of this endless series of changes upon changes,

which required expertise not only in typing but in cutting, pasting and collage as well.

A reference such as 3.4 (resp. 3.4(1)) refers to section 3.4 (resp. relation 3.4(1)) of the same chapter; if preceded by a capitalized Roman numeral it refers to the corresponding section or relation of the chapter denoted by that numeral.

A. BOREL, N. WALLACH[*]
July 1978

THE INSTITUTE FOR ADVANCED STUDY, Princeton, N. J. 08540
RUTGERS UNIVERSITY, New Brunswick, N. J. 08903

[*] The second named author did part of this work while enjoying the hospitality of Brandeis University. He also wishes to acknowledge partial support from NSF grant number MCS 77-04278 AO1.

116.

Stable and L^2-cohomology of arithmetic groups

Bull. Amer. Math. Soc., (N.S.) 3 (1980) 1025–1027

Introduction. In [1], [2] we gave a range of dimensions in which the real cohomology of an arithmetic or S-arithmetic subgroup Γ of a connected semi-simple group G over $\mathbf{Q}$ is naturally isomorphic to the space of harmonic forms on the quotient $X = G(\mathbf{R})/K$ of the group $G(\mathbf{R})$ of real points of G by a maximal compact subgroup K which are invariant under Γ and the identity component $G(\mathbf{R})^0$ of $G(\mathbf{R})$, and indicated some applications to the stable cohomology of classical arithmetic groups and to algebraic K-theory. In this note we first state an extension to nontrivial coefficients, since this has become of interest in topology and K-theory [7]. A chief tool in [2] was the proof that $H^*(\Gamma; \mathbf{C})$ could be computed using differential forms on $\Gamma\backslash X$ which have "logarithmic growth" at infinity. Theorem 2 extends this to more general growth conditions. This can be used to show that certain L^2-harmonic forms are not cohomologous to zero [9]. In §§3, 4, 5 we consider the L^2-cohomology space $H_{(2)}(\Gamma\backslash X)$ and relate it to the spectral decomposition of the space $L^2(\Gamma\backslash G)$ of square integrable functions on $\Gamma\backslash G$. Theorem 4 gives a sufficient condition under which it is finite dimensional, hence isomorphic to the space of square integrable harmonic forms, and §5 a series of examples in which it is not. For convenience, we assume G simple over $\mathbf{Q}$ and Γ torsion-free.

1   **1.** Let P_0 be a minimal parabolic Q-subgroup of G, S a maximal Q-split torus of P_0, N the unipotent radical of P and $\mathfrak{n}$ the Lie algebra of N. Let $X(S)$ be the group of rational characters of S and $\rho \in X(S)$ be such that $a^{2\rho} = \det \operatorname{Ad} a|_{\mathfrak{n}}$ for $a \in S$. For $\mu \in X(S)$ let $c(G, \mu)$ be the maximum of q such that $\rho - \mu - \eta > 0$, where η runs through the weights of S in $\Lambda^q\mathfrak{n}$. Let $c(G) = c(G, 0)$. If (r, E) is a finite-dimensional complex representation of $G(\mathbf{C})$, we let $c(G, r)$ be the minimum of $c(G, \mu)$, where μ runs through the weights of r with respect to S. It is easily seen that $c(G) \geqslant \Sigma_i c(G_i)$, where G_i runs through the simple factors of $G(\mathbf{C})$, and $c(G_i)$ is defined similarly, and that $c(G_i)$ is equal to $[(l-1)/2], l-1, l-2, l-1, 7, 13, 25, 5, 1$ if G_i is of type $A_l, B_l, C_l, D_l, E_6, E_7, E_8, F_4, G_2$.

THEOREM 1. *The natural homomorphism $H^*(\mathfrak{g}, \mathfrak{k}; E)^\Gamma \to H^q(\Gamma; E)$ is injective for $q \leqslant c(G, r)$, surjective if in addition $q < \operatorname{rk}_{\mathbf{R}} G$. If $E^G = (0)$, then $H^q(\Gamma; E) = 0$ for $q \leqslant c(G, r), (\operatorname{rk}_{\mathbf{R}} G - 1)$. If G is simply connected, these*

Received by the editors May 29, 1980.

1980 *Mathematics Subject Classification.* Primary 18H10; Secondary 20G10, 20G30, 53C39.

614

assertions remain true if Γ is replaced by an S-arithmetic subgroup or by $G(\mathbf{Q})$.

Here $\mathfrak{g}$ and $\mathfrak{k}$ stand for the Lie algebras of $G(\mathbf{R})$ and K. See [5] for relative Lie algebra cohomology. The proofs of these statements are similar in principle to those given or sketched in [1], [2] when r is the trivial representation, and moreover, take into account some results proved in [5]. If we have an inductive system of groups and representations without trivial constituents $(G_n, \Gamma_n, r_n, E_n)$ such as (G, Γ, r, E) and if $c(G_n, r_n) \to \infty$, then Theorem 1 implies that $H^q(\lim \Gamma_n, \lim E_n) = 0$ for $q > 0$.

2. On Siegel sets, we consider coefficients of differential forms with respect to special frames, as in [2]. For $\lambda \in X(S)$ we say that $\eta \in \Omega_{\lambda+}(\Gamma \backslash X)$ if the coefficients of η and of $d\eta$ satisfy a growth condition,

$$|f(x)| < a(x)^\lambda \, |P(\log a^{\alpha_1}, \ldots, \log a^{\alpha_l})|, \tag{1}$$

where $\alpha_1, \ldots, \alpha_l$ are the simple $\mathbf{Q}$-roots and P is a polynomial in l variables ($l = \dim S$). The proof of the following theorem is analogous to that of 7.4 in [2].

THEOREM 2. *If λ is dominant, then the injection $\Omega_{\lambda+}(\Gamma \backslash X) \to \Omega(\Gamma \backslash X)$ induces an isomorphism in cohomology. The elements of $\Omega^q_{\lambda+}$ are square integrable if $q \leqslant c(G, \lambda)$. The space of square integrable harmonic q-forms contained in $\Omega_{\lambda+}(\Gamma \backslash X)$ maps injectively into the cohomology of Γ for $q \leqslant c(G, \lambda) + 1$. If $\lambda < 0$, then $H^*(\Omega_{\lambda+})$ is canonically isomorphic to the complex cohomology with compact supports of $\Gamma \backslash X$.*

3. Let M be a Riemannian manifold. Let $\Omega_{(2)}(M)$ be the complex of differential forms η on M such that η and $d\eta$ are square integrable. By definition $H_{(2)}(M) = H^*(\Omega_{(2)}(M))$ is the space of L^2-cohomology of M. (See [6], where equivalent Hilbert space definitions are given.) Let $H_{(2)}(M)$ be the space of L^2-harmonic forms. It is known that if M is complete, then the natural map j: $H_{(2)}(M) \to H_{(2)}M$ is injective. If M is compact, then j is an isomorphism and $H_{(2)}(M) = H^*(M; \mathbf{C})$.

THEOREM 3. *There are canonical isomorphisms*

$$H_{(2)}(\Gamma \backslash X) = H^*(\mathfrak{g}, K; L^2(\Gamma \backslash G)^\infty)$$

and

$$H_{(2)}(\Gamma \backslash G) = H^*(\mathfrak{g}; L^2(\Gamma \backslash G)^\infty).$$

As usual, if (π, V) denotes a unitary representation of $G(\mathbf{R})$ then V^∞ denotes the space of C^∞-vectors in V. To establish Theorem 3, one proves first the second statement using a homotopy operator defined by the convolution by a compactly supported smooth function on G, and then deduces the first one by the comparison theorem for spectral sequences, applied to suitable spectral sequences in relative Lie algebra cohomology.

4. The space $L^2(\Gamma\backslash G)$ is the sum of the discrete spectrum $L^2(\Gamma\backslash G)_d$ and the continuous spectrum $L^2(\Gamma\backslash G)_{ct}$. By results obtained jointly with H. Garland [3], [4], $H_{(2)}(\Gamma\backslash X)$ is finite dimensional and is the direct sum of the spaces $H^*(\mathfrak{g}, K; H_i^\infty)$, where H_i runs through a set of irreducible constituents of $L^2(\Gamma\backslash G)_d$. By [8], $L^2(\Gamma\backslash G)_{ct}$ is a Hilbert direct sum of invariant subspaces, say V_i $(i \in I)$, each of which is a continuous integral of unitarily induced principal series (from parabolic **Q**-subgroups). By [4], $H^*(\mathfrak{g}, K; L^2(\Gamma\backslash G)_{ct}^\infty)$ is the sum of the $H^*(\mathfrak{g}, K; V_i^\infty)$ and can be nonzero only for finitely many terms. Those spaces can be computed as in [5, III] and can be nonzero only if the underlying parabolic subgroup is fundamental [5, IV] in $G(\mathbf{R})$. Together with Theorem 3, this proves

THEOREM 4. *The map $j\colon H_{(2)}(\Gamma\backslash X) \to H_{(2)}(\Gamma\backslash X)$ is an isomorphism if G has no proper parabolic **Q**-subgroup which is fundamental in $G(\mathbf{R})$, in particular if* rank G = rank K.

5. It is rather likely that if G has a proper fundamental parabolic subgroup P_1 defined over **Q**, then Γ has a subgroup Γ' of finite index such that $H_{(2)}(\Gamma'\backslash X)$ is infinite dimensional. This has been checked in a number of cases: (i) $G = \mathbf{SO}(n, 1)$ for $n \geqslant 3$ odd (with $\Gamma = \Gamma'$); (ii) the group P_1 is minimal over **R**; (iii) $G = \mathbf{SL}_n(\mathbf{R})$ and $\Gamma \subset \mathbf{SL}_n(\mathbf{Z})$. In those cases, infinite-dimensional cohomology occurs exactly in the dimensions q such that dim $X - l_0 < 2q \leqslant$ dim $X + l_0$, where l_0 = rank G − rank K.

REFERENCES

1. A. Borel, *Cohomologie réelle stable de groupes S-arithmétiques classiques*, C.R. Acad. Sci. Paris 274 (1972), 1700–1702.

2. ———, *Stable real cohomology of arithmetic groups*, Annales E.N.S. Paris (4) 7 (1974), 235–272.

3. ———, *Cohomology of arithmetic groups*, Proc. Internat. Congr. Math. Vancouver, 1974, vol. 1, pp. 435–442.

4. A. Borel and H. Garland, *Laplacian and discrete spectrum of an arithmetic group* (in preparation).

5. A. Borel and N. Wallach, *Continuous cohomology, discrete subgroups and representations of reductive groups*, Ann. of Math. Studies, No. 94, Princeton Univ. Press, Princeton, N.J., 1980.

6. J. Cheeger, *On the Hodge theory of Riemannian pseudomanifolds*, Proc. Sympos. Pure Math., vol. 36, Amer. Math. Soc., Providence, R.I., 1980, pp. 91–146.

7. T. Farrell and W. C. Hsiang, *On the rational homotopy groups of the diffeomorphism groups of discs, spheres and aspherical manifolds*, Proc. Sympos. Pure Math. vol. 32, part 1, Amer. Math. Soc., Providence, R.I., 1978, pp. 403–415.

8. R. P. Langlands, *On the functional equations satisfied by Eisenstein series*, Lecture Notes in Math., vol. 544, Springer-Verlag, Berlin and New York, 1976.

9. N. Wallach, L^2*-automorphic forms and cohomology classes on arithmetic quotients of* $SU(p, q)$ (to appear).

SCHOOL OF MATHEMATICS, INSTITUTE FOR ADVANCED STUDY, PRINCETON, NEW JERSEY 08540

117.

Commensurability classes and volumes of hyperbolic 3-manifolds

Ann. Sc. Norm. Super. Pisa, Cl. Sci., (4) **8** (1981) 1–33

The first purpose of this paper is to answer some questions raised by W. Thurston [24: 6.7.6] about families of commensurable hyperbolic 3-manifolds. Two hyperbolic 3-manifolds M, M' are commensurable if they have two diffeomorphic finite coverings. M is said to be minimal if it does not properly cover any other hyperbolic 3-manifold. We shall see that if $\mathcal{M}$ is a full commensurability class of orientable hyperbolic 3-manifolds of finite volume, then $\mathcal{M}$ contains infinitely many non-isomorphic minimal elements if $\pi_1(M)$ is definable arithmetically, but only one if it is not (and V-manifolds are allowed). Moreover the volumes of the elements in $\mathcal{M}$ are all integral multiples of some number (which is not necessarily one of the volumes, though).

An orientable hyperbolic 3-manifold M is the quotient of the hyperbolic 3-space H^3 by a discrete torsion-free subgroup Γ of the identity component $I(H^3)^0$ of the group if isometries of H^3 (which is isomorphic to $\boldsymbol{PGL}_2(\boldsymbol{C})$). We shall also allow Γ to have torsion, and then M is a V-manifold or an « orbifold » in the terminology of [24] (i.e., it looks locally as the quotient of euclidean space by a finite linear group). $M = H^3/\Gamma$ and $M' = H^3/\Gamma'$ are commensurable if Γ and Γ' are commensurable up to conjugacy, i.e., if Γ and a conjugate ${}^g\Gamma' = g\Gamma'g^{-1}$ of Γ' by some $g \in I(H^3)^0$ are commensurable (their intersection has finite index in both). Moreover, M is minimal if and only if Γ is maximal in its commensurability class. Given two commensurable subgroups Γ, Γ' of a group, let us define the generalized index of Γ' in Γ by

$$(1) \qquad [\Gamma:\Gamma'] = [\Gamma:\Gamma \cap \Gamma'] \cdot [\Gamma':\Gamma \cap \Gamma']^{-1} \,.$$

Pervenuto alla Redazione il 24 Maggio 1980.

Then clearly

$$(2) \qquad \mu(\Gamma') = [\Gamma : \Gamma'] \cdot \mu(\Gamma) ,$$

where $\mu(\Gamma)$ denotes the volume of H^3/Γ with respect to the hyperbolic metric. Therefore the above results are equivalent to the following theorem:

THEOREM. *Let Γ be a discrete subgroup of $PGL_2(C)$ of finite covolume and $\mathcal{A}_\Gamma$ the set of subgroups of $PGL_2(C)$ commensurable with Γ.*

(i) *If Γ is arithmetic, then $\mathcal{A}_\Gamma$ contains infinitely many nonconjugate elements which are maximal, or maximal among torsion-free elements of $\mathcal{A}_\Gamma$. If Γ is non-arithmetic, then $\mathcal{A}_\Gamma$ has a biggest element.*

(ii) *The indices $[\Gamma : \Gamma']$ ($\Gamma' \in \mathcal{A}_\Gamma$) are integral multiples of some number.*

In fact, we shall prove this more generally when $PGL_2(C)$ is replaced by a product

$$(3) \qquad G_{a,b} = PGL_2(R)^a \times PGL_2(C)^b , \qquad (a, b \in N, a + b \geq 1) ,$$

and, for convenience, Γ is assumed to be « irreducible » (cf. [17 : 5.20, 5.21]) [i.e., it is not possible to write $G_{a,b}$ as a direct product $G_{a,b} = H \cdot H'$ (H, H' closed connected, non-trivial) such that $(\Gamma \cap H) \cdot (\Gamma \cap H')$ has finite index in Γ. This is equivalent to $\Gamma \cap N = \{1\}$ for each proper normal closed subgroup N, or also to the fact that the projection of Γ on any infinite non-trivial quotient of $G_{a,b}$ is non-discrete]. Geometrically, this means that we consider irreducible discrete groups of automorphisms of the product $H_{a,b}$ of a copies of the upper-half plane H^2 by b copies of the hyperbolic 3-space H^3. Our initial case of interest is then $a = 0$, $b = 1$. Also included are Fuchsian groups ($a = 1, b = 0$) or Hilbert-Blumenthal groups. The notion of arithmetic groups in the present case will be recalled in § 3. One commensurability class of such groups is associated to a number field k with b complex places and a quaternion algebra B over k which is unramified at a set of a real places of k (3.3). We denote it by $C(k, B)$. If $a + b \geq 2$, Γ is automatically arithmetic (3.4).

In the non-arithmetic case, these results are trivial consequences of a theorem announced by G. A. Margoulis in [13], as will be seen in § 1, so that we shall be mainly concerned with the arithmetic case. There we shall get a hold of maximal elements in $\mathcal{A}_\Gamma$ by looking at their closures in the groups of p-adic points of the form of PGL_2 underlying the definition of Γ. For this we shall need some facts on the Bruhat-Tits building of SL_2 and on the maximal compact subgroups of SL_2 and PGL_2 over a p-adic field which are reviewed in § 2. The assertion (i) above is proved in § 4, and (ii) in §5. To a maximal order $\mathfrak{O}$ of B, we associate a maximal element $\Gamma_{\mathfrak{O}}$ of $C(k, B)$ such that the minimum of the volume $\mu(\Gamma)$ $(\Gamma \in C(k, B))$ is achieved on $\Gamma_{\mathfrak{O}}$. The g.c.d. of the volumes in the class are multiples of $2^{-c} \cdot \mu(\Gamma_{\mathfrak{O}})$ where c is at most equal to the number of primes dividing two in k. If for instance $a = 0$, $b = 1$, $k = \mathbf{Q}(\sqrt{-3})$, and H^3/Γ is not compact, then $c = 1$.

§ 7 gives an expression of the value of $\mu(\Gamma_{\mathfrak{O}}^1)$, where $\Gamma_{\mathfrak{O}}^1$ is the image of the group of elements of reduced norm 1 in $\mathfrak{O}$, in terms of data depending only on the field k and on the places of k at which the quaternion algebra B is ramified (7.3). This formula is deduced here from the fact that the Tamagawa number of a k-form of $\mathbf{SL}_2$ is one [29] and from local computations of volumes made in § 6. It includes formulae of G. Humbert (see [24: § 7]) when k is imaginary quadratic and of C. L. Siegel [23] and Shimizu [22] when k is totally real, (7.5). The volumes in the class $C(k, B)$ are all rational multiples of a number depending only on k and on the number of real places at which B is ramified. From this and a result of R. Baer [1], it follows that, given an arithmetic subgroup Γ_1 of $\mathbf{G}_{a,b}$, there exists an infinite set F of arithmetic subgroups of $\mathbf{G}_{a,b}$, which are not pairwise commensurable up to conjugacy, such that however $\mu(\Gamma)$ is a rational multiple of $\mu(\Gamma_1)$ for all $\Gamma \in F$ (7.6).

If $b = 0$, it is well-known that all volumes are commensurable. It is widely expected that this is not so when $b \neq 0$, but as far as I know, this has not been proved. This raises some questions on values of zeta functions at two (7.7).

If $a + b \geq 2$, it is known that the set of all volumes is discrete ([28], cf. 8.3), but this is not so for $a + b = 1$. However, we shall prove that the set of covolumes of arithmetic subgroups is discrete (8.2). For this, we shall use estimates of Odlyzko's on the discriminant of number fields [15] and the fact that, given a constant c, there exists an integer $n(c)$ such that if $\mu(\Gamma) \leq c$, then Γ is generated by $n(c)$ elements. (For $a = 1$, this is standard; for $b = 1$, this is proved in [24: Chap. 13], cf. 8.1) For $a = 0$, $b = 1$, the results of [24] imply then that the arithmetic subgroups are comparatively rare among all discrete subgroups of finite covolume of $\mathbf{PGL}_2(\mathbf{C})$. In 8.4 to 8.6, we give an arithmetic expression for the index $[\Gamma_{\mathfrak{O}} : \Gamma_{\mathfrak{O}}^1]$. It involves the class group of k, which makes it difficult to estimate it. Therefore we also single out a subgroup Γ_{R_f} intermediary between $\Gamma_{\mathfrak{O}}$ and $\Gamma_{\mathfrak{O}}^1$, whose covolume is independent of the class group, and which is equal to $\Gamma_{\mathfrak{O}}$ if k has class number one.

Finally, §§ 9 and 10 contain some remarks on, and examples of, groups operating on hyperbolic 3-space or products of upper half-planes.

The study of maximal arithmetic subgroups proceeds along rather standard lines and can be (has been) carried out in much greater generality.

In fact, part of what we prove here in the non-cocompact case is contained in [9; 18; 19]. However the case of forms of PGL_2 has some peculiar features. I have therefore preferred to limit myself to it, but give a rather complete treatment.

I thank E. Bombieri for his computations of small volumes in the arithmetic case, R. P. Langlands for some helpful remarks on the local measures and volumes and W. Thurston for suggesting 8.2 when $a = 0$, $b = 1$ and for many useful conversations.

1. – The non-arithmetic case.

Given a subgroup H of a group G, we let C_H be the *commensurability subgroup* of H in G, i.e., the set of elements $x \in G$ such that ${}^xH = xHx^{-1}$ is commensurable with H. It is obviously a subgroup, which contains all subgroups of G commensurable with H. A result announced by G. Margoulis [13: Theorem 9] implies that if $G = G_{a,b}$ and Γ (always assumed to be irreducible) has finite covolume in G, then either C_Γ is dense and Γ is arithmetically definable (see § 3 for this notion), or C_Γ is discrete and Γ is not arithmetically definable. [In deducing this from Margoulis' theorem, we also use the fact that if C_Γ is not discrete, then it is dense, as follows from [17: 5.13].] Therefore, in the latter case C_Γ is the biggest element in the commensurability class of Γ. The assertion (ii) is then clear in that case. Also, the commensurability class of Γ has only one maximal element, at any rate if we allow torsion. However, it is conceivable that $\mathcal{A}_\Gamma$ may contain infinitely many conjugacy classes of subgroups of finite index which are maximal among torsion-free subgroups, in which case there would again be infinitely many non-isomorphic minimal manifolds (in the strict sense, i.e., without V-singularities) in the commensurability class of H/Γ. It was pointed out to me by R. Griess and J-P. Serre, independently, that $L = = SL_2 Z/\{\pm 1\}$ indeed contains infinitely many non-conjugate subgroups of finite index which are maximal among torsion-free subgroups of L. However I do not know whether this is the rule or the exception in the case under consideration.

2. – Maximal compact subgroups and buildings for SL_2 and PGL_2 over a local field.

2.1. In this section F denotes a non-archimedean local field with finite residue field and $\mathfrak{o}_F$ its ring of integers. In fact, only the case where F is a finite extension of the field Q_p of p-adic numbers (p prime) will be needed, but the facts recalled here are also valid in the equal characteristic case. Let p be the characteristic and q the order of the residue field. Let $||$ be the normalized valuation of F and $v(\ )$ the order of an element. Thus

$$(1) \qquad |x| = q^{-v(x)}, \qquad \text{with } v(x) \in Z.$$

2.2. For the contents of the section, see e.g. [21: Chap. II] and [10. Prop. 2.30, 2.31]. Let $\mathcal{C}$ be the Bruhat-Tits building of $SL_2(F)$. It is a tree. $SL_2(F)$ is transitive on the edges, and the vertices form two orbits O_1, O_2 under $SL_2(F)$. The stability groups of the vertices are the maximal compact subgroups of $SL_2(F)$ and form two conjugacy classes, represented by $K_1 = SL_2(\mathfrak{o}_F)$, and the group K_2 of matrices of determinant 1 of the form

$$\begin{pmatrix} a & \pi \cdot b \\ \pi^{-1}c & d \end{pmatrix}, \qquad (a, b, c, d \in \mathfrak{o}).$$

There are $q + 1$ edges with a given vertex $P \in \mathcal{C}_v$, and the stability group of P operates transitively on them. In fact, by reduction mod v, the group K_1 identifies to $SL_2(F_q)$ and the set of edges incident to the fixed point of K_1 identifies to the projective line $P^1(F_q)$ over F_q. The stability groups of the edges are the Iwahori subgroups of $SL_2(F)$. They form one conjugacy class under $SL_2(F)$, represented by $K_1 \cap K_2$.

An automorphism of $\mathcal{C}$ either leaves O_1, O_2 stable or permutes them. We shall say that it is *even* in the former case, *odd* otherwise. Any (continuous) automorphism of $SL_2(F)$ induces an automorphism of $\mathcal{C}$. In particular $GL_2(F)$ operates on $\mathcal{C}$. The elements which induce even (resp. odd) automorphisms are those $x \in GL_2(F)$ such that $v(\det x)$ is even (resp. odd), whence the terminology. The center Z of $GL_2(F)$ acts trivially on $\mathcal{C}$ so that $PGL_2(F)$, which is the quotient $GL_2(F)/Z$ can (and will) be viewed as a group of automorphisms of $\mathcal{C}$. An element of $GL_2(F)$ or $PGL_2(F)$ will be said to be even (resp. odd) if it defines an even (resp. odd) automorphism of $\mathcal{C}$. The even elements in $PGL_2(F)$ form a subgroup $PGL_2(F)_0$ of index two. Thus $PGL_2(F)$ is transitive on the vertices and on the edges of $\mathcal{C}$. It has two conjugacy classes of maximal compact subgroups, the stability groups of the vertices and the stability groups of the edges (or, equivalently, of the middle points of the edges). A maximal compact subgroup is of the latter kind if and only it contains an odd element.

2.3. LEMMA. *Let $D \supset C$, $D' \supset C'$ be compact subgroups of $PGL_2(F)$. Assume that C fixes a vertex of $\mathcal{C}$, has no other fixed point in $\mathcal{C}$, and that C' contains an odd element. Then D and D' are not conjugate.*

In fact, since C fixes a vertex P_0 it consists of even elements. Moreover, any compact subgroup of $\boldsymbol{PGL}_2(F)$ must fix some point of $\mathfrak{C}$, therefore D also fixes P_0. Then D consists of even elements, hence is not conjugate to D'.

3. – Arithmetic subgroups of $\boldsymbol{G}_{a,b}$.

In this section we describe the discrete subgroups of $\boldsymbol{G}_{a,b}$ which are definable arithmetically, to be called arithmetic for short.

3.0. Before doing so, however, we would like to relate $\boldsymbol{G}_{a,b}$ to the full group of isometries of $\boldsymbol{H}_{a,b}$.

The group $\boldsymbol{PGL}_2(\boldsymbol{C})$ is connected, and is also the quotient $\boldsymbol{SL}_2(\boldsymbol{C})/\{\pm 1\}$. It is the group of orientation preserving isometries of H^3. On the other hand, $\boldsymbol{PGL}_2(\boldsymbol{R})$ has two connected components. Its component of the identity is $\boldsymbol{SL}_2(\boldsymbol{R})/\{\pm 1\}$. The group $\boldsymbol{PGL}_2(\boldsymbol{R})$ may be identified to the group of isometries of H^2, the elements of $\boldsymbol{SL}_2(\boldsymbol{R})/\{\pm 1\}$ are holomorphic, orientation preserving, while the others are antiholomorphic, orientation reversing. Therefore $\boldsymbol{G}_{a,b}$ is the group of all isometries of $\boldsymbol{H}_{a,b}$ which preserve each factor and the orientation of the three-dimensional ones. It has 2^a connected components and is of index $2^b \cdot a! \cdot b!$ in the full group of isometries of $\boldsymbol{H}_{a,b}$. We have singled it out since it turns out to be the most convenient to use for the discussion of arithmetic subgroups.

3.1. In the sequel, k is a number field, $\mathfrak{o}_k$ or simply $\mathfrak{o}$ the ring of integers of k, d the degree of $k/\boldsymbol{Q}$, V (resp. V_∞, resp. V_f) the set of places (resp. infinite places, resp. finite places) of k and r_1 (resp. r_2) the number of real (resp. complex) places of k. For $v \in V$, k_v denotes the completion of k at v and, if $v \in V_f$, $\mathfrak{o}_v$ is the ring of integers of k_v, $\mathfrak{p}_v$ the prime ideal at v and Nv the order of the residue field $\mathfrak{o}_v/\mathfrak{p}_v$.

A k-form of $\boldsymbol{PGL}_2$ (or $\boldsymbol{SL}_2$) is a linear algebraic group over k which is isomorphic to $\boldsymbol{PGL}_2$ (or $\boldsymbol{SL}_2$) over some extension of k. If G is a k-form of $\boldsymbol{PGL}_2$, then its universal covering $\tilde{G}$ is a k-form of $\boldsymbol{SL}_2$.

The group $\tilde{G}$ is the group of elements of reduced norm one in a quaternion algebra B over k. Either $B = \boldsymbol{M}_2(k)$ is the 2×2 matrix algebra over k and $\tilde{G} = \boldsymbol{SL}_2$ or B is a division quaternion algebra.

We let $\sigma\colon \tilde{G} \to G$ be the canonical projection. The group G can also be viewed as the quotient of a reductive k-group H with one-dimensional center, derived group $\tilde{G}$, by its center, namely the group defined by the invertible elements of B. We shall also denote by σ the canonical projection

$H \to G$. We let σ_v be the homomorphism $\tilde{G}_v \to G_v$ or $H_v \to G_v$ defined by σ $(v \in V)$, where, as usual, if M is a k-group, we denote by M_v the group $M(k_v)$ of points of M rational over k_v.

We recall that for any field $k' \supset k$, the map $\sigma \colon H(k') \to G(k')$ is surjective, because the kernel of σ is the center Z of H, which is isomorphic over k to the one-dimensional split algebraic torus $\boldsymbol{GL_1}$.

3.2. If H is an algebraic group over k, then a subgroup Γ of $H(k)$ is arithmetic if, given an embedding $\varrho \colon G \to \boldsymbol{GL_n}$ over k, the group $\varrho(\Gamma)$ is commensurable with $\varrho(G) \cap \boldsymbol{GL_n}(\mathfrak{o})$ (where, as usual, for any commutative algebra A, $\boldsymbol{GL_n}(A)$ is the group of $n \times n$ matrices with coefficients in A and determinant invertible in A).

3.3. Let Γ be a discrete subgroup of $\boldsymbol{G}_{a,b}$. It is said to be *definable arithmetically* if the following conditions are met: there exists a number field k with b complex places, at least a real places, a k-form G of $\boldsymbol{PGL_2}$, a set A of a real places such that

$$(1) \qquad G_w = \boldsymbol{PGL_2}(\mathbf{R}), \quad (w \in A), \quad G_w = \boldsymbol{SO_3} \quad (w \text{ real}, w \notin A),$$

and an isomorphism

$$(2) \qquad \iota \colon \boldsymbol{G}_{a,b} \xrightarrow{\sim} G_{S_1} = \prod_{w \in S_1} G_w ,$$

(where $S_1 \subset V_\infty$ is the union of A and of the complex places of k) which maps Γ onto an arithmetic subgroup of $G(k)$. Here, $G(k)$ is diagonally embedded in G_{S_1} by means of the natural inclusions $G(k) \subset G_w$. We note that, since G_w is compact for w real not in A, the arithmetic subgroups of G, viewed as subgroups of G_{S_1} via the diagonal embedding, are indeed discrete. There are two main cases:

(A) Γ is not cocompact in $\boldsymbol{G}_{a,b}$. Then $d = a + 2b$, and $S_1 = V_\infty$. The group G is just $\boldsymbol{PGL_2}$, viewed as a k-group.

(B) Γ is cocompact in $\boldsymbol{G}_{a,b}$. Then k may have any number $\geqslant a$ of real places. The group $\tilde{G}$ is the group defined by the elements of reduced norm one in a division quaternion algebra B over k which is ramified (at least) at all real places not contained in S_1, and H is the group defined by the invertible elements in B.

To simplify notation, identify G with $\varrho(G)$, with ϱ as in 3.2. For almost all (*i.e.*, all but finitely many) $v \in V_f$, the group $G(\mathfrak{o}_v) = G \cap \boldsymbol{GL_n}(\mathfrak{o}_v)$ is maximal compact in G_v. If Γ is arithmetic then its closure $Cl_v(\Gamma)$ in G_v is compact open, contained in $G(\mathfrak{o}_v)$ for almost all v's. Conversely, given a compact open subgroup L_v of G_v for each $v \in V_f$ which is equal to $G(\mathfrak{o}_v)$ for almost all v's, the group

$$(3) \qquad \Gamma_L = \{g \in G(k) \,|\, g \in L_v \text{ for all } v \in V_f\}\,, \qquad L = \prod_{v \in V_f} L_v\,,$$

is an arithmetic subgroup of G, and every arithmetic subgroup is contained in one of this type. The maximal ones are *among* the groups Γ_L, where L_v is maximal compact for all $v \in V_f$.

3.4. We recall that if $a + b \geq 2$, then every irreducible discrete subgroup of finite covolume of $G_{a,b}$ is arithmetic. This follows from results of G. A. Margoulis [7] (see also [25]).

4. – Maximal arithmetic subgroups of $G_{a,b}$.

4.1. We let $R(B)$ or R be the set of places at which B is ramified, R_∞ (resp. R_f) be the set of infinite (resp. finite) places in R, and $r_f = |R_f|$. Thus $B \otimes_k k_v$ is a division algebra if $v \in R$ and is isomorphic to $M_2(k_v)$ otherwise (and $|R|$ is even). Let $\mathfrak{O}$ be a maximal order in B. Then $\mathfrak{O}_v = \mathfrak{O} \otimes_k k_v$ is a maximal order of B_v. For $v \in R_f$ it is the unique maximal order of integral elements in B_v. If $\mathfrak{O}'$ is another maximal order, then $\mathfrak{O}_v = \mathfrak{O}'_v$ for almost all v's, $\mathfrak{O}'$ is the intersection of the $\mathfrak{O}'_v$ and the $\mathfrak{O}'_v$ can be prescribed arbitrarily at finitely many places. $\mathfrak{O}$ and $\mathfrak{O}'$ are said to have the same type if there exists $x \in B^*$ such that $x \cdot \mathfrak{O} = \mathfrak{O}' \cdot x$. The number of types of maximal orders is finite (and divides the class number of B). (For all this, see [4: §§ 8, 11].)

For $v \in V_f - R_f$, the group $\tilde{G}$ is isomorphic to SL_2 over k_v. We let $\mathfrak{T}_v$ be its Bruhat-Tits building and P_v the fixed point of $\tilde{K}_{1v} = \mathfrak{O}_v \cap \tilde{G}_v$. Furthermore, let e_v be an edge of $\mathfrak{T}_v$ incident to P_v; let Q_v be the middle point of e_v and P'_v the second end point of e_v. Let $\tilde{K}'_{1v}$ be the isotropy group of P'_v in $\tilde{G}_v$ and

$$(1) \qquad C_v = \{P_v, Q_v, P'_v\}\,.$$

We denote by K_{1v}, K_{2v} and K'_{1v} the isotropy groups of P_v, Q_v and P'_v in G_v. The groups K_{1v} and K'_{1v} are conjugate in G_v, the groups $\tilde{K}_{1v}$ and $\tilde{K}'_{1v}$ (resp. K_{1v} and K_{2v}) represent the two conjugacy classes of maximal compact subgroups in $\tilde{G}_v$ (resp. G_v) ($v \in V_f - R_f$) (cf. § 2). For convenience, we agree that for $v \in R_f$, the building $\mathcal{C}_v$ is reduced to a point and $G_v = K_{1v} = K_{2v} = = K'_{1v}$, $\tilde{G}_v = \tilde{K}_{1v} = \tilde{K}'_{1v}$.

4.2. The group $\tilde{G}$ is simple, simply connected, not compact at infinity, hence has the *strong approximation property*: the group $\tilde{G}(k)$, embedded diagonally in the restricted product $\tilde{G}(A_f)$ of the $\tilde{G}_v$ ($v \in V_f$) is dense. Without using the notion of restricted product, we can, in our case, express this as follows: let S be a finite subset of $V_f - R_f$. For $v \in V_f - S$, let L_v be a compact open subgroup of $\tilde{G}_v$ which is equal to $\tilde{G}(\mathfrak{o}_v)$ for almost all v, and put

$$(1) \qquad \tilde{G}(k)_L = \{g \in \tilde{G}(k) | g \in L_v \text{ for } v \in V_f - S\} \ .$$

Then, for any set of elements $g_v \in \tilde{G}_v$ ($v \in S$), there exists $g \in \tilde{G}(k)_L$ which is arbitrarily close to g_v for every $v \in S$.

This can also be formulated in the following way: let S be a finite subset of V_f; for $v \in S$, let D_v be a finite subset of $\mathcal{C}_v$ and $E_v = g_v \cdot D_v$ for some $g_v \in \tilde{G}_v$. Then there exists $g \in \tilde{G}(k)$ such that $g \cdot D_v = E_v$ for $v \in S$ and $g \cdot P_v = P_v$ for $v \in V_f - S$.

4.3. The group G has center reduced to the identity. Therefore the commensurability subgroup C_Γ (see § 1) of an arithmetic subgroup Γ is equal to $G(k)$ [2: Thm. 3]. It follows that if two arithmetic subgroups Γ, Γ' of G are conjugate in $G_{S_1} = \mathbf{G}_{a,b}$, then they are conjugate under an element of $G(k)$, hence $\mathrm{Cl}_v(\Gamma)$ is conjugate to $\mathrm{Cl}_v(\Gamma)$ in G_v for all $v \in V_f$. Also, if $\Gamma \subset \mathbf{G}_{a,b}$ is mapped onto an arithmetic subgroup of $G(k)$ under the isomorphism ι, then every subgroup of $\mathbf{G}_{a,b}$ commensurable with Γ is mapped by ι onto an (arithmetic) subgroup of $G(k)$. This then allows one to transfer the discussion of the commensurability class of Γ in $\mathbf{G}_{a,b}$ to that of $\iota(\Gamma)$ in $G(k)$.

4.4. PROPOSITION. *For two finite disjoint subsets S, S' of $V_f - R_f$ set*

$$(1) \qquad \Gamma_{S,S'} = \{g \in G(k) | g \in K_{1v}(\text{resp. } K_{2v}, \text{resp. } K'_{1v}) \text{ for } v \in V_f - (S \cup S');$$

$$(\text{resp. } S, \text{ resp. } S')\}.$$

(i) *For $v \in S'$ (resp. $v \in V_f$, $v \notin S \cup S'$), K'_{1v}(resp. K_{1v}) is the unique maximal compact subgroup of G_v containing $\Gamma_{S,S'}$.*

(ii) *Let Γ be an arithmetic subgroup of G containing an element which is odd at some $v \notin S$. Then Γ is not conjugate to a subgroup of $\Gamma_{S,S'}$.*

(iii) *Given an arithmetic subgroup Γ of G, let $S(\Gamma)$ be the set of v's such that Γ contains an element odd at v. Then there exists S' such that Γ is conjugate to a subgroup of $\Gamma_{S(\Gamma),S'}$, or to $\Gamma_{S(\Gamma),S'}$ itself if Γ is maximal.*

Given S, S' let for $v \in V_f$

$$(1) \quad L_v = K_{1v} \text{ (resp. } K_{2v}, \text{ resp. } K'_{1v}) \text{ if } v \notin S \cup S' \text{ (resp. } v \in S, \text{ resp. } v \in S').$$

(i) Amounts to asserting that P_v or P'_v, as the case may be, is the unique fixed point of $\Gamma_{S,S'}$ in $\mathfrak{C}_v$. For any $u \in V_f - \{v\}$, we may find a compact open subgroup M_u of $\tilde{G}_u$, equal to $\tilde{K}_{1u}$ for almost all u's, such that $\sigma(M_u) \subset L_u$ for $u \in V_f - \{v\}$. Set $M_v = \tilde{K}_{1v}$ (resp. $M_v = \tilde{K}'_{1v}$) if $v \notin S \cup S'$ (resp. $v \in S'$) and let

$$\tilde{\Gamma}_M = \{g \in \tilde{G}(k) | g \in M_u \text{ for } u \in V_f - R_f\} .$$

By strong approximation, $\tilde{\Gamma}_M$ is dense in M_v, hence P_v, or P'_v, is the unique fixed point of $\tilde{\Gamma}_M$ in $\mathfrak{C}_v$ $(v \in V_f - S)$. Since $\sigma(\tilde{\Gamma}_M) \subset \Gamma_{S,S'}$ by construction, (i) follows.

(ii) Is a consequence of (i) and 2.3.

(iii) By the generalities recalled in 3.3, there exists for every $v \in V_f$ a maximal compact subgroup J_v of G_v, equal to $K_{1,v}$ for almost all v's, consisting of even elements if $v \notin \Gamma(S)$, such that $\Gamma \subset \Gamma_J$, where J is the product of the J_v's. Let $T \subset V_f$ be the union of R_f and of the set of v's for which $J_v \neq K_{1,v}$. It contains $S(\Gamma)$. Using 4.1 and 4.2, we see that there exists $g \in \sigma(\tilde{G}(k))$ with the following properties:

$$g \in K_{1,v} \text{ if } v \notin T; \quad {}^gJ_v = K_{1,v} \text{ or } K'_{1,v} \text{ if } v \in T - S(\Gamma), \quad {}^gL_v = K_{2,v} \text{ if } v \in S(\Gamma) .$$

Then ${}^g\Gamma \subset \Gamma_{S(\Gamma),S'}$. There is then obviously equality if Γ is maximal. This proves (iii).

REMARK. A group $\Gamma_{S,S'}$ may be non-maximal. The point is that we cannot assert that for $v \in S$ the group $K_{2,v}$ is the unique maximal compact subgroup of G_v containing $\Gamma_{S,S'}$. It could happen that for some $v \in S$ no element of $\Gamma_{S,S'}$ is odd at v, and then $\Gamma_{S,S'}$ would fix pointwise the edge e_v containing P_v, Q_v (notation 4.1) and be a proper subgroup of $\Gamma_{S'',S'}$ where $S'' = S - \{v\}$. In order to show that there are indeed infinitely many non-conjugate maximal subgroups among the groups $\Gamma_{S,S'}$ we need therefore an existence statement. This is provided by the following lemma:

4.5. LEMMA. *Let $v \in V_f - R_f$. Then there exists a torsion-free arithmetic subgroup of G containing an element which is odd at v.*

Let n be such that G is embedded in $\boldsymbol{GL}_n$. If $g \in G(k)$ is of finite order, then its eigenvalues are roots of one, of degree $\leq d \cdot n$, hence there are only finitely many possibilities for the order of g. Therefore, for almost all $v \in V - R_f$ there exists a congruence subgroup K'_v of $K_{1,v}$ such that if $g \in K'_v \cap \cap G(k)$, then g has infinite order or $g = 1$. Choose one such place $v' \neq v$. Then any arithmetic subgroup contained in $K'_{v'}$ is torsion-free. Now we claim

(*) *There exists* $g \in G(k)$ *which is odd at* v *and contained in a compact open subgroup* L_u *of* G_u *for* $u \in V_f$, *where* $L_u = K'_{v'}$, *if* $u = v'$.

Assume this for the moment. We may take $L_u = G(\mathfrak{o}_u)$ for almost all u's. Then the arithmetic group Γ_L is torsion-free, since it is contained in $K'_{v'}$ and has an element odd at v, namely g. We are reduced therefore to proving (*).

By the Chinese remainder theorem, we can find an element $c \in k$ which is the square of a unit in $\mathfrak{o}_{v'}$, has order one at v, and is positive at all $v \in R_\infty$. In case (A), let

$$x = \begin{pmatrix} 0 & c \\ -1 & 0 \end{pmatrix} \in \boldsymbol{GL}_2(k) \, .$$

In case (B), let x be an element of reduced norm c in the quaternion algebra B which underlies the definition of G. Such an element exists by the norm theorem of Hasse-Schilling (see e.g. Prop. 3 in [28: XI, § 3]). The first condition implies the existence of an element $y_{v'} \in \tilde{G}_{v'}$ such that $\sigma_{v'}(x) = = \sigma_{v'}(y_{v'})$. Let T be the set of $v \in V_f$, $v \notin R_f \cup \{v'\}$ such that $\sigma(x) \notin G(\mathfrak{o}_v)$. It is finite. Using strong approximation, we can find $h \in \tilde{G}(k)$ such that

$$\sigma_{v'}(h \cdot y_{v'}) \in K'_{v'} \, , \quad h \cdot x(e_v) = e_v \ (v \in T), \quad h \in \tilde{G}(\mathfrak{o}_v) \quad \text{for } v \in V_f - T \, .$$

Then $\sigma_{v'}(h \cdot x) = \sigma_{v'}(h \cdot y_{v'}) \in K'_{v'}$ and $\sigma_v(h \cdot x)$ belongs to the stability group L_v of e_v for $v \in T$, to $G(\mathfrak{o}_v)$ for the other $v \in V_f - R_f$. Thus $\sigma(h \cdot x)$ satisfies the requirements imposed on g in (*).

4.6. THEOREM. *Let* Γ *be an arithmetically defined subgroup of* $\boldsymbol{G}_{a,b}$. *Then the commensurability class of* Γ *contains infinitely many non-conjugate elements which are maximal among discrete subgroups of* $\boldsymbol{G}_{a,b}$ *or maximal among torsion-free discrete subgroups of* $\boldsymbol{G}_{ab}$.

Let k and G be as in 3.3. Then, as pointed out in 4.3, it is equivalent to prove the same statement for the commensurability class of arithmetic subgroups of $G(k)$ and conjugacy by elements of $G(k)$. Let $\Gamma_1, \ldots, \Gamma_m$ be non-conjugate maximal (resp. maximal among torsion-free) arithmetic subgroups of $G(k)$. By 4.4, each one is conjugate to some subgroup of a group $\Gamma_{S,S'}$ and, by 4.2, 4.4, for almost all v's in $V_f - R_f$, the point P_v is the unique fixed point of Γ_i in $\mathcal{C}_v$ $(i = 1, \ldots, m)$. Fix one, say v_0. By 4.5, there exists a torsion-free arithmetic subgroup Γ' of G having an element which is odd at v_0. By 2.3, any arithmetic subgroup containing Γ' is not conjugate to any of the Γ_i's. Among those there is a maximal one and one which is maximal among torsion-free arithmetic subgroups, whence the theorem.

4.7. REMARK. The first assertion of 4.4 is contained in more general statements of [18; 19]. For $\boldsymbol{PGL_2}$ over a number field, the existence of infinitely many non-conjugate maximal arithmetic subgroups is already proved in [9].

4.8. In 4.7, we proved the existence of an element which is odd at a given place, but it may be odd at other places as well. The proof shows that, in order to produce a group for which $\Gamma(S)$ consists of just one given finite place v, it is enough to find $c \in k$ which has odd order at v, even order at all $u \notin R_f \cup \{v\}$, and is positive at all $u \in R_\infty$. This last condition is of course vacuous if $R_\infty = \emptyset$, in particular in case (A). The other two will be fulfilled if the prime ideal $\mathfrak{o} \cap \mathfrak{p}_v$ of $\mathfrak{o}$ has an odd power which is principal, in particular if $\mathfrak{o}$ is a principal ideal domain.

4.9. We shall write $\Gamma_{\mathfrak{O}}$ for $\Gamma_{S,S'}$ when S and S' are empty. Since $\sigma \colon H(k) \to G(k)$ is surjective (3.3) we have

$$(2) \qquad \Gamma_{\mathfrak{O}} = \Gamma_{\phi,\phi} = \sigma(\mathrm{Norm}\,\mathfrak{O}) \, ,$$

where

$$(3) \qquad \mathrm{Norm}\,\mathfrak{O} = \{x \in B^* | x \cdot \mathfrak{O} = \mathfrak{O} \cdot x\} \, .$$

Given S' finite in V_f, consider the set of points (R_v) where $R_v = P_v$ for $v \notin S'$ and $R_v = P'_v$ otherwise. There is a unique maximal order $\mathfrak{O}' = \mathfrak{O}(S')$ such that $\tilde{G}_v \cap \mathfrak{O}'_v$ fixes R_v for all v's. Thus we have

$$(4) \qquad \Gamma_{\phi,S'} = \Gamma_{\mathfrak{O}'} \, .$$

By strong approximation, the systems (R_v), for varying S', form a system of representatives for the conjugacy classes of maximal orders with respect to $\tilde{G}(k)$.

4.10. PROPOSITION. *Fix $S \subset V_f$. Then the groups $\Gamma_{S,S'}$ form finitely many conjugacy classes in $G(k)$, as S' varies through the finite subsets of $V_f - S$.*

Let $\mathfrak{O}'$ be as above and $\mathfrak{O}'' = \mathfrak{O}(S'')$ be similarly associated to S''. We claim that $\Gamma_{\mathfrak{O}'}$ and $\Gamma_{\mathfrak{O}''}$ are conjugate if and only if there exists $g \in B^*$ which is odd at exactly $S' \cup S'' - (S' \cap S'')$. Moreover, if there exists such an element, then there exists also $x \in B^*$ such that

(1) $\qquad x \cdot e_v = e_v \, (v \in S) \,, \quad x \cdot P_v = P_v \,, \quad x \cdot P'_v = P'_v \quad (v \in S' \cap S'') \,,$

(2) $\qquad\qquad x \cdot P_v = P_v \,, \quad (v \in V_f - (S \cup S' \cup S'')) \,,$

(3) $\quad x \cdot P'_v = P_v \quad (v \in S' - (S' \cap S'')) \,, \quad x \cdot P_v = P'_v \quad (v \in S'' - (S' \cap S'')) \,.$

The necessity of the condition is clear. If there exists such an element, say y, then by strong approximation (see end statement in 4.2), we can find $z \in \tilde{G}(k)$ such that $x = \sigma(zy)$ satisfies (1), (2), (3). But then $\Gamma_{S,S'}$ and $\Gamma_{S,S''}$ are conjugate under x. The proposition now follows from this and the finiteness of the type number of B (4.1).

5. – Comparison of the volumes in a commensurability class.

5.1. We consider the commensurability class $\mathcal{C}(k, B)$ of arithmetically defined subgroups of $\boldsymbol{G}_{a,b}$ defined by a k-form G of $\boldsymbol{PGL}_2$. We keep the notation of 3.3, 4.1, 4.4 and identify $\boldsymbol{G}_{a,b}$ with G_{S_1}. For $v \in V_f$, let Nv be the order of the residue field at v.

5.2. LEMMA. *Let S, S' be finite disjoint subsets of $V_f - R_f$ and S'' be a finite subset of $V_f - (R_f \cup S)$. For $v \in S''$ let $\mathcal{E}_v$ be the set of edges of $\mathcal{C}_v$ having P'_v (resp. P_v) as a vertex if $v \in S'$ (resp. $v \notin S'$). Then, given $f_v, f'_v \in \mathcal{E}_v$ $(v \in S'')$, there exists $\gamma \in \Gamma_{S,S'}$ such that $\gamma \cdot f_v = f'_v$ for $v \in S''$.*

This is again a consequence of strong approximation: Let M_v be the isotropy group in $\tilde{G}_v$ of P_v (resp. P'_v) if $v \in S''$, $v \notin S'$ (resp. $v \in S' \cap S''$). As recalled in 2.2, it is transitive on $\mathcal{E}_v$. Let then $g_v \in M_v$ be such that $g_v \cdot f_v = f'_v \, (v \in S'')$. By strong approximation, we can find $g \in \tilde{G}(k)$ which is arbitrarily close to g_v for $v \in S''$, fixes the point P_v for $v \notin S' \cup S''$, the point P'_v for $v \in S'$ and P_v, P'_v for $v \in S$. Then $\sigma(g) \in \Gamma_{S,S'}$ and $\sigma(g) \cdot f_v = f'_v$ for $v \in S''$.

5.3. THEOREM. *Let $\Gamma_{\mathfrak{O}}$ be as in 4.9. Let S, S' be finite disjoint subsets of $V_f - R_f$. Then there exists an integer m $(0 \leqq m \leqq |S|)$ such that*

(1) $\qquad\qquad [\Gamma_{\mathfrak{O}} : \Gamma_{S,S'}] = 2^{-m} \prod_{v \in S} (Nv + 1) \,.$

In particular $[\Gamma_{\mathfrak{D}} : \Gamma_{S,S'}] \geq 1$ and $[\Gamma_{\mathfrak{D}} : \Gamma_{S,S'}] = 1$ if and only if S is empty. Given $c > 0$, the groups $\Gamma_{S,S'}$ for which $[\Gamma_{\mathfrak{D}} : \Gamma_{S,S'}] \leq c$ are contained in finitely many conjugacy classes.

In this proof, it is understood that $v \in V_f$. We have

$$(2) \qquad [\Gamma_{\mathfrak{D}} : \Gamma_{S,S'}] = [\Gamma_{\mathfrak{D}} : \Gamma_{\phi,S'}] \cdot [\Gamma_{\phi,S'} : \Gamma_{S,S'}]$$

and (1) is equivalent to the following equalities

$$(3) \qquad [\Gamma_{\mathfrak{D}} : \Gamma_{\phi,S'}] = 1$$

$$(4) \qquad [\Gamma_{\phi,S'} : \Gamma_{S,S'}] = 2^{-m} \prod_{v \in R_f} (Nv + 1), \qquad (0 \leq m \leq |S|).$$

Let $\Gamma_1 = \Gamma_{\mathfrak{D}} \cap \Gamma_{\phi,S'}$. This is the subgroup of $G(k)$ which fixes the edges e_v $(v \in S')$ pointwise and the points P_v for $v \notin S'$. Lemma 5.2 implies:

$$(6) \qquad [\Gamma_{\mathfrak{D}} : \Gamma_1] = [\Gamma_{\phi,S'} : \Gamma_1] = \prod_{v \in S'} (Nv + 1),$$

which proves (3).

Let now $\Gamma_2 = \Gamma_{\phi,S'} \cap \Gamma_{S,S'}$. This is the subgroup of $G(k)$ which fixes P_v, P_v' for $v \in S$, the vertex P_v' for $v \in S'$ and P_v otherwise. By 5.2 we have

$$(7) \qquad [\Gamma_{\phi,S'} : \Gamma_2] = \prod_{v \in S} (Nv + 1).$$

On the other hand, if $g \in G_v$ stabilizes e_v, then g_v^2 fixes P_v and P_v'. As a consequence

$$(8) \qquad [\Gamma_{S,S'} : \Gamma_2] = 2^m \quad \text{for some } m \in [0, |S|].$$

(4) now follows from (7) and (8).

The right-hand side of (1) can be written as a product of factors indexed by $v \in S$, each of which is $\geq (Nv + 1)/2$, hence tends to infinity as v varies. Therefore, given $c > 0$, there exist only finitely many S such that

$$[\Gamma_{\mathfrak{D}} : \Gamma_{S,S'}] \leq c, \quad \text{for some } S'.$$

Since for fixed S, the groups $\Gamma_{S,S'}$ are contained in finitely many conjugacy classes (4.10) the last assertion is proved.

5.4. COROLLARY. *Let e be the number of places of k dividing 2 and not contained in R_f. Let Γ be a subgroup of $G_{a,b}$ commensurable with $\Gamma_{\mathfrak{O}}$. Then the volume $\mu(\Gamma)$ is an integral multiple of $2^{-e} \cdot \mu(\Gamma_{\mathfrak{O}})$. It is equal to $\mu(\Gamma_{\mathfrak{O}})$ if Γ is conjugate to a subgroup $\Gamma_{\phi,S'}$, and $> \mu(\Gamma_{\mathfrak{O}})$ otherwise.*

The group Γ is arithmetic in $G(k)$ (3.3). It is enough to consider the case where it is maximal. Γ is then conjugate to a group $\Gamma_{S(\Gamma),S'}$ (see 4.4). By 5.3, $[\Gamma_{\mathfrak{O}} : \Gamma_{S(\Gamma),S'}]$ is an integral multiple of the number

$$m_{S(\Gamma)} = \prod_{v \in S(\Gamma)} (Nv + 1)/2 \ .$$

If v divides the rational prime p, then Nv is a power of p, hence $Nv + 1$ is even except when v divides 2, Therefore all the factors in $m_{S(\Gamma)}$ are integers, except for those v which divide 2. All of them are > 1. Since $\mu(\Gamma) = [\Gamma_{\mathfrak{O}} : \Gamma]\mu(\Gamma_{\mathfrak{O}})$, the corollary follows.

5.5. Let S_2 be the set of primes of k dividing 2 and not contained in R. Let us denote by q_G the g.c.d of the numbers $\mu(\Gamma)$, where Γ runs through the arithmetic subgroups of G. We have just seen that

$$(1) \qquad\qquad q_G = 2^{-c} \cdot \mu(\Gamma_{\mathfrak{O}}) \ , \qquad \text{for some integer } c \in [0, |S_2|] \ .$$

If S_2 is empty, then $c = 0$. Assume now S_2 to be not empty. If there exists Γ such that $\Gamma(S)$ is not empty, contained in S_2, then $c \geq 1$. By 4.5, we know there exists Γ such that $\Gamma(S)$ contains any prescribed element of S_2 but this is not enough to insure that $c \geq 1$, because if $\Gamma(S)$ contains some v dividing an odd prime, then the factor $(Nv + 1)/2$ might still contribute a power of two which might compensate for the one stemming from a place dividing two at which Γ has an odd element. The remarks in 4.8 show that, in order to prove that $c = e$, it suffices to show that, given $u \in S_2$, there exists $c \in k$ which has an odd valuation at u, an even valuation at all other places in $V_f - R_f$ and is > 0 at the real places at which B is ramified. But such existence theorems do not seem easy to prove in general.

5.6. Assume we are in case (A) and that $a = 0$, $b = 1$. There exists then a square free negative rational integer m such that $k = Q(\sqrt{m})$. We have $c = 1, 2$ and more precisely $c = 2$ if and only if $m \equiv 1 \bmod 8$ (see e.g. [3]). For $-m = 1, 2, 3, 7, 19$ for instance, $\mathfrak{o}$ is a principal ideal domain. Therefore the exponent c in 5.4 (1) is given by:

$$(1) \qquad c = 1 \quad \text{if } -m = 1, 2, 3, 19; \qquad c = 2 \quad \text{if } -m = 7 \ .$$

6. – Some local computations of volumes.

6.1. Let $\mathfrak{g}$ be the Lie algebra of $\boldsymbol{SL_2(R)}$. Then $\mathfrak{g}_c = \mathfrak{g} \otimes_{\boldsymbol{R}} \boldsymbol{C}$ is the Lie algebra of $\boldsymbol{SL_2(C)}$. We view it as a 6-dimensional real Lie algebra. Then $\theta : x \mapsto - {}^t\bar{x}$ is an automorphism of $\mathfrak{g}_c$ whose fixed point set is the Lie algebra $\mathfrak{su}_2$ of $\boldsymbol{SU_2}$, i.e., the set of skew hermitian matrices. The orthogonal complement $\mathfrak{p}$ of $\mathfrak{su}_2$ with respect to the Killing form is the space of hermitian symmetric 2×2 matrices of trace zero. θ is the Cartan involution of $\mathfrak{g}_c$ associated to su_2. We have $H^3 = \boldsymbol{SL_2(C)}/\boldsymbol{SU_2}$ and the canonical projection identifies $\mathfrak{p}$ to the tangent space $T(H^3)_0$ to H^3 at the origin. On $\mathfrak{g}_c$ consider the hermitian form $g_0(x, y) = - 2Tr(x \cdot \theta(y))$, where Tr refers to the trace in the standard representation. It is hermitian positive non-degenerate, invariant under inner automorphisms of $\boldsymbol{SU_2}$. Then the hyperbolic metric is the left invariant Riemannian metric whose value at $\mathfrak{p} = T(H^3)_0$ is the inner product defined by the restriction of g_0 to $\mathfrak{p}$.

[To check this, note first that $2Tr = \frac{1}{4}B$, where B is the Killing form $B(x, y) = \mathrm{tr}\,(\mathrm{ad}\,x \circ \mathrm{ad}\,y)$, of $\mathfrak{g}_c$, viewed as real Lie algebra, and that, for the metric defined by the Killing form on $\mathfrak{p}$, the sectional curvature on the plane spanned by x, y is $B([x, y], [x, y]) \cdot A(x, y)^{-2}$, where $A(x, y)$ is the area of that plane. Then compute this expression for some choice of x and y, for instance h and u below.] Let

$$(1) \qquad h = \begin{pmatrix} 1 & 0 \\ 0 & -1 \end{pmatrix}, \; e = \begin{pmatrix} 0 & 1 \\ 0 & 0 \end{pmatrix}, \; f = \begin{pmatrix} 0 & 0 \\ 1 & 0 \end{pmatrix}, \; u = e + f, \; v = i(e - f).$$

Then $h/2, u/2$ and $v/2$ form an orthonormal base of $\mathfrak{p}$ with respect to g_0. In particular, if $d\mu$ denotes the volume element of the hyperbolic metric, then

$$(2) \qquad d\mu(h \wedge u \wedge v) = 8 .$$

In the real case, we have $H^2 = \boldsymbol{SL_2(R)}/\boldsymbol{SO_2}$. We identify $\mathfrak{p} = T(H^2)_0$ with the subspace of $\mathfrak{g}$ spanned by h and u. The computation sketched above also shows that the restriction of $- 2Tr(x \cdot \theta(y))$ to $\mathfrak{p}$ defines the hyperbolic metric, and that $h/2, u/2$ is an orthonormal basis of $\mathfrak{p}$.

6.2. Let $\omega_0^1, \omega_0^2, \omega_0^3$ be the left invariant 1-forms on $\boldsymbol{SL_2}$ whose values at the identity form a basis dual to the basis (h, e, f) of $\mathfrak{sl}_2$ given by 6.1(1), and $\omega_0 = \omega_0^1 \wedge \omega_0^2 \wedge \omega_0^3$. Let $\tilde{G}$ be a k-form of $\boldsymbol{SL_2}$. The group $\tilde{G}$ is then isomorphic to $\boldsymbol{SL_2}$ over some algebraic extension of k. Fix such an isomorphism φ and let $\omega = \varphi^* \omega_0$. Then ω is defined over k (see pp. 475-476 in [11]). It is then a « gauge form », which can be used to define first a local measure ω_v on $\tilde{G}_v$ for every v and then a Tamagawa measure $d\tau = = |D_k|^{-\frac{3}{2}} \prod_v \omega_v$ on the adelic group G_A. In this section, we are concerned with the local measures ω_v for $v \in V_\infty$. We have

$$(1) \qquad \omega_v = \omega \, , \qquad \text{if } v \text{ is real} \, ,$$

$$(2) \qquad \omega_v = \pm \, i^3 \cdot \omega \wedge \bar{\omega} \, , \qquad \text{if } v \text{ is complex} \, .$$

As usual, let ω_∞ be the product measure of the ω_v on $\tilde{G}_\infty$. Let again S_1 be the set of real v's such that $\tilde{G}$ is isomorphic to $\boldsymbol{SL}_2$ over k_v. Set

$$K_v = \boldsymbol{SO}_2 \qquad \text{if } v \text{ is real, } v \in S_1 \, ,$$

$$K_v = \tilde{G}_v = \boldsymbol{SU}_2 \qquad \text{if } v \text{ is real, } v \notin S_1 \, ,$$

$$K_v = \boldsymbol{SU}_2 \qquad \text{if } v \text{ is complex} \, .$$

Then $H_v = G_v/K_v$ is H^2 (resp, a point, resp. H^3) if v is real, $v \in S_1$ (resp. $v \notin S_1$, real resp. v complex). Let $d\mu_v$ be the hyperbolic volume elemnet on H_v in the first and last cases, the point measure in the second case.

LEMMA. *Let dk_v be the measure on K_v such that $\omega_v = d\mu_v \cdot dk_v$ ($v \in V_\infty$). Then the volume $v(K_v)$ of K_v with respect to dk_v is equal to π (resp. $4\pi^2$, resp. $8\pi^2$) if v is real in S_1 (resp. real not in S_1, resp. complex).*

Assume first v to be complex. Let

$$(1) \qquad \sigma^2 = \omega^2 + \omega^3 \, , \qquad \sigma^3 = \omega^2 - \omega^3 \, .$$

Then $\omega^1, \sigma^2, \sigma^3$ is the basis of $\mathfrak{g}^*$ dual to h, $u/2$, $-iv/2$. For a C-linear 1-form τ on $\mathfrak{g}_c$ let $R\tau$ and $I\tau$ be its real and imaginary part. We have

$$(2) \qquad \omega^1 \sigma^2 \sigma^3 = 2\omega \, ,$$

$$(3) \qquad 4\omega \wedge \bar{\omega} = \omega^1 \wedge \sigma^2 \wedge \sigma^3 \wedge \overline{\omega^1} \wedge \overline{\sigma^2} \wedge \overline{\sigma^3} = \pm \, 8i^3 \, R\omega^1 \wedge I\omega^1 \wedge R\sigma^2 \wedge I\sigma^2 \wedge R\sigma^3 \wedge I\sigma^3$$

$$(4) \qquad \omega_v = \pm \, (I\omega^1 \wedge I\sigma^2 \wedge R\sigma^3) \wedge (2R\omega^1 \wedge R\sigma^2 \wedge I\sigma^3) \, .$$

The elements ih, iu, iv form a basis of $\mathfrak{su}_2$, and h, u, v a basis of $\mathfrak{p}$. We have

$$(5) \qquad (2R\omega^1 \wedge R\sigma^2 \wedge I\sigma^3)(h \wedge u \wedge v) = 8 \, ,$$

which, in view of 6.1(2), shows that the second factor in the right-hand side of (4) is $d\mu_v$, up to sign.

Identify SU_2 to the standard unit sphere in $\mathbf{R}^4$ by using the real and imaginary parts of the matrix entries in the first row. The volume for the standard metric is then $2\pi^2$. It is readily seen that ih, iu, iv is an orthonormal basis for the standard metric. If dv_0 is the corresponding volume element on $\mathfrak{su}_2$, we have then

$$|2\omega| = |I\omega^1 \wedge I\sigma^2 \wedge R\sigma^3| = 4\,dv_0 \,,$$

and our first assertion for v complex follows. This also shows that volume of K_v for the positive measure defined by the restriction of ω is $4\pi^2$ if v is real, not in S_1.

In the first case, we have $d\mu_v = 2\omega^1 \wedge \sigma^2$, since $h/2$ and $u/2$ then form an orthonormal basis of $\mathfrak{p}$ in that case, as remarked above: therefore $dk_v = \sigma^3/4$, whence our assertion in that case.

6.3. LEMMA. *For $v \in R_f$, let $v(\tilde{G})$ be the volume of $\tilde{G}_v$ with respect to ω_v.* *Then*

$$(1) \qquad v(\tilde{G}_v) = (Nv + 1)\cdot Nv^{-2}\,, \qquad (v \in R_f)\,.$$

For comparison, let us recall that if $v \in V_f - R_f$, and K_v is a maximal compact subgroup of $\tilde{G}_v$, then ω_v is the standard measure on $SL_2(k_v)$ and K_v is conjugate, by an element of $GL_2(k_v)$, to $SL_2(\mathfrak{o}_v)$. We have then

$$(2) \qquad \omega_v(K_v) = Nv^{-3}\cdot\operatorname{Card} SL_2(F_v)\,,$$

where F_v is the residue field at v, hence

$$(3) \qquad \omega_v(K_v) = Nv^{-2}\cdot(Nv^2 - 1)\,, \qquad (v \in V_f - R_f)\,.$$

(cf. e.g. [16; 29]). In particular, the volumes given by (1) and (3) are rational numbers, but this follows from a general fact (see e.g. [16: 4.2.5]).

6.4. PROOF OF 6.3. This is a local statement, also valid in the equal characteristic case. We change notation and shift to a purely local situation. Let then E be a $\mathfrak{p}$-field [28: I, §3], F its unique unramified quadratic extension k_E and k_F the residue fields of E and F, and q the order of k_E. Then k_F is the quadratic extension of k_E and has order q^2. Let π be a uniformizing variable in E. It is then also one in F and $\mathfrak{p}_E = \pi \cdot \mathfrak{o}_E$, $\mathfrak{p}_F = \pi \cdot \mathfrak{o}_F$ are the maximal ideals of $\mathfrak{o}_E$ and $\mathfrak{o}_F$ respectively. We take as multiplicative representative system S' of k_F in F, the set $S = 0 \cup \langle w \rangle$ where w is a primitive $(q^2 - 1)$-st root of one in F. Let $x \mapsto x'$ be the non-trivial automorphism of F over E. We have $w' = w^q$.

Let B be the division quaternion algebra over E. It splits over F and contains F as a maximal subfield. We can write B as a cyclic algebra

$$(1) \qquad B = F \oplus F \cdot u$$

and may assume π to be equal to u^2. The map $x \mapsto x'$ extends to an involution of B which sends u to $-u$. We have

$$(2) \qquad (xy)' = y' \cdot x' \quad (x, y \in B), \qquad x \cdot u = u \cdot x' \quad (x \in F)$$

$$(3) \qquad (x + yu)' = x' - y \cdot u \quad (x, y \in F).$$

The reduced norm will just be denoted by N. We have

$$(4) \qquad Nb = xx' - \pi \cdot yy', \quad (b = x + yu, x, y \in F).$$

Let $\mathfrak{o}_B$ be the maximal order of B and $\mathfrak{p}_B$ its maximal ideal. We have

$$(5) \qquad \mathfrak{p}_B = \mathfrak{p}_F + \mathfrak{o}_F \cdot u, \quad \mathfrak{p}_B^2 = \pi \cdot \mathfrak{o}_B, \quad \mathfrak{o}_B/\mathfrak{p}_B = \mathfrak{o}_F/\mathfrak{p}_F = k_F.$$

We let B^1 (resp. K) be the subgroup of elements in B^* whose reduced norm is one (resp. a unit). The group K is the biggest compact subgroup of B^*. The reduced norm yields a surjective homomorphism of K onto $\mathfrak{o}_E^*$, with kernel B^1. It follows from the definitions that the measure ν, multiplied by the standard measure on E (which gives volume 1 to $\mathfrak{o}_E$) is the measure on B^* introduced in [11: p. 475]. Denote also by $\nu(\ )$ the corresponding volumes. We have then

$$(6) \qquad \nu(B^1) = \nu(K) \cdot \nu(\mathfrak{o}_E^*)^{-1} = \nu(K) \cdot q \cdot (q-1)^{-1}.$$

The natural projection of $\mathfrak{o}_B$ onto k_F, with kernel $\mathfrak{p}_B$, maps K onto k_F^*, hence

$$(7) \qquad \nu(K) = (q^2 - 1) \cdot \nu(\mathfrak{p}_B).$$

We write $F = E(\beta)$, where β is integral, and the reduction mod π of β generates k_F over k_E. Then

$$(8) \qquad \mathfrak{o}_F = \mathfrak{o}_E + \mathfrak{o}_E \cdot \beta, \quad \mathfrak{p}_F = \mathfrak{p}_E + \mathfrak{p}_E \cdot \beta.$$

We represent B as the set of matrices

$$(9) \qquad b = \begin{pmatrix} x & y \\ \pi y' & x' \end{pmatrix}, \qquad (x, y \in F) .$$

Then $Nb = \det b$. Write

$$(10) \qquad x = r + \beta s , \qquad y = u + \beta v .$$

On $\boldsymbol{GL}_2(F)$ we take as usual as coordinates the matrix entries a, b, c, d. On B^*, we use r, s, u, v. On $\boldsymbol{GL}_2(F)$ we have the standard invariant 4-form

$$(11) \qquad \omega = (\det)^{-1} \cdot da \wedge db \wedge dc \wedge dd .$$

On B^*,

$$(12) \quad a = r + \beta s , \quad b = u + \beta v , \quad c = \pi(u + \beta' v) , \quad d = r + \beta' s .$$

Therefore

$$(13) \qquad \omega|_{B*} = \pm (\beta - \beta')^2 \pi \cdot N^{-1} dr \wedge ds \wedge du \wedge dv .$$

$(\beta - \beta')^2$ is the discriminant of F over E, hence is a unit. This form is defined over E, and the measure v is the one associated to it. On K, the norm is one in absolute value, therefore our measure v is given by

$$(14) \qquad v = |\pi| \, dr \wedge ds \wedge du \wedge dv = q^{-1} \cdot dr \wedge ds \wedge du \wedge dv .$$

But now, by (5) and (8), the element b belongs to $\mathfrak{p}_B$ if and only if

$$(15) \qquad r, s \in \mathfrak{p}_E , \qquad u, v \in \mathfrak{o}_E .$$

Therefore

$$(16) \qquad v(\mathfrak{p}_B) = q^{-3} .$$

Together with (6) and (7), this yields

$$(17) \qquad v(B^1) = (q + 1) \cdot q^{-2} ,$$

as was to be proved.

6.5. REMARK. A different proof of 6.4 (17), or rather of an obviously equivalent equality, has been given by W. Casselman (Proc. Symp. Pur. Math. **33**, A.M.S. 1978, part 2, p. 155, lemma 5.2.1).

7. – Volumes and values of zeta functions at 2.

7.1. If Γ is a group, we let $\Gamma^{(2)}$ denote the subgroup generated by the squares of the elements of Γ. It is normal, and $\Gamma/\Gamma^{(2)}$ is a group of exponent 2. If Γ is finitely generated, then $\Gamma/\Gamma^{(2)}$ is a (finite) elementary abelian group of type $(2, 2, ..., 2)$. Its $\boldsymbol{F}_2$-rank is at most equal to smallest integer m such that Γ is generated by m elements. The group $\Gamma^{(2)}$ is the smallest normal subgroup of Γ such that $\Gamma/\Gamma^{(2)}$ has exponent 2.

7.2. We return to the commensurability class $\mathcal{C}(k, B)$. We let $N_{B/k}$ or simply N denote the reduced norm from B to k. Fix a maximal order $\mathfrak{O}$ of B, and let $\mathfrak{O}^*$ (resp. $\mathfrak{O}^1$) be the set of elements of $\mathfrak{O}$ whose reduced norm is a unit (resp. one). It is a group, and an arithmetic subgroup of H (resp. $\tilde{G}$). It is known that

$$(1) \qquad \mathfrak{O}^* = \{x \in \operatorname{Norm} \mathfrak{O} \,|\, Nx \in \mathfrak{o}^*\}, \qquad \mathfrak{O}^1 = \operatorname{Norm} \mathfrak{O} \cap \tilde{G}.$$

Not knowing of a good reference, we sketch the proof: For $v \in R_f$, we have $\mathfrak{O}_v^1 = \tilde{G}(k_v)$; for $v \in V_f - R_f$, we have $\mathfrak{O}_v^1 = \boldsymbol{SL}_2(\mathfrak{o}_v)$; hence in any case $\mathfrak{O}_v^1$ is equal to its normalizer in $\tilde{G}(k_v)$. The group $\mathfrak{O}_v^1$ is the v-adic closure of $\mathfrak{O}^1$ in $\tilde{G}(k_v)$. Consequently, if $x \in \operatorname{Norm} \mathfrak{O}$ has reduced norm one, it belongs to $\mathfrak{O}_v^1$ for all $v \in V_f$, hence to $\mathfrak{O}^1$. This proves the second equality of (1). By a theorem of Eichler [5], the map $x \mapsto Nx$ maps $\mathfrak{O}^*$ onto the group $\mathfrak{o}_{R_\infty}^*$ of units which are positive at R_∞. Let now $x \in \operatorname{Norm} \mathfrak{O}$ be such that $Nx \in \mathfrak{o}^*$. Since Nx has to be positive at R_∞, there exists then $y \in \mathfrak{O}^*$ such that $N(y \cdot x) = 1$. We have then $y \cdot x \in \mathfrak{O}^1$ and $x \in \mathfrak{O}^*$, whence the first equality of (1). Set

$$(2) \qquad \Gamma_{\mathfrak{O}^*} = \sigma(\mathfrak{O}^*) = \sigma(k^* \cdot \mathfrak{O}^*), \qquad \Gamma_{\mathfrak{O}}^1 = \sigma(\mathfrak{O}^1) = \sigma(k^* \cdot \mathfrak{O}^1).$$

Both are arithmetic subgroups of G, normal in $\Gamma_{\mathfrak{O}}$. Let $x \in \operatorname{Norm} \mathfrak{O}$. Then $N(x)^{-1} \cdot x^2$ has reduced norm 1, hence $x^2 \in k^* \cdot \mathfrak{O}^1$ by (1) and therefore

$$(3) \qquad \Gamma_{\mathfrak{O}}^1 \subset \Gamma_{\mathfrak{O}}^{(2)}. \qquad \textit{The group } \Gamma_{\mathfrak{O}}/\Gamma_{\mathfrak{O}}^1 \textit{ has exponent two}.$$

If $\tilde{G}$ is isomorphic to $\boldsymbol{SL}_2$ over k, we may choose a k-isomorphism which maps $\mathfrak{O}$ onto $\boldsymbol{M}_2(\mathfrak{o})$. We have then

$$(4) \qquad \begin{cases} \mathfrak{O}^1 = \boldsymbol{SL}_2(\mathfrak{o}), & \mathfrak{O}^* = \boldsymbol{GL}_2(\mathfrak{o}), \\ \Gamma_{\mathfrak{O}}^1 = \boldsymbol{SL}_2(\mathfrak{o})/\{\pm 1\}, & \Gamma_{\mathfrak{O}^*} = \boldsymbol{GL}_2(\mathfrak{o})/\{\pm 1\}. \end{cases}$$

7.3. THEOREM. *Let D_k denote the discriminant of k over $\mathbf{Q}$ and ζ_k the Dedeking zeta function of k. Let G be the k-form of $\mathbf{PGL}_2$ associated to a quaternion algebra B over k. Then*

$$(1) \qquad \mu(\Gamma_\mathfrak{D}^1) = \prod_{v \in R_f} (Nv - 1) \cdot \frac{2|D_k|^{\frac{3}{2}} \cdot \zeta_k(2)}{2^{2r_1 + 3r_2 - 2a} \cdot \pi^{2r_1 + 2r_2 - a}} \, .$$

In particular the volumes $\mu(\Gamma)$, where Γ is arithmetic in G, are all rational multiples of $\pi^{-d - r_1 + a} \cdot |D_k|^{\frac{3}{2}} \cdot \zeta_k(2)$.

(Cf. 3.1, 4.1 for the notation.) Let

$$(2) \qquad K_\infty = \prod_{v \in V_\infty} K_v \, , \qquad \nu(K_\infty) = \prod_{v \in V_\infty} \nu(K_v)$$

with $\nu(K_v)$ as in 6.2. We want to prove:

$$(3) \qquad \nu(\tilde{G}_\infty / \mathfrak{D}^1) = \mu(\Gamma_\mathfrak{D}^1) \cdot \nu(K_\infty)/2 \, ,$$

where $\nu(\ )$ on the left-hand side refers to the volume computed with ω_∞ and $\mathfrak{D}^1$ is diagonally embedded in $\tilde{G}_\infty$.

Let Γ be a torsion-free subgroup of finite index of $\mathfrak{D}^1$. Then $\sigma(\Gamma) \overset{\sim}{\to} \Gamma$ and $\sigma(\mathfrak{D}^1) = \mathfrak{D}^1/\{\pm 1\}$, hence

$$(4) \qquad [\mathfrak{D}^1 : \Gamma] = 2 \cdot [\sigma(\mathfrak{D}^1) : \sigma(\Gamma)] \, .$$

We have clearly

$$(5) \qquad \nu(\tilde{G}_\infty / \Gamma) = [\Gamma_\mathfrak{D}^1 : \Gamma] \cdot \nu(\tilde{G}_\infty / \Gamma_\mathfrak{D}^1) \, .$$

$$(6) \qquad \mu(\Gamma) = [\Gamma_\mathfrak{D}^1 : \sigma(\Gamma)] \mu(\Gamma_\mathfrak{D}^1) \, .$$

Since Γ is torsion-free, $\tilde{G}_\infty / \Gamma$ is fibered over H_∞ / Γ, with fiber K_∞, therefore

$$(7) \qquad \nu(\tilde{G}_\infty / \Gamma) = \nu(K_\infty) \cdot \mu(\Gamma)$$

and (3) follows from (4) to (7). By 6.2 we have

$$(8) \qquad \nu(K_\infty) = (8\pi^2)^b \cdot (4\pi^2)^{r_1 - a} \cdot \pi^a \, ,$$

whence

$$(9) \qquad \nu(\tilde{G}_\infty / \mathfrak{D}^1) = 2^{3r_2 + 2r_1 - 2a - 1} \pi^{2r_2 + 2r_1 - a} \mu(\Gamma_\mathfrak{D}^1) \, .$$

The Tamagawa number of $\tilde{G}$ is one [29]. This translates to

$$(10) \qquad \nu(\tilde{G}_\infty/\mathfrak{O}^1) = \left(\prod_{v \in R_f} (1 - Nv^{-2}) \cdot \nu(\tilde{G}_v)^{-1} \right) \cdot |D_k|^{\frac{3}{2}} \cdot \zeta_k(2) \,.$$

Together with 6.3 and (4), this yields (1).

7.4. *The case of matrix algebras.* 7.3 applies in particular to the case where R is empty, i.e., where B is the matrix algebra $\boldsymbol{M}_2(k)$ and $\tilde{G}$ is k-isomorphic to $\boldsymbol{SL}_2$. The proof of 7.4(1) is then slightly simpler, since 6.3 is not needed. 7.3(1) specializes to

$$(1) \qquad \mu(\boldsymbol{SL}_2\mathfrak{o}/\{\pm 1\}) = 2^{1-3b} \cdot \pi^{-d} \cdot |D_k|^{\frac{3}{2}} \cdot \zeta_k(2) \,.$$

7.5. REMARKS. (1) The formula 7.4(1) is due to G. Humbert for k imaginary quadratic $(d = 2, b = 1)$ (see [24: § 7]), and to C. L. Siegel [23] when k is totally real. The equality 7.3(1) for totally real fields in general follows from results of Shimizu [22: p. 193].

(2) To get the smallest covolume in $\mathcal{C}(k, B)$, we have to divide the right-hand side of 7.3(1) by the index of $\Gamma_\mathfrak{O}^1$ in $\Gamma_\mathfrak{O}$. At this point, all we know is that this index is $\leq 2^m$, where m is the smallest cardinality of a generating set for $\Gamma_\mathfrak{O}$ (7.1, 7.2(3)). In § 8, we shall give an expression for it in terms of data depending only on k and R.

7.6. PROPOSITION. *Let Γ_1 be an arithmetically defined subgroup of $\boldsymbol{G}_{a,b}$. Then there exist infinitely many commensurability classes of arithmetically defined subgroups of $\boldsymbol{G}_{a,b}$ such that the volumes $\mu(\Gamma)$ are all rational multiples of $\mu(\Gamma_1)$ when Γ runs through these classes.*

From 7.3, we see that, given a and b, the volumes $\mu(\Gamma)$ for the arithmetic subgroups defined by a k-form G of $\boldsymbol{PGL}_2$ (satisfying A) or B) of 3.3, of course) are all rational multiples of a number which depends only on k and a. Therefore, given k with b complex places and at least a real places, we need only to show that there are infinitely many k-forms of $\boldsymbol{PGL}_2$ associated to quaternion algebras over k which, at infinity, are ramified at exactly a places, such that two arithmetic subgroups of $\boldsymbol{G}_{a,b}$ associated to any two of them are not commensurable up to conjugacy.

Let A be the set of automorphisms of k. It is finite, of order $\leq d$. We can choose an infinite sequence of quaternion algebras B_i over k $(i = 1, 2, \dots)$

such that B_i is not isomorphic to any conjugate ${}^\sigma B_j$ $(\sigma \in A)$ of B_j for $i \neq j$ and, at infinity, B_i is ramified at exactly a places of k. [This follows immediately from the fact that a quaternion algebra is determined by its local invariants and that the only conditions imposed on those are to be zero almost everywhere and to have a sum $\equiv 0 \bmod 1$ (cf. [4: VII, § 5])]. Changing the notation slightly, we are then reduced to showing that if G and G' are k-forms of $\boldsymbol{PGL_2}$ associated to two such quaternion algebras B, B', where B' is not isomorphic to a conjugate of B, then an arithmetic subgroup Γ of G is not commensurable up to conjugacy to any arithmetic subgroup of $\boldsymbol{G}_{a,b}$ (3.3). Let $g \in \boldsymbol{G}_{a,b}$ be such that ${}^g\Gamma$ is commensurable with Γ'. Then ${}^g C_\Gamma = C_{\Gamma'}$ (where C_Γ, $C_{\Gamma'}$ denote the commensurability groups, cf. § 1). But $C_\Gamma = G(k)$, $C_{\Gamma'} = G(k')$. Therefore $G(k)$ would be isomorphic to $G'(k)$ as an abstract group. By a theorem of R. Baer [1: Thm. 2, p. 272] (see also [30: Thm. 4.1]), this would imply that G is isomorphic, as an algebraic k-group, to ${}^\sigma G'$ for some $\sigma \in A$, hence that B is isomorphic to ${}^\sigma B'$, a contradiction.

7.7. *Commensurability questions.* Let $b = 0$. In this case, all volumes are commensurable. In fact, if $\chi(\Gamma)$ is the Euler-characteristic of Γ, in the sense of C. T. C. Wall if Γ has torsion (cf. [20: p. 99]), then

$$(1) \qquad \chi(\Gamma) = \mu(\Gamma)/(-2\pi)^a ,$$

in agreement with the fact that $|D_k|^{\frac{1}{2}} \cdot \zeta_k(2)\pi^{-2d}$ is rational for k totally real. Let now $b \neq 0$. Then $\chi(\Gamma)$ is always zero. Although it is not expected that all volumes are commensurable, this has not been checked to far. In view of 7.3, to produce an example, it would be enough to exhibit two number fields k, k' of the same degree and the same non-zero number of complex places such that

$$(2) \qquad |D_k|^{\frac{1}{2}}\zeta_k(2) \notin \boldsymbol{Q} \cdot |D_{k'}|^{\frac{1}{2}} \cdot \zeta_{k'}(2) .$$

Apparently, nothing is known about this question. Of course, the truth of Milnor's conjectures about the Lobatshevski function [24: § 7] would provide many examples of quadratic imaginary fields k, k' satisfying (2).

8. – Discreteness of the set of arithmetic volumes.

In this section, we want to prove that the set of volumes $\mu(\Gamma)$, when Γ runs through the arithmetically defined subgroups of $\boldsymbol{G}_{a,b}$, is discrete (with finite multiplicities, see 8.2 for the precise statement). If $a + b \geq 2$, these subgroups are all the irreducible discrete subgroups of finite covolume, as already pointed out (3.4), and the discreteness has been proved by H. C. Wang ([27], see 8.3]). For the sake of uniformity we shall also include this case, although this is not really a new proof, since the idea of the proof of 8.1 in that case is taken from [26, 27].

8.1. LEMMA. *Let $a, b \in \mathbf{N}$ be given. Let $c > 0$. There exists an integer $m(c)$ such that if Γ is an irreducible discrete subgroup of $\mathbf{G}_{a,b}$ and $\mu(\Gamma) \leq c$ then Γ is generated by $m(c)$ elements.*

Let first $a = 1, b = 0$. Since Γ contains a subgroup Γ' of index two which preserves the orientation, we may assume $\Gamma \subset SR_2(\mathbf{R})/\{\pm 1\}$. In this case our assertion follows from the standard formula for $\mu(\Gamma)$: let m be the number of cusps of H^2/Γ, $\{\gamma_1, \ldots, \gamma_s\}$ a set of representatives of the classes of elliptic elements of Γ, e_j the order of γ_j $(1 \leq j \leq s)$ and g the genus of the standard compactification of H^2/Γ. Then Γ is generated by $2g + m + r - 1$ elements and we have

$$(1) \qquad \mu(\Gamma) = 2g - 2 + m + \sum_{j=1}^{j=s}(1 - e_j^{-1}).$$

Since $e_j \geq 2$, we see that $2g + m + r - 1 \leq 2\mu(\Gamma) + 2$.

If $a = 0$ and $b = 1$, 8.1 follows from the construction of all H^3/Γ with volume $\leq c$ by means of Dehn surgery applied to finitely many orbifolds, given in Chap. 13 of [24]; it will be proved explicitly in the final version of these Notes. For torsion-free Γ, all we shall need to know is that $H_1(\Gamma; \mathbf{Z}/2\mathbf{Z})$ has dimension bounded by some constant $n(c)$, and this follows directly from [24: Chap. 5]; in fact, it is shown there that the hyperbolic 3-manifolds of volume $\leq c$ are obtained by gluing some solid tori or cusps to finitely many compact manifolds with boundary a union of 2-dimensional tori.

Let now $a + b \geq 2$. Assume there is a sequence of irreducible subgroups Γ_n of $\mathbf{G}_{a,b}$ such that $\mu(\Gamma_n) \leq c$ and that the smallest cardinality g_n of a generating system of Γ_n tends to infinity. By the argument of [26: p. 137], recalled in [27: p. 480], there exists a subgroup Γ, which is a limit of the Γ_n, in the topology of the space of closed subgroups, such that $\mu(\Gamma) \leq c$, and moreover a homomorphism $r_n: \Gamma \to \Gamma_n$, for n big enough, defining a deformation of Γ which tends to the identity as $n \to \infty$. The group Γ is also irreducible, since r_n has to be trivial on any subgroup of Γ which is contained in a proper factor of $\mathbf{G}_{a,b}$. But then Γ is rigid (since $a + b \geq 2$), hence Γ_n is conjugate to Γ for n big enough, a contradiction with the assumption $g_n \to \infty$.

8.2. THEOREM. *Fix a and b. Let $c > 0$. Then there exist finitely many arithmetic subgroups $\Gamma_1, \ldots, \Gamma_{q(c)}$ of $G_{a,b}$ such that any arithmetic subgroup Γ of $G_{a,b}$ with covolume $\mu(\Gamma) \leq c$ is conjugate to one of the Γ_i's $(1 \leq i \leq q(c))$. In particular the set of volumes $\mu(\Gamma)$, where Γ runs through the arithmetically defined subgroups of $G_{a,b}$, is a discrete subset of the real line.*

In view of 5.3, 5.4, it suffices to prove this theorem for the set of arithmetic subgroups of the form $\Gamma_\mathfrak{O}$ defined in 4.9. We first show it for the groups $\Gamma_\mathfrak{O}^1$. Consider 7.3(1). Since $\zeta_k(2) \geq 1$, we have

$$(1) \qquad \mu(\Gamma_\mathfrak{O}^1) \geq 2 \cdot |D_k|^{\frac{3}{2}} \cdot (2\pi)^{-2r_1 - 2r_2} \cdot 2^{-r_2} .$$

$$(2) \qquad \mu(\Gamma_\mathfrak{O}^1) \geq 2|D_k|^{\frac{3}{2}} \cdot (4\pi^2)^{-r_1} \cdot (2\sqrt{2} \cdot \pi)^{-2r_2} .$$

Since there are only finitely many number fields with a given discriminant, it suffices to show that the right-hand side of (2) tends to infinity with the degree of k. But this follows from known estimates on the discriminant, e.g., from

$$(3) \qquad |D_k| \geq 50^{r_1} \cdot 19^{2r_2} , \qquad \text{for } d \text{ large enough}$$

which implies

$$(4) \qquad |D_k|^{\frac{3}{2}} \geq 353^{r_1} \cdot 82^{2r_2} , \qquad \text{for } d \text{ large enough} ,$$

and follows from 1.8 in [15].

Let now $c > 0$. By 8.1, there exists a constant $m(c)$ such that if $\mu(\Gamma_\mathfrak{O}) \leq c$, then $\Gamma_\mathfrak{O}$ has a generating set of cardinality $\leq m(c)$. Since $\Gamma_\mathfrak{O}/\Gamma_\mathfrak{O}^1$ has exponent two (7.2(3)), we have then

$$(5) \qquad [\Gamma_\mathfrak{O} : \Gamma_\mathfrak{O}^1] \leq 2^{m(c)} ,$$

(7.1), hence

$$(6) \qquad \mu(\Gamma_\mathfrak{O}^1) \leq c \cdot 2^{-m(c)} .$$

The possible $\Gamma_\mathfrak{O}^1$ form then finitely many conjugacy classes by the first part of the proof. In view of 5.3, the same is then true for the groups $\Gamma_\mathfrak{O}$.

8.3. REMARK. Let L be a connected semi-simple Lie group with center reduced to the identity and no compact factor. Theorem 8.1 in [27] asserts that the covolumes $\mu(L/\Gamma)$ (Γ discrete in L) form a discrete set (with finite multiplicities) if L has no three-dimensional factor. This should in particular apply to $L = \boldsymbol{PGL}_2(\boldsymbol{C})$, but there it is contradicted by the results of Thurston-Jorgensen [24: Chap. 5]. The mistake in [27] comes from a misunderstanding of rigidity in that group: the author uses a result he attributes to H. Garland and M. S. Raghunathan, but he misquotes it. However, the proof, as it stands, is valid for the irreducible subgroups of L, provided L is not locally isomorphic to $\boldsymbol{SL}_2(\boldsymbol{R})$ or $\boldsymbol{SL}_2(\boldsymbol{C})$, since in all those cases the rigidity theorem used by Wang is indeed available. In particular, this covers the case of our groups $\boldsymbol{G}_{a,b}$ for $a + b \geq 2$.

8.4. In this proof we have used discriminant estimates to handle the groups $\Gamma_{\mathfrak{D}}^1$ and then a geometric argument to go over to $\Gamma_{\mathfrak{D}}$. One can of course ask whether it would not be possible also to give an arithmetic proof for $\mu(\Gamma_{\mathfrak{D}})$, using a good estimate of $[\Gamma_{\mathfrak{D}}:\Gamma_{\mathfrak{D}}^1]$. We shall see that this is unlikely since this index depends in part on the class group of k. First we want to give an arithmetic description of it.

We denote by $\mathfrak{o}_{R_f}^*$ the group of R_f-units of k (elements which are integral at all finite places outside R_f) and by $\mathfrak{o}_{R_f,R_\infty}^*$ the group of elements of $\mathfrak{o}_{R_f}^*$ which are positive at R_∞.

We have

$$(1) \qquad [\mathfrak{o}_{R_f,R_\infty}^* : \mathfrak{o}_{R_f}^{*2}] \leq [\mathfrak{o}_{R_f}^* : \mathfrak{o}_{R_f}^{*2}] \leq 2^{r_1+r_2+r_f}, \qquad (r_f = |R_f|),$$

where the last inequality follows from the unit theorem. Let now

$$(2) \qquad B_{R_f}^* = \{b \in B^* \,|\, N(b) \in \mathfrak{o}_{R_f}^*\}, \qquad \Gamma_{R_f} = \sigma(k^* \cdot B_{R_f}^*).$$

We have $Nb \in \mathfrak{o}_{R_f,R_\infty}^*$ for $b \in B_{R_f}^*$. Moreover, the results of [5] imply that $B_{R_f}^* \in \mathrm{Norm}\,\mathfrak{D}$. We have then the inclusions

$$(3) \qquad \Gamma_{\mathfrak{D}} \supset \Gamma_{R_f} \supset \Gamma_{\mathfrak{D}^*} \supset \Gamma_{\mathfrak{D}}^1.$$

8.5. LEMMA. *The group $\Gamma_{R_f}/\Gamma_{\mathfrak{D}}^1$ is isomorphic to $\mathfrak{o}_{R_f,R_\infty}^*/\mathfrak{o}_{R_f}^{*2}$. In particular* $[\Gamma_{R_f}:\Gamma_{\mathfrak{D}}^1] \leq 2^{r_1+r_2+r_f}$.

Eichler's theorem implies that $b \mapsto Nb$ maps $B_{R_f}^*$ onto $\mathfrak{o}_{R_f,R_\infty}^*$. If now $Nb = c^2$, with $c \in \mathfrak{o}_{R_f}^*$, then $N(c^{-1} \cdot b) = 1$, hence $b \in k^* \cdot \mathfrak{D}^1$, and the first assertion is proved. The second follows from 8.4(1).

8.6. Let $D_\infty = D_\infty(B)$ (resp. $D_f = D_f(B)$) be the product of the primes in R_∞ (resp. R_f). Thus the ideal D_f is the square root of the discriminant of B. Let $I(k)$ (resp. $P(k)$) be the group of fractional (resp. principal) ideals of k and $P(k, D_\infty)$ the group of principal ideals generated by elements which are $\equiv 1 \bmod^* D_\infty$, i.e, which are positive at R_∞.

LEMMA. *Let M_1 (resp. M_2) be the subgroup of $I(k)$ generated by $P(k, D_\infty)$ (resp. $P(k)$) and the $\mathfrak{o} \cap \mathfrak{p}_v$ ($v \in R_f$). Let $J_1 = I(k)/M_1$ and let J_2 be the image of M_2 in J_1. Then $[\Gamma_{\mathfrak{D}} : \Gamma_{R_f}] = [{}_2 J_1 : J_2]$, where ${}_2 J_1$ is the kernel of the map $y \mapsto y^2$ in J_1. If k has class number one, then $\Gamma_{\mathfrak{D}} = \Gamma_{R_f}$.*

Let L_1 be the subgroup of $I(k)$ generated by the $\mathfrak{o} \cap \mathfrak{p}_v$ ($v \in R_f$) and the squares of all ideals. It follows from the description of an ideal as an intersection of local ideals that the elements of L_1 are the norms of the two-sided $\mathfrak{D}$-ideals. By a theorem of Eichler [5], an element of L_1 is the norm of a principal $\mathfrak{D}$-ideal if and only it belongs to $P(k, D_\infty)$.

Let now $x \in \operatorname{Norm} \mathfrak{D}$. There exists then a unique ideal $\mathfrak{m}(x)$ prime to D_f such that the ideal (Nx) is the product of $\mathfrak{m}(x)^2$ by a power product of the divisors of D_f. Let $\tau \colon \operatorname{Norm} \mathfrak{D} \mapsto J_2$ be the map which assigns to x the class of $\mathfrak{m}(x)$ in J_2. By the above, its image belongs to ${}_2 J_1$ and every element of ${}_2 J_1$ occurs in this way. For $c \in k^*$, we have $N(cx) = c^2 \cdot N(x)$, hence τ is constant on $k^* \cdot x$. Assume now that $\tau(x) \in J_2$. This means that we can write (Nx) as the product of a principal ideal (a^2) ($a \in k$) by a power product of the $\mathfrak{o} \cap \mathfrak{p}_v$ ($v \in R_f$). But then $N(a^{-1} \cdot x) \in \mathfrak{o}^*_{R_f}$, hence $x \in k^* \cdot B^*_{R_f}$. Thus τ defines an isomorphism of $\Gamma_{\mathfrak{D}}/\Gamma_{R_f}$ onto ${}_2 J_1/J_2$ and the first assertion is proved. If k has class number one, then $J_2 = J_1$, whence the second assertion.

8.7. The first part of the proof of 8.2 also shows that the $\mu(\Gamma_{R_f})$ form a discrete set, but I do not see how to go from there in the same way to $\mu(\Gamma_{\mathfrak{D}})$. An upper bound of ${}_2 J_1/J_2$ is the « narrow » class number $h_+(k)$. It may grow about as fast as $|D_k|^{\frac{1}{2}}$, which is too strong to be absorbed by 8.2(3), or even by the stronger estimates of [15]. Of course, for fields of a given degree, it is easy to show that the $\mu(\Gamma_{\mathfrak{D}})$ form a discrete set.

8.8. Another question raised by W. Thurston is whether the g.c.d. of the volumes in a commensurability class have a strictly positive lower bound. Since the volumes do by a well-known theorem of D. Kazdhan and G. A. Margoulis (see [17: XI, 11.9]), this is clear for non-arithmetically defined classes (cf. § 1). The first part of the proof of 8.2 also shows that the numbers $2^{-d} \cdot \mu(\Gamma_{R_f})$ have a strictly positive lower bound. In view of 5.4, this shows that the g.c.d. of the volumes in the commensurability classes attached to fields of class number one do have a strictly positive lower bound. I do not know whether this is true in general.

9. – Hyperbolic 3-folds.

9.1. In this and the next section, we consider the case of hyperbolic 3-space. We have then $a = 0$, $b = 1$, $r_2 = 1$, $d = r_1 + 2$, and 7.3(1) becomes

$$(1) \qquad \mu(\Gamma_{\mathfrak{D}}^1) = \prod_{v \in R_f} (Nv - 1) \cdot |D_k|^{\frac{3}{2}} \cdot \zeta_k(2)(2\pi)^{-2d+2} \,.$$

Since $a = 0$, the set R_∞ is the set of all real places of k, hence $\mathfrak{o}^*_{R_f, R_\infty}$ is the group $\mathfrak{o}^*_{R_f, +}$ all of totally positive R_f-units. Therefore, in view of 8.4, we get for the smallest volume in the given commensurability class

$$(2) \qquad \mu(\Gamma_{\mathfrak{D}}) = [\mathfrak{o}^*_{R_f, +} : \mathfrak{o}^{*2}_{R_f}]^{-1} \mu(\Gamma_{\mathfrak{D}}^1) \,, \quad \textit{if } k \textit{ has class number one}\,.$$

Assume now k to be imaginary quadratic. Then R_∞ is empty, $\mathfrak{o}^*_{R_f}$ is just the group of all R_f-units. It is the product of a cyclic group of even order by a free abelian group on r_f generators, hence

$$(3) \qquad [\mathfrak{o}^*_{R_f, +} : \mathfrak{o}^*_{R_f}] = [\mathfrak{o}^*_{R_f} : \mathfrak{o}^{*2}_{R_f}] = 2r_f + 1 \,, \quad \textit{if } k \textit{ is imaginary quadratic}\,.$$

k is necessarily quadratic imaginary in the non-cocompact case, and then R_f is also empty. We get

$$(4) \qquad \mu(\boldsymbol{SL}_2(\mathfrak{o})/\{\pm 1\}) = |D_k|^{\frac{3}{2}} \zeta_k(2)/4\pi^2 \,,$$

which is G. Humbert's formula. If k has class number one, $\mu(\Gamma_{\mathfrak{D}})$ is one-half of the right-hand side of (4).

9.2. It is not surprising from the general formula that small volumes should be tied up to fields of small discriminants and, in the compact case, to quaternion algebras which are as unramified as possible at the finite places. However, because of the factor $[_2 J_1 : J_2]$ we can confirm this only for group Γ_{R_f}, hence for $\Gamma_{\mathfrak{D}}$ if k has class number one. Remarks (a), (b), (c) below are due to E. Bombieri.

(a) Let $k = \boldsymbol{Q}(\sqrt{-3})$. Then the class number is one and

$$\mu(\boldsymbol{GL}_2(\mathfrak{o})/\{\pm 1\}) = 3^{\frac{3}{2}} \cdot \zeta_k(2)/8\pi^2 = 0.08458 \ldots \,.$$

Simple estimates show that this is the smallest value of $\mu(\Gamma_{R_f})$, when k runs through all the imaginary quadratic fields. It may also be the minimum of $\mu(GL_2(\mathfrak{o})/\{\pm 1\})$ for those cases. We now consider cocompact groups.

(b) Let $d = 2$. Then R_f has at least two elements. In this case the minimum of $\mu(\Gamma_{R_f})$ is realized when $D_k = -3$, $\prod\limits_{v \in R_f}(Nv - 1) = 6$, and then

$$(4) \qquad \mu(\Gamma_{R_f}) = \mu(\Gamma_{\mathfrak{D}}) = 0.126 \pm 0.0001 \ .$$

(c) Let $k = \boldsymbol{Q}(\theta)$, where θ is a root of $x^3 - x - 1$. Then $D_k = -23$, $[\mathfrak{o}_+^* : \mathfrak{o}^{*2}] = 4$. The prime $v = 2 - \theta$ divides 5 and $Nv = 5$. Take then $R_f = \{v\}$. The field k has also class number one. Then

$$(5) \qquad \mu(\Gamma_{\mathfrak{D}}) = 0.3536 \ldots \times \zeta_k(2) \leqq 0.47474 \ldots$$

and this $\Gamma_{\mathfrak{D}}$ seems a good candidate for the smallest volume when k has signature $(1, 1)$. At any rate 23 is the minimum of $|D_k|$ for these fields.

(d) Let $k = \boldsymbol{Q}(\theta)$ where $\theta = (3 + 2\sqrt{5})^{\frac{1}{4}}$. This is a quartic field of signature $(2, 1)$. Its discriminant is -275 and k is known to have the smallest discriminant in absolute value for fields of signature $(2, 1)$ [8; 14]. Take for B the quaternion algebra over k which is ramified at exactly the two real places of k. Then the group $\Gamma_{\mathfrak{D}}$ is the subgroup of orientation preserving transformations in the Coxeter group:

$$(6) \qquad \circ\!\!-\!\!-\!\!-\!\!\circ\!\!=\!\!=\!\!\circ\!\!-\!\!-\!\!-\!\!\circ \ ,$$

as was pointed out by W. Thurston. This appears so far to be the smallest volume known and it seems rather likely to be the smallest obtained from fields of signature $(2, 1)$. Eventually, for fields of high enough degree, the volumes have to become bigger, but it seems well possible that quaternion algebras over fields of relatively small degree with small discriminants might lead to smaller volumes. The next candidate would be the field of signature $(3, 1)$ with discriminant 4511, with R_f consisting of one place dividing a prime with small norm.

10. – Totally real fields. Fuchsian groups.

10.1. Assume now k to be totally real, i.e., $b = 0$. Then the functional equation yields

$$(1) \qquad \zeta_k(-1) = |D_k|^{\frac{3}{2}}\zeta_k(2)/(-2\pi^2)^d \ ,$$

and 7.3(1) can be written

$$(2) \qquad \mu(\Gamma_{\mathfrak{D}}^1) = (-1)^d \zeta_k(-1)\pi^a \cdot 2^{2a+1-d} \cdot \prod_{v \in R_f} (Nv - 1) .$$

The Euler-Poincaré characteristic $\chi(\Gamma_{\mathfrak{D}}^1)$ of $\Gamma_{\mathfrak{D}}^1$ is given by

$$(3) \qquad \chi(\Gamma_{\mathfrak{D}}^1) = (-2\pi)^a \mu(\Gamma_{\mathfrak{D}}^1) = (-2)^{a-d+1} \zeta_k(-1) \prod_{v \in R_f} (Nv - 1) .$$

The group $\Gamma_{\mathfrak{D}}^1$ is a quotient of $\mathfrak{D}^1$ by a group of order 2, hence

$$(4) \qquad \chi(\mathfrak{D}^1) = (-2)^{a-d} \zeta_k(-1) \prod_{v \in R_f} (Nv - 1) .$$

Note that if $a = d$, this is equal to $(-1)^d \chi(\mathbf{SL}_2 \mathfrak{o}_{R_f})$, in view of [20: p. 159].

10.2. Consider now the case of Fuchsian groups, where $a = 1$. Using 10.1(1) we can also write 7.3(1) as

$$(1) \qquad \mu(\Gamma_{R_f}) = 2^{d-3} \cdot \pi [\mathfrak{o}_{R_f,R_\infty}^* : \mathfrak{o}_{R_f}^{*2}]^{-1} \cdot |\zeta_k(-1)| \prod_{v \in R_f} (Nv - 1) .$$

This can be used in particular for the triangle groups which can be defined arithmetically. Some were already investigated by R. Fricke [6; 7] and a complete determination of those groups and of the associated quaternion algebras has been carried out by K. Takeuchi [12]. Moreover, it is shown there that the underlying groundfields have all class number one, so that $\mu(\Gamma_{R_f})$ realizes the minimum of the volume, and is the triangle group (recall that we have included orientation reversing isometries at the real places (3.0)).

As an example consider the group of the triangle $(2, 3, 7)$. Here $k = \mathbf{Q}(\cos 2\pi/7)$ is the maximal totally real subfield of the cyclotomic field of the seventh roots of 1. It is cubic, and we take for B a quaternion algebra ramified at exactly two infinite primes. Thus R_f is empty. It can be checked that $[\mathfrak{o}_{R_\infty}^* : \mathfrak{o}^{*2}] = 2$. Moreover, it is known that $\zeta_k(-1) = -1/21$ [20: p. 163]. We get indeed $\mu(\Gamma_{\mathfrak{D}}) = \pi/42$.

This group is denoted $\bar\Gamma_{(63)}$ in [6]. There Fricke also considers commensurable groups $\bar\Gamma_{(7)}, \bar\Gamma_{(14)}, \bar\Gamma$. The group $\bar\Gamma_{(7)}$ is the group of the triangle $(2, 4, 7)$, $\bar\Gamma_{(7)} = \bar\Gamma_{(14)} \cap \bar\Gamma_{(63)}$ has index 2 in $\bar\Gamma_{(14)}$ and 9 in $\bar\Gamma_{(63)}$. These groups can be described as follows in the set-up of §§ 4, 5.

Note first that 2 remains prime in k, and if v_0 is the corresponding place of k, then $Nv_0 = 8$. We may write $\bar{\Gamma}_{63} = \Gamma_{\mathfrak{O}}$, where $\Gamma_{\mathfrak{O}}$ is defined by the vertices (P_v) of the various Bruhat-Tits buildings. Then $\bar{\Gamma}_{(14)}$ is the group which fixes the P_v for $v \neq v_0$ and stabilizes the edge e for $v = v_0$. It indeed contains an element which is odd at exactly 2, namely $z \mapsto (z+1)/(1-z)$ [6: p. 456]. The reduction mod v_0 maps $\bar{\Gamma}_{63}$ onto the projective group of the projective line $\boldsymbol{P}^1(\boldsymbol{F}_8)$, and $\bar{\Gamma}_{(7)}$ on the stability group of a point, i.e., on the affine group of $\boldsymbol{F}_8$. The inverse image of the group of translations has then index 7 in $\bar{\Gamma}_{(7)}$, and is the group $\bar{\Gamma}$.

REFERENCES

[1] R. Baer, *The group of motions of a two-dimensional elliptic plane*, Compositio Math., **9** (1951), pp. 241-288.

[2] A. Borel, *Density and maximality of arithmetic subgroups*, J. Reine Angew. Math., **224** (1966), pp. 78-89.

[3] Z. I. Borevich - I. R. Shafarevich, *Number Theory*, Academic Press, New York, 1966.

[4] M. Deuring, *Algebren*, Erg. d. Math. u. i. Grenzgeb. **4**, Springer Verlag, 1935.

[5] M. Eichler, *Ueber die Idealtheorie hyperkomplexer Systeme*, Math. Z., **43** (1938), pp. 481-494.

[6] R. Fricke, *Ueber den aritmetischen Charaker der zu den Verzweigungen* (2, 3, 7) *und* (2, 4, 7) *gehörenden Dreicksfunctionen*, Math. Ann., **41** (1893), pp. 443-468.

[7] R. Fricke - F. Klein, *Vorlesungen über die Theorie der automorphen Functionen*, Band I, B. G. Teubner, Leipzig, 1893.

[8] H. J. Godwin, *On quartic fields with signature one with small discriminants*, Quart. J. Math. Oxford, **8** (1957), pp. 214-222.

[9] H. Helling, *Bestimmung der Kommensurabilitätsklasse der Hilbertschen Modulgruppe*, Math. Z., **92** (1966), pp. 269-280.

[10] N. Iwahori - H. Matsumoto, *On some Bruhat decompositions and the structure of the Hecke rings of p-adic Chevalley groups*, Publ. Math. I.H.E.S., **25** (1965), pp. 237-280.

[11] H. Jacquet - R. Langlands, *Automorphic forms on GL(2)*, Lecture Notes in Mathematics **114**, Springer-Verlag 1970.

[12] K. Takeuchi, *Commensurability classes of arithmetic triangle groups*, J. Fac. Sci. Univ. Tokyo, **24** (1977), pp. 201-212.

[13] G. A. Margoulis, *Discrete groups of isometries of manifolds of nonpositive curvature*, Proc. Int. Congress Math. 1974, Vancouver, Vol. 2, pp. 21-34.

[14] J. Mayer, *Die absolut-kleinsten Diskriminanten der biquadratischen Zahlkörper*, Sitzungsber. Akad. Wiss. Wien (IIA), **138** (1929), pp. 733-742.

[15] A. M. Odlyzko, *Some analytic estimates of class numbers and discriminants*, Invent. Math., **29** (1975), pp. 275-286.

[16] T. Ono, *On algebraic groups and discontinuous subgroups*, Nagoya Math. J., **27** (1966), pp. 297-322.

[17] M. S. Raghunathan, *Discrete subgroups of Lie groups*, Erg. d. Math. u. i. Grenzgeb., **68**, Springer Verlag, 1972.

[18] J. Rohlfs, *Ueber maximale arithmetisch definierte Gruppen*, Math. Ann., **234** (1978), pp. 239-252.

[19] J. Rohlfs, *Die maximalen arithmetisch definierten Untergruppen zerfallender einfacher Gruppen*, preprint.

[20] J-P. Serre, *Cohomologie des groupes discrets*, in Prospects in Math., Annals Math. Studies, **70**, Princeton U. Press 1970, pp. 77-168.

[21] J-P. Serre, *Arbres, amalgames, SL_2*, Astérisque, **46** (1977), Soc. Math. France.

[22] H. Shimizu, *On zeta functions of quaternion algebras*, Ann. of Math., (2) **81** (1965), pp. 166-193.

[23] C. L. Siegel, *The volume of the fundamental domain for some infinite groups*, Trans. A.M.S., **39** (1936), pp. 209-218.

[24] W. Thurston, *The geometry and topology of 3-manifolds*, mimeographed Notes, Princeton University.

[25] J. Tits, *Travaux de Margulis sur les sous-groupes discrets de groupes de Lie*, Sém. Bourbaki, Exp. 482, Février 1976, Springer L.N., **567**, pp. 174-190.

[26] H. C. Wang, *On a maximality property of discrete subgroups with fundamental domain of finite measure*, Amer. J. Math., **89** (1967), pp. 124-132.

[27] H. C. Wang, *Topics on totally discontinuous groups*, in Symmetric spaces, W. Boothby ed., M. Dekker 1972, pp. 460-487.

[28] A. Weil, *Basic Number Theory*, Grund. Math. Wiss., **144**, Springer-Verlag 1967.

[29] A. Weil, *Adeles and algebraic groups*, Notes by M. Demazure and T. Ono, The Institute for Advanced Study, 1961.

[30] B. Weisfeiler, *On abstract monomorphisms of k-forms of $PGL(2)$*, J. Algebra, **57** (1979), pp. 522-543.

The Institute for Advanced Study,
Princeton, New Jersey 08540

118.

Stable real cohomology of arithmetic groups II

Prog. Math., Boston **14** (1981) 21–55

Given a discrete subgroup Γ of a connected real semisimple Lie group G with finite center there is a natural homomorphism

$$j_\Gamma^q : I_G^q \to H^q(\Gamma;\mathbf{C}) , \qquad\qquad (q = 0,1,\dots) , \qquad\qquad (1)$$

where I_G^q denotes the space of G-invariant harmonic q-forms on the symmetric space quotient $X = G/K$ of G by a maximal compact subgroup K. If Γ is cocompact, this homomorphism is injective in all dimensions and the main objective of Matsushima in [19] is to give a range $m(G)$, independent of Γ, in which j_Γ^q is also surjective. The main argument there is to show that if a certain quadratic form depending on q is positive non-degenerate, then any Γ-invariant harmonic q-form is automatically G-invariant. In [3], we proved similarly the existence of a range in which j_Γ^q is bijective when Γ is arithmetic, but not necessarily cocompact. There are three main steps to the proof: (i) The cohomology of Γ can be computed by using differential forms which satisfy a certain growth condition, "logarithmic growth," at infinity; (ii) up to some range $c(G)$, these forms are all square integrable; and (iii) use the fact, pointed out in [16], that for $q \leq m(G)$, Matsushima's arguments remain valid in the non-compact case for square integrable forms.

The first purpose of the present paper is to generalize and sharpen the results of [3] in several ways. First of all we shall also consider the case of non-trivial coefficients, at any rate when they are defined by a finite dimensional complex representation E of G. This extension could already have been easily carried out in [3], but was not chiefly for lack of applications. However, recent work on the rational homotopy type of diffeomorphism groups [15] shows that it may be useful. In fact, it is done there for $\mathbf{SL}_n\mathbf{Z}$ and the adjoint representation. Second we shall consider other growth conditions and show (3.4) that

650

$H^*(\Gamma;E)$ can also be computed by using forms which, together with their exterior differential, are either of moderate growth or weakly λ-bounded (where λ is a dominant linear form on the Lie algebra of a maximal **Q**-split torus, see 3.2 for these notions). The proof is the same as that of the special case studied in [3: 7.4], and makes use of sheaf theory in the manifold with corners $\Gamma\backslash\bar{X}$ constructed in [8]. Those forms are square integrable up to a constant $C(G,\lambda,\tau^*)$ defined in §2, see (3.6). These growth conditions are expressed in terms of special frames in Siegel sets. In 3.10, we compare them with more usual notions of growth for functions on $\Gamma\backslash G$. Finally, as in [11], and following a development which has its origins in [20] and [22], we shall use relative Lie algebra cohomology and infinite dimensional unitary representations occurring in the spectrum of Γ, rather than Matsushima's original argument, and so can avail ourselves of some vanishing theorems recalled in 4.1. This leads again to the isomorphism of $I_G^{q\Gamma}$ and $H^q(\Gamma;\mathbb{C})$, but also to the vanishing of $H^q(\Gamma;E)$ in a certain range when E has no non-zero trivial subrepresentation (4.4). This range is the minimum of $C(G,\lambda,\tau^*)$ and of a constant defined by the vanishing of certain relative Lie algebra cohomology spaces (4.1). Since much information is known on the latter, this makes it worthwhile to study the former in more detail than in [3], and this is done in §2. Propositions 4.5 and 4.7 give two applications of these estimates.

Section 5 provides a counterpart to 3.4, which allows one to compute the cohomology with compact supports of $\Gamma\backslash X$ by means of forms which, together with their exterior differentials, are either fast decreasing (3.2) or weakly λ-bounded for $\lambda < 0$ (5.2). From this and 3.4 it follows that a non-zero fast decreasing Γ-invariant harmonic form is not cohomologous to zero in $H^*(\Gamma;E)$ and is cohomologous to a closed form with compact support (5.3). This applies in particular to harmonic cusp forms (5.5).

Finally, in §6, we show that 4.4 remains valid for S-arithmetic groups and groups of rational points, with essentially the same bounds (6.4). The main argument to effect this transition is contained in 6.2 and makes essential use of Bruhat-Tits buildings. As a result the stability theorems of [3] extend to S-arithmetic groups and groups of rational points and their consequences for the groups K_i of Quillen and $_\varepsilon L_i$ of Karoubi are also valid for rings of S-integers and number fields (6.5).

Some of the results proved here have been announced earlier, in particular in [2], [4: 6.1] and [6: Theorem 2].

<u>Notation and Conventions</u>. This paper is a sequel to [3], and we assume familiarity with it, in particular as regards Siegel sets and the representation of the invariant metric and differential forms with respect to special frames. However, we shall let K act on the right and G, Γ on the left, which causes some changes of signs and permutations of factors. For instance, A_t will now denote the subset of A where the simple roots are $\geq t$ (1.4). This is understood in the sequel. The Lie algebra of a Lie group H, U,... is denoted either by $L(H)$, $L(U)$,... or by the corresponding lower case German letter $\mathfrak{h}$, $\mathfrak{u}$,... .

The group G of this introduction will be replaced by the group $G(\mathbf{R})$ of real points of a (Zariski)-connected reductive $\mathbf{Q}$-group. It is not necessarily connected (in the ordinary topology); this is why we have to replace I_G by I_G^Γ in certain statements.

G *is a connected isotropic reductive $\mathbf{Q}$-group without non-trivial rational characters defined over* $\mathbf{Q}$, K *a maximal compact subgroup of* $G(\mathbf{R})$, θ *the Cartan involution of* $G(\mathbf{R})$ *with respect to* K [8], $X = G(\mathbf{R})/K$. *Moreover*, (τ, E) *is a finite dimensional complex rational representation of* G. *We assume* E *to be endowed with an admissible scalar product* [11: II, 2.2]. *In* §§1 *to* 5, Γ *is an arithmetic subgroup of* G.

1. Preliminaries on Reductive Groups

1.1 Let P be a parabolic $\mathbf{Q}$-subgroup of G, U its unipotent radical and $\sigma_P : P \to P/U$ the canonical projection. The Levi subgroups defined over $\mathbf{R}$ of P are then mapped isomorphically on P/U by σ_P. Let Z_d be the greatest central $\mathbf{Q}$-split torus of P/U. We let A_P denote the (topological) identity component of $Z_d(\mathbf{R})$. Any subgroup belonging to a Levi subgroup of $P(\mathbf{R})$ and mapped isomorphically onto A_P by σ_P will be called a *split component relative to* $\mathbf{Q}$ *of* P. In particular, if S is a maximal $\mathbf{Q}$-split torus in the radical of P, then $S(\mathbf{R})^\circ$ is a split component rel.$\mathbf{Q}$. However, such a choice may be too restrictive. There is one and only one split component rel.$\mathbf{Q}$ which is stable under the Cartan involution θ associated to K (see, e.g., [8]). It is not necessarily contained in a $\mathbf{Q}$-split torus, though. We shall refer to it as the split component rel.$\mathbf{Q}$ of P. Unless otherwise stated, Levi subgroups and split

components rel.**Q** are assumed to be θ-stable. The split components
rel.**Q** are conjugate under $N(R)$.

1.2 Let now P_o be a minimal parabolic **Q**-subgroup. The group P
is conjugate to a unique parabolic **Q**-subgroup containing P_o and, as
is well-known, this yields a unique monomorphism $A_P \to A_{P_o}$. Using the
projections σ_P and σ_{P_o}, we also get a unique monomorphism $A \to A'$,
where A (resp. A') is any split component rel.**Q** of P (resp. P_o),
whence also a canonical epimorphism $r_{PP_o}: X(A') \to X(A)$, where, as in
[3], $X(H)$ denotes the commutative group of continuous homomorphisms
of the real Lie group H into the multiplicative group $\mathbf{R}_+^*$ of strictly
positive real numbers. In particular, an element $\lambda \in X(A_{P_o})$ defines an
element $r_{P,P_o}(\lambda)$ of $X(A)$ for every parabolic **Q**-subgroup P, which
we shall also denote simply by λ.

1.3 We let $\Phi(P,A)$ denote the set of roots of P with respect
to A and $\Delta(P,A) = \{\alpha_1,\ldots,\alpha_\ell\}$ $(\ell = \dim A)$ the set of simple roots of
P with respect to A. An element

$$\lambda = \sum_{\alpha \in \Delta} c_\alpha(\lambda)\alpha \quad ,$$

is dominant (resp. dominant regular) if $c_\alpha \geqq 0$ (resp. $c_\alpha > 0$) for
all α. If $\lambda \in X(A_o)$ is dominant (resp. dominant regular), then
$r_{PP_o}(\lambda)$ is dominant (resp. dominant regular) for any proper parabolic
Q-subgroup P. For $\lambda,\mu \in X(A)$ we write $\lambda \geqq \mu$ (resp. $\lambda > \mu$) if
$\lambda - \mu \geqq 0$ (resp. $\lambda - \mu > 0$), and $\lambda \leqq 0$ (resp. $\lambda < 0$) if $-\lambda \geqq 0$
(resp. $-\lambda > 0$). As usual, ρ_P is defined by

$$a^{2\rho_P} = \det \operatorname{Ada}\big|_{\mathfrak{n}} \quad .$$

We have

$$\rho_P = r_{PP_o}\left(\rho_{P_o}\right) \quad .$$

Let $\mathfrak{h}$ be a Cartan subalgebra of $\mathfrak{g}$ containing $L(A_o)$. We
assume the root system $\Phi(\mathfrak{g}_c,\mathfrak{h}_c)$ be given an order compatible with
$\Phi(P_o,A_o)$. Let

$$\rho = \frac{1}{2} \sum_{\beta > 0} \beta \quad .$$

Then

$$\rho\big|_{\mathfrak{a}_o} \;=\; \rho_{P_o} \qquad\qquad .$$

1.4 In view of our shift from right to left for the action of Γ, the set A_t is defined here by

$$A_t \;=\; \{a \in A \mid a^\alpha \geqslant t \, , \qquad\qquad (\alpha \in \Delta(P,A)\} \qquad\qquad . \tag{1}$$

1.5 Given $\lambda \in X(A)$ we denote by $C(G,P,\lambda)$ the greatest integer q such that

$$\rho_P > \lambda + \mu$$

for every weight μ of A in $\oplus \Lambda^q \mathfrak{u}$. It would of course be equivalent to letting μ run through the weights of A in $\oplus_{j \leq q} \Lambda^j \mathfrak{u}$.

We write simply $C(G,\lambda)$ for $C(G,P,\lambda)$ when $P = P_o$. It is immediate that

$$C(G,P,r_{PP_o}(\lambda)) \;\geq\; C(G,\lambda) \qquad\qquad (\lambda \in X(A_o)) \qquad\qquad . \tag{1}$$

1.6 Let (σ,F) be a finite dimensional representation of G and $\lambda \in X(A_o)$. We let

$$C(G,\lambda,\sigma) \;=\; \inf_{\mu} C(G,\lambda + \mu) \quad ,$$

where μ runs through the restrictions to A_o of the weights of σ. It suffices of course to take the $\inf$ over the highest weights of the irreducible constituents of σ (for an ordering compatible with the one defined by P_o). For $\lambda = 0$, we denote this constant by $C(G,\sigma)$.

2. The Constant $C(G/k,\lambda)$

2.1 In this section we discuss the constant $C(G,\lambda)$, which we denote $C(G)$ or $C(G/\mathbf{Q})$ if $\lambda = 0$. It is equal to $C(\mathcal{D}G/\mathbf{Q},\lambda)$, where $\mathcal{D}G$ is the derived group of G, hence we may assume G to be semisimple. For the discussion it is also convenient to introduce this constant for more general groundfields than $\mathbf{Q}$. If H is a connected semisimple group defined over a field k, and $\lambda \in X(P_o)_k \otimes_{\mathbf{Z}} R$, we define then

$C(H/k,\lambda)$ as $C(G,P_o,\lambda)$ in 1.5. If k' is an extension of k, then any parabolic k-subgroup is a parabolic k' subgroup hence

$$C(H/k',\lambda') \leq C(H/k,\lambda) \tag{1}$$

where λ is obtained from λ' by restriction to a maximal k-split torus. If L is a k-group which is k-isogeneous to H, then $C(H/k,\lambda) = C(L/k,\lambda)$. If H is k-isogeneous to a product H_i ($1 \leq i \leq s$), then

$$C(H/k,\lambda) \;=\; \inf_i C(H_i/k,\lambda_i) \;, \tag{2}$$

where λ_i is obtained from λ by restriction to a maximal k-split torus of H_i. In view of its definition, $C(H/k,\lambda)$ depends only on the relative root system $_k\Phi(H)$ and on the multiplicities of the roots. Given a root system Φ, a set of natural integers $\{m_\alpha\}$ ($\alpha \in \Phi$), and $\lambda \in V[\Phi]$, where $V[\Phi]$ is the real vector space underlying the definition of Φ, fix an order on Φ, denote by ρ half the sum of the positive roots, each root α being counted m_α-times, and define $C(\Phi,\{m_\alpha\},\lambda)$ to be the greatest integer q such that $\rho > \mu + \lambda$ for any sum μ of q positive roots, where α occurs at most m_α times. Thus

$$C(H/k,\lambda) \;=\; C(_k\Phi(H),\{m_\alpha\},\lambda) \;, \tag{3}$$

where m_α is the dimension over k of the eigenspace in $L(H/k)$ corresponding to the root α and where $V[_k\Phi]$ is identified with $X(P_o)_k \otimes_{\mathbf{Z}} R$ in the usual way.

If the m_α are all equal to one, we write $C(\Phi)$ for $C(\Phi,\{m_\alpha\})$, $C(\Phi,\lambda)$ for $C(\Phi,\{m_\alpha\},\lambda)$. In particular, if H splits over k, then

$$C(H/k,\lambda) \;=\; C(\Phi(H),\lambda).$$

2.2 <u>Lemma</u>. *Let* Φ *be a finite set,* C, d, m_α ($\alpha \in \Phi$) *be strictly positive integers. Given a finite set* η *of elements belonging to* Φ, *let* $m_\eta(\alpha)$ *be the multiplicity of* α *in* η. *Let* ψ *be a finite set of elements of* Φ *of cardinality* $|\psi| \leq d.C$ *and such that* $m_\psi(\alpha) \leq d.m_\alpha$ *for all* $\alpha \in \Phi$. *Then* ψ *can be written as a disjoint union of subsets* ψ_i $(1 \le i \le d)$, *where* $|\psi_i| \le C$ *and* $m_{\psi_i}(\alpha) \le m_\alpha$ $(\alpha \in \Phi,\ 1 \le i \le d)$.

Proof by induction on d. There is nothing to prove if $d = 1$, so assume $d \ge 2$ and the lemma proved for $d - 1$. Let

$$\Phi_1 = \{\alpha \in \Phi \,|\, m_\psi(\alpha) > (d - 1)m_\alpha\} \quad , \tag{1}$$

$$r_\alpha = m_\psi(\alpha) - (d - 1)m_\alpha \quad , \qquad (\alpha \in \Phi_1) \ . \tag{2}$$

We have then

$$r_\alpha \le m_\alpha \quad , \qquad (\alpha \in \Phi_1) \tag{3}$$

and also

$$\sum_{\alpha \in \Phi_1} r_\alpha \le C \ . \tag{4}$$

Let θ be a subset of ψ which is maximal with respect to the following properties

$$|\theta| \le C; \quad m_\theta(\alpha) \le m_\alpha \quad (\alpha \in \Phi) \quad \text{and} \quad m_\theta(\alpha) \ge r_\alpha \ (\alpha \in \Phi_1). \tag{5}$$

Such subsets do exist in view of (3), (4). Set $\theta' = \psi - \theta$. By construction

$$m_{\theta'}(\alpha) \le (d - 1)m_\alpha \quad , \qquad (\alpha \in \Phi) \quad . \tag{6}$$

We claim moreover that $|\theta'| \le (d - 1)C$. This is clear if $\theta = |C|$. Assume now $|\theta| < C$. In view of the maximality assumption, we have then

$$m_\theta(\alpha) = \begin{cases} m_\alpha & , \quad (\alpha \in \Phi_1) \ , \\ \min(m_\alpha, m_\psi(\alpha)), & (\alpha \in \Phi - \Phi_1) \end{cases} \tag{7}$$

$$m_{\theta'}(\alpha) = \begin{cases} m_\psi(\alpha) - m_\alpha & (\alpha \in \Phi - \Phi_1, m_\psi(\alpha) > m_\alpha) \\ 0 & (\alpha \in \Phi - \Phi_1, m_\psi(\alpha) \le m_\alpha) \end{cases} \ . \tag{8}$$

It follows that

$$m_{\theta'}(\alpha) \leqq (d-1)m_{\theta}(\alpha) \quad , \qquad (\alpha \in \Phi) \quad , \qquad (9)$$

whence

$$|\theta'| \leqslant (d-1)|\theta| < (d-1)C \qquad . \qquad (10)$$

We then take $\psi_1 = \theta$ and apply the induction assumption to θ'.

2.3 <u>Lemma</u>. *Let Φ be a root system d, m_{α} ($\alpha \in \Phi$) strictly positive integers and let $\lambda \in V[\Phi]$. Then*

$$C(\Phi,\{dm_{\alpha}\},\lambda) \geqq d.C(\Phi,\{m_{\alpha}\},\lambda) \qquad . \qquad (1)$$

Given a set ψ of positive roots, let $\langle\psi\rangle$ be the sum of the elements in ψ and $m_{\psi}(\alpha)$ the multiplicity of α in ψ. Let $C' = C(\Phi,\{m_{\alpha}\},\lambda)$. Since $d\rho$ is half the sum of the positive roots with multiplicities dm_{α}, we have to prove

$$|\psi| \leqq d.C' \quad \text{and} \quad m_{\psi}(\alpha) \leqq d.m_{\alpha} \quad (\alpha \in \Phi) \Rightarrow d\rho > \lambda + \langle\psi\rangle \quad . \qquad (2)$$

By 2.2 we can write ψ as a disjoint union of subsets ψ_i ($1 \leqslant i \leqslant d$) such that $|\psi_i| \leqq C'$ and $m_{\psi_i}(\alpha) \leqq m_{\alpha}$ for all i and α. We have then

$$\rho > \lambda + \langle\psi_i\rangle \qquad (i = 1,\ldots,d) \qquad (3)$$

by assumption, whence

$$d\rho > d\lambda + \langle\psi\rangle \geqslant \lambda + \langle\psi\rangle \quad . \qquad (4)$$

2.4 <u>Remark</u>. This estimate is not sharp. Assume for instance that $\Phi = \{\pm\alpha\}$ is of type $\mathbf{A_1}$ and that $m_{\alpha} = 1$. Then $C(\Phi,\{m_{\alpha}\}) = 0$. However

$$C(\mathbf{A_1},d) = \left[\frac{d-1}{2}\right] \qquad .$$

This is due to the fact that we may have $d\rho > \langle\psi\rangle$ for sets ψ with more than dC' elements, for which the proof of the inequality does not reduce to the case $d = 1$.

2.5 <u>Proposition</u>. *Let k be a field, k' a finite separable extension of k. Let H^1 be a connected semi-simple k'-group and*

$H = R_{k'/k} H'$, *where* $R_{k'/k}$ *refers to the restriction of scalars* [25: 1].
Then $C(H/k) \geqq [k':k].C(H'/k')$.

Let $d = [k':k]$. There is a canonical isomorphism $_k\Phi(H) \to _{k'}\Phi(H')$
such that if α' corresponds to α, then $m_{\alpha'} = d.m_\alpha$ (cf. [10: 6.19,
6.21]). Our assertion then follows from 2.3, applied to H/k and
H'/k', and from 2.1(3).

2.6 <u>Remark</u>. If P'_o is a minimal parabolic k'-subgroup of H',
then $P_o = R_{k'/k} P_o$ is a minimal parabolic k-subgroup of H [10: loc.
cit.], and there is a canonical isomorphism $X(P'_o)_{k'} \otimes R \to X(P_o)_k \otimes R$
which is of course compatible with the isomorphism of relative root
systems used above. If λ in the latter space corresponds to λ' in
the former space by this isomorphism, then the same proof shows that
we have

$$C(H/k,\lambda) \geq [k':k]C(H'/k,\lambda') \qquad . \qquad (1)$$

2.7 <u>Theorem</u>. *Let* k *be a number field,* $\bar{k}$ *an algebraic closure
of* k, *and* H *an almost simple k-group. Let* G_i $(1 \leq i \leq s)$ *be the
simple factors of* $G/\bar{k}$. *Then* $C(H/k) \geq \Sigma_i C(G_i/\bar{k})$.

There exists a finite extension k' of k contained in $\bar{k}$ and
an absolutely almost simple k'-group H' such that H is isogeneous
to $R_{k'/k} H'$ [10: 6.21]. Since $C(H/k)$ is the same for two k-iso-
geneous k-groups, we may replace H by $R_{k'/k} H'$. By 2.5

$$C(H/k) \geq [k':k]C(H'/k') \qquad . \qquad (1)$$

We have $C(H'/k') \geq C(H'/\bar{k})$. Now $H'/\bar{k}$ is one of the simple factors
of $H/\bar{k}$. There are $[k':k]$ such factors, and they are isomorphic over
$\bar{k}$. The theorem now follows from (1).

2.8 <u>Remark</u>. As a sequel to 2.4, we note that this is not
necessarily sharp if H is not split over k. For instance, let
$k = Q$, $H' = SL_2/k'$ and $H = R_{k/Q} H'$. Then $C(H'/k') = 0$, but $C(H/k) =$
$[(d - 1)/2]$.

2.9 Using the tables of [12] one can compute $C(\Phi)$ for all
irreducible reduced root systems. We get the following list:

$$\Phi: \quad \mathbf{A}_\ell\,(\ell \geq 1), \quad \mathbf{B}_\ell\,(\ell \geq 3), \quad \mathbf{C}_\ell\,(\ell \geq 2), \quad \mathbf{D}_\ell\,(\ell \geq 4), \quad \mathbf{E}_6, \quad \mathbf{E}_7, \quad \mathbf{E}_8, \quad \mathbf{F}_4, \quad \mathbf{G}_2$$

$\Phi:$	$\mathbf{A}_\ell\,(\ell \geq 1)$	$\mathbf{B}_\ell\,(\ell \geq 3)$	$\mathbf{C}_\ell\,(\ell \geq 2)$	$\mathbf{D}_\ell\,(\ell \geq 4)$	$\mathbf{E}_6$	$\mathbf{E}_7$	$\mathbf{E}_8$	$\mathbf{F}_4$	$\mathbf{G}_2$
$C(\Phi):$	$[(\ell-1)/2]$	$\ell-1$	$\ell-2$	$\ell-2$	7	13	25	5	1

2.10   Let k, k', H and H' be as in 2.7. Let (τ',E') be an absolutely irreducible rational representation of H' which is defined over k'. Then (τ,E), where $\tau = R_{k'/k}\tau'$, $E = R_{k'/k}E'$ is a rational representation of H defined over k. It is not irreducible, but a direct sum of irreducible representations of the simple $\bar{k}$-factors of H. If S' is a maximal k'-split torus of H', then the greatest k-split subtorus of $R_{k'/k}S$ is a maximal k-split torus of H, whence an iso-morphism of $X(S')_{k'} \otimes \mathbf{R}$ onto $X(S)_k \otimes \mathbf{R}$ [10]. It maps the weights of τ' onto those of τ, but the multiplicity of the weight is multiplied by $[k':k]$. The restrictions to S of the weights of the constituents of τ are the same, with the original multiplicities. In particular there is only one highest weight. If $\lambda'_{\tau'}$ and λ_τ are the highest weights of τ' and τ, we have then with λ and λ' as before

$$C(H'/k',\lambda',\tau') \;=\; C(_{k'}\Phi(H'),\lambda' + \lambda'_{\tau'}) \tag{1}$$

$$C(H/k,\lambda,\tau) \;=\; C(_k\Phi(H),\lambda + \lambda_\tau) \quad . \tag{2}$$

Therefore 2.6(1) implies

$$C(H/k,\lambda,\tau) \;\geq\; [k':k]\,C(H'/k',\lambda',\tau') \quad . \tag{3}$$

2.11   Consider in particular the case where $\lambda' = 0$ and $\tau' = \mathrm{Ad}$ is the adjoint representation. Then τ is also the adjoint repre-sentation of H. Assume that H' is split over k'. The highest weight of τ' is the highest root δ_0. We have then

$$C(H/k,\mathrm{Ad}) \;\geq\; [k':k]\,.\,C(\Phi(H'),\delta_0) \quad . \tag{4}$$

The values of $C(\Phi,\delta_0)$ can also be computed for all types by using the tables in [12]. One finds

$\Phi:$	$\mathbf{A}_\ell$	$\mathbf{B}_\ell\,(\ell \geq 3)$	$\mathbf{C}_\ell\,(\ell \geq 2)$	$\mathbf{D}_\ell\,(\ell \geq 4)$	$\mathbf{E}_6$	$\mathbf{E}_7$	$\mathbf{E}_8$	$\mathbf{F}_4$	$\mathbf{G}_2$
$C(\Phi,\delta_0):$	$\left[\dfrac{\ell-3}{2}\right]$	$\ell-2$	$\ell-4$	$\ell-4$	6	12	23	4	0

3. Weak λ-boundedness of Differential Forms

3.1   If M is a manifold on which $G(\mathbf{R})$ operates (smoothly), then $\Omega(M;E)$ denotes the space of smooth differential E-valued forms on M. As usual, $G(\mathbf{R})$ acts on $\Omega^q(X;E)$ by the rule

$$(g \circ \omega)(x, Y_x) \;=\; \tau(g)(\omega(g^{-1}.x, g^{-1}.Y_x)),$$

$$(g \in G(\mathbf{R}), \; x \in M, \; Y_x \text{ a q-vector at } x) .$$

For a subgroup H of $G(\mathbf{R})$, we let $\Omega(M;E)^H$ be the space of H-invariant elements in $\Omega(M;E)$. Our main case of interest is $M = X$, but we shall also on occasion take $M = G(\mathbf{R})$.

3.2 <u>Growth Conditions for Differential Forms</u>. Let P be a parabolic $\mathbf{Q}$-subgroup and A its split component. As usual $M(\mathbf{R}) = {}^{O}M(\mathbf{R}) \times A$ is the (θ-stable) Levi subgroup of P, and $Z = {}^{O}M(\mathbf{R})/(K \cap M)$. Let 0 be a fixed point of K in X and

$$\mu_o: N(\mathbf{R}) \times Z \times A \;=\; Y \to X \tag{1}$$

be as in [3]. Let $\eta \in \Omega^j(X;E)$. With respect to a special frame, we write again

$$\mu_o^*(\eta) \;=\; \sum_I \eta_I \omega^I \qquad , \tag{2}$$

where, however, the coefficients η_I are smooth E-valued functions. Fix $\lambda \in X(A_o)$. We say that η is weakly λ-bounded if, given P and a Siegel set $\mathfrak{S}_{t,\omega} = \mu_o(\omega \times A_t)$, there exists a polynomial P in $\ell = \dim A$ variables such that

$$|\mu_o^*(\eta)_I (q,a)| \;\leqslant\; a^\lambda |P(\ell n \; a^{\alpha_1}, \ldots, \ell n \; a^{\alpha_\ell})| \;, \qquad (a \in A_t; \; q \in \omega) . \tag{3}$$

On the left-hand side $|\;|$ refers to the norm on E. on the right-hand side, $\alpha_1, \ldots, \alpha_\ell$ are the simple roots of P with respect to A. Since the precise nature of P will not intervene, we shall also write this

$$|\mu_o^*(\eta)_I (q,a)| \;\lesssim_w\; a^\ell \qquad , \qquad (a \in A_t; \; q \in \omega) . \tag{4}$$

We note that if $\lambda = 0$, and $E = \mathbf{C}$, this is the condition of logarithmic growth in [3].

The form η is λ-*bounded* if

$$|\mu_o^*(\eta)_{|}(q,a)| \lessdot a^\lambda , \qquad\qquad (a \in A_t ; \ q \in \omega) . \qquad\qquad (5)$$

It is of *moderate growth* if it is λ-bounded for some $\lambda \geqq 0$, and *fast decreasing* if it is λ-bounded for all λ's.

3.3 <u>Notation</u>. Let $\lambda \in X(A_o)$. We let $\Omega_\lambda^W(X;E)^\Gamma$ denote the spaces of smooth E-valued Γ-invariant forms η on X such that η and $d\eta$ are weakly λ-bounded. Moreover, $\Omega_{fd}(X;E)^\Gamma$ (resp. $\Omega_{mg}(X;E)^\Gamma$) will denote the complex of forms in $\Omega(X;E)^\Gamma$ which, together with their exterior differentials, are fast decreasing (resp. of moderate growth).

If $\lambda = 0$ and $E = \mathbb{C}$, this is just the complex C of Theorem 7.4 in [3], consisting of forms which, together with their exterior differentials, have logarithmic growth near the boundary.

3.4 <u>Theorem</u>. *Assume* $\lambda \in X(A_o)$ *to be dominant. Then the inclusions*

$$\Omega_\lambda^W(X;E)^\Gamma \rightarrow \Omega_{mg}(X;E)^\Gamma \rightarrow \Omega(X;E)^\Gamma \qquad\qquad (1)$$

induces isomorphisms in cohomology.

The proof differs only in minor details from that of Theorem 7.4 in [3] and we shall refer to the latter to the extent possible.

Let Γ' be a normal subgroup of finite index of Γ. Then we have

$$\Omega_\lambda^W(X;E)^\Gamma = (\Omega_\lambda^W(X;E)^{\Gamma'})^{\Gamma/\Gamma'} , \qquad\qquad (2)$$

$$H^*(\Omega_\lambda^W(X;E)^\Gamma = (H^*(\Omega_\lambda^W(X;E)^{\Gamma'})^{\Gamma/\Gamma'} , \qquad\qquad (3)$$

and similarly for $\Omega_{mg}(X;E)^\Gamma$ and $\Omega(X;E)^\Gamma$. We may therefore replace Γ by Γ', hence assume Γ to be torsion-free. Then E gives rise to a locally constant sheaf $\tilde{E}$ on the manifold with corners $\Gamma\backslash\bar{X}$. We define a presheaf F on $\Gamma\backslash\bar{X}$ by assigning to an open set $U \subset \Gamma\backslash\bar{X}$ the space of $\tilde{E}$-valued forms on $U \cap (\Gamma\backslash X)$ which, together with their exterior differentials, are weakly λ-bounded (resp. of moderate growth) near the boundary. This is clearly a sheaf.

In the corner which is the closure of the Siegel set of 3.2, we may take as local coordinates $\beta^i = a^{-\alpha_i}$ $(1 \leqq i \leqq \ell)$ and local coordinates x^j $(\ell < j \leqq n = \dim X)$ in ω. Hence the $d\beta^i$ $(i \leqq \ell)$ and the ω^j $(j > \ell)$

form a local frame of the cotangent bundle which extends smoothly to a local frame on the corner. If φ is a smooth function on $\Gamma\backslash\bar{X}$, then $d\alpha$ is a linear combination of the $d\beta^i$, ω^j with bounded coefficients. Since $d\beta^i/\beta^i = -da^{\alpha_i}/a^{\alpha_i}$, it follows *a fortiori* that the coefficients $(d\varphi)_j$ of $d\varphi$, expressed as linear combination of the ω^j, are bounded. We even have

$$(d\varphi)_j \prec a^{-\alpha_j} \quad , \qquad\qquad (1 \leq j \leq \ell) \quad , \qquad (4)$$

in $\mathcal{S}_{t,\omega}$. It follows then that if η and $d\eta$ are weakly λ-bounded (resp. λ-bounded, resp. of moderate growth), then so are $\varphi\eta$ and $d\varphi.\eta$. As a consequence, F is fine. There remains to see that it provides a resolution of $\tilde{E}$. Since λ is *dominant*, $\Gamma(U)$ contains the smooth E-valued functions, hence $H^o(F) = \tilde{E}$. The main point is then again to check that $H^q(F) = 0$ for $q \geq 1$. For this, it suffices to see that near a boundary point, the Poincaré lemma remains valid for weakly λ-bounded forms, or forms of moderate growth. This amounts to showing that the homotopy operator A of [3: p. 258] preserves these conditions. By (5), (7) *loc. cit.*, if η is of degree q, we have

$$A\eta = \sum_I c_I \omega^I \quad , \qquad\qquad (5)$$

where

$$c_I = \sum_j \pm x^j \int_0^1 t^{q-1} \eta_{I \cup \{j\}}(tx)\, dt \quad , \qquad (6)$$

the sum running through the indices j not belonging to I. By assumption,

$$\lambda = \sum_{\substack{1 \leq i \leq \ell \\ j}} d_i \alpha_i \quad , \qquad \text{with } d_i \geq 0 \text{ for } i = 1,\ldots,\ell. \qquad (7)$$

Set

$$d(x) = \sum_{1 \leq i \leq \ell} d_i . x^i \quad . \qquad\qquad (8)$$

We are interested only in $A\eta$ in some arbitrarily chosen smaller Siegel set, so we may assume $x^i \geq D$ for $i = 1,\ldots,\ell$ and some $D \geq 1$

and also $d(x) > 1$ unless $\lambda = 0$. Note also that the x^i for $i > \ell$ vary in a bounded interval. Assume now η to be weakly λ-bounded. There exist then $C, M > 0$ such that

$$|\eta_J(tx)| \leq C.e^{td(x)}.(x^1...x^\ell)^M , \qquad (x \in \mathfrak{S}, \ t \in [0,1]) . \qquad (9)$$

It is then elementary that we can find $C', M' > 0$ such that

$$\left| \int_0^1 t^{q-1}\eta_J(tx) \ dt \right| \leq C'.(x^1...x^\ell)^{M'}.e^{d(x)} , \qquad (x^i \geq D, \ i = 1,...,\ell) \qquad (10)$$

which implies that $A\eta$ is also weakly λ-bounded. It then also follows that $A\eta$ is of moderate growth if η is so, whence our assertion.

    3.5 <u>Scalar Product of Differential Forms</u>. The elements of $\Omega(X;E)^\Gamma$ are smooth sections of a Γ-bundle whose fibre at $x \in X$ is $\Lambda T_x^*(X) \otimes E$. On the latter, we put hermitian product which is the tensor product of the product stemming from the metric on X by the admissible product on E, to be denoted $(,)_x$.

    As is well-known (cf. e.g. [11; 21]), we may identify $\Omega(X;E)^\Gamma$ with the space of smooth cross-sections of K-bundle over $\Gamma \backslash X$ by means of the map

$$\eta \mapsto \eta^o , \qquad \text{where} \quad \eta^o(g) = \tau(g^{-1})(\eta \circ \pi)(g) , \qquad (1)$$

where $\pi: G(\mathbb{R}) \to X$ is the canonical projection. Since the hermitian product of the typical fibre is K-invariant, this allows one to define a scalar product on $\Omega(X;E)^\Gamma$. It is usually written

$$(\alpha,\beta) = \int_{\Gamma \backslash G(\mathbb{R})} (\alpha_g^o,\beta_g^o).dg , \qquad (\alpha,\beta \in \Omega(X,E)^\Gamma) , \qquad (2)$$

where dg is a Haar measure. The integrand is right invariant under K, so that the integral is in effect on $\Gamma \backslash X$. It is more convenient here for us to write it directly on $\Gamma \backslash X$, as

$$(\alpha,\beta) = \int_{\Gamma \backslash X} (\tau(g_x^{-1})\alpha_x,\tau(g_x)^{-1}\beta_x) \ dv_x , \qquad (3)$$

where dv_x is the Riemannian volume element on X and g_x denotes an element in $G(\mathbb{R})$ which brings 0 onto x.

We let $H^*(\Gamma;E)_{(2)}$ denote the space of elements in $H^*(\Gamma;E)$ which are representable by a square integrable closed form. By a theorem of Kodaira, recalled in [3: 2.4], these elements are also representable by square integrable harmonic forms.

3.6 <u>Lemma</u>. *Assume that (τ,E) is irreducible. Let $\lambda \in X(A_o)$. Then, if $j \leq C(G,\lambda,\tau^*)$, any weakly λ-bounded E-valued Γ-invariant j-form η is square integrable.*

It suffices to check that η is square integrable on any Siegel set $\mathscr{S}_{t,\omega}$, with respect to any proper parabolic **Q**-subgroup P. We use the notation of [3: 5.5]. We have then

$$(\eta,\eta)_{\mathscr{S}} = \int_{A_t \times \omega} \sum_{I,J} g^{IJ}(\tau(a^{-1}q^{-1})\eta_I, \tau(a^{-1}q^{-1})\eta_J) a^{-2\rho_P} \, dv_A dv_Z dv_N \ . \tag{1}$$

Let (e_ν) be an orthonormal basis of E consisting of eigenvectors of A, where $\nu \in X(S)$ is the weight of e_ν. We can write

$$\tau(q^{-1})\eta_I(q,a) = \sum_\nu \eta_{I,\nu}(q,a) e_\nu \tag{2}$$

$$\tau(a^{-1})\tau(q^{-1})\eta_I(q,a) = \sum_\nu \eta_{I,\nu}(q,a) a^{-\nu} e_\nu \qquad (q \in \omega, \ a \in A) \ . \tag{3}$$

Since ω is compact, we get from (2)

$$|\tau(q^{-1})\eta_I(q,a)|^2 = \Sigma|\eta_{I,\nu}(q,a)|^2 \bowtie |\eta_I(q,a)|^2 \ . \tag{4}$$

The linear form $-\nu$ is a weight of (τ^*,E^*), hence, if δ denotes the highest weight of τ^*:

$$\delta \geqslant -\nu \tag{5}$$

$$a^{-2\nu} \leqslant a^{2\delta} \qquad (a \in A_t) \ . \tag{6}$$

Together with (3), (4) this yields

$$|\tau(a^{-1}q^{-1})\eta_I(q,a)|^2 \quad a^{2\delta}|\eta_I(q,a)|^2 \ , \qquad (a \in A_t; \ q \in \omega) \ . \tag{7}$$

This reduces the estimate of the sum on the right-hand side of (1) to the scalar case considered in [3]. We get then, using (11), (12) in [3: 5.5]:

$$\left| \sum_{I,J} g^{IJ}(\tau(a^{-1}q^{-1})\eta_I, \tau(a^{-1}q^{-1})\eta_J \right| \leqq \sum |\eta_J|^2 a^{2\alpha(J)+2\delta} \tag{8}$$

$$\left| \sum_{I,J} g^{IJ}(\tau(a^{-1}q^{-1})\eta_I, \tau(a^{-1}q^{-1})\eta_J) \right| \prec_w \sum_J a^{2(\alpha(J)+\lambda+\delta)}$$

$$(a \in A_t, q \in \omega) \ , \tag{9}$$

which, by [3: 5.4], is integrable on $A_t \times \omega$ if

$$\rho_P > \alpha(J) + \lambda + \delta \tag{10}$$

for all J with j-elements. Since the $\alpha(J)$ are the weights of A in $\Lambda^j \mathfrak{u}$, this condition is indeed fulfilled whenever $j \leqslant C(G,\lambda,\tau^*)$.

3.7 **Proposition.** *Let* $\lambda \in X(A)$ *be dominant.*

(i) *Let* $j \leqq C(G,\lambda,\tau^*)$. *Then any j-form in* $\Omega_\lambda^W(X;E)$ *is square integrable and any cohomology class in* $H^j(\Gamma;E)$ *is representable by a square integrable harmonic form.*

(ii) *For* $j \leqq C(G,\lambda,\tau^*) + 1$, *the space of harmonic square integrable j-forms contained in* $\Omega_\lambda^W(X;E)^\Gamma$ *maps injectively into* $H^j(\Gamma;E)$.

(i) The first assertion is a special case of 3.6, applied to each irreducible constituent of E. In view of 3.4, any element of $H^j(\Gamma;E)$ is then representable by a closed square integrable form, hence by a harmonic one (3.5).

(ii) Let now $j \leqq C(G,\lambda,\tau^*) + 1$ and η be a harmonic square integrable j-form contained in $\Omega_\lambda^W(X;E)^\Gamma$. [Of course, by (i), the assumption of square integrability is redundant if $j \leqq C(G,\lambda,\tau^*)$.] Assume η is cohomologous to zero in $\Omega(X;E)^\Gamma$. Then, by 3.4, it is already cohomologous to zero in $\Omega_\lambda^W(X;E)$, hence there exists $\mu \in \Omega_\lambda^W(X;E)$ of degree $j-1$ such that $\eta = d\mu$. But μ is square integrable by (i), hence $\eta = 0$ [3: 2.5].

3.8 **Remark.** Theorem 3.7, for $\lambda = 0$, shows that $H^j(\Gamma;E) = H^j(\Gamma;E)_{(2)}$ for $j \leqq C(G,\tau^*)$. Similar criteria were already given in [24] for **Q**-rank 1 and announced in [17] in the general case. In fact the conditions given there are stronger. To compare, assume (τ,E) to be irreducible. Note first that our condition could be expressed by saying that $\rho > \mu + \nu$ for any weight μ of A in E^* and any weight ν of A in $\Lambda^j \mathfrak{u}$. These weights are the opposite of those of A in $\Lambda \mathfrak{u}^* \otimes E$, which is the complex used to compute $H^*(\mathfrak{u};E)$, so we can also phrase it by saying that $\rho + \mu > 0$ for any weight μ of A in $\Lambda^j \mathfrak{u}^* \otimes E$. Now the condition of [17;24] is that $s(\rho + \mu) > 0$ for certain elements s of $W(\mathfrak{g}_c, \mathfrak{h}_c)$. By a theorem of Kostant, these elements are such that $s(\rho + \mu) = \rho + \nu$ where ν runs through the weights of A in $H^*(\mathfrak{u};E)$. But $H^*(\mathfrak{u},E) = H(\Lambda \mathfrak{u}^* \otimes E)$. Thus the requirement in [17] is that $\rho + \mu > 0$ only for those μ which occur in $H^*(\mathfrak{u};E)$.

This is also contained in [27], where it is proved moreover that $H^*(\Gamma;E)_{(2)}$ is isomorphic to the space of L^2-harmonic forms up to that range.

3.9 We now relate the growth conditions introduced in 3.2 with conditions involving the coefficients of forms on $\Gamma \backslash G(\mathbf{R})$, written as linear combinations of Maurer-Cartan forms. Let then $\pi: G(\mathbf{R}) \to X$ be the canonical projection and (ω^i) be a basis of $L(\mathfrak{g}(\mathbf{R}))^*$. Any element $\eta \in \Omega(G;E)^\Gamma$ can be written

$$\mu = \sum_I \mu_I \omega^I \qquad . \tag{1}$$

We say that μ is *weakly λ-bounded* (resp. *λ-bounded*) if 3.2(3) (resp. 3.2(5)) is true on any Siegel set on $G(\mathbf{R})$. This is clearly independent of the choice of the basis of $L(\mathfrak{g}(\mathbf{R}))^*$. Furthermore, since $G(\mathbf{R}) = P(\mathbf{R}).K$, it suffices to check this condition for the restriction of μ to $P(\mathbf{R})$, if μ is K-finite. The form μ is *fast decreasing* (resp. of *moderate growth*) if it is λ-bounded for all (resp. some) λ.

3.10 <u>Proposition</u>. *Let $\eta \in \Omega(X;E)^\Gamma$ and $\lambda \in X(A_o)$.*

(i) *Assume that $\eta \circ \pi$ is (weakly) λ-bounded on $\Gamma \backslash G(\mathbf{R})$. Then η is (weakly) λ-bounded on $\Gamma \backslash X$.*

(ii) *$\eta \circ \pi$ is of moderate growth (resp. fast decreasing), if and only η is so on $\Gamma \backslash X$.*

In 3.2 we have used an isomorphism of manifolds $\mu_o: Y = U(\mathbf{R}) \times Z \times A \to X$. Since $X = P(\mathbf{R})/(K \cap P)$, we could equally well have lifted our forms to $P(\mathbf{R})$ and considered instead an isomorphism

$$\mu_o^*: Y' = U(\mathbf{R}) \times {}^o M(\mathbf{R}) \times A \to P(\mathbf{R}) \qquad , \tag{1}$$

given by $(u.m.a) \mapsto u.m.a$ $(u \in U(R),\ m \in {}^O M(R),\ a \in A)$.

On Y' we consider again special frames defined by the Maurer-Cartan forms ω^i on the three factors, those on $U(R)$ and A being chosen as before. The set of indices $I_{-1},\ I_0,\ I_\beta$ are as in [3: p. 250] except that I_0 refers to the Lie algebra of ${}^O M(R)$. For a set of indices I, we define the sum $\alpha(I)$ as in [3], the summands being contributed only by indices in the I_β's $(\beta \in \Phi(P,A))$. The elements ω^i also define a basis of the space of left-invariant 1-forms on $P(R)$.

Let us denote by $\bar{\omega}^i$ (resp. $\bar{\omega}^I$) the left-invariant form on $P(R)$ which is equal to ω^i (resp. ω^I) at the identity. We want to express the $\mu_o^*(\bar{\omega}^I)$ in terms of the ω^I, and conversely.

For a set I of indices, we denote by (I) the collection of sets of indices J such that the cardinalities of $J \cap J_{-1}$, $J \cap J_0$ and $J \cap J_\beta$ $(\beta \in \Phi(P,A))$ are the same as those of $I \cap I_{-1}$, $I \cap I_0$ and $I \cap I_\beta$. We note that $\alpha(J) = \alpha(I)$ for $J \in (I)$. Let $j \in I_\beta$, $k \in I_0$ and $\ell \in I_{-1}$. We have

$$\mu_o^*(\bar{\omega}^j + \bar{\omega}^k + \bar{\omega}^\ell)_{(u,m,a)} = \mu_o^*(uma(\bar{\omega}^j_1 + \bar{\omega}^k_1 + \bar{\omega}^\ell_1)) = \mu_o^*(uma(\omega^j_1 + \omega^k_1 + \omega^\ell_1))$$

$$= u.\mathrm{Adma}(\omega^j_1) + m.\omega^k_1 + a.\omega^\ell_1 \tag{2}$$

$(u \in U(R),\ m \in {}^O M(R),\ a \in A)$. Since

$$\mathrm{Ada}(\omega^j) = a^{-\beta}\omega^j \quad , \qquad (a \in A;\ j \in I_\beta) \ , \tag{3}$$

and ${}^O M(R)$ leaves $u(R)_\beta$ stable, there exist smooth functions c^j_i $(i,\ j \in I_\beta)$ on ${}^O M(R)$ such that

$$\mathrm{Adma}(\omega^j) = a^{-\beta} \sum_{i \in I_\beta} c^j_i(m)\omega^i , \qquad (m \in {}^O M(R);\ a \in A) \ . \tag{4}$$

From this it follows that we can write

$$\mu_o^*(\bar{\omega}^I)_{(u,m,a)} = a^{-\alpha(I)} \sum_{J \in (I)} c^I_J(m)\omega^J_{(u,m,a)} \quad , \tag{5}$$

where c^I_J is a smooth function on ${}^O M(R)$.

Conversely, we have

$$\mu_o^{*-1}(\omega_u^j + \omega_m^k + \omega_a^\ell) \;=\; \mu_o^{*-1}(u.\omega_1^j + m.\omega_1^k + a.\omega_1^\ell)$$

$$=\; uma\,(Ad(ma)^{-1}\omega_1^j + \omega_1^k + \omega_1^\ell)$$

and a similar computation shows the existence of smooth functions d_I^J on $^oM(\mathbf{R})$ such that

$$\mu_o^{*-1}(\omega^J)_{uma} \;=\; a^{\alpha(J)} \sum_{I\in(J)} d_I^J(m)\,\omega_{uma}^I \qquad . \tag{6}$$

For $\eta\in\Omega(X;E)^\Gamma$, let us write

$$(\eta\circ\pi)\big|_{P(\mathbf{R})} \;=\; \sum_I \bar\eta_I\bar\omega^I \,, \qquad \mu_o^*(\eta) \;=\; \sum_J \eta_J\omega^J \;. \tag{7}$$

It follows then from (5) that we have

$$\eta_J(u,m,a) \;=\; a^{-\alpha(J)} \sum_{I\in(J)} c_J^I(m)\,\overline{\eta_I}(uma) \qquad , \tag{8}$$

$$\overline{\eta_I}(uma) \;=\; a^{\alpha(I)} \sum_{J\in(I)} d_I^J(m)\,\eta_J(u,m,a) \tag{9}$$

$(u\in U(\mathbf{R})$, $m\in {}^oM(\mathbf{R})$, $a\in A)$. If now u and m vary in relatively compact sets, we get

$$|\eta_J(u,m,a)| \;\leqslant\; a^{-\alpha(J)} \sum_{I\in(J)} |\overline{\eta_I}(uma)| \tag{10}$$

$$|\overline{\eta_I}(uma)| \;\leqslant\; a^{\alpha(J)} \sum |\eta_J(u,m,a)| \qquad . \tag{11}$$

Then (i) follows from (10), and (ii) from (10) and (11). Note that the converse assertion to (i) need not be true since the right-hand side of (11) is not necessarily λ-bounded when the η_J's are so.

4. Some Vanishing and Isomorphism Theorems

4.1 Let H be a connected real semi-simple Lie group (with finite center, as usual), L a maximal compact subgroup of H and (μ,F) a finite dimensional complex representation of H. We let $M(H,\mu)$ or $M(H,F)$ denote the greatest integer such that $H^q(\mathfrak{h},L;V \otimes F) = 0$ for all $q \leq M(H.F)$ and all irreducible unitary representations of H with compact kernel, and $M(H,^*)$ be the minimum of $M(H,F)$ over all F's. The study of $M(H,^*)$ is one of the main objectives of [11].

Assume first H to be almost simple over $\mathbf{R}$. Then $M(H,^*) \geq \operatorname{rk}_{\mathbf{R}} H-1$ [11: V, 3.3] or [28], and $M(H,^*)$ is also at least equal to Matsushima's constant $m(H)$, which, in a few cases, may be $\geq \operatorname{rk}_{\mathbf{R}} H$ [11: II, 2.9]. If H is a product of non-compact $\mathbf{R}$-simple groups H_i $(1 \leq i \leq s)$ by a compact group H_o, then V decomposes into a Hilbert tensor product $\otimes_i V_i$, where V_i is an irreducible unitary representation of H_i, which is non-trivial for $i \geq 1$. If F decomposes similarly into a tensor product $\otimes F_i$, where F_i is a representation of H_i, then

$$M(H;F) + 1 = \sum_{1 \leq i \leq s} (M(H,F_i) + 1) \quad .$$

If H is a connected complex simple group, viewed as a real Lie group, then $M(H,^*) = r_H - 1$, where r_H is the constant given by T. Enright in [14]. Except for $\mathbf{SL}_n$, this constant is at least equal to the $\mathbf{R}$-rank and to $m(H)$. It is given by the following table [14]:

H:	$\mathbf{A}_\ell(\ell \geq 1)$,	$\mathbf{B}_\ell(\ell \geq 2)$,	$\mathbf{C}_\ell(\ell \geq 3)$,	$\mathbf{D}_\ell(\ell \geq 4)$,	$\mathbf{E}_6$,	$\mathbf{E}_7$,	$\mathbf{E}_8$,	$\mathbf{F}_4$,	$\mathbf{G}_2$
$r(H)-1$:	$\ell - 1$,	$2(\ell-1)$,	$2(\ell-1)$,	$2\ell-3$,	15,	26,	56,	14,	4

4.2 **Lemma.** *Let H, (μ,F) be as in 4.1, H' a connected semi-simple group and $\sigma: H' \to H$ a surjective homomorphism with finite kernel. Then $M(H',\mu) = M(H,\mu)$.*

Since any irreducible unitary representation with compact kernel of H is one of H' we have clearly $M(H',\mu) \leq M(H,\mu)$.

Let (π,V) be an irreducible unitary representation of H' with compact kernel. Let C be the kernel of σ. By assumption, it is contained in the kernel of every direct summand of (μ, F). It follows therefore from [11: I, 5.3] that if C is not in the kernel of π, then

$$H^*(\mathfrak{h}', L; V \otimes F) = 0 \quad .$$

As a consequence, the representations of H' which matter in the determination of $M(H', \mu)$ are only those which factor through H, whence the reverse inequality.

    4.3   Let G be connected, almost simple and isotropic over $\mathbf{Q}$. Then $G(\mathbf{R})^\circ$ is isogeneous to a direct product of non-compact simple real Lie groups. In fact, up to isogeny, G is of the form $R_{k/\mathbf{Q}}G'$, where k is a finite extension of $\mathbf{Q}$ and G' an almost absolutely simple k-group [10: 6.21], the group $(R_{k/\mathbf{Q}}G')(\mathbf{R})$ is the direct product of the groups $G'(k_v)$, where k_v runs through the archimedean completions of k, and the group $G'(k_v)$ is not compact and absolutely almost simple over k_v, hence almost simple over $\mathbf{R}$. If $H_1, \ldots, H_s$ are the normal simple subgroups of $G(\mathbf{R})^\circ$, then $G(\mathbf{R})^\circ$ is the quotient of $H = H_1 \times \ldots \times H_s$ by a finite subgroup. Any irreducible unitary representation V of $G(\mathbf{R})^\circ$ is one of H, hence is a Hilbert tensor product $V_1 \tilde{\otimes} \ldots \tilde{\otimes} V_s$, where V_i is an irreducible unitary representation of H_i $(1 \le i \le s)$.

    Let Γ be an arithmetic subgroup of G, contained in $G(\mathbf{R})^\circ$. It is then irreducible, so that 4.2 of [11: VII], applied to the inverse image of Γ in H, shows that if V occurs in $L^2(\Gamma \backslash G(\mathbf{R})^\circ)$, as a direct summand, then either V is trivial or no V_i is. In particular, *if V is not trivial, then its kernel is compact.*

    4.4 <u>Theorem</u>. *Let G be connected and almost simple over $\mathbf{Q}$.*

    (i) *Assume that (τ, E) does not contain any non-zero subspace on which $G(\mathbf{R})^\circ$ acts trivially. Then*

$$H^q(\Gamma; E) = 0 \qquad for \quad q \le M(G(\mathbf{R})^\circ, E), \ C(G, \tau^*); \qquad (1)$$

$$H^q(\Gamma; E)_{(2)} = 0 \qquad for \quad q \le M(G(\mathbf{R})^\circ, E) \qquad . \qquad (2)$$

    (ii) *The natural homomorphism $I_G^{q\Gamma} \to H^q(\Gamma; \mathbf{C})$ is injective for $q \le C(G) + 1$, surjective for $q \le M(G(\mathbf{R})^\circ, \mathbf{C}), C(G)$, and maps $I_G^{q\Gamma}$ onto $H^q(\Gamma; \mathbf{C})_{(2)}$ for $q \le M(G(\mathbf{R})^\circ, \mathbf{C})$.*

Let Γ' be a normal subgroup of finite index of Γ, contained in $G(\mathbf{R})^{\mathrm{o}}$. We have

$$H^*(\Gamma;E) \;=\; (H^*(\Gamma';E))^{\Gamma/\Gamma'} \;, \qquad I_G^{\Gamma} \;=\; (I^{\Gamma'})^{\Gamma/\Gamma'} \;, \tag{3}$$

and the first isomorphism obviously induces an isomorphism

$$H^*(\Gamma;E)_{(2)} \;=\; (H^*(\Gamma';E)_{(2)})^{\Gamma/\Gamma'} \;. \tag{4}$$

This reduces us to proving the theorem for Γ'. We may therefore assume $\Gamma \subset G(\mathbf{R})^{\mathrm{o}}$. In particular

$$I_G^{\Gamma} \;=\; I_G \;=\; H^*(\mathfrak{g},K^{\mathrm{o}};\mathbf{C}) \;. \tag{5}$$

Let us write the discrete spectrum $L^2(\Gamma\backslash G(\mathbf{R})^{\mathrm{o}})$ as a direct sum of irreducible invariant subspaces V_i $(i \in I)$. If V_i is not trivial, then it has compact kernel (4.3), hence

$$H^q(\mathfrak{g}(\mathbf{R}),K^{\mathrm{o}};V_i \otimes E) \;=\; 0 \quad \text{for} \quad q \leq M(G(\mathbf{R})^{\mathrm{o}},E),\ V_i \neq \mathbf{C} \;. \tag{6}$$

By [4; 7], $H^*(\Gamma;E)_{(2)}$ is a quotient of the sum of the spaces $H^*(\mathfrak{g}(\mathbf{R}),K^{\mathrm{o}};V_i \otimes E)$, hence $H^q(\Gamma;E)_{(2)}$ is a quotient of $H^q(\mathfrak{g}(\mathbf{R}),K^{\mathrm{o}};E)$ for $q \leq M(G(\mathbf{R})^{\mathrm{o}},E)$. But we have $H^*(\mathfrak{g}(\mathbf{R}),K^{\mathrm{o}};\mathbf{C}) = I_G$ and, in the case (i), $H^*(\mathfrak{g}(\mathbf{R}),K^{\mathrm{o}};E) = 0$ (see e.g. [11: II, 3.2]). This proves (2) and the last part of (ii).

By 3.7, for $\lambda = 0$, we have $H^q(\Gamma;E) = H^q(\Gamma;E)_{(2)}$, for $q \leq C(G,\tau^*)$ so that now (1) and the second assertion of (ii) follows. Finally, I_G is a space of harmonic forms contained in the complex $\Omega_0^w(X;E)^{\Gamma}$, hence 3.7(ii) yields the first assertion of (ii).

Remark. For $E = \mathbf{C}$, 4.4 is basically 7.5 of [3], with however somewhat better bounds. As an illustration, let us mention the following proposition:

4.5 Proposition. *Let* k *be a quadratic imaginary field,* $\mathfrak{o}$ *the ring of integers of* k, G' *an almost simple k-split k-group and* Γ *an arithmetic subgroup of* G'. *Then* $H^q(\Gamma;\mathbf{C}) = H^q(\mathfrak{g}';\mathbf{C})$ *for* $q \leq c(G')$, *where* c(G') *is given by the following table:*

$\mathfrak{g}'$	$A_\ell(\ell \geq 1)$,	$B_\ell(\ell \geq 3)$,	$C_\ell(\ell \geq 2)$,	$D_\ell(\ell \geq 4)$,	E_6,	E_7,	E_8,	F_4,	G_2
$c(G')$	$2[(\ell-1)/2]$,	$2(\ell-1)$,	$2(\ell-2)$,	$2(\ell-2)$,	14,	26,	50,	10,	2

Let $G = R_{k/\mathbf{Q}}G'$. We view Γ as an arithmetic subgroup of G. By 2.5, $C(G) \geq 2.C(\Phi(G'))$. Moreover $G(\mathbf{R}) = G'(\mathbf{C})$, viewed as a real Lie group, so that $M(G(\mathbf{R}),\mathbf{C})$ is given by Enright's vanishing theorem (4.1). Finally, in that case, $H^*(\mathfrak{g}(\mathbf{R}),K;\mathbf{C})$ may be identified to the Lie algebra cohomology $H^*(\mathfrak{g}';\mathbf{C})$. Our assertion now follows from 4.4(ii) and from the tables given in 2.10 and 4.1.

4.6 Consider now a sequence $(G_n,\Gamma_n,(\tau_n,E_n))$, of almost simple $\mathbf{Q}$-groups, arithmetic subgroups and non-trivial irreducible representations. If $M(G_n(\mathbf{R})^\circ,\tau_n)$ and $C(G_n,\tau_n^*)$ tend to infinity with n, then, given j, there exists $n(j)$ such that

$$H^j(\Gamma_n;E_n) = 0 \qquad \text{for all} \quad n \geq n(j) \quad . \tag{1}$$

For $M(G_n(\mathbf{R})^\circ,\tau)$ to tend to infinity, it suffices that the $\mathbf{Q}$-rank of G_n does so (4.1). We deal then with a sequence of classical groups from some n on. Then the coefficients of ρ, expressed as sum of simple roots, tend to infinity. Therefore $C(G,\tau^*) \to \infty$ if there exists a constant k such that the coefficients of the highest weight of τ_n^* are $\leq k$ for all n's. As an example, let us mention the following proposition:

4.7 <u>Proposition.</u> *Let k be a number field, d the degree of k over $\mathbf{Q}$, and $\mathfrak{o}$ the ring of integers of k. Let $G_n' = \mathbf{SL}_n$ (resp. $\mathbf{Sp}_{2n}$, resp. $\mathbf{SO}_{n,n}$) $(n \geq 4)$ and Γ_n' be a subgroup of finite index of $G_n'(\mathfrak{o})$. Let Ad_n be the adjoint representation of G_n'. Then*

$$H^j(\Gamma_n';Ad_n) = 0 \qquad for \quad j \leq d\left[\frac{n-4}{2}\right] \; (resp. \; d(n-4), \; resp. \; d(n-4)). \tag{1}$$

Let $G_n = R_{k/\mathbf{Q}}G_n'$. Then the canonical isomorphism $G_n'(k) = G_n(\mathbf{Q})$ maps Γ_n' onto an arithmetic subgroup Γ_n of G_n. Let (τ_n',E_n') be the adjoint representation of G'. Then the representation of (τ_n,E_n) of G_n obtained by restriction of scalars from (τ_n',E') is the adjoint representation of G_n. Over $\mathbf{C}$ it splits into d copies of (τ_n',E_n'), hence $H^*(\Gamma_n;E_n)$ is the direct sum of d copies of

$H^*(\Gamma_n';E_n)$. It suffices to prove (1) for Γ_n and the adjoint representation of G_n.

The adjoint representation is self-contragradient and its highest weight is the highest root δ_o. Using 2.11 one sees that the upper bound given for j in (1) is $\le C(G_n,\tau_n)$. The group $G_n(R)$ is a product of r_1 copies of $G_n'(R)$ and of r_2 copies of $G_n'(C)$, where r_1 (resp. r_2) is the number of real (resp. complex) places of k. We have therefore as a consequence of 4.1(1):

$$M(G_n(R)^o,{}^*) + 1 \ge r_1 M(G_n'(R)^o,{}^*) + r_2 M(G_n'(C){}^*) + r_1 + r_2 \quad . \tag{2}$$

From this and the results recalled in 4.1, one sees easily that the upper bounds for j in (1) are also majorized by $M(G_n(R)^o,{}^*)$, so that 4.7 follows from 4.4(i).

5. Decaying Forms and Cohomology with Compact Supports

5.1 We let $\Omega_c(X,E)^\Gamma$ denote the complex of forms in $\Omega(X,E)^\Gamma$ with compact support modulo Γ. Then $H^*(\Omega_c(X;E)^\Gamma) = H_c^*(\Gamma\backslash X;\tilde{E})$, where H_c^* refers to cohomology with compact supports. If Γ is torsion-free then $\tilde{E}$ is a locally constant sheaf and this follows from the de Rham theorem for cohomology with compact supports. If not, we can either reduce to that case by using a normal torsion-free subgroup of finite index of Γ, or by means of a mild extension of de Rham theorem. The discussion is the same as that given in [11: VII, 2.2].

5.2 **Theorem.** *Let* $\lambda \in X(A_o)$ *be* < 0. *Then the inclusions*

$$\Omega_c(X;E)^\Gamma \to \Omega_{fd}(X;E)^\Gamma \to \Omega_\lambda^w(X;E)^\Gamma \quad ,$$

induce isomorphisms in cohomology.

The reduction to the case where Γ is torsion-free is carried out exactly as in 3.4, so we assume Γ to be torsion-free.

In this proof, we let ν range over c, fd, λ and set $\Omega_\nu = \Omega_\nu(X;E)^\Gamma$ if $\nu = c$, fd, and $\Omega_\nu = \Omega_\lambda^w(X;E)^\Gamma$ if $\nu = \lambda$. Let F_ν be the presheaf which associates to U open in $\Gamma\backslash\bar{X}$ the restrictions to $U \cap (\Gamma\backslash X)$ of elements in Ω_μ. Again this is a sheaf and, as in 3.4, it is seen to be fine. The space of sections of F_ν is Ω_ν. Therefore, by the comparison theorem in sheaf theory [18: II, 4.6.2], to prove the theorem, it suffices to show that inclusions $F_c \to F_\nu$ $(\nu = fd, \lambda)$ induce isomorphisms of the derived sheaves. If $z \in \Gamma \backslash X$, then the stalk $F_{\nu,z}$ of F at z is just the space of germs of differential forms around z, for all values of ν, hence the stalks themselves are equal. Let now $z \in \Gamma \backslash \partial \bar{X}$. Clearly, $F_{c,z} = 0$. Hence there remains to prove

$$H^q(F_{\nu,z}) \; = \; 0 \qquad\qquad (\nu = fd, \; \lambda, \; q = 0,1,2,\ldots) \quad . \tag{1}$$

Let P be a parabolic $\mathbf{Q}$-subgroup such that z is contained in the face $e'(P)$ associated to P [8]. The point z is then in the closure of a Siegel set $\mathfrak{S} = \mathfrak{S}_{t,\omega}$ associated to P. In the latter we use the notation and conventions of 3.4. A function f in Ω_ν^o tends to zero as the argument in $\mathfrak{S}$ tends to z. If its differential is zero near z, it is then zero in $\mathfrak{S}$, near z, hence $F_{\nu,z}^o = 0$ and (1) is clear for $q = 0$. To prove (1) for $q \geq 1$ and $\nu = \lambda$, we have to show that if $\eta \in \Omega_\lambda^W(\mathfrak{S})$ is closed then its restriction to some smaller Siegel set $\mathfrak{S}'$ is the coboundary of an element in $\Omega_\lambda^W(\mathfrak{S}')$. We do this again by checking that the homotopy operator of the Poincaré lemma is compatible with our conditions. However, we have now to take the origin of our local coordinate system at z itself. We identify A_t to the strictly positive quadrant in the space with coordinates $\beta^i = a^{-\alpha_i}$ $(1 \leq i \leq \ell)$. Its closure $\bar{A}_t$ consists of the points with coordinates $\beta^i \in [0, 1/t]$. The canonical isomorphism of $A_t \times \omega$ onto $\mathfrak{S}$ extends to one of $\bar{A}_t \times \omega$ onto the corner containing z, and in fact z is in the image of $(0, \omega)$. We have

$$d \log a^{\alpha_i} \; = \; -d \ln \beta^i \; = \; - \omega^{i'} \qquad (1 \leq i \leq \ell) \quad . \tag{2}$$

We take now as local coordinates in the corner the $y^i = -(\log \beta^i)^{-1}$ $(1 \leq i \leq \ell)$ and a set of local coordinates $(y^i)_{\ell < j \leq n}$ in ω centered at z. We may assume that the coordinates y^j vary in $[-c, c]$ for $j > \ell$ and in $[0, c]$ for $j \leq \ell$, where c is some small strictly positive constant. The intersection D of $\Gamma \backslash X$ with this neighborhood of z in the given corner is then the set of points

$$D \; = \; \{y = (y^i) \mid y^j \in (0, c] \quad \text{if} \quad j \leq \ell, \; y^j \in [-c, c] \quad \text{if} \quad j > \ell\} . \tag{3}$$

By assumption

$$\lambda = - \sum_{1}^{\ell} d_i \cdot \alpha_i \qquad (d_i > 0; \; i = 1,\ldots,\ell) \quad . \tag{4}$$

The weak λ-boundedness of $\eta = \Sigma\, \eta_J \omega^J$ is originally expressed by the condition

$$|\eta_J(a,q)| \prec_w \prod_{1}^{\ell} (\beta^i)^{d_i} \qquad (a \in A_t,\; q \in \omega) \quad . \tag{5}$$

In D this is equivalent to requiring the existence of $d > 0$ and $M \in \mathbf{Z}$ such that

$$|\eta_J(y)| \leqq d \cdot \prod_{1}^{\ell} (y^i)^M \cdot e^{-d_i/y_i} \;, \qquad (y \in D) \qquad , \tag{6}$$

and, for all J's. Set

$$c(y) = \sum_{1 \leqq i \leqq \ell} d_i/y_i \quad . \tag{7}$$

Then, in D, and for c small enough

$$1 < c(y) < \infty \quad . \tag{8}$$

We shall now write (6) as

$$|\eta_J(y)| \prec_w e^{-c(y)} \;, \qquad (y \in D) \quad . \tag{9}$$

We have

$$d \log \beta^i = (y^i)^{-2} \cdot dy^i \;, \qquad (1 \leqq i \leqq \ell) \; . \tag{10}$$

On the other hand, the ω^j $(j > \ell)$ are linear combinations of the dy^j $(j > \ell)$ with bounded coefficients and *vice-versa*. Therefore if we write

$$\eta = \sum_J \eta_J \omega^J = \sum_J \eta'_J dy^J \;, \tag{11}$$

then the condition (9) for the η_J's and for the η'_J's are equivalent, i.e., η is weakly λ-bounded if and only if we have for all J's:

$$|\eta'_J(y)| \prec_w e^{-c(y)} \quad , \quad (y \in D) \quad . \tag{12}$$

Let $q \geq 1$ and $J = \{j_1,\ldots,j_q\}$. Consider the form $\eta = f.dy^J$. The homotopy operator A of the Poincaré lemma transforms η onto $\Sigma\, c_I\, dy^I$, where, for I equal to J with j_i erased:

$$c_I(y) = (-1)^{i-1}\, y^{j_i} \int_0^1 f(ty).t^{q-1}.\,dt \quad . \tag{13}$$

Assume that f satisfies (6). Then

$$|c_I(y)| \leq c.d \prod_1^{\ell} (y^i)^M \int_0^1 t^{M\ell+q-1}.e^{-c(y)/t}.\,dt \,, \,(y \in D) \,. \tag{14}$$

Going over to the variable $s = c(y)/t$, we see that the integral on the right-hand side of (14) is equal to

$$c(y)^{M\ell+q} \int_{c(y)}^{\infty} s^{-(M\ell+q+1)}.e^{-s}.\,ds \quad . \tag{15}$$

Let $N = M\ell+q+1$. It is easily seen that the integral in (15) is bounded by $C.c(y)^{-N}.e^{-c(y)}$, for some $C > 0$. Therefore c_I satisfies (9). As a consequence, $A(\sigma)$ is weakly λ-bounded if σ is so, for any form σ. If now σ is a closed form, then $\sigma = dA\sigma$, whence (1) for $\nu = \lambda$. But then it obviously follows also for $\nu = fd$ (in fact, the proof could be slightly simplified in that case since we need not keep the same exponent in the exponential).

    5.3 <u>Theorem</u>. *Let* H_{fd} *be the space of fast decreasing harmonic forms contained in* $\Omega(X;E)^\Gamma$. *Then the natural map of* $H_{fd} \to H^*(\Gamma;E)$ *is injective. If* $\eta \in H_{fd}$, *then* η *can be written in the form* $\eta = \mu + d\nu$, *where* μ *has compact support* mod Γ *and* ν *is fast decreasing.*

The second assertion follows from 5.2. The first one depends only on the fact that cohomology can be computed by means of a complex of forms with moderate growth (3.4). We could also use 7.4 of [3].

Let $\eta \in H^q_{fd}$. Assume that it is cohomologous to zero in $\Omega(X;E)^{\Gamma}$. Then it is already so in Ω_{mg}, e.g., hence there exists $\sigma \in \Omega^{q-1}_{mg}(X;E)$ such that $\eta = d\sigma$. In a Siegel set, the coefficients g^{IJ} associated to a special frame have all moderate growth (see [3: 5.5]). Therefore, for a differential form τ which has moderate growth (resp. is fast decreasing) the function $|\tau|_y$ has moderate growth (resp. is fast decreasing). It follows that the functions $|\eta|_y \cdot |\sigma|_y$ and $(\eta_y, d\sigma_y)$ are fast decreasing, in particular are integrable. Since η is harmonic, $\partial \eta = 0$ hence $(\partial \eta, \sigma_y) = 0$. We have then, by [3: 2.2]:

$$(\eta, \eta) \;=\; (\eta, d\sigma) \;=\; (\partial \eta, \sigma) \;=\; 0 \quad ,$$

hence $\eta = 0$.

5.4   Let $^{\mathrm{o}}L^2(\Gamma \backslash G)$ be the cuspidal spectrum of Γ, i.e., the subspace of $L^2(\Gamma \backslash G)$ consisting of cuspidal functions. The cohomology space $H^*(\mathfrak{g}(\mathbf{R}), K; {}^{\mathrm{o}}L^2(\Gamma \backslash G)^{\infty} \otimes E)$ may be identified with the space of harmonic E-valued forms whose coefficients are E-valued cusp forms. Its image in $H^*(\Gamma;E)$ is, by definition, the cuspidal cohomology $H_{cusp}(\Gamma;E)$ of Γ. Since a cusp form is fast decreasing, these harmonic forms belong to H_{fd} by 3.10, hence 5.3 has the following corollary, whose first part was already announced in [4: 6.1] for $E = \mathbf{C}$.

5.5   **Corollary.** *The natural map of* $H^*(\mathfrak{g}(\mathbf{R}), K; {}^{\mathrm{o}}L_2(\Gamma \backslash G)^{\infty} \otimes E)$ *onto* $H^*_{cusp}(\Gamma;E)$ *is injective. A cuspidal harmonic form* η *can be written in the form* $\eta = \mu + d\nu$, *where* μ *has compact support* mod Γ *and* ν *is fast decreasing.*

5.6   Assume Γ to be torsion-free and orientation preserving. Let n be the dimension of X. Then $H^n_c(\Gamma \backslash X;\mathbf{C}) \cong \mathbf{C}$ and this isomorphism is given by integration over $\Gamma \backslash X$. By 5.2, we have a canonical isomorphism $H^n(\Omega_{fd}(X;\mathbf{C})^{\Gamma}) = H^n_c(\Gamma \backslash X;\mathbf{C})$, whence also an isomorphism of $H^n(\Omega_{fd}(X;\mathbf{C})^{\Gamma})$ onto $\mathbf{C}$. We claim that this isomorphism is also defined by integration. To see this it is enough to show that

$$\int_{\Gamma \backslash X} d\eta \;=\; 0 \,, \qquad \text{if} \quad \eta \in \Omega^{n-1}_{fd}(X;\mathbf{C})^{\Gamma} \quad . \tag{1}$$

But η tends to zero at infinity so it extends to a smooth form on $\Gamma\backslash\bar{X}$ which is identically zero on the boundary $\Gamma\backslash\partial\bar{X}$. We can then apply Stokes' theorem to η on $\Gamma\backslash\bar{X}$, and (1) follows. Since the product of a fast decreasing form by a form of moderate growth is fast decreasing, it follows immediately that the following diagram, where the horizontal arrows are defined by integration over $\Gamma\backslash X$, and the vertical arrows are the isomorphisms given by 3.4 and 5.2, is commutative:

$$
\begin{array}{ccccc}
H^p(\Omega_{fd}(X;\mathbb{C})^\Gamma) & \times & H^{n-p}(\Omega_{mg}(X;\mathbb{C})^\Gamma) & \to & \mathbb{C} \\
\uparrow & & \downarrow & & \\
H^p_c(\Gamma;\mathbb{C}) & \times & H^{n-p}(\Gamma;\mathbb{C}) & \to & \mathbb{C} \qquad (0 \le p \le n) .
\end{array}
\tag{2}
$$

More generally, there are pairings

$$
\Omega^p_c(X;E)^\Gamma \times \Omega^q(X;E^*)^\Gamma \to \Omega^{p+q}_c(X;\mathbb{C})^\Gamma \quad ,
\tag{3}
$$

$$
\Omega^p_{fd}(X;E)^\Gamma \times \Omega^q_{mg}(X;E^*)^\Gamma \to \Omega^{p+q}_{fd}(X;\mathbb{C})^\Gamma \quad , \qquad (p,q \ge 0) ,
\tag{4}
$$

defined by exterior product and the trace map $E \otimes E^* \to \mathbb{C}$. Composed with integration over $\Gamma\backslash X$, when $p + q = n$, they yield a commutative diagram, where the horizontal arrows are gain perfect pairings:

$$
\begin{array}{ccccc}
H^p(\Omega_{fd}(X;E)^\Gamma) & \times & H^{n-p}(\Omega_{mg}(X;E)^\Gamma) & \to & \mathbb{C} \\
\uparrow & & \downarrow & & \\
H^p_c(\Gamma\backslash X;E) & \times & H^{n-p}(\Gamma;E^*) & \to & \mathbb{C} \qquad (0 \le p \le n) .
\end{array}
\tag{5}
$$

6. The Case of S-Arithmetic Groups

In this section, k *is a number field,* G' *a connected, almost* k-*simple isotropic* k-*group, and* $G = R_{k/\mathbb{Q}}G'$.

6.1 We let $\mathfrak{o}$ be the ring of integers of k, d the degree of k over $\mathbb{Q}$, V (resp. V_∞, resp. V_f) the set of (resp. archimedean, resp. finite) places of k and k_v the completion of k at v. For

$v \in V_f$, let $\mathfrak{o}_v$ denote the ring of integers of k_v.

If S is a finite set of places of k, set $S_\infty = S \cap V_\infty$ and $S_f = S \cap V_f$. Unless otherwise said, we assume that $S_\infty = V_\infty$. As usual, $\mathfrak{o}_S$ denotes the ring of S-integers of k (elements of k belonging to $\mathfrak{o}_v$ for all $v \in V - S$).

If H is a k-group, then we put

$$ H_v = H(k_v), \qquad H_\infty = \prod_{v \in V_\infty} H_v, \qquad H_S = \prod_{v \in S} H_v . \qquad (1) $$

A subgroup $\Gamma \subset H(k)$ is S-arithmetic if, for any faithful k-morphism $\mu: H \to \mathbf{GL}_m$, the group $\mu(\Gamma)$ is commensurable with $\mu(H) \cap \mathbf{GL}_m \mathfrak{o}_S$. The group Γ, diagonally embedded in H_S is discrete. If L is a compact open subgroup of H_{S_f}, then $\Gamma \cap (H_\infty \times L)$ is an arithmetic subgroup of H.

If $H' = R_{k/\mathbf{Q}}H$, then we have canonical isomorphisms $H'(\mathbf{Q}) \cong H(k)$ and $H'(\mathbb{R}) = H_\infty$. The first isomorphism maps arithmetic subgroups of H onto arithmetic subgroups of H'. It also allows one to view any $H'(R)$ module as a $H(k)$-module. This is tacitly understood in the sequel.

6.2 <u>Proposition</u>. *Assume G' to be simply connected. Let S be a finite set of places of V, Γ a S-arithmetic subgroup of G' and $N \in \mathbf{N}$.*

(i) Assume that for any pair $\Gamma_1 \subset \Gamma_2$ of arithmetic subgroups of G' contained in Γ the restriction map $H^q(\Gamma_2;E) \to H^q(\Gamma_1;E)$ is an isomorphism for $q \leqq N$. Then, for any arithmetic subgroup Γ_o of G' contained in Γ, the restriction map $H^q(\Gamma;E) \to H^q(\Gamma_o;E)$ is an isomorphism for $q \leqq N$.

(ii) If the assumption of (i) is satisfied for all finite $S \subset V$ and all S-arithmetic subgroups, then the restriction map $H^q(G'(k),E) \to H^q(\Gamma_o;E)$ is also an isomorphism for $q \leqq N$.

The proof uses Bruhat-Tits buildings [13]; its framework is the same as that of 3.7 in [5] or 6.9 in [9]. See [5;9] for more references concerning Bruhat-Tits buildings.

For $v \in V_f$, let X_v be the Bruhat-Tits building of G' over k_v, set $X_f = \prod_{v \in S_f} X_v$ and $G_f = G_{S_f}$. Let $\bar{X}$ be the completion by corners of X constructed in [8] and $\bar{X}_S = \bar{X} \times X_f$. The group G'_S operates naturally on $\bar{X}_S$. The group Γ, embedded diagonally in G'_S, operates properly on $\bar{X}_S$ and the quotient $\bar{X}_S/\Gamma$ is compact [9: 6.9]. The group Γ also operates on X_f via its projection on $G_f^!$ and the natural projection $X_s \to X_f$ commutes with Γ, whence a projection

$$\pi: \bar{X}_s/\Gamma \to X_f/\Gamma \quad . \tag{1}$$

The space X_f is a product of simplicial complexes, a "polysimplicial complex" in the terminology of [13]. The group G_f operates on X_f as a group of automorphisms of the polysimplicial structure and X_f/G_f is a polysimplex C, of dimension the sum of the k_v-ranks of G'. Since G' is simply connected, and G'_∞ is not compact, the projection of Γ on each G'_v ($v \in V_f$) is non-discrete hence Prop. 4.3 of [23] holds, and shows that Γ is *dense* in $G_f^!$. Since the isotropy groups of G_f on X_f are compact open, it follows that $X_f/\Gamma = X_f/G_f = C$.

We consider now the Leray spectral sequence (E_r) of π. We have $E_2^{p,q} = H^p(C;F^q)$ where F^q is the Leray sheaf of π whose stalk at $c \in C$ is $H^q(\pi^{-1}(c);E)$. For a face σ of C let L_σ be its isotropy group in G_f. It is compact open. The isotropy group Γ_σ of σ in Γ may be identified with $\Gamma \cap (G'_\infty \times L_\sigma)$, hence is an arithmetic subgroup. This is also the isotropy group in Γ of any point c in the interior $\mathring{\sigma}$ of σ. We have then a homeomorphism:

$$\pi^{-1}(c) \cong \bar{X}_\infty/\Gamma_\sigma \; , \qquad (c \in \mathring{\sigma}) \tag{2}$$

and therefore

$$H^q(\pi^{-1}(c);E) \cong H^q(\Gamma_\sigma;E) \quad , \qquad (c \in \mathring{\sigma}) \quad . \tag{3}$$

If σ' is a face of σ, then, $\Gamma_{\sigma'} \supset \Gamma_\sigma$ and we have a natural restriction map in cohomology. Thus the Leray sheaf is in this case just a system of coefficients, which assigns one graded space to each face, with natural mappings associated to inclusions of faces. By our assumption, there exists for $q \leq N$ a finite dimensional vector space F^q and isomorphisms $H^q(\Gamma_\sigma;E) \to F^q$ compatible with restrictions. Therefore in those dimensions, our sheaf is just an ordinary constant system of coefficients and we have

$$E_2^{p,q} = H^p(C;F^q) \quad , \qquad (p \in \mathbb{N}; \; q \leq N) \quad . \tag{4}$$

Since C is a polysimplex, it is acyclic, hence

$$E_2^{0,q} = F^q \qquad\qquad (q \leqq N) \qquad\qquad (5)$$

$$E_2^{p,q} = 0 \qquad\qquad (p > 0;\ q \leqq N)\ . \qquad\qquad (6)$$

We have therefore, in total degree $m \leqq N$:

$$F^m = E_2^{0,m} = {}^m E_2 = {}^m E_\infty = H^m(\Gamma;E) \qquad , \qquad\qquad (7)$$

and (i) follows.

Assume now this is true for all finite $S \subset V$ and all S-arithmetic groups. Fix an increasing sequence S_n $(n=1,2,\ldots)$ of subsets of V whose union is V. We can then view $G'(k)$ as the union of an increasing sequence of subgroups Γ_n, where Γ_n is S_n-arithmetic. There exists therefore an Eilenberg-MacLane space $K(G'(k),1)$ which is a union of subcomplexes K_n, where K_n is a $K(\Gamma_n,1)$ $(n=1,2,\ldots)$. By Theorem 2.10*, p. 273 of [26: VI], we have an exact sequence

$$\to \varprojlim{}^1 H^{q-1}(\Gamma_n;E) \to H^q(G'(k),E) \to \varprojlim H^q(\Gamma_n;E) \to 0, \qquad (q \in N)\ , \qquad (8)$$

where $\varprojlim{}^1$ is Milnor's first derived functor of $\varprojlim$ (*loc. cit.*). If $q \leqq N$, all the maps $H^q(\Gamma_{n+1};E) \to H^q(\Gamma_n;E)$ are isomorphisms, hence the $\varprojlim{}^1$ term is zero and the $\varprojlim$ term is just $H^q(\Gamma_1:E)$. The second assertion follows.

6.3 We let $\tilde{G}'$ denote the universal covering of G' and $\sigma: \tilde{G}' \to G'$ the canonical isogeny. Then $\tilde{G} = R_{k/G}\tilde{G}'$ is the universal covering of G. The canonical isogeny $\tilde{G} \to G$ will also be denoted by σ.

The group $\tilde{G}(R) = \tilde{G}'_\infty$ is always connected, while $G(R) = G'_\infty$ is not necessarily so. The homomorphism σ maps $\tilde{G}(R)$ onto $G(R)^0$ and has finite kernel.

As before, I_G denotes the space of $G(R)^0$ invariant forms on X and $j_\Gamma: I_G \to H^*(\Gamma;\mathbb{C})$ the natural homomorphism. If $G(R)$ is not connected, the space of $G(R)$-invariant harmonic forms may be of course $\neq I_G$, hence I_G^Γ may be $\neq I_G$, which introduces a minor complication. To state our next theorem we introduce still another constant, namely

$$i(G) = \max q: I_G^q = I_G^{G(R)} \qquad . \qquad\qquad (1)$$

If G(**R**) is connected, hence in particular if G is simply connected, then $i(G) = \infty$.

    6.4 <u>Theorem</u>. *Let* Γ *be equal either to* G'(k) *or to a S-arithmetic subgroup of* G'.

    (i) *Assume* E *does not contain any nonzero trivial* $G(\mathbf{R})^0$-*submodule. Then*

$$H^q(\Gamma;E) \;=\; 0\,, \qquad for \quad q \leq M(G(\mathbf{R})^0,\tau),C(G,\tau^*) \quad. \tag{1}$$

    (ii) *If* Γ *is S-arithmetic and contained in* $G(R)^0$, *then*

$$H^q(\Gamma;\mathbf{C}) \;=\; I_{G'}^q, \qquad for \quad q \leq M(G(R)^0,\mathbf{C}),C(G) \quad. \tag{2}$$

    (iii) *We have*

$$H^q(\Gamma;\mathbf{C}) \;=\; I_{G'}^q, \qquad for \quad q \leq M(G(\mathbf{R})^0,\mathbf{C}),C(G),i(G). \tag{3}$$

    [See 1.6, 4.1 and 6.3 for the definitions of $M(G(\mathbf{R})^0,\tau)$, $C(G,\tau^*)$, $C(G)$ and $i(G)$.]

    In view of 4.4, our assumptions imply in each case that if $\Gamma_1 \subset \Gamma_2$ are arithmetic subgroups of G' contained in Γ, then $H^q(\Gamma_2;E) \to H^q(\Gamma_1;E)$ is an isomorphism in the range indicated, and furthermore that $H^q(\Gamma_1;E) = 0$ in case (i), and $H^q(\Gamma_1;\mathbf{C}) = I_G^q$ in cases (ii), (iii). Therefore, if G is simply connected, the theorem follows from 6.2.

    In the general case, we first assume Γ to be S-arithmetic. As usual (see the beginning of 4.4), we may replace Γ by a subgroup of finite index. Since $\tilde{G}'(k) \to G'(k)$ maps S-arithmetic subgroups onto S-arithmetic subgroups [1: 8.12], we may then assume that $\Gamma = \sigma(\tilde{\Gamma})$, where $\tilde{\Gamma}$ is a torsion-free S-arithmetic subgroup of $\tilde{G}'(k)$. The map σ is then an isomorphism of $\tilde{\Gamma}$ onto Γ, hence $H^*(\tilde{\Gamma};E) = H^*(\Gamma;E)$. We are now back to the simply connected case. By our initial remark, this proves the theorem in this case, except that we have to replace G by $\tilde{G}$ in the constants giving the range for q. However $C(\tilde{G},\tau^*) = C(G,\tau^*)$ by definition, and $M(\tilde{G}(R),E) = M(G(R)^0,E)$, $M(G(R),\mathbf{C}) = M(G(\mathbf{R})^0,\mathbf{C})$ by 4.2. To prove (i) and (iii) when $\Gamma = G'(k)$, we then argue exactly as in the proof of 6.2(ii).

6.5   In [3], we applied 7.5 to sequences of arithmetic groups in classical groups. In view of 6.4, we can now do this for sequences of S-arithmetic groups, or also of groups of rational points. In particular, 11.1 remains valid if Γ_n stands for a S-arithmetic group or $G_n(\mathbf{Q})$. Similarly in 11.3, and in the examples which illustrate it, we may take for Γ_n' a S-arithmetic group or $G_n'(k)$ itself. Finally, the proofs of 12.2, and 12.3, giving the rank of the groups $K_i \mathfrak{n} \otimes \mathbf{Q}$ and $_\varepsilon L_i \mathfrak{n} \otimes \mathbf{Q}$ remain valid if $\mathfrak{n}$ is replaced by $\mathfrak{n}_S$ or by k. This then establishes the results announced in [2] and not proved in [3].

7.   References

[1] A. Borel, "Some finiteness properties of adele groups over number fields," *Publ. Math. I.H.E.S.* <u>16</u>, 5-30 (1963).

[2] A. Borel, "Cohomologie réelle stable de groupes S-arithmétiques," *C.R. Acad. Sci. Paris* <u>274</u>, 1700-1702 (1972).

[3] A. Borel, "Stable real cohomology of arithmetic groups," *Annales Sci. E.N.S. Paris* (4)<u>7</u>, 235-272 (1974).

[4] A. Borel, "Cohomology of arithmetic groups,"Proc. Int. Congress of Math. Vancouver, Vol. 1, 435-442 (1974).

[5] A. Borel, "Cohomologie de sous-groupes discrets et représentations de groupes semi-simples," *Astérisque* 32-33, 73-111 (1976).

[6] A. Borel, "Stable and L^2-cohomology of arithmetic groups," *Bull. A.M.S. (N.S.)* <u>3</u>, 1025-1027 (1980).

[7] A. Borel and H. Garland, "Laplacian and discrete spectrum of an arithmetic group" (in preparation).

[8] A. Borel and J-P. Serre, "Corners and arithmetic groups," *Comm. Math. Helv.* <u>48</u>, 436-491 (1973).

[9] A. Borel and J-P. Serre, "Cohomologie d'immeubles et de groupes S-arithmétiques," *Topology* <u>15</u>, 211-232 (1976).

[10] A. Borel and J. Tits, "Groupes réductifs," *Publ. Math. I.H.E.S.* <u>27</u>, 55-150 (1965).

[11] A. Borel and N. Wallach, "Continuous cohomology, discrete subgroups and representations of reductive groups," *Annals of Mathematics Studies* <u>94</u>; xvii + 387 p., Princeton University Press, 1980.

[12] N. Bourbaki, "Groupes et Algèbres de Lie," Chap. IV, V, VI, Act. Sci. Ind. 1337, Hermann, Paris, 1968.

[13] F. Bruhat and J. Tits, "Groupes réductifs sur un corps local I," *Publ. Math. I.H.E.S.* <u>41</u>, 1-251 (1972).

[14] T. Enright, "Relative Lie algebra and unitary representations of complex Lie groups," *Duke M. J.* <u>46</u>, 513-525 (1979).

[15] F.T. Farrell and W.C. Hsiang, "On the rational homotopy groups of the diffeomorphism groups of discs, spheres and aspherical manifolds," Proc. Symp. Pure Math. 32, Part 1 (1978), 403-415, A.M.S. Providence, RI.

[16] H. Garland, "A finiteness theorem for K_2 of a number field," *Annals of Math.* (2), 94, 534-548 (1971).

[17] H. Garland and W.C. Hsiang, "A square integrability criterion for the cohomology of an arithmetic group," Proc. Nat. Acad. Sci. USA, 59, 354-360 (1968).

[18] R. Godement, "Théorie des Faisceaux," Act. Sci. Ind. 1252, Hermann, Paris, 1958.

[19] Y. Matsushima, "On Betti numbers of compact, locally symmetric Riemannian manifolds," *Osaka Math. J.* 14, 1-20 (1962).

[20] Y. Matsushima, "A formula for the Betti numbers of compact locally symmetric Riemannian manifolds," *Jour. Diff. Geom.* 1, 99-109 (1967).

[21] Y. Matsushima and S. Murakami, "On vector bundle valued harmonic forms and automorphic forms on symmetric spaces," *Annals of Math.* (2) 78, 365-416 (1963).

[22] Y. Matsushima and S. Murakami, "On certain cohomology groups attached to hermitian symmetric spaces," *Osaka J. Math.* 2, 1-35, (1965).

[23] G. Prasad, "Strong approximation for semi-simple groups over function fields," *Annals of Math.* (2) 105, 553-572 (1977).

[24] M.S. Raghunathan, "Cohomology of arithmetic subgroups of algebraic groups II," *Annals of Math.* (2) 87, 279-304 (1968).

[25] A. Weil, Adeles and algebraic groups, Notes by M. Demazure and T. Ono, Institute for Advanced Study, Princeton, NJ, 1961.

[26] G. W. Whitehead, "Elements of homotopy theory," Grad. Texts in Math. 61, Springer Verlag, New York, 1978.

[27] S. Zucker, "L_2-cohomology of warped products and arithmetic groups," (to appear).

[28] S. Zuckermann, "Continuous cohomology and unitary representations of real reductive groups," *Annals of Math.* (2) 107, 495-516 (1978).

The Institute for Advanced Study
Princeton, NJ   08540   USA

(Received February 27, 1981)

119.

Mathematik: Kunst und Wissenschaft

Themen-Reihe der Carl Friedrich v. Siemens Stiftung XXXIII, München 1982

Es ist eine große Ehre, einen Vortrag in diesem Rahmen halten zu dürfen, aber eine, die für mich viele Schwierigkeiten mit sich bringt. Erstens die sprachliche: Seit vielen Jahren hat für mich Englisch den Vorrang unter fremden Sprachen angenommen. Ich fürchte, daß Französisch, meine Muttersprache, und Englisch unwillkürlich eingreifen werden, und bitte deshalb um Entschuldigung und um Gedult[0]. Zweitens sehe ich hier viele Mathematiker und ich bin mir bewußt, fast schmerzlich bewußt, daß eigentlich alles über mein Thema schon gesagt worden ist, und alle Dispute bereits geführt worden sind: Daß Mathematik nur Kunst sei, oder nur Wissenschaft – die Königin der Wissenschaften –, nur eine Dienerin der Wissenschaft, oder auch Kunst und Wissenschaft zugleich. Sogar mein Thema, in lateinischer Einkleidung: „Mathesis et Ars et Scientia Dicenda", tauchte als drittes Thema bei der Verteidigung einer Dissertation im Jahre 1845 auf. Der Opponent meinte, sie sei nur Kunst, keine Wissenschaft[1]. Es ist auch gelegentlich behauptet worden, daß Mathematik sehr trivial, fast tautologisch sei, also sicher unwürdig, entweder als Kunst oder als Wissenschaft betrachtet zu werden[2]. Jede Meinung kann man durch viele Zitate hervorragender Mathematiker stützen. Manchmal ist es auch möglich, durch geeignetes Zitieren ein und demselben Mathematiker ganz verschiedene Meinungen zuzuschreiben. Also möchte ich von vornherein betonen, daß die anwesenden beruflichen Mathematiker kaum etwas Neues hören werden.

Wenn ich mich aber an die Nichtmathematiker wende, treffe ich auf eine viel größere, sozusagen entgegengesetzte Schwierigkeit. Meine Aufgabe ist es, etwas über das Wesen, die Natur der Mathematik zu sagen; dabei darf ich jedoch nicht

---

0 In dieser Hinsicht möchte ich Dr. J. Schwermer für seine sehr nützliche Hilfe bei der Abfassung dieses Textes herzlich danken.

1 Es handelt sich um die Dissertation von L. Kronecker, cf. Werke, 5 Bde., Teubner, Leipzig 1895–1930, Bd. 1, S. 73. Der Opponent war G. Eisenstein.
Die mir bekannte Quelle für den Namen und die Stellungnahme des Opponenten ist eine Fußnote von E. Lampe zu einer Rede von P. du Bois-Reymond, „Was will die Mathematik und was will der Mathematiker?", die aus dem Nachlaß im Jahresbericht der Deutschen Mathematiker-Vereinigung 19 (1910) S. 190–198 von E. Lampe veröffentlicht wurde.

2 Für eine Diskussion einiger solcher Meinungen, siehe A. Pringsheim, „Über den Wert und angeblichen Unwert der Mathematik", Jahresbericht der Deutschen Mathematiker-Vereinigung 13 (1904), S. 357–382.

den Gegenstand dieser Äußerungen als bekannt voraussetzen. Sicher darf ich eine gewisse Vertrautheit mit griechischer Mathematik, etwa euklidischer Geometrie, vielleicht der Lehre der Kegelschnitte, sogar mit den Rudimenten von Algebra oder analytischer Geometrie, annehmen. Aber das hat wenig zu tun mit dem Gegenstand heutiger mathematischer Forschung. Von dieser mehr oder weniger geläufigen Grundlage ausgehend, haben die Mathematiker immer mehr abstrakte Theorien entwickelt, die immer weniger mit alltäglicher Erfahrung zu tun haben, sogar wenn sie später wichtige Anwendungen in den Naturwissenschaften haben. Der Übergang von einer Abstraktionsstufe zur nächsten war oft für die besten Mathematiker sehr schwierig und stellte damals einen äußerst kühnen Schritt dar. Es ist mir gar nicht möglich, eine befriedigende Übersicht dieser Anhäufung von Abstraktionen über Abstraktionen und deren Anwendungen in einigen Minuten zu geben. Doch finde ich es zu unbefriedigend, über Mathematik zu philosophieren und überhaupt nichts über deren Inhalt zu sagen. Ich möchte auch einen kleinen Vorrat von Beispielen zur Verfügung haben, um allgemeine Behauptungen über Mathematik oder über die Stellung der Mathematik gegenüber anderen Geistes- oder Naturwissenschaften illustrieren zu können. Deshalb werde ich versuchen, einige Beispiele von solchen Schritten zu beschreiben oder wenigstens anzudeuten. Ich kann dabei nicht alles definieren und genau einführen und ich erwarte auch nicht volles Verständnis. Aber das ist nicht wesentlich. Was ich mitteilen möchte, ist wirklich nur ein Gefühl für die Natur dieser Übergänge, vielleicht für ihre Kühnheit und für ihre Bedeutung vom Standpunkt einer Geschichte des Denkens. Ich verspreche auch, diesem Unternehmen nicht mehr als 20 Minuten zu widmen.

Ein Mathematiker strebt oft nach allgemeinen Lösungen. Man hat es gern, viele spezielle Probleme auf einen Schlag durch allgemeine Formeln zu lösen. Man kann dies Ökonomie des Denkens oder Faulheit nennen. Ein uraltes Beispiel ist die Lösung einer Gleichung zweiten Grades, etwa

$$x^2 + 2\,b\,x + c = 0.$$

Hier sind b und c reelle Zahlen. Gesucht wird eine reelle Zahl x, die dieser Gleichung genügt. Seit vielen Jahrhunderten drückt man x durch b und c mittels der Formel

$$x = -b \pm \sqrt{b^2 - c}$$

aus. Falls $b^2 > c$, kann man die Quadratwurzel ziehen und man bekommt zwei Lösungen. Falls $b^2 = c$, sagt man $x = -b$ ist eine doppelte Lösung. Falls aber $b^2 < c$, dann kann man die Quadratwurzel nicht ziehen und sagt, wenigstens am Anfang der Sekundarschule, es gebe keine Lösung. Im sechzehnten Jahrhundert wurden solche Formeln für Gleichungen dritten und sogar vierten Grades aufgestellt, z. B. für eine Gleichung wie

$$x^3 + a\,x + b = 0.$$

Eine solche Formel werde ich nicht angeben. Sie enthält Quadrat- und kubische Wurzeln, sogenannte Radikale. Aber ein sehr merkwürdiges Phänomen wurde in diesem Zusammenhang entdeckt, das „casus irreducibilis" genannt wurde. Wenn eine solche Gleichung drei verschiedene reelle Lösungen hat, z. B.

$$x^3 - 3\,x + 1 = 0,$$

und wenn man die Formel anwendet, die sie aus den Koeffizienten im Prinzip zu berechnen erlaubt, dann stößt man auf Quadratwurzeln negativer Zahlen, die zunächst keinen Sinn haben. Wenn man aber vergißt, daß diese nicht existieren, und sich nicht fürchtet, damit zu rechnen, dann eliminieren sie sich am Ende der Rechnung und man erhält doch die Lösungen, vorausgesetzt, daß man gewissen formalen Regeln sorgfältig folgt. Solche Wurzeln negativer Zahlen wurden dann „imaginäre Zahlen" genannt, im Unterschied zu den reellen Zahlen, und man stritt heftig darüber, ob es wirklich erlaubt war, solche nicht reellen Zahlen zu benutzen; Descartes zum Beispiel wollte damit nicht zu tun haben. Erst ungefähr um das Jahr 1800 wurde diese Frage in befriedigender Weise geklärt (wenigstens für einige). Man betrachtet die Gesamtheit der Punkte der Ebene, also Paare reeller Zahlen, und führt zwischen ihnen gewisse Verknüpfungen ein, die eine natürliche Verallgemeinerung der vier Grundoperationen sind. Man spricht dann von komplexen oder imaginären Zahlen. Man kann formal mit diesen mathematischen Objekten fast ebenso leicht wie mit den gewöhnlichen Zahlen rechnen und erhält zwei Lösungen, die manchmal reell, manchmal komplex sind. Im Fall der früheren Gleichung zweiten Grades wird man jetzt sagen, daß es zwei komplexe Lösungen gibt, falls $b^2 < c$.

Das ist natürlich zum Teil nur eine Konvention; aber es war gar nicht leicht, diesen komplexen Zahlen dasselbe Recht zur Existenz wie den reellen Zahlen zu geben und sie nicht bloß als ein Werkzeug zum Berechnen reeller Zahlen zu betrachten. Man hatte damals zwar keine strenge Definition der reellen Zahlen, aber die enge Verbindung mit Messungen, praktischen Rechnungen, gab den reellen Zahlen eine gewisse Realität, trotz Schwierigkeiten mit irrationalen oder negativen Zahlen. Jedoch mit den komplexen Zahlen verhielt es sich nicht so. Es war ein prinzipiell neuer Schritt, diese rein intellektuelle Schöpfung in den Vordergrund zu stellen. Als manche Mathematiker damit doch vertraut wurden, bemerkten sie, daß viele Operationen mit Funktionen, wie Polynomen, trigonometrischen Funktionen usw., noch einen Sinn hatten, wenn man komplexe Werte als Argumente zuließ. Dies war der Anfang der komplexen Analysis oder Funktionentheorie. Schon 1811 wies der Mathematiker Gauß auf die Notwendigkeit hin, eine solche Theorie an und für sich zu entwickeln:

„Es ist hier nicht von praktischem Nutzen die Rede, sondern die Analyse ist mir eine selbständige Wissenschaft, die durch Zurücksetzung jener fingierten Größen (der komplexen Zahlen) außerordentlich an Schönheit und Rundung verlieren würde"[3].

Anscheinend sah er nicht voraus, daß später diese komplexe Analysis auch viele praktische Anwendungen finden würde, zum Beispiel in der Elektrizitätslehre oder in der Aerodynamik.

---

3 Brief an F. W. Bessel, vom 18. 11. 1811. Siehe Briefwechsel zwischen Gauß und Bessel, G. F. Auwers Verlag, Leipzig 1880, S. 156.

Aber das ist nicht das Ende. Erlauben sie mir bitte, zwei weitere Schritte zur größeren Abstraktion zu erwähnen. Kehren wir zu unserer Gleichung zweiten Grades zurück. Jetzt kann man sagen, sie hat im allgemeinen zwei Lösungen, die vielleicht komplexe Zahlen sind. In ähnlicher Weise hat eine Gleichung n-ten Grades immer n Lösungen, wenn man komplexe Zahlen zuläßt. Seit dem sechzehnten Jahrhundert fragte man sich, ob es auch eine allgemeine Formel gibt, die eine feste Vorschrift liefern würde, um die Lösungen einer Gleichung fünften oder höheren Grades aus den Koeffizienten durch Radikale auszudrücken. Schlußendlich wurde bewiesen, es sei unmöglich. Ein Beweis (chronologisch eigentlich der dritte) wurde von dem französischen Mathematiker E. Galois im Rahmen einer allgemeineren Theorie gegeben, die damals gar nicht verstanden und dann vergessen wurde. Etwa fünfzehn Jahre später wurden seine Arbeiten wieder entdeckt und nur mit größter Mühe von recht wenigen verstanden, so neu war der Standpunkt. Galois betrachtete zu einer Gleichung die Menge derjenigen Permutationen der Wurzeln, die gewisse Relationen zwischen ihnen erhalten, und zeigte, daß gewisse Eigenschaften dieser Menge maßgebend sind. Das war dann der Anfang eines selbständigen Studiums von solchen Permutationsmengen, die von Galois Gruppen genannt wurden. Er zeigte, daß eine Gleichung dann und nur dann durch Radikale auflösbar ist, wenn die zugehörige Gruppe zu einer besonderen Klasse gehört, nämlich derjenigen der auflösbaren Gruppen, wie sie später genannt wurden. Der früher erwähnte Satz über Gleichungen, deren Grad wenigstens fünf ist, folgt dann aus der Tatsache, daß die Gruppe einer allgemeinen Gleichung n-ten Grades nur für $n = 1, 2, 3, 4$ auflösbar ist[4]. Die wichtigen Eigenschaften solcher Gruppen sind eigentlich unabhängig von der Natur der Objekte, die permutiert werden, und das führte zu dem Begriff einer „abstrakten Gruppe" und zu Sätzen von großer Tragweite, die in vielen Teilen der Mathematik anwendbar sind. Aber noch lange Zeit schien dies nur reine und sehr abstrakte Mathematik zu sein. Als ein Mathematiker und ein Physiker um das Jahr 1910 über das Curriculum für Physik an der Princeton University sprachen, sagte der Physiker, daß sie die Gruppentheorie sicher beiseite lassen dürften, da diese doch nie Anwendungen in der Physik haben werde[5]. Kaum zwanzig Jahre später erschienen drei Bücher über Gruppentheorie und Quantenmechanik, und seitdem sind Gruppen auch fundamental in der Physik.

Ein letztes Beispiel soll das folgende sein. Ich habe schon gesagt, wir können die komplexen Zahlen als Punkte der Ebene deuten. Ein irischer Mathematiker, W. R. Hamilton, fragte sich, ob man auch unter den Punkten des dreidimensionalen Raumes ein Analogon zu den vier Grundoperationen definieren und damit ein noch umfassenderes Zahlensystem bilden könne. Nach zehn Jahren ständiger Überlegung fand er die Antwort: Es ist nicht möglich im dreidimensionalen, aber

---

4 Eigentlich hatten schon einige Vorgänger von Galois Überlegungen angestellt, die man als die Anfänge der Gruppentheorie betrachten kann und die zum Teil Galois bekannt waren. Aber sein Standpunkt war so allgemein und abstrakt und noch dazu nur skizzenhaft beschrieben, daß er nur langsam assimiliert wurde. Für geschichtliche Information über die Gleichungstheorie und die Anfänge der Gruppentheorie, s. zum Beispiel N. Bourbaki, Eléments d'histoire des mathématiques, Hermann éd. Paris 1969, dritter und fünfter Artikel.
5 F. J. Dyson, "Mathematics in the physical sciences", Scientific American 211, September 1964, S. 129–146.

doch im vierdimensionalen Raum. Wir brauchen uns nicht den vierdimensionalen Raum hier anschaulich vorzustellen. Es ist einfach eine Sprechweise für Quadrupel von reellen Zahlen, anstatt von Tripeln oder Paaren von reellen Zahlen. Diese neuen Zahlen nannte er Quaternionen. Aber er mußte auf eine Eigenschaft der reellen und der komplexen Zahlen verzichten, die man bisher als selbstverständlich betrachtete, Vertauschbarkeit oder Kommutativität – nämlich daß $a \times b = b \times a$. Er zeigte auch, daß dieser Kalkül Anwendungen in der mathematischen Behandlung von Fragen der Physik oder Mechanik hatte. Nachher wurden viele andere algebraische Systeme mit einem nicht kommutativen Produkt definiert, insbesondere Algebren von Matrizen. Wieder schien dies ganz abstrakte Mathematik, ohne Verbindung mit der äußeren Welt. Doch Max Born, als er 1925 über gewisse Ansätze von W. Heisenberg nachdachte, entdeckte plötzlich, daß der geeignete Formalismus, um sie auszudrücken, gerade die Matrizenrechnung war, und daß man physikalische Größen durch algebraische Objekte darstellen muß, die nicht immer vertauschbar sind. Das führte zu den Unschärferelationen und war der Anfang der Matrizen-Quantenmechanik und der Zuordnung von Operatoren zu physikalischen Größen, die der Quantenmechanik zugrunde liegt[6].

Mit diesem Beispiel schließe ich diese Andeutungen zum Gegenstand der Mathematik ab. Sie sind natürlich sehr unvollständig und gar nicht für alle Teile der Mathematik repräsentativ. Die gegebenen Beispiele haben jedoch zwei Züge, die ich hervorheben möchte und die eine breite Gültigkeit haben. Erstens führen diese Entwicklungen in die Richtung größerer Abstraktion, weiter und weiter entfernt von der Natur. Zweitens, abstrakte Theorien, die an und für sich entwickelt wurden, fanden, manchmal ganz unerwartet, wichtige Anwendungen in den Naturwissenschaften. Eigentlich ist die Anpassung der Mathematik an die Bedürfnisse der Naturwissenschaften erstaunlich groß (ein Physiker sprach einmal von der „unvernünftigen Wirksamkeit der Mathematik"[7]) und wäre schon einer eingehenden Diskussion würdig, die ich leider übergehen muß.

---

6 Siehe B. L. van der Waerdens geschichtliche Einführung in "Sources in Quantum Mechanics", Classics of Science, Vol. 5, Dover Publ. New York 1967, insbesondere S. 36–38, 52–55. Cf. auch Diracs Bemerkungen über die Einführung der Nicht-Kommutativität in der Quantenmechanik in loc. cit.[7]

7 E. P. Wigner, "The unreasonable effectiveness of mathematics in the natural sciences", Communications on Pure and Applied Mathematics 13 (1960), S. 1–14.
Unter den vielen Aspekten dieser Wechselwirkung scheint mir insbesondere bemerkenswert, daß der mathematische Formalismus manchmal zu wesentlichen neuen rein physikalischen Ideen führt. Ein bekanntes Beispiel ist die Entdeckung des Positrons: 1928 stellte P. A. M. Dirac die quantenmechanischen relativistischen Gleichungen für die Bewegung des Elektrons auf. Diese Gleichungen ließen auch eine Lösung mit derselben Masse wie der des Elektrons, aber mit entgegengesetzter elektrischer Ladung zu. Versuche, diese Lösungen befriedigend zu erklären oder durch Umänderung der Gleichungen zu eliminieren, waren erfolglos. Das führte Dirac dazu, die Existenz eines entsprechenden Partikels zu vermuten, das später von Anderson nachgewiesen wurde. Siehe dazu P. A. M. Dirac, "The development of quantum theory" (J. R. Oppenheimer's memorial prize acceptance speech), Gordon and Breach, New York 1971.
Ein neueres und sogar umfangreicheres Beispiel wäre der Gebrauch der irreduziblen Darstellungen der speziellen unitären Gruppe SU(3) in drei komplexen Variablen, der zum sogenannten „eightfold way" geführt hat. Sehr auffallend war einer der ersten Erfolge dieser Theorie, nämlich die Entdeckung der Partikel Ω^-: Neun Baryonen wurden auf natürliche Weise, durch Betrachtung dreier ihrer charakteristischen Quantenzahlen, neun

Der Übergang zur größeren Abstraktion war nicht selbstverständlich, wie Sie schon dem Zitat von Gauß entnehmen können. Mathematik wurde zunächst für praktische Zwecke, wie Buchhaltung, Messungen, Mechanik, entwickelt und sogar die großen Erfindungen des siebzehnten Jahrhunderts, wie Infinitesimal- und Integralrechnung, waren zunächst hauptsächlich Werkzeuge, um Probleme der Mechanik, Astronomie und Physik zu lösen. Der Mathematiker Euler, der in allen Teilen der Mathematik und ihrer Anwendungen, bis hin zum Schiffbau, gewirkt hat, schrieb auch Arbeiten über reine Zahlentheorie und fühlte mehrmals die Notwendigkeit zu erklären, daß das ebenso berechtigt und wichtig sei wie mehr praktisch orientierte Arbeiten[8]. Mathematik war natürlich auch fast von Anfang an eine gewisse Idealisierung, aber für lange Zeit noch nicht so entfernt von der Realität oder, besser gesagt, von unserer Wahrnehmung der Realität wie die früher erwähnten mathematischen Begriffe. Als die Mathematiker diesem Wege folgten, wurden sie sich mehr und mehr dessen bewußt, daß ein mathematischer Begriff ein Recht zur Existenz schon hat, sobald er konsequent definiert ist, ohne notwendigerweise eine Verbindung zur physikalischen Welt zu haben, und er kann studiert werden, sogar wenn keine praktischen Anwendungen vorhanden zu sein scheinen. Kurz, man wurde zunehmend zu „reiner Mathematik" oder „Mathematik für Mathematik" gezogen. Aber wenn man manchmal diese Kontrolle der praktischen Anwendbarkeit ausläßt, entsteht sofort die Frage: Wie wird man den relativen Wert beurteilen? Sicher sind nicht alle Begriffe und Theoreme gleichberechtigt. Wie in G. Orwells „Animal Farm" müssen doch einige „gleicher" sein als andere.

Punkten einer ganz bestimmten aus zehn Punkten der Ebene bestehenden mathematischen Konfiguration [die zehn Gewichte einer irreduziblen zehndimensionalen Darstellung von SU (3)] zugeordnet; das veranlaßte M. Gell'man zu vermuten, es sollte auch dem zehnten Punkt ein Partikel entsprechen, das dann gewisse, wohl definierte Eigenschaften haben sollte. Dieses wurde etwa zwei Jahre später beobachtet. Eine weitere Entwicklung in diesem Gedankenkreis führte zur Theorie der „Quarks". Für die Anfänge dieser Theorie, siehe F. J. Dyson, loc. cit[5] und M. Gell-Mann und Y. Ne'eman, The eightfold way, W. A. Benjamin, New York 1964.

8 Siehe eine Anzahl von Arbeiten in L. Eulers Opera Omnia, insbesondere I.2, 62–63, 295, 461, 576, I.3, 5.2. Ich danke A. Weil, der mich darauf aufmerksam gemacht hat. Hier ist ein Beispiel, loc. cit. p. 62–63, 1747 veröffentlicht:

„Auch kümmert ihn [den Verfasser] nicht die Meinung der größten Mathematiker, die zuweilen behaupten, daß Erkenntnisse dieser Art geradewegs fruchtlos und es nicht wert sind, daß man auf ihre Untersuchung Mühe verwendet. Jede Erkenntnis der Wahrheit ist an sich etwas Herausragendes, auch wenn sie weit vom allgemeinen Gebrauch entfernt zu sein scheint; so sind auch alle Aspekte der Wahrheit, die uns zugänglich sind, so untereinander verbunden, daß keiner grundlos zurückgewiesen werden kann, auch wenn er geradewegs nutzlos scheint. Hinzu kommt, auch wenn irgendein bewiesener Satz nichts zu gegenwärtigem Nutzen beizutragen scheint, daß dennoch die Methode, vermittels der entweder Richtigkeit oder Falschheit herausgefunden wurde, meistens den Weg zu anderen Wahrheiten zu öffnen pflegt, die zu erkennen nützlicher sind.
Der Verfasser ist sich dessen sicher, daß er nicht nutzlos Mühe und Eifer bei der Erforschung der Beweise gewisser Sätze aufgewandt hat, in denen beachtenswerte Eigenschaften von Teilern von Zahlen enthalten sind.
Diese Lehre von den Teilern nämlich entbehrt nicht jeglicher Anwendung, sondern sie bezeugt sicherlich manchmal in der Analysis, daß sie einen nicht zu verachtenden Nutzen bietet. Weiterhin bezweifelt der Verfasser nicht, daß die angewandte Methode der Betrachtung irgendwann einmal bei anderen bedeutenderen Untersuchungen von nicht geringer Hilfe sein wird."

Gibt es denn innere Kriterien, die zu einer gewissermaßen objektiven Hierarchie führen? Sie merken, dieselbe Grundfrage stellt sich für Malerei, Musik, Kunst, wird also eine Frage der Ästhetik. In der Tat, eine übliche Antwort ist, daß Mathematik zum großen Teil eine Kunst sei, deren Entwicklung hauptsächlich von und mit ästhetischen Kriterien geleitet und beurteilt wird. Auf den Laien wirkt es oft überraschend, daß man von Schönheitskriterien in einem so abweisenden Gebiet wie Mathematik sprechen kann. Doch ist für den Mathematiker dieses Gefühl sehr stark, obwohl nicht leicht zu erläutern. Was sind die Regeln dieser Ästhetik? Worin besteht die Schönheit eines Theorems, einer Theorie? Freilich gibt es keine Antwort, die alle Mathematiker befriedigen würde, doch gibt es einen überraschenden Grad von Übereinstimmung, der mir viel größer scheint, als etwa in der Musik oder Malerei. Ohne behaupten zu wollen, ich könne das völlig erklären, werde ich doch sogleich versuchen, etwas Näheres darüber zu sagen. Vorläufig begnüge ich mich mit der Behauptung, daß die Analogie mit der Kunst eine ist, mit der viele Mathematiker übereinstimmen. Zum Beispiel meinte der englische Mathematiker G. Hardy, daß die Daseinsberechtigung der Mathematik, wenn überhaupt, nur als Kunst möglich sei[9]. Unsere Tätigkeit hat viel Gemeinsames mit der eines Künstlers: Ein Maler kombiniert Farben und Formen, ein Musiker Töne, ein Dichter Wörter und wir Ideen einer gewissen Art. Der Maler E. Degas hat auch gelegentlich Sonette geschrieben. Einmal, in einem Gespräch mit dem Dichter Stéphane Mallarmé, beklagte er sich, dies sei ihm so schwierig, obwohl er viele Ideen, sogar eine Überfülle von Ideen hätte. Mallarmé antwortete aber, daß Gedichte mit Wörtern, nicht mit Ideen gemacht werden[10]. Dagegen arbeiten wir mit Ideen. Dieses Gefühl einer Kunst wird noch verstärkt, wenn man daran denkt, wie der Forscher arbeitet und fortschreitet: Man soll sich gar nicht vorstellen, daß der Mathematiker ganz logisch und systematisch verfährt. Oft tappt er ganz im dunkeln, weiß gar nicht, ob er eine gewisse Behauptung beweisen oder widerlegen sollte, und wesentliche Ideen fallen ihm manchmal ganz unerwartet ein, sogar ohne daß er nachher einen klaren logischen Weg zurück zu früheren Überlegungen sehen kann. Wie im Fall eines Komponisten oder Künstlers sollte man dann von Inspiration sprechen[11]. Was der Mathematiker findet, ist auch manchmal so unerwartet, daß er fast

---

9 G. Hardy, "A mathematician's apology", Cambridge University Press 1940; Neudruck 1967 mit einem Vorwort von C. P. Snow, S. 139–140.

10 P. Valéry, «Degas, dense, dessin», A. Vollard éd Paris 1936; Oeuvres II, La Pléiade, Gallimard éd. Paris 1966, 1163–1240, insbesondere S. 1207–1209.

11 Als Illustration möge der folgende Auszug eines Briefes von C. F. Gauß an Olbers dienen, den Gauß am 3. 9. 1805 schrieb, kurz nachdem er ein Problem gelöst hatte, das ihn jahrelang beschäftigt hatte (das „Zeichen der Gaußschen Summen"):

„Endlich vor ein Paar Tagen ist's gelungen – aber nicht meinem mühsamen Suchen, sondern bloß durch die Gnade Gottes möchte ich sagen. Wie der Blitz einschlägt, hat sich das Räthsel gelöst; ich selbst wäre nicht im Stande, den leitenden Faden zwischen dem, was ich vorher wußte, dem, womit ich die letzten Versuche gemacht hatte, – und dem, wodurch es gelang, nachzuweisen . . ."

Siehe Gauß, Gesammelte Werke, Bd. 10 I, S. 24–25. Hier muß man auch H. Poincarés Beschreibung einiger seiner fundamentalen Entdeckungen über automorphe Funktionen erwähnen. H. Poincaré, «L'invention, mathématique» in Science et Méthode, E. Flammarion éd. Paris 1908, Chap. III.

mit Rodin sagen könnte, er wußte, was er suchte, erst, nachdem er es gefunden
hatte.

Andere Mathematiker aber widersetzen sich einer solchen Meinung und
behaupten, Beschäftigung mit Mathemtik, ohne durch die Bedürfnisse der Naturwissenschaften geführt zu werden, ist gefährlich und führt fast sicher zu Theorien,
die vielleicht sehr subtil sind, dem Geist einen besonderen Genuß verschaffen, aber
doch eine Art von geistigem Spiel darstellen, das völlig wertlos vom Standpunkt
der Wissenschaft oder Erkenntnistheorie ist. Zum Beispiel hat der Mathematiker
J. v. Neumann 1947 geschrieben[12]:

"As a mathematical discipline travels far from its empirical sources, or still more, if it is
second and third generation only indirectly inspired by ideas coming from 'reality', it is beset
with very grave dangers. It becomes more and more purely aestheticizing, more and more
purely l'art pour l'art ... there is a great danger that the subject will develop along the line of
least resistance ..., will separate into a multitude of insignificant branches ..."
"... In any event, ..., the only remedy seems to me to be the rejuvenating return to the
source: the reinjection of more or less directly empirical ideas."

Wieder andere nehmen eine Zwischenstellung ein: Sie erkennen völlig die
Wichtigkeit dieser ästhetischen Seite der Mathematik an, doch fühlen sie, es sei
gefährlich, die Mathematik zu weit an und für sich zu treiben. Zum Beispiel hatte
Poincaré geschrieben[13]:

«Et surtout leurs adeptes y trouvent des jouissances analogues à celles que donnente le
peinture et la musique. Ils admirent la délicate harmonie des nombres et des formes; ils
s'émerveillent quand une découverte nouvelle leur ouvre une perspective inattendue; et la
joie qu'ils éprouvent ainsi n'a-t-elle pas le caractère esthétique, bien que les sens n'y prennent aucune part? ...»
«C'est pourquoi je n'hésite pas à dire que les mathématiques meritent d'être cultivées
pour elles-mêmes et que les théories qui ne peuvent être appliquées à la physique doivent
l'être comme les autres.»

---

12 J. v. Neumann, "The mathematician" in Robert B. Heywood, The works of the mind,
University of Chicago Press 1947, 180–187. Collected Works, 6 Vol. Pergamon, New York
1961, I, S. 1–9:

„Wenn eine mathematische Theorie sich weit von ihren empirischen Quellen wegbe-
wegt, oder sogar mehr, wenn sie in ihrer zweiten oder dritten Phase, nur mehr indirekt
durch von der ,Realität' herrührende Ideen inspiriert wurde, dann läuft sie große Gefah-
ren. Sie wird mehr und mehr rein ästhetisieren, mehr und mehr rein ,l'art pour l'art' ... Es
besteht eine große Gefahr, daß sie längs der Linie kleinsten Widerstandes ... in eine Viel-
falt bedeutungsloser Unterfächer ... sich entwickeln wird.
... Auf alle Fälle, ... das einzige Heilmittel scheint mir eine verjüngende Rückkehr
zur Quelle: Eine Wiedereinführung mehr oder weniger direkt empirischer Ideen."
13 H. Poincaré, La Valeur de la Science, E. Flammarion, Paris 1905, Chap. 5, p. 139. Deutsche Übersetzung: Der Wert der Wissenschaft, Teubner, Leipzig 1906, S. 105. Eigentlich
ist dieses Kapitel eine Wiedergabe eines Vortrags, den Poincaré anläßlich des ersten
internationalen Mathematiker-Kongresses, Zürich 1897, gehalten hat:

„Überdies bereitet sie ihren Jüngern ähnliche Genüsse, wie die Malerei und die
Musik. Sie bewundern die zarte Harmonie der Zahlen und der Formen; sie bewundern
eine neue Entdeckung, die ihnen eine unerwartete Aussicht eröffnet; und hat die Freude,
die sie empfinden, nicht einen ästhetischen Charakter, obgleich die Sinne daran gar nicht
beteiligt sind? ...
Darum zögere ich nicht, zu sagen, daß die Mathematik um ihrer selbst willen gepflegt zu
werden verdient, und zwar die Theorien, die nicht auf die Physik angewendet werden können, ebensogut wie die anderen."

Aber doch einige Seiten weiter kommt er zu diesem Vergleich zurück, und fügt hinzu[14]:

«Si l'on veut me permettre de poursuivre ma comparaison avec les beaux-arts, le mathématicien pur qui oublierait l'existence du monde extérieur serait semblable à un peintre qui saurait harmonieusement combiner les couleurs et les formes, mais à qui les modèles feraient défaut. Sa puissance créatrice serait bientôt tarie.»

Dieses Leugnen der Möglichkeit einer abstrakten Malerei scheint mit besonders erwähnenswert, da wir uns in München befinden, wo ungefähr um dieselbe Zeit ein Künstler ganz tief mit diesen Fragen beschäftigt war, nämlich Wassily Kandinsky. Es war ungefähr 1903, als er bei der Betrachtung eines eigenen Gemäldes plötzlich schloß, daß der Gegenstand dem Gemälde schädlich sein kann, etwa weil er ein Hindernis zu einem direkten Zugang zu Formen und Farben, d.h. zu den eigentlichen malerischen Qualitäten des Werkes darstellt. Aber, wie er später schrieb[15], stellten sich dann „eine erschreckende Tiefe", eine Fülle von Fragen vor ihn, am wichtigsten: „Was soll den fehlenden Gegenstand ersetzen?" Kandinsky war sich der Gefahr einer Ornamentik, einer rein dekorativen Malerei, völlig bewußt und wollte sie unbedingt vermeiden. Entgegen Poincaré folgerte er aber nicht, daß eine Malerei ohne realen Gegenstand fruchtlos sein sollte. Wie Sie wissen, entwickelte er eine Theorie der „inneren Notwendigkeit" und des „geistigen Inhalts". Und seit etwa 1910 widmeten sich er und andere Maler in zunehmender Zahl der sogenannten abstrakten oder reinen Malerei, die wenig oder keinen Bezug zur Natur hat.

Will man aber die analoge Möglichkeit in der Mathematik nicht gestatten, dann wird man zu einer Auffassung der Mathematik geführt, die ich kurz so zusammenfassen möchte: Auf der einen Seite sei sie eine Wissenschaft, weil es ein Hauptzweck sei, der Naturforschung oder der Technik zu dienen. Dieses Ziel stehe eigentlich am Ursprung der Mathematik und sei ständig eine Hauptquelle von Problemen. Auf der anderen Seite sei sie eine Kunst, da sie eigentlich eine Schöpfung des Geistes sei, die mit geistigen oder intellektuellen Mitteln gefördert werde, viele davon aus der Tiefe des menschlichen Geistes herkommend, und worin ästhetische Kriterien in mancher Hinsicht maßgebend seien. Aber diese intellektuelle Freiheit, sich in einer Welt des reinen Gedankens zu bewegen, müsse doch in einem gewissen Maße begrenzt sein oder kontrolliert werden durch eventuelle Anwendungsmöglichkeiten in den Naturwissenschaften.

Aber solch ein Bild ist doch zu begrenzt, die letzte Klausel ist zu einschränkend, und viele Mathematiker haben sich für eine völlige Freiheit ihrer Tätigkeit eingesetzt. Zunächst, wie schon vorher hervorgehoben, wären viele Teile der Mathematik, die sich als wichtig für Anwendungen erwiesen, überhaupt nicht entwickelt worden, hätte man von vornherein auf Anwendungsfähigkeit beharrt. Trotz des

---

14 loc. cit. p. 147; deutsche Übersetzung S. 112:

> „Wenn ich meinen Vergleich mit den schönen Künsten fortsetzen darf, so wäre der reine Mathematiker, der die Existenz der äußeren Welt vergäße, dem Maler vergleichbar, der die Farben und Formen harmonisch zusammenzustellen verstünde, dem aber die Vorbilder fehlten. Seine schöpferische Kraft wäre bald versiegt."

15 W. Kandinsky, Rückblick 1901–1913, H. Walden Herausg. 1913. Neudruck vom W. Klein Verl. Baden-Baden 1955. Siehe S. 20–21.

vorigen Zitates hat auch v. Neumann darauf in einem späteren Vortrag ausdrücklich hingewiesen, wo er erklärte, es sei eigentlich in der ganzen Wissenschaft wahr, daß man oft erfolgreich ist, wenn man sich ausschließlich durch Kriterien intellektueller Eleganz leiten läßt und sich weigert, Sachen für Gewinn zu untersuchen [16].

"But still a large part of mathematics which became useful developed with absolutely no desire to be useful, and in a situation where nobody could possibly know in what area it would become useful: and there were no general indications that it even would be so ... This is true of all science. Successes were largely due to forgetting completely about what one ultimately wanted, or whether one wanted anything ultimately; in refusing to investigate things which profit, and in relying solely on guidance by criteria of intellectual elegance; ...
... And I think it extremely instructive to watch the role of science in everyday life, and to note how in this area the principle of laissez faire has led to strange and wonderful results."

Zweitens, und für mich wichtiger, gibt es Teile der reinen Mathematik, die bis jetzt keine oder nur wenige Anwendungen außerhalb der Mathematik gefunden haben, und die man doch einfach als große Leistungen hochschätzen muß. Ich denke zum Beispiel an die Theorie der algebraischen Zahlen, Klassenkörpertheorie, automorphe Funktionen, transfinite Zahlen usw.

Kehren wir einmal zum Vergleich mit der Malerei zurück und nehmen wir als Gegenstände die Probleme, die von der physikalischen Welt herrühren. Dann sehen wir zunächst, daß es Malerei nach der Natur ebenso wie reine oder abstrakte Malerei gibt.

Dieser Vergleich ist aber noch nicht völlig befriedigend, denn solch eine Beschreibung der Mathematik erfaßt doch gewisse ihrer wesentlichen Züge nicht, vor allem die Kohärenz und Einheit der Mathematik. In der Tat weist die Mathematik eine Kohärenz auf, die mir viel größer als die in der Kunst scheint. Man wird das schon ahnen, wenn man bedenkt, daß Theoreme oft von Mathematikern an weit entfernten Orten unabhängig bewiesen werden oder daß eine beträchtliche Anzahl von Arbeiten zwei, manchmal mehr Autoren haben. Es geschieht auch, daß Teile der Mathematik, die ganz unabhängig voneinander entwickelt wurden, unter der Einwirkung neuer Gesichtspunkte plötzlich tiefe Verbindungen aufweisen. Mathematik ist in großem Maße eine kollektive Arbeit. Vereinfachungen, Vereinheitlichungen halten das Gegengewicht zu der unaufhörlichen Entwicklung und Ausdehnung; sie bezeugen immer wieder eine merkwürdige Einheit, obwohl die

---

16 J. v. Neumann, "The role of mathematics in the Science and in Society", address to Princeton Graduate Alumni, June 1954. Cf. Complete Works, 6 Vol., Pergamon, New York 1961, Vol. VI, S. 477–490:

„Aber doch entwickelte sich ein großer Teil der Mathematik, der nützlich wurde, mit überhaupt keinem Verlangen, nützlich zu sein, und in einer Lage, wo niemand möglicherweise wissen konnte, in welchem Bereich er nützlich werden könnte: und es gab keine allgemeinen Anzeichen, daß es je so sein würde ...
... Das ist wahr in der ganzen Wissenschaft. Erfolge wurden größtenteils erhalten, indem man vollständig vergaß, was man letztendlich erreichen wollte oder daß man überhaupt etwas erreichen wollte, indem man sich weigerte, Sachen zu untersuchen, die einen Gewinn einbringen, und indem man sich ausschließlich durch Kriterien intellektueller Eleganz leiten läßt ...
... Und ich denke, es sei äußerst lehrreich, die Rolle der Wissenschaft im alltäglichen Leben zu beobachten, und zu bemerken, wie in diesem Gebiet das Prinzip des laissez faire zu sonderbaren und wunderbaren Resultaten geführt hat."

Mathematik viel zu groß ist, um von einem Individuum beherrscht zu werden. Es scheint mir schwierig, über solche Merkmale völlig Rechenschaft geben zu können durch bloße Zurückführung auf die bisher erwähnten Kriterien, nämlich entweder auf so subjektive Begriffe wie intellektuelle Eleganz und Schönheit, oder auf die Berücksichtigung der Bedürfnisse der Naturwissenschaften und der Technik. Man wird so zur Frage geführt, ob es noch andere Wegweiser und Kriterien als die vorigen gibt. Meiner Ansicht nach ist das der Fall, und ich möchte jetzt die Beschreibung der Mathematik mit einem wesentlichen Bestandteil vervollständigen und von einem dritten Standpunkt aus beleuchten. Als Vorbereitung dazu möchte ich abschweifen, oder scheinbar abschweifen, und mich mit der Frage befassen: „Hat Mathematik eine eigene Existenz? Schaffen wir Mathematik oder entdecken wir bloß, nach und nach, Theorien, die unabhängig von uns schon irgendwo existieren? Wenn dem so ist, wo ist der Ort dieser mathematischen Realität?" Es ist natürlich nicht so unbedingt klar, daß eine solche Frage wirklich sinnvoll ist. Doch ist dieses Gefühl, daß Mathematik auf irgendeine Weise präexistiert, sehr weit verbreitet. Es wurde zum Beispiel ganz scharf von G. Hardy ausgedrückt [17]:

"I believe that mathematical reality lies outside us, that our function is to discover or observe it, and that the theorems which we prove, and which we describe grandiloquently as our 'creations', are simply our notes of our observations. This view has been held, in one form or another, by many philosophers of high reputation, from Plato onwards, . . ."

Ist man ein gläubiger Mensch, dann wird man diese präexistierende mathematische Realität in Gott sehen. Eigentlich war das die Meinung Hermites, der einmal sagte [18]:

«Il existe, si je ne me trompe, tout un monde qui est l'ensemble des vérités mathématiques, dans lequel nous n'avons accès que par l'intelligence, comme il existe un monde des réalités physiques, l'un et l'autre indépendants de nous, tous deux de création divine.»

Es ist noch nicht lange her, daß ein Kollege in einer einführenden Vorlesung erklärte, die folgende Frage hätte ihn jahrelang beschäftigt: "Why has God created the exceptional series?" „Warum hat Gott die Ausnahmereihe geschaffen?"

Ein Verweis auf eine göttliche Herkunft kann zwar den Nichtgläubigen kaum befriedigen. Doch hat manch einer ein dunkles Gefühl, daß Mathematik irgendwo existiert, obwohl, wenn er darüber nachdenkt, er kaum dem Schluß entgehen kann, daß Mathematik ausschließlich eine menschliche Schöpfung sei.

Solche Fragen stellen sich auch für viele andere Begriffe, wie Staat, moralische Werte, Religion usw., und wären einen eigenen Vortrag wert. Aus Mangel an Zeit

---

17 Cf. G. Hardy, loc. cit. [9], S. 123–124:

> „Ich glaube, daß die mathematische Realität außerhalb unser liegt, daß es unsere Aufgabe ist, sie zu entdecken oder zu beobachten, und daß die Theoreme, die wir beweisen, und die wir schwülstig als unsere ,Schöpfungen' beschreiben, nur die Notizen unserer Beobachtungen sind. Diese Ansicht, in der einen oder anderen Form, ist von vielen Philosophen großen Rufes, von Plato an, vertreten worden."

18 G. Darboux, La vie l'Oeuvre de Charles Hermite, Revue du Mois, 10. 1. 1906, p. 46:

> „Es existiert, wenn ich nicht irre, eine ganze Welt, die die Gesamtheit der mathematischen Wahrheiten ist, zu der wir nur Zugang durch unseren Geist haben, wie auch eine Welt der physikalischen Realitäten existiert, die eine wie die andere unabhängig von uns, beide von göttlicher Schöpfung."

und Kompetenz muß ich mich schon mit einer kurzen, vielleicht zu einfachen Antwort zu diesem scheinbaren Dilemma begnügen und werde mich der These anschließen, daß wir dazu geneigt sind, all dem eine Existenz zu geben, was zu einer Kultur gehört, in dem Sinne, daß wir es mit anderen Leuten teilen und darüber Meinungen austauschen können. Etwas wird objektiv (im Gegensatz zu „subjektiv"), sobald wir davon überzeugt sind, daß es in der Gedankenwelt anderer Menschen in derselben Form wie in unserer existiert, und daß wir darüber gemeinsam nachdenken und diskutieren können[19]. Da die mathematische Sprache so präzis ist, ist sie ideal dazu geeignet, Begriffe zu definieren, worüber ein solcher Konsens existiert. Meiner Ansicht nach reicht das hin, um uns das Gefühl einer objektiven Existenz, einer Realität, der Mathematik zu geben, wie sie aus den oben zitierten Erklärungen von Hardy und Hermite hervorgeht. Man kann natürlich noch lange spekulieren, ob es einen anderen Ursprung hat, wie es Hardy und Hermite behaupten, aber das ist eigentlich nicht relevant für die Fortsetzung dieser Diskussion.

Bevor ich das weiterverfolge, möchte ich nebenbei bemerken, daß ähnliche Gedanken über unsere Auffassung der physikalischen Realität ausgedrückt worden sind. Beispielsweise hat Poincaré geschrieben[20]:

«Ce qui nous garantit l'objectivité du monde dans lequel nous vivons, c'est que ce monde nous est commun avec d'autres êtres pensants ...»
«Telle est donc la première condition de l'objectivité: ce qui est objectif doit être commun à plusieurs esprits et par conséquent pouvoir être transmis de l'un à l'autre ...» .

und Einstein[21]:

„Verschiedene Menschen können mit Hilfe der Sprache ihre Erlebnisse bis zu einem gewissen Grade miteinander vergleichen. Dabei zeigt sich, daß gewisse sinnliche Erlebnisse verschiedener Menschen einander entsprechen, während bei anderen ein solches Entsprechen nicht festgestellt werden kann. Jenen sinnlichen Erlebnissen verschiedener Individuen, welche einander entsprechen und demnach in gewissem Sinne überpersönlich sind, wird eine Realität gedanklich zugeordnet."

Jetzt wieder zur Mathematik. Die Mathematiker haben also eine geistige Realität gemein, eine riesige Menge von mathematischen Begriffen, Objekten, die teils bekannt, teils unbekannt sind, Theorien, Theoremen, gelösten und ungelösten Problemen, die sie mit geistigen Werkzeugen studieren. Diese Probleme und Begriffe wurden zum Teil von der physikalischen Welt suggeriert, zum größeren Teil jedoch direkt durch rein mathematische Überlegungen, zum Beispiel die Quaternionen. Diese Gesamtheit, obwohl vom menschlichen Geist herrührend, wirkt für

---

19 Siehe L. White, "The locus of mathematical reality: an anthropological footnote", Philosophy of Science 14 (1947), 289–303; auch in J. R. Newman, The World of Mathematics, 4 volumes, Simon and Schuster, New York (1956), Vol. 4, S. 2348–2364.
20 H. Poincaré, loc. cit.[13], p. 262, deutsche Übersetzung, S. 197:

„Was uns die Objektivität der Welt, in der wir leben, verbürgt, ist, daß wir diese Welt mit anderen denkenden Wesen gemeinsam haben ...
Das ist also die erste Bedingung der Objektivität: Was objektiv ist, muß mehreren Geistern gemein sein und folglich von einem dem anderen übermittelt werden können ..."
21 A. Einstein, Vier Vorlesungen über Relativitätstheorie, gehalten im Mai 1921 an der Universität Princeton, Fr. Vieweg und Sohn, Braunschweig 1922, S. 1. Englische Übersetzung in: The Meaning of Relativity, Princeton University Press, Princeton 1945.

uns wie eine Naturwissenschaft im üblichen Sinne, etwa Physik oder Biologie, und ist für uns ebenso konkret. Ich möchte sogar behaupten, daß sie nicht nur eine theoretische, sondern auch eine experimentelle Seite hat. Die erstere ist klar: Wir streben nach allgemeinen Sätzen, Prinzipien, Beweisen und Lösungsmethoden. Das ist die Theorie. Aber zuerst hat man oft keine Ahnung, was man erwarten soll und wie man vorgehen muß, und man gewinnt Verständnis und Intuition durch Experimentieren, d. h. durch Betrachtung von Spezialfällen. Man hofft so, erstens, zu einer vernünftigen Vermutung geführt zu werden, und zweitens, vielleicht auf eine Idee zu stoßen, die sich als verallgemeinerungsfähig erweist. Es geschieht natürlich auch, daß gewisse Spezialfälle an und für sich von großem Interesse sind. Das ist die experimentelle Seite. Daß wir mit intellektuellen Objekten operieren, eher als mit realen Objekten und Apparaten im Laboratorium, ist eigentlich unwesentlich. Das Gefühl, daß Mathematik in diesem Sinne eine experimentelle Wissenschaft ist, ist auch nicht neu. So schrieb Hermite an L. Königsberger um 1880[22]:

«Le sentiment exprimé dans ce passage de votre lettre où vous me dites ‹plus je réfléchis sur toutes ces choses, plus je reconnais que les mathématiques forment une science expérimentale aussi bien que toutes les autres sciences . . .› ce sentiment, dis-je, est aussi le mien.»

Traditionell wurden diese Experimente im Kopf oder mit Papier und Feder ausgeführt, und darum habe ich von intellektuellen Werkzeugen gesprochen. Ich sollte jedoch hinzufügen, daß seit etwa zwanzig Jahren echte Apparate, nämlich die elektronischen Rechenmaschinen, dabei eine zunehmende Rolle spielen. Sie haben eigentlich dieser experimentellen Seite der Mathematik eine neue Dimension gegeben. Das geht so weit, daß man schon von bedeutenden gegenseitigen und faszinierenden Einflüssen zwischen „Computer Science" und reiner Mathematik sprechen darf. Aber das ist nicht wesentlich für meine Diskussion und wird nicht weiterverfolgt werden.

Jetzt erwirbt das Wort „Wissenschaft" in meinem Titel eine erweiterte Bedeutung. Es bezieht sich nicht nur, wie früher, auf die Naturwissenschaften, sondern auch, und das in viel stärkerem Maße, auf die Auffassung der Mathematik selbst als experimenteller und theoretischer Wissenschaft, ich möchte fast sagen, als einer geistigen Naturwissenschaft.

Das macht es mir etwas leichter, über Motivierung und Ästhetik zu sprechen. Will man Anwendungen auf die Naturwissenschaften nicht besonders berücksichtigen, ist man doch nicht bloß auf intellektuelle Eleganz zurückgeworfen. Es bleiben noch fast praktische Kriterien, nämlich Anwendbarkeit in der mathematischen Wissenschaft selbst. Die Betrachtung dieser mathematischen Realität, der offenen Probleme, der Struktur, Bedürfnisse und Zusammenhänge verschiedener Teile, weist schon auf wahrscheinlich fruchtbare, wertvolle Richtungen hin und erlaubt dem Mathematiker, sich zu orientieren und Problemen wie Theorien relative Werte zu geben. Oft wird man es als Prüfstein für den Wert einer neuen

---

22 cf. L. Königsberger, „Die Mathematik eine Geistes- oder Naturwissenschaft", Jahresbericht der Deutschen Mathematiker-Vereinigung 23 (1914) 1–12:

„Das an der Stelle Ihres Briefes ausgedrückte Gefühl, an der Sie mir sagen: ‚Je mehr ich über alle diese Sachen nachdenke, desto mehr erkenne ich, daß die Mathematik eine experimentelle Wissenschaft wie alle anderen Wissenschaften ist‘, dieses Gefühl, sage ich, ist auch das meinige."

allgemeinen Theorie betrachten, ob sie alte Probleme lösen kann. Dies begrenzt de
facto die Freiheit des Mathematikers, vergleichbar etwa einem Physiker, der doch
nicht ganz willkürlich die Phänomene wählt, zu denen er eine Theorie zu erstellen
oder Experimente durchzuführen wünscht. Wie viele Beispiele zeigen, war es den
Mathematikern oft erlaubt, vorauszusehen, wie gewisse Teile der Mathematik sich
entwickeln würden, welche Probleme aufgegriffen werden sollten und wahrscheinlich bald gelöst werden würden. Äußerungen über die Zukunft der Mathematik
haben sich ziemlich oft als zutreffend erwiesen. Das ist nicht fehlerfrei, aber erfolgreich genug, um auf einen Unterschied mit der Kunst hinzuweisen. Analoge und so
verhältnismäßig erfolgreiche Betrachtungen über die Zukunft der Malerei z. B. gibt
es doch kaum.

Ich will das aber nicht zu weit führen. Die Auffassung der Mathematik als
einer „geistigen Naturwissenschaft" habe ich als einen von drei Bestandteilen, nicht
als das Ganze, vorgeschlagen.

Auf der einen Seite will ich gar nicht die Wichtigkeit der Wechselwirkungen
zwischen Mathematik und den Naturwissenschaften mindern. Erstens pflegt man,
wenigstens seit vielen Jahrhunderten, zu sagen, daß alle Disziplinen der Naturwissenschaften einer mathematischen Formulierung und Behandlung zustreben müssen; sogar daß eine solche Disziplin erst dann eine Wissenschaft im eigentlichen
Sinn ist, wenn das erreicht sei. Also ist es sicher wichtig, daß Mathematiker sich
bemühen, dabei zu helfen. Zweitens ist es zweifellos eine Leistung hohen Grades,
komplizierte Phänomene mathematisch zu formulieren und zu behandeln, und die
neuen Probleme, die somit eingeführt werden, stellen eine große Bereicherung der
Mathematik dar. Man denke etwa an die Wahrscheinlichkeitsrechnung. Ich behaupte nur, es ist einfach nicht nötig, diesem Gedanken der Anwendbarkeit den
Vorrang zuzubilligen, um wertvolle Mathematik zu treiben. Die Geschichte der
Mathematik zeigt, daß viele hervorragende Leistungen von Mathematikern stam-
men, die an äußere Anwendungen gar nicht dachten und durch rein mathematische Betrachtungen geleitet wurden. Und wie schon erwähnt und illustriert, haben
diese Beiträge oft wichtige Anwendungen in den Naturwissenschaften oder der
Technik gefunden, manchmal auf völlig unvorhergesehene Weise.

Auf der anderen Seite behaupte ich nicht, daß man alles ganz rational voraus-
sehen kann. Eigentlich ist dies auch in den Naturwissenschaften gar nicht der Fall;
zumal man auch nicht immer von vornherein weiß, welche Experimente interessant
sein werden. Hervorragende Mathematiker haben sich auch getäuscht und haben,
gerade im Hinblick auf Anwendbarkeit in der Mathematik, manchmal neue Ideen
als fruchtlos, belanglos, sogar gefährlich bezeichnet, die sich später als fundamental
erwiesen. Die Freiheit, nicht an praktische Anwendungen zu denken, die v. Neumann für die gesamte Wissenschaft verlangte, muß eben auch innerhalb der
Mathematik verlangt werden.

Dieser Analogie der Mathematik mit einer Naturwissenschaft könnte man
entgegensetzen, daß sie einen wesentlichen Unterschied übersieht: In den Naturwissenschaften oder der Technik stößt man manchmal auf Probleme, die man
unbedingt lösen muß, um überhaupt weiter fortschreiten zu können. In dieser
mathematischen Gedankenwelt hat man jedoch *de jure* die Freiheit, scheinbar
unlösbare, zu harte Probleme beiseite zu lassen und zu anderen, mehr verspre-
chenden, überzugehen, d.h., doch vielleicht Linien kleinsten Widerstandes zu fol-

gen, ganz wie v. Neumann fürchtete. Wäre das nicht eine Versuchung für einen
Mathematiker, der Mathematik als „die Kunst, Probleme zu finden, die man lösen
kann" definiert? Interessanterweise habe ich diese Definition von einem Mathe-
mathiker gehört, dessen Arbeiten besonders bewundernswert sind, weil sie so viele
Probleme behandeln, die zu der Zeit ganz speziell waren, aber sich später als fundamental erwiesen, und deren Lösungen neue Wege eröffnet haben, nämlich von
H. Hopf. Es ist jedoch nicht zu leugnen, daß oft Linien kleineren Widerstandes ver-
folgt werden, die zu recht trivialen oder bedeutungslosen Arbeiten führen. Es
kommt auch vor, daß eine erfolgreiche Schule später in eine fruchtlose Periode
gelangt und dann sogar in schlimmen Fällen einen schädlichen Einfluß hat. Aber
merkwürdigerweise kommt dann immer ein Gegenmittel, eine Reaktion, die das
neutralisiert und diese Irrwege und fruchtlosen Richtungen eliminiert. Bis jetzt
konnte Mathematik immer solche Wachstumskrankheiten überwinden, und ich bin
überzeugt, es wird immer so sein, solange es so viele begabte Mathematiker gibt. Es
ist doch sehr merkwürdig: Viele von uns haben dieses Gefühl einer Einheit der
Mathematik, aber es ist gefährlich, zu präzise Richtlinien im Namen dieser Einheit
vorzuschreiben. Es ist wichtiger, daß Freiheit herrscht, trotz gelegentlichen Miß-
brauchs. Warum das so erfolgreich ist, kann man nicht völlig erklären. Wenn man
etwa an Hopf denkt, kann man zu einem gewissen Maße seine Auswahl von Problemen sozusagen logisch begründen: Oft waren sie die ersten Spezialfälle eines
allgemeinen Problems, bei dem bekannte Beweismethoden nicht anwendbar waren.
Dessen war er sich natürlich bewußt. Aber das erklärt nicht alles. Wahrscheinlich
sah er nicht immer voraus, wie einflußreich seine Arbeiten sein würden, und gewiß
machte er sich auch darüber kaum Gedanken. Es ist einfach Teil der Begabung in
der Mathematik, zu guten Problemen hingezogen zu werden. Der Mathematiker ist
zu diesen teils durch rationale, wissenschaftliche Betrachtungen, teils durch bloße
Neugier, Instinkt, Intuition, rein ästhetische Betrachtungen geleitet. Das bringt
mich gerade zu meinem letzten Gegenstand, dem ästhetischen Gefühl in der
Mathematik.

Die Auffassung der Mathematik als einer Kunst, einer Poesie von Ideen habe
ich schon erwähnt. Davon ausgehend, wird man dann schließen, daß man, um
Mathematik zu schätzen, zu genießen, ein eigenartiges Gefühl für intellektuelle
Eleganz und Schönheit der Ideen in einer besonderen Gedankenwelt benötigt. Daß
dies dann den Nichtmathematikern kaum mittelbar ist, ist nicht überraschend:
Unsere Gedichte sind in einer recht speziellen Sprache geschrieben, der mathematischen Sprache; obwohl diese in vielen der üblichen Sprachen ausgedrückt
wird, ist sie doch einzigartig und in keine andere übersetzbar; und leider sind diese
Gedichte nur in der originalen Sprache zu verstehen. Die Ähnlichkeit mit einer
Kunst ist klar: Auch zur Wertschätzung der Musik oder Malerei muß man doch
eine gewisse Bildung haben, d. h. eine gewisse Sprache lernen.

Solchen Meinungen und Analogien habe ich lange zugestimmt. Ohne daß
meine grundsätzliche Stellung zur Mathematik sich geändert hat, möchte ich sie
doch etwas in die Richtung meiner vorherigen Äußerungen umformen. Ich glaube,
unsere Ästhetik ist nicht immer ganz so rein und esoterisch und bezieht auch ein
wenig mehr irdische Faktoren ein, wie Bedeutung, Tragweite, Anwendungsfähigkeit, Nützlichkeit, aber das innerhalb der mathematischen Wissenschaft. Unser
Urteil über ein Theorem, eine Theorie, einen Beweis, ist auch dadurch beeinflußt;

aber oft wird das einfach mit dem Ästhetischen gleichgesetzt. Ich möchte versuchen, es an dem Beispiel der früher erwähnten Galois-Theorie zu erläutern. Diese Theorie wird ganz allgemein als eines der schönsten Kapitel der Mathematik hochgeschätzt. Warum? Erstens löst sie eine sehr alte und zu jener Zeit die wichtigste Frage über Gleichungen. Zweitens ist sie eine äußerst umfassende Theorie, die weit über die ursprüngliche Frage der Lösbarkeit durch Radikale hinausgeht. Drittens beruht sie auf wenigen Grundsätzen großer Eleganz und Einfachheit, die in einem neuen Rahmen, mit neuen Begriffen formuliert werden, die größte Originalität aufweisen. Viertens haben diese neuen Gesichtspunkte und Begriffe, besonders der Gruppenbegriff, neue Wege eröffnet und einen tiefen Einfluß in der ganzen Mathematik gehabt.

Sie merken, daß unter diesen vier Punkten nur der dritte ein echtes ästhetisches Urteil ist, und eines, worüber man eine eigene Meinung nur haben kann, wenn man die technischen Einzelheiten der Theorie versteht. Die anderen haben doch einen davon verschiedenen Charakter. Solche Aussagen könnte man auch über Theorien in irgendeiner Naturwissenschaft aussprechen. Sie haben einen größeren objektiven Inhalt, und ein Mathematiker kann darüber seine eigene Meinung haben, sogar, wenn er die Theorie nicht völlig technisch beherrscht. Für die Zwecke dieser Diskussion habe ich diese vier Bestandteile getrennt, aber normalerweise würde ich das nicht immer so explizit machen, und alle vier tragen dazu bei, den Eindruck der Schönheit zu geben. Ich glaube schon, daß dieses Beispiel in solcher Hinsicht ziemlich typisch ist: Was wir als ästhetisch bezeichnen, ist eigentlich oft ein Verschmelzen verschiedener Betrachtungen. Zum Beispiel werde ich eine Beweismethode selbstverständlich als schöner empfinden, wenn sie neue unerwartete Anwendungen findet, obwohl die Methode sich nicht geändert hat. Sie ist vielleicht wichtiger geworden, aber in sich betrachtet, nicht schöner. Da alles sich in der Mathematik selbst abspielt, wird das dem Nichtmathematiker kaum helfen, in unsere ästhetische Welt einzudringen. Hoffentlich wird es ihm jedoch plausibel machen, daß unsere sogenannten ästhetischen Urteile einen größeren Konsens als in der Kunst aufweisen, einen Konsens, der sich in beträchtlichem Maße über geographische und zeitliche Grenzen erstreckt. Jedenfalls halte ich dies für einen Hauptgrund. Aber wieder muß ich mich hüten, das zu weit zu führen. Es ist eine Frage des Grades, nicht ein absoluter Unterschied. Ein ästhetisches Urteil über das Werk eines Komponisten oder Malers bezieht auch äußere Faktoren, wie Einfluß, Vorgänger, also die Stellung des Werkes im Gesamtbild, mit ein, aber doch schon in geringerem Maße. Auf der anderen Seite gibt es auch Meinungsverschiedenheiten und zeitliche Schwankungen in der Einschätzung mathematischer Werke, doch wieder nicht in so starkem Maße, würde ich hinzufügen. All diese Nuancen wären vieler Erläuterungen bedürftig, worauf ich aus Mangel an Zeit verzichten muß.

In der begrenzten Zeit, die mir zur Verfügung steht, wäre es natürlich bequemer, nur ganz scharfe Behauptungen über Mathematik zu äußern. Aber leider, oder glücklicherweise, wie andere menschliche Unternehmen, zu denen viele Menschen über viele Jahrhunderte beigetragen haben, weigert sich die Mathematik, mit einigen einfachen Formeln beschrieben zu werden. Fast jede allgemeine Behauptung über Mathematik muß irgendwie eingeschränkt werden. Eine Ausnahme doch, vielleicht die einzige, wäre diese Behauptung selbst. Ich hoffe, daß ich wenigstens den Eindruck hervorgerufen habe, daß Mathematik eine äußerst komplexe

Schöpfung ist, die so viele wesentliche gemeinsame Züge mit Kunst, experimentellen und theoretischen Naturwissenschaften aufweist, daß sie als alle drei zugleich betrachtet, und deswegen auch von allen drei unterschieden werden muß.

Ich bin mir dessen bewußt, daß ich mehr Fragen aufgestellt als beantwortet habe, und daß ich einige kaum oder gar nicht berührt habe, zum Beispiel den Wert dieser Schöpfung. Man kann natürlich auf die unzähligen Anwendungen in den Naturwissenschaften oder in der Technik, von denen viele einen großen Einfluß sogar auf unser tägliches Leben haben, hinweisen, und damit die Daseinsberechtigung der Mathematik sozial begründen. Doch muß ich gestehen, daß ich mich als reiner Mathematiker mehr für die Einschätzung der Mathematik an sich interessiere. Die Beiträge der verschiedenen Mathematiker kombinieren sich zu einer riesigen intellektuellen Konstruktion, die, meiner Ansicht nach, ein höchst imponierendes Zeugnis der Macht des menschlichen Denkens darstellt. Der Mathematiker Jacobi hat einmal geschrieben, daß der einzige Zweck der Wissenschaft die Ehre des menschlichen Geistes sei[23]. Ich glaube in der Tat, daß diese Schöpfung dem menschlichen Geist zur großen Ehre gereicht.

---

23 In einem Brief vom 2. Juli 1830 an A. M. Legendre, cf. K. G. J. Jacobi, Gesammelte Werke, G. Riemer, Berlin 1881–1891, Bd. 1, S. 453–455. Die betreffende Stelle lautet:

«Mais M. Poisson n'aurait pas dû reproduire dans son rapport une phrase peu adroite de feu M. Fourier, où ce dernier nous fait des reproches, à Abel et à moi, de ne pas nous être occupés de préférence du mouvement de la chaleur. Il est vrai que M. Fourier avait l'opinion que le but principal des mathématiques était l'utilité publique et l'explication des phénomènes naturels; mais un philosophe comme lui aurait dû savoir que le but unique de la science, c'est l'honneur de l'esprit humain et que sous ce titre, une question de nombres vaut autant qu'une question du système du monde."

Commentaires et corrections

Les Notes à chaque article sont numérotées consécutivement. Un symbole tel que XY.z dans la marge de gauche réfère à z, page XY de ce volume.

87. On periodic maps of certain $K(\pi, 1)$

57.1 Cette Note, non publiée antérieurement, répond à une question qui m'avait été posée par D. Sullivan en automne 1969. Elle a été communiquée un peu plus tard à F. Raymond et le Théorème 1, ou plutôt un corollaire de ce théorème, est démontré et utilisé dans un article de P. Conner et F. Raymond, Proc. of 2nd Conférence on compact transformation groups (1971), Springer Lecture Notes **299** (1972), p. 1−75.

89. Properties and linear representations of Chevalley groups

88.1 En fait, comme me l'ont signalé indépendamment J. Tits et D. Verma, la démonstration donnée ici présente une lacune. Une manière simple de la combler, suggérée par F. Bruhat et J. Tits, est de considérer la somme directe π' de π et de la représentation adjointe, munie de la base de Chevalley usuelle, et de *définir* $\mathbf{Z}(G_\pi)$ comme étant égal à l'anneau $\mathbf{Z}[G_{\pi'}]$ au sens de 3.4.

Par ailleurs, une démonstration de ces assertions sera contenue dans un article en préparation de F. Bruhat et J. Tits: «Groupes réductifs sur un corps local II», à paraître dans les Publ. Math. I.H.E.S.

98.2 Le fait que l'égalité $F_l = F_{l,\mathbf{z}} \otimes K$ soit écrite sans aucun commentaire semble impliquer qu'elle est considérée comme évidente. Il se trouve qu'elle est correcte, mais cela exige une démonstration (voir la Note 1 à [102]).

95. Some metric properties of arithmetic quotients of symmetric spaces and an extension theorem

169.1 La situation s'est éclaircie depuis. Tout d'abord, P. Kiernan (Bull. A.M.S. **80** (1974), 109−110) a montré que V^{**} est Hausdorff en utilisant le Théor. 3.7 du présent article et un analogue pour la compactification V^{**} de Piatetski-Shapiro, dû a P. Kiernan et K. Kobayashi (Inv. Math. **16** (1972), 237−348). De plus un article ultérieur de ces deux auteurs (J. Math. Soc. Japan **28** (1976), 577−580) montre comment déduire directement notre Théor. 3.7 de l'analogue relatif à V^{**} précité, dont la démonstration est plus simple. Je n'ai pas poussé plus loin mes efforts pour démontrer l'identité des deux topologies à partir de la théorie de la réduction, comme je l'envisageais ici, et ne sais si cela peut être mené à bien.

97. Homomorphismes «abstraits» de groupes algébriques simples
(avec J. Tits)

230.1   *Voir* J. Tits, Symposia Mathematica, Ist. Naz. di Alta Mat. **13** (1974), 479−499.

235.2   Une démonstration est donnée dans un article de G. Prasad, Bull. Soc. Math. France **100** (1982), 197−203.

98. Corners and arithmetic groups (with J-P. Serre)

285.1   La description donnée ici du voisinage d'une face n'est valable que si la face est associée à un sous-groupe parabolique minimal défini sur **Q**. Aussi retirons-nous toutes les assertion du § 10 ne se rapportant pas à ces faces. On peut rétablir une partie de ces énoncés en prenant comme voisinage d'une face $e(P)$ la réunion des transformés d'un ensemble de Siegel relatif à P par $\Gamma \cap P$. Une correction sera publiée dans les Comm. Math. Helv. L'erreur provient du fait que 7.3 (ii) est inexact. Mais ce résultat et le § 10 ne sont pas utilisés dans le reste de l'article.

99. Cohomologie de certains groupes discrets et Laplacien
p-adique [d'après H. Garland]

300.1   Pour G presque simple, et simplement connexe, la validité de ce théorème sans hypothèse sur le corps résiduel a été annoncée par W. Casselman (Bull. A.M.S. **80** (1974), 1001−1004), qui a publié sa démonstration, en supposant G déployé et semi-simple sur k (Jour. Fac. Sci. Univ. Tokyo Ser. IA, Math. Vol. **23** (1981), 907−928). Entre temps une démonstration générale a été donnée dans [115: XIII, 3.6 (i)].

301.2   Je se sais toujours pas si l'on peut s'affranchir de l'hypothèse sur le corps résiduel lorsque Γ n'est pas cocompact, en particulier si Γ est réduit à l'identité.

305.3   La cohomologie du complexe $L^*(X; \tilde{M})$ n'est autre que l'espace $H^*_{(2)}(X; \tilde{M})$ de cohomologie L^2 du complexe X, à coefficients dans $\tilde{M}$, au sens des cochaines simpliciales, complexe qui est supposé uniformément localement fini. La décomposition «de Hodge» 2.2(1) implique que l'espace $\mathbf{H}^r$ des r-formes harmoniques s'injecte naturellement dans $H^r_{(2)}(X; \tilde{M})$, et lui est égal si et seulement si dL^{r-1} est fermé dans L^r. Les inégalités locales, lorsqu'elles sont satisfaites, entraînent aussi cette condition, et pas seulement la nullité de $\mathbf{H}^r$. Elles impliquent donc la nullité de $H^*_{(2)}(X; \tilde{M})$. Il s'ensuit que dans la Proposition de 3.5, c'est non seulement $\mathbf{H}^r$ mais aussi $H^*_{(2)}(X; \tilde{M})$ qui est nul si le corps résiduel est suffisamment grand. Ici aussi, j'ignore si cette dernière hypothèse est nécessaire.

312.4   Dans le cas où G est de k-rang $\geqq 2$ et où Γ est «irréductible», le théorème vaut en effet pour toute représentation de dimension finie de Γ sur un corps de caractéristique zéro, conformément à ce qui est espéré ici (*voir* [115: XIII, 3.7]).

312.5 L'extension aux groupes semi-simples, obtenue tout d'abord en collaboration avec H. Garland et annoncée par lui dans son exposé au Proc. ICM Vancouver 1974, Vol. 1, 449–453) est établie dans [115], *loc. cit.*

313.6 Cela est démontré dans [106].

101. Cohomology of arithmetic groups

353.1 Les résultats du § 5 sont démontrés dans [108], et le Théor. 6.1 dans [118: § 5]. Les résultats obtenus en collaboration avec H. Garland (Théor. 6.2 et § 7) sont établis dans un article écrit en commun qui sera publié dans l'Amer. Math. Jour. (1983). Notons que le titre du § 7: «Square integrable cohomology», se réfère aux classes de cohomologie ordinaire représentables par une forme différentielle fermée de carré intégrable. Ce n'est pas la cohomologie de carré intégrable au sens devenu usuel ces dernières années, qui est celui de [116].

102. Linear representations of semi-simple algebraic groups

372.1 Le «basic principle» invoqué ici est l'invariance de la caractéristique d'Euler-Poincaré par changement de base, sous des hypothèses de platitude convenables. Cf. D. Mumford, Abelian Varieties, Oxford U. Press 1970, p. 50. L'égalité $F_l = F_{l,Z} \otimes K$ mentionnée plus haut dans la Note 2 [89] peut aussi s'écrire, dans le notations du § 6 considéré ici,

$$\dim H^0(G_k/B_k; \xi_r) = \dim H^0(G/B; \xi_r) \, .$$

Comme il y est remarqué, il suffit alors pour la démontrer de savoir que $\dim H^i(G_k/B_k; \xi_r) = 0$ pour $i \geqq 1$, ce que a été établi depuis en général par G. Kempf (Annals of Math. **103** (1976), 557–591). Des preuves plus simples ont été ensuite données par G. Kempf, W. Haboush, H. H. Andersen (*voir* un exposé d'ensemble de J. E. Humphreys (Preprint 1981) pour des références).

L'égalité des caractéristiques d'Euler sur **C** et sur k, interprétées comme G-modules virtuels, mentionnée ici comme assez probable, a été établie par W. Haboush (preprint, 1975).

372.2 *Voir* W. L. Griffith, Ill. J. Math. **24** (1980), 452–461. Beaucoup d'autres résultats ont été obtenus depuis sur ce problème. Cf. l'exposé de J. Humphreys cité ci-dessus ou aussi H. H. Andersen, Jour. Algebra **71** (1981), 245–258.

104. Cohomologie de sous-groupes discrets et représentations de groupes semi-simples

399.1 Pour un exposé systématique de tout ce qui concerne ici les sous-groupes discrets cocompacts et la cohomologie continue, *voir* [115].

414.2 Des exemples de sous-groupes discrets cocompacts où de covolume fini de
SU (2, 1) non définissables arithmétiquement ont été donnés par G. D. Mostow
(Pacific J. Math. **86** (1980), 171–276). Il en a ensuite fourni d'autres, dont le
principe de construction remonte à E. Picard. Une généralisation de cette méthode,
obtenue par Mostow et P. Deligne, permet de construire un exemple dans
SU (3, 1).

414.3 Cette conjecture n'est toujours pas démontrée, ni infirmée, en particulier pour
$n = 3$.

415.4 Ici, on a seulement besoin de savoir que si $H^*(\mathfrak{g},\mathfrak{k}\,;M_\pi \otimes V) \ne 0$ et si le poids
dominant de r est suffisamment régulier, alors π est dans la série discréte. Cela
résulte de 7.3 dans [115]. Plus généralement, sans supposer que G ait une série
discrète, ce théorème implique que π fait partie de la série fondamentale. De plus,
un travail de D. Vogan et G. Zuckerman (non encore publié) montre que l'on peut
supprimer le «suffisamment», autrement dit que si V a un poids dominant régulier
et si $H^*(\mathfrak{g},\,\mathfrak{k};M_\pi \otimes V) \ne 0$, alors M_π fait partie de la série fondamentale.

Sans référence à la cohomologie, il est aussi vrai que si π a un caractère
infinitésimal suffisamment régulier, alors π est dans la série fondamentale, mais,
à ma connaissance, ce résultat, bien connu de quelques experts, ne se trouve pas
dans la littérature.

105. Cohomologie d'immeubles et de groupes S-arithmétiques
(avec J.-P. Serre)

446.1 Dans le cas réel différentiable, on a encore la suite exacte du corollaire 3.3, *voir*
[115: V, 4.2], où l'on considère un analogue réel de la représentation de Steinberg.
Mais cette représentation n'est pas nécessairement irréductible et ses constituents
ne sont pas toujours unitaires.

111. On the development of Lie group theory

545.1 On suppose implicitement ici que G est connexe et simplement connexe (en tant
que groupe algébrique). Dans le cas général, les résultats sont essentiellement
semblables, mais plus compliqués à énoncer.

114. Symmetric compact complex spaces

599.1 L'expression «Zariski open» signifie ici: complémentaire d'un sous-ensemble
analytique fermé.

116. Stable and L^2-cohomology of arithmetic groups

614.1 Les résultats annoncés dans les §§ 1, 2 sont démontrés dans [118], le Théorème 3
(sous des hypothèses beaucoup plus générales) dans un article: «Regularization
theorems in Lie algebra cohomology. Applications» à paraître.

616.2 Après coup, j'ai découvert une erreur dans la démonstration de ce Théorème. W. Casselman a ensuite établi un lemme qui permettait de compléter la démonstration lorsque G et K ont même rang. Finalement, W. Casselman et moi avons montré que l'homomorphisme $\mathcal{H}_{(2)}(\Gamma \backslash X) \to H_{(2)}(\Gamma \backslash X)$ est un isomorphisme si G ne possède pas de sous-groupe parabolique propre défini sur $\mathbf{Q}$ qui contienne un sous-groupe parabolique fondamental, condition qui est un plus générale que celle du texte. Un travail commun à paraître («L^2 cohomology of locally symmetric manifolds of finite volume») contiendra notamment cet énoncé et les démonstrations des autres résultats des §§ 4, 5.

Bibliographie

La numérotation de gauche suit l'ordre chronologique de parution (de rédaction pour [30] et [87]). Un chiffre romain I, II ou III à droite d'un titre indique dans quel volume l'article en question est contenu. Les titres suivis du signe 0 sont ceux des travaux non reproduits ici.

1. (avec J. de Siebenthal) Sur les sous-groupes fermés connexes de rang maximum des groupes de Lie clos, C. R. Acad. Sci., Paris **226** (1948) 1662–1664   I, 1–2

2. Some remarks about Lie groups transitive on spheres and tori, Bull. Amer. Math. Soc. **55** (1949) 580–587   I, 3–10

3. (avec J. de Siebenthal) Les sous-groupes fermés de rang maximum des groupes de Lie clos, Comment Math. Helv. **23** (1949) 200–221   I, 11–32

4. Groupes d'homotopie des groupes de Lie, Espaces Fibrés et Homotopie, Sém. H. Cartan, E. N. S. Paris 1949–1950, Exp. 12, 13; Notes polycopiées, 2éme éd. Secrét. Mathématique, Soc. Math. France (1955)   0

5. Limites projectives de groupes de Lie, C. R. Acad. Sci., Paris **230** (1950) 1197–1199   I, 33–35

6. Sections locales de certains espaces fibrés, C. R. Acad. Sci., Paris **230** (1950) 1246–1248   I, 36–38

7. Le plan projectif des octaves et les sphères comme espaces homogènes, C. R. Acad. Sci., Paris **230** (1950) 1378–1380   I, 39–41

8. Groupes localement compacts, Séminaire Bourbaki, Exp. 29 (1949/50)   I, 42–54

9. (avec J-P. Serre) Impossibilité de fibrer un espace euclidien par des fibres compactes, C. R. Acad. Sci., Paris **230** (1950) 2258–2259   I, 55–56

10. Remarques sur l'homologie filtrée, J. Math. Pures Appl., (9) **29** (1950) 313–322   I, 57–66

11. Impossibilité de fibrer une sphère par un produit de sphères, C. R. Acad. Sci., Paris **231** (1950) 943–945   I, 67–69

12. Sous-groupes compacts maximaux des groupes de Lie, Séminaire Bourbaki, Exp. 33 (1950/51)   I, 70–76

13. Sur la cohomologie des variétés de Stiefel et de certains groupes de Lie, C. R. Acad. Sci., Paris **232** (1951) 1628–1630   I, 77–79

14. La transgression dans les espaces fibrés principaux, C. R. Acad. Sci., Paris **232** (1951) 2392–2394   I, 80–82

15. Sur la cohomologie des espaces homogènes de groupes de Lie compacts, C. R. Acad. Sci., Paris **233** (1951) 569–571   I, 83–85

16. (avec J-P. Serre) Détermination des p-puissances réduites de Steenrod dans la cohomologie des groupes classiques. Applications, C. R. Acad. Sci., Paris **233** (1951) 680–682   I, 86–88

17. Cohomologie des espaces homogènes, Séminaire Bourbaki, Exp. 45 (1950/51)   0

18. Cohomologie des espaces localement compacts d'après J. Leray, Notes. E. P. F. Zürich, 1951; 3ème édition: Lect. Notes Math. **2** (1964)   0

19. (avec A. Lichnérowicz) Groupes d'holonomie des variétés riemanniennes, C. R. Acad. Sci., Paris **234** (1952) 1835–1837   I, 89–91

20. (avec A. Lichnérowicz) Espaces riemanniens et hermitiens symétriques, C. R. Acad. Sci., Paris **234** (1952) 2332–2334   I, 92–94

21. Les espaces hermitiens symétriques, Séminaire Bourbaki, Exp. 62 (1951/52)   I, 95–103

22. Les fonctions automorphes de plusieurs variables complexes, Bull. Soc. Math. France **80** (1952) 167–182   I, 104–119

23. Sur la cohomologie des espaces fibrés principaux et des espaces homogènes de groupes de Lie compacts, (Thèse, Paris, 1952) Ann. Math., (2) **57** (1953) 115–207   I, 121–216

24. (avec J-P. Serre) Sur certains sous-groupes des groupes de Lie compacts, Comment. Math. Helv. **27** (1953) 128–139   I, 217–228

25. La cohomologie mod 2 de certains espaces homogènes, Comment. Math. Helv. **27** (1953) 165–197   I, 229–261

26. (avec J-P. Serre) Groupes de Lie et puissances réduites de Steenrod, Amer. J. Math. **75** (1953) 409–448   I, 262–301

27. Les bouts des espaces homogènes de groupes de Lie, Ann. Math., (2) **58** (1953) 443–457   I, 302–316

28. Homology and cohomology of compact connected Lie groups, Proc. Nat. Acad. Sci. USA **39** (1953) 1142–1146   I, 317–321

29. Sur l'homologie et la cohomologie des groupes de Lie compacts connexes, Amer. J. Math. **76** (1954) 273–342   I, 322–391

30. Représentations linéaires et espaces homogènes kähleriens des groupes simples compacts (inédit, Mars 1954)   I, 392–396

31. Kählerian coset spaces of semi-simple Lie groups, Proc. Nat. Acad. Sci, USA **40** (1954) 1147–1151   I, 397–401

32. Topics in the homology theory of fibre bundles, Univ. of Chicago 1954 (Notes by E. Halpern), Lect. Notes Math. **36** (1967)   0

33. Topology of Lie groups and characteristic classes, Bull. Amer. Math. Soc. **61** (1955) 397–432   I, 402–437

34. Nouvelle démonstration d'un théorème de P. A. Smith, Comment. Math. Helv. **29** (1955) 27–39   I, 438–450

35. (with C. Chevalley) The Betti numbers of the exceptional groups, Mem. Amer. Math. Soc. **14** (1955) 1–9 — I, 451–459

36. (with G. D. Mostow) On semi-simple automorphisms of Lie algebras, Ann. Math., (2) **61** (1955) 389–405 — I, 460–476

37. Sur la torsion des groupes de Lie, J. Math. Pures Appl., (9) **35** (1955) 127–139 — I, 477–489

38. Groupes algébriques, Séminaire Bourbaki, Exp. 121 (1955/56) — 0

39. Groupes linéaires algébriques, Ann. Math., (2) **64** (1956) 20–82 — I, 490–552

40. Transformation groups with two classes of orbits, Proc. Nat. Acad. Sci. USA **43** (1957) 983–985 — I, 553–555

41. Travaux de Mostow sur les espaces homogènes, Séminaire Bourbaki, Exp. 142 (1956/57) — I, 556–564

42. The Poincaré duality in generalized manifolds, Mich. Math. J. **4** (1957) 227–239 — I, 565–577

43. (with F. Hirzebruch) Characteristic classes and homogeneous spaces I, Amer. J. Math. **80** (1958) 458–538 — I, 578–658

44. (avec J-P. Serre) Le théorème de Riemann-Roch, d'après Grothendieck, Bull. Soc. Math. France **86** (1958) 97–136 — I, 659–698

45. (with F. Hirzebruch) Characteristic classes and homogeneous spaces II, Amer. J. Math. **81** (1959) 315–382 — II, 1–68

46. Fixed points of elementary commutative groups, Bull. Amer. Math. Soc. **65** (1959) 322–326 — II, 69–73

47. (with F. Hirzebruch) Characteristic classes and homogeneous spaces III, Amer. J. Math. **82** (1960) 491–504 — II, 74–87

48. On the curvature tensor of the hermitian symmetric manifolds, Ann. Math., (2) **71** (1960) 508–521 — II, 88–101

49. (with J. C. Moore) Homology theory for locally compact spaces, Mich. Math. J. **7** (1960) 137–159 — II, 102–124

50. Density properties for certain subgroups of semi-simple groups without compact components, Ann. Math., (2) **72** (1960) 179–188 — II, 125–134

51. Commutative subgroups and torsion in compact Lie groups, Bull. Amer. Math. Soc. **66** (1960) 285–288 — II, 135–138

52. Seminar on transformation groups, Ann. Math. Stud. **46** (1960), (with contributions by G. Bredon, E. Floyd, D. Montgomery, R. Palais) — 0

53. Sous groupes commutatifs et torsion des groupes de Lie compacts connexes, Tôhoku Math. J., (2) **13** (1961) 216–240 — II, 139–163

54. (with Harish-Chandra) Arithmetic subgroups of algebraic groups, Bull. Amer. Math. Soc. **67** (1961) 579–583 — II, 164–168

55. Some properties of adele groups attached to algebraic groups, Bull. Amer. Math. Soc. **67** (1961) 583–585 — II, 169–171

56. (avec A. Haefliger) La classe d'homologie fondamentale d'un espace analytique, Bull. Soc. Math. France **89** (1961) 461–513 — II, 172–224

57. (mit R. Remmert) Über kompakte homogene Kählersche Mannigfaltigkeiten, Math. Ann. **145** (1962) 429–439 — II, 225–235

58. (with Harish-Chandra) Arithmetic subgroups of algebraic groups, Ann. Math., (2) **75** (1962) 485–535 — II, 236–286

59. Ensembles fondamentaux pour les groupes arithmétiques, Colloque sur la Théorie des Groupes Algébriques, Bruxelles 1962, 23–40 — II, 287–304

60. Some finiteness properties of adele groups over number fields, Publ. Math., Inst. Hautes Etud. Sci. **16** (1963) 5–30 — II, 305–330

61. Arithmetic properties of linear algebraic groups, Proc. Int. Congr. Mathematicians, Stockholm 1962, Uppsala 1963, 10–22 — II, 331–343

62. Compact Clifford-Klein forms of symmetric spaces, Topology **2** (1963) 111–122 — II, 344–355

63. (with W. L. Baily Jr.) On the compactification of arithmetically defined quotients of bounded symmetric domains, Bull. Amer. Math. Soc. **70** (1964) 588–593 — II, 356–361

64. (avec J-P. Serre) Théorèmes de finitude en cohomologie galoisienne, Comment. Math. Helv. **39** (1964) 111–164 — II, 362–415

65. Cohomologie et rigidité d'espaces compacts localement symétriques, Séminaire Bourbaki, Exp. 265 (1963/64) — II, 416–423

66. (avec J. Tits) Groupes réductifs, Publ. Math., Inst. Hautes Etud. Sci. **27** (1965) 55–150 — II, 424–520

67. Statement of the index theorem. Outline of proof, Chap. I in: Seminar on the Atiyah-Singer index theorem by R. Palais et al., Ann. Math. Stud. **57** (1965) 1–11 — II, 521–531

68. A spectral sequence for complex analytic bundles, Appendix Two in: F. Hirzebruch, Topological methods in algebraic geometry, 3rd edition, 202–217, Springer 1966 — II, 532–547

69. (with W. L. Baily Jr.) Compactification of arithmetic quotients of bounded symmetric domains, Ann. Math., (2) **84** (1966) 442–528 — II, 548–634

70. Density and maximality of arithmetic subgroups, J. Reine Angew. Math. **224** (1966) 78–89 — II, 635–646

71. Opérateurs de Hecke et fonctions zêta, Séminaire Bourbaki, Exp. 307 (1965/66) — II, 647–661

72. Class invariants, Chap. III, IV in: Seminar on complex multiplication, with S. Chowla, C. S. Herz, K. Iwasawa, J-P. Serre, Lect. Notes Math. **21** (1966) — 0

73. Linear algebraic groups, Proc. Symp. Pure Math. **9,** Amer. Math. Soc. (1966) 3–19 — II, 662–678

74. Reduction theory for arithmetic groups, Proc. Symp. Pure Math. **9,** Amer. Math. Soc. (1966) 20–25 — II, 679–684

75. Introduction to automorphic forms, Proc. Symp. Pure Math. **9**, Amer. Math. Soc. (1966) 199–210   II, 685–696

76. (with T. A. Springer) Rationality properties of linear algebraic groups, Proc. Symp. Pure Math. **9**, Amer. Math. Soc. (1966) 26–32   II, 697–703

77. (with R. Narasimhan) Uniqueness conditions for certain holomorphic mappings, Invent. Math. **2** (1967) 247–255   II, 704–712

78. Sur une généralisation de la formule de Gauss-Bonnet, An. Acad. Bras. Cienc. **39** (1967) 31–37   II, 713–719

79. Ensembles fondamentaux pour les groupes arithmétiques et formes automorphes, Notes d'un cours à l'Inst. H. Poincaré 1964, rédigées par H. Jacquet, J.-J. Sansuq et B. Schiffmann, Ecole Normale Supérieure, Paris 1967   0

80. (with T. A. Springer) Rationality properties of linear algebraic groups II, Tôhoku Math. J., (2) **20** (1968) 443–497   II, 720–774

81. On the automorphisms of certain subgroups of semi-simple Lie groups, Proc. Inter. Colloquium on Algebraic Geometry 1968, Tata Institute, Bombay (1969) 43–73   III, 1–31

82. (avec J. Tits) On 'abstract' homomorphisms of simple algebraic groups, Proc. Colloquium on Algebraic Geometry 1968, Tata Institute, Bombay (1969) 75–82   III, 32–39

83. Injective endomorphisms of algebraic varieties, Arch. Math. **20** (1969) 531–537   III, 40–46

84. Introduction aux groupes arithmétiques, Actualités Sci. Ind. no 1341, Hermann, Paris (1969)   0

85. Linear algebraic groups (Notes by H. Bass), Math. Lecture Notes Series, Benjamin, Inc. New York (1969); Traduction russe Moscou MIR (1972)   0

86. Sous-groupes discrets de groupes semi-simples (d'après D. A. Kajdan et G. A. Margoulis), Séminaire Bourbaki, Exp. 358, (1968/69), Lect. Notes Math. **179** (1971) 199–216   III, 47–56

87. On periodic maps of certain $K(\pi, 1)$, (unpublished, 1969)   III, 57–60

88. Pseudo-concavité et groupes arithmétiques, Essays on Topology and Related Topics, Mémoires dédiés à G. de Rham, Springer (1970) 70–84   III, 61–75

89. Properties and linear representations of Chevalley groups, Seminar on algebraic groups and related finite groups, Lect. Notes Math. **131** (1970) 1–55   III, 76–108

90. (avec J-P. Serre) Adjonction de coins aux espaces symétriques; Applications à la cohomologie des groupes arithmétiques, C. R. Acad. Sci., Paris **271** (1970) 1156–1158   III, 109–111

91. (avec J-P. Serre) Cohomologie à supports compacts des immeubles de Bruhat-Tits; Applications à la cohomologie des groupes S-arithmétiques, C. R. Acad. Sci., Paris **272** (1971) 110–113   III, 112–115

92. (avec J. Tits) Eléments unipotents et sous-groupes paraboliques de groupes réductifs I, Invent. Math. **12** (1971) 95–104 — III, 116–125

93. Cohomologie réelle stable de groupes S-arithmétiques, C. R. Acad. Sci., Paris **274** (1972) 1700–1702 — III, 126–128

94. (avec J. Tits) Compléments à l'article: 'Groupes réductifs', Publ. Math., Inst. Hautes Etud. Sci. **41** (1972) 253–276 — III, 129–152

95. Some metric properties of arithmetic quotients of symmetric spaces and an extension theorem, J. Differ. Geom. **6** (1972) 543–560 — III, 153–170

96. Représentations de groupes localement compacts, Lect. Notes Math. **276** (1972) — 0

97. (avec J. Tits) Homomorphismes 'abstraits' de groupes algébriques simples, Ann. Math., (2) **97** (1973) 499–571 — III, 171–243

98. (with J-P. Serre) Corners and arithmetic groups. With an appendix by A. Douady and L. Hérault: Arrondissement des Variétés à coins. Comment. Math. Helv. **48** (1973) 436–491 — III, 244–299

99. Cohomologie de certains groupes discrets et Laplacien p-adique (d'après H. Garland), Séminaire Bourbaki, Exp. 437 (1973/74), Lect. Notes Math. **431** (1975) 12–35 — III, 300–314

100. Stable real cohomology of arithmetic groups, Ann. Sci. Ec. Norm. Super., (4) **7** (1974) 235–272 — III, 315–352

101. Cohomology of arithmetic groups, Proc. Int. Congr. of Mathematicians, Vancouver, 1974, Vol. 1 (1975) 435–442 — III, 353–360

102. Linear representations of semi-simple algebraic groups, Proc. Symp. Pure Math. **29**, Amer. Math. Soc. (1975) 421–439 — III, 361–373

103. Formes automorphes et séries de Dirichlet (d'après R. P. Langlands), Séminaire Bourbaki, Exp. 466 (1974/75), Lect. Notes Math. **514** (1976) 183–222 — III, 374–398

104. Cohomologie de sous-groupes discrets et représentations de groupes semi-simples, Astérisque **32–33** (1976) 73–112 — III, 399–438

105. (avec J-P. Serre) Cohomologie d'immeubles et de groupes S-arithmétiques, Topology **15** (1976) 211–232 — III, 439–460

106. Admissible representations of a semi-simple group over a local field with vectors fixed under an Iwahori subgroup, Invent. Math. **35** (1976) 233–259 — III, 461–487

107. (with B. M. Schreiber) p-adic linear groups with ergodic automorphisms, Isr. J. Math. **24** (1976) 199–205 — III, 488–494

108. Cohomologie de SL_n et valeurs de fonctions zeta aux points entiers, Ann. Sc. Norm. Super. Pisa, Cl. Sci., (4) **4** (1977) 613–636; Correction, ibid. **7** (1980) 373 — III, 495–519

109. (with G. Harder) Existence of discrete cocompact subgroups of reductive groups over local fields, J. Reine Angew. Math. **298** (1978) 53–64 — III, 520–531

110. (avec J. Tits) Théorèmes de structure et de conjugaison pour les groupes algébriques linéaires, C. R. Acad. Sci., Paris **287** (1978) 55–57   III, 532–534

111. On the development of Lie group theory, Proc. of the Bicentennial Congr. of the Dutch Math. Soc., Math. Centre Tract 100/101 (1979) 25–37 and Nieuw Archief voor Wiskunde (3) **27** (1979) 13–25; Math. Intell. **2.2** (1980) 67–72   III, 535–547

112. (with H. Jacquet) Automorphic forms and automorphic representations, Proc. Symp. Pure Math. **33,** Part 1, Amer. Math. Soc. (1979) 189–202   III, 548–561

113. Automorphic L-functions, Proc. Symp. Pure Math. **33,** Part 2, Amer. Math. Soc. (1979) 27–61   III, 562–596

114. Symmetric compact complex spaces, Arch. Math. **33** (1979) 49–56   III, 597–604

115. (with N. Wallach) Continuous cohomology, discrete subgroups and representations of reductive groups, Ann. Math. Stud. **94** (1980), Introduction   III, 605–613

116. Stable and L^2-cohomology of arithmetic groups, Bull. Amer. Math. Soc., (N.S.) **3** (1980) 1025–1027   III, 614–616

117. Commensurability classes and volumes of hyperbolic 3-manifolds, Ann. Sc. Norm. Super. Pisa, Cl. Sci., (4) **8** (1981) 1–33   III, 617–649

118. Stable real cohomology of arithmetic groups II, Prog. Math., Boston **14** (1981) 21–55   III, 650–684

119. Mathematik: Kunst und Wissenschaft, Themen-Reihe der Carl Friedrich von Siemens Stiftung XXXIII München 1982   III, 685–701

A paraître:

Cohomology and spectrum of an arithmetic group, Proc. of a Conference on Operator Algebras and Group Representations, Neptun, Rumania (1980), Pitman (1983)

On free subgroups of semi-simple groups, Enseign. Math. (2)

(with H. Garland) Laplacian and the discrete spectrum of an arithmetic group, to appear in the Amer. J. Math.

Regularization theorems in Lie algebra cohomology. Applications

(with W. Casselman) L^2-cohomology of locally symmetric manifolds of finite volume

L^2-cohomology and intersection cohomology of certain arithmetic varieties

Acknowledgements

Springer-Verlag would like to thank the original publishers of Armand Borel's papers for granting permission to reprint them here.

The numbers following each source correspond to the numbering of the articles in the bibliography at the end of each volume.

Reprinted from Amer. J. Math., © by Johns Hopkins University Press: 26, 29, 43, 45, 47

Reprinted from An. Acad. Bras. Cienc., © by Academia Brasiliera de Ciencias: 78

Reprinted from Ann. Math. Stud., © by Princeton University Press: 67, 115

Reprinted from Ann. Math., (2), © by Princeton University Press: 23, 27, 36, 39, 48, 50, 58, 69, 97

Reprinted from Ann. Sc. Norm. Super. Pisa, Cl. Sci., (4), © by Scuola Normale Superiore, Italy: 108, 117

Reprinted from Ann. Sci. Ec. Norm. Super., (4), © by Editions Bordas-Dunod-Gauthier-Villars: 100

Reprinted from Arch. Math.,© by Birkhäuser Verlag, Basel: 83, 114

Reprinted from Astérisque, © by Société Mathématique de France: 104

Reprinted from Bull. Am. Math. Soc., © by The American Mathematical Society: 2, 33, 46, 51, 54, 55, 63, 116

Reprinted from Bull. Soc. Math. France, © by Editions Bordas-Dunod-Gauthier-Villars: 22, 44, 56

Reprinted from C. R. Acad. Sci., Paris, © by Editions Bordas-Dunod-Gauthier-Villars: 1, 5, 6, 7, 9, 11, 13, 14, 15, 16, 19, 20, 90, 91, 93, 110

Reprinted from Comment. Math. Helv., © by The University of Zürich: 3, 24, 25, 34, 64, 98

Reprinted from Essays on Topology and Related Topics, © by Springer-Verlag Berlin Heidelberg New York: 88

Reprinted from Invent. Math., © by Springer-Verlag Berlin Heidelberg New York: 77, 92, 106

Reprinted from Isr. J. Math., © by Weizmann Science Press: 107

Reprinted from J. Differ. Geom., © by The Lehigh University: 95

Reprinted from J. Math. Pures Appl., (9), © by Editions Bordas-Dunod-Gauthier-Villars: 10, 37

Reprinted from J. Reine Angew. Math., © by Walter de Gruyter & Co.: 70, 109

Reprinted from Math. Ann., © by Springer-Verlag Berlin Heidelberg New York: 57

Reprinted from Mem. Amer. Math. Soc., © by The American Mathematical Society: 35

Reprinted from Mich. Math. J., © by The University of Michigan: 42, 49

Reprinted from Proc. Int. Congr. Mathematicians, Stockholm, 1962, © by The American Mathematical Society: 61

MIX
Papier aus verantwortungsvollen Quellen
Paper from responsible sources
FSC® C105338

If you have any concerns about our products,
you can contact us on
ProductSafety@springernature.com

In case Publisher is established outside the EU,
the EU authorized representative is:
Springer Nature Customer Service Center GmbH
Europaplatz 3, 69115 Heidelberg, Germany

Printed by Libri Plureos GmbH
in Hamburg, Germany